D0141632

FUNDAMENTALS OF ECOTOXICOLOGY

The Science of Pollution

FOURTH EDITION

FUNDAMENTALS OF ECOTOXICOLOGY

The Science of Pollution

FOURTH EDITION

Michael C. Newman

CRC Press is an imprint of the
Taylor & Francis Group, an **informa** business

CRC Press
Taylor & Francis Group
6000 Broken Sound Parkway NW, Suite 300
Boca Raton, FL 33487-2742

Printed in Canada on acid-free paper
Version Date: 20140320

International Standard Book Number-13: 978-1-4665-8229-3 (Hardback)

Library of Congress Cataloging-in-Publication Data

Newman, Michael C.
Fundamentals of ecotoxicology : the science of pollution / Michael C. Newman. -- Fourth edition.
pages cm
Includes bibliographical references and index.
ISBN 978-1-4665-8229-3 (hardcover)
1. Pollution--Environmental aspects. 2. Pollution--Health aspects. I. Title.

QH545.A1N49 2015
571.9'5--dc23 2014009747

Visit the Taylor & Francis Web site at
http://www.taylorandfrancis.com

and the CRC Press Web site at
http://www.crcpress.com

To Peg, Ben, and Ian

I pretend not to teach, but to inquire …

Locke (1690)

Contents

List of Vignettes

Preface

TENOR OF THIS BOOK

Nothing is new under the sun. Even the thing of which we say, "See, this is new!" has already existed in the ages that preceded us.

Ecclesiastes 1: 10–11

In contrast to the conceptual doldrums expressed above by Coheleth, we live in exciting times, rich in new discoveries and novel applications of familiar concepts: tremendous opportunity exists for intellectual growth. And there is now tremendous need by society for such growth. This is so obvious in fields such as molecular genetics and computer sciences that no elaboration is required. Advances in our understanding of the natural world come immediately to mind and range throughout the fields of planetary biogeochemistry, mathematics, and geology. Debates about Lovelock's **Gaia hypothesis*** (Lovelock, 1988; Margulis and Lovelock, 1989) have stretched our perspective beyond the ecosystem to consider contaminants in a global context. Recently, we learned that a **global distillation** moves volatile and persistent pesticides from warmer areas of use to cooler areas of the globe where they had been banned for decades or never used (Simonich and Hites, 1995). The imminent consequences of global warming have become an increasingly common topic in deliberations by decision makers from all countries of the planet (Kerr, 2007)—with some embarrassing political laggards. Indeed, Al Gore received the 2007 Nobel Peace Prize for focusing the world's collective attention on what he describes as a planetary emergency (Gore, 1992, 2006). Lovelock's (2003) observation about the scientific community's long-last acceptance of Gaia theory at a 2001 Amsterdam conference, "Then the ice began to melt," also reflects the present worldwide glacier meltdown and the unprecedented recent thinning of Arctic ice as a result of ignoring the Gaian context in environmental decision making. The global warming and future global scale crises require a Gaian framework for decision making.

Only a few decades ago, nonlinear dynamics and chaos theory laid out the limits of determinism. We now know that it is impossible to predict precisely the behavior of all but the simplest of systems. A more inclusive vantage emerged that allows us to better understand and make appropriate predictions while dealing with contaminants in our world. As an example involving individual organisms, the bioaccumulation models presented in Chapter 3 do not necessarily predict pollutant accumulation until a single equilibrium concentration is reached. In sharp contrast to current thinking, oscillations in body concentrations are expected under certain conditions (Newman and Jagoe, 1996). In some cases, similar oscillations are expected for many populations, including nontarget and target populations sprayed to pesticides (Newman and Clements, 2008; Newman, 2013). At an even higher level of ecological organization, new insights create anticipation that ecotoxicological tipping points are as plausible as gradual changes (Cairns, 2004; Scheffer et al., 2012). This includes responses to global warming, ocean acidification, and declining biodiversity. To accommodate complex temporal dynamics at different scales, new branches of ecology and associated techniques have emerged such as metacommunity (Wilson, 1992; Leibold et al., 2004), landscape (Foreman and Godron, 1986), and even planetary (Rambler et al., 1989) ecology. New techniques for making decisions and predictions about complex ecological systems have emerged and have been applied successfully to important issues (Boyd, 2012; Mougi and Kondoh, 2012). Similarly, the vantage of physical scientists has broadened to encompass global biogeochemistry (Butcher et al., 1992), atmospheric sciences (Bridgman, 1994; Lovelock, 2003), and observation systems (Butler, 2005).

Just a few decades ago, French scientists discovered that life's activities had so modified the geochemistry in Oklo (Gabon, Africa) 1.8 billion years ago that enough fissile uranium (^{235}U) was

* The earth's albedo, temperature, and surface chemistry are homostatically regulated by the biota.

brought together to reach critical mass (Lovelock, 1988, 1991). By creating the prerequisite conditions, Proterozoic life influenced the hydrological cycle and produced an oxidizing atmosphere. The consequence was algal **biomineralization** (biologically mediated deposition of minerals) of enough uranium to begin sustainable nuclear fission (Lovelock, 1988; Milodowski et al., 1990) and create Gabon's **natural Oklo reactors**. These natural fission reactors generated power for nearly a million years (Choppin and Rydberg, 1980), leaving clear evidence that life's activities had nuclear manifestations long before humans appeared on earth. The present distribution of residue from these natural reactors also provides environmental scientists with valuable clues about the long-term fate and migration of modern waste fission products. Such understanding becomes increasingly important as the transition begins from overdependence on petroleum-based energy sources to a mixture of energy sources, including nuclear fission.

My reasons for writing this book stem from the realization that new concepts and novel applications of existing ideas appear every day. This book is my effort to expose you to new and useful concepts of ecotoxicology. Not only are these concepts and facts fascinating in themselves, but they also provide us with the tools to avoid or solve environmental problems. As discussed in Chapter 1, our ecotoxicological problems become increasingly complex and encompass broader spatial and temporal scales with each passing year. Our practical understanding must evolve accordingly to maintain an acceptable quality of life.

> *... alert and healthy natures remember that the sun rose clear. It is never too late to give up our prejudices. No way of thinking or doing, however ancient, can be trusted without proof. What everyone echoes or in silence passes by as true to-day may turn out to be a falsehood to-morrow, mere smoke of opinion...*
>
> **Thoreau (1854)**

CONTENT AND ORGANIZATION OF THIS BOOK

This book is designed as a textbook for use in an introductory graduate or upper level undergraduate course, or as a general reference. It creates a basic understanding of the field that can be expanded further with *Ecotoxicology: A Comprehensive Treatment* (Newman and Clements, 2008), which was written to support a more advanced or intensive course. A third book, *Quantitative Ecotoxicology* (2013) also extends issues discussed in *Fundamentals of Ecotoxicology*, but frames them in a quantitative context.

The *Fundamentals of Ecotoxicology* provides a broad overview of ecotoxicology that ranges from molecular to global issues. Reflecting our present imbalance of knowledge and effort, this book does retain some bias toward lower levels of ecological organization, e.g., biochemical to organismal topics. Yet, it purposefully extends discussion further in each edition beyond the conventional ecosystem to include landscape, regional, and biospheric topics. The intent of this extension is to impart a perspective as encompassing as the problems facing us today: many of our most serious problems transcend the conventional ecosystem context. It follows that discussion of **ecotoxicants*** here will not be restricted to toxicants traditionally associated with direct poisoning of the individual, human or otherwise. The time has passed when it was sufficient for the ecotoxicologist to focus only on new human toxicants as they appear and then explore how they impact nonhuman species. An agent that eliminates an essential pollinator of a particular plant species (e.g., Devine and Furlong, 2007; Pettis and Deplane, 2010; Potts et al., 2010; Watanabe, 2008) will have as harmful an impact on the plant species' persistence as another agent that poisons the individual

* The classic toxicants or poisons are only a subset of agents that can harm or dissemble valued ecological entities. For example, a "nontoxic" agent can do harm by modifying key habitats, intra- or interspecies interactions, metapopulation or metacommunity dynamics, material cycling, or energy flow.

plants outright. An agent that modifies an organism's physical habitat, such as greenhouse gases, can have an adverse effect on that species. To omit discussion of some ecotoxicants for reasons of convention results in inadequate discussion of agents harmful to valued ecological entities.

> *We disagree with any suggestion that discussion include only conventional mammalian poisons in the environment that could do harm to other biota. Any ecotoxicant even nitrogen or CO_2, is relevant if present in sufficient amounts to perturb ecological entities.*
>
> **Newman and Clements (2008)**

Both human and ecological health issues are frequently interwoven here. The reason for this deviation for recent convention is simple*: to do otherwise would be inconsistent with our present approach to assessing hazards and risks from pollutants. Also, many mechanisms of action and bioaccumulation, and effect dynamics are common to all species. It is unreasonable to omit discussion of some particularly insightful materials simply because they were developed first for the human species. As examples, valuable insights are possible from human health studies that describe important biochemical mechanisms or useful epidemiological techniques. This is consistent with Newman and Clements (2008), who stated their rationale in this regards,

> *A more congruent treatment was also attempted [in* Ecotoxicology. A Comprehensive Treatment*] by including relevant human effects information rather than taking the contrived approach of "asking humans to step out of the picture" when discussing human influences on the biosphere.*

Before leaving this topic, one final reason can be given for discussing human health topics. Although we prefer the delusion of transcendence in environmental issues,† our decisions regarding ecological effects of contaminants are made with motives as profoundly anthropocentric as those involving human health. Why not discuss them together?

Topics are divided among 14 chapters. Vignettes written by expert guest authors are placed into each chapter to elaborate upon key themes or to highlight especially good examples.‡ The first chapter provides general history and perspective for the field. Chapter 2 details the essential features of the key contaminants of concern today, including their sources. Bioaccumulation and effects of contaminants are detailed at increasing levels of ecological organization in Chapters 3 through 12. The framework of these chapters is scientific, not regulatory. Regulatory aspects of the field are covered partially in Chapter 13, which addresses the technical issues of risk assessment. Although characterized by common themes, environmental laws and regulations vary with political setting and history so that it would be challenging to develop a chapter that does justice to all relevant laws. Instead, appendices illustrating important sets of laws and regulations are provided at the end of this book. Experts in each set of laws and regulations contributed discussions of key Australian, Chinese, Indian, South African, United States, and European Union legislation. It is hoped that

* Even a traditionalist argument exists for including humans in ecotoxicological discussions. The first coining of the term *ecotoxicology* (Truhaut, 1977) specifically included effects to humans.

† Decisions about ecological effects are often perceived as arising from selfless protection of the earth. In fact, our motives are based on the value we give to the heretofore free services provided by ecological systems. These services include generation of clean water and air, production of food, provision of biological materials for medical and genetic uses, and provision of pleasing settings for living and recreation. Decisions are based on the perceived value of these services relative to those of our technological services and goods. The delusion of selfless motivation in environmental stewardship and advocacy is sufficiently widespread as to be named the **Lorax Incongruity**. (Lorax is a character in the popular children's book by Dr. Seuss [Geisel and Geisel, 1971] who "speaks for the trees, for the trees have no tongues.") This well-intended, but intransigence-inducing, delusion is pervasive in society today.

‡ An instructor may wish to present the core materials in each chapter themselves and then assign vignettes for student-led discussions.

more such appendices will be added in future editions of this textbook. To provide a capstone for the textbook, the final chapter briefly summarizes and provides context for the volume.

There was a troublesome bias toward North American examples in past editions of this textbook. This reflected the author's limited background, not a quiet snub of non-North American interests. Recognizing this shortcoming, extra attention was paid with each new edition to gather information more broadly. The vignettes and appendices by a diverse team of guest authors contribute considerably to this end.

> *Except in their occasional introductions, science textbooks do not describe the sorts of problems that the profession may be asked to solve .... Rather, these books exhibit concrete problem solutions that the profession has come to accept as paradigms ... [however, students] must, we say, learn to recognize and evaluate problems to which no unequivocal solution has been given ....*
>
> **Kuhn (1977)**

Although most textbooks teach established science facts without much coverage of existing controversies, the youthfulness of ecotoxicology as a science (and the author's predilection) mandated that controversies be integrated into the chapters. This might annoy some instructors and puzzle some students; however, as expressed in the above quote, description of any science—especially a new one—as a static body of facts gives a false impression to students that little room remains for exploring novel themes. To safeguard against the author's mere opinion becoming confused with controversy, the term *incongruity* is used throughout the book to identify opinionative—perhaps even curmudgeonly—insights of the author that are not generally shared by most ecotoxicologists, e.g., the Lorax incongruity discussed earlier. The reader may decide to ignore most of them with no danger of becoming an uninformed ecotoxicologist.

To aid the reader, key terms not already identified by section headings are highlighted in the text and compiled in the glossary. Some important materials have been included in the footnotes so that the reader is asked not to unintentionally overlook these materials. Suggested readings are provided at the end of each chapter, and study questions are collected at the end of the book. Answers to the study questions are available for instructors from the publisher.

Acknowledgments

My sincerest thanks to the A. Marshall Acuff, Jr. Professorship of Marine Science for the support provided during the preparation of this edition. I am also very grateful to all vignette and appendix guest authors for their intelligent and thorough contributions to this book. These contributions greatly enrich the book and are much appreciated. Chapters written in previous editions by Thomas Hinton (first edition, Chapter 14) and Michael A. Unger (second edition, Chapter 2) contributed significantly to the reorganization and revisions of Chapters 2 and 13. The contributions of Dr. Eric Peters to numerous paragraphs discussing radionuclides became more ghostwriting than reviewing in places.

Author

Michael C. Newman is currently the A. Marshall Acuff, Jr. Professor of Marine Science at the College of William & Mary, Virginia Institute of Marine Science, where he also served as the dean of graduate studies for the School of Marine Sciences from 1999 to 2002. Previously, he was a faculty member at the University of Georgia's Savannah River Ecology Laboratory. His research interests include quantitative ecotoxicology, environmental statistics, risk assessment, population effects of contaminants, metal chemistry and effects, and bioaccumulation and biomagnification modeling. In addition to roughly 145 articles, he authored five books and edited another six on these topics. The English editions, and Mandarin and Turkish translations, of *Fundamentals of Ecotoxicology* have been adopted widely for introductory ecotoxicology courses. He taught full semester or short courses at universities throughout the world including the College of William & Mary, University of California—San Diego, University of Georgia, University of South Carolina, Jagiellonian University (Poland), University of Antwerp (Belgium), University of Hong Kong, University of Joensuu (Finland), University of Koblenz—Landau (Germany), University of Technology—Sydney (Australia), Royal Holloway University of London (United Kingdom), Central China Normal University, and Xiamen University (China). He served many international, national, and regional organizations including the Organisation for Economic Co-operation and Development, the U.S. EPA Science Advisory Board, the Hong Kong Areas of Excellence committee, and the U.S. National Academy of Science National Research Council. In 2004, the Society of Environmental Toxicology and Chemistry awarded him its Founder's Award, "the highest SETAC award, given to a person with an outstanding career who has made a clearly identifiable contribution in the environmental sciences."

Guest Authors

Jon A. Arnot
University of Toronto Scarborough
Toronto, Ontario, Canada

John Cairns, Jr.
Biology Department
Virginia Polytechnic Institute and State University
Blacksburg, Virginia

Edward J. Calabrese
School of Public Health and Health Sciences
University of Massachusetts Amherst
Amherst, Massachusetts

Peter Calow
School of Biological Sciences
University of Nebraska–Lincoln
Lincoln, Nebraska

Peter G. C. Campbell
INRS–Eau Terre Environnement Research Centre
Université du Québec
Québec City, Québec, Canada

Luisa E. Castillo
Institute for Central American Studies on Toxic Substances
Universidad Nacional
Heredia, Costa Rica

Ma Shan Cheung
Department of Ecology and Biodiversity
The University of Hong Kong
Hong Kong, China

William H. Clements
Department of Fish, Wildlife and Conservation Biology
Colorado State University
Fort Collins, Colorado

Mark Crane
AG-HERA Consulting
Oxfordshire, United Kingdom

Elba de la Cruz
Institute for Central American Studies on Toxic Substances
Universidad Nacional
Heredia, Costa Rica

Virya Bravo Durán
Institute for Central American Studies on Toxic Substances
Universidad Nacional
Heredia, Costa Rica

Valery E. Forbes
School of Biological Sciences
University of Nebraska–Lincoln
Lincoln, Nebraska

Robert W. Furness
Institute of Biomedical and Life Sciences
University of Glasgow
Glasgow, United Kingdom

John P. Giesy
Toxicology Center
University of Saskatchewan
Saskatoon, Saskatchewan, Canada

Bruce Grant
Biology Department
College of William & Mary
Williamsburg, Virginia

Albania Grosso
AG-HERA Consulting
Oxfordshire, United Kingdom

Mark E. Hahn
Biology Department
Woods Hole Oceanographic Institution
Woods Hole, Massachusetts

Landis Hare
INRS–Eau Terre Environnement Research Centre
Université du Québec
Québec City, Québec, Canada

Paul D. Jones
Toxicology Center
University of Saskatchewan
Saskatoon, Saskatchewan, Canada

James R. Karr
University of Washington
Seattle, Washington

Dmitry L. Lajus
Department of Ichthyology and Hydrobiology
St. Petersburg State University
St. Petersburg, Russia

Wayne G. Landis
Institute of Environmental Toxicology and Chemistry
Western Washington University
Bellingham, Washington

Helen Y. M. Leung
Department of Ecology and Biodiversity
The University of Hong Kong
Hong Kong, China

Kenneth M. Y. Leung
Department of Ecology and Biodiversity
The University of Hong Kong
Hong Kong, China

Samuel N. Luoma
Department of Zoology
Natural History Museum
London, United Kingdom

Donald Mackay
Trent University
Peterborough, Ontario, Canada

Karen McBee
Department of Zoology and Collection of Vertebrates
Oklahoma State University
Stillwater, Oklahoma

Theunis Meyer
Centre for Environmental Management
North-West University
Potchefstroom, South Africa

S. Bijoy Nandan
Department of Marine Biology, Microbiology and Biochemistry
Cochin University of Science and Technology
Kerala, India

James T. Oris
Department of Biology
Miami University
Oxford, Ohio

Eric L. Peters
Department of Biological Sciences
Chicago State University
Chicago, Illinois

Philip S. Rainbow
Department of Zoology
Natural History Museum
London, United Kingdom

Guritno Roesijadi
Marine Sciences Laboratory
Pacific Northwest National Laboratory
Sequim, Washington

Claudine Roos
Centre for Environmental Management
North-West University
Potchefstroom, South Africa

Clemens Ruepert
Institute for Central American Studies on Toxic Substances
Universidad Nacional
Heredia, Costa Rica

Mark Sandheinrich
Department of Biology
University of Wisconsin–La Crosse
La Crosse, Wisconsin

Glenn W. Suter, II
National Center for Environmental Assessment
U.S. Environmental Protection Agency
Cincinnati, Ohio

Kyle R. Tom
School of Marine Science
College of William & Mary
Gloucester Point, Virginia

Wolfgang K. Vogelbein
Virginia Institute of Marine Science
College of William & Mary
Gloucester Point, Virginia

Taiping Wang
Marine Sciences Laboratory
Pacific Northwest National Laboratory
Sequim, Washington

Michael StJ. Warne
Department of Science, Information
Technology, Innovation and the Arts
University of Queensland
Brisbane, Australia

CHAPTER 1

Introduction

On the day of the patients' victory at court, someone wrote a headline: "The Day that Tomoko Smiled." She couldn't possibly have known. Tomoko Uemura, born in 1956, was attacked by mercury in the womb of her outwardly healthy mother. No one knows if she is aware of her surroundings or not.

Smith and Smith (1975)

1.1 HISTORIC NEED FOR ECOTOXICOLOGY

It is natural and responsible to periodically reconsider the wisdom of our evermore complex and encompassing system of environmental regulations. Do United Nations treaties and European Union directives encroach too much on the sovereignty of nations? Have environmental regulations grown too costly for developed countries or too stifling for developing countries? It may be difficult at first glance to understand why national sovereignty should not be more respected or why significant amounts of the money now spent on environmental regulation should not be reallocated to the global economic crisis, critical social problems, medical research, technological innovation, education, reinvigorating space exploration, or other worthwhile endeavors, e.g., Lomborg (2001). But, just 50 years ago, it was easy to understand: Tomoko Uemura's mother understood.

Tomoko Uemura was born with severe and permanent neurological damage after her mother had unknowingly consumed mercury-laden fish. Tomoko was barely aware of her surroundings during her pain-filled 21 years of life. Although Tomoko's mother grew to understand the consequences of inattention to pollution, what she could not grasp as her personal tragedy unfolded was how the conditions leading to her daughter's agonizing life were allowed to come into existence in the first place.

Explanation of Tomoko's, and related, tragedies must begin with events that emerged a little more than half a century before she was born. At the close of the nineteenth century, complex changes were occurring unevenly across many countries. All grew out of the unprecedented shifts in human population size and distribution, and our singular talent for extracting resources and energy from the environment. This was a time of shortsighted exploitation of natural resources and cavalier attitudes toward worker health. Population expansion brought widespread land and soil degradation through farming, foresting, mining, smelting, and other activities. With expansion to fill all available frontier regions such as in the western United States, the option was no longer open to move to an unsullied area after despoiling local natural resources. Widespread degradation left the development of a sound knowledge base and practices for resource conservation as the only available alternative.

Perhaps soil degradation and eventual conservation is the clearest and most global illustration of this point. McNeill (2000) points out that two of the three historical surges in soil erosion overlapped with this period. The first did not, having occurred in the Middle East, India, and China circa 2000 BC–AD 1000. The second surge (1490s–1930s) did, starting with the European expansion into North America, South America, South Africa, Northern Africa, Australia, and New Zealand. The third, and ongoing, surge encompassed most of the world beginning in the 1950s. In addition to soil erosion, mining during the nineteenth and early twentieth centuries produced wide swaths of metal-contaminated soils in broad regions such as the Akita Prefecture of Japan, Silesia region of Poland, Ontario Province of Canada, and western United States, typically tainting local agricultural produce and waters.

Out of necessity, natural resource conservation encompassing land, water, wildlife, and fisheries resources became essential in developing countries. Typical of legislation that started to emerge during this period, the United States passed its first wildlife conservation act, the Lacey Act (1900). Society's ultimate embrace of a conservation ethic was clearly articulated in the 1905 revelation of U.S. President Theodore Roosevelt's head of the new Department of Forest Service, Gifford Pinchot:

> *Suddenly the idea flashed through my head that there was a unity in this complication – that the relation of one resource to another is not the end of the story.... All of [the] separate questions fitted into and made up the one great problem of the use of the earth for the good of man.*
>
> **Pinchot (1947)**

And another general movement materialized to cope with harm occurring to humans exposed to chemicals. Global, albeit uneven, trends in urbanization and industrialization brought with them harmful chemical consequences to human well-being that required redress.

People were gathering together in cities of a size never seen before in history. In addition to the infectious disease risks that emerged as large cities came into existence, urban air pollution–associated health risks appeared, becoming one of the first blatant pollution problems needing attention. As the twentieth century unfolded, coal burning was pervasive for industrial and domestic heating purposes. An archetypal consequence was the December 1952 London "fog" episode that killed 4000 Londoners outright (Anderson, 2009). Although this was an extreme case, appalling air pollution was being experienced in other cities including those in Europe (e.g., Athens, the Ruhr region of Germany, and numerous cities in Soviet-dominated countries), North America (e.g., Chicago, Mexico City, Pittsburgh, and St. Louis), and Asia (e.g., Calcutta). Large cities improved air quality temporarily by switching from coal to oil; however, the appearance of automobiles brought unhealthy air pollution back in the form of photochemical smog (McNeill, 2000).

Industries contributed substantially to city air pollution. Indeed, the term acid rain was first coined in 1872 by Angus Smith who identified it as the cause of extensive vegetation death around industrialized Newcastle and Liverpool (Markham, 1995). A 1930 air pollution episode precipitated by a brief inversion over an industrialized Belgian town in Meuse Valley increased death rates 10-fold and the sickened citizens with histories of respiratory illness (Anderson, 2009). This scenario played out yet again in Donora, Pennsylvania, when an October 1948 inversion held zinc smelter smoke close to the ground, increasing death rates by sixfold and sickening hundreds of residents. In Siberia, lung cancer rates of forced-labor residents skyrocketed when the Norilsk nickel mines and smelters came into existence in 1935 in support of Soviet industrial plans (McNeill, 2000).

Some of the first pieces of environmental pollution legislation (such as, the U.K. Clean Air legislation of 1956 and 1963) aimed at controlling health effects of air pollution in and around large cities and industries. In many cases, a local problem was resolved temporarily by building taller smoke stacks that spread pollutants over wider areas. They became a problem for another day.

On related fronts, labor rights and industrial hygiene movement leaders fought for a more evenhanded **corporatism**[*] during industrialized society's nonage.[†] An exemplary figure of this time was Alice Hamilton who founded the science of occupational health. Her career as an advocate, beginning circa 1910, included negotiation to control U.S. workplace poisons such as mercury (hatter industry), phosphorous (match manufacture), benzene (general industrial solvent), and radium (watch face painting) (Hamilton, 1985). She successfully added chemical agents to the list of workplace dangers needing resolution. As the industrial hygiene movement matured into the 1930s, employers, employee representatives, scientists, and government officials came together to resolve early industrial indiscretions and to assure future adherence to the principles of evenhanded corporatism. This coalescing of responsible and affected parties would eventually be adopted by those attempting later to cope with pollutants in the general environment. An expectation of a safe work environment was eventually established. As a final contributing social movement, awareness and political action emerged about harmful chemicals in foodstuffs and drugs. That movement established the expectation of safe foods and medicines, and in 1906, resulted in the creation of a new U.S. Food and Drug Administration.

To summarize, concepts, approaches, and institutions appeared during the first half of the twentieth century for addressing pressing problems of natural resource conservation, urban air quality, industrial workplace hygiene, and harmful chemicals in food and drugs. The associated social evolution established an approach and ethic that would next extend outward to address harmful chemicals in the general environment. Environmental pollutants became a serious social issue to resolve during the second half of the twentieth century.

Blended into these historical demographic and industrial trends midway through the century was the **Green Revolution**, which began in the 1940s and quickly spread throughout the world (Evenson and Gollin, 2003). The key goal of this revolution was to improve crop production through an integrated application of high-yield crop strains, chemical fertilizers, and chemical biocides. It too brought unique resource conservation, worker safety, food safety, and general pollution issues as the second half of the century began.

So an explanation can now be provided to Tomoko's mother about conditions that allowed her daughter's life to be so painful and brief. Tomoko was born just as society moved beyond the pale in its activities within natural systems that it depended on. Society was becoming aware of its mistakes and realizing that it had new responsibilities to carefully regulate toxicants in the general environment. Too late for Tomoko, beliefs, behaviors, and laws were poised for necessary change[‡] but had not yet changed. Change in our environmental ethic had not come soon enough for many such as Tomoko.

> *One of the newest fads in Washington – and elsewhere – is "environmental science." The term has political potency even if its meaning is vague and questionable. Lacking specific definition, it embraces every science – physical, natural, social – for all of them deal with man's surroundings and their influence and impact upon him.*
>
> **Klopsteg (1966)**

* Corporatism is "the belief that all parts of society [are] necessary to its harmonious functioning, and that therefore all parts should cooperate to see to the welfare of each part." (Clark, 1997)

† In the United States, these movements were embedded in what has been called the **Progressive Era** that occurred between the late 1890s until the United States' entry into the World War I in 1917. Occurring during the shift from agrarian to a more urban and industrialized state, the goal of Progressive Era was to transform democracy into a more just political system by replacing customs and dubious beliefs—self-evident intuitions—with modern ones, including innovations based on sound scientific reasoning and technology. A scientific lens was systematically focused on social and political issues such as those concerning race, women's voting rights, reasonable limits of capitalism, labor rights, and immigration.

‡ This objective explanation is likely very unsatisfactory to anyone who, if only for a moment, imagines themselves in place of Tomoko or her mother. For the interested reader, a subjective narrative for the Minamata poisonings is provided in one of the best photojournalism works to date, *Minamata* (Smith and Smith, 1975). The Smiths' book and Carson's *Silent Spring*, were major catalysts for what would become a global movement to control environmental pollutants.

The vastness of the earth has fostered a tradition of unconcern about the release of toxic wastes. Billowing clouds of smoke are diluted to apparent nothingness; discarded chemicals are flushed away in rivers; insecticides "disappear" after they have done their job; even the massive quantities of radioactive debris of nuclear explosions are diluted in the apparent infinite volume of the environment ... [But] we have learned in recent years that dilution of persistent pollutants even to trace levels detectable only by refined techniques is no guarantee of safety. Nature is always concentrating substances that are frequently surprising and occasionally disastrous.

Woodwell (1967)

After World War II, the **dilution paradigm** (the solution to pollution is dilution) was gradually replaced by the **boomerang paradigm** (what you throw away can come back to hurt you). Two widely publicized epidemics of heavy metal poisoning from contaminated food had occurred in Japan. By the 1950s, enough organic mercury was transferred through the marine food web to poison hundreds of people in Minamata Prefecture. Nearly a thousand people, including Tomoko Uemura, fell victim to **Minamata Disease** before Chisso Corporation halted mercury discharge into Minamata Bay. In a major mining region of Japan (Toyama Prefecture), citizens were slowly being poisoned from 1940 to 1960 by cadmium in their rice. This outbreak of what became known as **itai-itai disease** was linked to irrigation water contaminated with mining wastes.*

In 1945, open-air testing of nuclear weapons began at Alamogordo, New Mexico, and nuclear bombs exploded over Hiroshima and Nagasaki later that same year. Nine years later, the Project Bravo bomb exploded at Bikini Atoll, dropping fallout over thousands of square kilometers of ocean including several islands and the ironically named fishing vessel, *Lucky Dragon* (Woodwell, 1967). The Marshall Islands of Ailinginae, Rongelap, and Rongerik received radiation levels of 300–3000 rem† within 4 days of this detonation (Choppin and Rydberg, 1980). Fourteen years later, Hempelmann (1968) would report elevated prevalence of nodular thyroids in Marshallese children (78% of the 19 exposed children vs. 0.36%–1.7% of unexposed children) notionally caused by the 1954 detonation radiation.

Fallout radioactivity in the air here increased sharply yesterday to 17.4 measured units, almost twice Sunday's 8.25 micromicrocuries per cubic meter of air. A micromicrocurie is one millionth of a millionth of the radiation strength of a gram of radium.

***Washington Post* (1961)**

On a broader scale, the hemispheric dispersal and unexpected accumulation of fission products in foodstuffs from these and subsequent detonations eventually created concern about possible long-term health effects. Initially, fallout had elicited only brief comment such as the snippet above taken from the December 12, 1961 weather page of the *Washington Post*. But concern grew quickly as the public read more thoughtful articles such as the 1963 *Washington Post* front-page exposé reporting "Persons living within 400 miles of the Nevada nuclear test site have been exposed during the last dozen years to far more radiation from radioactive iodine

* The name, itai-itai, which literally means "ouch-ouch," reflects the extreme joint pain associated with the disease. Doctors gave the disease this moniker based on the exclamations of patients as they came into clinics and hospitals for help.

† A **rem** or **R**oentgen **e**quivalent **man** is a measure of radiation that takes into account the differences in biological effects of various types of radiation. It relates the radiation dose received to its potential biological damage. As such, it is a convenient unit for defining allowable radiation exposures, e.g., the average person receives approximately 0.360 rem (360 mrem) of radiation annually. The rem has been replaced as the official unit by the sievert (Sv). (1 rem = 0.01 Sv.) In contrast, the **curie** used later is a straightforward measure of radioactivity. One curie is 2.2×10^6 disintegrations per minute (dpm). Although still used widely as in this book, the curie has been replaced as the official unit of radioactivity by the **becquerel** (Bq). One curie is 3.7×10^{10} Bq.

than hereto realized" (Simons, 1963).[*] From 1960 to 1965, human body burdens of ^{137}Cesium increased rapidly worldwide and then slowly decreased as the United States, former Soviet Union, France, United Kingdom, and China bowed to public pressure to cease open-air testing (Shukla et al. 1973).

Unreported discharge of radionuclides occurred in addition to these overt releases. Most were kept from the general public for reasons of national security. On the northwest coast of England, a fire in the Windscale plutonium-processing unit released 20,000 Ci of radioactive iodine (^{131}I) to the surrounding area (Dickson, 1988). Release of ^{131}I is particularly disconcerting because it concentrates in the thyroid, greatly increasing cancer risk. After atmospheric release, ^{131}I in fallout can contaminate local vegetation, be ingested by dairy cattle, and accumulate in thyroids of dairy product consumers. At the secret Soviet Chelyabinsk 40 military plant in the Urals, plutonium processing had furtively discharged 120 million curies to a nearby lake and enough to the Techa River to induce radiation poisoning in citizens living downriver (Medvedev, 1995). A 1957 storage tank explosion at this same facility released 18 million curies of radioactive material, forcing approximately 11,000 people to evacuate a 1,000-km^2 area (Trabalka et al. 1980; Medvedev, 1995). From 1944 to 1966, knowledge of releases from the U.S. Atomic Energy Commission's Hanford Site in Washington State was kept from the general public. The complex released 440,000 curies of ^{131}I into the atmosphere between 1944 and 1947 (Stenehjem, 1990). An estimated 20,000 curies were released to the Columbia River on May 12, 1963, from the Hanford K-East reactor (Stenehjem, 1990).

Concern about pollutant effects to nonhuman species was also growing. Recollect from our discussion of trends during the first half of the century that a natural resource conservation ethic had come into being and institutions were established to ensure adherence to basic conservation principals. Pesticides such as the then widely used DDT[†] (dichlorodiphenyltrichloroethane or 1,1,1-trichloro-2,2-di-(4-chlorophenyl)-ethane) began accumulating in wildlife to alarming concentrations, resulting in direct toxicity and sublethal effects. The indisputable success of DDT in combating insect vectors of human disease such as malaria, yellow fever, and encephalitis, had encouraged its broadened use to control agricultural and nondisease-related pests. Responsible government agencies—the U.S. Department of Agriculture and Public Health Service, for instance—had found little evidence for concern in the early 1940s as DDT use spread and intensified. However, U.S. Department of Interior did express concern by the mid-1940s about widespread DDT application and possible harm to wildlife (Nelson and Surber, 1946). Then, in 1954, Professor Wallace at Michigan State University noticed dead and dying robins after DDT spraying on campus to control Dutch elm disease (Carson, 1962). Hunt and Bischoff (1960) and Dolphin (1959) documented from 1957 to 1960 the overt death of Western grebes (*Aechmophorus occidentalis*) after bioaccumulation of the DDT-related pesticide, DDD (1,1-dichloro-2,2-bis[*p*-chlorophenyl] ethane) through a freshwater food web (Clear Lake, California). Following excessive DDD application to this lake, enough accumulated in the grebes' brains to cause axonic dysfunction and death. Dolphin (1959) described a 1949 administration of DDD to control the nonbiting gnat, *Chaoborus astictopus*, of Clear Lake as "involving introducing approximately 40,000 gallons of a 30% DDD formulation … from drum-laden barges"! In 1962, Rachel Carson brought these and many other events together in a remarkably well-crafted and factually accurate exposé, *Silent Spring*.

* The basis for this newspaper article was a 1962 *Science* article by Lapp who examined the risk of infant thyroid cancer as a consequence of open-air testing. Estimates were done for the area surrounding the Nevada test site and also for a distant location in Troy, New York, that received considerable fallout (circa 2–4 μCi$\cdot{}^{131}$I sq. mi.$^{-1}$) on April 26, 1953, from a Nevada detonation.

† DDT was an extremely important tool for disease control throughout the world. Indeed, Paul Müeller was awarded the 1948 Nobel Prize in medicine for discovering its value as an insecticide. Its importance in the context of disease vector control is often overshadowed by our present understanding of its adverse effects on nontarget species if used indiscriminately. Indeed, UN Environmental Programme delegates at the May 17, 2004 Stockholm Convention on POPs conceded that a complete DDT ban was unreasonable in malarial regions of the world. The World Health Organization now prescribes careful reintroduction of DDT for malaria control in developing countries (Lubick, 2007b).

The men who make pesticides are crying foul. "Crass commercialism," scoffs one industrial toxicologist. "We're aghast," says another. "Our members are raising hell," reports a trade association.... Statements are being drafted and counter-attacks plotted.

Lee (1962)

Notwithstanding attacks upon the work of this "quiet women author" (Lee, 1962), Rachel Carson succeeded in drawing the public's attention to the consequences of pesticide accumulation in wildlife (Souder, 2012). Although relatively nontoxic to humans, it had become clear that DDT and DDE (dichlorodiphenyldichloroethylene or 1,1-dichloro-2,2-bis-(*p*-chlorophenyl)-ethylene) inhibited Ca-dependent ATPases in the shell gland of birds, resulting in shell thinning and increased risk of breakage of eggs after being laid (Cooke, 1973, 1979). Birds at higher trophic levels were particularly vulnerable because DDT and DDE were resistant to degradation and accumulated in tissue lipids. The result was an increase in concentration with each trophic exchange in a food web. Reproductive failure of raptors and fish-eating birds became widespread. For instance, the average number of offspring per pair of osprey (*Pandion haliaetus*) nesting on Long Island Sound dropped from 1.71 young/nest (1938–1942) to only 0.07–0.40 young/nest by the mid-1960s (Spitzer et al. 1978).* Reproductive output of raptor populations decreased similarly in Alaska (Cade et al. 1971) and other regions of North America (Hickey and Anderson, 1968). Reproduction of brown pelicans (*Pelecanus occidentalis*) on the South Carolina coast from 1969 to 1972 fell below that needed to maintain viable populations (Hall, 1987). Ratcliffe (1967, 1970) reported the same downward trends for falcons (*Falco peregrinus*) and other raptors in the United Kingdom.

As the adverse effects of DDT became apparent, and perhaps more importantly, insect pests began developing resistance to DDT, agricultural chemists began synthesizing a complex arsenal of new pesticides (Figure 1.1). A wide array of pesticides differing in mode of action, toxicity, and environmental persistence came into existence soon after the 1939 introduction of DDT.

From among all of these instances of harm from pollutants in the general environment, the Minamata poisonings in Japan and DDT accumulation in raptors and fish-eating birds became the two events that most captured the public's attention and accelerated a paradigm shift (dilution paradigm to boomerang paradigm) (Figure 1.2). Together, they drew attention away from giddy industrialization and the well-intended Green Revolution to the consequences of inattention to pollutants in the environment. They were among the first issues to give impetus to the science of ecotoxicology.

1.2 CURRENT NEED FOR ECOTOXICOLOGY EXPERTISE

Today, we are hearing and seeing dire warnings of the worst potential catastrophe in the history of human civilization: a global climate crisis....

Gore (2006)

Pollution is not in the process of undermining our well-being. On the contrary, the pollution burden has diminished dramatically in the developed world.

Lomborg (2001)

* Osprey populations have rebounded. Ambrose (2001) reports that fewer than 8,000 breeding pairs existed in the United States in 1981 but that number increased to 14,246 pairs by 1994.

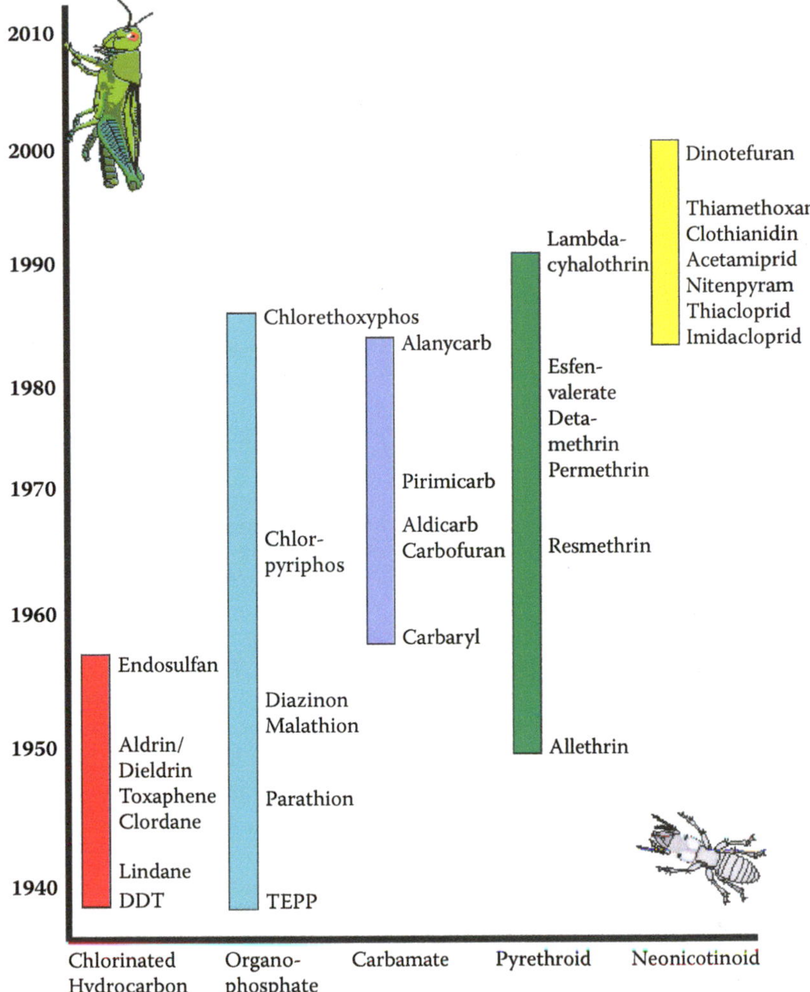

Figure 1.1 The general development chronology of the five major types of organic insecticides. The information in this figure was derived from Casida and Quistad (1998) (Figure 1), Jeschke and Nauen (2008), and Jeschke et al. (2011). The neonicotinoids are plotted with patent year, i.e., imidaclopid (1985), thiaclopid (1985), nitenpyram (1988), acetamiprid (1989), thiamethoxam (1992), and dinotefuran (1994).

Everyone would like to feel that the problems just described reflect early mistakes in our global techno-industrial revolution that will not be repeated.* That would be an unequivocal mistake. Our population size and technological ingenuity have created a world containing harmful pollutants that can no longer be diluted to harmless levels: the existence of real risk from pollutants now requires due diligence. Techno-industrial progress and environmental legislation proceed unevenly within and among countries, creating ample opportunity for repetition of past mistakes. And novel problems continue to emerge despite our increased diligence and complex regulations. It would be

* This premise is sufficiently prevalent to warrant a label, the **Bad Old Days Incongruity**. The assumption is often made that the worst environmental issues are now past and present issues are being handled adequately with existing legislation and technology. This assumption does not stand up to evidence such as that described herein. The adverse impacts of present and emerging issues are as serious—arguably much more serious in many instances—than those of the past. Increased attention and sophistication in dealing with these issues is required as the world's population increases and our techno-economic systems become more complex and far reaching.

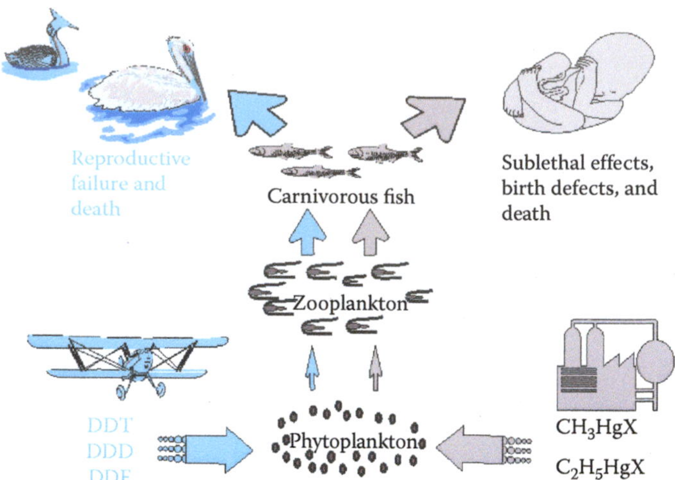

Figure 1.2 Two of the first pollutants to draw attention to the inadequacies of the dilution paradigm were DDT and methylmercury. They accelerated the emergence of the boomerang paradigm. Both chemicals were returned to humans or to valued wildlife species by transfer through food webs.

absurd to argue that, because early problems have been solved, attention and resources can now move from environmental pollution* to other important issues. As we approached and then entered the present millennium, problems extended more and more frequently to transnational and global scales (Figure 1.3). Many now are imbedded in **laggard systems** (Kerr, 2007), that is, systems with adverse effects that only slowly approach a critical state after conditions become established for their emergence. Adverse consequences of our present problems are equivalent to, or often more serious than, those of historic problems although they may manifest more subtly. This makes wise decisions difficult to reach for the public and mandates an increasingly more subtle and granular understanding of ecotoxicological phenomena. It also requires a level of international cooperation that is unfamiliar, and intermittently discomforting, for citizens of many countries. Effective attention must be given to a wide range of contaminants.

Nuclear materials still require our attention and resources. The core of Three Mile Island Reactor Unit 2 (Harrisburg, Pennsylvania) melted on March 28, 1979, releasing 3 Ci of radiation (Booth, 1987). In 1986, nearly 30 years after the Chelyabinsk 40 explosion in the Urals, the Chernobyl Reactor 4 core melted down in the Ukraine, creating the largest radioactive release in history (301 million curies as estimated by Medvedev [1995]). Fallout from Chernobyl spread rapidly across the Northern Hemisphere. The estimated 350,000 m^3 of waste from uranium (and rare earth element) mining and processing operations near the Estonian coastal city of Sillamäe remains and continues to release high levels of radon gas decades after Estonia won back its independence from the Soviet Union (Raukas, 2004). Three reactors in the Japanese Fukushima Daiichi nuclear power facility melted down on March 11, 2011 (Dauer et al. 2011), releasing enough radioactive material to eventually kill a projected 130 humans via fatal cancers (Ten Hoeve and Jacobson, 2012). The International Atomic Energy Agency (2007) estimates that the world's output of nuclear power will double by 2030 as countries' fossil fuel–dominated energy strategies shift to encompass more options. Increased presence of nuclear power stations in our landscape seems inevitable.

* Even a careful reading of Lomborg's book from which the above, superficially contrary quote was taken reveals that his appraisal embraces a central theme of increased thoughtfulness in balancing the benefits and adverse consequences of our activities, that is, "… it is absolutely vital for us to be able to prioritize our efforts in many different fields, e.g., health, education, infrastructure and defense, as well as the environment" (Lomborg, 2001).

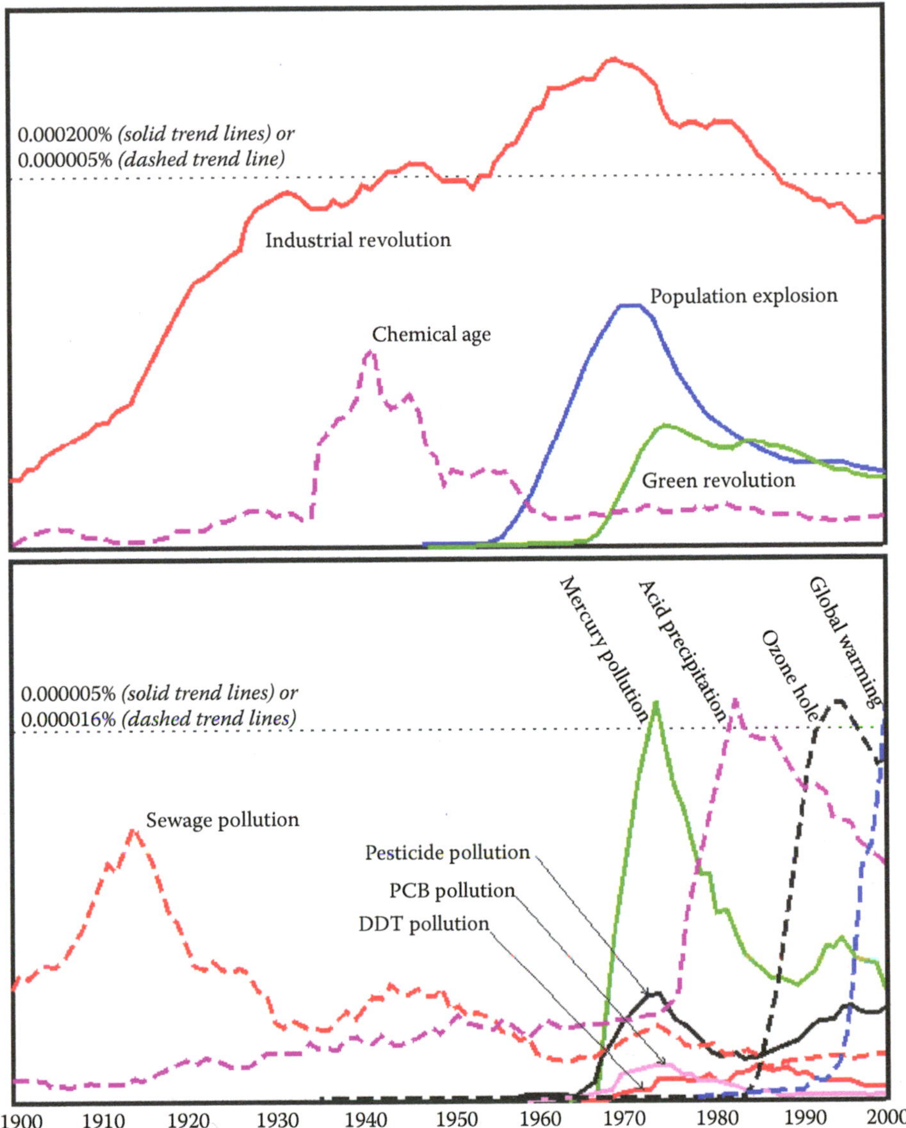

Figure 1.3 Results of an NGram analysis of several bigrams relevant to ecotoxicology. The top panel is the prevalence of bigrams related to social trends that were found in the millions of English books scanned to date by Google (see ngrams.googlelabs.com). The vertical axis reflects the occurrence of the bigram as a proportion of all of the bigrams in these books. The bottom panel depicts the same for bigrams related to impacts of environmental pollution. Clearly, a sequence of events occurring in the twentieth century established an awareness of a sequence of pollution types. An important theme to note is the expending scale for each in the temporal sequence of pollution types. For example, sewage pollution was usually associated with a water body. The more recent pollution types, e.g., the ozone hole or global warming, encompass scales of a continent and the entire biosphere. (Modified from Figure 8.1 in Newman, M. C., *Quantitative Ecotoxicology,* 2nd Edition, Taylor & Francis/CRC Press, Boca Raton, FL, 2013.)

Chemical wastes require continued attention and funding. A myriad of environmental issues remain in affiliated countries after the collapse in 1991 of the Soviet Union (Tolmazin, 1983; Edwards, 1994). Wastes from Soviet-era metal mining and smelting activities in Poland remain extensive and only partially remediated (Figure 1.4). Tributyltin (TBT), an antifouling agent in

Figure 1.4 Pervasive metal contamination exists in Poland's Olkusz mining region. Lead and zinc have been mined here since the medieval ages; however, industrial activities surged through the 1960s and 1970s in a manner typical of the times. Similar, widespread metal contamination was produced around the Avonmouth (UK), Gusum (north Sweden), Copperhill (Tennessee), and Flin Flon (Canada) smelters. In the early 1980s, releases at the Polish mining region decreased as production demand abated, and modern electrofilters and scrubbers were installed (top panel). (See Łagisz et al. [2002], Stone et al. [2001, 2002], and Niklińska et al. [2006] for details about ecotoxicological consequences.)

marine paints, has decimated coastal mollusk populations throughout the world (Bryan and Gibbs, 1991; Leung et al. 2006) yet has only recently been effectively regulated through international agreement. Mercury in fish and game remains a concern as new sources appear such as mercury used in South American gold mining (e.g., de Lacerda et al. 1989; Branches, 1993; Reuther, 1994; Alho and Vieira, 1997; Malm, 1998; Wade, 2013) (Figure 1.5). Subsurface agricultural drainage in the San Joaquin Valley of California brought selenium in the Kesterson Reservoir and Volta Wildlife Area to concentrations causing avian reproductive failure (Ohlendorf et al. 1986). Such high selenium concentrations from irrigation and other sources remain a global issue (Lemly, 2004). Efforts to avoid harm to humans (Ember, 1980; Settle and Patterson, 1980; Millar and Cooney, 1982) and wildlife by reducing lead in products such as gasoline and lead shot have been effective only since the late 1970s. Even well into the 1980s, debate continued about lead's effects and the

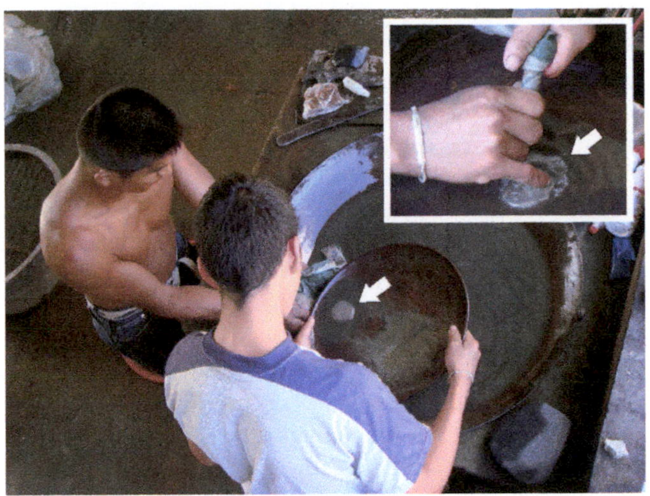

Figure 1.5 Mercury being used to extract gold from ore mined in Portovelo, Ecuador (June 2007). Pulverized ore is agitated with a puddle of mercury (arrows) to form a gold–mercury amalgam, which in this picture, is being placed into a piece of cloth. Mercury is hand-squeezed from the amalgam and remaining mercury is driven off by heating with a propane torch. This process results in substantial human exposure and mercury release into the environment. (Note that the person wringing the mercury from the amalgam has a silver bracelet into which the mercury will also readily dissolve to form a silver–mercury amalgam.) (Courtesy of Lane, K., College of William & Mary.)

need for federal regulation (Anderson, 1978; Marshall, 1982; Anonymous, 1984a; Ember, 1984; Putka, 1992).* Regardless, lead acid battery fabrication and recycling plants in Kenya still result in abnormally high blood lead levels in workers (Were et al. 2012). In late 2007, concern about lead emerged again as high concentrations were measured in toys manufactured by China: China's booming economy overran its capacity to maintain adequate quality control on these imports.

There is little reason to suspect that pollution events such as those that emerged in Europe, North America, and Asia during the 1970s and 1980s will not continue to emerge. The need remains for vigilance and accurate assessment of associated risk. As an infamous instance of confused risk communication, the controversy about the Hooker Chemicals and Plastics Corporation's waste dump sites at Hyde Park and Love Canal became hysterical as the public waited for a clear statement of risk from authorities and scientists (Culliton, 1980; Anonymous, 1981, 1982; Smith, 1982). Similar instances involving high perceived risk and uncertainty will inevitably appear in the future, requiring accurate estimation and communication of real risk. As an example that no one wishes to risk repeating, poor risk estimation and communication resulted in the December 2, 1984, explosion of a storage tank at a Union Carbide pesticide plant (Bhopal, India) and release of a methyl isocyanate cloud that killed 2,000 people and harmed another estimated 200,000 (Anonymous, 1984b; Heylin, 1985; Lepkowski, 1985). The impact on human life from this overnight catastrophe far exceeded that of the 1956 Minamata poisonings.

The intense activities associated with petroleum production, transport, and consumption also require our close attention. Recent examples to support this statement are easily found. On March 16, 1978, the *Amoco Cadiz* supertanker ran aground at Portsall (France) and released

* Relative to our slow acceptance of lead's adverse effect, Tackett (1987) provides a revealing quote by Benjamin Franklin (July 31, 1786). "This my dear Friend is all I can at present recollect on the Subject. You will see by it, that the Opinion of this mischievous Effort from lead is at least above Sixty Years; and you will observe with Concern how long a useful Truth may be known and exist, before it is generally receiv'd and practis'd on." Indeed, Michaels (2008) describes acute lead toxicity in hundreds of workers only a few decades ago that was related to lead-based gasoline additive production. As recently as one decade ago—200 years after this quote was made, the value of reducing lead in gasoline was being actively questioned.

roughly 209,000 m^3 of crude oil (Ellis, 1989). On March 24, 1989, March 24, 1989, the *Exxon Valdez* spilled 41,340 m^3 of crude oil into Prince William Sound. The oil covered an estimated 30,000 km^2 of Alaskan shoreline and offshore waters (Piatt et al. 1990). Marine bird populations are still recovering from this spill (Lance et al. 2001). From August 2, 1990 until February 26, 1991, the largest oil release to have ever occurred at that time was deliberately spilled by Iraqi troops occupying Kuwait. Half a million tons (roughly equivalent to 522,000 m^3) of crude oil from the Mina Al-Ahmadi oil terminal were pumped into the Arabian Gulf (Sorkhoh et al. 1992). Plumes of contaminating smoke from the intentional ignition of Kuwaiti oil wells by the Iraqi troops were visible from space (Figure 1.6). Beginning on April 20, 2010, and lasting for 84 days (Atlas and Hazen, 2011), the *Deepwater Horizon* drilling rig blowout released between four and five million barrels (approximately 715,442 m^3) of oil into the Gulf of Mexico, exceeding the 1990 Kuwaiti spill in size and easily displacing the 1989 *Exxon Valdez* spill as the largest oil spill in U.S. history (Camill et al. 2010; Kerr et al. 2010).* No fundamental change has occurred since the twenty-first century began that would exclude these kinds of releases happening at the current frequency into the near future.

Other smaller or more diffuse, but incrementally as damaging, events also require expertise in ecotoxicology. Beyond the intentional release described above, the Arabian Gulf receives 67,000 m^3 of oil annually from smaller leaks and spills (Sorkhoh et al. 1992). Before the *Exxon Valdez* spill in Alaska, a 1978 act of sabotage to the trans-Alaska pipeline had released 2540 m^3 of oil onto land near Fairbanks. In October 2001, 1081 m^3 of oil gushed from a hole shot in the trans-Alaska pipeline by an intoxicated man. The average number of oil spills and volume per spill in or around U.S. waters from 1970 to 1989 were 9,246 and 47,000 m^3, respectively, with no obvious downward

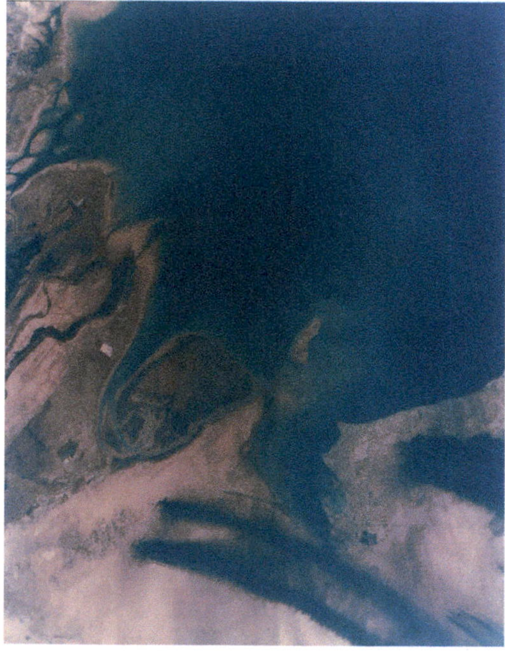

Figure 1.6 Kuwaiti oil wells set afire by Iraqi troops as seen from a U.S. space shuttle flight. Oil wells are seen burning north of the Bay of Kuwait and immediately south of Kuwait City. (Courtesy of NASA.)

* It also exceeded the June 1979 Mexican IXTOC 1 well oil spill that lasted 9 months and released three and a half million barrels (556,430 m^3) into the southern Gulf of Mexico (Kerr et al. 2010).

trend in either through time (Table 8 in Gorman, 1993). Three more events seem typical of those we should expect into the foreseeable future. On November 8, 2007, an estimated 220 m^3 of bunker fuel was spilt into San Francisco Bay from a container ship that struck the Bay Bridge due to a navigational mix-up. A week later, 4921 m^3 of oil was spilt near the Strait of Kerch when the Russian tanker, *Volganeft-139*, broke up in a north Black Sea storm. Reports by MSNBC suggested inadequate ship maintenance as a significant contributor to what the Krasnodar region's governor, Alexander Tkachyov, referred to as an "ecological catastrophe" (MSNBC News Services, 11/12/2007). In contrast to the previously discussed 1957-to-1960 loss of a few hundred grebes on Clear Lake to DDD, this one wreck quickly killed 30 thousand oiled seabirds and damaged critical habitat along a major bird flyway. One month after these two oil spills (December 7, 2007), an estimated 12,520 m^3 of crude oil spilled from the single-hull supertanker, *Hebei Spirit*, after it struck a crane on a barge off the west coast of Korea, resulting in Korea's largest spill. Clearly, real risk for harmful oil spills will be present into the near future and require our full attention.

Chemicals from our agricultural activities also require continued diligence in the twenty-first century. At the time Rachel Carson was writing *Silent Spring* (*circa* 1960), annual production of synthetic organic chemicals was 43.9 billion kilograms. Worldwide production had climbed to 145.1 billion kilograms by 1970 (Corn, 1982). By 1985, U.S. use of pesticides roughly doubled from the 227 million kilograms used in 1964 (Figure 7 in Gorman [1993]). Growing dependency on a complex array of pesticides has now become a global trend that is particularly problematic for countries with rapidly developing economies.

> *As we approached the* [Chinandega, Nicaragua] *airport the now familiar stench of chemicals became overpowering. As we walked down the airstrip to the health clinic in the airport complex, I glanced across the adjacent fields toward the nearby dwellings of town. Nothing was moving in the open space, not birds, not insects, none of the creatures normally found in such abundance in the tropical climates. Here indeed, Rachel Carson's prophecy seemed all too real.*
>
> **Murray (1994)**

Many persistent pesticides banned in developed countries are still used in developing countries (Simonich and Hites, 1995). Less persistent, but more toxic, pesticides are also used under inadequate regulation in many developing countries, inflicting more harm to humans and ecological entities than the harm from DDT described by Rachel Carson (Murray, 1994; Roth et al. 1994; Castillo et al. 1997, 2000; Henriques et al. 1997; Ecobichon, 2001; Murray et al. 2002; Wesseling et al. 2005).

> *In light of the growing evidence that many chemicals disrupt hormones, impair reproduction, interfere with development, and undermine the immune system, we must now ask to what degree contaminants are responsible for dwindling animal populations.*
>
> **Colborn et al. (1996)**

Unsuspected effects of conventional contaminants are being discovered and brought to the public's attention by diverse means such as the bestselling book, *Our Stolen Future* (Colborn et al. 1996). Reflecting a welcome maturation of our collective environmental ethic, *Our Stolen Future* was taken seriously by industries instead of receiving the indignation and counterattacks that greeted *Silent Spring*. This is important because much uncertainty remains about pollutant effects to endocrine systems. An important instance underscoring the need for increased diligence involves one of the most widely used agrochemicals today, atrazine. Despite decades of controversy, it has now been determined to harm aquatic organisms with key effects involving endocrine system dysfunction (Rohr and McCoy, 2010).

And unsuspected movements of conventional pollutants are also being elucidated. **Persistent organic pollutants** (POPs)* are particularly disconcerting because they are now known to disperse widely and accumulate to high concentrations in wildlife (e.g., Weber and Goerke, 2003; Tanabe, 2004; Lubick, 2007a) and humans (e.g., Landrigan et al. 2002; Lorber and Phillips, 2002; Pronczuk et al. 2002; Solomon and Weiss, 2002) (Figure 1.7) in regions far removed from their points of release. In response to this new understanding, the 2004 Stockholm Convention called for a gradual, international elimination of the most prominent, including polychlorinated biphenyl, dioxins, furans, and nine pesticides. Also the European Union's Registration, Evaluation, and Authorization of Chemicals (REACH) directive has included as one important aim the reduction of POPs (Tanabe, 2004).

China is choking on its own success. The economy is on a historic run, posting a succession of double-digit growth rates. But the growth derives, now more than at any time in the recent past, from a staggering expansion of heavy industry and urbanization that requires colossal inputs of energy, almost all from coal, the most readily available, and dirtiest, source.

Kahn and Yardley (2007)

New situations of concern continue to emerge in which conventional contaminants cause undeniable harm. Often, the stage of a country's economic development is a major contributor. Certainly, the unethical shipping of toxic waste from developed nations to underdeveloped African countries that occurred in the 1980s is one blatant example in which this was the case (Vir, 1989; Lipman, 2002; Simpson, 2002). Another is China's current economic growth as described in the aforementioned quote. China has begun to conscientiously balance economic growth and environmental stewardship (Liu and Diamond, 2005; Aunan et al. 2006; Fu et al. 2007; Zhang, 2007), but major obstacles remain. Features of the current Chinese historic economic growth, such as the **Township**

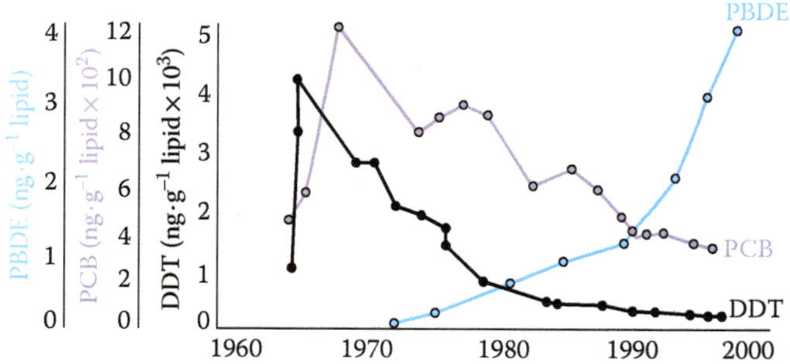

Figure 1.7 Trends in three persistent organic pollutants (POPs) in Swedish breast milk. DDT was banned in Sweden in the early 1970s, and accordingly, concentrations drop through the decades after peaking in the mid-1960s. A less obvious decline is seen for polychlorinated biphenyl (PCB) concentrations, resulting in continued concern about adverse PCB effects on humans and wildlife. Concentrations of brominated POPs used as fire retardants (polybrominated diphenyl ethers [PBDE]) increased rapidly through the decades; however, considerable effort is now being spent in the European Union to move away from PBDE use (Birnbaum and Staskal, 2004; Betts, 2007; Blum, 2007). Note that the scales for DDT, PCB, and PBDE are ng·g^{-1} of milk lipid ×10^3, ×10^2, and ×10^0, respectively. (Data extracted from Figures 3, 9, and 11 in Solomon, G. M. and Weiss, P. M., *Environ. Health Perspec.*, 110, A339–A347, 2002.)

* POPs are those organic pollutants that are long-lived in the environment. Many tend to increase in concentration as they move through food webs. Often, **Persistent, Bioaccumulative Toxicants** or Persistent, Bioaccumulative Chemical is the specific term used for POPs that tend to biomagnify in food webs.

and Village Enterprise (TVE) sector,* create circumstances with the potential for creating much more environmental damage than typical during such an economic surge. Other challenges to environmental compliance need resolution as China's economy booms (McElwee, 2011). One issue is the current tendency for regulators responsible for enforcement to be paid through, and consequently answerable to, local officials who might favor development over environmental protections. A third, less apparent, issue is deeply rooted guanxi† that may obligate a regulator to local affiliates. This obligation might take precedence over, or sway interpretation of, loosely written central government regulations. Despite these impediments, effective environmental concern is now crucial in many regions of China that have unique problems. For instance, the extensive pollution in Baotou (Inner Mongolia) is one linked to China's growth to dominate global rare earth element mining‡ (e.g., Yongxing et al. 2000; Zhang et al. 2000; Li, 2006; Wen et al. 2006) (Figure 1.8).

Yet another recent instance involved economically driven changes in conventional pesticide use in Central America (Figure 1.9). For several decades preceding the 1980s, agro-economic forces moved Nicaraguan agriculture away from a modest importation of arsenate-based (1920s cotton production) and then organochlorine pesticides. The amounts of pesticide imported and the toxicities of those imported pesticides increased dramatically when Nicaraguan agriculture shifted from large cotton plantations to smaller farms growing a range of export crops (Murray, 1994). Lax regulation based on ineffective pesticide use registration was pervasive. The Pan American Health Organization (2002) estimated that 97.5% of all pesticide use in Nicaragua from 1992 to 2000 was unregistered. The surge in acute human poisonings shown in the top panel of Figure 1.9 resulted from this regulatory failure and unsafe handling by small farmers who were accustomed to less toxic pesticides. A more recent, but similar, trend is clear in the bottom panel.

PBDEs, used as fire retardants in furniture, are structurally similar to the known human toxicants PBBs, PCBs, dioxins, and furans. In addition to having the similar mechanisms of toxicity in animal studies, they also bioaccumulate and persist in both humans and animals.

Blum (2007)

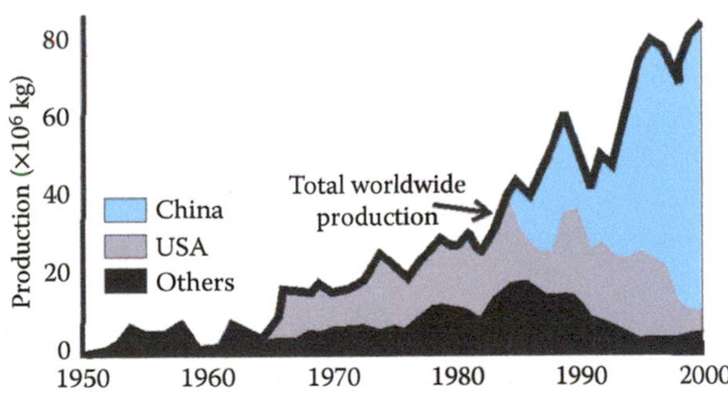

Figure 1.8 China's rapid rise to dominance in rare earth element production (redrawn from Figure 1 of the USGS Fact Sheet 087-02 [USGS, 2002]). These elements are essential components of modern color television screens, liquid-crystal displays of computer screens, optical networks, glass polish systems, and magnet technologies.

* The **TVE sector** comprised 20 million small factories throughout the Chinese countryside (Tilt, 2006). These small factories are responsible for much of the economic upswing in China but also are the most difficult to control relative to pollution. Drawing on World Bank analysis of China, Tilt (2006) states that the TVE sector accounts for 60% of the "air and water pollution, endangering human health and posing a serious threat to agro-ecosystems."

† As applied here, guanxi (关系) refers to a tradition of behaviors within a personal network of relationships in which members perceive a mutual obligation and capacity to ask for a favor in return. Guanxi is important and widespread in China.

‡ The **rare earth elements** include La, Ce, Pr, Nd, Pm, Sm, Eu, Gd, Tb, Dy, Ho, Er, Tm, Yb, Lu, and Y. They are the lanthanides with atomic numbers from 57 to 71 plus scandium (atomic number 21) and yttrium (atomic number 39).

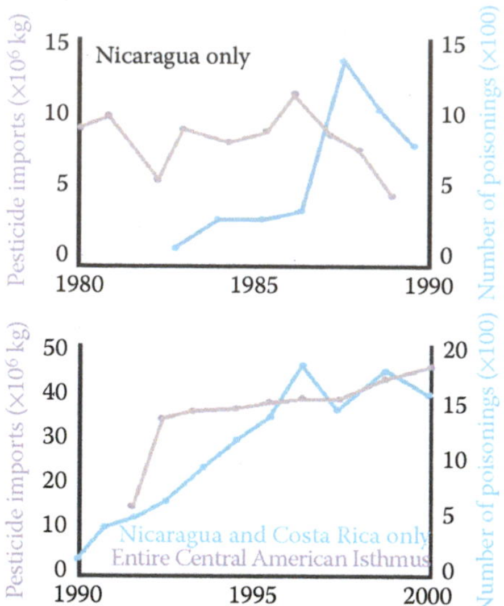

Figure 1.9 Pesticide use and effects in Central American countries. Exemplary of trends during the 1980s are those shown in the top panel for Nicaragua (from Figure 3-3 and Table 4-3 of Murray [1994].) The bottom panel depicts the same themes today of increasing pesticide importation into the Central American isthmus (Pan American Health Organization, 2002) and consequent acute human poisoning, e.g., that for Costa Rica and Nicaragua. (From Wesseling, C. M. et al. *Toxicol. Appl. Pharm.*, 207, S697–S705, 2005.)

And releases of new chemicals are becoming a concern. Brominated fire retardants, chlorination by-products, fluorinated chemicals, synthetic estrogens, hormones from livestock production, alkylphenol ethoxylates and their degradation products, manufactured antimicrobial products, pharmaceuticals, and constituents of personal care products* are important examples of heretofore ignored contaminants that are presently discharged in large quantities (Hale et al. 2001; Meyer, 2001; Daughton, 2004; Chapman, 2006). Being unconventional contaminants, most are inadequately regulated (Daughton, 2004). And novel wastes occasionally emerge such as those associated with nanotechnologies (Daughton, 2004; Chapman, 2006) and electronic wastes (Widmer et al. 2005). Only during the last few years have ecotoxicologists begun to take adequate notice of the potential, widespread impacts of these materials in natural systems.

Contaminants amenable to atmospheric transport have become especially disconcerting because of the spatial scale over which they have an impact. This is certainly a major concern with many POPs. Acid rain is now a transnational problem (Likens and Bormann, 1974; Likens, 1976; Cowling, 1982), damaging both aquatic (Glass et al. 1982; Baker et al. 1991) and terrestrial (Cowling and Linthurst, 1981; Ellis, 1989) ecological systems. With an estimated 69% of China's energy derived from coal, acid rain has become extensive in Asia (Larssen et al. 2006). Chlorofluorocarbons (CFCs) used as propellants and coolants have been linked to ozone depletion in the stratosphere (Zurer, 1987, 1988; Kerr, 1992), and efforts are being made to greatly reduce their use (Crawford, 1987). But, despite the milestone 1987 Montreal Protocol call for complete elimination of CFC use by 2000, efforts by lawmakers were still underway into the mid-1990s to delay, and even avoid, any U.S. reduction of CFC emissions (Lee, 1995). Such common delay tactics result in unnecessary harm later.

* **Pharmaceuticals and Personal Care Products** released into the environment, especially from wastewater treatment facilities, are often referred to collectively as PPCPs.

The above monotonous litany of problems is not intended to convince the reader that techno-industrial advancement is incompatible with environmental integrity and human well-being. Rather, it is intended to establish two simple truths. First, by the mid-twentieth century, the dilution paradigm failed with clearly unacceptable consequences to human health and ecological integrity. Second, expertise in ecotoxicology is now critical to our well-being. The Minamata poisonings and eggshell thinning by DDT were the first blatant signs—not the high water marks—of unacceptable environmental pollution. The approaches and ethics established during the resource conservation, workplace health, and food safety movements were modified to address pollution issues faced by society.

Major environmental problems remain and new ones appear annually that are as, or more, significant than historical problems. Appropriately, environmental themes are so interwoven into our culture that associated concepts are communicated to the general public under catchy titles such as *Silent Spring, Our Stolen Future*, or *An Inconvenient Truth*. Expertise in ecotoxicology is essential for weighing the costs and benefits of the innumerable technological and industrial decisions affecting our lives. Nonmarket goods and services, and natural capital (Odum, 1996; Prugh, 1999) must be included in decision making. Such services provided pro bono by nature are estimated to be in the range of $33 trillion annually, twice the annual gross domestic product of the countries of the earth (Rousch, 1997). Investment of time, thought, and resources to avoid damage to service-providing natural systems is economically, as well as ecologically, wise behavior. Environmental regulations reduce human suffering, foster sustainable economic prosperity, and allow responsible environmental stewardship.

> *... growth and the environment are not opposites – they complement each other. Without adequate protection of the environment, growth is undermined; but without growth it is not possible to support environmental protection.*
>
> **Lomborg (2001)**

VIGNETTE 1.1 Emergence and Future of Ecotoxicology

John Cairns, Jr.
Virginia Institute and State University

Quite naturally, toxicology began with concerns about adverse effects of chemical substances on humans. The focus of toxicology gradually extended to organisms domesticated by humans and then to wild organisms of commercial, recreational, or aesthetic value. Unfortunately, these toxicity tests are all homocentric, commendable but inadequate when human society is dependent on an ecological life-support system. The close linkage between human health and the health of natural systems makes an ecocentric toxicity component essential. Figure 1.10 illustrates the dimensions of this challenge and shows that, despite remarkable progress in the field of ecotoxicology over the last four decades, there is still a long way to go.

In its earliest stages, environmental toxicology depended on short-term laboratory tests with single species that were low in environmental realism but with satisfactory replicability. However, such tests used neither endpoints nor levels of complexity characteristic of ecosystems. Neither did they include cyclic phenomena and many types of variability that are the norm for the complex, multivariate systems known as ecosystems. In short, the "eco" was seriously deficient in the field of ecotoxicology. Although simple tests were often used as surrogates for more important properties, toxicity tests at lower levels of biological organization, such as single species, were not readily validated at higher levels of biological organization, such as communities, ecosystems, or landscapes.

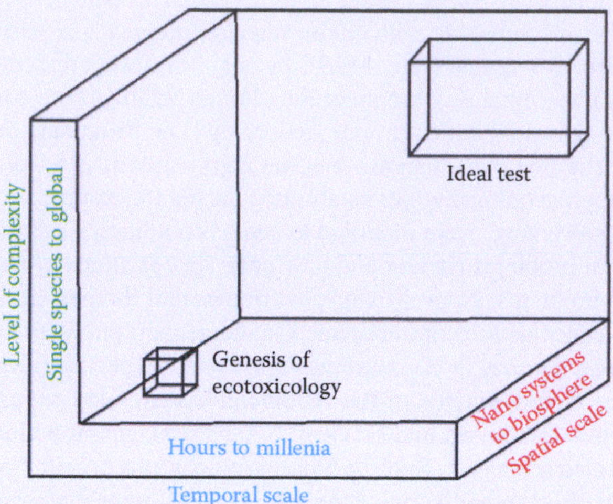

Figure 1.10 Schematic depiction of the genesis of ecotoxicology and the ideal goal about which one can,at present, only speculate. All three major components should be matched to the problem of interest to achieve the goal.

RESHAPING THE PLANET

Arguably, the impetus for the development of ecotoxicology resulted from the unmistakable environmental transformations that occurred in the twentieth century on a scale unique in human history. The world has become increasingly humanized, and ecosystems have become more fragmented and diminished in aggregate size. Consequently, ecosystems have begun to lose resilience and need greater protection from threats to their integrity. This situation is an uncontrolled "experiment" on a planetary scale for which the outcome is uncertain. Although environmental change has been the norm for 4 billion years, the planet has been altered by the human species for 4 million years; however, the rate and intensity of change that occurred in the last century are a cause for deep concern. One might reasonably assert that ecotoxicology is an attempt to provide some rules for the planetary game human society is playing. Almost every sport has elaborate rules that are discussed in great detail by fans. The human, therefore, needs to recognize the natural laws that determine the outcome of the game of life in which all are participants.

FUNCTIONING DESPITE UNCERTAINTY

Both the human condition and the "tools of the trade" (ecotoxicology) are constantly changing. Theories and practices once thought to be sufficient have been shown to be inadequate, often with stunning rapidity. Ecotoxicology can make major contributions in reducing the frequency and intensity of environmental surprises by (1) determining critical ecological thresholds and breakpoints, (2) developing ecological monitoring systems to verify that previously established quality control conditions are being met, (3) establishing protocols for the protection and accumulation of natural capital, (4) providing guidelines for implementing the **precautionary principle**,* (5) developing guidelines for anthropogenic wastes that contribute to ecosystem health, and (6) responding to environmental changes with prompt remedial ecological restoration measures when evidence indicates that an important threshold has been crossed.

* The cautious policy or position that, even in the absence of any clear evidence and in the presence of high scientific uncertainty, action should be taken if there is any reason to think that harm might be caused.

NATURAL CAPITAL AND INDUSTRIAL ECOLOGY

Natural capital consists of resources, living systems, and ecosystem services. Natural capitalism envisions the use of natural systems without abusing them, which is essential to sustainable use of the planet. Sustainable use of the planet requires a mutualistic relationship between human society and natural systems, and affirms that a close relationship exists between ecosystem health and human health. Natural capitalism deals with the critical relationship between natural capital—natural resources, living systems, and the ecosystem services they provide—and human-made capital (Hawken et al. 1999).

Industrial ecology is the study of the flows of materials and energy in the industrial environment, and the effects of these flows on natural systems (White, 1994; Graedel and Allenby, 1995). The essential idea of industrial ecology is the coexistence of industrial and natural ecosystems. Properly managed, industrial ecology would enhance the protection and accumulation of natural capital in areas now ecologically degraded or at greater risk than necessary. However, the most attractive feature of industrial ecology may be that it would involve temporal and spatial scales greater than those possible with even the most elaborate microcosms or mesocosms. To optimize the quality and quantity of information generated by these hybrid systems, some carefully planned risks must be taken, and regulatory agencies must be sufficiently flexible to permit them. Because some ecologic damage is inevitable under these circumstances, ecotoxicologists must be knowledgeable of the practices commonly used in ecological restoration. Even industrial accidents can be a valuable source of ecotoxicological information if they are immediately studied by qualified personnel and the information widely shared. Regulatory and industrial flexibility in assessing experimental remedial measures would also enhance the quality of the information base. The obstacles to achieving this new relationship between industry and regulatory agencies boggle the mind, yet there seems to be no comparable, cost–effective means of acquiring the needed ecotoxicological information over such large temporal and spatial scales. Convincing the general public and its representatives of the values inherent in this approach will be a monumental task, but the consequences of making ecotoxicological decisions with inadequate information are appalling.

SPECULATIONS

Failure to react constructively to unsustainable practices does not always lie in not knowing what is wrong or even in not knowing what to do about it but, rather, in the failure to take this knowledge seriously enough to act on it. Thus, ecotoxicologists have a responsibility to raise public literacy about their field to ensure that the information they generate is taken seriously and used effectively. Ecotoxicologists and other environmental professionals must be aware that their data, predictions, estimates, and knowledge will be used in a societal context that is embedded in an environmental ethos (or set of guiding beliefs). Because science is often idealized as independent of an ethos, there is some level of tension between the essentially scientific task of estimating risk and uncertainty, and the value-laden task of deciding what level and type of risks are acceptable. If sustainable use of the planet becomes a major goal for human society, it is difficult to visualize how the mixture of science and value judgment can be avoided. One hopes that the process of science, with its priceless quality control component, can remain intact for data generation and analysis while producing better and more relevant information to be used with intelligence and reason in making value judgments. However, sustainable use of the planet will require a major shift in present human values and practices. At the same time, ecotoxicology will be evolving, possibly rapidly, so that keeping system and order in the mixture will be difficult.

Although some of the trends in ecotoxicology briefly described here seem probably inevitable, the direction of the field will almost certainly be determined by environmental surprises, which are ubiquitous and unlikely to occur in convenient places at convenient times. Worse yet, the temptation will be strong to study them entirely with existing methodology, which will probably be inadequate.

A multidimensional research strategy is needed that emphasizes ecosystem complexity, dynamics, resilience, and interconnectedness, to mention a few important attributes. However, major obstacles exist to the development of such a program in the educational system, governmental agencies and industry, and with a citizenry increasingly suspicious of science and academe in general. Society depends primarily on its major universities for the generation of new knowledge; however, this function has become a commodity produced for sale, which means the research direction is all too often a function of marketability. An unfortunate consequence is that writing grant proposals (now often termed contracts) consumes an ever-increasing proportion of the time of ecotoxicologists and other environmental professionals. Ecotoxicologists may often postpone visionary, long-term projects whose outcomes are highly uncertain for short-term projects of severely limited scope determined by the perceived needs of the funding organization rather than being truly exploratory undertakings.

Some counter trends exist to these discouraging developments, often occurring outside of "mainstream" science. A number of new journals are challenging the fragmentation of knowledge, and publications are espousing the consilience ("leaping together") of knowledge. Increasingly, within academe, the focus is on issues that transcend the capabilities of a single discipline or even a few disciplines. In addition, environmental professionals, such as ecotoxicologists, are finding ways to minimize the effects of budgetary constraints. However, it still seems likely that one or more major environmental catastrophes will be needed to persuade decision makers that a major shift in approach is needed to cope more effectively with the ecotoxicological and other uncertainties, which human society now faces and are likely to increase substantially in the future.

Much evidence is also available on other major issues affecting ecotoxicology, including the following:

1. Climate change, including global warming: Rainfall patterns are changing and mountain snow, the source of water for many rivers, is decreasing. Markedly reduced flow in major rivers means far less dilution of the wastes discharged into them. Major floods displace contaminated sediments and probably the ecological partitioning of the chemicals associated with them. Unstable climate conditions will weaken many species, making them more vulnerable to toxic stress. Drought-stressed terrestrial plants and animals will also often be more sensitive to toxicants. Finally, because ecosystems are structured to optimize energy and nutrient flow, loss of keystone species will alter both ecosystem function and structure. All these factors must be carefully considered when designing and interpreting ecotoxicological tests at different levels of biological organization.
2. Peak oil and "tough oil" (e.g., tar sands): Once oil supplies have peaked globally, petroleum availability will rapidly decline. Efforts will be made to get the remaining oil from fields that have had declined output. However, the major challenge will be to study the ecotoxicological effects of the attempts to convert the tar sands into some type of fuel. The Athabasca site will furnish much useful information, but as usual, each site will be ecologically unique in some characteristics, and, as a consequence, each will require somewhat different ecotoxicological methods and procedures. Validation of the ecotoxicological predictions will also require site-specific modifications.

3. Ecological overshoot: **Ecological overshoot** means using resources faster than they can be regenerated. August 22 was Earth Overshoot Day for 2012, marking the date when humanity had exhausted nature's budget for the year (Global Footprint Network, www.footprintnetwork.org). The most obvious solution to the ecological overshoot is to curtail resource extraction and use. However, with global human population increasing by 230,970 persons per day (2012 World Population Data Sheet, Population Reference Bureau), plus adverse economic impacts, this approach is unlikely. The most obvious short-term solution is to restore or rehabilitate damaged ecosystems. Because toxic substances were probably involved in many such situations, ecotoxicological tests will be essential to determine when the site will be suitable for recolonization.

ACKNOWLEDGMENTS

I am indebted to Darla Donald for editorial assistance meeting the requirements of the publisher.

Recognition of our increasing influence on the biosphere since the late eighteenth century has led a growing group of scientists to the weighty proposition that the Holocene epoch has ended and a new epoch begun. Their argument begins by restating that we divide the history of our planet into periods based on profound geological, atmospheric, biological, or climate changes during its evolution. For instance, the present Holocene epoch is agreed to have started as the last ice age ended. The argument continues by pointing out that several human-induced changes have taken place that are similar in impact to many used in the past to distinguish among geological epochs. Steffen et al. (2011) identify four such changes: (1) climate change; (2) significant alteration of elemental cycles such as those of nitrogen,* phosphorus, and sulfur; (3) substantial modification of the terrestrial water cycle; and (4) precipitation of the sixth mass extinction event in the history of life. The name, **Anthropocene**, was proposed for this new epoch by E. F. Stoermer of the University of Michigan, and since then, has been advocated by Nobel Laureate, Paul Crutzen,† in many publications, e.g., Crutzen (2002), Rockström et al. (2009a,b), Steffen et al. (2011), and Zalasiewicz et al. (2010). They propose the late eighteenth century as the beginning of this epoch dominated by human activity (Crutzen, 2002; Zalasiewicz et al. 2010). Suggestive of the value of accepting that a new epoch has emerged, the Anthropocene context is already being used to gauge our impact on the oceans (e.g., Vidas, 2011; Tyrrell, 2011), climate (Kellie-Smith and Cox, 2011), and land resources (Ellis, 2012).

One important theme coming out of initial assessments of the Anthropocene is the comparison of current use of resources to the estimated planetary boundary values. As long as humanity does not exceed these planetary boundary limits, it will be unencumbered in its pursuit of social and economic evolution progress (Figure 1.11). Currently, our activities have exceeded the values for climate change, nitrogen cycling, and species loss. It is disconcerting to read these analyses and realize that boundary values are not yet established for chemical pollutants. How can ecotoxicologists "[respond] to environmental changes with prompt remedial ecological restoration … when an important threshold has been crossed" as John Cairns advocates in Vignette 1.1 if we have no idea if an important threshold has been crossed yet? We remain uncertain whether we are or are not exceeding the biosphere's capacity to cope with our chemical wastes.

* For example, Law (2013) describes recent findings that atmospheric deposition of nitrogen from agricultural fertilizers use has increased so much that its influence can be seen on global evergreen forests.

† Crutzen received the 1995 Nobel Prize in Chemistry with Mario Molina and F. Sherwood Rowland based on their work on atmospheric ozone chemistry and the impact of anthropogenic releases on the formation of the ozone hole.

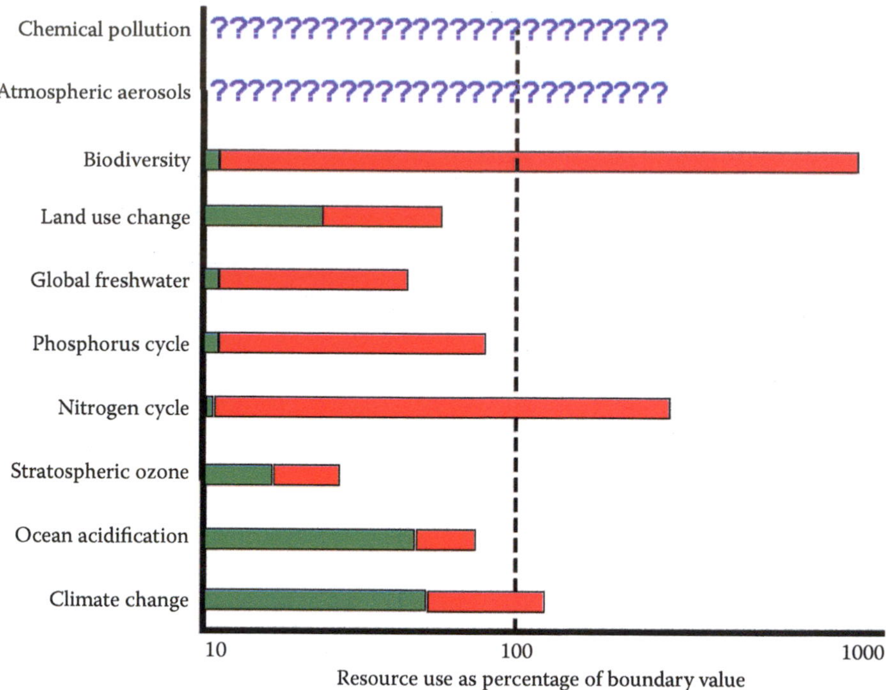

FIGURE 1.11 **Planetary boundary values** and estimates of preindustrial (green) and current values (red). Information extracted from Table S2 of Rockström et al. (2009b). According to these authors, humanity "has the freedom to pursue long-term social and economic development" only if these boundary values are not exceeded. Climate change units: carbon dioxide concentration; Ocean acidification: aragonite saturation ratio; Stratospheric ozone: ozone expressed in Dobson unit; Nitrogen cycle: Metric tonnes removed from the atmosphere for human use annually; Phosphorus cycle: Metric tonnes of flow into oceans annually; Global freshwater use: consumptive use of runoff as km^3 annually; Land use change: percentage of land cover converted to cropland; Biodiversity: number of species lost per million species annually.

1.3 ECOTOXICOLOGY

The subject of this book could have been labeled environmental toxicology or ecotoxicology because the definitions of both terms are rapidly converging. Often, use of the term, environmental toxicology, implies that only effect of environmental contaminants to humans will be discussed. Such an implication would be inappropriate for this book. Some definitions of ecotoxicology seem to exclude discussion of humans except as the source of contaminants (Table 1.1), but the original definition given to ecotoxicology by Truhaut (1977) includes effects to humans. Truhaut's rationale, as paraphrased by Bacci (1996), is quite sound: "human health cannot be protected unless in conjunction with wildlife protection…." Ecotoxicology was selected here with the intention of emphasizing effects to ecological entities without ignoring complementary knowledge associated with effects to humans. Ecotoxicology was chosen after some introspection because a strong and confining conventional ecosystem-or-lower emphasis is implied in many of its definitions. For example, the textbook by Connell (1990) specifies that the scope of ecotoxicology includes "organisms, populations, communities, and ecosystems" (Table 1.1). More and more, this conventional individual → population → community → discrete ecosystem context is necessarily being extended to metacommunity → ecosystem → landscape → ecoregion → continent → hemisphere → biosphere. There is a simple reason of this: the conventional framework has gradually grown insufficient to encompass all germane

Table 1.1 Definitions of Ecotoxicology and Environmental Toxicology

Definition	Reference
Environmental toxicology	
1. The study of the effects of toxic substances occurring in both natural and manmade environments	Duffus (1980)
2. The study of the impacts of pollutants upon the structure and function of ecological systems (from molecular to ecosystem)	Landis and Yu (1995)
Ecotoxicology	
1. The branch of toxicology concerned with the study of toxic effects, caused by natural and synthetic pollutants, to the constituents of ecosystems, animals (including human), vegetable and microbial, in an integrated context	Truhaut (1977)
2. The natural extension from toxicology, the science of poisons on individual organisms, to the ecological effects of pollutants	Moriarty (1983)
3. The science that seeks to predict the impacts of chemicals upon ecosystems	Levin et al. (1989)
4. The study of the fate and effect of toxic agents in ecosystems	Cairns and Mount (1990)
5. The science of toxic substances in the environment and their impact on living organisms	Jørgensen (1990)
6. The study of toxic effects on nonhuman organisms, populations and communities	Suter (1993)
7. The study of the fate and effect of a toxic compound on an ecosystem	Shane (1994)
8. The field of study, which integrates the ecological and toxicological effects of chemical pollutants on populations, communities and ecosystems with the fate (transport, transformation and breakdown) of such pollutants in the environment	Forbes and Forbes (1994)
9. The science of predicting effects of potentially toxic agents on natural Hoffman et al. (1995) ecosystems and nontarget species	Hoffman et al. (1995)
10. The study of the pathways of exposure, uptake and effects of chemical agents on organisms, populations, communities, and ecosystems	Connell (1990)
11. The study of harmful effects of chemicals on ecosystems; the harmful effects of chemicals (toxicology) within the context of ecology	Walker et al. (2001)
12. The study of harmful effects of chemicals upon ecosystems and includes effects on individuals and consequent effects at the levels of population and above	Walker et al. (2012)

subjects. As applied here, **ecotoxicology** is the science of contaminants in the biosphere and their effects on constituents of the biosphere, including humans.

1.4 ECOTOXICOLOGY: A SYNTHETIC SCIENCE

1.4.1 Introduction

Ecotoxicology is a synthetic science in that it draws together insights and methods from many disciplines (Figure 1.12). Questions about effect are posed at the molecular (e.g., enzyme inactivation by a contaminant), to the population (e.g., local extinction), and to the biosphere (e.g., global warming) levels of biological organization. Questions of fate and transport are addressed from the chemical (e.g., dissolved metal speciation), to the habitat (e.g., contaminant accumulation in depositional habitats), and to the biosphere (e.g., global distillation of volatile pesticides) levels of physical scale. Sometimes, this can produce a confusing complex of scales and associated specialties. The key to maintaining conceptual coherency in this complex of interwoven and hierarchical topics was articulated by Caswell (1996), "… processes at one level take their mechanisms from the level below and find their consequences at the level above.… Recognizing this principle makes it clear that there are no truly 'fundamental' explanations, and make it possible to move smoothly up and down the levels of the hierarchical system without falling into the traps of naive reductionism or pseudo-scientific holism." Understanding fates and effects at all levels is essential for effective environmental stewardship (Newman, 1996). Equally important is the integration of our collective understanding of fate and effects at all levels into a coherent and self-consistent body of knowledge (Newman and Clements, 2008).

Although all levels are equally important, they contribute differently to our efforts and understanding (Figure 1.13). Questions dealing with lower levels of the conceptual hierarchy, such as, biochemical effects, tend to be more tractable and have more potential for easy linkage to a specific cause than do effects at higher levels such as the biosphere. Changes in δ-aminolevulinic acid dehydratase (ALAD) activity in red blood cells can be assayed inexpensively and quickly linked to lead exposure of humans or wildlife. The general loss of fish species from Canadian lakes was much more challenging to document and to link to sulfur and nitrogen oxide emissions from a distant city that produced acid precipitation. As a consequence, effects at lower levels of the ecological hierarchy are used more readily in a proactive manner than are those at higher levels. They can indicate the potential for emergence of an adverse ecotoxicological effect, whereas effects at higher levels are useful in documenting or prompting a regulatory reaction to an existing problem. Although highly tractable and sensitive, the ecological relevance of effects at lower levels is much more ambiguous than effects at higher levels of organization. Most reasonable biologists would agree that a 50% reduction in species richness is a clear indication of diminished health of an ecological community. But a 50% increase in metallothionein in adults of an indicator species provides an equivocal indication of the health of species populations contributing to the associated community. Relative

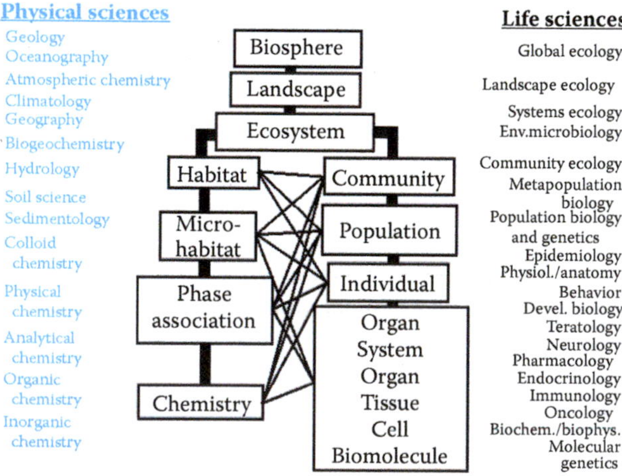

Figure 1.12 Hierarchical organization of topics addressed by ecotoxicology. Disciplines contributing to understanding abiotic interactions are listed on the left side of the diagram and those contributing to understanding biotic interactions are listed on the right. Important interactions, denoted by lines connecting components, occur between biotic and abiotic components.

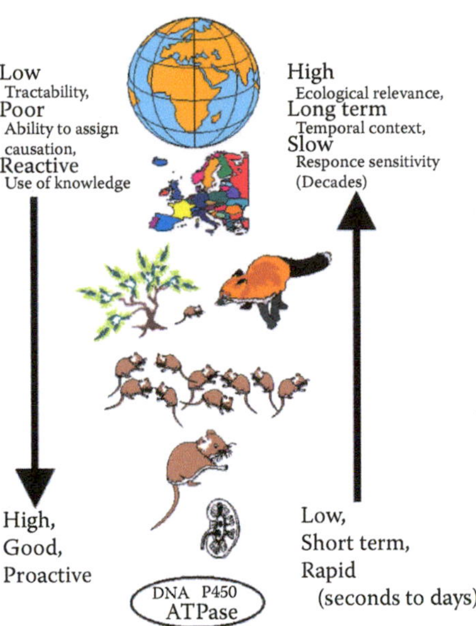

Figure 1.13 Hierarchical organization of topics in ecotoxicology relative to ecological relevance, general tractability, ability to assign causation, general use of knowledge, temporal context of consequence, and temporal sensitivity of response.

to those at higher levels of biological organization, effects at lower levels tend to occur more rapidly after the stressor appears and to disappear more quickly after it is removed. Considering all of these points, it is clear that information from all relevant levels of biological organization should be used together as described by Newman (2001, 2013) and Newman and Clements (2008).

1.4.2 Science, Technology, and Practice

There is a general lack of unanimity in ecotoxicology that leads to much confusion: there seems often to be no single voice indicating how one should conduct oneself. The main objective of this section is to describe how the outwardly inconsistent activities of ecotoxicologists come together to address three intermeshing and equally important goals.

Ecotoxicology has diverse goals (scientific, technological, and practical) in addition to diverse contributing disciplines (Slobodkin and Dykhuizen, 1991; Newman, 1995). The diversity of subjects prompts most ecotoxicologists to specialize in particular areas and to look only peripherally at other information. Subsets of ecotoxicologists must block out of consideration major portions of our collective knowledge structure to move effectively toward their respective and more focused goals. For example, distinct, but overlapping, subsets of information and methodologies are used by the scientist, analyst, and regulator. Whereas a scientist might relay on the conventional hypothetico-deductive method during laboratory experimentation, this approach would be an impediment to a regulator who might use a weight-of-evidence or expert opinion approach instead to expeditiously assess the need for remediation at a particular site despite high uncertainty. The use of a standard method with a suboptimal detection limit might hinder achievement of a scientific goal such as accurately quantifying rates at which a microcontaminant moves between ecosystem components, yet be a necessary compromise for achieving the regulatory goal of generating a consistent water quality database from which decisions and actions can be derived.

Slobodkin and Dykhuizen (1991) observed that much confusion is generated if the distinct intentions and approaches are not understood and respected by the professionals in any applied science. Difficulties arise if methods of one ecotoxicologist are judged unacceptable by another without recognition of differing goals. Because the goals of ecotoxicologists are tightly intermeshed and boundaries are not easily drawn, there is sometimes the appearance of inconsistency or confusion in the field. Again, the purpose of this section is to dispel some of this apparent inconsistency by delineating these three goals or sets of intentions (scientific, technological, and practical), and generally describing the means by which they are pursued. One extremely important aspect of practical ecotoxicology, environmental law, cannot be discussed adequately in the available space. As a partial remedy that provides a general impression of existing legislation in different countries, short appendices covering key Australian, Central American, Chinese, Indian, European Union, South African, and U.S. laws and regulations are offered to the reader by guest authors.

1.4.2.1 Scientific Goal

The goal of any science is to organize knowledge based on explanatory principles (Nagel, 1961). It follows that the scientific goal of ecotoxicology is to organize knowledge, based on explanatory principles, about contaminants* in the biosphere and their effects. The approaches used to reach this scientific goal are well established. But they are worth reviewing because they are taught informally and consequently, many unthoughtful opinions exist regarding the conduct of science. The discussion below is condensed from Newman (1995, 1996, 2001) who reviewed the works of Sir Karl Popper, and extraordinary articles by Platt (1964) and Chamberlin (1897) from the context of ecotoxicology. It is modified at the end by extending these features based on materials in Newman and Clements (2008) and Newman (2013).

In the early history of science, untested or weakly tested theories were used to explain specific phenomena (Chamberlin, 1897). A question was presented to an acknowledged expert and explanation given based on some prevailing or ruling theory. This was enough to fit the phenomena or observation into the existing knowledge structure. Facts gradually accumulate around the ruling theory, fostering a sense of consistency that enhances belief. Such uncritical acceptance of an explanation based on a ruling theory (**precipitate explanation**) is considered unacceptable in modern science. In fact, the first of Descartes' four rules of proper scientific reasoning is the following:

> *... never to accept anything as true that I did not know to be evidently so: that is to say, carefully to avoid precipitancy and prejudice.*
>
> **Descartes (1637)** (Translated by Sutcliffe [1968])

Credibility based solely on conventional wisdom is no longer an acceptable approach to science. The foundation premise of science is now *nullius addictus judicare in verba magestri*, that is, "inquiry is not obliged to the word of any authority." Unfortunately, precipitate explanation can reappear periodically in scientific disciplines, especially applied sciences such as ecotoxicology. Consequently, it is important to recognize precipitate explanation in any field and avoid it in your own work.

Modern sciences have replaced the ruling theory with the working hypothesis. The **working hypothesis** is never accepted as true and only serves to enhance the development of facts and their relations by functioning as the focus of the falsification process (Chamberlin, 1897). One decides to conditionally accept the working hypothesis based on current evidence. Experiments and less-structured experiences are used to test the working hypothesis. The falsification process

* Terms such as pollutant, contaminant, xenobiotic, and stressor have specific and distinct connotations. A **pollutant** is "a substance that occurs in the environment at least in part as a result of man's activities, and which has a deleterious effect on living organisms" (Moriarty, 1983). A **contaminant** is "a substance released by man's activities" (Moriarty, 1983). There is no implied adverse effect for a contaminant although one might exist. A **xenobiotic** is "a foreign chemical or material not produced in nature and not normally considered a constitutive component of a specified biological system. [It is] usually applied to manufactured chemicals" (Rand and Petrocelli, 1985). A **stressor** is that which produces a stress. **Stress** "at any level of ecological organization is a response to or effect of a recent, disorganizing or detrimental factor" (Newman, 1995). As will be discussed, the terms have slightly different legal definitions too.

is often conducted using the null hypothesis–based statistics developed by Ronald A. Fisher in the 1920s for the Popperian falsification of hypotheses. The working hypothesis approach still has a proclivity toward precipitate explanation because a central theory or hypothesis tends to be given favored status during testing. Chamberlin (1897) suggested application of the **method of multiple working hypotheses** to reduce this tendency. The method of multiple working hypotheses reduces precipitate explanation and subjectivity by considering all plausible hypotheses simultaneously so that equal amounts of effort and attention are provided to each. In fields where multiple causes or interactions are common, it also lessens the tendency to stop after "the cause" has been discovered.

In any modern science, a hypothesis is never assumed to be true regardless of the approach used. But it can gain enhanced status after repeated survival of rigorous testing. Status is not legitimately enhanced unless tests also have high powers to falsify. Unfortunately, consistent application of weak testing can lead to the progressive dominance of an idea by repetition alone. Weak testing is occasionally used to promote an idea or approach; consequently, members of any science such as ecotoxicology must be able to recognize false paradigms that emerge from weak testing and to avoid weak testing in their own work. Further, tests involving imprecise or biased measurement should be avoided because they frequently generate false conclusions and foster confusion.

Gradually, observational and experimental methods produce a framework of explanatory principles or paradigms about which facts are organized. These **paradigms** (generally accepted concepts in a healthy science that withstood rigorous testing and, as a result, hold enhanced status as causal explanations) are learned by members of a discipline and define the major context for inquiry in the field. They act as nuclei around which ancillary concepts are formulated and as a framework for further testing and enrichment of fact. Unlike ruling theories, these paradigms remain subject to future scrutiny, revision, rejection, or replacement. They are explanations that are currently believed to be the most accurate and useful reflections of truth, but they are not absolute truths. This is an important distinction to keep in mind. For example, the paradigm of matter conservation (matter cannot be created nor destroyed) was an adequate explanation of phenomena until Einstein demonstrated its conditional nature. It was then incorporated into a more inclusive paradigm (relativity theory) with the qualification that relativistic mass (mass + mass equivalent of energy) is constant in the universe, but mass may be converted to energy and vice versa ($\Delta E = \Delta m (c^2)$). In a field with mixed goals such as ecotoxicology, the conditional status of scientific explanations or paradigms is sometimes forgotten.

Two general and interdependent types of behavior occur in any science: normal and innovative science (Kuhn, 1970). **Normal science** works within the framework of established paradigms, and increases the amount and accuracy of our knowledge within that framework. The contribution of normal science is the incremental enhancement of facts and articulation of ideas with which paradigms can be reaffirmed, revised, or replaced by new paradigms (Kuhn, 1970). Most scientific effort is normal science and the collective work of ecotoxicologists is no exception. In contrast, **innovative science** questions existing paradigms and formulates new paradigms. Innovative science is completely dependent on normal science and can only occur after the incremental enrichment of knowledge brought about by normal science has uncovered inconsistencies between facts and an established paradigm. The opposite is also true. Normal science without enough innovative science results in stagnation and dubious inferences.

Although normal science tends to be more important in a young field, an excessive preoccupation with details ("tyranny of the particular" [Medawar, 1967]) or measurement (*idola quantitatus* [Medawar, 1982]) can slow the maturation and progress of a science. Conversely, insistence on rigorous hypothesis testing before the accumulation of enough facts and establishment of accurate measurement techniques can lead to premature rejection of a hypothesis that might otherwise be accepted. Both normal and innovative science must be balanced in a healthy science. In ecotoxicology, many areas still require more normal science before innovative science can be applied effectively. In many other areas of this maturing science, the tyranny of the particular and *idola quantitatus* exist at the expense of much needed innovative science (Newman, 1996). A balance between normal and innovative science is essential to effectively achieve the scientific goal of ecotoxicology. The long-term benefit of such a healthy balance will be optimal efficiency and effectiveness in environmental stewardship.

1.4.2.2 Technological Goal

The technical goal of ecotoxicology is to develop and then apply tools and methods to acquire a better understanding of contaminant fate and effects in the biosphere. Often, some activities in technology are indistinguishable from normal science; however, their goals are distinct. Relative to plainly scientific endeavors, the benefits to society of technology are more immediate but slightly less global. Although analytical instrumentation is an obvious component of ecotoxicological technology, other components include standard procedures and approaches, and computational methods. Many of these technologies can also become pertinent to the practical goals of ecotoxicology *when used to address specific problems.* Consequently, the distinction between technology and practice is also based on context.

The development of analytical instrumentation able to detect and quantify low concentrations of contaminants in complex environmental matrices has been essential to the growth of knowledge. For example, the number of commercial atomic absorption spectrophotometers (AAS) increased exponentially in the 1950s and 1960s, making possible the rapid measurement of trace element contamination in diverse environmental materials (Price, 1972). Flameless AAS methods lowered detection limits for most elements and enhanced analytical capabilities even further. Now, a wide range of atomic emission, atomic absorption, atomic fluorescence, and mass spectrometric techniques are available for the study of elemental contaminants at levels ranging from $mg{\cdot}g^{-1}$ to $pg{\cdot}kg^{-1}$ concentrations. Gas chromatography (GC) techniques allowed study of the more volatile organic contaminants. Techniques including GC coupled with a mass spectrometer (GC–MS), more effective columns for separation, and improved detectors have all enhanced our understanding of fate and effects of organic contaminants. For organic compounds less amenable to GC-related techniques, innovations such as advanced separation columns and high-pressure pumps have quickly improved high-pressure liquid chromatographic (HPLC) methods. Overarching all of these advances have been computer-enhanced sample processing, analytical control, and signal processing. These, and a myriad of instrumental techniques, have appeared in the last few decades and allowed rapid advancement of scientific ecotoxicology.

Again, procedures and protocols are also important components of environmental technology. Pertinent procedures vary widely. As an example, they may include such activities as the mapping of **ecoregions**—relatively homogeneous regions in ecosystems or associations between biota and their environment—as a practical means for defining sensitivity of U.S. waters and lands to contaminants (Omernik, 1987; Hughes and Larsen, 1988). These naturally similar regions of the country are grouped for development of a common study or management strategy. Another important example is a crucial technology created through seminal papers such as the series defining the generation and analysis of aquatic toxicity data (e.g., Sprague, 1969, 1970; Buikema et al. 1982; Cherry and Cairns, 1982; Herricks and Cairns, 1982). The establishment of a procedural paradigm for ecological risk assessment (e.g., Environmental Protection Agency [EPA], 1991a) constitutes a technological advance as well as a contribution to ecotoxicology's practical goals. General methods for **biomonitoring** (use of organisms to monitor contamination and to imply possible effects to biota or sources of toxicants to humans) (e.g., Goldberg, 1986; Phillips, 1977) and applying **biomarkers** (cellular, tissue, body fluid, physiological, or biochemical changes in extant individuals that are used quantitatively during biomonitoring to imply presence of significant pollutants or as early warning systems for imminent effects) (e.g., McCarthy and Shugart, 1990) are also important technologies developed in the last several decades. Most biomonitoring programs are only possible now because of the advances in analytical instrumentation described above.

Experimental design schemes, statistical methods, and computer technologies are also important here. Valuable descriptions of experimental designs and statistical methods are provided by professional organizations (e.g., American Public Health Association, 1981) and government agencies (e.g., EPA 1985a, 1988a, 1989a,b), facilitating effective data acquisition to enhance our understanding of contaminant fate and effects. These often have easily implemented computer programs associated with them (e.g., EPA, 1985a, 1988a, 1989b). Other computer programs have been developed by EPA to enhance scientific progress. An example is the MINTEQA2 program (EPA, 1991b), which predicts speciation and phase association of inorganic toxicants such as transition metals. Numerous programs for statistical analysis of toxic effects data are available from EPA (e.g., EPA, 1985a, 1988a, 1989b) and commercial sources. Geographic information system (GIS) technologies discussed in later chapters have been developed to study nonpoint source contamination over large areas such as watersheds (e.g., Adamus and Bergman, 1995).

Some technology-related approaches are difficult to understand if an inappropriate context is forced upon them. Some are focused primarily on supporting scientific goals while others are designed to support the practical goals of ecotoxicology. Unfortunately, the complex blending of goals in ecotoxicology makes this a common source of confusion. The designed use of any technology must be kept clearly in mind to avoid confusion and generation of misinformation. For example, standard or operational definitions (e.g., acute vs. chronic effect, sublethal vs. lethal exposure) may have dubious scientific value relative to predicting the actual impact of toxicants in an ecological context. The 96-hour duration of the conventional acute toxicity test was selected because it fits conveniently into the workweek, not because it has any particular scientific underpinnings. The operational distinction made between acute and chronic exposures is also partially arbitrary. An acute "sublethal" exposure could eventually produce a fatal cancer. Regardless, when appropriately and thoughtfully applied, such standard definitions and associated tests are invaluable in applying our technology to various scientific subjects, such as to determining if the free metal ion is the most toxic species of a metal by using standard acute toxicity endpoints.

Qualities valued in technologies are effectiveness (including cost effectiveness), precision, accuracy, appropriate sensitivity, consistency, clarity of outcome, and ease of application. As discussed below, several of these qualities are also important in practical ecotoxicology.

1.4.2.3 Practical Goal

The practice of ecotoxicology has as its goal the application of available scientific knowledge and technologies to document or solve specific problems. During the process, some scientific knowledge will be marginalized to expeditiously resolve the problem at hand. What might seem from a rigid scientific vantage to be an incomplete depiction of reality is, in fact, the most expeditious means of defining and resolving the problem. Many technologies are relevant to practical ecotoxicology; however, the goal of their application is also to solve or document a particular environmental situation. Techniques appropriate for the practical ecotoxicologist may be general such as methods for determining contaminant leaching from wastes (e.g., Anonymous, 1990) or the application of microarray techniques to infer the cause of a particular adverse effect. Predictive software such as the QUAL2E program (EPA, 1987a), which estimates stream water quality under specific discharge scenarios, may also be important tools in achieving practical goals. Other tools include specific steps to take during implementation of a method, e.g., biomarker-based biomonitoring on U.S. Department of Energy sites (McCarthy et al. 1991). They might involve guidelines for the practice of risk assessment on hazardous waste sites (EPA, 1989c) or for waste basin closure. In each of these instances, the goal is not to understand the ecotoxicological phenomena more completely or to develop a technology to better study a system. The goal is to address and resolve a specific problem. Indeed, attempts to conduct scientific work in such efforts or expend resources only to develop a novel technology could delay progress toward the practical goal—solving the immediate problem and removing potentially harmful pollutants.

Practical tools may also include criteria and standards for regulation of specific discharges or water bodies. For example, water quality **criteria** are estimated concentrations of toxicants based on current scientific knowledge that, if not exceeded, are considered protective for organisms or a defined use of a water body (or some other environmental media). Criteria are developed for individual contaminants, e.g., aluminum (EPA, 1988b), cadmium (EPA, 1985b), copper (EPA, 1985c), lead (EPA, 1985d), and zinc (EPA, 1985d) using a standard approach (i.e., EPA, 1985e). On the basis of scientific knowledge, they are used to recommend toxicant concentrations not to be exceeded as a result of discharges into waters. Although the example here is that for water, criteria are also defined for other media such as air (see U.S. Clean Air Act in the Appendix) and sediments (Shea, 1988; Di Toro et al. 1991).

On the basis of the criteria and the specified use of a water body, water quality standards may be set for contaminants. **Standards** are legal limits permitted by each U.S. state for a specific water body and thought to be sufficient to protect that water body. Both criteria and standards are designed with the intent to avoid specific problems, but they are based partially on existing scientific knowledge. Consequently, a healthy growth of scientific knowledge in ecotoxicology is essential to improving our progress toward the practical goals of ecotoxicology. Indeed, criteria and standards (EPA, 1983) are revised periodically to accommodate new knowledge.

Effectiveness, precision, accuracy, sensitivity, consistency, clarity, and ease of application are valued in practical ecotoxicology as well as in technical ecotoxicology as discussed above. Also important to practical ecotoxicology are the following: unambiguous results, safety, and clear documentation of progress during application.

1.5 SUMMARY

At the close of the World War II, the dilution paradigm failed with clearly unacceptable consequences to human health and ecological integrity. The influence of humans on the biosphere has grown so extensive that scientists now propose that we have entered into a new geological epoch, the Anthropocene. Expertise in ecotoxicology is now critical to our well-being in the Anthropocene. Ecotoxicology, the science of contaminants in the biosphere and their effects on constituents of the biosphere including humans, has emerged to provide such expertise.

Ecotoxicologists have overlapping yet distinct scientific, technological, and practical goals that must be understood and respected. Our current knowledge available for achieving these goals (Figure 1.14, upper panel) requires further expansion and more integration (Newman and Clements, 2008). Although the knowledge applied to these goals overlaps, there remain many instances of inappropriate or inadequate integration. For example, present regulations remain biased toward single species tests done in the laboratory, yet our scientific knowledge clearly indicates that results from multiple species tests are at least as valuable to understanding risk. Recognizing the continual need for reintegration of knowledge, lawmakers have wisely incorporated periodic review and revision into major legislation and associated regulations. Further, new or improved technologies are continually drawn into our scientific efforts, e.g., new molecular technologies applied to assay genetic damage and identify causes of adverse effects. The scientific foundations of the field should also expand and come into balance with technology and practice. This is done most effectively using the methods described in this chapter and explained more fully by Newman and Clements (2008) and Newman (2013). Passé and flawed behaviors such as precipitate explanation, overdominance of normal science, the tyranny of the particular, and *idola quantitatus* should be avoided as impediments to scientific progress. An understanding and respect for the different goals of ecotoxicologists must also prevail to appropriately apply science and technology to practical problems of environmental stewardship.

Ecotoxicology is rapidly becoming a mature discipline and, hopefully, will soon achieve an effective balance of knowledge to best address its scientific, technical, and practical goals (Figure 1.14, lower panel). Scientific understanding must expand in all directions, especially toward

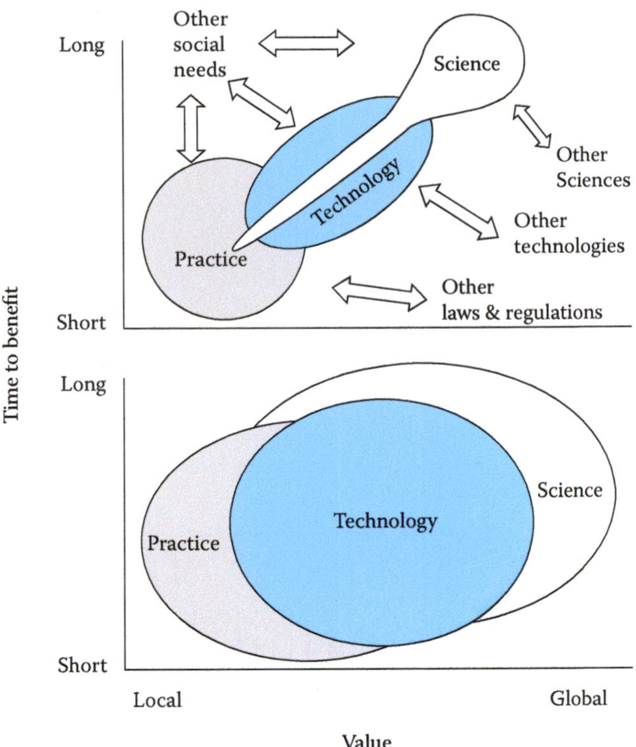

Figure 1.14 Present (top panel) and ideal (bottom panel) balance among scientific, technological, and practical components of ecotoxicology. The relative amount of effort in each is reflected by area on the plots of scale of value (local to global) and time to realize (short to long term) benefit from these components of the field.

more global and long-term phenomena. Technology and practice must do the same. The sound laws presently implemented in most countries should, or are rapidly evolving, to accommodate the continual improvements to technology and science.

SUGGESTED READINGS

Carson, R., *Silent Spring*. Houghton-Mifflin Co., Boston, 1962.

Chamberlin, T. C., The method of multiple working hypotheses. *J. Geol.,* 5, 837–848, 1897.

Colborn, T., D. Dumanoski, and J. P. Myers, *Our Stolen Future*, Penguin Books, New York, 1996.

Kline, B. *First along The River. A Brief History of the U.S. Environmental Movement,* 4th Edition. Rowman & Littlefield Publishers, Inc., Lanham, MD, 2011.

Marco, G. J., R. M. Hollingworth, and W. Durham, *Silent Spring Revisited*, American Chemical Society, Washington, DC, 1987.

McNeill, J. R. *Something New under the Sun. An Environmental History of the Twentieth-Century World*, W. W. Norton & Company, New York, 2000.

Newman, M. C., Ecotoxicology as a science, in *Ecotoxicology: A Hierarchical Treatment*, Newman, M. C. and C. H. Jagoe, Eds., CRC/Lewis Publishers, Boca Raton, FL, 1996.

Newman, M. C., and W. H. Clements, Chapters 1 and 36, in *Ecotoxicology. A Comprehensive Treatment*, Taylor & Francis/CRC Press, Boca Raton, FL, 2008.

Newman, M. C., *Quantitative Ecotoxicology,* 2nd Edition, Taylor & Francis/CRC Press, Boca Raton, FL, 2013.

Platt, J. R., Strong inference. *Science,* 146, 347–353, 1964.

Smith, W. E. and A. M. Smith, *Minamata*, Holt, Rinehart and Winston, New York, 1975.

Woodwell, G. M., Toxic substances and ecological cycles. *Sci. Am.,* 216, 24–31, 1967.

CHAPTER 2

Major Classes of Contaminants

2.1 INTRODUCTION

Which contaminants presently concern us the most? Many have been recognized and given close attention for decades, whereas others have only recently emerged as concerns. Only the most prominent conventional or emerging contaminants are introduced here for the sake of space.

Following convention, the chemical contaminants are divided grossly into those organic and inorganic. The terms organic and inorganic were applied originally to indicate whether the chemical came from living organisms (organic) or mineral sources (inorganic); however, this distinction is blurry.* Fortunately, the distinction is clearest with organic compounds of most interest here, which are composed of carbon chains or rings. Organic compounds are composed predominantly of the nonmetal elements from groups 1, 4–7A of the periodic table (Figure 2.1); however, there are important instances in which metals are integrated into biomolecules such as the cytochromes and hemoglobin, or contaminants such as methylmercury and organotins.

Organic contaminants include intentional poisons such as insecticides, herbicides, fungicides, and wood preservatives. They become a problem when nontarget species come into contact with sufficiently high concentrations of them. Other organic contaminants are unintentional poisons, e.g., degreasers, solvents, and various industrial by-products. Some organic compounds such as those in personal care products (e.g., detergents and musks) or pharmaceuticals (e.g., drugs, antibiotics, and birth control substances) are designed to be directly beneficial to an individual's well-being yet still cause problems after release into the environment. Similarly, inorganic contaminants are composed of intentional and unintentional poisons. Some are released for a very specific purpose (e.g., sodium arsenate as a pesticide), but others are released by a wide range of human activities such as the lead used in batteries, plumbing, gas additives, and many other products. Inorganic contaminants, such as nitrite in drinking waters, become a problem only if our activities elevate their concentrations to abnormal levels. Indeed, several metals are essential to life but, above certain concentrations, become harmful.

A mammalian toxicologist would not consider all of the chemicals discussed in this chapter to be toxicants because they do not directly poison individuals. However, contaminants† such as excessive amounts of phosphorus or nitrogen species in lakes can have harmful consequences to the associated ecological community. Similarly, global changes in atmospheric gases such as carbon dioxide (CO_2) and methane have a pronounced influence on the earth's ecosystems at concentrations that are not directly toxic to humans. Putrescible compounds might not kill aquatic biota outright but can kill large numbers of organisms indirectly during their microbial degradation and consequent

* The interested reader is referred to Larson and Weber (1994) for discussion of this ambiguity.

† The term, **contaminant**, simply refers to a substance released by human activities. No harmful effect is necessarily implied in its release. However, the term, **pollutant**, refers to a contaminant that has or likely will have a deleterious effect on humans or ecological entities.

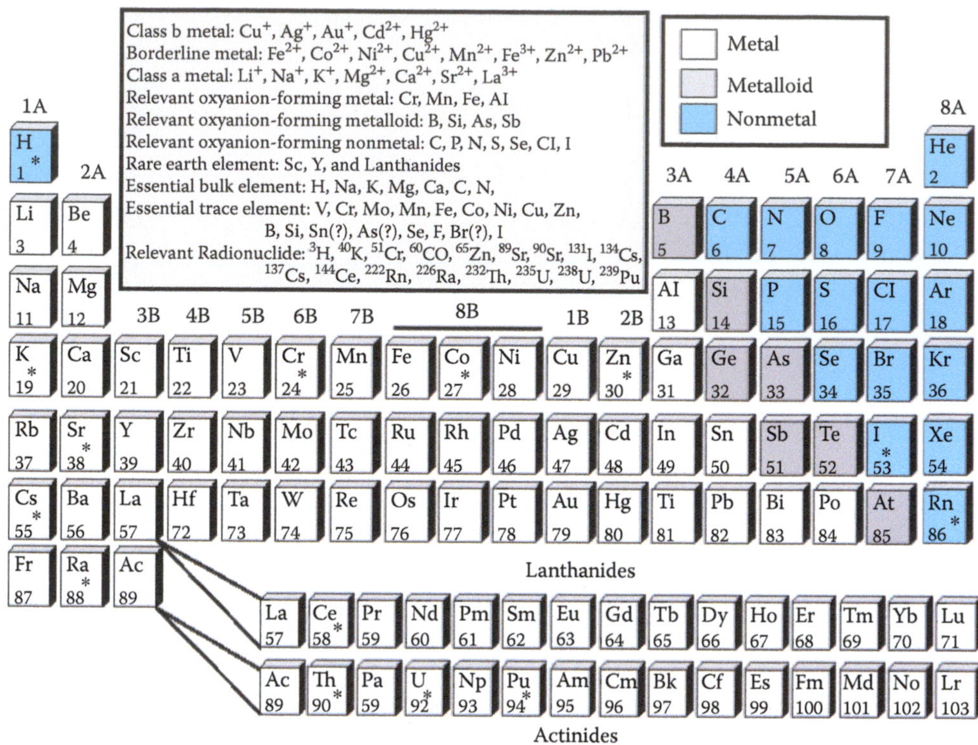

Figure 2.1 The periodic table is modified slightly here by the addition of an inset that highlights different clusters of elements relative to ecotoxicological features. The metals (white boxes) make up the majority of elements. The nonmetals (blue boxes) are fewer in number than the metals but include the major elements of life. The metalloids (gray boxes) have properties intermediate between the metals and nonmetals. Those elements for which environmentally relevant radioactive isotopes are discussed later in this chapter are indicated with an asterisk (*).

depletion of dissolved oxygen of receiving waters. Changes in thermal conditions can disrupt ecological communities by influencing innumerable biological rates at the level of the individual organism. Consequently, these contaminants are discussed along with the more conventional poisons as **ecotoxicants**.

2.2 MAJOR CLASSES OF CONTAMINANTS*

2.2.1 Inorganic Contaminants

The periodic table is a reasonable tool for framing our discussion of different inorganic contaminants (Figure 2.1). Elements within each of its 18 groupings have similar outer orbital configurations, resulting in common themes in their environmental chemistries and toxicologies. The elements in this table can be divided generally into metals, nonmetals, or metalloids (semimetals).

* Respecting the diverse backgrounds of students who might use this book, brief summaries of some chemistry concepts are inserted into Section 2.4 toward the end of this chapter. The intent is to provide sufficient insight to those who do not have much chemistry training without distracting those who do. Topics covered in Section 2.4 are indicated within the text as (Section 2.4., Topic).

Most elements are metals (Section 2.4, "Metals"). Pure metals are characteristically solid at room temperature although mercury is the exception. They are described as good electrical and thermal conductors with high luster and malleability. They emit electrons readily on heating. However, these conventional descriptors of pure metals are not very helpful to the environmental scientist who deals with metal compounds, complexes, and ions at minute concentrations. More useful to the environmental scientist are characteristics that influence metal movement among environmental phases and interaction with components of the environment including toxicological interactions. A major characteristic that contributes to these metal behaviors for the environmental scientist is chemical bonding as determined by outer orbital electron behavior. Generally, metals tend to lose electrons when reacting with nonmetals. For an important example, metals are electron donors in many normal biological reactions and act as cofactors within coenzymes (vitamins). The extent to which a metal shares its electrons during bonding varies (Section 2.4, "Bonds").

Because most of the relevant metals in our environment do not exist prominently in the pure elemental state, the effects of metal compounds, complexes, and ions are often much more relevant to the ecotoxicologist. Normal biological processes can be negatively affected by uptake of metals from the environment. Ecotoxicologists should recognize the characteristics that influence their changes in state (e.g., their oxidation, ionization, and incorporation into organic form) in the environment, and subsequent movements and interactions of the metals.

Nonmetals are a minority cluster of elements in the upper right side of the periodic table (plus hydrogen). They include the noble gases that, by definition, do not readily react with other elements. With the exception of radioactive radon, the noble gases will not hold our attention beyond this brief mention. Although a minority in the table, the remaining nonmetals constitute the majority of the bulk elements making up living organisms. They, including the halogens* in period 7A, tend to be very electronegative (Section 2.4, "Electronegativity"). In contrast to the metals, they have a strong tendency to attract electrons during chemical interactions. The strength of that attraction depends on several factors including their outer orbital electron configuration, which, in turn, partially determines how polar their covalent bonds might be with other elements such as metals. Generally, the larger the difference in electronegativity between two interacting elements, the more the bond between them will tend toward ionic and away from covalent.

Metalloids, the remaining elements in the periodic table, are intermediate in their properties, having some qualities of metals. For example, they can be semiconductors with their electrical conductance increasing with temperature. They can also form strong covalent bonds with some other elements.

2.2.1.1 *Metals and Metalloids*

The conventional light/heavy or transition metal characterization of metals is unhelpful to ecotoxicologists and chemists in general (Duffus, 2002) and will not be applied here. Other metal properties are much more relevant than elemental density. Such being the case, how should metals best be discussed by environmental scientists? The widely used and more relevant approach is to consider their bonding tendencies (Shaw, 1961; Pearson, 1963; Nieboer and Richardson, 1980; Williams and Turner, 1981). It applies Hard Soft Acid Base (HSAB) Theory (Section 2.4) to classify metals and the chemical moieties with which they interact.

* The group 7A halogens are prominent in later discussions of organic contaminants in which they are frequently found. Particularly prominent in this regard are chlorine and bromine, but also fluorine and iodine in a few cases. Halogens are highly electronegative, that is, electrons involved in covalent bonding between a halogen and another atom tend to be more attracted toward the halogen than the other atom. Halogens also have a strong tendency to be present as anions in solution, e.g., Cl^-.

The top panel of Figure 2.2 is a **complexation field diagram** (based largely on HSAB theory) that describes how the stability of bonds changes among dissolved metals in freshwater and seawater systems. Progressing across the diagram from class A (left) to B (right) metals, the stability of metal covalent bonds with soft ligands increases. Relevant soft ligands in the aquatic environments are Cl^-, S^{2-}, and HS^-. As an example relevant to biomolecules, the sulfur atoms in the amino acids, methionine and cysteine, are soft donor atoms. Bond stability with intermediate ligands increases diagonally from the top left corner to the bottom right corner of the diagram. Relevant ligands that tend toward an intermediate/borderline hard status include the dissolved anions, CO_3^{2-} and OH^- in natural waters and organic carboxyl and phenolic groups in biomolecules. The stability of ionic bonds with hard ligands increases as you move down the diagram. Relevant hard ligands in natural waters include F^-, SO_4^{2-}, and NH_3. Examples in biomolecules are the phenolate side chains of tyrosine, phosphate groups in DNA or RNA, and carboxylate groups (R-COOH). These

Covalent bond stability

Ionic bond stability	a	Borderline a	Borderline b	b
I	Cs, Li, K Na, Rb		Cu(I) TI(I)	Ag Au(I)
II	Ba, Ca Mg, Sr	Co, Cu(II) Fe(II), Mn(II), Ni, Pb, Zn	Cd	Hg
III	Rare Earth Elements		Bi	TI(III)
IV	AI,Be,Cr(III) Fe(III),U(IV)	In		

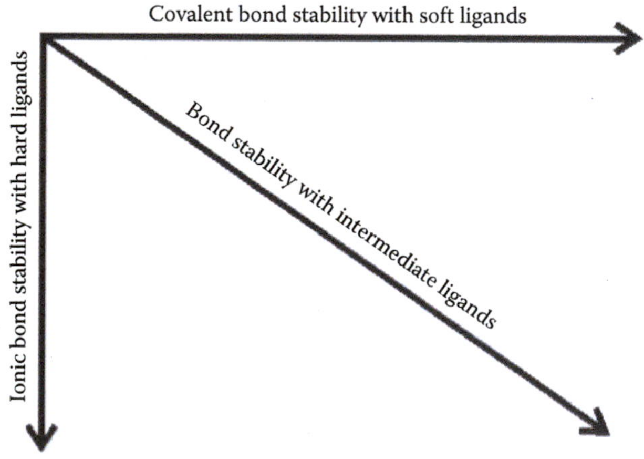

Figure 2.2 A modified complexation field diagram for cationic metals. The diagram is based on Figures 1 and 2 in Newman and Jagoe (1994), which, in turn, drew on Figure 12 and the associated text from Turner et al. (1981), which dealt with cation-ligand complexes in salt and fresh water bodies. The headings across the top refer to an element's Hard Soft Acid Base status and the I-IV down the right side of the diagram refer to single charged (I), double charged (II), triple charged (III), or hydrolysis-dominated (IV) cations. Roman numerals after an elemental symbol indicates the oxidation state if relevant. Solid blue, weakly complexed cation; striped blue, cations with widely variable complexation dependent on ionic conditions; striped gray, chloro-dominated cations; and solid gray, cations with hydrolysis-dominated complexation.

biomolecular ligand examples illustrate how the described binding trends are relevant for predicting biological effects as well as chemical form in the environment.

The Class A, B, and intermediate metals can be placed into these metal–ligand binding trends as depicted in Figure 2.2. The panel at the bottom of this figure shows trends in ionic and covalent bonding of metals with different classes of ligands (Section 2.4, "Binding Trends"). Such a depiction allows one to quickly understand how a dissolved metal might speciate in natural waters or interact with biomolecules. For example, Na^+ in natural waters does not form strong covalent or ionic bonds with dissolved ligands. On the other hand, dissolved cadmium speciation in seawater is dominated by complexes with Cl^- such as $CdCl^-$ and $CdCl_2^0$. Similarly, the strong covalent bonds formed by Hg^{2+} or Ag^+ with many functional groups of biomolecules correctly suggest how poisonous these metals ions are. Similarly, Ba^{2+} forms much more stable ionic bonds than K^+ at membrane K^+ channels of excitable tissues.* High Ba^{2+} doses can cause death by interfering with normal K^+ passage in cells of nerves and muscles. Of course, if a metal forms an oxyanion (Section 2.4, "Oxyanion/Oxycation"), its interactions are also determined by features of that polyatomic anion. Table 2.1 is one compilation of hard, soft, and intermediate metals (Lewis acids) and corresponding ligands (Lewis bases).

Another quality to understand is whether or not an element is essential to life (Figure 2.1). Essential elements include bulk and trace components of biological entities. The bulk essential elements include only four Class A metals. Many more trace essential elements are metals. Too low as well as too high concentrations of essential elements in organisms can be harmful. If another element is very similar in its chemistry to an essential element, that essential **element analog** can enter into biochemical processes in place of the essential element and interfere with normal biological functioning if present at too high a concentration relative to that of the essential element.

Metalloids are also important to understand relative to their binding chemistries and this same scheme can be applied to understand their fate and effects. Probably, the most environmentally relevant is arsenic, which in many ways acts like its neighboring nonmetal on the periodic table, selenium. Both arsenic and selenium form oxyanions that often influence their movements in the environment, biochemical transformations, and effects on biota. In the case of arsenic, both arsenite (AsO_2^-) and arsenate (AsO_4^{3-}) are common anionic forms. The biochemical similarity of arsenate

Table 2.1 Compilation of Class A, Intermediate, and B Metals (Acids) and Ligands (Bases)

Class A

Metals: Cs^+, K^+, Li^+, Na^+, Be^{2+}, Ba^{2+}, Ca^{2+}, Mg^{2+}, Sr^{2+}, Al^{3+}, Fe^{3+}, Co^{3+}, rare earth elements (+3)

Ligands: NH_3 (also placed in Intermediate), H_2O, OH^-, R-NH_2 (amine), R-COOH, R-OH, R-O^-, R_2-C = O (ketone), R-O-R (ether), CO_3^{2-}, SO_4^{2-}, F^-, HPO_4^{2-}, NO_3

Intermediate

Metals: Co^{2+}, Cu^{2+}, Fe^{2+}, Mn^{2+}, Ni^{2+}, Pb^{2+}, Sn^{2+}, Zn^{2+}, Ga^{3+}, In^{3+}

Ligands: R-NH_2, Br^-, Cl^- (also placed in Class B), NO_2^-, SO_3^{2-}

Class B

Metals Ag^+, Au^+, Cu^+, Tl^+, Cd^{2+} (also placed in intermediate), Hg^{2+}, Pt^{2+}, Pd^{2+}, Bi^{3+}, Tl^{3+}, Pt^{4+}

Ligands: CN^-, CO, SCN^-, R_2-S, R-SH, RS^-, S^{2-}, I^-

Sources: Table 2.8 of Fraüsto da Silva, J.J.R. and Williams, R.J.P., *The Biological Chemistry of the Elements: The Inorganic Chemistry of Life*, Oxford University Press, Oxford, 1991; Figure 6 of Pearson, R.G., *J. Am. Chem. Soc.*, 85, 3533-3539, 1963; Table 1 of Nieboer, E. and Richardson D.H.S., *Environ. Pollut.*, 1B, 3-26, 1980.

Note: Classifications vary among authors, so those of Fraüsto da Silva and Williams (1991) were tabulated if differences existed.

* Notice from the Section 2.4 discussion of Binding Trends that the stability of a metal's ionic bond with a hard ligand increases with the square of the metal ion's charge, that is, Z^2/r. The K^+ and Ba^{2+} have very similar radii of 1.33 and 1.35 Å, respectively. Therefore, the square of their charges indicates that Ba^{2+} will have ionic bonds four times as stable as those of K^+ with a soft ligand. This results in the poisonous effects of Ba^{2+} on excitable tissues such as those making up nerves and muscles.

to the phosphates that are involved in essential biochemical processes and molecules contributes to arsenate's poisonous tendencies. As an oxyanion, arsenic can also form compounds with toxic metals.

Now that the general trends in metal and metalloid qualities have been described, our discussions can turn to individual elements. Each of the most prominent elements is described briefly in Table 2.2 relative to its uses, sources to the environment, and general effects. For convenience, organometals discussed below are included in the table.

Table 2.2 Description and Sources of Environmentally Important Metals and Metalloids

Metal	Description
Aluminum	A naturally abundant metal in the environment that, under low pH conditions such as those resulting from acid precipitation or mine drainage, can increase to unusually high dissolved concentrations that can kill aquatic species.
Arsenic	This metalloid and its compounds are used in numerous products including metal alloys, pesticides (e.g., $Pb_3(AsO_4)_2$), wood preservatives (chromated copper arsenate preserving), plant desiccants, and herbicides (e.g., $Na_3As_3O_3$). It is associated with coal fly ash and is also released during gold and lead mining. It is often present as an oxyanion, e.g., AsO_4^{3-}, $HAsO_4^{2-}$, and $H_2AsO_4^{-}$. It is toxic and carcinogenic.
Cadmium	This metal is used in alloy production, electroplating, and galvanizing. It is also used in pigments, batteries, plastics, and numerous other products. It is generated during zinc ore processing. Smokers are exposed to high levels of cadmium in cigarettes. It is toxic and carcinogenic.
Chromium	Chromium is used in alloys, catalysts, pigments, and wood preservatives. It also is used in tanning and as an anticorrosive. It may be present as Cr (VI) or Cr (III). These are referred to as hexavalent and trivalent chromium, respectively. Cr (VI), but not Cr (III), is carcinogenic and is the more toxic of the two forms. Chromium is often present as an oxyanion, e.g., CrO_4^{2-}, $HCrO_4^{-}$, and CrO_7^{2-}.
Copper	Copper is used extensively for wiring and electronics, and plumbing. It is also used to control growth of algae, bacteria, and fungi. It is an essential element that is toxic at high concentrations.
Lead	This poisonous metal is ubiquitous due to its widespread use in gasoline, batteries, solders, pigments, piping, ammunition, paints, ceramics, caulking, and numerous other applications. Use of antiknock compounds containing lead was discontinued in many countries. Although used until the early 1980s to solder tin cans, this use also has been banned after concerns about health effects became apparent. It causes anemia and neurological dysfunction with chronic exposure. Poisoning of birds can occur by ingestion of lead shot or sinkers.
Mercury	Used in electronics, dental amalgams, chlorine-alkali production, gold mining, and paints. It is an excellent industrial catalyst and, because it is liquid at ambient temperatures, as a component in electrolysis. It was used extensively as a biocide, involving seed treatments to prevent fungal growth and growth inhibition in numerous industries such as pulp mills. Today, burning of fossil fuels, especially coal, is a major source of mercury throughout the world. This highly toxic metal can accumulate to high concentrations in biota, primarily in the organometallic form.
Nickel	This metal is used in alloys such as stainless steel and for nickel plating. It also has innumerable other uses including battery production (Ni-Cd batteries). At sufficiently high concentrations, nickel is both toxic and carcinogenic.
Rare earth metals	The rare earth elements have emerged as important metals to consider in ecotoxicology. They are essential components of modern color television screens, displays of computer screens and cell phones, optical networks, glass polishing systems, and magnet technologies. As such, their mining, processing, and eventual disposal increase as the use of the technologies they support expand.
Selenium	Selenium is a nonmetal that can behave similar to arsenic. It is used in the production of electronics, glass, pigments, alloys, and other materials. It is a by-product of gold, copper, and nickel mining. It is also associated with coal fly ash.
Zinc	This essential metal is used extensively in protective coatings and galvanizing to prevent corrosion. It is also used in alloys. It is less toxic than most metals listed here.

Table 2.2 (*Continued*)

Metal	Description
Organotin	Tributyltin (TBT) was used as an antifouling paint on hulls of ships and boats until recently banned by the International Maritime Organization's "International Convention on the Control of Harmful Anti-fouling Systems in Ships." At very low concentrations, it can cause extensive damage to molluscan populations, resulting in shell abnormalities in oysters and modification of sexual characteristics of snails. Organotins such as trimethyltin (TMT) and triethyltin (TET) are neurotoxicants.
Organolead	Organic compounds of lead such as tetraalkyllead have been used extensively as antiknock additives to gasoline. At high concentrations, these compounds (e.g., trialkyllead produced via liver metabolism from tetraalkyllead) can cause neurological dysfunction and other problems.
Organomercury	Methylmercury and other organomercury compounds are used or produced as a consequence of industrial and microbial processes. They are used as biocides as in seed coatings. Methylmercury is found in many fish species consumed by humans. These compounds cause neurological and cardiovascular damage.

2.2.1.2 Organometallic Compounds

Some metals combine with organic structures to produce useful products (Figure 2.3). Others such as organomercury species are produced naturally by microbial processes in anoxic sediments and other environmental phases. Consequently, they have qualities associated with both metals and organic compounds. For example, the metal atom can still be an electron donor in toxicological reactions and the uncharged part of the molecule can increase the organometallic compound's solubility in lipids. The lipids into which the organometallic compound dissolves can be those in living tissues, or those released into the environment on death from organisms and now associated with detrital material. As a specific example, dimethylmercury and monomethylmercury will partition much more readily into lipid-rich organisms or organic material in sediments than the hydrated Hg^{2+} ion (Section 2.4, "Partitioning"). Another feature of dimethylmercury that enhances its ability to diffuse into lipid-rich biological membranes is its linear form: the molecule has lower steric hindrance of its movement between lipid molecules in membranes than if its C-Hg bond angles were more acute and its resulting molecular profile larger.

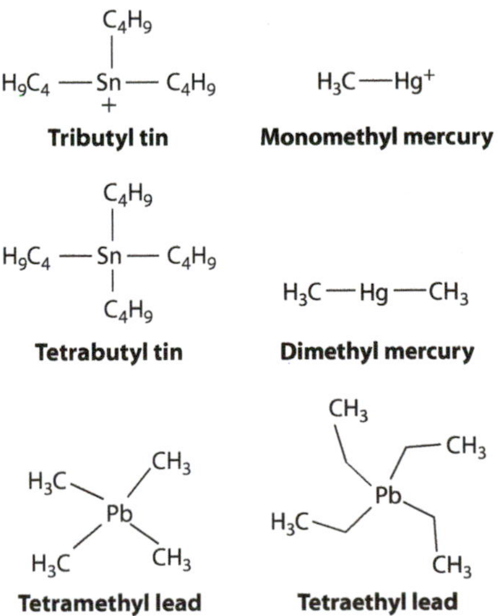

Figure 2.3 Chemical structures of the organometallic compounds that are currently of most concern to ecotoxicologists. Note that the four branches of the tetraethyl lead are C_2H_5.

2.2.1.3 *Inorganic Gases*

Carbon dioxide (CO_2) from combustion concerns environmental scientists because concentrations in the atmosphere are slowly increasing and are linked to global climate changes. Global warming as a consequence of increased atmospheric CO_2 and other gases is responsible for diverse consequences including the reduction in polar ice coverage (Figure 2.4). Nitrogen oxides (NO_x) and sulfur dioxide (SO_2) are also produced at high volumes by combustion in stationary (e.g., coal power plants) and mobile (e.g., motor vehicles) sources. NO_x and SO_2 react in the atmosphere to produce low pH precipitation popularly referred to as acid rain. There is also epidemiological evidence of adverse health effects of these gases such as linkage to various human lung and cardiovascular diseases, and even death (see review by Kampa and Castanas [2008]). High levels of SO_2 can cause plant leaf necrosis.

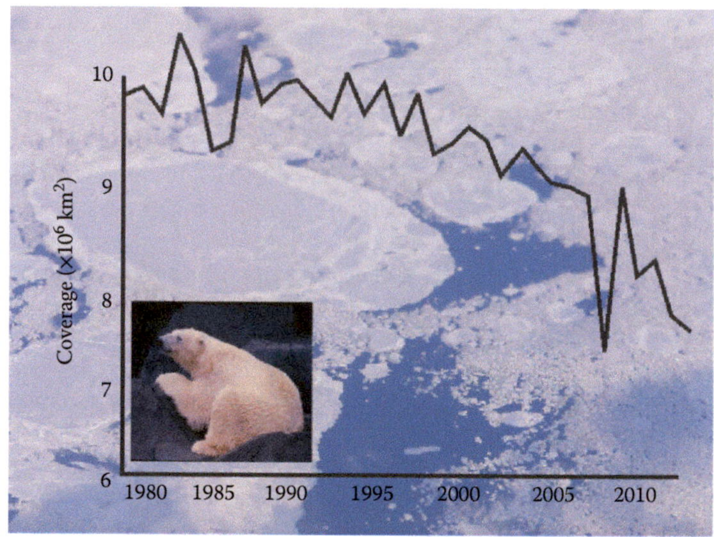

Figure 2.4 Decrease in October Arctic ice extent from 1979 to 2012 as documented by the U.S. National Snow and Ice Data Center (http://nsidc.org/arcticseaicenews/). The background photograph is pack ice in Ungava Bay of the Arctic Ocean (northeast Canada) as photographed on July 1, 2007 during a historically thin summer polar ice condition. Pack ice melt is normally slow until mid-July when it then accelerates in this part of the Arctic Ocean. Because feeding by polar bears on seals is strongly influenced by pack ice dynamics, scientist and Native American Inuit groups have begun to focus on potential impacts of global warming trends on the persistence of polar bear populations (Derocher et al. 2004; Wiig, 2005; Stirling and Parkinson, 2006). Some predictions indicate the potential for substantial thinning (Bluhm and Gradinger, 2008; Walsh, 2008) or even complete disappearance of summer ice (Kerr, 2007; Serreze et al. 2007; Stroeve et al. 2007). Like the penguins shown later in Figure 12.1, polar bears can accumulate high concentrations of persistent organic pollutants (Kumar et al. 2002), which suggests a complex situation exists for polar bears at this time. (Photograph by M. C. Newman.)

2.2.1.4 *Anionic Contaminants Including Nutrients*

An excess of nitrogen and phosphorus nutrients in aquatic and terrestrial systems change the structure and functioning of associated ecological communities. **Cultural (accelerated) eutrophication** is the classic example of such an adverse effect (see review by Smith et al. [1999]). Figure 2.5 depicts blatant accelerated eutrophication in a nuclear reactor cooling water lake. The top photograph shows a plume of cool, nutrient-rich Savannah River (South Carolina) water that was being pumped through a nuclear reactor into the cooling water reservoir, L Lake, during a prolonged reactor down period. As the cool river water entered the receiving reservoir, it sank below the warmer reservoir water, which

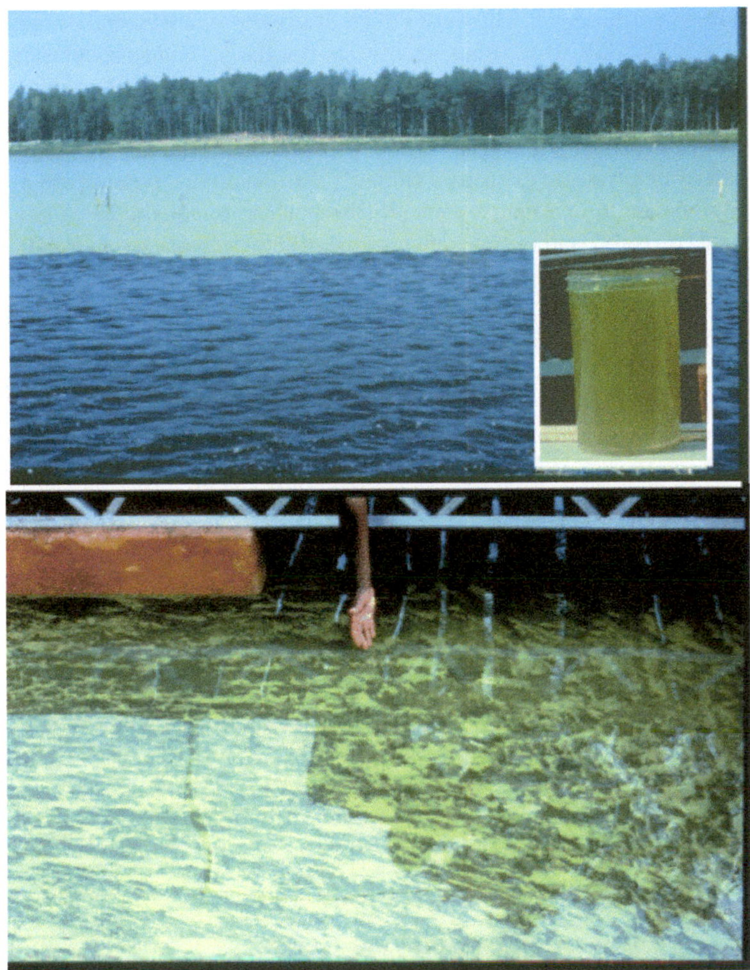

Figure 2.5 Blue-green algal bloom in a nuclear reactor cooling reservoir resulting from high phosphorus loadings. The top panel shows cool, nutrient-rich river water entering the lake from the reactor when nuclear production was stopped (foreground). The river water is cooler than the lake water so it sank below the resident lake water within a 100 m of the point of discharge. The lake water in the background is distinctly greener in color because of the dense blue-green algal growth realized after the nutrient-rich water was allowed to reside for a few days in the summer sunlight. The top panel inset and bottom panel are closer views of the dense blue-green algal growth.

appears much lighter in the photograph due to an extensive and persistent *Anabaena* blue-green algal bloom. The combination of high nutrients concentrations in the river water plus high light conditions in the shallow reservoir caused so much growth that the demarcation was very clear between the newly arrived river water mass and the warmer reservoir water. The bottom photograph is a closer view of the floating clumps of *Anabaena*. The insert shows a clear plastic cup containing *Anabaena*-rich lake water. Such an aquatic system is rendered ecologically dysfunctional despite the reactor engineers' position at the time that the ecological community of the cooling reservoir was biologically balanced.* For example, nocturnal episodes of very low dissolved oxygen in the reservoir caused episodic fish kills. This eutrophic lake will be discussed again in Chapter 11.

* "Biologically balanced" is a feature of a receiving water's biological integrity thought to be critical for meeting the requirements of the U.S. Federal Clean Water Act. (See the discussion of the Clean Water Act given in the Summary of U.S. Laws and Regulations, Appendix 3.)

Nitrogen species can cause other adverse effects if present at sufficiently high concentrations. Nitrate can enter water bodies from runoff or sewage discharge. High concentrations in drinking water cause **methemoglobinemia**. Methemoglobinemia in newborn infants is also called blue-baby syndrome because of the baby's skin coloring resulting from the reaction of nitrite with hemoglobin that converts it to the methemoglobin form, which is incapable of normal transport of oxygen. The nitrite is produced from nitrate in the baby's stomach. Nitrosamines, potent carcinogens, can form from nitrogen compounds in drinking waters. Nitrite is toxic to aquatic biota and can also cause methemoglobinemia directly. Both nitrate and nitrite can also decrease oxygen affinity of hemocyanin in the hemolymph of aquatic invertebrates (Cheng and Chen, 2002).*

Sulfate can become problematic as when SO_2 released into the atmosphere from combustion sources reacts with moisture to produce sulfuric acid. Exposure of sulfide ores from metal mines to oxygen and chemoautotrophic bacteria such as *Thiobacillus ferrooxidans* generates low pH drainage with very high sulfate concentrations. This also occurs with pyrite in coal mines or piles. These low pH, high sulfate effluents are called **acid mine drainage** or **acid rock drainage**, depending on the source. Sluiced ash from coal power plants can also release sulfate in settling basin waters and produce discharge waters with high sulfate concentrations.

Much more subtle human impacts on dissolved anions can harm ecosystems. A recent example is the effect of increasing atmospheric CO_2 concentrations on coral reefs. Increased CO_2 diffusing into the ocean and the increasing global temperatures cause ocean acidification by influencing the equilibrium concentrations of the CO_2–bicarbonate–carbonate buffering system:

$$
\begin{aligned}
&CO_{2(air)} \leftrightarrow CO_{2(dissolved)} + H_2O \\
&CO_{2(dissolved)} + H_2O \leftrightarrow H_2CO_3 \\
&H_2CO_3 \leftrightarrow H^+ + HCO_3^- \\
&HCO_3^- \leftrightarrow H^+ + CO_3^{2-} \\
&HCO_3^- + H_2O \leftrightarrow H_2CO_3 + OH^- \\
&CO_3^{2-} + H_2O \leftrightarrow HCO_3^- + OH^- \\
&H_2CO_3 \leftrightarrow H_2O + CO_2
\end{aligned}
$$

The resulting shift in equilibrium concentrations reduces the bioavailability of carbonate (CO_3^{2-}) to biota such as corals that require it to build hard, carbonate-based structures. Predictions of the consequences include functional collapse of coral reef ecosystems (Hoegh-Guldberg et al. 2007). These same shifts in ocean CO_2 and pH are also predicted to cause direct effects on individual organisms such as impairing skeleton development of purple urchin (*Strongylocentrotus purpuratus*) larvae (Kintisch and Stokstad, 2008).

The perchlorate anion (ClO_4^{-1}) from munitions and rocket solid propellant production or use has become a local concern, especially in contaminated groundwaters of the United States. Concern about perchlorate escaping from military sources was heightened by the discovery that perchlorate can modify thyroid function (Urbansky, 2002). More recently, its interference with iodine transport into milk during human lactation created concern about effects to nursing infants (Dohán et al. 2007).†

* Although not an inorganic anion, ammonia is another common inorganic contaminant that can adversely influence biota. Ammonia can be toxic to aquatic biota near sources such as sewage discharges. Ammonia toxicity is very dependent on the pH of the receiving waters.

† Perchlorate's interference with thyroid and breast function has a common mechanism. The Na^+/I^- symporter (NIS) protein facilitates I^- uptake by thyroid and breast cells. Perchlorate interferes with NIS functioning in the cell's plasma membrane. Consequences of sufficiently high perchlorate exposure can be modified synthesis of iodine-containing hormones in the thyroid and reduced iodine incorporation into milk (Dohán et al. 2007).

2.2.2 Organic Contaminants

Because of their diversity, organic substances are unquestionably among the most useful and problematic compounds generated by humans. A general chapter such as this one cannot possibly do justice to organic contaminants. Several books would be needed to adequately explore the many ways that elements combine to form organic compounds, the many congeners and isomers of important compounds, the diverse ways that organic compounds affect biological systems, and the bewildering ways that they can be broken down. Only a broad-brush description of the most prominent organic contaminant classes will be attempted.

Organic compounds are crudely divided into aliphatic and aromatic compounds with some compounds having parts that are aliphatic and others that are aromatic in nature. Aromatic compounds include benzene or similar compounds in their structure (Section 2.4, "Aromatic Compounds"). Aliphatic compounds are composed of carbon chains with various bonds and other nonaromatic structures (Section 2.4, "Aliphatic Compounds").

Some organics have ionizable groups that allow them to be quite water soluble and to react in specific ways with biomolecules. The extent to which an organic compound is ionized in solution depends on the pH and the pK_a of its ionizable group(s) (Section 2.4, "Ionization"). Other compounds do not ionize, but are polar because the electrons involved in their bonds are unequally shared by the atoms within the molecule (Section 2.4, "Polar Bond"). Still other organic molecules are nonpolar and uncharged, making them very soluble in lipids (lipophilic) and sparingly soluble in water (hydrophobic). The compounds in Figures 2.6, 2.7, and 2.12 are typical lipophilic contaminants (Section 2.4, "Lipid Solubility").

Finally, organic compound partitioning that involves a gas phase can be important to understand. Henry's Law (Section 2.4, "Partitioning") predicts that an equilibrium will be established eventually in which a compound's concentration in solution is proportional to its partial pressure in the gas phase. Relevant situations of such partitioning include partitioning between a water body and the atmosphere, a plant and air, and the exchange of a substance in an organism's lipid-containing blood and its exhaled air.

With this general context established, descriptions of the major organic contaminants can be provided below. More details regarding their accumulation in ecological systems and effects will be covered in following chapters.

2.2.2.1 *Hydrochlorofluorocarbons and Chlorofluorocarbons*

This general grouping of compounds includes the chlorofluorocarbons (CFCs), which are composed of C, F, and Cl atoms, and the hydrochlorofluorocarbons (HCFCs), which contain H in addition (Figure 2.6). These compounds were used widely for refrigeration; air-conditioning; firefighting; foam blowing in Styrofoam and polyurethane production; various solvent applications such as electronics cleansing, propellants like those in aerosol cans; and in the case of 1-chloro-1,1-difluoroethane, as intermediates in fluoropolymer production. Generally, they are nontoxic, odorless gases at room temperature that can be made liquid easily with mild compression (Bunce, 1991). This last quality makes them ideal for the uses just listed. They are moderately soluble in water and have low lipophilicity. The HCFCs were introduced as substitutes for CFCs because they have much less capacity to deplete the ozone layer of the atmosphere (Chapter 12). This is largely a consequence of the hydrogen atom, which makes the HCFCs more reactive in the troposphere than the very inert CFCs. For example, the potential of HCFC-141b to deplete stratospheric ozone [ozone depletion potential (ODP) = 0.1] is only 1/10th that of CFC-11, which has an ODP of 1 by convention (Bunce, 1991).

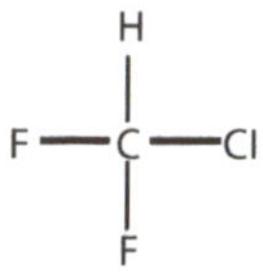

Chloro-difluoro-methane (CFC-22)

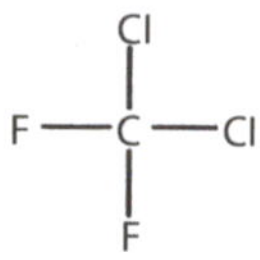

Dichloro-difluoro-methane (CFC-12)

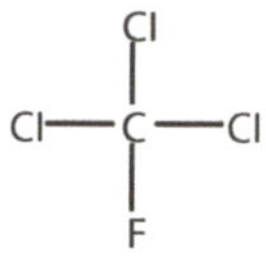

Trichloro-fluoro-methane (CFC-11)

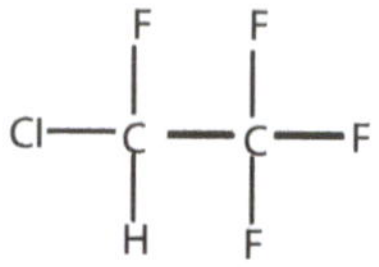

1-Chloro-1,2,2,2-tetrafluoroethane (HCFC-124)

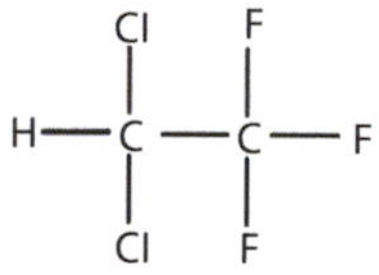

1,1-Dichloro-2,2,2-trifluoroethane (HCFC-123)

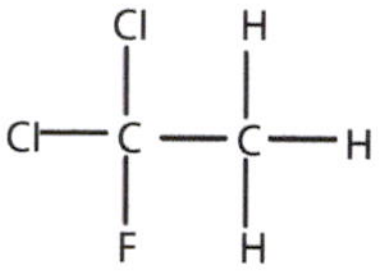

1,1-Dichloro-1-fluoroethane (HCFC-141b)

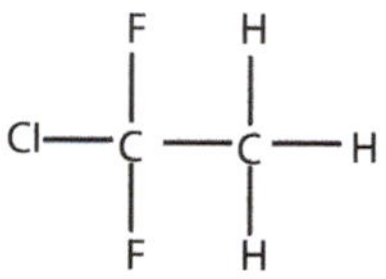

1-Chloro-1,1-difluoroethane (HCFC-142b)

Figure 2.6 Examples of chlorofluorocarbons (CFCs) and hydrochlorofluorcarbons (HCFCs) including their commercial designations, e.g., CFC-22, which is also called R-22.

2.2.2.2 *Organochlorine Alkenes*

There is no practical way to completely explain the rich nomenclature associated with the many organic compounds described here. Of necessity, brief and general qualities will be defined instead. For example, an alkene is distinguished here simply as a compound with a carbon = carbon double bond (covalent sharing of four electrons) as is evident in the compounds depicted in Figure 2.7. Their names end with ⁻ene as in ethylene or propylene.* (Remember that the C = C bonds drawn by convention in an aromatic ring do not qualify as this kind of sharing of two electron pairs by adjacent C atoms because the electrons are actually in resonance within the entire C ring.)

Organochlorine alkenes (Figure 2.7) were used in large volumes as solvents and degreasers. An important example is tetrachloroethylene, which is also called tetrachloroethene, perchloroethylene, PCE, and Perc. It was used in large amounts as a dry cleaning solvent and a degreaser for metal parts. Another important example is trichloroethylene (also called trichloroethene, TCE, and Triclene), a degreaser and solvent used in many processes. Trichloroethylene was even used to remove caffeine from coffee. Both of these organochlorine alkenes are immiscible and denser than water, making them prominent members of an important class of contaminants known as **Dense, Nonaqueous Phase Liquids (DNAPLs)** that often become associated with groundwater and drinking water contamination.

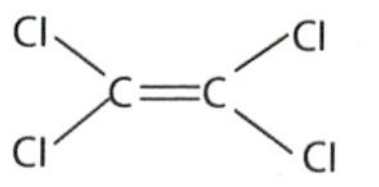

1,1,2,2-Tetrachloroethylene (Perchloroethylene)

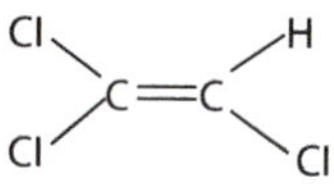

1,1,2-Trichloroethylene

Figure 2.7 Two important organochlorine alkenes contributing to Dense, NonAqueous Phase Liquids (DNAPLs) contamination.

* Aliphatic compounds with just single C–C bonds (covalent sharing of two electrons) are called alkanes and their names end with –ane as in methane, ethane, propane, or 1-chloro-1,1-difluroethane. Compounds with a triple carbon bonds (sharing six electrons) such as acetylene are alkynes.

2.2.2.3 Polycyclic Aromatic Hydrocarbons

Polycyclic (or polynuclear) aromatic hydrocarbons (PAHs) are composed of fused aromatic rings (Figure 2.8). They are produced by incomplete burning of wood, coal, or petroleum products such as that occurring in internal combustion engines of vehicles. The simplest (naphthalene) has only two rings but PAHs can have up to seven rings (coronene) (Albers, 2003). Alkyl and aromatic groups might be added to their basic fused-ring structures (e.g., 2,6-dimethylnaphthalene). Many, such as benzo[a]pyrene, can be precursors of potent carcinogens. They are abundant and widespread in our modern environment, being associated with air, soil, and water pollution. **Pyrogenic (pyrolytic) PAHs** are formed along with soot during combustion, that is, brief, high temperature combustion events. Sources of pyrogenic PAH to the environment include creosote, which is used widely to treat wood. Forest or grassland fires also release PAHs. Other PAHs, called **petrogenic PAHs**, originate from diverse petrochemical sources such as oil spills. They form slowly under low temperature conditions during petrogenesis and are often alkylated, that is, have added alkane groups as on the 2,6-dimethylnaphthlene in Figure 2.8. These PAHs can comprise 0.7%–7% of crude oils by weight (Kennish, 1997). Because their release is less pervasive, the petrogenic PAHs do not appear to the extent that the pyrogenic PAHs do in the sediment record (Burgess et al. 2003). **Biogenic PAHs** are also produced naturally by plants, fungi, and bacteria (Albers, 2003).

The signature abundance patterns of the various PAHs found in environmental samples are frequently used to infer the relative contributions of combustion and petroleum sources to any contamination. In general, petrogenic PAH mixtures tend to have more PAH compounds with alkyl substitutions than do pyrogenic PAH mixtures (Burgess et al. 2003). Petrogenic PAHs tend to be more prone than pyrogenic PAHs to alteration after entering the environment, especially the lower molecular weight PAHs that are more readily biodegraded (Burgess et al. 2003).

Other generalizations can be made about PAHs. The low molecular weight PAHs tend to be more toxic than the higher molecular weight PAHs (Kennish, 1997). The larger PAHs tend to be more carcinogenic than the smaller PAHs. The low molecular weight PAHs tend to readily move into the air and the larger PAHs become associated with solids (Bunce, 1991).

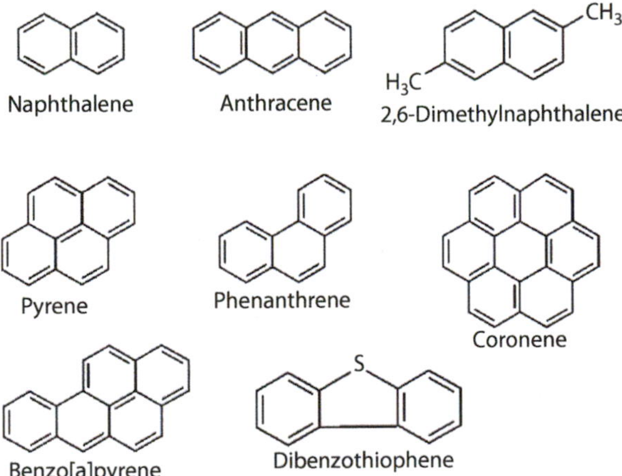

Figure 2.8 Examples of polycyclic aromatic hydrocarbons ranging from the two-ringed naphthalene to the seven-ringed coronene. These hydrocarbons can possess alkyl groups as shown for 2,6-dimethylnaphthalene. Although not a polycyclic aromatic hydrocarbon, the sulfur heterocyclic aromatic compound, dibenzothiophene, is also shown here because it commonly contributes to the aromatic hydrocarbon-related toxicity of spilt oils.

The PAHs have received much deserved study because of their abundance, ubiquity and diversity of sources, toxicity, teratogenicity, and carcinogenicity. They will appear numerous times in the chapters that follow.

2.2.2.4 Polyhalogenated Benzenes, Phenols, and Biphenyls

2.2.2.4.1 Polychlorinated Benzenes and Phenols

Polychlorinated benzenes such as the hexachlorobenzene (HCB) depicted in Figure 2.9 or 1,2,4-trichlorobenzene can be environmental contaminants of concern. HCB is a persistent contaminant used extensively as a fumigant, component of wood preservatives, and precursor in chlorinated pesticide synthesis. The polychlorinated phenols (bottom of Figure 2.9) are used as wood preservatives [e.g., trichlorophenol (TCP) and pentachlorophenol (PCP)] and fungicides (e.g., PCP). TCP was also used as a precursor in herbicide production. The polychlorinated phenols can be generated during kraft pulp mill operations when chlorine in the bleaching step reacts with natural phenolic compounds.

2.2.2.4.2 Polychlorinated Biphenyls (and Terphenyls)

The polychlorinated biphenyls (PCBs) were used as or in lubricants, heat conductors in electrical transformers, plasticizers, printing, and many other applications (Figure 2.10). The PCBs were produced commercially as mixtures of congeners* such as Aroclor 1221. Their accumulation in and toxic effects to humans and wildlife are a major concern with some biota being very sensitive to these effects, such as mink that consume tainted fish. The PCBs degrade very slowly in the environment and are soluble in fats and oils. Consequently, they can accumulate to very high concentrations in tissues. As an example, tissues of *Orca* whales off the coast of the U.S. Pacific Northwest continue to have very high PCB concentrations despite the U.S. ban on PCB production several decades ago (Hickie et al. 2007; Lubick, 2007a).

Polychlorinated terphenyls (PCTs) are similar in structure to PCBs but have a third aromatic ring to which chlorine atoms can be added. Produced as commercial mixtures such as Aroclor 5432 or 5460, they had similar uses in industry as the PCBs (Hale et al. 1990). U.S. production of PCTs was banned in the early 1970s.

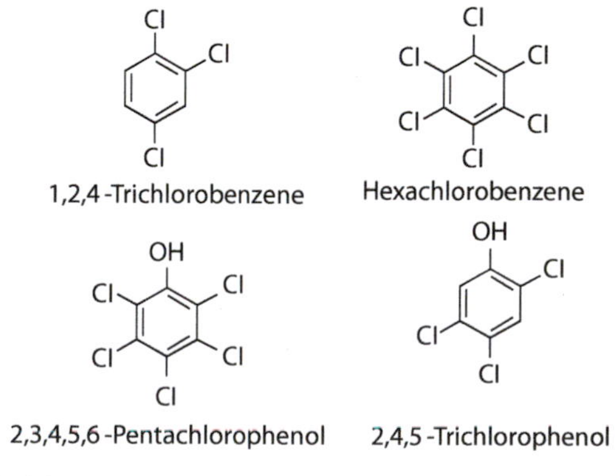

Cl Cl Cl 1,2,4-Trichlorobenzene

Cl Cl Cl Cl Cl Cl Hexachlorobenzene

OH Cl Cl Cl Cl Cl 2,3,4,5,6-Pentachlorophenol

OH Cl Cl Cl 2,4,5-Trichlorophenol

Figure 2.9 Examples of polychlorinated benzenes and phenols.

* The term, **congener**, is relevant to mixtures of similar compounds such as PCBs that differ along a common theme. Here, the PCB congeners in Aroclor 1221 are all of the various biphenyls in the mixture with different numbers and positions of the substituted chlorine atoms.

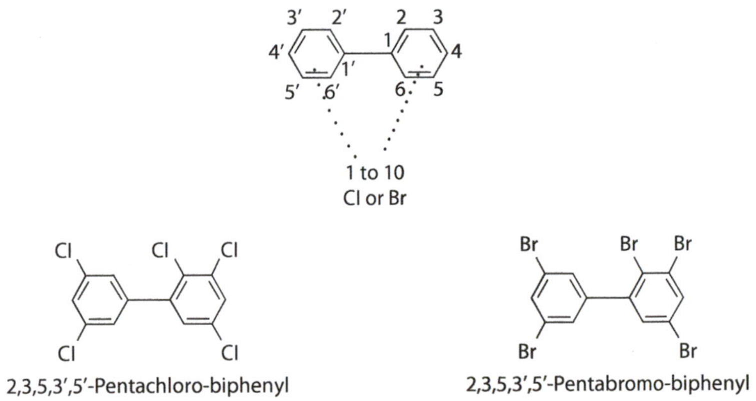

1 to 10 Cl or Br

2,3,5,3′,5′-Pentachloro-biphenyl

2,3,5,3′,5′-Pentabromo-biphenyl

Figure 2.10 The general structure of polychlorinated or polybrominated biphenyls is shown at the top of this figure. One to 10 of either chlorine or bromine atoms are placed along the two rings. Two examples are provided below this general structure.

2.2.2.4.3 Polybrominated Biphenyls

The polybrominated biphenyls (PBBs) share many qualities with PCBs (Figure 2.10). They are slow to degrade and tend to accumulate to high concentrations in biota. Members of the general category of brominated fire retardants, the PBBs, were removed from the market in the 1970s after a poisoning incident in Michigan in which a commercial fire retardant product was accidentally packaged as an animal feed supplement (Birnbaum and Staskal, 2004).

2.2.2.5 Polychlorinated Naphthalenes

Prior to cessation of their production, the polychlorinated naphthalenes (PCNs) (Figure 2.11) were incorporated into diverse electrical devices, wires, and plastics. They were used commercially for "impregnation of wood, paper and textiles to attain waterproofness, flame resistance and protection against insects, molds and fungi" (Jakobsson and Asplund, 2000). They were produced as commercial mixtures such as Halowax 1031. Now, PCN congeners are released in landfill leachate, waste incinerator ash and fumes, and chlor-alkali plant sludge. They are also by-products of the

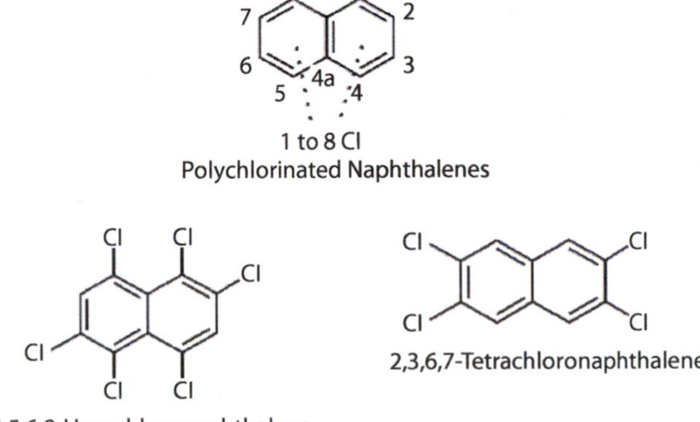

Figure 2.11 The general structure of polychlorinated naphthalenes in which one to eight chlorine atoms are substituted into the molecule. Two examples of polychlorinated naphthalenes are also shown.

production of some metals and are present as minor contaminants in PCB products. Currently, they are considered among the most widespread and persistent of the organic contaminants.

2.2.2.6 Polychlorinated Dibenzodioxins and Dibenzofurans

The general structures of polychlorinated dibenzodioxins and furans are provided in Figure 2.12. The furans have a characteristic single O-containing ring stabilized by electron resonance. The dioxins have a heterocycle with two oxygen atoms. Both polychlorinated dibenzodioxins and dibenzofurans have many congeners. Neither was manufactured intentionally, but appeared as contaminants during incineration or the synthesis of other compounds such as PCBs and some herbicides (Bunce, 1991).

Polychlorinated dibenzofurans (PCDFs) are released as contaminants of other commercial products such as PCB mixtures and chlorophenols. They can be generated during combustion or bleaching associated with kraft pulp mills.

Dioxins enter the environment as contaminants in herbicides (e.g., Agent Orange) and wood preservatives. Polychlorinated dibenzo-p-dioxins (PCDDs) are contaminants in commercial products such as PCBs and chlorophenols. Dioxins are also formed as combustion by-products and during the bleaching process of kraft pulp mills. Some dioxins are extremely toxic.

2.2.2.7 Pesticides

Most organochlorine pesticides and many of the organic chemicals already discussed (examples being HCB, PCBs, PCDFs, and PCDDs) are classified as persistent organic pollutants (POPs) because they are resistant to degradation in the environment and tend to increase in concentration with movement up through food webs. They can persist at alarming concentrations in biota and many can disperse globally as will be discussed later (Weber and Goerke, 2003; Tanabe, 2004; Lubick, 2007a). Sometimes, POPs are called persistent toxicants that bioaccumulate (PTB) or persistent bioaccumulative toxic (PBT) chemicals. The most widely recognized organochlorine pesticide members of the POPs are aldrin, chlordane, 1,1-dichloro-2,2-bis-(p-chlorophenyl) ethane (DDD), dichlorodichloroethylene or 1,1-dichloro-2,2-bis-(p-chlorophenyl)- ethene (DDE), dichlorodiphenyltrichloroethane or 2,2-bis-(p-chlorophenyl)-1,1,1-trichloroethane (DDT), dieldrin/endrin, heptachlor, mirex, and toxaphene. Chlordecone (kepone) is also a POP that can cause significant

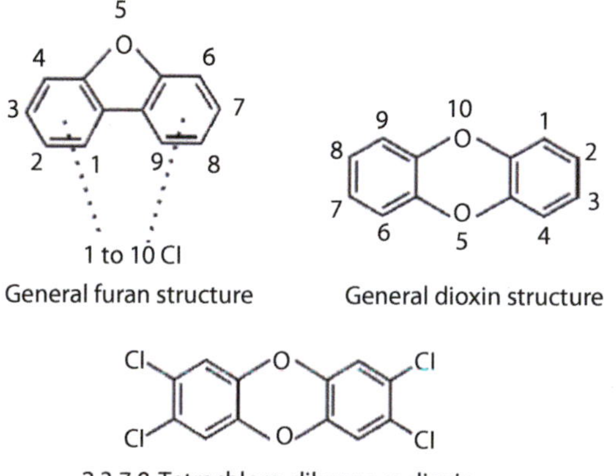

Figure 2.12 The general structures of polychlorinated dibenzofurans and dibenzodioxins. The dioxin shown (2,3,7,8-tetrachloro-dibenzo-p-dioxin) is one of the most toxic organic contaminants known and is often simply referred to as "dioxin" in regulatory and popular publications.

contamination associated with pesticide manufacturing (Kennish, 1992). Of course, being so structurally similar to PCBs, the PBBs will behave as POPs.

Numerous national laws and international treaties ban production of members of the POPs. Recognizing the global context of the problem posed by POPs, the United Nations Stockholm Convention on Persistent Organic Pollutants was established in 2001, eventually signed by 50 countries, and became binding in 2004. Its aim was to eliminate 12 POPs*; however, some use of DDT in developing countries that are still combating endemic malaria was permitted until a suitable replacement can be found (Lubick, 2007b).

2.2.2.7.1 *Organochlorine*

The organochlorine pesticides degrade very slowly and tend to be soluble in lipids such as those of organisms. This results in bioaccumulation and possible increase in concentrations with passage through food webs. Important organochlorine pesticides are shown in Figures 2.13 through 2.15.

The DDT depicted in Figure 2.13 was an extraordinarily valuable insecticide that was banned in the 1960s when its adverse effects to wildlife became apparent. In addition to being a degradation product of DDT, DDD was used widely as a pesticide and was banned along with DDT after lethal and sublethal poisonings of wildlife occurred (Dolphin, 1959). DDE can be produced from DDT in the environment or as a DDT metabolite in organisms (Bunce, 1991). All (DDT, DDD, and DDE) are present as mixtures of isomers in the environment. Often, concentrations of DDT, DDD, and DDE in samples are summed and reported as ΣDDT, total DDT, or tDDT.

Other prominent organochlorine pesticides are shown in Figure 2.14. These chlorinated pesticides share with DDT an acute toxicity mode of action involving disruption of neural transmission. They inhibit ATPases that are crucial for membrane ion transport in nervous tissues. Aldrin, chlordane, and dieldrin/endrin are cyclodiene† pesticides and tend as a class to be among the more toxic organochlorine pesticides to taxa such as fish (Kennish, 1997). Toxaphene (camphechlor), classified as a polychloroterpene‡ or polychlorinated camphene pesticide, acts like a cyclodiene pesticide. Lindane, primarily one isomer (gamma) of hexachlorocyclohexane (HCH), tends to be metabolized more readily in organisms and to degrade more rapidly in the environment than the other organochlorine pesticides just described.

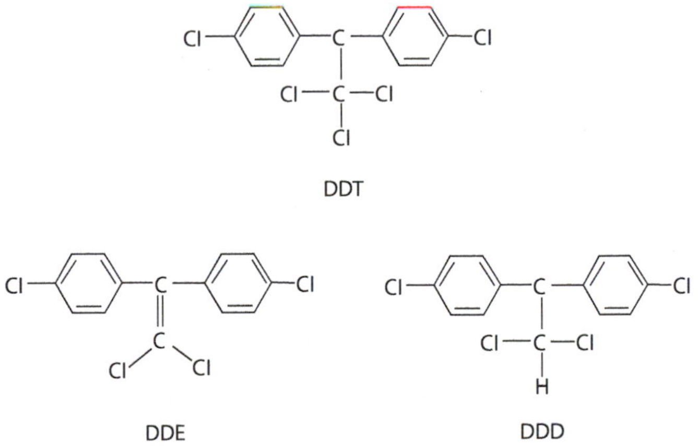

DDT

DDE

DDD

Figure 2.13 DDT and two closely related chlorinated contaminants with aromatic (diphenyl) and aliphatic (ethane or ethene) components.

* Aldrin, chlordane, DDT, heptachlor, endrin/dieldrin, HCB, mirex, toxaphene, PCBs, PCDD, and PCDFs.

† A **diene** is an alkene with two double bonds in its structure. These pesticides are produced by reactions of hexachlorocyclopenta<u>diene</u>.

‡ A **terpene** is made up of isoprene units, i.e., $H_2C{=}C(CH_3){-}CH{=}CH_2$ that are often inserted head to tail in a compound. In this case of toxaphene, chlorine atoms have been bonded to the basic camphene structure.

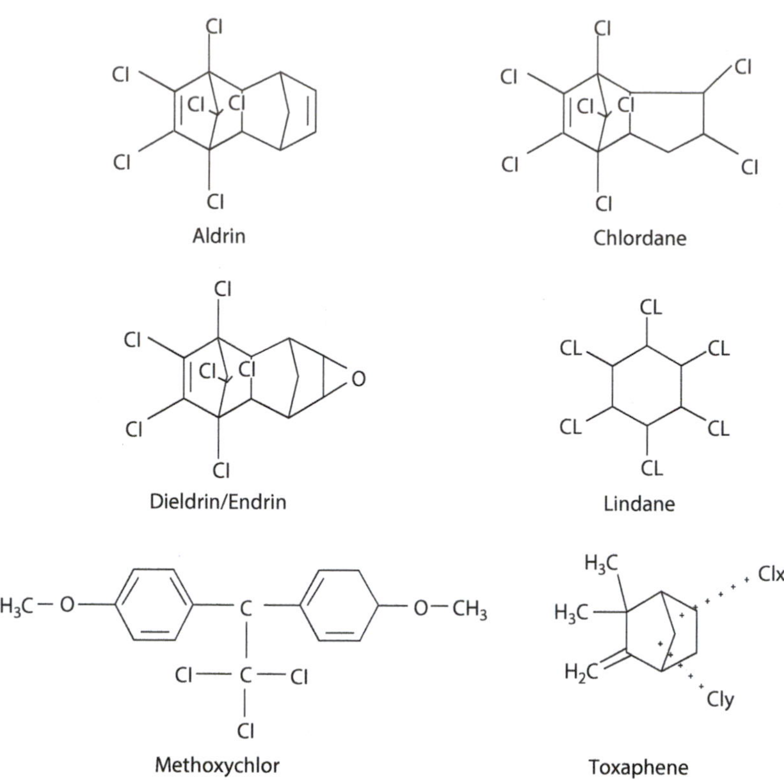

Figure 2.14 Seven prominent organochlorine pesticides including dieldrin and endrin, which are isomers of each other. Notice how similar the structure of methoxychlor is to that of dichlorodiphenyltrichloroethane. Toxaphene has different numbers of chlorine atoms attached to the ring shown on the plane of the page (x) and the ring extending upward from the plane of the page (y). Toxaphene (along with strobane) is a polychloroterpene insecticide. Aldrin and chlordane also are shown with rings flat on and perpendicular to the plane of the page. Note that the carbon ring of lindane is not an aromatic ring. For an organochlorine pesticide, lindane degrades relatively rapidly in the environment (Kennish, 1997). Note that these pesticides are actually used as mixtures of isomers.

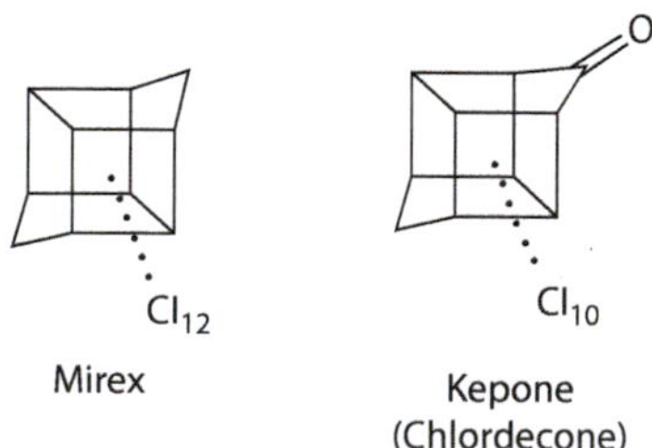

Figure 2.15 Two similar organochlorine pesticides, mirex (dodecachlorooctahydro-1,3,4-methano-1H-cyclobuta [cd]pentalene) and kepone (chlordecone).

The pesticides in Figure 2.15 had wide application including mirex's use in fire ant baits and as a fire retardant (Kennish, 1997). Chlordecone was used in fire ant bait and was generated as a degradation product during mirex production (Kennish, 1997; Blus, 2003). Both were banned in the United States in the late 1970s.

2.2.2.7.2 *Organophosphorus*

These pesticides degrade faster in the environment than organochlorine pesticides. However, they tend to be more acutely toxic to mammals than organochlorine pesticides and are often involved in human pesticide poisonings such as those described in Chapter 1. They can also accumulate in fats and oils of organisms. The mode of action for these compounds is the inhibition of acetylcholinesterase activity. So, like the organochlorine pesticides, organophosphorus pesticides are neurotoxicants. However, they act by blocking acetylcholinesterase, an enzyme essential in breaking down the neurotransmitter, acetylcholine. Important examples are shown in Figure 2.16. Others include fenitrothion, fethion, fonofos, and parathion.

2.2.2.7.3 *Carbamate*

Like organophosphate insecticides, carbamate insecticides degrade rapidly in the environment and cause neural dysfunction by inhibiting acetylcholinesterase. Carbamates have high acute toxicity to mammals but their toxicities tend to be lower than those of the organophosphate pesticides. Examples include those shown in Figure 2.17. They are derived from carbamic acid (H_2NCOOH), having the general structure of R_1–NH–COO–R_2 (Manahan, 1993).

2.2.2.7.4 *Pyrethrum and Pyrethroid*

Along with rotenone, toxaphene, and the neonicotinoid insecticides,* these toxicants are themselves, or are similar to, natural pesticides produced by plants. Pyrethrins extracted from pyrethrum flowers were used as pesticides in the early 1800s (Tomizawa and Casida, 2005). The neurotoxic effects of pyrethrum and pyrethroid pesticides result from their interference with normal sodium channel functioning in cells of nervous and other excitable tissues. All are similar to pyrethrum

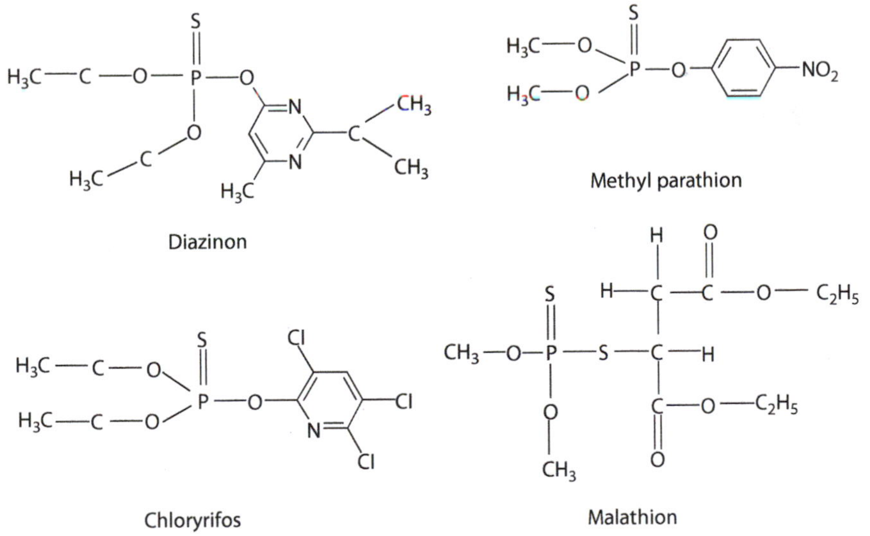

Diazinon

Methyl parathion

Chloryrifos

Malathion

Figure 2.16 Four common organophosphate pesticides.

* The pesticides described in this section are those that are themselves, or are similar to, natural plant products. **Toxaphene** is produced by adding chlorine atoms to the plant-produced camphene molecule. The **neonicotinoid insecticides** are also modifications on the naturally produced insecticide, nicotine, structure. **Rotenone** is a **natural plant product pesticide**. It was extracted from the roots of plants (*Derris* sp.) as early as the 1820s and used as a broad spectrum pesticide. It acts as a respiratory inhibitor by inhibiting the oxidation of NADH to NAD.

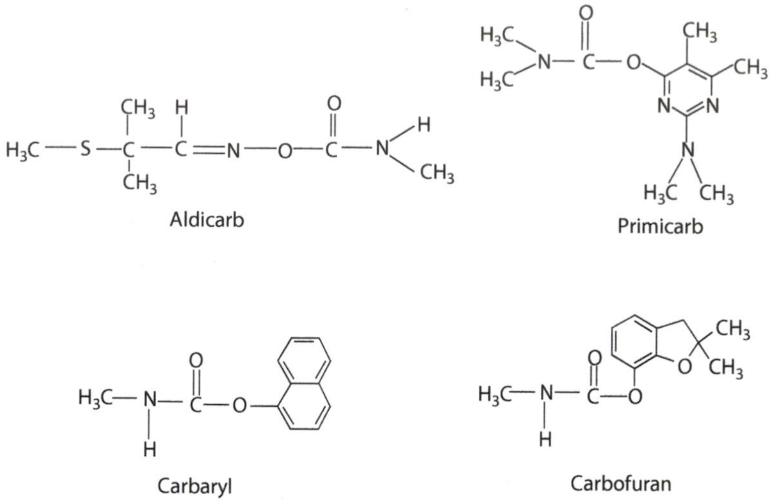

Figure 2.17 Representative carbamate pesticides including the widely used carbaryl (Sevin).

produced by the plants, *Chrysanthemum cineum* and *C. cinerariaefolium*. Some are natural products although they might now be produced synthetically. Others are analogs of or similar to the natural compounds. The natural compounds are called **pyrethrins** and their synthetic analogs or derivatives are called **pyrethroids**. Pyrethrin 1 in Figure 2.18 is a pyrethrin. Allethrin and permethrin are analogs of natural compounds (i.e., are pyrethroids) that do not incorporate a cyano group (triple bonded C and N) in their structures. They are classified as Type I pyrethroids for this reason.

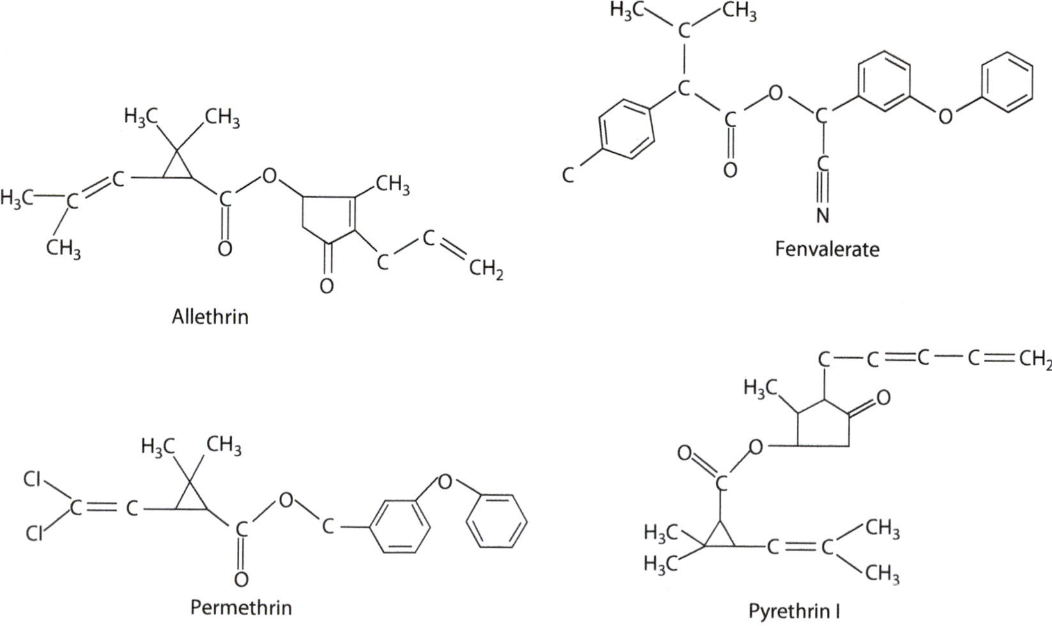

Figure 2.18 Representative pyrethrin (pyrethrin 1) and pyrethroid (allethrin, fenvalerate, and permethrin) pesticides. The cyano group in fenvalerate distinguishes it as a Type II pyrethroid and the absence of this group from allethrin and permethrin make them Type I pyrethroids.

Fenvalerate, which contains a cyano group, represents the Type II pyrethroids in the figure. Acute poisoning symptoms manifest differently for Type I and II pyrethroids. Most of these pesticides are used as mixtures of isomers. Because they degrade quickly, most related environmental problems are associated with acute exposures.

2.2.2.7.5 Neonicotinoid

Nicotine has a long history of use as a pesticide and the **neonicotinoid insecticides** (Figure 2.19) are modifications of this naturally produced alkaloid. The development of neonicotinoids began in the 1970s, and since then, nicotine derivatives have emerged as the most recent class of insecticides to be developed (Figure 1.1). They currently account for 17% of all worldwide insecticide sales (Tomizawa and Casida, 2005; Jeschke and Nauen, 2008) with one of the most widely used (imidacloprid) having been in the market only since 1992 (Kagabu, 1997). They act as agonists to the postsynaptic nicotinic acetylcholine receptors of insects (Tomizawa and Casida, 2005). The toxicity of these systemic insecticides to mammals, birds, and fish is relatively low. This combined with their low lipophilicity (Log K_{ow} values in the range of $^{-}0.66$ to 1.26) makes them ideal as systemic insecticides. An especially attractive feature of these insecticides is their high **pesticide safety factor**, that is, the quotient of $LD50_{mammal}/LD50_{insect}$ due to their specific mode of action. Kagabu (1997) reports a safety factor of 0.08 for

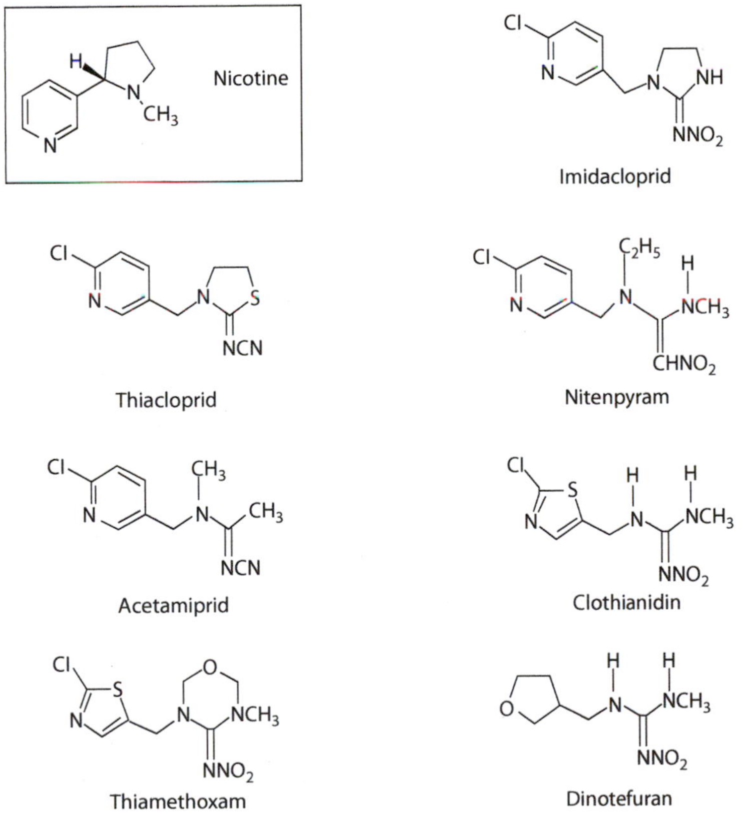

Nicotine

Imidacloprid

Thiacloprid

Nitenpyram

Acetamiprid

Clothianidin

Thiamethoxam

Dinotefuran

Figure 2.19 Members of the neonicotinoid class of pesticides are shown here in addition to the naturally produced nicotine molecule. The nicotine structure is shown boxed on the top left of the figure.

nicotine and contrastingly high safety factors for organophosphate (5–37) and carbamate (17) pesticides. The rate of insect resistance development to neonicotinoid insecticides also appears to be slow (Nauen and Denholm, 2005). They are gradually replacing many organophosphate and carbamate pesticides because of their relatively low risk to nontarget species (Jeschke and Nauen, 2008).*

This trend might end soon, however, due to growing concerns about their adverse impact on insect pollinators. Nectar and pollen consumers appear to be exposed to harmful levels of these (and other) systemic pesticides, notionally reducing the efficiency of foraging and also rearing of young bees (Thompson, 2001; Dicks, 2013). The European Food Safety Authority (EFSA) recently published its concerns about risk to bees resulting from current uses of clothianidin (EFSA, 2013a), imidacloprid (EFSA, 2013b), and thiamethoxam (EFSA, 2013c). Concern has also arisen recently about the impact of neonicotinoid insecticides on cerebellar neurons of mammals (Kimura-Kuroda et al. 2012).

A few points require mention before we leave the topic of pesticides. The progression from organochlorine to carbamate pesticides involved an increase in degradability in the environment. However, the replacement of POP pesticides with the more readily degradable pesticides came at a cost of increased mammalian toxicity. Figure 2.20 plots LD50† values for pesticides fed orally to rats. There is a clear shift toward pesticides with higher mammalian toxicity (lower LD50 values) with the movement from organochlorine to organophosphate and carbamate pesticides. The mammalian toxicity lessened considerably with the use of pyrethroids and neonicotinoids.

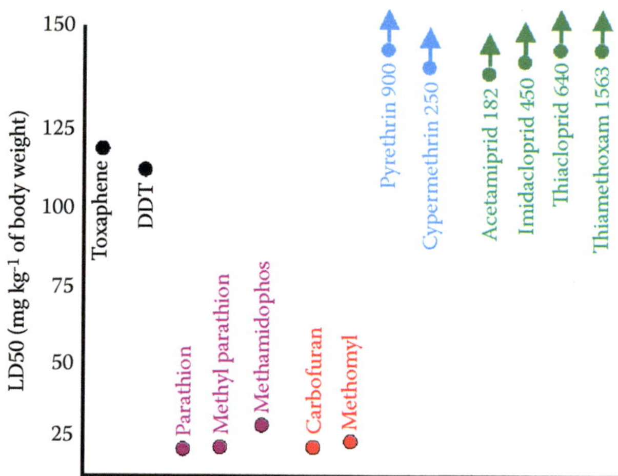

Figure 2.20 LD50 values for orally dosed rats. (Organochlorine [black], organophosphate [purple], and carbamate [red] pesticide data taken from Table of Murray [1994]; pyrethroid [blue] pesticide data taken from Kagabu [1997]; and neonicotinoid [green] pesticide data were obtained from Table 3 of Tomizawa and Casida [2005]. The actual LD50 values for the pyrethroid and neonicotinoid pesticides are given after the insecticide name. The arrow indicates an LD50 that is higher than the maximum value for the *y* axis.)

* Concern about their use remains because of potential impacts on honey bee and other pollinator insects.

† As discussed later, the LD50 is the estimated dose that would kill 50% of exposed individuals.

VIGNETTE 2.1 Endosulfan and Cashew Production in Southern India

S. Bijoy Nandan
Cochin University of Science and Technology

ENDOSULFAN

Endosulfan, a cyclodiene organochlorine pesticide, is a broad-spectrum insecticide and acaricide* that was first registered for use in the United States in 1954 to control agricultural pests on a variety of crops. It was used on vegetables, fruits, paddy, cotton, cashew, tea, coffee, tobacco crops, and also to control tsetse flies and mites. Endosulfan is a mixture of two stereo isomers (Figure 2.21) of approximately 70% α-endosulfan and 30% β-endosulfan. Endosulfan sulfate is a major, equally toxic degradation product of these isomers (Extoxnet, 1993; EPA, 2001b).

Endosulfan is currently classified as a highly hazardous chemical by the United States Environmental Protection Agency (EPA), World Health Organization (WHO), and Indian Institute of Toxicology Research (IITR); as a persistent toxic substance by the European Union (EU); and as a category II-Moderately Hazardous chemical by the United Nations Environmental Program (UNEP). In 2002, U.S Fish and Wildlife Service studies found endosulfan residues in food and water to pose serious health hazards, and accordingly, restrictions were tightened on its agricultural use. To date, it has been banned in over 80 countries, including India (UNEP, 2002a).

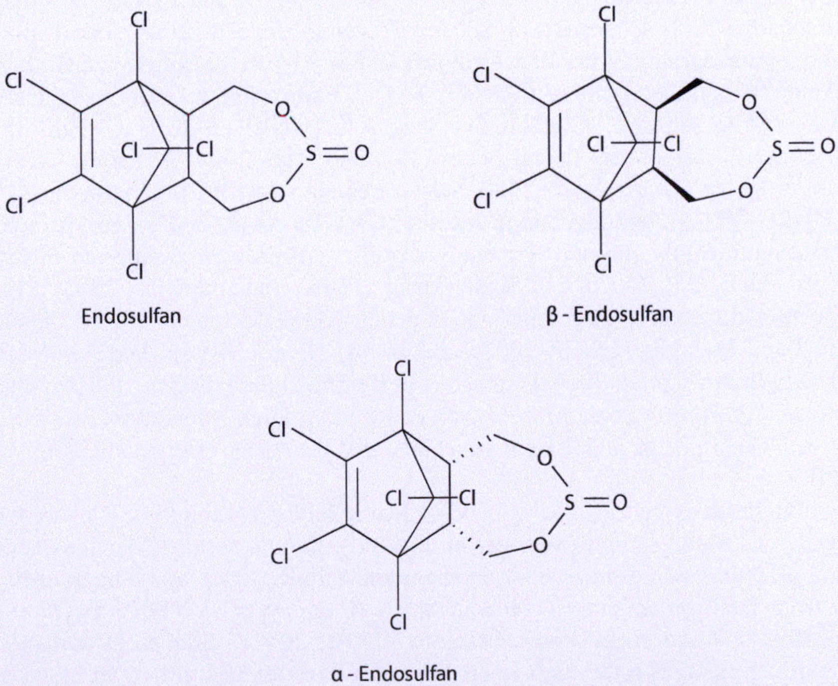

Endosulfan

β - Endosulfan

α - Endosulfan

Figure 2.21 Molecular structures of endosulfan, α-endosulfan, and β-endosulfan.

* An **acaricide** is a chemical agent used intentionally to kill mites or ticks.

ENDOSULFAN USE IN SOUTHERN INDIA CASHEW PRODUCTION

India occupies an important position in the world production of cashew nut. Production in India during 2012–2013 was about 7.3 lakh* tonnes in a cultivated area of approximately 9.8 lakh hectares (Ragunathan, 2012). The leading cashew producing states of India are Kerala, Maharashtra, and Andhra Pradesh with the Kannur and Kasargod districts in Kerala accounting for roughly 70% of the total area under cashew cultivation.

The major insects that decrease cashew yield are the tea mosquito, flower thrip, and stem, root, fruit, and nut borers. The nymphs and adults of tea mosquito (*Helopeltis antonii*) suck sap from new leaves, shoots, inflorescence, and even young nuts and fruit of the cashew plant. To eradicate the tea mosquito in about 4860 ha of cashew plantations in nine Kasargod villages of Kerala, the Plantation Corporation of Kerala (PCK) had resorted to two or three aerial sprayings of endosulfan annually (1978–2001). Residents were informed 2 days prior to spraying by megaphone announcements, and notices posted in public and government institutions. Nonetheless, many remained unaware of the spraying as well as any need to take precautions. Some water sources were covered with coconut leaves during PCK spraying; however, preventing contamination of many water sources like ponds, rivulets, streams, and wells was impossible. The congenital deformities in cattle, mass death of bees, fishes, and birds that occurred with aerial spraying were largely ignored by the PCK and government agencies (Government of Kerala, 2003).

GENERAL ENVIRONMENTAL LEVELS AND HEALTH EFFECTS

Endosulfan has been detected in surface and groundwater in India (Rao and Pillale, 2001; Akhil and Sujatha, 2012), Ghana (Ntow, 2001), Pakistan (Tariq et al. 2004), Portugal (Cerejiera et al. 2003), South Africa (Fatoki and Awofolus, 2004), Mexico (Miglioranza et al. 2004a,b), Central America, and the Caribbean (UNEP, 2002b). Endosulfan residues were detectable in marine waters of India (UNEP, 2002a), Southeast Asia (UNEP, 2002c), and Gulf of Mexico, and other coastal areas of the United States (Hoh and Hite, 2004). Residues were reported from rivers in China (Zhang et al. 2004) and Guatemala (UNEP, 2002b), mountain lakes in Europe (UNEP, 2002d), and lagoons in Mexico (Carvolho et al. 2002) and Spain (Cerejiera et al. 2003). Significantly elevated concentrations of endosulfan were reported in Colombian drinking waters (UNEP, 2002b) and South Africa (Fatoki and Awofolus, 2004). The coastal and estuarine sediments of Argentina and Jamaica (Gonzalez et al. 2002), Ghana (Ntow, 2001), Israel and Malaysia (UNEP, 2002c), Benin and Nigeria (Fatoki and Awofolus, 2004), Honduras (Miglioranza et al. 2004a), and soils in Benin, Nigeria, Sudan, and Zambia (Fatoki and Awofolus, 2004) were contaminated with endosulfan. High levels of endosulfan residues were even reported in fauna and flora of Greenland and the Arctic (Herrmann, 2003; Vorkamp et al. 2004).

Endosulfan residues have also been found in diverse biological samples. They were detected in umbilical cord blood samples (Cooper et al. 2001), human serum (Martinez et al. 2002; Younglai et al. 2002), adipose tissue (Amaraneni and Pillale, 2001), and human milk samples collected from healthy lactating women in Spain (Campoy et al. 2001), Egypt (Sang and Petrovic, 1999), and cotton packers in Pakistan (UNEP, 2002a). Studies of human exposure in Kasargod in Kerala reported high levels of endosulfan residues in human blood and milk [Centre for Science and Environment (CSE), 2001; Joshi, 2001; Thanal, 2001; Vankar et al. 2001; Sanghi et al. 2003].

* A lakh is 100,000 in South Asian counting systems.

Health impacts of endosulfan residue exposure are well documented. Endosulfan exposure of adult rats resulted in reduced intratesticular spermatid counts, sperm abnormalities, and changes in the marker enzymes of testicular activities (lactate dehydrogenase, sorbitol dehydrogenase, γ-glutamyl transpeptidase, and glucose-6-phosphate dehydrogenase). There have been reports from Spain of undescended testes in human populations due to endosulfan exposure (Saiyed et al. 2003). Endosulfan has also been reported to have inhibited testicular androgen biosynthesis, caused renal and testicular damage, and inhibited leucocyte and macrophage migration resulting in adverse effects on humoral- and cell-mediated immune system in laboratory animal experiments (Sang and Petrovic, 1999).

Relative to acute occupational exposure, endosulfan is highly toxic and can be fatal if inhaled, swallowed, or absorbed through the skin. Ingestion or inhalation of high levels of the pesticide can lead to convulsions and death. It directly affects the central nervous system and may cause recurrent epileptic seizures (Extoxnet, 1996; Thanal, 2002). Its inhibition of brain acetylcholinesterase can cause detrimental effects in the brain (Naqvi and Vaishnavi, 1993). Autopsy of an intentional ingestion case revealed damage to liver, lungs, and brain (Lu et al. 2000). Although it is a proven endocrine disruptor causing irreparable changes in gonads and thyroid gland, its long-term effects have not been fully studied (Soto et al. 1994; CDC, 1997; Saiyed et al. 2003). Endosulfan has estrogenic properties comparable to those of DDT (Soto et al. 1994). Endosulfan can also promote tumor formation (Fransson-Steen et al. 1992). For example, it can induce proliferation of human breast estrogen sensitive cells (Preziosi, 1998), promoting breast cancer development (Ibarluzea et al. 2004). α- and β-endosulfan have the capacity to alter the genetic material, as documented for chromosomal changes in mammalian cell cultures (Jamil et al. 2004). The data from a study conducted by the U.S. National Cancer Institute revealed that endosulfan exposure induced malignant neoplasms in many tissues of male and female rats, including the endocrine organs of males (Reuber, 1981). Although no direct evidence is available to link carcinogenicity of endosulfan in humans, surveys have created the suspicion of a linkage between human cancer and endosulfan chronic exposure (Quijano, 2000).

The National Institute of Occupational Health (NIOH) reported higher prevalence of neurobehavioral disorders, congenital malformations in female children in Kasargod district chronically exposed to endosulfan spraying (NIOH, 2002). NIOH surveys of school children from the Kasargod villages of Padre and Meenja revealed serious health problems such as delay in sexual maturity and establishment of secondary sexual characters, early menarche among girls, and higher incidence of goiter (NIOH, 2002). Similarly, studies conducted by the CSE in Padre village in Kasargod reported blood samples containing endosulfan residues several hundred times higher than the residue limits for water (CSE, 2001).

In response to the various health and environmental problems arising from aerial spraying of endosulfan in the villages of Kasargod, the Kerala State Council for Science, Technology and Environment (KSCSTE) was entrusted by the government of Kerala to conduct detailed investigations on the environment, health, and social aspects of the affected villages of the district (Figure 2.22).

ISSUES IN DIRECTLY AFFECTED VILLAGES

Studies conducted by the KSCSTE with the active support of the Centre for Water Resources Development and Management (CWRDM, Government of Kerala) noted a pervasive, noxious odor in villages after endosulfan aerial spraying. People reported difficulty breathing, eye

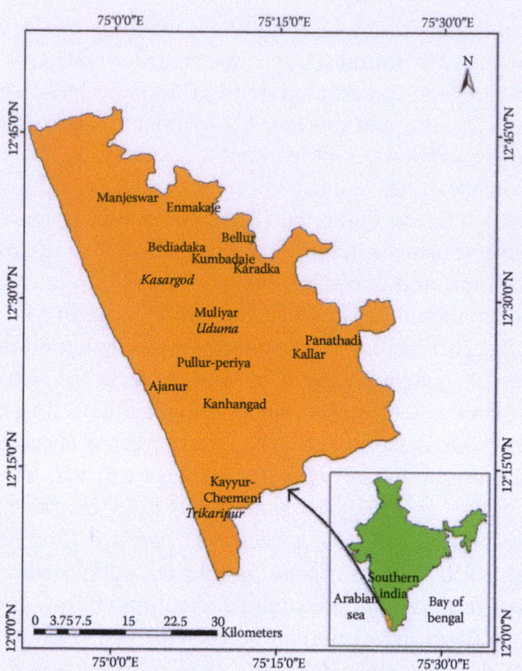

Figure 2.22 The endosulfan-impacted panchayaths of the Kasargod district of Kerala are shown in the left map. The map of southern India on the right shows the general location of Kerala in orange with the box around the general location of the Kasargod district. (Map of panchayaths modified from the Kerala State Council for Science, Technology and Environment report of monitoring of endosulfan residues in the panchayaths of Kasargod.)

irritation, headache, dizziness, and skin itching with swelling when scratched. Some grazing animals had seizures that were often followed by death. Uterine bleeding resulting in abortion and sometimes death was reported in pregnant cows and dogs.

The pesticide residue in the water samples collected from different wells and water bodies in the affected panchayaths* was below the detection limit and within the permissible limits set by the U.S. EPA and WHO (KSCSTE, 2011). The highest soil concentration of 16.91 $\mu g \cdot kg^{-1}$ was recorded from the Pullur–Periya panchayath (α-endosulfan: 7.38 $\mu g \cdot kg^{-1}$, β-endosulfan: 9.53 $\mu g \cdot kg^{-1}$) whereas samples from most other affected villages of Kasargod were near the detection limit. The highest pesticide levels in sediment were from the site used for cleaning the helicopter sprayer of and filling it with pesticide. Endosulfan sulfate, a persistent degradation product of endosulfan, was detected only in one soil sample from Bellur and one sediment sample from Karadukka village. Endosulfan sulfate has high toxicity and a half-life in the environment of 100–150 days, that is, several times longer than the parent compounds, α- and β-endosulfan. The pH values of soil, sediment, and water samples in the selected panchayaths were acidic, and under these acidic conditions, the endosulfan isomers are likely to persist longer in soil. Akhil and Sujatha (2012) reported α-endosulfan concentration of 7.8 $\mu g \cdot L^{-1}$ in Parappa and 3.1 $\mu g \cdot L^{-1}$ in Rajapuram area water during the premonsoon period, whereas Panathur panchayath registered the highest concentration of α-endosulfan (58 $\mu g \cdot L^{-1}$) followed by Peria village (37 $\mu g \cdot L^{-1}$) during the postmonsoon period. However, the premonsoon level

* A local small village or town in India that is self-governed.

of α-endosulfan was maximum in Panathady (56 $\mu g{\cdot}L^{-1}$) followed by Rajapuram (40 $\mu g{\cdot}L^{-1}$) panchayath. The increase in α-endosulfan in soil even after its ban in 2004 may be attributed to illegal and clandestine usage for agricultural purpose. Based on the scientific investigations and literature discussed previously, it seems well established that endosulfan aerial spraying caused serious detrimental effects on the environment and humans in selected villages of Kasargod, Kerala.

Studies by KSCSTE also reported a decline in plant diversity from 40% to 70%, particularly involving native species in the endosulfan sprayed areas. Studies have also reported mass death of honeybees, fishes, frogs, snakes, squirrels, hares, peacocks, crows, and butterflies in early 1978 on or near the cashew plantations during aerial spraying. The common jackal, porcupine, wild boar, civet cats, and bats completely disappeared during the spraying period; however, there were signs of their return 5–6 years after aerial spraying ended. The fishes, *Mesonoemacheilus triangularis* and *M. guentheri,* reported earlier were rarely seen from the rivers of the area whereas the frogs, *Rana temporalis* and *Micrixalus* sp., were not observed during the KSCSTE study. The major bird species that were missing in the plantations as a result of pesticide spraying included flycatchers, babblers, and other endemics such as small sunbird, crimson-throated barbet, and white-bellied tree pie.

According to the PCK, there were 292 houses located inside the plantation area whereas 2026 houses were outside the area. The estimated numbers of people living in these houses were about 1,058 and 10,738, respectively (USNIEHS, 2003). Residents living in and around the plantation areas lived mostly on the hills, valleys, and slopes that were exposed to the hazardous chemical for 24 years (1978–2001). The endosulfan-contaminated water from the hills drained into streams and drinking water wells of households in the valley. The PCK workers who prepared and mixed pesticides resided on and near the plantations. No protective devices over face or body were provided to these staff, even during spraying operations. Some developed skin problems, lymph node enlargement, chronic asthma, and cancer due to the pesticide exposure.

A farmer in Padre village (Kasargod) reported the birth of three calves with deformed limbs and stunted growth. This prompted a local journalist, Shree Padre, to write a December 1981 article "Life Cheaper than Cashew" in the newspaper, *The Evidence*. This was the first printed report on the possible environmental and health problems in Kasargod due to aerial spraying of endosulfan in the cashew plantations. A 2002 health survey organized by the Government of Kerala on the adverse effects to people from endosulfan-sprayed (Padre village) and nonsprayed (Meenja village) areas* of Kasargod identified cases like congenital abnormalities (95 in sprayed area and 70 in nonsprayed area), mental retardation (871 in sprayed and 83 in nonsprayed), cancers (58 in sprayed and 34 in nonsprayed), infertility (104 in sprayed and 75 in nonsprayed), growth retardation (25 in sprayed and 11 in nonsprayed), and psychiatric problems (46 in sprayed and 33 in nonsprayed). There were reports of infertility in men, whereas women and children were found to be anemic and subject to frequent fevers and diseases, suggesting immunotoxic effects. The doctors practicing in the area reported an alarming number of cases of breast enlargement in boys (**gynaecomastia**). Women were suffering from serious gynecological problems, requiring treatment to correct hormonal abnormalities and some

* Respectively, these counts were derived from a survey of 80,275 and 80,281 people in the sprayed and nonsprayed areas (Government of Kerala, 2003). Because the numbers of individuals were nearly identical, these prevalence estimates can be used directly without normalization to a conventional sample size.

hysterectomies. Organochlorine pesticides are known to cause endometriosis and breast cancer. Infertility and miscarriages were also high in women, and in some cases, endometriosis has also been reported from the area (Saiyed et al. 2003). Swelling of legs and hands was specifically reported by many of the plantation workers involved in pesticide mixing and spraying operations. It is well known that long-term exposure to organochlorines can affect the nervous system affecting hearing and vision. Some plantation workers reported that they experienced eye irritation, burns, chronic headaches, and sometimes, loss of vision (Thanal, 2002).

A brief survey conducted in the Periya–Pullur, Muliyar, Cheemeni, and Rajapuram areas reported that several people were suffering from oral, throat, stomach, prostate, or intestinal cancer (Soto et al. 1994). According to Dr. Y. S. Mohankumar, a practicing physician in Enmakaje panchayath, there were 51 cancer deaths mostly due to liver and blood cancer in 26 households near the Kodenkeri stream of the panchayath. Cancer deaths recorded in Enmakaje panchayath alone had increased from 37 (1982–1987), to 49 (1988–1993), and to 71 (1994–1999). This reflects an increase of 33% in just 6 years and 92% in 12 years. Comparatively, residents of the unsprayed Meenja panchayath, which was away from the PCK plantations, did not show a significant increase in cancer deaths over the same period (Thanal, 2002).

Thanal, a nongovernment organization, reported from the villages of Belur, Enmakaje, and Muliyar congenital anomalies including cerebral palsy, slowed mental development, epilepsy, and physical deformities such as staghorn limbs and deformed or incompletely grown limbs (Figure 2.23). Psychiatric problems also arose in the area with growing evidence that suicides and suicidal tendencies may have been caused by prolonged exposure to neurotoxins. Deaths due to rheumatism, paralysis, and arthritis were also on the rise in the pesticide-affected areas (Thanal, 2002).

Published studies documenting endosulfan mutagenic effects to bacteria and yeast cells suggest that exposure to the pesticide may be the cause of the mutagenic effects such as those described previously to exposed residents. Endosulfan is also an endocrine disruptor, and during early developmental stages, very low levels might trigger malfunctioning of thyroid gland

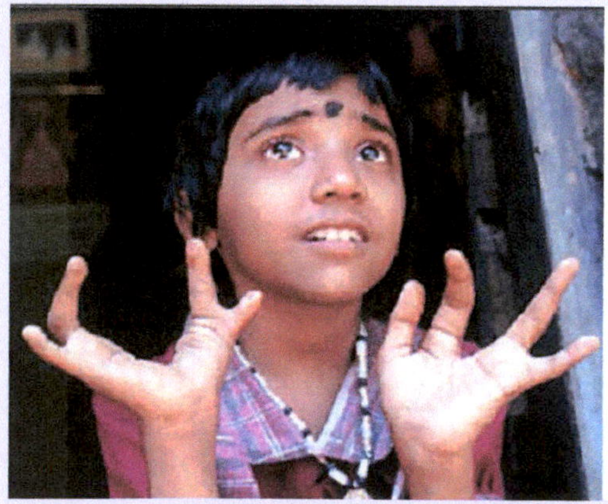

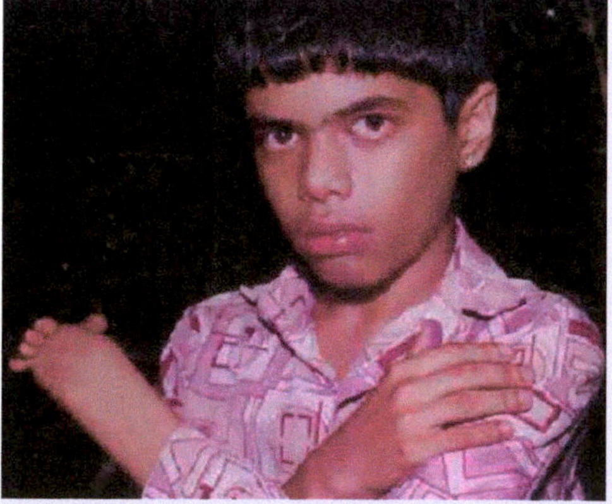

Figure 2.23 Photographs of children from Kasargod affected by the aerial spraying of endosulfan. The young man was born with deformed hands and the young woman was born with what are called staghorn hands and also a deformed leg. (Photographs reproduced with permission from the materials. Photographs of children in Kasargod living around plantations where endosulfan was sprayed for 24 years since 1976, developed and circulated by the Thanal Conservation Action and Information Network, Trivandrum, Kerala, India.)

leading to failure in brain development and complex developmental disorders in the fetus (Naqvi and Vaishnavi, 1993; EPA, 2001).

LEGAL AND SOCIAL ISSUES

The pioneering efforts of journalist Shree Padre were followed by that of Dr. Y. S. Mohan Kumar (1997) and later Leelakumari Amma (1999). Together they drew attention to and created a widespread awareness of the environmental and health issues associated with aerial spraying of endosulfan. Consequent to their efforts, the Hosdurg Munsiff Court endosulfan ban in February 2001 stopped further aerial spraying on PCK plantations of Kasargod.

But the controversy did not end there. In another brief analysis by the Fredrick Institute of Plant Protection and Toxicology (FIPPAT) reported that endosulfan residues were not detected in water, soil, milk, or blood samples, and that the health issues in Kasargod villages were not due to the pesticide. Later, the CSE contradicted the FIPPAT report. The CSE extensively analyzed the endosulfan residues information and established that all the environmental samples contained traces of endosulfan, and that the pesticide was responsible for the health and environmental problems in Kasargod. The Government of Kerala then imposed a ban on the use of endosulfan in the cashew plantations in August 2001 and also appointed an expert committee headed by Dr. Achuthan to study and analyze the effects of aerial spraying of endosulfan in cashew plantations of Kasargod. The committee strongly recommended that the government ban aerial spraying and establish a 5-year pesticide moratorium in Kasargod plantation area. In November 2001, as an initiative of the Thanal Conservation Action and Information Network and the Pesticide Action Network Asia and Pacific, Dr. Romeo F. Quijano (Department of Pharmacology and Toxicology, College of Medicine, University of Philippines) visited Kasargod to assess what role endosulfan might play in the health problems among the people of the affected panchayaths. Quijano's fact-finding team strongly recommended a permanent ban on endosulfan and a comprehensive health and environmental survey in affected areas to determine the magnitude of harm (Quijano, 2002). Further, the NIOH, Hyderabad established in 2002 that the inhabitants of Padre village in Kasargod had enough endosulfan residues in their serum to potentially disrupt normal endocrine functioning. Based on these reports, the Honorable High Court of Kerala banned the use of endosulfan in Kerala in August 2002. In contradiction of these reports and findings, Dr. O. P. Dubey headed a committee constituted by the Ministry of Agriculture (Government of India) that concluded in March 2003 that endosulfan should be permitted and that there were no links between the use of endosulfan and health issues. Following the Dubey Committee report, there was widespread public protest against the Indian government. Another committee headed by Dr. C. D. Mayee was constituted to study the entire issue in December 2004 (KSCSTE, 2011). Taking into account the apprehensions of the Kerala residents, this Committee strongly argued immediate ban on the use of endosulfan. In December 2004, the Kerala State Pollution Control Board directed all District offices of the Board to enforce the endosulfan use restriction imposed by the Honorable High Court of Kerala and to initiate action against violators. In November 2010, the KSPCB issued necessary notification banning endosulfan under the provisions of the Water (Prevention and Control of Pollution) Act, 1974 and Air (Prevention and Control of Pollution) Act, 1981. The Department of Agriculture (Government of Kerala) issued a gazette notification in December 2005 withholding the use of endosulfan in Kerala. In 2007, the endosulfan victim's relief and remediation cell was constituted by the Government of Kerala, comprising the local government officials; health, social, and agricultural departments; and citizenry. In consultation with the cell, the state government allocated 50 lakh rupees (circa US$80,000) for the relief, remediation, and rehabilitation of the people affected by endosulfan spraying. The National Human Rights Commission in December 2010 conducted an enquiry by the Indian Council of Medical Research (ICMR) in Kasargod regarding the health aspects due to endosulfan. As an outcome of the enquiry, the commission recommended the government of India initiate proper administrative and legislative action to ban the use of endosulfan and to conduct a nationwide survey of populations witnessing aerial spraying of endosulfan. On May 13, 2011, a three-judge bench of the Honorable Supreme Court of India responded to public interest generated by the Democratic Youth Federation of India—Kerala wing to ban the manufacture, sale, use, and export of endosulfan throughout the country. Authorities were directed to seize the permits issued to the manufacturers of endosulfan, citing its adverse health impacts on humans. The Honorable Court in its judgment stated that "right to life, guaranteed under the Article 21 of the Constitution of India, is the most fundamental of all human rights, and any decision affecting human life, or which may put an individual's life at risk, must call for the most anxious scrutiny." The Honorable Court has also appointed a joint committee headed by the director general of the ICMR and the commissioner for agriculture, Government of India to initiate a detailed scientific study to understand the ill effects of endosulfan on life and the environment. The Honorable Court also asked the Committee to suggest a "safe" substitute for endosulfan.

CONCLUSION

It is evident from the existing literature that endosulfan has impacted the ecology and human health of the Kasargod district. It is equally evident that this persistent pesticide will remain in the region for some time after the ban on its use.

Regular environmental monitoring of the soil, water, biodiversity, and community health is necessary into the near future to determine the effectiveness of remediation measures. Medical assistance and economic compensation should be provided to victims. To this end, a Government of Kerala committee recommended formation of a tribunal to settle disputes over compensation eligibility and also recommended formation of an Endosulfan Relief Fund with contributions from the government, PCK, and others.

VIGNETTE 2.2 Ecotoxicology and Pesticides in Central America

Luisa E. Castillo, Clemens Ruepert, Elba de la Cruz, and Viria Bravo Durán
Universidad Nacional

Pesticide use in Central America is one of the highest in the world. Pesticides were introduced in the 1950s for use on agriculture and for vector control. By 1970, pesticide imports amounted to 5.18 million kilograms in Costa Rica (Hilje et al. 1987), and in Nicaragua and Guatemala the insecticides used on cotton alone during 1977 surpassed 8.5 million kilograms (Murray, 1994). The importation of insecticides, fungicides, herbicides, and other related products in Honduras doubled from 1992 to an approximate value of 11,000 tn in 1996 (Kammerbauer and Moncada, 1998). Costa Rica, the largest user of agricultural pesticides in Central America, imported nearly 150,000 tn of active ingredients between 1977 and 2000 (Valcke et al. 2005). The major crops in Central America include bananas, pineapple, rice, sugarcane, coffee, oil palm trees, and vegetables (Food and Agriculture Organization of the United Nations, 2012).

Most organochlorines were restricted in the region in the early 1980s, almost a decade later than in the developed world. Also, in some countries, the use of DDT continued for vector control for at least an additional decade. In Costa Rica, it was used until 1985, in Nicaragua until 1989, and in Belize until 1990. Importation and use of pesticides in Central America shifted during the 1980s to other less persistent but more toxic pesticides. Currently an estimate of 39,000 tn of active ingredients of pesticides per year are imported annually for agricultural use in Central America (Tables 2.3 and 2.4), the main groups of pesticides used are organophosphates, carbamates, and pyrethroids among the insecticides; the dithiocarbamic fungicides; and phenoxyacids, phosphonoglycines, dipyridyls, and triazines as herbicides. The organochlorines are still represented in the Central American region by endosulfan—an insecticide used in most countries with the exception of Belize in crops such as rice, melons, pineapples, and vegetables. This compound has been included recently in the Stockholm Convention and a decrease in

Table 2.3 Average Annual Consumption of Pesticides in Central America for the Period of 2005–2009

Country	Pesticide Imports (kg a.i.·ha^{-1} of Cultivated Land)	Average Annual Imports (t a.i. × 10^{3})
Central America[a]	4.5	39
Belize[a]	4.5	0.5
Costa Rica[a]	24.8	12.5
El Salvador[a]	2.0	1.8
Guatemala[a]	5.8	13.6
Honduras[a]	3.7	5.7
Nicaragua[a]	1.4	3.1
Panama[a]	3.6	2.4
EU (average for 25 countries)[b]	2.1	219.7

[a]Bravo et al. in press.
[b]Nadin (2007).

Table 2.4 Most Imported Pesticides in Costa Rica during the Period 2005-2009 and Their Toxicity to Aquatic Organisms

More Imported Pesticide	2005–2009 (Tons)	Amphibia	Fish	Crustacea	Algae	Aquatic Fern
Methamidophos	24,917	Moderate	Moderate	High	Low	No data
Endosulfan	4,748	High	High	High	Moderate	No data
Carbaryl	661	High	High	High	Moderate	Low
Cypermethrin	954	High	High	High	Moderate	No data
Terbufos	1,385	No data	High	High	Moderate	No data
Oxamyl	22,561	No data	High	High	Moderate	No data
Mancozeb	36,779	High	High	High	High	No data
2,4-D	1,761	Moderate	Moderate	Moderate	Moderate	Low
Glyphosate	1,234	Moderate	Moderate	Moderate	Low	Moderate
Paraquat	17,800	High	Moderate	High	High	Moderate
Atrazine	2,540	Moderate	High	High	High	High

Toxicity symbol: High (red), Moderate (yellow), Low (green), No data (purple)

the quantities imported is observed in most Central American countries since 2007 and in Honduras since 2009 (Bravo et al. 2013).

Among the most imported compounds in Central America during the period of 2005–2009 are 2,4-D, amethrine, atrazine, chlorothalonil, diuron, glyphosate, mancozeb, methyl bromide, paraquat, terbufos, and tridemorf (Bravo et al. in press).

In Costa Rica, a percentage greater than 90% of the pesticide imported is moderate to extremely toxic to fish and crustaceans, and 50% to algae. About 49% of the total imported in the country during 2000–2009 is highly toxic to amphibians and 37% is moderately to highly toxic to aquatic plants (De la Cruz et al. 2014).

In addition to high volume and toxicity of pesticides in use in Central America, tropical conditions and usage practices can also increase environmental risks. Among the later, aerial applications, use of pesticide-impregnated plastic bags, and uncovered applications to the soil of granular nematicides can increase emissions to the environment. The lack of regulatory control related to pesticide use is an additional problem in the Central American countries.

Most knowledge of pesticide behavior and distribution in the environment comes from studies conducted in the drier and colder temperate zones (Arbeli and Fuentes, 2007); however, there are important differences with tropical ecosystems. Factors that can contribute to the shorter period of disappearance in the soil substrate of humid tropical areas are high temperatures, which favor microbial and chemical degradation, and also volatilization and solarization, which enhance photodecomposition. High biomass content has also been mentioned as a factor contributing to the decrease in environmental residues in tropical aquatic ecosystems (Calero et al. 1993).

A factor limiting the interpretation of data related to degradation and fate of pesticides in the environment is that most studies conducted in the tropics include only parent compounds and not degradation products. The importance of this fact can be illustrated by a study of degradation of chlorothalonil in tropical conditions (Chaves et al. 2007). Half-lives in soil under field conditions were lower than those reported in temperate environments. However, residues of the main metabolite, 4 hydroxy-chlorothalonil, were detected even 85 days after application. Also background concentrations of all chlorothalonil compounds, including degradation products, of up to 15 $ng \cdot g^{-1}$ (dry weight) were found at the beginning of the study, approximately 4 months after the previous application. Chlorothalonil is one of the most imported fungicides in Costa Rica and is applied in banana, coffee, and vegetable plantations.

The risk of toxicity appears to be increased at higher temperatures (Brecken-Folse et al. 1994; Howe et al. 1994; Rand, 1995); however, biochemical detoxification and degradation of the chemical may also increase with temperature. Factors that can increase toxicity of pesticides to aquatic biota at higher temperatures are higher solubility of toxicants in water, an augmented rate of uptake and blood flow, and increased bioconcentration. Also oxygen availability diminishes at higher temperatures and the combined effect of this decrease with that of a toxicant could become more serious than the effect of the toxicant alone (Viswanathan and Krishna Murti, 1989).

VULNERABILITY OF TROPICAL ECOSYSTEMS

The contaminants released into the environment generally end up in aquatic ecosystems. Runoff and groundwater leaching will be favored by rainy conditions prevailing in some of the tropical areas. In addition, common practices such as the cultivation and use of pesticides in slopes and in close proximity to streams and rivers, as well as the extensive drainage systems in crops such as bananas increase pesticide runoff. Considering also the high-use pattern in the Central American countries, it is not surprising that agrochemical residues originating from terrestrial applications are ubiquitous in streams and rivers (Castillo et al. 2000, 2006; Polidoro et al. 2009; Echeverría-Sáenz et al. 2012; Diepens et al. 2014; De la Cruz et al. 2014), eventually draining into estuaries and coastal seas.

Tropical ecosystems have greater species richness and ecological diversity than temperate areas, which means than the occurrence of species that are more sensitive cannot be discarded. There are not many studies regarding this issue, and some studies have not found differences in the sensitivity of temperate and tropical species (Rico et al. 2011). However, Rico et al. (2011) only addresses two compounds and a small number of Amazonian freshwater organisms.

Also, it is not known whether tropical ecosystems such as coral reefs and mangroves are more or less sensitive than temperate systems. Mangroves are considered especially susceptible to herbicides and it has been suggested that this susceptibility is possibly related to the physiological stress of living in a saline environment.

Corals could also be more susceptible to herbicides due to their association to algae in their zooxanthellae. Herbicides can penetrate coral tissues and reduce photosynthetic efficiency of the zooxanthellae; bleaching of recruits or adult coral branches is a common reaction to chronic exposures or high concentrations of PSII herbicides (Van Dam et al. 2011). Central America has important coral reef areas, including the world's second most important barrier reef; however, studies on pesticide occurrence and their impact are scarce (Van Dam et al. 2011).

Hydrodynamics of these coastal tropical ecosystems, coastal lagoons, mangroves, and coral reefs, can trap pollutants increasing their residence time, which in turn can increase exposure times for endemic organisms. But also mangrove systems and seagrass beds can help to trap sediments and play a role in the active removal of dissolved agrochemicals from the water (Burke et al. 2011), thus protecting coral reefs.

Several studies have documented the occurrence of pesticide residues in coastal lagoons of Nicaragua (Carvalho et al. 2002) and Costa Rica (Castillo et al. 2000, 2012), and coral reefs in Costa Rica (Abarca and Ruepert, 1992; De la Cruz and Castillo, 2003) and Panama (Glynn et al. 1984).

Cotton was grown for several decades in the region when persistent and very toxic organophosphates were heavily used in agriculture. Large areas of the Pacific Coast were deforested and large quantities of pesticides were applied without any environmental concern (Murray, 1994). The cotton production ceased in the 1990s, but exposure to the compounds used will remain for a long time. The first ecotoxicological study in Central America can be traced back to 1973 when cotton was the main export crop in the region. This study reported the presence of organochlorine residues in estuarine and marine fish and invertebrates of the Guatemalan coastal area, influenced by cotton plantations in the area (Keiser et al. 1973).

In the Cholutecan River Basin of Honduras, an area where different agricultural production systems exist, environmental samples collected throughout 1995–1997, contained detectable organochlorine, organophosphate, and other pesticides (Kammerbauer and Moncada, 1998). Organochlorines were the main residues in fish tissue and lagoon sediments. The organochlorine residues in the Cholutecan river basin were linked to earlier intensive pesticide applications associated with cotton production, other crops, and public health purposes. The highest concentrations were found associated with an area of more intensive agricultural production and the lowest concentrations were found in the small watershed associated with traditional agricultural production. The river samples from the Choluteca melon production area contained 20 different compounds. The effluents from this watershed drain directly into the estuaries of the Gulf of Fonseca where shrimp farmers might be affected by pesticide residues in the water.

Toxaphene residues were identified in 100% of sediment and 81% of fish muscle samples collected during 1991 at Lake Xolotlán, Nicaragua, where the only organochlorine production plant in the region was located. Sediment and fish muscle samples had maximum concentrations of 187 $\mu g \cdot kg^{-1}$ and 1131 $\mu g \cdot kg^{-1}$ fresh weight (fw) respectively. DDTs were detected at concentrations ranging from 2 to 102 $\mu g \cdot kg^{-1}$ fw in sediments and 2 to 114 $\mu g \cdot kg^{-1}$ fw in fish muscle tissue (Calero et al. 1993). Mayer and Mehrle (1977) reported developmental defects in two fish species with concentrations of toxaphene in muscle of 400 $\mu g \cdot kg^{-1}$. No such problems have been identified in Nicaragua.

The intensive use of agrochemicals in cotton-growing areas of the Pacific Coast of Nicaragua caused an extensive contamination on the coastal lagoons of Chinandega with organochlorine and organophosphate pesticides (Castilho et al. 2000; Carvalho et al. 2002; Lacayo et al. 2005).

Studies of atmospheric transportation of organochlorines in the region show air and soil residues present in pristine areas of Costa Rica and Belize (Alegria et al. 2000; Daly et al. 2007b; Shunthirasingham et al. 2011). Also legacy organochlorines continue to be found in wildlife of the region as will be commented on later.

CURRENT-USE PESTICIDES IN RIVERS AND STREAMS

Costa Rica is well known for its environmental conservation efforts. Currently, Costa Rica has over 25% of its total land area under some kind of protection status and ranks among the 20 countries with the highest biodiversity in the world. However, extensive agricultural areas, with high input of agrochemicals, are located upstream of important wetland ecosystems and other protected area. Major crops grown in the Atlantic coast of the country include bananas and pineapples.

Bananas are grown in about 42,000 ha and use approximately 49 kg a.i.$\cdot ha^{-1}$,* whereas the area grown with pineapple has expanded more than 500% with respect to 1999, reaching more than 45,000 ha in 2010, with a use of approximately 30 kg a.i.$\cdot ha^{-1}$ (Castillo et al. 2012; Echeverría-Sáenz et al. 2012). Most of the bananas and pineapples produced in Costa Rica are grown in the Atlantic zone. Thus the study of the impact of pesticides in these areas is considered of high priority.

Studies carried out in these areas since 1993 have demonstrated extensive surface water contamination, presence of pesticides in sediments, and biological impacts, especially near the banana and pineapple plantations. The frequency of detection and the concentrations found were highest close to the agricultural areas. In the banana plantation, area peak concentrations in surface waters were observed following application of ground and aerially applied pesticides. The fungicides used in packing plants (imazalil and thiabendazole), as well as several insecticide/nematicides (chlorpyrifos, diazinon, cadusafos, carbofuran, ethoprofos, and terbufos) were detected in concentrations that potentially could damage aquatic life (Castillo et al. 1997, 2000). Other fungicides frequently found in surface waters in the past years include propiconazole and chlorothalonil and more recently also difenoconazole. Several of the pesticides mentioned have been identified in recent years as the probable cause of fish mortalities occurring in the area, specifically ethoprofos, terbufos, and chlorothalonil. These compounds have been identified in tissues of the dead fish and in the water collected at the time of the die-offs. In one case linked to ethoprofos, a cholinesterase assay was carried out, showing inhibition of the enzyme. Values of ethoprofos concentrations in the water ranged from 0.1 to 2.9 $\mu g \cdot L^{-1}$, over the maximum limit set by The Netherlands to protect aquatic organisms

* a.i. is the abbreviation for "active ingredient," such as, the 49 kg of pesticide active ingredient per hectare referred to here.

(0.063 μg·L^{-1}). The high toxicity of nematicides, their solubility, and their mode of application (i.e., applied and left uncovered on the ground) increase the risk for water organisms. In addition, the drainage systems of banana plantations may increase pesticide runoff to the creeks and rivers. A follow-up study of nematicide applications showed peak concentrations following applications. Macroinvertebrate communities at these sites showed significant changes in community composition following the pesticide applications in the banana plantations (Castillo et al. 2006).

The herbicides, bromacil, diuron, and amethryn, were found in three different studies in surface waters of a pineapple-growing area in the Atlantic zone of Costa Rica (Castillo and Ruepert, 2005; Ugalde, 2007; Echeverría-Sáenz et al. 2012). The first study reported concentrations in surface waters of the persistent and highly mobile herbicide, bromacil ranged from 1 to 55 μg·L^{-1}, 65% of samples had over 5 μg·L^{-1} of bromacil. Parallel studies in the laboratory with waters collected at the same sites showed inhibition of growth of lettuce seeds and algae (*Pseudokirchneriella subcapitata*), the percentage of inhibition was well correlated with the concentration of bromacil in the samples. Other herbicides present in the samples were diuron (2.2–5.6 μg·L^{-1}) and amethryne (0.3–3.9 μg·L^{-1}). The correlation with lettuce and algae growth inhibition increased when all three herbicides were considered. In addition, several insecticides, carbaryl, chlorpyrifos, diazinon, and ethoprofos, were detected in 10%-17% of the surface water samples in concentrations ranging from below detection levels to 32 μg·L^{-1} in the case of diazinon. Other insecticides had maximum concentrations of up to1.6 μg·L^{-1}. Daphnia 48-hour mortalities of 73%–100% were observed in the presence of these compounds.

The second study (Ugalde, 2007) reported lower concentrations of herbicides (0.3–2.3 μg·L^{-1}) and the occurrence of several insecticides and fungicides such as diazinon, ethoprofos, chlorothalonil, and triadimefon. Diazinon is toxic to fish and aquatic invertebrates (Tomlin, 2003) and is one of the four most frequently detected pesticides in streams in the United States. (USGS, 1997). A parallel study of the phytoplankton showed lower diversity and higher abundance in the contaminated streams than in the pristine streams used as control. Higher abundance in contaminated sites could be related to herbicide concentrations below threshold levels affecting phytoplankton and to the concentration of insecticides above toxicity thresholds for crustaceans and other predators of phytoplankton. Also the higher availability of nutrients in the contaminated sites could have favored the growth and reproduction of some phytoplankton species. Tolerant species were present in higher numbers in the contaminated site, whereas sensitive species were present in the reference site but not in the contaminated site (Ugalde, 2007).

In a more recent study, Echeverría-Sáenz et al. (2012) characterized environmental hazards of pesticides from pineapple production in riparian communities. Pesticides detected along the Río Jiménez watershed in concentrations above 1 μg·L^{-1} included herbicides such as amethryn, bromacil, and diuron; the insecticides such as carbaryl and diazinon; and fungicides such as triadimefon and epoxiconazole. The study showed clear relationships among high levels of herbicides and poor plant growth, high levels of organophosphate pesticides and anticholinesterase effects on fish, mortality of crustaceans, and deterioration of macroinvertebrate communities.

The frequent presence of herbicides could eventually have an impact on primary production in the downstream wetland as well as indirect effects on the valuable species they protect such as the manatee, an endangered species that feeds exclusively on macrophytes.

The use of pesticides remains high for export crops such as bananas and pineapples and the studies of the water bodies downstream of these agricultural lands show consistent results of pesticide concentrations and effects on aquatic organisms and communities.

ORGANOCHLORINE COMPOUNDS IN WILDLIFE

Organochlorine residues are still detected worldwide in abiotic and biotic samples (UNEP, 2003). Studies in the Central American region also document their presence. A study of organochlorine compounds in vertebrates in a conservation area in northwestern Costa Rica (Klemens et al. 2003) showed widespread contamination both geographically and across taxonomic boundaries. The most frequently detected organochlorine compounds in the area included p,p'-DDE, benzene hexachloride, heptachlor, and dieldrin. Turtles showed the highest frequency of contamination of the four groups analyzed, which included also amphibians, rodents, and birds. Turtles are longer lived species and thus would have more time to accumulate organochlorine compounds. They might also feed higher in the food chain than the granivorous rodent species analyzed, which was the least contaminated group in the study. The levels documented in snapping turtles in this study have been linked to developmental abnormalities (Klemens et al. 2003). Atmospheric long-distance transport was considered as the likely major source of organochlorine pesticide contamination, because the study area has been protected as of the mid-1960s. Atmospheric transport of organochlorines has been documented in numerous studies in the region (Standley and Sweeney, 1995; Daly et al. 2007a,b; Gouin et al. 2008; Shunthirasingham et al. 2011).

The occurrence of organochlorine compounds has also been reported in crocodile species in Belize and Costa Rica (Wu et al. 2000; Pepper et al. 2004; Rainwater et al. 2007; Grant et al. 2013) and dolphins in Panama (Borrell et al. 2004). The most prevalent of these compounds in crocodile eggs was p,p'-DDE, which occurred in 96% of the eggs examined in concentrations up to 372 $ng{\cdot}g^{-1}$, in 69% of chorioallantoic membranes of crocodile eggs, and in all samples of caudal scutes. The likely primary route of organochlorine pesticide contamination in eggs is believed to be maternal transfer but transfer from nest material has also been documented in reptile eggs (Cañas and Anderson, 2002). Caudal scutes collected in Costa Rica from a polluted river showed average concentrations of p,p'-DDE and p,p'-DDT of 340 and 254 $ng{\cdot}g^{-1}$ (wet weight [ww]), respectively. Crocodiles are the highest predator of this river, which receives pollutant inputs from a variety of sources including urban, industrial, and agricultural areas. A recent study of pesticides in blood from spectacled caiman in the Caribbean region of Costa Rica (Grant et al. 2013) reported 7 of the 9 compounds present were organochlorines; all except endosulfan have not been used in Costa Rica since the late 1980s or early 1990s. Their presence is probably linked to the long half-lives of POPs and reflects the degree to which their prey was contaminated by current and past use of pesticides. Two pyrethroids compounds detected, permethrin and cypermethrin, are currently used in vector control.

Blubber and skin samples from dolphins from the western Pacific coast of Panama showed mean levels of HCB (hexachlorobenzene) and tDDT (dichlorodiphenyltrichloroethane) of 0.064 and 2.30 $mg{\cdot}kg{-}^{1}$, respectively. These levels are considered low, and the high observed ratio of p,p'-DDE/tDDT reflects the local reduction of DDT use in the area since the 1970s.

ATMOSPHERIC TRANSPORTATION

Pesticides not only move into the aquatic environment, they are also transported atmospherically from their sites of application and deposited in distant locations (Alegria et al. 2000; Shen et al. 2005; Daly et al. 2007a,b).

Dispersal by air of pesticides can lead to exposure of the terrestrial natural areas nearby or downwind from the agricultural areas. Large-scale dispersion and deposition of organochlorine pesticides to areas far from the original site of application is well documented (UNEP, 2003; Shen et al. 2005). Transport of current-use pesticides into high elevation ecosystems has been documented in North America (LeNoir et al. 1999). In general, the possible effects of the air dispersal of pesticides on organisms, populations, or ecosystems in the tropics are not well known. High rain rates, persistent cloud cover, steep temperature gradients, and soils rich in organic matter should favor the accumulation in tropical mountains of organic chemicals that are readily scavenged by rain and fog.

Several authors have reported the presence of persistent organochlorine pesticides in vertebrate and invertebrate species collected from a conservation area in northwestern Costa Rica (Standley and Sweeney, 1995; Klemens et al. 2003) and linked it to possible long-distance atmospheric transport of these compounds, but without clear evidence. Research in Costa Rica by Daly et al. (2007a) indicates that current-use pesticides used in the lowlands are being transported and accumulated to protected areas such as national parks and volcanoes. Soil and air samples for pesticide residue testing were taken from a broad network of 23 sampling stations with different climate and environmental characteristics across Costa Rica (Daly et al. 2007a,b). Most of the sites were located in protected areas with no real pesticide use in the past. Air samples were taken during a year using passive air samplers. The authors concluded that the current-use pesticides are of greater concern than the POPs that were found in much lower concentrations. The insecticide, endosulfan, was found in air and soil in relative high concentrations compared to studies done in other parts of the world. The highest soil concentrations were found at high altitudes (>2500 m) up to 3 $\mu g \cdot kg^{-1}$. The highest air concentrations were found in the highly populated Central Valley area. Endosulfan is used in several crops in the vicinity of the Central Valley, among others in ornamental plants and coffee; it is also used in pineapple plantations in the lowlands. A more recent study by Shunthirasingham et al. (2011) shows endosulfan sulfate as the most abundant pesticide residue in water, with concentrations ranging from 0.4 to 9.4 $ng \cdot L^{-1}$. Its levels were highest in water sampled from bromeliads. Consistent with calculations of cold trapping* in tropical mountains, concentrations of endosulfan sulfate increased with altitude.

High concentrations of the widely used fungicide, chlorothalonil, were found in air at sites near banana- and vegetable-growing areas. The highest soil concentrations of this compound were also found in the same sites at high altitude. Pesticide air concentrations in protected mountain areas were sometimes almost 10-fold higher than in areas adjacent to farms. The results indicate that the air masses can carry certain pesticides from lowland agriculture up mountainsides, where the low temperatures and high precipitation rates will cause their deposition. Daly et al. (2007a) used a simulation model describing pesticide transfer, fate, and distribution along elevation gradients in which a mountain contamination potential (MCP) for pesticides was developed. Based on the partition coefficients of pesticides between air and

* Shunthirasingham et al. (2011) describe **mountain cold trapping** as the enrichment of persistent organic compounds in various media (such as water, snow, soil, and tissues) with an increase in elevation up mountain slopes. The mechanism for cold trapping, as implied in its name, is compound transition out of the mobile atmospheric gaseous phase at the lower temperatures that prevail at higher elevations and into less mobile solids phases such as particulates, water droplets (such as rain and fog), snow, soil, and other components of the land surface. The increased amount of water moving from gaseous to liquid or solid phases with increased altitude also contributes to the movement of persistent organic compounds out of the atmosphere and onto the land and biota at higher elevations.

water, and octanol–air, it is expected that insecticides like diazinon and chlorpyrifos will show a similar fate as endosulfan and chlorothalonil. In fact, a study in the Central Valley of Costa Rica found high chlorpyrifos air concentrations with an annual level of 2.2 $ng{\cdot}m^{-3}$ (Gouin et al. 2008).

Fog might be another relevant deposition pathway to high-altitude tropical cloud forest as the study by Shunthirasingham et al. (2011) showed the fog samples to be relatively enriched in some of the analyzed pesticides, such as dacthal and chlorothalonil.

The deposition of pesticides can probably be expected in the mountain areas of the other countries of Central America; however, the situation in Costa Rica with its extensive agricultural activities and high pesticide use in the northern and eastern lowlands could result in higher impact on its central highlands.

These findings might help explain the occurrence of amphibian declines in tropical montane forests. Although fungal infections in combination with climate change have been implicated in the decline of amphibian populations in the neotropics (Carey et al. 2001; Pounds et al. 2006; Whitfield et al. 2007), this does not exclude a potential role of airborne contaminants especially considering that most declines have occurred in upland areas where deposition of pesticides was known to occur and amphibian declines in the mountains of California have been linked to pesticide deposition (Sparling et al. 2001; Davidson, 2004). In recent years, there is increasing research and evidence of the toxicity of pesticides to amphibians (Mann et al. 2009; Belden et al. 2010; McMahon et al. 2011; Brühl et al. 2013). Further research is needed to elucidate the ecotoxicological risk of the atmospheric dispersion of current-use pesticide to altitude ecosystems.

This short overview of pesticide studies in the Central American isthmus shows that pesticide residues are present and strongly associated with areas of intensive agriculture, although even in the areas practicing more traditional agriculture, pesticide residues were not absent. They are also present in pristine areas as coastal lagoons located downstream of agricultural areas and montane forests as the atmospheric transport studies have shown. Further studies are needed to improve the knowledge of environmental risks and impacts of pesticides in tropical ecosystems; at the same time it is necessary to develop strategies and effective policy framework to assist in minimizing adverse impact on environment and human health.

2.2.2.8 *Herbicides*

These biocides include products such as the bipyridiums, paraquat and diquat (Figure 2.24), which have nitrogen heterocyclic rings in their structures. Note that the rings are aromatic in that the associated electrons stabilize the structures by being in resonance within the rings. The herbicides also include triazines (Figure 2.25) such as atrazine that can be problematic on entering water bodies adjacent to agricultural fields. A recent meta-analysis of atrazine identified consistent indirect adverse effects to amphibians and fish exposed to atrazine (Rohr and McCoy, 2010). Phenoxy herbicides (e.g., 2,4-D) are also important in the control of dicots and function by disrupting plant growth regulation.

2.2.2.9 *Oxygen-Demanding Compounds*

Putrescible materials possess high biochemical oxygen demands and can, in aquatic systems, reduce oxygen concentrations to stressful or lethal minima. Sewage-associated effluents are one,

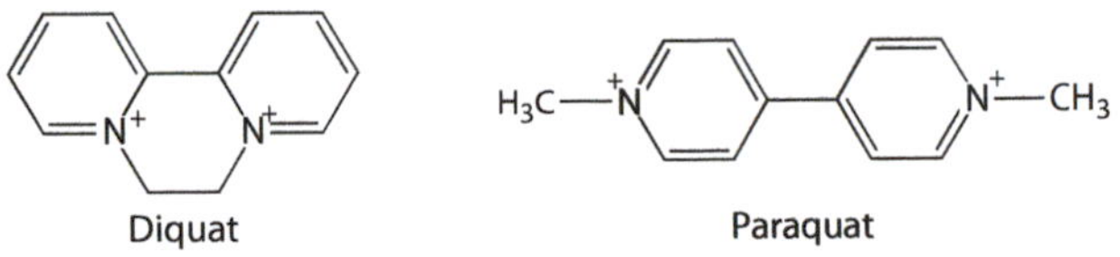

Diquat Paraquat

Figure 2.24 Two important bipyridium herbicides. Notice that both carry positive charges associated with the nitrogen atoms in their rings.

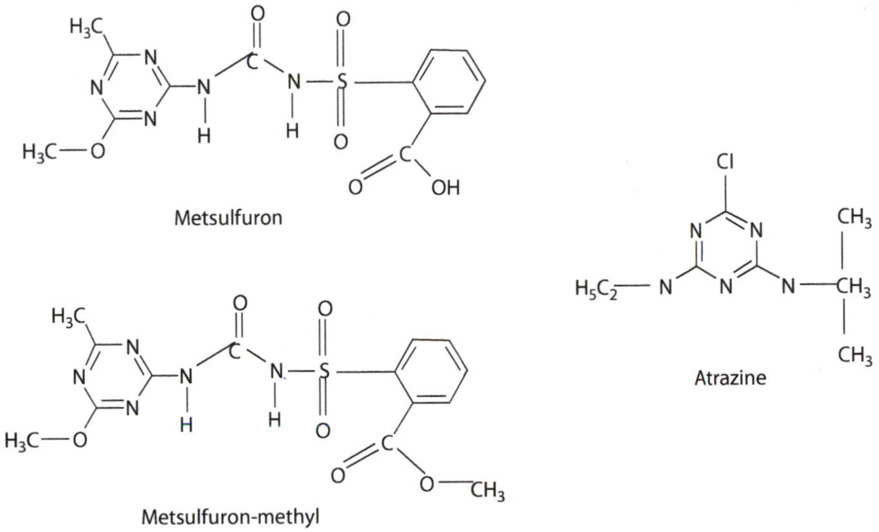

Metsulfuron

Atrazine

Metsulfuron-methyl

Figure 2.25 Three triazine herbicides including atrazine, which is a very widely used herbicide.

but certainly not the only, important source of such materials. Poultry or meat processing facilities, and wood pulp operations can be locally damaging to waters into which they discharge putrescible wastes. Excess algal productivity in eutrophic waters can generate high levels of putrescible dead algal matter and contribute to anoxic events occurring during the night when the ratio of microfloral respiration to photosynthesis is high.

2.2.2.10 *Other Important Compounds*

Numerous other contaminants deserve mention but only a few will be highlighted here due to space limitations.

The alkylphenols, p-nonylphenol, are generated as mixtures of many isomers with branching occurring at different carbon atoms in its structure, such as the one isomer shown at the top of Figure 2.26. They find their way into the environment from diverse industrial activities. They are used in plastics, as adjuvants in pesticides, and as surfactants in cleansers and detergents. They can be produced during detergent synthesis, but importantly, are breakdown products of the nonylphenol ethoxylate (NPE) nonionic surfactants. The NPE has had diverse uses such as surfactants, cleaning agents, detergents, and pesticide adjuvants. Hale et al. (2000) estimates that most nonylphenols in the environment are derived by the breakdown of 4-nonylphenol polyethoxylates (NPEOs).

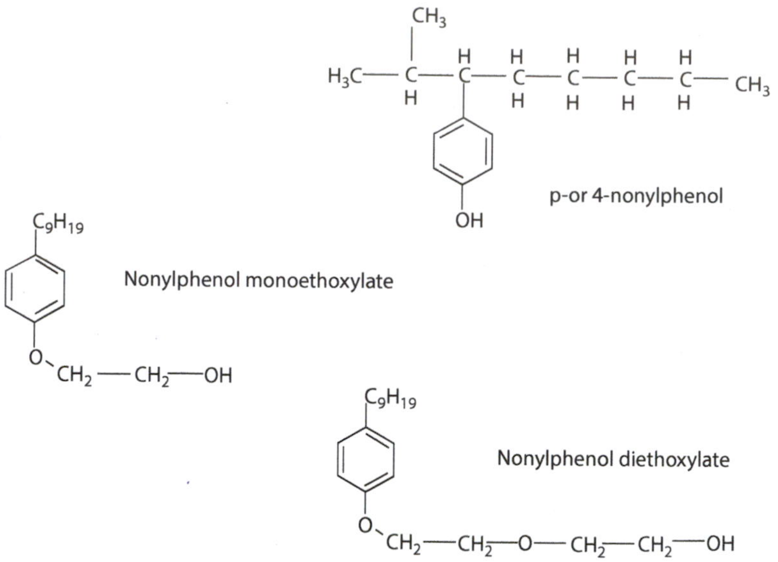

p-or 4-nonylphenol

Nonylphenol monoethoxylate

Nonylphenol diethoxylate

Figure 2.26 The structure of p- or 4-nonylphenol (top) and also two nonylphenol ethoxylates, nonylphenol mono- and diethoxylate. (The C_9H_{19} in these two nonylphenol ethoxylates represent the nine-carbon alkane part of a nonylphenol such as p-nonylphenol.) The commercial nonylphenol ethoxylates degrade to relatively persistent, degradation products such as these three compounds (Cox, 1996). The depicted mono- and diethoxylates degradation products have one and two ethylene oxide groups, respectively. Much of the nonylphenol present in the environment comes from the degradation of 4-nonylphenol polyethoxylates with nine or more ethylene oxide groups in their structures. Generally, the more highly ethoxylated 4-nonylphenol polyethoxylates are the most readily degraded. (From Hale, R.C. et al. *Environ. Toxicol. Chem.*, 19, 946-952, 2000.)

Concern about p-nonylphenol contamination stems from its ability to act like estrogen. As an example of such activity, male sheepshead minnow (*Cyprinodon variegatus*) exposed to p-nonylphenol produce a protein (**vitellogenin**) normally synthesized only by females during egg production (Hemmer et al. 2001, 2002).

Surfactants in cleansers and detergents released into the environment have hydrophilic and hydrophobic regions to their structures to facilitate their function. Widely used in the 1960s, the commercial surfactants known as alkyl benzene sulfonates (ABSs) (Figure 2.27) were very resistant to environmental breakdown and caused widespread problems on release into aquatic systems. Examples of ABS in Figure 2.27 are the branched tetrapropylene benzene sulfonate and the ABS with the aromatic group on the terminal carbon of the aliphatic chain. Slow ABS biodegradation prompted their replacement with the more easily degraded, linear alkyl benzene sulfonates (LASs) that lack the branching on their aliphatic chains (Manahan, 1993). The LASs have the general structure shown at the bottom of Figure 2.27. The α-benzene sulfonate in Figure 2.27 is an LAS.

The LASs are used widely in detergents including household and personal care products. They are present as "a complex mixture of homologues with different alkyl chain lengths and phenyl isomers with different points of attachment of the benzene ring to the alkyl chain" (Larson and Woltering, 1995) with an average alkyl chain length of 12 carbons. They are more readily broken down than the ABS by microbes in the environment, and consequently, many of the problems encountered in the 1960s with surfactants in aquatic systems have been reduced. They can cause problems in aquatic systems below sources where they can be present at high concentrations. They are released from such diverse and widespread sources as municipal sewage treatment plants, septic systems, and solid wastes (Larson and Woltering, 1995).

The general structure of perfluoroalkyl sulfonates (PFASs) and an example of a PFAS are shown in Figure 2.28. These compounds are generated for many purposes, especially surface treatment of paper, fabric, carpet, and leather. They are also used in firefighting foams. Perfluorooctane sulfonate (PFOS) is the most widely detected PFAS in the environment. It has

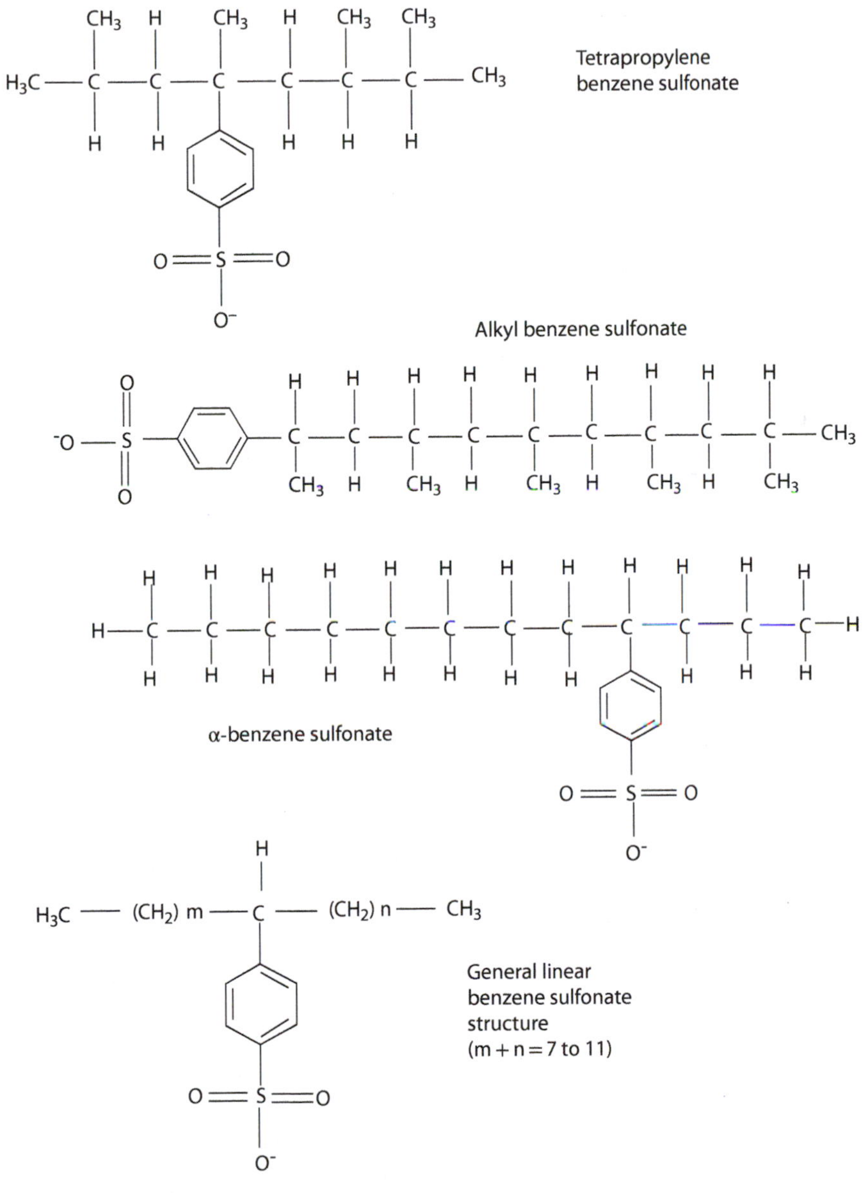

Tetrapropylene benzene sulfonate

Alkyl benzene sulfonate

α-benzene sulfonate

General linear benzene sulfonate structure (m + n = 7 to 11)

Figure 2.27 Anionic surfactants, alkyl benzene sulfonate (ABS) and linear alkyl benzene sulfonate (LAS), used in cleansers and detergents. The tetrapropylene benzene sulfonate and alkyl benzene sulfonate with the aromatic group on the terminal carbon are more resistant to degradation than the LAS, α-benzene sulfonate shown here. The general structure of LAS is shown at the bottom of this figure.

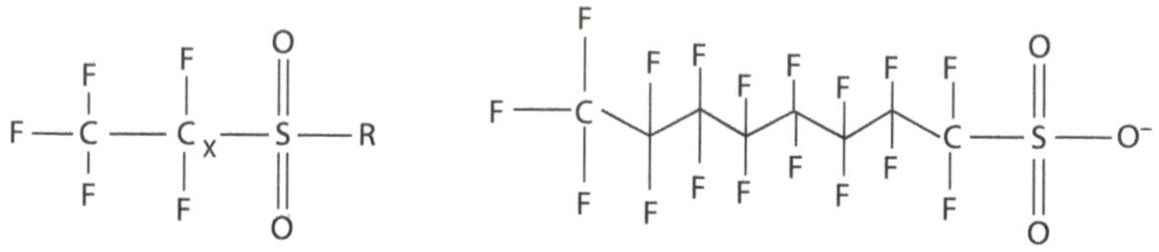

General perfluoroalkyl sulfonate (PFAS) structure

Perfluorooctane sulfonate (PFOS)

Figure 2.28 The general structure of a perfluoroalkyl sulfonate and an example (PFOS).

been used in insecticides as an adjuvant and in metal plating, floor polishes, and numerous other applications. A 2001 global survey (Giesy and Kannan, 2001) found widespread PFOS accumulation in wildlife tissues and the tendency to increase in concentration with movement through food webs to piscivorous species. There was a drop in production and use at that time (2000–2002) (Verreault et al. 2007). In a study of temporal trends (Verreault et al. 2007), PFAS concentrations in Herring gull (*Larus argentatus*) eggs from northern Norway appear to have increased from the early 1980s into the early 1990s. They then appear to have leveled off. The persistence and widespread distribution of PFAS in the environment prompted restriction of their use in EU countries, and the addition of PFOS and related compounds to the United Nations Stockholm Convention POPs list.

VIGNETTE 2.3 Perfluoroalkyl Substances in the Environment: History of an Environmental Issue

John P. Giesy and Paul D. Jones
University of Saskatchewan

ISSUE

Perfluoroalkyl acids (PFAs) are synthetic, perfluorinated, straight- or branched-chain organic acids with a variety of reactive moieties, including among others, carboxylate or sulfonate groups (Giesy and Kannan, 2001, 2002). The PFAs are manufactured, and in some cases, released directly to the environment but can also be formed through transformation of many precursor molecules containing a perfluoroalkyl moiety (Martin et al. 2004a). These include, among others, perfluorinated alcohols and various sulfonamides, fluorotelomer alcohols (FTOHs), and perfluoroalkylsulfonamides (Figure 2.29). Collectively, this family of chemicals will be referred to in this vignette as perfluoroalkyl substances (PFSs). Many of the environmentally relevant compounds are fatty acid analogues and have been classified as perfluorinated fatty acids (PFFAs). PFSs are synthetic products that have been manufactured for over 50 years (Giesy and Kannan, 2002). They have been used in many commercial products including refrigerants, surfactants, polymers, pharmaceuticals, wetting agents, lubricants, adhesives, pesticides, corrosion inhibitors, and stain resistant treatments for leather, paper, and clothing (Sohlenius et al. 1994; Key et al. 1997; Giesy and Kannan, 2002). Other uses have included aqueous film forming foams (AFFF) for firefighting, mining and oil well surfactants, acid mist suppressants for metal plating and electronic etching baths, alkaline cleaners, floor polishes, photographic film, denture cleaners, shampoos, and as an ant insecticide (Moody and Field, 2000). Beginning

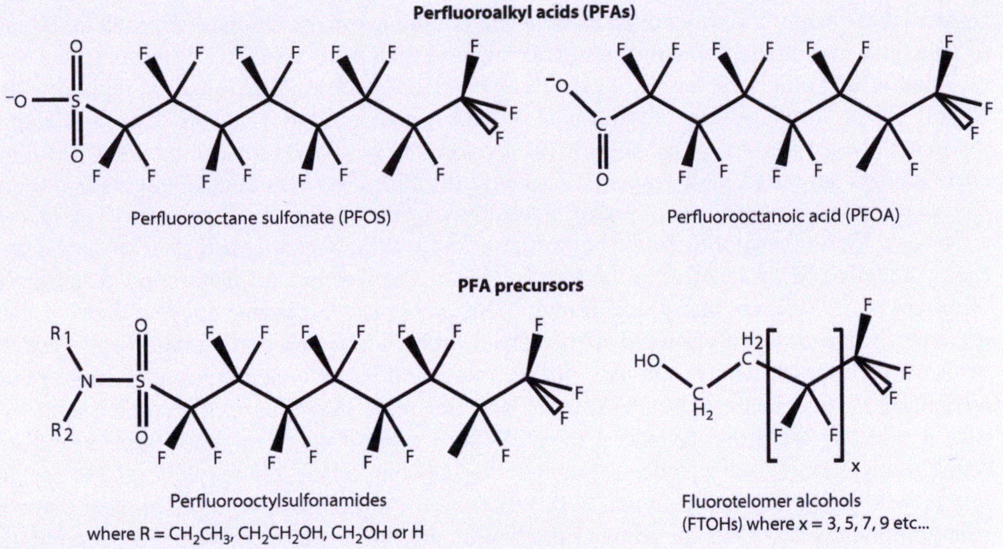

Perfluoroalkyl acids (PFAs)

Perfluorooctane sulfonate (PFOS)

Perfluorooctanoic acid (PFOA)

PFA precursors

Perfluorooctylsulfonamides
where R = CH_2CH_3, CH_2CH_2OH, CH_2OH or H

Fluorotelomer alcohols
(FTOHs) where x = 3, 5, 7, 9 etc...

Figure 2.29 Structures of some representative perfluoroalkyl substances.

in the late 1990s, we had the opportunity to participate in research that lead to actions that rapidly reduced the potential risks posed by these chemicals. Here we provide this narrative as a positive example of chemists and biologists, academics, industries, and government agencies working together to avoid an environmental problem without economic dislocation or disruptions of useful products. Furthermore, this is an example that demonstrates that application of monitoring and risk assessment processes can lead to rapid solutions.

Even though relatively large quantities of PFSs were being used in commerce, they did not become an environmental issue until the late 1990s. The reason for this was several fold. First, until that time identification and quantification of PFAs were limited by a lack of available instrumentation that was both sensitive and applicable to nonvolatile anionic compounds (Giesy and Kannan, 2002; Martin et al. 2004b). The emergence of liquid chromatography instruments coupled to relatively inexpensive and reliable electrospray ionization mass spectrometry systems provided the perfect tool with which to measure PFAs in complex environmental samples. The first of such methods was published in 2001 (Hansen et al. 2001). Since that time, the number of scientific studies reporting on the environmental concentrations, behavior, or toxicology of PFAs has grown immensely, such that by the time this chapter was written it was difficult to keep this narrative up-to-date with emerging reports. Historical measurements used cumbersome and nonspecific methods, such as total organic fluorine, and thus were not suitable for environmental research. A historical perspective of PFA methods is provided briefly here, but is also available in more detail in Giesy and Kannan (2002) and a more critical evaluation of modern methods is provided in Martin et al. (2004b).

In addition to the lack of analytical methods and standards, PFSs were not thought to be a likely environmental issue because they were thought to be bound tightly into polymers. This was because the primary use of the most abundant PFSs was to impart water and soil repellency to textiles, particularly carpeting. The need and desire to look for PFSs in the environment was also limited because, from a regulatory perspective, they fell under the polymer exclusion clause of the Toxic Substances Control legislation in the United States. Basically, if a compound is produced as or bound into a high molecular weight polymer, it was thought to not be available to enter the wider environment and not thought to be a potential problem. Because of this, less information on environmental fate and toxicity is required by government agencies than might otherwise have been the case. For all of these reasons, the first report of PFSs, mainly PFAs, in the general environment did not appear until 2001 when Giesy and Kannan published the results of a global survey of a few PFSs in wildlife from around the world. These early studies surprised many scientists because the PFSs were thought to not enter the environment, and if they did, were thought not to be very mobile or very persistent. The fact that they were found in wildlife from the northern and southern hemispheres, urban and remote areas including Midway Atoll in the Pacific Ocean, the Arctic, and Antarctic was surprising (Kannan et al. 2001a,b; Giesy and Kannan, 2002). Geospatial trends showed that the greatest concentrations were observed in industrial areas where the compounds were produced or used, with lesser concentrations in urban areas, and lesser concentrations in more remote terrestrial environments in North America and Europe and the least concentrations in more remote oceanic areas. Nevertheless, the PFSs were ubiquitous in the environment and were found to biomagnify, not as much as organochlorine compounds, but they did tend to accumulate in wildlife (Giesy and Kannan, 2002). However, unlike the many classes of organochlorine compounds that made up most of the historically important persistent, bioaccumulative, and toxic chemicals on which most of the paradigms of environmental fate and accumulation were founded, the PFSs did not tend to partition into lipids. Instead they were bound tightly to proteins (Jones et al. 2003a). Thus, previously developed methods and predictive relationships were inappropriate.

ENVIRONMENTAL FATE AND EXPOSURE

Because of the high-energy carbon–fluorine (C–F) bond and their highly oxidized state, PFAs are largely resistant to biotic and abiotic degradation mechanisms such as hydrolysis, photolysis, microbial degradation, and metabolism and are thus environmentally persistent (Giesy and Kannan, 2001, 2002). Due to their fluorinated tails and moderate to high surface activity, the phase-partitioning behavior of PFSs is different than that of chlorinated hydrocarbons. Longer perfluoroalkyl chains are amphiphobic, that is, they are both oleophobic* and hydrophobic and thus do not mix well with either water or oil (Kissa, 2001). This and the presence of anionic functional groups result in some PFSs having surfactant properties and so they migrate to the interface of solutions due to the competing action of hydrophobic and hydrophilic moieties. These properties are considerably different from those of chlorinated and brominated hydrocarbons, which are hydrophobic, but comparably lipophilic. The hydrophilic nature of some PFSs is due to the presence of a charged moiety, such as carboxylates, sulfonates, or a quaternary ammonium, but may also be due to polar, yet neutral, moieties such as an alcohol group. PFSs tend to be water soluble and not accumulated into fat. They have a relatively low vapor pressure and thus are not transported in air once they are in the ionic form. However, they have been distributed globally in the atmosphere because some compounds, such as the alcohols, amides, and amines are more volatile (Giesy and Kannan, 2002).

HAZARD ASSESSMENT

Subsequent to the initial reports of PFSs in wildlife by Giesy and Kannan (2001, 2002) interest of government agencies in PFSs increased dramatically. Subsequently, studies were funded

* An oleophobic compound is one with a very low affinity for oils.

and supported by industry to conduct national surveys in wildlife, and then in water, air, soils, and food. Thus, a substantial body of knowledge was accumulated about the distribution of the PFSs in the environment, but little was known about the potential toxicity of these compounds. Because PFSs are chemically stabilized by strong covalent C–F bonds, they were historically considered to be metabolically inert and nontoxic (Sargent and Seffl, 1970). Accumulating evidence has demonstrated that PFSs are actually biologically active and can cause peroxisomal proliferation,* increased activity of lipid and xenobiotic metabolizing enzymes, and alterations in other important biochemical processes in exposed organisms (Sohlenius et al. 1994; Obourn et al. 1997). In wildlife, the most widely distributed PFSs, perfluorooctane sulfonic acid (PFOS), accumulates primarily in the blood and in liver tissue (Giesy and Kannan, 2001). Therefore, although the major target organ for PFSs is presumed to be the liver, this does not exclude other possible target organs such as the pancreas, testis, kidney, and even blood itself (Olson and Anderson, 1983). In fact, PFSs were not only accumulating in the tissues of wildlife, they were found at relatively great concentrations in the blood (So et al. 2006) and breast milk (Yeung et al. 2006) of humans.

Until recently, most toxicological studies have been conducted on perfluoroctanoic acid (PFOA) and perfluorodecanoic acid (PFDA), rather than the more environmentally prevalent PFOS. However, PFOS appears to be the ultimate degradation product of a number of commercially used perfluorinated compounds and the concentrations of PFOS found in wildlife are greater than the concentrations of other perfluorinated compounds (Kannan et al. 2001a,b; Giesy and Kannan, 2002). However, more recent studies have found that there are many different PFSs in the environment. The toxicity of the primary PFSs in the environment, PFOS, was recently reviewed by Beach et al. (2006) and Water Quality Criteria for the Protection of Aquatic Organisms were estimated. Subsequently, the toxicity of PFOS to birds was determined *in vivo* (Newsted et al. 2005) so that toxicity reference values (TRVs) could be determined.

To be able to assess the potential risk of complex mixtures of PFSs, it was necessary to determine the critical mechanism of action. This was done by the use of advanced molecular techniques, including gene chips (Hu et al. 2002a,b; Guruge et al. 2006). It was learned that there were several critical mechanisms of action of the PFSs compounds. First, it affected the properties of membranes (Hu et al. 2002a,b) and second, it disrupted metabolism of fatty acids. This is not surprising, because the structure of the PFSs is analogous to that of fatty acids, but they cannot be metabolized (Guruge et al. 2006). Thus, accumulation of PFSs can lead to a cascade of different effects. However, using simple in vitro bioassays based on this knowledge allowed the development of quantitative structure activity relationships (QSARs) (See Chapter 4) that revealed that the toxicity of the various PFSs was directly proportional to the length of the perfluorinated tail and not related to the terminal moiety. The most toxic compounds were those with 8-carbon tails and toxicity was affected by whether the tail was straight chained or branched.

When the fact that PFSs, in particular PFOS, were ubiquitous in the environment was combined with the fact that the dose–response relationship was very steep and the endpoint assessed in rats, lethality, was severe, it was quickly decided by the principle manufacturers

* **Peroxisomes** are membrane-bound organelles in the cytoplasm that can generate but also consume hydrogen peroxide. Catalytic conversion of some organic compounds in the peroxisomes produces hydrogen peroxide which, in turn, can be consumed by catalase in the peroxisome. Thus, they can serve in detoxification but also hydrogen peroxide production can create oxidative stress (see Chapter 6). Some compounds can cause the cell's peroxisomes to increase (**peroxisome proliferation**), elevating oxidative stress and associated cell damage or death.

of PFOS-based products, the 3M company and the United States, in consultation with the U.S. EPA agreed to voluntarily phase out most of the production of PFOS-based products. The cessation of production could have resulted in disruption of many PFSs use areas, many of high value, or in some cases, critical to the production of high-value products. However, the decision was made easier by the fact that by understanding the mechanism of action and the QSARs that had been developed, it was possible to develop replacement products that were not accumulated and were essentially not toxic.

When current concentrations of PFSs in water (So et al. 2004) and wildlife have been compared to thresholds for toxic effects, it has been found that the critical levels above which toxicity would have been expected to have occurred were not exceeded, even in those locations where the greatest concentrations had been observed (Rostkowski et al. 2006). Thus, although the concentrations in the tissues of some animals came near, they did not exceed concentrations that would have been expected to cause adverse effects on wildlife populations.

Currently, monitoring of concentrations of PFSs is ongoing and the PFSs have been added to the lists of compounds for which many agencies now monitor. In addition, several PFSs were added in 2009 to the Stockholm Convention list of persistent, bioaccumulative, and toxic compounds (UNEP, 2010). Since the phase out of the manufacture of PFOS-based products, measurements of concentrations of PFOS in tissues of wildlife in the Arctic of eastern Canada have been confirmed to be decreasing. This indicates that the voluntary actions taken by 3M were effective and that the environmental response has been rapid.

CONCLUSIONS AND LESSONS LEARNED

Regulatory agencies had thought that they had put in place screening procedures and were requesting sufficient information to make informed decisions about the types of chemicals that could be manufactured, used, and released to the environment safely. The discovery in the late 1990s of an entirely new class of compounds was surprising. How did this happen? First, the perceptions that the PFSs were inert and bound tightly into polymers and were not available to enter the environment led to complacency about the potential to cause adverse effects. Second, chemists find what they look for. Current monitoring programs are not well suited to looking for novel compounds. The nature of the compounds and a lack of awareness or urgency in looking for them in the environment mean that there were no instrumental methods available and certainly no certified standards or reference materials (Martin et al. 2004b). Once the issue was recognized by working together, experts from industry and academia were able to develop and apply methods of identification and quantification, and to conduct a global survey to answer the most critical questions about exposure. Furthermore, by using modern molecular techniques, it was possible to discern some of the critical mechanisms of action, develop predictive models, and design replacement products all within a period of less than 1 year. This allowed industry and governmental agencies to make effective decisions that removed a potentially hazardous material from use. This has resulted in rapid reductions in environmental concentrations and consequent risks while maintaining useful products. With close cooperation, understanding, and trust between academia, industry, and regulatory agencies, this rapid alleviation of a potentially dangerous environmental situation would not have been possible.

Phthalates (Figure 2.30) are used widely in plastics, especially polyvinyl chloride. They are also incorporated into adhesives and paints. Their widespread use and ability to modify endocrine function* have prompted concern about phthalate contamination.

2.2.2.11 Additional Emerging Organic Contaminants of Concern

Many more contaminants have recently emerged as compounds of concern. The coverage here includes important contaminants from diverse origins and significant potential effects. Predictably, this list will continue to grow and the perceived importance of contaminants on the list will shift through time so the few key emerging classes of contaminants described here should be seen as current examples of contaminants of increasing concern.

Disinfection by-products result from reactions of halogen-based disinfectants with natural compounds of waters. By-products include trihalomethane, haloacetic acids, haloacetonitriles, and haloketones. These chlorination by-products have emerged as compounds of particular concern. Chlorine gas or compounds used to disinfect drinking waters produce chlorinated organic compounds with potentially toxic or carcinogenic effects. Chlorination produces trihalomethanes such as the carcinogen, chloroform ($CHCl_3$). As one sublethal effect example, chlorination by-products in drinking water are suspected of modifying the menstrual cycle of women (Windham et al. 2003). Trihalomethane in drinking water might also modify semen quality in healthy men (Fenster et al. 2003). In addition to trichloromethanes, two other chlorination by-products of toxicological concern are dichloroacetic acid and dichloroacetonitrile. Disinfection via bromination can produce compounds such as the trihalomethanes, dibromochloromethane ($CHClBr_2$), and tribromomethane (bromoform, $CHBr_3$).

Brominated fire retardants include the PBBs already discussed and the polybrominated biphenyl ethers (PBDEs), tetrabromobisphenol A (TBBPA), and hexabromocyclododecanes (HBCD). These brominated fire retardants are depicted in Figure 2.31.

The PBDE have two aromatic rings joined by an ether bond (R-O-R). They differ in the number of bromine atoms bound to their aromatic rings. The general grouping of PBDEs is based on whether the associated commercial mixture has an average of five (pentaBDE), eight (octaBDE), or ten (decaBDE) bromine atoms. Technical mixtures of these PBDEs are composed of many congeners.

The PBDEs have been used in a wide array of polymer products including back coating on fabrics, electronics, and numerous household products. PentaBDE is used in polyurethane foams such as those in furniture cushions. It can make up to 30% by weight of such foams (Birnbaum and Staskal, 2004). It is slowly released by volatilization through time (Blum, 2007) or as the foam itself disintegrates. Like pentaBDE, decaBDE is a mixture of congeners but is mostly (97%) BDE209. It is used in plastics and back coating of fabrics.

The PBDEs have high lipophilicity and slowly degrade in the environment (Birnbaum and Staskal, 2004), warranting the present attention of mammalian and ecological toxicologists to

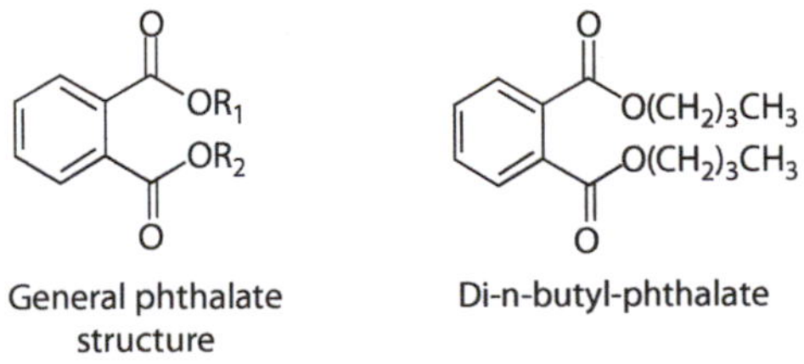

Figure 2.30 The general phthalate structure and the example of di-n-butyl-phthalate.

* The **endocrine system** is composed of several tissues. It broadly includes any tissues or cells that release a chemical messenger (hormone) intended to signal or induce a physiological response in some target tissue.

the possibility of adverse effects of PBDE exposure (Birnbaum and Staskal, 2004). For example, very high concentrations of PBDEs have been measured in house cats and a notional link made to increased feline hyperthyroidism (Dye et al. 2007). Similar to results for Asians and North Americans, high concentrations of PBDEs were found in umbilical cord serum and breast milk sampled from women of Madrid, Spain. Indeed, Solomon and Weiss (2002) found that PBDEs are the only POPs in their survey that were currently increasing each year in human milk (Figure 1.7). These studies indicate that significant routes for prenatal and perinatal exposure exist, especially exposure to the highly brominated congeners (hepta- and decaBDE) (Gónara et al. 2007). The potential for endocrine disruption including thyroid dysfunction, reports of neoplasm-induction in rats, and developmental neurotoxicity in mice makes the possibility of high PDBE exposure worrisome (Birnbaum and Staskal, 2004). In 2004, penta- and decaBDE were banned in Europe. U.S. manufacture of penta- and octaBDE ended in 2004, but decaBDE is still produced in high volumes in the United States. Their use is greatly curtailed in the United States and elsewhere (Betts, 2007).

Tetrabromobisphenol A (Figure 2.31) is incorporated by reaction into polymers such as those of printed circuits. After release into the environment, it can accumulate in biota due to its high lipophilicity. However, TBBPA does degrade in the environment and is eliminated from biota after gaining entry into tissues (Birnbaum and Staskal, 2004). TBBPA is a suspected disruptor of endocrine function. The cycloalkane fire retardant, hexabromocyclodecane (HBCD) shown in Figure 2.31, was incorporated into plastics and other polymeric materials. Its use in the United States was minor compared to TBBPA: HBCD was used more in Europe (Birnbaum and Staskal, 2004).

Enough pharmaceutical and veterinary agents enter the environment to warrant mention of their possible ecotoxicological side effects (Jørgensen and Halling-Sørensen, 2000; Daughton, 2002; Boxall et al. 2003, 2004; Jones et al. 2003b; Derksen et al. 2004; Ankley et al. 2007). Although too diverse a class of contaminants to permit completely accurate generalizations, these agents tend to be less persistent than the POPs discussed earlier. Yet their continuous release from sewage treatment plants, stockyards, areas of **biosolids** (defined as stabilized sewage sludge) application, and other sources prompted Daughton (2002) to refer to them as pseudopersistent pollutants.

Many such agents are intended to kill or control infectious viruses, bacteria, fungi, protozoa, and metazoan parasites. Still other relevant human pharmaceuticals are intended as anti-inflammatory

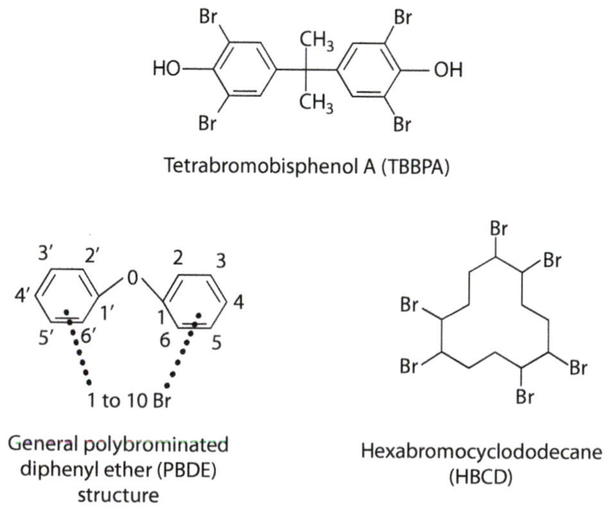

Tetrabromobisphenol A (TBBPA)

General polybrominated diphenyl ether (PBDE) structure

Hexabromocyclododecane (HBCD)

Figure 2.31 The brominated fire retardants including the polybrominated biphenyl ether general structure, tetrabromobisphenol A (TBBPA), and hexabromocyclododecane (HBCD). Note that another brominated flame retardant (polybrominated biphenyl in Figure 2.10) has already been discussed.

drugs (Schwaiger et al. 2004; Triebskorn et al. 2004), β-blockers, antiepileptics, analgesics, antidepressants, and hormonally active agents (Carballa et al. 2004; Derksen et al. 2004). Veterinary medicines also include growth promoters (Meyer, 2001; Boxall et al. 2003). Many of these substances pose risk in ecosystems under certain combinations of administration route, metabolism within exposed organisms, environmental degradation, and partitioning among environmental phases. Their concentrated application in intensive livestock production or aquaculture can result in large amounts being present in the immediate environment, e.g., intensive ivermectin use for control of endoparasites in stockyards (Boxall et al. 2003). The land application of manures or biosolids can result in chronic release into still other environments.

Perhaps among the most relevant classes of such pharmaceuticals and veterinary agents are hormone analogs and their metabolites (Figure 2.32). Our earliest concern about the adverse effects of such chemicals emerged in the 1970s (Naz, 1999). For years, pregnant women at risk of spontaneous abortion were provided diethylstilbestrol (DES), an agent that acts like the hormone, estradiol. This practice was based on the assumption that at-risk women were less capable than others of maintaining hormonal conditions conducive to pregnancy and DES improved their hormonal state (National Research Council, 1999). Its use on humans stopped when certain cancers began appearing in daughters of women who took DES. Concerns about other hormone analogs have emerged more recently. Now, many women take the 17 β-estradiol (Figure 2.32) analog, 17 α-ethynylestradiol (EE_2), for birth control purposes. Still other hormones and their analogs are presently administered to livestock for purposes of reproductive control, growth promotion, or increasing milk production.

These hormone analogs and their metabolites find their way into the environment via discharge from sewage treatment plants, drainage from stock or poultry facilities, and run-off from land onto which biosolids had been applied (Carballa et al. 2004). Diverse ecotoxicological effects have been reported including impaired reproduction of fish exposed to ethynylestradiol (Nash et al. 2004) and modified sexual development of fish exposed to the synthetic androgen, 17 α-methyltestosterone (Pawlowski et al. 2004). Trenbolone acetate, an anabolic steroid used to promote livestock growth, is metabolized in cattle to 17 β-trenbolone. Considerable amounts of this metabolite move from feedlots into the natural environment where it is reported to have a relatively long half-life of approximately 260 days (Schiffer et al. 2001). Concern is growing about the potential adverse impacts of 17 β-trenbolone on nontarget species (Ankley et al. 2003) because it has androgenic and antiglucocorticoid* effects, and can cause developmental abnormalities in mammals (Wilson et al. 2002).

A final group of emerging contaminants is those associated with personal care products. They include components of cosmetics, medicines, and household products such as fragrances, musks, and bactericides (Bester, 2007). The contaminants composed of the pharmaceuticals discussed previously and these products are collectively known as PPCPs (**Pharmaceuticals and Personal Care Products**). Concentrations of many are reduced during sewage treatment processes but others are more resilient. Carballa et al. (2004) report removals of approximately 70%–90% for fragrances, 40%–65% for anti-inflammatory drugs, 65% for 17 β-estradiol, and 60% for the antibacterial sulfamethoxazole during sewage treatment. The musks shown in Figure 2.33 are used widely in personal care products and released globally (e.g., Osemwengie and Steinberg, 2001; Zeng et al. 2008), resulting in concern about their effects within ecosystems. They have been found to bind to hemoglobin in fish species (Mottaleb et al. 2004a,b,c). Because of their persistence and potential to bioaccumulate, musk xylenes were banned in Japan in the 1980s and are under consideration of banning in Europe (Mottaleb et al. 2004a).

* The **glucocorticoids** such as cortisol are pivotal in glucose metabolism. Among their other roles, they function in the body's general response to stress, serve as modifiers of glucose metabolism, and modify immune response including inflammation.

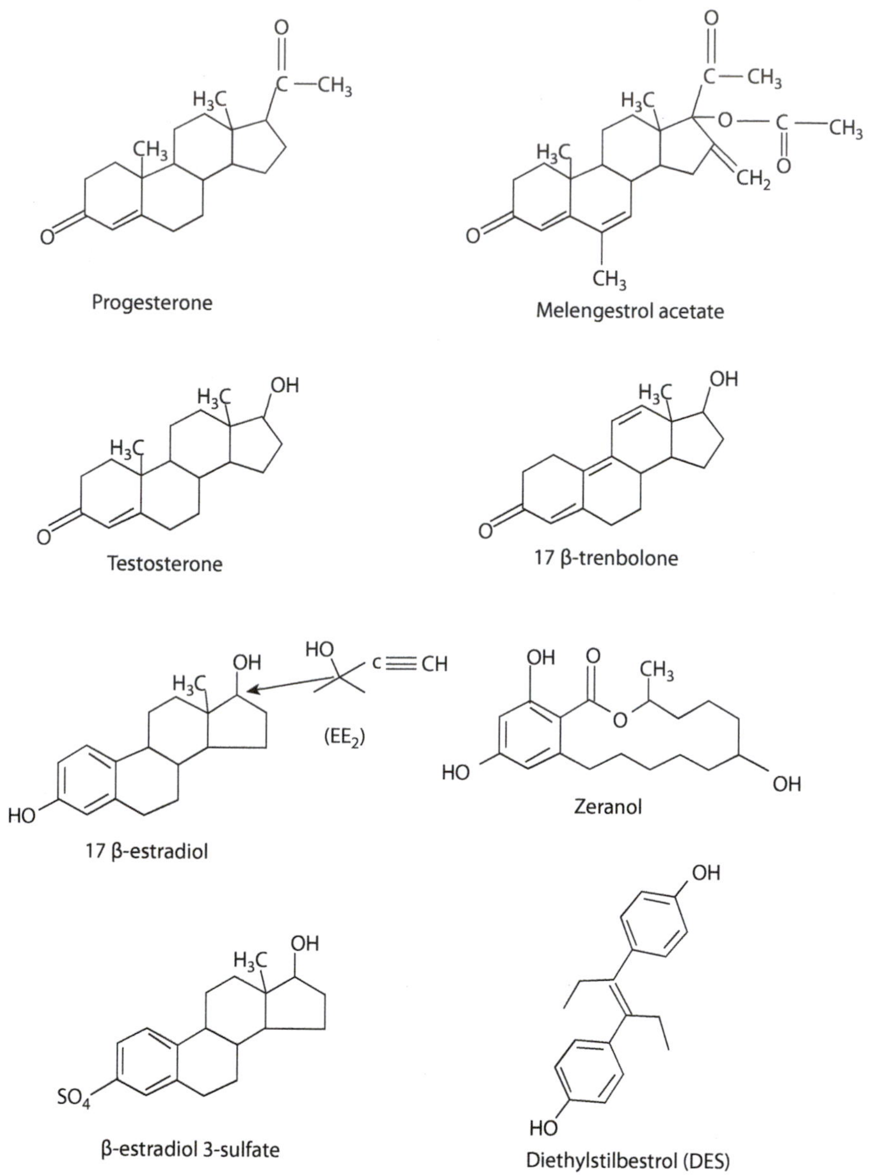

Progesterone

Melengestrol acetate

Testosterone

17 β-trenbolone

17 β-estradiol

(EE$_2$)

Zeranol

β-estradiol 3-sulfate

Diethylstilbestrol (DES)

Figure 2.32 Hormones and their synthetic analogs and conjugates. Diethylstilbestrol, a synthetic estrogen, was used until the 1970s to reduce the risk of spontaneous abortions in pregnant women with unanticipated detrimental consequences. Progesterone, testosterone, and 17 β-estradiol are natural hormones. Examples of their synthetic analogs are shown to their right (modified from Figure 1 of Meyer [2001]). Zeranol, a synthetic estrogen, and trenbolone, a synthetic androgen, are used to enhance livestock growth. Similarly, melengestrol acetate is provided for growth enhancement of female livestock as well as for regulating reproduction (Meyer, 2001). 17 β-ethynylestradiol (EE2 or EE$_2$), a 17 β-estradiol analog used for birth control purposes by women, is similar to 17 β-estradiol but has an additional -C≡CH group as shown in this figure. Both, 17 β-ethynylestradiol and 17 β-estradiol are conjugated with -SO_4 in the body and eliminated in the bile as discussed in Chapter 3.

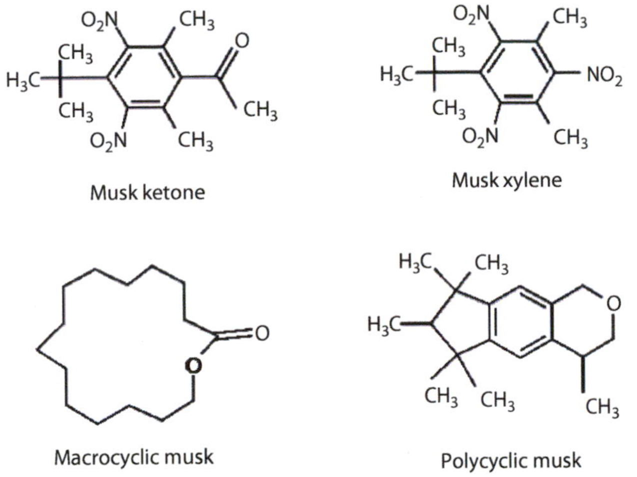

Musk ketone

Musk xylene

Macrocyclic musk

Polycyclic musk

Figure 2.33 Large amounts of synthetic musks are produced annually and used widely in products like toiletries, laundry products, cosmetics, and perfumes. Osemwengie and Steinberg (2001) quote a 1999 worldwide musk production estimate of 6 million kilograms. The musk ketone and xylene shown at the top are nitro musks. At the bottom are examples of macrocyclic and polycyclic musks. Many synthetic musks and their metabolites are not readily biogradable and are lipophilic enough to raise concerns about bioaccumulation in biota including humans (Osemwengie and Steinberg, 2001). (Note that xylene refers to the presence of a benzene ring with two attached methyl groups. Ketone refers to the presence of a carbon atom with a double bond to an oxygen atom and its remaining two bonds with other organic structures, that is, RR'CO.)

2.2.3 Radiations

To this point, we have discussed contaminants based on their binding chemistries and resulting compounds; that is, properties involving primarily the behavior of the outer orbital electrons of atoms and also structural properties of compounds. Now, we will broaden the discussion to include changes involving the state of inner orbital electrons and components of atomic nuclei.

The term **radiation** refers to the propagation of energy through space. This can be in the form of photons, ejected atomic nuclei or their fragments, or subatomic particles. Heat energy, visible and ultraviolet (UV) light, x-rays, and gamma (γ) rays are all examples of radiations that are photons of electromagnetic energy. Radiations are also generated within the nuclei of unstable elements: such nuclei are called radionuclides, and their emitted radiations include alpha (α) particles (helium nuclei with a $^+2$ charge), beta (β) particles (electrons or **positrons** with a $^-1$ or $^+1$ charge), and γ photons. Neutrons (which have no charge) and protons (which have a $^+1$ charge) might interact with other atomic nuclei but are not emitted from radionuclides (although they are emitted if other atoms are excited by radiations). Neutrons might create radionuclides when generated by anthropogenic processes. Cobalt-60 (^{60}Co) is an example of a radionuclide manufactured using neutrons in nuclear reactors. It is used as a γ radiation source in irradiation facilities.

There are many sources of radiations that can affect organisms. Radiations are generated naturally through the decay of radioactive elements on earth. All of these radionuclides were originally formed in the cores of stars other than our sun, or through the decay series of elements such as uranium or thorium (which slowly decay, and in turn, produce radioactive elements such as polonium, radium, and radon). Radiations from outside the earth include **cosmic rays**, which are predominantly the electron-free nuclei of iron atoms (Fe^{23+}), protons emitted from the sun, and subatomic muons and neutrinos that impinge on the earth's atmosphere (or are produced by extraterrestrial radiations in the atmosphere). Although these radiations rarely interact directly with living tissue, they can be important because they can react with atmospheric gases and produce cosmogenic radionuclides (especially ^3H and ^{14}C) that can become incorporated into organisms. The radioactive emissions from the earth's crust combine with atmospheric radiation such that the total radiation dose at sea level is about evenly divided between the two. At increasing altitude, radiation exposure from the cosmos increases by approximately two-fold for every mile increase in altitude. Latitude also affects radiation exposure. The earth's magnetic field deflects charged particles near the equator but channels them into the atmosphere at the magnetic poles.

In addition to these natural sources, radiation is generated from nuclear weapons production and testing, nuclear energy production, and numerous medical, research, and industrial applications. For example, technetium exists only in short-lived radioactive forms and does not occur naturally on earth. Technetium, as ^{99}Tc is a major component of nuclear waste, and Tc isotopes are used in medical diagnostic tests of lung damage. Human activities can also influence natural electromagnetic radiation reaching ecosystems. For example, the intensity of UV radiation reaching the earth's surface has increased due to our CFC emissions.

This short section will sketch out rudimentary concepts and details about potentially harmful kinds of radiation in the environment. More details about their associated effects and risks will be discussed in Chapters 12 and 13.

Terrestrial radiations can be characterized as either electromagnetic photons or particles emitted during the decay of unstable atomic nuclei. The more unstable atomic nuclei decay the most rapidly and energetically. Often this involves a **decay series** in which a particular radionuclide transforms into multiple elements or isotopes in sequence. Photons can also be emitted during orbital electron transitions. Photons have no mass or charge whereas particles have mass and can be charged. Infrared and UV radiations, γ-rays and x-rays are all electromagnetic radiations. Gamma-(γ) and x-rays have higher energies (shorter wavelengths) than infrared and UV radiation, and their wavelengths and energies can overlap. They differ solely in their mode of generation: **γ-rays** are emitted during nuclear decay and **x-rays** are emitted during shifts of inner orbital electrons. X-rays can be generated when electrons produced during nuclear decay (or from sources outside the atom) are captured and cause a consequent shift in inner orbital electrons (Wang et al. 1975). The characteristic energy associated with any particular γ- or x-ray photon strongly influences its capacity to transfer energy to and damage living tissue. The energies associated with various kinds of electromagnetic radiations fall roughly into the following ranges (Bureau of Radiological Health, 1970):

Infrared: 4.1×10^{-4} to 1.6 eV
UV: 3.3 to 410 eV
X-ray: 10 to 3×10^{10} eV
γ-ray: 8×10^3 to 10^7 eV

In contrast to the electromagnetic radiation just described, α and β radiations involve particle emission from the nucleus. An **α-particle** is composed of a helium nucleus (2 neutrons and 2 protons) ejected during radioactive decay. As such, it carries a 2^+ charge and is massive compared to β-particles. Because of this higher mass and charge, α emissions penetrate only micrometers into tissue. This is the reason that α-emitting radionuclides create negligible risk outside of the body. But they can produce extensive damage in even minute quantities if inhaled or ingested. This explains the reputation of plutonium isotopes as being the most deadly of toxicants on earth.

The **β-particles** are either an electron (negatron or β^-) or positron (β^+). Either can be released during nuclear decay. A β-particle has only 1/7300 the mass of an α-particle and does not have a discrete, characteristic energy like an α-particle. Instead, a β-particle has a distribution of energies characterized by a maximum (E_{max}) and an average ($\bar{E}$) energy. The shape of the energy distribution curve varies among β-emitters. The $\bar{E}$ is approximately 1/3 E_{max} for β^- decay and 2/3 E_{max} for β^+ decay (Wang et al. 1975). Having both mass and charge, the α- and β-particles interact much more with living tissues than electromagnetic photons. However, these same qualities also restrict their penetration into tissues. Even the highest energy α-particles are generally stopped within micrometers after interacting in living tissues. The much less massive and lower charged β-particles penetrate living tissues to several centimeters if they are highly energetic, but most are stopped at the surface.

Some radiations originate directly from radionuclides (Section 2.4, "Nuclides") present in our environment. Other types of radiation originate from radioactive decay occurring outside of our immediate environs, such as γ-rays from cosmic sources or the tritium (^{3}H) formed as atmospheric nitrogen interacts with cosmic rays. Still other radiation such as x-rays or UV radiation can be generated during shifts in energy states of inner (x-rays) and outer (UV) orbital electrons of elements. Radiation can be ionizing in which case ion pairs are generated when the radiation interacts and gives up energy to matter. It only takes approximately 34 eV of energy to form ion pairs in air (Wang et al. 1975) and only 3.9 eV to remove the first electron from a cesium atom (Considine, 1995). A double bond between carbon atoms can be broken with only approximately 5 eV. So, a general rule of thumb might be proposed for our discussion that radiation with energies of roughly 10 eV or higher might be considered ionizing. For example, γ- and x-rays, and perhaps some energetic UV photons in our environment would be ionizing radiation. Of course, α- and β-particles also produce ion pairs in matter, including the matter making up living tissue.

Production of ions along tracks traveled by radiation through tissues is the primary cause of radiation-induced damage. The nature of radiation's interactions with living tissues depends on the radiation's energy (wavelength), mass, and charge. A massless, high-energy x-ray photon passes easily through tissue but the massive and charged α-particle will have limited penetration. Yet, the amount of energy that an α-particle can transfer to tissue during interaction will be very high relative to that of an x-ray.

In addition to the nature and source of a radioactive material, the ability to manifest harm depends on the radionuclide's movement into appropriate proximity to sites of effect. The movement of a radionuclide in the environment, and in organisms, depends on its conventional chemistry and decay characteristics, and its similarity to other stable nuclides such as essential elements. For example, naturally occurring radiopotassium (^{40}K) and chemically similar radiocesium (^{137}Cs) will move through biological systems like stable potassium.

Radon gas can be used to illustrate that the level of radiation exposure is influenced by the nature and amount of a source, radioactive decay rate, and possible generation in a decay series of the radioactive element (Section 2.4, "Radioactive Decay Series"). The radioactive gas, ^{222}Rn, is released into the biosphere as one product from the natural uranium–radium decay series. The amount of various natural parent radionuclides (uranium and radium) present in an area depends on its local geology and the amount of radon gas released is dictated by the amount of these parent radionuclides. The ^{222}Rn has a relatively short half-life and produces solid α-emitting metals such as polonium that become associated with fine particles in the home. Inhalation of these α-emitting, fine particulates creates an avenue for deep penetration into the lungs. Local deposition of high amounts of energy from the α-particle emission results in tissue damage and potentially cancer. So, generation of ^{222}Rn from a decay series, the gaseous nature of the ^{222}Rn that allows it to move into the home, the production of α-emitting metals that associate with fine dust that can penetrate deeply into the lungs, and the deposition of large amounts of energy in lung tissue via α-particles make ^{222}Rn exposure a significant concern.

2.2.3.1 Expressing Radioactivity

As mentioned briefly in Chapter 1, measures of radioactivity and possible radiation effects are easily described. The original, and still widespread, unit for expressing radioactivity is the **curie**. By convention, a curie is 2.2×10^{6} nuclear disintegrations per minute. The **becquerel** (Bq) replaced the curie as the official unit of radioactivity. There are 3.7×10^{10} Bq in 1 curie.

Measures also exist for quantifying radiation-induced damage in biological tissues. These measures adjust for the different capacities for various kinds of radiation to penetrate and cause damage to living tissues. Radiation exposure was originally expressed in Roentgens (R), which is 87.6 ergs of energy imparted to a gram of air. Of course, we are more interested here in the amount of energy imparted to living tissues, that is, in radiation dose. Radiation dose was expressed in **rads**, which is 100 ergs of ionizing radiation energy in a gram of material. The official radiation dose is the **Gray** (Gy). The old unit of a rad is 0.01 Grays. A **rem** or Roentgen equivalent man accounts for differences in the various types of radiation for potentially producing biological damage. Again, to reconcile the units to the current international standard, the rem has been replaced as the official unit by the sievert (Sv). One rem is equal to 0.01 Sv.

Also important to consider are the differences among radiation types and tissues. An ionizing radiation must penetrate and deposit energy in a tissue to cause damage. Charged particles such as α- and β-particles penetrate poorly because they deposit their energy by ionizing atoms near their point of emission. Uncharged neutrons or photons can pass considerable distances into tissues before interacting with the electron cloud, or much less probably, the nucleus of an atom. Neutrons produce an effect only during interactions with the nucleus but, when this occurs, the atom becomes unstable and emits its newly acquired energy as a β-particle, proton, or γ photon. These in turn can interact and potentially damage tissue.

The amount of energy that a particular photon or particle can transfer is characteristic of that radiation. **Quality factors** (Q) are given to various kinds of radiation to account for their different capacities to damage tissues. A quality factor of 1 was given to β, γ, and x-rays and 10 to α-particles. More precise measure of energy transfer to matter (such as tissue) during passage is the **Linear Energy Transfer (LET)**, which has units of amount of energy per unit of path length (eV per micrometer).

2.2.3.2 *Radionuclides**

The themes explored previously can now be applied with some of the more environmentally relevant natural and human-produced radiation sources. Eisler (1994) divides radionuclides into four convenient groups. In the first group are those with very long half-lives (more than 10^9 years) such as ^{235}U, ^{238}U, ^{232}Th, and ^{40}K that are generally associated with natural sources. The second group (e.g., ^{222}Rn and ^{226}Ra) is progeny (or daughter) radionuclides produced by the decay series of such long-lived parent radionuclides. The third group includes ^{3}H and ^{14}C, which are formed by the continual cosmic ray bombardment in the upper atmosphere mentioned previously. The fourth group includes radionuclides formed during nuclear reactor operation, nuclear weapons detonations, and other human activities, for example, ^{60}Co in irradiators. Members of each of these groups will be discussed in the order they appear in a periodic table (Figure 2.1).

The most relevant examples from the group 1A elements are hydrogen, potassium, and cesium. The first two are essential bulk elements. As already mentioned, tritium is produced continuously in the atmosphere by cosmic ray bombardment of nitrogen atoms. Having an unstable combination of one proton and two neutrons, ^{3}H decays with a half-life of approximately 12.3 years by emitting β radiation (β^- particles) with a E_{max} of 0.0186 MeV (MeV = 10^6 eV). Potassium-40 (^{40}K) has a very long half-life (1.26×10^9 years), so the majority of this radionuclide in living organisms and their environs came from primordial sources. Approximately 0.0117% of the potassium on the earth is ^{40}K. ^{40}K produces β^--particles (1.314 MeV E_{max}), β^+-particles (0.483 MeV E_{max}), and γ-rays (1.46 MeV) on decaying to ^{40}Ca. Radionuclides of cesium are all anthropogenic in origin. Both ^{134}Cs and ^{137}Cs are generated or released accidentally from nuclear production facilities or nuclear weapon detonations. Carlton et al. (1992) report that 3.5×10^7 Ci (1.3 TBq) of ^{137}Cs were released by atmospheric testing of nuclear weapons before the mid-1960s ban. The Chernobyl reactor accident released approximately 1.3×10^6 Ci of ^{134}Cs and approximately 9.7×10^5 Ci of ^{137}Cs (IAEA, 2006). And other ^{137}Cs sources exist. For example, Carlton et al. (1992) note that two Soviet satellites powered by nuclear

* Constants for these radionuclides are taken from the U.S. Federal Food and Drug Administration's *Radiological Health Handbook* (Bureau of Radiological Health, 1970) and Kocher (1981).

reactors reentered the atmosphere, releasing ^{137}Cs as they disintegrated. The ^{134}Cs decays away more rapidly (half-life of 2.04 years) than ^{137}Cs (half-life of 30.0 years), making ^{137}Cs more of a long-term concern after release. The ^{134}Cs produces β^--particles (0.662 MeV E_{max}) and γ-rays (0.87 MeV). The ^{137}Cs emits β^--particles (1.176 MeV E_{max}) and γ-rays (0.662 MeV). Concern about ^{137}Cs is heightened by its behavior as a potassium analog: it behaves similarly to potassium in biochemical pathways and processes. It is taken up at comparable rates to potassium and, like potassium, it accumulates in cells. However, it is eliminated at a slower rate than potassium and, as a consequence, can increase in concentration with movement through food webs, that is, to biomagnify.

In group 2A are two elements warranting discussion here. Radiostrontiums (^{89}Sr and ^{90}Sr) are produced by fission sources. Strontium-89 decays more quickly (half-life of 52.7 days) than ^{90}Sr (half-life of 27.7 years). Strontium-90 emits a β^--particle with an E_{max} of 0.546 MeV. While doing this, it produces a radioactive daughter (^{90}Y) that decays quickly (with a 64 hour half-life), producing another β^--particle with a much higher E_{max} of 2.27 MeV. The most common radium isotope ^{226}Ra (half-life of 1600 years) is the second group 2A radionuclide of concern. It is formed in the uranium decay series and is much more radioactive than uranium. It emits high-energy α-particles (4.7 or 4.6 MeV), x-rays of several energies, and several shorter lived radioactive daughters. Radium was once widely used as a self-luminous paint for watch faces and instrument panels. It is still used for medical treatments.

Although their radiations are short range and relatively harmless on the exterior of the body, harm from ^{90}Sr and ^{226}Ra is greatly increased because they are calcium analogs in biological systems. In invertebrates, ^{90}Sr concentrates in molluscan shells and anthropod cuticles (Polikarpov, 1966; Hinton et al. 1992). Eisler (1994) indicates that coconut crab (*Birgus latro*) present on the Bikini Atoll after nuclear bomb detonations experienced high chronic radiation dosing from ^{90}Sr accumulated in their exoskeletons and also from the ^{137}Cs in their muscle tissues. Strontium and radium accumulate in vertebrate skeletons, and are often referred to as bone-seeking radionuclides. The point about bone-seeking behavior of some radionuclides can be further illustrated with two examples. The first is the tragic radium poisoning of young women who painted glowing radium onto clock and watch faces (Clark, 1997). They unwittingly ingested ^{226}Ra that became incorporated into their bones and, as it decayed, the 4.7 MeV, massive and charged α-particles deposited substantial energy in local bone and marrow, causing extensive bone tissue damage. These radiations also damaged DNA in marrow stem cells that resulted in fatal cancers (Macklis, 1993; Clark, 1997). Indicative of our ignorance at that time (before the 1930s) about radionuclide effects, ^{226}Ra ingestion from licking a sharp point to paint brushes was not the only exposure route. As one survivor recollects, "The girls [sneaked] the radium to paint their toenails to make them glow" (Adams, 2005).

^{90}Sr movement in food webs is also influenced by its being a calcium analog. It becomes incorporated into minimally digestible insect cuticle in terrestrial ecosystems, creating what is called a **calcium sink**. Because these exoskeletons are not digested efficiently by predators, the amount of ^{90}Sr available to be transferred to the next trophic level is only a small portion of that present (Reichle et al. 1970).

Three intermediate metal radionuclides (6B to 1B groups), ^{51}Cr, ^{60}Co, and ^{65}Zn, are environmentally relevant. All can be produced by **activation**, that is, by the bombardment of a stable nuclide with neutrons to produce a radioactive nuclide. The production of these products is generally limited, but they can find their way into the environment through accidents. Of particular concern is ^{60}Co that is produced by neutron activation in fission reactors or for use as an irradiation source. Compared with ^{137}Cs, ^{60}Co has a relatively short half-life (5.2 years). It is much more radioactive, emitting both a β^--particle (1.48 MeV E_{max}) and two high-energy γ photons (1.17 and 1.33 MeV). The two other nuclear reactor activation products, ^{65}Zn and ^{51}Cr, can be discharged in reactor cooling waters. Both of these activation products were discharged into the Columbia River (Washington State) by the Department of Energy's Hanford nuclear reactors (Cutshall, 1974; Dauble and Poston, 1994). During the period of plutonium production, they were the primary activation products reaching the mouth of the Columbia River (Osterberg,

1975). Both are isotopes of essential elements, and as such, enter into specific biochemical pathways. Fortunately, both have relatively short half-lives (27 days for ^{51}Cr and 245 days for ^{65}Zn).

Two nonmetal radionuclides that will be described here besides ^{3}H are ^{131}I of group 7A and ^{222}Rn of group 8A. The first, ^{131}I has a short half-life (8.05 days) and emits a β^--particle (0.806 MeV E_{max}), γ-ray (0.364 MeV), and a ^{131m}Xe daughter. Infamously, there have been historic, large ^{131}I releases from the U.K. Windscale/Sellafield processing facility (1957), U.S. Hanford facility (1944–1947), and Chernobyl reactor in the Ukraine (1986). Being an essential element isotope, ^{131}I accumulates to high levels in and damages the thyroid gland. Harley (2008) reports a 2002 childhood thyroid cancer incidence rate for Belarus of 7.5 children per 100,000 due to the ^{131}I released from the Chernobyl accident. As already mentioned, radioactive radon gas is slowly generated from decay of natural radionuclides. It has a short half-life (3.8 days) and emits a 5.49 MeV α-particle, a γ-ray (0.510 MeV), and ^{218}Po daughter. It can be a significant lung cancer risk factor in areas having geologies that favor high outgassing of ^{222}Rn and an exposure situation that tends to retain the radon gas such as a well-sealed house above a cellar (Pawel and Puskin, 2004).

One lanthanide and three actinide series radionuclides can be environmental concerns. Radioactive cerium (^{144}Ce) is a β^- emitter (0.31 MeV E_{max}, half-life of 284 days) that also releases a 0.080 MeV γ-ray and a rapidly decaying daughter radionuclide (^{144}Pr, 2.99 MeV E_{max} β^- and 0.695 MeV γ-ray). It has been released into the environment from nuclear bomb tests, fission reactors, and nuclear accidents. Although not a calcium analog, it is not surprising given its position on the periodic table, that ^{144}Ce also can be incorporated in exoskeleton of invertebrates and made less available for transfer to higher trophic levels (Reichle et al. 1970). The long-lived ^{232}Th (half-life of 1.41×10^{10} years) is present in our environment from natural sources. It produces a 4.01 MeV α-particle, β^--particle (0.042 MeV E_{max}), γ-ray (0.058 MeV), and daughter radionuclides. The remaining radionuclides, ^{235}U, ^{238}U, and ^{239}Pu, can be introduced into the environment from fission reactors and bombs. Low levels of uranium radioisotopes can also be introduced in the depleted uranium used to add mass to military projectiles and armoring. Both ^{235}U and ^{238}U have long half-lives (7.1×10^8 and 4.51×10^9 years, respectively). Both produce α-particles (4.20 and 4.58 MeV), γ-rays, and radioactive daughters. Thorium and uranium have a history of being incorporated into consumer products. As examples, thorium was used as mantles of kerosene lamps and uranium was used in ceramic glazes. ^{239}Pu also has a long half-life (2.4×10^4 years) and emits an α-particle (5.16 MeV) and γ-rays. As α-emitters, they can be very potent inducers of cancer if brought into appropriate close contact with living tissue.

2.2.3.3 Ultraviolet Radiation*

Initially, it may seem out of kilter to discuss this relatively benign form of electromagnetic radiation here. Granted, our release of CFCs into the atmosphere has thinned the ozone layer with a concomitant increase in UV radiation reaching the earth's surface, but what could possibly elevate a modest increase in UV irradiation to the level of attention paid to effects of strongly ionizing radiations? The clear answer is that UV irradiation is pervasive and has a significant impact to ecosystems, especially to primary producers (Newman and Clements, 2008). Increased UV radiation reaching the earth's surface is a particular concern for the earth's southern regions for reasons to be discussed in Chapter 12. Also, UV radiation can enhance adverse effects of other contaminants. As an important example, UV radiation is strongly absorbed by PAHs† such as anthracene that, in the resulting excited state, generate reactive

* Relevant UV radiation is categorized as **UVA** (wavelengths of 320–400 nm, energies of 3.87–3.1 eV), **UVB** (290–320 nm, 4.3–3.87 eV), and **UVC** (less than approximately 290 nm, >4.3 or >4.4 eV). Generally, UV radiation of energies of 4.7 eV or greater is absorbed in the atmosphere. Particularly crucial, ozone absorbs UV in the range of 240–320 nm (5.17–3.87 eV) so most of the UV radiation reaching the earth's surface is the less energetic UVA range.

† Those structures of organic molecules that absorb energy from the radiation are called **chromophores**. As radiation is absorbed, the chromophore goes from a stable to an excited state. Excitation is possible if the difference in energies of the involved molecular orbitals of the chromophore is in the range of that of the absorbed radiation.

oxygen species. These reactive oxygen species cause biomolecular damage (Chapter 6) (Toyooka and Ibuki, 2007) and even lethality (Bowling et al. 1983). In the case of PAHs, it is possible that some UV radiation can photomodify them, producing more toxic products. **Photomodification** might involve photooxidation or photolysis of the molecule (Huang and Adamson, 1993; El-Alawi et al. 2002).

2.2.3.4 Infrared Radiation

The present global climate change is a result of modified infrared radiation retention by the earth. The seriousness of the environmental consequences remains actively debated by many scientists; however, the vast majority of scientists agree that significant, adverse consequences are presently appearing (Kerr, 2007). Some effects are more subtle than others. A subtle effect of CO_2-induced global warming on coral reef and purple urchin sustainability has already been provided as an example.

2.2.4 Genetic Contaminants

The contamination of ecosystems with introduced genes is a topic that could easily be relegated to another discipline—perhaps crop sciences or conservation biology. But only a decade ago, one might have made a similar judgment that hormones and their analogs would be best discussed by endocrinologists and livestock scientists, not ecotoxicologists. That judgment would have deterred the progress of ecotoxicologists in understanding and making wise decisions about hormonally active compounds in the biosphere. Like the past emergence of hormonally active substances as relevant ecotoxicants, genes engineered into genomes of valued species are now emerging as ecotoxicologically relevant. Minimally, it is prudent to take the position that genetic contamination might create some instances of genetic pollution (Sharples, 1991; Beringer, 2000). Some products of genetic engineering also influence how conventional pesticides and herbicides are used. So, brief mention is warranted about the potential ecotoxicological consequences of **genetically modified organisms (GMO)** and their products in the environment.

There are several key points worth summarizing relative to **GMO biosafety** (i.e., "the effects of [genetically modified] crops and their products on human health and the environment" [Lu and Snow, 2005]). First, some genetically modified (GM) crop species produce toxins* that could find their way to nontarget species in the biosphere. As an example, insect resistance or herbicide tolerance genes incorporated into domesticated rice strains might find their way into noncrop *Oryza* species (Lu and Snow, 2005). Second, some GM crop strategies make the host plant more resistant to herbicides, allowing more herbicide to be applied to kill more weeds. Canola that is genetically modified to be more tolerant of herbicides is used in the United States despite the consequences of heavier herbicide applications that this permits and the potential for tolerance gene transfer to noncrop species (Rieger et al. 2002). Third, products of introduced genes in transgenic crops are released to nearby ecosystems habitats and could potentially harm nontarget species. The best example is the incorporation of the *Bacillus thuringiensis* (Bt) toxin gene into crop species. The soil bacterium, Bt, produces an endotoxin that is effective in controlling Lepidoptera. This bacterium has been applied to plants for this purpose for many years but, with the advent of modern genetic techniques, the Bt gene has now been inserted into numerous crop species that then express the Bt gene in most of their cells. This creates the potential for some ecotoxicological harm if nontarget species exposure to Bt toxin becomes pervasive. Hansen Jesse and Obrycki (2000) note that Bt corn pollen can settle on milkweed plants near fields and kill monarch butterfly (*Danaus plexippus*) larvae feeding on the milkweed. Fourth, their use should inevitably become part of integrated pest management programs in which pesticides, herbicides, and other agricultural tools are carefully balanced to maximize sustainable yield and minimize adverse impacts.

* Toxin and toxicant are not synonymous terms. A **toxin** is a toxicant of animal, plant, fungal, or microbial origin such as toxicants in animal venom or belladonna alkaloids of plant origin.

Many such programs rely on pesticide-free refuges to reduce the rate at which pesticide resistance is acquired by target insects. Introduced genes are potential avenues for inadvertent introduction of pesticide Bt into refuges (Glaser and Matten, 2003; Chilcutt and Tabashnik, 2004). Fifth, some GMO could modify processes in the biosphere, and in doing so, alter ecosystems. An example is Frostban, a bacteria-associated gene intended to reduce crop frost damage (Naimon, 1991). At-risk crop plants such as strawberry would be sprayed with two bacteria having inserted genes that inhibit ice nucleation. However, ice nucleation genes in natural bacteria can be crucial for rainfall formation (Naimon, 1991). Heavy spraying of these GM bacteria and their subsequent dispersal into the atmosphere might create the risk of modified precipitation patterns. Sixth, some genes can move to other species and shift the competition among species in ecological communities. Certainly, the incorporation of the Bt gene into endemic species could significantly alter herbivory and the competition among herbivores in ecological communities. Creeping bentgrass (*Agrostis stolonifera*) modified for high herbicide resistance is a good example of a species prone to rapid flow of inserted genes to other species. Planted on golf courses, it is a wind pollinated and highly outcrossing species with many closely related endemic species throughout the world. Surveys of genetic markers integrated into this GMO show rapid and widespread flow of incorporated genes into sentinel *A. stolonifera* plants and native *Agrostis* species growing tens of kilometers from GMO pollen origins (Watrud et al. 2004; Reichman et al. 2006).

2.2.5 Nanomaterials

Nanomaterials are natural and manmade materials having at least one dimension of 100 nm or less (Peralta-Videa et al. 2011): **nanoparticles** are nanomaterials with minimally two dimensions being between 1 and 100 nm. Although nanoparticles can be natural in origin as in the cases of silica and asbestos fibers, the term is most commonly used for manufactured particles. Natural nanoparticle sources include fires, volcanoes, biologically produced magnetite, and ferritin (Oberdörster et al. 2007). Anthropogenic sources include unintentional ones such as internal combustion engines, smelters, and electric motors. Nanoparticles also include intentionally engineered particles.

Interest in the harmful effects of nanomaterials broadened rapidly from a justifiable concern about a few natural (e.g., asbestos fibers), industry-associated (e.g., silica) and combustion-derived (e.g., diesel soot) materials to encompass a wide range of engineered materials. Engineered nanomaterials are used in products including electronics, optics, textiles, medical devices, pharmaceuticals, cosmetics, packaging, water treatment technologies, fuel cells, catalysts, biosensors, and bioremediation materials (Handy et al. 2008, Ju-Nam and Lead, 2008). These new materials include single- and double-walled carbon nanotubes, metallic, metal oxide, metal hydroxide, quantum dot, and polystyrene nanoparticles that can be broadly classified as follows (Peralta-Videa et al. 2011):

- Organic
  - Fullerenes (C_{60} and C_{70})*
  - Carbon nanotubes
    - Single-walled carbon nanotubes (SWCNTs)
    - Multiwalled carbon nanotubes (MWCNTs)
- Inorganic
  - Metal oxides (e.g., CeO_2, Fe_3O_4, Fe_2O_3, TiO_2, ZnO_2, and ZnO)
  - Metals (e.g., Au and Ag)
- Quantum dots (CdSe)†

* A **fullerene** is a large molecule composed of 60 or more carbon atoms with the form of a hollow sphere, ellipsoid, or tube. In the present discussion of nanoparticles, nanotubes are separated from the other fullerenes. The C60 and C70 refer to the number of carbon atoms in the molecule with C60 being the well-known and widely used spherical buckminsterfullerene.

† **Quantum dots** are semiconductor nanocrystals whose electronic properties are dictated by their crystalline structure and size. They are commonly colloidal nanocrystals of cadmium selenide, cadmium telluride, cadmium sulfide, or zinc selenide with diameters of approximately 2–10 nm.

Some nanomaterials can be mixtures of these classes.

The same concerns about health effects of asbestos fibers have been expressed for carbon nanotubes due to their high aspect ratios (Pacurari et al. 2010). The SWCNT has especially high and C_{60} fullerene especially low alveolar macrophage cytotoxicity with MWCNT being intermediate in its toxicity (Jia et al. 2005). As an ecotoxicological example, SWCNT acted as respiratory toxicants of rainbow trout (*Oncorhynchus mykiss*), causing oxidative damage and osmoregulatory changes (Smith et al. 2007). Gold and silver nanoparticles can be a source of dissolved metals and cause oxidative stress as discussed in Chapter 6. Indeed, one of the major applications of silver nanoparticles is as a bactericide incorporated into various products. Gold nanoparticles are more inert than those of silver. Metal ions can also be released from quantum dots once they enter the environment. Ju-Nam and Lead (2008) also postulate that they could cause oxidative stress (see Chapter 6) by transferring energy to nearby oxygen atoms.

The movement and phase association of nanoparticles after release into the environment depend on their physical and chemical features. The same is true regarding their fate on entering an organism. As mentioned, dissolution of some can result in harmful concentrations of dissolved metals. A human health example is **metal fume fever**, a disease caused by breathing in metal-rich fumes that are produced during smelting, metal casting, welding, and similar activities. Metals inhaled deep into the lungs are released from the metal-rich particles, often zinc oxide, and gain entry to the bloodstream. Another example is the silver ion released from silver nanoparticles (Figure 2.34) that can be toxic to aquatic or soil organisms. Metal oxides can have negative surface charges and, as a

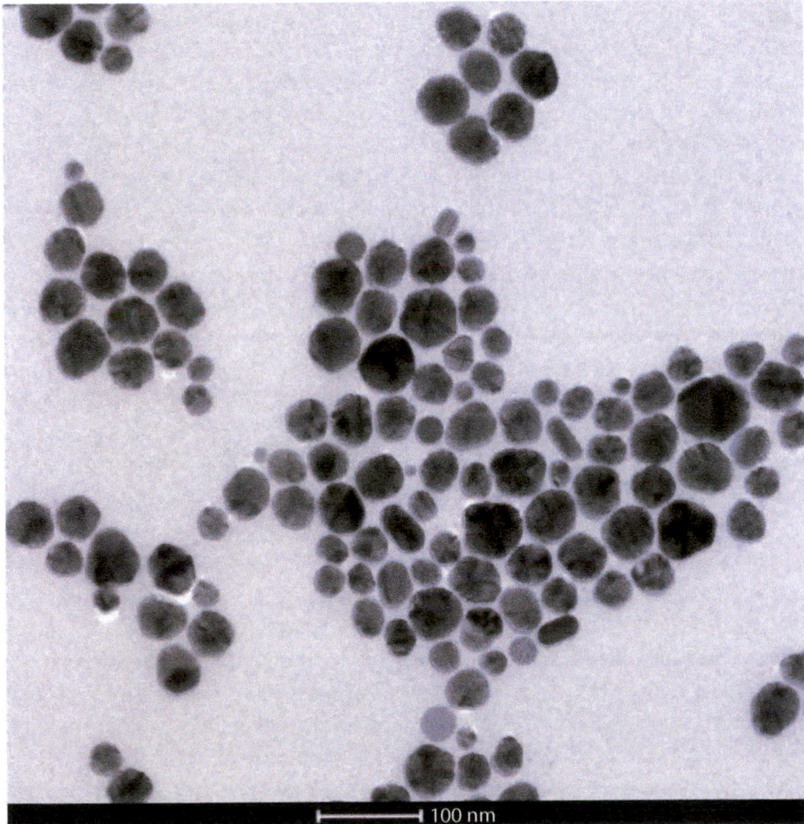

Figure 2.34 Silver nanoparticles that are increasingly used as antimicrobial agents in medical products. Yang et al. (2012) exposed the soil nematode, *Caenorhabditis elegans*, to these particles, finding that oxidative dissolution of silver contributed to nanoparticle toxicity. (Transmission electron micrograph courtesy of S. Marinakos, Duke University, Durham, NC.)

consequence, can become associated with other ions and coated with natural dissolved organic materials. Aggregation dynamics can be important for nanoparticles such as carbon nanotubes that tend to be hydrophobic and form aggregates that influence their movement in aquatic systems. Many can associate with surfaces by adsorption. The dynamics of colloidal aggregates into which nanoparticles have become incorporated then determines movement within the environment. Physical association with other molecules can influence nanoparticle movement into and their effects on organisms. Once inside an organism, a nanoparticle can become surrounded by proteins (and other macromolecules) that form a **protein corona**. The corona influences nanoparticle movement and possible effects inside the organism. The nature of the corona depends on nanoparticle size and surface properties (Lundqvist et al. 2008). Some elements of the corona will change as the nanoparticle moves from the circulatory system into the cell cytoplasm and between subcellular compartments (Lynch and Dawson, 2008). Lundqvist et al. (2011) envision a corona hard core that re-equilibrates with its surroundings as the nanoparticle complex moves among biological compartments.

2.2.6 Thermal Pollution

Thermal pollution is any shift in environmental thermal conditions of sufficient magnitude to adversely impact human or ecological well-being. The scale of thermal pollution can be local as in the case of a discharge from a power production plant to global as in the case of global warming. Localized thermal pollution of the atmosphere and hydrosphere has a long history of study. More recently the scale of consideration for thermal pollution has expanded to encompass continents and even the entire biosphere due to global climate change.

The nineteenth century British meteorologist, Luke Howard (Howard, 1820), was the first to document the influence of cities on regional air temperature. This phenomenon was later labeled the **urban heat island effect**. He noted that temperatures in cities are typically higher than those of the surrounding rural/suburban region because the structures and materials comprising the city more effectively capture heat than do those of the surrounding countryside. An epiphenomenal advective cell can be created when the warmer air rises above the city, cools, and then eventually returns to the land surface where it is drawn back into the city again. Pollutants such as particulates tend to be retained in urban areas by this circulation cell. Elevated thermal conditions have been monitored around large cities such as New York City (Bornstein, 1968), cities of Quebec (Oke, 1973), and more recently China's large megacities (Wang et al. 1990; Li et al. 2004). Local warming of the urban atmosphere can also contribute to thermal pollution in water bodies receiving runoff from paved surfaces (Herb et al. 2008).

Thermal pollution in the hydrosphere has a long history too (Hoak, 1961; Davidson and Bradshaw, 1967); however, most concerns emerged in the 1940s when the number of U.S. fossil fuel power plants doubled with a corresponding increase in discharge of large amounts of heated waters into water bodies (Davidson and Bradshaw, 1967; Walker, 1989). Similar thermal discharges occurred from nuclear power and production reactors (Figure 2.35). The elevated temperatures shift chemical equilibria (e.g., concentrations of chemical components of the carbonate buffering system), concentrations of dissolved gases (e.g., the dissolved oxygen concentrations noted in Figure 2.35), and biogeochemical transformations (e.g., nitrogen transformation). They also can adversely impact the bioenergetics of individual organisms (e.g., body condition of bass [Esch and Hazen, 1980]), relationships between hosts and their pathogens (e.g., the pathogenic amoebae, *Naegleria fowleri* [Griffin, 1972; Marciano-Cabral, 1988] and parasites (e.g., parasites of bream, *Abramis brama* [Pojmańska and Dzika, 1987]), and also relationships among members of aquatic communities (e.g., the shift from diatom to green to blue-green algal dominants as temperatures increase in receiving waters [Cairns, 1971]). Reservoir management can include discharge of cold hypolimnetic waters from the bottom of reservoirs into the water body below the dam, causing harm to aquatic biota. Finally, warm and cold plumes injected into water bodies can disrupt or block fish migration patterns.

Global warming is also occurring and has drawn much deserved attention recently. Most focus is on greenhouse gas effects on global climate change; however, Nordell (2003) noted that direct heat dissipation from fossil fuel use has also contributed to global temperature rise. Regardless, global climate change continues to modify ecosystems worldwide. Massive changes are occurring such as the loss of global ice-sheet mass summarized as follows (from Shepherd et al. [2012]):

Ice-sheet	Change from 1992 to 2011(±standard deviation) (gigatonnes/year)
Greenland	$^{-}152 \pm 49$
East Antarctic	$^{+}14 \pm 43$
West Antarctic	$^{-}65 \pm 26$
Antarctic Peninsula	$^{-}20 \pm 14$
Net change per year	$^{-}223$

Reactor state
Hot Cold
Temperature (°C)
61 16
(30–68) (8–24)
Dissolved O_2 (mg L^{-1})
3.9 9.3
(2.7–6.7) (7.1–11.5)
Temperature (°C)
52 17
(31–60) (7–28)
Dissolved O_2 (mg L^{-1})
4.4 9.0
(3.3–6.7) (7.0–11.0)
Temperature (°C)
39 20
(25–46) (8–25)
Dissolved O_2 (mg L^{-1})
6.7 9.5
(5.5–8.7) (6.4–12.3)

Figure 2.35 Thermal pollution associated with a nuclear production reactor (K-Reactor on Pen Branch, U.S. Department of Energy Savannah River Site, South Carolina, 1985–1986). Arithmetic means and ranges of water temperature (°C) and dissolved oxygen concentration (mg $O_2 \cdot L^{-1}$) are provided for times when the reactor was active (Hot) and inactive (Cold) (Newman, 1986). Due to pump engineering issues, water flow was maintained through the system regardless of reactor production status. Photographs and water quality information are those from approximately 2 km below the K-reactor discharge into Indian Grave Creek (top), 9 km below the discharge (middle), and 12 km below the reactor discharge where the discharged waters entered the extensively damaged 40.5 km^2 bald cypress-tupelo river swamp adjoining the Savannah River (bottom). (Photograph by M. C. Newman.)

2.3 SUMMARY

This chapter provides a broad-brush description of the important environmental contaminants, including their natures, behaviors in ecological systems, and underlying qualities. They include inorganic and organic contaminants, radiation, nanoparticles, and thermal pollution. Discussions of the organic contaminants included emerging contaminants that have specific biological activities such as modification of endocrine functions. The different kinds of and human influence on environmental radiation were explored. The new issue of nanoparticles as contaminants and old topic of thermal pollution were sketched out with the intention of adding more detail in later chapters. Also at the end of the chapter, biological macromolecules with deliberate specific biological functions—genes—were explored using GM crop species as examples.

2.4 BACKGROUND CHEMISTRY CONCEPTS AND DEFINITIONS

Aliphatic Compounds: These nonaromatic compounds (i.e., containing no rings of six carbon atoms that share resonating electrons) are extraordinarily diverse. They can incorporate single, double, or triple bonds, or branches in their structures. As examples, all the compounds in Figures 2.6, 2.7, 2.15, 2.28, and 2.29 are aliphatic compounds. So are lindane (Figure 2.14), malathion (Figure 2.16), pyrethrin, and allethrin (Figure 2.18). Figures 2.25 and 2.26 show large organic compounds with aliphatic and aromatic regions. More involved structures are possible including heterocyclic structures (i.e., those with more than one kind of atom in the cyclic structure such as diazinon in Figure 2.16 and atrazine in Figure 2.25).

Aromatic Compounds: These compounds have one (or more) six carbon ring(s) (C_6H_6) in which high stability is imparted because carbons bonds of the ring are intermediate between single and double bonds. The involved electrons are in resonance: they move amongst the six carbons to produce this state. This is depicted as shown below with alternating single and double bonds or with a circle inside a six carbon ring of single bonds.

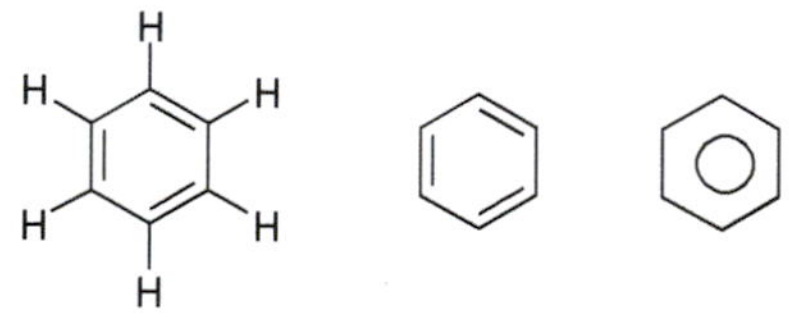

Many substitutions can, and do, occur in this ring such as the addition of –OH to produce a phenol, –Cl to produce chlorobenzene, or $–NO_2$ to produce nitrobenzene.

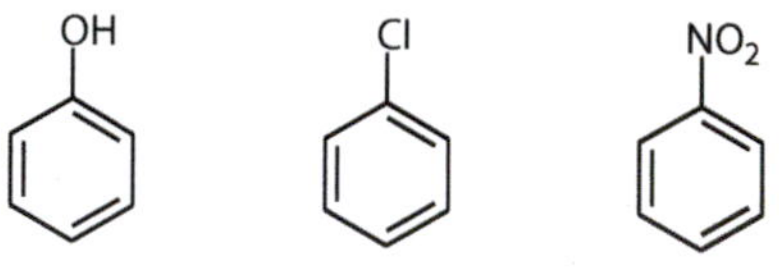

All of the compounds in Figures 2.8 through 2.13 are aromatic compounds.

Binding Trends: The ionic bond stability of a metal with a hard ligand is estimated with the metal ion charge (Z) and radius (r): polarizing power = Z^2/r. The covalent bond stability can be quantified by $\Delta\beta$. It is calculated with the stability constants (β_{MX}) of the metal fluoride and chloride complexes: $\Delta\beta = \log \beta_{MF} - \log \beta_{MCl}$. Different binding trends are used to categorize metals, e.g., Class A, B, or intermediate. A metal's categorization into one or another HSAB-based class can vary among publications because their Z^2/r and $\Delta\beta$ are continuous, not discrete categorical variables.

Bonds: Bonds can be classified generally as covalent or ionic depending on how equitably electrons are shared by the interacting atoms. An electron pair can become so predominately associated with one of the atoms that only electrostatic forces hold the charged ions together after transfer of electrons. The resulting bond is ionic. An atom that donates an electron pair during interaction is called a ligand or donor atom whereas the atom accepting the electron pair is called the receptor atom. The electron pairs are shared to differing degrees with covalent bonds. A nonpolar covalent bond results if the electrons are shared equitably between the two atoms. The covalent bond becomes increasingly polar as the density of the electron cloud becomes more and more asymmetric between the two interacting atoms.

Electronegativity: Electronegativity, the atom's attraction for electrons during reaction with another atom, dictates many qualities of an atom during chemical reaction such as bond ionic/covalent nature, strength, or length. Very electronegative elements such as fluorine or chlorine have a strong tendency to become negative ions. The very electropositive group 1A metals have the opposite tendency of readily forming positive ions. Various measures of electronegativity exist such as the Mulliken, Pauling, Allred-Rochow coefficients (Barrett, 2002). Often, an **electronegativity coefficient** is designated as χ. The value of this coefficient reflects the atom's tendency (in a molecule) to attract electrons during chemical reaction. As examples, fluorine and cesium have very different Pauling electronegativities of 4.0 and 0.7, respectively, whereas carbon (χ = 2.5) and nitrogen (χ = 3.0) have similar electronegativities. The relative electronegativities of two reacting atoms determine such important bond qualities as the degree to which the bond displays ionic versus covalent characteristics. The greater the difference in electronegativities, the greater the ionic nature of the resulting bond. The predominant nature of chemical bonding between cesium and fluorine would be ionic but that between carbon and nitrogen would be covalent.

Hard Soft Acid Base Theory (HSAB): HSAB theory provides a general scheme for quantifying metal binding tendencies. The hard-soft label has to do with the propensity for the metal's outer electron shell to deform (polarize) during interactions with ligands. The Class A metals in the periodic table (1A, 2A, 3A) tend to be hard relative to the softer Class B metals (in periods from Mo to W, and Sb to Bi) (Jones and Vaughn, 1978; Fraústo da Silva and Williams, 1991). Hard metals are not as polarizable as soft metals. Other metals are intermediate to the soft and hard metals. Acid–base refers to the Lewis acid (accepting an electron pair during interaction) or base (donating an electron pair during interaction) context for predicting the nature and stability of the metal interaction with ligands. The atom accepting the electron pair is called an electrophilic atom, whereas that donating the electron pair is called the nucleophilic, donor, or ligand atom. Ahrland (1968) has a fuller discussion of this theme.

Ionization: The ionizable group of an organic compound might be completely, partially, or sparingly ionized depending on the solution pH and also the group's dissociation constant (K_a), e.g., acetic acid:

$$CH_3CO_2H \leftrightarrow CH_3CO_2^- + H^+$$

$$K_a = \frac{[H^+][CH_3CO_2^-]}{[CH_3CO_2H]} = 1.75 \times 10^{-5}$$

The pK_a is the negative of the $\log_{10}$ of the ionization constant (K_a): $-\log_{10}(K_a)$. The pH and pK_a can be entered into the Henderson–Hasselbalch relationship to estimate the fraction of the compound that is present in ionized form. In the case of acetic acid, log ([$CH_3CO_2^-$]/[CH_3CO_2H]) = pH–pK_a.

Ligand: Ligand is an anion or molecule that forms a coordination compound or complex with metals. A complex is simply the combined metal–ligand and can be charged ($FeCl^+$) or neutral

($Cd(CN)_2$). Some ligands, especially organic ligands, can be quite complex with several of their atoms sharing different electron pairs. Such multidentate ligands are also called chelates, chelating agents, or chelators. Natural dissolved organic compounds can act as chelators. Like metals, ligands (Lewis bases) can be classified as hard or soft depending on how readily the donor atom's outer electron cloud distorts during interaction with a metal (Lewis acid).

Lipid Solubility: Lipophilicity of a compound is quantified most often using a coefficient (K_{ow} or P) for its partitioning between a lipid (often n-octanol) and water phase. The compound's equilibrium concentrations in the n-octanol ($[X]_o$) and aqueous ($[X]_w$) phases are used to determine the K_{ow}:

$$K_{ow} = \frac{[X]_o}{[X]_w}$$

Often, due to the magnitude of the K_{ow} values, the logarithm of the K_{ow} is the preferred metric of compound lipid solubility.

Metals: Distinction is sometimes made between heavy and light metals. By convention, the term **heavy metal** is reserved for metals with specific gravities of five or greater (Nieboer and Richardson, 1980), but the literature contains many deviations from this definition. Light metals are defined as those in groups 1A and 2A; heavy metals or transition metals are those in groups 3B–8B, 1B, and 2B. (The lanthanide and actinide metals are broken into separate series within 3B.) In general, heavy or transition metals tend to form various complexes with other elements with predictable trends as described below. This classification of metals has substantial shortcomings (Nieboer and Richardson, 1980; Duffus, 2002) and will not be discussed further in this textbook as a consequence.

Nuclides: A nuclide is a nucleus with a specified number of neutrons and protons (and amount of energy). The number of protons and neutrons in a nuclide is designated by a superscripted number in front of its symbol (^{15}N) or a number separated from its name by a hyphen (nitrogen–15). The electrons around the nucleus might change, e.g., ^{131}I and $^{131}I^-$, but the nuclide remains the same. Some nuclides are stable but others are unstable. The natural isotopes of nitrogen, ^{14}N and ^{15}N, are stable nuclides. (**Isotopes** are atoms with the same number of protons but different numbers of neutrons.) Other nuclides (radionuclides or radioisotopes) such as ^{131}I have particular complements of neutrons and protons that are less stable than some other combination. Generally, a neutron/proton quotient of one for lighter nuclides or slightly less than one for heavier nuclides is the most stable state (Wang et al. 1975). Radionuclides emit particles and photons to reach a more stable state. Some do this quickly but others decay slowly, resulting in differences in radioactive decay rates. Some decay through a sequence of unstable or meta-stable nuclides (decay series) until the most stable nuclide is produced.

Oxyanion/Oxycation: Many elements combine with oxygen to form an oxide-containing, negative (oxyanion or oxoanion) or positive (oxycation or oxocation) ion. Figure 2.1 provides environmentally important examples of metals, metalloids, and nonmetals that can form oxyanions. Examples of oxycations are FeO^{2+}, MoO_2^{2+}, UO_2^{2+}, and VO^{2+}.

Partitioning: A charged or uncharged chemical substance will preferentially associate with different environmental phases (partition) depending on its properties. The water molecules surrounding a simple, charged metal ion will form a hydration sphere, enhancing its tendency to remain in the aqueous phase and not to associate with a relatively nonpolar organic phase such as the lipids in cell membranes. In contrast, a nonpolar organic contaminant will tend to partition more into a lipid phase than an aqueous phase. The relative tendency of a substance to associate with one phase or another eventually produces an equilibrium distribution of concentrations in the available phases.

The distribution or **partition coefficient** (or its logarithm) is used to quantify partitioning between phases, e.g., $K_d = [X_{\text{organic phase}}]/[X_{\text{aqueous phase}}]$.

The partitioning involving a gaseous phase is handled similarly. **Henry's Law** states that the compound's concentration in the aqueous phase (C_c) will be proportional to its partial pressure in the gaseous phase (P_c).

$$K = \frac{C_c}{P_c}$$

The assumption is that (practical) equilibrium exists for a dilute solution of the compound. Another metric similar to the K_{ow} (see Section 2.4, "Lipid Solubility") exists if the focus is on air and a lipid phase such as that associated with a biological membrane. The K_{oa}, a measure of lipophilicity, is the partition coefficient for a compound between n-octanol and overlaying air.

Polar Bond: A covalent bond can be a polar bond if the electron cloud associated with the bound atoms is denser around one atom than the other. The polarity is depicted as partial charges in the chemical structure as shown for water below:

δ+ δ+
H H
O
δ−

Polarity will increase as the electronegativity of the paired atoms becomes increasingly different. The electronegativities of C (2.5), Br (2.8), and H (2.1) are similar, so the polarity of their bonds will be low. However, those of F (4.0), O (3.5), Cl (3.0), and N (3.0) are less similar to those of C and H. Extending this simple discussion of polarity to more complicated molecules, regions of complex organic molecules can have differences in partial charges due to other atoms in their vicinities. A contrasting pair of compounds with which this can be illustrated is mirex and kepone (Figure 2.15). The kepone molecule is more polar than mirex because of the O atom. The polarity of a molecule is expressed with its dipole moment. The water molecule shown previously has a dipole moment of 1.85; the nonpolar, carbon tetrachloride (CCl_4) molecule has a dipole moment of 0. Generally, compound lipophilicity decreases as the dipole moment increases (Boethling and Mackay, 2000).

Radioactive Decay Series: There are three natural and one artificial series of radionuclides in which one (parent) radionuclide decays to another (progeny) in a series of steps until the most stable element is formed. The series include the uranium–radium, thorium, actinium, and neptunium series (Wang et al. 1975; Considine, 1995). The neptunium series is artificial, and as such, is irrelevant here. Some radionuclides in a series might be short lived such as ^{222}Rn (half-life of 3.82 days) in the uranium–radium decay series, but others might be very long lived such as ^{238}U in that same series (4.51×10^9 years). The uranium–radium series begins with ^{238}U, decays through 14 radionuclides, and ends with the stable nuclide, ^{206}Pb. It generates gaseous ^{222}Rn in the process, which can create a cancer risk in certain parts of the world. The thorium and actinium series end with the stable nuclides, ^{208}Pb and ^{207}Pb, respectively. (As an aside, the relative amounts of stable, radiogenic lead isotopes from these decay series and the nonradiogenic ^{204}Pb are often used to trace the origins of lead because their amounts depend on a mining area's geology, e.g., Newman et al. (1994). Often they are used to date sediments that have accumulated lead in the past from natural and leaded gasoline sources.)

SUGGESTED READINGS

Day, T. A. and P. J. Neale, Effects of UV-B radiation on terrestrial and aquatic primary producers. *Annu. Rev. Ecol. Syst.*, 33, 371–396, 2002.

Harley, N. H., Chapter 25, Health effects of radiation and radioactive materials, in *Casarett & Doull's Toxicology: The Basic Science of Poisons,* 7th Edition, Klaassen, C. D., Ed., McGraw Hill, New York, NY, 2008.

Körner, C., Biosphere responses to CO_2 enrichment. *Ecol. Appl.,* 10, 1590–1619, 2000.

Manahan, S. E., *Fundamentals of Environmental Chemistry*, Taylor & Francis/CRC Press, Boca Raton, FL, 2000.

Newman, M. C. and W. H. Clements, *Ecotoxicology: A Comprehensive Treatment*, Taylor & Francis/CRC Press, Boca Raton, FL, 2008.

Wania, F. and D. Mackay, Tracking the distribution of persistent organic pollutants. *Environ. Sci. Technol.,* 30, 390A–396A, 1996.

CHAPTER 3

Uptake, Biotransformation, Detoxification, Elimination, and Accumulation

Models are, for the most part, caricatures of reality, but if they are good, then, like good caricatures, they portray, though perhaps in distorted manner, some of the features of the real world.

Kac (1969)

3.1 INTRODUCTION

It is important to understand and to be able to predict the accumulation of contaminants in biota because contaminant effects are a consequence of concentrations in target organs or tissues. Also, because human exposure often occurs through the consumption of tainted food, the ability to predict the amount of contaminant accumulating in these potential vectors of exposure is crucial to avoiding human poisonings. The basic processes of uptake, biotransformation, and elimination that result in bioaccumulation will be discussed in this chapter. Chapters 4 and 5 will then add detail within this framework.

Bioaccumulation is the net accumulation of a contaminant in, and some special cases on, an organism from all sources including water, air, and solid phases in the environment. Solid phases might include food, soil, sediment, or fine particles suspended in air or water. By convention in ecotoxicology, bioaccumulation is not the same as the more restricted term, **bioconcentration** that has come to mean the net accumulation of a contaminant in an organism from water only. These two terms and their distinctions arise from an earlier, somewhat-dated debate in aquatic toxicology about the relative importance of water and food sources of contaminants to aquatic organisms.

The study of bioaccumulation has always relied heavily on mathematical models. This has greatly enhanced progress: Understanding of these tools has become essential to prediction of bioaccumulation and consequent effects. However, as expressed in the quote above, it is important to understand that these models are only useful caricatures and understanding mathematical models does not constitute a sufficient and accurate understanding of the associated phenomena.

Figure 3.1 depicts the common rendering of bioaccumulation processes to a simple mathematical model. As with any mathematical model, the only incorporated details are those essential to understanding and predicting the general behavior of the system with sufficient accuracy. More details might be needed if the model produces inaccurate predictions. At the top of this figure, a fish is exposed to a contaminant through the water passing over its gills and the food passing through its gut. Dermal exposure might also be important depending on the contaminant qualities

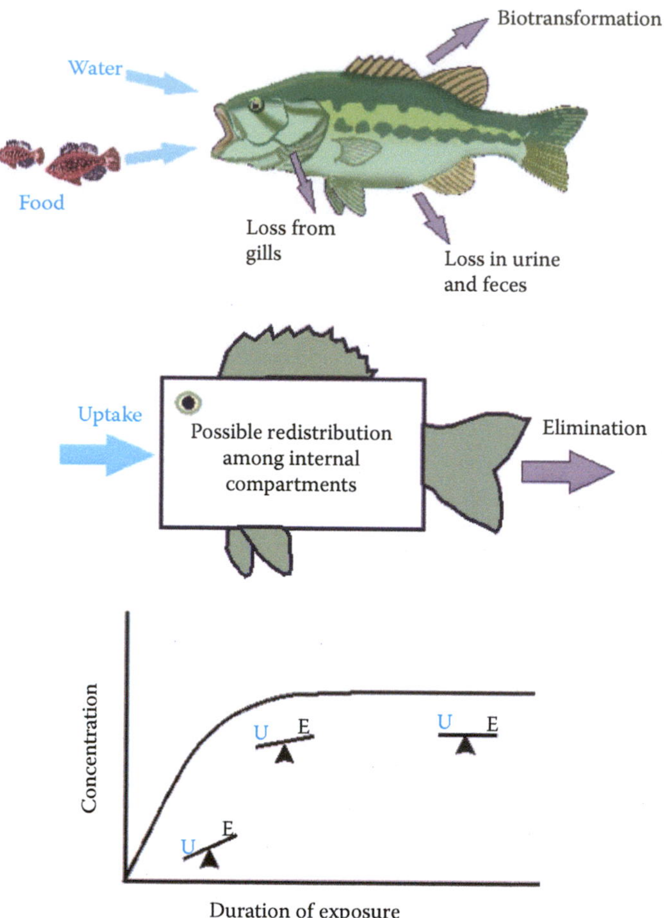

Figure 3.1 A simplified conceptualization of bioaccumulation. At the top of this figure, the fish (largemouth bass) is thought to potentially take in a substance from its food and water, and lose it through the gills, urine, and feces. There might be internal redistribution or biotransformation of the contaminant. This process is rendered to a simple box and arrow diagram (middle graph). Here, only uptake from water is assumed to be significant, and all elimination processes are described by one elimination component. The most common mathematical description of this model predicts a gradual increase in contaminant in the fish until a steady-state concentration is obtained as depicted in the graph at the bottom of this figure. (U, uptake; E, elimination.)

(Landrum et al. 1996) and the surface-to-volume ratio of the organism (Barron, 1995). In this example of a largemouth bass, only one important source of constant concentration is assumed for the sake of numerical expediency (center panel). All other sources are considered insignificant or adequately reflected in the one uptake coefficient. Some contaminant might be lost from the gills or pass over the gills without being taken up. It could also be excreted in the urine or eliminated in the feces after passage through the gut or entering the gut through biliary excretion. The contaminant might be transformed and redistributed to various compartments within the fish, e.g., from the gill into the blood plasma and then to the kidney. In this example, only one elimination coefficient is included. Multiple processes are mathematically represented in this one coefficient or some are assumed trivial relative to a single, dominating elimination process. Also, no biotransformation or effects to the organism are prominent here. Kinetics and rate coefficients are assumed to remain constant over the time of accumulation. These details may be rendered to a box and arrow model (center of figure) with associated simplifying assumptions regarding the mathematics of uptake, elimination, and internal compartment exchanges. In this example, uptake and elimination take place for one compartment in which the contaminant is mixed homogeneously and instantly. (The practical validity of these assumptions and simplifications should be tested during model development.) Mathematical expression of this simplest of models predicts a time course for bioaccumulation such as that shown at the bottom of Figure 3.1. This curve of a gradual, monotonic increase to a maximum concentration is a result of the change in relative influences of uptake (U) and elimination (E) processes on the change in internal concentration during the course of exposure. At the beginning of the exposure, little contaminant is contained in the fish and available for elimination: uptake dominates relative to elimination in the initial dynamics, and internal concentrations increase. But, as more and more contaminant accumulates in the fish, more contaminant becomes available within the fish for elimination. Elimination becomes increasingly important and the rate of increase in internal concentration begins to decline. Eventually, a balance between uptake and elimination results in a steady-state* concentration in the fish that will be maintained as long as conditions remain constant.

Although described here in terms of concentration (amount of contaminant per amount of organism such as 25 μg of lead per g of tissue), such models are also developed in terms of **body burden**, the mass or amount of contaminant in the individual (e.g., 2500 μg of lead per individual). The details and mathematical expression of this general model and models derived from it will be developed primarily in terms of concentration in this chapter.

3.2 UPTAKE

3.2.1 Introduction

Uptake (the movement of a contaminant into an organism) can occur by several mechanisms and involve the dermis, gills, pulmonary surfaces, or the gut. In all cases, the process begins with interactions with cell surfaces in tissues. Simkiss (1996) categorized uptake by a cell into three general routes: (1) lipid, (2) aqueous, and (3) endocytotic routes. The lipid route encompasses the passage of lipophilic contaminants through the bilayer of membrane lipids. Some small, uncharged polar molecules such as CO_2, glycerol, and H_2O might also diffuse to varying degrees through the lipid bilayer (Alberts et al. 1983). The aqueous route uses two general types of **membrane transport proteins** that either form channels (**channel proteins**) or act as **carrier proteins** in the membrane, which transfer hydrophilic contaminants into cells. Some channels (**porins**) are nonspecific; others are quite specific relative to the substances passing through them. A porin is composed of linked β-strand proteins that combine to form a structure with nonpolar components of amino acids facing outward to the lipid membrane and polar components facing into the porin core. Porin channels have internal features (eyelets) that can dictate the size of solutes that pass through them. A notable porin is the **amiloride-blockable sodium channel** responsible for the sodium readsorption that takes place within distal nephrons of the vertebrate kidney. Sodium in the tubule lumen diffuses passively through these sodium channels on the apical surface of nephron cells whereupon an Na^+, K^+-ATPase on the basolateral surface then actively pumps the sodium from the cell into the

* Landrum and Lydy (1991) comment that the terms steady state and equilibrium are often used incorrectly as synonyms. Steady state refers to a constant concentration in an organism resulting from processes (e.g., uptake, elimination, and internal exchange among compartments) including those requiring energy. However, equilibrium concentrations resulting from chemical equilibrium processes do not require energy to be maintained. Steady-state concentrations resulting from bioaccumulation can be considerably higher than those predicted for chemical equilibrium.

serosal compartment (Eaton et al. 1995).* The functioning of channels such as this sodium channel can be influenced by the presence of other chemical substances including other contaminants (Simkiss, 1996). In contrast to porins, carrier proteins such as the Na^+, K^+-ATPase just described, facilitate the transport of a specific ion or molecule, or specific class of molecule across the membrane. **Calmodulin**, a membrane-associated messenger protein that functions to transport calcium, is another good example of a carrier protein. Calmodulin is involved in many processes including inflammation and apoptosis, which will be described in later chapters.

Possible mechanisms for uptake include adsorption, passive diffusion, active transport, facilitated diffusion or transport, exchange diffusion, and endocytosis (Newman, 1995). Several of these processes can be important for any particular combination of contaminant, exposure scenario, and species.

However, before a contaminant can be taken up, it must first interact with some surface of the organism. Movement onto the organism can involve and be modeled as adsorption. **Adsorption** is the accumulation of a substance at the common boundary of two phases such as that between a liquid solution and solid. An adsorption example might be metal ion exchange with a hydrogen ion associated with a ligand on the surface of an insect's integument. The more general term **sorption** will be used instead of adsorption if the specific mechanism by which a compound in solution becomes associated with a solid surface is unknown or poorly defined.

Two equations are widely applied to quantitatively define adsorption, the **Freundlich** and **Langmiur isotherm equations**. The Freundlich equation (Equation 3.1) is an empirical relationship and the Langmuir equation (Equation 3.2) is a theoretically derived relationship

$$\frac{X}{M} = KC^{\frac{1}{n}} \tag{3.1}$$

where X = amount adsorbed, M = the mass of adsorbent, K = a derived constant, C = the concentration of solute in the solution after adsorption is complete, and n = a derived constant, and

$$\frac{X}{M} = \frac{abC}{1+bC} \tag{3.2}$$

where X, M, and C are as defined above, a = the adsorption maximum (amount), b = affinity parameter reflecting bond strength. A plot of C (abscissa or x axis) versus X/M (ordinate or y axis) will result in a curve shaped like the one at the bottom of Figure 3.1. Such relationships can be linearized with equations such as Equation 3.3 (Freundlich isotherm) or Equation 3.4 (Langmiur isotherm) to facilitate data fitting by linear regression:

$$\log\frac{X}{M} = \log K + \frac{\log C}{n} \tag{3.3}$$

* A Na^+, K^+-ATPase or sodium–potassium pump exchanges K^+ from outside the cell for Na^+ inside the cell, expending an ATP in the process: it facilitates active transport. It is classified as an **antiporter** because it is a protein integrated into a membrane that transports one (or more) molecule/ion across the membrane in exchange for another molecule/ion. Three Na^+ leave the cell as two K^+ enter. It contrasts with a **symporter** or cotransporter that transports two or more molecules/ions together in the same direction across a membrane. Returning to the renal tubules for a symporter example, the Na^+/K^+/$2Cl^-$ symporter active in the loop of Henle is important in ion regulation. The K^+/I^+ symporter protein mentioned in Chapter 2 is another. A **uniporter** is a protein integrated into the membrane that facilitates transport of a single molecule/ion across the membrane without any exchange as in the case of an antiporter. **Valinomycin**, a carrier protein that transports K^+ across membranes by passive diffusion, is a uniporter.

$$\frac{C}{X/M} = \frac{1}{ab} + \frac{1}{a}C \tag{3.4}$$

Adsorption theory and these equations have been used successfully to define toxicant movement onto diverse biological surfaces such as those of unicellular algae (Crist et al. 1988), fish gills (Pagenkopf, 1983; Janes and Playle, 1995), periphyton (Newman and McIntosh, 1989), and zooplankton (Ellgehausen et al. 1980). Crist et al. (1988) found that uptake of H^+ by algae involved rapid adsorption followed by a slow diffusion into the cell. Langmuir equation constants *a* and *b* (Equation 3.2) were also used by Crist and coworkers to define relative metal interactions with algal cell surfaces.

Diffusion is the movement of a contaminant down a chemical or electrochemical gradient.* It might be simple diffusion of a charged ion through a channel protein or passage of a lipophilic molecule through the lipid route (Figure 3.2). Simple diffusion does not require expenditure of energy. If diffusion involves a channel protein, passage through a channel is influenced by ion charge and size, and features of the hydration sphere around an ion. Channels can be gated and respond to various chemical or electrical conditions. Diffusion could also be facilitated by a carrier

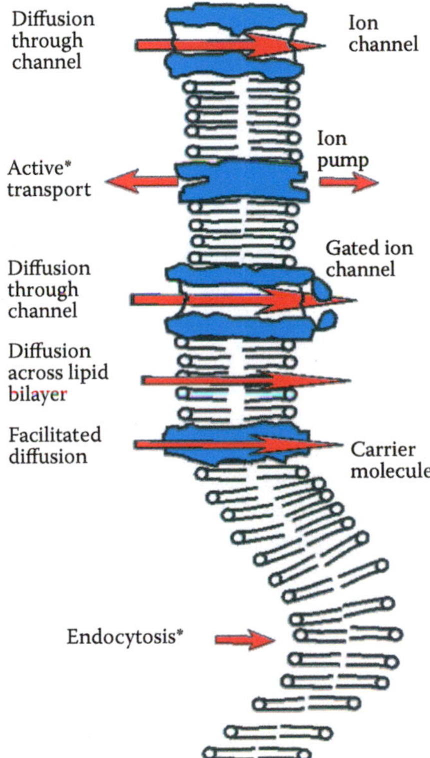

Figure 3.2 Mechanisms of uptake of contaminants into cells. Simple diffusion can occur across the lipid bilayer or through an ion channel formed by a channel protein. Channels may be gated and their functioning influenced by chemical and electrical conditions. Facilitated diffusion occurs through a carrier protein. Active transport moves the solute up an electrochemical gradient. Here, the Na^+, K^+-ATPase pump is illustrated. Potassium is pumped into the cell as sodium is pumped out. The last mechanism for cellular uptake is endocytosis. As indicated by an asterisk, endocytosis and active transport require energy.

* The term electrochemical gradient can refer to a concentration, activity, or electrical gradient.

protein. **Facilitated diffusion** occurs down an electrochemical gradient, requires a carrier protein, does not require energy, but is faster than predicted for simple diffusion. Because a carrier protein is involved, facilitated diffusion can be subject to saturation kinetics and competitive inhibition. As already mentioned, some facilitated diffusion involves the exchange of ions across a membrane (**exchange diffusion**). Diffusion accurately describes uptake of many nonpolar organic compounds or uncharged inorganic molecules such as ammonia (NH_3) (Fromm and Gillette, 1968; Thurston et al. 1981) or $HgCl_2^0$ (Simkiss, 1983) by the lipid route (Spacie and Hamelink, 1985; Barber et al. 1988; Erickson and McKim, 1990), and many charged moieties such as dissolved metals and protons (H^+) through the aqueous route. Facilitated diffusion is expected for some metals and organic contaminants (Landrum and Lydy, 1991). Diffusion back and forth across a membrane can be complicated if the chemical is changed after crossing that membrane. For example, pentachlorophenol can pass through a membrane into the blood stream but then be converted to the charged pentachlorophenate. This compound is much less capable of passing back across the membrane and out of the organism.

Diffusion is described most often with Fick's Law (Equation 3.5):

$$\frac{dS}{dt} = -DA\frac{dC}{dX} \tag{3.5}$$

where dS/dt = rate of contaminant movement across the surface, D = a diffusion coefficient, A = the surface area through which diffusion is occurring, and dC/dX = the concentration gradient across the boundary of interest, e.g., the difference in concentration between the two sides of a cell membrane. This equation is incorporated into many models of bioaccumulation.

Energy is required for the **active transport** of a contaminant up an electrochemical gradient. Because it involves a carrier molecule, active transport might be subject to saturation kinetics and competitive inhibition. Perhaps, the best example of active transport is cation transport by membrane-bound ATPases. These ATPases, using energy from the hydrolysis of ATP, act as a coupled ion pump to simultaneously remove some ions from the cell (e.g., sodium cation) while moving others (e.g., potassium cation) into the cell. Radiocesium, a chemical analog of potassium, can be taken up by this mechanism (Newman, 1995). Some metals (e.g., cadmium taken up as an analog of calcium) and some large, hydrophilic compounds can also be subject to active transport (Cockerham and Shane, 1994).

Endocytosis (pinocytosis or phagocytosis) may be important in uptake, especially for contaminants entering an organism in food. Simkiss (1996) details an excellent example of contaminant metals being taken up by the transferrin route normally involved in iron assimilation. Iron and other metals are bond to the membrane-associated transferrin protein. The metal–transferrin complex moves to a specific region of the cell surface where it becomes engulfed and incorporated into a vesicle. In the cell, the vesicle fuses with a lysosome and the associated metal is released.

3.2.2 Reaction Order

The specific kinetics of uptake is often defined in terms of reaction order; therefore, a short review of reaction order concepts is required here. In the context of a reaction involving only one reactant, reaction order refers to the exponent (n) to which the reactant concentration (C) is raised in the equation describing the reaction rate, $dC/dt = kC^n$. In the case of a zero-order reaction, n is 0 and $dC/dt = kC^0$. This reduces to $dC/dt = k$ and k has units of $C \cdot h^{-1}$. The concentration of the product will increase independent of reactant concentration with zero-order kinetics. For first-order kinetics ($dC/dt = kC$), the rate will change with reactant concentration and the units for k are h^{-1}. First-order reaction kinetics is the most commonly observed and applied kinetics for bioaccumulation modeling. Higher-order reaction kinetics is occasionally warranted. However, also common are saturation kinetics such as Michaelis–Menten kinetics. Saturation kinetics is relevant with active or facilitated transport. Saturation kinetics also occurs for enzyme-mediated processes such as those associated with detoxification. Above a certain reactant concentration, a system is saturated and transport cannot precede any faster than a maximum velocity (V_{max}), i.e., pseudo-zero-order kinetics. However, the kinetics shifts to first order as reactant concentrations drop below saturation conditions. For example, Mayer (1976, as reported in Spacie and Hamelink (1985)), saw evidence of saturation for the elimination of di-2-ethylhexylphthalate from fathead minnows (*Pimephales promelas*). Differential Equation 3.6 describes the change in reactant concentration by Michaelis–Menten kinetics:

$$-\frac{dC}{dt}=\frac{V_{\max}C}{K_m+C} \tag{3.6}$$

where V_{max} = the maximum rate of reactant change (C·h^{-1}), and K_m = half-saturation constant. A plot of the velocity of concentration change (*V*, *y* axis) against concentration (*C*, *x* axis) would look very much like the plot at the bottom of Figure 3.1. The velocity of concentration change increases with reactant concentration only to a certain point. There would be a maximum velocity (V_{max}) that could not be exceeded regardless of any further increase in concentration. The concentration at which V was equal to $V_{max}/2$ would be the K_m in Equation 3.6. Again, although zero, first, or saturation kinetics are applied to uptake, first-order uptake is the most commonly applied in modeling.

3.3 BIOTRANSFORMATION AND DETOXIFICATION

3.3.1 General

Once a contaminant enters the organism, it becomes available for possible **biotransformation**, the biologically mediated transformation of a chemical compound to another. Most biotransformations involve enzymatic catalysis and, as a consequence, can be subject to saturation kinetics and competitive inhibition. Biotransformation can lead to enhanced elimination, detoxification, sequestration, redistribution, or activation. It can enhance the rate of loss from the organism as is often the case if a lipophilic xenobiotic is converted to a more reactive or hydrophilic compound, e.g., naphthalene oxidation to naphthalene diol. The contaminant might be rendered to a nontoxic form. Some contaminants may be transformed to a form that is retained within the organisms but is sequestered away from any site of possible adverse effect. With **activation**, the adverse effect of a contaminant is made even worse by biotransformation or an inactive compound is converted to one with an adverse bioactivity. As an example of activation, the organophosphorus pesticide, parathion,* undergoes oxidative desulfuration to form the very potent paraoxon (Hoffman et al. 1995). General processes associated with biological transformations (Figure 3.3) are described below.

3.3.2 Metals and Metalloids

Although the term biotransformation is used most appropriately for the transformation of organic compounds, metals and metalloids are subject to biological transformation of a sort, resulting in their elimination from or sequestration within the individual (Figure 3.3a). However, some

* Parathion is similar in structure to the methyl parathion shown in 2.16 except the two methyl (CH_3^-) groups attached to the oxygen atoms on the left of the methyl parathion structure are replaced by ethyl ($H_5C_2^-$) groups. Methyl parathion will also be biotransformed by oxidative desulfuration to methyl paraoxon. The paraoxon of either of these compounds simply has a double-bonded oxygen atom inserted in place of the double-bonded sulfur atom.

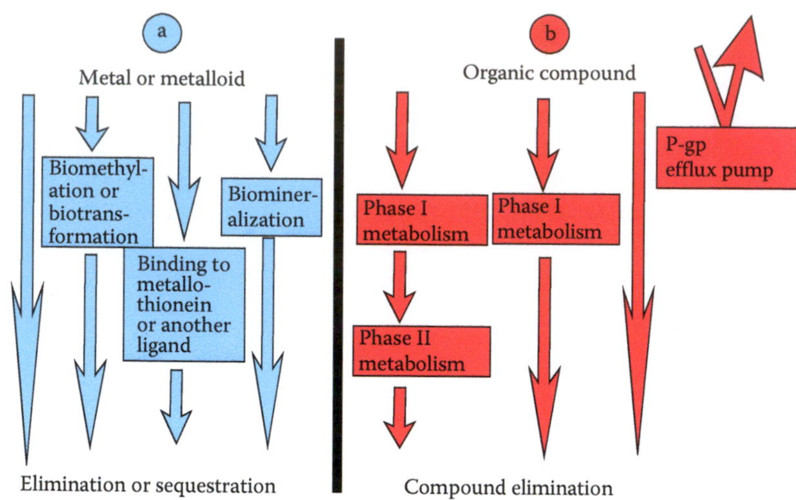

Figure 3.3 General mechanisms of biotransformation and detoxification of inorganic (a, blue) and organic (b, red) contaminants (see text for explanation).

inorganic ions can quickly bind to a plasma-associated ligand and become available for removal from the organism without any transformation. Lyon et al. (1984) demonstrated the role of biological ligand binding to metal ions in the blood of crayfish in determining the relative rates of such elimination for a series of metal ions.

Microbes genetically adapted to metal contaminated environments can have an enhanced ability to add methyl or ethyl groups to a metal ion as in the case of ionic mercury conversion to methylmercury (Wood and Wang, 1983). Trivalent (As^{3+}) and pentavalent (As^{5+}) arsenic entering plants or animals might be rendered less toxic by methylation to monomethylarsonic and dimethylarsinic acids (Nissen and Benson, 1982; Peoples, 1983). Arsenic may also be converted to an arsenosugar (i.e., trimethylarsonium lactate), arsenobetaine, or a phospholipid such as *O*-phosphatidyl trimethylarsoniumlactate (Cooney and Benson, 1980; Edmonds and Francesconi, 1981). Selenium entering a plant as selenite (SeO_3^{-2}) can be reduced and integrated into selenocysteine.* If incorporated into proteins, such selenoamino acids result in protein dysfunction, providing one mechanism for selenium toxicity (Brown and Shrift, 1982). In selenium tolerant plants, large amounts of the nonprotein amino acids (Se-methylselenocysteine, selenocystathionine, and others) are produced, suggesting that high rates of synthesis of these amino acids serve as a means of selenium detoxification (Brown and Shrift, 1982).

Metals may be bound and sequestered from sites of toxic action by metallothioneins and similarly functioning molecules. **Metallothioneins** are a class of relatively small (~7000 Da) proteins with approximately 25%–30% of their amino acids being sulfur-rich cysteine and possessing the capacity to bind six to seven metal atoms per molecule (Hamilton and Mehrle, 1986; Shugart, 1996) (Vignette 6.2). They are commonly induced by metals, including cadmium, copper, mercury, and zinc.† Silver, platinum, and lead also have been reported to induce

* Note that selenate (SeO_4^-) can be converted to selenite and then undergo these transformations. Selenate can behave as an analog of the oxyanion of its periodic table neighbor, sulfur, and be converted to selenite by sulfate-reducing enzymes (Brown and Shrift, 1982; Fraüsto da Silva and Williams, 1991).

† As pointed out by Simkiss and Taylor (1981), metallothioneins are involved primarily with the detoxification of Class B metals. (The HSAB classification of metals was covered in Chapter 2.) **Class B metal cations** (Nieboer and Richardson, 1980) have filled d orbitals of 10–12 electrons and low electronegativity. They form "*soft*" spheres readily deformed by adjacent ions. They easily form covalent bonds with donor atoms such as sulfur. Class B metals include Cu^+, Ag^+, Au^{+1}, Cd^{2+}, and Hg^{2+}. **Class A metal cations** (e.g., Li^+, Na^+, K^+, Mg^{2+}, Ca^{2+}, and Sr^{2+}) have inert gas electron configurations, high electronegativity, and hard spheres. Intermediate between Class A and B metals are **intermediate or borderline metal cations** such as Fe^{2+}, Co^{2+}, Ni^{2+}, Cu^{2+}, Mn^{2+}, and Fe^{2+} (Stumm and Morgan, 1981; Brezonik et al. 1991).

metallothioneins (Garvey, 1990). In addition to their role in essential metal homeostasis, they can be induced by elevated concentrations of toxic metals, bind these metals, and reduce the amount available to cause a toxic effect (Fraüsto da Silva and Williams, 1991; Knapen et al. 2007). Metallothioneins and metallothionein-like proteins are found in many vertebrates and invertebrates, and also have been reported in higher plants (Grill et al. 1985; Gonzalez-Mendoza et al. 2007). A class of metal-binding polypeptides, **phytochelatins,** are induced by metals in plants, including algae (Grill et al. 1985; Clemens et al. 2001; Pawlik-Showrońska et al. 2007; Seth et al. 2008). Grill et al. (1985) found that phytochelatins were induced in plants by the metal ions, Cd^2, Cu^2, Hg^2, Pb^2, and Zn^{2+}. They can also function in regulation and detoxification of metals. Surprisingly, Clemens et al. (2001) found a phytochelatin synthase gene expressed in the nematode, *Caenorhaditis elegans*, suggesting that phytochelatins might also function in metal homeostasis in some animals.*

Finally, metals and other cations may be sequestered or eliminated through biomineralization. Metals such as lead (e.g., Bercovitz and Laufer, 1992; Newman et al. 1994) and radionuclides such as radiostrontium or radium (Grosch, 1965) can be incorporated into relatively inert shell, calcareous exoskeleton, or bone. This could become a concern in human cases of prolonged bed rest or microgravity of space flight in which lead deposited in bones is released back into circulation as bone mass slowly decreases with time (Kondrashov et al. 2005; Smith et al. 2005) Some metals are incorporated into exoskeleton (including the gut lining) of soil invertebrates and lost on molting (Beeby, 1991). Other metals such as lead can be incorporated into molluscan shell (Beeby, 1991; Newman et al. 1994) and rendered unavailable for interaction at target sites. Radiostrontium from atmospheric fallout accumulates in bone along with calcium. Children whose bodies are actively building and reworking bone during development are particularly vulnerable to harm due to radiostrontium accumulation. The consequence of such sequestration in bone may be extensive damage, such as in the case of another bone-seeking radionuclide, radium. In 1925, an outbreak of radium poisoning was reported among young woman employed at brushing radium-laced paint onto watch faces.† They ingested radium while licking the paintbrush bristles to a fine point as they worked. The radium accumulated in bone causing extensive bone lesions or fatal anemia by damaging the bone marrow (Grosch, 1965).

Metals may be sequestered by incorporation into a variety of granules or concretions in addition to sequestration in structural tissues (Mason and Nott, 1981; Simkiss and Taylor, 1981; Pynnönen et al. 1987; Vijver et al. 2004). Such granules are usually associated with the midgut, digestive gland, hepatopancreas, Malpighian tubules, and kidneys of invertebrates (Roesijadi and Robinson, 1994). They are also found in other specialized cells of invertebrates, and connective tissues of vertebrates and invertebrates (Roesijadi and Robinson, 1994). Mercuric selenide granules (calculi) are also found to sequester mercury in tissues of cetaceans (Martoja and Viale, 1977; Frodello et al. 2000; Mackey et al. 2003).

Hopkin (1986) described four general categories of granules in invertebrates. Type A granules are intracellular granules with 0.2–3 μm in diameter built as concentric layers of calcium (and magnesium) pyrophosphate and an organic matrix of lipofuscin. Metal precipitation with phosphate is mediated by pyrophosphatase (Howard et al. 1981). These Type A granules are found in most invertebrate phyla (Roesijadi and Robinson, 1994). Type A granules accumulate Class A and some

* As will be discussed in Chapter 6, phytochelatins are now classified as a class of metallothioneins and the definition of metallothionein has been expanded to include peptides as well as proteins.

† Our collective ignorance of the effects of radionuclides such as radium was staggering at this time. Even more shocking an example is the intentional ingestion of radium in the patent medicine, Radithor. Its consumption was responsible for numerous painful ailments and unnecessary deaths. Its use was banned only after the fatal poisoning of a prominent New York socialite (Macklis, 1993). The reader interested in the history of industrial radium poisoning is directed to Clark (1997).

intermediate metals such as manganese and zinc. Type B granules are also intracellular granules of roughly the same size as Type A granules and have high concentrations of sulfur. Consequently, they have high concentrations of Class B and intermediate metals (copper, cadmium, mercury, silver, and zinc) that avidly bind to sulfur-containing groups. Type C granules are intracellular granules rich in the iron-containing products of ferritin. The last granule (Type D) is extracellular, can be 20 μm in diameter, is composed of calcium carbonate, and functions to buffer the hemolymph (blood) of mollusks.

The various types of granules differ in their locations in the organism. In the marine snail, *Littorina littorea*, Type A granules are found in basophil cells of the digestive diverticulum and Type D granules are located in calcium cells of connective tissues of the foot and other tissues.

Type A, B, and C granules may be storage sites of metals and can be important in the elimination of metals by their discharge from cells into the gut lumen (Hopkin, 1986; Beeby, 1991). Elimination can involve **exocytosis** (fusion of intracellular vesicles with the cell membrane and emptying of vacuole content to the cell exterior) or cell lysis with release of associated granules.

3.3.3 Organic Compounds

Depending on their properties, organic contaminants can be eliminated rapidly or be subjected to metabolism* with subsequent excretion of metabolites. During biotransformation, lipophilic compounds are often, but not always, made more amenable to excretion by conversion to more hydrophilic products. Biotransformations of organic contaminants can be separated into Phase I and II reactions (Figure 3.3b). Generally, reactive groups such as $-COOH$, $-OH$, $-NH_2$, or $-SH$ are added or made available by **Phase I reactions**, increasing reactivity, and often hydrophilicity. The more reactive product can engage in further transformation. Although predominantly oxidation reactions, hydrolysis and reduction reactions are also important Phase I reactions (George, 1994). One of the most common Phase I reactions involves addition of oxygen to the xenobiotic by a **monooxygenase** (Hansen and Shane, 1994). After formation, the products of Phase I reactions can be eliminated or enter into **Phase II reactions** (Figure 3.3b). Conjugates are formed by Phase II reactions, which inactivate and foster elimination of the compound. Compounds conjugated with xenobiotics or their biotransformation products include acetate, cysteine, glucuronic acid, sulfate, glycine, glutamine, and glutathione (Hansen and Shane, 1994; Landis and Yu, 1995). As an example, the hormone, 17 β-estradiol, is conjugated with sulfate to produce the 17 β-estradiol 3-sulfate shown in Figure 2.32.

These Phase I and II reactions can be illustrated with the metabolism of naphthalene (Figure 3.4). Phase I oxidation (naphthalene → naphthalene epoxide) and hydrolysis (naphthalene epoxide → naphthalene 1,2-diol) are shown in this example to produce more reactive and water-soluble metabolites of naphthalene. Subsequent Phase II conjugation with glucuronic acid is shown for naphthalene 1,2-diol. Many more reactions than shown here are involved in the metabolism of xenobiotics as discussed in more detail in Chapter 6. Many involve inducible enzyme systems subject to the complex regulation typical of biochemical pathways. Subsequently, the modeling of bioaccumulation and elimination of the original xenobiotic can be complicated by the dynamics of associated metabolites, especially if a radiotracer is used in the experimental design. In the case of an activated compound, the dynamics of a metabolite can be more important to predict than that of the original compound.

* Lech and Vodicnik (1985) and others object to using the term, metabolism, in this context and suggest biotransformation as the preferred word for biochemically mediated conversion of xenobiotics. They feel that the term metabolism should be restricted to biochemical reactions of "carbohydrates, proteins, fats, and other normal body constituents."

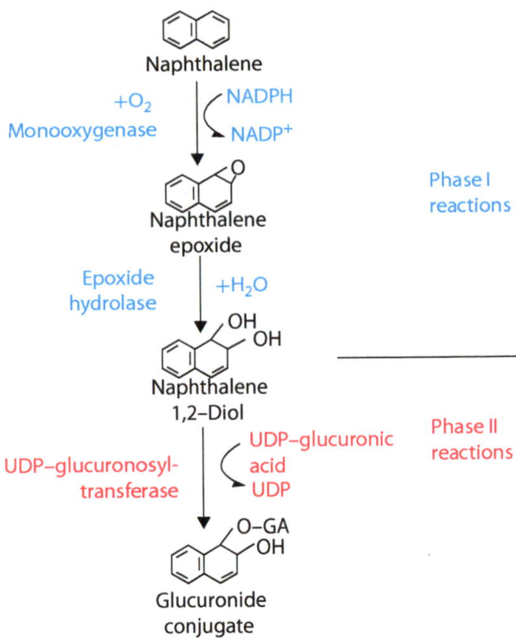

Figure 3.4 Metabolism of naphthalene including Phase I (blue) and Phase II (red) reactions. The first Phase I reaction (naphthalene → napthalene epoxide) is an oxidation reaction and the second Phase I reaction (naphthalene epoxide → naphthalene 1,2-diol) is a hydrolysis reaction. UDP-glucuronic acid is formed by condensation of UTP (high-energy nucleotide, uridine triphosphate) with glucose-6-phosphate. UDP, uridine diphosphate; GA, glucuronic acid. (Composite Figures 1, 5, and 14 of Lech, J.J. and M.J. Vodicnik, *Fundamentals of Aquatic Toxicology*, Rand, G.M. and S.R. Petrocelli, Eds., Hemisphere Publishing Corp., Washington, DC, 1985.)

3.4 ELIMINATION

3.4.1 Elimination Mechanisms

Elimination is contaminant loss from an organism resulting from its biotransformation and/or excretion. Although often used synonymously, depuration and clearance do not have the same meaning as elimination (Barron et al. 1990). **Depuration** is a term associated with a particular experimental design in which the organism is placed into a clean environment and allowed to lose contaminant through time. It is associated with a specific set of laboratory or field conditions in which the amount or concentration of contaminant in the organism decreases. **Clearance**, as will be discussed shortly, is a term used when modeling bioaccumulation kinetics in a certain way and reflects the rate of substance movement between compartments normalized to concentration. Clearance has units of volume·time^{-1}, such as mL·h^{-1}.

If one monitors concentration in individuals over time, another phenomenon, **growth dilution,** might produce a decrease in concentration in an organism. Contaminant concentration can decrease in a growing organism because the amount of tissue in which the contaminant is distributed is increasing. Growth dilution is not a component of elimination because the total amount of contaminant, or **body burden**, has not changed as a result of growth.

Cellular elimination can involve several of the transport and transformation mechanisms already described. Some mechanisms can be specific to a certain class of compounds but others can operate in a more general manner. The glutathione conjugates formed with various electrophilic compounds or metabolites during Phase II detoxification can be removed by a specific ATP-dependent **glutathione**

S-conjugate export pump (Ishitawa, 1992). Contaminants or their metabolites (including many glutathione conjugates) that are organic anions can be eliminated from cells by the **organic anion transporters (OATs)**. For example, OATs are involved in the removal from the cell of the phenoxy herbicide, 2,4-D (Sweet, 2005). Even mercury conjugated with cysteine, *N*-acteylcysteine, or glutathione can be transported from kidney proximal tubule cells by OATs (Sweet, 2005). Another important example of a cellular elimination mechanism is the **p-glycoprotein (P-gp) pump** in membranes. This energy-requiring pump eliminates metabolites from the cell and also is the first line of defense against contaminants that could potentially gain entry to a cell and associated tissues (Bard, 2000) (Figure 3.3). For example, the P-gp on the lumen surface of the mammalian intestinal epithelium actively pumps some compounds back into the lumen during their initial steps of entering of the epithelial cells (Renwick, 2008). The functioning of P-gp is usually described with the "flippase" model: A metabolite coming into contact and binding with the P-gp on the inner cell membrane surface is "flipped" back out to the outer membrane surface by the P-gp. The P-gp is involved in **multixenobiotic resistance (MXR)** in which increased P-pg activity due to exposure to one compound increases resistance to similar (moderately hydrophobic, low molecular weight, and planar) compounds (Segner and Braunbeck, 1998). Other mechanisms can be more general in function. For example, Na^+/K^+-ATPases can be responsible for the movement of various metals such as silver (Bury et al. 1999) out of cells.

The relative importance of specific elimination mechanisms varies among unicellular organisms and metazoan plants, vertebrates, and invertebrates as well as among contaminants. Plants lose contaminant by transformation, leaching, evaporation from surfaces, leaf fall, exudation from roots, or herbivore grazing (Duffus, 1980; Newman, 1995). Animals may eliminate contaminants by transport across the gills, exhalation, bile secretion, secretion from the hepatopancreas, secretion from the intestinal mucosa, shedding of granules, loss in feces, molting, excretion through the kidney or an analogous structure, egg deposition, or loss in hair, feathers, milk, and skin. Although a reasonable argument could be made that aluminum bound to fish gill mucus has not been taken up by the fish yet, such aluminum is rapidly lost by mucus sloughing (Wilkinson and Campbell, 1993). In higher animals, elimination may involve loss in milk, sweat, saliva, and genital secretions (Duffus, 1980). The liver, gills or lungs, and kidney are often the primary elimination routes for animals.

Elimination from the gills can be rapid for xenobiotics with low lipophilicity, i.e., those with log K_{ow}* values of 1–3 (Barron, 1995). Nonpolar organic compounds resistant to biotransformation tend to diffuse slowly across the gill. Spacie and Hamelink (1985) provide DDT, di-2-ethylexylphthalate (DEHP), phenol, and pentachlorophenol as examples of compounds that can be eliminated in significant amounts this way.

Large, nonpolar molecules and associated metabolites can be incorporated into bile in the liver and lost in feces (Duffus, 1980; Spacie and Hamelink, 1985; Barron, 1995). In humans, compounds with molecular weights exceeding approximately 300 Da are eliminated in significant amounts in the bile (Gibaldi, 1991). Many metals (e.g., aluminum, cadmium, cobalt, mercury, and lead) and metalloids (e.g., arsenic and tellurium) complexed with proteins or other biochemical compounds in the plasma are incorporated into bile (Camner et al. 1979).

After passage into the liver, a contaminant enters the hepatic sinusoids where it may be absorbed by parenchymal cells and then biotransformed. The compound or its metabolites either return to the sinusoids or become incorporated into bile. Those incorporated into bile can be eliminated in feces; however, some compounds entering the small intestine in bile may be subject to reabsorption and repeated passage through the liver. This **enterohepatic circulation** increases the persistence of some compounds in the body, and in so doing, may increase damage to the liver (Duffus, 1980; Gibaldi, 1991). Although a minor complication of most elimination processes, compounds incorporated into saliva

* Recall from Section V in Chapter 2 that $\mathbf{K_{ow}}$ is the partition coefficient for a compound between *n*-octanol and water. It reflects the lipophilicity of a compound and is used to imply relative partitioning of a xenobiotic between aqueous phases of the environment and lipids in an organism. It is sometimes designated by the capital letter, P.

can also establish a similar cycle (Wagner, 1975). A Phase II reaction metabolite can be reabsorbed after being deconjugated, e.g., hydrolyzed, in the intestine (Wagner, 1975) and also become involved in cycling. As a relevant example, Schultz et al. (2001) found that conjugated 17 α-ethynylestradiol in male rainbow trout (*Onchoryncus mykiss*) underwent considerable enterohepatic circulation. Metals can exhibit enterohepatic circulation depending on the metal complex size. Trivalent arsenic and methylmercury exhibit significant enterohepatic circulation (Camner et al. 1979). For arsenic, this involves active transport. Competition between compounds during incorporation into bile, and factors affecting bile formation and enterohepatic circulation will dictate the effectiveness of bile elimination.

Kidney excretion tends to be important for compounds with molecular weights less than approximately 300 Da. It is also the primary route of metal (e.g., cadmium, cobalt, chromium, magnesium, nickel, tin, and zinc) excretion by mammals (Roesijadi and Robinson, 1994); however, there are exceptions in which **gastrointestinal excretion** dominates. Metals such as cadmium and mercury can be excreted directly through the intestinal mucosa by active or passive processes (Camner et al. 1979). Gastrointestinal excretion can also involve loss by normal cell sloughing of the intestine wall.

Renal elimination involves three processes: (1) glomerular filtration, (2) active tubular secretion, and (3) passive reabsorption. Filtration through capillary pores allows passage of most xenobiotics except those bound to plasma proteins. Elimination of some metals and lipophilic compounds by this passive filtration mechanism is inhibited by such binding. In contrast to filtration, renal secretion is an energy-requiring movement up a concentration gradient that depends on the contaminant concentration, concentrations of competing compounds, urine pH, delivery rate to carrier proteins in the proximal tubule, and the relative affinity of the compound for the carrier proteins of the tubule and the plasma proteins (Gibaldi, 1991). Because secretion involves carriers, it can be subject to competitive inhibition and saturation kinetics (Gibaldi, 1991). Weak organic acids and bases can be actively secreted to the urine or reabsorbed depending on urine pH. Beryllium elimination involves tubular secretion (Camner et al. 1979).

After a high concentration gradient is established, compounds can undergo passive reabsorption into the blood: this process favors lipid-soluble compounds and is less favorable to ionized or water-soluble compounds (Gibaldi, 1991). Cadmium bound to metallothionein can undergo reabsorption in renal tubules (Camner et al. 1979). Some compounds can be actively reabsorbed (Gibaldi, 1991). Obviously, pH of the urine in the distal tubule will strongly influence reabsorption of weak acids and bases by influencing the amount of the compound that is ionized (Spacie and Hamelink, 1985), and metals such as lead and uranium (Camner et al. 1979).

Other routes of elimination can be important depending on the organism and contaminant. Volatile organic compounds can be lost in expired breath, and some lipophilic compounds can be lost from species such as fish by deposition in lipid-rich eggs. Arthropods have the capacity to eliminate contaminants such as metals by molting (Lindqvist and Block, 1994), peritrophic membrane sloughing from the midgut, and hindgut cuticle loss during molting (Hopkin, 1989), or by discharge of metal-rich granules. Birds may incorporate metals such as lead (Amiard-Triquet et al. 1992; Burger et al. 1994) and mercury (Becker et al. 1994) into feathers. Mammals can eliminate lipophilic compounds (e.g., DDT) or calcium analogs (e.g., strontium) in milk. Fin whales transfer some of their body burden of polychlorinated biphenyls and DDT to their young by normal placental processes before giving birth, and after birth, transfer more to the calf in milk (Aguilar and Borrell, 1994). Some metals and metalloids are incorporated into skin and hair (e.g., Roberts et al. 1974), facilitating removal from the body.*

* Incorporation of elements into hair is used to monitor exposure as typified by two studies of historical significance. Analysis of hair from Napoleon Bonaparte (Maugh, 1974) indicates that, sometime during his exile on Elba or in Saint Helena, he was slowly being poisoned with arsenic, a common political tool of the time. In the second case, hair from Sir Isaac Newton was found to have extremely high levels of mercury, a common element used in his experiments. Newton habitually tasted his chemicals during experimentation. Poisoning is now identified as the most probable reason for his "year of lunacy" (1693), a year filled with abnormal behaviors characteristic of mercury's neurological effects (Broad, 1981).

3.4.2 Modeling Elimination

Given the variety of elimination mechanisms, it is surprising that first-order kinetics is used in most models of elimination. Zero-order and saturation kinetics are used less frequently and tend to be used more in complex model formulations.

Compartment models of elimination similar to that occurring in Figures 3.1 and 3.5 can be formulated as rate constant, clearance volume, or fugacity models. All three approaches are equivalent for the simple, single source models shown here and associated constants may be interconverted (Newman, 2013). Statistical moments methods for quantifying elimination without formulation of a specific compartment model are also common in pharmacology (Yamaoka et al. 1978) and have application to ecotoxicology (Newman, 2013).

Rate constant–based models use constants such as the first-order rate constants described earlier to quantify the rate of change of concentration (or amount) of toxicant in one or more compartments. Frequently, compartments are not physical (e.g., excretion from the liver) but are mathematical compartments (e.g., elimination from the "fast" component). But some interpretations of mathematical compartments imply associated physical compartments; consequently, it is important to determine exactly the type of compartment model being described when assessing this type of research.

A simple, first-order rate constant model can be used to describe elimination of contaminant from an organism after it has been moved to a clean environment and allowed to depurate (left of Figure 3.5). The model can be expressed in terms of concentration (C) or body burden (amount in the individual or X):

$$\frac{\mathrm{d}C}{\mathrm{d}t} = -kC \tag{3.7}$$

where C = concentration in the compartment, k = rate constant for the concentration-based formulation (h^{-1}), and t = time or duration of elimination:

$$\frac{\mathrm{d}X}{\mathrm{d}t} = -kX \tag{3.8}$$

where X = amount in the compartment, k = rate constant for the amount-based formulation (h^{-1}), and t = time or duration of elimination. Here (Equation 3.8), k is the fractional loss of the amount of

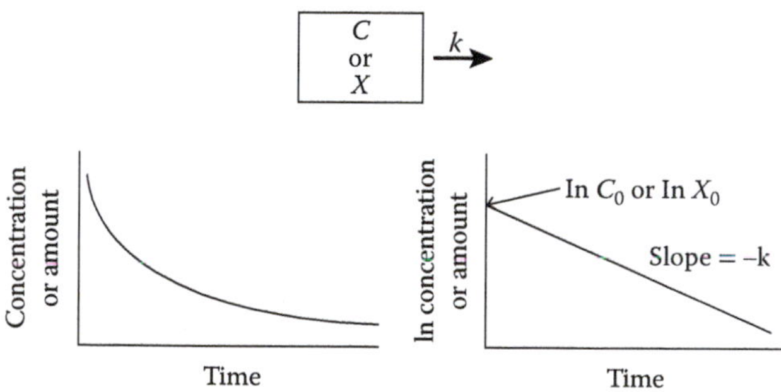

Figure 3.5 Elimination of a contaminant under a depuration scenario including linearization to extract k and C_0 or X_0.

substance, or more explicitly, the amount lost in a unit of time divided by the amount of substance in the compartment. Equations 3.7 and 3.8 are integrated to Equations 3.9 and 3.10 to allow prediction of elimination from an initial concentration (C_0) or amount (X_0) to any time (t) during elimination:

$$C_t = C_0\, e^{-kt} \tag{3.9}$$

$$X_t = X_0\, e^{-kt} \tag{3.10}$$

Figure 3.5 (lower left) shows the exponential elimination of a substance described by these equations. To fit depuration data to these equations, the natural logarithm (ln) of concentration or body burden could be plotted against time (lower right of Figure 3.5). The resulting line has a y intercept equal to the natural logarithm of the initial concentration or body burden. The slope is an estimate of $-k$. If linear regression of the logarithm of concentration or logarithm of body burden versus time is used to estimate C_0 or X_0, a commonly ignored bias exists in intercept estimates that can and often should be corrected (Newman, 1995, 2013).

The time required for the amount or concentration to decrease by 50% ($t_{1/2}$, **biological half-life**) is (ln 2)/k. The **mean residence time** of a molecule of compound (τ) in the compartment is $1.44t_{1/2}$ or, more directly, $k = 1/\tau$. If two or more elimination mechanisms are responsible for elimination from a compartment, they can be easily included in the models:

$$C_t = C_0\, e^{\Sigma - k_i t} \tag{3.11}$$

$$X_t = X_0\, e^{\Sigma - k_i t} \tag{3.12}$$

where k_i = the ith individual elimination rate constants.

For models described by Equations 3.11 and 3.12, the **effective half-life** (k_{eff}) is (ln 2)/Σk_i. Equations 3.11 and 3.12 are particularly useful if a radiotracer is used to quantify elimination because the (first-order) radioactive decay rate constant (λ) can be included in the formulation, i.e., $\Sigma k_i = k + \lambda$ for a one component elimination model with radioactive decay of the tracer.

Rate constant–based models can be developed for biexponential or multiexponential elimination. One such conceptual model (top and bottom left of Figure 3.6) has elimination from two

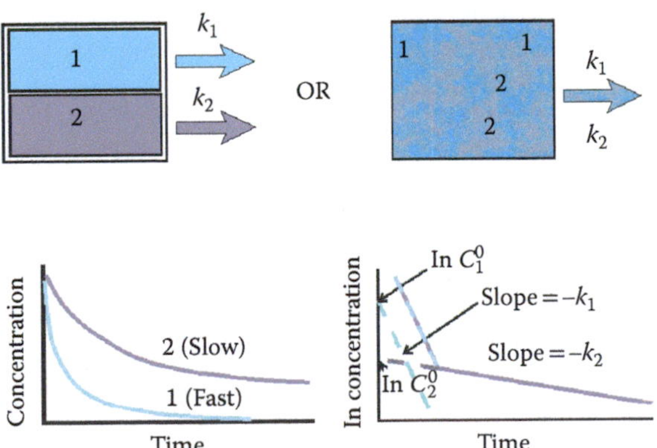

Figure 3.6 Elimination involving two compartments and two elimination constants including backstripping to calculate k_1, k_2, C_1, and C_2. The upper left is the most idealized and least realistic visualization: the upper right is a more realistic visualization involving two discrete compartments (blue and gray) that are interspersed as might occur for contaminants in discrete bone compartments.

different compartments within an organism displaying distinct ("fast" and "slow") elimination kinetics. The two compartments might be discrete but interspersed (top right) as in the case of some contaminants in various discrete bone structures. The integrated forms of this model are Equations 3.13 and 3.14:

$$C_t = C_1^0 e^{-k_1 t} + C_2^0 e^{-k_2 t} \tag{3.13}$$

where C_t = total concentration assuming equal volumes or masses for compartments 1 and 2, C_1^0 = initial concentration in compartment 1, C_2^0 = initial concentration in compartment 2, k_1 = elimination rate constant for compartment 1, and k_2 = elimination rate constant for compartment 2. Obviously, the assumption of equal sizes for the two compartments is an inconvenient assumption of this model. This constraint will be resolved later using clearance-volume models. However, the assumption of equal compartment sizes is not a difficulty in the case of the model based on amount in the various compartments:

$$X_t = X_1^0 e^{-k_1 t} + X_2^0 e^{-k_2 t} \tag{3.14}$$

where X_t = total amount of contaminant in the organism, X_1^0 = initial amount in compartment 1, X_2^0 = initial amount in compartment 2, k_1 = elimination rate constant of compartment 1, and k_2 = elimination rate constant for compartment 2.

With the linearizing method just described for monoexponential elimination, biexponential elimination produces a line with a distinct break in slope (lower left of Figure 3.6, solid line). A **backstripping (backprojection** or **residual) procedure** can be used to extract the model parameters from such a multiexponential elimination curve. First, the region of the ln concentration versus time curve where predominantly slow elimination occurs (the straight part of the line after the break in slope in Figure 3.6, lower left.) is used to fit a line for component 2. The slope of this line is $-k_2$ and the y intercept estimates ln C_2^0. The linear equation just derived for this second component (ln C_2 = ln C_2^0 $k_2 t$) is then used to predict concentrations of contaminant in compartment 2 for all sampling times during depuration. These predicted concentrations in compartment 2 are subtracted from the original data observed for all times prior to the break in the line to estimate the concentrations present in compartment 1 during elimination; these "stripped away" data for component 1 produce a straight line (Figure 3.6, lower left panel, dashed line) with a slope of $-k_1$ and y intercept of ln C_1^0. Wagner (1975) and Newman (2013) detail this and more accurate means of backstripping exponential curves of elimination. Computer programs, including the R packages PK and PKfit, are readily available to implement these methods and to assess the results statistically.

One of the most common instances in which a plasma or blood concentration–time "biphasic curve" such as the one being backstripped at the bottom of Figure 3.6 manifest is the situation with an initial period of distribution among compartments (**distribution phase**) because the introduced dose does not instantaneously reach distribution equilibrium in the organism (Gibaldi, 1991; Renwick, 2008). The first linear portion of the plot reflects this initial stage.

Other multiple compartment models can be fit with this rate constant–based approach. One of the more common, two compartment elimination models (Figure 3.7, top) can be fit in a similar manner. In this model, a contaminant is introduced as a bolus or single mass into a central compartment (e.g., a single injection of the dose into the blood) and the compound is then subject to passage into and out of another peripheral or storage compartment. The compound can only be eliminated from the central compartment. The resulting elimination curves for these two compartments are illustrated at the bottom of Figure 3.7. Initially, the concentration in the central compartment (1) is high and there is no compound in the peripheral compartment (2).

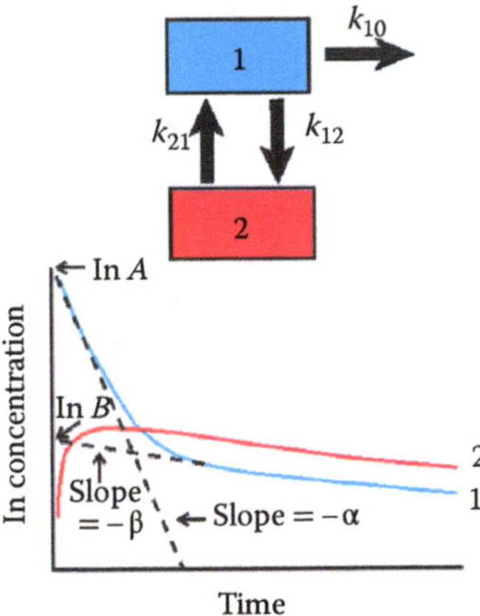

Figure 3.7 Elimination from a two compartment model after bolus introduction of the contaminant into compartment 1. A, B, α, and β are used to estimate k_{12}, k_{10}, and k_{21} as described in the text.

The concentration in the central compartment decreases and the concentration in the peripheral compartment begins to increase. The concentration in compartment 1 eventually describes a biexponential curve and the curve for compartment 2 eventually runs parallel to the end component of compartment 1 because elimination is occurring from compartment 1 only. If the two dashed lines shown in the bottom panel of Figure 3.7 are used to obtain intercepts and slopes for the apparent biexponential elimination dynamics from compartment 1, the microconstants (k_{10}, k_{21}, and k_{12}) can be estimated with the following equations. For example, one could monitor blood concentrations through time and fit the results to Equation 3.13 with backstripping methods. The resulting model is the same as Equation 3.13 except A, B, α, and β are substituted into the equation. Obviously, these constants do not retain the same physical interpretation as C_1, C_2, k_1, and k_2:

$$k_{21} = \frac{A\beta + B\alpha}{A + B} \tag{3.15}$$

$$k_{10} = \frac{\alpha\beta}{k_{21}} \tag{3.16}$$

$$k_{12} = \alpha + \beta - k_{21} - k_{10} \tag{3.17}$$

where A and B = antilogs of the y intercepts for components 1 and 2, respectively, and α and $\beta = -1$ times the slopes of components 1 and 2, respectively. The biological half-life for a contaminant using this model is $(\ln 2)/\beta$ (Barron et al. 1990).

Many similar multiple compartment models can be developed under the rate constant–based modeling scheme. However, modeling can often be made easier as more compartments are added if a **clearance volume–based** formulation is applied instead. In this approach, the substance is distributed in and cleared from compartments of different volumes. Clearance (Cl) is expressed as a flow, the volume of compartment completely cleared of the substance per unit time (volume·time^{-1}).

The apparent volume of distribution concept is used in clearance volume–based models. It can be envisioned with the example of the introduction of a dose of compound into a compartment. After introduction, the dose distributes itself within the compartment and a compartment concentration is realized. The estimated compartment volume (V) is dose/concentration. For example, a dose of a chemical is injected into the blood and allowed to distribute in the circulatory system. The **apparent volume of distribution** (V_d) for blood would be dose/concentration. Volume estimation becomes more involved as the compound distributes among many compartments, but methods for estimation of apparent volumes and clearance rates in such a situation have been developed in pharmacology (Newman, 2013). If other compartments are involved, their apparent volumes of distribution are derived mathematically, and importantly, *expressed in units of volume of the reference compartment*. Often the reference compartment is the blood or plasma compartment. Consequently, these volumes are not physical volumes *per se*, but mathematical volumes. The total dose (D_T) in the organism is distributed among compartments as defined by the concentrations in the compartments (C) and the compartment V_d values (V). For example, if blood is the reference compartment and there are n additional compartments,

$$D_T = C_b V_b + \sum_{i=1}^{n} C_i V_i \tag{3.18}$$

where subscripts of b and i denote blood and the nonblood compartments, respectively. The k_i values of the rate constant formulation are equal to Cl_i/V_i for the clearance volume–based formulation.

Because clearance volume–based formulation allows development of more complex models and parameterization of these models, they have been used in pharmacology to describe the internal kinetics of drugs (**pharmacokinetics**) and poisons (**toxicokinetics**) for many years. Consequently, a wealth of data, techniques, software, and expertise has grown up around this approach. Clearance volume–based models also allow direct incorporation of many physiological parameters into models, and thus, enhance predictive capabilities of models under different conditions and for different species. (Pharmacokinetic models that include physiological and anatomical features in describing internal kinetics are called **physiologically based pharmacokinetics** or **PBPK models**.)

A final compartment model that uses different units but is equivalent to the rate constant–based and volume-based formulations is the fugacity model. It is based on the escaping tendency of a substance from a compartment, that is, its **fugacity**. Fugacity (f) is expressed as a pressure [Pascal (Pa)] and is related to substance concentration in a phase, $C = fZ$, where Z is the fugacity capacity (mol·m^{-3} Pa^{-1}) of the phase. The rate of transport between two compartments (1→2) is N (mol·h^{-1}). This N is equal to $D(f_1 - f_2)$, where D is a transport constant with units of mol·h^{-1} Pa^{-1}. The k_i values for the rate constant–based model shown above are equivalent to the D_i/V_iZ_i of fugacity models, where V is the compartment volume. The major advantage of the fugacity model is that units are the same regardless of the phases being considered. Consequently, wide differences in concentrations are easily accommodated in complex models such as those including water, sediment, food, and biological compartments (Mackay and Paterson, 1982; Gobas and Mackay, 1987; Mackay, 1991) (Vignette 3.1).

Finally, there exists in the pharmacology literature a body of methods for calculating elimination qualities such as mean residence time for substances in individuals that do not require the assumption of a specific model. Yamaoka et al. (1978) introduced a statistical moments approach to pharmacokinetics that uses only the area under the curve (AUC) of a plot of concentration (y axis) versus time (x axis). The mean residence time, its associated variance, and other parameters can be estimated with the AUC. Although still underused in ecotoxicology (Newman, 1995, 2013; Newman and Clements, 2008), these methods can be very useful if an exact model is judged to be unnecessary, impractical, or impossible to define. Also, because of their general form, these methods may be used if the exact model is known.

3.5 ACCUMULATION

Bioaccumulation is the net consequence of uptake, biotransformation, and elimination processes within an individual. The simplest, rate constant–based model includes first-order uptake from one source into a single compartment and first-order elimination from that compartment:

$$\frac{dC}{dt} = k_u C_1 - k_e C \qquad (3.19)$$

where C_1 = concentration in the source (e.g., 1 = water), C = concentration in the compartment (e.g., fish), k_u = uptake clearance (mL·g^{-1} h^{-1}),* and k_e = elimination rate constant (h^{-1}). This equation integrates to Equation 3.20:

$$C_t = C_1\left(\frac{k_u}{k_e}\right)(1 - e^{-k_e t}) \qquad (3.20)$$

The concentration in organisms can be predicted at any time (t) based on Equation 3.20. The resulting bioaccumulation curve is shown in Figure 3.1 (bottom panel) and in the accumulation phase of Figure 3.8. The clearance volume–based and fugacity equivalents of Equation 3.20 are the following:

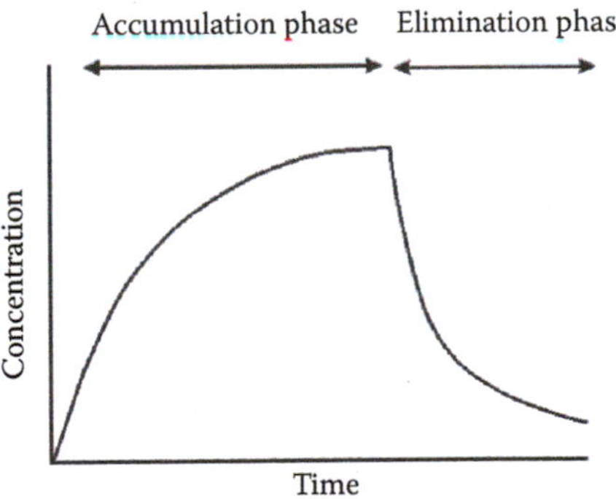

Figure 3.8 Simple bioaccumulation through time as described by Equation 3.20 followed by elimination as described by Equation 3.9. Data of this type are generated with an accumulation–elimination experiment. In the accumulation phase, kinetics is dictated by both uptake and elimination. The organism is then allowed to depurate and kinetics in this elimination phase is solely a consequence of elimination.

* The units of k_u are often expressed as h^{-1} under the assumption of equal density of the source (e.g., water) and compartment (e.g., tissue). A milliliter of source volume and a gram of tissue then cancel each other out and units become h^{-1}. This makes k_u appear as a rate constant. However, k_u reflects a clearance from the source (Landrum et al. 1992), and consequently, should retain units of flow normalized to mass. Details supplied by Peter Landrum are provided as Appendix 9 of this book for the student interested in the derivation of these units for k_u.

$$C_t = V_d C_1 (1 - e^{-\frac{Cl}{V_d}t}) \tag{3.21}$$

$$C_t = (\frac{Z}{Z_1}) C_1 (1 - e^{-\frac{D}{VZ}t}) \tag{3.22}$$

where V_d and Cl = V_d and Cl for the organism, respectively, and Z and Z_1 = fugacity capacities for the organism and source, respectively; D = transport constant for the organism, and V = the volume of the organism.

As the concentration in the organism approaches steady state (i.e., the rightmost, bracketed term $(1 - e^x)$ in Equations 3.20 through 3.22 approaches 1), the relationship between the concentration in the organism (C at steady state or C_{ss}) and the source (C_1) is described as the following:

$$\frac{C_{ss}}{C_1} = \frac{k_u}{k_e} = V_d = \frac{Z}{Z_1} \tag{3.23}$$

These terms define the steady-state concentration in the organism relative to the source. If the environmental source is water, the k_u/k_e, V_d, or Z/Z_1 estimate the **bioconcentration factor** (BCF). This ratio is defined by many workers as the **bioaccumulation factor** (BAF or BSAF)* if the contaminant source is sediment. Normalization may be done for some nonpolar contaminants in these models. The ratio is often referred to as the **accumulation factor** (AF) if the concentrations are normalized to amount of lipid in the organism and to the amount of organic carbon in sediments.

Often bioaccumulation models are fit to the curve of concentration in the organism versus time of exposure using nonlinear regression methods. For the rate constant–based formulation, this method estimates k_u and k_e simultaneously. Although often adequate, this can lead to some difficulty in computations (i.e., poor convergence on a solution, low estimate precision, and an undesirable degree of covariance between estimates of k_u and k_e). A combined accumulation–elimination design (Figure 3.8) is occasionally used to avoid some of these difficulties as in the studies of Yamada et al. (1994) and Murphy and Gooch (1995). In this design, organisms are exposed to a contaminant and allowed to accumulate until internal concentrations approach practical steady-state concentration (i.e., 95% of C_{ss}). Then, the organisms are removed from the contaminated arena and allowed to depurate in a clean setting. The drop in internal concentration is measured with time during this elimination phase. With these data, the k_e can then be estimated independent of k_u from the elimination phase data. This estimate of k_e can then be used in the nonlinear fitting for the accumulation phase data. Now, the nonlinear regression need only estimate k_u from the accumulation data.

Other formulations of these bioaccumulation models allow inclusion of more detail. If there had been an initial concentration of contaminant in the organism before the trial exposure, the contaminant concentration can be predicted through time by combining Equations 3.9 (elimination of initial concentration, C_0) and 3.20 (bioaccumulation from source 1):

$$C_t = \frac{k_u}{k_e} C_1 (1 - e^{-k_e t}) + C_0 e^{-k_e t} \tag{3.24}$$

* As discussed in Chapter 5, the term, bioaccumulation factor is also used in a more general context to mean accumulation from other sources as well. It is often used if the exact source remains undefined or poorly defined as in a field survey. Some workers will use the more specific term, **biota-sediment accumulation factor (BSAF)** to make clear that the factor relates accumulated contaminant to that in sediments (Spacie et al. 1995).

Multiple elimination components can be included (Equation 3.25):

$$C_t = \frac{k_u}{k_{e1} + k_{e2}} C_1 (1 - e^{-(k_{e1} + k_{e2})t}) \tag{3.25}$$

Under the expedient, but sometimes overly optimistic, assumption that growth dilution can be described as e^{-gt} (where g = a rate constant akin to the elimination rate constant), growth dilution can be included as a component in a model such as Equation 3.25.

Uptake from food and water can be incorporated into these models using an estimated **assimilation efficiency** (α, amount assimilated per amount ingested in food) and specific ration (R, amount of food consumed per amount of organism). Assuming concentrations in water (C_1) and food (C_2) are constant:

$$C_t = \frac{k_u C_1 + \alpha R C_2}{k_e} (1 - e^{-k_e t}) \tag{3.26}$$

A general model incorporating multiple sources (k_{uj}), many elimination components (k_{ei}) and an initial concentration (C_0) can be defined:

$$C_t = \frac{\sum_{j=1}^{m} C_j k_{uj}}{\sum_{i=1}^{n} k_{ei}} \left(1 - e^{-\left(\sum_{i=1}^{n} k_{ei}\right)t}\right) + C_0 e^{-\left(\sum_{i=1}^{n} k_{ei}\right)t} \tag{3.27}$$

Data requirements become increasing difficult to meet as models become more complex. Consequently, most pharmacokinetic models are useful compromises between realism and expediency. This is important to understand to extract the fullest understanding from modeled systems without making the unintentional transition to naïve overinterpretation.

VIGNETTE 3.1 Fugacity and Bioaccumulation

Jon A. Arnot
University of Toronto Scarborough

Donald Mackay
Trent University

In this vignette, we outline why the fugacity concept is useful when predicting or evaluating the bioaccumulation of organic chemicals in both water and air respiring organisms and in food webs. We first briefly review fugacity as a thermodynamic criterion of phase equilibrium and its relationship to concentration. We show that fugacity can be readily applied in nonequilibrium situations to calculate fluxes by processes such as diffusion, food uptake, fecal egestion, and biotransformation. Fugacity enables us to view the bioaccumulation phenomena in a new and often more insightful light. We conclude by suggesting that the fugacity concept can be entirely complementary to concentration-based predictions, especially when compiling mass balance models of biological uptake and when interpreting biomonitoring data.

FUGACITY: SOME FUNDAMENTALS

Fugacity, as a thermodynamic quantity, was devised by G. N. Lewis in 1901 as a more convenient criterion of phase equilibrium than the more fundamental concept of chemical potential (Lewis, 1901). It has units of pressure, preferably using the SI unit of pascal, Pa (1 standard atmospheric pressure is 101,325 Pa). Fugacity can be viewed as an "escaping tendency." The word is derived from fugacious, a now rarely used adjective meaning fleeting or transient as in the smell of flowers in a gentle breeze. Fugacity is widely used in chemical engineering practice when calculating the compositions of fluid phase mixtures that are in contact and may approach equilibrium partitioning, as applies for example in petroleum distillation columns. It has been successfully applied to environmental partitioning phenomena in which organic chemicals approach or reach equilibrium between diverse phases, media or compartments such as air, water, soil, and organisms (Mackay, 2001). For most environmental purposes fugacity can be regarded as equal to partial pressure, thus the fugacity of oxygen in the atmosphere and in water at equilibrium with air is about 0.2 atmospheres or 20,000 Pa. In this atmospheric situation the relationship between fugacity f (Pa) and concentration C ($\text{mol}\cdot\text{m}^{-3}$) is readily deduced from the Ideal Gas Law by equating f to partial pressure P as follows, where n is moles, V is volume (m^3), R is the gas constant (8.314 Pa $\text{m}^3\cdot\text{mol}^{-1}$ K^{-1}), and T is absolute temperature (K).

$$C = \frac{n}{V} = \frac{P}{RT} = \frac{f}{RT} = Z_A f \tag{3.28}$$

The proportionality constant or fugacity capacity for air (Z_A) thus has a value of about 4×10^{-4} $\text{mol}\cdot\text{m}^{-3}\cdot\text{Pa}^{-1}$ and at a defined temperature is constant for all nonassociating organic vapors. Z values can be calculated for other phases if the equilibrium partition coefficient K_{12} or C_1/C_2 is known since K_{12} is Z_1/Z_2. Calculation of fugacity capacities usually starts in air with Z_A. Using the air–water partition coefficient Z_W is calculated for water because K_{AW} is Z_A/Z_W. This can be extended to octanol if K_{OW} is known because K_{OW} is Z_O/Z_W. For lipid phases, Z_L is usually assumed to be equal to Z_O. A set of Z values thus enables concentrations in a variety of phases to be converted to fugacities and vice versa.

There is a common misconception that the fugacity concept can only be applied in equilibrium conditions. In practice, any calculation of rates of transport and transformation that can be done using concentrations and the corresponding rate parameters can also be done in terms of fugacity. Rates are conventionally expressed as Df ($\text{mol}\cdot\text{h}^{-1}$), where D is a fugacity "rate coefficient". For example, a reaction rate may be CVk ($\text{mol}\cdot\text{h}^{-1}$), where V is volume and k is a rate constant (h^{-1}). The corresponding D value is VZk ($\text{mol}\cdot\text{h}^{-1}\cdot\text{Pa}^{-1}$). D values can also be determined for flows of chemical as in food uptake by organisms or in respiratory uptake or loss.

One area in which fugacity has proven to be valuable is understanding and modeling bioaccumulation phenomena (Connolly and Pedersen, 1988; Gobas, 1993; Gobas et al. 1999; Mackay, 1982; McLachlan, 1996). Here, we show the equivalence between fugacity and concentration-rate constant models and illustrate why fugacity is so useful in this context.

BIOCONCENTRATION IN AQUATIC ORGANISMS (UPTAKE ONLY BY RESPIRATION FROM WATER)

The conventional concentration-rate constant expression first advanced by Neely and colleagues (Neely et al. 1974; Branson et al. 1975) for bioconcentration in fish takes the form of the differential equation

$$\frac{dC_O}{dt} = k_1C_W - k_2C_O \tag{3.29}$$

where C_O and C_W are the concentrations in the organism and in water and k_1 and k_2 are uptake and clearance (or loss) rate constants. When a steady state is reached

$$C_O = \left(\frac{k_1}{k_2}\right) \cdot C_W = \text{BCF} \cdot C_W \tag{3.30}$$

where BCF is a bioconcentration factor, i.e., C_O/C_W. It can be regarded as a fish–water partition coefficient. This equation, thus, contains three possible parameters, k_1, k_2, and BCF (which is k_1/k_2) but only two can be defined independently resulting in some controversy about which are the two "fundamental" parameters. The corresponding fugacity version is

$$\frac{V_OZ_O df_O}{dt} = D_R\left(f_W - f_O\right) \tag{3.31}$$

where V_O is the organism volume, Z_O its Z value, D_R is the single respiration uptake and Cl rate parameter, and f_W and f_O are the fugacities of the chemical in the water and in the fish. At steady state f_O equals f_W and equilibrium applies. As a result,

$$\frac{C_O}{C_W} = \frac{Z_O \cdot f}{Z_W \cdot f} = \frac{Z_O}{Z_W} = \text{BCF} \tag{3.32}$$

where f is the common fugacity, Z_W depends on the substance's K_{AW} and Z_O can be expressed simply as $Z_W{\cdot}L{\cdot}K_{OW}$, where L is the lipid content of the organism and K_{OW} is the octanol–water partition coefficient, which is usually assumed to equal the lipid–water partition coefficient.

When loss processes other than respiration are negligible (e.g., growth and biotransformation), bioconcentration is simply an equilibrium partitioning phenomenon and the BCF is $L{\cdot}K_{OW}$. The rate parameter D_R can be shown to be both k_1Z_W/V_O and k_2Z_O/V_O.

It can now be argued that there are two fundamental bioconcentration parameters. The first, in fugacity terms is $L{\cdot}K_{OW}$ (or Z_O/Z_W), or in concentration terms BCF (or k_1/k_2), the thermodynamic partitioning quantities. Second is the kinetic quantity D or k_1 that is controlled by the ventilation or water flow rate to the respiring fish gills. In an aquatic system, the simplest first-order assumption is that equilibrium applies and all organisms reach the same fugacity as the water; however, their concentrations vary depending on their lipid contents. It is for this reason that lipid normalization is so useful when interpreting biomonitoring data, because this is effectively a calculation of fugacity (Burkhard et al. 2012).

In natural waters some of the chemical present is sorbed to various forms of organic matter and may thus not be "available" for uptake, particularly for hydrophobic chemicals (log $K_{OW} > 5$) (Parkerton et al. 2008). The total chemical concentration present and typically measured in the water may then exceed the dissolved concentration that is "available" for uptake by the organism. Often a "bioavailable fraction" is estimated. Bioavailability is automatically included in the fugacity calculation because it is only the truly dissolved chemical that can exert a fugacity in the pure water phase (f_W). In other words, the Z value of the bulk water phase (Z_{WT}) containing the sorbent exceeds than that of the pure water phase (Z_W).

The reader is cautioned that in this vignette to avoid the complication of different units in concentration and in rate constants, we have assumed no biotransformation and that equilibrium is controlled only by lipid content. In practice these elements require careful consideration.

BIOACCUMULATION IN AQUATIC ORGANISMS (UPTAKE FROM FOOD AND WATER)

The fugacity mass balance equation is obtained by introducing D values for respiration D_R (as before and containing an absorption efficiency, AE), dietary intake D_D (also containing a parameter for AE) and the loss processes of fecal egestion D_E, biotransformation D_B and possibly growth dilution D_G as shown in Figure 3.9. The differential equation now expands to

$$\frac{V_O Z_O df_O}{dt} = D_R f_W + D_D f_D - f_O \left(D_R + D_E + D_B + D_G\right) \tag{3.33}$$

where f_D is the fugacity of the food or diet. At steady state, this becomes

$$f_O = \frac{D_R f_W + D_D f_D}{D_R + D_E + D_B + D_G} \tag{3.34}$$

This formulation was first proposed by Mackay, Gobas, and colleagues (Gobas et al. 1988; Clark et al. 1990). The relative importance of each process is immediately apparent from its D value. This general equation reduces to several simpler versions under limiting conditions. The bioaccumulation factor is then $f_O Z_O / f_W Z_W$ in fugacity terms or again C_O/C_W, but equilibrium does not necessarily apply because f_O can be greater or less than f_W.

When D_R is large compared to the other D values (as often applies to persistent substances of relatively low K_{OW}), then f_O approaches f_W as in bioconcentration. When D_D and D_E are large (as applies to persistent substances of high K_{OW}), then f_O approaches f_D (D_D/D_E). This ratio, D_D/D_E, is usually about 3–10 for fish and represents a limiting biomagnification factor (BMF), i.e., a maximum fugacity increase from prey to predator. If biotransformation D_B is substantial, f_O will be less than f_W or f_D, and the extent of bioaccumulation is reduced. An equivalent set of bioaccumulation mass balance equations can be compiled in which each process is expressed as a rate constant and phase equilibrium is expressed using partition coefficients. The equations can be shown to be ultimately algebraically equivalent (Powell et al. 2009). Interpretation can, however, be more difficult especially when describing in detail the food uptake process in the gastrointestinal tract.

BIOMAGNIFICATION

Of particular concern in ecological and human risk assessments is biomagnification, the increase in concentration from prey to predator and with trophic level or position in food webs. This applies to persistent, hydrophobic chemicals with K_{OW} values exceeding 10^5. The

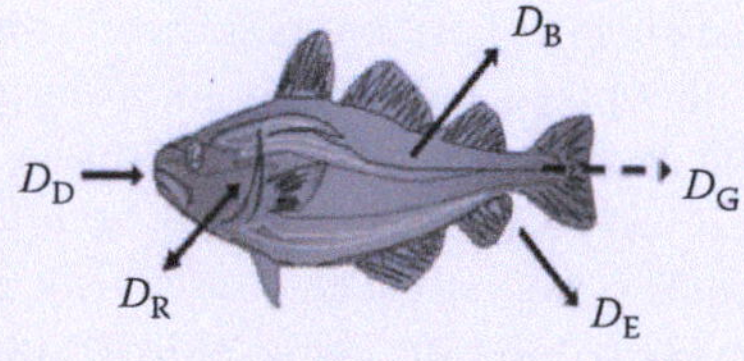

Figure 3.9 An illustration of chemical transport and transformation processes (D values) for bioaccumulation in fish.

biomagnification factor (BMF) is the ratio of the predator (C_2) to prey or diet (C_1) concentration. Viewed in terms of fugacity this ratio is also $(Z_2/Z_1)/(f_2/f_1)$. The first group depends on the differences in lipid content of the organisms and may be greater or less than 1.0. The second group reflects the increase in fugacity from diet to consumer resulting from gastrointestinal magnification and is caused by the loss of lipids and mass as the food is digested. It is also D_D/D_E. In view of the importance of biomagnification, there is a continuing effort to develop improved models of gastrointestinal assimilation for fish, birds, and mammals. Fugacity-based models have proved indispensable in this context by expressing the diversity of complex processes in rigorous and understandable equations as the food is digested and excreted (Gobas, 1993; Gobas et al. 1999; Moser and McLachlan, 2002; Drouillard et al. 2012).

FOOD WEB MODELS

If organism-specific Z and D values are available, it is relatively straightforward to compile models of food webs in which organisms consume each other, i.e., f_O for the prey becomes f_D for the predator. For nonmetabolizing substances the organism fugacities throughout the food web are often similar, but with possible increases in fugacity due to biomagnification and reductions in fugacity due to faster growth. A fugacity-based food web model has been compiled by Campfens and Mackay (1997), that is analogous to concentration-based food web formulations (Gobas, 1993; Morrison et al. 1996, 1997; Arnot and Gobas, 2004).

AIR-BREATHING ANIMALS

In principle, the same equations used for fish can be applied to animals that respire air (birds and mammals) by replacing f_W in water by f_A the fugacity in air. The respiration D value now depends on Z_A rather than Z_W. Loss by respiration is slower for air breathing organisms because for most chemicals Z_A is usually much smaller than Z_W. Homeotherms generally have higher energy demands than fish; therefore, they consume more food and D_D is larger. Digestion efficiencies can also be higher, increasing the D_D/D_E ratio. These differences between avian and mammalian species and aquatic species (fish) can result in more significant biomagnification for air-breathing organisms. The limiting BMF D_D/D_E can reach values of 30–100 representing a considerable increase in prey-to-predator fugacity and concentration (Kelly et al. 2004, 2007). It is for this reason that birds are often particularly vulnerable to contamination by persistent organic chemicals from consumption of fish and terrestrial organisms. Humans are also subject to these fugacity increases and resulting toxic effects providing an additional incentive to evaluate and improve bioaccumulation models using concentrations or fugacity or preferably both.

ROLE OF FUGACITY IN INTERPRETING MONITORING DATA

When examining monitoring data for concentrations of a variety of organisms in a food web, it is generally found that the concentrations in water, sediments, and organisms vary greatly in magnitude and units. Lipid normalization of organism concentrations is often useful because it reveals the equilibrium status of the organisms. By converting all concentrations to fugacities, the variation in values is greatly reduced and the equilibrium status of the entire system becomes apparent. Often, fish fugacities are similar to that of the water, and fugacities in benthic organisms are similar to that of sediments. Biomagnification, or trophic magnification, becomes apparent as does trophic dilution in the form of a fugacity reduction from prey to predator, usually caused by rapid relatively rapid biotransformation in the predator. The utility of the fugacity approach for converting various bioaccumulation metrics (e.g., BCF, BAF, BMF, trophic magnification factor, etc.) into fugacity ratios for bioaccumulation hazard assessment is discussed elsewhere (Burkhard et al. 2012; Mackay et al. 2013).

A CASE STUDY OF BIOACCUMULATION AND BIOMAGNIFICATION IN FISH

In this case study, we model the bioaccumulation of a hypothetical substance in a 1 kg (1 L) fish with uptake from both water by respiration and diet by contaminated food. The numerical values of the chemical, the fish and the exposure quantities have been rounded off to enable the calculations to be more easily followed. The values are, however, similar to those suggested by Arnot and Gobas (2004) in their model of bioaccumulation in aquatic systems.

The chemical has a molar mass of 250 $g{\cdot}mol^{-1}$, a Henry's law constant H of 0.05 $Pa{\cdot}m^3{\cdot}mol^{-1}$ (dimensionless $K_{AW} = 2 \times 10^{-5}$), a log K_{OW} of 6, and a biotransformation rate constant (k_B) of 0.002 day^{-1} corresponding to a half-life of 346 days. The majority of organic chemicals are subject to biotransformation, but estimates of biotransformation rates and half-lives have historically been very limited. Recently, databases and models for biotransformation half-life and rate constant estimates have become available (Arnot et al. 2008, 2009; Brown et al. 2012); therefore, we later illustrate the implications of increasing and decreasing the assumed biotransformation half-life by a factor of 4 to highlight the sensitivity of the bioaccumulation models to this parameter.

The fish has a respiration rate of 200 $L{\cdot}day^{-1}$ and an uptake assimilation efficiency (AE) at the gill of 0.5, thus the fish effectively clears 100 $L{\cdot}day^{-1}$ or 0.1 $m^3{\cdot}day^{-1}$ of dissolved chemical (G_R). It has a feeding rate of 0.05 $L{\cdot}day^{-1}$ with an uptake assimilation efficiency of 0.4. The dietary uptake rate (G_D) is thus 0.02 $L{\cdot}day^{-1}$ or 20×10^{-6} $m^3{\cdot}day^{-1}$. The fish's lipid content is 10% (assumed equivalent to octanol). The growth rate constant is 0.001 day^{-1}.

The total chemical concentration in water is 250×10^{-6} $mg{\cdot}m^{-3}$ or 250 $ng{\cdot}m^{-3}$; however, due to the chemical's hydrophobicity and the assumed organic matter in the water in this case study (sorbent), only 80% of the chemical is truly dissolved, i.e., 200 $ng{\cdot}m^{-3}$ or 0.2 $ng{\cdot}L^{-1}$ or 0.8×10^{-9} $mol{\cdot}m^{-3}$. The diet (prey) contains 0.025 $g{\cdot}m^{-3}$ of chemical, i.e., 10^{-4} $mol{\cdot}m^{-3}$, and has a lipid content of 5%.

We first calculate the Z values in water, octanol, fish and diet, all with units of $mol{\cdot}m^{-3}{\cdot}Pa^{-1}$.

Water	$Z_W = Z_A/K_{AW} = 1/H = 20$
Octanol and lipid	$Z_O = Z_W K_{OW} = 20 \times 10^6$
Fish	$Z_F = 0.1\ Z_O = 2 \times 10^6$ (lipid content is 10%)
Diet	$Z_D = 0.05\ Z_O = 10^6$ (lipid content is 5%)

The bioconcentration factor (BCF) is Z_F/Z_W or 10^5. The fugacities in the water and diet can then be deduced from concentrations dissolved in water and in the diet in units of $mol{\cdot}m^{-3}$.

Water	$f_W = C_{WD}/Z_W = 0.8 \times 10^{-9}/20 = 0.04 \times 10^{-9}$ Pa or 0.04 nPa.
Diet	$f_D = C_D/Z_D = 10^{-4}/10^6 = 10^{-10}$ Pa or 0.1 nPa

The diet is thus at a higher fugacity than the water by a factor of 2.5, presumably because of bioaccumulation and biomagnification at lower trophic levels of the food web.

In the Arnot and Gobas model, the fecal egestion rate constant is calculated from the composition and feces loss rates which in turn are calculated from digestive efficiencies for lipids and nonlipid components of the ingested diet (prey). In this example, we assume for simplicity that the D value for egestion is a factor (Q) less than the diet uptake D value where Q is 5. Q reflects the reduced lipid content and quantity of the feces as compared to the ingested diet (prey).

The D values for uptake and loss processes can then be calculated, all with units of $mol{\cdot}Pa^{-1}{\cdot}day^{-1}$.

Respiration	$D_R = G_R Z_W = 2$ (both uptake and loss)
Dietary uptake	$D_D = G_D Z_D = 20$
Egestion	$D_E = D_D/Q = 5$
Biotransformation	$D_B = V_F Z_F k_B = 4$
Growth dilution	$D_G = V_F Z_F k_G = 2$

The total D value for the loss process $D_T = (D_R + D_E + D_B + D_G)$ is thus 13. The steady-state fugacity of the fish can now be calculated as

$$f_F = \frac{f_W D_R + f_D D_D}{D_T} = \frac{0.08 + 2.0}{13} = 0.16 \text{ nPa} \tag{3.35}$$

This fugacity in the predator is four times that of the water and 1.6 times that of the prey. The net BMF in terms of concentration is 3.2 consisting of the ratio of Z values of 2.0 and the fugacity ratio of 1.6. The concentration in the fish is $f_F{\cdot}Z_F$ or 0.32×10^{-3} mol·m^{-3} or 0.08 g·m^{-3} or 80,000 ng·kg^{-1}. The bioaccumulation factor (BAF), expressed as the ratio of fish to total water concentration, is 320,000 (log BAF = 5.5) as would be expected for a hydrophobic chemical (log K_{OW} = 6) that is poorly biotransformed. If there was no biotransformation and growth, the fish fugacity would almost double 0.3 nPa and the BMF would increase to nearly 6.0.

The fluxes to and from the fish are, in units of nmol·day^{-1}.

Respiration uptake	$D_R f_W = 0.08$
Dietary uptake	$D_D f_D = 2.00$
	Total uptake = 2.08
Respiratory loss	$D_R f_F = 0.32$
Egestion loss	$D_E f_F = 0.80$
Biotransformation loss	$D_B f_F = 0.64$
Growth loss	$D_G f_F = 0.32$
	Total loss = 2.08

Aspects of this mass balance (that is clearly satisfied by the above fluxes) are worthy of note.

The fugacity in the fish exceeds that of the water by a factor of 4; therefore, the rate of respiratory loss is 4 times that of the uptake. There is thus net loss by respiration. Uptake is primarily by food ingestion and that rate is 25 times that of respiratory uptake. The loss processes are approximately 39% by egestion, 31% by biotransformation, and 15% each by respiration and growth dilution.

The fish contains 0.32×10^{-6} mol of chemical or 80 μg. Because the loss rate is 2.08×10^{-9} mol·day^{-1} or 520×10^{-9} g·day^{-1} or 0.520 μg·day^{-1}, the residence time of the chemical in the fish is 80/0.52 or 154 days. This is also $V_F Z_F / D_T$. It will thus take 154 days for an increase or decrease in concentration to e^{-1} or 37% of the steady-state value. The corresponding half-time for uptake or clearance is 107 days.

Although biotransformation is slow with a half-time of 346 days it is a significant loss process and thus has an effect on bioaccumulation. Adjusting the parameters in the above equations shows that increasing the biotransformation rate by a factor of 4 causes the fish fugacity to fall to 0.08 nPa or half the value above. Decreasing the biotransformation rate by a factor of 4 causes the fish fugacity to rise by a factor of 1.3. It is thus important to obtain a reliable estimate of the biotransformation rate as well as of fecal egestion and growth, especially for hydrophobic chemicals for which respiratory exchanges are relatively slow.

The rate constants corresponding to the D values and the rates (ng·day^{-1}) are as follows:

Respiratory uptake	$k_1 = 100$ L·kg^{-1}·day^{-1}	rate $= k_1C_W = 20$
Dietary uptake	$k_D = 0.02$ kg·kg^{-1}·day^{-1}	rate $= k_DC_D = 500$
Respiratory loss	$k_2 = 0.001$·day^{-1}	rate $= k_2C_F = 80$
Egestion loss	$k_E = 0.0025$·day^{-1}	rate $= k_EC_F = 200$
Biotransformation loss	$k_B = 0.002$·day^{-1}	rate $= k_BC_F = 160$
Growth dilution	$k_G = 0.001$·day^{-1}	rate $= k_GC_F = 80$

The concentrations are as follows: C_W is 0.2 ng·L^{-1}, C_D is 25,000 ng·kg^{-1}, and C_F is 80,000 ng·kg^{-1}. Inspection of these rates and concentrations shows that both approaches are equivalent.

CONCLUSIONS

We have sought to present some fundamental aspects of fugacity and demonstrate that fugacity and conventional concentration formulations of the biological uptake equations are entirely equivalent. The use of fugacity can simplify the equations, provide additional insights into the phenomena and assist the interpretation of biomonitoring information by comparing fugacities in organisms that comprise food webs.

3.6 SUMMARY

In this chapter, the methods and mathematics associated with the uptake, biotransformation, and elimination of contaminants were described. The general mechanisms of uptake were described: adsorption, passive diffusion, active transport, facilitated diffusion, exchange diffusion, and endocytosis. Biological transformations of metals, metalloids, and organic compounds were described briefly. Binding proteins and peptides, and biotransformation of and sequestration of metals and metalloids were described in detail. Phase I and Phase II reactions in the transformation of organic compounds were discussed. Elimination by a variety of mechanisms was discussed for plants and animals. Details of elimination through liver, kidney, and gill were highlighted.

Models were developed based on three formulations: (1) rate constant, (2) clearance volume, and (3) fugacity. All are equivalent in their basic forms but each has its own advantages and disadvantages. Also, statistical moments methods exist that do not require a specified model. Although rate constant–based models have the longest history in ecotoxicology, clearance volume–based models have much promise in allowing linkage to the pharmacokinetics literature and techniques. They also are most amenable to generation of PBPK models. Fugacity models have a great advantage for including extremely different phases/compartments because they express contaminant levels in identical units for all compartments. Statistical moments methods require the least amount of information, and no model is required.

SUGGESTED READINGS

Barber, M. C., A review and comparison of models for predicting dynamic chemical bioconcentration in fish. *Environ. Toxicol. Chem.*, 22, 1963–1992, 2003.

Barber, M. C., Dietary uptake models used for modeling the bioaccumulation of organic contaminants in fish. *Environ. Toxicol. Chem.*, 27, 755–777, 2008.

Connell, D. W., *Bioaccumulation of Xenobiotic Compounds*, CRC Press, Boca Raton, FL, 1990.

Gibaldi, M., *Biopharmaceutics and Clinical Pharmacokinetics*, Lea and Febiger, Philadelphia, PA, 1991.

Hansen, L. G. and B. S. Shane, Xenobiotic metabolism, in *Basic Environmental Toxicology*, Cockerham, L. G. and B. S. Shane, Eds., CRC Press, Boca Raton, FL, 1994.

Himmelstein, K. J. and R. J. Lutz, A review of the applications of physiologically based pharmacokinetics modeling, *J. Pharmacokinet. Biop.*, 7, 127–145, 1979.

Mackay, D., *Multimedia Environmental Models. The Fugacity Approach*, Lewis Publishers, Boca Raton, FL, 2001.

Newman, M. C., *Quantitative Ecotoxicology,* Second Edition, Taylor & Francis/CRC Press, Boca Raton, FL, 2013.

Newman, M. C. and W. H. Clements, *Ecotoxicology: A Comprehensive Treatment,* CRC Press, Inc., Boca Raton, FL, 2008.

Parkinson, A. and B. W. Ogilvie, Biotransformation of xenobiotics, in *Casarett & Doull's Toxicology: The Basic Science of Poisons,* 7th Edition, Klaassen, C. D., Ed., McGraw-Hill, New York, NY, 2008.

Renwick, A. G., Chapter 4. Toxicokinetics, in *Principles and Methods of Toxicology,* 5th Edition, Hayes, A. W., Ed., CRC/Taylor & Francis Press, Boca Raton, FL, 2008.

Shen, D. D., Toxicokinetics, in *Casarett & Doull's Toxicology: The Basic Science of Poisons,* 7th Edition, Klaassen, C. D., Ed., McGraw Hill, New York, NY, 2008.

Vijver, M. G., C. A. M. Van Gestel, R. P. Lanno, N. M. Van Straalen, and W. J. G. M. Peijnenburg, Internal metal sequestration and its ecotoxicological relevance: A review. *Environ. Sci. Technol.,* 38, 4705–4712, 2004.

CHAPTER 4

Factors Influencing Bioaccumulation

Without question!

The chemical company called Chisso poisoned the fishing waters of Minamata, poisoned the aquatic food chain, and eventually poisoned a great number of inhabitants. Chisso poured industrial poisons through waste pipes until Minamata Bay was a sludge dump, the heritage of centuries destroyed.

Smith and Smith (1975)

4.1 INTRODUCTION

4.1.1 General

The degree to which a contaminant accumulates in biota depends on the qualities of the contaminant, the organism, and the environmental conditions under which the organism and contaminant are interacting. The qualities of the contaminant determine the physical and chemical form in which it is present in the environment and the degree to which it is available to be taken up, biotransformed, and eliminated. Physiological, biochemical, and genetic qualities determine an organism's ability to take up, biotransform, and eliminate the contaminant. Developmental or sex-related changes can influence bioaccumulation. For example, age- and sex-correlated lipid content will influence accumulation of a lipophilic contaminant. Ecological and behavioral characteristics of an organism determine routes of exposure and efficiency of uptake from each potential source: a predator is exposed through its prey but a pelagic open-ocean species is not exposed directly to sediment-associated contaminant. The *milieu* in which interaction between a contaminant and organism takes place can affect speciation and phase association of the contaminant and, consequently, its availability for accumulation. Environmental conditions can also directly modify the functioning of an organism. For example, temperature has a clear influence on rates of pertinent physiological and biochemical processes taking place within the organism. Other factors such as salinity and pH strongly modify ion regulation and osmoregulation and, in so doing, influence the uptake of many contaminants. Qualities of the microenvironment at the site of interaction may be as, or more, important than those of the general environment. The water chemistry at the surface microlayer of the gill (Janes and Playle, 1995) or that of interstitial waters immediately surrounding an infaunal species (Campbell et al. 1988) strongly influence uptake.

Basic qualities of chemicals, organisms, and the environment that have the strongest influence on bioaccumulation are sketched out in this chapter. They are discussed first for inorganic contaminants and then for organic contaminants. Finally, general biological processes that transcend this organic/inorganic dichotomy are discussed.

4.1.2 Bioavailability

Bioavailability is the extent to which a contaminant in a source is free for uptake (Newman and Jagoe, 1994). This is gauged most often as the extent to which the substance can or does reach systemic circulation—the blood stream (Paustenbach and Madl, 2008; Shen, 2008). In many definitions or specific studies, especially those associated with pharmacology or pharmacokinetics, bioavailability implies the degree to which the compound is free to be taken up *and to cause an effect at the site of action** (Gibaldi, 1991). In such studies, one is focused on determining the **biologically effective dose**, the delivered amount actually available to have an adverse or clinical effect on the relevant target cells, organs, or organ systems (Paustenbach and Madl, 2008). This is a reasonable qualification in the context of many environmental and clinical studies. But the most general definition of bioavailability seems warranted in ecotoxicology because contaminant availability for bioaccumulation in a food or prey species for which no effect is expected could be as much a concern as the availability to an organism that is directly affected. Even with adoption of this broader definition, the term can still be used in the context of the degree to which a toxicant is available to have an effect, e.g., availability of metal in sediments to kill benthic species (Carlson et al. 1991).

Bioavailability is measured or implied in many ways. Relative bioavailability of a contaminant in two types of food may be implied from measured differences in bioavailability for individuals exposed to similar concentrations of contaminant in the different foods. This qualitative approach has been used for contrasting different foods (Newman and McIntosh, 1983; Reinfelder and Fisher, 1991), soils (Zagury et al. 2006), sediments (Luoma and Bryan, 1978; Langston and Burt, 1991), and water chemistries (Wright and Zamuda, 1987; Driscoll et al. 1995). Landrum and Robbins (1990) suggested that comparisons of bioavailability for different sediments can be done by measuring the amount of a contaminant in sediments and the amount in organisms inhabiting the sediments, or by measuring uptake clearance rates of organisms placed in the various sediment types. Using this approach, Hickey et al. (1995) compared the impact of feeding modes on bioaccumulation of polychlorinated biphenyls (PCBs) and polynuclear aromatic hydrocarbons (PAHs) in the deposit feeder, *Macomona liliana*, and filter feeder, *Austrovenus stutchburyi*.

In Chapter 3 (Equation 3.26), assimilation efficiency was used to quantify bioavailability from food. Assimilation efficiency or percentage retention (efficiency expressed as a percentage) may be determined by feeding known amounts of contaminant, often using a radiotracer, and measuring the increase in contaminant within the fed individual. For example, bioavailabilities from food were quantified in this manner for the organotin antifouling agents, tributyltin and triphenyltin, in Red seabream (*Pagrus major*) (Yamada et al.1994), and chlordane *trans* and *cis* isomers in channel catfish (*Ictalurus punctatus*) (Murphy and Gooch, 1995). The mass incorporated into tissue divided by the mass fed to the individual is calculated assuming that all of the contaminant measured in the organism has become interspersed in or incorporated into its tissues. Much effort is made in designing this type of experiment to minimize the amount of unassimilated contaminant in the gut, hepatopancreas, or other similar sites. Without this precaution, the estimated assimilation efficiency would be biased upward from the true assimilation efficiency.

* Together, the contaminant concentration and bioavailability in a source are very often all that are needed to determine the exposure that an organism will experience. This exposure then dictates the probability and intensity of adverse effects to the organism, that is, the risk associated with exposure.

Bioavailability is also measured another way in drug and contaminant pharmacokinetics studies. Such studies are concerned with determining the biologically effective dose of a drug, that is, the amount entering the blood and available to have an effect. To do this, a dose might be administered orally and the amount appearing in the blood compared to that in the blood after the same dose is injected intravenously. Bioavailability quantified in this context involves both the amount of and rate at which the drug or toxicant enters the organism (Gibaldi, 1991). This can be done by comparing the areas under the curves (AUCs) of concentration versus time plots obtained for the different routes of administration (Figure 4.1). The **absolute bioavailability** is estimated from the AUC for any route or form of the compound divided by the AUC for the compound after direct injection into the bloodstream. By definition, the dose injected directly into the bloodstream was completely bioavailable and that administered by the route of interest is compared to that 100% bioavailable injected dose. For example, Equation 4.1 might be used to estimate the bioavailability of a dose of an ingested compound

$$F = \frac{AUC_{oral}}{AUC_{iv}} \tag{4.1}$$

where AUC_{oral} = AUC for dose D administered orally and AUC_{iv} = AUC for dose D administered intravenously. Figure 4.1 (upper panel) shows two curves from which AUCs would be generated with methods such as trapezoidal estimation (Renwick, 2008; Newman, 2013). **Relative bioavailability from two different sources** can be estimated in a similar fashion by choosing one source to serve as a reference. Figure 4.1 (lower panel) shows the blood concentration–time curves used to estimate relative bioavailability of a dose (D) administered in food 1 and then in food 2.

With some modification, bioavailability can be estimated if different doses were used for the various routes of introduction or administered forms (Newman, 1995). Assuming that AUC

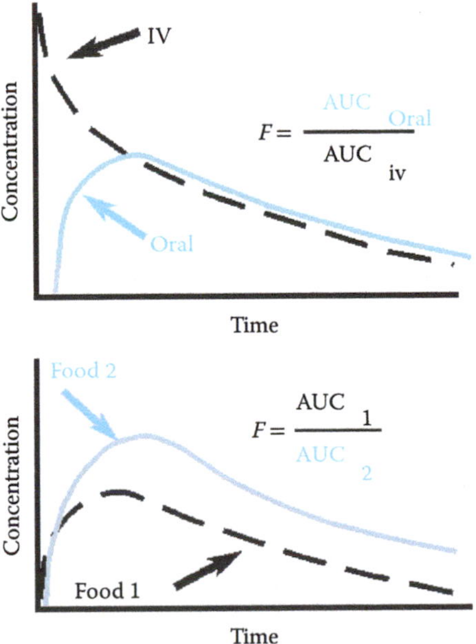

Figure 4.1 Estimation of absolute bioavailability of an ingested dose by comparison of AUCs for ingestion and intravenous injection (top panel), and relative bioavailability by AUC comparison for the same dose administered in two different foods (bottom panel).

is a linear function of dose within the range of applied doses, i.e., not displaying saturation kinetics,

$$F = \frac{\text{AUC}_{\text{oral}} / D_{\text{oral}}}{\text{AUC}_{\text{iv}} / D_{\text{iv}}} = \frac{\text{AUC}_{\text{oral}}\, D_{\text{iv}}}{\text{AUC}_{\text{iv}}\, D_{\text{oral}}} \tag{4.2}$$

where D_{iv} and D_{oral} = the doses administered by intravenous injection and orally, respectively (Gibaldi, 1991; Shen, 2008). However, if half-lives were different for the two routes of administration, the associated difference in clearance is usually corrected by using the following equation instead of Equation 4.2 (Gibaldi and Perrier, 1982):

$$F = \frac{\text{AUC}_{\text{oral}}\, D_{\text{iv}}\, (t_{1/2})_{\text{iv}}}{\text{AUC}_{\text{iv}}\, D_{\text{oral}}\, (t_{1/2})_{\text{oral}}} \tag{4.3}$$

Other AUCs may be applied to estimate bioavailability under various conditions. For example, AUCs for concentration–time curves of urine or bile could be used effectively for some compounds. With statistical moments analysis of the AUCs or parameter estimation for the models described in Chapter 3, **mean residence times** (**MRT**, the mean time of drug or compound residence in a compartment similar to τ described in Chapter 3) can be calculated (Yamaoka et al. 1978). The rate of drug or contaminant absorption is estimated with MRTs for various routes of introduction as the difference between MRTs. The **mean absorption time** (MAT) for a drug or contaminant in food is $MRT_{food} - MRT_{iv}$ (Gibaldi, 1991). Assuming first-order kinetics, an **absorption rate constant** (k_a) can be calculated to be MAT = k_a^{-1} (Gibaldi, 1991; Renwick, 2008). It can also be generated using the methods of backstripping or residuals illustrated in Figure 4.2 (Renwick, 2008). Parameters such as the absorption rate constant are meaningful for estimating how quickly an ingested, inhaled, imbibed, or otherwise introduced toxicant becomes available to interact at a target site within the organism.

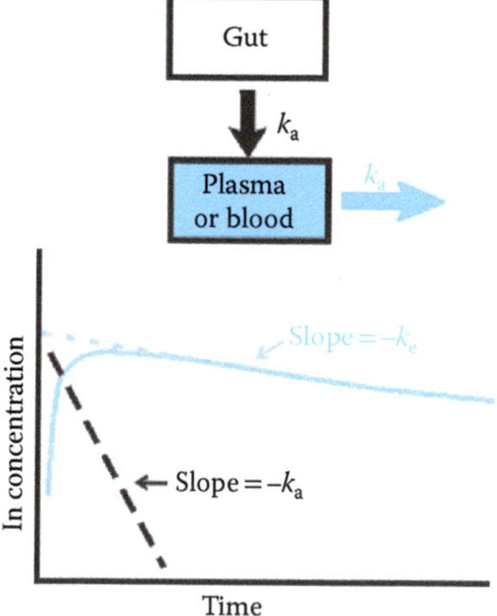

Figure 4.2 A visual illustration of absorption rate constant estimation using the method of residuals (bottom panel is a modification from Figure 4.12 of Renwick [2008]). Assuming first-order elimination and absorption from the gut, the concentration in the blood or plasma is plotted after dosing via ingestion (solid blue curve). As done in Figure 3.7, a linear model is fit to the terminal, linear part of this curve (dashed blue line). The differences are then plotted between measured concentrations and corresponding predicted concentrations for all points taken before the linear portion of the (blue) plasma/blood concentration–time curve (black dashed line). The absorption rate constant ($-k_a$) is derived from the slope of this derived line.

4.2 CHEMICAL QUALITIES INFLUENCING BIOAVAILABILITY

4.2.1 Inorganic Contaminants

4.2.1.1 Bioavailability from Water

Water chemistry affects bioavailability by changing the chemical species present and the functioning of uptake sites. For example, pH has an obvious effect on the equilibrium, $NH_3 + H^+ \leftrightarrow NH_4^+$. The resulting distribution of ammonia species is important to understand because the neutral NH_3 passes much more readily through the cell membrane than does the charged NH_4^+ (Lloyd and Herbert, 1960), and consequently, is the more bioavailable or poisonous form of ammonia. Similarly, the bioavailability of cyanide (HCN), a weak acid but extremely potent inhibitor of cytochrome oxidase, is influenced by the effect of pH on the equilibrium, $H^+ + CN^- \leftrightarrow HCN$ (Broderius et al. 1977). Bioavailability of HCN may also be further modified by the presence of iron (Fe (II)), which combines with HCN to form a less toxic ferrocyanide, $Fe(CN)_6^{-4}$ (Manahan, 1993). Prediction of the amount of bioavailable toxicant is complicated for both HCN and dissolved sulfide (H_2S, HS^-, and S^{2-}) because more than one species contributes simultaneously to toxicity (Broderius et al. 1977).

The bioavailability of a dissolved metal and metalloid can also be affected by chemical speciation. Metal cations compete with other cations for dissolved **ligands**,* that is, anions or molecules that form coordination compounds and complexes with metals. Ligands forming complexes with metals include dissolved organic compounds and inorganic species. Natural organic ligands such as humic and fulvic acids have a wide range of relevant functional groups. Among the most important in complexation are carboxylic and phenolic groups. The major inorganic species involved in metal complexation in saltwater include BOH, BOH_4^-, Cl^-, CO_3^{2-}, HCO_3^-, F^-, $H_2PO_4^-$, HPO_4^{2-}, NH_3, OH^-, Si $(OH)_4$, and SO_4^{2-} (Öhman and Sjöberg, 1988; Newman and Jagoe, 1994). Relevant major inorganic species for freshwaters are Cl^-, CO_3^{2-}, HCO_3^-, F^-, OH^-, and SO_4^{2-}. The ligands, NH_3, HS^-, and S^{2-} are important for anoxic conditions. Of course, H_2O is also an important ligand that forms a hydration sphere around cations and, in so doing, can influence bioavailability. As discussed in Chapter 3, size and charge of a hydrated cation can influence its passage through membrane protein channels.

At thermodynamic equilibrium, the distribution of a particular dissolved cation among its various species can be estimated as a function of concentrations of competing cations, pH, Eh, ligand concentrations, temperature, and ionic strength. These predictions, or directly measured concentrations of the free (aquated) ion, can be used to normalize metal concentrations under a variety of conditions to better estimate bioavailable metal.

As a general rule, bioavailability or toxicity is correlated with the free metal concentration (Andrew et al.1977; Dodge and Theis, 1979; Allen et al. 1980; Borgmann, 1983). This concept is sufficiently prevalent as to be given a name, the **free-ion activity model**† (FIAM). Indeed, in formulating the free-ion-activity model (FIAM) for the first time, Morel (1983) referred to "the universal importance of free metal ion activities in determining the uptake, nutrition, and toxicity of all cationic trace metals." It must be kept in mind that the concentrations of other species are also correlated with that of the free-ion concentration (or activity), and their bioavailability is often difficult to define independently as a consequence. Further, it is not always the case that the free ion is the most, or only, available form of a dissolved metal. Simkiss (1983) suggested 30 years ago that neutral complexes of some

* Ligands share electron pairs with metals. The ligand is a **monodentate ligand** if only one electron pair is shared and multidentate if more than one pair is shared. **Multidentate ligands** are also called **chelates**. Multidentate ligands are prefixed, e.g., bi- or tridentate, to indicate the number of electron pairs involved.

† The activity is used here instead of concentration to encompass situations where there is significant nonideal behavior of ion concentrations due to interionic interactions. In very dilute solutions, the distinction is not as important as in more concentrated solutions such as seawater where the activity coefficients are necessary to relate concentration and activity. By convention, the activity coefficient (activity/concentration) is one with infinite dilution.

Class B metals (e.g., $HgCl_2^0$) could be much more lipophilic than charged species (e.g., Hg^{2+}), and this high lipophilicity combined with the dominance of neutral chloro complexes in marine systems might enhance mercury bioavailability.* Also, the platinum group metals (Pt, Pd, and Rh), which are becoming a concern due to their use in automobile catalytic converters, can form complexes with organic chelators. Zimmermann et al. (2003) examined the octanol–water distribution coefficients of some of these metal complexes and demonstrated that the greatest transfer into the octanol was observed with those chelators associated with fish bile. These findings and the increasing concentrations of these metals in road dust suggested to Zimmerman et al. (2003) that these metals, especially Pd, could display significant accumulation in biota.

Competition among cations at uptake sites is also modified by water chemistry. Crist et al. (1988) described the competition of H^+ on the initial adsorption of metals to sites (notionally carboxylic groups of pectin) on algal cells. They also showed the interaction of dissolved metals with Ca^{2+}, Mg^{2+}, and Na^+ at algal cell–binding sites. The influences of pH and major cations on silver (Janes and Playle, 1995) and aluminum (Wilkinson and Campbell, 1993) uptake on gills have been quantified based on competition for binding to gill surface sites (surface-associated ligand groups). Campbell and Tessier (1996) discussed several studies of competition of metals with hardness cations (Ca^{2+} and Mg^{2+}) at various biological surfaces. Although not the complete explanation, competition between hardness cations and metals is suggested as one reason for the decrease in bioavailability or toxicity of metals often measured with an increase in freshwater hardness.

The water chemistry of the microlayer at a biological surface can be extremely important. Excretion of NH_4^+, NH_3, HCO_3^-, and CO_2 from the gills can rapidly modify the chemistry of water as it passes over gill surfaces (Newman and Jagoe, 1994). Depending on the bulk water chemistry, the shift in water chemistry at surfaces can be enough to modify bioavailability of metals such as aluminum (Neville and Campbell, 1988; Exeley et al. 1991; Playle and Wood, 1989, 1991).

VIGNETTE 4.1 Metal Speciation: A Continuum (Exposure Medium → Interface between the Living Organism and Its Environment → The Intracellular Environment)

Peter G. C. Campbell and Landis Hare
Université du Québec

SOME DEFINITIONS

The term, speciation, was originally used in a scientific context to refer to the evolutionary process by which new biological species arise. However, environmental chemists adopted the term in the 1970s and have adapted its meaning for their own purposes. The International Union of Pure and Applied Chemistry, the official arbitrator of chemical terminology, defines the speciation of an element as "its distribution amongst defined chemical species in a system," where the term, species, refers to a specific form of an element that is defined as to its isotopic composition, electronic or oxidation state, and/or complex or molecular structure (Templeton et al. 2000). This is the definition that we will retain, and the system of interest to us will correspond to a lake or river, i.e., a freshwater aquatic environment. Examples of metal species in fresh waters are shown on the left-hand side in Figure 4.3 and include the following:

- The "free metal ion," M^{z+} (e.g., Cu^{2+} or Al^{3+}), to which water molecules are bound transiently (e.g., $t_{½}$ less than a nanosecond for Cu^{2+}) by electrostatic forces.
- Inorganic metal complexes ($L_{inorganic}$–M), involving the interaction of the metallic cation with inorganic ligands such as the hydroxide, carbonate, chloride, or fluoride anions.

* Although impurities in the olive oils used in this early study were eventually found to have produced an artifact (Personal Communication: K. Simkiss, P. Campbell, P. Rainbow, and R. Blust), high lipid solubility of the neutral mercury chloro complex was subsequently demonstrated by Mason et al. (1996).

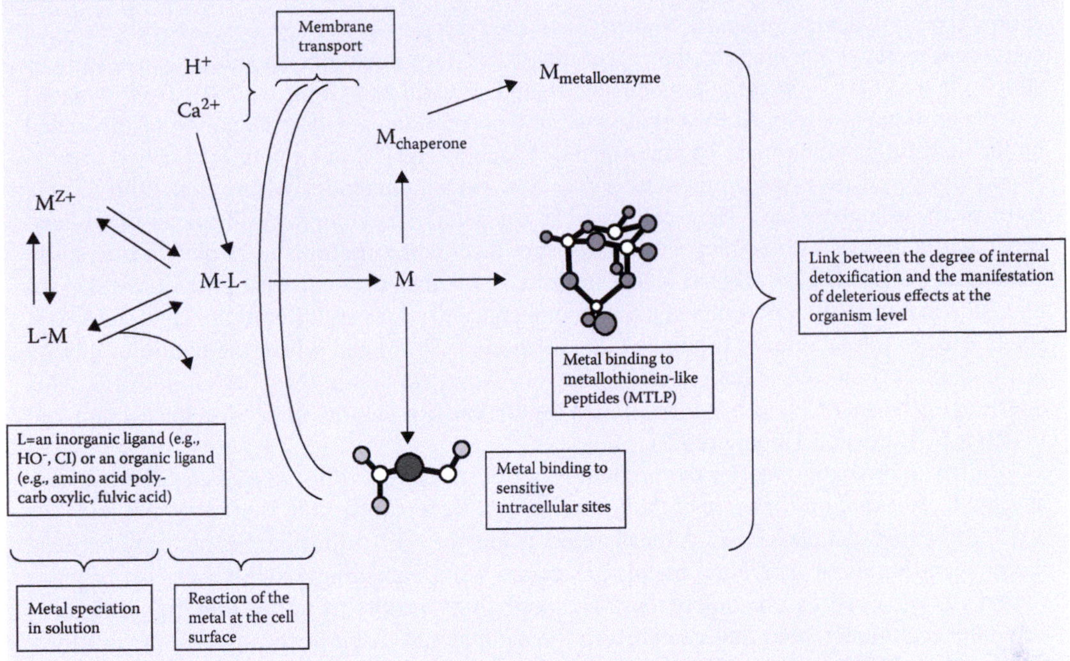

Figure 4.3 Metal (M) speciation in water (left), the interaction of the metal at an external biological membrane, and metal transport into the cell (right) where it is partitioned among various cellular fractions.

- Simple organic metal complexes ($L_{organic}$–M), involving monomeric organic ligands of natural or industrial origin such as alanine, citrate, or ethylenediamine tetra-acetate (EDTA).
- Organic metal complexes (L_{humic}–M), involving natural organic matter, e.g., fulvic or humic acids.

Exchange of metal ions among ligands involves changes in weak coordinate bonding. Unlike the case for transformation of organic molecules, these changes are normally reversible, with the important consequence that the speciation of a metal varies as a function of the chemistry of the medium in which it is found. In contrast, the speciation of an organic contaminant is largely determined by the form in which the contaminant entered the environment and by any subsequent degradation processes.

How Does One Determine the Speciation of a Metal in Fresh Water?

Metal speciation is dynamic, not static: For most metals and under most environmental conditions, the kinetics of the complexation and dissociation reactions shown on the left-hand side in Figure 4.3 are quite rapid and the various metal species will tend to be in equilibrium. Under such conditions, one can calculate the metal's speciation using a chemical equilibrium model, e.g., *MINEQL+* (Schecher and McAvoy, 2001) or *Visual MINTEQ* (Gustafsson, 2012). Necessary input data for such models include the total concentrations of the dissolved cations and anions present in the sample, the total concentrations of the dissolved ligands present in solution, the pH, and the temperature.

For inorganic species of common metals in dilute waters, these models yield accurate predictions (Turner, 1995). However, natural systems normally also include variable amounts of fulvic and humic acids. Because these ligands can bind successively to more than one metal cation (i.e., polydentate ligands), and because their properties change progressively as each additional metal cation binds to the parent molecules, they cannot be modeled successfully as simple, low molecular weight, monodentate ligands (Dudal and Gerard, 2004). Two advanced models have been developed recently to take into account the polydentate nature of fulvic and humic acids: the Windermere Humic Aqueous Model or WHAM (Tipping et al. 1998) and the Nonideal Competitive Adsorption–Donnan or NICA–Donnan model (Milne et al. 2001, 2003). Both of these models have been calibrated using data from laboratory titrations of isolated fulvic and humic acids, and they are able to reproduce these titration curves quite successfully (Dudal and Gerard, 2004). What is less obvious is whether they can also predict metal complexation in natural waters, where the total metal concentrations are often at or below the lowest levels used in the laboratory titrations (Unsworth et al. 2006) and where the nature or quality of the natural dissolved organic matter may vary from site to site (Mueller et al. 2012). This question of the accuracy of metal speciation predictions for natural waters has been examined in detail by Lofts and Tipping (2011).

Rather than simply rely on the predictions of chemical equilibrium models, one can also attempt to measure one or more of the metal species present (often the free metal ion itself, or labile inorganic complexes). Such measurements are very difficult to make, not only because the concentrations of individual metal species are often vanishingly small, but also because the results obtained can be ambiguous. This ambiguity results from the sampling and measurement techniques used that can perturb the sample and displace the speciation equilibria (Batley et al. 2006

Why Should We Be Interested in Metal Speciation?

The speciation of a metal will affect its geochemical behavior and its bioavailability, both of which are of interest in ecotoxicology. Let us consider the link between the speciation of a metal and its bioavailability. To elicit a biological response from a target organism and/or to accumulate within the organism, a metal must first interact with and traverse a cell membrane. For hydrophilic metal species, this interaction of the metal with the cell surface can be represented as the formation of a surface complex, M-X-membrane, where X-membrane is a cellular ligand present at the cell surface. In chemical terms, we have simply added a biological surface as an additional ligand on the left-hand side in Figure 4.3. The metal-accepting ligand on the biological surface (X-membrane) may be a metal-sensitive site, or a membrane protein involved in transporting the metal into the intracellular environment. Transport of a metal across a membrane is assumed to be slower than the initial surface complexation step. In such cases, the biological surface reaches equilibrium with the exposure medium, and the concentration of the M-X-membrane species proves to be a good predictor of the biological effect. In simple systems such as those with constant pH and [Ca^{2+}], the concentration of this surface complex will vary as a function of the free metal ion concentration in the exposure medium and thus [M^{z+}] will be a good predictor of metal bioavailability (Campbell, 1995).

As indicated in Figure 4.3, other cations (e.g., H^+, Ca^{2+}, Mg^{2+}) that compete with M^{z+} for binding at the metal-accepting ligand on the biological surface can afford some protection against metal toxicity. Indeed, it is well recognized that many metals are less toxic in hard waters than in soft waters (Meyer et al. 1999). In recent years, these various effects have been integrated into a modeling framework, the **Biotic Ligand Model** (BLM), which takes these competitive interactions into account (Paquin et al. 2002; Niyogi and Wood, 2004). The BLM has proved successful in predicting the acute toxicity of metals to model aquatic organisms and is gaining widespread acceptance among the scientific and regulatory communities, notably in the United States and within the European Union, where it has been used extensively as a tool in environmental risk assessments for metals (International Council on Mining and Metals, 2007). Extension of the BLM approach to assess the chronic impacts of metals is, however, not straightforward. For example, the BLM currently assumes that for a given aquatic species the binding properties of the metal-accepting ligand on its biological surface, i.e., the biotic ligand, are constant. However, recent research suggests that these properties can be affected by prior exposure to metal, i.e., by acclimation (Kamunde et al. 2002; Klinck et al. 2007; Lavoie et al. 2012). Since metal accumulation and toxicity are governed by these key epithelial properties, applications of the BLM to real-world cases of chronic exposure will have to take acclimation into account.

Future Perspectives

From a chemistry perspective, metal speciation should be important not only in the exposure medium and at the biological interface, but also within living cells (Figure 4.3, right-hand side). However, the intracellular environment differs greatly from that which prevails outside the cell. For example, the pH may vary over 4 or more pH units in the exposure environment, but within the cell, the pH is normally close to 7 and remains virtually constant (except in digestive vacuoles). Likewise, the concentrations of metal-binding ligands within the cell are much higher than in a typical lake or river, and the resulting high ligand:metal ratios will favor metal complexation. The chemical nature of intracellular ligands is also very different from that of the ligands found externally. For example, complexation by thiol groups is far more important in the intracellular environment than outside the cell. One of the important consequences of this ligand-rich environment is that the free metal ion concentration within the cell is calculated to be vanishingly low ($<10^{-15}$ M) (Finney and O'Halloran, 2003). Metal exchange reactions that proceed by dissociation of the original metal complex followed by binding of the free metal to the new ligand will be so slow as to be ineffective in the intracellular environment. Instead, it has been suggested that incoming (essential) metals are bound to metal chaperone ligands, which then bind to the appropriate receptor ligand and relinquish their metal (O'Halloran and Culotta, 2000; Finney and O'Halloran, 2003). Metallothionein (Figure 4.3), an inducible thiol-rich peptide found in most animal cells, likely plays such a role for essential metals such as copper or zinc (Otvos et al. 1987). In addition, metallothionein is also involved in the detoxification of nonessential metals such as cadmium and silver.

The intracellular environment represents a new frontier for metal speciation research, one that has recently acquired its own descriptor, metallomics (see Lobinski et al. 2010). In a limited number of cases, researchers have been successful in quantifying defined chemical species (Garcia et al. 2006), in which case use of the term, intracellular metal speciation, is appropriate, but in most cases, the researchers have determined intracellular metal partitioning. Partitioning refers to the distribution of an element amongst various fractions that are separated according to their physical (e.g., size, solubility) or chemical (e.g., bonding, reactivity) properties (Templeton et al. 2000). To determine metal partitioning within living organisms, one must normally gently homogenize the biological tissue in a cold, isotonic medium so as to avoid disrupting membrane-bound cellular organelles, such as nuclei, mitochondria, microsomes, and lysosomes. Subsequent steps involve the separation of the different metal-binding ligands and the quantification of the amount of metal associated with each ligand pool. Differential centrifugation or high-performance liquid chromatography (HPLC), or some combination of the two, can be used to perform the separation step (Wallace et al. 2003; Giguère et al. 2006; Lobinski et al. 2010).

Intuitively, based on the clear demonstration that metal speciation outside a living cell affects how available the metal is to the cell, one might expect to find similar links between subcellular metal partitioning and metal-induced effects. Indeed, one of the tenets of classical metal toxicology is that metals cause toxicity if they bind to inappropriate, metal-sensitive intracellular sites, and that there is a threshold concentration below which exposed organisms manage to prevent metals from binding to these metal-sensitive sites (Mason and Jenkins, 1995). One of the goals of research in this area is thus to be able to distinguish between metals that have been successfully detoxified, for example, by sequestration in metal-rich granules or by binding to metallothionein, and those that have, on the contrary, been able to bind to inappropriate, metal-sensitive sites (Figure 4.3). When applied to wild aquatic organisms, living under conditions of chronic metal exposure, such a distinction would enable one to determine whether or not the organisms were subject to metal-induced stress (Campbell et al. 2008; Campbell and Hare, 2009).

Subcellular metal partitioning is important not only for predicting the effects of metals on organisms, but also for determining the ease with which metals in prey organisms will be assimilated when they are eaten by those above them in the food chain. Pioneering work in the 1990s showed that the assimilation of metals by marine herbivores depended on the form in which the metals were present in their algal food (Reinfelder and Fisher, 1991; Reinfelder et al. 1997). Thus elements present in the liquid portion of algal cells (cytosol) tended to be assimilated more efficiently by copepods than those associated with algal cell walls. Subsequent studies have shown that the efficiency with which a variety of herbivores and carnivores assimilate metals can indeed be predicted from measurements of metal partitioning in the cells of their prey (Wallace and Luoma, 2003; Zhang and Wang, 2006; Dubois and Hare, 2009a,b). Although metals in some prey fractions (e.g., proteins) are more likely to be assimilated than others (e.g., granules), the availability of a given prey fraction has been shown to vary among predators depending on the rate at which they ingest prey and their digestive physiology (Rainbow et al. 2011). Given this fact, it can be misleading to assume that a given prey fraction is either entirely available or unavailable for transfer to predators (Dubois and Hare, 2009a). Still, the overall approach has proven very useful for explaining why some metals move up food chains more easily than other metals and why animals at the top of some food chains have higher metal concentrations than do those of an equivalent trophic level but dependent on a different food chain (Stewart et al. 2004).

The purpose of this vignette has been to demonstrate the importance of metal speciation at various levels, starting outside the living organism in the exposure medium, including the cell membrane that separates the abiotic from the biotic environment, and finishing within the cell. The recurring message has been that since various metal species behave differently in each of these realms, attempts to predict metal bioavailability and toxicity without considering metal speciation are clearly foolhardy!

4.2.2 Bioavailability from Solid Phases

The bioavailability of inorganic contaminants in aerosols, food, sediments, soils, and other solid phases of the environment is difficult to predict accurately; however, some general themes do emerge from the literature. The direct availability from solid phases is only one part of the story. For example, availability of a metal from a particular solid phase such as sediments can be determined by its capacity to partition into the interstitial waters. These general phenomena will be discussed with examples here.

The bioavailability of metals or metalloids in solid aerosols* suspended in air is determined not only by their chemical forms in the solid but also by the size of the particulates and the distribution

* An **aerosol** is a collection of liquid or solid particles suspended in a gas such as air (Valentine and Kennedy, 2008).

of the element within the particulates. As an example, arsenic tends to condense onto outer layers of smaller coal fly ash particles as they move up the smoke stacks of coal burning power plants and, because of this surface deposition, the arsenic is more available than if it were uniformly distributed throughout small to large ash particles (Hulett et al. 1980; Wangen, 1981). Lead halides in automobile exhaust dissolves more readily in the lung after inhalation than lead in road dust that has weathered to compounds such as lead sulfate (Laxen and Harrison, 1977). Also, lead is present at highest concentrations in small particles of road dust that gain deeper access to the lungs than large particles (Biggins and Harrison, 1980). In humans, larger particles are removed by nasal hair, and associated contaminants are unavailable as a consequence. Particles with diameters of approximately 5–10 μm gain entry only into the nasopharyngeal region although some <1 μm particles can also be trapped there. More particulates with diameters of 1 μm or less can go much deeper into the terminal bronchioles, alveolar ducts, and alveoli (Cordasco et al. 1995) with substantial deposition possible for those <0.01–0.1 μm in the terminal bronchioles (Valentine and Kennedy, 2008; Witschi et al. 2008). Clearly, the depth of passage and consequent bioavailability of contaminants in inhaled particles are related to particle size and other factors. Boethling and MacKay (2000) suggest that absorption of contaminants in lungs is highest for those compounds with water solubilities higher than their lipid solubilities. They also indicate that polar compounds tend to be more bioavailable in inhaled aerosols than nonpolar compounds.

Bioavailability of contaminants in food is a function of many factors. Just as with inhaled particulates, the size of a food particle can determine bioavailability as well as the chemical form of the affiliated contaminants. Particle size of materials passing though the human gut can modify bioavailability of some contaminants and drugs (Gibaldi, 1991). Bivalve mollusks have complex sorting mechanisms on the gills and palps, in the gut, and in the digestive diverticula that are strongly affected by particle size. The size of the food particle will determine the degree to which it participates in these different digestive processes. Some small particles such as iron oxide and iron saccharate can even be taken into phagocytes while still on the gill of oysters (Galtsoff, 1964). This process of sorting and digestion is further complicated by environmental factors such as temperature and tidal rhythm, which modify feeding behavior and digestive processes of bivalves (Morton, 1970). Nanoparticles ingested by insects can also bioaccumulate as in the case of tobacco hornworms fed gold nanoparticles (Figure 4.4).

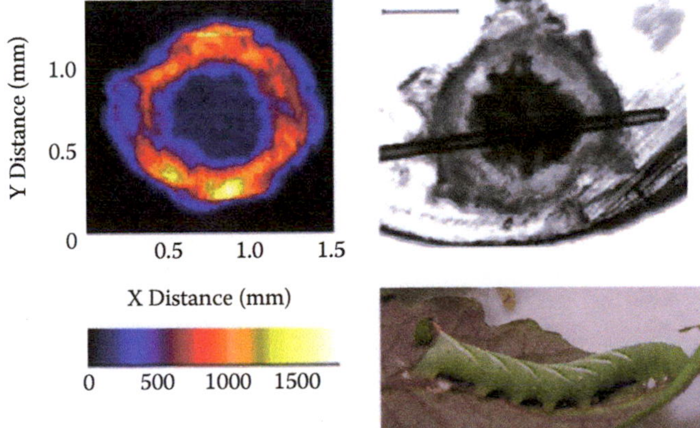

Figure 4.4 Synchrotron x-ray fluorescence microprobe (μXRF) map of fluorescence from the Au L-α edge of gold (upper left) and corresponding light micrograph of a cross-section (upper right) from a tobacco hornworm (*Manduca sexta*, lower right) fed plants exposed to 5-nm nanoparticles. The fluoresence intensity legend (counts per second, lower left) indicates the relative amounts of gold in areas of the gut cross-section. (Images courtesy of Jonathan Judy. Study details in Judy, J.D., et al. *Environ. Sci. Technol.*, 45, 776–781, 2011; Judy, J.D., et al. *Environ. Sci. Technol.*, 46, 12672–12678, 2012.)

VIGNETTE 4.2 Bioavailability of Metals to Aquatic Biota

Philip S. Rainbow and Samuel N. Luoma
Natural History Museum

A metal is bioavailable only if it is in a form that can be taken up by an organism. Not all forms of metal encountered by any organism in its environment will be bioavailable. The term *bioavailability*, therefore, refers to that fraction of the total ambient metal that an organism actually takes up when encountering or processing environmental media, summated across all possible sources of metal, including water and food as appropriate (after Rand et al. 1995; Luoma and Rainbow, 2008). For example, the metal bioavailable to an aquatic alga would be a fraction of the dissolved metal in the surrounding medium, whereas the metal that is bioavailable to an aquatic invertebrate animal would include fractions of both dissolved metal and metal derived from ingested food. There is a necessary correlation between the bioavailability of a metal to an organism and its uptake rate. Since high uptake rates correlate with toxicity, high bioavailabilities correlate with potential ecotoxicological effects.

An understanding of the bioavailability of metals to aquatic biota depends on an understanding of the processes of uptake. Whether taken up from solution across a permeable external epithelium or taken up in the alimentary tract of an animal, trace metals need to cross the apical cell membrane of a body cell (Figure 4.5) (see also Figure 3.2 and Section 3.2).

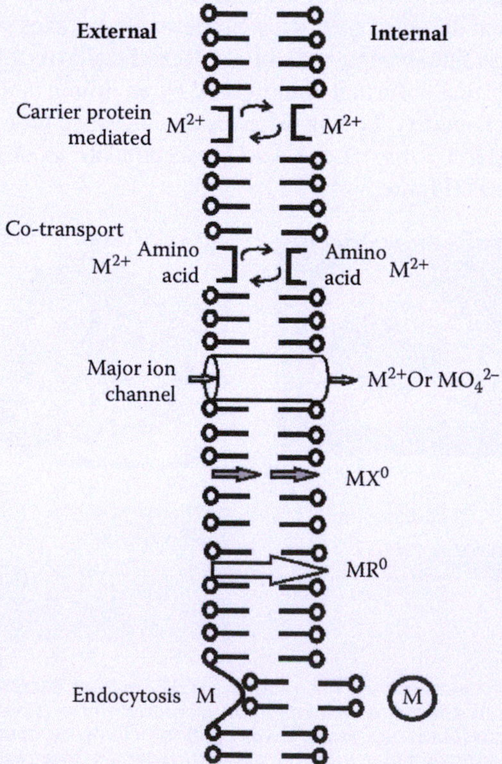

Figure 4.5 Means by which metals can move into cells.

The eukaryotic cell membrane consists of a lipid bilayer traversed by many proteins. The lipid bilayer acts as a barrier to the passage of water and water-soluble chemicals into the cell. It is the embedded proteins that typically allow (and control) the transfer of soluble chemicals and particularly ions across the membrane. As shown in Figure 4.5, there are several mechanisms by which a trace metal may cross the apical membrane.

Carrier Proteins (Transporters)

Specific carrier (transporter) proteins exist for specific (expectedly essential) trace metals, as well as for major ions and nutrients. Transport via a carrier from the outside to the inside of a cell does not involve the need for energy, proceeding by passive facilitated diffusion. A concentration gradient is required but not a concentration gradient of total metal. A concentration gradient of the specific chemical form of the trace metal that binds with the carrier protein externally can be sustained by rapid transformation of the metal once it reaches the cell interior, for example by binding to intracellular proteins. For dissolved metals that behave as cations, experimental results are best explained if the form bound to the carrier is considered to be (or rather, is best modeled by) the free metal ion, a concept referred to as the FIAM (Campbell, 1995). In most natural waters, only a commonly small proportion of the total metal concentration may be in the form of the free metal ion, much of the remaining dissolved metal being complexed by inorganic or organic complexing agents (Luoma and Rainbow, 2008).

Major Ion Channels

Protein channels also exist in the membrane, forming temporary aqueous pores of a specific diameter to carry a specific major ion like calcium or sodium across the cell membrane. Trace metals may trespass into these ion channels, e.g., cadmium into calcium channels. The free cadmium ion radius (0.92 Å) is very similar to that of the calcium ion (0.94 Å), and it is chemically inevitable that some cadmium from the external medium will enter epithelial cells through a calcium channel. Whether this route of entry for cadmium is as significant as that of cadmium entry via facilitated diffusion on trace metal carrier proteins (perhaps zinc) depends on the specific animal in question and its physiological requirement for high calcium uptake. In teleost fish, zinc uptake in the gills may also occur at least partly via a calcium channel in gill chloride cells, supplementing zinc uptake via a zinc transporter. A proportion of the uptake of dissolved copper and silver by teleost fish gills occurs via a sodium channel; because these metals are transported as monovalent cations (Luoma and Rainbow, 2008). Because it is the free metal ion that enters the channel, the free metal ion can again be considered to reflect the bioavailable form for this uptake route too, in accord with the FIAM.

Some trace metals and metalloids will not follow the FIAM because their commonly dissolved ionic forms in seawater are anionic, e.g., arsenate, chromate, molybdate, selenate, and vanadate, offering potential for their entry into cells via major anion routes for sulfate or phosphate.

Cotransport

Cotransport of trace metals with amino acids is likely to be more important in the alimentary tract than at external surfaces. For example, cotransport is the major mechanism for alimentary uptake of methylmercury, because its complex with cysteine mimics another amino acid (L-methionine) (Luoma and Rainbow, 2008).

Endocytosis

Endocytosis is another process by which a metal may enter a cell. Particulate material is engulfed by the membrane of the cell, as in the case of uptake of particulate iron oxides by the gill cells of mussels or particulate vanadium in the pharynx of tunicates. Endocytosis will be the mechanism of uptake of metallonanoparticles that are likely to be increasingly important as aquatic contaminants.

Other Possible Routes

Nonionic, nonpolar chemical species are more soluble in lipid than their ionic counterparts, and so it is possible that some trace metals may enter the cell in this form. Many organometallic compounds are also typically lipophilic and therefore theoretically able to cross lipid bilayers relatively easily.

Trace metals may, therefore, potentially enter an epithelial cell and thence the body of an organism by a variety of mechanisms and indeed by several mechanisms simultaneously. All mechanisms are not equally important at all times, nor are more than one mechanism always quantitatively significant in the total uptake of trace metals. Indeed the relative importance of different mechanisms of entry will change with the species of organism and its physiological state, the metal and its chemical form, and the location and type of epithelium and its physiological role. Such variation defines what metal is bioavailable to what organism under what conditions. The FIAM adjusted for competition from major ions (the BLM) will adequately model some, but not all, mechanisms. Arguably, the BLM will account for most of the uptake from solution by most organisms for most cationic metals under most environmental conditions.

Concentration is the geochemical factor that always influences uptake rates, if the proportion of different physicochemical metal forms is constant. Other external factors affect uptake rates when they influence the abundance of the bioavailable form, even in the absence of changes in total dissolved concentration. Complexation of free metal ions with organic or inorganic complexing agents usually will reduce bioavailability from solution. Changes in salinity also will affect the rate of uptake by marine or estuarine organisms of particular trace metals such as zinc and cadmium, which are predominantly complexed with chloride in seawater. In freshwater, decreasing pH reflects increasing concentrations of hydrogen ions, which can occupy ligands in solution, replacing metals and causing increased free metal ion activity. The extra H^+ ions at low pH may, conversely, compete with metal ions at the membrane-binding site and decrease the rate of metal uptake. The balance between these two contradictory effects determines the uptake rate from solution, for example in the phantom midge larva *Chaoborus* (Hare and Tessier, 1998).

It is increasingly apparent that the diet is an important route for the uptake of trace metals by animals, which accounts for a great deal of the interspecific variation in total metal uptake rates. Uptake from food is strongly determined by what an organism eats and this differs very widely. Mussels and other filter feeders remove suspended material from the water column to obtain their food. That material includes plant cells, detritus, and inorganic particles, including those with an organic coating. Even superficially similar species differ in the degree to which they separate these materials; but most filter feeders ingest some proportion of all types of particles. Typically the addition of living material such as phytoplankton to suspended material increases the efficiency with which a bivalve can obtain metal from its food (Luoma and Rainbow, 2008).

Uptake from food is ultimately a function of how much an animal ingests (feeding rate), the concentration and form of the metal in the food, and how much of that metal is extracted and assimilated by the animal. Feeding rates differ widely among species, as dictated by life history. The AE is a measure of the percentage of the ingested metal that is assimilated into (initially) the alimentary epithelium of the feeding animal. The AE can be measured for a particular species in any set of circumstances. Metals are typically ingested in an innocuous solid form, but are released for transport across the membrane of the digestive tract by powerful solubilizing forces in the digestive tract. Metals released by extracellular digestion are concentrated into the fluids of the gut medium by complexation with high concentrations of small organic molecules. Gut epithelial cells are adapted for the uptake of many of the organic molecules released by digestion such as amino acids, and metals may be cotransported with amino acids such as histidine and cysteine. The chemical form of the metal in the food is also an important consideration. An early proposal was that only metal bound to the soluble fraction in prey is available to higher trophic levels, but it is now appreciated that many animals have digestive processes that can assimilate metal from more fractions than just the metal in cell solution. At the other end of the spectrum it was also suggested that trace metals are not trophically bioavailable after being detoxified and incorporated into trace metal-rich granules in prey. It is now known, however, that predatory gastropod mollusks can assimilate cadmium, silver and zinc from metal-rich granules in their prey (Luoma and Rainbow, 2008). In general, it appears that there is a progression of the dietary bioavailability of a metal to a particular feeding animal from metal bound in soluble forms > metal bound to insoluble intracellular materials > insoluble metal-rich granules in the food species. It is likely that the absolute AE among these fractions varies among the consumers (Rainbow et al. 2011). The differences in availability of different fractions and differences in fractionation among food types, together with the assimilative powers of the predator, combine to cause different AE from different foods.

The AE is driven by specific traits, such as digestive processing, that determine how long the food is retained in the digestive tract, how much metal is released, the form of metal that is released by digestion, and whether that form can be taken up by gut epithelial cell. Given the range of gut structures and digestive processes found in animals, it is not surprising that the AE varies considerably among species. Thus, as stressed for bioavailability in general, the bioavailability of a metal in a particular diet depends on the feeding animal and is not a generalized feature of the diet.

Biological factors are thus influential. Bioavailability is strictly applicable to a single species according to its functional ecology and physiological traits. This combination of traits determines the proportional uptake of metal from different sources, such as solution or diet. Thus, different organisms take up trace metals at different rates under identical physicochemical conditions. These interspecific biological differences are often large.

Literature describing the bioavailability of elements ingested by humans provides several telling examples of additional factors modifying assimilation. Diet can have a strong impact on bioavailability; we have all been instructed about the do's and don'ts of eating while taking various medications. Chronic zinc deficiency and consequent dwarfism found in regions of the Middle East are linked to the poor zinc availability in the predominantly cereal diet of these peoples and the custom of eating clay onto which zinc can adsorb (Sandstead, 1988). Zinc is more available in meats relative to cereals, and clay in the diet sequesters zinc during its passage through the intestine. Protein in the diet increases zinc availability, likely due to enhanced absorption of zinc that is chelated by histidine and cysteine (Sandstead, 1988). In contrast, a protein rich diet reduces calcium bioavailability.

Bioavailability of ingested lead to humans and nonhuman species is a topic of much deserved attention. The lead in paint chips is all too bioavailable to small children ingesting them. Many avian poisonings also involve lead. The bird species of concern are those feeding in wetlands or fields spattered with shotgun pellets from sporting activities. Lead sinkers from the bottom of fishing ponds may also be a source of lead to large waterfowl such as swans or some geese species (Pain, 1995). Raptors feeding on birds containing lead shot are exposed too (Pain, 1995). Indeed, 338 of 4300 dead bald eagles that were examined in a survey by the U.S. Department of Interior were found to have been poisoned by lead in shot or bullet fragments from their prey (Franson et al. 1995). Shot is retained in the gizzard of birds as gizzard stones and lead is slowly released under the associated grinding and acidic conditions (Amiard-Triquet et al. 1992). Carrion scavengers are especially prone to lead poisoning through the inadvertent ingestion of spent lead bullets and shot in carcasses. Efforts to reintroduce the endangered California condor, *Gymnogyps californianus*, are hampered by this scavenger's susceptibility to such poisoning (Wiemeyer et al. 1988; Cade, 2007) that has in some cases required wildlife conservationists to administer chelation therapy to debilitated condors (Wynne and Stringfield, 2007).

Bioavailability of metals and metalloids in sediments is difficult to estimate (Luoma, 1989); however, it is thought to be determined by concentrations in interstitial water and concentrations in different solid phases. The solid phase concentrations influence bioavailability by dictating the concentrations of metal in the interstitial waters surrounding the biota in addition to defining the amounts of solid phase metal available for direct ingestion by benthic species.* Because total metal concentration in sediments can be a poor indicator of available metal (Tessier et al.1984), many estimates of bioavailable metals depend on partial extractions such as a 1 N HCl extract (Krantzberg, 1994) or a series of sequential extractions of sediments thought to grossly separate particular metal-binding fractions of the solid sediments (e.g., Tessier et al. 1979; Babukutty and Chacko, 1995). As an example using a single extractant, sediment-bound metals extracted into an EDTA solution were correlated with bioavailability to several marine invertebrates (Ray et al. 1981). Relative to sequential extractions, Tessier et al. (1984) found that bioaccumulation of sediment-associated metals in a freshwater mussel was best correlated with metal concentrations in the more easily extracted fractions in the extraction series shown in Figure 4.6.

For oxic sediments, several general trends can be identified regarding metal bioavailability. First, easily extracted (1 N HCl) iron, notionally reflecting iron hydrous oxides, tends to inhibit metal bioavailability (Newman, 1995). Presumably this reflects the avid binding of metals to oxides in oxic sediments. Consequently, metal concentrations in a 1 N HCl sediment extract may be normalized to the simultaneously extracted ("easily extracted") iron to account for this effect (e.g., Luoma and Bryan, 1978). Although less consistent than this effect of iron, an increase in sediment organic carbon can diminish bioavailability for some metals (Crecelius et al. 1982) and metal concentrations can be normalized to sediment organic carbon content (Ogendi et al. 2007). Finally, more easily extracted fractions in sequential extractions tend to be more bioavailable than more tightly bound metals (Tessier et al. 1984; Rule and Alden, 1990; Young and Harvey, 1991; Newman, 1995).

For anoxic sediments, bioavailabilities of some metals (e.g., cadmium, chromium, lead, mercury, and nickel) are correlated with sulfide concentrations as reflected by the **acid-volatile sulfides** (AVS), sulfides extracted with 1 N HCl believed to be predominately iron sulfides, especially metastable amorphous FeS, mackinawite (FeS), and greigite (Fe_3S_4) (Morse and Rickard, 2004; Campbell et al. 2006). The presence of sufficient amounts of AVS sequesters the sediment-associated metals as highly insoluble metal sulfides. Equilibrium between the extremely insoluble metal sulfides and the large amounts of iron sulfides is established, favoring metal precipitation (right side of the equation): $Cd^{2+} + FeS_{(S)} \leftrightarrow CdS_{(S)} + Fe^{2+}$. This maintains very low interstitial water concentrations of toxic metals and, according to Di Toro et al. (1990), renders them unavailable, especially those metals that bind avidly with sulfur. For anoxic sediments, a 1 N HCl extract may be analyzed for

* Based on this premise, Campbell et al. (1988) classified sediment-associated organisms as **Type A** (those in contact with sediments but unable to ingest particles) and **Type B** (those capable also of ingesting particles) **organisms**. Examples of Type A organisms may be benthic algae or rooted macrophytes. Examples of Type B organisms are detritivores or many suspension feeders. With this distinction, bioavailability is discussed for the two classes relative to interstitial water concentrations only (Type A) or solid phase plus interstitial water (Type B).

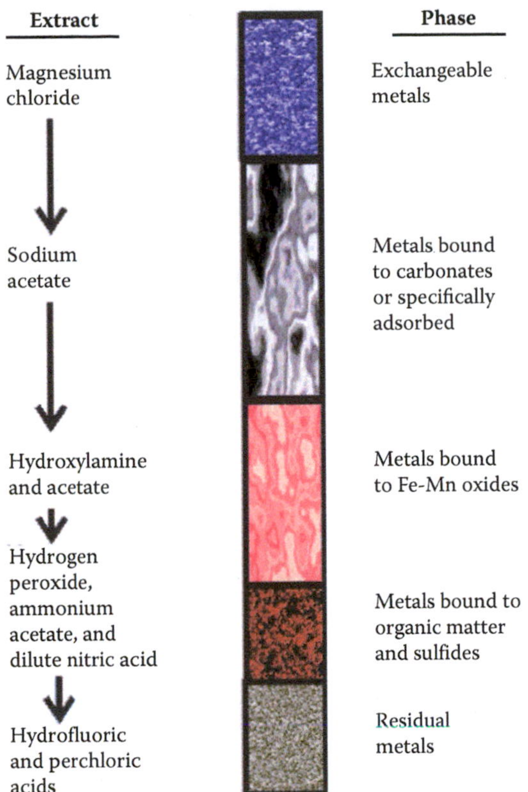

Figure 4.6 Fractionation of sediment-associated metals by sequential extraction. In the scheme shown here (Tessier et al. 1984), an aliquot of sediment is sequentially extracted with magnesium chloride, sodium acetate, hydroxylamine and acetate, hydrogen peroxide, ammonium acetate and nitric acid, and, finally, strong acid to produce five different extracts. These extracts are thought to grossly reflect the amount of metal in the forms noted at the right side of the figure. These are operational definitions and the phase descriptions often do not accurately reflect the true phase association of the extracted metals.

AVS and simultaneously extracted metals (SEMs). The metal concentrations (SEM) are normalized to AVS, e.g., Ankley et al. (1991, 1996), or Carlson et al. (1991). Di Toro et al. (1990) argued initially that, for values of SEM/AVS less than 1, the metal is precipitated as sulfide and relatively unavailable to have a toxic effect on associated benthic species. Recent applications of this approach use SEM-AVS instead of SEM/AVS with SEM-AVS > 0 suggesting the presence of bioavailable metals that might cause an adverse impact on biota. Di Toro et al. (2005) extended the SEM-AVS method to include normalization of the lethal concentration to organic carbon content of sediments, finding good prediction for the onset of toxicity in sediments containing increasing amounts of cadmium, copper, nickel, lead, and zinc. However, controversy exists about the general utility of this approach as it is presently applied. For example, some benthic invertebrates create an oxic microenvironment around themselves as in the case of a burrow microenvironment and prediction based on anoxic geochemistry might be inadequate for such species (Campbell et al. 2006). Also, many species ingest sediment, and consequently, particulate-associated metals gain entry to organisms via ingestion in addition to uptake from interstitial waters. Ogendi et al. (2007) also recently explored metal bioavailability to midge (*Chironomus tentans*) larvae using SEM-AVS and SEM-AVS normalized to the organic carbon content of freshwater sediments. In their study, the organic carbon normalization did not appear to enhance predictions. The metals to which this method has been applied have been Class B and some intermediate metals, and generalizations are not yet warranted about all

Class B and intermediate metals. Long et al. (1998) indicated that SEM-AVS normalization was not more accurate than simply normalizing amounts of metal to dry weight of sediment. DeJonge et al. (2009) also found metal accumulation in chironomids and tubificid worms from field locations to be most correlated with total metal concentration in sediments or total metal concentration normalized to organic carbon or clay content, but not AVS. Lee et al. (2000) published an article in *Science*, concluding that "[The] evidence refutes the prevailing view that the bioavailability of toxic metals is regulated by metal interactions with pore water and reactive sulfides." Morse and Rickard (2004) argue for caution by describing the associated chemical complexities of iron sulfide diagenesis and spatial variability, and the toxicity of dissolved sulfides themselves in sediments. Campbell et al. (2006) conclude that, despite legitimate counterarguments to the general application of this method, "tests of the model on field-collected sediments and on metal-spiked sediments have consistently shown excellent agreement (>90%) between predictions and observations of the *absence* of toxicity … Predictions of the occurrence of toxicity are admittedly much less accurate …" The conclusion at the moment can only be that application of the SEM-AVS approach is warranted if done with a careful understanding of its limitations and appropriateness of its application for specific metals.

A **biomimetic approach** to estimating bioavailability of contaminants can also be used for sediments and other ingested solids. A portion of the solid material containing the contaminant is placed into a synthetic digestive juice solution, extracted for a period of time, and the amount of released contaminant measured. This was the approach taken by Rodriguez et al. (1999) for estimating bioavailable arsenic in soils. Another version of this approach would involve extraction with actual gastric juices drawn from organisms as was done by Yan and Wang (2002) to estimate bioavailable metals in sediments ingested by an annelid. Comparison of the biomimetic method by Chen and Mayer (1999) and Fan and Wang (2003) to the SEM-AVS approach showed moderate agreement. However, results for some metals showed poor agreement with the SEM-AVS approach or the influence of other solid phases.

4.2.3 Organic Contaminants

4.2.3.1 Bioavailability from Water

Bioavailability of organic compounds from water and other sources has been described with **structure–activity relationships** (SARs) that use molecular qualities of the compounds to predict activity or availability. Often such qualitative relationships predict changes in activity of a drug or toxicant with structural changes such as the addition of a chloride atom to or removal of a methyl group from a parent molecular structure. If expressed quantitatively, SARs become **quantitative structure–activity relationships** (QSARs). A QSAR is a quantitative—often statistical—relationship between molecular qualities and bioactivity such as bioavailability or toxicity. Molecular qualities include measures of lipophilicity, steric conformation, molecular volume, ionization, polarity, or reactivity. But, in ecotoxicology, the most commonly used are measures of lipid solubility of organic compounds, such as K_{OW} (Figure 4.7). Partitioning between *n*-octanol and water has been the most common procedure used to reflect partitioning between water and lipids in organisms, e.g., Mackay (1982). Chiou (1985) used triolein (glyceryl trioleate)—water partitioning (i.e., K_{TW}) to better reflect partitioning between water and triglycerides in organisms (Figure 4.7c). The more widely applied *n*-octanol—water partitioning was found to accurately reflect partitioning between triolein—water to a log K_{OW} of 6. Above 6, the K_{OW} predictions dropped slightly below expectations from perfect linear concordance between K_{TW}- and K_{OW}-based predictions. This suggested that application of log K_{OW} to predict partitioning with lipids might be slightly biased above a log K_{OW} in the range of 6.

In the simple K_{OW} approach, the organism is envisioned as a membrane enveloped pool of emulsified lipids. In this conceptual model, uptake and elimination are controlled by permeation of the membrane and/or permeation through aqueous phases (Connell, 1990) with the predominant

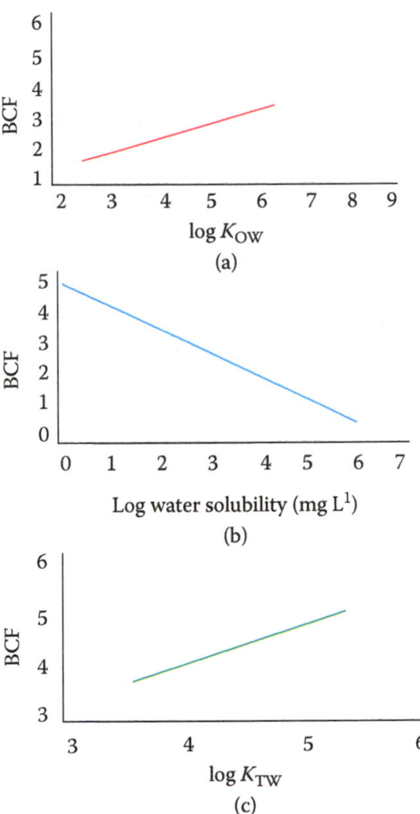

Figure 4.7 Relationships between bioconcentration factors for a variety of organic compounds and log K_{OW} (trout muscle), log water solubility (marine mussel), and K_{TW} or log of triolein–water partition coefficient (log K_{TW}, pooled data for rainbow trout and guppies). (Panels a, b, and c are modified from Figure 1 of Neely et al. (1974), Figure 1 of Geyer et al. (1982), and Figure 3 of Chiou (1985), respectively.)

process being dictated by the qualities of the specific compound in question. Uptake of small, hydrophilic molecules is strongly influenced by membrane permeation but large, hydrophobic compounds are controlled more by permeation through aqueous phases. Very large molecules (e.g., molecular size >9.5 Å) (Landrum and Lydy, 1991) will not pass through the lipid membrane. Uptake across the membrane is dictated by Fick's Law (Equation 3.5).

The consequence of these behaviors of organic compounds that vary in size and lipophilicity can be illustrated with the work of Connell and Hawker (1988) (Figure 4.8). In the upper panel in Figure 4.8, log of the uptake constant is plotted against log K_{OW} for a series of chlorinated hydrocarbons accumulating in three fish species. Among log K_{OW} values of approximately 3–6, uptake increases linearly with log K_{OW} and is thought to be controlled primarily by membrane permeation. Above a log K_{OW} value of 6, the rate of increase slows and eventually uptake begins to decrease with increasing K_{OW} as the large molecular size of the most lipophilic compounds begin to impede diffusion in aqueous phases of the fish. Elimination (log k_e^{-1}) is restricted by membrane diffusion at low K_{OW} values and becomes linear (diffusion controlled, see Equation 3.5) between log K_{OW} values of approximately 2.5–6.5. Above this point, the lipid solubility begins to deviate from a perfect correlation with *n*-octanol-based predictions: K_{OW} becomes an increasingly poorer surrogate of lipid-water partitioning as compounds become more and more lipophilic. Chiou (1985) explained this deviation quantitatively with the **Flory–Huggins Theory** relating solubility to solvent molecular size. The discrepancy between molecular size of *n*-octanol and lipids of organisms becomes increasingly important for larger, more lipophilic compounds. The net result of these factors on uptake and elimination is shown in the bottom panel in Figure 4.8. Below approximately log K_{OW} of 3, bioaccumulation of the most water soluble compounds is controlled by permeation of the membrane. Bioaccumulation is determined primarily by diffusional processes between log K_{OW} values of 3–6. Above this range, the relationship is strongly influenced by inhibiting effects of increased molecular size on diffusion, and accordingly, bioconcentration factors drops with increasing log K_{OW}.

Other factors contribute to bioavailability of organic compounds from water. The above model would not apply for compounds undergoing extensive biotransformation. Also, bioavailability of ionizable organic compounds would be influenced by pH, as already described and soon to be described in more detail for ingested, ionizable contaminants. Models (QSARs) can be developed, which include lipophilicity and ionization (e.g., Lipnick, 1985) as well as other factors. Classic models based on the Hansch equation relate bioactivity to hydrophobicity, electronic, and steric qualities of molecules (Lipnick, 1995). **Linear solvation energy relationships** (LSERs) are based on molecular volume, ability to form hydrogen bonds, and polarity or ability to become polarized (Blum and Speece, 1990).

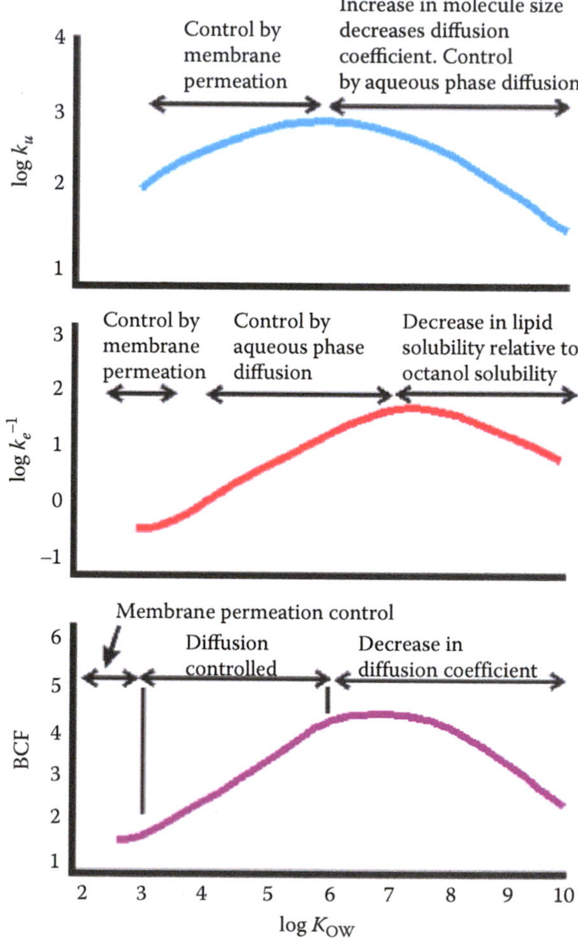

Figure 4.8 A summary of the changes in uptake, elimination, and bioaccumulation (bioconcentration factor) as related to K_{OW}. (Modified from Figures 3, 4, and 5 of Connell, D.W. and D.W. Hawker, *Ecotox. Environ. Safe.*, 16, 242–257, 1988.)

4.2.3.2 *Bioavailability from Solid Phases*

For ionizable contaminants, gastric pH can influence availability with the direction of effect being determined by the pK_a of the contaminant.* (The pK_a is $-\log_{10}$ of the ionization constant (K_a) for a weak, Brønsted acid, where $K_a = ([H^+][X^-])/[HX]$.) Weakly acidic, organic compounds with pK_a values greater than 8 are unionized in the human gastrointestinal tract but bioavailability of those with pK_a values between 2.5 and 7.5 is pH sensitive (Gilbaldi, 1991). According to the **pH-partition hypothesis** (Shore et al. 1957; Wagner, 1975), bioavailability is determined by diffusion of the unionized form from the gastrointestinal lumen across the "lipid barrier" created by the gut lining and into the tissues as determined by pK_a and pH. The proportion of the compound remaining unionized can be estimated with the **Henderson–Hasselbalch relationship** for monobasic acids (donors of single protons) (Equation 4.4) and monoacidic bases (acceptors of single protons) (Equation 4.5) (Wagner, 1975):

$$f_u = \frac{1}{1+10^{\mathrm{pH}-\mathrm{p}K_a}} \tag{4.4}$$

$$f_u = \frac{1}{1+10^{\mathrm{p}K_a-\mathrm{pH}}} \tag{4.5}$$

However, the ionized form may also contribute to bioavailability (Wagner, 1975; Gibaldi, 1991; Palm et al. 1999). In such cases, the diffusion rates and concentrations of both unionized and ionized forms of the contaminant contribute to estimation of bioavailability.

The K_{OW} influences bioavailability of lipophilic compounds in food. Spacie et al. (1995) noted a maximum availability of contaminant in food at log K_{OW} values of 6 with uptake being lower for very hydrophobic and large compounds. They discuss several studies indicating low bioavailability to fish of several lipophilic organic contaminants in food. Donnelly et al. (1994) indicated that organic compounds with log K_{OW} values of 4–7 such as PCBs are quickly absorbed to soils and, consequently, are not readily available to terrestrial plants. In contrast, compounds such as many pesticides with log K_{OW} values of approximately 1–2 are taken up more easily by plants. An important example would be the neonicotinoid pesticides that have generally low lipid solubility and high water solubility. This allows them to enter plants from soils and contributes to their value as systemic insecticides (Tomizawa and Casida, 2005).

If one views oral bioavailability from the biologically effective dose vantage, **Lipinski's rule of 5** becomes handy for gauging bioavailability of relatively small and lipophilic compounds (Lipinski et al. 2001).† These four rules of thumb indicating poor absorption or permeation are the following:

1. More than five hydrogen bond donors, that is, the sum of NH and OH groups exceeds five
2. More than 10 hydrogen bond acceptors, that is, the sum of N and O atoms exceeds 10
3. A molecular weight greater than 500 Da
4. A calculated log K_{OW} value above 5

Molecular weights beyond 500 generally result in poor intestinal permeability as does a calculated log K_{OW} above five. In addition, Lipinski et al. (2001) found that compound bioavailability was related to

* The bioavailability as influenced by gut pH is more complicated than this because the pH conditions of various regions of the gut are quite different. Weakly basic compounds will be more rapidly absorbed in the small intestine than in the acidic stomach. However, many acidic compounds can also be absorbed more effectively in the small intestine than the stomach because of the large amount of surface area of the small intestine relative to the stomach. Consequently, **gastric-emptying rates** (the rate at which the contents of the stomach are emptied into the small intestine) can influence bioavailability for many substances, e.g., paracetamol (Heading et al. 1973) by modifying the time that a compound remains under different pH conditions.

† Unfortunately, one Lipinski's rule of 5 involves 10 instead of 5!

the number of NH and OH bonds, and the number of N and O atoms it possesses. The primary reason for this impact was that hydrogen bonding inhibits membrane permeation by influencing partitioning between strongly hydrogen-bonding water and nonhydrogen-bond-accepting membrane phases. Poor bioavailability is anticipated if more than one of the above rules is violated. As a pertinent example, aldicarb (Figure 2.17) would not violate these rules given its 190.26 mol wt, 1.359 log K_{OW}, ΣO+N of 4 and Σ NH+OH bonds of 1. Of course, substrates with specific transporter mechanisms are the exceptions.

In sediments, bioavailability to benthic species usually—but not always—decreases with increasing log K_{OW} (Landrum and Robbins, 1990). This is likely due to the enhanced partitioning of nonpolar organic compounds to the sediment solid phases with consequent low concentrations in the interstitial waters. Any increase in sediment organic carbon content can diminish the bioavailability of nonpolar organic compounds much as AVS decreases bioavailability of metals in anoxic sediments.* A maximum bioavailability has been noted at a log K_{OW} of approximately 6 for some series of chlorinated hydrocarbons (Landrum and Robbins, 1990).

Bioavailability of organic contaminants in sediments can be estimated by a range of techniques, many of which have already been described. For example, a semipermeable membrane placed into sediments was used by Lyytikäinen et al. (2003) to reflect the sediment-associated polychlorinated dibenzofurans, polychlorinated dibenzodioxins, and polychlorinated diphenyl ethers (PCDE)† available to the annelid, *Lumbriculus variegatus*. Results suggested that lipophilicity, dipole moment, molecular size, and structural configuration all played a part in determining bioavailability of these contaminants in sediments. Lipophilicity alone was inadequate for prediction. As a final example, a digestive fluid extract approach like the biomimetic approach described earlier for metals was used by Weston and Maruya (2002) to screen for sediment-associated contaminants of concern. They found PCBs and high molecular weight PAHs to be sufficiently bioavailable to be a concern for the two marine test species, *Macoma nasuta* and *Arenicola brasiliensis*.

4.3 BIOLOGICAL QUALITIES INFLUENCING BIOACCUMULATION

4.3.1 Temperature-Influenced Processes

Temperature is perhaps the most widely studied and important factor affecting the general physiology of individual organisms. This being the case, it should be no surprise that temperature influences biochemical and physiological processes associated with bioaccumulation. Indeed, the strong positive relationship noted in the 1960s between metal (zinc or radiocesium) excretion rate and temperature-dictated metabolic rates of poikilotherms (Mishima and Odum, 1963; Williamson, 1975) and homeotherms (Pulliam et al.1967; Baker and Dunaway, 1969) lead researchers to explore ^{65}Zn elimination as a way of measuring metabolic rates of free-ranging individuals. Unfortunately, such use was compromised because free-ranging animals and laboratory-maintained animals often differed in other important ways that influenced elimination rates (e.g., Pulliam et al. 1967). These early studies identified much inexplicable variability and bias in results. It should be remembered that temperature also determines important rates such as those for feeding, growth, and egestion in addition to key cellular qualities such as membrane fluidity and lipid composition.

Generally, increases in temperature within normal physiological ranges have been shown to increase bioaccumulation, e.g., mercury in mayfly nymphs (Odin et al. 1994), cadmium and mercury in mollusks (Tessier et al. 1994), cadmium in Asiatic clams (Graney et al. 1984), and DDT in rainbow trout (Reinert et al. 1974). Cesium (^{134}Cs) uptake was highest at temperatures optimal for food

* This general partitioning of contaminants between solid and dissolved phases of sediments with consequent effects on bioavailability and toxicity (bioactivity) is the foundation for the equilibrium partitioning approach to sediment criteria development. See Shea (1988) and Di Toro et al. (1991) for more details.

† As described in Chapter 2, the **polychlorinated diphenyl ethers** (PCDE) are similar in structure to the PBDE except they are polychlorinated instead of polybrominated.

consumption and growth of rainbow trout (Gallegos and Whicker, 1971). The biological half-life ($t_{1/2}$) for elimination of ^{134}Cs from rainbow trout increased as water temperatures increased according to an exponential relationship*, $t_{1/2} = (\text{Constant})\, e^{-0.106t}$ where t = temperature in °C (Ugedal et al. 1992). For the rainbow trout, retention of methylmercury was approximately 1.5 times longer at 0.5°C to 4.0°C than at water temperatures of 16°C to 19°C (Ruohtula and Miettinen, 1975).

Some studies report no effect of temperature changes on bioaccumulation kinetics. Smith and Green (1975) reported that methylmercury uptake and elimination by freshwater clams were not significantly affected by temperature. Clearly, effects of temperature are sometimes complex and can deviate from the general trends discussed here.

Watkins and Simkiss (1988) gave a fascinating example of such a complication. They, like others, found enhanced accumulation of a contaminant (zinc in the marine mussel, *Mytilus edulis*) with an increase in water temperature (10°C increase to 25°C). But they also examined the effect of fluctuating temperatures between 15°C and 25°C, and found that zinc bioaccumulation was even higher than its bioaccumulation in the constant 25°C treatment. They hypothesized that this result was linked to shifts among various zinc pools within the mussels. Both free and ligand-associated zinc are present in the mussels: $Zn^{2+} + L^{2-} \leftrightarrow ZnL$. When water temperature increases, the equilibrium for the complexation reaction shifts to favor more zinc existing free of the biochemical ligands. This zinc then becomes more available for incorporation into granules. On cooling, zinc entering from outside the animal establishes equilibrium with ligands again, thus replenishing the zinc lost to granules. As temperatures increase and decrease, the process is repeated with enhanced accumulation of zinc in granules. The cyclic association with ligands, shift toward dissociation from ligands, and sequestering in granules ratchets zinc concentrations higher than if the mussel were left at a higher temperature of 25°C.

4.3.2 Allometry

Allometry, the study of size and its consequences (Huxley, 1950), can also be important to consider for bioaccumulation. Metabolic rate and a myriad of other anatomical, physiological and biochemical qualities of organisms change with size (see Adolph [1949] and Heusner [1987] for more detail) and, in so doing, modify uptake, transformation and elimination rates. The commonly observed consequence is size-dependent bioaccumulation. However, because age and size are correlated in most species, allometric effects are often confused with age or exposure duration effects in surveys of bioaccumulation. Regardless, many studies detail allometric effects on contaminant uptake (Newman and Mitz, 1988; Schultz and Hayton, 1994), biotransformation (Walker, 1987) (Figure 4.9), elimination (Mishima and Odum, 1963; Reichle, 1968; Gallegos and Whicker, 1971; Ugedal et al. 1992; Hendricks and Heikens, 2001), and general bioaccumulation (Boyden, 1974, 1977; Landrum and Lydy, 1991; Warnau et al. 1995; Hendricks and Heikens, 2001; Hendricks et al. 2001). These have been reviewed for metal bioaccumulation by Newman and Heagler (1991) and Hendricks and Heindricks (2001), for linkage to metabolic rate by Fagerström (1977), and for scaling of pharmacokinetic parameters by Hayton (1989). Most resort to the classic power model for **scaling**, that is, the manipulation of allometric data (size versus some physiological, morphological, or biochemical quality) to produce quantitative relationships.

* Both power and exponential relationships will be discussed in this chapter. A **power relationship** is a mathematical relationship in which the dependent variable (Y) is related to the independent variable (X) raised to some power. For example, $Y = aX^b$. A power relationship can be linearized and conveniently fit by linear regression by taking the log X and log Y in the regression ($\log Y = b \log X + \log a$). In contrast, an **exponential relationship** is a mathematical relationship in which the Y variable is related to some constant raised to the X variable, i.e., $Y = a10^{bX}$. An exponential relationship is transformed to $\log Y = bX + \log a$ as done earlier to fit exponential (first-order) elimination kinetics. Newman (1993, 2013) provides details on fitting these models using linear regression.

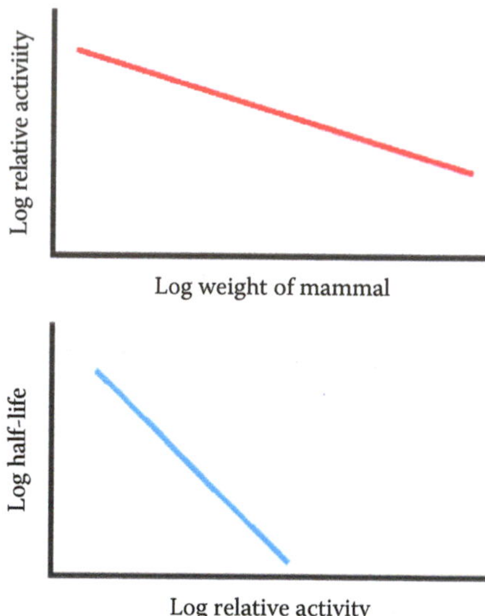

Figure 4.9 The scaling of monooxygenase activity (top panel) and the relationship between relative monooxygenase activity and biological half-life of xenobiotic metabolism (bottom panel). (Composite and modified from Figures 2 and 3 of Walker, C. H., *Drug Metab. Rev.*, 7, 295–323, 1978.)

The general power function used in classic allometric scaling is the following:

$$Y = aX^{b} \tag{4.6}$$

where Y = some quality being scaled to size, X = some measure of individual (or species) size, and a and b are constants often derived by regression analysis. This equation may be used to link size to morphological (e.g., gill surface area), physiological (e.g., blood flow rate or gill ventilation rate), or biochemical (e.g., monooxygenase activity) qualities. It has also been used to model contaminant body burden (amount per organism) to animal size. Equation 4.6 can be converted by dividing both sides by mass to generate Equation 4.7, which expresses the relationship for bioaccumulation in concentration units instead of body burden units

$$Y = aX^{1-b} \tag{4.7}$$

where X, a and b = same as in Equation 4.6 but Y = concentration, not body burden. Much has also been made of the estimated b-values for scaling bioaccumulation because the b-values are a major theme in the classic physiological literature, e.g., Hendricks and Heikens (2001). This important scaling issue remains a fruitful one for study because precipitate explanation from the classic allometry, physiological, and anatomical literature is still frequently invoked without sufficient scrutiny of the situation being explained.

In the mid-1970s, Boyden (1974, 1977) compiled metal body burden data for mollusks and established the power model (Equation 4.6) for scaling contaminant accumulation. He identified three classes of models based on their associated b-values (Figure 4.10). One class had b-values of 1 and reflected a simple proportionality. There was a constant number of binding sites in the tissues regardless of size. Body burden increased linearly with weight of the organism so concentration was independent of weight. Another relationship with b-values in the general range of 0.77 had body burdens and concentrations that changed in a nonlinear fashion with weight. The concentration was higher for small individuals than large individuals. The b-value of 0.77 suggested to Boyden a linkage to metabolic rate that, in general, has a b-value of approximately 0.75 during scaling.

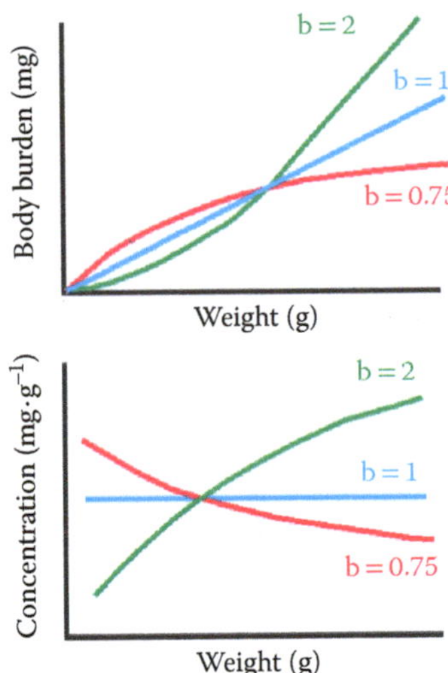

Figure 4.10 The relationships between organism size (weight) and body burden (amount per individual) (top panel) and concentration (amount per gram of tissue) of metal. Modified from Figure 13 of Newman (1995) with slight exaggeration of curvature of the concentration versus size plots to make clear the nonlinear nature of two of these curves ($b = 2$ and $b = 0.75$). The point of intersection for both plots occurs, where $X = 1$. (Reprinted with permission from Newman, M.C., *Quantitative Methods in Aquatic Ecotoxicology*, Lewis Publishers, Boca Raton, FL, 1995. Copyright Lewis Publishers, an imprint of CRC Press, Boca Raton, FL.)

However, Fagerström (1977) was quick to point out that b-values for states (body burden) and fluxes (metabolic rate) cannot be compared directly. He demonstrated mathematically that a b-value of 1 would be expected for contaminant burdens driven by metabolic rate. Boyden's simple hypothesis was inadequate. The final class noted by Boyden had a b-value of 2 or greater (burden related to the square of body weight) and reflected a gradual increase in concentration with an increase in size. He suggested that elements such as cadmium that are subject to rapid removal from circulation and very avid binding in some tissues conform to this class of relationships. He suggested incorrectly that these relationships (b-values) would be constant for species–element pairs. Soon after Boyden proposed this constancy of b-values, Cossa et al. (1980) and then Strong and Luoma (1981) demonstrated considerable spatial and temporal variation in b-values. Further, Newman and Heagler (1991) added to and reanalyzed Boyden's original data set, and found no clear evidence for distinct classes of relationships based on b-values. They found a generally skewed distribution of b-values with medians in the range of 0.80–0.83. They also identified several sources of bias in Boyden's approach. Regardless, Boyden made a major contribution by clearly establishing the power model for scaling body burdens, and hypothesizing plausible and testable underlying mechanisms. His approach is used often to normalize body burden/concentration data taken during surveys of organisms of differing sizes. In such cases, the empirically derived constants are adequate for normalization of data. Regardless of its value, it should not be forgotten that the foundation for such scaling is empirical.

Another equally important use of scaling is computer modeling to predict general behaviors of bioaccumulation dynamics and to isolate important factors controlling bioaccumulation kinetics. As an early example, a power model was used to model ^{137}Cs half-life in humans of different sizes (Eberhardt, 1967). In some such cases, estimation of the exact values for parameters may not be critical and the range of expected values may be sufficient. Today, more complicated allometric explanations and relationships than Boyden's, usually embedded in physiologically based pharmacokinetics (PBPK) models, are used to predict size effects on bioaccumulation (Barber et al. 1988; Hayton, 1989; Schultz and Hayton, 1994; Hendricks et al. 2001; Weijs et al. 2010). Examples of such complicated models are given in Figure 4.11. Temperature, scaling, and other

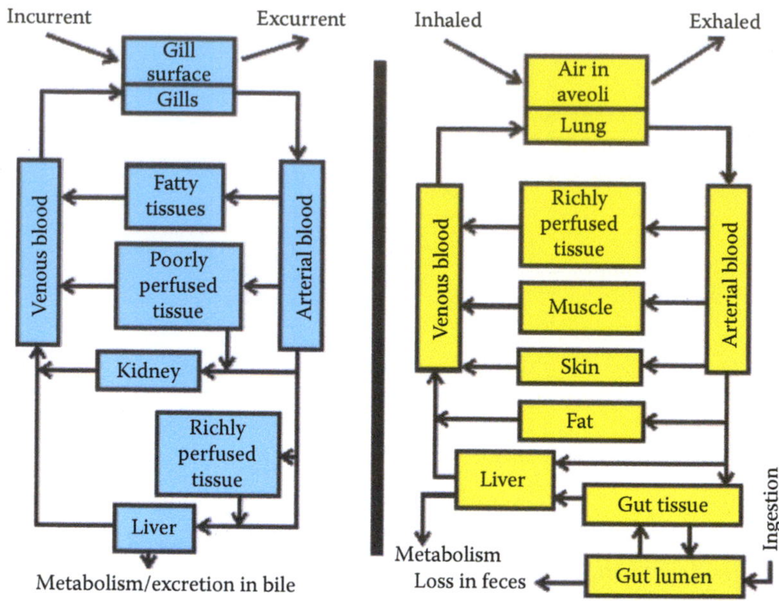

Figure 4.11 Two PBPK models. The model to the left involves the fish uptake of a xenobiotic across the gills, distribution among five compartments and blood, and loss via the liver (Nichols et al. 1990). The model on the right involves styrene inhalation and ingestion by a mammal, distribution into a series of compartments, and loss from liver metabolism and defecation (Paterson and Mackay, 1987). In the original publications, the fish model was formulated as a clearance volume-based model and the mammalian model was formulated as a fugacity-based model.

important factors can be included into these PBPK models. However, as models become more complicated, they require increasingly more and more estimated parameters for the system being modeled.

4.3.3 Other Factors

Many other factors can complicate prediction of bioaccumulation. Bioaccumulation may differ between sexes. The female fin whales studied by Aguilar and Borrell (1994) eliminate organochlorine compounds via transfer to young before birth and during nursing; this is obviously not an avenue of elimination for males. Consequently, older female whales have lower concentrations than older male whales. Diet, as alluded to above, can also influence bioavailability. Shifts in diet associated with ecological or developmental qualities of an organism have consequences relative to bioaccumulation. Acclimation of euryhaline species can change their ability to cope with a contaminant; killifish uptake of pentachlorophenol and excretion of a glucuronide conjugate of pentachlorophenol decreased with the lowering of salinity (Tachikawa and Sawamura, 1994). Lipid content of an individual may also influence bioaccumulation. Rainbow trout with high total body lipid content (21%) accumulated more pentachlorophenol and had lower elimination rate constants than those with lower (13%) lipid content (van den Heuvel et al. 1991).

Inorganic toxicants can be influenced by normal regulatory processes associated with essential chemicals. For example, potassium body burden of humans has a strong influence on the accumulation of its analog, cesium (Leggett, 1986). Other essential elements* such as zinc are carefully regulated within the body. For example, a **cysteine-rich intestinal protein (CRIP)** exists to enhance

* The **essential elements** such as H, Na, K, Mg, Ca, V, Cr, Mo, Mn, Fe, Co, Ni, Cu, Zn, B, C, Si, N, P, O, S, Se, F, Cl, and I. It is presently unclear if Sn, As, and Br are also essential (Mertz, 1981; Fraüsto da Silva and Williams, 1991)

the uptake of zinc by cells in the intestine wall of mammals (Roesijadi and Robinson, 1994). This regulation of essential elements influences availability or effects of nonessential analogs. An excess of an essential element analog such as cadmium can cause symptoms of an apparent (zinc) deficiency (Neathery and Miller, 1975): an excess of zinc may lessen toxic effects of cadmium (Leland and Kuwabara, 1985). Similarly, copper and zinc interact relative to effects on algal cell function (Rueter and Morel, 1981). Many essential elements and some of their analogs are defined as **biologically determinant**; their concentrations in the organisms remain relatively constant over a wide range of environmental concentrations (Reichle and Van Hook, 1970). Other elements are **biologically indeterminant** and their concentrations in organisms are directly proportional to environmental concentrations.

Perhaps less direct, but as telling, is the role of phosphate biochemistry on the accumulation of arsenate in marine species. In the phosphate deficient waters of the Great Barrier Reef, arsenate enters into the normal biochemical pathways involved in phosphate assimilation and is incorporated into zooxanthellae that live in the mantel of the giant clams, *Hippopus hippopus*, *Tridacna maima*, and *T. derasa* (Benson and Summons, 1981). Arsenic then accumulates to extraordinary levels, gaining entrance into the clam via the biochemistry of the symbiotic algae.

4.4 SUMMARY

In this brief chapter, factors influencing bioaccumulation and bioavailability of contaminants were discussed along with several methods of measuring bioavailability. Solid and dissolved sources of organic and inorganic contaminants were detailed relative to bioavailability. Finally, the importance of temperature- and size-dictated changes in biological functions and structures were discussed. Other important factors that influence bioaccumulation were identified including sex, diet, lipid content, and elemental essentiality. Many details associated with accumulation from trophic exchange were omitted because they are discussed thoroughly in Chapter 5.

SUGGESTED READINGS

Campbell, P. G. C., A. G. Lewis, P. M. Chapman, A. A. Crowder, W. K. Fletcher, B. Imber, S. N. Luoma, P. M. Stokes, and M. Winfrey, *Biologically Available Metals in Sediments*, NRCC No. 27694, NRCC/CNRC Publications, Ottawa, Canada, 1988.

Hamelink, J. L., P. F. Landrum, H. L. Bergman, and W. H. Benson, Eds., *Bioavailability: Physical, Chemical and Biological Interactions*, CRC Press, Boca Raton, FL, 1994.

Luoma, S. N., Can we determine the biological availability of sediment-bound trace elements? *Hydrobiolia,* 176/177, 379–396, 1989.

Newman, M. C., *Quantitative Ecotoxicology,* Second Edition, Taylor & Francis/CRC Press, Boca Raton, FL, 2013.

Newman, M. C. and W. H. Clements, *Ecotoxicology: A Comprehensive Treatment*, Taylor & Francis/CRC Press, Boca Raton, FL, 2008.

Newman, M. C. and M. G. Heagler, Allometry of metal bioaccumulation and toxicity, in *Metal Ecotoxicology: Concepts and Applications*, Newman, M. C. and A. W. McIntosh, Eds., Lewis Publishers, Chelsea, MI, 1991.

CHAPTER 5

Bioaccumulation from Food and Trophic Transfer

Far too often food chains have been envisioned as mechanisms operating solely to concentrate pollutants as they move from prey to predator. Less often are they objectively recognized as ecological processes, with the net effect of concentration or dilution of materials during their transport along the food chain being dependent upon a complex of biological variables.

Reichle and Van Hook (1970)

5.1 INTRODUCTION

As evidenced by Minamata disease and dichlorodiphenyltrichloroethane (DDT) poisoning of birds, the transfer of contaminants through trophic webs* can have undesirable consequences to top predators. Some contaminants such as mercury and DDT display **biomagnification**, an increase in contaminant concentration from one trophic level (e.g., primary consumer) to the next (e.g., secondary consumer) because of accumulation from food. The possibility of biomagnification must be considered in any thorough assessment of ecological or human risk. However, because biomagnification played a pivotal role in our awakening to environmental issues, it is sometimes invoked as a ruling theory when equally plausible, alternate explanations are present (Moriarty, 1983; Beyer, 1986; Laskowski, 1991). For example, predators tend to live longer than prey species and have more time to accumulate some contaminants than do prey. The result may be higher concentrations in predators than prey (Moriarty, 1983). Predators are often larger than prey and allometric effects on bioaccumulation can result in higher concentrations of some contaminants in predators relative to prey (Moriarty, 1983). Also, for lipophilic contaminants, higher lipid content in predators than prey can also result in increases in contaminant concentrations with trophic level. Lower food web organisms tend to grow faster than those higher in the food web so growth dilution might be more pronounced at lower levels than at higher levels (Huckabee et al. 1979). Difficulties in defining trophic status, especially if species change their feeding habits with age or life stage (Huckabee et al.1979), can confound and render subjective conclusions of biomagnification. Also, most field studies of bioaccumulation do not distinguish between water and food sources, pressing many researchers toward unjustified speculation about biomagnification. Finally, some communities show such wide variation in concentrations among species of a particular trophic level that trends noted among trophic levels are often questionable (Beyer, 1986; Laskowski, 1991). Such calculations and

* The term food web is often preferable to food chain in discussions of trophic transfer. The concept of a web of interactions is often more accurate than that of an orderly transfer from one distinct level to the next highest level only. Many species feed on an array of prey that, in turn, have equally complex feeding strategies. Such **omnivory** is best accommodated in the context of a trophic web.

their interpretations are frequently biased toward biomagnification. Reports of biomagnification must be read carefully to peel away unintentional biases of investigators. Yet, because of the important consequences of biomagnification for many contaminants and ecological communities, the concept is well worth investigating.

Before any further discussion of biomagnification, it is necessary to mention that the terms, **bioamplification** and **trophic enrichment** are infrequently used instead of biomagnification. The general term, bioaccumulation is occasionally used incorrectly as a synonym for biomagnification. Sometimes, biomagnification is used to describe field observations of increasing concentrations with increasing trophic level regardless of the ambiguity about the magnitude of uptake from food relative to water. It is not used in that context here as the historical literature clearly associates trophic transfer of contaminant with biomagnification. Results of biomagnification studies using this term otherwise can be needlessly confusing and should be scrutinized carefully to determine the inferential strength of associated conclusions.

The term bioamplification has recently been given a distinct meaning by MacDonald et al. (2002) who envision two processes contributing to bioamplification, solvent switching, and solvent depletion. Solvent switching takes place when a contaminant partitions preferentially into one solvent or phase (e.g., lipids of a prey species) relative to another (e.g., its surrounding media or food). There is an increase in concentration in the prey but the final media–prey fugacity does not change with solvent switching. When the prey is eaten, solvent depletion occurs with the digestion of the lipids containing the contaminant. The contaminant then partitions into the lipids of the predator, increasing concentration and fugacity of the predator. Of course, bioenergetics also contributes because consumers must take in more biomass than they themselves possess to survive. The dynamics are such that, in terms of the fugacity transport constant discussed in Chapter 3, the contaminant D_{in} is greater than D_{out} for members of the food web. So biomagnification of a persistent hydrophic compound involves **concentration** and **fugacity amplification,** which, in turn, involve a series of solvent switching and depletion events. Daley et al. (2009, 2011) used the concepts of concentration and fugacity amplification to describe bioamplification of polychlorinated biphenyl (PCB) in fish eggs and ephemeral emergent insects. The consumption of stored nutrients in fish egg during larval development resulted in bioamplification. Stream insects were found to bioamplify PCBs because of weight and lipid loss during emergence. This potentially increased PCB consumption by terrestrial predators of emergent insects.

Returning to the context of biomagnification, it is not required to have adverse effects as a consequence of trophic transfer. If contaminant concentrations are very high in a food item, species farther up the trophic web might still be exposed to concentrations sufficient to produce an adverse effect. This was the case with Japanese afflicted with itai-itai disease as discussed in Chapter 1. Another example involves metals and metalloids that accumulate to extremely high concentrations in materials coating submerged surfaces in lentic and lotic systems (Newman et al. 1983; 1985). These materials (Figure 5.1) including the associated *aufwuchs* (**periphyton**) are ingested by an important grazer/scrapper guild of freshwater organisms and have the potential to cause adverse effects even in the absence of biomagnification (Newman and Jagoe, 1994). Also, the flux into an organism, not only the net concentration in that organism, may contribute to determining the adverse outcome of toxicant exposure.

Biomagnification is one of three possible outcomes for trophic transfer of contaminants. Concentrations may be similar in both predator and prey: There might be no statistically significant upward or downward trend in concentrations. Alternatively, as is commonly the case, the contaminant concentration may decrease as trophic position increases. During each transfer, the required balance among ingestion rate, uptake from food, internal transformations, and elimination does not exist for the conservative transfer of a contaminant: concentrations decrease with each exchange. Diminution with increasing trophic level is called **trophic dilution** (Reichle and Van Hook, 1970) or **biominification** (Campbell et al. 1988). The trophic dilution of lead noted in an experimental

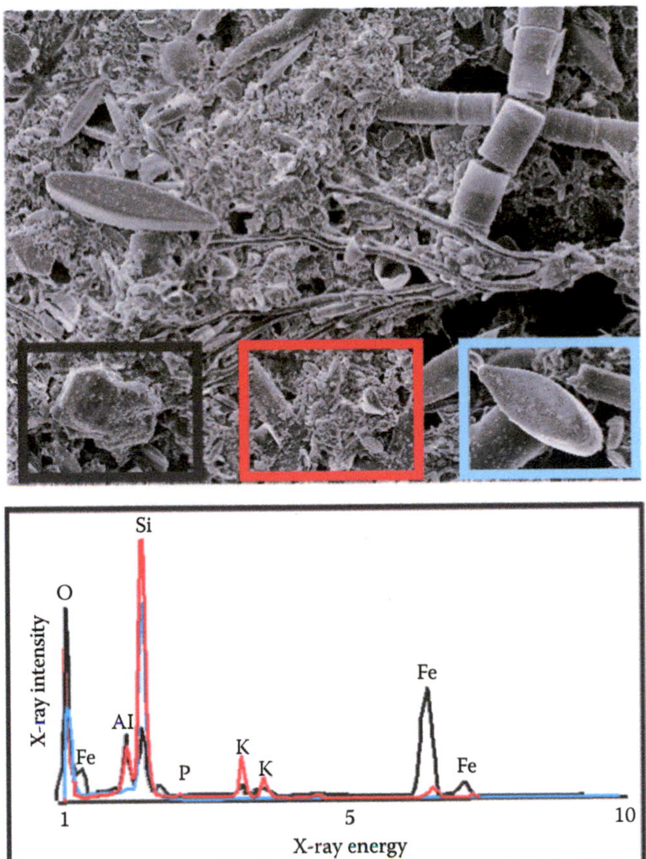

Figure 5.1 A scanning electron microscope image of procedurally defined periphyton removed from submerged surfaces of the South River (Virginia). The inserts are higher magnification images of selected areas. X-ray spectra obtained from three contrasting areas of peryphyton were generated during electron bombardment of the samples in the scanning electron microscope and are depicted at the bottom of this figure. The elements associated with each peak of characteristic x-rays are also shown and correspond in color to the framed micrograph regions. The spectra drawn with a black line was rich in Al, O, Si, K, and particularly iron, indicating mineral composition with considerable iron oxides. The spectrum drawn in red was from an area that was rich in Al, O, Si, and K, suggesting primarily mineral composition again with perhaps considerable quartzite and clays. The last (blue line) was taken from a diatom that has a frustule composed of silicon dioxide as is clear from the large Si and O x-ray peaks. (M. C. Newman, Unpublished micrographs and spectra.)

multilevel food web by Soto-Jiménez et al. (2011) is one recent example. **Bioreduction** (Nott and Nicolaidou, 1993) has also been used in describing trophic dilution but the meaning seems to shift slightly to focus on rendering the contaminant less bioavailable in the prey biomass with a consequent ineffective assimilation in the predator (Vijver et al. 2004). For example, Nott and Nicolaidou (1993) noted that, after being ingested by the carnivorous gastropod, *Nassarius reticulatus*, half of the zinc bound in intracellular phosphate granules of the marine gastropod prey species, *Littorina littorea*, was egested without being assimilated. The terrestrial food web calcium sink described in Chapter 2 for ^{90}Sr is another process resulting in bioreduction.

This chapter was developed to specifically explore bioaccumulation from food because of the importance of this phenomenon (e.g., Vignette 5.1), especially in terrestrial systems where food can be the predominant source of contaminants to biota. The theme brought out by the aforementioned quote that bioaccumulation from food is a complex process that has been viewed too simplistically in the past, will be enriched in this chapter. Unlike Chapter 4, the focus will be on the sequential transfer of contaminants among species rather than bioaccumulation within an individual. Details about differential transfer of radioisotopes and organic isomers will be provided. Means of measuring trophic status will be discussed too because of its crucial role in bioaccumulation studies.

VIGNETTE 5.1 Birds as Monitors of Mercury Pollution

Robert W. Furness
University of Glasgow

Mercury is discharged into the environment in large quantities by volcanic activity, and so has always been present in food webs. One challenge is to identify elevated levels of mercury in food chains because of pollution against a background of highly variable natural levels. The natural process of mercury cycling results in highest concentrations tending to occur in aquatic ecosystems rather than in terrestrial ones. Mercury is highly toxic, but there is evidence that some aquatic animals are able to cope with levels of mercury that would harm terrestrial animals (Monteiro and Furness, 2001a; Finkelstein et al. 2007; Kenow et al. 2007; Ackerman etal. 2008; Burgess and Meyer, 2008; Evers et al. 2008), suggesting that mercury toxicity has led to the evolution of tolerance of mercury in aquatic biota to an extent that has not occurred in terrestrial ecosystems (Thompson et al. 1991). Anthropogenic sources of mercury pollution have increased considerably with industrial development, and particularly with the increase in combustion of fossil fuels (Slemr et al. 2003), but include a wide range of both local- and wide-scale sources. Mercury pollution may occur locally, for example, where mercury is used in gold mining, and may contaminate a river downstream from the mining activity (Camargo, 2002), or where an industrial plant discharges mercury in its waste (e.g., see Vignette 5.2). But most mercury pollution is diffuse, with burning of fossil fuels and incineration of waste resulting in mercury entering the atmosphere and being spread over large areas (Pacyna et al. 2006). Indeed, the high volatility of elemental mercury can lead to extremely long-range transport of this metal by the jet stream, with deposition occurring on dust particles and rainfall right around the planet (Mason and Sheu, 2002). Anthropogenic emissions of mercury occur mostly in the northern hemisphere. The separate atmospheric circulation patterns between the two hemispheres result in much higher contamination in the northern than in the southern hemisphere, with very little transport of mercury across the equator (Lamborg et al. 1999). Modeling of mercury discharges from industrial pollution into the environment, and its transport through atmospheric systems, has suggested that levels may have increased in the environment of the northern hemisphere by a factor of about 4 or 5 over the last 150 years with most mercury entering marine systems, but has probably changed rather little in the southern hemisphere (Pirrone et al. 1998; Lamborg et al. 1999; Mason and Sheu, 2002; Pacyna et al. 2006). In recent years, technological developments have reduced the proportion of mercury reaching the atmosphere from industrial processes in North America and Europe, but increasing use of fossil fuels in these regions, and the rapid industrial development in parts of Asia have continued to add to mercury discharges to the environment. Wildlife provides an opportunity to monitor mercury pollution by sampling animal tissues to assess trends and spatial patterns of mercury pollution (Monteiro and Furness, 1995; Evers et al. 2007). In particular, whereas it is very difficult to monitor historical levels of mercury that used to be present in the atmosphere or in water bodies many decades ago, we can use animal specimens stored in museum collections to investigate historical changes in mercury.

Elemental and inorganic mercury tend not to enter food chains, as assimilation of inorganic mercury by animal digestive systems and through lungs or gills is poor. Most elemental or inorganic mercury ends up in sediments. However, inorganic mercury is converted by certain bacteria into methylmercury, and this organic form is very efficiently assimilated from water or food. Thus, the conversion of mercury into the organic form is a critical step in determining the uptake of mercury into food chains (Monteiro and Furness, 1995). Methylation of mercury tends to occur in environments with low oxygen levels, low pH, and high levels of dissolved organic compounds, environments favored by sulfate-reducing bacteria that are largely responsible for methylation. These conditions are sometimes found, for example, in deep sea environments, in coastal marine sediments, and in some freshwater lakes. One particular example is the bottom of newly constructed freshwater reservoirs, where decay of organic matter as a new reservoir is filled with water can lead to conditions favoring methylation of mercury.

Organic mercury is lipid soluble, and tends to bioaccumulate in animals and to biomagnify up the food chain (Evers et al. 2005; also Vignette 5.2). As a result, almost all mercury in vertebrates, and especially in top predators, is in the form of methylmercury. Organic mercury is also much more toxic to vertebrates than inorganic mercury. Some marine mammals, and probably also a few seabirds, can transform organic mercury back into inorganic mercury, which can then be stored in a nontoxic form in the liver or kidney as crystals of mercury selenide. This demethylation process may be important for animals that have little ability to excrete methylmercury, such as cetaceans. However, virtually all mercury present in fish is methylmercury. This is also the case in most birds and in mammals that grow hair. Indeed, the processes of excretion of methylmercury are key to understanding the patterns of mercury concentrations found in different kinds of animals. In the case of fish, mercury excretion appears to be extremely limited, and mercury concentrations (almost all of which is methylmercury) increase with the age of the fish. This trend means that larger fish tend to have higher levels of mercury, and longer lived fish tend to have more mercury than short-lived fish. In addition, fish that feed at higher trophic levels tend to accumulate more mercury, as do fish that live in locations affected by conditions favoring methylation (for example, fish that live in the deep ocean). In contrast, birds and mammals have more opportunity to excrete mercury as it naturally tends to bind to growing hair and feathers, and to enter eggs.

Methylmercury has a strong tendency to bind to sulfur-containing amino acids. Sulfur-containing amino acids are present in large amounts in keratin, the main protein constituent of hair and feathers. As a result, methylmercury molecules circulating in blood tend to enter into growing feathers or hair. Birds and mammals have concentrations of mercury in their blood in dynamic equilibrium with levels of mercury in their diet and with amounts of mercury sequestered in lipid-rich internal tissues. Thus, birds and mammals with high mercury loads, or mercury intake, grow feathers or hair containing correspondingly high levels of mercury. It is rather easy to cut small samples of feathers or hair from live birds and mammals and use these samples to measure mercury exposure. This opens the opportunity for sampling from large numbers of healthy wild animals, avoiding the need to kill animals to measure mercury in organs such as the liver. Furthermore, the fact that museums hold large historical collections of bird skins (with particularly large numbers collected in the "Victorian era" around 1880–1900) permits the analysis of mercury levels in feathers from specimens that can indicate how levels of mercury in food webs have changed over the last 150 years, and it can be established that mercury levels in feathers do reflect mercury levels in the diet of the birds.

Cetaceans that demethylate mercury accumulate increasing amounts of inorganic mercury, so show increasing levels with age. The same might be true of birds, such as albatrosses, that may also demethylate mercury and store it in the liver and kidney, but rather little is known about this. It seems that most birds do not demethylate mercury because they can lose methylmercury into growing feathers. However, albatrosses are unusual among birds. Large albatrosses can take more than a year to reproduce, and only molt at the end of this, replacing only a proportion of their plumage. This low frequency of molt may require them to demethylate mercury to avoid accumulating potentially toxic levels. Other birds molt more often and more extensively, and can lose most of their mercury burden each year in their main annual molt. As a result, in most birds the level of mercury in their internal tissues does not increase with age (Furness et al. 1990). The exception to this is that young birds (recently fledged) tend to have lower levels of mercury than do adults (Monteiro and Furness, 1995). But adult birds show no further accumulation across the years of adult life. There is a slight tendency for female birds to have marginally lower levels of mercury in their plumage than that found in males, and this relates mainly to the fact that female birds excrete some mercury into the eggs they lay (eggs are also rich in sulfur-containing amino acids and so naturally tend to bind mercury circulating in the blood as the eggs are formed). But this difference between the sexes is very small, and is often obscured because mercury levels vary considerably among individuals (Lewis et al. 1993).

Bird feathers were first used to monitor changes in mercury pollution when they were used to study a wildlife crisis caused by mercury seed dressings used to protect newly planted cereal seeds. It was shown that granivorous birds accumulated high levels of mercury through feeding on these seeds, and that birds of prey accumulated even higher levels by feeding on poisoned granivorous birds. Many granivorous animals were poisoned, but of particular concern was that many birds of prey died as a consequence of mercury poisoning. Birds of prey populations were greatly reduced, and so the use of mercury seed dressing was halted. These studies used analysis of mercury in feathers as a convenient way of demonstrating the high levels of mercury transferring from seed dressings to birds of prey. But they were only partly aware that mercury levels vary considerably from feather to feather. More recent studies have had to investigate these patterns in more detail to assess what mercury levels in feathers mean, and to determine the best sampling methods to use. Complementary techniques, such as stable isotope analysis, can also help in the interpretation of patterns of mercury contamination (Thompson et al. 1998; Bearhop et al. 2000; Tavares et al. 2008).

Several experiments have been carried out to test the validity of using feathers to monitor mercury in the diet of birds. Giving hand-reared birds a diet with increased or reduced levels of mercury leads to higher, or lower, levels of mercury in their growing feathers (Lewis and Furness, 1991). Giving oral doses of mercury to wild seabirds also alters levels in their blood, and in their growing feathers, following a standard linear dose–response model (Monteiro and Furness, 2001a,b). Data from wild birds also support the results obtained from experiments. A study of the levels of mercury in a variety of species of seabirds in the Azores, and in their diets, demonstrated that there was a strong positive correlation between feather mercury and diet mercury levels (Figure 5.2), indicating that samples of feathers can provide a valuable integrative measure of mercury levels in the food web. However, there are important patterns of mercury level in relation to feather type. It seems that birds have little ability to excrete methylmercury when they are not growing new feathers. Mercury levels increase in the bird's liver and muscle tissues when birds are between molts (for example, while breeding in most species). Most bird species have a molt after the breeding season, and mercury levels are highest in the first-grown feathers of the seasonal molt, and tend to decrease as molt progresses because, by the end of the molt, most of the mercury in the internal tissues has gone into growing feathers. So it is important to understand the molt pattern of the bird and to sample appropriate feathers (Furness et al. 1986). If we take single body feathers, mercury levels can vary considerably from feather to feather. This may be because many birds will grow some new feathers at odd times.

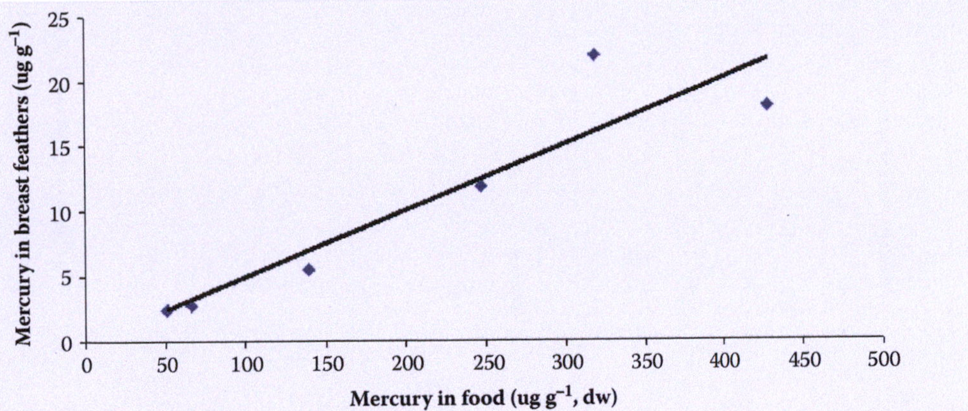

Figure 5.2 Relationship between mercury in body feathers and mercury in the diet for six species of seabirds in the Azores. (Redrawn from Monteiro, L. R. et al. *Mar. Ecol.- Prog. Ser.*, 166, 259–265, 1998.)

If, for example, a body feather is lost outside the normal molting period, most birds will regrow that feather immediately. It may get a high dose of mercury if it happens to grow just before the normal molt starts, or a very low dose if it grows at the end of the molt period. By sampling and pooling about six to ten small body feathers, a mercury measurement can be obtained that is a good measure for the individual bird, because it averages out these variations. As we get to understand bird molt patterns better, the possibility arises to select particular feathers to get a more detailed picture for an individual bird. For example, some birds molt some feathers in their breeding area, and others in their wintering area.

Historical changes in mercury levels in marine food chains can be seen very clearly from studies of mercury levels in body feathers of seabirds. There has been about a fourfold increase in mercury levels in feathers of Atlantic puffins (Figure 5.3) and other seabirds from the British Isles over the past 150 years (Thompson et al. 1992), almost exactly the amount predicted by models of mercury emissions and atmospheric transport during the industrial period since 1900, and closely similar to time series of mercury concentrations measured in appropriate sediment cores

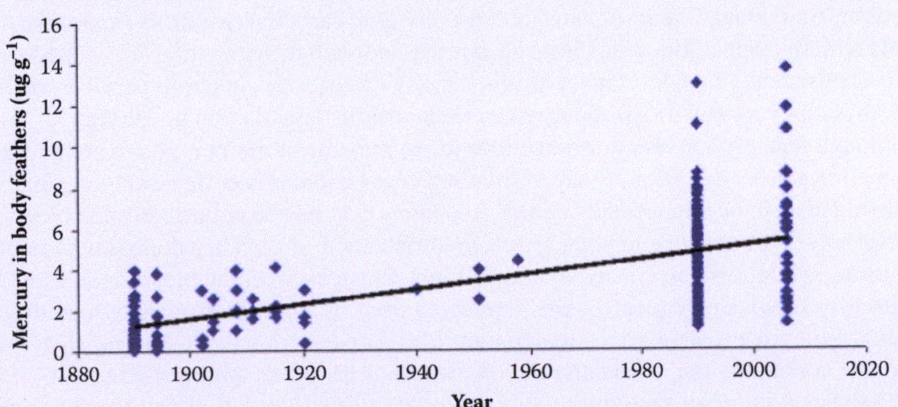

Figure 5.3 Mercury concentrations in pooled samples of several body feathers from Atlantic puffins from Scotland. (Data from Thompson, D. R. et al. *J. Appl. Ecol.*, 29, 79–84, 1992, updated with new data from 2006.)

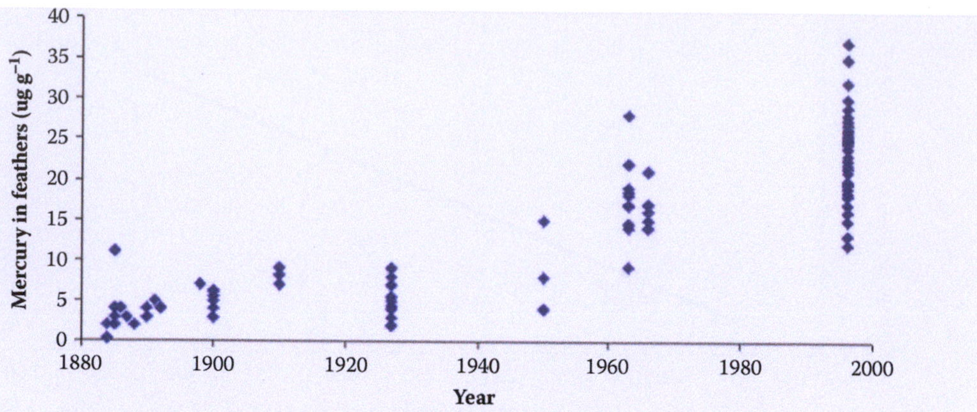

Figure 5.4 Mercury concentrations in pooled samples of several body feathers from Bulwer's petrels in the Azores. (Redrawn from Monteiro, L. R. and R. W. Furness, *Environ. Toxicol. Chem.*, 16, 2489–2493, 1997.)

(e.g., Pirrone et al. 1998). In the central North Atlantic, there have been similar large increases in mercury in seabirds, and especially in seabirds that feed on mesopelagic prey such as Bulwer's petrel (Figure 5.4). The large increase in birds feeding on deep-water fish relates to the conversion of inorganic mercury to methylmercury in the deep sea where low oxygen conditions prevail (Monteiro and Furness, 1997). Such patterns show no evidence of global contamination by mercury used in gold mining, where high levels might have been predicted during gold rushes in the preindustrial period (Camargo, 2002). Perhaps mercury used in gold mining was transported in aquatic medium and incorporated into sediments, rather than entering the atmosphere. Some time series of bird feathers show other patterns. In the southern North Sea there was a huge increase in mercury levels in Herring Gulls during the Second World War, which may indicate industrial production without pollution controls (Lewis et al. 1993) with some further increase during the latter twentieth century but evidence of reduced mercury pollution in the most recent decades. In contrast, there has been no increase, or only a relatively small increase, in mercury levels in seabirds from the southern hemisphere (Thompson et al. 1993; Becker et al. 2002; Gauthier-Clerc et al. 2005).

One feature of studies of mercury in birds has been the finding that levels vary considerably from bird to bird, even taking account of factors such as trophic status, age, sex, and breeding status of the individual. Causes of much of this individual variation (see for example, Figures 5.3 and 5.4) remain unclear. But it has been shown that individuals with a high level one year tend to have a consistently high level in other years; high or low levels appear to be a feature of individual physiology as well as a response to levels in the diet (Bearhop et al. 2000).

Although feathers are very convenient for measurements of mercury, there are alternatives and sometimes these may be preferable. Sampling eggs is often easier than sampling feathers as birds do not need to be trapped and handled, and many eggs can be collected from colonial nesting birds. Levels of mercury in eggs give a good reflection of mercury contamination of food in the local area around the colony where the birds forage early in the breeding season (Lewis and Furness, 1993). Consequently, eggs have been used to monitor changes in mercury levels over periods of a few years or decades (Braune, 2007) and to investigate geographical patterns of mercury contamination (Evers et al. 2003; Braunc et al. 2005; Anthony et al. 2007). Blood can be used as a sampling medium to indicate mercury levels over very short time scales, and so can be more useful when seasonal patterns are of interest, or at times of year when feathers are not growing (Bearhop et al. 2000). Other keratin structures accumulate mercury in the same way as feathers, and it has been suggested that claws may be as good as, or possibly better than feathers for measuring mercury exposure (Hopkins et al. 2007).

Analysis of mercury in bird feathers can also inform us about unknown features of the biology of birds. Variation in mercury levels between feathers may help to explain molt sequences in birds with complex molt cycles. Arcos et al. (2002) showed that seabirds feeding on fishery discards in the Mediterranean accumulated higher levels of mercury than seabirds feeding on pelagic fish, because bottom-living fish feeding on benthic invertebrates tend to accumulate more mercury than those feeding within the water column on zooplankton. A new species of storm-petrel was discovered recently because birds that nested at one time of year had much higher levels of mercury than those nesting in a different season but in the same colony (Monteiro et al. 1995). Further investigation identified that the high-mercury birds not only fed on a different diet (Monteiro et al. 1998) but were also genetically distinct and that this storm-petrel should be split into two sibling species (Friesen et al. 2007).

5.2 QUANTIFYING BIOACCUMULATION FROM FOOD

5.2.1 Assimilation from Food

Assimilation of a contaminant in food by individuals is one metric applied to estimate contaminant transfer between members of different trophic levels. It can be quantified as already described for bioavailability from food and used to estimate the magnitude of trophic transfer between the levels represented by the food item (e.g., prey or primary producer) and consumer (e.g., predator or primary consumer). The amount of contaminant in the organism after a specific time duration, usually the time to reach a practical steady state concentration, is divided by the total amount of contaminant fed to the organism to estimate assimilation efficiency (see Chapter 3 for details). Preferably, the amount ingested and amount egested over a period can be used to estimate assimilation efficiencies. The area under the curve (AUC) method is used less frequently despite its obvious usefulness.

Assimilation can be measured by a **twin tracer technique**, a technique that introduces a radiotracer of the substance to be assimilated simultaneously with an inert radiotracer to which assimilation is compared (Weeks and Rainbow, 1990). The inert tracer is not assimilated to any appreciable amount after ingestion and the retention of the assimilated tracer is quantified when all (or most) of the inert tracer has passed through the organism. For example, the assimilation of a ^{14}C-labeled organic compound may be compared to the amount of notionally inert ^{51}Cr fed simultaneously to a zooplankton species (Bricelj et al. 1984). This technique is based on the assumption that the two radiotracers pass through the gut at approximately the same rate and that the inert tracer is not absorbed to any significant extent. Both are incorporated into the same food items together to foster nearly identical movement through the animal's gut.* (As with all such uses of radiotracers, the **specific activity concept** is central to accurate implementation: For practical purposes, the radionuclide (e.g., ^{14}C) that is being used to reflect movement of the stable nuclide (e.g., C in an organic contaminant) is assumed to behave identically to its nonradioactive nuclide in chemical and biological processes.) In the absence of such an effective pairing of isotopes for exposures, a radioisotope may be fed to an organism and the difference between the amount of that isotope ingested and egested over a time course is used to estimate assimilation efficiency. The time to pass through the gut is estimated and the amount remaining after gut

* With sediments, it can be difficult to incorporate dual tracers identically. The radiotracers may be distributed differently among sediment fractions. It then becomes important to account for feeding selectivity as some fractions (e.g., fine, organic carbon-rich particles) may be processed differently from others (e.g., large particles).

evacuation is assumed to be assimilated. The advantage of the pairing with an inert tracer is that it provides a clear indication of when sampling of egested materials can be stopped—when all or most of the inert tracer has been egested. After that point, any radioactive compound remaining is assumed to have been taken up into the organism.

The assimilation efficiencies for various foods and members of a food web are pieced together to predict trends in trophic transfer. They are used to complement field surveys and more complex laboratory experiments as described in the following sections.

5.2.2 Trophic Transfer

5.2.2.1 Defining Trophic Position

Challenges to identifying trophic status of species exist so a few paragraphs are required to describe existing methods for identifying trophic structure.

One means of reducing ambiguity is to conduct a laboratory experiment so that the trophic structure is imposed by the experimental design. The disadvantage of this approach is that the artificial context imposed by the design might result in an unrealistic depiction of trophic dynamics occurring in natural communities. Recognizing this flaw, field surveys are often conducted to complement or to test conclusions from laboratory experiments.

In field surveys, the prevalent method for determining trophic status is gathering and integrating information from the natural history literature. This is adequate for interpreting results of field surveys only to a degree. Much uncertainty can remain because species in communities have complex feeding strategies that change with age, life stage, time, and community composition. Some of this uncertainty can be further reduced with visual observations of species interactions and analysis of gut content (Kling et al. 1992). Regardless, most such renderings and associated quantification simplify trophic status to a discrete state for specific species, i.e., primary producer, primary consumer, secondary consumer, and so forth. This would be an inadequate description for omnivorous species that feed at several levels.

Natural biochemical tracers within the food web can also provide information in some cases. Insight about food sources to members of marine pelagic, and less often benthic, food webs can be obtained using fatty acids trophic markers. Synthesized fatty acids are diverse and different taxonomic groups produce distinct kinds of fatty acids. Many are incorporated into consumer tissue without transformation of any sort.* Fatty acid composition in consumers has been applied to indicate vascular plant, microalgal, and macroalgal food sources (Lillebø et al. 2012). Zooplankton trophic studies have taken advantage of the distinct fatty acid profiles of diatoms and dinoflagellates (Dalsgaard et al. 2003). Diatoms are rich in 16:1*n*-7, 16:4*n*-1, and 20:5*n*-3 fatty acids whereas dinoflagellates are rich in 22:6*n*-3 and 18:4*n*-3 fatty acids (Kelley and Scheibling, 2012). Various taxonomic groupings of animals, vascular plants, macrophytes, and bacteria can also have distinct fatty acid signatures useful in exploring trophic relationships.

* A short discussion of fatty acid nomenclature might be needed to follow this discussion. Most **fatty acids** are chains with even numbers of carbons, commonly 14 and 24 carbons. There is a methyl (CH_3–) group on one end and a carboxyl (–COOH) group on the other. They can have varying numbers of double bonds in the chain. If the fatty acid has no 1, ≥2, ≥4 double bonds, they are categorized as saturated fatty acids (SFA), monosaturated fatty acids (MUFA), polyunsaturated fatty acids (PUFA), or highly unsaturated fatty acids (HUFA), respectively. Fatty acids with ≥24 carbons are long-chain fatty acids (LCFA). A fatty acid might be identified as $A{:}Bn{-}x$ where A = number of carbon atoms, B = number of double bonds, and x = the position of the first double bond relative to the terminal methyl group (Budge et al. 2006; Kelly and Scheibling, 2012). Under this system, double bonds in the molecule are assumed to be separated by a –CH_2– group ("methylene-separated") except in uncommon instances (Budge et al. 2006).

An accurate and more general method of quantifying trophic status has become readily available during the last 15 years (Cabana and Rasmussen, 1994). Enough **isotopic discrimination*** of light elements such as C, N, and S occurs during trophic transfers to allow quantification of trophic status in natural communities (Fry, 1988; Hesslein et al. 1991). Isotopic discrimination tends to reduce the amount of lighter isotopes (^{12}C, ^{14}N, or ^{32}S) in organisms relative to the heavier isotopes (^{13}C, ^{15}N, or ^{34}S) during trophic exchange. The lighter isotopes are eliminated from the organisms more readily than the heavy isotopes. Carbon isotopic changes with trophic transfer are clear but smaller in magnitude than those of nitrogen isotopes (Hesslein et al. 1991; Cabana and Rasmussen, 1994). The differences in carbon isotopes can be influenced by isotopic discrimination in the photosynthetic pathways of C_3 and C_4 plants,† and the differential dissolution of carbon isotopes (in atmospheric CO_2) in natural waters and subsequent equilibrium among dissolved carbon dioxide, carbonate ion, and bicarbonate ion in natural water bodies (Peterson and Fry, 1987). Carbon isotopic ratios vary more within a trophic level too, making them a less effective indicator of trophic status than nitrogen isotopes (Hesslein et al. 1991). However, carbon isotopes do provide information about carbon sources (e.g., C_3 and C_4 photosynthetic pathways) as well as trophic structure. Sulfur isotopic ratios may change only slightly with trophic structure (Hesslein et al. 1991) or not at all (Fry, 1988) with trophic exchange but can reflect sulfur source, making it a better indicator of food source than trophic status. For example, Hesslein et al. (1991) used sulfur isotopes to identify food bases for broad whitefish (*Coregonus nasus*) and lake whitefish (*Coregonus clupeaformis*).

Changes in stable nitrogen isotope composition seem to be the best indicator of trophic status, e.g., Rau (1981), Fry (1988), Hesslein et al. (1991), and Kling et al. (1992).

$$\delta^{15}N = 1000\left[\frac{(^{15}N_{sample})/(^{14}N_{sample})}{(^{15}N_{air})/(^{14}N_{air})} - 1\right] \quad (5.1)$$

Changes in nitrogen isotopes (^{14}N and ^{15}N) in organisms within a trophic web are quantified relative to the isotopic ratio of air (Equation 5.1).‡

The units for $\delta\ ^{15}N$ are ‰ ("per mill" or "per millage"). The $\delta\ ^{15}N$ increases with each trophic exchange because the lighter isotope (^{14}N) is more readily excreted. Isotopic discrimination of nitrogen does not appear to be associated with uptake, metabolic breakdown to amino acids, or deamination (Minagawa and Wada, 1984). Nor is there any evidence of discrimination in the urea cycle or in uric acid formation: The discrimination is relatively independent of the diverse processes of urine formation (Minagawa and Wada, 1984).

Changes in $\delta\ ^{15}N$ at each trophic exchange can range widely. Minawaga and Wada (1984) describe a range from 1.3‰ to 5.3‰ per trophic exchange. However, the average increase is 3.4‰

* Isotopic discrimination is the effect of isotope mass on the rate at or extent to which it participates in some biological or chemical process. Isotopic discrimination or the isotope effect results from differences in the kinetic energy of associated molecules with slightly different masses because they contain different isotopes of an element, for example, ^{14}N instead of ^{15}N. Discrimination can also result from distinct vibrational and rotational qualities of molecules (Wang et al. 1975). Discrimination between isotopes is measured as a **discrimination ratio** with a ratio of one indicating that no discrimination was occurring.

† Photosynthesis occurs differently in **C_3 and C_4 plants**. The C_3 and C_4 refer to the number of C atoms of the first compound into which the CO_2 is initially incorporated by a plant: C_4 plants combine CO_2 to form oxaloacetate and C_3 plants combine CO_2 to form glycerate 3-phosphate. The C_3 plants tend to dominate ecosystems and to have a higher $\delta\ ^{13}C$ than C_4 plants. Examples of C_4 plants are tropical and saltmarsh grasses, and the crop species such as corn, sugarcane, and sorghum (Peterson and Fry, 1987).

‡ The $\delta\ ^{13}C$ is similar except the two carbon isotopes, ^{13}C and ^{12}C are inserted into this relationship instead and the reference is a mineral standard material (most often, belenmite from the Pee Dee Formation) instead of air.

per trophic exchange (Minagawa and Wada, 1984; Cabana and Rasmussen, 1994). The δ ^{15}N can also change with animal age (Minagawa and Wada, 1984). Despite these complications, δ ^{15}N can reveal the general trophic structure of diverse field communities (Figure 5.5) and enhance interpretation of contaminant movement in ecological communities. Indeed, it can also be used with human populations to identify diet, e.g., δ ^{15}N for human populations with different but unquantified

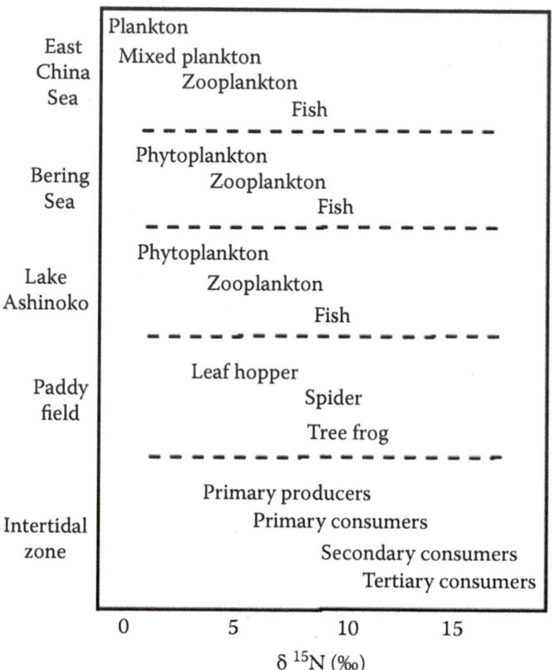

Figure 5.5 The change in δ ^{15}N (‰) with increase in trophic level for the diverse communities of the East China Sea (an oligotrophic sea with significant amounts of nitrogen fixation at the time of phytoplankton sampling, which brought the δ ^{15}N to near atmospheric levels), the Bering Sea (North Atlantic Ocean), Lake Ashinoko (freshwater lake), a paddy field (terrestrial system) in Konosu, Japan, and an intertidal zone. (A composite and modification of Figures 1 and 2 in Minagawa, M. and E. Wada, *Geochim. Cosmochim. Ac.*, 48, 1135–1140, 1984.)

amounts of diet coming from marine and terrestrial systems (Scheninger et al. 1983). Finally and importantly, this approach has the advantage of quantifying intermediate trophic positions so that the simplistic assignment of a species to a single trophic level can be avoided. It can be augmented with a dual isotope (δ ^{15}N and δ ^{13}C) approach to also imply major vegetable sources of biomass (Fry, 1991) to members of the ecological community.

Trophic level can still be used to translate relative δ ^{15}N values for biota within a food web into convenient units (Jardine et al. 2006),

$$TL_{Consumer} = \frac{\delta^{15}N_{Consumer} - \delta^{15}N_{Primary\ Producer}}{\Delta^{15}N} + 1 \tag{5.2}$$

where $TL_{Consumer}$ = estimated trophic level of the consumer organism of interest, $\delta\,^{15}N_{Consumer\ or\ Primary\ Producer}$ = the δ ^{15}N for the consumer of interest or primary producers, and $\Delta^{15}N$ = the increase in δ ^{15}N from a trophic level to the next higher trophic level (the **enrichment factor**).

5.2.2.2 Estimating Trophic Transfer

Perhaps the most abstract approach to quantifying biomagnification is to divide the contaminant concentration at some trophic level n (C_n) by that at the next lowest trophic level (C_{n-1}) (e.g., Bruggeman et al. 1981; Laskowski, 1991). This **biomagnification factor (B)** may involve individual organisms of discrete known or assumed trophic status:

$$B = \frac{C_n}{C_{n-1}} \tag{5.3}$$

Such a biomagnification factor is based on the assumption that concentrations have reached steady state in the sampled individuals and that the capacity is the same for them (e.g., individuals have the same lipid content if lipophilic compounds are being studied). This B can also be expressed in terms of the rate constant–based bioaccumulation model.

$$B = \frac{C_n}{C_{n-1}} = \frac{\alpha f}{\mathrm{k_e}} \tag{5.4}$$

where α = assimilation efficiency for the ingested species, f = the feeding rate (mass of food mass of individual^{-1} time^{-1}), and k_e = the elimination rate constant (Bruggeman et al. 1981). Note that assimilation efficiency often declines as feeding rate increases (Clark and Mackay, 1991; McCloskey et al. 1998) so this parameterization is conditional.

Biomagnification factors could be estimated for samples of many organisms from two trophic levels by using a body mass-weighted mean concentration for the two trophic levels,

$$B' = \frac{\left(\sum_{i=1}^{x} C_{n.i} w_{n.i}\right)\left(\sum_{j=1}^{z} w_{n-1.j}\right)}{\left(\sum_{j=1}^{z} C_{n-1.j} w_{n-1.j}\right)\left(\sum_{i=1}^{x} w_{n.i}\right)} \tag{5.5}$$

where w = the weight of individuals sampled from the n or $n-1$ trophic levels.

Definitions for some bioaccumulation indices acknowledge that water might also contribute to differences in concentrations measured for individuals at different trophic levels. This is necessary in field surveys that do not differentiate between water and food sources to individuals. The **bioaccumulation factor (BF)*** is such an index that has the same form as Equation 5.3 except the source is not necessarily food alone.

$$\mathrm{BF} = \frac{C_{\mathrm{organism}}}{C_{\mathrm{source}}} \tag{5.6}$$

where C_{organism} = concentration in the organism resulting from uptake from food and water sources, and C_{source} = concentration in the reference source of contaminant. For lipophilic organic compounds,

* Note that bioaccumulation factor designated BSAF is used in a more restricted sense in this book to mean the ratio of concentration in an organism and concentration in sediments (Chapter 3). Also, BAF is the same as BF throughout the literature. For the sake of clarity only, BF will be used here to designate the ratio derived under the assumption that both water and food sources may be contributing to body concentrations of contaminants.

concentrations may be expressed as a mass per mass of lipid basis. If the BF is greater than 1, biomagnification might be occurring. If it is less than 1, trophic dilution is suggested, although other factors such as allometric processes or growth dilution could be contributing to the changes. Experimental designs may employ tandem exposures of two subsets of individuals to either contaminant in water alone or contaminant in both food and water. The differences in the treatment results (bioconcentration factor [BCF] derived from the water-alone treatment and BF derived from the food plus water treatment) can then be used to assess the significance of trophic exchange relative to accumulation from water alone (Bruggeman et al. 1981). Assuming that uptake from water dominates at the primary producer level only and discrete trophic levels with negligible overlap, the transfer up the trophic structure can be calculated with knowledge of the concentration at the lowest trophic level and estimates of assimilation efficiencies and feeding rates at each trophic level (Ramade, 1987; Newman, 2013).

Commonly, the concentrations at trophic levels are referenced to that of the primary or some lowest defined source of contaminant. For example, a **concentration factor (CF)** could be estimated for all trophic levels with the concentration in the water used in the denominator, $CF = C_n/C_{water}$. Reichle and Van Hook (1970) took this approach to estimate concentration factors for radionuclides in a terrestrial food chain with concentrations in plant leaves as the reference concentration. The increase or decrease in concentration relative to that of the source is expressed as a multiple of the source concentration.

The methods described to this point require assignment of species to discrete trophic levels, but this is often an unrealistic simplification of trophic dynamics because omnivory is common in most food webs. It is more effective to quantify trophic transfer by relating concentrations in members of the community to corresponding $\delta\ ^{15}N$ values. Figure 5.6 depicts the results of such an exercise documenting the increase in mercury from zooplankton → a zooplanktivorous shrimp (*Mysis relicta*) → pelagic forage fish → lake trout (*Salvelinus namaycush*) of Canadian lakes (Cabana and Rasmussen, 1994; Cabana et al. 1994). In a survey of seven Canadian shield lakes with various trophic structures leading to lake trout (Figure 5.7), data for mercury concentration and $\delta\ ^{15}N$ fell between the predicted values for a discrete trophic structure. The diagram on the left side of Figure 5.7 shows the expected values for $\delta\ ^{15}N$ based on three possible, discrete structures. (Structure is specified as the number of levels above zooplankton that the lake trout occupy.)

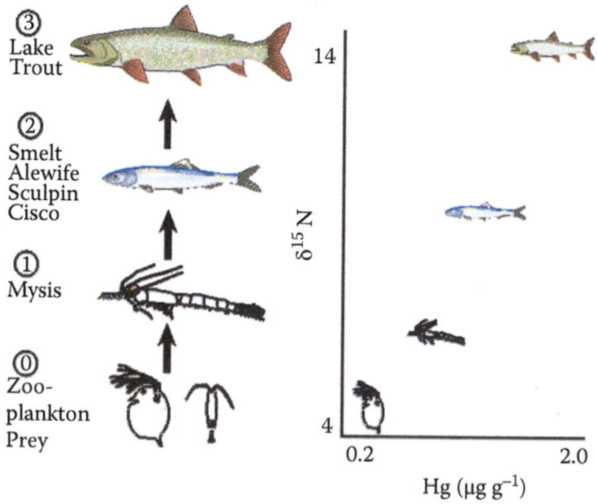

Figure 5.6 The increase in mercury from zooplankton to a zooplanktivorous shrimp (*Mysis relicta*) to pelagic forage fish to lake trout (*Salvelinus namaycush*). The trophic structure is quantified with $\delta\ ^{15}N$: its value increasing roughly 3.4‰ at each trophic exchange. A clear relationship is evident between mercury concentration and trophic status of individuals, indicating biomagnification of mercury. (Constructed using data and Figure 2 of Cabana, G. and J. B. Rasmussen, *Nature,* 372, 255–257, 1994.)

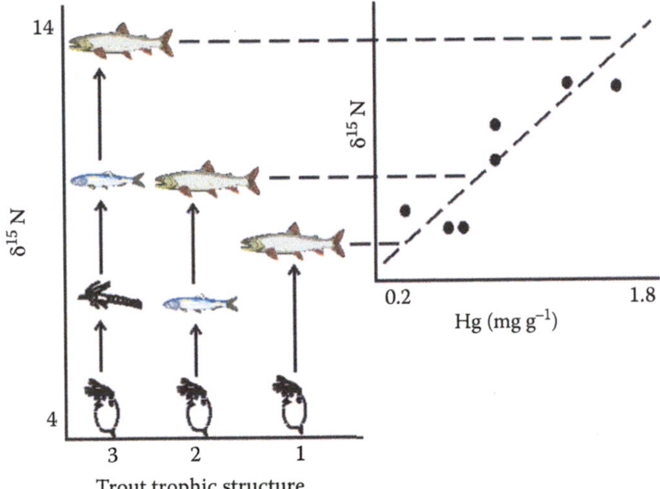

Figure 5.7 Correlation between mercury concentration and trophic structure (δ ^{15}N) for Canadian lakes with different trophic structure relative to lake trout (*Salvelinus namaycush*). In some lakes, shrimp and/or forage fish are missing, giving rise to various trophic structures relative to lake trout. Using a discrete assignment of trophic structure, 3 = same structure as shown in Figure 5.6 (zooplankton → shrimp → forage fish → lake trout), 2 = shortened structure without shrimp (zooplankton → forage fish → lake trout), and 1 = a simple structure with both shrimp and forage fish being insignificant. (Forage fish were predominantly alewife [*Alosa pseudoharengus*], cisco [*Coregonus artedii*], whitefish [*Coregonus clupeaformis, Prosopium cyclindraceum*], sculpin [*Myoxocephalus thompsoni* ?], and smelt [*Osmerus mordax*].) The panel on the upper right shows the actual mercury concentrations and δ ^{15}N values for lake trout from seven Canadian lakes. Note that trout from these lakes are shifted slightly to positions intermediate between the discrete trophic structure levels. (Composite and modification of Figures 2 and 3 of Cabana, G. and J. B. Rasmussen, *Nature*, 372, 255–257, 1994.)

Regression models can be developed if a continuous variable such as δ ^{15}N was used to quantify trophic transfer. Figures 5.6 and 5.7 suggest that a simple linear regression technique could be used to model mercury increase with trophic transfer. Studies of bioaccumulation of organic compounds suggested to Broman et al. (1992) and Rolff et al. (1993) that an exponential model may be more appropriate in many cases. Such exponential models are the norm (Jardine et al. 2006) and can involve base 10 or, less often, base e:

$$C = ae^{b(\delta^{15}\text{N})} \text{ or } a10^{b(\delta^{15}\text{N})} \quad (5.7)$$

where C = concentration in the organism within the food web, and a and b are estimated parameters. The parameter b is the **biomagnification power**. A positive b indicates a proportional increase in concentration with increase in position within a trophic web (biomagnification) and a negative b indicates a proportional decrease in concentration with increase in position (trophic dilution). Broman et al. (1992) found that decreases and increases in concentrations of various polychlorinated dibenzo-*p*-dioxins and dibenzofurans in a pelagic food web could be modeled with this relationship.

If trophic levels are employed in models in lieu of δ ^{15}N, the food web magnification factor (FWMF, also called trophic magnification factor [TMF]) can be estimated as the b in Equation 5.7 when TL is substituted for δ ^{15}N (Jardine et al. 2006), i.e.,

$$C = a_{\text{TL}}e^{b_{\text{TL}}(\text{TL})} \text{ or } a_{\text{TL}}10^{b_{\text{TL}}(\text{TL})} \quad (5.8)$$

5.3 INORGANIC CONTAMINANTS

5.3.1 Metals and Metalloids

Assimilation studies of metals and metalloids have used the twin tracer technique and single isotope differences between ingested and egested element. Bricelj et al. (1984) fed $^{14}C/^{51}Cr$ labeled algae to clams (*Mercenaria mercenaria*) for 30 to 45 minutes and then measured the amount of both tracers in feces after the clams were transferred to water containing unlabeled algae. The amounts lost after nearly total recovery of the notionally inert ^{51}Cr label in the feces were used to estimate ^{14}C assimilation. Chromium was assimilated in very small amounts but carbon assimilation was approximately 80%. With a similar design (^{60}Co used as an inert tracer for ^{65}Zn assimilation), zinc assimilation from tissue of the macroalga, *Laminaria digitata*, fed to the amphipod, *Orchestia gammarellus*, was estimated to be nearly 100% (Weeks and Rainbow, 1990). Reinfelder and Fisher (1991) used this design to determine assimilation efficiencies for a sequence of elements when calanoid copepods were fed labeled diatoms. For ^{14}C assimilation, ^{51}Cr was used as the inert tracer. For ^{75}Se assimilation, ^{241}Am was the inert tracer in the twin tracer technique. Assimilation of other elements (^{110m}Ag, ^{109}Cd, ^{32}P, ^{35}S, and ^{65}Zn) was calculated without an inert isotope marker by dividing the amount retained after the presumed complete gut evacuation by the amount ingested. Assimilation was estimated for sulfur and zinc using algae from cultures in the log phase or stationary phases of growth (Figure 5.8). Given the short residence time in the gut, the elemental content of the algal cytosol was assumed to reflect that available for "liquid" digestion by the zooplankton. This assumption was supported by a strong correlation between assimilation efficiencies for the various elements and the fraction of each present in the algal cytoplasm. Based on these results and earlier work with thorium, lead, uranium, and radium (Fisher et al. 1987), Reinfelder and Fisher (1991) suggested that Class B metals and those borderline metals that have greater affinities for sulfur than nitrogen or oxygen (e.g., cadmium, silver, and zinc) will be present in the cytoplasm in higher proportions than Class A metals (e.g., americium, plutonium, and thorium) that have greater affinities for oxygen than nitrogen and sulfur. Assimilation efficiencies are generally higher for the Class B and borderline metals than Class A metals because of their higher availability in the cytoplasm for "liquid" digestion during rapid passage through the zooplankton gut.

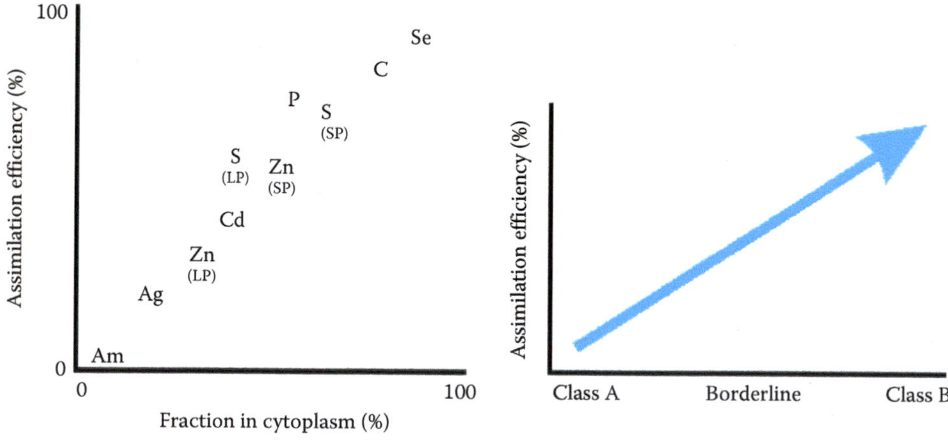

Figure 5.8 Assimilation efficiencies of elements contained in diatoms (*Thalassiosira pseudonana*) fed to zooplankton (*Acartia tonsa*, *Acartia hudsonica*, or *Temora longicornis*) plotted against the amount of each element associated with the cytoplasm of the diatoms (left panel). SP = data derived using stationary phase (senescent) diatom cultures and LP = results derived from diatom cultures in the log phase of growth. (Modified from Figure 1 of Reinfelder and Fisher [1991].) The right panel depicts the general trend in assimilation efficiency hypothesized relative to the Class B—borderline—Class A classification of metals.

Additional factors can influence assimilation efficiencies of metals in other feeding interactions. As mentioned, mollusks can sequester metals in granules and lower their potential for damage. The winkle, *Littorina littorea,* does so with zinc. When tissues of this gastropod are fed to the predatory gastropod, *Nassarius reticulatus*, only a small proportion of the zinc in the associated granules is available for assimilation in the predator (Nott and Nicoladou, 1993). In contrast, there are considerable decreases in magnesium, phosphorus, and potassium in granules during passage through the predator's gut. The detoxification of zinc in the winkle tissue by incorporation into granules decreases the assimilation efficiency of zinc during predation. The assimilation of other metals associated with invertebrate intracellular granules is also probably influenced to varying degrees. Despite the widespread detoxification of metals via sequestration in granules, much work remains to be done regarding their influence on the availability of various classes of metals.

Trophic transfer studies suggest that biomagnification can be more the exception than the rule for metals and metalloids. Wren et al. (1983) studied a series of elements (aluminum, barium, beryllium, boron, calcium, cadmium, cobalt, iron, lead, mercury, magnesium, manganese, molybdenum, nickel, phosphorus, sulfur, strontium, titanium, vanadium, and zinc) at various trophic levels of a Canadian Precambrian Shield lake and found evidence of biomagnification for only mercury. Cushing and Watson (1971) found no evidence for zinc biomagnification in a food chain of water → periphyton → carp. Terhaar et al. (1977) found that mercury, but not silver, showed evidence of biomagnification in an aquatic food chain. Using *in situ* enclosures placed in an aquatic system, Gächter and Geiger (1979) found no evidence for biomagnification of inorganic mercury, copper, cadmium, lead, or zinc. They speculated that alkylated mercury, but not inorganic mercury, increased with increasing trophic status of species. That is consistent with many studies, including the study depicted in Figure 5.9 and Vignette 5.2. Numerous authors report biomagnification of mercury in aquatic systems (Wren and MacCrimmon, 1986; Cabana et al. 1994; Kidd et al. 1995)

VIGNETTE 5.2 Models of Mercury Trophic Transfer Using Nitrogen Isotopes

Kyle R. Tom and Michael C. Newman
College of William & Mary

This brief vignette illustrates the contrasting methylmercury biomagnification and inorganic mercury trophic dilution in a typical contaminated lotic system, the South River in Virginia. Between 1929 and 1950, inorganic mercury was released into the river from a plant where it was a catalyst for polymer production. More than 50 years after intentional release ceased, mercury remains at high concentrations in the 100 mile reach below the historic point of discharge.

In 2007, mercury trophic transfer was modeled in the most contaminated part of this reach as part of a larger effort to assess resource damage and predict consequences of various remediation scenarios. Models encompassed trophic transfer from periphyton to a piscivorous fish that is prized by local fisherman (largemouth bass, *Micropterus salmoides*). Samples of periphyton, invertebrates, forage fish, and largemouth bass were taken in the reach extending from the historic source at Waynesboro, Virginia, to 23 miles (approximately 14.4 km) downriver. Samples of diverse biota were taken at five locations within the study reach and analyzed for total mercury and methylmercury concentrations, and $\delta\ ^{15}N$. These data were fit to the general model depicted by Equation 5.7 except sampling location was incorporated as a covariate by adding river mile (RM) below the historic source:

$$C = a\, e^{b(\delta^{15}N) + b_2(RM)}$$

For statistical reasons and ease of fitting, linear transformation was done on Equation 5.7 before fitting. The results are summarized in Table 5.1. The intercept can be converted to the a in the aforementioned equation, that is, $a = e^{\text{Intercept}}$.* Methylmercury concentration clearly increases with trophic position ($\delta\ ^{15}$N) (Figure 5.9) as indicated by the positive and statistically significant coefficient for trophic position (Table 5.1). This coefficient is an estimate of the biomagnification factor (b in Equation 5.7). Methylmercury concentration also increases with distance downriver from the source. In contrast, there was a decrease in inorganic mercury with trophic position ($\delta\ ^{15}$N) as indicated by the negative and statistically significant coefficient for trophic position, i.e., a negative biomagnification factor (Table 5.1). As Figure 5.10 shows, the percentage of the total mercury present as methylmercury increases with trophic position from periphyton to piscivorous fish until essentially all mercury in most fish is methylmercury.

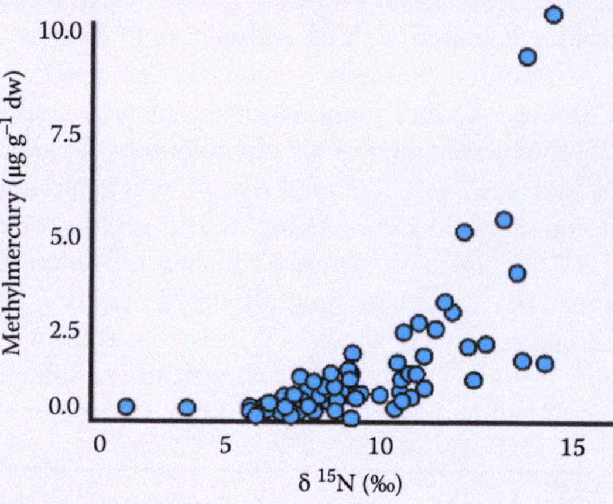

Figure 5.9 Methylmercury biomagnification in the South River (Virginia) is clear using $\delta\ ^{15}$N values to indicate trophic position of biota in this river. (Data from Tom, K. R. et al. *Environ. Toxicol. Chem.*, 29, 1013–1020, 2010.)

Table 5.1 Mercury Trophic Transfer Models for South River (Virginia). The Regression r^2 Values for the ln of Methylmercury and ln of Inorganic Mercury Models were 0.78 and 0.17, Respectively (n = 65)

Regression Parameter	MethylMercury	Inorganic Mercury
Intercept		
Estimate	−5.2	−0.18
Estimate Standard Error	0.3	0.73
p value for H_o of 0	<.0001	0.8093
Coefficient for $\delta\ ^{15}$N (b)		
Estimate	0.45	−0.16
Estimate Standard Error	0.03	0.08
p value for H_o of 0	<.0001	0.0354
Coefficient for River Mile (b_1)		
Estimate	0.05	0.06
Estimate Standard Error	0.01	0.02
p value for H_o of 0	<.0001	0.0044

* Although not directly relevant to discussions here, there is a bias in this back transformation that can be corrected as mentioned in Chapter 4 and detailed in Chapter 2 of Newman (2013).

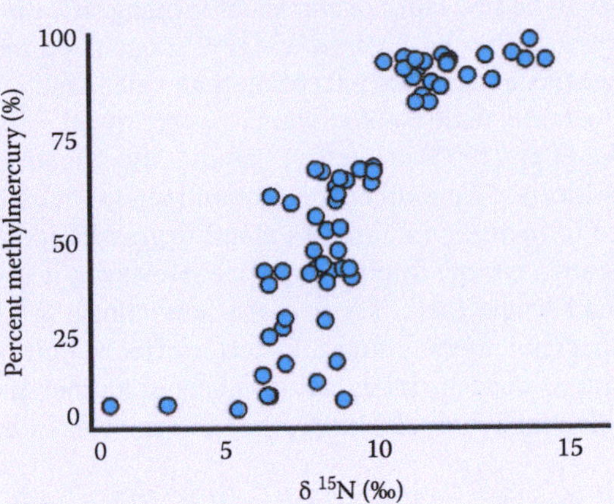

Figure 5.10 Increase in the percentage of the total mercury that is methylmercury in members of the aquatic food web of the South River below a source of inorganic mercury (Virginia). Biota sampled ranged from periphyton to invertebrates to forage fish to piscivorous fish. (Data from Tom, K. R. et al.*Environ. Toxicol. Chem.*, 29, 1013–1020, 2010.)

A major goal of this sampling and modeling was to develop a predictive tool with which to judge how a reduction in mercury entering the trophic web (i.e., via procedurally defined periphyton) would lower concentrations in largemouth bass that are valued and consumed by anglers in this region of the United States. Most of the mercury in this species is methylmercury so predictions of methylmercury movement through the trophic web would fulfill this goal: A good predictive model for total mercury trophic transfer is unnecessary for this contaminated reach of the South River. The methylmercury model was judged to be adequate to such purposes based on an associated prediction r^2 value of 0.76.* This is fortunate because, although the biomagnification factor for total mercury was positive and statistically significant (0.17 with a standard error of 0.02, $p < .0001$), the regression r^2 for the total mercury model was poor (0.32). Good predictions were unlikely to be obtained for total mercury. Given the successful generafi-cients that relate methylmercury concentration in procedurally defined periphyton to dissolved or fine sediment phases of the river.

For comparison, we (Newman et al. 2011, Wang et al. 2013) recently built ^{15}N-based trophic models for several locations on the adjoining terrestrial floodplain. The methylmercury biomagnification factor for the floodplain food web was substantially higher and more variable than that for the river.

* A regression r^2 such as that shown in Table 5.1 is a biased metric for assessing adequacy of a model *for predictive purposes*. A prediction r^2 is used here for that reason. The interested reader is referred to Neter et al. (1990) or Newman (2013) for more detail regarding this cross-validation metric.

although Huckabee et al. (1979) cautioned that, even with mercury, other factors correlated with trophic status of an organism may confound the identification of biomagnification. Mance (1987) reviewed the literature and found only a few cases suggesting metal or metalloid biomagnification in aquatic systems. In addition to mercury, there was an occasional report of arsenic increase in food webs. Campbell et al. (2005) recently noted that the alkali metal, rubidium, can biomagnify in freshwater and marine food webs. As mentioned below and by Campbell et al. (2005), another alkali earth metal, cesium, is also thought to biomagnify.

Similarly, there appear to be few clear examples of biomagnification of metals or metalloids in terrestrial systems. Beyer (1986) and Laskowski (1991) concluded independently that biomagnification in terrestrial systems is more the exception than rule. Zinc, an essential and internally regulated metal, might increase with trophic status in terrestrial ecosystems deficient in this element (Beyer, 1986). Wu et al. (1995) suggested that biomagnification occurred in a selenium-contaminated area of California. Selenium concentration factors relative to water extractable soil concentrations were noted to increase in a soil → plant → grasshopper (*Dissosteria pictipennis*) → praying mantis (*Litaneutria minor*) trophic sequence. However, it is difficult to determine the accuracy of this conclusion because there was no consistent pattern of increase with each transfer and covariates such as species longevity and allometric effects were not considered. Only the grasshopper to praying mantis concentration factor indicated an increase in selenium concentration but, according to Table 9 of Wu et al. (1995), the increase was quite variable among sample sites.

5.3.2 Radionuclides

This separate consideration of radionuclides is admittedly arbitrary and creates some overlap with the metals and metalloids discussed in the earlier section. However, the unique sources and effects of radioactive contaminants provide some justification for this separation.

Assimilation efficiencies of radionuclides vary widely, e.g., 80% for ^{134}Cs (algae → carp) to nearly 0% for ^{144}Ce (food → fish), which becomes essentially unavailable after being incorporated into structural tissue of prey (Reichle et al. 1970). Amphipods-fed brine shrimp had assimilation efficiencies of 6.2%, 9.4%, and 55% for ^{144}Ce, ^{46}Sc, and ^{65}Zn, respectively. Cesium-137 and ^{134}Cs, widespread fission products with long half-lives (circa 30 years for ^{137}Cs and 2 years for ^{134}Cs), have high assimilation efficiencies (65% to 94%) from a wide range of food sources in terrestrial and aquatic systems. Similarly, ^{47}Ca assimilation efficiencies might be quite high in diverse systems (69% to 98%). Almost 100% of ^{86}Rb and ^{187}W were assimilated by grazers of plant foliage (Reichle et al. 1970). Reichle et al. (1970) and Blaylock (1982) provide a good summary of this type of data for radionuclides in aquatic and terrestrial systems.

Some (radionuclides of nitrogen, phosphorus, potassium, and sodium) increase with passage up the food web but others (^{47}Ca) can be diluted (Reichle and Van Hook, 1970). The increase or decrease is a consequence of their availability in the environment relative to the physiological need for each. Davis and Foster (1958) observed that the most common trend for many radionuclides was accumulation to relatively high concentrations at the lowest trophic level (primary producers) with subsequent diminution at each transfer thereafter. For example, ^{32}P displayed this behavior in aquatic systems examined by Kahn and Turgeon (1984). Polikarpov (1966) summarized and described processes contributing to concentration factors for many radionuclides in aquatic biota.

Essential elements in short supply and their radioactive analogs tend to increase during trophic exchange (Reichle and Van Hook, 1970).* However, the calcium analog, ^{90}Sr, does not increase with trophic exchange because it becomes incorporated into bone or other structural tissues and, in so doing, has a very low bioavailability to predators. The consequence is trophic dilution for ^{90}Sr (Woodwell, 1967). Reichle et al. (1970) identified **calcium sinks** such as arthropod cuticles as a mechanism for trophic dilution for calcium and its analogs in terrestrial communities.

The radiocesium isotopes, ^{134}Cs and ^{137}Cs, are potassium analogs that have received much deserved attention because of their release from fission-related processes and their relatively long half-lives. Radiocesium and potassium are taken up with similar, high efficiencies but radiocesium

* Recollect that an elemental **analog** is an element that behaves like, but not necessarily identical to, another element in biological processes, e.g., cesium is an analog of potassium and strontium is an analog of calcium.

tends to be eliminated more slowly than potassium (McNeill and Trojan, 1960).* Consequently, the potential exists for a net increase in cesium with each trophic exchange. Radiocesium (^{137}Cs) from nuclear weapons fallout did biomagnify with passage from plant tissue to mule deer (*Odocoileus hemionus*) to cougar (*Felis concolor hippolestes*), suggesting that biomagnification could also increase ^{137}Cs activities in humans who consume tainted mule deer (Pendleton et al. 1964). There was a 3.4-fold increase of ^{137}Cs from deer to cougar. Similar increases have been described for atmospheric fallout–related ^{137}Cs in a food chain leading from lichens to caribou to Alaskan Eskimos, wolves, and foxes (French, 1967; Woodwell, 1967). Further evidence for the increase in ^{137}Cs concentrations with trophic transfer is the observation that ^{137}Cs activities are often higher in piscivorous fish than nonpiscivorous fish (Blaylock, 1982; Rowan and Rasmussen, 1994).

5.4 ORGANIC COMPOUNDS

A wide range of behaviors is exhibited by organic contaminants relative to trophic transfer but, in general, biomagnification of compounds not subject to metabolism seems to be predictable in aquatic food webs with log K_{ow}. Connolly and Pedersen (1988) suggested that biomagnification is possible if log K_{ow} values were approximately above 4, whereas Kelly et al. (2007) suggested log K_{ow} values greater than 8 as an approximate upper limit for biomagnification in aquatic food webs. Earlier, Thomann (1989) indicated a range from 5 to 7 for compounds prone to biomagnification. Above 7, biomagnification begins to become hampered by diminished assimilation efficiencies and BCF for lower trophic levels such as phytoplankton. Below 5, decreased uptake and increased elimination rates begin to limit the capacity for biomagnification. Russell et al. (1995) found that PCBs with log K_{ow} values of 6.1 or above were subject to biomagnification when white bass (*Morone chrysops*) consumed emerald shiners (*Notropis atherinoides*) (Figure 5.11).

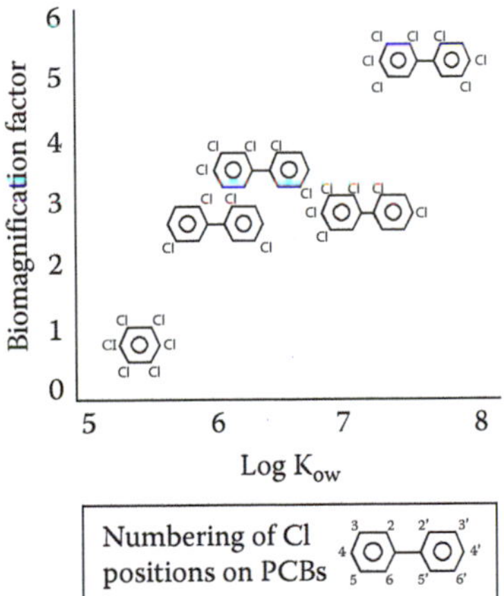

Figure 5.11 The increase in biomagnification factor (concentration in white bass/concentration in emerald shiner) with increasing log K_{ow} values for hexachlorobenzene (bottom left) and four polychlorinated biphenyls (PCBs), 2,2′,5,5′-tetrachlrobiphenyl, 2,2′,3,4,5′-pentachlorophenyl, 2,2′,3,4,4′,5-hexachlorobiphenyl, and 2,2′,34,4′,5,5′-heptachlorobiphenyl. (Modified from Figure 2 in Russell et al. [1995].) Note that a key to the numbering of chloride atoms attached to the biphenyl rings is provided at the bottom of this figure. The prefixes, tetra-, penta-, hexa-, and hepta-, refer to the total number of chloride atoms in the PCB molecule.

* Like isotopic discrimination described earlier, a discrimination factor can be estimated for analogs. Discrimination between an element and its analog may be quantified as a **discrimination factor or ratio**, e.g., $([Cs]_{food}/[K]_{food})/([Cs]_{body}/[K]_{body})$. The discrimination ratio for cesium and potassium in humans is approximately 0.33 (McNeill and Trojan, 1960).

PCBs often exhibit biomagnification (e.g., Hickie et al. 2007) although Clayton et al. (1977) argued otherwise in cases involving zooplankton. Using nitrogen isotopes to quantify trophic status, PCB concentrations in lake trout from 83 Canadian lakes were shown to be determined by biomagnification processes and species lipid content (Rasmussen et al. 1990). Using the lake trout trophic classification discussed earlier for mercury, concentrations of PCBs were shown to increase 3.5-fold for every trophic exchange.* Lake Michigan food web studies also demonstrated PCB biomagnification (Evans et al. 1991). Clark and Mackay (1991) found the PCB, 2,2',3,3',4,4',5,6'-octachlorophenyl† to biomagnify in guppies (*Poecilia reticulata*) exposed in the laboratory. Bruggeman et al. (1981) fed PCBs to goldfish (*Carassius auratus*) and found that biomagnification occurred, seemingly as a consequence of water solubility–correlated elimination rates.

DDT, DDE, DDD, or the sum of these pesticides may increase with each trophic transfer in field communities (Evans et al. 1991; Kidd et al. 1995). (These pesticides may be studied individually or together as ΣDDT.) Toxaphene may display biomagnification (Sanborn et al. 1976; Kidd et al. 1995) but to a lesser extent than DDT (Evans et al. 1991). Dibenzo-*p*-dioxins and dibenzofurans also were found to increase with trophic position in a Northern Baltic Sea food web (Broman et al. 1992).

All isomers and congeners do not behave similarly during trophic exchanges in aquatic food webs. For example, PCB congeners behaved distinctly in Arctic trophic levels: Arctic cod (*Boreogadus saida*) → ringed seal (*Phoca hispida*) → polar bear (*Urus maritimus*) (Muir et al. 1988). The tri- and tetrachloro PCBs were the dominant congeners in cod. Penta- and hexachloro PCB congeners dominated in the ringed seal, and hexa- and heptochloro PCBs were the dominant congeners in polar bears. Older seals tended to have more of the highly chlorinated PCBs than younger seals. The reason for this age difference was likely a combination of the slower elimination of the more highly chlorinated congeners, and the difference in diet between young and adult seals. Young seals tend to eat more amphipods than do adults. In studies of trophic transfer of dibenzo-*p*-dioxins and dibenzofurans in a Baltic food web, the total concentration of these contaminants did not increase but the most toxic of the isomers did exhibit biomagnification to high levels in eider ducks (*Somateria mollissima*) (Broman et al. 1992; Rolff et al. 1993). This study underscores the importance of considering individual as well as summed concentrations of isomers or congeners. Catfish (*Ictalurus punctatus*) preferentially accumulate *cis* more readily than *trans* chlordane because the *cis* isomer appears most resistant to metabolism (Murphy and Gooch, 1995). Likely for this same reason, the *cis* form of chlordane increased relative to the *trans* form in a freshwater trophic chain (Sanborn et al. 1976).

The situation differs when air-breathing species are involved as in the case of terrestrial food webs or marine food webs leading to a marine mammal (Kelly et al. 2007). Kelly et al. (2007) surveyed data for water-breathing and air-breathing animals from Canada to ascertain the best way of predicting persistent organic pollutants (POPs) biomagnification. In contrast to the rules given above, some compounds with log K_{ow} values less than 5 did display high biomagnification in a lichen → caribou → wolf and a marine mammal-focused food web involving ringed seals and beluga whales. These authors suggested that air-breathing species tend to be more effective in absorbing contaminants from their prey and show higher biomagnification than their

* Interestingly, these authors make the point that fisheries management strategies commonly thought to enhance sports fishing can do damage instead. Forage species are added to enhance the trophic base of a lake and produce more trout. This is seen as a positive action. But each addition to the trophic structure enhances accumulation of PCBs in trout. If the addition of forage species brings trout tissue concentration above a certain level, a fish consumption advisory or ban can result. The forage fish stocking may damage the sports fishery in some cases.

† Recall from earlier discussions that PCBs were manufactured and used as mixtures of chlorinated biphenyls with different numbers and positions of associated Cl atoms. The individual PCBs which share a common form but have different numbers of Cl atoms at different positions are called **congeners**. The numbering of Cl atom positions on the biphenyl common structure of PCB congeners is shown in Figure 5.11.

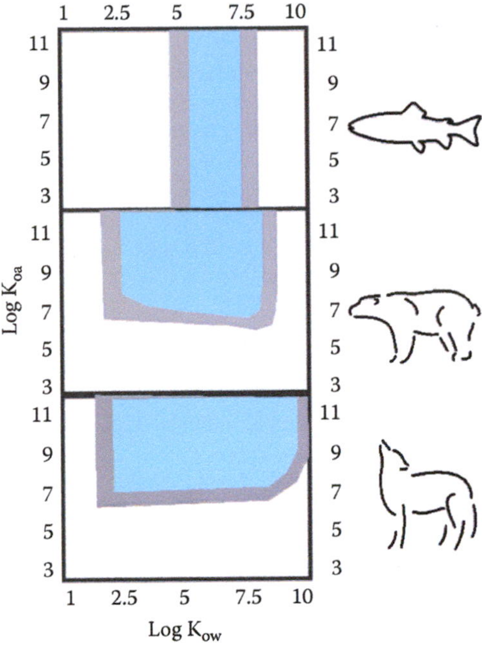

Figure 5.12 The influence of K_{ow} and K_{oa} on POP biomagnification in aquatic piscivore-focused (top panel), marine mammal-focused (middle panel), and terrestrial mammal-focused (bottom panel) food webs. Combinations of log K_{ow} and log K_{oa} associated with a biomagnification factor in the final web predator are shown: 2 or less (white areas), approximately >2 to 25 (grey area), and >25 (blue area). (This figure was drawn using information from three panels in Figure 3 of Kelly, B. C. et al. *Science,* 317, 236–239, 2007).

water-breathing counterparts. They also found that accumulation in air-breathing species was influenced by the K_{oa}-dependent respiratory elimination. The K_{oa} for a compound is similar in concept to the K_{ow} but the partitioning being quantified is between the air and the lipid surrogate, *n*-octanol. So, for an aquatic food chain involving only water-breathing species, the K_{ow} is generally adequate to predict biomagnification of POPs. The range of K_{ow} for which biomagnification was noted was $>10^5$ to $<10^8$. For a terrestrial food web such as that just described for lichen → caribou → wolf, both K_{ow} and K_{oa} are required to predict biomagnification. Generally, POPs with log K_{ow} values between 2 and 10 and log K_{oa} of 6 or greater were found to biomagnify in the studied terrestrial food web. The marine mammal-focused food web (with polar bear as the top predator) included water-breathing species so it was intermediate between the aquatic and terrestrial food webs (Figure 5.12).

5.5 SUMMARY

In this chapter, the transfer of contaminants during trophic interactions was discussed. Methods of estimating such transfer were provided for discrete and continuous trophic structures. Conditions conducive to and examples of biomagnification and trophic dilution were outlined for metals, metalloids, radionuclides, and organic compounds.

This chapter ends Section II of the book, which provides description of mechanisms leading to bioaccumulation of contaminants. Section III of the book details the consequences of such bioaccumulation. Effects will be discussed from the biochemical to the global level of ecological organization.

SUGGESTED READINGS

Broman, D., C. Näf, C. Rolff, Y. Zebühr, B. Fry, and J. Hobbie, Using ratios of stable nitrogen to estimate bioaccumulation and flux of polychlorinated dibenzo-*p*-dioxins (PCDDs) and dibenzofurans (PCDFs) in two food chains from the northern Baltic. *Environ. Toxicol. Chem.,* 11, 331–345, 1992.

Jardine, T. D., K. A. Kidd, and A. T. Fisk, Applications, considerations, and sources of uncertainty when using stable isotope analysis in ecotoxicology. *Environ. Sci. Technol.,* 40, 7501–7511, 2006.

Kelly, B. C., M. G. Ikonomou, J. D. Blair, A. E. Morin, and F. A. P. C. Gobas, Food web-specific biomagnification of persistent organic pollutants. *Science,* 317, 236–239, 2007.

McNeill, K. G. and O. A. D. Trojan, The cesium-potassium discrimination ratio. *Health Phys,*. 4, 109–112, 1960.

Newman, M. C., *Quantitative Ecotoxicology,* 2nd Edition, Taylor & Francis/CRC Press, Boca Raton, FL, 2013.

Newman, M. C. and W. H. Clements, Ecotoxicology*: A Comprehensive Treatment*, Taylor & Francis/CRC Press, Boca Raton, FL, 2008.

Reinfelder, J. R. and N. S. Fisher, The assimilation of elements ingested by marine copepods. *Science,* 251, 794–796, 1991.

Rowan, D. J. and J. B. Rasmussen, Bioaccumulation of radiocesium by fish: the influence of physicochemical factors and trophic structure. *Can. J. Fish. Aquat. Sci.,* 51, 2388–2410, 1994.

Vijver, M. G., C. A. M. Van Gestel, R. P. Lanno, N. M. Van Straalen, and W. J. G. M. Peijnenburg, Internal metal sequestration and its ecotoxicological relevance: A review. *Environ. Sci. Technol.,* 38, 4705–4712, 2004.

Weeks, J. M. and P. S. Rainbow, A dual-labeling technique to measure the relative assimilation efficiencies of invertebrates taking up trace metals from food. *Funct. Ecol.,* 4, 711–717, 1990.

CHAPTER 6

Molecular Effects and Biomarkers

Pollutants affect biological systems at many levels, but all chemical pollutants must initially act by changing structural and/or functional properties of molecules essential to cellular activities.

Jagoe (1996)

... if effects on the ecosystem are to be predicted and understood, it is necessary to identify the effects of a toxicant on lower levels of biological organization, such as the subcellular level. Specific and sensitive biochemical methods can, therefore, serve as early warning indicators and adverse effects on the ecosystem can be avoided by taking protective measures.

Haux and Förlin (1988)

6.1 INTRODUCTION

Two key concepts relative to biochemical effects of toxicants are exemplified by the above quotes. First and most fundamental, all toxicant effects begin by interacting with biomolecules. Effects can then cascade through the biochemical → subcellular → cellular → tissue → organ → individual → population → community → ecosystem → landscape → biosphere levels of organization.* Consequently, an understanding of effects at the biochemical level may provide some insight into the root cause of effects seen at next few higher levels. Also, by understanding biochemical mechanisms, we can better predict effects of untested contaminants based on their similarity of biochemical mode of action to that of well-understood contaminants. And, if several contaminants are present, specific biochemical changes can provide valuable clues about which contaminant is having an effect. In some histochemical methods, tissue localization of biochemical changes can also provide information relative to exposure at target organs or sites. For example, high Phase I monooxygenase activity in a specific tissue may be measured after animals are exposed in the laboratory to a toxicant† known to be metabolized to a potent carcinogen. This finding may be linked to survey results indicating high incidence of cancers in the same tissue of individuals from a contaminated

* It is also true that some causal sequences might be better studied from a top-down vantage (Newman and Clements, 2008). For example, herbicide alteration of vegetation in a forbs and shrub–grass community can adversely impact a resident rodent population's dominance structure and ultimately the overall fitness of individual rodents (Johnson and Hansen, 1969).

† A compound that is converted to a carcinogen is a **procarcinogen.** For example, 2-acetylaminofluorene is a procarcinogen that is converted to a carcinogen by Phase I enzymes (*N*-hydroxylation by monooxygenases) (Stegeman and Hahn, 1994). Similarly, cytochrome P-450 monooxygenases transform BaP to potent carcinogens. Remember from Chapter 3 that these are cases of activation.

site. So, understanding the biochemical mode of action enhances our grasp of causal structure and our ability to predict effects at higher levels of biological organization. This understanding is greatly enhanced by biochemical information and methods developed first in medical sciences.

Second, a technical advantage is gained by understanding toxicant effects on biomolecules and molecular responses to contaminants. Changes in biomolecules or suites of biomolecules can indicate exposure to bioavailable contaminants in field situations. The biochemical quality that is changing can be used as a biomarker. As discussed in Chapter 1, a biomarker is a biochemical, physiological, morphological, or histological quality used to imply exposure to or effect of a toxicant. Biomarkers can also indicate that sufficient toxicant was available for enough time to elicit a response or effect (Melancon, 1995; van der Oost et al. 2003). Because biochemical changes generally are detectable before adverse effects are seen at higher levels of biological organization, the biochemical marker approach is often applied as an early warning or proactive tool. This is a great advantage because responses at higher levels such as the ecosystem are usually measurable only after significant or permanent damage has occurred. It follows that biochemical markers are also useful to monitor the shift back to a normal state after cleanup of a contaminated site. Regardless of their proactive or retroactive utility, the ecological realism (ability to accurately reflect an ecologically meaningful effect or response) is lower for molecular biomarkers than for indicators based on higher level changes such as species richness or reproductive failure.

The following qualities enhance the usefulness of a biochemical change as a biomarker. First, it should be measurable before any adverse, material consequences occur at higher levels of biological organization (Haux and Förlin, 1988; Campbell and Tessier, 1996). This enhances its value as a proactive tool. Second, measurement of the ideal biomarker should be rapid, inexpensive, and sufficiently easy so as to be amenable to widespread use by ecotoxicologists. Third, its measurement should accommodate standard quality control/quality assurance practices. Fourth, the ideal biomarker should be specific to a single toxicant or class of toxicants (Haux and Förlin, 1988; Campbell and Tessier, 1996), although nonspecific biomarkers have value too (Monserrat et al. 2007). Fifth, a clear concentration–effect relationship must exist for the toxicant and biomarker (Haux and Förlin, 1988). Sixth, the ideal biomarker should be applicable to a broad range of **sentinel species** (feral, caged, or endemic species used in measuring and indicating the level of contaminant effect during a biomonitoring exercise) so that it might have the widest possible application (Sanders, 1990). Seventh, established linkage of biomarker changes with some toxicant-related decrease in individual fitness is desirable (Sanders, 1990) but not always necessary. This enhances ecological relevance in any discussions of change in a biomarker. Finally, the system should be sufficiently well understood so that other qualities of the organism or its environment that influence the biomarker can be accommodated in the experimental design and data interpretation (Campbell and Tessier, 1996). For example, ambient temperature, age, and sex can influence monooxygenase activity and should be considered in the design of biomonitoring studies using this biomarker (Kleinow et al. 1987).

The remainder of this chapter examines biochemical aspects of ecotoxicology with the dual goals of understanding toxicant mode of action and describing current molecular biomarkers. Enzyme activities, conjugates, and products of Phase I and II breakdown of organic compounds are discussed first. Focus then shifts to metallothioneins and stress proteins. Enzymes and products associated with oxidative stress, toxicant effects on nucleic acids, and symptoms of enzyme dysfunction are detailed. Aspects of these topics are explored again during discussions of cancer (Chapter 7).

6.2 ORGANIC COMPOUND DETOXIFICATION

A variety of biochemical processes can transform a xenobiotic present in the cell. The overarching theme to these transformations is either to render the xenobiotic more amenable to further reaction or to make it more easily eliminated from the cell. Integrated with these biochemical transformation mechanisms are complementing mechanisms designed to either prohibit initial xenobiotic entry into the cell, or to remove the xenobiotic or its metabolites from the cell. The membrane-associated P-glycoprotein (P-gp) pump is a member of the ATP-binding cassette (ABC) family of transporters that can reject xenobiotics as they begin entry to the cell, often acting as the cells first line of defense (Bard, 2000; Dean and Annilo, 2005). The ABC transporters can also remove contaminants from the cell as in the case of tributyltin export from oysters (*Saccostrea forskali*) (Kingtong et al. 2007). Once a xenobiotic has been transformed in the cell, cellular mechanisms such as the ATP-dependent glutathione S-conjugate pump (Ishitawa, 1992) can remove Phase II reaction products from the cell. The ABC transporters including P-gp can also remove products of Phase I and II reactions (Bard, 2000; Hoffmann and Kroemer, 2004). Anionic metabolites can be removed by organic anion transporters (OATS) (Sweet, 2005). This array of transport and detoxification mechanisms works in concert to minimize the amount of xenobiotic available to cause damage at cellular sites of action.

6.2.1 Phase I Transformations

As will be detailed shortly, Phase I reactions involve hydrolysis, reduction, and oxidation of xenobiotics (Parkinson and Ogilvie, 2008). Functional groups are produced or exposed during Phase I reactions, rendering the compound more reactive and perhaps producing a biotransformation product that could be acted on in Phase II detoxification reactions. Many Phase I products might also be more water soluble (Parkinson 1996; Parkinson and Ogilvie, 2008) than the original xenobiotic. Phase I products can enter into Phase II reactions but some are eliminated effectively without further Phase II biotransformation. "A commonality of biotransformation reactions is the conversion of hydrophobic xenobiotics into more polar, more easily excreted compounds" (Kemper et al. 2008). Cells of different tissues differ in their capacities for Phase I and II reactions.

Cytochrome P-450 monooxygenase is the major Phase I system. Biochemical shifts associated with **cytochrome P-450** (CYP) monooxygenase induction are often used to document a response to organic contaminants such as polycyclic aromatic hydrocarbons (PAHs) (Wirgin et al. 1994; Buesen et al. 2002; Abrahamson et al. 2007), chlorinated hydrocarbons (Walker et al. 1987), organic compounds in pulp and paper mill effluents (Soimasuo et al. 1995), polychlorinated biphenyls (PCBs) (Brumley et al. 1995), hydrocarbons, dioxins, and dibenzofurans (Goksøyr and Förlin, 1992). These monooxygenases are involved in the conversion of a very wide range of xenobiotics as well as eicosanoids (icosanoids), fatty acids, sterols including cholesterol, vitamins, steroid hormones, and many drugs (Guengerich, 2008). When our understanding of this system was less detailed, the cytochrome P-450 monooxygenase system was often called the **mixed function oxidase** (MFO) system (Di Giulio et al. 1995); however, more precise terms referring to the monooxygenase function or specific CYP isoforms are now preferred. Although most Phase I oxidations are associated with the CYP monooxygenase system, the cytochrome b_5 and NADH-cytochrome b_5 reductase system is also important in xenobiotic metabolism (Di Giulio et al. 1995). Kemper et al. (2008) reviewed recent work with cytochrome b_5 reductase that indicates it could also pass electrons to cytochrome b_5 and then to the CYP monooxygenase system. **Epoxide hydrolases**, which add water to an epoxide to produce a dihydrodiol (Timbrell, 2000), are another group of Phase I enzymes found mostly in the microsomal fraction (Franklin and Yost, 2000). The hydrolyzed epoxide might have been produced originally by CYP monooxygenase oxidation as in the case of the naphthalene shown in Figure 3.4 (Parkinson and Ogilvie, 2008). Other examples include **esterases** and **amidases** that facilitate ester and amide hydrolysis, respectively (George, 1994; Timbrell, 2000). Esterases and amidases act on xenobiotics that tend to be similar in structure to endogenous compounds (Kemper et al. 2008). Although both are found in microsomes, esterases are found mostly in the cytoplasm (Franklin and

Yost, 2000). Another important Phase I monooxygenase is the **flavin monooxygenase** or **flavin-containing monooxygenase (FMO)**. This microsomal flavin-containing Phase I monooxygenase is capable of oxidizing nucleophilic (electron-deficient) nitrogen, sulfur, and phosphorus groups in xenobiotic compounds. Like the CYP monooxygenases about to be described in detail, the FMO system requires nicotinamide adenine dinucleotide phosphate (NADPH) and molecular oxygen and has different reaction specificities depending on the isoform composition of the complex (Kemper et al. 2008, Parkinson and Ogilvie, 2008).

The CYP monooxygenase system is associated primarily with and assayed in the microsomal fraction, the cell fraction composed of membrane vesicles (**microsomes**) derived from the endoplasmic reticulum (ER) during routine separations by tissue homogenization followed by ultracentrifugation. The name P-450 is derived from P for pigment and 450 nm, the wavelength at which they have maximum light absorption when bound to CO (Haux and Förlin, 1988; Di Giulio et al. 1995). It is present at high levels in livers of many vertebrates and can be induced by xenobiotics (Parkinson and Ogilvie, 2008).

The CYP monooxygenases are hemoproteins present in the highest amount with ER membranes although they are also present in eukaryote mitochondria (Omura, 1999; Parkinson and Ogilvie, 2008) and prokaryote cytoplasm (Omura, 1999). More specifically, the CYP monooxygenase system is an assemblage of **isoforms*** [i.e., distinct proteins (enzymes) differing in their substrate specificity]. The widely varying nature of the isoforms influences the CYP monooxygenase system's ability to act on such a wide range of substrates (Kemper et al. 2008). Each isoform has a molecular weight of approximately 45–55 kDa (Goksøyr and Förlin, 1992; Timbrell, 2000) and contains an iron protoporphyrin IX (heme) prosthetic group (Timbrell, 2000). In the membrane, the isozymes are associated with the 78-kDa **NADPH cytochrome P-450 reductase**, a flavoprotein that transfers electrons to the isoenzyme assemblage (Timbrell, 2000; Kemper et al. 2008). The P450 isozymes, reductase, and membrane phospholipids form the functional unit responsible for many Phase I oxidations. The reaction normally depicted as facilitated by this system involves the incorporation of an oxygen atom into the organic compound

$$RH + NADPH + O_2 + H^+ \rightarrow ROH + NADP^+ + H_2O \tag{6.1}$$

where RH is an organic compound undergoing hydroxylation. It involves two enzymes (CYP isozymes and NADPH-CYP reductase), NADPH, and molecular oxygen. Although hydroxylation is shown here, CYP monooxygenases also catalyze epoxidation, (N-, O-, and S-) dealkylation, oxidative deamination, (S-, P-, and N-) oxidation, desulfuration reactions, and oxidative and reductive dehalogenation (Stegeman and Hahn, 1994; Di Giulio et al. 1995). Timbrell (2000) describes these reactions in detail, using toxicant and drug transformations as examples.

There is a rich nomenclature to be understood to make sense of studies of these diverse CYP monooxygenases. The genes that code for CYP monooxygenases are grouped into families and subfamilies based on similarities in DNA sequences. Names of genes and gene products are based on the root CYP (cytochrome **P**-450). Numbers and letters are added to the CYP root to designate the particular family (e.g., CYP1 or CYP2), subfamily (e.g., CYP1A or CYP1B), and gene (e.g., CYP1A1 or CYP1A2) (Goksøyr and Förlin, 1992). A standard denotation is used for the protein, mRNA, and DNA associated with any particular gene. The protein and mRNA are most often designated as described above, e.g., CYP1A1, but in some publications, might be designated as P-450 1A1 instead. The DNA designation is always italicized, e.g., *CYP1A1* (Goksøyr and Förlin, 1992). This is important to keep in mind because protein, mRNA, and

* Mark Hahn (personal communication) indicates that **isozyme** or **isoenzyme** was used instead of the term, isoform, in the literature but isoform is now the preferred term.

DNA are all used in studies of CYP monooxygenases. Proteins are often assayed with specific antibodies and the mRNA with DNA probes. For example, Haasch et al. (1993) measured immunoreactive CYP1A1 protein and hybridizable CYP1A1 mRNA to suggest fish liver CYP response to PAH and PCB contamination. As a second example, tomcod (*Microgadus tomcod*) response to xenobiotics was monitored along the North American coast with CYP1A mRNA (Wirgin et al. 1994).

Members of the CYP1, CYP2, and CYP3 families are very important in xenobiotic transformation (Parkinson and Ogilvie, 2008). As examples, the CYP1A and CYP1B are involved in the general transformation of diverse xenobiotics and also estrogens. Kemper et al. (2008) indicate that CYP1A is involved in PAH transformation and CYP1B is active in the conversion of molecules such as aromatic amines and caffeine. Because of their prominent roles in xenobiotic transformations, the CYP1 and CYP2 tend to be expressed in tissues that come into contact with xenobiotics, e.g., lung, small intestine, skin, and liver (Omura, 1999) and tissues associated with detoxification (liver and also kidney) (Franklin and Yost, 2000). Expression can often be tissue specific. For example, Parkinson and Ogilvie (2008) note that CYP1A1 and CYP1A2 activity expressions are so different in liver and nonliver tissues of humans that CYP1A1 and CYP1A2 can be considered extrahepatic and hepatic enzymes, respectively.

Additional nomenclature appears if enzymatic activities are used as measures of CYP monooxygenase activity. CYP monooxygenase activity can be measured as hydroxylation of benzo[a]pyrene (BaP) in which case results are expressed as **AHH** (**a**ryl **h**ydrocarbon **h**ydroxylase) activity. Another common assay involves O-deethylation of ethoxyresorufin with activity being expressed as **EROD** (**e**thoxy**r**esorufin **O-d**eethylase) activity (Goksøyr and Förlin, 1992; Sarkar et al. 2006; Abrahamson et al. 2007). CYP monooxygenase activities measured as AHH or EROD are common biomarkers reflecting response (induction) to significant amounts of bioavailable xenobiotic. For example, Soimasuo et al. (1995) used EROD activity in whitefish (*Coregonus lavaretus*) liver to measure CYP1A response to pulp and paper mill effluent (Figure 6.1). Abrahamson et al. (2007) noted elevated gill EROD activity in caged trout reared near sources of PAH. Haasch et al. (1993) used EROD activity in tandem with assays of CYP1A1 protein and mRNA to monitor fish (catfish, *Ictalurus punctatus*; largemouth bass, *Micropterus salmoides*; and killifish, *Fundulus*

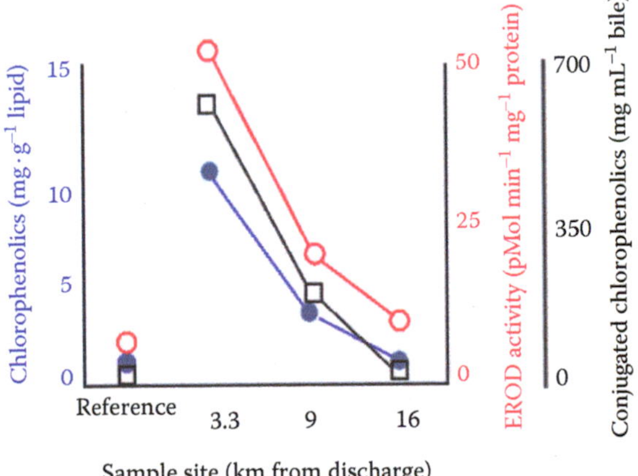

Figure 6.1 The response of biomarkers in juvenile whitefish (*Coregonus lavaretus*) to pulp and paper mill effluents. The concentration of total chlorophenolics in the gut lipids of this sentinel species (●) were elevated in whitefish taken 3.3 km below the discharge relative to the reference samples and rapidly decreased with distance from the discharge. Similarly, two biomarkers, EROD activity (○) and conjugated chlorophenolics in the bile (□), were highest near the discharge and decreased with distance from the paper and pulp mill effluent. (Generated by combining information from Figures 3, 4, and 5 of Soimasuo, R., et al. *Aquat. Toxicol.*, 31, 329–345, 1995.)

heteroclitus) response to PAH and PCB exposure. Other enzyme assays that reflect P-450 activity include ethoxycoumarin O-deethylase (ECOD), aflatoxin B1 2,3-epoxidase (AFBI), lauric acid ω-1 hydrolase (LA), testosterone hydroxylase (TH), and phenanthrene hydroxylase (AH) assays (Goksøyr and Förlin, 1992, Brumley et al. 1995). An important point to keep in mind when interpreting data from activity assays is that any one "monooxygenase activity" (e.g., EROD) might be catalyzed jointly by multiple CYP isoforms, and each isoform might act on multiple substrates, i.e., contribute to multiple activity measurements.

Induction of CYP monooxygenase is influenced by numerous factors including ambient temperature (Kleinow et al. 1987), the particular toxicant and species combination being studied (Haux and Förlin, 1988), body weight (Parke, 1981), animal sex (Haux and Förlin, 1988), hormone titers (Haux and Förlin, 1988), and tissue oxygen tension (Parke, 1981). When these modifying factors are taken into account, CYP monooxygenase induction has proved to be a reliable biomarker. As discussed already, it was used by Haasch et al. (1993), Wirgin et al. (1994), Soimasuo et al. (1995), and Abrahamson et al. (2007) as an indicator for significant exposure to a variety of contaminants. Van Veld et al. (1990) measured CYP protein (immunoassay) and activity (EROD) in spot (*Leiostomus xanthurus*) to indicate PAH contamination in the Chesapeake Bay region. Brumley et al. (1995) used induction of these same biomarkers in sand flatheads (*Platycephalus bassensis*) as indicators of response to PCB exposure.

VIGNETTE 6.1 Cytochrome P-450 Monooxygenases and Their Regulation

Mark E. Hahn
Woods Hole Oceanographic Institution

GENERAL

Cytochromes P-450 are heme-containing enzymes that catalyze the oxidative (and occasionally reductive) biotransformation of organic compounds. Collectively, CYPs are the primary group of proteins catalyzing Phase I biotransformation reactions, and thus represent a key part of the "chemical defensome" that has evolved in response to the presence of toxic xenobiotic and endogenous chemicals. CYPs are notable for several features that make them of great interest in toxicology, including ecotoxicology: (1) the large intraspecific and interspecific diversity of CYP genes and isoforms, (2) the broad range of compounds that can be substrates for CYP-mediated biotransformation, (3) the important roles of CYPs in both detoxication and toxication (activation) processes, and (4) the induction of selected CYPs in response to chemical exposure, making them potentially useful as biomarkers of exposure, effect, or susceptibility.

Much of our fundamental understanding of the roles and regulation of CYPs in toxicology comes from studies in mammalian species (primarily mice, rats, and humans). One objective of research in environmental toxicology and ecotoxicology is to determine the extent to which results obtained in mammalian systems (and other model systems) can be extrapolated to other phylogenetic groups, such as nonmammalian vertebrates and various invertebrate taxa. Increasingly, with the expanding number of whole genome sequences and associated technologies, information on CYP diversity, function, and regulation in organisms of interest in ecotoxicology is becoming more widely available.

CYP DIVERSITY AND EVOLUTION

The CYP superfamily is highly diverse with respect to the numbers of genes and their variability within and among species. CYP nomenclature, which is based on evolutionary relationships among the proteins (Nelson, 2006), does not coincide precisely with functional characteristics but provides a useful framework within which to consider—and predict—functional and regulatory properties of the CYP genes. The most up-to-date information on CYP nomenclature and diversity can be found at the P450 website, http://drnelson.uthsc.edu/CytochromeP450.html.

Genome sequencing projects involving model species have provided insight into CYP diversity and how it varies among taxa. The focus here is on metazoans, but similar patterns can be seen in other organisms. Table 6.1 illustrates the number of CYP subfamilies and genes in the CYP families most prominently involved in metabolizing toxic xenobiotic and endogenous compounds. In deuterostomes (chordates and echinoderms), these are families 1–4. In protostomes, families 1–3 (as defined in deuterostomes) are not present (Goldstone, 2008), although CYP1-like, CYP2-like, and CYP3-like genes have been identified (Baldwin et al. 2009; Zhang et al. 2012; Zanette et al. 2013). CYPs in families 4, 6, and 9 appear to carry out xenobiotic biotransformation in protostomes.

A common feature revealed by genome sequencing is the species-specific expansion of certain CYP families. For example, mice have more than three times as many CYP2 genes as do humans (Nelson et al. 2004). The CYP2 family is expanded also in sea urchin (Goldstone et al. 2006), whereas in *Drosophila melanogaster*, CYP families 4 and 6 contain the most genes (Strode et al. 2008). The exact numbers of identified CYP genes in animal genomes (Table 6.1) will change somewhat with further study. The key point is that there is tremendous CYP diversity and that it differs among animal groups both in terms of the numbers of CYP genes involved in detoxication and the specific CYP families or subfamilies that have undergone expansion.

Table 6.1 CYP Diversity in Metazoans: A Phylogenetic Perspective

Group	Representative	Total Number of CYP Genes	CYP1	CYP2	CYP3	CYP4	CYP6	CYP9
Mammals	Human	57	3[a] (A, B)	16 (A, B, C, D, E, F, J, R, S, U, W)	4 (A)	12 (A, B, F, V, X, Z)	NP[b]	NP
	Mouse	102	3 (A, B)	50 (A, B, C, D, E, F, G, J, R, S, T, U, W)	8 (A)	20 (A, B, F, V, X)	NP	NP
Reptiles/birds	Chicken	41	4 (A, B, C)	(C, G, H, J, R, U, AC)				
Amphibians	*Xenopus tropicalis*		3 (A, B, C)			(F, V)		
Bony fishes	*Takifugu rubripes* (pufferfish)	54	4 (A, B, C)	17 (K, N, P, R, U, X, Y, Z)	5 (A, B)	4 (F, T, V)	NP	NP

(Continued)

Table 6.1 *(Continued)*

Group	Representative	Total Number of CYP Genes	CYP1	CYP2	CYP3	CYP4	CYP6	CYP9
	Danio rerio (zebrafish)	94	5 (A, B, C, D)	47 (K, N, P, R, U, X, Y, AA, AD, AE)	5 (A, C)	4 (F, T, V)	NP	NP
Echinoderms	Sea urchin	120	11[c]	73[c] (U)	10[c]	10[c] (F, V)	NP	NP
Arthropods	*Drosophila melanogaster*	84	NP[d]	NP[d]	NP[d,e]	32	22[e]	5[e] (B, C, F, H)
	Daphnia pulex	75	NP[d]	NP[d]	NP[d]	38 (C, AN, AP, BX, BY)	NP	NP
Molluscs	*Lottia gigantea*	67	NP[d]			6		
	Crassostrea gigas	136	NP[d,f]	NP[d,f]	NP[d,f]	~20	NP	NP
Cnidarians	*Nematostella*	82	NP[d]	NP[d]	NP[d,e]	3	NP[d,e]	NP[d,e]

Note: Data were obtained from various sources, including P450 home page (http://p450.sophia.inra.fr/index_link.html), published articles (Nelson, 2003; Nelson et al. 2004; Goldstone et al. 2006, 2007, 2010; Goldstone, 2008; Strode et al. 2008; Baldwin et al. 2009; Zhang et al. 2012), and J. V. Goldstone (personal communication).

[a] Numbers indicate the number of genes in the family; letters indicate the subfamilies present.

[b] NP = not present.

[c] In phylogenetic analyses, sea urchin CYP genes usually cluster outside of the known vertebrate subfamilies; subfamily designations have not yet been established. (For more detailed discussion, see Goldstone, J.V., et al. *Dev. Biol.*, 300, 366–384, 2006; Goldstone, J.V., et al. *Mol. Biol. Evol.*, 24, 2619–2631, 2007.)

[d] Many CYP genes in protostomes cannot be classified into typical CYP families. A higher-order classification system, in which multiple CYP families are represented by "Clans," is sometimes used. Clan 2 includes CYP families 1, 2, 17, 18, and others; Clan 3 includes families 3, 5, 6, 9, and others; Clan 4 includes CYP family 4 and others. *Nematostella* has 39 genes in Clan 2 that are most closely related to CYP17 genes, and 20 genes in Clan 3 that are most closely related to CYP5 genes. (Goldstone, J.V., *Cell Biol. Toxicol.*, 24, 483–502, 2008.)

[e] CYP6 and CYP9 families are closely related to the CYP3 family within the "CYP3 clan." (From Nelson, D.R., *Methods Mol. Biol.*, 320, 1–10, 2006.)

[f] For molluscs, CYP1-like, CYP2-like, and CYP3-like genes have been identified (Zhang et al. 2012; Zanette et al. 2013). These do not meet the thresholds of sequence identity (>40%) needed for inclusion in these CYP families but could share some functional or regulatory properties with genes in those families. In oyster, a large number of CYP genes have been annotated as CYP2-like.

Diversity in CYP gene sequences implies differences in function or regulation, which can take many forms. A single amino acid difference can change the substrate specificity of a CYP protein, so functional diversification following gene duplication can be rapid. Changes in CYP catalytic function could involve changes in rates of substrate turnover, altered substrate specificity from one substrate to another, or broadening of a narrow substrate range to one that can accommodate a wider variety of chemicals.

Some (perhaps most) of the diversity in xenobiotic-metabolizing CYPs in some animals may have evolved in response to the presence of toxic natural products in their food. For example, many terrestrial plants are known to contain high concentrations of allelochemicals* that may act as feeding deterrents. Some insects have evolved the ability to detoxify these chemicals via specific CYP proteins that either are highly expressed or are induced by their allelochemical substrates (Li et al. 2007). Similar examples of such chemical ecology involving CYP proteins and allelochemicals can be found in freshwater and marine environments (e.g., Whalen et al. 2010).

CYP REGULATION AND INDUCIBILITY

The roles of CYPs as xenobiotic detoxification enzymes and as biomarkers are intimately tied to the phenomenon of inducibility, that is, the increased synthesis of CYP proteins following exposure to environmental chemicals. Consequently, there is great interest in understanding how CYPs are regulated, both positively (induction) and negatively (repression). Although CYP induction is widespread in animals, plants, and microbes, the specific CYP forms that are induced and the underlying mechanisms of induction vary. In vertebrate animals, and likely also in invertebrates, the induction of xenobiotic-metabolizing CYPs occurs through binding of inducers to receptor proteins that act as ligand-activated transcription factors† to increase transcription of selected CYP genes. CYP induction can also occur through posttranscriptional mechanisms such as enhanced stability of CYP proteins, but such mechanisms are not as well understood.

CYP1 Induction. The classic example of CYP induction in animals is that of CYP1A. The CYP1A induction response was discovered more than 40 years ago as the increase in aryl hydrocarbon hydroxylase (AHH) activity following exposure to polynuclear aromatic hydrocarbons (PAHs) such as BaP or planar halogenated aromatic hydrocarbons (HAHs) such as 2,3,7,8-tetrachlorodibenzo-*p*-dioxin (TCDD) or non*ortho*-substituted polychlorinated biphenyls (PCBs) (Okey, 2007). CYP1A induction has been studied extensively and is the basis for the widely used measurement of EROD activity as a marker of exposure to PAHs or HAHs (Bucheli and Fent, 1995). Although most of the early assessments of CYP1A induction occurred through measurement of its catalytic activities (AHH and EROD), other CYP forms can contribute to these activities. Thus, quantification of CYP1A mRNA [by real-time reverse transcription polymerase chain reaction (RT-PCR)] or CYP1A protein (by western blot or immunohistochemistry) provides a more sensitive and specific measure of CYP1A expression.

CYP1A genes are found in deuterostomes, but CYP1A induction has so far been demonstrated only in jawed vertebrates: cartilaginous and bony fishes and more recently evolved vertebrate groups (amphibians, reptiles, birds, and mammals) (Stegeman and Hahn, 1994). In these vertebrate animals, CYP1A is highly induced, often by more than 100-fold, providing a robust response that is easily measured as described above.

The mechanism of CYP1A induction involves high-affinity binding of inducers to the **aryl hydrocarbon receptor** (AHR), a basic helix–loop–helix-PAS family protein that is found in animals from cnidarians to humans. Although the AHR is found in nearly all animals, it appears to function in CYP regulation only in vertebrates (Hahn, 2002). There are reports of induction of CYP1A-like proteins in invertebrates, such as molluscs, based on changes in EROD or AHH activities or proteins recognized by antibodies to vertebrate CYP1A after chemical exposure. However, molecular and genomic studies indicate that CYP1 genes are not found in protostomes (Goldstone et al. 2007), although there are CYP1-like genes (Zhang et al. 2012; Zanette et al. 2013). Nevertheless, there is

* Toxic allelochemicals are produced by plants as a defense mechanism against grazers or to inhibit competing plant species.

† A **transcription factor** is a protein involved in regulation of gene expression that binds with specific promoter elements of a gene.

no strong evidence for an AHR-CYP1 signaling pathway in these animals. This does not preclude the existence of other inducible CYP forms in these animals (e.g., Menzel et al. 2005).

CYP2 and CYP3 Induction. The induction of CYP2 genes, originally referred to as the phenobarbital-like pattern of induction because of this classical CYP2 inducer, has been known for as long as that of CYP1A. Other CYP2 inducers include *ortho*-substituted PCBs, dichlorodiphenyltrichloroethane, and other organochlorine pesticides. CYP3 genes are induced by steroids and steroid-related drugs such as pregnenolone 16a-carbonitrile (PCN). Unlike the case for CYP1 induction, the mechanism by which CYP2 and CYP3 genes are induced was an enigma until recently. Part of the mystery surrounding CYP2 regulation concerned the apparent lack of CYP2 induction in fishes, in contrast to the well-conserved CYP1A induction response in these animals. It is now known that induction of CYP2 and CYP3 genes occurs largely through activation of nuclear receptors known as the **constitutive androstane receptor** (CAR) and **pregnane-X-receptor** (PXR), respectively (Handschin and Meyer, 2003). Both CAR and PXR are found in mammals, whereas most other vertebrates possess a single PXR-like receptor. Fish, in particular, appear to lack CAR, providing a molecular explanation for a long-standing mystery concerning the absence of a CYP2 induction response like that seen in mammals.

Induction of Other CYP Forms. Other CYP genes also are inducible by xenobiotics, and thus have potential as biomarkers. CYP4 genes involved in fatty acid metabolism are induced by compounds that cause peroxisome proliferation, including some hepatocarcinogens, through a mechanism involving another nuclear receptor known as peroxisome proliferator-activated receptor (PPAR). Similarly, CYP19 genes, which encode aromatase enzymes (estrogen synthases), can be inducible by exposure to estrogens or xenoestrogens acting through estrogen receptors (ERs). CYP19 (aromatase) genes are of particular interest in bony fishes, which have two forms—a brain-specific form and an ovarian form—that are regulated independently (Sawyer et al. 2006).

Although most of the information about CYP inducibility in animals comes from studies in vertebrates, CYP induction is also seen in invertebrates. Inducible CYPs are well known in insects, especially in relation to insecticide resistance and chemical ecology. For example, CYP4, CYP6, and CYP9 genes are often upregulated in insecticide-resistant strains or are induced by plant toxins (allelochemicals) in the diet of specialist consumers (Li et al. 2007; Strode et al. 2008). Additional examples of CYP inducibility will no doubt emerge from studies of other invertebrate groups.

CYP FUNCTIONS IN BIOTRANSFORMATION AND AS BIOMARKERS

The importance of CYPs in ecotoxicology is multifaceted, involving their catalytic activity with xenobiotic substrates, secondary products generated during those reactions, and the use of CYP expression as a biomarker. CYP enzymes catalyze biotransformation reactions that result in detoxication as well as activation, sometimes performing both types of reactions on a single substrate. Because of this, it can be difficult to predict the specific outcome (harmful or protective) of any one CYP acting on a xenobiotic. An example is the PAH, BaP. CYP1A enzymes catalyze the conversion of BaP to several products, including less toxic hydroxylated metabolites and more toxic bioactivation products such as epoxides that go on to form adducts with DNA. Because of the well-known bioactivation reactions, it had been predicted that loss of CYP1A would be protective against the toxicity of BaP or other PAHs. However, knockout of the CYP1A1 gene enhanced the sensitivity of mice to BaP toxicity and DNA damage, probably through reduced elimination leading to increased accumulation of BaP, coupled with its bioactivation by other enzymes (Uno et al. 2004). Similar results have been obtained in PAH-exposed fish embryos in which CYP1A expression has been inhibited or knocked down by morpholino **antisense oligonucleotides** (Billiard et al. 2007).

Another potential CYP-dependent mechanism of toxicity is through CYP induction leading to altered metabolism of endogenous signaling chemicals, such as hormones. In addition, CYP enzymes such as CYP1A can release reactive oxygen species, which can be stimulated by the binding of xenobiotic substrates (Schlezinger et al. 1999). Uncoupling of CYP reactions in this way has been suggested as an important mechanism of embryotoxicity. For example, a mechanistic role for CYP1A was suggested by the identification of CYP1A as the single most highly induced gene in fish embryos exposed to TCDD (Handley-Goldstone et al. 2005). Although recent studies have shown that CYP1A is not involved in embryotoxicity of TCDD in zebrafish (Carney et al. 2004), for other compounds or other species CYP-generated reactive oxygen could still be highly relevant.

The role of CYPs as biomarkers is a complex and somewhat controversial issue, perhaps best illustrated using CYP1A as an example. The induction of CYP1A (AHH, EROD, CYP1A protein, or mRNA) may be the most widely used biomarker in ecotoxicology. But biomarker of what? There is no doubt that CYP1A induction is a sensitive and specific *biomarker of exposure* to compounds capable of acting as agonists for the AHR, such as chlorinated dioxins, planar PCBs, PAHs, and others. However, CYP1A also can be induced by natural compounds, so caution must be used in interpreting elevated CYP1A levels in the absence of confirming data from chemical analysis of xenobiotic concentrations. To the extent that it indicates that the inducers are present in the animal at levels sufficient to elicit a biochemical response, CYP1A induction can also be considered a *biomarker of effect*. Whether it is a *biomarker of adverse effect* will depend on the demonstration that the induced CYP1A itself is involved in a mechanism of toxicity, or that there are parallel pathways of toxicity (e.g., other AHR-regulated target genes) that are induced at the same dose levels. To date, such evidence is rare. On another cautionary note, there are instances where even use of CYP1A induction as a biomarker of exposure provides misleading information. This would occur, for example, when CYP1A catalytic activity is inhibited by high levels of xenobiotics (in which case it would be better to measure CYP1A protein or mRNA) (Hahn et al. 1996), or at sites in which populations of animals have evolved resistance or tolerance to PAHs or dioxin-like compounds (Van Veld and Nacci, 2007; Nacci et al. 2010). In these cases, low expression of CYP1A would mask the fact that the animals are highly exposed. However, the lack of CYP1A inducibility could be used as a *biomarker of susceptibility*, signaling resistance to the toxicity of the chemicals at those sites.

SUMMARY

CYP proteins have central roles in toxicology, including ecotoxicology. CYP genes are highly diverse, with multiple functional characteristics and complex regulatory patterns. The investigation of CYPs as biomarkers and as mediators of chemical effect requires careful consideration of knowledge obtained in model systems, placed in a phylogenetic context with the species of concern. The emergence of molecular and genomic information and technologies is facilitating a more complete understanding of the roles of CYP proteins in environmental toxicology and ecotoxicology.

SUGGESTED READINGS

Bucheli, T. D., and K. Fent, Induction of cytochrome P450 as a biomarker of environmental contamination in aquatic ecosystems. *Crit. Rev. Environ. Sci. Technol.*, 25, 201–268, 1995.

Goldstone, J. V., A. Hamdoun, B. J. Cole, M. Howard-Ashby, D. W. Nebert, M. Scally, M. Dean, D. Epel, M. E. Hahn, and J. J. Stegeman, The chemical defensome: Environmental sensing and response genes in the *Strongylocentrotus purpuratus* genome. *Dev. Biol.*, 300, 366–384, 2006.

Goldstone, J. V., A. G. McArthur, A. Zanette, T. Parenta, M. E. Jonsson, D. R. Nelson, and J. J. Stegeman, Identification and developmental expression of the full complement of cytochrome P450 genes in zebrafish. *BMC Genomic,* 11, 643, 2010.

Hahn, M. E., Aryl hydrocarbon receptors: Diversity and evolution. *Chem. Biol. Interact.*, 141, 131–160, 2002.
Nebert, D. W. and T. P. Dalton, The role of cytochrome P450 enzymes in endogenous signalling pathways and environmental carcinogenesis. *Nat. Rev. Cancer*, 6, 947–960, 2006.
Nelson, D. R., J. V. Goldstone, and J. J. Stegeman, The cytochrome P450 genesis locus: The origin and evolution of animal cytochrome P450s. *Philos. Trans. R. Soc. Lond. B. Biol. Sci.*, 368, 20120474, 2013.
Stegeman, J. J. and M. E. Hahn, Biochemistry and molecular biology of monooxygenases: Current perspectives on forms, functions, and regulation of cytochrome P450 in aquatic species, in *Aquatic Toxicology: Molecular, Biochemical and Cellular Perspectives*, D. C. Malins, and G. K. Ostrander, Eds., CRC/Lewis Publishers, Boca Raton, FL, 1994.

6.2.2 Phase II Transformations

Phase II enzymes might also be induced by xenobiotics and used in biomarker studies. They involve **conjugation**, that is, "the addition to foreign compounds of endogenous groups which are generally polar and readily available in vivo" (Timbrell, 2000). The readily available compounds that are bound to the toxicants (or their metabolites) can be carbohydrate derivatives, amino acids, glutathione, or sulfate (Timbrell, 2000). The result is production of a conjugated compound that is more polar and consequently more readily eliminated. In some cases, the conjugation permits ready recognition by specific elimination processes of the conjugated toxicant or metabolite, i.e., the glucuronic acid portion of the conjugate is recognized by the transport process (Parkinson and Ogilvie, 2008).

As mentioned already, Phase I biotransformations render a toxicant more readily available for further reaction, and often, less lipophilic as in the cases of many xenobiotics acted on by CYP monooxygenases. Conjugation produces a more polar molecule from the original toxicant or, very often, the product of a Phase I reaction. The resulting increase in water solubility and decreased lipid solubility makes the molecule more prone to elimination. However, this is not always the case. Conjugation of a highly lipophilic, volatile organic compound that is usually eliminated via exhalation could diminish the rate at which it is eliminated from the body (Parkinson, 1996). Some xenobiotics are made less toxic by conjugation (Parkinson and Ogilvie, 2008).

Uridinediphospho glucuronosyltransferases [UDP-glucuronosyltransferase (UDP-GT)] are the most prominent of the Phase II group of enzymes, and in humans, are found at highest levels in the liver (Parkinson and Ogilvie, 2008). These ER-associated 55-kDa enzymes catalyze **glucuronidation**, that is, the transfer of **glucuronic acid** from uridine diphosphate glucuronic acid to electrophilic xenobiotics or their transformation products to produce a polar or often charged conjugate (Di Giulio et al. 1995; Franklin and Yost, 2000). The transfer usually involves a carboxyl, hydroxyl, phenol, amine, and free sulfhydryl groups on the compound (Franklin and Yost, 2000; Parkinson and Ogilvie, 2008). Such a glucuronidation is shown of naphthalene 1,2-diol in Figure 3.4. Different UDP-GTs exist such as those that conjugate planar (GT1) and nonplanar (GT2) phenols (Franklin and Yost, 2000). For example, Phase II GT2 glucuronidation converts morphine to morphine-3-glucuronide directly without an initial Phase I biotransformation. UDP-GT also binds glucuronic acid covalently with electrophilic compounds such as PAHs without any initial Phase I transformation (George, 1994).

Glutathione S-transferases (GSTs) that attach **glutathione** (GSH, a tripeptide made of cysteine, glutamate, and glycine [glu–cys–gly]) to the xenobiotic or its transformation products are another group of Phase II enzymes in the cytosol. The GSTs are composed of approximately 25 kDa subunits that combine in various combinations that allow GSTs to react with a wide range of substrates (Franklin and Yost, 2000). Conjugates with glutathione are more hydrophilic and amenable to elimination. They can also be recognized by specific transporter systems.

Another Phase II enzyme is **sulfotransferase**, an enzyme in the cytosol and Golgi apparatus membranes that conjugates a sulfate with compounds including some endogenous hormones. The sulfate is ionized under physiological conditions, making the conjugates readily eliminated. This sulfonation (sulfation) occurs to alcohols, phenols, and less frequently, aromatic amines (Franklin and Yost, 2000; Parkinson and Ogilvie, 2008), using 3-phosphoadenosine-5'-phosphosulfate (PAPS) as a substrate. The PAPS can be present in limited supply in the cell, and this being the case, can be depleted during detoxification. Sulfonation can be important at low doses, but glucuronidation will become increasingly dominant as dose increases in humans (Franklin and Yost, 2000; Parkinson and Ogilvie, 2008).

There are Phase II processes that contribute less to enhancing elimination than they do to masking reactive parts of a compound (Franklin and Yost, 2000). **Amino acid conjugation**, conjugation of a xenobiotic possessing a carboxylic acid group with an endogenous amino acid, is one such Phase II category of reactions. Another is **acetylation** by *N*-acetyltransferases in which amides are formed in xenobiotic compounds (Parkinson and Ogilvie, 2008).

Both the induced activities of these enzymes and concentrations of conjugated products can serve as biomarkers. Di Giulio et al. (1995) suggested that, in general, these enzymes were less valuable biomarkers than Phase I enzymes; however, work by Kobayashi et al. (2002) suggest good potential as biomarkers. As discussed above, Soimasuo et al. (1995) examined GST and UDP-GT induction, concentrations of conjugates in bile, and EROD in whitefish. Induction of EROD was clearly demonstrated for these fish during exposures to pulp and paper mill effluent; however, induction of GST and UDP-GT was not as clear. Conjugated metabolites were sensitive biomarkers that had an obvious trend with distance from discharge (Figure 6.1). Activity of UDP-GT, and conjugates with glucuronic acid and sulfate did increase after laboratory exposure of sand flatheads to PCBs, suggesting to Brumley et al. (1995) that these qualities might be acceptable biomarkers of PCB exposure.

6.3 METALLOTHIONEINS

Metallothioneins were first isolated from horse kidneys by Margoshes and Vallee (1957) who described them as proteins with a high cadmium content and correspondingly high number of sulfhydryl residues. It was proposed at the time that a potential function of metallothionein was cadmium detoxification. Metallothioneins are now found to be ubiquitously distributed among both prokaryotes and eukaryotes, and to occur in organisms as diverse as bacteria, fungi, invertebrates, and vertebrates. In animals ranging from lower invertebrates to humans, they are found in highest concentrations in the liver or functionally equivalent organs, (e.g., the hepatopancreas), kidneys, lung or gills, and intestines (Roesijadi, 1992). They have been of considerable interest for their role in the ecotoxicology of metals and, particularly, for application as a biomarker for environmental metal contamination.

Metallothioneins are defined as "polypeptides resembling equine metallothionein in several of their features" (Kojima et al. 1999). The general features, summarized below, were based on those of the proteins originally isolated from horse kidneys:

1. Low-molecular weight with a high metal content.
2. High cysteine content, and no aromatic amino acids or histidine.
3. Unique amino acid sequence, especially regarding cysteine placement.
4. Metal-thiolate clusters.

Based on early mammalian studies, none were believed to contain aromatic amino acids or histidine (e.g., Hamilton and Mehrle, 1986); however, there are enough exceptions in nonmammalian metallothioneins that the lack of histidine is no longer seen as a distinguishing characteristic

(Blindauer, 2008). They have high heat stability, a quality often used during their isolation (Winge and Brouwer, 1986). Poorly characterized proteins or proteins not conforming precisely to the above qualities have been called **metallothionein-like proteins** in the past (Hamilton and Mehrle, 1986).

Metallothioneins are typically small proteins, ranging from 2 to 7 kDA depending on the species. However, in 1985, nonproteinaceous plant polypeptides with biochemical similarities to metallothionein were first reported (Grill et al. 1985). They were named **phytochelatins*** and classified as a nonprotein form of metallothionein (Fowler et al. 1987). As discussed briefly in Chapter 3, phytochelatins are inducible, metal-binding peptides isolated from plants and more recently from some animals (Clemens et al. 2001; Brulle et al. 2008).

Reflecting the diversity of organisms from which metallothioneins have been identified, the molecules are diverse in fine structure, and metallothioneins are now considered a superfamily that is subdivided into 15 subfamilies that generally correspond to the taxonomic origin of the molecules (Binz and Kägi, 1999).

Metallothioneins function in the uptake, internal compartmentalization, sequestration, and excretion of essential (e.g., copper and zinc) and nonessential (e.g., cadmium, mercury, and silver) metals. Their involvement in the normal homeostasis of essential metals results in their presence at basal levels in the absence of toxic metal exposure. Fluctuations in metallothionein levels can also occur with normal biological processes such as molting and reproduction (Roesijadi, 1992), or changes in glucocorticoid hormone levels (Karin and Herschman, 1981; Haq et al. 2003). Reflecting their roles in metal detoxification and sequestration, they are induced by elevated levels of the above metals. This increases the capacity of the organism to bind and effectively sequester toxic metals away from molecular sites of toxic action. Enhanced levels of metallothioneins have been linked to enhanced fitness during metal exposure of individuals (Bouquegneau, 1979; Sanders et al. 1983; Hobson and Birge, 1989; Chesman et al. 2007). Further, enhanced capacity to produce metallothionein and the consequent lessening of toxic metal effects have been linked to population adaptation to chronic metal exposure. For example, Maroni et al. (1987) found that metallothionein gene duplication in *Drosophila melanogaster* populations is associated with enhanced production of metallothionein and higher survival during copper or cadmium exposure.

Metallothioneins have been used as specific biomarkers for metal contamination under a variety of exposure scenarios. For example, metallothionein levels in whelk (*Bullia digitalis*) consuming metal-tainted grasses reflected the extent of site contamination (Hennig, 1986). Hepatic metallothionein levels in juvenile trout (*Salmo gairdneri*) (Roch et al. 1982) and metallothionein in the freshwater mussel, *Anodonta grandis* (Couillard et al. 1993) were effective biomarkers of metal contamination (Figure 6.2). More recent work has focused on using metallothionein gene expression as a biomarker for metal contamination.

Such biomarker studies can be explored even further based on the spillover hypothesis (Hamilton and Mehrle, 1986; Campbell and Tessier, 1996). Under the assumption that binding by metallothionein sequesters toxic metals away from sites of potential adverse action, the **spillover hypothesis** states that toxic effects will begin to emerge after exceeding the capacity of metallothioneins to bind metals. The unbound metals then "spill over" to interact at sites of adverse action. The amount of metal in cells in excess of that bound to metallothionein is correlated with some measure of adverse effect. One example of such use of the spillover hypothesis is the study by Klaverkamp et al. (1991) of white sucker (*Catostomus commersoni*) inhabiting lakes around the Flin Flon smelter (Manitoba, Canada).

* If free of metals, **phytochelatin** has the form of (γ-glutamic acid-cysteine)$_n$-glycine with *n* being 3, 5, 6, or 7 (Grill et al. 1985).

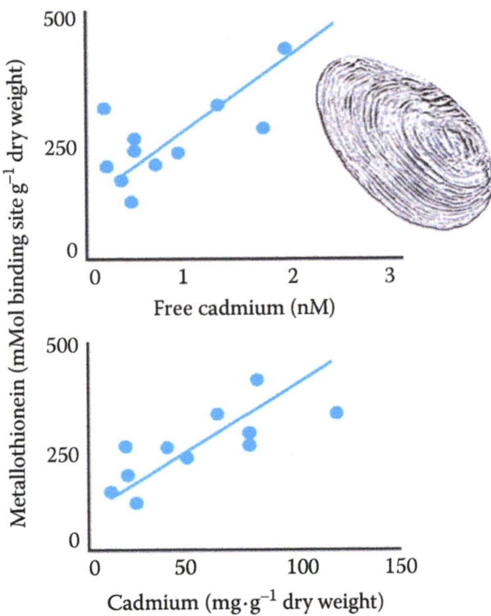

Figure 6.2 The relationships between free cadmium (Cd^{2+}) concentration in lake water (sediment-water interface) and metallothionein concentration in the freshwater mussel, *Anodonta grandis* (top panel), and cadmium concentration in the entire mussel and metallothionein concentration in this mussel (bottom panel). Each point represents samples from a different lake in the Rouyn-Noranda mining area of Quebec, Canada. (Modified from Figures 3 and 4 of Couillard, Y., et al. *Limnol. Oceanogr.*, 38, 299–313, 1993.)

VIGNETTE 6.2 Metallothioneins

Guritno Roesijadi
Pacific Northwest National Laboratory

GENERAL

Metallothioneins, a family of low-molecular-weight, metal-binding proteins discovered and initially characterized approximately four decades ago, have since been shown to be ubiquitously distributed in diverse taxa. Metallothioneins are known to be involved in regulation of essential metals, specifically of zinc and copper; protection against the toxicity of these and nonessential metals such as cadmium, silver, and mercury; and protection against free radical toxicity (Cherian and Chan, 1993). Although there is still debate over assignment of specific functions, metal detoxification, particularly of cadmium, remains a central function of this protein (Klaassen et al. 1999). The first reports of the occurrence of metallothioneins in species such as fish and molluscs in the mid-1970s had ecotoxicological contexts, and these reports were closely followed by proposals for the use of metallothioneins in pollution monitoring, a focus that prevails to the present time.

BIOCHEMISTRY AND MOLECULAR REGULATION

The amino acid sequence of metallothionein is characterized by the abundance and location of cysteine residues (Kojima et al. 1999), which form cys–cys, cys–X–X–cys, and cys–X–cys motifs, where X represents another amino acid. Metals bind cooperatively to the sulfhydryl groups of the cysteines, resulting in the formation of characteristic metal clusters, typically two, in metallothioneins of diverse taxa ranging from mammals (Braun et al. 1992) to invertebrates such as the sea urchin (Wang et al. 1995), crustaceans (Otvos et al. 1982), snail (Dallinger et al. 2001), and nematode (You et al. 1999). It has recently been shown for metallothioneins found to contain histidine and that this amino acid is involved in metal coordination as well (Blindauer, 2008).

Metallothioneins are normally present at low, basal levels and can be induced by a variety of substances and physicochemical conditions (Kägi, 1991). For this reason, they have been considered by some to be general stress-related proteins (Ryan and Hightower, 1996). Zinc is usually associated with metallothioneins under basal conditions, and the most thoroughly studied signal transduction pathway for expression of metallothioneins involves transcriptional regulation via the zinc-dependent metal transcription factor **MTF-1** and its binding to metal-response elements (MRE) in the promoter of the metallothionein gene (Andrews, 2000). This pathway for metallothionein expression is conserved in organisms as evolutionarily distant as the mouse, human, fish, and fruit fly (Laity and Andrews, 2007). MTF-1 controls the expression of metallothioneins and other genes directly involved in the intracellular zinc sequestration and transport. However, other metals known to induce metallothionein use signal-transduction pathways that do not involve MTF-1. Induction by copper in yeast occurs via the copper-specific transcription factor for metallothionein induction ACE1 (Szczypka and Thiele, 1989), and induction of cadmium in cells of higher animals occurs by way of the upstream stimulatory factor (USF) and its interaction with a composite USF/antioxidant response element (Daniels and Andrews, 2003). The existence of direct pathways for these metals strengthens arguments that metal detoxification and protection against metal toxicity are specific functions of metallothionein.

Metals are not the sole inducers of metallothioneins and the metallothionein gene also contains promoter elements* that are responsive to other signals (Haq et al. 2003). Glucocorticoid-response elements and antioxidant-response elements, for example, exist in the promoter sequences of metallothioneins in diverse organisms, enabling free radicals and steroid hormones to induce metallothionein.

Over the years, a fundamental role for metallothionein in metal trafficking through the metallothionein–apothionein redox couple has emerged. It has been known from cell-free studies with metallothionein and target metalloproteins that Zn-metallothionein is capable of donating zinc to apo-zinc-metalloproteins such as zinc transcription factors and metalloenzymes (Kang, 2006). Zn-metallothionein has also been shown to be capable of rescuing cadmium-bound proteins through abstraction of cadmium and donation of zinc (Roesijadi et al. 1998; Ejnik et al. 1999). Recent work has extended observations of direct interactions between Zn-metallothionein and apo-zinc-metalloenzymes to living cells (Maret and Krezel, 2007).

METALLOTHIONEINS AND TOXICOLOGY

Overexpression of metallothionein and the consequent binding of inducing metals are believed to be the basis for increased resistance to metals such as cadmium. This overexpression is usually due to induction by metal exposure. Gene amplification (Beach and Palmiter, 1981; Koropatnick, 1988) and gene duplication (Otto et al. 1986) have been shown to increase the potential for induction above that for cells or organisms that have not undergone such

* **Promoter elements** are regulatory sequences upstream of the coding region of a gene.

phenomena. Transgenic mice* that overexpress metallothionein also exhibit greater resistance to cadmium (Liu et al. 1995).

Loss-of-function and gain-of-function experiments represent the most convincing examples for a protective function of metallothionein against metal toxicity. Deletion of the metallothionein gene in yeast and mice, for example, results in loss of protection against cadmium, copper, mercury, or zinc toxicity (Hamer et al. 1985; Michalska and Choo, 1993; Masters et al. 1994a; Satoh et al. 1997). In oyster blood cells, disruption of metallothionein expression with antisense oligonucleotides results in susceptibility to cadmium toxicity at concentrations not otherwise toxic (Butler and Roesijadi, 2001). The loss of copper resistance in metallothionein-disabled yeast is restored by insertion of a human metallothionein gene (Thiele et al. 1986).

Several types of intracellular interactions can be envisioned for the role of metallothionein in protection against metal toxicity. These include binding and sequestering newly taken-up metal ions (Roesijadi and Klerks, 1989), metal abstraction from structures that have bound toxic metal ions (Huang and Adamson, 1993), and metal exchange with metal ions that have replaced zinc in zinc-binding sites (Roesijadi et al. 1998; Ejnik et al. 1999). In these interactions, the interception of newly taken-up metal ions by metallothionein forestalls toxic interactions, and molecules whose structure and function have been compromised by binding toxic metals are rescued by metallothionein. The latter recognizes an active role for metallothionein in repairing structures already affected by binding certain toxic metals. If metallothionein expression is not sufficient to keep up with exposure, the protection conferred by metallothionein can be compromised and result in toxicity.

That factors other than metals can induce metallothioneins introduces a broader toxicological context for the function of metallothionein. The presence of antioxidant response elements in the metallothionein gene and the ability of metallothionein to scavenge free radicals imply a function in protection against oxidative stress (Andrews, 2000). In addition, induction of metallothionein through pathways targeting glucocorticoid response elements in promoter sequences implicates metallothionein expression in hormonally mediated processes associated with general stress responses. The ability to respond to other stressors has implications for natural populations that are responding to complex and variable environmental conditions, because metals alone are not responsible for modulating metallothionein gene expression. Induction of metallothionein is not, by itself, diagnostic of metal exposure.

Gene expression profiling is currently being used to elucidate the functional relationship between metallothionein gene expression and expression of other genes in response to metal and other stressors. The ability to screen for expression of a multitude of genes simultaneously is greatly facilitating the analysis of complex interactions that involve metallothionein. Its applications in ecotoxicology will also be of value in understanding mechanisms of response in natural environments impacted by multiple chemical agents. Microarrays for analyzing gene expression in nonmammalian animal species used as biomedical models, e.g., the nematode, *Caenorhabditis elegans*; the zebrafish, *Danio rerio*; and the fruit fly, *Drosophila melanogaster*, are commercially available and applicable to related species of environmental interest. Moreover, full genome sequences are becoming increasingly available for a variety of species of environmental relevance. Coupled with global proteomic approaches that analyze proteins by mass spectrometry, it should be possible through bioinformatic approaches to elucidate the complex subcellular interactions underlying response to contaminant exposures.

* A **transgenic organism** is simply an organism modified by genetic engineering.

METALLOTHIONEINS AND ECOTOXICOLOGY

Sensitive and sophisticated techniques to measure metallothionein, its expression, or its metal-binding status are now routine. Analyses directed to different steps in the gene-expression pathway that leads to formation of metallated metallothionein can provide complementary information for evaluating response of metallothionein in natural environments:

1. Metallothionein mRNA as an indicator of metallothionein gene expression (estimated from northern blot, ribonuclease protection assay, quantitative RT-PCR).
2. Metallothionein as an indicator of mobilization of detoxification pathways (estimated from metal content, differential pulse polarography, radionuclide substitution assay, immunoassay, electrophoresis with fluorescent probes, and capillary electrophoresis).
3. Metals bound to metallothionein as an indicator of detoxified toxic metals (estimated from metal analysis of isolated metallothionein or metallothionein-containing intracellular fractions by atomic absorption spectrometry and inductively coupled plasma-mass spectrometry).

Moreover, global approaches associated with genomic or proteomic profiling, e.g., cDNA arrays, two-dimensional gel electrophoresis of proteins, and mass spectrometry of proteins are capable of detecting metallothionein or its transcript in a complex background of proteins or mRNA.

In general, exposure to certain metals—notably cadmium, zinc, or copper—results in elevated levels of metallothionein mRNA, metallothionein, and metallothionein-bound metals. This has been amply demonstrated in laboratory experiments and has been extended to natural populations inhabiting metal-contaminated environments. However, of these measures, only that for metallothionein-bound metals provides a direct reflection of the contribution that detoxification by metallothionein makes to metal bioaccumulation. Although metallothionein mRNA and metallothionein should, in principle, reflect changes in levels of inducing metals, actual measurements can be confounded or obscured by simultaneous response to other endogenous and exogenous factors that can modulate metallothionein gene expression. Factors such as season, temperature, reproductive condition, and sex can affect levels of either metallothionein or metallothionein mRNA (Hylland et al. 1998; Mouneyrac et al. 2001; Van Cleef-Toedt et al. 2001). The complexity of factors that can influence induction needs to be accounted for when interpreting the significance of changes in any of the measures associated with metallothionein induction or binding of metals (Amiard et al. 2006). Furthermore, elevations in metallothionein or the pool of metallothionein-bound metals do not by themselves signify induction, because increases in the amount of metallothionein and its bound metals can occur in response to metal exposure without induction of metallothionein. For example, cadmium can displace zinc from metallothionein or compete with zinc in binding to newly synthesized metallothionein, resulting in increased stability of metallothionein, reduced turnover of both metallothionein and cadmium, and accumulation of metallothionein and its bound metals without benefit of induction. This appears to have been the case in a population of cadmium-exposed oysters in which metallothionein-bound cadmium levels were elevated as a result of environmental exposure while metallothionein mRNA levels were not (Roesijadi, 1999). From the perspective of developing specific biomarkers based on responses associated with metallothionein; however, a focus on response variables that exhibit the least ambiguous relationship to metal exposure and bioaccumulation would be the most desirable.

The existence of metal-specific isoforms of metallothionein that respond preferentially to specific toxic metals would be of value in hazard assessment. To date, a copper-specific metallothionein in yeast (Fürst et al. 1988) and a cadmium-specific form in an earthworm (Sturzenbaum et al. 2004) exhibit such properties. Metallothionein isoforms with similar promise have recently been reported in oysters (Tanguy and Moraga, 2001; Jenny et al. 2004) and a species of mussel (Geret and Cosson, 2002; Banni et al. 2007). These could have utility in identifying specific metals that are eliciting the biological response.

Our current understanding of metallothionein in the diverse species inhabiting natural environments does not allow universal generalizations regarding its utility as a biomarker applicable to all species despite about three decades of research on the topic. In some species, other pathways and biological compartments are quantitatively more important for sequestering metals, and changes in the response of metallothionein are difficult to detect. It is clear, though, that natural populations in metal-enriched environments have responded in many instances through increased mobilization of metallothionein; in some cases, such populations have adapted and evolved resistance to metal toxicity (Klerks and Levinton, 1989; Laurent et al. 1991). It had been speculated for some time that such events can have ecological implications apart from changes in fitness of adapted populations, since detoxification would enable organisms to accumulate higher concentrations of metals and facilitate trophic transfer. Recent studies demonstrate that (1) the cell fraction containing metallothionein in prey is the most significant contributor to the transfer of cadmium to predators (Wallace and Lopez, 1997; Seebaugh and Wallace, 2004) and (2) this cadmium transfer can result in toxic effects in the predator (Wallace et al. 1998, 2000). Thus, it appears that the binding of toxic metals to metallothionein has a direct relationship with transfer of metals and their toxic effects up the food chain. Although the evolution of metal resistance can provide advantage to the resistant organisms by enhancing their viability in metal contaminated environments, broader ecological costs may be associated with such adaptation.

The increasing application of genomic and proteomic profiling in ecotoxicological studies is enabling the analysis of complex interactive responses that occur as a result of contaminant exposure. Whereas earlier approaches have required *a priori* assumptions regarding what response or suite of responses would be appropriate to analyze, global expression analysis of transcripts or proteins has the potential of providing information on identifying patterns or networks of responses, which can be confirmed with more targeted experimentation. It is becoming increasingly apparent (Snell et al. 2003; Monsinjon and Knigge, 2007) that analysis of a single response as a sole biomarker is unlikely to provide the information necessary for informative risk assessment and decision making due to the complexity of responses. This is true even for a single contaminant, let alone the complexity associated with multicontaminant exposures in many natural environments. Expression profiling in both fish (Krasnov et al. 2005; Williams et al. 2006; Auslander et al. 2008) and invertebrates (Custodia et al. 2001; Dondero et al. 2006) has shown changes in metallothionein expression in response to both single metals such as cadmium or copper, and mixtures that include organic compounds. Although still at a descriptive phase, coupled with targeted analysis of specific responses, the use of these global approaches adds a powerful tool to elucidating environmentally relevant response at the cellular level.

SUMMARY

The progress made in understanding the biochemistry, molecular biology, physiology, and toxicology of metallothioneins has greatly increased our understanding of its role in toxicological functions. This progress has been instrumental in refining approaches to the study of metallothionein in natural populations. An understanding of metallothionein function in exposure to metals and other contaminants will improve our understanding of ecotoxicological relationships, as well as cellular and organismal function. One can expect increased application of genomic and proteomic profiling to contribute to a better appreciation of the role metallothionein plays in ecotoxicological processes.

SUGGESTED READINGS

Amiard, J. C., C. Amiard-Triquet, S. Barka, J. Pellerin, and P. S. Rainbow, Metallothioneins in aquatic invertebrates: Their role in metal detoxification and their use as biomarkers. *Aquat. Toxicol.*, 76, 160–202, 2006.

Andrews, G. K., Regulation of metallothionein gene expression by oxidative stress and metal ions. *Biochem. Pharmacol.*, 59, 95–104, 2000.

Haq, F., M. Mahoney, and J. Koropatnick, Signaling events for metallothionein induction. *Mutat. Res.*, 533, 211–226, 2003.

Klaassen, C. D., J. Liu, and S. Choudhuri, Metallothionein: An intracellular protein to protect against cadmium toxicity. *Annu. Rev. Pharmacol. Toxicol.*, 39, 267–294, 1999.

Laity, J. H. and G. K. Andrews, Understanding the mechanisms of zinc-sensing by metal-response element binding transcription factor-1 (MTF-1). *Arch. Biochem. Biophys.*, 463, 201–210, 2007.

Monsinjon, T. and T. Knigge, Proteomic applications in ecotoxicology. *Proteomics*, 7, 2997–3009, 2007.

Roesijadi, G., Metallothionein and its role in toxic metal regulation. *Comp. Biochem. Physiol.*, 113C, 117–123, 1996.

Snell, T. W., S. E. Brogdon, and M. B. Morgan, Gene expression profiling in ecotoxicology. *Ecotoxicology*, 12, 475–483, 2003.

Vergani, L., M. Grattarola, C. Borghi, F. Dondero, and A. Viarengo, Fish and molluscan metallothioneins: A structural and functional comparison. *FEBS J.*, 272, 6014–6023, 2005.

Wallace, W. G., T. M. H. Brouwer, M. Brouwer, and G. R. Lopez, Alterations in prey capture and induction of metallothioneins in grass shrimp fed cadmium-contaminated prey. *Environ. Toxicol. Chem.*, 19, 962–971, 2000.

6.4 STRESS PROTEINS

The **cellular stress response** is an "orchestrated induction of key proteins that form the basis for the cell's protein protection and recycling system" (Sanders and Dyer, 1994).* A cellular stress response can be elicited by heat, anoxia, some metals, some xenobiotics, ethanol, sodium arsenate, or UV radiation (Craig, 1985; Thomas, 1990; Di Giulio et al. 1995; Lewis et al. 1999; Geist et al. 2007). Stress-induced proteins were first studied in organisms that experienced abrupt changes in temperature (5–15° C). These **heat shock proteins (hsp)** were distinguished from one another according to their molecular weights. Respectively, there are 90, 70, 60, 16–24, and 7 kDa groupings designated as hsp90, hsp70, hsp60, low-molecular weight (LMW), and ubiquitin. These proteins are now known to be induced by other stressors and, consequently, have been renamed **stress proteins** to reflect this more general role. The terminology originally based on heat stress has been modified. Now the stress protein groupings are also called stress90, stress70, chaperon 60 (cpn60), LMW, and ubiquitin. The term **chaperon**, used collectively for stress90, stress70, and cpn60, reflects their role of associating with and directing the proper folding and coming together of proteins. They also protect proteins from denaturation and aggregation, and enhance refolding of damaged protein to a functional conformation. Their final role is in the transport of proteins to their intercellular location where they then are folded to a functional conformation (Craig, 1993).

Stress proteins are induced by physical and chemical agents that have significant **proteotoxicity** (toxicity due to protein damage) (Hightower, 1991). Enhanced production of stress90, stress70, cpn60, LMW, and ubiquitin is initiated by denatured protein to protect, repair, or vector for breakdown the proteins of the cell. Some stress proteins such as stress70 are always present (Welch, 1990), but others (LMW) appear only under stressful conditions (Di Giulio et al. 1995). Stress90 is present at high concentrations under unstressed conditions and is induced to even higher levels

* This is one of the most often discussed characteristics of the cellular stress response; however, Sanders (1990) also discussed as part of this response the induction of **glucose-regulated proteins** (grps) under low glucose or oxygen conditions. The grps are structurally similar to hsp, are present at basal levels in unstressed cells, and are induced in glucose- or oxygen-deficient cells exposed to toxicants, which modify calcium metabolism, e.g., lead (Sanders, 1990).

by stress. Stress70 is present at lower concentrations during normal conditions and is induced by stress. Cpn60, which is found in the mitochondria and facilitates protein movement and folding, is present at low levels under normal conditions and is inducible. Stress70 and cpn60 are good candidates as biomarkers because they are inducible by stress, are highly conserved proteins, and are not normally present at high levels as is stress90 (Sanders, 1990). The LMW proteins are more variable in structure and in inducibility among species: these qualities detract from the utility of LMW as biomarkers. Ubiquitin is induced by stress and its structure is very conservative evolutionarily, making it another potential biomarker (Sanders, 1990).

Stress proteins recognize and bind to exposed regions of denatured proteins that are rich in hydrophobic peptides (Agard, 1993). With the aid of stress70 and stress90, denatured or aggregated proteins are unfolded and refolded properly to restore their functioning. Proteins damaged beyond repair are bound by ubiquitin, which then helps move them to lysosomes for breakdown (Di Giulio et al. 1995).

Increased concentrations of the various stress proteins are used as biomarkers of general cellular stress response. For example, Sanders and Martin (1993) correlated general contamination with cpn60 and stress70 levels in archived tissues of marine species. For plants (broad bean, *Vicia faba*), stress70 was found to increase with chronic lead exposure (Wang et al. 2008). Because different stressors induce the various stress proteins to different degrees, Sanders and Dyer (1994) suggested that the patterns of stress protein induction could be used to suggest the particular toxicant inducing the response. Patterns from field samples can be compared to those obtained with single candidate toxicants in the laboratory. They referred to this approach as **stress protein fingerprinting**. Because the stress proteins have evolved in a very conservative manner and the cellular stress response is so universal, Sanders and Dyer (1994) suggested that this approach has more universal application than many other biomarker systems.

6.5 OXIDATIVE STRESS AND ANTIOXIDANT RESPONSE

Molecular oxygen is both benign and malign. On one hand it provides enormous advantages and on the other it imposes a universal toxicity. This toxicity is largely due to the intermediates of oxygen reduction, i.e. O_2^-, H_2O_2, and $OH^\bullet$, and any organism that avails itself of the benefits of oxygen does so at the cost of maintaining an elaborate system of defenses against these intermediates.

Fridovich (1983)

Combining molecular oxygen reduction with energy generation during aerobic metabolism (upper panel, Figure 6.3) creates the potential for **oxidative stress**, i.e., damage to biomolecules from free oxyradicals. Oxyradical-generating compounds such as hydrogen peroxide (H_2O_2) are also produced by aerobic metabolism and other processes, contributing to oxidative stress. Free radicals* such as the superoxide radical ($O_2^{\cdot-}$) and hydroxyl radical ($^\bullet OH$) damage proteins, lipids, DNA, and other biomolecules. Oxyradicals are produced during electron transport reactions in mitochondria and microsomes, photosynthetic electron transport, phagocytosis, and normal catalysis by prostaglandin synthase, guanyl cyclase, and glucose oxidase (Di Giulio et al. 1989). Complex regulatory processes such as those associated with the hypoxia-inducible factor 1 (HIF-1) are needed to finely regulate cellular oxygen (Semenza, 2007).†

* A **free radical** is a molecule having an unshared electron. The unshared electron is usually designated by a dot, $^\bullet$ Free radicals are very reactive. **Oxyradicals** are free radicals with an unshared electron of oxygen, e.g., $RO^\cdot$, where R is some oxygen-containing compound.

† Although not discussed here, HIF-1 regulation involves innumerable shifts such as those in glycolytic enzyme and glucose transporter expression.

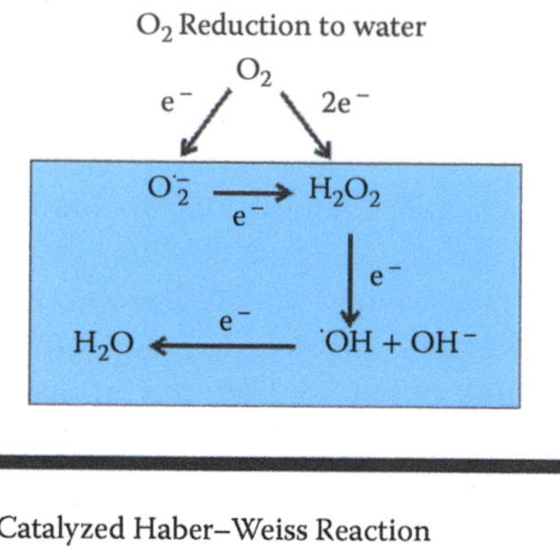

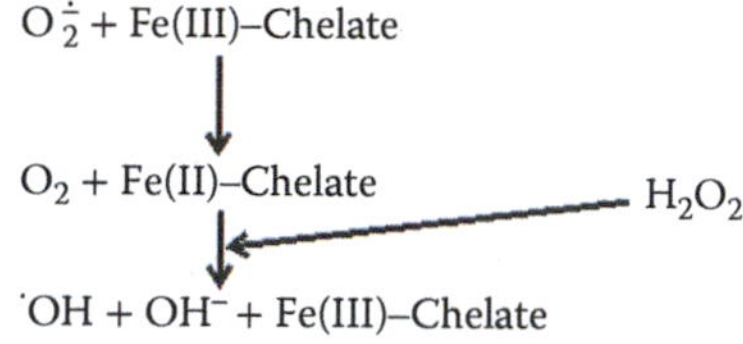

Figure 6.3 Reduction of O_2 to water during aerobic respiration ($O_2 + 4e^- \rightarrow H_2O$). Molecular oxygen can be converted to the superoxide radical ($O_2^{\dot{-}}$) by adding one electron (e^-) or to hydrogen peroxide (H_2O_2) by adding two electrons. The superoxide radical can be converted to hydrogen peroxide by the addition of another electron. Hydrogen peroxide can then be reduced to the hydroxyl radical ($^{\bullet}OH$), producing a hydroxyl anion (OH^-). Water is produced with the addition of a fourth electron. Two highly reactive, free radicals are generated along this reaction sequence. Hydrogen peroxide is also produced and, through the **Haber–Weiss reaction** ($O_2^{\dot{-}} + H_2O_2 \rightarrow {}^{\bullet}OH + OH^-$ as shown in the top panel in the box), generates oxyradicals. The **catalyzed Haber–Weiss reaction** is a greatly accelerated Haber–Weiss reaction catalyzed by metal chelates as shown in the bottom panel. Metals can also generate free radicals directly. As one example, chromium reacts with cysteine to produce thiol radicals (Leonard et al. 2004) which, in turn, can generate $O_2^{\dot{-}}$.

Organisms cope with oxidative stress in two fundamental ways. They produce antioxidants that react with oxyradicals. These antioxidants include Vitamin E, Vitamin C, β-carotene, catecholamines, glutathione, and uric acid (Winston and Di Giulio, 1991). For example, Pawlik-Skowrońska et al. (2007) detected lowered glutathione concentrations in seaweeds sampled from sites with elevated metal levels.* Recently, it became clear that metallothionein can also act as an oxyradical scavenger because of its high number of sulfhydryl groups (Monserrat et al. 2007) (Vignette 6.2). In addition to antioxidants, enzymes that reduce the amount of oxyradicals present at any instant are involved in lowering oxidative damage. **Superoxide dismutase (SOD)** decreases the amount of the superoxide radical in the cell by catalyzing reaction 6.2:

$$2\,O_2^{\bullet -} + 2\,H^+ \rightarrow H_2O_2 + O_2 \tag{6.2}$$

Catalase (CAT) and **glutathione peroxidase** reduce levels of hydrogen peroxide via reactions 6.3 and 6.4, respectively.

$$2\,H_2O_2 \rightarrow H_2O + O_2 \tag{6.3}$$

* Notice here that glutathione has roles as a substrate for Phase II conjugation and as an antioxidant. It is also a precursor for phytochelatin synthesis (Pawlik-Skowrońska et al. 2007).

$$2\ \text{Reduced Glutathione} + H_2O_2 \rightarrow \text{Oxidized Glutathione} + H_2O \quad (6.4)$$

Xenobiotics can cause oxidative damage indirectly by interfering with these mechanisms for coping with oxidative stress. As an example, high lead concentrations disrupt metabolic pathways, resulting in elevated reactive oxygen species concentrations in plants (Wang et al. 2008). Adverse changes to peroxisomes due to contaminants can contribute to oxidative stress. As briefly mentioned already (footnote in Vignette 2.3), **peroxisomes** are single membrane bound organelles generated from the ER that contain enzymes associated with oxidative reactions. They are also involved in hydrogen peroxide production and decomposition. They function in respiration dependent on hydrogen peroxide and oxidation of fatty acids (Lazarow and Fujiki, 1985). Their formation can be induced by contaminants (Lazarow and Fujiki, 1985). Bhattacharya and Bhattacharya (2007) found that some oxidative stress associated with catfish (*Clarias batrachus*) exposed to arsenic was partially due to arsenic modification of hydrogen peroxide metabolizing enzymes in peroxisomes.

Xenobiotics may also participate directly in reactions leading to oxidative stress. Some xenobiotics form oxyradicals such as **alkoxyradicals** ($RO^{\bullet}$) and **peroxyradicals** ($ROO^{\bullet}$) and, in doing so, cause damage (Di Giulio et al. 1995). As an example of free radical formation from a xenobiotic, carbon tetrachloride can undergo the following reaction to generate the trichloromethyl radical, $CCL_4 + e^- \rightarrow CCl_3^{\bullet} + Cl^-$. This reaction, involving the NADPH-cytochrome P-450 system, contributes to carbon tetrachloride damage (necrosis, cancer) to the human liver where the free radical reacts with lipids, proteins, and DNA (Slater, 1984). Xenobiotics can also form free radicals during interactions with oxyradicals. For example, promethazine reacts with the hydroxyl radical to produce OH^-, becoming a free radical itself in the process. PAHs are activated by biotransformation to their free radical forms (Slater, 1984). Still other contaminants (quinones, aromatic nitro compounds, aromatic hydroxylamines, bipyridyls such as paraquat and diquat, and some chelated metals [Di Giulio et al. 1989]) are reduced to radicals and then enter into **redox cycling** to produce superoxide radicals from molecular oxygen. The initial reduction of the xenobiotic to a free radical may be facilitated by one of several reductases (Di Giulio et al. 1989). Because the xenobiotic enters the redox cycle as a free radical but exits in its original form at the end of a cycle, it is available to recycle many times and produce considerable amounts of oxyradicals.

Because contaminants can interfere with the normal mechanisms of coping with oxidative stress and can contribute to free radical concentrations themselves, considerable effort has been spent in studying contaminant-related changes in oxidative stress. Elevated levels of superoxide dismutase, catalase, or glutathione peroxidase in exposed individuals can suggest elevated oxidative stress. Changes in these enzymes might be observed for individuals in contaminated sites (Regoli and Principato, 1995) or those exposed to contaminants in the laboratory (Víg and Nemcsók, 1989; Liu et al. 2007; Wang et al. 2008). As an example, Prevodnik et al. (2007) assessed xenobiotics and catalase activity in mussels (*Mytilus edulis*) from the Baltic Sea. As a plant example, guaiacol and ascorbate peroxidases are upregulated during lead exposure of broad bean (*Vicia faba*) seedlings (Wang et al. 2008). Changes in antioxidant pools may also be examined. As additional examples involving mussels, *M. galloprovincialis* exposed to metals (Regoli and Principato, 1995) and *Geukensia demissa* exposed to paraquat (Wenning et al. 1988) showed significant shifts in glutathione concentrations.

Increased concentrations of free radicals can cause membrane dysfunction due to **lipid peroxidation** (oxidation of polyunsaturated lipids). **Malondialdehyde**, a breakdown product from lipid peroxidation, is indicative of oxidative damage of lipids from a variety of toxicants (Thomas, 1990). For example, Atlantic croaker (*Micropogonias undulatus*) exposed to PCBs (Aroclor 1254) or cadmium had elevated concentrations of microsomal malondialdehyde (Wofford and Thomas, 1988). Also, minnow (*Poeciliopsis lucida*) cells had elevated malondialdehyde in their microsomes after anthracene exposure (Choi and Oris, 2000).

Free radicals can form covalent bonds with a variety of biomolecules. Free radicals covalently bond to membrane components such as enzymes and receptors, changing their structures and functions (Slater, 1984). Free radicals cause damage by reacting with sulfhydryl groups of proteins and other biomolecules and, in doing so, influence their function (Slater, 1984). A final, important reason exists for the interest paid to contaminant-influenced oxidative stress. The formation of free radicals near DNA can lead to mutations and, as a consequence, increased risk of cancer or other genotoxic consequences.* Malins (1993) found DNA (guanine and adenine) lesions induced by the hydroxyl radical in fish exposed to carcinogens. He suggested that these types of lesions in fish and humans lead to increased misreading of the DNA template and increased cancer risk.

6.6 DNA MODIFICATION

As suggested in the last paragraph, toxicants and their products can be discussed relative to **genotoxicity**, i.e., the damage by a physical or chemical agent to genetic materials such as chromosomes or DNA. At the molecular level, agents such as free radicals can produce breaks in one or both strands of the DNA molecule (Figure 6.4). Oxyradicals also can oxidize bases. Xenobiotics and their metabolites may bond covalently to a base or, less frequently, to another portion of the DNA molecule to form an **adduct**.

Metals can bind to phosphate groups and heterocyclic bases of DNA and, in doing so, change the stability and normal functioning of the DNA (Eichhorn et al. 1970). Magnesium bonds to phosphate groups in the DNA backbone to stabilize the DNA structure, but copper can bond between bases,

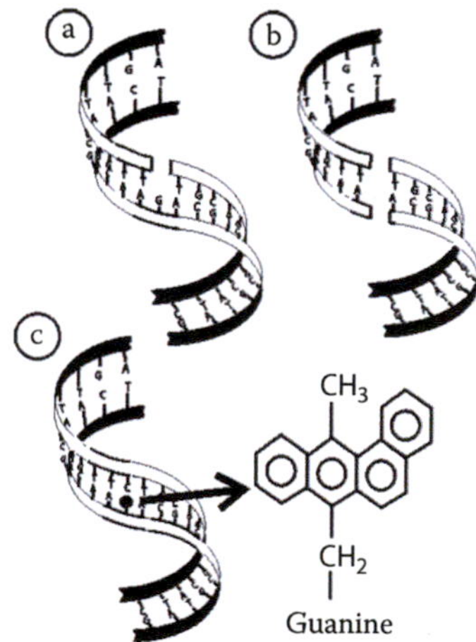

Figure 6.4 Various types of damage can occur to DNA due to toxicants. Toxicants or free radicals can produce (a) single- or (b) double-strand breaks. (c) Xenobiotics or their metabolites may react with bases to form adducts. Here, metabolism of 7,12-dimethyl-benz[a]anthracene leads to covalent bonding of its metabolite to guanine to form a DNA adduct. Interactions such as those with free radicals can also modify bases, i.e., oxidize bases such as thymine and guanine to thymine glycol and 8-hydroxyguanine, respectively. (From Di Giulio, R.T., et al. in G.M. Rand, [Ed.], *Fundamentals of Aquatic Toxicology: Effects, Environmental Fate, and Risk Assessment*, 2nd Edition, Washington, DC, Taylor & Francis, 1995.)

* The formation of free radicals is a key to radiation effects too. Free radicals formed from water and molecular oxygen during irradiation cause damage to cells. However, free radicals can also be used to our advantage in radiation treatments of cancers where our intent is to kill cancer cells. Damage to cancer cells can be enhanced during radiation treatment by administering a radiosensitizer such as derivatives of nitro imidazoles (Slater, 1984). **Radiosensitizers** enhance the production of free radicals in the area receiving radiation that leads to more effective destruction of cancer cells.

compete with the normal hydrogen binding, and destabilize the DNA structure (Eichhorn, 1975). Because the matching of specific base pairs is a matter of degree of attraction for complementary versus noncomplementary pairs, the modification of hydrogen bonding by metals can contribute to base mispairing. Also, mercury can form strong cross-links between the strands of the DNA molecule. Normal DNA repair mechanisms are overwhelmed if alterations occur too frequently, resulting in high mutation rates.

The expected diploid chromosome number (2*N*) can be disrupted due to chromosome breakage and result in a deviation from the usual number of chromosomes (**aneuploidy**) or structural aberrations in chromosomes. Agents that cause chromosome damage in living cells are classified as **clastogenic**. All of these genotoxic effects can have **mutagenic** (causing mutations), **carcinogenic** (causing cancers), and **teratogenic** (causing developmental malformations) consequences (Jones and Parry, 1992).

Various methods are used to assess genotoxicity. Chromosome damage may be determined by microscopic examination after appropriate staining. Aneuploidy and other clastogenic effects may be quantified using flow cytometry after staining the DNA in cells with the appropriate fluorescent dye. With flow cytometry, fluorescence is used to measure the amount of DNA in individual cells of a sample as each cell flows through an excitation light beam. The distribution of DNA concentrations in the population of cells is examined for notable numbers of cells with atypical amounts of DNA (Shugart, 1995). Clastogenic activity may also be reflected in the number of **micronuclei** (membrane bound masses of chromatin) in cells. The presence of many micronuclei suggests impairment of the cell's ability to divide properly (Jones and Parry, 1992).

DNA adducts may be assayed using a ^{32}P-labeling method (Jones and Parry, 1992). Deoxyribonucleoside 3'-monophosphates are produced by hydrolysis of the DNA molecule. Then, these deoxyribonucleoside 3'-monophosphates are labeled with ^{32}P. Most of the deoxyribonucleoside 3'-monophosphates contain one of the four normal bases in DNA but some will have bases covalently bound to an adduct. The labeled deoxyribonucleoside 3'-monophosphates are separated chromatographically to the four normal bisphosphates of adenosine, cytidine, guanosine, and thymidine, plus bases with adducts. The amount of radioactivity associated with the base-adduct fraction reflects the number of DNA adducts and the potential for genotoxic effects. The occurrence of adducts has been correlated with cancer risk (e.g., Gaylor et al. 1992) or used to simply imply general genotoxicity (e.g., Ericson and Larsson, 2000; Jönsson et al. 2004; Østby et al. 2005; Gagné et al. 2006).

The amount of DNA breakage can also be estimated and correlated with exposure to genotoxic contaminants. Shugart (1988,) described an alkaline unwinding assay to estimate the degree of single-strand breakage. After a specified time of exposure to alkaline conditions, the DNA strands from samples unwind to different degrees depending on the relative amounts of double- and single-stranded DNA in the sample. Differences in fluorescence intensity of single- and double-stranded DNA are used to measure the relative amounts of each in the samples. Identical amounts of isolated DNA from control and exposed individuals are placed under alkaline conditions, allowed to unwind for a set time, and then the fraction of DNA that is double stranded in each sample estimated via fluorescence. This fraction is used to imply the relative amount of single-strand breaks in the various samples. For example, this fraction dropped within 10 days of bluegill sunfish (*Lepomis macrochirus*) exposure to BaP (Shugart, 1988), suggesting an increase in single-strand breaks with exposure. Meyers-Schöne et al. (1993) correlated environmental contamination with single-strand breaks in DNA of freshwater turtles using this alkaline unwinding assay.

A technique related to the alkaline unwinding assay, **single-cell gel (or comet) electrophoresis**, also is used to correlate genotoxicant damage to DNA strand breakage (Figure 6.5). Cells from exposed individuals are embedded in a gel under alkaline conditions so that unwinding can occur. How much unwinding will depend on the number and kinds of breaks in the double-stranded DNA molecules (Fairbairn et al. 1995; Rojas et al. 1999; Brendler-Schwaab et al. 2005). The DNA moves

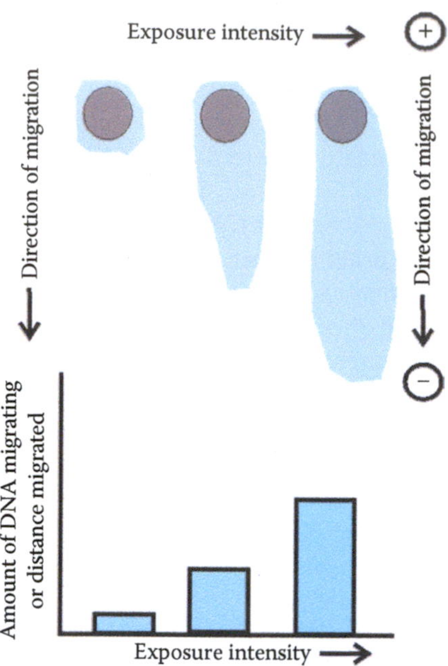

Figure 6.5 Comet electrophoresis for the detection of DNA strand breakage in individual cells. Cells are embedded in a gel, subjected to alkaline conditions, subjected to electrophoresis under alkaline conditions, stained to visualize DNA, and the amount of DNA in the resulting "comet" tail (blue) or the distance migrated by the DNA making up the tail is used as a measure of the amount of DNA breaks present. "Comets form as broken ends of the negatively charged DNA molecule become free to migrate in the electric field toward the anode. The ability of DNA to migrate is a function of both the size of the DNA and the number of broken 'ends' which may be attached to larger pieces of DNA but which can still migrate a short distance from the comet head" (Fairbairn et al. 1995). DNA patterns are depicted here for three cells that had different intensities of exposure and consequent differences in DNA breakage.

from the position of the embedded cell at amounts and rates dependent on the extent of breakage. Simple measures of damage might include the proportion of cells with comet tails or the ratio of tail width to length (Rojas et al. 1999). Other common metrics are the proportion of the total amount of DNA in the tail or the tail moment. **Tail moment** is the tail length times the fraction of the DNA that is in the tail as opposed to the comet head. Applications of the comet assay range widely from earthworms (*Eisenia fetida*) exposed to metals (Reinecke and Reinecke, 2004), to white storks (*Ciconia ciconia*) living near metal-contaminated habitat (Pastor et al. 2001), to humans exposed to uranium compounds (Prabhavathi et al. 2003).

6.7 ENZYME DYSFUNCTION AND SUBSTRATE POOL SHIFTS

Contaminants can have significant effects on enzymes and these effects are used routinely as biomarkers. Metals can influence protein-mediated catalysis, transport, and gas exchange by modifying protein structure (secondary, tertiary, or quaternary) and stability (Ulmer, 1970). Metals bind to a wide range of electron donor groups of proteins such as imidazole, sulfhydryl, hydroxyl, carboxyl, amino, quanidinium, and peptide groups (Eichhorn, 1975). Consequent changes in secondary and tertiary structure can lead to lowered or elevated enzyme activities. Normally, metals stabilize quaternary structure of many proteins: substitution of another metal for the usual stabilizing metal can interfere with the coming together of peptide chains to form stable and functional oligomers such as those in many multimeric enzymes. Some enzymes are stabilized by metals (e.g., lysozyme by Mg^{2+}) and substitution of other metals for these stabilizing metals can enhance denaturation (Ulmer, 1970). Metals are also present at active sites of biomolecules such as carboxypeptidase, alkaline phosphatase, carbonic anhydrase, cytochrome c, and hemoglobin (Eichhorn, 1975). Displacement of the appropriate metal by another can change the functioning of these proteins.

Several steps in the synthesis of heme (Figure 6.6) are modified by toxicants, and associated enzyme activities and substrate pools are used as biomarkers to reflect sublethal poisoning. **Porphyrins** are produced as intermediates in heme synthesis. Porphyrins with four to eight carboxyl groups tend to be produced in excess and are excreted in urine. The relative amounts of excreted porphyrins tend to be consistent among individuals. However, mercury poisoning interferes with normal heme synthesis, promotes oxidation of reduced porphyrins, and shifts porphyrin pools in the urine as a consequence. For example, rats exposed to mercury have increased levels of porphyrins with four and five carboxyl groups (Woods et al. 1993). Woods et al. (1993) used such shifts to indicate sublethal exposure to mercury in dentists who use mercury in silver–mercury amalgam fillings. Male dentists categorized as having either no detected mercury, or greater than 20 $\mu g \cdot L^{-1}$ of mercury in their urine had pentacarboxylporphyrin concentrations of 0.76 and 3.07 $\mu g \cdot L^{-1}$, respectively.

Polyhalogenated aromatic hydrocarbons such as PCBs may also interfere with heme synthesis, possibly by inhibiting uroporphyrinogen decarboxylase or by cytochrome P-450 monooxygenase generation of oxyradicals that oxidize porphyrinogens (Peakall, 1992). Hepatic porphyrins have been used as biomarkers of polyhalogenated aromatic hydrocarbon exposure to species such as pike (*Esox lucius*) (Koss et al. 1986) and herring gulls (*Larus argentatus*) (Fox et al. 1988).

Other biomarkers can indicate effect to the respiratory pigments of animals. Cadmium exposure of flounder (*Pleuronectes flesus*) depressed blood hematocrit, hemoglobin titers, and red blood cell counts (Johansson-Sjöbeck and Larsson, 1978). Lead depresses the activity of **δ-aminolevulinic acid dehydratase** (δ-ALAD or ALAD), an enzyme catalyzing the conversion of δ-aminolevulinic

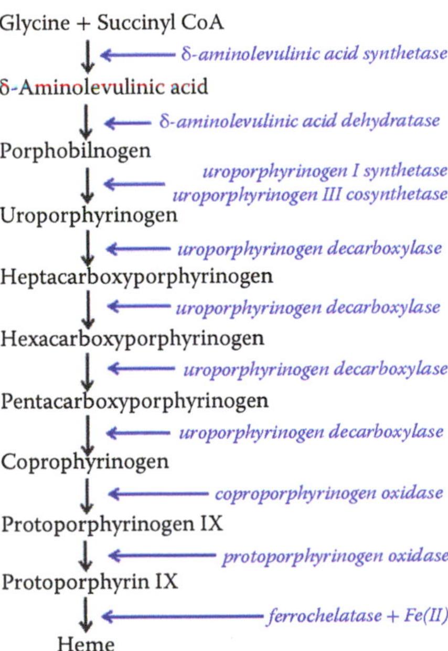

Figure 6.6 Steps in the synthesis of heme. Enzymes catalyzing each conversion are italicized and placed to the right of the reaction sequence. (Modified from Figure 1 of Woods, J.S., et al. *J. Toxicol. Environ. Health*, 40, 235–246, 1993.)

acid to porphobilinogen during heme synthesis (see Figure 6.6). Exposure of rainbow trout (*Salmo gairdneri*) to lead lowered ALAD activities and caused anemia (Johansson-Sjöbeck and Larsson, 1979). Fish exposed to lead-containing mine drainage also had low blood ALAD activities (Dwyer et al. 1988; Schmitt et al. 1993). This effect of lead can be modified by the presence of metals such as cadmium and zinc (Berglind, 1986; Schmitt et al. 1993).

Because of their importance in animal osmoregulation and cell water regulation, ATPases (adenosine triphosphatases) have been the subject of considerable study (Thomas, 1990). Their presence on gill surfaces makes them particularly vulnerable to contact with toxicants. Yap et al. (1971) exposed bluegill sunfish (*Lepomis machrochirus*) to PCBs and noted depressed Mg^{2+}-, Na^+-, and K^+-ATPase activities. Rock crab (*Cancer irroratus*) gills exposed to cadmium or lead also showed depressed Na^+- and K^+-ATPase activity.

6.8 SUMMARY

Understanding molecular effects enhances our ability to assign causal linkage to effects at higher levels of organization and to predict effects of untested chemicals based on similar molecular interactions with biomolecules. It also provides biomarkers as tools to proactively measure effects in field situations. Some biomarkers are specific (lead's effect on ALAD activity), specific to a class of toxicants (metallothioneins), or general (stress proteins). All have value if the temptation to attach too much ecological relevance to a biochemical response is avoided.

SUGGESTED READINGS

Dean, M. and T. Annilo, Evolution of the ATP-binding cassette (ABC) transported superfamily in vertebrates. *Annu. Rev. Genomics Hum. Genet.,* 6, 123–142, 2005.

Lewis, S., R. H. Handy, B. Cordi, Z. Billinghurst, and M. H. DePledge, Stress proteins (HSP's): Methods of detection and their use as an environmental biomarker. *Ecotoxicology,* 8, 351–368, 1999.

Newman, M. C. and W. H. Clements, *Ecotoxicology: A Comprehensive Treatment*, Taylor & Francis/CRC Press, Boca Raton, FL, 2008.

Sarkar, A., D. Ray, A. N. Shrivastava, and S. Sarker, Molecular biomarkers: Their significance and application in marine pollution monitoring. *Ecotoxicology,* 15, 333–340, 2006.

Timbrell, J., *Principles of Biochemical Toxicology*, 3rd Edition, Taylor & Francis, Philadelphia, PA, 2000.

van der Oost, R., J. Beyer, and N. P. E. Vermeulen, Fish bioaccumulation and biomarkers in environmental risk assessment: A review. *Environ. Toxicol. Pharmacol.,* 13, 57–149, 2003.

CHAPTER 7

Cells, Tissues, and Organs

... the problem took the form of habitat pollution → DDE accumulation in prey species → DDE in predators → decline in brood size → potential extermination. The same phenomenon can [be written as] lipid soluble toxicant → bioaccumulation in organisms with poor detoxification systems... → vulnerable target organ [shell gland] → inhibition of membrane-bound ATPases... → potential extermination. Ecologists would claim a decline in population recruitment, biochemists an inhibition of membrane enzymes.

Simkiss (1996)

7.1 INTRODUCTION

Two superficially opposing views of how to deal with hierarchical subjects exist.* At one end of the spectrum of approaches is **reductionism (microexplanation),** which attempts to understand the behavior of the simplest units at the lowest levels of organization, and then uses this understanding to explain phenomena at all higher levels. By analogy, a clock is understood completely if one understands the workings of all of its parts, or the fate of a species is predictable if one knows the impact of the pollutant of concern on its survival, reproduction, growth, and development. In seeming contrast, **holism** holds that, because unique properties emerge at higher levels of organization and complex interactions among parts beget complex dynamics, a descriptive understanding of higher order processes is much more useful for prediction than building a mechanistic structure from the lowest or most fundamental level to the highest. Continuing with the exposed species' fate example, one must conduct field studies to gain knowledge about the species in its ecosystem, including how the pollutant influences its population dynamics, community interactions, and the pollutant's impacts on the biogeochemistry of the species' habitat. In reality, both approaches are successful only when used in a mutually supportive fashion (see Newman and Clements, 2008). Alone, neither works beyond a limited scope. Prediction from a holistic study is limited to the immediate object of study or very similar ones if a researcher does not understand the components of the system but, instead, only describes phenomena at the highest level of organization. Limited prediction makes movement toward the scientific goal of ecotoxicology extremely ineffective. The

* **Macroexplanation** is a third, less obvious but important, vantage (Newman and Clements, 2008). With macroexplanation, the properties of the whole are used to explain the nature or behavior of the parts. For example, the observation of a population shift to fewer older individuals might be explained by stating that the young individuals were more sensitive to a toxicant. Or allele frequencies measured in a population might be used to imply the lower fitness of one genotype versus another in a contaminated environment. As discussed later, the information aggregated in the population qualities that is required to make such inferences is unavailable, making explanation prone to error. The name for the uncertainty inherent in macroexplanation from aggregated information is the **problem of ecological inference**.

reductionist approach alone is also ineffective because our limited knowledge makes prediction of some emergent properties impossible based solely on mechanisms at lower levels. Also causal structure does not always proceed up the artificial, hierarchical structure of biological organization. Effects at the tissue level can be influenced by physiological (one level up) in addition to biochemical (one level down) mechanisms. Arguably, the physiological causes could be traced back to their biochemical mechanisms. However, our incomplete understanding of cascading events, the network of intermeshed cause–effect phenomena, and the uncertainty associated with the magnitude of each phenomenon in the causal web would impart to such an exercise a high likelihood of failure beyond a limited number of links.

In some cases, causal structure can be defined clearly for several levels of organization. One such causal sequence from the biochemical → cell → individual that can be quickly evoked is activation of a xenobiotic by monooxygenases → production of oxyradicals and consequent DNA damage → increased risk of liver cancer → increased chance of death to an individual. However, extrapolation of this sequence to the population level would be difficult, and to the level of an ecological community level highly speculative. Causal linkage fails almost immediately for other cases such as that between δ-ALAD activity in red blood cells and population viability. Future work will eventually allow clearer definition for some of these linkages, but it is unlikely that clear linkage will ever be made in all instances.

Success in ecotoxicology is enhanced by keeping holistic goals and limitations in mind while applying reductionistic methods to questions as much as reasonable. Regardless of the level being examined, one should apply a creed of studying the "simplest system you think has the properties you are interested in" (Levinthal quoted in Platt, 1964). This should be done in full anticipation that important properties might emerge at higher levels that will require accommodation in predictions at higher levels.

As discussed in Chapter 1 and reflected again in the quote earlier, conceptual coherency is maintained by understanding that "… processes at one level take their mechanisms from the level below and find their consequences at the level above … Recognizing this principle makes it clear that there are no truly 'fundamental' explanations,* and makes it possible to move smoothly up and down the levels of the hierarchical system without falling into the traps of naive reductionism or pseudo-scientific holism" (Caswell, 1996; Bartholomew, 1964). Thus, coherency is maintained in a conceptual "relay race" in which all legs (levels of biological organization) are equally crucial to achieving the goal. Attempts to fulfill the goal of ecotoxicology using either the reductionistic or holistic approaches alone are equally absurd attempts to "swallow the ocean". Whether attempted in one impossibly large gulp or in an impossibly large number of small gulps, the ocean will not be swallowed.

7.2 GENERAL CYTOTOXICITY AND HISTOPATHOLOGY

The study of effects at the cell, tissue, and organ level of organization is invaluable for several reasons. In the context of the hierarchical relay race just described, these effects are consequences of biochemical mechanisms that provide interpretative insight about potential harm to individuals. They integrate damage done at the molecular level (Hinton and Laurén, 1990a), perhaps modulated by higher level processes such as those of the endocrine or immune systems. These biomarkers can

* A related incongruity arises as a consequence of not fully appreciating this context for ecotoxicological study. Often, statements are made that a particular level of organization is the most important to understanding ecotoxicological phenomena. Such statements are reminiscent of the Ptolemaic theory (the Earth is the center of the universe) and reflect a false paradigm which can be called the **Ptolemaic incongruity** here. This incongruity that any particular level of biological organization holds the central role in the science of ecotoxicology leads to confusion, narrowness of vision, and wasteful intransigence.

be used as an early warning system for potential effects at the level of individual and, sometimes, population. **Histopathology**, the study of change in cells and tissues associated with communicable or noncommunicable disease, provides a cost–effective way to verify toxicant effect as well as exposure (Hinton et al. 1987; Hinton, 1994; Handy et al. 2003).

Currently, there are two shortcomings of histopathological biomarkers to be overcome. First, the normal histology and variations in normal histology with season, diet, reproductive cycle, and other processes can be poorly understood for sentinel species (Hinton and Laurén, 1990a). Much more descriptive work—normal science—is required in this area to improve the application of histopathological biomarkers. Second, although methods exist to routinely quantify effects, most histopathological studies are qualitative in nature and needlessly neglect quantification (Jagoe, 1996). Both of these limitations are slowly being overcome by more normal science and a change in emphasis to quantitative methods.

7.2.1 Necrosis and Apoptosis

A wide range of **lesions** (pathological alterations of cells, tissues, or organs) can indicate exposure to toxicants and suggest mechanisms of action. Often, they are associated with a particular **target organ** as a result of preferential toxicant transport to, accumulation in, or activation within that organ. As an important example of xenobiotic activation within a specific organ, the high biotransformation activity and consequent activation of carbon tetrachloride in the liver is responsible for its hepatotoxicity. Also, cadmium damage to the kidney proximal tubules is linked to its transport to and association with metallothionein in the kidney during oral cadmium poisoning. Cadmium associated with metallothionein enters the kidney, is removed from circulation in the glomeruli, and is reabsorbed in the proximal tubules where lysosomal activity can release it (Goldstein and Schnellmann, 1996). Cadmium damages kidney cells once free of metallothionein.

Cytotoxicity (toxicity causing cell death*) may be detected in a tissue or target organ as localized or diffuse **necrosis,** cell death from disease or injury (Figure 7.1). Pyknosis, one of the most obvious indicators of necrosis, involves the cell nucleus. With **pyknosis** (**pycnosis**), the distribution of chromatin in the nucleus changes with this material condensing into a strongly staining, basophilic mass. The nucleus becomes irregular in shape (Sparks, 1972). Often pyknosis is followed by **karyolysis**, the disintegration of the nucleus. The cytoplasm of necrotic cells tends to become more acidophilic† (eosinophilic) than that of viable cells. Mitochondria swell and more granules can appear in the cytoplasm. Necrosis might also be indicated by the displacement or separation of the cell from its normal location in a tissue (Meyers and Hendricks, 1985), e.g., cell sloughing from the gill epithelium or arterial wall. Necrosis is often associated with inflammation, which will be described in Section 7.2.2.

Different types of necrosis manifest after various insults. **Coagulation necrosis** is characterized by extensive coagulation of cytoplasmic protein, making the cell opaque. The cell's outline and position within its tissue are retained for some time after cell death (Sparks, 1972; Hinton, 1994). As instances of its appearance upon poisoning, coagulation necrosis occurs in the mammalian alimentary tract after ingestion of phenol or after acute inorganic mercury exposure

* In everyday use, terms such as death lack the distinctions that are important in ecotoxicology. The distinction between cell death and somatic death is a good example. Cell death or necrosis occurs in living as well as dead individuals. Death of an individual is **somatic death**. As we have already seen, stress is another term requiring unusual attention in its use. In Chapter 6, we discussed cellular and oxidative stress. Later, we will discuss Selyean (individual) and ecosystem stresses. Each describes a different phenomenon.

† An **acidophilic component** of a cell (e.g., often much of the cytoplasm) is one that is readily stained by an acidic dye such as eosin and a **basophilic component** (e.g., the DNA-rich nucleus or cytoplasmic RNA) is one that is readily stained by a basic dye such as hematoxylin (hematoxylin). In general, preparations such as most of those shown in this chapter, hematoxylin and eosin are used to stain the nuclear and cytoplasmic components.

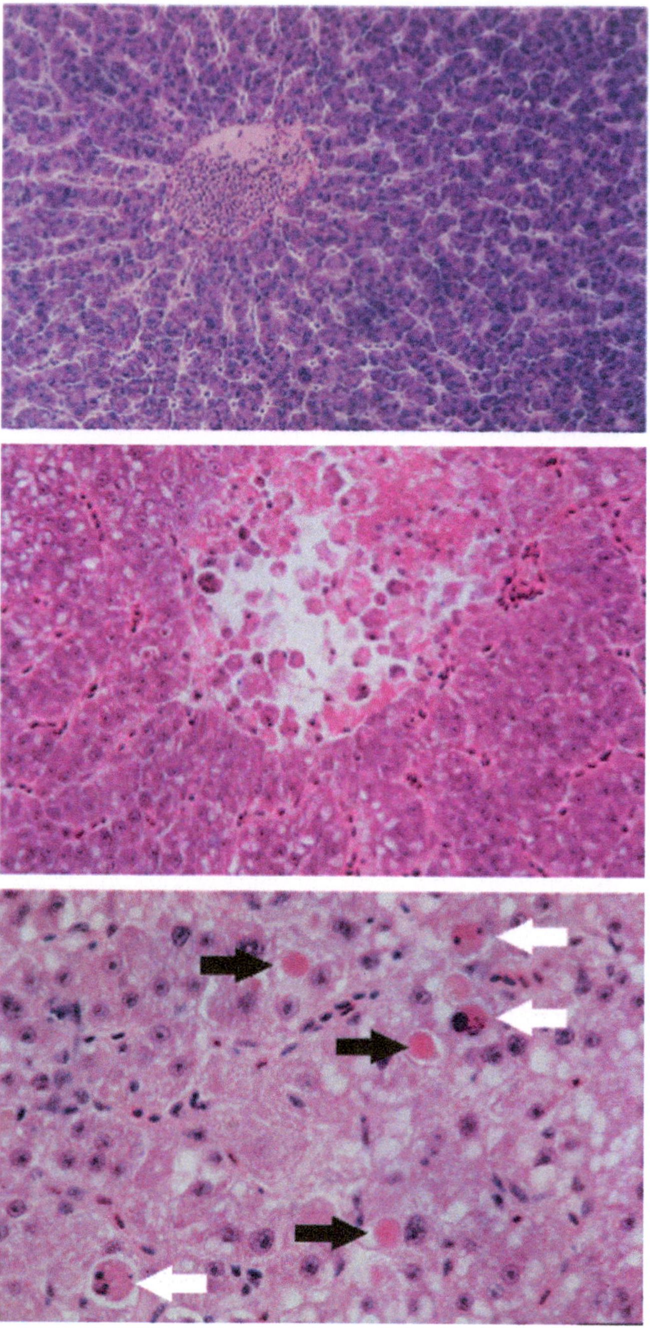

Figure 7.1 The top panel is a section through a normal *Fundulus heteroclitus* liver with branching hepatic tubules lined with hepatic sinusoids. Note that the hepatocytes are relatively uniform in size and shape. The middle panel is an example of necrosis in the liver. Notice the difference in staining between the living and dead cells. Dead cells show nuclear pyknosis and karyolysis, and loss of cell adherence. The bottom panel shows necrosis of individual cells, not a localized area as seen with the necrosis shown in the middle panel. Three necrotic cells are at the tips of the dark arrows. They are round or oval remnants that stain strongly with eosin. The basophilic chromatin remnants are visible in dead cells identified by the white arrows. (Color rendering of Figure 4.1 in Newman and Clements, 2008. Photomicrographs and general descriptions provided by W. Vogelbein, Virginia Institute of Marine Science.)

because both poisons denature and precipitate proteins (Sparks, 1972). Renal failure may result from accumulation and eventual spillover of metallothionein-associated metals in kidneys (the target organ) and subsequent coagulation necrosis (Hinton and Laurén, 1990a) of kidney cells. Other diverse cases involving coagulative necrosis include hepatic necrosis of children accidentally poisoned with boric acid (Chao et al. 2005), hepatic necrosis in rabbits after copper ingestion (Cooper et al. 1996), hepatic necrosis in mice exposed to the algal toxin, cylindrospermopsin 1 (Murphy and Thomas, 2001), and fish gill mucus cell necrosis following crude oil exposure (Prasad, 2006). This type of necrosis can also occur with an abrupt cessation of blood flow as in the case of a myocardial infarction. Hinton and Laurén (1990b) suggest that coagulation necrosis can be a good biomarker for cytotoxicity. **Liquefactive (cytolytic) necrosis** occurs with rapid breakdown of the cell as a consequence of the release of cellular enzymes. Fluid-filled, necrotic spaces in tissues might appear if many cells undergo liquefactive necrosis, especially if the involved tissue has considerable enzymatic activity (Meyers and Hendricks, 1985). Hinton and Laurén (1990b) suggest that liquefactive necrosis is less useful as a biomarker than coagulation necrosis because liquefactive necrosis is also often associated with infections. Several other forms of necrosis have been described but also are less useful than coagulative necrosis as biomarkers of cytotoxicity. With **caseous necrosis**, cells disintegrate to form a mass of fat and protein. **Gangrenous necrosis** is a combination of coagulation and liquefactive necrosis (Sparks, 1972) often resulting from puncture (with an associated lack of blood supply to the damaged tissue*) and subsequent infection. Meyers and Hendricks (1985) describe **fat necrosis** that involves deposits of saponified fats in dead fat cells. **Zenker's necrosis** occurs only in skeletal muscle and is similar to coagulation necrosis. All reflect cell death but coagulation necrosis seems to best reflect toxicant effect.

Apoptosis, often called programmed cell death (PCD), is another useful biomarker of toxicant effect (Sweet et al. 1999). The appearance of cells that have undergone apoptosis can be distinct from necrotic cells (Figure 7.1). With apoptosis, the chromatin condenses and the cytoplasm shrinks. The shrinking cell detaches from adjacent cells and its shape becomes irregular (Plaa and Charbonneau, 2008). The apoptotic cell can break into membrane-bound fragments called **apoptotic bodies**. In contrast to necrosis, apoptosis is not normally accompanied by inflammation. Apoptosis occurs for single cells but, as already detailed, necrosis can involve clusters or foci of many dead cells in affected tissue (Plaa and Charbonneau, 2008). Apoptosis occurs if a damaged or otherwise compromised cell begins a programmed sequence of biochemical steps intended to result in its death and removal from the tissue. Central to this process is activation of a particular class of proteases, **caspase enzymes**. Apoptosis begins by activation of caspases and the disruption of the mitochondrial membrane potential (Sweet et al. 1999; Jaeschke, 2008; Plaa and Charbonneau, 2008). According to Sweet et al. (1999), the cell then undergoes changes in calcium, potassium, and water fluxes; loss of intercellular junctions; and surface extensions. Finally, the cell's chromatin condenses, DNA fragments, and apoptotic bodies form.

Apoptosis can appear at lower toxicant concentrations than necrosis, and as a consequence, might be a very good biomarker for detecting low-level toxicant effects (Sweet et al. 1999; Plaa and Charbonneau, 2008). Chabicovsky et al. (2004) provide a relevant example of toxicant-induced apoptosis that occurred in a terrestrial snail (*Helix pomatia*) exposed to cadmium. Apoptotic cells were phagocytized by epithelial cells of this snail and subsequently cleared from tissues. Interestingly, a cadmium overloading phenomena described earlier in Chapter 6 as the metallothionein spillover hypothesis also occurred in these chronically exposed snails.

Before leaving this discussion of apoptosis, the point needs to be made that PCD also occurs normally in many instances such as in the developing organism to facilitate formation

* The resulting inadequacy of blood supply to surrounding tissues is called **ischemia**.

of anatomical features such as fingers or to obtain the proper positioning of connecting neurons. It also occurs with natural cell replenishment such as that occurring continuously for skin cells or neutrophils. However, apoptosis resulting from toxicant exposure is distinct because it is intended specifically to remove compromised cells that might otherwise foster tissue dysfunction or even cancer. Indeed, failure of normal apoptosis can contribute significantly to carcinogenesis.

7.2.2 Inflammation

Inflammation can be useful as a biomarker of toxicant effect (Figures 7.2 and 7.3). It is a response to cell injury or necrosis that isolates and destroys the offending agent or damaged cells (Sparks, 1972). As such, inflammation is often associated with toxicant-induced necrosis. For example, hepatic necrosis due to toxicant action can be accompanied by inflammation (Hinton and Lauŕen, 1990b). The process of inflammation continues through to lesion healing. The net result of inflammation is healed tissue with damaged cells being replaced by functioning cells (La Via and Hill, 1971).

The four **cardinal signs of inflammation** are heat, redness, swelling, and pain, although heat is irrelevant for poikilotherms and redness is relevant to red-blooded animals. Regardless, the underlying processes remain universal in all animals. Blood vessels in the damaged region dilate to increase blood flow to the area, causing redness and heat in red-blooded poikilotherms. Pain is caused by the pressure of the resulting tissue swelling as fluid from the blood passes through blood vessel walls and into the inflamed tissues. Leucocytes leave the blood vessels and enter the area of damaged tissue. Consequently, the infiltration of such cells (neutrophilic* granulocytes or neutrophils, mononuclear cells, and lymphocytes) also indicates inflammation is occurring. Later in the process, small blood vessels begin to form and connective tissue begins to grow in a mass called the **granulation tissue** (La Via and Hall, 1971). Scar tissue may be formed due to fibroblast and collagen infiltration of the damaged tissue (Sparks, 1972). If inflammation continues for too long as a consequence of chronic damage or infection, dense collections of collagenous connective tissue form

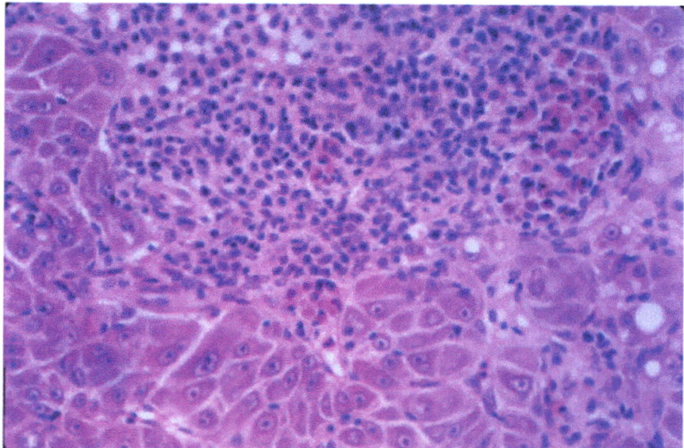

Figure 7.2 Inflammation in the liver of the estuarine fish, *Fundulus heteroclitus*. At the top center of the top photomicrograph is a focus of inflammation. (Color rendering of Figure 4.1 in Newman and Clements, 2008. Photomicrograph provided by W. Vogelbein, Virginia Institute of Marine Science.)

* The term neutrophilic means simply not having a strong, or preferential, staining with acidic or basic dyes.

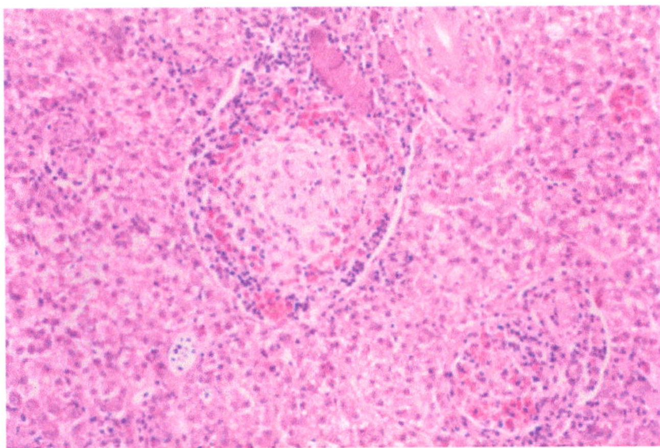

Figure 7.3 Macrophage aggregates are clumps of macrophages containing large amounts of ceroid/lipofuscin. Aggregates are surrounded by a fine capsule of connective tissue, are often associated with sites of inflammation, and, in mummichog (*F. heteroclitus*) from contaminated habitats, are greatly elevated in number and in size. Vogelbein and co-researchers suggest that this reflects the terminal stage of an inflammatory process in response to specific and nonspecific degenerative necrotic changes. (Photomicrograph and general description provided by W. Vogelbein, Virginia Institute of Marine Science.)

and might produce tissue dysfunction. Such accumulations can serve as good biomarkers of chronic inflammation.

Although the inflammatory response is common to all metazoan phyla, some important details vary such as the specific cells involved or the role of environmental factors. Students interested in studying nonmammalian species are urged to consult books describing the pathology and immunology of invertebrates and lower vertebrates (e.g., Sparks, 1972; Cooper, 1976) rather than relaying solely on mammalian pathology books.

7.2.3 Other General Effects

Several other general effects of toxicants are observed in cells and tissues. Swelling of mitochondria has been correlated with metal contamination (Aloj Totaro et al. 1986) or poisoning (Squibb and Fowler, 1981). A few of the more prominent effects will be highlighted in this section.

Accumulated **lipofuscin** (**ceroid***) is a common biomarker of toxicant effect (Figure 7.3). Lipofuscin is a mixture of degradation products of lipid peroxidation mixed with proteins that accumulates in cell vacuoles called **residual bodies** (La Via and Hall, 1971). It is composed of oxidized, cross-linked lipid and proteins (Jung et al. 2007) that appears as light yellow to brown granules in the cells' cytoplasm. Lipofuscin accumulates with age in some cell types (e.g., neurons) giving it the common moniker, **age pigment** (La Via and Hall, 1971; Aloj Totaro et al. 1986). It also increases with exposure to some toxicants. For example, lipofuscin granules in squid (*Torpedo marmorata*) neurons increased with copper exposure (Aloj Totaro et al. 1985, 1986), suggesting

* Ceroid, **ceroid pigment**, age pigment, and lipofuscin are often treated as synonymous terms as done here. Lipofuscin and ceroid have the same composition but form under different pathological conditions; thus, the distinction (Terman and Brunk, 1998). These pigmented degradation products are not metabolized further by proteolytic mechanisms nor removed from the cell, leading to their accumulation with time. They often are linked to aging or pathologies such as toxicant chronic exposure. Jung et al. (2007) provide a recent review of these materials although their review emphasizes aging.

elevated levels of lipid peroxidation due to oxyradical production by copper. Supporting this suggestion, that part of the squid nervous system with the most lipofuscin (the electric lobe) had the lowest activities of superoxide dismutase of all central nervous system tissues examined. As another example, the cadmium-exposed snails discussed earlier relative to apoptosis had elevated levels of residual bodies in their excretory cells (Chabicovsky et al. 2004). In some cases of metal intoxication, metals can be present in high concentrations in residual bodies (Domouhtsidou and Dimitriadis, 2000).

Metal-rich granules are another biomarker that were already discussed. Granules are found in almost all animal phyla (Simkiss, 1981a). Beeby (1991) summarized the different types of granules found in terrestrial invertebrates. Calcium-rich granules include calcium carbonate granules associated with pH buffering and calcium homeostasis. Calcium phosphate granules are involved in metal storage and detoxification: their levels can increase with increasing metal exposure. **Cloragosomes** of earthworms are similar in composition and function to the calcium phosphate granules of arthropods and mollusks except they have higher organic content. Soil isopods have also been found to have copper storage granules (**cuprosomes**) rich in sulfur. Several studies suggest that metals exchange between metallothionein or metallothionein-like protein and granules (e.g., Figure 7.4).

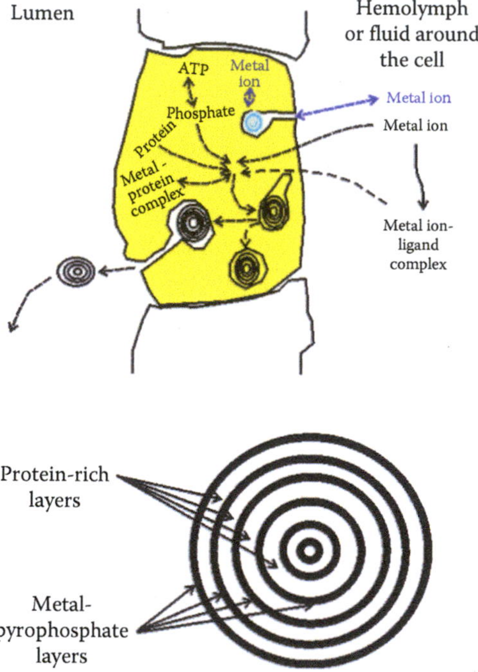

Figure 7.4 A simplified rendering of the role of intracellular granules in metal storage, detoxification, and elimination in invertebrate tissues. Two general types are shown here: calcium carbonate (blue) and calcium pyrophosphate (black/white laminated) granules. These granules are approximately 1–100 μm in diameter and are built up by concentric layers (Simkiss, 1981a). The calcium carbonate-based granules are involved in pH and calcium regulation, and are contained in vacuoles that can be connected to the cell's exterior via pores. The calcium (and magnesium) phosphate granules are usually smaller than 5 μm (Hopkin and Nott, 1979) and have concentric phosphate-rich layers separated by layers with much more protein. Metals such as cobalt, iron, lead, manganese, and zinc can accumulate to high concentration in these granules by combining with the phosphate, and in some cases, the proteins (Simkiss, 1981b). The presence and abundance of these granules can be very tissue-specific. Toxic metals can remain sequestered in calcium pyrophosphate granules within vacuoles or released from the cell. Some metals such as cadmium and mercury tend to remain more associated with proteins than granules in invertebrate tissues. (Simkiss, 1981b.)

Unusually, high densities of metal-rich granules in invertebrate tissues such as the hepatopancreas, digestive gland, arthropod midgut, gills, and kidney might suggest a general detoxifying response to metal exposure (Brown, 1978; Hopkin and Nott, 1979; Simkiss and Taylor, 1981; Mason et al. 1984; Hopkin et al. 1989; Correia et al. 2002; Bonneris et al. 2005). **Calculi**, mercuric selenide granules thought to sequester mercury, can also be found in liver connective tissues of toothed whales (Mackey et al. 2003).

7.3 GENE AND CHROMOSOME DAMAGE

Contaminants can produce changes in DNA, and in so doing, affect cells and tissues. These changes can increase both somatic and genetic risk. **Somatic risk** is the risk of an adverse effect to the exposed individual resulting from genetic damage to somatic cells (e.g., damage leading to cancer) and **genetic risk** is the risk to progeny of an exposed individual as a consequence of heritable genetic damage (e.g., damage to germ cells or gametes leading to a nonviable fetus or an offspring with a birth defect). These effects are manifested in a variety of cellular structures or processes (Figure 7.5). Several effects are routinely used as biomarkers and provide a mechanism for consequences at the individual level such as mutation or cancer.

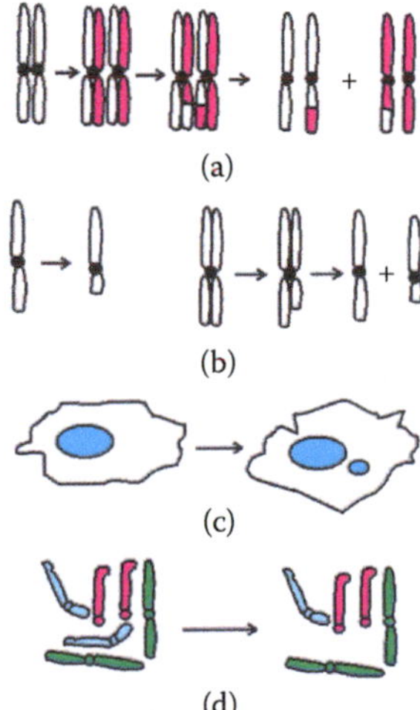

Figure 7.5 Damage to genetic materials can be assessed with a variety of cellular qualities. Increased rates of sister chromatid exchange (a) can suggest the rate of DNA damage although the exchange itself is not injurious. Beginning as a pair of homologous chromosomes (a **homolog**), four chromatids are formed. Sister chromatids in this **tetrad** (four chromatids paired as two homologs at metaphase) may exchange material. In the sister chromatid exchange assay, DNA is stained in the first round of cell division with 5-bromodeoxyuridine and the cells then synthesize unstained DNA in subsequent divisions. This results in stained and unstained chromatids that may exchange DNA in future cell divisions. This exchange is seen as chromatids with both stained and unstained segments. The rate of exchange is correlated with the frequency of DNA breakage. Chromosomes may be damaged (b, left) producing a chromosomal aberration. Viewed under the light microscope, these aberrations may appear as breaks or gaps in chromosomes. If only one chromatid in a homolog is broken, a chromatid aberration occurs (b, right). With a chromatid aberration, only one chromosome in one of the two daughter cells would be damaged. Failure of mitotic processes can produce micronuclei (c) or aneuploidy, a deviation from the usual ploidy (e.g., the usual 2*N* diploid number of chromosomes) (d).

Toxicant-mediated chromatid breakage can be correlated with the incidence of **sister chromatid exchange (SCE)*** of DNA (Figure 7.5a). Dixon and Clarke (1982) showed a clear dose-SCE response after mutagen† (mitomycin C) exposure of the blue mussel, *Mytilus edulis*. A similar relationship was found for *M. edulis* larvae exposed to mutagens (Harrison and Jones, 1982). Consequently, SCE has been proposed as an assay of DNA damage (Tucker et al. 1993). In this assay, one chromatid in a pair is stained with 5-bromodeoxyuridine. After two cycles of cell division, tissue samples are examined for DNA exchange of labeled segments between chromatids. With no exchange, chromatids would remain either completely stained or unstained after cell division. If exchange between chromatids has occurred, chromatids with both unstained and stained segments would be produced. The number of SCE per metaphase or SCE per chromosome is used as a measure of DNA damage.

Structural **chromosomal aberrations** include breakage and loss of segments of DNA, or chromosomal rearrangements. Chromosomal breaks involve double-strand breaks as shown in Figure 7.5b (left). Fragments of chromosomes might also be apparent. Also visible may be gaps in chromosomes, small discontinuities in the chromosome. A **chromatid aberration** occurs if only one strand (chromatid) is broken (Figure 7.5b, right).

There have been numerous studies using aberrations as biomarkers. Chromosomal aberrations in bone marrow cells were higher in mice (*Peromyscus leucopus*) and cotton rats (*Sigmodon hispidus*) from petroleum-, heavy metal-, and polychlorinated biphenyl (PCB)-contaminated sites than in mice and rats from uncontaminated sites (McBee et al. 1987). Similarly, chromosomal aberrations in blood lymphocytes of petroleum refinery workers (0.023–0.037 breaks per lymphocyte) were elevated relative to those of control populations (0.015–0.021 breaks per lymphocyte) (Khalil, 1995). Lead poisoning increased the incidence of chromosomal aberrations in mouse bone marrow cells (Forni, 1980). Chromosomal and chromatid aberrations in workers at a lead oxide factory were elevated by sublethal lead exposure (18.7% vs. 5.1% abnormal metaphases per lymphocyte for factory workers and controls, respectively) (Forni, 1980). More recently, workers exposed to uranyl compounds were found to have elevated levels of leucocyte chromosomal aberrations (Prabhavathi et al. 2003) as were downtown Prague police officers notionally exposed to high levels of PAH (Srama et al. 2007).

Anomalies during cell division, such as chromosome damage and/or spindle dysfunction, can produce micronuclei (MN), that is, nuclear segments isolated in the cytoplasm from the nucleus (Nikinmaa, 1992) (Figure 7.5c). Red blood cells of fish exposed to toxicants have been reported to have elevated numbers of MN (Nikinmaa, 1992). Soft-shell clams (*Mya arenaria*) exposed to PCB-contaminated sediments of the New Bedford Harbor (Massachusetts) had a threefold higher incidence of MN in blood cells than clams from a clean site (Martha's Vineyard) (Dopp et al. 1996). Further, the number of leukemic cells per milliliter of hemolymph was correlated with the number of blood cells with MN, suggesting a common mechanism between production of MN and severity of leukemia in clams from contaminated sites.

Failure of chromosomes to properly segregate during cell division can result in aneuploidy (Figure 7.5d). With a chromosome preparation, such a condition would appear as an atypical number of chromosomes in the cell, e.g., $2N-1$ or $2N+1$ chromosomes. As discussed in Chapter 6, flow cytometry (FCM) has been used to quantify aneuploidy in cells taken from individuals exposed to genotoxic agents. Lamb et al. (1991) used flow cytometric methods to document aneuploidy in the red blood cells of turtles (*Trachemys scripta*) from a reservoir with elevated levels of radiocesium (^{137}Cs) and radiostrontium (^{90}Sr).

* Recollect that, before cells divide, the DNA in each chromosome is duplicated to produce two identical chromatids. As the chromosomes condense before division, each appears as a pair of **chromatids** connected at a common centromere. At the metaphase plate, these **sister chromatids** come together with those of the other homologous chromosome to form a **tetrad**. The four sister chromatids (two per chromosome) may exchange segments of homologous DNA by the breaking and rejoining with crossing over of DNA segments.

† A **mutagen** is a physical or chemical agent capable of producing mutations.

VIGNETTE 7.1 Chromosome Damage

Karen McBee
Oklahoma State University

Chromosomes have been analyzed for evidence of structural damage that results from exposure to clastogenic compounds since early in the twentieth century, but effects of environmental contaminants on chromosome structure have been investigated in field settings only for about 30 years (Shugart, 1999; Bickham et al. 2000; Dmitriev and Zakharov, 2001). Almost all routine laboratory protocols and techniques have been adapted for use with a wide array of organisms including plants (Al-Sabti and Kurelec, 1985a; Kumar et al. 1989), invertebrates (Al-Sabti and Kurelec, 1985b; Bolognesi et al. 1996), fish (Al-Sabti, 1985; Al-Sabti and Hardig, 1990), and mammals (Peakall and McBee, 2001). The most commonly used techniques are metaphase chromosome aberration (CA) analysis, sister chromatid exchange (SCE), flow cytometry (FCM), and micronucleus (MN) formation.

Visualization of lesions in metaphase spreads (Figure 7.6) historically was the most commonly used and intuitively obvious way to document chromosome damage. Assays to detect induction of CAs typically have involved microscopic examination of chromosomal spreads extracted from blood, bone marrow, spleen, or testis from animals and root tip squashes from plants. Cells are scored for the presence of several categories of chromosomal aberrations including breaks, deletions, gaps, and translocations of fragments of one chromosome onto one or more nonhomologous chromosomes. Among the first studies in wild mammals exposed to environmental contaminants was that of Thompson et al. (1988) in which chromosome and chromatid aberrations in cotton rats (*Sigmodon hispidus*) living close to two hazardous waste dumps were compared to animals from a control site. When all classes of aberrations were lumped together, the frequency of occurrence was higher in animals from hazardous waste sites than in animals from the control site. When individual classes of aberrations, such as chromatid breaks, were considered separately, animals from hazardous waste sites showed higher frequencies, but these differences were not statistically significant. Chromosomal aberrations were examined in bone marrow of white-footed mice (*Peromyscus leucopus)* and cotton rats (*S. hispidus)* collected over a 2-year period at a site contaminated

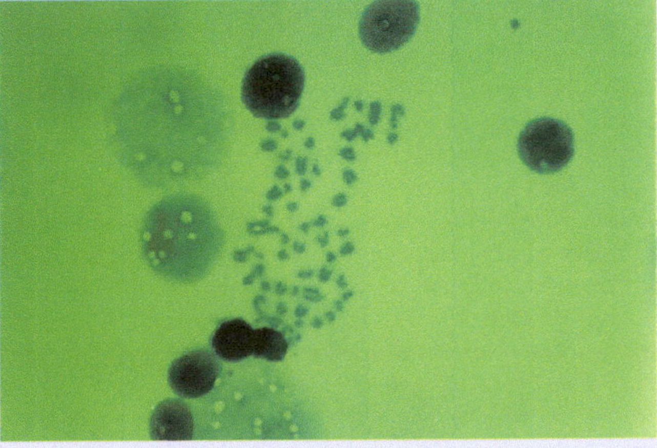

Figure 7.6 Metaphase chromosome spread from *Sigmodon hispidus* (hispid cotton rat) showing multiple chromatid breaks. (Photograph by Karen McBee.)

with a complex mixture of petrochemical refinery wastes, diesel fuel, heavy metals, and PCBs. Animals of both species from the contaminated site showed significant increases in number of aberrant cells per individual and mean number of lesions per cell compared to animals from two reference sites; however, there were no statistical differences for either parameter when animals from the two reference sites were compared (McBee et al. 1987). *Peromyscus* from the contaminated site had significantly higher levels of lead and chromium but not cadmium and zinc in muscle tissue than did animals from the reference sites; however, there were only weak correlations (r = 0.12–0.48) between level of lead or chromium and number of aberrant cells or lesions per cell suggesting that something other than metals were responsible for induced genetic lesions (Tull-Singleton et al. 1994). Shaw-Allen and McBee (1993) examined bone marrow metaphase chromosomes from *P. leucopus*, *S. hispidus*, and fulvous harvest mice (*Reithrodontomys fulvescens*) collected from a site contaminated with the PCB mixture Aroclor 1254 and two uncontaminated reference sites, and found no statistical differences for aberrant cells per individual or number of lesions per cell for any species among the three sites. Eckl and Riegler (1997) collected feral *Rattus rattus* and *Rattus norvegicus* in the vicinity of a waste disposal site near Salzburg, Austria, and found that the level of chromosomal damage in hepatocytes was directly related to age (as determined by weight) for *R. rattus* but not for *R. norvegicus*. Bueno et al. (1992) collected individuals of two genera (*Akodon* and *Oryzomys*) of South American rodents from an uncontaminated island near Florianpolis and from three sites in the Sango River Basin, which receives wastes from "coal washing" processes that result in the area being heavily contaminated with mercury, lead, cadmium, copper, and zinc. Both *Oryzomys* and *Akodon* showed statistically significant increases in percent cells with chromosomal aberrations at the sites along the Sango River compared to the site near Florianpolis. Mexican free-tailed bats (*Tadarida brasiliensis*) from Carlsbad Caverns and from a maternity colony located in a cave in northwestern Oklahoma were examined for levels of chromosome aberrations in bone marrow. No statistically significant differences were observed between males and females or between individuals from the two caves over a 4-month period even though animals from the Carlsbad Caverns colony consistently had higher levels of DDT and its clastogenic metabolites in their tissues than did animals from Oklahoma (Thies et al. 1996). Topashka-Ancheva et al. (2003) analyzed CA in the Macedonian mouse (*Mus macedonicus*), sibling vole (*Microtus epiroticus*), yellow-necked field mouse (*Apodemus flavicollis*), and bank vole (*Myodes glareolus*) collected from within a large industrial site contaminated with Cu, Pb, Cd, and Zn. Individuals of all species from sites with high levels of Pb and Cd had the highest percentage of chromosomal aberrations. Alibi et al. (2006) found dose-dependent increases in number of structural aberrations and significantly lower mitotic indices in bone marrow cells of *R. norvegicus* dosed with leachates containing heavy metals from municipal landfills. *P. leucopus* collected at Tar Creek Superfund Site, a 40 sq mi area heavily contaminated with lead, cadmium and zinc, showed no significant difference for number of aberrant cells per individual or mitotic index compared to mice collected from uncontaminated reference sites (Hays, 2010).

Sister chromatid exchange is the reciprocal interchange of DNA at homologous loci between sister chromatids. Chromosomes that have undergone SCE are not damaged in the conventional sense in that they are structurally intact, but increased strand breakage during S-phase leads to increased SCE. Nayak and Petras (1985) examined SCE levels in house mice (*Mus musculus*) from several locations in southern Ontario and found significant negative correlations between SCE level and both distance to the Windsor-Detroit urban complex and distance to the nearest industrial complex. Tice et al. (1987) found that *P. leucopus* collected from an EPA Superfund site had significantly higher levels of SCE in bone marrow compared to animals from uncontaminated sites in the northeastern United States. Furthermore, animals collected at the Superfund site during the winter had significantly fewer SCEs compared to animals collected during the summer even though levels of SCEs in animals collected from the uncontaminated sites were not different between the two collecting periods. Ellenton and McPherson (1983) examined SCE in Herring Gull (*Larus argentatus*) eggs collected from five breeding colonies in polluted areas around the Great Lakes and found no difference in levels of SCEs compared to those in a colony from a pristine area on the Atlantic coast.

Micronuclei (MN) are cytoplasmic nuclear bodies that are formed when whole or fragmented chromosomes are not incorporated into the nuclei of daughter cells or when small fragments of chromatin are retained from polychromatic erythrocytes after expulsion of the nucleus in the process of erythrocyte maturation in mammals. Al-Sabti and Hardig (1990) showed that the frequency of MN formation in perch (*Perca fluviatilis*) decreased with increasing distance from wastewater discharge points on the Baltic Sea. Tice et al. (1987) showed increased levels of MN in animals from a Superfund site compared to pristine sites in the northeastern United States. Cristaldi et al. (1990) demonstrated increased levels of MN formation in house mice (*M. musculus*) collected in northern Italy 6–12 months after the nuclear incident at Chernobyl, Ukraine, compared to animals collected at the same sites 5 years before the incident. However, Rodgers and Baker (2000) and Rodgers et al. (2001) examined bank voles (*M. glareolus*) from near the Chernobyl reactor and found no difference in frequencies of MN in animals from inside the exclusion zone where radiation doses were estimated to be as high as 15–20 rads per day compared to animals from a reference population outside the exclusion zone where radiation doses were negligible. Theodorakis et al. (2001) used MN to investigate induction of chromosome damage in kangaroo rats (*Dipodomys merriami*) inhabiting two atomic blast sites at the Nevada Test Site but were not able to demonstrate significant differences among the blast sites and two reference sites.

Microscope-based assays of chromosome damage can be time consuming and difficult to score accurately and consistently if the scorer is not carefully trained, so there has been considerable interest in development of quicker, easier methods to determine levels of chromosome damage. Flow cytometry (FCM) can be a fast and cost-effective method to detect chromosomal damage. Variation in amount of nuclear DNA within the cells of an individual is measured as the coefficient of variation (CV) around the mean DNA content for all cells in the G_0/G_1 stage of the cell cycle. Increased CVs result from unequal amounts of DNA being distributed to daughter cells at mitosis when the mitotic cell contained broken chromosomes (Shapiro, 2003). Bickham et al. (1988) reported increased CVs and increased levels of aneuploidy in *T. scripta* collected from ponds contaminated with radioactive wastes, but Lamb et al. (1995) working at the same ponds a few years later found no evidence of increased CVs or aneuploidy levels. Matson et al. (2005) documented a significant positive correlation between concentrations of three-ring PAHs in sediments and increased whole peak CVs in two species of turtles (*Emys orbicularis* and *Mauremys caspica*) among several sites in Azerbaijan but did not see a similar trend for micronucleus assays. Hays and McBee (2007) found that red-eared slider turtles (*T. scripta*) from Tar Creek Superfund Site did not have increased CV but did have significantly higher levels of aneuploidy when compared to turtles caught at a national wildlife refuge. Jung et al. (2011) demonstrated increased half-peak CVs for killifish (*Fundulus heteroclitus*) collected from the Elizabeth River, Virginia. They also found increased levels of sediment PAH concentrations and DNA adducts formed by benzo [*a*] pyrene metabolites but did not see damage arising from oxidative DNA adducts.

Although, these assays are relatively easy and time- and cost-effective to perform, we still know little about their meaning in terms of the overall health of a population or community or even their relation to each other. Husby et al. (1999) found no difference in levels of CA in *Peromyscus* from abandoned strip mine sites and reference sites in eastern Oklahoma, but did find significantly greater levels of DNA strand breaks (Husby and McBee, 1999) in the same animals. Hausbeck (1995) found only weak correlations between levels of heavy metal bioaccumulation in tissues of these same animals and levels of chromosome damage found by Husby et al. (1999). Tull-Singleton et al. (1994) found only weak relationships between levels of CA and lesions measured by FCM and levels of lead, cadmium, chromium, and zinc in livers of *P. leucopus*. Although Hays (2010) found no evidence of increased genetic damage in *P. leucopus* from within Tar Creek Superfund Site, Phelps and McBee (2010) showed that *P. leucopus* from the same sites were larger in size, showed a higher frequency of dental anomalies, and had higher incidence of botfly infestations. They also showed that the small mammal community at sites within Tar Creek Superfund Site had reduced diversity compared to reference sites and was dominated by a single species, *P. leucopus* (Phelps and McBee, 2009). No other species was collected in large enough numbers to conduct chromosomal analyses. Tice et al. (1987) suggest potential for seasonal variability in sensitivity to or level of exposure to environmental genotoxicants. A series of studies of CA, tissue accumulation, population demographics, immunological response, and enzyme induction (McMurry, 1993; McBee and Lochmiller, 1996; Peakall and McBee, 2001) conducted on *S. hispidus* at three locations within a large Superfund site and three matched reference sites resulted in evidence of significant responses in several endpoints among the six sites, but the pattern of responses was not consistent across sites or seasons or among endpoints. Summed together, these studies indicate that, just as in a controlled laboratory setting, contaminants can induce chromosomal damage in wild populations, but accurate interpretation of effects of chromosomal damage and its relationships with other toxic effects may be far more difficult than its documentation. As more rapid molecular techniques are becoming available, analysis of chromosomal damage in wild populations that have been exposed to clastogenic pollutants may be abandoned before it is understood.

7.4 CANCER

Normal cells have the capacity to multiply and increase in tissues. This increase in the numbers of cells in a tissue or organ via mitosis that results in an increase in tissue or organ size is called **hyperplasia**. Hyperplasia in response to a variety of normal stimuli such as that involved in the tissue repair process is one type of **physiologic hyperplasia** (La Via and Hill, 1971). This particular form of physiologic hyperplasia is called **compensatory hyperplasia** and occurs to compensate for lost or damaged tissue, e.g., the enlargement of a remaining kidney after removal of a diseased, second kidney. **Hormonal hyperplasia** is a second form of physiologic hyperplasia that occurs when hormone-signaled changes in organ or tissue function occur, such as the increase in numbers of human breast or uterus cells associated with childbearing. Sometimes, **excessive hyperplasia** might occur during response to injury or irritation, resulting in one type of **pathologic hyperplasia**. This excessive hyperplasia is seen with hyperthyroidism such as that suspected by Dye et al. (2007) and Guo et al. (2012) to be linked to polybrominated diphenyl ether concentrations in domestic cats. Such abnormal hyperplasia involves an inappropriate response to growth factors or hormonal stimuli. A second form of pathologic hyperplasia is **neoplastic*** **hyperplasia,** which is distinct from other forms because it involves a hereditary change to cells such that they no longer respond properly to chemical signals that would normally control their growth. Neoplastic hyperplasia provides a mechanism for

* **Neoplasia** are "hyperplasia which is caused, at least in part, by an intrinsic heritable abnormality in the involved cells" (La Via and Hill, 1971).

cancerous growth. Neoplasia that grow slowly and tend to remain differentiated in their morphology are not as invasive of neighboring tissues as other neoplasia and are classified as **benign neoplasia**. Those that take on undifferentiated forms, tend to grow rapidly, and invade other tissues are classified as **malignant neoplasia**. Malignant neoplasia or cancers are more life-threatening than benign cancers. Pieces of the original malignant cancerous growth can dislodge, move to another tissue via the circulatory or lymphatic system to establish other foci of cancerous growth. This process of **metastasis** spreads a malignant cancer from the site of origin throughout the body.

Cancer is a result of a heritable change in cells. A chemical (e.g., numerous carcinogens), physical (e.g., ionizing radiation), or biological (e.g., a retrovirus) agent modifies a gene or its normal relation to other genes, which results in neoplasia. This may occur through point mutations, deletions, additions, rearrangements as described above, or by gene insertion by a virus. A retrovirus might insert DNA into a cell's chromosome and transform that cell so it responds improperly to growth regulating signals. A chemical or physical agent may interact with the cell's genome such that a gene involved with cell signaling or the regulation of normal growth and differentiation of cells (**proto-oncogene**)* is changed to an **oncogene**, an altered gene causing cancer. Change to an oncogene results in loss of normal growth dynamics and/or differentiation of the cell. Cancer results from such inappropriate activation or activity of a gene that would normally be involved in cellular growth and/or differentiation (Moolgavkar, 1986). Other genes called **suppressor genes** function normally to suppress cell growth and might inhibit abnormal growth. Like proto-oncogenes, suppressor genes are thought to function in the process of growth and differentiation in normal cells, but are most often explored in the context of suppressing growth—and even stimulating apoptosis—of cells with a mutated gene. A cancerous tumor can also develop if some agent or event affects the functioning of the suppressor gene. Modifications to proto-oncogenes and/or suppressor genes facilitate carcinogenesis.

It follows from the earlier discussion that some agents initiate the neoplastic process and others promote the development of the tumor. This fact has led to the classification of two groups of agents: **initiators**, which convert normal cells to latent tumor cells, and **promoters**, which enhance the growth and continued expansion of latent tumor cells (Figure 7.7). Cancers are thought to pass through an initiation stage in which the DNA is permanently changed and then a promotional stage in which the modified cell then undergoes clonal increase. Cancer can be promoted by cell proliferation consequent to lesion formation (e.g., those associated with asbestos fiber in the lung or possibly carbon nanotubes [Poland et al. 2008]), high-hormone levels-associated hyperplasia (**hormonal oncogenesis**) (La Via and Hill, 1971), and chemical agents. Often, inappropriate inactivation of suppressor genes is associated with the promotion process (Aust, 1991), resulting in unregulated growth (Van Beneden and Ostrander, 1994). After initiation and promotion, the final step of carcinogenesis occurs, that is, cancer progression. **Cancer progression** is the change in the biological attributes of neoplastic cells over time that leads to a malignancy: it involves an additional genetic mutation in the neoplastic cell line to produce cancer cells able to sustain themselves in the body (Beck et al. 2008). There can be a long **latent** (or **latency**) **period** between exposure to a carcinogen and the appearance of an observable cancer because carcinogenesis involves several stages, each having an associated probability of occurring. This makes assignment of cause and effect difficult because the exposure might have occurred years prior to effect manifestation.

The latent period and multistage process leading to cancer also make prediction of the exact shape of the dose–cancer response curve challenging. Two theories exist and are supported by different data. The **threshold theory** holds that there is no effect below a certain low dose (e.g., Downs and Frankowski, 1982; Cohen, 1990). Above this threshold, the slope of the response versus dose curve increases, resulting in a dose–response curve with the general appearance of a hockey stick (Figure 7.8). The **linear-no threshold theory** is based on several radiation-induced cancer studies

* Gregus (2008) specifies that the protein products of the proto-oncogenes are either growth factors, intracellular signal transducers, cell surface receptors of growth factors, or nuclear transcription factors. (A nuclear transcription factor is a protein that binds to DNA regions, facilitating DNA to RNA transcription.)

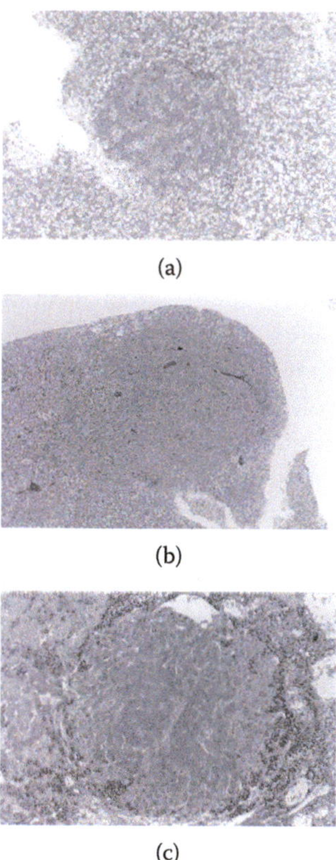

(a)

(b)

(c)

Figure 7.7 Progression of neoplasms in the medaka (*Oryias latipes*) initiated with the carcinogen, diethylnitrosamine. Fish had more rapid development of foci, and shorter times between exposure and realized effect (latent period) if they were fed the promoter, 17-β-estradiol in their diet. The top micrograph shows a basophilic focus of cellular alteration in the liver 25 weeks after exposure to diethylnitrosamine and provision of a diet containing estradiol (hematoxylin and eosin, X150, courtesy of Janis Brencher Cooke, University of California, Davis). The middle micrograph shows a solid basophilic adenoma in medaka liver exposed as already described for the top micrograph. (hematoxylin and eosin, X75, Courtesy of Janis Brencher Cooke, University of California, Davis). The bottom micrograph shows a hepatocellular carcinoma in medaka liver at 12 weeks of exposure. Note the distinct architecture of the cancer relative to the surrounding tissue (hematoxylin and eosin, X125, courtesy of Swee The, University of California, Davis).

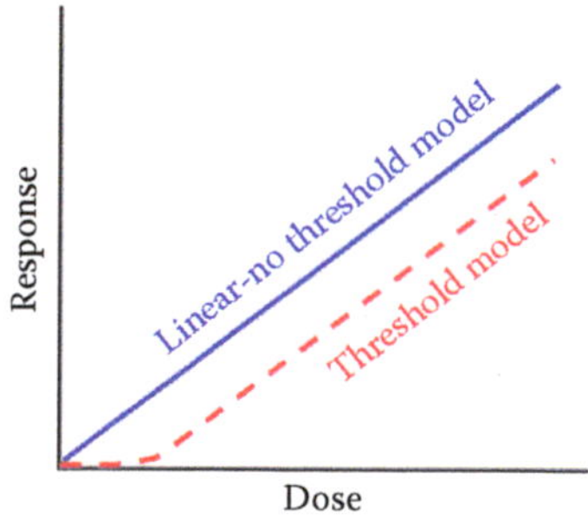

Figure 7.8 The linear-no threshold and threshold theories describe two dose-response models differing from each other at lower doses. The linear-no threshold model is described by a straight line and the threshold model is described by a "hockey stick" curve.

where no apparent threshold seemed to exist for effect. Cancer initiation being the mechanism for radiation-induced carcinogenesis lends support to this theory because even a single mutation has a chance of producing a cancer. This theory carries the assumption that any lack of detected cancer below a certain dose reflects our inability to measure low incidences of cancers at these exposure levels. This dose–response model describes a line with no threshold.

Chemical and physical agents can act by initiation or promotion of the process leading to cancer. Relevant mutagenic mechanisms are those described previously for alterations of DNA and chromosomes. Further, agents that inhibit DNA repair may also increase the probably of cancer. Differences in DNA **repair fidelity** (accuracy in repairing and returning the DNA to its original state after damage) associated with various agents can also result in differences in carcinogenicity. For example, Robison et al. (1984) suggested that, although chromate, nickel, and mercury damage DNA, there is a difference in the nature of the damage and its repair. Generally, mercury produces more single-strand breaks than nickel and chromate, which cause more protein–DNA cross-linking. The difference in the repair fidelity between these two types of damage leads to nickel and chromate (lower repair fidelity) being more carcinogenic than mercury (higher repair fidelity).

The presence of tumors and various tumor cell qualities are used as biomarkers for environmental carcinogens. In the study of micronuclei in soft-shell clams discussed earlier (Dopp et al. 1996), leukemic cells were measured using an antibody assay that recognized cell surface alterations. Enzymatic and histochemical alterations associated with cancer cells are also used (e.g., Moore and Myers, 1994). Often, the incidence of cancers is correlated with the level of contaminant to which a population is exposed. For example, Martineau et al. (2002) noted a correlation between pollution and cancer in St. Lawrence River beluga whales (*Delphinapterus leucas*). Liver neoplasia in sole (*Parophyrs vetulus*) from Puget Sound were correlated to sediment contamination by PAHs (Stein et al. 1990). The incidence of cervical neoplasia in Czech women increased 2 years (latent period) after radionuclide release from the Chernobyl disaster (Borovec, 1995) and a 1980–2004 survey of recovery workers and other local Ukrainians (Prysyazhnyuk et al. 2007) found elevated thyroid cancer incidence rates.

7.5 GILLS AS AN EXAMPLE

Gills are often damaged or changed by toxicant exposure (Evans, 1987; Hinton et al. 1987), and will be discussed here as an important example that incorporates many of the nongenetic cellular effects described in this chapter.

The fish gill (Figure 7.9, top panel) is composed of **primary lamellae** (filaments) that extend outward at right angles from the branchial arches. On the dorsal and ventral sides of each primary lamella are parallel rows of **secondary** (respiratory) **lamellae** that are the major sites of gas exchange. The epithelium covering the lamellae is a double layer of cells with intercellular lymphoid spaces between the cell layers. **Chloride cells** (Figure 7.10), which function in ion regulation, are found predominantly on the epithelium of primary lamellae but also on the secondary lamellae.

After exposure to such diverse toxicants as ammonia, metals, detergents, high or low pH, nitrophenols, and pesticides, the outer epithelium of the secondary lamellae often lifts away to produce an enlarged, fluid-filled space between itself and the inner epithelial cell layer (Skidmore and Tovell, 1972; Mallatt, 1985; Evans, 1987; Cengiz and Ünlü, 2003; Lease et al. 2003;) (Figure 7.11). This, for example, occurred for gills of zebrafish (*Danio rerio*) larvae after exposure to contaminated sediments from Brazilian reservoirs (Fracácio et al. 2003). Granulocyte densities can increase in the space between the outer and inner epithelial layers, indicating inflammation. With acute exposure to toxicants such as zinc, chloride cells can separate from the epithelium or ruptures can occur in the outer epithelial layer. Cells might appear swollen with distended mitochondria, indicating necrosis.

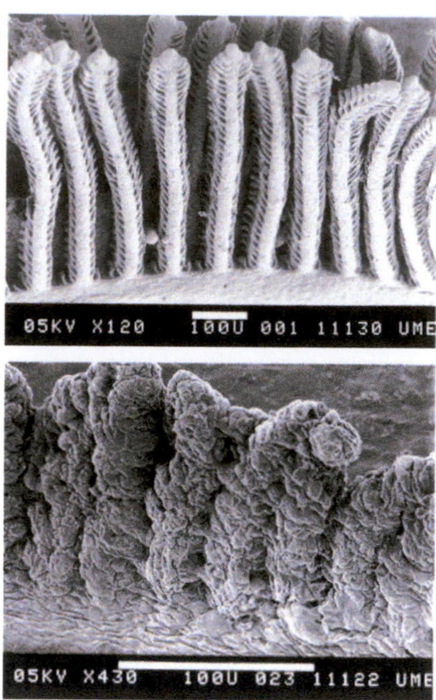

Figure 7.9 Electron micrographs of gills from Atlantic salmon (*Salmo salar*) fry. The top panel shows the normal gill morphology with primary lamellae extending (vertically here) from the branchial arch. Perpendicular to and on both sides of the main axis of each primary lamellae are the secondary lamellae. The bottom micrograph shows the gills of salmon fry after 30 days exposure to 300 $\mu g \cdot L^{-1}$ of aluminum. Note the extensive fusion of the secondary lamellae. (Courtesy of C. H. Jagoe, Florida A&M University.)

The number of chloride cells on the primary lamellae increased with exposure of rainbow trout (*Salmo gairdneri*) to dehydroabietic acid, a component of Kraft mill effluent (Tuurala and Soivio, 1982), and mosquitofish (*Gambusia holbrooki*) exposed to inorganic mercury (Jagoe et al. 1996). Also, the size of mosquitofish chloride cells increased with inorganic mercury exposure. This chloride cell size change is an example of **hypertrophy**, an increase in cell size (and function) resulting from an increase in the mass of cellular components (La Via and Hill, 1971; Meyers and Hendricks, 1985). The hyperplasia and hypertrophy of chloride cells are believed to be a compensatory response to ion imbalance resulting from gill damage (Jagoe et al. 1996). Individually or together, hyperplasia and epithelial lifting can produce a fusion of secondary lamellae with a consequent reduction in gaseous exchange capacity across the gill surface. Exposure of salmon fry to aluminum (Figure 7.9, bottom) produced extensive fusion of lamellae. The hyperplasia and hypertrophy of chloride cells of the secondary lamellae caused similar fusion of lamellae as the spaces between primary lamellae were filled with cells from the primary lamellae (Figure 7.10, bottom panel). **Thrombosis** (a clot formed in the circulatory system that could cause infarction of local tissue) can also occur with toxicant exposure (Figure 7.12). The consequences at the physiological and individual levels were a reduction in oxygen exchange across gill surfaces and death at lethal exposures.

7.6 LIVER AS AN EXAMPLE

The selection of gills as the first example does not suggest that the gill is the most useful organ relative to detecting cellular effects. The midgut epithelium of insects can also be an important target tissue (Beaty et al. 2002). The organs and tissues of the endocrine system also manifest

Figure 7.10 Micrographs of mosquitofish (*Gambusia holbrooki*) gills in cross section. The top panel shows the normal gill with secondary lamellae extending outward from the primary lamellae. Note the large chloride cells on the primary lamellae between the secondary lamellae. After exposure for 14 days to 60 μg·L^{-1} of inorganic mercury (bottom panel; see Jagoe et al. 1996), the primary lamellar epithelium began filling in the spaces between adjacent secondary lamellae. The chloride cells (large cells with lightly stained cytoplasm) were involved to a large extent in this hyperplasia, becoming larger (hypertrophy) and more abundant on the primary lamellae. The secondary lamellae appeared to shorten or disappear as a consequence. Although reported elsewhere in response to gill irritation with toxicants (e.g., Tuurala and Soivio, 1982), no necrosis, separation of epithelium from the secondary lamellae, or inflammation were noted. (Toluidine blue; distance between secondary lamellae at their base is circa 20 μm; Courtesy of C. H. Jagoe, Florida A&M University.)

important changes associated with general stress (e.g., Tsigos and Chrousos, 2002), receptor-modifying agents (Volz et al. 2008),* or specific endocrine-disrupting substances (e.g., Hontela, 1998; Colborn, 2002; Brown et al. 2004a,b; Mayne et al. 2005). Among the most common organs studied relative to lesions are the liver and analogous organs of invertebrates. They tend to be involved with detoxification, and therefore, are often sites of activation. They also display more active cell proliferation than many other organs. For these reasons, liver lesions will be discussed here as a second

* As explored in this citation, some agents influence liver condition by disrupting bile formation and movement. **Cholestasis**, physical blockage of bile secretion, can manifest with environmental contaminant exposure. Such effects can be complicated ones involving interactions with receptors such as the ah receptor (AHR). As an example, fish exposed to the biliary toxicant, α-naphthylisothiocyanate, displayed liver swelling and apoptosis of bile preductular epithelial cells (BPDECs) (Volz et al. 2008). This was followed by gall bladder discoloration, suggesting an adverse hepatobiliary alteration. There was evidence that ah receptor activation and cytochrome P-450 monooxygenase induction where involved. Presence of an ah receptor agonist reduced acute hepatobiliary alterations.

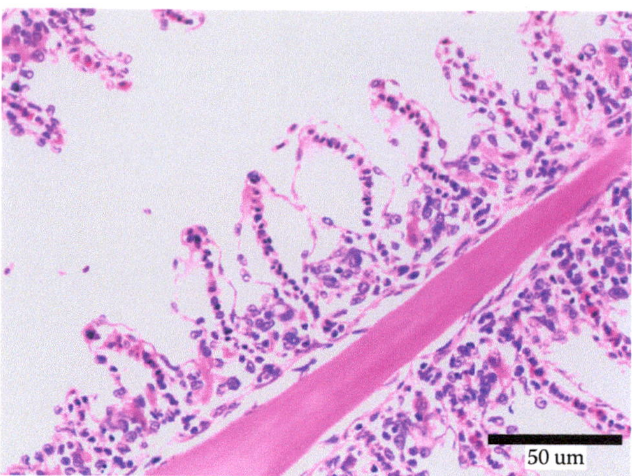

Figure 7.11 Gill epithelial lifting, indicative of edema between the outer epithelial layer and underlying blood sinusoid. (Photomicrograph and general description provided by W. Vogelbein, Virginia Institute of Marine Science.)

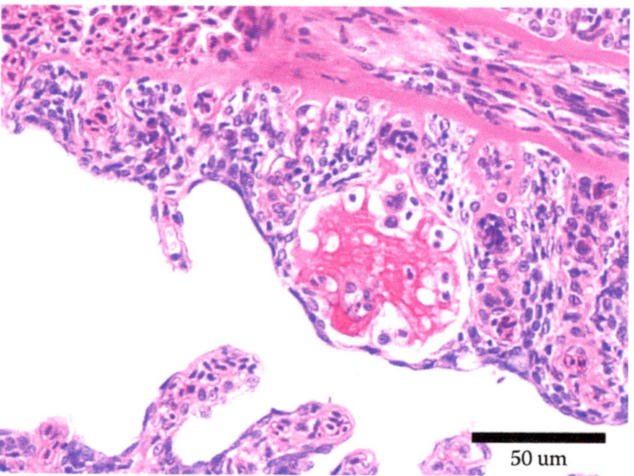

Figure 7.12 Mild hyperplasia and thrombosis if a lamellar sinusoid of mummichog (*Fundulus heteroclitus*) gill. (Photomicrograph and general description provided by W. Vogelbein, Virginia Institute of Marine Science.)

important example that includes some of the nongenetic and genetic (cancer-associated) cellular effects described earlier.

As one good example of using histological biomarkers of the liver, Stentiford et al. (2003) examined the qualities shown in Table 7.1 of flounder (*Platichthys flesus*) from four British estuaries that differed in intensity of pollution. The Alde site was relatively uncontaminated but the remaining three sites were described as industrially impacted, having high sediment PAH concentrations. The presence of EROD and PAH adducts were also noted in fish from the three polluted estuaries.

Table 7.1 Example of Histopathological Biomarkers

Lesion Type	Brief Description
Non-neoplastic	
Phospholipidosis	Excess accumulation of phospholipids in cell
Fibrillar inclusions	Inclusion of fibrils
Cell/nuclear polymorphism	More than one distinct form of cell or nucleus
Spongiosis hepatic	Lesion involving perisinusoidal liver cells with a distinct, mesh-like morphology (spongiotic structures)
Benign Neoplastic	
Hepatocellular adenoma	Adenoma[a] involving liver cells
Cholangioma	Benign neoplasm involving bile duct cells
Hemangioma	Benign neoplasm involving blood vessels
Pancreatic acinar cell adenoma	Benign exocrine pancreatic neoplasm (see Fournie and Hawkins, 2002 for a detailed fish example)
Malignant Neoplastic	
Hepatocellular carcinoma[b]	Malignant neoplasm involving liver cells
Cholangiocarcinoma	Malignant neoplasm of the liver involving bile ducts
Pancreatic acinar cell carcinoma	Malignant neoplasm of the exocrine pancreas
Mixed hepatobilary carcinoma	Malignant neoplasm involving a mixture of cell types
Hemangiosarcoma[c]	Malignant neoplasm directly supplied with blood from vessels and often filled with blood
Hemangiopericytic sarcoma	Malignant vascular neoplasm originating in the **pericytes**[d] of the capillaries
Altered Cell Foci	
Clear altered cell foci	Areas (foci) of altered cells that appear clear
Vacuolated altered cell foci	Foci of altered cells containing numerous vacuoles
Eosinophilic altered cell foci	Foci of eosinophilic altered cells
Basophilic altered cell foci	Foci of basophilic altered cells
Mixed altered cell foci	Foci of mixtures of the above altered cells

Note: These lesions were chosen by Stentiford et al. (2003) in their study of contaminant effects to flounder (*Platichthys flesus*) liver. Other biomarkers not listed were general necrosis, apoptosis, lymphocyte or monocyte infiltration, granuloma, and melanomacrophage centers.

[a] An **adenoma** is a benign epithelial tumor of glandular tissue. An adenoma that becomes malignant is called an **adenocarcinoma**.

[b] A **carcinoma** is a cancer involving epithelial cells that tends to be malignant, i.e., spread over or into the organ or nearby organs, and to metastasize to more distant organs.

[c] A **sarcoma** is a malignant cancer involving mesodermal cells such as those of muscle, cartilage, blood vessels, fat, or bone.

[d] **Pericytes** are a type of cell associated with capillaries or small blood vessels.

Non-neoplastic, and both benign and malignant neoplastic lesions were examined as histopathological biomarkers (Table 7.1 and Figure 7.13).

Several similar examples have been published such as those associated with the contaminated Elizabeth River estuary (Mulvey et al. 2002, 2003; Ownby et al. 2002). The primary research focus of the works of Mulvey and co-investigators was the impact of contamination on genetic qualities of population of the fish, *F. heteroclitus*. However, liver lesion data were also collected for *F. heteroclitus* in this same estuary (Vogelbein and Unger, 2006) that suggested linkage between sediment PAH concentrations and preneoplastic and neoplastic liver lesions. Although PAH was probably the cause of these lesions, no serious exploration of other possible factors contributing to lesion manifestation was done for Elizabeth River *F. heteroclitus* such as the careful field surveys by Myers et al. (1994) of liver lesions of fish taken along the Pacific coast of the United States.

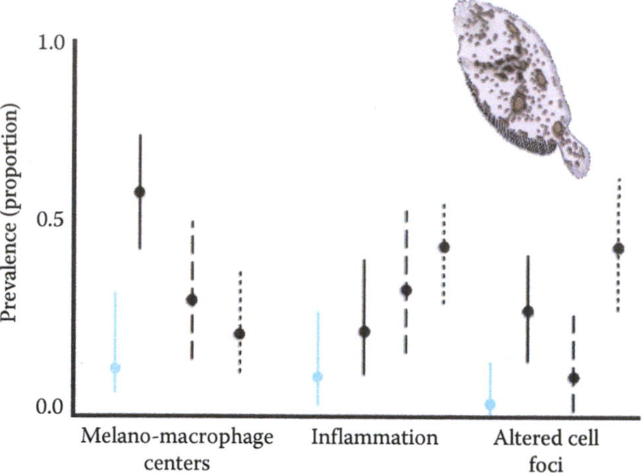

Figure 7.13 Prevalence of flounder liver lesions sampled in autumn from four British estuaries that differ in levels of pollution (Stentiford et al. 2003). Prevalence (proportion of examined fish livers) and sample sizes were extracted from Table 2 of the original publication to generate the 95% (asymmetric) confidence intervals via the Wilson method (Newcombe and Altman, 2000) for the observed proportion. Inflammation and altered cell foci have already been described as biomarkers. The melano-macrophage centers referred to here are centers of pigment-containing macrophages and leucocytes associated with chronic inflammation (Agius and Roberts, 2003; Leknes, 2007). (Alde: Solid blue 95% confidence intervals; Tyne: Solid black 95% confidence intervals; Tees: Long black dash 95% confidence intervals; Merser: Short black dash 95% confidence intervals.)

VIGNETTE 7.2 Polycyclic Aromatic Hydrocarbons and Liver Cancer in Fish

Wolfgang K. Vogelbein
College of William & Mary

Polycyclic aromatic hydrocarbons are ubiquitous environmental contaminants derived largely from the incomplete combustion of organic matter. These large complex organic molecules comprised fused aromatic ring structures, can be divided into two broad classes: the low-molecular weight (LAH: 1–3 benzene rings) and high-molecular weight compounds (HAH: 4–6 benzene rings). Major sources of environmental contamination include industrial dischargers such as metal smelting facilities; wood treatment plants using creosote; oil refining operations; and accidental spills of fuel oil, crude oils, and other petroleum products. Atmospheric emissions from incineration and internal combustion engines (e.g., automobiles) and their associated deposition and urban runoff represent another major source of PAH contamination. In heavily industrialized sites, PAHs often co-occur with a variety of other chemical pollutants, including heavy metals, pesticides, and polychlorinated biphenyls (PCBs).

PAHs, especially the high-molecular weight compounds, are highly lipophilic and rapidly sorb to organic and inorganic matter. Thus, in aquatic environments, PAHs tend to bind to detritus and become immobilized in bottom sediments. Concentrations in surface waters are generally very low. Despite extensive and tenacious adsorption to sediments, PAHs appear to be readily bioavailable to aquatic organisms. Fish are exposed to PAHs from their diet, by aqueous exposure and by direct contact with sediments (Meador et al., 1995a). Many benthic invertebrates, especially those that ingest sediment, can bioaccumulate PAHs, and, therefore, may constitute an important dietary source of PAH in fish (Meador et al. 1995b).

Although sediment PAHs are generally bioavailable and tend to bioaccumulate in some aquatic invertebrates, they do not normally accumulate to any appreciable degree within tissues of aquatic vertebrates such as teleosts. In addition, there does not seem to be any significant biomagnification of PAHs across trophic levels in fish (Suedel et al. 1994). Thus, measured PAH concentrations are generally very low in fish tissues and are not a useful method of assessing environmental exposure. The primary reason for these low tissue concentrations is that vertebrates rapidly and efficiently biotransform hydrophobic contaminants into polar, more water-soluble forms that can then be more easily excreted. Xenobiotic metabolizing enzyme systems such as the cytochrome P-450 monooxygenases and glutathione-*S*-transferases are responsible for biotransformation of these chemicals. Thus, these enzymes serve an important detoxification function and provide a mechanism whereby large hydrophobic and potentially toxic xenobiotics are eliminated from the tissues and the body. In vertebrates, the primary organ responsible for the metabolic detoxification of hydrophobic organic contaminants such as PAH is the liver.

Although they do not generally bioaccumulate in vertebrate tissues, PAHs have been documented to cause significant adverse health effects in exposed organisms. Although metabolic conversion of PAHs serves mainly in detoxification, some intermediate metabolites produced during this process exhibit potent cytotoxic, immunotoxic, mutagenic, and carcinogenic properties. Recent investigations have shown that some PAHs including benzo[a]pyrene, benzfluoranthene, benz[a]anthracene, dibenz[a,h]anthracene, indeno[1,2,3-c,d]pyrene, and dibenzo[a,l]pyrene are highly carcinogenic in laboratory rodents (e.g., National Toxicology Program, 1999). Carcinogenic potential of some of these PAHs has been confirmed in fishes (e.g., Hawkins et al. 1990, 1995; Hendricks et al. 1985; Reddy et al. 1999). In addition, most of these carcinogens have been documented at biologically significant concentrations in PAH contaminated aquatic sediments (Reddy et al. 1999).

Over the past 25 years, numerous studies have been conducted on the adverse health effects of contaminant exposure in wild fish. These studies have shown that some fish populations inhabiting highly industrialized environments exhibit many of the same health impacts documented by PAH exposures of laboratory rodents and fish. An important focus of these studies has been the relationship between chemical exposure, in particular the PAHs, and development of liver cancer in wild fish. In North America, cancer epizootics have been documented in fish from over 40 freshwater and estuarine waterways (Clark and Harshbarger, 1990). These include liver cancers and associated lesions in winter flounder, *Pleuronectes americanus*, from Boston Harbor, Massachusetts (Moore and Stegeman, 1994; Moore et al. 1996), brown bullhead, *Ictalurus nebulosus*, from tributaries of the southern Great Lakes region (Baumann, 1989; Baumann et al. 1991), English sole, *P. vetulus*, and other bottom fish species from Puget Sound, Washington (Myers et al. 1987, 1998) and mummichogs, *F. heteroclitus*, from the Elizabeth River, Virginia (Vogelbein et al. 1990, 1997). PAHs are implicated as causative agents in all of these epizootics. However, the most thorough investigations to date are those conducted on the English sole in the Pacific Northwest. In general, the prevalence of fish liver disease has been found to increase with increasing industrialization and urbanization. Liver lesion prevalences as high as 90% have been documented in fish inhabiting waterways where sediment PAH concentrations are high, whereas lesion prevalences have been very low or inconsequential (<1%) in fish from relatively uncontaminated habitats. Further, PAH levels in benthic prey organisms and fish gut contents (Malins et al. 1984; Myers et al. 1987, 1993), induction of cytochrome P-450-mediated xenobiotic metabolizing enzymes (Van Veld et al. 1992, 1997), DNA adduction in fish tissues (Varanasi et al. 1986), and PAH metabolite levels in fish bile (Krahn et al. 1984) all are reported to be much higher in fish from PAH-contaminated environments than in fish from relatively uncontaminated habitats. These findings strongly support the view that PAHs are bioavailable and that they are the causative agents responsible for the development of liver disease, including cancer in wild fish. However, direct experimental evidence that sediment PAHs cause liver cancer in wild fish is scant. To date, the most direct evidence for this association has been provided by Metcalfe et al. (1988) who induced hepatocellular carcinomas in rainbow trout, *Oncorhynchus mykiss* by microinjecting fry with sediment extracts from Hamilton Harbor, Ontario, and by Schieve et al. (1991) who induced altered hepatocellular foci (AHF) in English sole exposed to an extract of a chemically contaminated marine sediment. Vogelbein and Unger (2006) induced altered hepatocellular foci and hepatic neoplasms in mummichogs following long-term (year-long) laboratory exposure to PAH-contaminated sediments and diet.

The adverse effects of PAH exposure in wild fishes generally comprise a complex spectrum of liver lesions represented here in mummichogs from PAH-contaminated sites in the Elizabeth River, Virginia. Although livers of fish from uncontaminated reference sites appear texturally homogeneous (Figure 7.14a), livers of fish from creosote-contaminated sites in the Elizabeth River exhibit multiple small focal lesions, pigment foci, large tumorous nodules, and fluid-filled cysts (Figure 7.14b). Histologically, livers of fish from uncontaminated sites exhibited typically normal teleost tissue structure (Figure 7.14c). In contrast, mummichogs from the creosote-contaminated sites exhibited a spectrum of lesions that included cytotoxic and specific degenerative cellular changes and a suite of lesions indicative of hepatocarcinogenicity. These included preneoplastic altered hepatocellular foci (Figures 7.14d and 7.14e), large benign neoplasms called hepatocellular adenomas (Figure 7.14f) and hepatocellular carcinomas (Figures 7.14g and 7.14h) (Vogelbein et al. 1990, 1999). In addition, these fish exhibited extrahepatic neoplasms including tumors of the exocrine pancreas, bile ducts, vascular system, kidney, and lymphoid tissues (Fournie and Vogelbein, 1994; Vogelbein and Fournie, 1994; Vogelbein et al. 1997).

Recent field surveys in the Elizabeth River provide additional support for the association between toxicopathic liver disease in mummichogs and sediment PAH concentrations. High liver lesion prevalences in mummichogs corresponded with high sediment PAH concentrations in the most industrialized portions of the southern and eastern branches of the river. In contrast, low prevalences of altered hepatocellular foci and hepatic neoplasms and low sediment PAH levels were observed in the less industrialized, more residential portions of the river, including the western branch, portions of the eastern and southern branches and the Lafayette River. Several major sediment remediation efforts are underway in the Elizabeth River. To document the effectiveness of these site remediation efforts, biological and chemical monitoring efforts are planned. Because the mummichog is nonmigratory, with a summer home range of about 30–40 m (Lotrich, 1975), liver histopathology in this species, in conjunction with sediment PAH analyses, are proposed as an effective method for tracking environmental recovery following

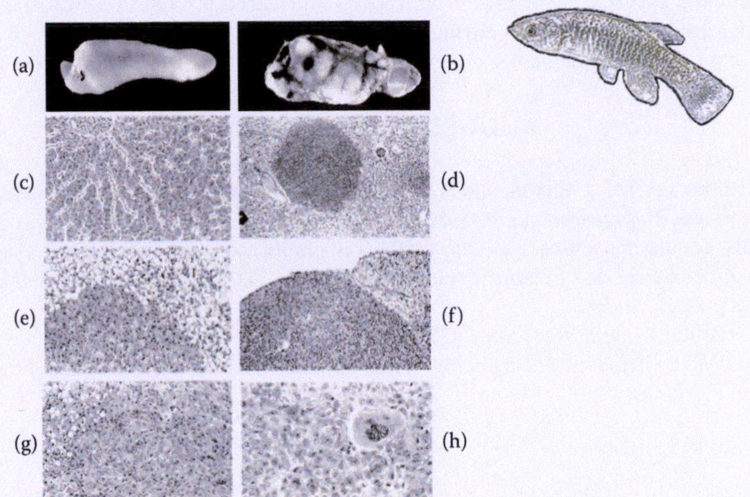

Figure 7.14 Gross and histopathologic anatomy of the liver of mummichogs, *Fundulus heteroclitus*, inhabiting uncontaminated and PAH-contaminated environments in the Elizabeth River, Virginia, United States. (a) Normal healthy liver of mummichog from uncontaminated environment. (b) Liver of mummichog from a PAH-contaminated habitat exhibiting multiple focal lesions, tumorous nodules and cystic lesions. (c) Histologic section of normal healthy mummichog liver showing typical tubulosinusoidal architecture (hematoxylin and eosin; X200). (d) Altered hepatocellular focus (eosinophilic focus) in liver of exposed mummichog (hematoxylin and eosin; X40). (e) High magnification of eosinophilic focus in Figure (d) showing subtle blending of the lesion border with surrounding normal tissue (hematoxylin and eosin; X150). (f) Large hepatocellular adenoma showing swelling of liver capsule and well-demarcated border with normal liver tissue (hematoxylin and eosin; X75). (g) Hepatocellular carcinoma showing sharp locally invasive border with normal liver tissue and cellular and nuclear pleomorphism of the tumor cells (hematoxylin and eosin; X200). (h) High magnification of a hepatocellular carcinoma showing cellular and nuclear pleomorphism typical of poorly differentiated highly malignant neoplasms (hematoxylin and eosin; X400).

specific site remediation efforts. If proposed remediation efforts are effective in cleaning up heavily contaminated portions of the river, then mummichog liver lesion prevalences should be seen to decline significantly over time.

7.7 SUMMARY

In this chapter, effects at the cellular, tissue, and organ levels from exposure to toxicants were described. They included necrosis, inflammation, and specific changes including increased amounts of lipofuscin deposits and metal-laden granules. Damage to genes and chromosomes was described along with several associated assays for damage. Cancer development as a consequence of exposure to chemical and physical agents was discussed briefly and linked to damage due to somatic mutations. Cancer is a clear example of consequences to individuals of cellular damage. Finally, gill and liver responses to toxicants were used as examples integrating several of the nongenetic and genetic cellular changes discussed in this chapter. These examples linked changes at the cellular level to consequences at the cellular (hyperplasia and cancer) or physiological (ion imbalance and decreased oxygen diffusion) to the individual (somatic death) level of biological organization.

The changes described in this chapter are often used as biomarkers and many good reviews provide guidance for their application as such, e.g., Handy et al. (2003). Some reviews are quite specific. Au (2004) and Monserrat et al. (2007) recently reviewed their use as biomarkers for estuarine and marine species. Stentiford et al. (2003) and van der Oost et al. (2003) provide excellent and extensive reviews for fish biomarkers.

SUGGESTED READINGS

Doxon, D.R., A.M. Pruski, L.R.J. Dixon, and A.N. Jha, Marine invertebrate eco-genotoxicology: A methodological overview, *Mutagenesis,* 17, 495–507, 2002.

Hinton, D.E, Cells, cellular responses, and their markers in chronic toxicity of fishes, in *Aquatic Toxicology: Molecular, Biochemical and Cellular Perspectives*, Malins, D.C. and G.K. Ostrander, Eds., CRC Press Inc., Boca Raton, FL, 1994.

Li, A.P. and R.H. Heflich, *Genetic Toxicology*, CRC Press, Boca Raton, FL, 1991.

Newman, M.C. and W.H. Clements, *Ecotoxicology: A Comprehensive Treatment*, Taylor & Francis/CRC Press, Boca Raton, FL, 2008.

CHAPTER 8

Sublethal Effects to Individuals

Our notions of law and harmony are commonly confined to those instances which we detect; but the harmony which results from a far greater number of seemingly conflicting, but really concurring laws, which we have not detected, is still more wonderful.

Thoreau (1854)

8.1 GENERAL

In the last chapter, we discussed effects at the cellular to organ levels including cell death. Death of individuals (somatic death) will be detailed in the next chapter. Sandwiched between these two categories of effects are sublethal effects to individuals. Emphasis in the last chapter was placed on the use of sublethal effects to indicate mode of action or as biomarkers. The emphasis here will be on adverse consequences. Although these sublethal effects are often more difficult to measure than lethal effects, they are likely as, or more, important in determining the ultimate consequences to species under many pollution scenarios.

Sublethal effects are effects occurring at concentrations or doses below those producing direct somatic death. They are most often recognized as some change in an important physiological process, growth, reproduction, behavior, development, or a similar quality. Nearly always, they are adverse or putatively adverse effects that diminish an individual's fitness.

The concept of sublethal effects was formalized first in pharmacology or mammalian toxicology. Although useful and widely applied, the meaning of the term sublethal effect is more ambiguous in ecotoxicology. Some sublethal effects will have lethal consequences in an ecological context, i.e., an arena in which the individual must successfully compete with other species, avoid predation, find food and mates, and cope with multiple stressors. For example, an individual may be able to survive a sublethal exposure but have reduced ability to evade predators or effectively forage for food. In a natural setting, the consequence of a sublethal exposure might be death. The concept of **ecological mortality or death** is used to describe the toxicant-related diminution of fitness of an individual functioning in an ecosystem context that is of a magnitude sufficient to be equivalent to somatic death (Newman, 1995, 2013). It is important to remain open to the very real possibility of ecological mortality at exposures that are judged sublethal based on results from laboratory assays. Also, a sublethal effect that results in an individual that is incapable of producing viable offspring could be considered a lethal effect because the individual's Darwinian fitness (that being, its ability to contribute offspring to the next generation) could be equivalent to that of a dead individual (Rand and Petrocelli, 1985). On the other hand, complex behaviors in the wild such as avoidance of contaminated habitat may ameliorate sublethal consequences of contamination.

8.2 SELYEAN STRESS

The concept of stress is frequently invoked when dealing with sublethal effects. The implication is that sublethal exposure causes individuals to function suboptimally. The concept of stress to individuals must be clarified here because it does not always have the same meaning to different people. When applied to individuals in a medical or scientific sense, stress has a precise meaning that does not correspond directly with its general meaning. This confusion of the scientific/medical and general meanings of stress in the ecotoxicology literature has led to many problems.

Hans Selye formulated the medical concept of stress as applied to individuals about 60 years ago. As a medical student, he noted a nonspecific response of the human body if extraordinary demands were made on it (Selye, 1950, 1973). He referred to this nonspecific response as stress and defined it as "the state manifested by a specific syndrome which consists of all the nonspecifically induced changes within a biological system" (Selye, 1956).* Selyean stress is a specific suite of changes constituting the body's attempt to reestablish or maintain homeostasis while under the influence of a stressor (Adams, 1990).

Selye defined the **General Adaptation Syndrome (GAS)** associated with stress as based on "the tenet that all living organisms can respond to stress as such, and that in this respect the basic reaction pattern is always the same, irrespective of the agent used to produce stress" (Selye, 1950). The GAS is often superimposed on the specific effects of a toxicant or some other agent. It has three phases: (1) the alarm reaction, (2) adaptation or resistance, and (3) exhaustion phases (Figure 8.1). All phases of the GAS serve to resist deviation from, or to regain, homeostasis (Selye, 1950, 1956; Newman and Clements, 2008). In the alarm component, human blood pressure and heart rate increase as short-term responses to compensate for an immediate stressor. Also in humans, adrenal cortex secretory cells release their granules and hypoglycemia (low blood glucose levels) can appear. Because these short-term, energy-intensive mechanisms to resist the effect of the stressor cannot

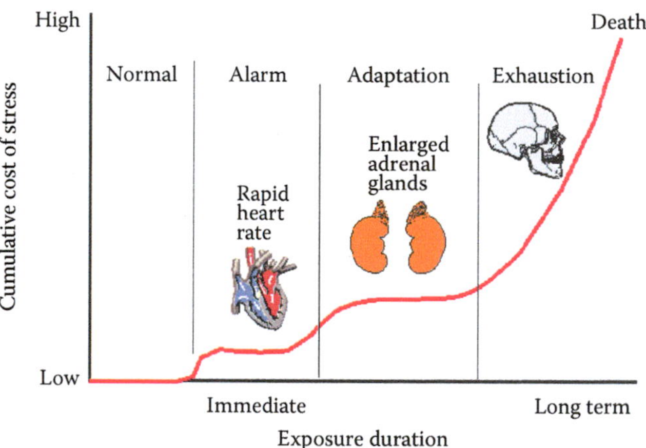

Figure 8.1 The three stages of the General Adaptation Syndrome (GAS). In the initial alarm phase, immediate responses to stress include increases in blood pressure and heart pumping rate. Cells of the adrenal cortex also discharge their granules into the blood. With longer durations of exposure, adrenal enlargement occurs and cells of the adrenal cortex may become rich in granules again (Selye, 1973). Finally, after sufficient exposure, the ability of the body to compensate for the stressor's effects is exceeded and the individual slowly becomes exhausted. If the stress continues, the individual will die.

* We will further qualify stress to an individual as **Selyean stress** so as to distinguish it from other applications of the term in this book and throughout the literature.

be maintained indefinitely, the organism may enter the second stage if stress persists. Responses that enhance tissue level compensation such as enlargement of the adrenal cortex are typical of the adaptation stage; however, generally adverse responses such as shrinkage of the thymus, lymph nodes, and spleen, and the appearance of gastric ulcers can also occur for mammals stressed to this phase (Selye, 1973; Chrousos, 1998). The granules that disappeared from adrenal cortex cells will reappear at this stage. If exposure to enough stressor continues for sufficient time, the individual's ability to resist change is exceeded and it gradually moves into an exhaustion phase. In this last stage, the individual slowly fails in its efforts to compensate for the effects of the stressor and will eventually die if exposure continues.

The complex neuroendocrine dynamics associated with stress are too involved to be discussed here (see Newman and Clements, 2008, pp 138–140). The interested reader is directed to Chrousos (1998) and Tsigos and Chrousos (2002) who produced excellent summaries of the general neuroendocrinology of stress. Regardless of its complexity, stress-related changes are clear in reports of toxicant-induced shifts in the hypothalamic–pituitary–adrenal or hypothalamic–pituitary–interrenal axes (e.g., Gendron et al. 1997; Hontela, 1998; Benguira and Hontela, 2000; Koch et al. 2006; Newman and Clements, 2008).

It is important to understand that a definitive characteristic of Selyean stress is that it is a specific syndrome. Phenomena such as necrotic damage were not included within the stress concept as originally proposed by Selye: the term, Selyean stress, should not be used in describing these effects. Necrosis constitutes damage, not a response to a stressor. It is important to keep clear these distinctions among Selyean stress, damage, specific responses to, and effects of toxicants. Most of the sublethal effects described in the remainder of this chapter are not cases of Selyean stress, yet they might be discussed as such elsewhere in the ecotoxicological literature. And, to emphasize an important point, nonstress effects of toxicants might coexist with stress-related effects (Selye, 1950). Because of the central role of the endocrine system in Selyean stress, this certainly can be the case for endocrine disruption* that manifests from exposure to toxicants.

8.3 GROWTH

Growth is often chosen as the response variable to measure sublethal effects. Not only is it easy to measure but it also integrates a suite of biochemical and physiological effects (Mehrle and Mayer, 1985) into one quality that is often associated with individual fitness. A few examples can show how effectively growth can be used for diverse species. Growth is reduced for fish held under acidic conditions or green tree frog (*Hyla cinerea*) tadpoles exposed to a combination of low pH and elevated aluminum concentrations such as those found in water bodies impacted by acid rain (Jung and Jagoe, 1995). Growth of the aquatic macrophyte, *Juncus effuses*, is altered by mixtures of agrochemicals (Lytle and Lytle, 2005). Woltering et al. (1978) measured reduced growth of largemouth bass (*Micropterus salmoides*) exposed to ammonia and attributed the reduced growth to decreased feeding rate. Growth was reduced for redbreast sunfish (*Lepomis auritus*) inhabiting a stream contaminated with mixed waste (Adams et al. 1992). In contrast, white sucker (*Catostomus commersoni*) from a metal-contaminated lake had higher growth rates but reduced longevity relative to white suckers from a nearby, uncontaminated lake (McFarlane and Frazin, 1978). Some toxicants reduce growth in fish but differences in body size between exposed and nonexposed individuals may disappear later due to compensatory growth (Sprague, 1971).

The dose–response relationship for toxicant-influenced growth often conforms to a threshold model (Figure 7.8), but this is not always the case. Sometimes, a stimulatory effect is shown with

* Environmental pollutants that interfere with the normal functioning of human and nonhuman animal endocrine systems are called **endocrine disruptors** (or **modifiers**).

exposure to low, subinhibitory levels of toxicants or physical agents. Such **hormesis*** is not usually a toxicant-specific response. Instead, it is a general phenomenon that appears during exposure to a variety of stressors (Calabrese, 2008, 2010) (Vignette 8.1). For example, peppermint (*Mentha piperita*) exposed to the growth inhibitor, phosfon, grew fastest at the lower concentrations of 2.5×10^{-12} to 2.5×10^{-8} M (Calabrese et al. 1987). Faster growth of peppermint at these low concentrations than in the control group yields a biphasic dose–response. The **biphasic dose-effect model** is shaped like the threshold model in Figure 7.8, but the curve dips down from the controls before increasing with dose. In this case, the downward dip would reflect enhanced growth of plants exposed to low concentrations but, in other cases, it might reflect decreased mortality or increased fecundity at low concentrations relative to controls. Stebbing (1982) and Sagan (1987) suggest that the mechanism for hormesis may be regulatory overcompensation or an over response of organisms after subinhibitory challenge. Regardless of the mechanism, hormesis has been documented for survival, growth, seed germination, cancer incidence, antibody titer, and fertility in individuals exposed to a variety of agents including toxicants and radiation (Stebbing, 1982; Sagan, 1987; Doust et al. 1994; Fagin, 2012) (Vignette 8.1).

Interest in hormesis is not restricted to nonhuman species. Indeed, hormesis is a foundation concept of **homeopathic medicine**, a branch of medicine not given full attention by many physicians in North America but practiced to varying degrees throughout the world. Homeopathic medicine, founded by Samuel Hahnemann, is based on the **law of similars** (a drug that induces symptoms similar to those of the disease will aid the body in defending itself by stimulating the body's natural responses). By stimulating these responses (symptoms), a drug is thought to enhance the processes that the individual uses in resisting the disease.

VIGNETTE 8.1 Dose Response: Comparing Hormesis with the Threshold and Linear No-Threshold Models

Edward J. Calabrese
University of Massachusetts Amherst

Throughout the twentieth century, the threshold dose–response has been accepted as the most fundamental and reliable model by which to predict responses across the entire dose–response continuum. This has been the case for the broad spectrum of the biological and biomedical sciences. It was supported by voluminous data and consistent with common experience. The only regulatory exception to this near universal acceptance of the threshold dose–response was in the case of how regulatory agencies, such as the U.S. Environmental Protection Agency (EPA), assessed risk from exposures to carcinogens, whether chemical or radiation. In this case, the EPA made a decision in the late 1970s following the recommendations of the Safe Drinking Water Committee of the U.S. National Academy of Sciences (NAS) to reject the use of the threshold model to assess risks from exposures to carcinogens and to replace it with the use of a linear at low-dose modeling strategy (NAS, 1977; Calabrese, 2013a). This dose–response dichotomy for assessing risks to noncarcinogens and carcinogens through different dose–response models has continued to the present.

During the past three decades, important developments have occurred that suggest that dose–response strategies and methods for assessing carcinogens, and even noncarcinogens may be incorrect. First, technological advances have been profound especially in the area of

* Calabrese (2008) recently defined hormesis generally as "a biphasic dose–response phenomenon characterized by a low-dose stimulation and high-dose inhibition."

chemical analysis with detection limits being progressively lowered far below that to which exposure standards were designed to protect against. This has raised questions about whether very low doses of contaminants to which humans and other species are commonly exposed could have adverse biological effects and has led to extensive toxicological testing over a much broader range of doses than in previous decades. The data from such studies have revealed that reproducible stimulatory responses to doses below toxicological thresholds commonly occur. These biphasic dose–responses had been previously identified in the early 1940s by Southam and Ehrlich (1943) and referred to as hormesis. Second, in addition to interest in lower doses as a result of analytical advances, the early 1980s introduced the concept of cell culture and *in vitro* toxicology, leading to test systems involving 96-well plate assays in which a compound could be assessed across 10 to 11 concentrations and compared to its unexposed control and/or sham control. The *in vitro* toxicology methodology introduced a practical means to also assess many doses in the below toxic threshold zone. Consistent with the above findings, many of these experiments revealed the reliable occurrence of hormetic-like biphasic dose–responses, contradicting expectations of the threshold dose–response model. These findings have been reported for numerous endpoints including a wide range of immune responses (Calabrese, 2005a), human tumor cell line responses (Calabrese, 2005b), proliferative responses in numerous yeast cell lines (Calabrese et al. 2006) as well as a broad range of responses in other biological models and tissues (Calabrese and Blain, 2005, 2009, 2011). Third, findings also were consistent with similar developments that were unfolding in the pharmacological sciences in which biphasic dose–responses were becoming ever more commonly reported with endogenous and exogenous agonists across essentially all receptor systems. These numerous observations lead Szabaldi (1977) to propose a receptor-based mechanistic model to account for the occurrence of the biphasic dose–responses that could be broadly generalized. Fourth, the concepts of adaptive (Samson and Cairns, 1977) and preconditioning (Murry et al. 1986) responses were independently being discovered during this period within the toxicological and biomedical sciences, respectively. Careful evaluation of the dose–response relationships of the adapting and preconditioning doses also revealed the occurrence of biphasic dose–responses consistent with the hormetic dose–response model (Calabrese, 2007, 2008; Calabrese et al. 2007, 2013).

These converging and collective findings were reflected in the need to reevaluate the nature of the dose–response, especially in the low-dose zone. Although the data clearly indicated that hormetic dose–responses were real and reproducible, it was not evident how frequently they might occur in the toxicological and pharmacological literature. This lead to an assessment of over 21,000 articles from leading toxicological journals from the mid-1960s to the present using highly rigorous *a priori* entry and evaluative criteria to evaluate the frequency of hormesis in the toxicological literature (Calabrese and Baldwin, 2001). Hormetic dose–responses occurred nearly 40% of the time, making it a dose–response of considerable significance. Because the evaluative criteria were rigorous, it was thought that the actual frequency may be considerably higher. In a follow-up evaluation with the same data set, the hormetic dose–response was far more common than the threshold dose–response, occurring some 2.5 times more often than its rival threshold model (Calabrese and Baldwin, 2003).

These findings provided a strong challenge to the threshold model, but also to basic toxicological thinking. The most fundamental principle in toxicology is the dose–response relationship and these data indicated that the field had somehow made a mistake on its most fundamental and central tenet (Calabrese, 2011). Of particular significance is that even more extensive follow-up head-to-head comparisons of the hormetic dose–response model with the threshold model have confirmed the initial findings that not only did the hormetic dose–response accurately predict responses in the low-dose zone, far outperforming the traditional threshold and linear no-threshold (LNT) models, but that these two models performed quite poorly and in an unacceptable manner (Calabrese et al. 2006).

While the term **hormesis**, which means to excite in Greek, had its origin with the 1943 paper of Southam and Ehrlich, this concept was originally proposed nearly 100 years earlier in the work of Rudolph Virchow (Henschler, 2006). Subsequently, it was made more visible, yet extremely controversial, by Hugo Schulz (1887, 1888) (Figure 8.2) who associated this dose–response concept with the medical practice of homeopathy, thinking it provided its underlying explanatory principle. Unfortunately, the association of this biphasic dose–response with homeopathy led to Schulz and his dose–response theory becoming the objects of considerable criticism and frequent ridicule (Clark, 1927, 1937). Thus, at a critical time of concept consolidation in the pharmacological and toxicological disciplines, the hormetic biphasic dose–response, as promoted by Schulz, became marginalized, being supplanted by the threshold dose–response model, which became the mainstay of food and drug and environmental governmental regulatory activities. The censoring of the hormetic dose–response became nearly complete as it became excluded from all major textbooks in pharmacology and toxicology, was never included as a session at major professional society meetings, was excluded from funding priorities by federal agencies, and was deemed as functionally nonexistent by initial biostatistical modeling that constrained dose responses to be forced through the origin even when data were clearly below control values, a process continuing to the present day (Calabrese, 2005c,d, 2011).

Figure 8.2 Hugo Schulz (1853–1932). Pioneer in the study of hormesis.

The **hormetic dose–response**, as noted above, is a biphasic dose–response characterized by a low-dose stimulation and a high-dose inhibition. The dose–response is frequently plotted as an inverted U-shaped dose–response (Figure 8.3). This is typically the case when the endpoints measured are ones such as growth, cognition, and longevity. However, the hormetic dose–response can also be plotted as J-shaped as seen when assessing disease incidence. The hormetic stimulation in the low-dose zone is believed to represent a modest overcompensation to a disruption in homeostasis following the induction of stress or injury, making it a dose–time–response relationship. Unless the temporal component is incorporated into the study design, the hormetic stimulation could be easily missed. The requirement of the use of numerous doses and the incorporation of the temporal elements places heightened resource and time constraints on investigators, often resulting in the reduced likelihood of discovering and assessing hormesis.

The hormetic dose–response, including its quantitative features, has been found to be very general, being independent of biological model, endpoints measured and the chemical class/physical agent (Calabrese and Blain, 2005). The fact that hormetic dose–responses are independent of biological model and have similar quantitative features indicates that it is adaptive in nature and highly conserved. The quantitative features of the hormetic dose–response indicate that the maximum stimulation is consistently modest being only 30%–60% greater than the control. The reasons for the modest magnitude of the stimulatory response have not been assessed in detail but suggest that this process reflects an efficient regulation of biological resources. It also indicates that modest and properly timed and integrative adaptive responses can be biologically leveraged to protect against massive exposures subsequent to the initial adapting dose. The width of the stimulatory response, while usually ranging from 1/10 to 1/20 of the threshold dose, can at times extend over a very broad dose range approaching and exceeding 1000-fold, although this is usually no more than several percent of the many thousands of examples studied. The reason for the greater variability in the width of the stimulatory response is likely related, at least in part, to the degree of heterogeneity of the study population.

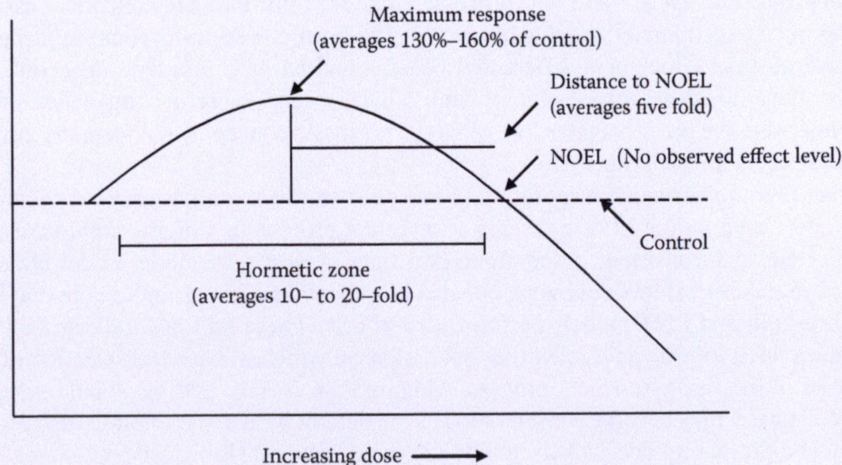

Figure 8.3 Dose–response curve depicting the quantitative features of hormesis.

Hormesis occurs in essentially all members of the population, ranging from the very susceptible to the highly resistant. The only difference is that the hormetic dose–response is shifted either to the left for the very susceptible and to the right on the graph for the highly resistant. The quantitative features of the dose–response, however, remain similar, regardless of susceptibility.

Hormesis also occurs across agents regardless of their toxicological potency. It has been well-established that even when agents differ by many orders of magnitude with respect to toxic potential, the quantitative features of the hormetic dose–responses are likewise similar in their quantitative features.

The concept of chemical interactions, including additivity and synergism, has been assessed with respect to hormesis. The key to understanding chemical interactions with hormesis is that the response relates to increases in performance, not toxicity, as is the case in the traditional toxicological models of interactions. The magnitude of the stimulation in hormetic chemical interactions is still constrained to be within the 30%–60% increase range over the controls. Thus, the "major" change seen in hormetic synergy occurs with dose but not with response. This is a novel finding and one that has important toxicological and biomedical implications.

The concept of hormesis can have important implications for the risk assessment process (Cook and Calabrese, 2006). First, the low-dose stimulation has the potential to be a beneficial or an adverse effect. If the response were deemed an adverse effect it could result in the toxic threshold being lowered. An example of an adverse effect in the hormetic stimulatory zone might be the enhancement of an immune response that would be considered adverse, such as an increase in autoimmune response or the stimulation of proliferation of an organ weight/size such as with the prostate. Examples of potential beneficial effects could be an increase in growth or longevity, or a decrease in disease incidence. Sometimes a low-dose stimulation might not be of sufficient magnitude to have known biological consequences for the individual or the population. If the low-dose stimulation were considered a beneficial effect it should have the potential to affect the risk assessment process. At the present time the EPA (2004) does not take into account the occurrence of adaptive and/or hormetic responses in their risk assessment methodology, being only concerned with potential adverse effects. At the same time, EPA has a long history of conducting cost–benefit analysis of exposure assessment standards. In the case of beneficial hormetic effects, these would have the potential to significantly affect assessment of costs and benefits. Thus, these two processes that are built into EPA chemical assessment procedures need to be more effectively integrated. The hormetic dose–response model provides an objective means to document differential benefits and harm within heterogeneous populations across the entire dose–response continuum. This is a very powerful contribution to the risk assessment evaluative process that is lost when the hormesis concept is not formally considered as is the case at the present time.

The fact that the hormetic dose–response has been shown to outperform the current set of default models used by the EPA in the risk assessment process in a highly significant manner is an important and consistent observation. Not only does the hormetic model make better predictions in the critical low-dose zone but also it is even more important to note that the long revered threshold and LNT models perform very poorly. These findings indicate that regulatory agencies need to provide a scientific foundation on which to base the selection of default models used in the risk assessment process. Although the current findings would suggest that the hormetic model may deserve consideration as the default model, the findings also reveal that the threshold and LNT models lack the ability to provide accurate low-dose predictions, thereby failing to deliver on a critical societal need.

The hormetic dose–response model is finally receiving the chance to be studied and evaluated by the broader scientific community after nearly a century of suppression, as a result of its being unnecessary collateral damage in the long and often bitter conflict between traditional medicine and homeopathy. In fact, numerous examples of hormetic mechanisms have been documented in the toxicological and pharmacological literature (Calabrese, 2013b). One must wonder what the history of the dose–response might have been in the twentieth century if Hugo Schulz had simply reported his findings on the occurrence of biphasic dose–responses and not have linked it to homeopathy, thereby bringing an avalanche of criticism, ridicule, and lost opportunity. Society will benefit from a fresh look at hormesis in these initial decades of the twenty-first century.

8.4 DEVELOPMENT

8.4.1 Developmental Toxicity and Teratology

Some contaminants can adversely impact the developing embryo, fetus, or immature individuals such as larvae and juveniles. This can result in death (embryo lethality), anatomical malformation, functional deficiencies, or slowed growth. In some cases, there are critical periods in embryo or fetus development when a particular developmental effect can occur but this is not always the case. The *in utero* effect of mercury on humans was mentioned briefly in our earlier discussion of the Minamata disease. *In utero* effects of mercury include macrocephali (abnormally large head), asymmetrical skull, depressed optical region of the skull, lowered IQ, poor muscular coordination, hearing loss or impairment, poor speech, poor walking skills, and mental retardation (Khera, 1979). Any physical or chemical agent such as mercury that is capable of causing developmental malformations is called a **teratogen. Teratology** is the science of fetal or embryonic abnormal development of anatomical structures. Notice that some of the qualities mentioned above for mercury involve functional deficiencies, i.e., lowered IQ, that may not be considered in classic teratology. Also, some contaminants slow growth of the developing organism but do not produce an anatomical abnormality. The broader term, **developmental toxicity** is often used to include altered growth, functional deficiencies and even death in addition to teratogenic effects (Weis and Weis, 1989a; Rogers and Kavlock, 2008).

Teratogenic effects are often thought to have a threshold: a critical amount or concentration of teratogen is needed before an effect manifests (Weis and Weis, 1989a). This is due to "the high restorative growth potential of the … embryo, cellular homeostatic mechanisms, and maternal metabolic defenses" (Rogers and Kavlock, 2008).

Although some teratogens, such as the infamous thalidomide,* are specific in their action, most teratogenic contaminants are believed to be relatively nonspecific. According to **Karnofsky's law**, any agent will be teratogenic if it is present at concentrations or intensities producing cell toxicity during development (Bantle, 1995). Teratogens act by interfering with genomic imprinting,† disrupting mitosis, interfering with transcription and translation, inter- and intracellular signaling

* In the early 1960s, the sedative, thalidomide was given to pregnant European women to treat nausea. Babies were born without limbs **(amelia)** or with limbs having reduced bone lengths **(phocomelia)**. It soon became apparent that abnormalities occurred if it was taken during a critical 2-week period of active limb morphogenesis (McBride, 1961; Lenz, 1968, 1996; La Via and Hill, 1971). Roughly, 10,000 children were born with severely malformed limbs. This event stimulated revision of drug testing procedures. Although it has become symbolic of the tragic consequences of drug use prior to extensive testing, thalidomide is prescribed more judiciously today as an immunosuppressant and is considered for treatment of some symptoms of AIDS including associated cancers (Richardson et al. 2002).

† **Genomic imprinting** occurring during gametogenesis provides the mechanism by which an individual only inherits one functional copy of specific genes from the two parents, that is, so that either the paternal or maternal gene is expressed, but not both (Villar et al. 1995).

or enzyme activity, disturbing metabolism, or producing nutritional or substrate deficits (Weis and Weis, 1987; Rogers and Kavlock, 2008). Consequences of these disruptions include abnormal cell interactions, migration and growth, and excessive or inadequate cell death. In general, effects early in development tend to be more deleterious than those occurring later because early damage affects cells that will go on to differentiate and to become involved in a wider range of organs and tissues (Bantle, 1995). The dose can also influence the nature of the effect: low doses of a toxicant might cause growth retardation but higher doses might cause malformation (Rogers and Kavlock, 2008).

Many developmental toxicology studies focus on effects occurring during or after exposure of the egg to contaminants. Exposure may occur across the placenta, from contaminants deposited in the yolk, from egg exposure before fertilization, between egg shedding and elevation of the chorion, or after elevation of the chorion (Weis and Weis, 1989a). However, evidence suggests that some congenital diseases and birth defects in humans might also be linked to male exposure to contaminants (Gardner et al. 1990; Stone, 1992). For example, men working at the Sellafield nuclear fuel processing plant on the Southwest coast of England had a higher incidence of offspring with leukemia than control groups (Gardner et al. 1990), notionally due to a chromosomal translocation that activated proto-oncogenes (Evans, 1990; Kondo, 1993). Previously, such **male-mediated toxicity** (disease and birth defects produced by paternal exposure to a physical or chemical agent) had been judged to be insignificant based on epidemiological studies of atomic bomb survivors of Hiroshima. However, careful study has shown several instances of male-mediated toxicity in humans (Anderson, 2005). Stone (1992) reports that males in some professions (painters, mechanics, and farmers) may be at higher risk of fathering children with teratogenic problems. Spouses of pesticide applicators (Petrelli et al. 2000) and stainless steel welders (Hjollund et al. 2000) experienced elevated incidence rates of spontaneous abortions.

A wide range of developmental problems has been described for organisms exposed to contaminants. Metamorphic delay, thyroid gland, and other morphological abnormalities can occur in developing invertebrates, amphibians, and fishes exposed to xenobiotics that disrupt endocrine function (Omura et al. 2001; Marcial et al. 2003; Boudreau et al. 2004; Degitz et al. 2005; Fenske et al. 2005; Tietge et al. 2005; Wollenberger et al. 2005). For fish, the most common developmental problems are those of the skeletal system and associated musculature, circulatory system, optical system, and retardation of growth (Weis and Weis, 1989a; Moreels et al, 2006). Skeletal system problems include **scoliosis** (the lateral curvature of the spine) and **lordosis** (the extreme, forward curvature of the spine) (Figure 8.4). Another problem reported with exposure to contaminants such as oil spill-related polycyclic aromatic hydrocarbons (PAH) is **pericardial edema** (fluid accumulation in the pericardium, i.e., the sac surrounding the heart) (Incardona et al. 2006, 2011) (Figure 8.5). Slowing of growth may increase the time a developing individual remains in a critical stage and, consequently, the likelihood of a problem becoming manifest (Weis and Weis, 1989a). Birds exposed to contaminants produced embryos with a variety of eye, limb, beak, heart, and brain abnormalities (Ohlendorf et al. 1986; DeWitt et al. 2006) or had increased incidence of egg failure (Henny and Herron, 1989). Weis and Weis (1995) showed that **behavioral teratology** (behavioral abnormalities in otherwise normal appearing individuals arising after exposure to an agent as an embryo) can also be important. They showed that exposure of mummichog (*Fundulus heteroclitus*) embryos to methylmercury lowered the ability of this fish to capture prey after hatching.

There are standard assays for measuring developmental effects of environmental contaminants. The widely accepted frog embryo teratogenesis assay-*Xenopus* (**FETAX**) uses embryos of the clawed frog, *Xenopus laevis*. Although the test species is not native to North America nor Europe, the convenience of producing ample amounts of eggs and sperm, well-established procedures for the assay (e.g., Bantle and Sabourin, 1991; ASTM, 1993), and the large data base for this species make it an appealing tool (Bantle, 1995). In the assay, eggs are exposed to different concentrations of the contaminant for a set time, e.g., 96 hours. The proportions of exposed eggs showing mortality and the proportion of living embryos with developmental abnormalities are scored for each

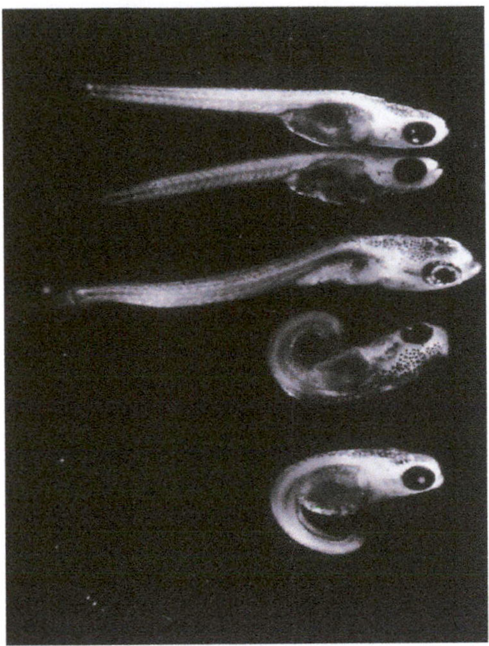

Figure 8.4 Fish (*Fundulus heteroclitus*) exposed to no (top three individuals) or 10 mg·L^{-1} of Pb^{2+} (bottom two individuals) during development. The effect threshold is typically 1 mg·L^{-1} of Pb^{2+}. Note the failure of the exposed individuals to uncurl after hatching. (Courtesy of P. Weis, UMDNJ—New Jersey Medical School.)

treatment concentration. Using formal methods described in the next chapter, the concentrations producing 50% mortality of eggs (LC50) and producing 50% of embryos with abnormalities (EC50 or TC50) are estimated. (EC stands for effective concentration and TC stands for teratogenic concentration.) A **teratogenic index** (TI) is calculated as the LC50 divided by the EC50. It reflects the developmental hazard of a contaminant. Higher values indicate increased developmental hazard of the tested contaminant. Bantle (1995) opines that TI values less than 1.5 indicate little developmental hazard. For example, a TI of 2 was derived with fertilized toad (*Bufo arenarum*) eggs exposed to the flavinoid, naringenin, suggesting high developmental hazard (Pérez-Coll and Herkovits, 2004). The most prominent effects were stunted growth, underdeveloped gills and tail, and neurotoxity.

Osano et al. (2002) provide a good example of applying the FETAX assay to understanding environmental risks of chloroacetanilide herbicides and their aniline degradation products. Atypically, *Xenopus laevis* was endemic to the study area about which the risk statements were being made, i.e., agricultural areas of Kenya. (This coincidence inspired the atypically droll line in the author's acknowledgments, "And thank you God for providing the *Xenopus* frogs at my doorstep"!) The agricultural chemicals, alachlor and metolachlor, are widely applied in Kenya. Alachlor and its degradation product, 2,6-diethylaniline, and also metolachlor and its degradation product, 2-ethyl-6-methylaniline were used in assays. Embryotoxicity, growth, and teratogenicity were noted after 96 hour of exposure (midblastula to early gastrula stages of development).* Although alachlor and metolachlor had TI of 1.7 and 0.2, respectively, their degradation products

* Recollect that the solid ball of cells (morula) initially formed by a developing zygote keeps cleaving until the blastula stage is reached. The blastula is a ball of cells surrounding a cavity. The animal gastrula is formed in the next stage (gastrulation) with the invagination of the blastula to form the endoderm, mesoderm, and ectoderm cell layers around a central cavity (archenteron) that becomes the gut.

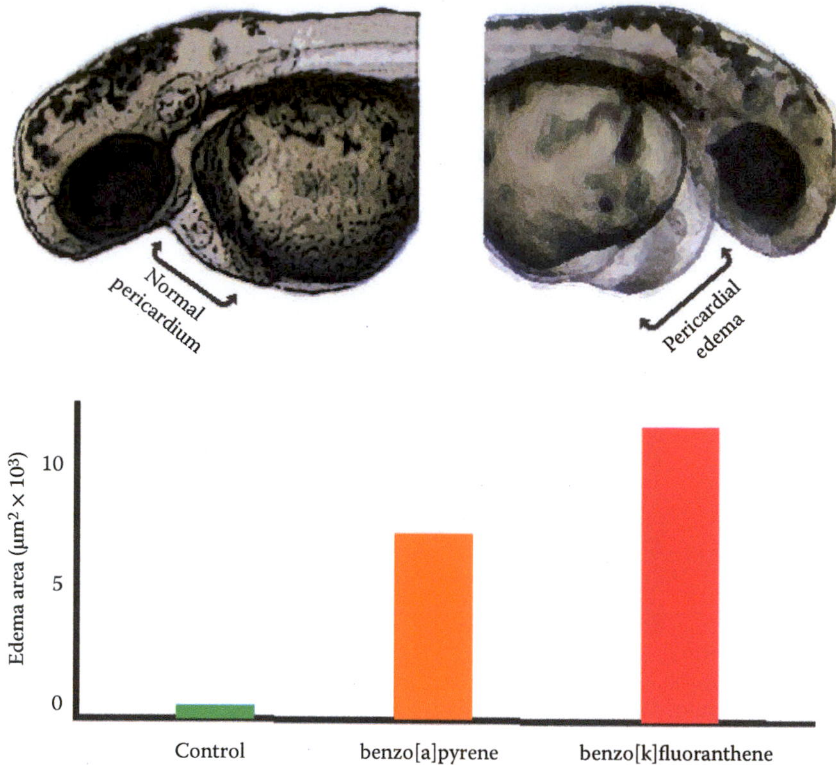

Figure 8.5 Pericardial edema of zebrafish (*Danio rerio*) larvae expressed as increased pericardial area for dosed larvae relative to controls (Incardona et al. 2011). Edema was measured after larvae were exposed from 4 to 48 hours postfertilization to control (green, dimethyl sulfoxide solvent only), 40 μM benzo[a]pyrene (orange), or 40 μM benzo[k]fluoranthene (red). Pencil sketch images show a normal pericardium (left) and an extremely edemic pericardium of a larva that had been exposed to benzo[k]fluoranthene. The difference in toxicity between the two 5-ring PAHs is a consequence of their interactions with the aryl hydrocarbon receptor discussed in Vignette 6.1 as being pivotal to cytochrome P450 induction in vertebrates. Exposure to these PAHs results in different tissue-specific inductions of CYP1A. Functional consequences were noted such as embryo heart rate that decreased from control (128 beats/minutes, bpm) to benzo[a]pyrene-exposed (120 bpm) to benzo[k]fluoranthene (93 bpm) treatment. (Original photographs from Figure 1A and D of Incardona et al. (2011) were digitized, modified electronically, and rendered to pencil sketches to produce the upper illustrations. The data shown in the lower histogram were visually extracted from their Figure 1E.)

had TI of 2.1 (2,6-diethylaniline) and 2.7 (2-ethyl-6-methylaniline). The results for metolachlor illustrated that a chemical with no teratogenic risk can be the source of a degradation product that poses a significant teratogenic risk.

8.4.2 Sexual Characteristics

Fish and reptiles exposed in the field to contaminants can manifest changes in secondary sex characteristics that are thought to potentially diminish individual fitness (e.g., Guillette et al. 1996, 1999; Milnes et al. 2002; Parrott et al. 2003). **Estrogenic chemicals** (xenobiotic estrogens or xenoestrogens) mimic estrogen and can cause changes in the sexual characteristics of individuals. Like estrogen, these chemicals regulate the activity of estrogen responsive genes by binding to estrogen receptors (Jobling et al. 1996). They disrupt hormonal systems, affecting sex organ development, behavior, and fertility. For example, affected male sea gulls might ignore nesting colonies, and females may pair and nest together as a consequence of modified behavior by estrogenic chemicals (Hunt and Hunt, 1977; Luoma, 1992). Indeed, gull egg treatment with DDT results in feminization of the reproductive system of hatched males (Fry and Toone, 1981). Agonistic (estrogenic) substances bind to and have an effect at estrogen receptors. Males might develop female traits or female endocrine-related features might change. Antagonistic or antiestrogenic substances block receptors, preventing normal binding of estradiol (Leung et al. 2002). Estrogenic chemicals include xenobiotics such as DDT, DDE, dioxin, PCB, alkylphenols (e.g., p-nonyl-phenol and the surfactant, alkylphenol polyethoxylate), and pharmaceuticals released in sewage effluent. Some xenobiotic estrogens bind to hormone receptors and induce an abnormal, elevated response but some with minimal estrogenic activity block the normal hormone's action as a consequence of binding (McLachlan, 1993). Kelce et al. (1995) found that some estrogenic chemicals such as DDE can also block androgen receptor-mediated processes, and in doing so, act as **androgen receptor antagonists**.

Bergeron et al. (1994) provided a straightforward demonstration of PCB congeners acting as xenobiotic estrogens. Eggs of the red-eared slider turtle (*Trachemys scripta*) were exposed to either: no PCB, different concentrations of PCB, or the hormone, 17β-estradiol (Figure 8.6). Because this species displays temperature-dependent sex determination, researchers can manipulate the sexes of hatchlings by controlling incubation temperatures. In the experiment, eggs were incubated at a temperature (26 C to 28°C) that would normally produce all males. But the application of 200 μg of 2′, 4′, 6′ -trichloro-4-biphenylol to the eggshell surface resulted in 100% of the hatchlings emerging as females despite the incubation temperature. The estrogenic activity of a second PCB (2′, 3′,4′, 5'-tetrachloro-4-biphenylol) was much lower than that of 2′, 4′, 6′-trichloro-4-biphenylol.

Masculinization of females can also result from exposure to contaminants. **Imposex** (the imposition of male characteristics on females, e.g., a penis or vas deferens) occurred for mosquitofish (*Gambusia* sp.) exposed to Kraft mill effluent (Howell et al. 1980). Females displayed male reproductive behavior and developed gonopodia. (A gonopodium is a modified anal fin that functions as an intromittent organ in males.) Bortone et al. (1989) suggested that mosquitofish

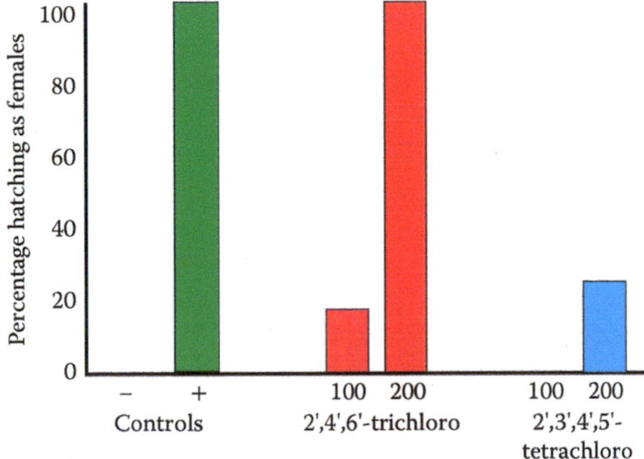

Figure 8.6 The percentage of hatchling turtles that are females after exposure to PCBs. Eggs were incubated at 26°C to 28°C that will cause all hatchlings to be males (negative control). Eggs in the negative control treatment (–) were spotted on their surfaces only with the solvent used for the other treatments (95% ethanol) and those in the positive control treatment (+) were spotted with the hormone, 17β-estradiol The two PCBs spotted onto eggs were 2′, 4′, 6′-trichloro-4-biphenylol (labeled 2′, 4′, 6′-trichloro) and 2′, 3′,4′, 5′- tetrachloro-4-biphenylol (labeled 2′,3′,4′,5′-tetrachloro). Both were dosed at 100 and 200 μg per egg as indicated above the PBC labels. (Modified from Figure 1 of Bergeron, J.M., et al. *Environ. Health Perspect.*, 102, 780–781, 1994. With permission.)

exposed to Kraft mill effluent take in sterols (stigmastanol and β-sitosterol) that have been modified by *Mycobacterium smegmatis* to compounds having androgenic effects. Tributyltin (TBT) compounds used in antifouling paints have also been implicated in high incidence of imposex in marine snails inhabiting coastal regions of North America and England (Saavedra Alvarez and Ellis, 1990; Bryan and Gibbs, 1991). In areas of heavy boat traffic, the occurrence of a large proportion of reproductively incompetent individuals resulted in decimated populations of ecologically important whelk species and consequent shifts in the composition of the ecological community (Bryan and Gibbs, 1991).

VIGNETTE 8.2 Use of Neogastropods as an Indicator of Tributyltin Contamination along South China Coast

Ma Shan Cheung, Helen Y. M. Leung and Kenneth M. Y. Leung
The University of Hong Kong

Over the last 40 years, organotin compounds such as tributyltin (TBT) and triphenyltin (TPT) have been widely used as disinfectants and biocides because of their effective biocidal properties (Bennett, 1996). Among them, TBT is commonly used worldwide as a wood preservative and an antifouling agent in boat paint. Due to its high effectiveness and relative low cost, the production and usage of TBT as an antifouling agent have increased dramatically since the 1970s (de Mora, 1996). Unfortunately, beginning in the late 1970s, TBT has been found by many scientists to be highly toxic to nontarget marine species (Alzieu, 1996; Antizar-Ladislao, 2008). For instance, TBT leached from antifouling paints was corroborated to cause shell deformation, growth reduction, and reproductive failure in the oyster *Crassostrea gigas*; widespread TBT contamination in coastal areas thus resulted in a massive financial loss to the shellfish industry in Europe between the late 1970s to 1980s (Alzieu, 2000).

Astonishingly, TBT is also an endocrine disruptive chemical that can cause female neogastropods (e.g., the dogwhelk, *Nucella lapillus*, and the rock shell, *Thais clavigera*) to develop male sexual characters, i.e., the penis and vas deferense (Gibbs et al. 1987, 1988; Horiguchi et al. 1998). Such an abnormal development in the female neogastropods is called imposex, which is defined as superimposition of male characters onto the female (Gibbs et al. 1987). Over the last 20 years, TBT-mediated imposex has been observed in 150 species of gastropods (Matthiessen et al. 1999). Recent studies have discovered that imposex induction is mediated through the direct binding of TBT or TPT onto the Retinoid X Receptor (RXR) and hence triggering the RXR signaling pathway in *N. lapillus* and *T. clavigera* (Nishikawa et al. 2004; Castro et al. 2007; Horiguchi et al. 2007). The RXR is well-known in mammals as a key factor involved in the mediation of several hormone response systems. However, the effects of imposex on the reproductive activity of the female gastropod vary among different gastropod species. In *Nassarius reticulates* and *Ilyanassa obsolata*, for example, development of imposex does not affect the reproduction of the female (Horiguchi et al. 1994; Gibbs and Bryan, 1996). In other species like *N. lapillus* and *T. clavigera*, the development of vas deferens tissues can block the opening of the vulva, i.e., opening of oviduct, at the late stage of imposex, resulting in failure in egg capsule release. This causes sterilization and leads to mortality of imposex females (Bryan et al. 1987). In severe cases, such a lethal consequence can significantly reduce the number of sexually mature females in the wild population, and even lead to local extinction; the latter is commonly observed in areas nearby marinas, harbors, ports, and major navigation channels where the ambient TBT concentration is high (Matthiessen et al. 1999; Leung et al. 2006).

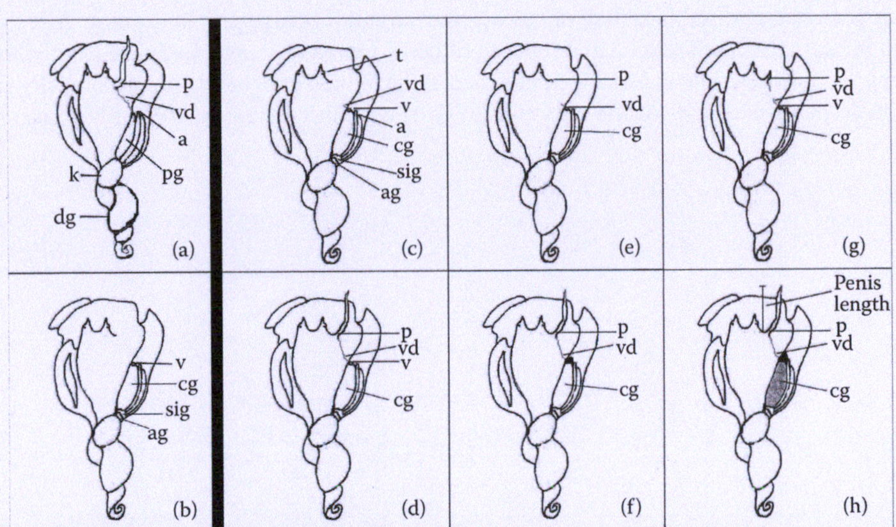

Figure 8.7 Anatomical features of (a) a normal male, (b) a normal female, and (c–h) the effects of imposex on the reproductive systems of female *Thais clavigera* at different vas deferens sequence index (VDSI) stages: (c) develops vas deferens at the region ventral to the genital papilla (stage 1); (d) penis development starts behind the bottom of right tentacle (stage 2) ; (e) develops distal region of vas deferens at the base of small penis (stage 3); (f) development vas deferens completes and the penis enlarges to the size similar to male (stage 4); (g) vulva is displaced by the overgrowth tissue of vas deferens and become invisible, blister like protuberances may appear around the site of papilla (stage 5); (h) lumen of capsule gland contains material of aborted egg capsule and formed a translucent or brown mass (stage 6). a, anus; ag, albumen gland; cg, capsule gland; dg, digestive gland; k, kidney; p, penis; pg, prostate gland; t, tentacle; v, vulva; and vd, vas deferens. (Modified from Gibbs, P.E., et al. *J. Mar. Biol. Assoc. UK*, 68, 715–731, 1988; Li,Z., The Incidence of Imposex in Hong Kong and the Value of *Thais clavigera* (Gastropoda: Muricidae) as a Bioindicator of TBT Pollution, PhD Diss., The University of Hong Kong, Hong Kong, 2000. With permission.)

Given that the severity of imposex often correlates well with the level of ambient TBT contamination in seawater, determination of imposex stages in female neogastropods is proven to be an excellent, sensitive biomarker to gauge TBT contamination in coastal environments, and to reflect its bioavailability and adverse impacts to the health and fitness of the animals (Hagger et al. 2006). In *Nucella lapillus*, imposex can be induced by TBT at concentrations as low as 0.5 ng Sn·L^{-1} (Bryan et al. 1987). The intensity of imposex can be readily quantified by measuring the relative penis size in female gastropods (i.e., the relative penis size index [RPSI]) and determining the developmental stage of vas deferens (i.e., vas deferens sequence index [VDSI]). Sex of a neogastropod can be determined by identification of the sperm ingesting gland and the capsule gland in females and the presence of the prostate gland in males (Figure 8.7). The RPSI is determined based on a comparison of the female penis size with the male penis size and can be calculated as the ratio of (length of the female penis^3/length of male penis^3) × 100, which provides the relative mass of the gastropod penis (Gibbs et al. 1987). The penis length can be measured to the nearest 0.1 mm for nonnarcotized males and females under a stereomicroscope. The VDSI, which was first developed by Gibbs et al. (1987, 1988) for *N. lapillus* and later adopted for other neogastropod species like *Thais clavigera* (Li, 2000; Leung et al. 2006), has seven stages (0–6), with stage 0 indicating no effect. Using *T. calvigera* as an illustration, various VDSI stages are summarized and shown in Figure 8.7. At stage 1, a proximal infolding first appears at the bottom of the right tentacles of female gastropods, i.e., the initiation of vas deferens development. At stage 2, in addition to the appearance of distal infolding of vas deferens, a small penis starts to develop. After which, the two infolding sections extend and the size of penis increases (stage 3). Subsequently, the two infolding sections fuse into one and the penis size of the female gastropod is approximately similar to the male penis (stage 4). At the next two VDSI stages, the female gastropod becomes sterile

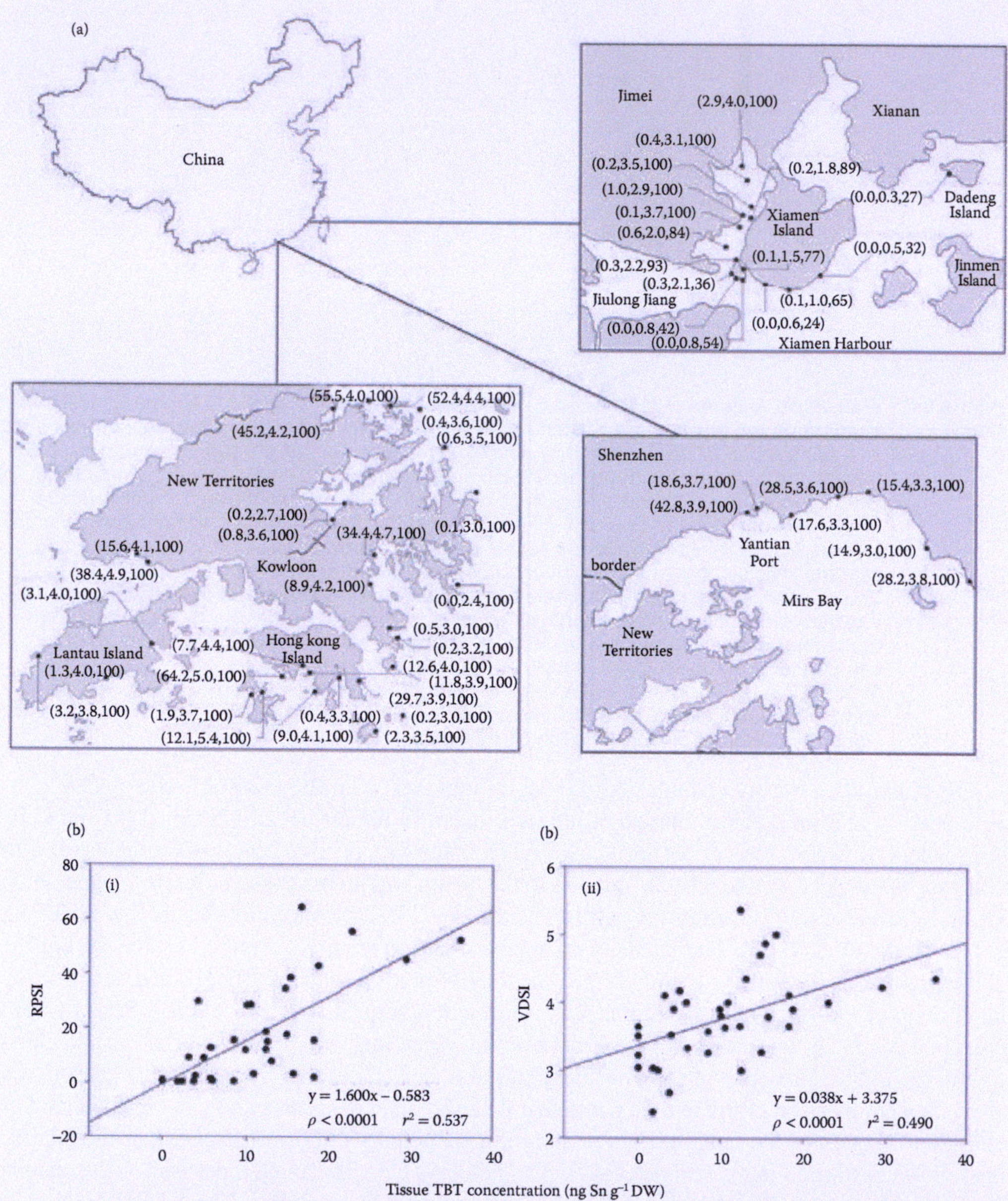

Figure 8.8 (a) Map of China showing the study sites in Hong Kong, Shenzheng and Xiamen. The figures in brackets are RPSI, VDSI, and percentage of imposex (%), respectively, in *Thais clavigera*. Data were extracted from Leung et al. (2006) (Hong Kong), Chan et al. (2008) (Shenzheng), Tang et al. (2009), and Wang et al. (2008) (Xiamen). (b) The relationship between (i) RPSI or (ii) VSDI and the tissue concentration of TBT in *T. clavigera*. Results of the regression analysis are also shown.

due to the blockage of the opening of oviduct. At stage 5, the vulva is displaced by the overgrowth tissue of vas deferens and becomes invisible. Finally, the aborted egg capsules are retained in the capsule gland and appear as a translucent or brown mass (stage 6). In general, both RPSI and VDSI increase with increasing TBT concentrations in ambient seawater and in body tissues of the neogastropod, and increase as the proportion of female neogastropods in the population decreases (Bryan et al. 1987; Leung et al. 2006). In general, the VDSI is usually less affected by seasonality and can provide a better resolution when compared with the RPSI (Chan et al. 2008).

While serving as *the factory of the world* and becoming one of the most important global economic entities, China is facing unprecedented challenges in combating pollution and upholding proper environmental management. For instance, the United Nations estimated that the total annual use of antifouling paints in China reaches to 65,000 metric tons (MT) in which around 20,000 MT (ca. 31%) are TBT-based paints (United Nations Development Program: http://goo.gl/5GLsKE). Therefore, TBT contamination is common and widespread along the Chinese coastline (e.g., Jiang et al. 2001; Zhang et al. 2002; Gao et al. 2004; Shi et al. 2005; Leung et al. 2006; Yang et al. 2006; Chan et al. 2008). On the basis of biomonitoring studies in Hong Kong, Shenzhen, and Xiamen between 2004 and 2007, TBT contamination has been quite serious along the South China coastline, especially in areas with heavy marine traffic and intense shipping activities as indicated by the high values of the RPSI, VDSI and percentage of imposex in each population of *Thais clavigera* (Figure 8.8a). Living in the busiest cargo ports in the world, *T. clavigera* populations in Hong Kong and Shenzhen showed the highest percentage of imposex (i.e., 100%) with the VDSI beyond the stage 3 in most cases. Such pervasive TBT contamination was likely attributable to intense shipping activities and TBT-leaching from a great number of idling and active vessels. Although all imposex indexes were comparatively lower in Xiamen, there was a clear trend of TBT contamination being generally higher in the inner bay (i.e., east of Xiamen Island) than the west coast of Xiamen Island (Figure 8.8a). This might be related to the poor flushing rate and more intense marine traffic in the inner bay of Xiamen. By pooling all available data from Hong Kong, Shenzhen, and Xiamen, it is clear that the imposex index (RPSI or VDSI) significantly increases with increasing tissue concentration of TBT in *T. clavigera* (Figure 8.7b), indicating a dose-dependent effect of TBT on the abnormal development in the female *T. clavigera*. Also, it has been shown that the VDSI often decreases with increasing distance from the major port, marina, harbor, or pier (Leung et al. 2006; Chan et al. 2008).

To rectify the environmental impacts of TBT contamination, many developed nations and regions have partially banned the use of TBT on small vessels (<25 m in length) since the early 1990s (Antizar-Ladislao, 2008). Nonetheless, leaching of TBT from large vessels still poses considerable threat to many marine organisms (Jörundsdóttir et al. 2005; Leung et al. 2006). In September 2007, a complete ban of TBT was approved by the majority of member states of the International Maritime Organization (IMO) through a global implementation of the Convention on the Control of Harmful Antifouling Systems on Ships (AFS Convention; IMO 2005). This ban entered into force on 17 September 2008 (http://goo.gl/135GGC). Under this AFS Convention, all vessels in China will no longer be permitted to apply or reapply organotin compounds in their antifouling systems. The United Nations Development Program has also provided some financial support through its Global Environmental Facility to support the long-term phase out of TBT-based paints in China. It is, therefore, anticipated that the situation of TBT contamination in China's coasts will improve in the near future. The effectiveness of the implementation of the AFS Convention and its enforcement can be evaluated by monitoring the imposex status in *Thais clavigera* and other neogastropod species through temporal comparison with data from previous studies. The study of imposex in marine neogastropods is thus essential to environmental management of organotin contamination and merits further studies to cover more monitoring locations along the Chinese coastline and provide additional baseline data for meaningful temporal comparison in the future.

8.4.3 Developmental Stability

Developmental stability, the capacity of an organism to develop into a consistent phenotype in a specific environment, may also be influenced by contaminants and used to suggest contaminant impact. Developmental stability is notionally related with fitness although Lens et al. (2002) question whether the linkage between these two is generally true. Individuals that fail to achieve a consistent phenotype often show lower survivorship or reduced reproductive output than those that do. As will be seen, developmental stability can be inexpensively measured in many field or laboratory studies, and is relatively universal in its applicability (Zakharov, 1990; Graham et al. 1993a). However, fluctuations are often small in magnitude so particular attention is required to conduct a reliable study of developmental stability (Leung et al. 2000) (Vignette 8.3). If studied carefully, developmental stability has advantages as an indicator of contaminant effects.

Developmental stability can be examined by measuring deviations from perfect form. For bilaterally symmetrical organisms, this is often calculated as deviations from perfect symmetry. Bilateral characters, such as the lengths of the right and left eye stalks of a crab, can be measured and the unsigned difference ($d = |\text{Length}_{\text{right}} - \text{Length}_{\text{left}}|$) or signed difference calculated. Measured characteristics can be **meristic** (phenotypes are expressed in discrete, integral terms such as number of bristles) or continuous such as eye stalk length. Usually, many characters are measured or counted. Deviations from perfect bilateral symmetry (**fluctuating asymmetry [FA]**) measured within a population are thought to reflect perturbations of normal developmental processes. If the mean of the d values from a population is 0 and the distribution of d is normal, the variance of d within the population is one measure of FA. Using the notation of Zakharov (1990), the mean (M_{d}) and variance (σ_d^2) of d are simply the following:

$$M_{\text{d}} = \frac{\Sigma d_i}{n} \tag{8.1}$$

$$\sigma_{\text{d}}^2 = \frac{\sum (d - M_{\text{d}})^2}{n - 1} \tag{8.2}$$

where n = the number of individuals measured. Bilateral organisms can also show directional asymmetry or antisymmetry. (But see additional details in Vignette 8.3.) Directional asymmetry exists if the mean was not 0. For example, measurement of d for weights of left and right arms of right-handed humans would display directional asymmetry. Antisymmetry is indicated if the distribution of d was bimodal. Graham et al. (1993b) give the example of male fiddler crab claws as antisymmetry: some males have larger right claws but others have larger left claws. Although directional asymmetry and antisymmetry can be used to measure contaminant effects on developmental stability, FA is most often used.

Minimal FA is expected under benign conditions. FA is expected to increase as conditions become increasingly different from some optimal range as in the case of increasing concentrations of a contaminant. This increase in FA suggests movement away from developmental stability.

Several field and laboratory studies have measured FA to assess toxicant effects. For example, Graham et al. (1993a) measured traits of morning glory (*Convolvulus arvensis*) growing at various distances from an ammonia plant in the Ukraine and found that FA was highest for leaves taken nearest the factory. In a laboratory study in which flies (*Drosophila melanogaster*) were fed food containing either lead or benzene, FA increased with exposure to these toxicants, suggesting a diminished ability to maintain a consistent phenotype (Graham et al. 1993b).

VIGNETTE 8.3 Developmental Instability and Fluctuating Asymmetry

Dmitry L. Lajus
St. Petersburg State University

DEVELOPMENTAL INSTABILITY, FLUCTUATING ASYMMETRY, AND THEIR RELATION TO STRESS

Any one organism regulates its phenotype. That is why we can recognize species, subspecies, and sometimes populations. This stabilized flow has two components: (1) canalization and (2) developmental stability (Waddington, 1957; Zakharov, 1989). **Canalization** is an ability to produce a consistent phenotype under different conditions. Its converse is **phenotypic plasticity**. Developmental stability in contrast, refers to the ability of a developing organism to produce a consistent phenotype in a given environment. Very often as a counterpart of the term "developmental stability" one uses the term "developmental instability" (DI). Thus DI causes variation around the expected (target) phenotype that should be produced by a specific genotype in a specific environment.

Initially, small deviations from ideal development on the molecular level accumulate and result in higher level deviations, such as FA (Palmer, 1994; McAdams and Arkin, 1999; Fiering et al. 2000; Graham et al. 2010). FA represents random deviations from perfect symmetry (Ludvig, 1932), can be found in the majority of symmetric structures (Figure 8.8), and is most often used for measuring DI. Higher similarity between structures and thus lower FA means lower DI.

Advantages of this approach are its universal characters in terms of both kinds of organisms (e.g., inclusive of virus, plant or animal, and kinds of stress), comparatively low cost, and possibility of application with noninvasive techniques. Much attention has been paid to DI as a measure of stress and fitness during the last several decades (Mather, 1953; Palmer and Strobeck, 1986; Leary and Allendorf, 1989; Zakharov, 1989; Parsons, 1990; Graham et al. 1993a; Leung and Forbes, 1996; Møller and Swaddle, 1997; Lens et al. 2002; Leung et al. 2003; Hoffmann and Woods, 2003). Since the early 1990s, the approach has attracted increasing attention that is manifested, for instance, in the steady growth of citations of papers with the keyword "fluctuating asymmetry" in the SCI database. Citations have increased from less than 100 to more than 4000 citations for the period spanning 1990–2007.

It has been found that FA can change due to environmental and genetic factors. Potentially, any one environmental factor can cause DI. DI can increase due to nonoptimal temperature in experimental (Leary et al. 1992; Ruban, 1992; Campbell et al. 1998) and natural situations (Alados et al. 1993), or high densities of individuals (Wiener and Rago, 1987; Leary et al. 1991). The level of FA is associated with growth rate (Zakharov, 1989; Lajus, 2001), susceptibility to parasites (Vrijenhoek, 1994; Escos et al. 1997; Reimchen, 1997), and heterozygosity (Blanco et al. 1984; Leary et al. 1992; Kozhara, 1994). Inbreeding often leads to reductions of developmental stability due to expression of deleterious recessive alleles (Leary et al. 1983; Bongers et al. 1997).

Effects of pollutants were documented in a number of studies. For instance, Valentine et al. (1973) observed higher levels of FA in waters affected by industrial pollution along the U.S. coast. Valentine and Soulé (1973) found in experiments that DDT can cause DI. There are also other examples of effect of pollutants on fish, (e.g., Ames et al. 1979; Ostbye et al. 1997). Analysis of museum specimens showed that the present-day Baltic seals are more asymmetric than the museum specimens from the 1940s to1950s, presumably due to the heavy pollution of the Baltic Sea (Zakharov and Yablokov, 1990). FA has been shown to increase in several species due to radiation (Zakharov and Krysanov, 1996).

MEASURING OF FLUCTUATING ASYMMETRY

Before analysis of FA, it is necessary to test whether the trait actually manifests this type of asymmetry. FA is unintentional asymmetry: it is not programmed in the developmental process. There are two types of intentional (or adaptive) asymmetry: (1) **Directional asymmetry** is very common and occurs when the character on the one side is normally greater than on the other, i.e., both presence and direction of asymmetry are predetermined. (2) **Antisymmetry** occurs when structure is greater on one side than on the other, but it is not predetermined on which side. Examples of antisymmetry are the left- and right-sided individuals in some populations of flatfish, left- and right-coiling snails. Structures with antisymmetry and directional symmetry also possess FA.

Statistically, FA is characterized by normal distribution of left and right (L–R) values of trait with a mean equal to zero. Normal distribution of L–R with a mean differing from zero is a characteristic of directional asymmetry, and bimodal or platicurtic distribution of L–R with mean equal to zero is a property of antisymmetry (Palmer and Strobeck, 1986). Tests for the distribution shape of L–R values allow one to distinguish these types of symmetry and then to select characters manifesting only FA or to apply corrections to remove directional symmetry.

There are many FA indices (Palmer, 1994; Palmer and Strobeck, 2003). The simplest index is FA = L–R. But, because differences between paired structures usually increases with growth, it is common to apply a simple size-correction, using the index $FA = |L-R|/0.5(L + R)$. Despite such size standardization, an association between size and asymmetry often remains. Therefore, tests and further corrections for such association must be performed. Measuring FA in terms of variance, i.e., var (L–R) (Leamy, 1984; Kozhara, 1994; Palmer, 1994; Lajus et al. 2003a) is necessary when one wishes to compare the magnitude of FA with other variances, such as total phenotypic variance and variance associated with measurement error (ME).

Measurement error is a great concern in FA studies because levels of FA are typically in the order of 1% or less of the size of the trait. Because the asymmetry is so small and characters frequently cannot be measured with high accuracy, ME can account for a large fraction of the between-side variance. Empirical studies confirm this prediction. Contribution of ME to FA very much depends on the measured character and sometimes is rather high, far exceeding 50% (Palmer and Strobeck, 1986; Hubert and Alexander, 1995; Merila and Bjorklund, 1995; Lajus and Alekseev, 2000).

Control over the ME is a standard procedure in FA studies now. To quantify the ME and thus to assess the true FA, it is necessary to perform the replicate measurements of the same specimen and then to apply analysis of variance (ANOVA). Sources of the ME may be instrument error, alterations of position of specimen during analysis, mistakes of recording, and alterations in the operator's way of taking measurements. Often ME decreases with increased operator experience.

It is important to keep in mind that correct quantification of ME is possible only when repeated measures are taken from the very beginning of analysis. For instance, if, as is frequently done, analysis of digital images only involves replicated measuring of the images themselves to control ME, alterations during preparation of specimens for imaging will not be included. Such preparation errors can contribute substantially to total measured FA.

Because it is impossible to completely avoid ME, the task of a researcher is to minimize it, to correctly quantify it, and to avoid its effect on sample comparison. Note that main criterion while selecting characters based on ME is not its absolute, but relative, magnitude, i.e., the amount of ME relative to the total FA. For this reason, characters with moderate ME and high FA can be preferable for analysis than those with low ME and low FA.

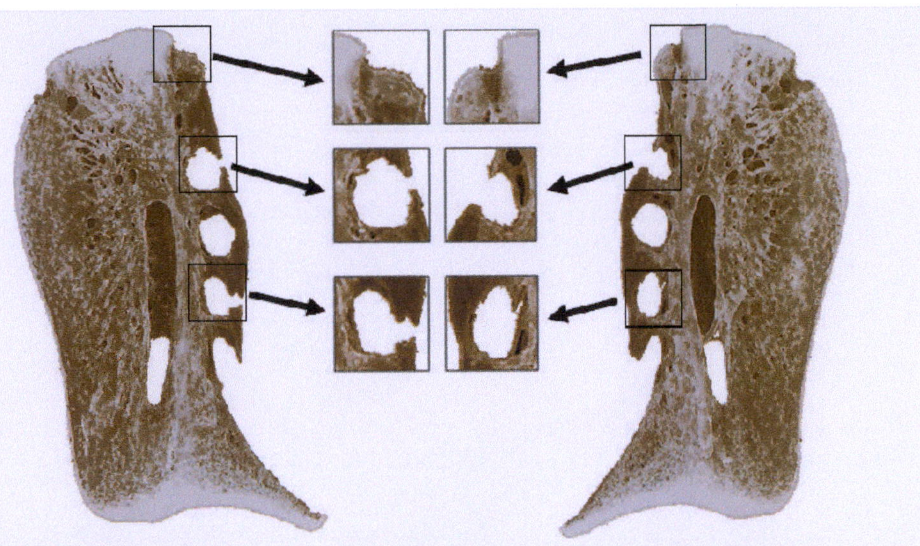

Figure 8.9 Example of fluctuating asymmetry: paired cranial bone *Ceratohyale* from Pacific herring, *Clupea pallasii*.

Figure 8.10 Leaf of seagrass *Halophila ovalis* (Alismatales, Hydrocharitaceae). Distances between vein interceptions and length of veins can be used for measuring fluctuating asymmetry. (From Ambo-Rappe, R., et al. *Ecol. Indic.*, 8, 100–103, 2008; Ambo-Rappe, R., et al. *J. Environ. Chem. Ecotoxicol.*, 3, 149–159 2011.)

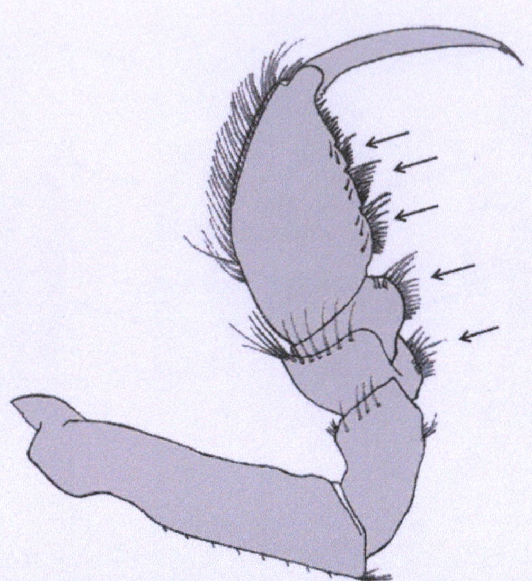

Figure 8.11 A thoracic leg of *Saduria entomon* (Crustacea, Isopoda) with characters (groups of chaeta on internal surface shown by arrows) convenient for analysis of fluctuating asymmetry. (Modified from Lajus, D.L., et al. *Ann. Zool. Fennici*, 40, 411–419, 2003. With permission.)

It is also important to equalize ME among samples. Because ME can be different for different operators, or even for the same operator at different times, this may result in between-sample differences in observed FA, even if in the absence of a true difference. To avoid such artifacts, the analyses must be performed not sample by sample, but specimen by specimen after randomization.

An ideal trait for analysis of FA must be easily measured, i.e., does not require complicated preparations before taking measurements, manifests ideal FA (i.e., normally distributed R–L values around zero), does not correlate with size, and does not possess substantial ME. Examples of actual traits used for FA analysis are the number of fin rays and sensory pores in fish; number of setae in crustaceans; width of plant leaves and length of veins in leaves; position of interceptions of veins on wings and number of bristles in insects; number, position, and size of cranial foramina in vertebrates; and muscle prints in mollusk shells. The most popular in all taxa are measures of external structures. Figures 8.9 through 8.13 show some characteristics convenient for measuring FA in various organisms.

One of the most important questions arising while assessing DI is whether or not results of the study depend on the traits being used, i.e., whether asymmetry of different traits varies concordantly across samples. Review of available literature shows that, in general, the answer is yes. FAs of different traits manifest significant concordance across samples, and in lesser extent, across individuals, i.e., there is population-wide and organism-wide DI. At the same time, the magnitude of concordance is usually quite low. There are statistical and biological explanations for this.

The first reason is purely statistical. Because FA is a variance and it is calculated from only two sides (one degree of freedom) of a bilateral trait in a given individual, it has high sampling error. Thus, we should expect that more sampling effort would be needed to study FA. Other reasons come from biology. For instance, traits developed at different stages of ontogenesis can be affected by different environmental conditions that result in differences in response to these conditions in terms of FA. Also, departure from concordance of FAs across samples can be

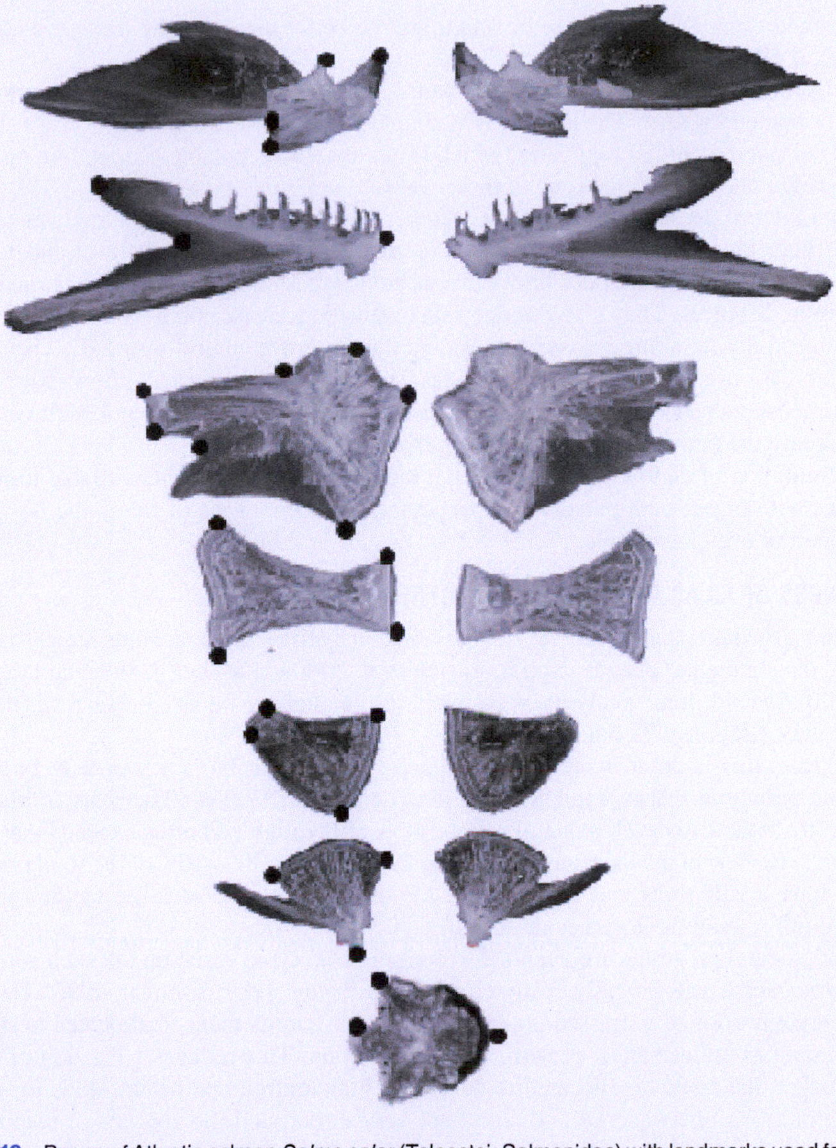

Figure 8.12 Bones of Atlantic salmon *Salmo salar* (Teleostei, Salmonidae) with landmarks used for analysis of fluctuating asymmetry. (From Yurtseva, A., et al. *J. Appl. Ichthyology*, 26, 307–314, 2010.)

Figure 8.13 Basal part of *Zostera capricorni* (Liliopsida, Zosteraceae) leaf. Distances between interceptions of median vein and lateral veins (arrowed) can be used for measuring translational fluctuating asymmetry. (From Ambo-Rappe, R., et al. *Environ. Bioindic.*, 2, 99–116, 2007.)

explained by different genetic properties of different traits, in particular, involvement of epistatic interaction of genes responsible for trait development (Leamy and Klingenberg, 2005). In some cases, different samples are characterized by different mean—asymmetry relationships that also can result in departures from concordance (Lajus, 2001). All of these examples show that, to some extent, DI is trait-specific and might be better measured by traits, which already showed their efficiency, than by others.

Usually, in FA analysis several or many traits are used. Multiple traits can be analyzed individually or pooled together. The latter way is simpler, but it ignores specifics of traits. Different traits are considered only as replicates, which is not always acceptable. Depending on goals of the study researcher may choose one of these ways.

During the last decade, more and more studies on FA have used digital analysis of images for identifying the position of landmarks, i.e., distinctive location structure, and to handle their coordinates. Such techniques have obvious advantages compared with traditional manual measuring of distances: They allow easier data acquisition, avoid manual data entry mistakes, reduce subjectivity in taking measurements, and thus improve control over ME. They are also more suitable for use of noninvasive techniques. Digital image techniques allow shape analysis using geometric morphometry methods (Smith et al. 1997; Klingenberg and McIntyre, 1998), which increase the power of morphometric analysis.

One limitation of digital techniques is that they limit the number of available traits due to difficulties with object manipulation. In particular, digital analysis of many meristic traits is very difficult or even impossible.

USEFULNESS OF FA AS AN INDICATOR OF STRESS

Despite growing interest in FA, doubts about its utility for assessing stress have been expressed during the last decade (e.g., Bjorksten et al. 2000a,b; Møller, 2000; Van Dongen and Lens, 2000). Indeed, some analyses of FA yield results that are far from researcher's expectations and may disagree with other indicators.

To address this issue, it is necessary to keep in mind that DI is a source of one type of phenotypic variation, that is, random variation (Lajus et al. 2003a). That part of phenotypic variation attributable to developmental instability is rather high and often exceeds that of other components. Review of the literature shows that DI comprises 30%–50% of the total phenotypic variance for meristic traits and 10%–40% of the total phenotypic variance for morphometric traits used in FA analysis (Lajus et al. 2003).

Developmental programs are intended to control phenotypic variation but such control over growth processes is energetically costly (Koehn and Bayne, 1989; Sommer, 1996). Because of this, the larger portion of a stressed organism's available energy must be allocated to important functions such as maintenance, growth, and reproduction. This reduces a fraction of the total energy budget that could be allocated to developmental control, and hence, leads to increased DI. Therefore, links between DI and stress are very direct, and the absence of response of a stressed population in terms of DI does not suggest that DI is an inappropriate measure of stress. Rather, it suggests that this specific factor does not cause detectable effect on the development of a particular trait.

Moreover, under natural conditions, different factors affect developmental instability in different ways and these effects may be superimposed. *Barnacle*, *Tetraclita japonica*, from locations with different levels of heavy metal contamination in Hong Kong showed no differences in FA. Further analyses revealed, however, higher level of proteinase in animals from more polluted locations, arguing that animals here have better feeding conditions. Most likely, absence of differences between barnacles from clear and polluted locations is due to superimposing of two factors influencing DI in the opposite ways: decrease caused by heavy metal contamination, and increase due to more favorable feeding conditions. It is important to emphasize here that barnacles from lower zone of littoral showed lower asymmetry than those from upper zone, revealing that structures used for FA measuring are sensitive to environmental stress

(Leung, Lajus, Chan, Yurtseva, unpublished data). Similar results were obtained on seagrass, *Halophila ovalis*, from Australia (Ambo-Rappe et al. 2008, 2011).

Random phenotypic variation may manifest itself not only in departures from bilateral symmetry but also in departures from other types of morphological asymmetry such as radial, rotational, dihedral, translational, and fractal (Graham et al. 2010). In general, random phenotypic variation can be seen in all cases when nonrandom phenotypic variation is negligible.

This nonrandom, or among-individual, variation is caused by genotypic heterogeneity and phenotypic plasticity, and may be neglected if both of these components are small enough. This takes place, for instance, when studying within-clone or within-individual variation associated with a homogenous environment. And not only morphological traits, but any phenotypic traits, such as life-history, behavioral, or physiological traits, manifest random variation that can change under stress (Sukhotin et al. 2003). Wide application of bilateral asymmetry for measuring DI takes place not because this is the only possible way to analyze DI, but rather because not many researchers are aware about other approaches.

Techniques for analysis of FA on traits that are not bilateral were developed for translationally symmetrical structures of several plant species (Alados et al. 2001, 2006; Sinclair and Hoffmann, 2003; Tan-Kristanto et al. 2003; Ambo-Rappe et al. 2007) and showed high potential. It is important to note that translational and rotational FA symmetry have one important advantage over bilateral FA because the number of structures in these cases is usually more than two; therefore, their analysis provides statistically more robust measures of variance.

CONCLUSION

The use of FA as an indicator of environmental stress has a number of attractive features. It is based on a simple idea that stress disturbs the control of morphogenesis for all living organisms. However, in practice, the analysis is not simple and includes a number of steps. Researchers differ in their opinion about the utility of this approach. Advocates often ignore the problems while critics highlight them. Moreover, critics often forget that an ideal indicator of a population's stress is still not available and all indicators have shortcomings.

It is necessary keep in mind while studying FA that it is a general measure of stress, i.e., it responds on various stresses including those of genetic and environmental origins. This is both advantageous and disadvantageous. On one hand, the joint effects of different factors may cause a mixed effect on development, especially in the wild, that is not easy to interpret. FA could not be used to infer causality of stress. On the other hand, FA may reflect synergistic interaction among stressors and assess the overall quality of organisms in a comparatively easy way. Probably, FA is the most effective, initial indicator that a population might be under stress (Leung et al. 2003). In this respect, FA is best used in combination with more specific biomarkers of environmental stress, particularly that arising from chemical pollution.

8.5 REPRODUCTION

Lowered fitness of individuals due to reproductive impairment is arguably among the most useful of sublethal effects measured by ecotoxicologists (Sprague, 1971, 1976). In the first chapter, reproductive failure traced to DDT/DDE inhibition of Ca-dependent ATPase in the eggshell gland (Kolaja and Hinton, 1979) and consequent eggshell thinning was discussed as one of the earliest events leading to our increased awareness of pesticide impacts to nontarget species. Already in this chapter, we discussed adverse impacts to the sea gull reproductive system due to an estrogenic xenobiotic and population failure of whelks experiencing high prevalence of imposex as a consequence of TBT exposure. Reproduction remains one of the most frequently measured qualities in both field and laboratory studies of contaminants (Newman and Clements, 2008).

The following field studies further highlight the value of measuring reproductive variables. White suckers (*Catostomus commersoni*) from a lake near the Flin Flon metal smelters (Manitoba, Canada) had smaller eggs, and reduced egg and larval survival than suckers sampled from a clean lake (McFarlane and Franzin, 1978). *Catostomus commersoni* had a high incidence of reproductive failure in a lake contaminated with metals (Munkittrick and Dixon, 1988); however, this failure was attributed to modification of the food base for the sucker, not to a direct effect of the metals. Western mosquitofish (*Gambusia affinis*) taken from a selenium-contaminated water body showed lowered fry survival and more stillborn fry than mosquitofish from a clean site (Saiki and Ogle, 1995). Starry flounder (*Platichthys stellatus*) from a highly urbanized region of the San Francisco Bay had more previtellogenic oocytes and lower embryo success than flounder sampled from a less urbanized region (Spies et al. 1989). Purple sea urchin (*Strongylocentrotus purpuratus*) egg fertilization was diminished by oil production effluent (Krause, 1994).

Numerous laboratory experiments show the impact of toxicants on reproductive performance and augment field studies such as those just mentioned. In one such study, reproductive impairment caused by metals was estimated by using a target 16% reduction in the number of young produced per female *Daphnia magna* (Biesinger and Christensen, 1972). Mosquitofish (*G. holbrooki*) in mercury-spiked mesocosms had low numbers of late stage embryos (Mulvey et al. 1995). Standard laboratory methods also exist for measuring various reproductive qualities as affected by contaminants. Weber et al. (1989) outline these methods for reproduction of zooplankton species and fathead minnow (*Pimephales promelas*). Chapman (1995) describes a standard method for quantifying fertilization success of purple urchin eggs.

All of these methods are useful for predicting effects on field populations if combined with an adequate understanding of the reproductive biology and ecology of the species in question. Effectiveness of identifying consequences of the sublethal effect is uncertain in the absence of such information. For example, a delay in the onset of reproduction may have trivial consequences for one species but serious consequences for another (Newman and Clements, 2008; Newman, 2013).

8.6 PHYSIOLOGY

Deviations from homeostasis associated with sublethal exposure often manifest as physiological alterations. Such physiological alterations may be used to infer a mode of action of the toxicant as well as to document a lowered capacity to maintain homeostasis or normal functioning (Brouwer et al. 1990). For example, acetylcholinesterase inhibitors* such as many organophosphate and carbamate insecticides affect feeding, respiration, swimming, and social interactions of nontarget species by impairing the senses and neuromuscular activity (Mehrle and Mayer, 1985). Such toxicants may be straightforward in their mode of action but others may involve a complex series of causal linkages. For example, contaminant-modified functioning of the endocrine system may impact numerous processes under hormonal control (Brouwer et al. 1990; Willingham, 2004). Currently, the ecotoxicology of endocrine modifying chemicals is an very active area of research that addresses effects linked mechanistically to phenomena already described such as effects to the hypothalamic–pituitary–adrenal or hypothalamic–pituitary–interrenal axes and to general reproduction. They are

* **Acetylcholinesterase inhibitors** are compounds that inhibit the normal functioning of acetylcholinesterase, an enzyme which breaks down the neurotransmitter, acetylcholine. After release from the presynaptic neuron, acetylcholine diffuses across the synapsis to bind at a receptor on the postsynaptic neuron (or on the neuromuscular junction), resulting in nerve impulse transmission. At the receptor site, it must then be hydrolyzed by acetylcholinesterase to choline and acetic acid to facilitate subsequent normal transmission of nerve impulses.

also relevant to discussions in the next section about immunology. Examples not covered elsewhere include thyroid dysfunction (Colborn, 2002), which have been reported in many exposed species including birds (e.g., Mayne et al. 2005), reptiles (e.g., Gunderson et al. 2002), and fish (e.g., Brown et al. 2004a,b).

Some physiological effects have characteristic threshold concentrations. Ammonia toxicity to rainbow trout (*Oncorhynchus mykiss*) is a classic example of a physiological effect with a threshold (Lloyd and Orr, 1969). As ammonia concentration increases, it causes increased water flux into trout that is counterbalanced by increasing urine production. Ammonia only becomes lethal above a threshold reflecting the maximum rate of urine production.

Physiological changes often studied for nonhuman animals include impaired performance (e.g., swimming speed or stamina), respiration, excretion, ion regulation, osmoregulation, and bioenergetics (e.g., food conversion efficiency) (Sprague, 1971). Common sublethal effects to plants include water status, stomatal function, root growth, respiration, transpiration, nitrogen or carbon fixation, chlorophyll content, and photosynthesis (Baker and Walker, 1989; Marwood et al. 2003). Most of these sublethal effects are assumed to provide insight about important physiological functions for which diminished efficiency lowers the fitness of affected individuals.

Studies of toxicant effects on respiration may describe oxygen consumption directly under normal, resting, or maximum exertion regimes. Also used is the **scope of activity** or **metabolic scope**, the difference between the rates of oxygen consumption under maximum and minimum activity levels. The scope of activity reflects the respiratory capacity or amount of energy available to meet the diverse demands on or activities of an organism. The oxygen consumption rate may be combined with the nitrogen (ammonia) excretion rate to suggest the relative dependence of respiration on carbohydrate and lipid resources versus the deamination of amino acids. For example, white mullet (*Mugil curema*) exposed to benzene and shrimp (*Macrobranchium carcinus*) exposed to metals had significant shifts in the O:N ratio (Correa, 1987; Correa and Garcia, 1990).

More complicated indicators of metabolism may be needed to address specific problems. For example, energetic analysis may reveal a disruption of energy balance associated with toxicant exposure (Dillon and Lynch, 1981; Olsen et al. 2007). The **scope of growth** (P = production) may be calculated from an energy budget of exposed individuals. The scope of growth is the amount of energy taken into the organism in its food (A) minus the energy used for respiration (R) and excretion (U): $P = A - R - U$ (Cockerham and Shane, 1994). It reflects the energy available for growth and production of young. For organisms exposed to toxicants, energy must be expended in response to the toxicant and the prediction is that scope of growth could decrease as a consequence.

The **adenylate energy charge (AEC)** (Equation 8.3) can also reflect the balance of energy transfer between catabolic and anabolic processes.

$$AEC = \frac{ATP + 1/2\ ADP}{ATP + ADP + AMP} \tag{8.3}$$

where ATP, ADP, and AMP = concentrations of adenosine tri-, di-, and monophosphate, respectively. Giesy et al. (1981) reported a drop in AEC in crayfish (*Procambarus pubescens*) and freshwater shrimp (*Palaeomonetes paludosis*) exposed to cadmium, and suggested a diminished energy status for these exposed crustaceans. Application of this assay, used mostly with studies of microbiota and cell cultures, has become uncommon in ecotoxicology during the last few decades.

Sublethal effects on respiratory activity or respiratory organs may be determined by examining movements associated with respiratory organs. In a common assay, fish are placed into a chamber receiving a toxicant solution or an effluent suspected of containing toxic components. The chamber is equipped with electrodes that measure the change in low-voltage electrical fields

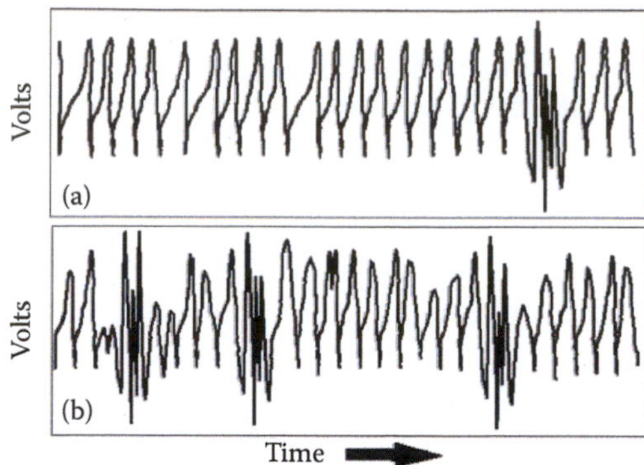

Figure 8.14 Patterns for ventilation and coughing for fish. In panel a, ventilation (small evenly spaced peaks) and coughing (strong, rapid cluster of peaks toward the right-hand side) are traced through time for a control individual. Ventilation frequency increases and amplitude decreases, and cough frequency increases often with exposure to contaminants (panel b).

produced by respiratory movements. In the absence of contaminant, changes in potential over time (Figure 8.14, panel a) show regular and uniform, respiratory movements with occasional **coughs** or **gill purges**. The coughs are abrupt, periodic reversals of water flow over the gills that function to dislodge and eliminate excess mucus from the gill surfaces. Cough frequency may increase during exposure to various contaminants, often because the irritant causes excess mucus production on the respiratory surfaces (Figure 8.14, panel b). As examples, bluegill sunfish (*Lepomis macrochirus*) exposed to heavy metals or chlorinated hydrocarbons (Bishop and McIntosh, 1981; Diamond et al. 1990) and brook trout (*Salvelinus fontinalis*) exposed to very low concentrations of methyl- or inorganic mercury (Drummond et al. 1974) had increased cough rates. The frequency and amplitude of the signals from ventilation also are sensitive indicators of sublethal effect. Ventilation frequency increased with exposures to cadmium (Bishop and McIntosh, 1981), copper (Thompson et al. 1983), zinc (Thompson et al. 1983), chlorinated hydrocarbons (Diamond et al. 1990), and other organic compounds (Kaiser et al, 1995). Diamond et al. (1990) and Kaiser et al. (1995) showed that the amplitude of the ventilation signal can also decrease with exposure to sublethal concentrations of toxicants.

Osmotic and ion regulation capacities can be compromised with contaminant exposure. Atlantic salmon (*Salmo salar*) in ammonia-spiked seawater experienced an increase in plasma osmolality* (Knoph and Olsen, 1994). Eel (*Anguilla rostrata*) osmoregulation was disrupted by DDT inhibition of Na^+, K^+- and Mg^{2+}-ATPase activity in the intestine where the flux of water is linked with the flux of these ions (Janicki and Kinter, 1971). Ion balance was disrupted in situations such as Atlantic salmon (*S. salar*) exposed to ammonia, flounder (*Platichthys flesus*) exposed to cadmium, and rainbow trout (*O. mykiss*) exposed to copper (Larsson et al. 1981; Laurén and McDonald, 1985; Knoph and Olsen, 1994). Acid conditions, alone or in combination with elevated aluminum concentrations, also alter ion regulation by fish (Fromm, 1980).

A wide range of physiological qualities are also measured for plants under the influence of contaminants. Metals (cobalt, nickel, and zinc) modify the water balance, stomatal closure, and leaf orientation of bean (*Phaseolus vulgaris*) seedlings (Rauser and Dumbroff, 1981). Exposure to PCBs (Doust et al. 1994) and heavy metals (Baker and Walker, 1989) can reduce photosynthetic activity of plants. Foy et al. (1978) provide an excellent review of plant physiological responses to metals.

* Osmolality is the number of osmoles of solute per kilogram of solvent. An osmole is the number of moles of dissolved ions that contribute to the osmotic pressure of a solution.

Air pollutants causing oxidative damage such as ozone and sulfur dioxide (SO_2) cause leaf **chlorosis**, a blanching of green color due to the lack of chlorophyll production or the increased destruction of chlorophyll (Temple, 1972; Malhotra and Hocking, 1976). Some changes are associated with defensive stomatal changes that attempt to restrict sulfur dioxide absorption (Madamanchi and Alscher, 1991). Heavy metals and sulfur dioxide can affect general plant respiration, and carbon and nitrogen fixation in addition to disrupting photosynthesis (Malhotra and Hocking, 1976; Baker and Walker, 1989).

8.7 IMMUNOLOGY

Toxicant influence on immunological competence has already been alluded to in discussions of GAS and toxicant-modified endocrine systems. Toxicant effects on immunocompetence or disease resistance are being reported with increasing frequency (Anderson, 1990; Luebke et al. 1997; Newman and Clements, 2008). Humoral features of the immune system might be impacted as in the case of the humoral responsiveness of great tit, *Parus major*, living in the vicinity of a metal smelter (Snoeijs et al. 2004). Cellular immune features are also susceptible to change by toxicants. Organochlorine contaminant concentration in the blood of glaucous gulls, *Larus hyperboreus*, was correlated with heterophil and lymphocyte levels in the blood (Bustnes et al. 2004). Phagocytic capacity of pulmonary macrophages of toads (*Bufo marinus*) was reduced at 1–24 hours after exposure to ozone (Dohm et al. 2005). As one example involving an invertebrate, malathion exposure altered phagocytosis by lobster phagocytes (De Guise et al. 2004). This lobster study was prompted by a possible link between the massive 1999 lobster die-off in Long Island Sound (United States) and malathion spraying for West Nile virus vector control. Lobsters at that period were found to have high prevalence of a *Paramoeba* sp. infection and the authors speculated that the disease outbreak might have occurred because lobsters were immunocompromised. As a final example, the hepatopancreatic digestive cells of blue mussels (*Mytilus edulis* and *M. galloprovincialis*) exposed to pollutants that produce oxidative stress had lysosomal alterations (Moore et al. 2007). This suggested an increase in **autophagy**[*] to remove damaged molecules and organelles, and also, that similar alterations likely were occurring with other phagocytic cells involved in cellular immunological processes.

8.8 BEHAVIOR

Various animal activities are studied in **behavioral toxicology**, the science of abnormal behaviors produced by exposure to chemical or physical agents. In addition to those already discussed relative to ventilation movement and developmental toxicology, behavioral abnormalities include changes in preference or avoidance, activity level, feeding, performance, learning, predation, competition, reproduction, and a variety of social interactions such as aggression or mutual grooming (Table 8.1) (Henry and Atchison, 1991). Most often, these effects are measured in a laboratory setting but some studies measure *in situ* changes in behavior (e.g., Gray, 1990). Unfortunately, behavioral effects still are underused in assessments of risk for three reasons (Giattina and Garton, 1983): (1) some behaviors are believed to be difficult to objectively score; thus, leaving open the possibility of generating biased information, (2) considerable variability can exist in behavioral data, and (3) it is often difficult to extrapolate accurately from highly structured laboratory experiments of

* **Autophagy** is the destruction of organelles or damaged proteins in the cell's lysosomes. This cellular process eliminates damaged cell components.

Table 8.1 Behaviors Commonly Used to Reflect Sublethal Effects of Contaminants

Behavior	Examples
1. Preference or avoidance	Change in response to light, temperature, salinity, or current. May avoid or move toward a stimulus differently after toxicant exposure. Salmon exposed to DDT shift their preferred water temperature from 19.1°C to 23.4°C (Ogilvie and Miller, 1976).
2. Activity level	Fatigue (lethargy) of workers with chronic lead poisoning (Bornschein and Kuang, 1990). Hyperactivity of fiddler crabs (*Uca pugilator*) exposed to TBT (Weis and Perlmutter, 1987) or Arctic char exposed to chlorine (Jones and Hara, 1988).
3. Feeding	Cessation of or diminished feeding by fish after exposure (Jones and Hara, 1988); Deviation from predictions of optimal foraging theory (Sandheinrich and Atchison, 1990).
4. Performance	Ability to swim against a current or maintain proper orientation to a current (rheotaxis) (Little and Finger, 1990). Critical swimming speed of fish lowered by exposure (Schreck, 1990).
5. Learning	Memory impairment of humans exposed to excess metals or metalloids, and memory loss due to mercury poisoning (Bornschein and Kuang, 1990); poorer response and higher error rate of exposed mammals in "lever pulling" learned behavior experiments (Gad, 1982).
6. Respiratory activity	See examples given in the chapter section on physiological effects of sublethal exposures.
7. Predation	Lowered ability to avoid predator (largemouth bass) by mosquitofish (*G. holbrooki*) exposed to radiation or mercury (Kania and O'Hara, 1974; Goodyear, 1972); suboptimal predator foraging or prey-switching (Atchison et al. 1996).
8. Competition	Zooplankton species grazing and filtration rates modified by toxicants (general details discussed by Atchison et al. 1996).
9. Reproductive behavior	Decreased libido after occupational lead exposure (Bornschein and Kuang, 1990); masculization of mosquitofish discussed earlier in this chapter.
10. Social interactions	Grooming in mammals after exposure (Gad, 1982); emotional lability of humans exposed to excess manganese, or increased irritability or depression associated with lead poisoning of humans (Bornschein and Kuang, 1990).

behavior to behavior in field situations. However, the first two problems can be minimized by careful design and execution of experiments and the third point is no more true for behavioral assays than for many assays used today in risk assessment. Regardless, the results of behavioral studies are most effectively used to assess contaminant impact if combined with results from a suite of other lethal and sublethal effects studies (Atchison et al. 1987).

Recent studies of changed behavior after toxicant exposure are easily found in the literature. General activity changes are often the most obvious nonsocial behavioral effect. As a teleost example, general activity of Japanese medaka (*Oryzias latipes*) was reduced by pesticide exposure (Teather et al. 2005). PCB-exposed, benthic worms (*Lumbriculus variegatus*) had reduced sediment burial rates, reflecting consequent reduced sediment bioturbation (Landrum et al. 2004). Predator avoidance was changed in fish exposed to mercury (*Notemigonus crysoleucas*) (Webber and Haines, 2003) or cadmium (*Salvelinus fontinalis*) (Riddell et al. 2005). Even learning can be affected as was the case with honeybees (*Apis mellifera*) exposed to insecticide (Decourtye et al. 2004). Impaired predator avoidance is also a common sublethal effect of toxicant exposure (e.g., Preston et al. 1999; Wojtaszek et al. 2005). Changes in social behavior range widely from sexual behavior of mosquitofish (*Gambusia holbrooki*) in pulp mill effluent (Toft et al. 2004), breeding behavior of tetrachlorodibenzo-p-dioxin-exposed osprey (*Pandion halieatus*) (Woodford et al. 1998), fish aggressive behavior after metal exposure (*Oncorhynchus mykiss*) (Sloman et al. 2003a,b), and schooling of PCB-exposed fish (*O. latipes*) (Nakayama et al. 2005).

VIGNETTE 8.4 Role of Behavior in Ecotoxicology

Mark Sandheinrich
University of Wisconsin–La Crosse

Behavior integrates genetic, biochemical, physiological, and environmental attributes that influence the evolutionary fitness of organisms. Consequently, behavior transcends single levels of biological organization and provides a link between subcellular processes that may be measured in the laboratory and ecological responses to contaminants observed in the field. For example, Weis et al. (2001) showed that altered levels of brain neurotransmitters and thyroid hormones of killifish (*Fundulus heteroclitus*) from a contaminated site were correlated with altered behavior, specifically reduced locomotor activity, impaired feeding, and increased vulnerability of killifish to predators. These altered behaviors explained reduced growth, condition, and longevity of the fish in the field as well as population changes in a major prey, the grass shrimp (*Palaemonetes pugio*). The role of behavior in ecotoxicology has been defined primarily by studies of (1) the sublethal effects of chemicals on organisms, (2) behavioral modification of organism exposure and chemical toxicity, (3) the use of behavior in identifying toxic modes of action, and (4) the incorporation of behavior into routine monitoring of water quality.

Numerous reviews have concluded that behavior is an ecologically important and sensitive indicator of toxicant stress in aquatic organisms (e.g., Westlake, 1984; Atchison et al. 1987; Beitinger, 1990; Henry and Atchison, 1991; Birge et al.1993; Little et al. 1993; Kasumyan, 2001; Dell'Omo, 2002; Scott and Sloman, 2004). Contaminants affect a variety of behaviors (Table 8.1), many that influence growth, reproduction, and survival—the classic endpoints of standard acute and chronic toxicity tests. For example, Little and Finger (1990) reported that swimming behavior of fish exposed to a variety of chemicals was altered at concentrations as low as 0.7% to 5% of the median lethal concentration. Swimming activity was frequently affected at concentrations that subsequently reduced growth. Sandheinrich and Atchison (1990) noted that feeding behavior of fish was disrupted by toxicants at or near concentrations that decreased growth and the altered behavior was often observed within hours of exposure. Moreover, changes in feeding and predatory behavior or altered behavioral susceptibility to predation can have indirect effects on community composition and species abundance (Fleeger et al. 2003; Relyea and Hoverman, 2006). Behavior is especially susceptible to modification by endocrine disrupting chemicals (EDCs) (Clotfelter et al. 2004) and, in some instances, may be more predictive of the effects of EDCs on reproduction than changes in physiological or hormonal endpoints (Zala and Penn, 2004; Sandheinrich and Miller, 2006).

Despite its sensitivity as an indicator of contaminant stress, behavior is not routinely used in hazard assessment or in establishment of water quality criteria. Lack of test standardization and field verification of behavioral responses are major challenges that have limited the acceptance of behavior as regulatory endpoints (Little, 1990). However, avoidance responses to contaminants, particularly by fish, do occur in the field and are ecologically important because the response can alter aquatic communities through emigration of organisms (Atchison et al. 1987). Sprague and colleagues showed that Atlantic salmon (*Salmo salar*) avoided low concentrations of copper and zinc in the laboratory (Sprague, 1964) and subsequently reported that Atlantic salmon migrating upstream during spawning runs avoided areas contaminated with a mixture of zinc and copper (Sprague et al. 1965; Saunders and Sprague, 1967). Geckler et al. (1976) reported that production of fish within a stream was reduced by avoidance of copper. Standardized acute and chronic toxicity tests conducted on site failed to predict this response. Avoidance of contaminated areas has been legally accepted as evidence of injury for Natural Resource Damage Assessments under proceedings of the Comprehensive Environmental

Response, Compensation, and Liability Act of 1980 (Natural Resource Damage Assessments, 1986; Little et al. 1993).

Although little studied, one of the more intriguing areas of ecotoxicology is the effect of behavior on rate of uptake and sensitivity of organisms to toxicants. Feeding, avoidance or attraction to contaminated areas, and social behaviors may alter exposure of fish to contaminants. For example, hierarchial positions within social groups of freshwater fish, such as salmonids and centrarchids, are established through aggressive or agonistic behaviors that may include nudging, biting, chasing, and flaring fins. Consequently, subordinate individuals within the group often respond by producing elevated levels of the stress hormone cortisol (Fox et al. 1997), which subsequently alters osmoregulation and transport of ions (e.g., metals) by the gills (Jobling, 1995). Sloman et al. (2002a) found that the position of a fish within a social hierarchy affected uptake of copper. Sublethal concentrations of copper did not alter established hierarchies within laboratory populations of rainbow trout (*Oncorhynchus mykiss*), but subordinate fish had significantly higher concentrations of copper in gills and liver than dominant fish due to increased uptake of copper from water. Behaviorally altered rates of toxicant uptake may influence susceptibility to toxicants. Sparks et al. (1972) found that the subordinate fish of a pair of bluegill (*Lepomis macrochirus*) succumbed more quickly to a lethal concentration of zinc than the dominant fish. When provided with a shelter, the frequency of aggressive interactions decreased between the fish and there was no difference in their sensitivity to zinc.

Behavior of aquatic organisms may also be used to identify the mode of action of chemicals. For example, Diamond et al. (1990) showed that the amplitude and frequency of ventilation, number and type of gill purges, and frequency of erratic movement in bluegill could be used to differentiate between different groups of chemicals. The heavy metals, zinc and cadmium, decreased the ventilatory amplitude and increased the frequency of gill purges within minutes of exposure. The chlorinated hydrocarbons, dieldrin and trichloroethylene, increased erratic movement, number of gill purges, and ventilatory frequency of the bluegill.

Drummond and coworkers used behavioral and morphological changes in 30-days-old fathead minnows (*Pimephales promelas*) to classify organic chemicals according to toxic modes of action (Drummond et al. 1986). They initially developed a system for visual observation of 40 different behavioral and morphological characteristics from 10 general categories (Drummond et al. 1986). These categories included equilibrium, locomotor activities, schooling and social behavior, body movement and coloration, ventilatory patterns, and general pathology and mortality. Fathead minnows were observed during acute exposure to 139 single chemicals. Loss of schooling was usually the first major symptom to occur and was caused by 96% of the chemicals. Discriminant function analysis, a statistical method for identification of groups and patterns, showed that behavioral data could be used to identify and separate chemicals into four categories based on probable cause of action. Narcosis-inducing chemicals (e.g., alcohols, ethers, ketones, phthalates) caused loss of equilibrium, depressed swimming activity, altered ventilation, and darkened body color. Chemicals that disrupt metabolism (e.g., benzenes, phenols) induced hyperactivity and increased ventilatory rates and amplitude. Neurotoxins (e.g., carbamate and organophosphate insecticides) depressed swimming activity but increased ventilation, tetany, deformities, and sensitivity to outside stimulation. Skin irritants (e.g., acrolein) produced symptoms similar to disrupters of metabolism, but fish were extremely hyperactive and exhibited agonistic and other unusual behaviors. Drummond and Russom (1990) refined these categories and classified more than 300 organic compounds based on three behavioral toxicity syndromes that represent a different mode of toxic action: (1) hypoactivity syndrome (narcosis), (2) hyperactivity syndrome (metabolic dysfunction), and (3) physical deformity syndrome (neurological dysfunction). Subsequently, Rice et al. (1997) used similar methods to assess sublethal toxicity to Japanese medaka (*Oryzias latipes*) of five chemicals with different modes of action. Behavioral toxicity syndromes may be useful, in conjunction with other endpoints, for predicting modes of actions of unknown compounds, identifying toxicants in complex effluents, and for testing wastewater before discharge into the environment.

One of the most practical uses of behavior is for the continuous monitoring of water quality at wastewater or drinking water treatment facilities and for providing a method of detection of toxicants from episodic events, such as storm water runoff or accidental chemical spills (Gerhardt et al. 2006). Instrumentation is commercially available to identify changes in behavior of fish, mussels, and water fleas (*Daphnia*) due to a toxic event and to signal an alarm to facility personnel. Water from lakes, rivers, or treatment plants is continuously passed through single or multiple chambers containing aquatic organisms and their behavior is monitored electronically. Biomonitoring systems are used throughout Europe to protect surface waters but have received only limited use in North America (Mikol et al. 2007).

The amplitude and frequency of rhythmic ventilatory movement, gill purges, and erratic locomotor behaviors of captive fish are sensitive indicators of environmental stress (Diamond et al. 1990). Neuromuscular activity of the fish generates a microvolt bioelectric signal. Electrodes within the chamber capture the signals, which are subsequently amplified and transmitted to a computer system. Significant departures from baseline activity due to abnormal behavior are indicative of a toxic condition (Shedd et al. 2001).

Monitoring systems that measure changes in position of the valves (i.e., gape) of mussels have also been developed (e.g., Sluyts et al. 1996). Small sensors are attached to the shell of the mussel and measurements of shell gape are made by an electromagnetic induction technique. Shell gape and the number of valve movements change under stressful conditions, such as those elicited by exposure to chemicals.

Video imaging software and complex algorithms are used to evaluate changes in the swimming behavior of groups of *Daphnia*. Measurements of mean swimming speed, distribution of swimming speeds among individuals, turns and circling movements, height of the *Daphnia* in the water column, distance between organisms, and number of moving *Daphnia* are combined to calculate a toxic index. Statistically significant changes in the toxic index trigger an alarm (Lechelt et al. 2000).

In conclusion, behavior is a sensitive indicator of sublethal toxicant stress and provides an ecologically relevant measure of contaminant effects. In developing methods for measuring chemically altered behavior, Little et al. (1993) proposed that endpoints should be evaluated based on (1) the sensitivity and ability of the behavioral response to measure effect or injury from exposure, (2) the utility of the response in identifying the chemical causing toxicity, and (3) the capacity of the behavioral response to increase predictability of the ecological consequences of toxicant exposure. Behavioral endpoints of greatest use in ecotoxicology will be those that provide linkages between the endpoints of growth, reproduction, and survival derived in the laboratory and changes in population and community structure predicted to occur in the field.

8.9 DETECTING SUBLETHAL EFFECTS

8.9.1 Conventional Approach

Sublethal and chronic lethal effects are detected and quantified in several ways. For regulatory testing, data are often generated with experimental designs similar to that shown at the top of Figure 8.15. Effects are measured at one, or perhaps, several time intervals in replicates receiving various toxicant concentrations. For example, fathead minnow (*Pimephales promelas*) larvae may be exposed to a series of toxicant concentrations (four replicate tanks containing 10 embryos each per exposure concentration) and the growth of the larvae measured after 7 days. The results are then analyzed using either a hypothesis testing or regression method.

Hypothesis testing most often involves a classic one-way ANOVA and post-ANOVA approach (Figure 8.15), although other methods are applicable. Generally, in **analysis of variance (ANOVA)**, the total variance (total sum of squares) is broken down into the variance among and within treatments, i.e., associated with differences among concentration treatments and differences for replicates within each concentration treatment. The mean variance within treatments (mean sum of $squares_{within}$) is assumed to reflect the sampling or error variance, and that among treatments (sum of $squares_{among}$) estimates the mean error variance plus any additional variance that might be associated with the treatment. According to conventional thinking, these two estimated measures of variance would be equal to some specific value if there were no differences among treatment means. According to classic hypothesis testing, this statement leads to a conventional statistic for testing the null hypothesis of equal means among treatments. The ratio of these two variance estimates ($F =$ mean sum of $squares_{among}$/mean sum of $squares_{within}$) is compared to tabulated critical F statistics to test for significant deviation from the null hypothesis of no difference among treatments.

The ANOVA approach has two statistical assumptions that require attention: (1) equal variances among treatments and (2) normally distributed data. These two assumptions are assessed

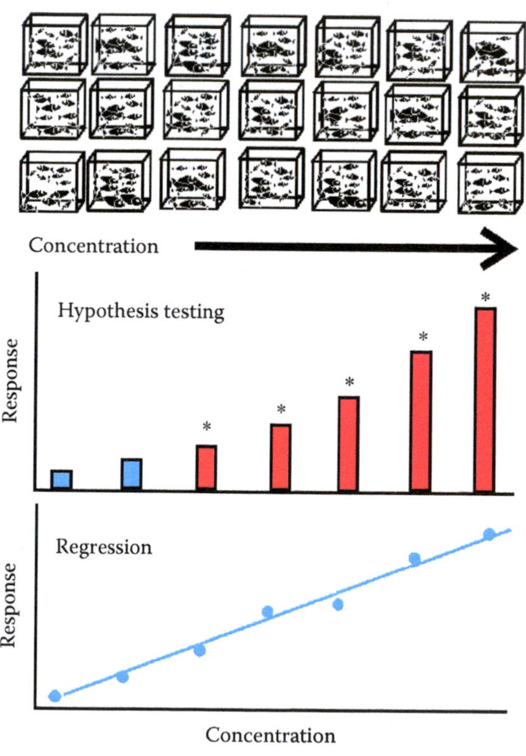

Figure 8.15 For regulatory testing, sublethal effects are most often analyzed using hypothesis testing methods. A series of tanks (seven sets of triplicates here) receiving increasing concentrations of contaminant are used to estimate the response for each concentration, including a control set of tanks. Data are then tested to determine if the response at each concentration was significantly different from that of the control. In the middle panel,* is used to indicate exposure concentrations with effects that are significantly different from the control. Here, the responses of the five highest concentrations (red) were significantly different from the control: the lowest exposure concentration was not significantly different from the control. These data can also be used in regression models to develop a predictive relationship between concentration and effect (bottom panel).

before performing the ANOVA.* Normality is conveniently tested with a formal method such as the **Shapiro–Wilk's test** although graphical methods are also adequate (Newman, 2013). The assumption of homogeneity of variance is tested using one of several tests (**Bartlett's test** is the most commonly applied) for data that, according to some test such as the Shapiro–Wilk's test, appear to satisfy the assumption of normality. The ANOVA methods are relatively robust to violations of these two formal assumptions (Miller, 1986; Salsburg, 1986). Newman (2013) states that "… probabilities derived from ANOVA are close to the real probabilities if the underlying distribution is at least symmetrical and the variances for the treatments are within three-fold of each other."

Often transformations of the data aid in meeting these formal requirements of ANOVA. The most common transformation is the **arcsine square root transformation,** which is particularly helpful in meeting the assumption of homogeneous variances for proportions of exposed individuals responding,

$$\text{Transform } P = \arcsin\sqrt{P} \tag{8.4}$$

where P = the measured effect such as the proportion of the exposed organisms responding.

If ANOVA leads to the rejection of the null hypothesis, we infer by convention that the treatment (i.e., exposure concentration) had a significant influence on mean response but we do not know which treatment means were significantly different. Extending the fathead minnow growth example, we infer that mean growth rate was different among the different exposure concentrations but we do not know which concentrations had mean growth rates that were significantly higher or lower than the others. Notionally, a series of post-ANOVA methods can help identify the specific treatments that are different from each other (Figure 8.16). Some methods test for significant difference between all possible pairs with the control (control mean versus each treatment concentration mean individually). For example (Figure 8.15), six pairs of means are tested because there was a

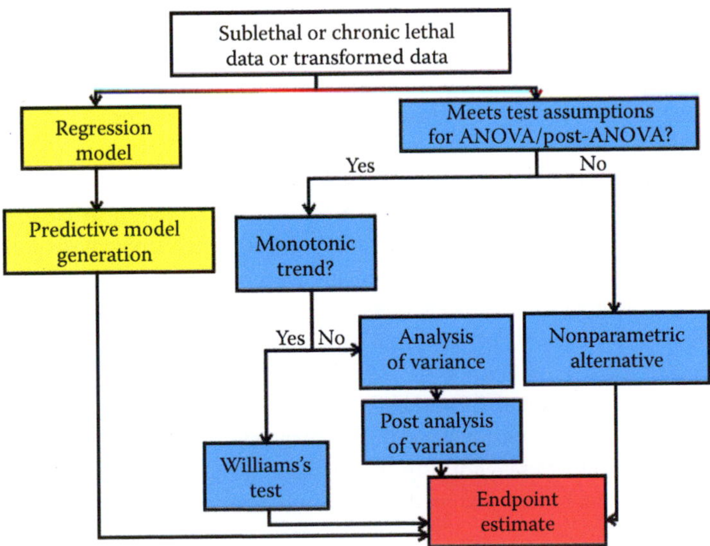

Figure 8.16 A flow diagram of conventional methods applied to analyze sublethal (and chronic lethal) effects.

* A third and very important assumption of ANOVA is that observations are independent. It is usually satisfied with a careful experimental design in which subjects are randomly assigned to treatments.

control treatment and six exposure concentration treatments: one pair for each concentration tested against the control. Other methods are also applied as will be discussed.

If the data or transformed data violate one or both assumptions of ANOVA, some statistical power may be sacrificed on order to use nonparametric, post-ANOVA tests. **Steel's many-one rank test** and the **Wilcoxon rank sum test with Bonferroni's adjustment** are the most commonly used (Weber et al. 1989). Steel's many-one test is often recommended by U.S. regulators if there are equal numbers of observations for all treatments (e.g., duplicate tanks of fish for all concentrations including the control treatment). The Wilcoxon rank sum test with Bonferroni adjustment is recommended often for unequal number of observations (Weber et al. 1989) although the slightly more powerful Steel's many-one rank test can be used in designs with unequal observation numbers (Newman, 2013).

If the data or their transformations are acceptable for parametric ANOVA, U.S. EPA test manuals indicate that several post-ANOVA methods are available and have more statistical power than the nonparametric methods just described. **Dunnett's test** or ***t* tests with a Bonferroni adjustment** can be applied although Dunnett's test is the slightly more powerful. Newman (2013) suggests that, although widely used, the *t* test with a Bonferroni adjustment is also slightly less powerful than the equally convenient but less commonly applied ***t* test with a Dunn–Šidák adjustment**.

If one can assume that a monotonic trend (i.e., a consistent increase or decrease in response) will occur with increasing toxicant concentration, the even more powerful **Williams's test** can be applied. Unlike the tests described above, the alternate hypothesis is no longer inequality of means between treatment pairs. Instead, the alternate hypothesis becomes "there is a monotonic trend in effect with treatment concentration." The test is done in two steps. In the first, significant deviation from the null hypothesis (equal mean responses among all treatments) is tested. If there is a significant deviation, a second step of the test is completed to identify the lowest concentration treatment having a significantly different mean from the control. Obviously, Williams's test would not be appropriate if one suspected hormesis in the data because the underlying assumption of monotonicity could not be met.

The U.S. EPA methods most often applied to these kinds of studies were just described to illustrate the ANOVA-based approach to analyzing sublethal effects. However, the Organisation for Economic Co-operation and Development (OECD) has recently published a detailed document that deviates somewhat from the recommendations just described. The reader—especially the non-North American reader—is urged to examine OECD (2006) before conducting such analyses of sublethal effects data.

Any of these methods could produce a data summary as depicted in the middle panel of Figure 8.15. A mean response of the control (leftmost bar) is compared to mean responses for the treatment concentrations and those significantly different from the control response are identified. An asterisk (*) was placed above treatment means judged to differ significantly from the control.

With results from analyses of the kind just described, biological effects of various concentrations of chemicals are inferred for sublethal and chronic lethal effects. However, the process of using tests of statistical significance to infer biological significance is more difficult than it may first appear. Even statistical hypothesis tests done in the most appropriate manner can only show that something is present in the data set that deviates significantly from that expected under the null hypothesis: they say nothing about the biological significance of that deviation. Considerable judgment must be applied to statistical results to successfully predict ecotoxicological consequences.

To assist in the extrapolation of statistical results to material ecotoxicological consequences, many descriptive concepts and terms have been developed. The **no observed effect concentration (or level) (NOEC or NOEL)** is the highest test concentration for which there was no statistically significant difference from the control response. To emphasize that the effect is an adverse one, the term is sometimes expanded to no observed adverse effect concentration or NOAEC. In Figure 8.15, the lowest experimental concentration would be the NOEC (second blue bar from the left).

The **lowest observed effect concentration (or level) (LOEC or LOEL)** is the lowest concentration in a test with a statistically significant difference from the control response (red, third bar from the left). Again, the word, adverse, might be added to the term. The **maximum acceptable toxicant concentration** (**MATC**) is "an undetermined concentration within the interval bounded by the NOEC and LOEC that is presumed safe by virtue of the fact that no statistically significant adverse effect was observed" (Weber et al. 1989). Notice that the boundaries for statistical significance are represented by NOEC and LOEC, and the dubious presumption is made that biological "safety" is to be found within those boundaries. Weber et al. (1989) define the term, **safe concentration** to mean "The highest concentration of toxicant that will permit normal propagation of fish and other aquatic life in receiving waters. The concept of a 'safe concentration' is a biological concept, whereas the 'no observed effect concentration' is a statistically defined concentration."

The above definitions and their implied applications in assessing risk have several shortcomings. First, the values that can be taken by the NOEC and LOEC are totally dependent on the particular concentrations chosen for the experiment: they are tied as much to the experimental design as to any toxicological reality. Next, the process produces higher than optimal NOEC and LOEC values if one uses a suboptimal experimental design (low statistical power) or poor technique (high error variance). Consequently, inferior design and technique can be rewarded with higher NOEC and LOEC values than would be calculated with superior design and technique. The concentrations identified as having an effect would be higher with inferior methods. The presumption of a "safe" MATC between the NOEC and LOEC cannot be extended beyond the design, species, and exposure durations used in the specific test without much additional supportive information. The MATC has no statistical confidence interval because the LOEC and NOEC are used to define it. Finally, such data may have dubious predictive value for estimating a safe concentration ("... that [permitting] propagation of fish and other aquatic life in receiving waters"). A statistically significant reduction in reproduction (e.g., 50%) may be much higher than that which will eventually lead to local extinction of some species populations (e.g., 20%) but not others. Regardless, these types of data, augmented with supportive data and professional judgment, are used extensively in ecological risk assessments today.*

Another approach to analyzing concentration–response data is statistical regression (Figure 8.15, bottom panel) (Stephan and Rogers, 1985; Hoekstra and Van Ewijk, 1993). Data may be fit to a specific concentration–effect model by least squares or maximum likelihood methods. Concentrations (and their associated confidence intervals) having some biologically meaningful level of effect, such as a 10% reduction in fecundity, are calculated through interpolation with the model. The ability to extrapolate downward from results may be another advantage of this approach if one is confident in the shape of the concentration–response model. However, if one incorrectly assumes a linear model when another model is more appropriate, predictions from regression models will lead to false conclusions. Also, if so much variation exists in the data that a good model cannot be identified, the regression approach becomes compromised and the ANOVA approach may be required (Stephan and Rogers, 1985).

8.9.2 Fundamental Issue to Resolve

A dilemma emerges at the end of our discussions of conventional ways of quantifying sublethal effects. On one hand, not all students reading this book will have taken a statistics course. Yet, the ANOVA-based methods just described are—there is no nice way to put it—simply wrong

* Statistical significance is often used directly to assign ecological significance in ecological risk assessment. This unwise practice will be labeled the **maulstick incongruity** here. (A maulstick is a stick used by artists to simply steady the brush-hand while painting.) Just as an inferior painting would be expected from an artist who used the maulstick instead of the brush to apply paint to canvas, use of statistical methods alone instead of biological data supported by statistical methods to determine biological significance leads to an inferior decision. The misuse of an otherwise effective tool leads to an inferior product and misinformation.

and a certain level of statistical training is needed to understand why. If no mention were made of fundamental flaws in current practices so as to avoid frustrating the reader without a statistics background, all readers would leave this section feeling that all is well. All is not well. Yet including a very detailed discussion would not provide a reader lacking a statistical background with much useful insight. The only course of action seems to be to give some general details for all readers and then point to numerous references for the reader with more training.

The current procedures were formulated based on our historic understanding of hypothesis testing that many informed statisticians now realize was wrong. The generally practiced "reject the null hypothesis and confirm the theory" (Cohen, 1994) or "null ritual" (Gigerenzer, 2004) approach is an inconsistent blending of Fisher's significance testing and Neyman and Pearson's hypothesis testing that emerged as a compromise among statisticians of the early twentieth century (Krueger, 2001; Sterne and Davey Smith, 2001; Ziliak and McCloskey, 2008). The result has been decades of a collective illusion that seemed to function "to make the final product, a significant result, appear highly informative, and thereby justify the ritual" (Gigerenzer, 2004). Detailed descriptions of the problems with the current approach are given in Bakan (1966), Cohen (1994), Fidler et al. (2004), Gigerenzer (2004), Krantz (1999), Krueger (2001), McCloskey (1995), and Sterne and Davey Smith (2001). Newman (2008), Newman and Clements (2008), and Newman (2013) provide detailed discussion of the problems and potential solutions relative to sublethal effects testing.

The bare essentials of the problem can be summarized as follows. The p value associated with a calculated test statistic is the probability of getting the observed data or more extreme data if the null hypothesis were true. It is not the probability of the null hypothesis being true. Confusion about this point has caused considerable misinterpretation of sublethal effects test data. Further, the common inference that a low p value (e.g., $p \leq 0.05$) supports the alternate hypothesis (an effect is present) is incorrect (Sterne and Davey Smith, 2001). This pervasive mistake is called the **inverse probability error** by Cohen (1994). It is based on a misunderstanding that only two options exist: there is and is not an effect. This is untrue from either the Fisher or Neyman–Pearson context of testing. In the context of testing as advocated by Fisher, the options are the null hypothesis is rejected (falsified) or no conclusion can be reached about the null hypothesis (not falsified). No "there is an effect" option exists. From the decision-making context of Neyman and Pearson, there are four decision states: (1) true positive, (2) false positive, (3) true negative, and (4) false negative.* When the inverse probability error is made, two of these decisions states are ignored. Also, the common assumption is invalid that the size of the p value indicates how significant a difference is (Bakan, 1966). Another common and equally incorrect practice is to test for a statistical difference without regard for whether that difference is biologically or materially significant (Bakan, 1966; Gigerenzer, 2004), i.e., the **nil hypothesis error**. These and other aspects of current hypothesis test practices constitute a fundamental inferential incongruity that we will label the **immaterial significance incongruity**.

Sound decisions from these kinds of tests require more rigor but only small changes to test designs. Defensible decisions using these kinds of tests about whether or not a treatment probably has a material effect requires *a priori* assignment of appropriate Type I (α) and II (β) error rates, a specified effect size (ES) judged to be materially significant, and some estimate of the prior probability of the hypothesis being true (e.g., R, the number of "true relationships" divided by the number of "no relationships" estimated before testing). As illustrated recently for ecotoxicity testing by Bundschuh et al. (2013), this allows one to calculate the **positive predictive value** (PPV) (the probability of the alternate hypothesis being true given a significant hypothesis test result) and **negative predictive value** (the probability of the null hypothesis being true given a nonsignificant

* Respectively, the true positive, false positive, true negative, and false negative decisions mean the following: the hypothesis was true and your test indicated it was true, the hypothesis was untrue but your test indicated it was true, your hypothesis was untrue and your test indicated it was untrue, and the hypothesis was true but your test indicated it was untrue.

hypothesis test result) (Wacholder et al. 2004; Rizak and Hrudey, 2006; Newman, 2008; Newman and Clements, 2008; Newman, 2013),

$$\text{PPV} = \frac{(1-\beta)R}{R-\beta R+\alpha} \tag{8.4}$$

$$\text{NPV} = \frac{1-\alpha}{1+\beta R-\alpha} \tag{8.5}$$

If one reflects for a moment on the definition just given for PPV, it will be clear that most researchers falsely believe that they are dealing with PPV when they judge significance based on p values. Again, a p value is the estimated probability of getting the observed or more extreme data if the null hypothesis were true, not the PPV.

8.10 SUMMARY

In this chapter, sublethal effects were described including the general adaptation syndrome and effects to growth, development, reproduction, physiology, immunology, and behavior. Many effects overlap these artificial categories, e.g., developmental toxicology. The substantial uncertainty is emphasized about whether or not such sublethal effects may result in death in an ecological arena. Conventional statistical methods used to detect, model, and predict sublethal response to toxicants were described along with the difficulties of using the associated results to predict ecotoxicological impact or risk. Regardless of the difficulties in prediction, measures of sublethal effects are likely to be as important, or more important, than the measures of acute or chronic lethal effect described in the next chapter to accurately assess the consequences of contamination.

Provided at the end of the chapter is discussion of an emerging crisis about how to measure sublethal effects. From a scientific point of view, the current approach is based on flawed significance testing methods that are slowly being replaced in statistics and scientific fields that depend on statistics (Newman, 2013). On the basis of this knowledge and the importance of making the best possible judgments about environmental effects of pollution, it could be argued that ecotoxicologists should immediately abandon the old approach and adopt a new valid one, regardless of recommendations in regulatory guidance documents. As Bacon (1620, reprinted in 1944) states in his *Novum Organum*, "... truth is rightly named the daughter of time, not of authority." Surely, a sound understanding of our current environmental problems requires that we must apply the best available tools and abandon conventional, but delusive, opinions. But, as discussed in Chapter 1, this would cause considerable disruption of the activities of technical and practical ecotoxicologists who rely on consistent application of agreed-upon methods and the enormous databases built over decades with the now questioned methods. The tremendous challenge is to avoid perpetuating a fundamental error while simultaneously respecting the technical and practical needs of any applied field. Bacon's quote continues,

> *It is not wonderful, therefore, if the bonds of antiquity, authority, and unanimity, have so enchained the power of man, that he is unable (as if bewitched) to become familiar with things themselves.*

More recently, Otto Nuerath used the analogy of a ship at sea to describe how difficult it is to change a scientific approach that has outlasted its time. Quine (1960) describes what has become known as the Neurathian bootstrap,

> *Neurath has likened science to a boat which, if we are to rebuild it, we must rebuild it plank by plank while staying afloat in it...Our boat stays afloat because at each alteration we keep the bulk of it intact as a going concern.*

The **Neurathian bootstrap incongruity** seems an appropriate moniker for the continued retention of the NOEC/LOEC approach despite its fundamental shortcomings relative to the intended task. So much regulatory structure depends on this approach that its immediate elimination would cause a crisis. Yet understanding the reason why this approach has been retained beyond its useful lifetime does not trump the fact that environmental issues now facing humankind require rapid and clear-headed progress in finding the best solutions. As I stated in Newman (2013),

> *Clearly, we are forced to work with the planks available onboard while requiring some major fitting at a shipyard...Ecotoxicologists would benefit at this time by bringing their ship* [in] *for* [a] *refitting...*

SUGGESTED READINGS

Atchison, G. J., M. G. Henry and M. B. Sandheinrich, Effects of metals on fish behavior: A review. *Environ. Biol. Fish.*, 18, 11–25, 1987.

Calabrese, E. J., Hormesis: Why it is important to toxicology and toxicologists. *Environ. Toxicol. Chem.*, 27, 1451–1474, 2008.

McLachlan, J. A., Functional toxicology: A new approach to detect biologically active xenobiotics. *Environ. Health Perspec.*, 101, 386–387, 1993.

Newman, M. C., "What exactly are you inferring?" A closer look at hypothesis testing. *Environ. Toxicol. Chem.*, 27, 1013–1019, 2008.

Newman, M. C., *Quantitative Ecotoxicology*, 2nd Edition, Taylor & Francis/CRC Press, Boca Raton, FL, 2013.

Newman, M. C. and W. H. Clements, *Ecotoxicology: A Comprehensive Treatment*, Taylor & Francis/CRC Press, Boca Raton, FL, 2008.

Palmer, A. R., Waltzing with asymmetry. *Bioscience*, 46, 518–532, 1996.

Selye, H., The evolution of the stress concept. *Am. Sci.*, 61, 692–699, 1973.

Sprague, J. B., Measurement of pollutant toxicity to fish—III. Sublethal effects and "safe" concentrations. *Water Res.*, 5, 245–266, 1971.

Stebbing, A. R. D., Hormesis—The stimulation of growth by low levels of inhibitors. *Sci. Total Environ.*, 22, 213–234, 1982.

Weis, J. S. and P. Weis, Effects of environmental pollutants on early fish development. *CRC Crit. Rev. Aquat. Sci.*, 1, 45–73, 1989a.

CHAPTER 9

Acute and Chronic Lethal Effects to Individuals

Pollutants matter because of their effects on populations, and so, indirectly, on communities too, but pollutants act by their effects on individual organisms.

Moriarty (1983)

9.1 GENERAL

9.1.1 Overview

Most methods used by ecotoxicologists to determine lethality have their origins in mammalian toxicology. In the developing field of ecotoxicology, this adoption allowed very rapid, initial advancement since established techniques and concepts could be quickly incorporated. It also led to some inconsistencies that must be resolved as ecotoxicology progresses. Also, some very worthwhile techniques in mammalian toxicology have yet to be assimilated into ecotoxicology. This chapter will explore these borrowed concepts and techniques, discuss those remaining underexploited, and identify inconsistencies between application in mammalian toxicology and ecotoxicology.

9.1.2 Acute, Chronic, and Life Stage Lethality

Early in ecotoxicology, a broad distinction was made between acute and chronic lethality. **Acute lethality** refers to death following a brief, and often intense, exposure. The traditional duration of an acute exposure in ecotoxicity testing is generally 96 or fewer hours (Sprague, 1969). Although death is assumed to occur during or immediately after exposure, there are exceptions in which acute exposure results in death over a substantial period. Perhaps, the most obvious example is the acute radiation syndrome following an intense and very brief exposure to deeply penetrating, ionizing radiation. Death does not occur immediately but, depending on the received body dose and damage to the gastrointestinal and marrow cells, within days to weeks of the exposure. Truhaut (1970) gives the additional delayed mortality after brief exposure examples of respiratory failure after paraquat ingestion, kidney tumor formation in rats after ingestion of dimethylnitrosamine, edema of the lungs because of pyrrolizidine exposure, and nervous tissues damage from organophosphorous compounds. Citing Barnes (1968), he refers to such toxicants as **hit-and-run poisons**. This issue emerges again from a different vantage in ecotoxicity testing that attempts to predict lethal consequences to populations exposed for a set time to a toxicant. Tests such as those described below measure only death that occurred up to the end of the exposure but, depending on the toxicant, substantial numbers of the survivors can die in the time immediately following exposure cessation

(Zhao and Newman, 2004, 2006).* **Chronic lethality** refers to death resulting from a more prolonged exposure. By recent convention in ecotoxicology, a chronic test should be at least 10% of the duration of the species' life span (Suter, 1993), but this is not always the case. Sometimes a test of shorter duration is discussed as a chronic test. The distinction of convenience between acute and chronic is often blurred.

Another important distinction can be made for lethality testing based on life stages. An elaborate **life-cycle study** might determine lethality, growth, reproduction, development, or other important qualities at all stages of a species' life, e.g., Mount and Stephan (1967). **Critical life stage testing** focuses on a particular life stage such as neonates. Often, the most critical life stage is, or is assumed to be, an early life stage, leading to the development of **early life stage (ELS)** tests (McKim, 1985; Weber et al. 1989). The critical life stage approach is based on the sound assumption that protection of the most sensitive stage will ensure protection of all life stages: The most sensitive stage of an individual's life cycle will determine that individual's fate under lethal challenge. A dubious extension of this concept (**weakest link incongruity**) is often made in which one assumes that exposure of field populations to concentrations identified in testing as causing significant mortality at a critical life stage will result in significant impact on the field population. However, loss of individuals at certain life stages may or may not have much bearing on population growth or the risk of local extinction for species populations.† For example, a 10% reduction in the number of larvae during an oyster spawn because of toxic effect may have minimal impact on the likelihood of an oyster population becoming extinct. Very high mortality is expected for the planktonic larvae under normal conditions and oyster populations accommodate widely varying annual recruitment. Also, oyster populations often are reseeded with larvae from nearby populations. It is important to keep in mind that the term *critical life stage* refers to the life stage of an individual most sensitive to poisoning, not necessarily the life stage most critical to population viability. More will be added to this point in the Chapter 10.

9.1.3 Test Types

Tests used to quantify lethality vary depending on the medium of concern, i.e., water, effluent, food, sediment, or soil. The nature of the toxicant or toxicant mixture may also influence the test design. For example, exposure solutions might not be aerated for tests involving volatile compounds. For TBT-based antifouling paints, dosing may be done by submerging painted surfaces (e.g., discs or rods) in the feed water flowing into the different exposure tanks (Bryan and Gibbs, 1991). Different amounts of TBT leach from these surfaces into the tanks, producing a range of exposure concentrations. Exposure duration, test species, resources, and expendable time also influence method selection. Chronic exposure of a suite of endemic species in the laboratory might be highly desirable yet impractical. Instead, a representative species that is easily cultured in the laboratory might be exposed for only the most sensitive time in its life cycle. The results may then be used to imply the risk to endemic species over their entire life cycles.

A series of exposure designs has been established for tests quantifying lethal effects of toxicants in waters. In **static toxicity tests**, individuals are placed into one of a series of exposure

* Zhao and Newman (2004) suggested that toxicity testing protocols can be changed so that latent mortality occurring after cessation of exposure ends is also noted and combined with that occurring during exposure. With these exposure and postexposure mortality data, a **complete median lethal concentration** can be calculated. Similar adjustments for postexposure effects are also important to consider with pulsed or intermittent exposures of populations as might occur with periodic runoff after pesticide spraying events (Zhao and Newman, 2006; Newman, 2013).

† This situation results because of the difference in emphasis taken by medical/mammalian toxicologists and ecotoxicologists. For a toxicologist dealing with humans, the individual is justifiably the focus of decisions, concepts, and associated methodologies. The ecotoxicologist tends to focus on population and community viability instead. This distracting inconsistency is a consequence of the rapid infusion of methods from mammalian toxicology into ecotoxicology.

concentrations and the exposure water is not changed during the test. The advantage of this design is that it is easy to perform and inexpensive. Also, minimal volumes of toxic solutions are generated (Peltier and Weber, 1985). But toxicant concentrations often change during exposures because of sorption to the container walls and other solid phases, volatilization, bacterial transformation, photolysis, and other processes. Waste products of the test organisms may build up during the test and oxygen concentrations may drop to undesirable levels. For these reasons, most static tests are used to measure acute lethality, not chronic lethality. Static tests are becoming uncommon because of the shortcomings just mentioned. A **static-renewal test** can reduce some of these problems to tolerable levels. Test solutions are completely or partially replaced with new solutions periodically during exposures, or organisms are periodically transferred to new solutions. A **flow-through test** uses continuous flow or intermittent flow of the toxicant solutions through the exposure tanks. A flow-through system eliminates or greatly minimizes the problems just discussed for static tests. However, flow-through tests can generate large volumes of toxicant solutions that must be treated. They also require more time, space, and expense (Peltier and Weber, 1985). Although individual containers of various concentrations can be used as sources of test waters, often a special apparatus called a **proportional diluter** (Figure 9.1) mixes and then delivers a series of dilutions of the contaminant solution to the test tanks. The solution being diluted may be a toxicant solution or an effluent suspected of having an adverse effect on aquatic biota. With effluent testing, exposures levels are expressed as percentages of the exposure water volume made up of the effluent (e.g., 45% effluent blended with 55% diluent by volume) resulting in the toxic response.

There are several methods associated with solid-phase testing. Organisms may be placed into spiked or contaminated soils. This type of test has been used with soil invertebrates such as nematodes (Donkin and Dusenbery, 1994) and earthworms (Gibbs et al. 1996). Similarly, sediment toxicity tests may involve spiked or contaminated sediments. The **spiked bioassay (SB) approach** generates a concentration–response model for, or tests hypotheses regarding, effects to individuals placed into sediments spiked with different amounts of toxicant (Giesy and Hoke, 2001).

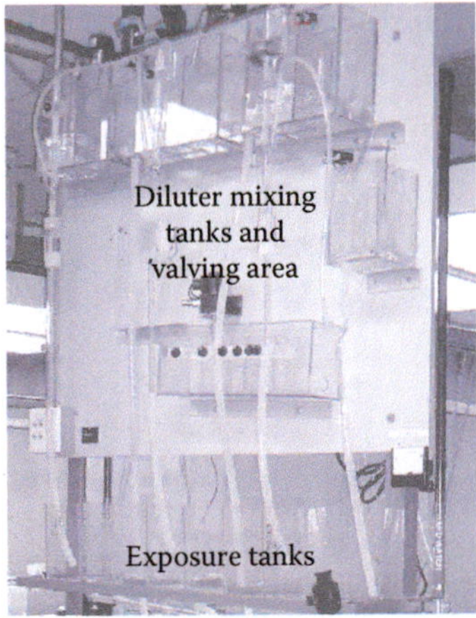

Figure 9.1 Proportional diluters, such as the one shown here, are commonly used to deliver a test water flow of a predetermined series of toxicant or effluent concentrations to replicate test tanks containing experimental organisms. Various proportions of dilution water and toxicant solution (or effluent) are mixed in the dilutor tanks and valving area. Waters with the resulting series of toxicant (or effluent) concentrations are then delivered to tanks (Exposure Tanks). Each exposure tank contains a specific number of experimental organisms.

Concentration may be based on total concentration in the sediment, interstitial water concentration, or the concentration in some notionally bioavailable fraction of the sediments. Sediment toxicity can also be implied using an elutriate test with a nonbenthic test species. In an **elutriate test**, a nonbenthic species such as *Daphnia magna* is exposed to an elutriate produced by mixing the test sediment with water and then centrifuging the mixture to remove solids (McIntosh, 1991). Exposure to various dilutions of the elutriate allow an amount-versus-response analysis as described earlier for effluent tests. The results of such a test would seem most appropriately used to assess lethality for situations such as those involving plumes produced during sediment dredging activities or perhaps runoff events.

Still other tests focus on effects of ingested toxicants. Studies involving ingestion often employ replicate cages of animals (Figure 9.2) fed different doses of the agent of interest. The agent is introduced into the animal's feed or might be introduced via **gavage**, that is, fed to the individual through a small tube inserted down its throat. The amount of toxicant ingested becomes the dose that is related to lethal or sublethal response. In these kinds of tests, an **up-and-down experimental design** might be used to obtain lethality metrics while minimizing the number of individuals subjected to adverse toxicant effects (Dixon and Mood, 1948; Dixon, 1965, 1991; Dixon and Massey, 1969; Bruce, 1985,1987). The individuals are not all dosed simultaneously with this method. Instead, they are dosed sequentially. Using the example of grebes in Figure 9.3, the first grebe is dosed and it dies. A second animal is given a lower dose and it also dies. A third animal is given a still lower dose and it survives. The next grebe is given a slightly higher dose and it dies. This process continues until a stopping rule as described in the above citations is met. This kind of sequential dosing allows calculation of conventional lethal metrics while killing fewer animals.*

Figure 9.2 An avian toxicity testing facility in which a series of doses are fed to subsets of mallard ducks so that lethal or nonlethal effects can be quantified.

* The reader interested in knowing more about the up-and-down approach should read Robert Bruce's original descriptions of the approach (Bruce, 1985, 1987), the Dixon citations given above, and more recent methods descriptions such as those of Auletta (2002).

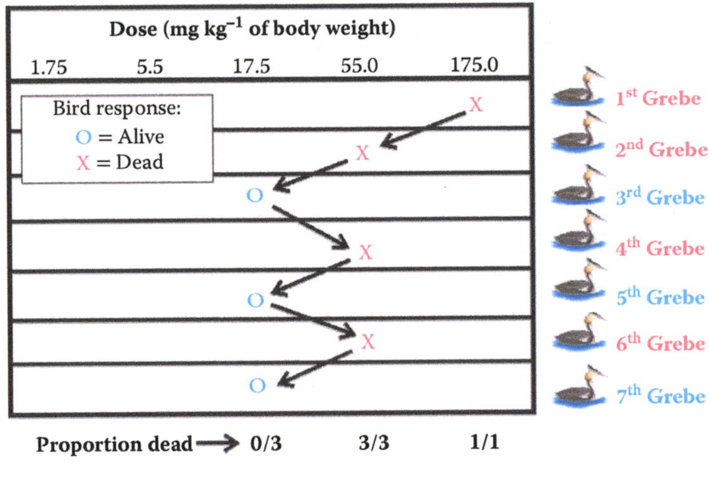

Figure 9.3 Up-and-down experimental design for estimating LD50.

9.2 DOSE–RESPONSE

9.2.1 Basis for Dose–Response Models

More as a consequence of the early history of ecotoxicology than through careful comparison to all alternatives, most lethality tests in ecotoxicology involve the dose–response* approach. This approach and associated quantitative methods were taken directly from mammalian toxicology and remain the cornerstone of toxicity testing in ecotoxicology. However, an alternative, time-to-response design to be described in Section 9.3 is being utilized with increasing frequency with every passing year.

With the **concentration–response approach**, a series of toxicant concentrations is delivered to containers as illustrated earlier in Figure 8.15. There are replicate containers for each treatment (concentration) that allows estimation of variation within each treatment and at least one control treatment receiving no toxicant.† Individuals are randomly placed into the containers until a predetermined number of individuals is present in each, e.g., 10 fish per exposure tank. Mortality is tallied in each tank as the number dying of the total number exposed after a certain time, or sometimes, a series of times such as 24, 48, 72, and 96 hours. The paired data (proportion dying and exposure concentration) for the tanks are used to calculate lethality.

The predominant methods for analyzing dose– or concentration–response data are based on the nearly forgotten **concept of individual effective dose (IED)** or **individual tolerance**. According to this concept, there exists a smallest dose (or concentration) needed to kill any particular individual

* To improve readability, dose–response and concentration–response are used interchangeably here. The analyses of these two models are identical: only the delivery of toxicant differs. A dose–response model is associated with a delivered amount of toxicant, e.g., 5 mg of toxicant is injected into the organism. A concentration–response model is associated with exposure to a concentration in the organism's environment, e.g., 5 mg L^{-1} of toxicant is present in the water into which a fish is placed.

† A solvent may be used to produce the toxicant solution as required for some sparingly soluble, organic compounds. In such a case, a solvent-control treatment is included too and methods for quantifying toxic effect are modified slightly to include this control in the experimental design. The modifications are made to ensure that any effect of the solvent on response does not unintentionally bias the results.

and this IED is a characteristic of that individual. This concept and its applicability to dose–response data analysis is illustrated in Figure 9.4. The upper panel shows the skewed, log normal distribution of IED values thought to be typical of populations. A sigmoidal dose–response curve would be produced if seven random samples of 10 fish each from this same population were given doses corresponding to the six IED groupings in the top panel and a control dose. This presumptive log normal curve is the basis for the probit method, the most common approach to analyzing dose–response data (lower panel).

Although the IED concept is presented almost exclusively as the foundation of dose–response models, other concepts are invoked to support various methods of analysis. The log logistic model has been suggested for years as an alternative to the log normal model, leading to a protracted controversy about the relative values of logit versus probit methods. The log logistic model also predicts a sigmoidal curve like that shown in the lower panel of Figure 9.4. The foundation for the logistic model is its linkage to processes such as enzyme kinetics, autocatalysis, and adsorption phenomena (Berkson, 1951).

Berkson (1951) questioned the IED concept and advocated the use of the log logistic model instead of the log normal model. He based his argument on an experimental screening of combat pilots for tolerance to high altitude conditions. Candidate aviators were placed into decompression chambers and their individual tolerances measured. Assuming the IED concept was correct, those failing to exhibit symptoms of the "bends" to a certain critical pressure passed the test and low tolerant individuals were rejected from further consideration as pilots. Berkson broke from the standard test protocol and asked that a set of candidates be retested to see if individual responses remained the same between tests. The results showed poor agreement between repeated tests: The IED concept had failed to explain the dose–response phenomena.

As a consequence, counter argument to the IED concept has been made based on the idea that individuals do not have unique tolerances in some instances. The argument is made that, in such cases, the probability of death for any individual is a consequence of a random process or set of processes

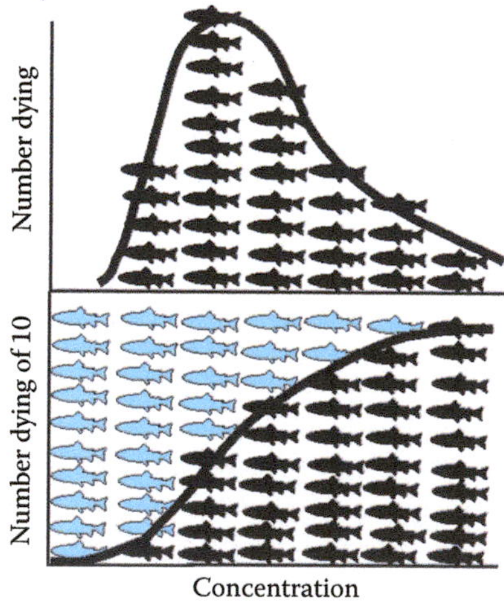

Figure 9.4 The individual effective doses (IED) concept and the analysis of dose–response lethality data. The top panel depicts the distribution of IEDs among 35 fish taken randomly from a population. Each individual's response was placed into a category (column) based on whether its IED fell into one of six different ranges of IED values. The distribution of such IED values within a population is thought to be log normal as evidenced by an asymmetric curve with a few very tolerant individuals. The bottom panel shows the results of exposing sets of 10 individuals from this same population to a series of doses for a set time. Surviving fish (blue) had IED values greater than the exposure dose and dead fish (black) had IED values less than or equal to the exposure dose. The result is a typical, sigmoidal dose–response curve.

that may conform to a log normal, log logistic, or another model. Berkson (1951) and Finney (1971) argued that some processes, such as those described in Chapter 7 for cancer risk, are based on probabilities that a specific sequence of events will occur and then lead to death or cancer. The distribution of differences in occurrence of the event (i.e., appearance of a clinical cancer or death) is not related to differences among unique individuals: the distribution is a consequence of probabilities associated with events taking place in all individuals. In this model, which particular individual responds quickly ("sensitive") or slowly ("tolerant") is a matter of chance alone. Individual response is a consequence of random events that are described by probability distributions. Certainly, the observation that the log normal model seems to work for tests with microbes and zooplankton composed of cloned individuals casts some doubt on the IED concept as the sole underlying explanation for the log normal model. Gaddum (1953) suggested that a random process in which several "hits" are required to produce death could also form the basis for the log normal model. Remarkably, whether one or both of these concepts is the basis for the majority of dose–response relationships has remained poorly tested until recently. Newman and McCloskey (2000) formally tested these concepts and found that neither was universally valid. Continuing this assessment of the IED general explanation for sigmoidal dose–response curves, Zhao and Newman (2007) also found with a different set of organisms and toxicants that the IED could not be assumed to be the sole explanation for sigmoidal dose–response curves.

Although this point may appear trivial, the consequences of repeated exposure of a population are different under these two concepts (Newman, 2001; Zhao and Newman, 2007). This can be illustrated with a thought problem in which all covariates affecting lethal impact such as animal sex, size, and age are identical for all individuals. If the IED concept were correct, survivors of the first exposure would be the most tolerant and the impact on the population of survivors would be less in the subsequent exposure. In the second case, where probability of death is the same for all individuals, the survivors will not be inherently more tolerant and the impact of a second exposure would be as large as that of the first.

A Weibull model also provides good fit for the dose–response curve but is seldom used (Christensen, 1984; Christensen and Nyholm, 1984; Newman, 1995; Newman and Dixon, 1996). The Weibull model can describe a multistage process such as that discussed in Chapter 7 for carcinogenesis. When applied, it seems to fit dose–response data as well as the generally accepted log normal and log logistic models.

Many dose–response relationships for lethality have threshold concentrations below which no discernible increase in mortality occurs. Most of the models described here can and should be modified to include lethal thresholds if required. Further, with chronic exposures, natural or spontaneous mortality may be occurring simultaneously with toxicant-induced mortality. Such spontaneous mortality can also be included in dose–response models if necessary. Similarly, models incorporating hormesis are possible although applied infrequently as discussed in Vignette 8.1.

9.2.2 Fitting Data to Dose–Response Models

Methods have been developed to analyze dose–response data based on the concepts and models just discussed. Data (proportions responding paired with doses or concentrations) might be used directly or after transformation. If done, the objective of transformations is most often to make linear the relationship between dose and response. Measurements or transformations of measurements used in the analysis of biological tests are termed **metameters**. Both dose or concentration metameters, and effect metameters can be used for dose–response data. The most common dose or concentration metameter is the logarithm of dose or concentration. Which effect metameter is appropriate depends on whether the log normal, log logistic, or another model is assumed. For example, the log normal model is assumed if the log dose or log concentration transformation is paired with the probit metameter of the proportion dying. The log logistic model is assumed if the log dose or concentration is paired with the logit metameter of the proportion dying.

The probit transformation is derived from the **normal equivalent deviates (NED)**, the proportion dying expressed in units of standard deviations from the mean of a normal curve. For example, a proportion corresponding to the mean (50% of exposed individuals are dead) would have an NED of 0. A proportion below the mean by one standard deviation (approximately 16% of exposed individuals are dead) would have a NED of approximately −1.* In introducing the probit method when mechanical adding machines were the technology of the day, Bliss (1935) viewed negative NED values as inconvenient and added five to NED values to avoid negative numbers.† The resulting metameter is the **probit**. Probit analysis (log normal model) is performed using the log dose or log concentration versus probit (or NED) of the proportion dead,

$$\text{Probit }(P) = \text{NED }(P) + 5 \tag{9.1}$$

where P = proportion of exposed individuals that died by the end of the exposure and NED = the normal equivalent deviation.

The **logit** or log-odds metameter is based on the log logistic model and has the form

$$\text{Logit }(P) = \ln\left[\frac{P}{1-P}\right] \tag{9.2}$$

A transformed logit is more commonly employed than that calculated by Equation 9.2 because values of this transformed logit are nearly the same as probit values except for proportions at the extreme ends of the curves

$$\text{Transformed Logit} = \left[\frac{\text{Logit }(P)}{2}\right] + 5 \tag{9.3}$$

where logit(P) = logit value estimated by Equation 9.2. Other effect metameters are used much less commonly as discussed in Newman (2013). The Weibull transformation is one metameter that could be used more often (Christensen, 1984; Christensen and Nyholm, 1984). Equation 9.4 gives the form of the **Weibull metameter**.

$$U(P) = \ln(-\ln\{1-P\}) \tag{9.4}$$

All of these metameters will generate a straight line for appropriate dose–response data (Figure 9.5). Except at the tails of these lines, the probit and transformed logit metameters have nearly identical values for any set of data. Alternatively, the nonlinear models can be fit with many software packages and, as a consequence, the linearizing transformations just described become unnecessary.

The **median lethal dose (LD50)** and its associated confidence limits are the most common statistics derived from dose–response models. The LD50 is the predicted dose resulting in death of 50% of exposed individuals by a predetermined time such as 48 or 96 hours. Similarly, a **median lethal concentration (LC50)**‡ and its confidence limits are commonly derived from

* The NED value of any proportion (P) can be estimated quickly with the Excel function, NORMINV(P,0,1). For example, NORMINV(0.16,0,1) returns −0.99446.

† Our present need to avoid negative numbers is doubtful. Analysis with the NED instead of the probit produces the same results. The NED used in this way is often called the **normit** metameter and the associated method, normit analysis.

‡ For sublethal or ambiguously lethal effects, the term **median effective concentration (EC50)** may be used instead of LC50. It is often applied to species such as invertebrates for which death is difficult to score and events such as cessation of ventilation or general movement are scored. It is also the term used if sublethal events are being analyzed.

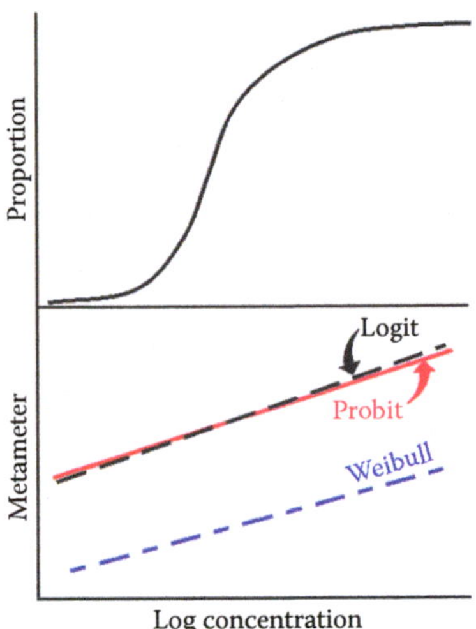

Figure 9.5 The typical sigmoid curve for concentration–response data (top panel) and lines resulting from the probit (solid black), logit (dashed black), and Weibull (dashed blue) transformations (bottom panel).

concentration–response data. It is the concentration predicted to result in death of 50% of exposed individuals by a predetermined time. Median values were adopted as benchmarks instead of low values (e.g., 1% or 5%) because medians tend to be more consistent and to have narrower confidence intervals than lower percentiles. Median values also have an advantage because models such as the probit and logit produce very similar results toward the median. Although these advantages are quite real, it is important to keep in mind that the median was not chosen because it has any particular biological significance. In many cases, concentrations killing lower proportions would be much more meaningful for the ecotoxicologist attempting to determine risk upon toxicant release to the environment.

Many ecotoxicologists estimate values other than the median. Such LCX statistics (e.g., 96 hours LC5 = the concentration predicted to kill 5% of exposed individuals by 96 hours) are more helpful in assessing adverse effects, but they have generally wider confidence intervals and results are more model dependent. This last point is often forgotten because of the misconception, based on our preoccupation with the median, that all models give similar statistics. Model independence is true for practical purposes at the median for the probit and logit models but it becomes progressively less so toward the tails of the model. Careful model selection (e.g., probit vs. logit analysis) becomes important as attention moves away from the LC50. Methods of comparing candidate models will be discussed toward the end of this section.

Numerous methods for estimating the LC50 are available (Table 9.1). Indeed, interpolation from a simple line produced with different sets of metameters, such as the probit of P versus log of concentration (Figure 9.6), could be used to graphically estimate the LC50. However, the 0% and 100% mortality treatments would not be easily plotted on such a graph, and visual fitting would be subjective. For these reasons, more formal methods are applied for estimation of the LC50, its 95% confidence interval, and the slope of the concentration–response line. The last two parameters are as important as the LC50 itself. The LC50 has little utility without some measure of confidence in its calculated value. Also, the slope of the line provides valuable information as can be illustrated easily with Figure 9.6. Imagine a second line intersecting the drawn line at the LC50 but give this second line a much steeper slope. Although the LC50 would be the same for both lines, a small change in concentration has much more of an effect with one toxicant (steep slope) than the other (shallow slope). This is an important piece of information that is often neglected.

Table 9.1 Established Methods for Estimation of LC50

Method	Advantages	Disadvantages	References
Litchfield–Wilcoxon	Quick, semigraphical method	Results are dependent on the individual who is fitting the data "by eye."	Litchfield and Wilcoxon, 1949; Stephan, 1977; Peltier and Weber, 1985; Newman, 2013
Maximum Likelihood Estimation (MLE) or χ^2 Fitting of Specific Model	Powerful, parametric method that can use any of a series of possible models	Requires a specific model such as the log normal or log logistic models; iterative method that is tedious to do manually; MLE results are slightly biased; MLE methods may not converge properly.	Armitage and Allen, 1950; Berkson, 1955; Stephan, 1977; Peltier and Weber, 1985; Newman, 2013
Spearman–Karber	A robust, nonparametric method not requiring a specific model; can trim data to minimize the undue influence of extreme values.	Toxicity curve must be symmetrical; not as powerful as parametric methods.	Hamilton et al. 1977; Stephan, 1977; Peltier and Weber, 1985; Newman, 2013
Binomial	Can be used for data with no partial kills.	Estimation of confidence interval ignores sampling error (Newman, 2013)	Stephan, 1977; Peltier and Weber, 1985; Newman, 2013
Moving Average	Easily implemented	Simple equations for this method require specific progression of toxicant concentrations and numbers of replicates	Stephan, 1977; Peltier and Weber, 1985

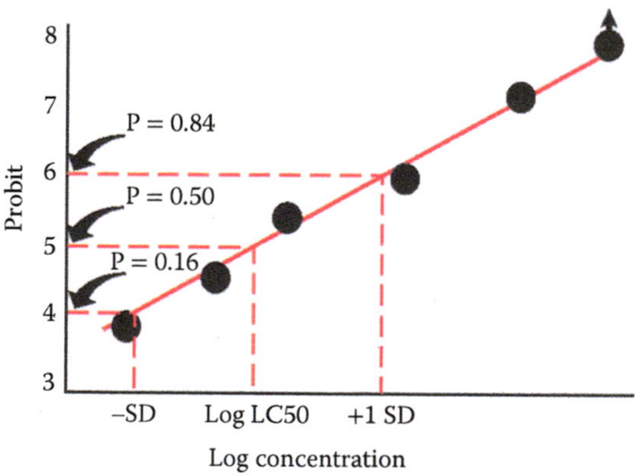

Figure 9.6 Based on a log normal model, probit methods can be used to generate a line from concentration-effect data. An estimate of the log LC50 ± one standard deviation can be approximated from such a graph. Note that a point with 100% mortality is often indicated by an arrow attached to the data point. This indicates that 100% mortality would likely have occurred before the time endpoint and some lower concentration likely would have produced 100% at exactly that time endpoint. Other, more rigorous methods discussed in the text provide better estimates.

The easiest, but most subjective, method for estimation of an LC50 is the **Litchfield–Wilcoxon method**. It is a semigraphical method in which data are graphed first as in Figure 9.6 and the points on the line corresponding to proportions 16%, 50%, and 84% mortality are used to estimate the LC50 and its 95% confidence interval. The antilogarithm of the log concentration corresponding to the $P = 0.50$ is taken as the LC50. A slope factor (S) is calculated from the concentrations corresponding to 16% (LC16), 50% (LC50), and 84% (LC84) mortalities.

$$S = \frac{(LC84/LC50) + (LC50/LC16)}{2} \tag{9.5}$$

This S and the total number of animals exposed in treatments within the range of LC16 and LC84 (N') are used to generate the upper and lower 95% confidence limits.

$$f_{LC50} = S^{2.77/\sqrt{N'}} \tag{9.6}$$

$$\text{Upper Limit} = LC50 \cdot f_{LC50} \tag{9.7}$$

$$\text{Lower Limit} = LC50/f_{LC50} \tag{9.8}$$

As easy as this method is, it is an inconsistent tool because visual fitting of the line will vary among individual practitioners.

Maximum likelihood estimation (MLE) for the log normal, log logistic, or other models is a parametric method that avoids subjectivity but takes on the assumption of a specific model. Probit, logit, and other approaches are most often applied with MLE methods if there were two or more **partial kills** (treatments in which some, but not all, exposed individuals are killed). The MLE results have very good precision relative to those of other methods. The MLE estimates can be slightly biased for small sample sizes; however, this bias is often in the range of those of the other methods. The MLE method is an iterative one and, for this reason, it is most often done with a computer. Data are fit to the model, a goodness-of-fit statistic is calculated, then the estimated parameters are changed slightly and the process is repeated until the goodness-of-fit statistic indicates that the maximum likelihood method found ("converged on") a set of model parameter estimates that best fit the data. Occasionally, the MLE method will fail to converge after many iterations or will converge on an inappropriate (local) solution.

Because various models can be fit to data, questions arise about goodness-of-fit among candidate models. The χ^2 values estimated for each model fit to the data can be used for this purpose. Because the χ^2 value will decrease as fit improves, the ratio of χ^2 values for the different models will reflect relative goodness-of-fit for each model to the data. If the **χ^2 ratio** is less than 1, the model whose χ^2 value is in the numerator (e.g., χ^2_{probit} for $\chi^2_{probit}/\chi^2_{logit}$) fits the data better than the model whose χ^2 value is in the denominator. More formal tests can be carried out with the χ^2 ratio to determine if the fit for one model is significantly better than that of the other.

If it is difficult or unnecessary to justify a specific model for the dose- or concentration-effect data, the nonparametric **Spearman–Karber method** is available to estimate the LC50 and its confidence intervals. The technique only requires a symmetrical toxicity curve. The technique has many steps but can be applied with a hand calculator if necessary. During application of this method, values at the extreme tails may or may not be trimmed to minimize undue influence of extreme values on the estimate. Trimming rules are provided by Hamilton et al. (1977). If trimming is done, the influence of anomalous values will be reduced; however, the standard error of the estimate will increase if done without careful consideration of the data quality.

Two other methods are commonly discussed in ecotoxicology. The **binomial method** allows estimation of the LC50 if there were no partial kills although sampling error is ignored with this method (Newman, 1995, 2013). The **moving average method** may be implemented with straightforward equations if the toxicant concentrations are set in a geometric series, and there are equal numbers of individuals exposed in each treatment.

Occasionally, a model cannot be unambiguously fit to data for chronic toxicity tests. An analysis of variance design as described earlier might be applied instead. The lowest concentration at which mortality is significantly higher than the control is determined via hypothesis testing.

9.2.3 Incipiency

Incipiency when applied to lethality of contaminants is the lowest concentration (or dose) at which an increase in toxicant concentration (or dose) begins to produce an increase in the measured effect. It is often measured with the **incipient median lethal concentration (incipient LC50)**, the concentration below which 50% of exposed individuals will live indefinitely relative to the lethal effects of the toxicant. This concentration is also called the asymptotic, ultimate, or threshold LC50 by various authors. It may be determined graphically in several ways. Figure 9.7 shows one approach taken by van den Heuvel et al. (1991) for rainbow trout (*Oncorhynchus mykiss*) exposed to pentachlorophenol. The LC50 values were determined for a series of exposure times and the reciprocals of the LC50 values plotted against duration of exposure. Alternatively, a double logarithm plot may be used (Newman, 2013). The concentration at which the curve becomes parallel to the *x* axis is an estimate of the incipient LC50. Although widely used and convenient to estimate incipiency, these incipiency measures are difficult to assign ecotoxicological significance. Recall that the median value has ambiguous meaning relative to the continued viability of an exposed population. It follows that any measure of incipiency based on the median will also carry the same ambiguity relative to ecological significance. Statistical limitations also exist for the graphical approach described above (Chew and Hamilton, 1985). For instance, concentration is set in the design yet treated as a dependent variable in subsequent analysis. Applying conventional regression models under such conditions is statistically inappropriate. Newman (2013) explores more complicated models incorporating incipiency.

9.2.4 Mixture Models

Many contaminants of concern are introduced to the environment as mixtures and manifest joint effects in potentially complex ways. Consequently, estimation of lethal effect is more difficult than described to this point for exposure to single toxicant. Regardless of the difficulty, it is important to understand joint effects of chemicals. Such understanding begins by distinguishing among four different conditions that can arise with mixtures: potentiation, additivity, synergism, and antagonism.

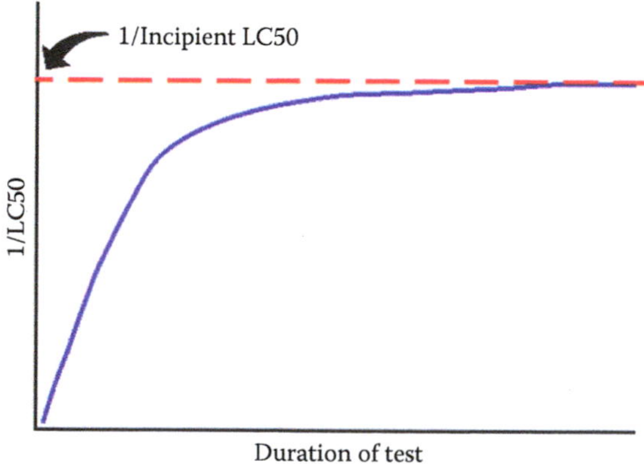

Figure 9.7 The incipient lethal concentration is estimated as the point at which the curve of 1/LC50 versus duration of exposure begins to run parallel to the *x* axis.

Potentiation might occur if one chemical, not toxic itself at the exposure concentration or dose, enhances the toxicity of a second chemical in a mixture.* For example, sublethal concentrations of isopropanol greatly enhance the toxic effects of carbon tetrachloride to the mammalian liver (Klaassen et al. 1987). Disulfiram (also called tetraethylthiuram disulfide or Antabuse) greatly enhances ethanol toxicity in humans (Timbrell, 2000). This ethanol potentiation that occurs at non-toxic doses of disulfiram led to its use as a medication to reinforce drink abstinence by alcoholics. After consumption, ethanol is converted by alcohol dehydrogenase to acetaldehyde, which is then converted to acetate by aldehyde dehydrogenase. Disulfiram inhibits acetaldehyde breakdown, producing an acetaldehyde buildup. The resulting acetaldehyde syndrome includes headache, nausea, vomiting, and dizziness. Potentiation is also used to improve the effectiveness of some insecticide formulations. Piperonyl butoxide, added to insecticide formulations, potentiates pesticide action by inhibiting pesticide breakdown by the cytochrome P450 monooxygenase system. Less pesticide need be released to the environment during application.

Joint effects of chemical mixtures can also result in effect additivity, synergism, or antagonism. Simply put, **additivity** exists if the measured mixture effect was the sum of the expected effects for the individual toxicants. **Synergism** occurs if the observed effect level of the mixture was higher than the sum of the predicted effects for the individual toxicants in the mixture. Toxicant **antagonism** occurs if the observed effect level of the mixture was lower than that predicted by summing the predicted effects for the individual toxicants in the mixture. If antagonism is framed in the special context of antidotes, the chemical whose effect is thought to be reduced is the **agonist** and the chemical that reduces that effect is the **antagonist** or antidote.

Combined effects of chemicals have been illustrated using the simple concept of **toxic units (TUs)**, amounts or concentrations of different toxicants expressed in units of lethality such as units of LD50 or LC50. However, this approach is slowly being replaced by more sophisticated ones. At the onset of its use, the TU was most often expressed as a fraction of the incipient LC50. An example in which TUs are based on the incipient LC50 might be chemical A with an incipient LC50 of 20 mg L^{-1} being present at 0.5 TU in a 10 mg L^{-1} solution. Similarly, chemical B (LC50 = 100 mg L^{-1}) would be present as 0.5 TU in a 50 mg L^{-1} solution. If the toxicity of two toxicants in combination were (concentration) additive, the simple sum of the TUs of the two toxicants would equal the actual toxicity measured for the mixture, e.g., 0.5 TU of A + 0.5 TU of B should equal 1.0 TU of effect when combined as a mixture. In the above example, a mixture of 10 mg L^{-1} of A plus 50 mg L^{-1} of B should result in 50% of the exposed individuals dying. If the mixture actually results in less than 1 TU of lethality, chemicals A and B are said to be less than additive or antagonistic. They are synergistic if their combined effect is more than additive.

The **isobole approach** to mixture analysis is another way of exploring mixture effects (Figure 9.8) (see Chen and Pounds [1998] for further discussion). In Figure 9.8, the solid line connecting 1.0 on both axes reflects additivity and deviation from additivity if a point lies to the lower left (synergism) or upper right (antagonism) of this line. The dashed line represents a hypothetical hyperbole for two toxicants displaying synergistic joint action. In isobole analysis, the observed line produced for different proportions of the toxicants is compared to the theoretical line for strict additivity.

Kinds of antagonism can be identified based on underlying mechanisms (Klaassen et al. 1987). **Functional antagonism** results from two chemicals eliciting opposite effects on physiological or biochemical functions and, as a consequence, counterbalancing each other. An example involving neurotransmitter influence on the heart would be the counterbalancing effects of norepinephrine released from sympathetic nerves (increases contraction rate) by acetylcholine released from parasympathetic nerves (decreases contraction rate). With **chemical antagonism**, two toxicants react

* The term *potentiation* is also used by some authors (e.g., Thompson, 1996) in the context of synergism as discussed below.

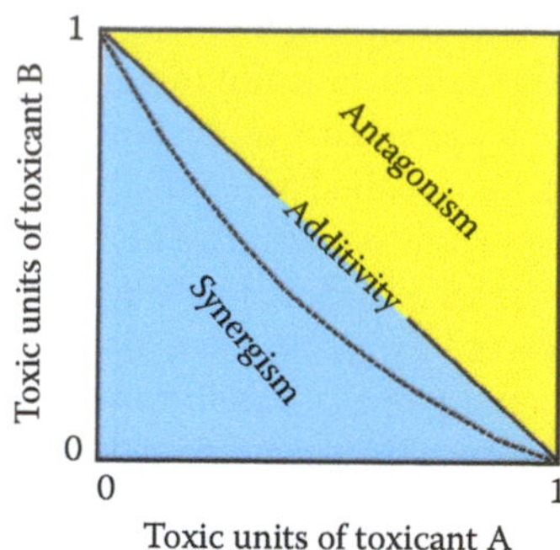

Figure 9.8 An illustration of the isobole approach. The solid line connecting 1.0 on both axes reflects additivity and deviation from additivity if a point lies to the lower left (synergism) or upper right (antagonism) of the figure. The dashed line represents a hypothetical hyperbole for two toxicants A and B displaying synergistic joint action.

with one another to produce a less toxic product. For example, cyanide and a toxic metal may combine in mixture to form a less toxic complex. The undesirable consequence of chemical antagonism can be illustrated by iron complexation with heparin that can interfere with immediate heparin treatment of individuals with deep vein thrombosis or pulmonary embolism.* Protamine sulfate is a chemical antagonist of warfarin, a drug taken after heparin treatment by such patients with a blood clotting disorder. **Dispositional antagonism** involves the absorption, transformation, uptake, movement within the organism, deposition at specific sites, and elimination of the toxicants. The presence of the two toxicants together shifts one or more of these processes to lower the impact of the toxicants on the site(s) of action or target organ(s). This might involve lowered chances of interaction with a target, e.g., less time or lowered concentration available to interact. For example, ethanol enhances mercury elimination in mammals (Hursh et al. 1980; Khayat and Shaikh, 1982) and could modify the toxic effect of mercury as a consequence. The last type, **receptor antagonism**, occurs where two or more toxicants bind to the same receptor and each toxicant blocks the other from fully expressing its toxicity. Klaassen et al. (1986) gave the example of using oxygen to counter the effects of carbon monoxide poisoning based on receptor antagonism. Newman and McCloskey (1996a) found evidence of receptor antagonism for binary mixtures of metals' effects on response of the **Microtox® assay**, a rapid, bacterial assay in which a decrease in bioluminescence is thought to reflect toxic action. A final example of receptor antagonism is atropine administration to block orthophosphate pesticide binding to nerve receptors.†

The TU approach described above is based on the concept of **concentration additivity**: after adjustment for relative potency, concentrations of toxicants can be added together to predict effects under the assumption of additivity. However, as we saw earlier in this chapter, toxicant concentration (or dose) is often related to lethal effect by a sigmoid, not a linear, relationship. Some relationships are pseudolinear over a range of concentrations of interest and concentration additivity can be used to approximate effects. But summation of potency-adjusted concentrations or doses to predict combined effect is not always appropriate. (The interested reader is directed to Berenbaum [1985] for a mathematically explicit explanation of this point that the **summation rule** based on concentration or dose is not generally appropriate.) Instead, to explore additivity, effect levels predicted with models such as the probit or logistic model for each toxicant in a mixture can be added together, not

* Long-term drug therapy of individuals prone to such blood clots involves administration of warfarin, a vitamin K production antagonist that acts by blocking an epoxide reductase production of the vitamin K essential to blood coagulation.

† Although more relevant to medical toxicology, another type of antagonism is possible. **Induction antagonism**, the administration of one agent that induces enzymes before exposure to a second agent, resulting in lowered toxic effect of the second chemical.

the toxicant concentrations in the mixture.* For example, the effect predicted for a certain exposure concentration of toxicant A is predicted from a probit model and added to the effect predicted from another probit model for a certain concentration of toxicant B. The sum of the predicted effects could then be compared with the observed effect of the actual combination of the two concentrations of A and B to assess conformity to or deviations from effect additivity. For example, the separate effects of two toxicants on the proportion of exposed individuals dying (P_A and P_B) are estimated with the following equations:

$$\text{Probit}(P_A) = \text{Intercept}_A + \text{Slope}_A(\log \text{Concentration}_A)$$

$$\text{Probit}(P_B) = \text{Intercept}_B + \text{Slope}_B(\log \text{Concentration}_B)$$

To progress from this approach to a more involved treatment of mixtures, the distinction must be made between toxicants that act independently or similarly. Toxicants can have **similar joint action** in which case, they act by the same mechanism (i.e., share identical modes of action) and "one component can be substituted at a constant proportion for the other & toxicity of a mixture is predictable directly from that of the constituents if their relative proportions are known" (Finney, 1947). **Independent joint action** of toxicants exists if each toxicant produces an effect independent of the other and by a different mode of action (Finney, 1947). Relative to quantifying joint action of such chemicals, Finney (1947) makes an important distinction, "In mixtures whose constituents act similarly any quantity of one constituent can be replaced by proportionate amount of any other without disturbing the potency, but for mixtures whose constituents act independently the mortalities, not the doses, are additive." This distinction is important to keep in mind in the following discussions.

A group of individuals is exposed to a mixture of Concentration_A and Concentration_B, and the proportion dying (P_{A+B}) measured. The three effect proportions (P_A, P_B, and P_{A+B}) can then be used to estimate any deviation from predictions of simple independent joint action. The predicted effect level for the mixture would be the following if the mixed toxicants A and B acted independently in producing the effect (e.g., death) (Finney, 1947):

$$\text{Predicted } P_{A+B} = P_A + P_B(1 - P_A) \tag{9.9}$$

This often is rearranged to the following (e.g., Berenbaum, 1985):

$$\text{Predicted } P_{A+B} = P_A + P_B - P_A P_B \tag{9.10}$$

Using Equation 9.9 or 9.10, the predicted P_{A+B} can be compared to the observed proportion dying after exposure to the mixture of A and B in this case of independent action of the toxicants in the mixture. They should be equivalent with simple independent, joint action.

Likely, it is not obvious to the reader why the right side of Equation 9.9 is not simply $P_A + P_B$. The reason that the right side of this equation contains the $(1 - P_A)$ term can be understood if one envisions the proportion responding as being the probability of an exposed individual dying. If the independently acting toxicants A and B are mixed at the specified concentrations, the probability of dying from A is estimated as P_A. However, an individual must have survived the effects of A to be available to die from B. The probability of surviving the exposure to A is simply $1 - P_A$, so the P_B must be multiplied by $1 - P_A$ instead of any presumed 1 to accommodate for this fact. This relationship can also be rearranged to the following,

$$P_{A+B} = 1 - (1 - P_A)(1 - P_B) \tag{9.11a}$$

* The details provided here are only the most rudimentary for understanding this complex and important topic of joint mixture effects. For a fuller understanding, the reader is urged to review Finney (1947), Plackett and Hewlett (1952), Berenbaum (1985), Chen and Pounds (1998), Eide and Johnsen (1998), Groen et al. (1998), and Mumtaz et al. (1998).

Finney (1947) expands this equation to calculate mixture effects for more than two toxicants with independent action to the following:

$$P_{\text{A+B+C+...}} = 1 - (1 - P_{\text{A}})(1 - P_{\text{B}})(1 - P_{\text{C}})... \tag{9.11b}$$

A slightly different approach is used for toxicants that have similar action. Toxicants with similar action often show parallel slopes in their probit models so a simple measure of relative potency can be calculated to predict the effect of two similarly acting toxicants in mixture (Finney, 1947). Using Finney's equations converted to the form shown above:

$$\text{Probit}(P_{\text{A}}) = \text{Intercept}_{\text{A}} + \text{Slope (log Concentration}_{\text{A}})$$
$$\text{Probit}(P_{\text{B}}) = \text{Intercept}_{\text{B}} + \text{Slope (log Concentration}_{\text{B}})$$

The log of the relative potency can be calculated as follows:

$$\log \rho_{\text{B}} = \frac{(\text{Intercept}_{\text{B}} - \text{Intercept}_{\text{A}})}{\text{Slope}} \tag{9.12}$$

So the predicted effect for the mixture of A and B can be calculated with the equation:

$$\text{Probit}(P_{\text{A}} + P_{\text{B}}) = \text{Intercept}_{\text{A}} + \text{Slope} \cdot \log(\text{Concentration}_{\text{A}} + \rho_B\{\text{Concentration}_{\text{B}}\}) \tag{9.13}$$

Other approaches can be used. As a more involved example, Carter et al. (1988) applied the regression methods for the logistic isobologram model:

$$\log\left[\frac{P}{1-P}\right] = \beta_0 + \beta_{\text{A}} C_{\text{A}} + \beta_{\text{B}} C_{\text{B}} + \beta_{\text{AB}} C_{\text{A}} C_{\text{B}} \tag{9.14}$$

where P = proportion of exposed individuals dying, C_{A} = the concentration metameter for toxicant A, C_{B} = the concentration metameter for toxicant B, β_0 = the y-intercept, β_{A} = regression coefficient for the effect of toxicant A, β_{B} = regression coefficient for the effect of toxicant B, and β_{AB} = regression coefficient for the interaction term, $C_{\text{A}}C_{\text{B}}$. Deviations from additivity would be suggested by the β_{AB} coefficient. The equivalent probit model would be the following:

$$\text{Probit}(P) = \beta_0 + \beta_{\text{A}} C_{\text{A}} + \beta_{\text{B}} C_{\text{B}} + \beta_{\text{AB}} C_{\text{A}} C_{\text{B}}. \tag{9.15}$$

Perhaps, the best example of coping with mixtures of similarly acting toxicants is the approach taken for mixtures of compounds having aryl hydrocarbon (AH) receptor interaction as the first stage of their toxic action. Compounds with very similar modes of action of this kind include dioxins, dibenzofurans, and dioxin-like polychlorinated biphenyls (PCBs). These chemicals are often present in mixtures so assessment of their net effect must be made. Based on empirical laboratory evidence, **toxic equivalency factors (TEFs)** are calculated for each such compound in the mixture and the concentrations of individual chemicals in the mixture are added together after multiplication of each by its toxic equivalency factor. One of the most toxic AH receptor-binding compounds is 2,3,7,8-tetrachlorodibenzo-p-dioxin (TCDD). It is given a toxic equivalency value of 1 and other similarly acting compounds are given experimentally derived TEF scaled to the TCDD TEF of 1. For example, a moderately active PCB might be given a TEF of 0.001. The **toxic equivalent (TEQ)** of this PCB would be its concentration multiplied by its TEF, i.e., its effect scaled to that of TCDD. Assuming that these compounds are similarly acting, the joint effect of a mixture can be expressed as the sum of the TEQs for all of the constituent dioxins, dibenzofurans, and dioxin-like PCBs (Van den Berg et al. 1998).

Let us return for a moment to assessments of mixture additivity. Many mixture studies in ecotoxicology use the simple additive index (AI) approach to determine whether additive or nonadditive action is occurring. For example, Thomulka and Lange (2003) used this approach with bacterial response to mixtures of dinitrobenzene and trinitrobenzene. Marking and Dawson (1975) generated an **additive index** for assessing the joint action of toxicants in mixtures. Letting, A_m and B_m = the toxicity (e.g., incipient LC50) of toxicants A and B when present in mixture, and A_i and B_i = toxicity of A and B when they are tested separately,

$$\frac{A_m}{A_i} + \frac{A_m}{B_i} = S. \tag{9.16}$$

The toxicants are antagonistic if $S < 1$, synergistic if $S > 1$, or additive if $S = 1$ (Figure 9.9, "sum of toxic contributions" scale). Although shown here for a binary mixture, several toxicants can be added to Equation 9.16, if desired. The units are not linear to the right and left sides of 1 on this scale so a change from +1 to +2 is not of the same magnitude as a change from −1 to −2. Units can be made linear by making values to the left of additivity equal to $-S + 1$ and values to the right of additivity equal to $1/S - 1$:

$$\text{Additive Index (AI)} = -S + 1 \text{ for } S > 1.0 \tag{9.17}$$

$$\text{Additive Index (AI)} = \frac{1}{S} - 1 \text{ for } S \leq 1.0 \tag{9.18}$$

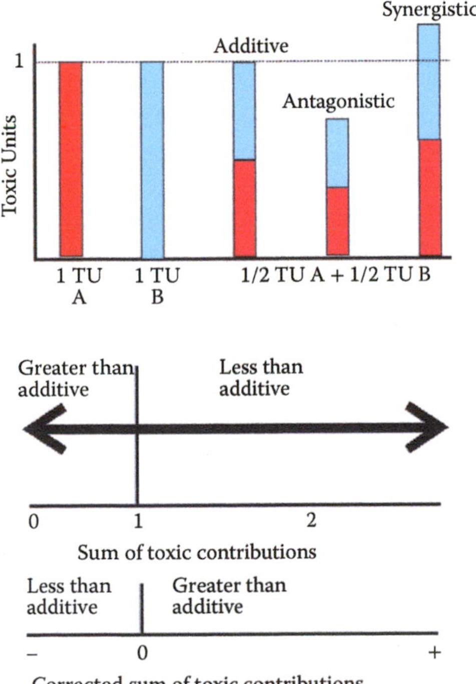

Figure 9.9 The combined effect of toxicants may be quantified by expressing toxicant concentrations in mixtures as toxic units (TU) (upper panel). If the realized effect expressed in terms of TU is less than the calculated sum of TU for both toxicants A and B in a mixture, the chemicals are said to be antagonistic. Their effect is synergistic if the realized effect is greater than the calculated effect based on their individual actions. The two scales include the nonlinear scaling of Marking and Dawson (1975) (middle) and additive index (lower), which is a linear scaling of combined toxicant effect.

Notionally, this additive index is linear on both sides of additivity (0) (Figure 9.9, "corrected sum of toxic contributions"). Negative and positive numbers indicate less than and greater than additivity, respectively.

In common ecotoxicology practice, the TU and additive index are the most widely used tools for quantifying joint toxicant effects. Although the mathematical models provided for independently acting and similarly acting toxicants are more inclusive, Marking's additive index (Marking and Dawson, 1975; Marking, 1985) is a straightforward and pragmatic means in many instances for visualizing combined effects of mixtures. The concept of concentration additivity also forms the foundation for estimation of combined effects of contaminants in solutions or solids. For example, Di Toro et al. (1990) assumed additivity in estimating the combined effects of metals in contaminated sediments and performed mathematical modeling accordingly. The current use of the TU and additive index is a consequence of their ease of application and historical role in ecotoxicology, not their direct linkage to underlying theory, or power to quantify and test for significant deviations from additivity. More thorough analysis of mixtures can be done with the general linear model approach including interaction terms (e.g., Equations 9.9 through 9.15; see Neter et al. [1990] for details of regression methods). There are now many convenient software packages for personal computers, which implement such general linear modeling procedures that should allow more general implementation of these methods.

9.3 SURVIVAL TIME

9.3.1 Basis for Time–Response Models

The **time–response approach** is becoming an increasingly common alternative to the dose– or concentration–response approach. In the conventional dose– or concentration–response approach, exposure time is held constant although results (numbers dying) are sometimes acquired at a series of time intervals. This approach provides a gross indication of the influence of exposure duration; however, full consideration of temporal dynamics is sacrificed to generate estimates of chemical toxicity expressed as amounts or concentrations.

Accurate measurement of the effect of exposure duration is also extremely important to assessing ecotoxicological consequences of contamination: not all toxicant releases into the environment are 48 or 96 hours in duration. To accomplish this, times-to-death (TTD) are measured in the time–response approach. More data are generated for each exposure tank, resulting in enhanced statistical power. For example, 10 TTD values can be generated instead of one proportion from an exposure treatment involving 10 fish. Also advantageous, important qualities influencing toxic consequences such as individual sex or weight can be more easily included in the analysis because the endpoint (TTD) is associated with individuals. The enhanced power allows more effective inclusion of other qualities such as temperature, toxicant concentration, or nutritional history. The inclusion of concentration in the survival time approach provides a highly desirable model that predicts effect as a function of both exposure duration and concentration (Figure 9.10). Finally, as we will discuss in the Chapter 10, results of time–response analysis can be readily incorporated into demographic analysis.

Despite these clear advantages, ecotoxicologists still underutilize time–response methods. Traditions including those taught to students and formalized in regulations seem to have a strong hold on ecotoxicologists in this area. Kooijman (1981), Chew and Hamilton (1985), Dixon and Newman (1991), Newman and Aplin (1992), Newman et al. (1994), Newman (1995), Roy and Campbell (1995), Newman and Dixon (1996), Newman and McCloskey (1996b), Crane et al. (2002),

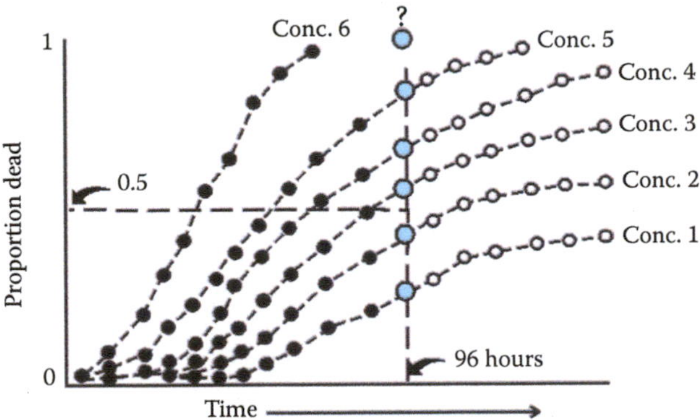

Figure 9.10 An idealized time course of mortality occurring at six different concentrations. Note that, if only data from 96 hours were used, only five points (blue) with perhaps some estimate of a sixth for Concentration 6 (blue with question mark) would be available to calculate the LC50. In contrast, many more data points (approximately 70 points) would be gathered by 96 hours if times-to-death for individuals were noted instead of the final proportion dying in each tank. For a more complete data set, observations could be made that noted mortality occurring at intervals before (black) and after (white) 96 hours.

and Newman (2013) provide general discussion and examples of time–response methods applicable to ecotoxicology.

9.3.2 Fitting Survival Time Data

The **Litchfield method** is a very simple method for analyzing survival time data (Litchfield, 1949). It was developed during the same year as the Litchfield–Wilcoxon method for estimating LC50 and the methods differ only in minor details. First, the proportion of exposed individuals dying is tabulated for a series of exposure durations. The probit of the proportion responding is plotted against the logarithm of exposure duration to produce a straight line, e.g., Figure 9.11. (Originally log-probability paper was used instead of these transformations.) A line is fit by eye to these data and the time corresponding to the probit for 50% mortality is the estimated **median lethal time** (LT50).* To calculate the 95% confidence interval, a slope factor is derived as the following:

$$S = \frac{(\text{LT84}/\text{LT50}) + (\text{LT50}/\text{LT50})}{2} \tag{9.19}$$

where LT16, LT50, and LT84 = exposure durations corresponding to when 16%, 50%, and 84% of exposed individuals were dead, respectively. An f_{LT50} is calculated to estimate the 95% confidence interval for the LT50.

$$f_{\text{LT50}} = S^{1.96/\sqrt{N}} \tag{9.20}$$

$$\text{Upper Limit} = \text{LT50} \cdot f_{\text{LT50}} \tag{9.21}$$

$$\text{Lower Limit} = \text{LT50}/f_{\text{LC50}} \tag{9.22}$$

* As discussed for LC50, a **median effective time** (ET50) can be used instead of LT50 if the effect is sublethal (e.g., time-to-stupefication) or ambiguously lethal (e.g., time-to-valve gap for bivalves).

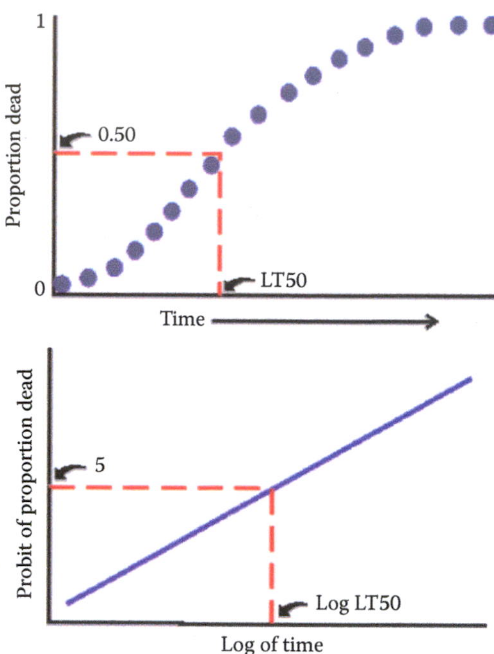

Figure 9.11 Linear transformation of time-to-death data assuming a log normal model. A sigmoidal curve is generated if the cumulative proportion of the exposed individuals that have died is plotted against exposure time (upper panel). Under the assumption of a log normal model, a straight line is produced (lower panel) if log time is plotted against the probit of the proportion dead. A curve which is not straight, but instead has a break in its slope, is called a **split probit**. Traditionally, a split probit is assumed to reflect either two distinct mechanisms of toxicity at the beginning and later in the exposure, or two distinct subpopulations of individuals in the exposed group that differ in tolerance to the toxicant. A third possibility exists, the log normal model is an inappropriate model for the data, but is rarely assessed.

where N = the total number of individuals exposed during the test if all individuals died during the test. If there were survivors, the N in Equation (9.20) is adjusted as described originally by Litchfield (1949) or more conveniently as described by Newman (2013).

Although ecotoxicology has traditionally considered only this log normal model, other models for TTD data are commonly assumed or explored in other fields. The most commonly employed include the exponential, gamma, Weibull, normal, log normal, and log logistic models. Fit to these models may be visually assessed with a series of linear transformations (e.g., Table 9.2). The model with a set of transformations that result in a straight line (e.g., Figure 9.11 for the log normal model) is selected as the most appropriate.

Like dose–response data, TTD data can be analyzed with several methods, ranging from the simple Litchfield method to more involved nonparametric, semiparametric, and fully parametric methods. The nonparametric **product-limit (Kaplan–Meier) methods** do not require a specific model for the survival curve. Product-limit methods allow estimation of survival through time and associated computations can be used to test for significant differences among treatments. At the other extreme, fully parametric techniques assume a specific model for the survival curve (e.g., log normal or Weibull model) and a specific function relating survival to covariates (e.g., to exposure concentration or animal size). The underlying distribution of mortality through time can be selected from a series of candidate models such as those in Table 9.2. These models are usually fit using a maximum likelihood method as described earlier for dose–response data.

Table 9.2 Transformations of Times-to-Death Data Used to Select from among Candidate Models

Candidate Model	Transformation of Mortality	Transformation of Time
Exponential	ln $S(t)$	t
Weibull	ln [–ln $S(t)$]	ln t
Normal	Probit [F(t)]	t
Log Normal	Probit [$F(t)$]	ln t
Log Logistic	ln [$S(t)/F(t)$]	ln t

t, duration of exposure; $S(t)$, the cumulative survival to time t expressed as the proportion of exposed individuals still alive at time t; and $F(t)$, the cumulative mortality to time t which is 1 – $S(t)$.

Depending on which model is selected to describe the shape of the survival curve, the fully parametric model takes one of the two general forms: proportional hazard or accelerated failure time. Selection of an exponential or Weibull model produces a proportional hazard model. A **proportional hazard model** can be used to conveniently relate the hazard (proneness to or risk of dying at any time, t) of one group (e.g., toxicant exposed) quantitatively to that of a reference group (e.g., not exposed). Results of proportional hazard models for human mortality are often expressed as easily understood **relative risks**, risks of one group expressed as a multiple of that of another. For example, the risk of dying from lung cancer may be X times higher for smokers relative to nonsmokers or the risk of surviving a heart attack is Y times higher if one exercises relative to that for someone who does not exercise regularly. Proportional hazard models can be expressed in the general form:

$$h(t,x_i) = \mathrm{e}^{f(x_i)}\, h_0(t) \tag{9.23}$$

where $h(t, x_i)$ = the hazard at time, t, as modified by the value (x_i) of the covariate x, $h_0(t)$ = the hazard of some reference group or type, which is described by a specific model such as a Weibull model, and $f(x_i)$ = some function of the covariate (x) making the hazards proportional. For example, $f(x)$ may be a simple function making hazards proportional among different animal sizes, e.g., $f(x) = a + b$ (log animal weight).

Use of the log normal, log logistic, normal, or gamma model results in an **accelerated failure time model**, a model in which the time-to-death (ln TTD_i) of a particular type of individual (e.g., smoker) is changed ("accelerated") as a function of some covariate (e.g., classification relative to smoking habit). For example, an increase in toxicant concentration may accelerate the expected time until death of an individual. The simple form of the accelerated failure time model is the following:

$$\ln \mathrm{TTD}_i = f(x_i) + \varepsilon_i \tag{9.24}$$

The error term (ε) is fit using an assumed model such as a log normal model. Some function of the covariate modifies the ln TTD. It can be any function of the covariate such as that given above for the effect of animal weight, e.g., $f(x) = a + b$ (log animal weight).

Most applications fitting parametric survival time models allow parameter estimation so that predictions can be made about a **median time-to-death** (MTTD) or some proportion other than 50% mortality. They also allow one to test for statistically significant effects of different covariates on lethal effect.

Often, especially in clinical studies comparing various treatments, it is impossible or unnecessary to model the underlying survival curve. Sometimes, the exact underlying model is much less important than estimating the influence (hazard or relative risk) of some covariate on survival. For example, the focus of a study may be on determining if a postsurgical treatment improves patient survival relative to the standard treatment (reference treatment). In such a study, knowledge about the exact form of the survival curve is not required. A semiparametric method (**Cox proportional hazard model**) is designed to allow examination of proportional hazards without taking on the assumption of any specific model for the baseline or reference hazard.

9.3.3 Incipiency

The LT50 or MTTD may be used to estimate lethal incipiency, the concentration at which 50% of exposed individuals will live indefinitely relative to the toxicant effects. The point at which the line for toxicant A begins to run parallel to the log MTTD axis is an estimate of this lethal threshold in Figure 9.12. Note that there may be no apparent threshold for some toxicants (e.g., Toxicant B). Also, there may be a minimum time required before an effect can be expressed. Toxicant B illustrates such a **minimal time to response**.

9.3.4 Mixture Models

Mixture effects can also be quantified for the time–response approach as done with the dose–response approach. For example, the lethal threshold estimated with the MTTD or some other estimate is used instead of the incipient LC50. However, more involved treatments become possible because more data can be extracted from a toxicity test with these methods. As an example, Roy and Campbell (1995) took advantage of this enhanced power of survival time models to quantify the combined effect of aluminum and zinc to young Atlantic salmon (*Salmo salar*).

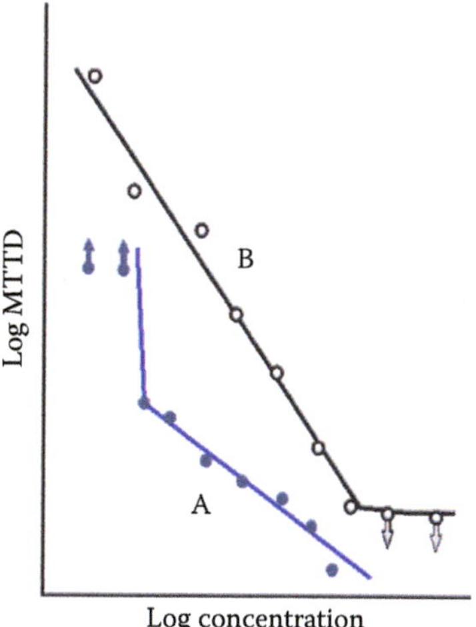

Figure 9.12 Incipiency for the time-to-death approach. Chemical A (blue circles) has a distinct incipient median time-to-death (MTTD). Below a certain concentration, 50% of exposed individuals will live indefinitely relative to the effect of toxicant A. Note that the two lowest concentrations resulted in less than 50% mortality regardless of how long the individuals were exposed. Some toxicants (B, black circles) may show no evidence of incipiency. There may also be a minimum time to get a response, e.g., individuals can only die so fast regardless of the toxicant concentration. Toxicant B in the diagram has such a threshold. (Arrows attached to a point signify that the "true" value is probably in the indicated direction from the point; e.g., the true log MTTD for the last point on curve B is less than the value of the last point.) (Modified from Figure 5 in Sprague, J. B., *Water Res.*, 3, 793–821, 1969 and Figure 4 in Newman, M. C., *Quantitative Methods in Aquatic Ecotoxicology,* Lewis Publishers, Boca Raton, FL, 1995.)

9.4 FACTORS INFLUENCING LETHALITY

9.4.1 Biotic Qualities

Many biological qualities influence toxicant effects. Some have already been discussed, as examples, developmental stage, lipid pools within the individual, feeding behavior, and induction of detoxification mechanisms. Life stage is also important (e.g., Campbell, 1926). Because of their widespread occurrence, two general biological qualities that can influence lethality, acclimation, and allometry will be discussed briefly in this section.

Acclimation* is the modification of biological functions, especially those physiological, or structures to maintain or minimize deviations from homeostasis despite change in some environmental quality such as temperature, salinity, light, radiation, or toxicant concentration. It is an expression of phenotypic plasticity of individuals in response to a sublethal change in some environmental factor. Often the distinction is made in the literature between acclimation (shifts taking place in a controlled or laboratory setting) and **acclimatization** (shifts taking place under natural conditions). It is important to understand that acclimation is not the shift in population qualities, i.e., not a change in genetic composition or demographic structure in response to a toxicant. Such adaptation or genetic change in a population as a consequence of selection will be discussed in the Chapter 10.

Acclimation can occur after a preexposure to sublethal concentrations of toxicants so that survival of individuals is enhanced during a subsequent, more intense exposure. However, some toxicants cause damage regardless of their concentrations, e.g., cyanide preexposure does not enhance survival of rainbow trout because of long-lasting kidney damage that occurred during the first exposure (Dixon and Sprague, 1981a,b). In contrast, preexposure of human lymphocytes to low levels of radiation leads to reduced chromatid aberrations during a subsequent exposure to X-rays (Olivieri et al. 1984). Dixon and Sprague (1981a) found that, above a certain concentration, preexposure of rainbow trout to copper enhanced their tolerance to otherwise lethal concentrations of copper. It is important to note that the enhanced tolerance increased with copper concentration only to a certain point; beyond that point, preexposure caused damage and consequent, diminished tolerance in preexposed fish. Enhanced tolerance was a complex function of both acclimation concentration and time (Dixon and Sprague, 1981b). For example, both strongly influenced lethal consequences to salmon exposed to high temperatures after varying intensities and durations of preexposure (Elliott, 1991).

Allometry can also influence toxicant effect on individuals (Newman and Heagler, 1991; Newman, 2013). The classic work of Bliss (1936) used arsenic intoxication of silkworm larvae to estimate the influence of size on toxic impact. First, he took TTD data and transformed it to rate of toxic action (rate of toxic action = 1000/TTD). Next, he performed a multiple regression to generate the model:

$$\text{Rate} = a + b_1(\log \text{dose}) + b_2(\log \text{weight}) \tag{9.25}$$

where a, b_1, and b_2 are constants derived during model fitting. This relationship was transformed to generate an adjustment factor (weight^h) for the influence of animal weight, where $h = b_2/b_1$.

$$\text{Rate} = a + b_1 \quad \log \frac{\text{dose}}{\text{weight}^h} \tag{9.26}$$

* Acclimation can also be used to identify the time allowed for an organism, population, community, or ecosystem to stabilize to a set of conditions before testing. For example, the term might be used to describe the establishment of mesocosms for ecotoxicological testing. Acclimation will not be used in that context (relative to an experimental protocol) here.

This approach has been adopted during the last 60 years as the primary approach to incorporating size effects into TTD data. The approach (Bliss, 1936) was later modified to dose- or concentration-effect models (Anderson and Weber, 1975),

$$\log \text{LC50} = \log a + b \log \text{weight} \quad (9.27)$$

or

$$\text{LC50} = a \text{ weight}^b \quad (9.28)$$

Newman et al. (1994) and Newman (2013) opined that a more general approach exploring the various models described for survival time models leads to a better fit to these data.

9.4.2 Abiotic Qualities

Numerous physical and chemical factors modify the toxic action of contaminants. As discussed, a compound's lipid solubility can enhance its accumulation in organisms and the extent of consequent harm. For example, Figure 9.13 shows a simple QSAR for 48 hour LC50 for grass shrimp exposed to a series of oil spill-related polycyclic aromatic hydrocarbons (PAHs) with a range of lipid solubilities. Ambient and acclimation temperatures modify the impact of some toxicants. For example, toxic impact (LC50) of cadmium was modified by acclimation and ambient temperature during exposures of a snail (*Potamopyrgus antipodarum*) (Møller et al. 1994). Ambient light can also influence the action of photolabile chemicals by changing the rate at which they break down to more toxic products. **Photoinduced toxicity**, toxicity of a chemical in the presence of light because of the production of toxic photolysis products or activating other molecules via energy captured during interactions with light, was found for bluegill sunfish (*Lepomis macrochirus*) exposed to the PAH, anthracene (Bowling et al. 1983) (Vignette 9.1). Some chemicals can also enhance the **photosensitivity** (sensitivity of cutaneous tissues to the effects of light evoked by a chemical) of individuals. Channel catfish (*Ictalurus punctatus*) treated with the antibiotic, oxytetracycline, become

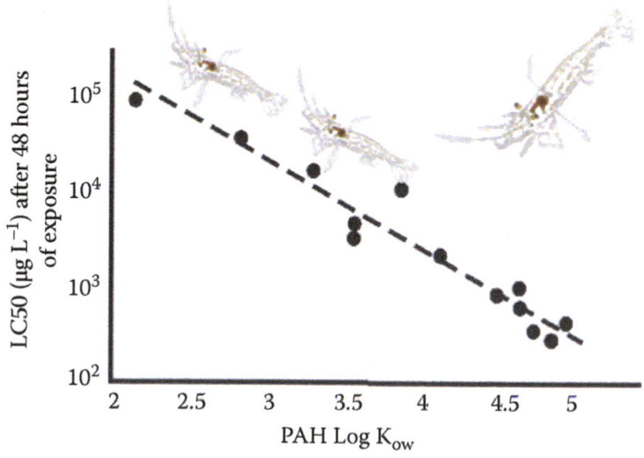

Figure 9.13 An illustration of lipid solubility effect on LC50 of grass shrimp (*Palaemonetes pugio*) exposed to a series of PAH, commonly associated with oil spills. (Data from Unger et al. *Environ. Toxicol. Chem.* 27, 1802–1808, 2008.)

extremely sensitive to sunlight, resulting in skin (sunburn) and eye (lesion) damage (Stacell and Huffman, 1994).*

Organic and inorganic constituents of waters might modify metal toxicity (e.g., Andrew et al. 1977; Bradley and Sprague, 1985; Azenha et al. 1995). Freshwater **hardness** (sum of the concentrations of dissolved calcium and magnesium) is often related to toxic effect of metals such as cadmium (Carroll et al. 1979), copper (EPA, 1985c), lead (Davies et al. 1976; EPA, 1985d), and zinc (EPA, 1985f). Several mechanisms have been advanced, including competition of hardness cations (Ca^{2+} and Mg^{2+}) with toxic metals for binding sites on biomolecules, modification of biological processes, and modification of metal speciation in the exposure waters. These effects on metal toxicity are predicted in EPA water quality criteria documents (e.g., EPA, 1985e) with a simple, empirical model:

$$\log \text{Toxicity Endpoint (e.g., LC50)} = \log a + b\,(\log \text{Hardness}) \tag{9.29}$$

or

$$\text{Toxicity Endpoint} = a\ \text{Hardness}^{b} \tag{9.30}$$

where log a and b = estimates from linear regression of the intercept and slope. Although Equation 9.30 produces a biased prediction of toxic effect (Newman, 1991, 1993, 2013), it is used extensively in the ecotoxicological literature and water quality criteria documents without consideration of the easily corrected bias.

Sediment and soil qualities also influence toxicity as described in earlier discussions of factors modifying bioavailability. Further, QSARs are readily developed for lethality as well as bioaccumulation. The reader is encouraged to review information covered in Chapter 4 as topics discussed there are relevant to this discussion of abiotic factors influencing toxicity.

VIGNETTE 9.1 Photoinduced Toxicity of Polycyclic Aromatic Hydrocarbons in Aquatic Systems

James T. Oris
Miami University

INTRODUCTION

The interaction of solar radiation with organic and inorganic chemicals has shaped the composition of the atmosphere and life on earth. Over evolutionary periods, molecules that absorb wavelengths of sunlight were incorporated into living organisms, converting solar energy into chemical reactions that split water into hydrogen ions, high-energy electrons, and molecular oxygen. These photochemical reactions were used to convert carbon dioxide into reduced organic chemicals that could be used to power the biochemical reactions of cells and, as a by-product, released molecular oxygen into the atmosphere. Photochemical reactions of atmospheric oxygen with solar radiation led to the formation and maintenance of high concentrations of ozone in the upper atmosphere, absorbing high-energy ultraviolet radiation (UVR) and preventing DNA damaging reactions from occurring at the surface of the earth. The rest, they say, is history.

* This is also the case for humans taking certain antibiotics. Physicians and labels on some prescription antibiotics warn patients not to spend much time in the sun because of increased sensitivity to sunburn.

Thus, throughout evolutionary history, organisms have utilized the energy from the sun to make food and essential nutrients, and to power the chemical reactions of life. However, not all photochemical reactions are beneficial. Complex organic molecules with multiple, conjugated covalent bonds such as polycyclic aromatic hydrocarbons (PAHs) have outer orbital electrons that absorb solar radiation in the wavelength range that reaches the earth's surface. Energy absorbed by these electrons elevates them to excited state orbitals. That energy can be sufficient to break bonds (photolysis), can be released in the form of light and heat (fluorescence or phosphorescence), or can be passed on to other molecules in proximity to the excited state chemical (**photosensitization**). In particular, if the absorbed energy is passed to molecular oxygen, that transfer creates highly reactive oxygen species that degrade cellular molecules and lead to organismal damage, dysfunction, and death (photoinduced toxicity).

HISTORICAL PERSPECTIVES

In their interactions with the environment, humans have known about the detrimental effects of photoinduced toxicity for thousands of years. Perhaps, the first recorded description of photoinduced toxicity came from Egyptian papyrus writings dating back to 1550 BC (Bennedetto, 1977). Those writings describe severe irritation to the skin after ingestion of specific plant materials, now known to contain polycyclic aromatic photosensitizing compounds called psoralens, and subsequent exposure to sunlight. In the last century, as the use of fossil fuels became prevalent for transportation and energy production, medical researchers began to explore the relationship between combined exposure to PAH and UVR, and adverse outcomes such as irritation, edema, severe sunburn, and cancer of the skin (e.g., Findlay, 1928; Doniach and Mottram, 1937; Burkhardt, 1939). Polycyclic aromatic hydrocarbons in coal, oil, and coal tar such as anthracene, methylanthracene, and fluoranthene were identified as common contributors to UVR-induced phototoxicity (Burkhardt, 1939; Kochevar et al. 1982). In practical terms, workers in occupations closely associated with PAHs, like chimney sweeps and those who work with asphalt or creosote, know from experience that they need to avoid sunlight after exposure to these chemicals—all you have to do is ask. Once, while installing old railroad ties in my yard for landscaping purposes, a good neighbor of mine who was a rail yard worker made a point to come over and warn me not to sit or lay bare skin on the ties because "everyone knows that you will get a hell of a sunburn."

Although photoinduced toxicity of PAH could be considered common knowledge, this knowledge was largely relegated to the medical literature and terrestrial biologists until the latter part of the twentieth century. Because of limitations in analytical techniques and instrumentation and, perhaps, a lack of interdisciplinary collaboration with physical oceanographers and photochemists, aquatic biologists largely ignored the potential for photoinduced toxicity in fish and aquatic plants or invertebrates. The notion that UVR could cause adverse effects in aquatic systems began to emerge in the early to mid-1900s with reports by marine and fisheries biologists describing sunburn and mortality of aquatic organisms exposed to sunlight (e.g., Klugh, 1929; Crowell and McCay, 1930; Bell and Hoar, 1950). One report described mortality in trout held in "black asphaltum painted" troughs after 2 days of exposure to bright sunlight (Dunbar, 1951). The report's author suspected that UVR may have contributed to the mortality, but he did not connect that to the combined exposure to PAH's leaching from the asphalt painted troughs with the UVR exposure. From this time into the late 1970s biologists began to appreciate the importance of UVR in aquatic systems and began to explore how PAH and other compounds may impact the health of aquatic organisms.

In the late 1970s and early 1980s, a group of researchers at the Savannah River Ecology Laboratory (Aiken, SC) were studying anthracene in freshwater systems. Anthracene is a linear, three-ring PAH with a water solubility of approximately 30 μg L^{-1} at 20°C in freshwater. It is a common component of coal, oil, coal tar, and asphalt and was long known in the medical literature, but never in the ecological or aquatic literature, to be a photosensitizing compound. The studies being conducted were aimed at the use of laboratory experiments to develop, and a large outdoor mesocosm study to help validate, a model of the fate and dynamics of anthracene in streams. A concentration of one-half the water solubility of anthracene was chosen to conduct the experiments, and nothing from the lab studies indicated that this concentration of compound was toxic to any of the organisms tested. However, within a few hours of initiating the outdoor mesocosm study, caged fish in the anthracene streams showed signs of severe stress and quickly died. Instead of writing off the experiment as a failure, the investigators did additional experiments and discovered that fish were killed only when they were exposed simultaneously to both anthracene and sunlight (Bowling et al. 1983). This study became the first to show that photoinduced toxicity of PAHs could occur at environmentally relevant concentrations in aquatic environments (Giesy et al. 2013).

Further studies since that time have elucidated the chemical/physical characteristics of PAHs necessary to cause photoinduced toxicity, the biochemical and physiological mechanisms of action, and environmental factors that serve to attenuate the toxicity in natural aquatic systems.

CHEMICAL/PHYSICAL CHARACTERISTICS

Many types of molecules can act as photosensitizers. However, for simplicity sake, this essay will focus solely on PAHs in the aquatic environment. The first and second laws of photochemistry dictate the necessary conditions for photoinduced toxicity (Diamond, 2003). To be phototoxic, a PAH must absorb light energy (first law) and that energy must be exactly equal to the difference in the molecule's outer shell electrons' ground state and excited state orbitals (second law). Thus, PAHs that have the potential to be phototoxic must have absorption spectra within the wavelength range of solar radiation present in a particular body and depth of water. In the aquatic environment, this implies that phototoxicity may occur throughout the range of environmental solar radiation (UVB: 280–320 nm; UVA: 320–400 nm; PAR: 400–700 nm*). PAHs that fall within this category typically have three to five benzene rings.

If the absorbed energy does not break bonds via photolysis, excited state electrons with paired spin (singlet excited state) return to ground state by emitting light and heat in the form of fluorescence. Fluorescence lifetimes are very short (approximately 10^{-9} seconds) and there is little chance of the energy from those electrons interacting with other molecules. Some PAHs have a propensity to use some of the absorbed energy to unpair the spin of excited state electrons, dropping to an intermediate excited state—the triplet excited state. Because the electron spin is unpaired in the triplet state, it takes longer for these electrons to return to ground state. Energy from the triplet excited state is typically released as light and heat in the form of phosphorescence, with lifetimes $>10^{-3}$ seconds. Because phosphorescence lifetimes are so long, there is much greater chance of interaction between the excited state PAH and other molecules, and that energy can be transferred to generate excited state forms of those molecules. In particular, when this energy is transferred to molecular oxygen the energy state of the molecule is converted from its normal triplet ground state to its extremely reactive singlet excited state.†

Thus, PAHs capable of causing photoinduced toxicity must have outer electron orbital characteristics that enhance the absorption of ambient solar radiation leading to the formation of triplet excited state molecules that transfer their energy to oxygen. In attempts to predict the potential for photoinduced toxicity, empirical studies in aquatic organisms demonstrated that three- to five-ring compounds that absorbed radiation in the 310–400 nm range (primarily

* PAR is the abbreviation for photosynthetically active radiation portion of the light spectrum.

† Molecular oxygen is atypical in that it exists in its ground state as a less reactive, triplet state; if molecular oxygen were more typical and existed in the highly reactive singlet state, our lives would be very different.

UVA) and that had long phosphorescence lifetimes tended to be more phototoxic (Newsted and Giesy, 1987; Oris and Giesy, 1987). Further study of photochemical properties of phototoxic PAH indicated that compounds with an energy difference between the Highest Occupied Molecular Orbital (HOMO) and the Lowest Unoccupied Molecular Orbital (LUMO) in the range of 6.8–7.4 eV would generate the reactive oxygen species necessary to cause photoinduced toxicity (Mekenyan et al. 1994). Given a combination of absorption spectra, phosphorescence lifetimes, and HOMO–LUMO gap information, one can predict which PAH may act as phototoxic compounds.

BIOCHEMICAL AND PHYSIOLOGICAL MECHANISMS OF ACTION

Phototoxic PAHs are lipophilic and partition into membranes of aquatic organisms. When those membranes are subsequently exposed to UVA and singlet oxygen is generated via photosensitization, subsequent reactions of singlet oxygen lead to the formation of other reactive oxygen species (ROS) such as superoxide anion and hydroxyl radical. These ROS interact with biomolecules and cause disruption, dysfunction, or destruction of membrane and membrane components. The most commonly observed effect in aquatic organisms is abstraction of hydrogen from membrane lipids, leading to lipid peroxidation (Choi and Oris, 2000). In vivo studies demonstrate histopathologies consistent with lipid peroxidation in gill and skin membranes (Oris and Giesy, 1985; Weinstein et al. 1997), and organisms exhibit overt signs of respiratory and osmoregulatory stress (McCloskey and Oris, 1993), with cause of death typically attributed to asphyxiation (Oris and Giesy, 1985).

ENVIRONMENTAL ATTENUATION FACTORS

In order for photoinduced toxicity to occur, photoxic PAHs must be taken up by the organism. Once PAHs have been accumulated, the organism must be exposed to the appropriate wavelength and intensity of solar radiation. Thus, any environmental factor that alters bioavailability of PAHs, changes the intensity or spectral quality of light, or affects the formation of ROS can potentially attenuate the photoinduced toxicity of PAHs. The natural attenuation of light in water decreases intensity of radiation with depth, and dissolved and particulate materials can greatly enhance that attenuation (Rose et al. 2009). Organisms that live in deep waters, migrate diurnally to live at depth during daylight hours, or inhabit shaded or turbid waters, are therefore less susceptible to UVR-related impacts (Williamson et al. 1999). Dissolved organic carbon (DOC) has been shown to both decrease bioavailabilty of phototoxic PAHs absorb UVA/B wavelengths in water, and strongly attenuate photoinduced toxicity in laboratory studies (Oris et al. 1990; Weinstein et al. 1999). Chemicals or factors that enhance bioavailability have been shown to enhance photoinduced toxicity (e.g., methyl tertiary butyl ether [MTBE]) (Cho and Oris, 2003).

RELEVANCE IN NATURAL AQUATIC SYSTEMS

Over the past 30+ years, laboratory studies and shallow outdoor mesocosm studies have clearly demonstrated that a range of PAHs are strongly phototoxic to aquatic organisms. There has been much speculation, however, about the relevance of PAH-photoinduced toxicity in natural aquatic systems (e.g., McDonald and Chapman, 2002). Criticisms of laboratory studies have been that water sources have been largely devoid of natural dissolved and suspended materials that decrease bioavailability of PAH and decrease penetration of UVB and UVA into the water column. In addition, laboratory and most mesocosm studies have been conducted using concentrations of PAH that, although generally below water solubility limits (ranges 10–1000 μg L^{-1}), are significantly higher than those typically observed in natural waters (ranges 10–1000 ng L^{-1}). Counter arguments for the extrapolation of laboratory and mesocosm results to natural field conditions rely on a photochemical principle referred to as the **Brunson–Roscoe Law of Reciprocity*** (Dworkin, 1958). In photochemistry, the rate of a reaction is dependent on both the concentration of the photosensitizer (i.e., PAH) and the intensity of the actinic radiation (i.e., UVA or UVB). The greater the concentration of PAH and the higher the intensity of UV, the greater the rate of ROS production will be. These rate-determining factors are reciprocal. In other words, at low PAH concentrations, a higher UV intensity is required to elicit the same rate of ROS production compared with at high PAH concentrations. As laboratory experiments cannot mimic the intensity of the sun (the best we can achieve is around 10% of ambient sunlight), the concentrations of PAHs used in laboratory experiments need to be at least 10× greater than in natural systems to see the same effect.

The principle of reciprocity has been used to develop models that predict the expected level of phototoxicity in a body of water based on the measured or predicted body residue of the sum of phototoxic PAHs and the intensity of UV radiation (Sellin Jeffries et al. 2013). We use body residue instead of water concentration to account for bioavailability differences because of water quality parameters. In addition, because each PAH converts radiation energy to ROS at different rates, the model expresses PAHs in molar units multiplied by a quantum efficiency factor scaled to a common PAH. In our models, we scale all quantum efficiency factors to the PAH anthracene to calculate the relative phototoxicity of all PAHs to a common scale. Predictions of toxicity levels can then be made at any combination of PAH concentration and UV intensity. In a risk assessment context, the model can be used to estimate levels of predicted toxicity given a measure of PAH and ambient UV values. That model can be used to estimate toxicity at any depth or water column habitat (e.g., pelagic or nearshore) if UV attenuation data are also available.

The reciprocity model along with field toxicity tests have been used to provide support to management decisions concerning PAH in natural aquatic environments. In Lake Tahoe (border of California and Nevada), field and outdoor experiments were used to show that heavy use of carbureted two-cycle engines in personal watercraft (that is, jet skis) could cause toxicity to zooplankton at concentrations of PAHs as low as 50 ng L^{-1} (Miller et al. 2003). This information was used in part during the decision-making process to ban this engine type from the Tahoe Basin (Sward and Doyle, 1998). This approach has also been used in the study of establishment and expansion of nonnative fish species in Lake Tahoe (Gevertz et al. 2012; Tucker et al. 2012), with special emphasis on the need to control nearshore water quality.

Large-scale, accidental releases of PAHs have also been the subject of study in the field. The reciprocity model, toxicity test data, water quality data, ambient UV monitoring and attenuation data, and fish life history data have been combined to provide a current and retrospective risk assessment of phototoxic PAHs released from the 1989 *Exxon Valdez* Oil Spill (Sellin Jeffries et al. 2013). That assessment was able to predict the extent of impact in a target fish population shortly after the spill and again more than 10 years after the spill. Because the waters of Prince William Sound are relatively productive, with significant amounts of suspended and dissolved organic materials, it was predicted that only a small percentage of fish under study would have been impacted as these environmental factors both reduce PAHs bioavailability and UV penetration into the water column (Sellin Jeffries et al. 2013). More recently this approach has been incorporated into the impact assessment of the *Deepwater Horizon* Oil Spill. The less productive, clear open waters of the Gulf of Mexico, combined with its subtropical latitudes and, thus, higher ambient levels of UV will likely make a significant difference in some of the phototoxic impact predictions.

* The general Brunson–Roscoe law of photochemistry refers to the combined influence of the intensity and duration of light on the generation of product such as might occur with photographic film development or reactive oxygen species production by PAHs. Cumulative irradiance estimated with light intensity and exposure duration (intensity × time = cumulative irradiance) determines the amount of product generated.

On the basis of the reciprocity model, empirical toxicity data, and knowledge of the range of PAHs present in a variety of aquatic habitats—from relatively pristine to highly contaminated—the relevance of PAH-photoinduced toxicity in natural aquatic environments depends on a specific body of water and its location. We have shown that photoinduced toxicity is highly relevant in clear, oligotrophic systems with a high intensity UV regime (low latitudes or high elevations). Depending on the habitat depth and location of the organisms, moderately productive systems may only experience significant PAH-photoinduced toxicity under high PAHs conditions (e.g., spills, industrial inputs, or intense boating activity). Turbid, eutrophic, or dystrophic systems are not likely to be impacted by PAH-photoinduced toxicity. In general, we would not predict PAH-photoinduced toxicity in the field when DOC is greater than about 3 mg L^{-1} or turbidity prevents UV from penetrating more than about 1 m.

9.5 SUMMARY

In this chapter, lethality was described under acute and chronic exposure scenarios. The possibility of differences in toxic impact at different life stages of an individual was explored along with associated implications. Basic toxicity test designs were outlined and their relative advantages and disadvantages highlighted. The predominant approach to measuring toxicity (dose– or concentration–response design) was detailed and methods of analyzing dose–response data were contrasted. An alternative approach, the survival time design, and methods for analyzing TTD data were compared with the dose–response methods. Toxic incipiency and ways of quantifying effects of toxicant mixtures were provided for the dose–response and survival time designs. Finally, biotic and abiotic factors modifying toxicity were explored only briefly because many had already been discussed in the context of factors modifying bioavailability (Chapter 4).

SUGGESTED READINGS

Anderson, P. D. and L. J. Weber, Toxic response as a quantitative function of body size. *Toxicol. Appl. Pharmacol.,* 33, 471–483, 1975.

Christensen, E. R., Dose–response functions in aquatic toxicity testing and the Weibull model. *Water Res.,* 18, 213–221, 1984.

Dixon, D. G. and J. B. Sprague, Acclimation to copper by rainbow trout (*Salmo gairdneri*)—A modifying factor in toxicity. *Can. J. Fish. Aquat. Sci.,* 38, 880–888, 1981a.

Newman, M. C., *Quantitative Ecotoxicology,* 2nd Edition, Taylor & Francis/CRC Press, Boca Raton, FL, 2013.

Newman, M. C. and M. S. Aplin, Enhancing toxicity data interpretation and prediction of ecological risk with survival time modeling: An illustration using sodium chloride toxicity to mosquitofish (*Gambusia holbrooki*). *Aquat. Toxicol.,* 23, 85–96, 1992.

Newman, M. C. and W. H. Clements, *Ecotoxicology: A Comprehensive Treatment*, Taylor & Francis/CRC Press, Boca Raton, FL, 2008.

Newman, M. C. and J. T. McCloskey, Time-to-event analyses of ecotoxicology data. *Ecotoxicology,* 5, 187–196, 1996.

Sprague, J. B., Measurement of pollutant toxicity to fish. I. Bioassay methods for acute toxicity. *Water Res.,* 3, 793–821, 1969.

Stephan, C. E., Methods for calculating an LC50, in *Aquatic Toxicology and Hazard Evaluation.* ASTM STP 634, Mayer, F. L. and J. L. Hamelink, Eds., American Society for Testing and Materials, Philadelphia, PA, 1977.

CHAPTER 10

Effects on Populations

The plethora of ecological problems generated by both industrial and domestic contaminants calls for an effective means to reliably assess population- and community-effects.

Emlen and Springman (2007)

10.1 OVERVIEW

Sometimes, the focus of our efforts is protection of individuals, particularly in studies of humans or legally protected species such as endangered species. The Migratory Bird Treaty involving the United States, Canada, and United Kingdom is one such case that prohibits the taking of individual birds of specific listed species. More often, effects to individuals are measured with the intent of predicting consequences to populations: the primary goal of most ecotoxicologists is assuring persistence and vitality of populations within ecological communities. Such has been the context for recent studies of population declines or collapses of invertebrates (e.g., Chan et al. 2008; EFSA 2013a–c), birds (Donald et al. 2001; Rattner et al. 2004), and amphibians (Sparling et al. 2001). As an example, now drawing much-deserved interest are the declines in insect pollinator populations worldwide, e.g., Figure 10.1. Another recent example is the drop in populations of oriental white-backed vulture (*Gyps bengalensis*) in India notionally due to the use of the drug diclofenac on cattle. In this instance of drug pollution, vultures feeding on treated cattle die of kidney failure (Balmford, 2013).

Qualities used to assess population-level effects include many already discussed, but they are interpreted in a different manner. For example, cancer may now be examined in the context of relative incidences in populations occupying different microhabitats or having different demographic qualities. Age-dependent effects of toxicants may be woven into explanations of demographic change. Differences in individual effective doses might be considered in the context of selection for tolerant genotypes in populations. Knowledge of activation mechanisms for carcinogens may be sought to support inferences about causal structure in studies of disease prevalence in field populations.

Explored in this chapter are the ways of determining the status of exposed populations. Also discussed are population responses that could lessen adverse effects. First, approaches are detailed for describing the occurrence of toxicant-related disease in extant populations. Associated epidemiological information helps us to assess the imminence of failure of the afflicted population in addition to estimating the probability of an individual with certain qualities being adversely affected. Second, impacts on population demography are described. The advantages are put forward for generating toxicity test data amenable to demographic analysis. Third, we expand discussions to consideration of a network of subpopulations in a landscape mosaic that are linked by migration,

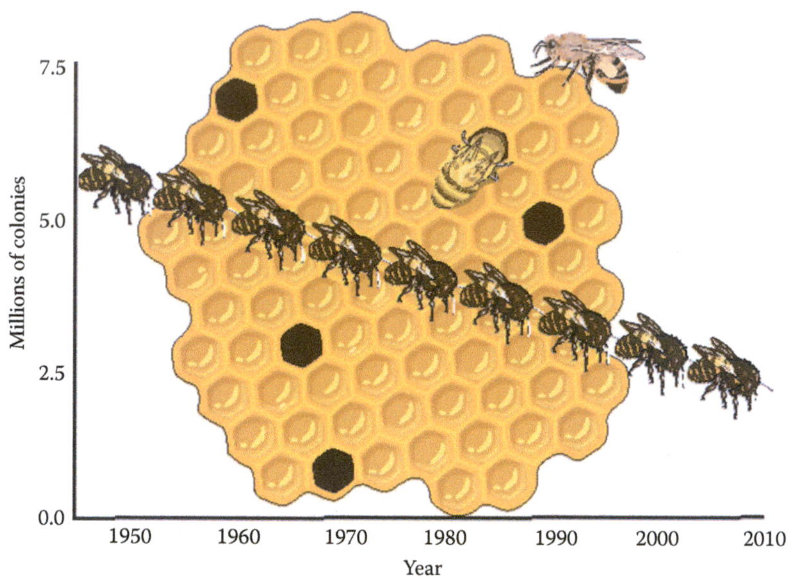

Figure 10.1 The decline in honey bee colonies in the United States as estimated by Pettis and Delaplane (2010) from National Agricultural Statistics Services records. (The line depicted with honey bees is the downward trend approximated from 1945 to 2008 data points in Figure 1 of Pettis, J.S. and K.S. Delaplane, *Apidologie,* 2010, 1–8, 2010.)

that is, metapopulations. The final issue is population genetics including the evolution of tolerance that might occur after prolonged population exposure to contaminants.

10.2 EPIDEMIOLOGY

Epidemiology is the science concerned with the cause, incidence,* prevalence, and distribution of infectious and noninfectious diseases in populations. Most often, disease is linked through correlation with risk factors† such as the qualities of individuals and **etiological agents** (an agent responsible for causing, initiating, or promoting the disease [Rench, 1994]). An example of an etiological agent might be the high-mercury concentrations in seafood taken from Minamata Bay. Two definitions are relevant to studies of disease resulting from chemical (e.g., pollutants or chemicals in the workplace) and physical (e.g., radiation, UV light, high temperatures, or asbestos fibers) etiological agents. Here, **environmental epidemiology** is defined as that subdiscipline of human epidemiology concerned with diseases caused by chemical or physical agents in the environment (Rench, 1994). **Ecological epidemiology**, frequently associated with retrospective ecological risk assessments, is the name often given to epidemiological methods applied to determining the cause, incidence, prevalence, and distribution of adverse effects to nonhuman species inhabiting contaminated sites (Suter, 1993).

* Incidence and prevalence have slightly different meanings in epidemiology. Disease **incidence** is the number of new individuals having the disease in a certain time interval, e.g., incidence rate = 10 cases per 1000 person-years of exposure. **Prevalence** is simply the total number or proportion of individuals with the disease at a particular time, e.g., 157 cases in New York City (Ahlbom, 1993). Often prevalence is expressed as a proportion, percentage, or ratio, e.g., 157 cases per 10,000 people in 1957.

† A **risk factor** is any quality of an individual (e.g., age or dietary habits) or an etiological factor (e.g., chronic exposure to high levels of a toxicant) that modifies an individual's risk of developing the disease.

A range of straightforward metrics can be generated during epidemiological analyses. Those from life tables, as described in Section 10.3.3, were among the first to be applied to human epidemiology. Disease **incidence rate** (I, expressed in units of individuals or cases per unit of exposure time being considered in the study) for a nonfatal condition can be calculated as the number of individuals with the disease divided by the total time that the population had been exposed, e.g., 10 new cases per 1000 person-years:

$$I = \frac{N}{T} \tag{10.1}$$

where N = number of diseased individuals or cases, and T = the total time at risk of contracting the disease (Ahlbom, 1993). The T may be expressed as the total number of time units that individuals were exposed to disease risk during the study period, e.g., per 1000 person-years of exposure. As already mentioned, prevalence can be expressed in several ways such as a straightforward proportion as in Figure 10.2. In other cases, **prevalence** (P) is expressed as the incidence rate (I) times the amount of time (t) that individuals were at risk:

$$P = I \times t \tag{10.2}$$

If there were 2 cases per 1,000 person-years, the prevalence in 10,000 person-years (e.g., in a population of 1,000 people exposed for 10 years) would be (2 cases/1,000 person-years) (10,000 person-years) or 20 cases.

Occurrence of disease in a population can also be expressed relative to that in another population. Often one is a control or reference population. The simple difference in incidence rates can be used to compare disease in two populations, e.g., 25 more cases per year in population A than in population B. The difference often is expressed in terms of a standard size (10,000 individuals) because the populations will likely differ in size. Also, the ratio of

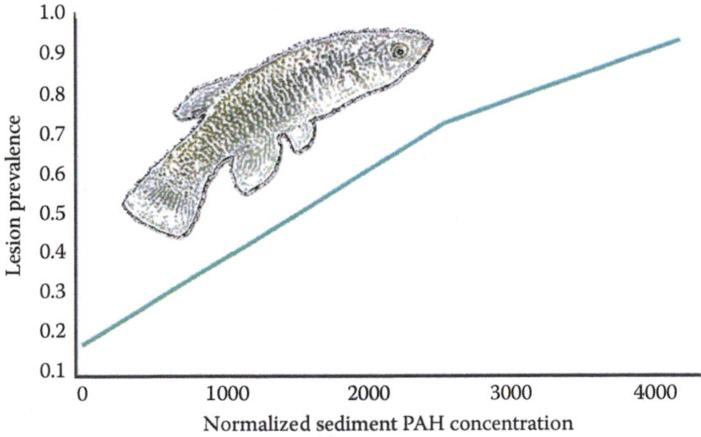

Figure 10.2 Prevalence of liver lesions for mummichog (*Fundulus heteroclitus*) inhabiting sites on the Elizabeth River (Virginia, United States) with different concentrations of polynuclear aromatic hydrocarbon (PAH) in sediments. Prevalence is expressed as the proportion of sampled individuals with lesions and PAH concentration was expressed as (milligram PAH per kilogram of dry sediment)/(sediment total organic carbon content/100). (See Vignette 7.2 for details.) Figure derived from Figure 6.1 in Newman, M.C., *Quantitative Ecotoxicology, 2nd Edition*, Taylor & Francis/CRC Press, Boca Raton, Florida, 2013.)

occurrences of the disease in the two populations (**relative risk [RR]**) can be expressed as the ratio of incidence rates (**rate ratio**):

$$RR = \frac{I_A}{I_0} \tag{10.3}$$

where I_A = incidence rate in population A, and I_0 = incidence rate in the reference or control population. For example, 23 diseased individuals occurring per year in a standard sample size of 10,000 individuals for a heavily industrialized city may be compared to an annual incidence rate of 0.5 individuals per 10,000 individuals in a small town. The RR would be expressed as a rate ratio of 46. Notice that RR calculated with survival time models described earlier can be used in such epidemiological analyses too.

Relative risk can also be expressed as an **odds ratio** in case-control studies. The number of individuals with the disease that were (a) or were not (b) exposed, and the number of reference individuals free of the disease that were (c) or were not (d) exposed to the risk factor are used to estimate the odds ratio (Ahlbom, 1993):

$$\text{Odds Ratio} = \frac{a/b}{c/d} = \frac{ad}{bc} \tag{10.4}$$

Say, for example, that 750 cases of a fatal disease were documented with 500 of them associated with individuals who had been previously exposed to an etiological agent (a) and 250 of them (b) were associated with individuals who had never been exposed to this agent. In another sample of 500 control individuals showing no signs of the disease, 60 had been exposed (c) and 440 (d) had not been exposed to the agent. The odds ratio would be (500)(440)/(250) (60) or roughly 14.7. This high odds ratio suggests that exposure does influence proneness to the disease. An example study using odds ratios is provided by Rench (1993) in which a disproportionately high number of soil conservationists fell victim to non-Hodgkin's lymphoma relative to a control group, notionally because of high pesticide exposures associated with their jobs.

There exists a wealth of methods such as proportional odds and logistic regression models (SAS Institute, 1990) for analyzing odds data from epidemiological studies. Easily assessable textbooks such as those written by Ahlbom (1993), Marubini and Valsecchi (1995), and Rosner (2006) describe these and other statistical methods applicable to epidemiological data. They detail the calculation of confidence intervals for estimates and statistical tests of significance. Although most draw their examples from human epidemiology and clinical studies, there are no inherent obstacles to application of these techniques to ecological epidemiology. For example, the lesion prevalence information shown in Figure 10.2 was fit to a logistic model in Newman (2013). These powerful methods are gradually increasing in use by ecotoxicologists.

Rules have been developed to enhance inferential soundness of epidemiology because most approaches in this field relay heavily on inferentially weak correlations of disease with risk factors. Two such sets of rules, Hill's (1965) nine aspects of noninfectious disease association and Fox's (1991) **rules of practical causal inference**, are commonly applied in ecotoxicology. Only Hill's rules will be described, as both sets of rules are related and similar. **Nine aspects of disease association** have been identified by Hill (1965) for assigning linkage of a risk factor and disease in environmental epidemiology: strength of association, consistency of association, specificity of association, temporal association, biological gradient (dose–response), biological plausibility, coherence of the association, experimental support, and analogy. The *strength of association* between some risk factor and disease is important to consider. For example, the 200-fold higher prevalence of scrotal cancer for chimney sweeps relative to men with other occupations added strength to the supposition that the cancer was initiated or promoted by some occupational risk factor (carcinogenic polynuclear aromatic hydrocarbons [PAHs] in soot and tar). *Consistency of association*, the consistent observation of the association under numerous, varying conditions, also strengthens conclusions. For example, soundness of the conclusion associating asbestos fibers with lung cancer is enhanced if the incidence of lung cancer is high for either male and female workers in many, diverse occupations sharing one common factor—high asbestos fiber densities in workplace air. Another example would be the linkage of lung cancer incidence in several European cities to high levels of the air pollutant, nitrogen dioxide (Vineis et al. 2006). However, consistency may also be generated by a bias in the data set, so caution must be exercised in applying this rule. As an example of potential bias, concentrations of other air pollutants were correlated with those of nitrogen dioxide in these cities. Identification of an association under very specific situations (*specificity of association*) can enhance one's ability to assign causation or linkage. For example, only a very specific type of behavior or occupation may have the associated linkage between the risk factor and disease. Indeed, Hill (1965) observed that it was the lack of specificity in the association between lung cancer and smoking that allowed counterarguments to any causal linkage to persist for so long. A *temporal association* might be considered to reinforce causation. The cause or promoter of the disease should be present before the disease occurs. There is potential for considerable uncertainty and bias here also. Obviously, a clear *biological gradient (dose–response) in the association* strengthens evidence also for a relationship. Although the knowledge is often not available, *biological plausibility* (a plausible, underlying mechanism for the association) can enhance the probability of making a correct linkage between a disease and risk factor. *Coherence of the association* with what is already known about the disease is also very helpful. For example, findings of liver lesions after chronic exposure of laboratory rats to a particular toxicant could support inference of an association between human exposure to that toxicant in the workplace and an increased incidence of liver cancer. *Experimental support of association* (manipulation of the association with measured change in the disease response) can help if practical. This type of support in human studies is appropriately restricted for ethical reasons and, consequently, tends to be more applicable for nonhuman species. However, these data are available for humans under special circumstances. Hill (1965) provides one example of lung and nasal cancer incidences in nickel refinery workers before and after workplace exposure routes were controlled. The incidence of cancer dropped significantly after control measures were taken to reduce worker exposure. Figure 10.3 shows an analogous situation with the effect of DDE on nest productivity of osprey (*Pandion haliaetus*). The nestling production gradually increased after a ban on the application of such pesticides. This recovery supports causal linkage between DDE and reproductive damage. As discussed in Chapter 1, the association between DDE and raptor population decline was also supported by biological plausibility, i.e., eggshell thinning. *Analogy* is the final quality of associations enhancing the accuracy with which risk factors are identified. Similarity of the association to another well-documented association fosters accuracy. Hill (1965) suggests that the demonstration of thalidomide's effect on fetuses enhanced the credibility of subsequent suggestions of birth defect linkage to other drugs taken by pregnant women.

None of these nine factors alone leads to unambiguous identification of a true association. Rather, the goal is to accumulate enough information with all of the nine factors to identify associations as either sufficiently plausible or implausible. Of course, another possible outcome of an epidemiological study is the conclusion that insufficient evidence exists to judge plausibility.

These same nine factors can enhance plausibility of association between a risk factor and some adverse consequence in ecological epidemiology. Suter (1993, Table 10.4) outlined Hill's nine factors, providing a slightly more ecological explanation for each. The only change was associated with a greater capacity to perform and to emphasize controlled experiments with nonhuman species. These and similar logical tools will be discussed again in more detail in Chapter 13.

Figure 10.3 The gradual increase in active nest productivity (average number of young fledged per active nest) of Long Island Sound osprey. Nesting success was extremely low before the widespread banning of pesticides such as dichlorodiphenyldichloroethylene (DDE). The nest productivity slowly recovered (top panel) as DDE concentrations decreased in eggs from osprey nests (bottom panel), suggesting that DDE was a significant risk factor in nest failure. (Modified from Figure 1 of Spitzer et al. (1978), background images drawn from photographs of M. Newman and NASA.)

10.3 POPULATION DYNAMICS AND DEMOGRAPHY

10.3.1 Overview

Precisely because exposure to a toxicant can result in mortality as well as multiple sublethal effects, the use of simplistic toxicity metrics often results in underestimates of the total effects of toxicants.

Stark and Banks (2003)

... ecologically relevant effects assessment should involve endpoints based on population-level responses.

Forbes et al. (2001a)

A **population** is a group of individuals of the same species occupying a defined space at a particular time. As studied extensively in ecology and demography, external factors influence the size, nature, and distribution of nonhuman* and human† populations. Growth under limitation in artificial systems such as agricultural plots and species tolerance ranges as determined in the laboratory were used to predict presence and size of populations in the field. Ecotoxicologists have extended

* Perhaps the best known examples are **Liebig's law of the minimum** and **Shelford's law of tolerance**. Liebig's law states that a population's size will be limited by some essential factor in the environment that is scarce relative to the amounts of other essential factors, e.g., phosphorus will limit algal growth in many lakes. Shelford's law states that a species' tolerance along an environmental gradient (or a series of environmental gradients) will determine its realized population distribution and size in the environment. For example, salinity and temperature gradients may define the location and abundance of an oyster species along the east coast of the United States.

† The Englishman, Thomas R. Malthus (1766–1834) established a series of assumptions and observations regarding limitations on human populations. **Malthusian theory** was first published in 1798 as the profoundly influential essay, *An Essay on the Principle of Population as it Affects the Future Improvement of Society with Remarks on the Speculation of Mr. Godwin, M. Condorcet and Other Writers.*

this traditional ecological approach to predicting population sizes and probabilities of local extinction of populations exposed to contaminant gradients based on laboratory, mesocosm, small-field plot, and enclosure studies.

Often predictions of population viability are made from results of toxicity tests, and ANOVA-based NOEC and LOEC values. Concentrations that affect such qualities as individual growth, larval and adult survival, and number of young produced per female are combined to make predictions. Decisions may be based on whether or not change was statistically significant. As we discussed, the major difficulty with this approach is that statistical significance does not dictate biological significance. For this and other reasons already discussed, a movement away from such an approach has begun. An illustrative sampling of publications urging abandonment of such use of measures like the NOEC includes Chapman et al. (1996), Jager (2012), van Dam et al. (2012), and Van der Vliet et al. (2012). A more thoughtful approach might be to decide on a magnitude of change in some relevant quality beyond which the population is assumed to be adversely impacted (Forbes et al. 2001a; Stark and Banks, 2003; Emlen and Springman, 2007). The rationale is the following: (1) to move forward in environmental stewardship, consensus must be reached about the level of change required to trigger concern or action by regulators, and (2) consensus opinion should focus on a level of biologically meaningful change that one can detect with reasonable confidence in most cases (e.g., Hoekstra and Van Ewijk, 1993). Both of these pragmatic approaches (NOEC-LOEC and magnitude of change for some population quality) are potentially compromised as they do not directly answer the question of whether or not the population will remain viable despite the presence of the toxicant. A 20% reduction of a particular quality may be catastrophic for one species population but trivial for another species population: a 20% change in one quality may be trivial but a similar change in another quality may lead to imminent population extinction. Fortunately, traditional population and demographic analyses can be used to predict the possible outcomes of exposure and their probabilities of occurring. Although most toxicity testing methods do not produce information directly amenable to demographic analysis, more and more ecotoxicologists have begun to design tests and interpret results in this context (Newman and Clements, 2008; Newman, 2013), e.g., Kammenga and Laskowski (2000). Further, current EPA documents describing ecological risk assessment methods clearly recognize and articulate this inconsistency between traditional toxicity tests and data needs.

> *During the past two decades, toxicological endpoints (e.g., acute and chronic toxicity) for individual organisms have been the benchmarks for regulations and assessments of adverse ecological effects. ... The question most often asked regarding these data and their use in ecological risk assessments is, 'What is the significance of these ecotoxicity data to the integrity of the population?' More important, can we project or predict what happens to a pollutant-stressed population when biotic and abiotic factors are operating simultaneously in the environment? Protecting populations is an explicitly stated goal of several Congressional and Agency mandates and regulations. Thus it is important that ecological risk assessment guidelines focus upon protection and management at the population, community, and ecosystem levels...*
>
> **EPA (1991a)**

The focus has begun to shift to population vital rates. This approach, as described here, has much promise for improving prediction of population effects of contaminants.

10.3.2 General Population Response

The simplest models of population response treat all individuals identically and predict change in total number or density of individuals over time. Surveys of widespread population trends such as that by Sarokin and Schulkin (1992) may treat populations in this general manner. Changes in total numbers of individuals might be correlated with epidemics of pollutant-linked cancers in the nonhuman population, or **epizootics** (outbreaks of disease in a large number of individuals) caused by a biological agent to pollution-weakened populations.

Unrestrained, exponential growth of a population can be predicted as a function of the population size (N) and its **intrinsic** (or **Malthusian**) **rate of increase** (r) with a simple differential equation:

$$\frac{dN}{dt} = rN \tag{10.5}$$

With knowledge of the initial population size (N_0), its size at any time (t) in the future can be predicted.

$$N_t = N_0 e^{rt} \tag{10.6}$$

Doubling times ($t_d = (\ln 2)/r$) can also be estimated for a population if r is known.

A difference equation can be used instead of Equation 10.5 if population size is measured at discrete intervals such as might be done with a population with nonoverlapping generations. An annual plant or insect population could have nonoverlapping generations:

$$N_{t+1} = \lambda N_t \tag{10.7}$$

where λ is the **finite rate of increase***†, and N_{t+1} and N_t are population size at times $t + 1$ and t, respectively. If population size is measured initially (N_0) and at some time in the future (N_t), r can be estimated from λ.

$$\lambda = \frac{N_t}{N_0} = e^r \tag{10.8}$$

All three parameters, λ, r, and t_d, have been used as meaningful metameters for adverse population effects. As an early example, Marshall (1962) calculated the effects of γ radiation on the intrinsic rate of increase for *Daphnia pulex*. Rago and Dorazio (1984) detailed means to estimate the influence of toxicants on zooplankton population finite rate of increase. These qualities have the advantage over measures such as a simple percentage reduction in reproduction in that they are easily fit into predictive ecological models. They also are readily incorporated into results from more complex, life table methods as discussed in Section 10.3.3.

Obviously, exponential population growth cannot continue indefinitely. In the simplest form, these equations are modified so that growth rate decreases as the population size approaches some **carrying capacity** of the environment (K, the maximum population size expressed as total number of individuals, biomass, or density that a particular environment is capable of sustaining). For populations with overlapping generations, Equation 10.9 is relevant. The classic **Ricker model** (Equation 10.10) is relevant to populations with nonoverlapping generations or experimental designs with discrete intervals of population growth.

$$\frac{dN}{dt} = rN\left[1 - \frac{N}{K}\right] \tag{10.9}$$

* This term is widely used; however, May (1976a) believes that a better term is the **multiplicative growth factor per generation**. For further discussion of this point, please refer to May (1976a) or Newman (2013).

† Metapopulation models that focus on whether or not patches are occupied are called **occupancy models**. In contrast, **structured metapopulation models** focus on the distribution of population sizes among patches (Hanski and Gilpin, 1991).

$$N_{t+1} = N_t \, e^{r(1-N_t/K)} \tag{10.10}$$

Even these simplest of models (Equations 10.9 and 10.10) predict very complex behavior for population dynamics under certain conditions (May, 1974, 1976b). According to these models, populations may increase to and remain at K, oscillate around K with gradual oscillation dampening until settling at K, oscillate indefinitely around K, or fluctuate chaotically. Consequently, field observations of population densities other than those near a K or of population densities that fluctuate widely, do not necessarily reflect an adverse consequence of contamination. Also, recovery of a population after a toxicant-related decrease in population size might not always involve a simple, monotonic increase back to K. Oscillations may occur and influence the probability of successful recovery or extinction. For example, Simkiss et al. (1993) found that sublethal exposure of blowflies (*Lucilia sericta*) to cadmium modified population dynamics in a food-limited environment.

10.3.3 Demographic Change

... ten times twelve solar years were the term fixed for the life of man, beyond which the gods themselves had no power to prolong it; that the fates had narrowed the span to thrice thirty years, and that fortune abridged even this period by a variety of chances, ...

Niebuhr's *History of Rome* as cited in Deevey (1947)

A more detailed analysis can be made by examining population **vital rates**, rates at which important life-cycle events or processes such as birth, migration, and death occur for individuals in populations. Vital rates can be considered for age classes or life stages to obtain a rich understanding of population qualities such as rate of change and stable age structure. Life tables are constructed with these age-/stage-specific vital rates* to predict population qualities influencing persistence. As suggested by Bezel and Bolshakov (1990), such information is critical for accurate assessment of population effects of pollutants because irreversible shifts in population structure are more often the cause of population extinction than outright death of individuals.

Life tables may include only survival data similar to that modeled in Chapter 9. The conventional age-specific notation includes x as the unit of age and $\mathbf{l}_x$ as the number of individuals in a cohort that are alive at x. Such **$\mathbf{l}_x$ tables** or **schedules** summarize information just as the survival models described earlier might; however, no specific underlying distribution need be assumed in the l_x schedule (Table 10.1). The **age-specific number of individuals dying** ($d_x = l_x - l_{x+1}$) and the **age-specific death rate** (proportion dying or probability of dying in interval x or $q_x = d_x/l_x$) can also be derived from l_x.

Some ecologically meaningful statistics can be generated from l_x tables alone as illustrated in Table 10.2. This table is a rendering of the high-dose survival data in Table 10.1; however, l_x is now expressed as the number of survivors. The average years lived (L_x) is estimated as $(l_x + l_{x+1})/2$ for each age class. The total years lived (T_x) for the age class is estimated by summing the L_x values from the bottom of the chart (e.g., for $x = 21$ here) up to the pertinent age class (x). The **expected life span** for individuals of age x can then be estimated as $e_x = T_x/l_x$. Numerous recent studies have used these types of metrics to better understand toxicant effects to populations, e.g., Chandler et al. (2004), Laskowski (2001), Mauri et al. (2003), Moe et al. (2001), Raimondo and McKenney (2005a,b; 2006), and Tanaka and Nakanishi (2001). As an early example, Bechmann (1994) measured changes in expected life span of a marine copepod exposed to copper and found an increase

* Discussions here illustrate the age-specific approach to demographic analysis; however, a stage-specific approach is also commonly used. Instead of an individual's age, the stage in the individual's life cycle is used in computations for the stage-specific approach. For example, computations might be done using various instars and the adult stages of an arthropod, or the various stages of a mammal's life (neonate, weanling, pre-adult, adult). See Newman (2013) for details.

Table 10.1 Survival (l_x) for *Daphnia pulex* Exposed to 0 and 75.9 R·h^{-1} of Radiation

Age Class or x (days old)	Control (0 R·h^{-1})	High Dose (75.9 R·h^{-1})
0	1.00	1.00
1	1.00	0.98
2	1.00	0.96
3	1.00	0.96
4	1.00	0.96
5	1.00	0.96
6	1.00	0.96
7	0.98	0.96
8	0.98	0.96
9	0.98	0.96
10	0.98	0.96
11	0.98	0.96
12	0.98	0.94
13	0.98	0.94
14	0.98	0.94
15	0.98	0.94
16	0.98	0.94
17	0.98	0.81
18	0.98	0.67
19	0.98	0.29
20	0.98	0.17
21	0.98	0.02
22	0.91	0.00
23	0.81	0.00
24	0.58	0.00
25	0.49	0.00
26	0.35	0.00
27	0.28	0.00
28	0.19	0.00
29	0.14	0.00
30	0.14	0.00
31	0.14	0.00
32	0.07	0.00
33	0.05	0.00
34	0.05	0.00
35	0.00	0.00

Source: Derived from Table I of Marshall J.S., *Ecology,* 43, 598–607, 1962.
Note: In this table, the proportion of the original cohort surviving in each treatment is shown instead of raw counts of individuals. These proportions are identical to the cumulative survivals or S (t) described for survival data in Chapter 9.

in life span at low copper concentrations, suggesting a compensatory response by the copepod population.

Age-specific birth rates (m_x, the mean number of females born to a female of that age class) can be added to produce an **l_xm_x life table**. Much more information can then be extracted from the population. Newman (1995, 2013) gave the fictitious example (Table 10.3) of an l_xm_x table for a population living in a contaminated mesocosm. The expected number of females to be produced by a newborn female during her lifetime (R_0 or **net reproductive rate**) can be estimated from this table

Table 10.2 Estimation of Expected Life Span (e_x) from the High Dose in Table 10.1 Assuming 100 Individuals in the Original Cohort

Age (x)	l_x	d_x	q_x	L_x	T_x	e_x (days)
0	100	2	0.02	99.0	1774.0	17.7
1	98	2	0.02	97.0	1675.0	17.1
2	96	0	0.00	96.0	1578.0	16.4
3	96	0	0.00	96.0	1482.0	15.4
4	96	0	0.00	96.0	1386.0	14.4
5	96	0	0.00	96.0	1290.0	13.4
6	96	0	0.00	96.0	1194.0	12.4
7	96	0	0.00	96.0	1098.0	11.4
8	96	0	0.00	96.0	1002.0	10.4
9	96	0	0.00	96.0	906.0	9.4
10	96	0	0.00	96.0	810.0	8.4
11	96	2	0.02	95.0	714.0	7.4
12	94	0	0.00	94.0	619.0	6.6
13	94	0	0.00	94.0	525.0	5.6
14	94	0	0.00	94.0	431.0	4.6
15	94	0	0.00	94.0	337.0	3.5
16	94	13	0.14	87.5	243.0	2.6
17	81	14	0.17	74.0	155.5	1.9
18	67	38	0.57	48.0	81.5	1.2
19	29	12	0.41	23.0	33.5	1.2
20	17	15	0.88	9.5	10.5	1.1
21	2	2	1.00	1.0	1.0	1.0
22	0	-	-	-	-	-

Note: The l_x in this table is l_x in Table 10.1 multiplied by 100.

Table 10.3 $l_x m_x$ Life Table for a Fictitious Population in a Contaminated Mesocosm

Age Class Midpoint of Class (x)	l_x	m_x	$l_x m_x$	$x l_x m_x$
0 to <1 year, 0.5	1.000	0.000	0.000	0.000
1 to <2 year, 1.5	0.312	2.238	0.698	1.047
2 to <3 year, 2.5	0.095	1.390	0.132	0.330
3 to <4 year, 3.5	0.037	0.410	0.015	0.053
4 to <5 year, 4.5	0.003	0.400	0.001	0.005
5 to <6 year, 5.5	0.000	-	$\Sigma = 0.846$	$\Sigma = 1.435$

Source: Modified from Newman, M.C., *Quantitative Methods in Ecotoxicology*, Lewis Publishers, Boca Raton, FL, 1995.

as the sum of the $l_x m_x$ products in the table. In this case ($R_0 = 0.846$), each female is not replacing herself and the population size will likely decline if conditions do not change. The **mean generation time** (T_c) is estimated as the sum of the $x l_x m_x$ column divided by R_0. (The x used here is the midpoint of the age class, e.g., 0.5 for the 0- to 1-year age class.) The mean generation time for this population is 1.435/0.846 or 1.7 years. An approximation of the intrinsic rate of increase is $\ln R_0/T_c$ or -0.098, indicating again a decline in population size through time. The intrinsic rate of increase is more accurately estimated with the **Euler–Lotka equation,**

$$\sum_{x=0}^{\text{infinity}} l_x m_x e^{-rx} = 1. \tag{10.11}$$

The estimate of r generated above may be used initially in the Euler–Lotka equation. The estimated r is then adjusted up- or downward until the solution to the left side of the equation is sufficiently close to one. The r resulting in such a condition is deemed adequate for most purposes. Use of the Euler–Lotka equation carries the assumption that the population is stable. If conditions do not change with time, a population with a particular r will eventually establish a stable distribution of individuals among the various age classes. Such a population is called a **stable population.**

For a stable population, the expected contribution of offspring during the remaining life of an individual (**reproductive value** or V_A) is easily calculated for each age class. This V_A is the reproductive value for the x age class, divided by that for a neonate, i.e., V_x/V_0.

$$V_A = \sum_{x=A}^{\text{infinity}} \frac{l_x}{l_A} m_x \tag{10.12}$$

As examples, a neonate, reproductive adult, and post-reproductive adult might be expected to produce 1.2, 5.5, and 0, respectively, during their remaining lifetime.

This V_A can be used to reinforce the point discussed previously that it is not always appropriate to estimate the consequences to population viability based solely on reductions in most sensitive life stage survival and NOEC values for lowered reproduction. Reproductive value can also be used to assess such consequences to populations. Indeed, Kammenga et al. (1996) demonstrated with such an approach that the most sensitive life-cycle trait of the nematode, *Plectus acuminatus,* was not the most critical demographic quality impacted by cadmium exposure. For example, a toxicant that eliminated the age class or set of age classes with the highest reproductive value could have strong, adverse consequences to the population's viability. Conversely, a toxicant might have very minor consequences relative to population viability if it eliminated large numbers of individuals in an age class that had very low-reproductive value. This approach was underutilized for assessments of effect but has recently enjoyed much broader appreciation, e.g., Akçakaya, et al. (2008) and Barnthouse et al. (2008). Barnthouse et al. (1987) did effectively include reproductive value in their models of fish population response to contaminants and Martínez-Jerónimo et al. (1993) did estimate V_A for *Daphnia* populations exposed to increasing concentrations of Kraft mill effluent.

In one of the earliest ecotoxicological studies applying simple demographic methods, Marshall (1962) exposed female *Daphnia pulex* to various doses* of ^{60}Cobalt γ radiation (Figure 10.4) and estimated age-specific changes in birth and death rates. The changes in intrinsic rate of increase for the exposed populations were then calculated from these vital rates. Stable age structure of populations can be estimated if the rate of increase (r or λ) and l_x values are known, and used to suggest contaminant effects. The proportion of all individuals in age class x (C_x) is estimated to be the following:

$$C_x = \frac{\lambda^{-x} l_x}{\sum_{i=0}^{\text{infinity}} \lambda^{-i} l_i} \tag{10.13}$$

* Radiation is expressed here as the exposure dose rate, i.e., Roentgen per hour. A **Roentgen (R)** is a measure of the amount of energy deposited in some material by a certain amount of radiation. By convention, it is expressed relative to energy dissipation in 1 cc of dry air. Use of R to express dose allows one to normalize for the different amounts of energy that are deposited by different types of radiation. You will remember from Chapter 1 that R was incorporated into the rem, Roentgen equivalent man. The rem is the measure of radiation dose expressed in units of potential effect to humans.

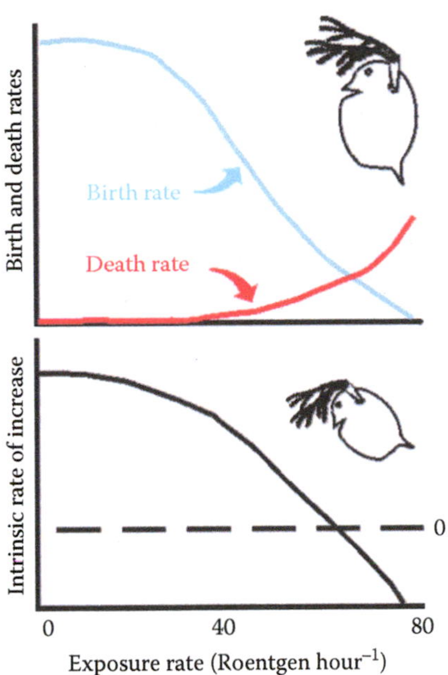

Figure 10.4 Birth, death, and intrinsic rate of increase for *Daphnia pulex* exposed continuously to different amounts of ^{60}Cobalt γ radiation. Age-specific survival (l_x) and fertility (m_x) rates were measured for females exposed to different levels of radiation, and the intrinsic rate of increase estimated by iterative solution of the Euler–Lotka equation. (Modified from Figures 1 and 2 of Marshall, J.S., *Ecology,* 43, 598–607, 1962.)

Marshall (1962) applied this approach to predict stable age structures for *Daphnia pulex* exposed to radiation (Figure 10.5). The stable age structure slowly shifted away from a preponderance of neonates at 0 R/day to a structure composed primarily of older individuals at 75.9 R/day.

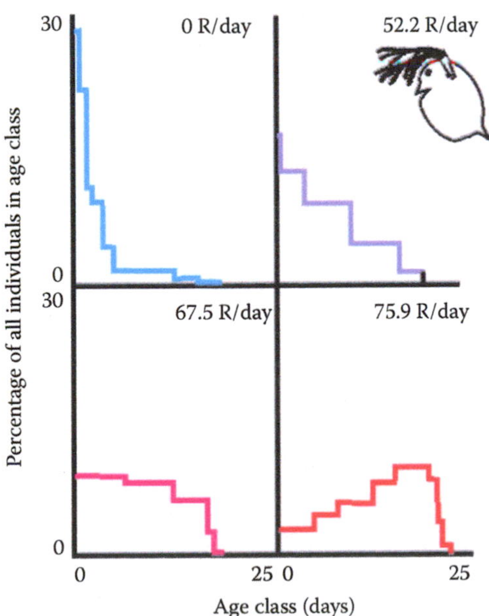

Figure 10.5 Stable age structures of *Daphnia pulex* exposed to 0–75.9 R·h^{-1} of ^{60}Cobalt γ radiation. Note the gradual trend toward a structure with few neonates as dose rate increases. (Adapted from Figure 3 of Marshall, J.S., *Ecology,* 43, 598–607, 1962.)

The life table approach to demographic analysis just presented was used because it involves relatively straightforward computations. A much more powerful approach based on matrix algebra was developed by Leslie (1945, 1948) and has been widely applied to demography (e.g., Caswell, 2001). A **Leslie matrix*** can be constructed that has the probability (P_x) of a female alive in period x_i to x_{i+1} being alive in period x_{i+1} to x_{i+2} in the matrix subdiagonal and the number of daughters (F_x) born in the interval t to $t + 1$ per female of age x to $x + 1$ in the top row of the matrix. The Leslie matrix is multiplied by a vector of the number of individuals alive at each age class at some time (t) to estimate the number of individuals expected in each period at some future time, $t + 1$. This multiplication can be repeated for many time steps to project the population size and structure through time.

$$\begin{bmatrix} F_0 & F_1 & F_2 & F_3 & \ldots & F_\omega \\ P_0 & 0 & 0 & 0 & \ldots & 0 \\ 0 & P_1 & 0 & 0 & \ldots & 0 \\ 0 & 0 & P_2 & 0 & \ldots & 0 \\ \ldots & \ldots & \ldots & \ldots & \ldots & \ldots \\ 0 & 0 & 0 & 0 & P_{\omega-1} & 0 \end{bmatrix} \bullet \begin{bmatrix} n_{0,t} \\ n_{1,t} \\ n_{2,t} \\ n_{3,t} \\ \ldots \\ n_{\omega,t} \end{bmatrix} = \begin{bmatrix} n_{0,t+1} \\ n_{1,t+1} \\ n_{2,t+1} \\ n_{3,t+1} \\ \ldots \\ n_{\omega,t+1} \end{bmatrix}$$

Demographic methods are being used to great advantage in an increasing number of ecotoxicological studies of copepods (Daniels and Allan, 1981; Bechmann, 1994; Green and Chandler, 1996; Chandler et al. 2004), mysids (Raimondo and McKenney, 2005a,b, 2006), and cladocerans (Schober and Lampert, 1977; Winner et al. 1977; Daniels and Allan, 1981; Hatakeyama and Yasuno, 1981; Van Leeuwen et al. 1985; Day and Kaushik, 1987; Wong and Wong, 1990; Martínez-Jerónimo et al. 1993; Tanaka and Nakanishi, 2001). Sibly (1996) tabulated a literature search of such studies that included studies primarily of arthropods, but also of algae, gastrotrichs, nematodes, and humans.

Caswell (1996) developed a demographic assay for the effects of pollutants and advocated more attention to demography. Ferson and Akçakaya (1991) produced software that allows risk assessors to estimate population persistence probabilities based on pollution-induced changes in demographic qualities. Alternatively, the PopTool Excel® add-in shareware from the CSIRO can be downloaded from www.cse.csiro.au/CDG/poptools at this time. A significant literature and theory base are beginning to accumulate on alterations of life history traits due to contamination (e.g., McFarlane and Franzin, 1978; Neuhold, 1987; Sibly and Calow, 1989; Bezel and Bolshakov, 1990; Holloway et al. 1990; Schnute and Richards, 1990; Adams, 1992; Mulvey et al. 1995; Postma et al. 1995; Sibly, 1996; Akçakaya et al. 2008; Barnthouse et al. 2008). Indeed, McGraw and Caswell (1996) have applied such life history analysis to estimate an overall fitness for individuals. Such an overall fitness for organisms under different pollutant exposure conditions would be an excellent response variable for concentration/dose-effect models.

10.3.4 Energy Allocation by Individuals in Populations

Responses of individuals making up a population have been described relative to energy allocation. These responses can produce significant changes in population demographics. Sibly and coworkers (Sibly and Calow, 1989; Holloway et al. 1990; Sibly, 1996) describe an **optimal stress response** for species exposed to toxicants. The optimal stress response involves a shift in the balance in energy allocation between somatic growth rate and longevity (survival) to optimize Darwinian fitness under stressful conditions. Sibly (1996) gives several examples of toxicants influencing these crucial

* If a life stage-based approach is warranted, an equivalent **Lefkovitch matrix** approach (Lefkovitch, 1965; Caswell, 2001) can be applied to analyze population effects of toxicants (Newman and Clements, 2008).

demographic qualities. Kooijman (Kooijman, 1993; Kooijman and Bedaux, 1996) provides a theory-rich approach using energy budgeting for individuals as the central theme through which survival, growth, and reproduction are affected by toxicants. This **Dynamic Energy Budget (DEB) approach** has been formalized into the DEBtox model (Kooijman and Bedaux, 1996; Kooijman et al., 2008).

Toxicant-related effects can be interpreted in the context of the more encompassing **principle of allocation**: there exists a cost or trade-off to every allocation of energy resources. Also discussed as the **concept of strategy**, this principle is used to interpret responses ranging from immediate responses to stress (e.g., early sexual maturity or delayed growth) (Sibly, 1996) to evolutionary responses (e.g., enhanced metal tolerance at the cost of impaired growth rate) (Wilson, 1988). There exists a limited amount of energy available that must be optimally allocated among different processes and functions by an individual to enhance Darwinian fitness. Energy spent producing defense proteins (e.g., cytochrome P-450 monooxygenase isoforms) cannot be spent for reproduction or growth.

Related to this concept is the **limited lifespan paradigm** that a genetically defined maximum lifespan is an inherent quality of an individual.* Parsons (1995) extends this to the more germane **rate of living theory of ageing,** the total metabolic expenditure of a genotype is generally fixed and longevity depends on the rate of energy expenditure. He notes that the immediate response to stress, an increase in metabolic rate, diminishes longevity. He further advocates a **stress theory of ageing** that selection takes place for resistance to stress, and as an epiphenomena, individuals resistant to stress will predominate in extreme age classes of a population. The diminution of homeostasis under stress should be lowest in individuals with highest longevity. He illustrates the concept with populations of *Drosophila*, correlating longevity with reduced metabolic rates under stress and increased antioxidant activity.† He refers to the work of Koehn and Bayne (1989) in which high stress resistance was associated with efficient use of metabolic resources. In discussing genetic evidence supporting this stress theory of ageing, he notes that different genetically determined forms of the glycolytic enzyme, glucosephosphate isomerase (abbreviated as GPI or PGI) seem to impart an advantage to individuals in a variety of species under stressful conditions. As we will soon see, this observation is reinforced relative to ecotoxicology by genetic studies.

VIGNETTE 10.1 Effects of Contaminants on Population Dynamics

Valery E. Forbes and Peter Calow
University of Nebraska–Lincoln

INTRODUCTION

In assessing the risks of chemicals for ecological systems, in most cases we are concerned with protecting populations rather than just the individuals within them (Forbes et al. 2008). It is therefore important to consider the capacity for populations to recover after impact (i.e., to consider responses in terms of population growth rate, [pgr]), and to consider impacts of toxicants on long-term population size (i.e., population abundance or biomass over time). In the first scenario, it is assumed that population dynamics are independent of population density, whereas in the second scenario population dynamics are assumed to be under density-dependent control. In what follows, we first consider pgr responses to chemical exposure, and then we consider impacts on long-term population size, taking account of the influence

* Curtsinger et al. (1992) disagree with this concept.

† The observation of increased longevity with increased antioxidant activity could also be used to support the **disposable soma theory of ageing**. This theory suggests that ageing is a consequence of a gradual accumulation of cellular damage via random molecular defects (Parsons, 1995).

of population density. Finally, we address complications arising from life-cycle variability, temporal and spatial influences, and multiple stressors.

SCENARIO 1: EFFECTS OF CONTAMINANTS IN DENSITY-INDEPENDENT POPULATIONS

This scenario is important when considering the recovery of populations following the impacts of chemicals, as well as in assessing likely impacts of chemical insults on declining populations. Typical tests performed for chemical risk assessment provide information on individual-level traits, most notably survival, reproduction, and growth. Thus the key question is, "To what extent are these measurements on individual-level traits likely to be more or less sensitive to chemicals than is pgr itself?" In other words, are risk assessments based on individual-level responses likely to under- or overestimate risks to populations?

Forbes and Calow (1999) addressed this question by performing a review of the relevant literature that included 41 studies, 28 species and 44 toxicants. They found that out of 99 species by chemical cases considered there were only five in which chemical effects on pgr were detected at lower exposure concentrations than those resulting in statistically detectable effects on any of the individual-level traits. In 81.5% of the cases considered, the percentage change in pgr was less than the percentage change in the most sensitive of the individual-level traits, 2.5% in which the percentage change in pgr was equal to that of the most sensitive trait and 16% in which the percentage change in pgr was greater than the percentage change in the most sensitive individual-level trait. Although proportional changes in pgr were significantly correlated with proportional changes in fecundity and with time to first reproduction, these correlations were weak, and trend analysis indicated that the relationships were nonlinear. Surprisingly, the correlation between proportional reduction in overall survival (i.e., juvenile and adult survival probabilities pooled) and proportional reduction in pgr was not statistically significant. Overall, there was no consistency in which of the measured individual-level traits was the most sensitive to toxicant exposure, and none of them, considered individually, was a very precise predictor of toxicant effects on pgr.

Taking a somewhat different approach, Forbes et al. (2001a) used a generalized form of elasticity analysis on a two-stage life-cycle model to consider the extent to which small percentage changes in individual life-cycle variables result in greater or lesser changes in pgr. When population size was unchanging over time, small changes in time to first reproduction, time between reproductive events, juvenile and adult survival, and fecundity resulted in, at most, the same percentage changes in pgr. However, if pgr was allowed to increase above one (growing population), small changes in life-cycle variables could lead to proportionally greater changes in pgr. Subsequent analyses, based on a review of ecotoxicological studies, demonstrated an inverse relationship between the elasticity of life-cycle traits and their sensitivity to toxicants under density-independent conditions (Forbes et al. 2010). This analysis again demonstrates that changes in single individual-level responses (i.e., survival, reproduction, growth), which are the most common endpoints measured in ecological risk assessments, are not robust indicators of population-level responses. Interestingly, fecundity tended to be most sensitive to toxicant exposure (showing the largest percentage decreases of all traits), whereas it tended to have the lowest elasticity, indicating that impacts on this trait had a relatively small effect on pgr. Thus, chronic toxicity tests that frequently use reproduction as a test endpoint (Walker et al. 2012) may be substantially overestimating effects on pgr.

Conclusions

From the available evidence it would seem that the relationships between the life-cycle traits typically measured in toxicity tests and pgr are nonlinear, differ among traits, and depend on the state of the population (i.e., growing, shrinking, stable). It is difficult to make generalizations, and so if population (rather than individual) impacts are of primary concern, risk assessments should either measure impacts on population growth directly or integrate measured responses in survival, growth, and reproduction through the use of appropriate population models.

SCENARIO 2: EFFECTS OF CONTAMINANTS IN POPULATIONS UNDER DENSITY-DEPENDENT CONTROL

This scenario is important when considering the impacts of chemicals on stable populations. It is presumed that most natural populations are under some form of regulation at some point in their life cycle (Moe, 2008). Under such stable conditions, pgr will tend to zero and population size may be a better indicator of population-level impacts of chemical exposure (Grant et al. 1998; Hayashi et al. 2009). Because most standard tests used for ecological risk assessment are performed under density-independent conditions, this raises the question of whether and to what extent observations/predictions of chemical impacts on populations that ignore density dependence under- or overestimate responses of populations subject to density-dependent control.

Forbes et al. (2001b) reviewed experimental studies exploring density by toxicant interactions. Results were mixed, with some studies showing additive interactions between density and chemical effects on pgr, whereas others found less-than-additive effects or more-than-additive effects. There is even some indication that the form of the interaction may vary across a chemical concentration gradient, with effects shifting from less-than-additive at low-toxicant concentrations to more-than-additive at higher toxicant concentrations (Linke-Gamenick et al. 1999). In addition, it appears that the kind of interaction that may be observed is to an important extent constrained by the kind of experimental design used. This same study explored hypothetical interactions between toxicity and density using a general two-stage demographic model (Forbes et al. 2001b). In these analyses, the key determinant of toxicant–density interactions was the shape of the functional relationship between pgr and the life-cycle variables contributing to it. Depending on these shapes, various outcomes in terms of the additivity, less-than-additivity, or more-than-additivity between density and toxicant effects on pgr were possible. Similarly, Raimondo (2013) combined a density-dependent matrix population model with chronic toxicity concentration–response models for three hypothetical and four real chemicals and confirmed that the functional form of density dependence is a critical mechanism driving the interactions between chemicals and density. Density-dependent compensation appears to be greatest when there are strong density-dependent pressures for a life-cycle variable impacted by the chemical, and is less when the majority of chemical impacts are for life-cycle variables with no or weak density dependence. Also critical is the life-cycle variable(s) on which density dependence is acting, and compensation appears more likely when there is density dependence for survival versus that for sublethal traits such as growth and reproduction (Moe, 2008; Raimondo, 2013).

In contrast to the results from density-independent scenarios showing that pgr is often less sensitive to chemicals than are the individual life-cycle traits contributing to it (described earlier), Hayashi et al. (2009) found that the percentage reduction in equilibrium population size often exceeded the percentage reductions in individual life-cycle traits in density-dependent populations. Likewise, Wang and Grimm (2010) found that population density was more sensitive for detecting population-level effects than population growth rate for simulated stable populations.

Conclusions

The main messages from this scenario are that density needs to be taken into account in ecotoxicological assessments when predicting impacts of chemicals on populations under density-dependent control. Since density × toxicant interactions are difficult to generalize, density needs to be introduced more explicitly into ecotoxicological studies. The analyses described in this scenario rely heavily on a variety of population models and demonstrate how such models can yield important insights into the mechanisms of population-level impacts of chemicals under complex and realistic conditions.

COMPLICATIONS

From Life Cycles

It is implicitly assumed in ecotoxicological analyses that the relationship between effects of chemicals on individual-level variables means the same for different taxa. Thus, 50% mortality (LC50) is treated as implying the same in demographic terms for all species. Yet, analyses of demographic models very clearly show that this is not the case and that the demographic consequences of similar changes in individual-level variables can vary widely among species with different life cycles. For example, Calow et al. (1997) considered a series of simplified but plausible scenarios to explore the complications introduced by life cycles. A number of general conclusions arose out of this analysis. (1) As expected, the effect on pgr of a toxicant that reduces juvenile survival or fecundity will be greater for **semelparous** (i.e., species that reproduce once) as compared with **iteroparous** (i.e., species that reproduce more than once) **species**, and the reverse will be the case for effects of toxicants on adult survival. (2) **Iteroparous species** with life cycles in which the time to first reproduction is shorter than the time between broods will be more susceptible to toxicant impacts on survival or fecundity than will species in which time to first reproduction is longer than the time between broods. (3) Anything that shortens time to first reproduction relative to the time between broods (e.g., increased temperature or increased food availability) is expected to increase the population-level impact of toxicant-caused impairments in survival or fecundity. (4) Lengthening of the time to first reproduction should lessen the population-level impact of toxicant-caused impairments in survival or fecundity.

These results have important consequences for the interpretation of species sensitivity distributions, which are typically constructed from responses of individual-level variables. Such distributions reflect physiological (toxicological) variability of individuals of different species rather than species (population) variability. In theory, life-cycle variability can magnify, ameliorate, or reverse individual toxicological differences, and this can cause problems for carrying out risk assessments using sensitivity distributions based only on toxicological variability of individual-level traits (Forbes and Calow, 2002; Stark et al. 2004; Spromberg and Birge, 2005).

From Temporal and Spatial Influences

Standard tests used for ecological risk assessment generally require that constant exposure concentrations are maintained for the duration of the test. However, for some exposure scenarios (e.g., accidents, spills, or pesticide exposure following application to agricultural fields), exposures will be temporary and/or repeated. Toxicokinetic/Toxicodynamic (TKTD) models provide a powerful tool to assess the impacts of time-varying exposure scenarios on individual-level responses (Nyman et al. 2012) and have also been able to aid in extrapolating individual-level responses among species (Kretschmann et al. 2012).

Under realistic field conditions, exposure to chemicals is likely to vary in space, as well as in time, at scales that may be relevant for exposed populations, e.g., providing refuges or hotspots of contamination. Likewise, habitat quality may vary spatially and interact (positively or negatively) with the impacts of chemicals. To assess ecological risks for such situations requires the use of spatially explicit models such as metapopulation matrix models (see Vignette 10.2) or individual-based models (Railsback and Grimm, 2012).

From Multiple Stressors

It has long been recognized that organisms in the field are rarely exposed to single chemicals in isolation, but that contaminated sites more often contain complex mixtures of toxic chemicals. Likewise, natural populations are subject to a variety of biotic (e.g., diseases, predation, competition) and abiotic (e.g., environmental fluctuations, habitat change) stressors that may influence their susceptibility to chemical impacts. For site-specific risk assessments, it may be possible to perform field studies that adequately capture the relevant complexities. However, for assessing the risks of chemicals in general, it would be impractical to test all possibly relevant scenarios empirically. Mechanistic effect models provide the only way to systematically and efficiently explore the influence of multiple stressors, including mixture effects, on chemical risk. Because such models can easily become complex, as well as difficult to parameterize, validate, and communicate, development of accepted principles of Good Modeling Practice is critical (Grimm et al. 2009).

Conclusions

An important conclusion from the above is the realization that life-cycle variability can play an important part in differential sensitivity to chemicals among populations of different species. Likewise, inclusion of ecological complexities such as spatiotemporal variability in chemical exposure or in habitat features and the influence of multiple stressors may be important but rarely considered explicitly in standard risk assessments. A variety of mechanistic models exist that can help to address these issues, but their use is still limited.

WAY FORWARD

The main methods for assessing the ecological effects of chemicals rely on tests on individual organisms. As this vignette has shown, single individual-level responses may either over- or underestimate population-level impacts. There remain serious challenges in developing experimental and modeling approaches that can assess population-level impacts on the basis of the kinds of data that are usually available in practice (Forbes et al. 2008). These include having the necessary data to inform sufficiently realistic models, knowing how complex to build the models, knowing how specific to make the models, and knowing the extent to which other ecological interactions should be incorporated. But there are many benefits to be gained by developing and applying such mechanistic effect models in ecological risk assessments. They add value by incorporating better understanding of the mechanistic links between individual responses and population size and structure and by incorporating greater levels of ecological complexity so that the assessment of contaminant effects can be made more scientifically sound and thus a better basis for environmental protection. Progress is being made on this front, particularly in Europe. Key to this success has been the engagement of multiple stakeholders from academia, industry, and regulatory authorities in various workshops; support from key authorities (i.e., the European Food Safety Authority) in making it possible to consider appropriate models in regulatory risk assessments; and substantial funding from the European Seventh Framework Programme to support the development of research in this area (CREAM, PITN-GA-464 2009-238148; http://cream-itn.eu/).

SUGGESTED READINGS

Akçakaya, H.R., J.D. Stark and T.S. Bridges, Eds. *Demographic Toxicity: Methods in Ecological Risk Assessment,* Oxford University Press, Oxford, UK, 2008.

Barnthouse, L.W., W.R. Munns, Jr. and M.T. Sorensen, Eds. *Population-Level Ecological Risk Assessment*, Taylor & Francis/CRC Press, Boca Raton, FL, 2008.

10.4 METAPOPULATIONS

Populations often unevenly occupy areas with patches of superior and inferior habitat (Figure 10.6). Subpopulations within a superior habitat may have high rates of increase but those within inferior habitats may have negative growth rates (r). Some subpopulations will act as a source of individuals due to the surplus of offspring produced while others will act as sinks with surplus individuals moving in from outside the patch. This source-sink or patch structure results in dynamics distinct from those predicted by the simple models discussed so far (Pulliam and Danielson, 1991; Ares, 2003; Hanski and Ovaskainen, 2003; Zartman and Shaw, 2006). As an example, a habitat so contaminated that reproduction is impossible can still have individuals in it if there was a nearby source of individuals. Maurer and Holt (1996) illustrate the relevance of this concept to ecotoxicology with the example of pesticide impact on wildlife inhabiting a patchy habitat.

Interest in the dynamics of populations inhabiting a mosaic of habitats has spawned the relatively new field of metapopulation ecology. The complex of populations occupying a habitat mosaic within a landscape that can exchange individuals by migration is called a **metapopulation**, and the goal of metapopulation ecology is to explain and predict the dynamics of such systems. Equation 10.14 is a simple metapopulation model* that is thought to include some of the more relevant aspects of metapopulation dynamics to ecotoxicology. Other information and examples relevant to ecotoxicology can be found in Angeler and Alvarez-Cobelas (2005), Chaumot et al. (2003), O'Connor (1996), Newman and Clements (2008) and Newman (2013).

$$\frac{dp}{dt} = m(1-p) - ep(1-p) \tag{10.14}$$

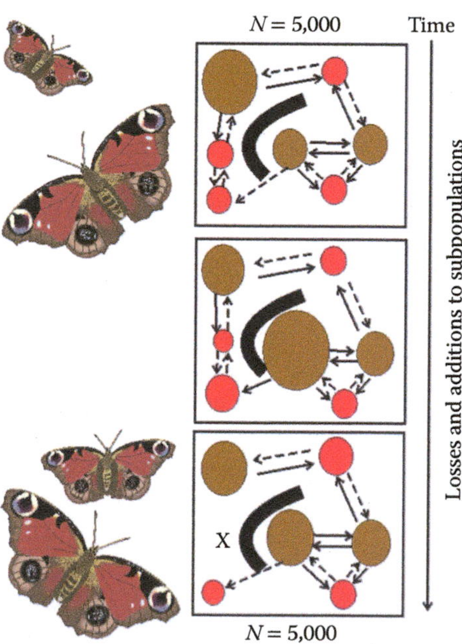

Figure 10.6 An illustration of metapopulation dynamics. A mosaic of population patches are present at the onset (upper panel) with some of the 5000 individuals in the composite metapopulation being in inferior patches (red) and others in good or superior patches (brown). Barriers to migration may exist such as that depicted with a black arch in the center of the panel. Some patches are net source (solid arrows) and sinks (dashed arrows) of individuals. Numbers of individuals in the subpopulation might fluctuate (or even drop to zero as depicted with a red X in the lower panel) but the general metapopulation size remains relatively constant here.

* Metapopulation models that focus on whether or not patches are occupied are called **occupancy models**. In contrast, **structured metapopulation models** focus on the distribution of population sizes among patches (Hanski and Gilpin, 1991).

where p = the proportion of available patches that are occupied, e = the probability of local extinction in a patch, and m = the probability of population reappearance in a vacant patch. Two important themes are embedded within this model: the rescue effect and propagule rain effect. The **rescue effect** is simply the increased probability of a vacated patch reoccupation if nearby patches are occupied: the probability of a patch extinction is lowered as p increases. In contrast, the **propagule rain effect** is independent of the occupation density of nearby patches. The likelihood of a patch population extinction is decreased and that of patch population reappearance is increased if there is a seed bank or dormant stage that acts as a constant source of propagules to the patch regardless of the proportion of nearby patches that are occupied. Both of these effects are relevant to ecotoxicology. If a patch was vacated due to the action of a stressor, patch density-dependent and -independent mechanisms can reestablish the patch population. Effective migration corridors among patches are also crucial for effective recovery of a toxicant-stressed patch.

Three additional features of metapopulations have ecotoxicological consequences. It is possible that an individual could be exposed in one patch but, after migration to another, have the effect of toxicant exposure manifest in the uncontaminated patch. Spromberg et al. (1998) call this spatial separation between exposure and effect, the **effect at distance hypothesis** but it might be categorized in some cases as a manifestation of the antirescue effect. An **antirescue effect** is the increased risk of extinction of a patch due to increased immigration.* Another issue arises from the fact that individuals in different patches have different fitnesses. Some habitat patches can be of such high quality that they are **keystone habitats**, habitats essential to maintaining the vitality of the entire metapopulation. The loss of a hectare of keystone habitat to contamination or to habitat remediation can be much more damaging than the loss of an equivalent area of marginal habitat. Finally, metapopulation biologists have defined the related concepts of minimum viable metapopulation (MVM) and minimum amount of suitable habitat (MASH) (Hanski et al. 1996). The **minimum viable metapopulation** is the minimum number of interacting subpopulations needed to ensure long-term persistence of the metapopulation, e.g., a 95% chance of survival for 100 years. The **minimum amount of suitable habitat** refers to the minimum density of habitat patches needed to ensure metapopulation persistence. Clearly, lowered habitat quality and habitat fragmentation due to pollution could lower MASH, bringing metapopulations closer to their MVM.

Other metapopulation issues require mention before leaving the metapopulation context. The first is to note that species are not always passive conformers to the state of the habitat mosaic. For example, intertidal mussels move to create optimum patches relative to feeding and survival (Grünbaum, 2011; De Jager et al. 2011). Also mathematical analyses of metapopulation models suggest the **core-satellite hypothesis**, that is, a collection of species metapopulations in an ecological community will display a bimodal distribution of patch occupancy frequencies (p): most of the species metapopulations will occupy either a large portion or a very small portion of the habitat patches with few species metapopulations in between these extremes. Formal treatment of this concept as detailed in Gotelli (1991) and Hanski and Gyllenberg (1993) is beyond this discussion, but the point can be made that the states of metapopulations will influence higher order features of ecological communities. Any influence of pollution on metapopulations should be expected to impact ecological communities.

* A similar antirescue scenario would be the immigration of individuals carrying an infectious disease into a patch. Another example involves periwinkle (*Littorina saxatilis*) metapopulations that are characterized by intertidal zone patches occupied by different ecotypes. Heavy wave action experienced in the upper-middle intertidal zone favors ecotypes with smooth shells but heavy crab predation in the lower-middle intertidal favors those with heavy, ridged shells. Hydrids of these two ecotypes live between these two zones of ecomorphs. If the number of one ecotype drops and ecomorphs not well suited to the microhabitat migrate into it, the chance of a patch extinction increases because the individuals now occupying it have lowered fitness under the associated conditions (Rolán-Alvarez et al. 1997). A similar effect might be expected for metapopulation composed of pesticide resistant and susceptible subpopulations. In fact pesticide-free refuges are often a component of integrated pest control programs for this reason.

VIGNETTE 10.2 Action at a Distance: The Impacts of Chemicals on the Dynamics of Spatially Explicit Populations

Wayne G. Landis
Western Washington University

One of the primary challenges of environmental toxicology is the extrapolation of what happens in the laboratory to effects at the scale of the population. Populations exist in space and time, and the world is notoriously patchy. Populations are the providers of numerous ecological services including forestry, fishing, recreation, and the survival of culturally significant species. Some populations are purposefully suppressed by introducing toxicants; these include pest and disease species, invasive species, and native species that interfere with commercially important species. This brief note describes a way of analyzing the effects of toxicants on populations at broad spatial scales and with a number of spatial arrangements. Then we will pick a more interesting case, one with an invasive and native species competing for similar resources, with a toxicant in the environment.

For the purposes of this vignette, a population is a group of potentially interacting organisms of the same species. A population comprises organisms of different ages, different genders, and only part of the population is reproductively active. The organisms share a great deal of genetic material, enough so that successful breeding can occur.

The remainder of this section demonstrates that understanding the spatial distribution and migration of organisms is critical to understanding the impacts of xenobiotics on populations, and the ecological systems that contain them.

POPULATIONS IN A LANDSCAPE

Organisms that make up a population are not evenly distributed in the environment, and migrate to other patches of habitat. There are several different spatial arrangements that populations may take (Figure 10.7). Populations may exist as a single patchy population, an isolated population, a metapopulation, or a mainland-island metapopulation. A metapopulation is defined as a "population of populations" (Levins, 1969) connected through immigration and emigration. In a metapopulation, most organisms spend a majority of their life span in a single patch, but occasional migration does occur. Not all available habitats that can successfully maintain the species are always occupied. A single patchy population is characterized by frequent migration by individuals between the various habitat patches. In an isolated population structure, no migration takes place. A population in a mainland-island pattern exists where one population is much larger than the populations in surrounding patches. The properties of different types of distributions affect the types of impacts that a toxicant can have on one patch and how those patches affect populations in other patches.

In creating simulation models of metapopulation or patch populations, there are several typical assumptions. First, there is an assumption that there is a **minimum viable population (MVP)** size below which extinction of the population in the habitat patch will occur. Second, an assumption is made that a carrying capacity exists, that is a population size that can be maintained without a tendency to increase or decrease.

A population in a habitat patch serves as a sink if it is below the MVP and is recruiting emigrants. A subpopulation serves as a source for nearby habitat patches by providing immigrants to them. The **rescue effect** is where a population that is below the MVP is rescued by immigrating organisms from a source population or patch. If a population within one patch becomes extinct due to a stressor, it may be rescued by other nearby patches.

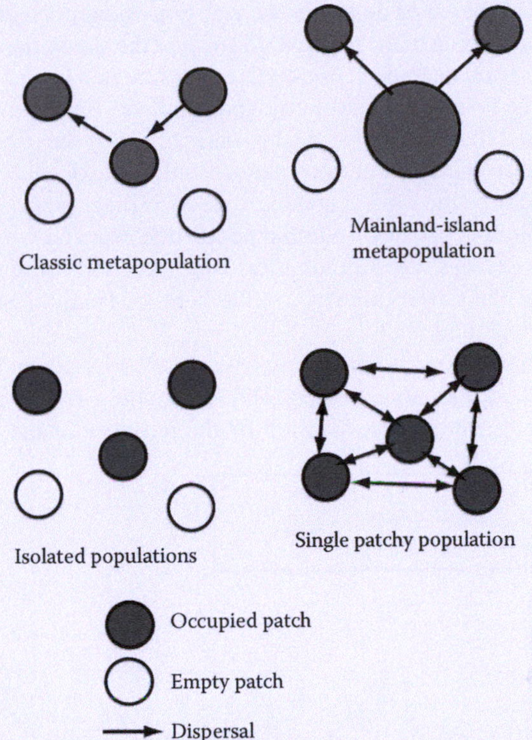

Figure 10.7 Kinds of patch structure in a patchy environment. The models described here generally fit the single patchy population with varying arrangements and distances between patches.

IMPACTS OF CHEMICALS ON SPATIALLY EXPLICIT POPULATIONS

Using a template from Wu et al. (1993), my colleagues and I developed a computer model of a generic animal patchy population that has at least one contaminated patch (Spromberg et al. 1998). The basic framework of the simulation model is presented in Figure 10.8. The single-species

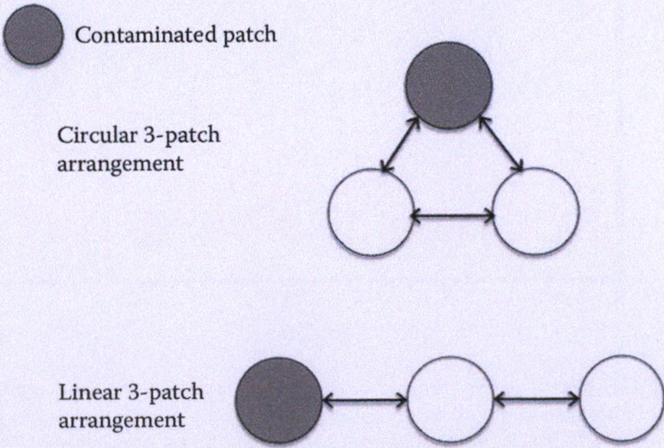

Figure 10.8 Arrangement of habitat patches in the patchy population models used in our studies. The contaminated patch could be in any position within the model landscape. Distances between the habitats could also be altered, changing the rate of migration between each habitat patch.

patchy population model is based on deterministic equations for growth of the population, migration between patches and the fate of the toxicant. In some of the simulations, the toxicant is persistent and in others the toxicant degraded. To estimate exposure in a habitat patch the models use a statistical distribution, the Poisson, in a stochastic (probabilistic) function. The amount of toxicant the organisms are exposed to in the habitat patch is determined by the persistence of the chemical in the patch and the chance encounter of the organism with the toxicant.

The simulations were begun with a standard set of initial conditions with an amount of toxicant in one or more of the patches, an initial population size and a set distance between the habitat patches. The model was then run for a one-time step, and those results are used for the next iteration. Typically, the models are run for 300 iterations and the sizes of the populations in each habitat patch are plotted.

In our first simulations a persistent toxicant was placed in the end patch, and all of the patches had the same initial population size. The first finding is that populations in patches removed from the contamination were affected by the presence of the toxicant (Figure 10.9).

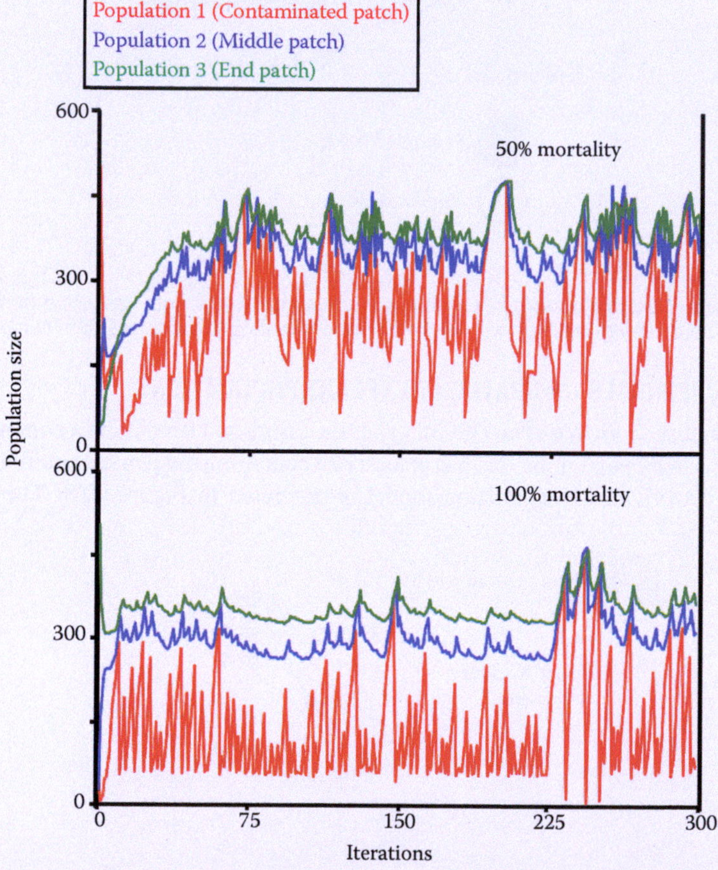

Figure 10.9 Action at a distance. In the three patch simulations where the end contaminated patch has a toxicant concentration equal to an EC50, (top panel) the dosed patch has wide fluctuations. However, there is an occasional overlap among all three populations. The populations in non-dosed habitat patches are below the carrying capacity of 500 and exhibit fluctuations. Even at an EC100 (bottom panel) in the dosed patch, organisms are still extant and occasionally reach numbers comparable to the non-dosed patches. In both simulations, the population in the dosed habitat patch is rescued by the other patches. The initial population size for each habitat patch was 100, and the distances between the patches are equal.

The effects were the reduction of the population below carrying capacity and a fluctuation in population size in all of the patches. The reduction in number and the fluctuations in the non-dosed patches resulted in population sizes that were equal to those in the dosed habitat patch, even with a dose equivalent to the EC50. In the simulations when the dosed patch was at EC100, organisms could still be found in the dosed patch due to the rescue effect from other patches. Because of the stochastic nature of the exposure between the toxicant and the organism, the simulations are repeatable only in type of outcome, not in the specific dynamics.

We next performed simulations with a toxicant that degrades. The finding was that many of these simulations had several discrete outcomes from the same set of initial conditions. The range and types of outcomes depended on the specifics of toxicant concentration, initial population size, and distance between patches. The outcomes were as varied as all three populations reaching carrying capacity to all three becoming extinct with associated probabilities of occurrence.

In the first example, the simulation had three patches that were at a specified distance from each other. Only one patch contained the toxicant that was degraded by half through the simulation. With an initial population size of 100 in each of the patches, 80% of the simulations resulted in all three of the populations in the patches reaching the MVP. All three populations reached carrying capacity in 20% of the simulations.

In the second example, the initial populations' sizes were increased to 140 for each habitat patch. In no instance did the populations decline to the MVP. In 82% of the simulations the outcome was all three populations reaching the carrying capacity. However, 18% of the simulations resulted in a stable oscillation, or bifurcation, of all three populations. A very different outcome, yet only the initial population sizes were altered.

In the previous examples, all of the habitat patches had the same initial population size. In the next series of simulations, the initial population sizes were set so that one habitat patch had a source population, and the others have initial populations at the MVP or lower. The patch that acts as the source was at the end or in the middle patch in the simulations.

The finding from these simulations is that the habitat patch that is dosed is important in determining the potential range of outcomes (Table 10.4). In a series of simulations, the source patch on the end had an initial population size of 200, and all of the other patches started at 50, near the MVP. When the source was habitat patch one and was dosed with a non-persistent toxicant, the four outcomes were (1) in 50% of the simulations only the population in patch 2 survived and it was at the MVP; (2) in 26% of the runs the populations in patches 1 and 2 survived at the MVP; (3) in 10% of the simulations all three populations survived at the MVP; and (4) in only 14% of the runs all three populations reached carrying capacity. Placing the toxicant in non-source patches 2 and 3 in the middle and at the far end with population densities at 200,

Table 10.4 Outcomes of Various Simulations

Outcome by Patch Number	End Dosed (source patch) (%)	Middle Dosed (source patch) (%)
2 MVP	50	0
1, 2 MVP	26	0
1, 2, 3 MVP	10	28
1, 2, 3 cc	14	16
1, 3 MVP	0	56
Source Patch Not Dosed		
1, 2, 3 cc	100	100

MVP, minimal viable population; cc, carrying capacity.
Note: If not explicitly mentioned, the population in that patch is extinct.

50, and 50 for patches 1, 2, and 3, respectively, resulted in all populations reaching carrying capacity.

In another series of simulations, the arrangement was that habitat patch 2, in the middle, was the source (population 200) and dosed with the toxicant while the other patches had population sizes of 50. In these simulations, there were only three possible outcomes observed. (1) In 56% of the runs the populations in patches 1 and 3 existed at the MVP. (2) In 28% of the cases all three patches reached MVP. (3) In 16% of the cases all three populations reached carrying capacity. As with the first set of simulations applying the contaminant to non-source populations (in this case at the ends of the landscape), all patches reached carrying capacity.

In both series of simulations, the alteration in possible outcomes and outcome frequency depended on the location of the contaminated patch in the context of that specific landscape arrangement. Dosing of the source population had a much greater impact than dosing the sink populations.

The simulations summarized above clearly demonstrate that a relationship exists between patch arrangement, initial population size, and the placement and amount of the toxicant in determining the number and frequencies of discrete outcomes. Changes can lead to new outcomes and alter the probabilities of each of these outcomes. Several of these findings have been confirmed by Johnson (2002) using an individual-based modeling in a two-patch landscape.

The findings of these modeling efforts have led Spromberg et al. (1998) to hypothesize "Action at a Distance." The basic premise of action at a distance is that the impact of a toxicant can be transmitted to populations in other habitat patches by changes in the rate and direction of migration between patches by the individual organisms. Action at a distance does not require the direct contamination of a patch or of any organism that resides in a patch. The patterns in the resulting dynamics can be varied and nonlinear. The question is "are these results artifacts of the numerical simulations or can they be expressed in simulated populations and ecological systems?"

Experimental confirmation of action at a distance has been provided by the research of Louis Macovsky (1999). This study created a novel laboratory metapopulation model of the flour beetle, *Tribolium castaneum*. Arranged linearly, habitat patches were linked by density-dependent dispersal of the adult morph. Patches were monitored for the indirect effects on population demographics beyond the patch that received a simulated adulticide over the period of approximately one and one half egg-to-adult cycles. It was demonstrated that indirect effects do occur in patches beyond the patch where adulticide occurred. The indirect effects were dose related and correlated with distance from the directly disturbed patch.

INTERACTING POPULATIONS IN A PATCHY ENVIRONMENT

In two instances, the models developed by Spromberg have been expanded to include interacting populations. The first model (Landis et al. 2000) investigated the potential effects of novel genetic elements being introduced into bacterial populations. The second (Deines et al. 2005) examined the results of competition between a native and an invasive species. In both cases, terms had to be added to describe the interactions of the species within each patch and migration between patches.

In Landis et al. (2000), we found that the introduction of a genetic element that altered the fitness of a host within a landscape resulted in a number of different dynamics. As the rate of infection increased, the dynamics could be forced into severe oscillations that could later be damped. It also was not necessary for the movable genetic element to increase the fitness of the host: what was important was the rate of infection. The pattern of patches was important in determining the specifics of the interactions. We also found that although the specifics of the rate of infection were highly variable, it was inevitable that the infectious genetic element would be spread to the host population.

Similar in many respects to the situation for movable genetic elements is the question of the risk of invasive species. In Deines et al. (2005), we reported the results of our modeling efforts to describe the patch dynamics of non-indigenous (invasive) species within three patch systems. Our models included an invasive organism, multiple patches, a difference in sensitivity to the toxicant (fitness) between the competitors, and interaction with a toxicant. As in our earlier models, certain initial conditions would result in multiple outcomes in the colonization of the invasive. One surprise was the importance of having the invasive established in a habitat patch somewhat isolated from the other patches so that the invasive could establish a large population without the native being rescued. We termed this the "beach-head effect." The beach-head allows the invasive population to build to sufficient numbers so that it now acts as a source population for other invasions to other regions of the landscape. We found that with our models there were clear optima between distance between patches, the competitive ability of the invasive, and the percent spread of the invasive to the patches. In one example (Figure 10.10), the optimal distances between patches were 10 to 50 model units. Distances longer or shorter than this resulted in no invasion. Also contrary to our expectations, at the longer distances an increase in competitive ability of the invasive was not important.

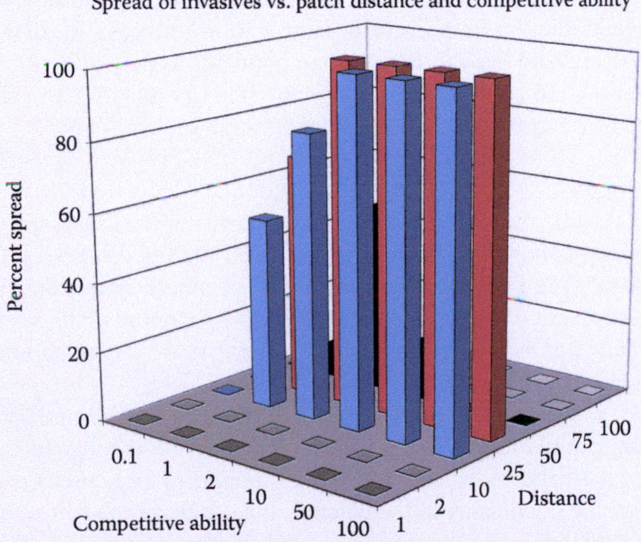

Figure 10.10 Interaction between distance between patches, competitive ability, and spread of the invasive. Note that a range of outcomes can occur depending on the specific distances and competitive ability in this system. Even at a competitive ability far above the native species invasion does not occur if the patches are either too close or too far away. At a distance of 50 units, the optimal competitive ability is actually twice that of the native.

These and other studies of how different species interact depending on spatial relationships are the beginning of the incorporation of metacommunities into an assessment of toxicological impacts. Given that streams, forests, and many other environments are constructed of a variety of habitat patches, this requires a consideration of special relationships when multiple species are interacting.

IMPORTANCE OF PATCH DYNAMICS

In the 15 years since publication of Spromberg et al. (1998), the importance of the spatial relationships of habitat patches has become recognized. First, there are now multiple examples of populations of regulatory interest that exist in patchy or metapopulations. Weakfish (Thorrold et al. 2001) and the Pacific herring along the British Columbian coast have been identified as populations existing as interacting patches. Recent studies of common loons (Walters et al. 2008); mummichogs (Nacci et al. 2008); and the amphipod, *Leptocheirus plumulosus* (Bridges et al. 2008) have all incorporated metapopulation modeling into the assessment of toxicant effects over large spatial scales.

Spromberg and Scholz (2011) have used patch dynamics to understand how stocks of Coho salmon are impacted by toxicant inputs that result in pre-spawn mortality. They found a relationship between the number of source patches (streams) and the likelihood of extinction of the species. As the number of sinks (spawning streams with high mortality) is increased, so does the probability of extinction of the entire population. Again, action at a distance is revealed. An entire population can become extinct although the direct toxicant effects are localized to one area.

Wilson and Hopkins (2013) have modeled the metapopulation dynamics of pond-breeding toads (*Bufo americanus*) when mercury is the toxicant. They have extensive data on the effects of mercury in field populations in the South River watershed of Virginia and have used those data in the modeling efforts. Dispersal from source populations can maintain other populations that would decline to extinction. However, an increase in distance between the patchy populations decreased the ability of source populations to support sink areas. Mercury contamination of the source population dramatically impairs the ability of the source to support the sinks, causing potential extinction in those patches.

IMPLICATIONS

Agriculture and urbanization have fractured the patterns of the landscape producing a variety of habitat patches. Aquatic environments have been modified, altering the movement of organisms by the introduction of dams, modification of channels, or by contamination of water and sediment. Terrestrial environments are fragmented by shopping malls, roads, and suburbia. In several studies, action at a distance appears as a property of patchy populations. There are four clear implications.

First, the spatial context of the population being studied must be understood. Impacts on one part of the patchy population can have effects on other parts through changes in migration patterns across the landscape. A change in migration patterns or sources and sinks within a landscape can change the occupancy of the patches. Placement of the same amount of toxicant in different parts of the landscape changes the ranges of potential outcomes for the populations.

Second, multiple outcomes are likely from the same set of initial conditions. Our simulations often resulted in more than one outcome being realized by the same set of initial conditions. This is due to the fact that a probabilistic function was incorporated to describe the dosing of the individuals within a patch. Outcomes as divergent as possible for a population, from extinction to reaching carrying capacity, can be possible from the same set of initial conditions. Only a clear knowledge of the properties of the organisms as individuals, of the distribution of the toxicant, and the spatial arrangement of the populations will allow an accurate prediction.

Third, indirect effects can cause extinction over an area broader than the occurrence of the contaminant. Action at a distance dictates that there does not have to be exposure to a contaminant by individuals of a habitat patch for impacts to be realized upon the population. Measurement of contaminant levels in organisms of that patch will not indicate any exposure, although there is an impact due to contamination in another part of the landscape. Exposure and effect need to be understood in a landscape context.

Fourth, the idea of a reference site is archaic when patch dynamics are understood. If one habitat or patch is connected to another contaminated site by migration, then it cannot serve as a reference site. So what about using areas clearly not linked by migration? Then there is the problem of ensuring sufficient genetic and community similarity. Bottom line, there is no such thing as a reference site when it comes to populations and landscapes.

10.5 POPULATION GENETICS

Kettlewell (1955) provided a telling investigation of **industrial melanism** (the gradual increase to predominance of melanic forms in industrialized regions) in peppered moths (*Biston betularia*) of Great Britain (Majerus, 2009; Rudge, 2010). Peppered moths are active at night but remain still on surfaces during the day to avoid the notice of visual predators, especially birds. The fitness of rare dark morphs quickly increased relative to the light morphs as surfaces darkened with soot. Although not thought of as such, this premier example of natural selection in the wild is an equally good example to ecotoxicologists of contaminant effects on population genetics. It also reinforces the important point that physical or chemical contaminants do not have to kill or impair individuals outright to influence populations. Rapid shifts in population genetics occurred as a consequence of shifting Darwinian fitness relative to a species interaction, predation.

VIGNETTE 10.3 Industrial Melanism: Genetic Adaptation to Pollution

Bruce Grant
College of William & Mary

Industrial melanism refers to the evolution of dark body colors in animal species that live in habitats blackened by industrial soot. Kettlewell (1973) and Majerus (1998) list some of approximately 100 examples that have been reported in a variety of species. Most are insects, and of these the vast majority are Lepidoptera, especially night-flying moths that rest by day. By far, the most thoroughly documented example is the peppered moth, *B. betularia.*

The common name derives from the appearance of the typical phenotype, a pale moth "peppered" with white and black scales. Melanic phenotypes are effectively solid black, and are often called carbonaria (Figure 10.11). Intermediates (or *insularia*) also occur. The range of phenotypes is genetically determined by multiple alleles at a single locus (Lees and Creed, 1977; Grant, 2004). Most population studies focus on temporal and geographic frequency distributions of the fully melanic *carbonaria*, but see Cook and Grant (2000) for an analysis of *insularia.*

The first melanic specimen was captured in 1848 near Manchester, England. By 1895, about 98% of the specimens near Manchester were melanic, and this once rare phenotype had spread across the industrial regions of Britain. Recent DNA analysis by Saccheri and his colleagues (Van't Hof et al. 2011) reveal significant linkage disequilibrium between the carbonaria-determining region and nonallelic marker loci within haplotypes indicating that melanism in British populations had a single mutational origin. With only one generation per year, the nearly complete reversal in phenotype frequency was astonishingly rapid and the limited recombination between the ancestral carbonaria haplotype and the more diverse typical haplotypes is the genetic signature of a strong selective sweep (Van't Hof et al. 2011).

Figure 10.11 Typical and melanic peppered moths, *Biston betularia cognataria*. At rest, the wings span 4–5 cm. (Photograph by Bruce Grant.)

But selection for what? Tutt (1896) proposed the answer to this question: *camouflage*. Peppered moths are active at night and remain still during daylight hours, resting or hiding on the surfaces of trees. Most insectivorous birds hunt by day, and locate prey primarily by vision. Moths that are well concealed against the backgrounds on which they rest are more likely to escape detection by birds than are conspicuous moths. Tutt proposed that "speckled" peppered moths in undisturbed environments gain protection from predators by their resemblance to lichens. In manufacturing regions where lichens have been destroyed by pollution and the surfaces of trees blackened by soot, the speckled forms fall victim to bird predators, whereas the melanic forms escape detection and pass their genes on to their progeny. Tutt recognized the impact of pollution in disrupting the effectiveness of the protective coloration the speckled pattern lent to peppered moths. Blackening of habitats by industrial soot shifted the survival advantage to the once rare melanic phenotype resulting in the evolution of a new camouflage pattern better adapted to the perturbed local conditions.

Tutt's camouflage hypothesis was not tested until the 1950s when Kettlewell (1973) initiated a series of experiments to determine whether or not birds actually ate melanic and typical peppered moths selectively. His experiments included three main components: (1) quantitative rankings of conspicuousness (to the human eye) of typical and melanic phenotypes resting on various backgrounds, (2) direct observations of predation by birds on moths placed onto tree trunks, and (3) recapture rates of marked moths released onto trees in polluted and unpolluted woodlands in different regions of Britain. His results showed that conspicuous moths were the first to be found and eaten by birds, and that the recapture rates of released moths were in agreement with Tutt's predictions about protective camouflage.

The success and publicity of Kettlewell's experiments made industrial melanism the classic textbook example of natural selection that could actually be observed as it occurred. As part of the normal process of science, the design and execution of these pioneering experiments came under intense scrutiny over the years. Some criticisms of the work were valid and some were trivial, and many came from Kettlewell himself. Unfortunately, these routine scientific discussions were intentionally distorted by creationists who sought (and still seek) to discredit evidence for evolution. A legitimate critique of the field, in a book by Majerus (1998), and its subsequent reviews, drew attention to perceived deficiencies in the evidence supporting bird predation: do peppered moths actually rest on the *trunks* of trees, and do birds selectively eat peppered moth phenotypes under more natural conditions than those employed by Kettlewell? Ultimately all of the criticisms were met in the largest single predation study ever undertaken in which Majerus (Cook et al. 2012) released 4864 peppered moths over a 6-year period at controlled densities and phenotype frequencies. In addition, his own painstaking observations revealed that wild moths routinely rest on tree trunks (validating previous experiments inexpertly disqualified for having used trunks), as well as among upper branches where he conducted his own experiments. He also directly observed nine different bird species consume peppered moth phenotypes selectively. The selection coefficients calculated from his data are consistent with the current widespread declines in melanic frequency across the United Kingdom. For an analysis and description of Majerus's work, see Cook et al. (2012).

Extensive experimental work over the past half century has confirmed that bird predation is the selective agent driving the evolution of melanism in peppered moth populations. Although other potential agents have been proposed, none have the experimental backing that selective bird predation has accumulated. But whether or not bird predation is the only *agent* of selection, the evidence for natural selection itself rests solidly on the long-term changes in natural populations of the frequency of the alleles producing melanic phenotypes, well documented at numerous locations by independent workers over many decades. A change in allele frequency precisely meets the operational definition of genetic evolution, and of those known evolutionary forces capable of producing such change (mutation, migration, genetic drift, and selection), only natural selection can account for the steady trajectory and velocity of the allele-frequency changes that have been recorded. Criticisms about the evidence identifying bird predation as the *agent* of selection, valid or otherwise, should not be conflated with the fundamental population genetic evidence for the occurrence of natural selection.

Because reflectance from the surface of tree bark is strongly negatively correlated with atmospheric levels of suspended particles (Creed et al. 1973), the testable prediction from Tutt's camouflage hypothesis is that melanic phenotypes should be more common in sooty, polluted regions than they are in unpolluted regions. This prediction was clearly met by the frequency of melanic peppered moths in populations surveyed across Britain during the 1950s (Kettlewell, 1973).

Several studies have shown that the geographic distribution of melanism in British peppered moths is more strongly correlated with sulfur dioxide (SO_2) concentration than with smoke (Steward, 1977). Sulfur dioxide (SO_2), as a gas, is more widely dispersed in the atmosphere than particulate matter that tends to settle locally as soot. Although selection may operate locally, gene flow from the migration of individuals among populations has contributed to more gradual clines in melanism in peppered moths, a relatively mobile species, than in the sedentary *Gonodontis* (= *Odontopera*) *bidentata*, in which the frequency distribution of melanic phenotypes is sharply subdivided over its range (Bishop et al. 1978). The power of gene flow to obscure genetic adaptation to local conditions resulting from selection was too little appreciated by early workers who entertained various forms of nonvisual selection in attempts to account for apparent anomalies in clines. Since that time, Saccheri et al. (2008) have shown low levels of genetic differentiation at nine microsatellite loci from population samples along a 130-km transect in Britain that indicate median dispersal distances to be five times greater than earlier estimates reported by Bishop (1972).

In 1956, Britain initiated the Clean Air Act to establish so-called smokeless zones in heavily polluted regions. Following significant reductions in atmospheric pollution, particularly SO_2, melanic peppered moths have declined in frequency as the typical form recovered (Clarke et al., 1985). Indeed, the decline in melanism in the latter half of the twentieth century is much better documented than the increase in melanism in the latter half of the nineteenth century. A continuous record begun in 1959 by Clarke near Liverpool includes over 18,000 specimens that show a drop in the frequency of melanics from 93% to 8% by 1996 (Grant et al. 1998). The most recent national survey taken in 1996 shows that marked declines in melanism have occurred everywhere in Britain (Grant et al. 1998). The predicted correlation between changes in the levels of pollution and in the frequencies of melanic phenotypes in peppered moth populations has been firmly established.

Parallel evolutionary changes have also occurred in the American peppered moth, *B. betularia cognataria* (Grant et al. 1996) in which its melanic phenotype, indistinguishable in appearance from British carbonaria, is produced by an allele at the same genetic locus (Grant, 2004). Owen (1961) showed that the rise and spread of melanism started in America about 50 years later than in Britain. Museum collections do not include any melanic specimens before 1929 in the vicinity of Detroit, Michigan, but by 1959, when Owen began sampling moth populations there, the frequency of melanics exceeded 90%. In 1963, clean air legislation was inaugurated in the United States. Records for southeastern Michigan show that atmospheric SO_2 and suspended particulates have declined significantly, and by 1994, the frequency of melanic peppered moths had fallen below 20% (Grant et al. 1996). The decline in melanism has continued with only 15 melanic moths (5.3%) among 283 specimens taken in 2001 (Grant and Wiseman, 2002). The significance of a sharp decline in melanism in American peppered moths coincident with reductions in atmospheric pollution is complemented by comparisons of SO_2 concentrations from southeastern Michigan to western Virginia, where melanism has not exceeded 5% over the same time interval. The Michigan location recorded higher concentrations of SO2 than the Virginia location in 23 of the 25 years that measurements were taken at both places (Grant and Wiseman, 2002).

The recent decline in melanism in Michigan, and its near absence in Virginia might be mistaken for *thermal* melanism, a phenomenon not uncommon in the Lepidoptera (Majerus, 1998). However, there is direct evidence against thermal melanism in *B. betularia*; namely, the absence of latitudinal clines. This is clear from latitudinal variation in melanic frequencies in Britain (Kettlewell, 1973) and in Scandinavia (Douwes et al. 1976).

Both American and British peppered moth populations are now converging on monomorphism for their respective typical phenotypes correlated with reduced levels of atmospheric pollution on both sides of the Atlantic. These phenotypic changes reflect genetic changes as the populations continue to adapt by natural selection to environments modified by human activity.

10.5.1 Change in Genetic Qualities

Toxicants can influence population genetics in many ways although ecotoxicologists tend to focus on selection-associated changes. By mechanisms already discussed, chemicals and radionuclides can change the genetic qualities of individuals within a population directly. But, toxicants can also influence population genetics in less direct ways. A toxicant can reduce the **effective population size** (the number of individuals contributing genes to the next generation) and result in a net loss of genetic variation. A **genetic bottleneck** occurs if there are too few individuals available to ensure an allele makes it into future generations. Under less severe conditions, reduction of effective population size may accelerate the rate of loss of a rare allele from a population. **Genetic drift** (random change in allele frequencies in a population) is accelerated at low effective population sizes. The net result of a toxicant's effect on a population, even in the absence of selection, could be the loss of a specific allele or an overall decrease in genetic variability. As genetic variation is the raw material for evolutionary change, this could reduce the ability of a population to adapt to and survive future changes in its environment (Derycke et al. 2007; Nowak et al. 2007). Murdoch and Hebert (1994) found reduced variability in mitochondrial DNA of the brown bullhead (*Ameiurus nebulosus*) from the Great Lakes and attributed it to pollutant-induced reductions in effective population size. Kopp et al. (1992) demonstrated a reduction in genetic heterozygosity in stressed populations of the central mudminnow (*Umbra limi*).

Under the presumption that individuals possessing the most genetic variation tend to be the most robust,* the argument has been made that a decrease in genetic variability could occur in populations impacted by toxicants and reduce the overall fitness of associated individuals (Mulvey and Diamond, 1991). The individuals most able to survive and reproduce in the polluted environment may be the most heterozygous. For example, Kopp et al. (1992) noted for laboratory assays that central mudminnows (*U. limi*) with highest genetic diversity had enhanced abilities to survive stressful conditions (low pH and high aluminum concentrations). Diamond et al. (1989) found that mosquitofish (*Gambusia holbrooki*) with high numbers of heterozygous loci tended to have longer times-to-death during mercury exposure than mosquitofish with fewer heterozygous loci. However, Newman et al. (1989) demonstrated later that the particular heterozygosity effect described by Diamond et al. (1989) was an artifact. Schlueter et al. (1995) found no effect of heterozygosity on time-to-death of fathead minnows (*Pimephales promelas*) exposed to high concentrations of copper. Although plausible mechanisms exist (e.g., multiple heterosis, inbreeding depression, or overdominance), presumptions cannot be made at this time about any specific field situation that multiple locus heterozygosity does or does not influence fitness relative to toxicant effects.

10.5.2 Acquisition of Tolerance

Toxicants can, and often do, act as selective agents for exposed populations. **Natural selection**, the process by which genes from the most fit individuals are overrepresented in the next generation,† can result in enhanced tolerance to toxicants as is amply demonstrated by the adaptation of pests to pesticides. Considerable human effort is made to counterbalance tolerance acquisition by insect (e.g., Comins, 1977; Mallet, 1989), rodent (Webb and Horsfall, 1967; Partridge, 1979), and other target species of pesticides. Populations of nontarget species can also adapt and become more tolerant of toxicants. The probability of obtaining tolerance or the rate at which tolerance increases is influenced by many factors (Table 10.5).

The likely, but sometimes overlooked, consequence of exposure is local extinction of the exposed population, not enhanced tolerance (Klerks and Weis, 1987; Mulvey and Diamond, 1991). Pollution as a selection force is extreme in its rate of change relative to many other environmental factors to which organisms must adapt (Moriarty, 1983). Populations successfully adapting to so rapid a change in environmental conditions are probably exceptional. Therefore, the occasional arguments made to ease regulations based on the premise that adapted field populations will be more tolerant than predicted from laboratory assays using nonadapted individuals are flawed because they would ignore this possibility (Klerks and Weis, 1987).

* Measures of fitness have been correlated with the number of loci found to be heterozygous in individuals (Samallow and Soule, 1983; Koehn and Gaffney, 1984; Danzmann et al. 1986). The mechanism for the enhanced fitness is often assumed to be **multiple heterosis**, a generally higher fitness as a consequence of combined advantages of being heterozygous for each individual locus (heterosis). (**Heterosis** is the general term used to describe the superior performance of heterozygotes.)

† Evolution via natural selection carries several assumptions. It is assumed that surplus numbers of individuals are produced by populations. In a particular environment, individuals vary in their abilities to survive and reproduce, i.e., their fitness. All or a portion of these differences in fitness are heritable. The net result is natural selection.

Table 10.5 Qualities Modifying the Rate of Tolerance Acquisition

Quality	Specific Influence on Tolerance Acquisition
1. Genetic Qualities	
Dominance	Most rapid in early generations if tolerance is controlled by a dominant allele
Single gene versus many genes involved	Most rapid if determined by a single gene
Two or more selection components	Opposing selection components can balance each other or slow tolerance acquisition
Relative differences in fitness	Most rapid if the differences in fitness among tolerant and sensitive individuals is large
2. Reproductive Qualities	
Rate of increase and generation time	Most rapid with high population growth rate and short generation time
Size of population	In general, smaller populations will have less variation than larger populations
3. Ecological Qualities	
Migration	Influx of nontolerant individuals due to immigration could slow tolerance acquisition
Presence of refugia	The presence of refugia such as uncontaminated areas will slow tolerance acquisition
Life stage	Sensitive life stage will have large influence on tolerance acquisition

Source: Modified from Table 1 in Mulvey, M. and S.A. Diamond, in *Metal Ecotoxicology: Concepts and Applications,* Lewis Publishers, Chelsea, Michigan, 1991.

Differences in tolerance* of target and nontarget species may be controlled by a single gene (**monogenic control**) or several genes (**polygenic control**) (Figure 10.12). Monogenic control of tolerance to endrin and other cyclodiene pesticides was found for mosquitofish (*Gambusia affinis*) populations exposed during agricultural spraying (Wise et al. 1986; Yarbrough et al. 1986). Wirgin et al. (2011) found base deletions to the aryl hydrocarbon receptor 2 gene of tomcod (*Microgadus tomcod*) were the basis of PCB in Hudson River (New York, United States) tomcod populations. Cadmium-resistant Foundry Cove (New York, United States) populations of the oligochaete, *Limnodrilus hoffmeisteri*, were also believed to be a result of strong selection for one allele. In contrast, Posthuma et al. (1993) found polygenic control of heavy metal tolerance in populations of the springtail, *Orchesella cincta*. Enhanced tolerance of this soil insect was associated with differences in metal excretion efficiency among populations.

Tolerance acquisition varies relative to its cost, that is, the allocation of metabolic resources (Figure 10.13). Hickey and McNeilly (1975) found that metal-tolerant plants are at a disadvantage relative to nontolerant plants if grown in a non-contaminated soil. Postma et al. (1995) found that midges (*Chironomus riparius*) from cadmium-tolerant populations had poorer survival, growth, and reproductive success than nontolerant populations if reared in low cadmium conditions. The lowered fitness was attributed to an apparent zinc deficiency in tolerant individuals living under low cadmium conditions. The rapid loss of cadmium resistance in Foundry Cove populations of the oligochaete, *Limnodrilus hoffmeisteri*, was thought by Mackie et al. (2009) to indicate a cost for resistance.

There may be cross-resistance among toxicants depending on the mechanism underlying enhanced tolerance. **Cross-resistance** or **co-tolerance** is the condition in which enhanced tolerance

* Distinction is made by many authors between the terms, tolerance and resistance. **Tolerance** is often reserved for enhanced ability to cope with a factor due to physiological acclimation. **Resistance** is used if the enhanced abilities are associated with genetic adaptation. Following the lead of Weis and Weis (1989b), tolerance is used here for both acclimation and genetic adaptation.

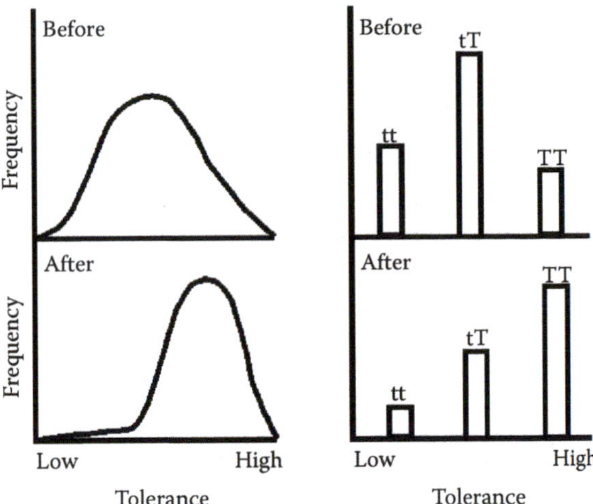

Figure 10.12 Shifts in tolerance expected under polygenic (left side of figure) or monogenic (right side of figure) control. With polygenic control, differences in tolerance will appear continuous as shown in the "before" selection panel on the left side of the figure. With selection, the mean tolerance will shift upward and the variation about this mean will narrow. With tolerance determined by a single gene (right side), the distribution of genotypes among homozygous for intolerance (tt), ance allele (T). In this illustration, the T allele is dominant to the t allele. (Modified from Figure 1 of Mulvey, M. and S.A. Diamond, in *Metal Ecotoxicology: Concepts and Applications,* Lewis Publishers, Chelsea, Michigan, 1991.)

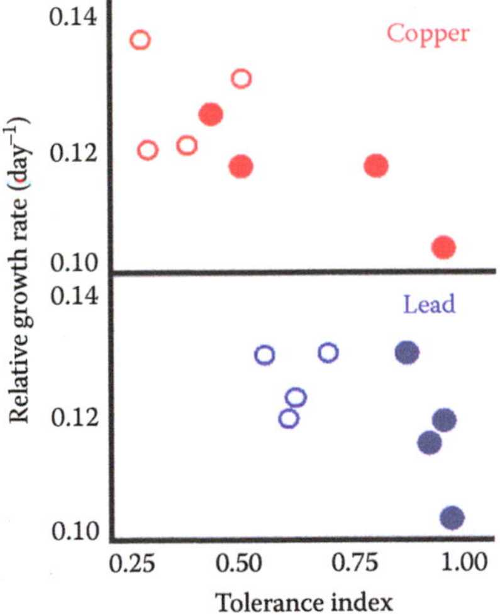

Figure 10.13 The cost in terms of growth rate for plant (*Agrostis capillaris*) strains with varying degrees of heavy metal tolerance. Plants were taken from reference locations (open circles) and areas with long histories of heavy metal mining in Wales (closed circles). Growth was measured in near-optimal media with heavy metal concentrations below those causing stress. Tolerance was measured as [2 (root length in heavy metal solution)]/[root length in a heavy metal solution + root length in a control solution]. Plants adapted to high metal concentrations (high-tolerance index) had generally slower growth under ideal conditions than nontolerant strains, indicating a trade-off associated with tolerance. (Modified from Figures 1 and 2 of Wilson, J.B., *Evolution,* 42, 408–413, 1988.)

to one toxicant also enhances tolerance to another. For example, plants tolerant to one *s*-triazine herbicide display cross-resistance to other *s*-triazine herbicides due to elevated levels of an herbicide-binding protein (Erickson et al. 1985). Lead detoxification is higher in isopod populations tolerant to copper (Brown, 1978). Metallothionein gene duplication as an adaptation to elevated levels of one metal (Lange, 1989) could enhance tolerance to other metals. Enhanced rotenone tolerance imparted by elevated cytochrome P-450 monooxygenase activity (Fabacher and Chambers, 1972) would likely enhance tolerance to other pesticides detoxified by this mechanism.

10.5.3 Measuring and Interpreting Genetic Change

Allozymes (allelic variants of an enzyme coded for by a particular locus) were first applied in the 1970s by Nevo and coworkers (Nevo et al. 1977, 1978, 1981; Lavie and Nevo, 1982, 1986; Baker et al. 1985) to measure changes in population genetics as a consequence of environmental pollution. They remain useful for certain types of assessments of population response to toxicants (Gillespie, 1996) although modern DNA techniques have justifiably supplanted them in most studies. Allozymes can rapidly be scored in large numbers of individuals using starch gel or some other type of electrophoresis. The different forms of an enzyme notionally coded for at a locus are separated from tissue homogenates of individuals in an electrical field during electrophoresis and biochemical staining is then used to visualize enzyme activity on the electrophoretic medium. Genotype for each sampled individual is then implied from the pattern of staining on that medium.

Currently, flourishing DNA- and RNA-based technologies have greatly accelerated the study of contaminant effects on population genetics. Techniques particularly relevant at the moment include restriction fragment length polymorphism (RFLP), random amplification of polymorphic DNA (RAPD), microsatellite DNA analysis, and DNA sequencing but others are quickly coming into common usage (see Newman and Clements, 2008, pages 311–313 for more details). These technologies have fostered the emergence of the new field of **ecotoxicogenomics**, the study of toxicant effects on organism genomes and related changes in biological functions. Snape et al. (2004) first coined this term, ecotoxicogenomics to mean "the study of gene and protein expression in non-target organisms that is important in responses to environmental toxicant responses." This field tends more toward the laudable goal of enriching our understanding of how toxicant-gene interactions influence functions than directly enriching our understanding of population-level processes. However, an ever-increasing number of related studies do use the associated new technologies to directly measure population-level processes. Mulvey et al. (2002, 2003) successfully applied sequencing of mitochondrial DNA as well as allozymes to show correspondence between habitat PAH contamination and fish (*Fundulus heteroclitus*) population structure within a landscape. Reichman et al. (2006) used these technologies to document the flow of an herbicide resistance gene from genetically modified crop species into bent grass (*Agrostis stolonifera*) populations. Microsatellite loci were studied in cattails (*Typha angustifolia* and *T. latifolia*) near the Chernobyl reactor site (Tsyusko et al. 2006). Although the brief discussion below tends to draw on allozyme-based studies to introduce concepts, the reader should understand that many more excellent nucleic acid-based studies are now being conducted that greatly enrich our understanding of toxicant effects on population genetics.

Allele and genotype frequencies in field populations have suggested pollutant-related loss of genetic diversity (Kopp et al. 1992) or selection (Battaglia et al. 1980; Gillespie and Guttman, 1989; Heagler et al. 1993). Interpretation of field results is frequently supported by laboratory studies suggesting differential mortality among allozyme genotypes (Battaglia et al. 1980; Diamond et al. 1989; Newman et al. 1989; Heagler et al. 1993; Keklak et al. 1994; Schlueter et al. 1995). For example, Heagler et al. (1993) interpreted changes in the frequency of alleles associated with a glucosephosphate isomerase locus (GPI-2) of a field population of mosquitofish (*Gambusia holbrooki*) using results of survival analysis of GPI-2 genotypes exposed to high concentrations of mercury in the laboratory.

Too often, the enzyme itself is assumed in allozyme-based studies to be responsible for the observed differences in fitness among genotypes without due consideration of alternative explanations. For example, the different allozymes are thought to have different availabilities of sites to bind with metals, and consequently, susceptibilities to inactivation by the metals. Although this is a reasonable explanation, it is seldom tested rigorously. It was untrue in one case in which it was tested. Differences in GPI-2 genotype sensitivity under acute mercury exposure of mosquitofish (*Gambusia holbrooki*) were not a consequence of differential inactivation of allozymes by mercury (Kramer et al., 1992; Kramer and Newman, 1994). Results suggested that differences among genotypes were more readily interpreted in the context of optimal energy resource allocation under general stress. Further, a scored enzyme locus might only be acting as a marker for a closely linked gene that is actually responsible for the difference in tolerance among genotypes. Such **genetic hitchhiking*** is very often given inadequate consideration as a mechanism underlying the observations.

Alleles can be unevenly distributed throughout a structured population, e.g., among lineages. If differences in tolerance exist within the structure, this would result in correlations between tolerance and genotype that falsely suggest that an allele itself is directly linked to tolerance. Lee et al. (1992) reinforced this point by demonstrating a strong family effect relative to mosquitofish (*Gambusia holbrooki*) tolerance to mercury.

Population structuring may also produce an apparent deficit of heterozygotes relative to Hardy–Weinberg expectations.† These deficits may be mistakenly attributed to selection against heterozygotes. However, a deficit of heterozygotes can arise if two genetically distinct groups of individuals are mixed as in the case of unknowingly sampling a highly structured population under the assumption of uniformity in the sample. It can also occur if significant amounts of migration have occurred as in the case of an influx of individuals into a population recently decimated by a pollution event. The **Wahlund effect** predicts that there will be a net deficit of heterozygotes if two populations, each in Hardy–Weinberg equilibrium but possessing different allele frequencies, are mixed and their combined genotype frequencies quantified. This effect has been ignored in most pollution-related studies to date. Woodward et al. (1996) demonstrated such a Wahlund effect during a population study of midges (*C. plumosus*) inhabiting a mercury-contaminated lake.

Genotype-related differences in pollutant tolerance are almost exclusively examined relative to survival differences. However, selection occurs at several stages in the life cycle of an individual (Figure 10.14). There are several such **selection components**: viability selection, sexual selection, meiotic drive, gametic selection, and fecundity selection. **Viability selection** or selection based on differential survival begins at the zygote and continues throughout the life of the individual. This component can be further broken down into viability at different ages or stages of development. **Sexual selection** involves differential mating success of individuals. It might be associated with females or males, i.e., female or male sexual selection. **Meiotic drive** is the differential production of gametes by different heterozygous genotypes. One allele may be underrepresented in the gametes produced by a heterozygous individual. **Gametic selection** involves differential success of gametes produced by heterozygotes. The last component, **fecundity selection**, is the production of more offspring by matings of certain genotype pairs than by other genotype pairs. Selection

* Endler (1986) defines genetic hitchhiking as "a situation in which a given allele changes in frequency as a result of linkage or gametic disequilibrium with another selected locus … [it can] give a false impression of selection at a particular locus … Similarly, if there is genotypic correlation among quantitative traits, then selection will appear to affect a trait directly, although it is actually only affected through its correlation with another selected character."

† At **Hardy–Weinberg equilibrium**, the frequency of genotypes will remain stable through time. For a two allele locus (i.e., T and t), the frequencies of the genotypes will be q^2 for TT, 2pq for Tt, and p^2 for tt, where q and p are the allele frequencies for T and t, respectively. The following conditions are assumed: (1) the population is large (effectively infinite) and composed of randomly mating, diploid organisms with overlapping generations, (2) no selection is occurring, and (3) mutation and migration are negligible.

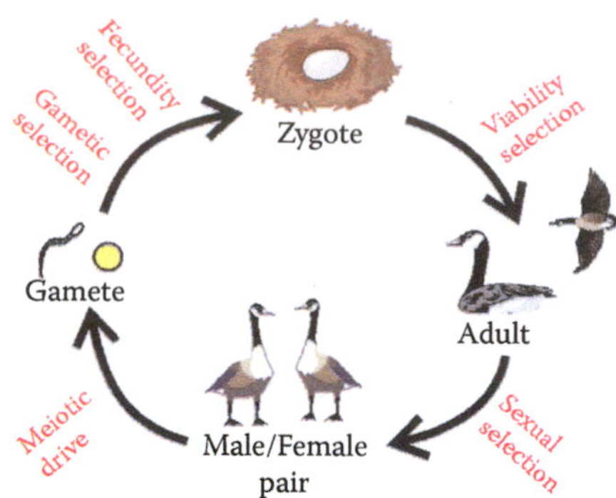

Figure 10.14 Components of the life cycle of an individual in which natural selection (selection components) can occur. Although rarely considered, selection components can be acted upon by contaminants. Please see the text for a detailed explanation.

components can act in opposite and balancing directions (Endler, 1986); therefore, measurement of only one, such as viability selection, may result in inaccurate predictions of changes in allele frequencies under selection pressures from pollutants. For example, Mulvey et al. (1995) found this to be the case with mercury-exposed mosquitofish (*Gambusia holbrooki*). The GPI-2 genotype that was at a disadvantage during acute exposure (viability selection) was not the same as that genotype at a disadvantage relative to female sexual selection. Unfortunately, selection component analysis is ignored in most studies.

10.5 SUMMARY

In this chapter, the importance of assessing effects to populations and metapopulations is emphasized. A brief sketch of epidemiological metrics and logic is provided as applicable to ecotoxicology. Demographic approaches are described that greatly improve our ability to predict the population consequences of toxicant exposure. Examples of their increasing use by ecotoxicologists are provided. The potential influence of toxicants on population genetics is outlined in addition to the possible consequences of changes. Acquisition of tolerance, factors influencing the rate of tolerance acquisition, and other related processes are described.

SUGGESTED READINGS

Daniels, R.E. and J.D. Allan, Life table evaluation of chronic exposure to a pesticide, *Can. J. Fish. Aquat. Sci.,* 38, 485–494, 1981.

Mulvey, M. and S.E. Diamond, Genetic factors and tolerance acquisition in populations exposed to metals and metalloids, in *Metal Ecotoxicology: Concepts & Applications,* Newman, M.C. and A.W. McIntosh, Eds., Lewis Publishers, Chelsea, MI, 1991.

Newman, M.C., *Quantitative Ecotoxicology*, 2nd Edition, Taylor & Francis/CRC Press, Boca Raton, FL, 2013.

Newman, M.C. and W.H. Clements, *Ecotoxicology: A Comprehensive Treatment*, Taylor & Francis/CRC Press, Boca Raton, FL, 2008.

Rench, J.D., Environmental Epidemiology, in *Basic Environmental Toxicology*, Cockerham, L.G. and Shane, B.S., Eds., CRC Press, Boca Raton, FL, 1994.

Sibly, R.M. and P. Calow, A life-cycle theory of responses to stress, *Biol. J. Linn. Soc.,* 37, 101–116, 1989.

CHAPTER 11

Effects to Communities and Ecosystems

The accumulation of persistent toxic substances in the ecological cycles of the earth is a problem to which mankind will have to pay increasing attention ... What has been learned about the dangers in polluting ecological cycles is ample proof that there is no longer safety in the vastness of the earth.

Woodwell (1967)

11.1 OVERVIEW

11.1.1 Definitions and Qualifications

A classically defined ecological **community** is "an assemblage of populations living in a prescribed area or physical habitat: it is an organized unit to the extent that it has characteristics additional to its individual and population components ... [it is] the living part of the ecosystem" (Odum, 1971). This community—perhaps of a locality such as a lake or meadow—is made up of species that interact and form an organized unit although some species might interact only loosely (Magurran, 1988).

Acts in what Hutchinson (1965) has called the 'ecological theatre' are played out on various scales of space and time. To understand the drama, we must view it on the appropriate scale.

Wiens (1989)

Toxicant exposure often changes a species' niche, that is, its role in the "ecological theatre." Niche can be defined more succinctly as the **Hutchinsonian niche,** "... the certain biological activity space in which an organism exists in a particular habitat. This space is influenced by the physiological and behavioral limits of a species and by effects of environmental parameters (physical and biotic, such as temperature and predation) acting on it" (Wetzel, 1982). A species in a particular area has a **fundamental niche** in which it could exist based on its physiological and other limitations, and a **realized niche**, which is that portion of its fundamental niche that it actually occupies. The realized niche is often smaller than the fundamental niche due to the influence of other species, such as predators and competitors.

The impracticality of studying all species, or even all important species, in any community results in studies that focus on some taxonomic or functional subset of the community such as the fur bearers of a woodland or fish in a lake. Pielou (1974) suggests that the term **taxocene** should be used to distinguish these taxonomically defined subsets from true communities. Magurran (1988) suggests the term **species assemblage** for any operationally defined grouping such as planktonic filter feeders. Many models and indices are framed in the community context but applied pragmatically to taxocenes or species assemblages. Although such an approach remains valuable and necessary, interpretation of associated results should be tempered with the understanding that a species assemblage or taxocene is not the complete community.

Similarly, the **ecosystem** concept is a useful abstraction or simplification that should not be taken for a completely accurate depiction of reality (Newman, 2013). The **ecosystem** concept combines the biota (community) and abiotic environment into an organized system. (See Golley (1993) for a detailed discussion of the ecosystem concept.) Species interact with each other and loosely interact with their physical environment. Biotic and abiotic components act together to direct the flow of energy and cycling of materials. Obviously, application of this concept to real situations is highly dependent on the scale (time and space), distinctiveness of system boundaries (e.g., a distinct, spring-fed lake vs. a diffuse bottomland hardwood ecosystem between an upland and a river), and the particular qualities under study (e.g., cation flux from a watershed ecosystem vs. oxygen dynamics of a dimictic* lake). This concept must be applied intelligently to avoid unsound conclusions about qualities of an operationally defined ecosystem using ideal characteristics of the ecosystem abstraction.

The classic ecosystem framework just described has an implied emphasis on a smaller scale than that of the MacArthur–Wilson approach discussed in Section 11.1.2. The classic ecosystem focuses more on species interactions and the MacArthur–Wilson approach considers primarily movement of species between distant ecosystems. The metacommunity concept integrates the species interactions within communities with the migration of individuals of different species within a network of communities. A **metacommunity** is defined as "a set of local communities that are linked by dispersal of multiple potentially interacting species" (Leibold et al. 2004) (Figure 11.1). A richer understanding results from the metacommunity context for phenomena at different scales, allowing explanation of observations inexplicable from either a classic community or MacArthur–Wilson island biogeography vantage alone. Embellishment of the metacommunity concept, like that to the metapopulation concept described in the last chapter, enriches our understanding of many issues about to be described including predator–prey interactions (Taylor, 1990), competitive coexistence (Amarasekare and Nisbet, 2001; Mouquet and Loreau, 2002; Amarasekare, 2003), trophic cascades (Shurin et al. 2002), species diversity (Loreau, 2000), symbiosis (Klausmeier, 2001), community closure (Gilbert et al. 1998; Lundberg et al. 2000) and connectivity (Roberts, 1997), experimental approaches in community ecology (Skelly, 2002), and food web structure (Holyoak, 2000).

11.1.2 Context

Community, metacommunity, and ecosystem qualities are all affected by abiotic factors including pollutants (Dunson and Travis, 1991). Despite this, pollutant effects at these levels have been addressed in less detail than warranted in the past (Taub, 1989). Clements and Kiffney (1994) noted that only 12% of 699 environmental toxicology articles published from 1980 to 1982 dealt with populations, communities, or ecosystems. They further noted a disappointingly low percentage (18% of all articles) in more recent (1992) publications of the journal *Environmental Toxicology and Chemistry.* A quick review of the 2012 issues of that same journal indicates again a modest 10% of all publications on these topics. This initial neglect probably reflects the field's historical roots in mammalian toxicology, a field that emphasizes effects to individuals. A better balance is needed and hopefully will emerge soon in ecotoxicology. To contribute to such a balance, more than half

* Many lakes become thermally stratified with warm, less-dense water toward the surface and cold, more-dense waters at their bottom. As temperatures change with season, lake layers often mix as these temperature and density differences break down. A dimictic lake is one that mixes completely from surface to bottom twice each year (often spring and fall). The dissolved oxygen dynamics are distinct for the top (epilimnion) and bottom (hypolimnion) layers.

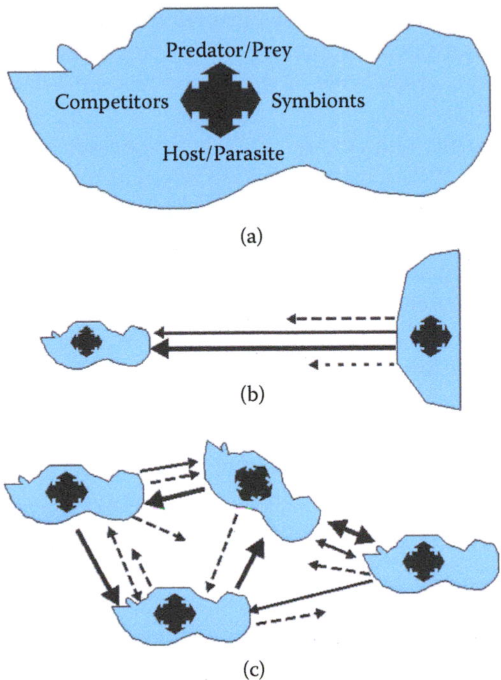

Figure 11.1 The metacommunity context includes species interactions and species migration dynamics among communities. Classic community ecology emphasizes species interactions such as those shown as (a) in influencing community structure and function. Classic MacArthur–Wilson theory emphasizes the likelihood of individuals of different species establishing or disappearing from an area such as a physical island (b). Metacommunity ecology combines both interspecies interactions and movements in a network of communities to suggest influence on structure and function (c).

of the pages in the sister textbook to this one (Newman and Clements [2008], *Ecotoxicology. A Comprehensive Treatment*) were devoted to higher order ecological themes.

Causal mechanisms for community change are often to be found at the next lower level of organization, i.e., at the population or metapopulation level. For example, change can occur because a particular species population's viability was lowered sufficiently by a toxicant's effect on growth, survival, or reproduction. This scenario is consistent with the approach advocated in Chapter 1 for maintaining conceptual coherency in any hierarchical science. However, **emergent properties** must also be considered carefully at higher levels of organization. Properties emerge in hierarchical systems that cannot be predicted solely from our limited understanding of a system's parts or components.* The counterexample to that just given is the indirect loss of several species because an important keystone species was killed directly by the toxicant. (A **keystone species** is one that influences the community by its activity or role, not its numerical dominance.) A species resistant to the direct action of a toxicant would disappear because another species performing a crucial role in the community was eliminated. Another example is industrial melanism in which community processes (i.e., predator–prey interactions) influenced population genetics (i.e., predominance of melanism). A third example might be the loss of a migration corridor in a metacommunity with a consequent loss of a species in an adjacent community. Causal structure is reversed with interactions among species populations in the community (higher level) producing an impact to

* This concept is central to the tedious holistic–reductionistic debate in ecology. Time wasted debating this obvious point distracts ecotoxicologists from the real challenge, enhancing the inferential strength of their science regardless of the conceptual vantage taken.

a population (lower level). Some final examples involve trophic webs. **Trophic cascades** exist in which predators influence plants of a community as will be discussed shortly. Also, weak interactions within a food web play a crucial role in stabilizing resource–consumer dynamics (McCann et al. 1998; Neutel, et al. 2002).

This chapter deals primarily with communities, but processes occurring in whole ecosystems are discussed toward the end of the chapter. Many ecosystem topics have already been described and some additional properties associated with ecosystems will be addressed in Chapter 12. This chapter begins by describing simple species interactions relative to the influence of toxicant action. Then community qualities, including structure and function, are described in the context of laboratory, mesocosm, and field research. Although much of the ecotoxicological work done with communities has been descriptive, the emphasis in this chapter will be explanatory principles derived primarily from experimental efforts. Field studies and methods are detailed toward the end of the chapter. Most field methods focus on structural changes observed in species assemblages such as soil arthropods or stream macroinvertebrates. Conventional community indices (such as species richness) or more specialized indices (the index of biological integrity [IBI]) are applied to species assemblages from and around contaminated areas. Community functions are assessed less often, but with increasing frequency during the last decade. These functions are discussed briefly toward the end of the chapter.

11.1.3 General Assessment of Effect

A wide range of practical experimental approaches has been taken to determine the concentration of toxicants below which the community is protected. The **most sensitive species approach** takes the results for the most sensitive of *all tested species* as an indicator of that concentration most likely to protect *all species in the community.* Despite the great advantage of its simplicity, several difficulties arise with this notionally cost-effective approach (Cairns, 1986). One must make the dubious assumption that the tested species and measured effects truly reflect the most sensitive within the community. The most sensitive species approach might not be cost-effective if one considers the high costs of a bad decision based on a compromised approach (Cairns, 1986). Remember also that the biological significance of the effect remains ambiguous if effect metameters such as NOEC are used, i.e., the maulstick incongruity.

In the last two decades, an embellishment of this most sensitive species approach has appeared for notionally determining concentrations protective of a community. A collection of available NOEC values or some other common measure of effect are pooled for a collection of species to estimate the concentration "protective" of all but a predetermined percentage of all species (e.g., 5%) (Figure 11.2) (Van Straalen and Denneman, 1989; Wagner and Løkke, 1991; Posthuma et al. 2002). It was originally presented as a threshold measure of the concentration generally protective of the community and some current applications still seem to adhere to this underlying assumption when interpreting results. But Hopkin (1993) questioned the assumption that a loss of 5% of all species is generally acceptable for protection of an ecological community. Hopkin's concerns seem justified given the compromised performance of the technique in a recent examination of pesticide effects on soil organisms (Frampton et al. 2006). Arguing from the context of soil communities, Hopkin noted that elimination of one keystone species such as the earthworm would dramatically influence the community even if 95% of all species were protected. Other examples in which the removal of species with crucial community functions are easy to find. In the case of birds, removal of the hornbill species from its community would disrupt seed dispersal and the elimination of a hummingbird species would disrupt pollination of plant species (Sekercioglu, 2006). The pollination crisis occurring now throughout North America and Europe (Biesmeijer et al. 2006; Holden, 2006; Stokstad, 2006) involves critical plant–pollinator interactions that could be treated peripherally in this general approach. The 2000–2008 drop in oriental white-backed vulture (*Gyps bengalensis*) populations

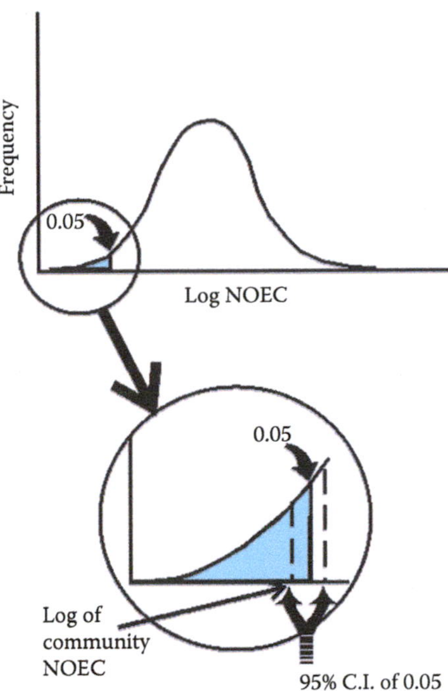

Figure 11.2 A pragmatic method of estimating a community-level NOEC. The 5% quantile is used to estimate the log concentration at which all but 5% of species would be protected. The concentration corresponding to the 5% quantile or the upper limit of the 95% confidence interval around this estimate could be used as the community-level no observed effect concentration. (From Van Straalen, N. M. and Denneman, C. A., *Ecotox. Environ. Safe.*, 18, 241–251, 1989; Wagner, C. and Løkke, H., *Water Res.*, 25, 1237–1242, 1991.)

to 1% of their former size due to exposure to the veterinary drug, diclofenac, is another example in which community function was central to predicting consequences.* Exposure resulted from the vulture's role as a scavenger feeding on dead livestock treated with the anti-inflammatory drug (Oaks et al. 2004; Balmford, 2013).† Species sensitivity analysis using conventional ecotoxicity testing with recommended species would have missed this connection. Given these examples, it is difficult to justify the assumption that the collection of NOEC or LCX values derived from laboratory tests would accurately reflect harmful concentrations for a community (Jagoe and Newman, 1997; Forbes and Calow, 2002; Newman et al. 2002). Often, values are derived from those of standard test species and are biased toward certain convenient taxa. For example, NOEC data for the drug, diclofenac, might be available for a domestic chicken, pigeon, Japanese quail, and mynah species (Hussain et al. 2008), but not vultures. Beyond this bias toward laboratory species and tendency to overlook ecological interactions, it is difficult to know how many NOEC values are needed to effectively capture the differences among species in an entire community or even a species assemblage.

* Like the Long Island osprey recovery that took place after DDT banning (see Figure 10.3), the oriental white-backed vulture populations of India, Nepal, and Pakistan are also recovering after the 2006 banning of diclofenac for veterinary uses (Balmford, 2013).

† The neglect of the ecological qualities of species and focus on a percentage of species is so pervasive that it deserves a name, the **pound-of-flesh incongruity**. Settling on a 5% of species cut-off is like Antonio's unthoughtful agreement with Shylock of a "pound of flesh" bond to secure a loan in Shakespeare's *Merchant of Venice*. It matters very much which pound of flesh is taken.

11.2 INTERACTIONS INVOLVING TWO OR A FEW SPECIES

11.2.1 Predation and Grazing

An adverse effect on predator–prey interactions can lead to local extinction of a species population even if toxicant concentrations are below those causing diminished growth, reproduction, or survival of individuals in the population. This premise prompted laboratory experiments quantifying such effects. A simple predator–prey arena (Figure 11.3) was used to demonstrate the influence of γ irradiation on the ability of mosquitofish (*Gambusia holbrooki*, formerly *G. affinis holbrooki*) to avoid predation by largemouth bass (*Micropterus salmoides*) (Goodyear, 1972). Mosquitofish were provided with a shallow refuge to simulate normal mosquitofish behavior of avoiding bass predation by staying in the shallows close to the water's edge. The influence of irradiation on predator avoidance over 10 days was dose-dependent with more mosquitofish failing to stay in the refuge as radiation dose increased. This approach was applied again by Kania and O'Hara (1974) to demonstrate that sublethal concentrations of inorganic mercury increased predation in a concentration-dependent fashion. Mosquitofish previously exposed to low concentrations of mercury were incapable of maintaining the most effective orientation relative to predator location in the test chamber.

Some predator–prey laboratory studies involve field-exposed predators and measure their effectiveness in taking prey. One such example is a study of field-exposed shrimp (*Palaemonetes pugio*) in which effectiveness of predation upon brine shrimp (*Artemia franciscana*) was measured (Perez and Wallace, 2004). Another involved predation by a fish (*Fundulus heteroclitus*) taken from sites with different contamination histories (Weis et al. 2000).

Still other predator–prey studies incorporate more complexity. In an arena including artificial plants as both prey refugia and predator cover, fathead minnows (*Pimephales promelas*) exposed to cadmium were more vulnerable to largemouth bass predation than were unexposed minnows (Sullivan et al. 1978). Concentrations producing a significant increase in vulnerability were lower than those measured for any other sublethal effect. Increased predation was explained in the context of the abnormal schooling behavior of the exposed minnows. Similar studies examined the influence of a fire ant bait pesticide (mirex) on pinfish (*Lagodon rhomboides*) predation upon grass shrimp (*Palaemonetes vulgaris*) (Tagatz, 1976). More recently, turbellarian predation on isopods as influenced by cadmium (Ham et al. 1995), and *Hydra* predation on *Daphnia* after lindane

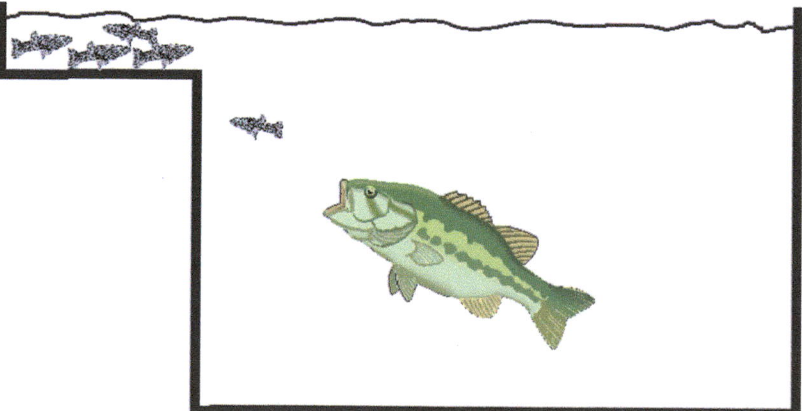

Figure 11.3 Experimental arena used to measure the effect of radiation (Goodyear, 1972) or inorganic mercury (Kania and O'Hara, 1974) on mosquitofish avoidance of predation by largemouth bass. The stressors diminished the ability of the mosquitofish to avoid predation by remaining in a shallow refuge.

(γ-hexachlorocyclohexane) exposure (Taylor et al. 1995) were quantified after exposure of both predator and prey.

Insight gleaned from such simple studies can be enriched greatly by drawing on the principle of allocation (see Chapter 10). Organisms must effectively allocate energy among many activities to optimize their Darwinian fitness. For example, a predator may change its prey consumption rate as prey densities increase. Such a **functional response** (a change in some predator function, such as prey consumption rate, in response to changes in prey density) was studied for a largemouth bass-mosquitofish system (Woltering et al. 1978). Both predator and prey were exposed to ammonia, and changes in prey consumption and bass weight were monitored at different ammonia concentrations. Increases in prey consumption rate with an increase in prey density were slowed at high ammonia concentrations, as was the increase in bass weight. In fact, because the mosquitofish were more tolerant of ammonia than bass, the mosquitofish harassed the bass in high ammonia and prey density treatments, resulting in a weight loss of the predator. Clearly, the influence of toxicants on predator–prey interactions is important and complicated.

Atchison and coworkers (Sandheinrich and Atchison, 1990; Henry and Atchison, 1991; Atchison et al, 1996) provide a sound description of predator–prey interactions from the context of energy allocation. They argue from the extensive literature on **optimal foraging theory** (the ideal forager achieves a maximum net rate of energy gain by optimally allocating its time and energy to the various components of foraging) that many important components of foraging are influenced by toxicants. For example, a predator must optimize time and energy spent in prey searching, identification, choice, pursuit and capture, handling, and ingestion. Any of these might be altered by the presence of toxicants. Atchison and coworkers cite numerous studies in which predator foraging activities are modified by toxicant exposure.

Toxicants can influence herbivore grazing too. Lürling and Scheffer (2007) describe one instance involving pollution effects on green algal (*Scenedesmus obliquus*) defense against grazing by zooplankton. To make it more difficult to be grazed upon when *Daphnia* sp. is present, *S. obliquus* form multicellular clusters instead of remaining single cells. The surfactant, FFD-6, interferes with the normal chemical signaling used by the algae to sense the presence of the grazer. Drawing analogy to disruption by pollutants of endocrine signaling within the body, the authors refer to this grazer-to-alga signal disruption as **info-disruption**. Relyea (2003) and Sih et al. (2004) also found pesticide info-disruption for amphibians exposed to pesticides. Atrazine caused info-disruption in fish (Moore and Waring, 1998; Saglio and Trijasse, 1998; Moore and Lower, 2001; Tierney et al. 2007) but not toad (*Bufo americanus*) tadpoles (Rohr et al. 2009).

Kersting (1984) measured pesticide (Dichlobenil) influence on *Daphnia* grazing on a green alga by estimating the deviation in normal herbivore–plant (*Daphnia magna–Chlorella vulgaris*) dynamics in a microecosystem (Figure 11.4). He allowed the *Daphnia–Chlorella* system to come to steady state and plotted grazer density versus algal density through time. A lag function was incorporated in the plot because there was a delay of approximately 7 days before the grazer density* could respond fully to any change in algal density. A 95% tolerance ellipse was drawn to define the limits of the system's behavior in the absence of the pesticide. The implication was that it would be unusual for a point to be outside this ellipse under normal dynamics. The pesticide was introduced and the consequent grazer density versus algal density points plotted. As these points fell outside the tolerance ellipse, the grazer–plant system was judged to have changed as a consequence of the pesticide exposure. A normalized ecosystem strain index (S) was used to quantify the degree to which the system had changed relative to its normal state.

Another species interaction concept to grasp is the influence of predators on other members of the community, that is, trophic cascades. Trophic cascades are prominent features of many

* In contrast to a functional response, change in predator or grazer number through increased reproductive output, decreased mortality, or increased immigration in response to changes in prey or food densities is called a **numerical response**.

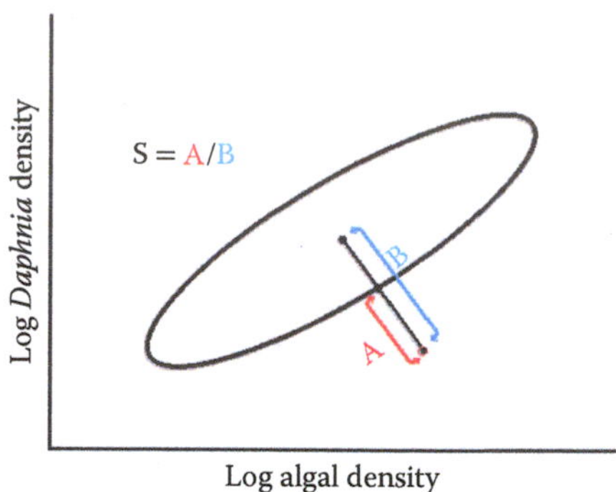

Figure 11.4 Calculation of normalized ecosystem strain as described by Kersting (1984). Strain is equal to A/B in this diagram. Ecosystem strain is measured relative to the normal behavior of the system as reflected by the 95% tolerance ellipse. These particular dimensions may be replaced by variables other than algal and grazing *Daphnia* densities. (Modified from Figure 4 in Kersting, K., *Ecol. Bull.,* 36, 150–153, 1984.)

ecological communities in which predator presence depresses the prey species density (or some function/behavior) enough that the next lower trophic levels/positions are released from the negative influence of the prey species. The importance of trophic cascades often becomes apparent when predator densities are changed, perhaps due to human intervention or pollutants. As a recent example, reduced numbers of apex predatory sharks along the U.S. Atlantic coast released prey species such as cownose ray, *Rhinoptera bonasus*, a mesopredator that feeds heavily on bivalves. The increased abundance of rays has slowed recovery of several important bivalve fisheries (Myers et al. 2007). Knight et al. (2005) describe another trophic cascade involving freshwater ponds and adjacent terrestrial species. Fish feeding on dragonfly larvae influence the number of emerging adult dragonfly that prey on bee pollinators of pond edge plants: changes in pond fish population density influenced terrestrial plant reproduction. These kinds of top-down processes can even influence toxicant movement. Jones et al. (2013) found during a mesocosm study that a decrease in fish density increased the flux of methylmercury from the water to the land by increasing the number of methylmercury-laden emergent insects produced in the water and moving onto land. They also noticed a bottom-up effect in which increasing nutrients in a mesocosm increased emergent insect production, and in so doing, again increased methylmercury flux onto land. So bottom-up and top-down processes influenced methylmercury flux. Such processes are common in nature so they should not to be treated as marginally relevant to ecotoxicology. Shurin et al. (2002) found trophic cascades to be important in a wide range of ecosystems albeit to varying degrees (Figure 11.5). In their analyses, top-down influences were consistently larger on herbivores than plants. Predator presence depressed herbivore biomass or density in all systems with the change being greatest for the marine and lentic* benthos, and weakest for stream and land systems. Predator presence increased plant biomass with the strongest influence in lentic and benthic systems, and weakest in terrestrial and marine planktonic systems.

The existence of trophic cascades is one illustration of the importance of considering community interactions in ecotoxicology. Another popular one in ecology textbooks is the Operation Cat Drop anecdote (O'Shaughnessy, 2008). To control mosquito vectors for malaria during the early 1950s, the World Health Organization sprayed Sarawak and Sabah villages (Borneo) with DDT, benzene hexachloride, and later, deildrin. Unfortunately, sprayed cockroaches that were fed upon by geckos became a vector for pesticide accumulation in geckos. The geckos were fed upon by the village cats that received a lethal insecticide dosing.† The release from cat predation allowed the resident rat population to explode, eventually forcing authorities to parachute in replacement cats, hence the name "Operation Cat Drop." As a further complication, caterpillars that fed on thatched

* **Lentic** waters are standing waters such as lakes or ponds. In contrast, **lotic** waters are flowing water bodies such as streams and rivers.

† Pesticide ingestion by cats also resulted by licking of their DDT-dosed fur and paws during grooming.

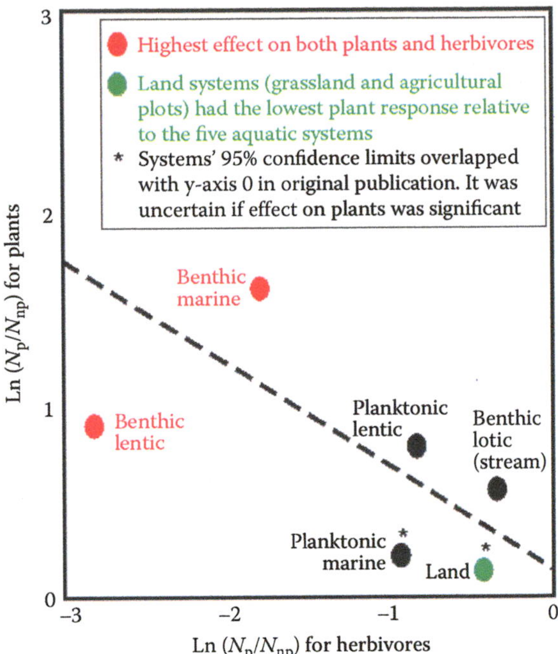

Figure 11.5 The influence of predator presence on plant biomass was expressed as the natural logarithm of the ratio of amount of biomass with (N_p) and without (N_{np}) the predator being present, ln (N_p/N_{np}). The same was calculated for herbivore biomass or density with and without the predator being present. Primary producers were periphyton (benthic lentic and stream), macroalgae (benthic marine), or grasses and forbs (land). Logarithm of quotients are plotted here for plants (*y* axis) versus herbivores (*x* axis). No predator influence of plant or herbivore biomass would be suggested if a point was 0 on the appropriate axis. The line passing through the points would have a slope of −1 if the influence was the same for plants and herbivores. The slope is actually shallower than −1. (Derived from Figure 1 in Shurin et al. *Ecol. Lett.*, 5, 785–791, 2002.)

roofs were less sensitive to DDT than their parasites (chalcid wasps). The caterpillars thrived in the absence of their parasites, causing many thatched roofs to rot and collapse.

Examples of toxicant-related shifts in community interactions are being published with increasing frequency. Zvereva and Kozlov (2006) documented increased damage by larvae of the leaf-mining moth (*Eriocrania* sp.) to mountain birch (*Betula pubescens* subsp. *czerepanovii)* growing adjacent to nickel–copper smelters in Monchegorsk, Russia. Neither the birch themselves nor their predators were directly harmed by the pollution; but a parasitoid of the moth larvae was. Ecological release from the parasitoid allowed densities of the herbivorous leaf-miners to increase in polluted birch stands. As another example, McMahon et al. (2011) noted that the fungicide, chlorothalonil, induced top-down (trophic cascade) and bottom-up effects that modified ecosystem functions. Clearly, pollutant's adverse effects on species interactions are as important to understand and evaluate as are direct effects to individuals. A recent publication of Clements et al. (2012) reinforces this conclusion, but warns that community responses are often context specific. The state of communities might vary along a gradient (such as along a river) or among patches, and the community state will influence response to toxicants.

11.2.2 Competition

Interspecies competition, the interference with or inhibition of one species by another, can be influenced by toxicants and also contribute to changes in community structure. This could involve **interference competition** in which one species interferes with another as might occur with territoriality. A toxicant-induced change in aggressive behavior, such as that displayed by bluegill exposed to cadmium (Henry and Atchison, 1979), could easily shift the balance of interference competition. **Exploitation competition**, where species compete for some limiting resource such as food, can also be affected by toxicants. For various freshwater zooplankton under toxicant exposure, Atchison et al. (1996) noted differences in filtration rates and suggested that toxicants can produce shifts in exploitation competition among these potentially competing filter feeders.

Atchison and coworkers (Atchison et al. 1996) also lament the low number of studies in behavioral ecotoxicology and attribute this deficiency to the indirect, and relatively complicated, means by which interspecies competition is measured, that is, as changes in population dynamics of competing species. In the absence of direct information, indirect information was used as evidence to suggest that exploitation competition could be affected by contaminants. Newman (2013) details some methods for quantifying the effects of toxicants on interspecies competition. Fortunately, an increasing number of pollutant-influenced competition studies are emerging in the ecology and ecotoxicology literatures.

Again, it is implied in this and the previous discussion that realized niches change with toxicant-induced shifts in competition and foraging behavior. If balanced competition among species in the community is assumed to enhance species packing by fostering optimal niche separation among species, competitive dysfunction might decrease species diversity. As we will see shortly, shifts associated with species abundance curves support this conclusion. Interestingly, Chattopadhyay (1996) suggests that there are conditions under which toxicants can also stabilize variations in competing species populations.

11.3 COMMUNITY QUALITIES

11.3.1 General

Several community or species assemblage qualities are measured routinely to assess toxicant effect. The number of species inhabiting a contaminated site may be compared to the number at a reference site. The presence or absence of indicator species may be noted. For example, a species that is extremely sensitive to the pollutant might be used much as the proverbial canary in the coal mine. Its presence suggests that a site is likely unimpacted. The absence of a particularly sensitive species might suggest impact; however, the species might not be there for another reason. The population declines of the Long Island Sound osprey and oriental white-backed vulture of India have already been mentioned as examples of such change for sensitive species. Another example is the disappearance of pH-sensitive species from lakes undergoing acidification. The mysid shrimp, *Mysis relicta*, disappeared from an experimentally acidified lake when pH was lowered to 6 (Schindler, 1996). Not only was it a sensitive species that provided an early warning of deteriorating conditions, it disappearance was noteworthy also because it was functioning as a keystone species in the lake. A final example includes mayflies (e.g., *Baetis* sp.) that tend to be sensitive indicators for a variety of pollutants in freshwater systems (Ford, 1989).

Alternatively, a rise to predominance of pollution-tolerant species might suggest a deteriorating community. Benthic communities found below high BOD (biochemical oxygen demand) outfalls from sewage treatment plants are typically dominated by heterotrophs tolerant of low dissolved oxygen concentrations. The oligochaete, *Tubifex tubifex*, is a common benthic species at such polluted sites. The *Sphaerotilus* bacterium is an indicator, which forms extensive, filamentous mats below sewage discharges. Indeed, based on these and other community features, Kolkwitz and Marsson (1908) described a **saprobien spectrum**, a change in community composition expected at different distances below a discharge of putrescible organic waste into a river or stream. Characteristic species define the polysaprobic, mesosaprobic, and oligosaprobic zones below a sewage discharge depending on the oxygen concentrations, amounts of putrescible organic material, and stage of stream recovery.

Several qualities apparently influence community (or ecosystem) **vulnerability** (susceptibility to irreversible damage) to toxicants (Cairns, 1976). Low **elasticity** (the ability to return to a prestressed condition), **inertia** (ability to resist change), and **resilience** (the number of times a community can return to its normal state after perturbation) all contribute to vulnerability. Elasticity is enhanced by the ease with which new individuals can move back into the affected area, i.e., qualities of the associated metacommunity. This was demonstrated in a mesocosm study of deltamethrin impact in which zooplankton species, except *Chaoborus* larvae, recovered primarily due to propagules in the mesocosm, not immigrants from outside sources (Hanson et al. 2007). Inertia may be influenced by the structural redundancy in the community and previous adaptation of the community to environmental variability. Resilience is influenced by the elasticity of the community and the frequency of perturbation. Too frequent perturbation of a community with low elasticity gradually ratchets the community downward toward a degraded state.

Implicitly, all of this discussion is based on the assumption that a community is in some kind of balance and can be expected to return to that balanced state after a perturbation. Pratt and Cairns (1996) point out that this concept of community steady state is pervasive in ecotoxicology. It extends into regulations such as those setting environmental criteria that seek to protect "balanced biological communities." In contrast to this concept of a community deviating from and then returning to a steady state condition after the stressor is removed, Matthews et al. (1996) suggest that disturbed communities might not return to their original states. This suggestion is consistent with the increasingly expressed view of ecologists that communities are not steady state systems (Pratt and Cairns, 1996). They retain important features determined by historical events and can have several possible stable states (Chase, 2003). Matthews et al. (1996) argue that communities retain information about occurrences in their past, dubbing this argument the **community conditioning hypothesis**. Any dynamics back toward some norm will also reflect the history of the community: one cannot assume without justification that the community will return to its original state. The implication here is that pollution effects to communities can be present long after the toxicant is gone and that any anticipation of a return to an original state is presumptuous. It is this author's opinion that both views (steady state and community conditioning hypotheses) are useful if applied appropriately. One might expect a general recovery of some community qualities to a near "normal," but unique, state after community disruption by toxicants. Chase (2003) indicates that communities are most likely to have several possible stable states if they exist within "systems with large regional species pools, low rates of connectance, high productivity and low disturbance."

VIGNETTE 11.1 Ecological Resilience as a Measure of Recovery in Aquatic Communities

William H. Clements
Colorado State University

THEORETICAL BACKGROUND

Since their introduction into the ecological literature over 30 years ago, the concepts of ecological resistance and resilience have been subjects of considerable controversy (Holling, 1973). This debate was fueled in part because of inconsistent terminology and because of difficulty defining specific measures of recovery. I will adopt terminology consistent with Van Nes and Scheffer (2007) to define ecological **resilience** as the level of perturbation a community can withstand before it is pushed to an alternative stable state and **recovery rate** as the rate at which a community returns to equilibrium following disturbance (Figure 11.6). **Recovery rates** in aquatic communities are influenced by many factors, including the type of perturbation (e.g., physical disturbance vs. chemical stressor), the amount of natural variation that a community experiences, the presence of other stressors, availability of colonists and life history characteristics of dominant species (Rapport et al. 1985; Niemi et al. 1990; Bellwood et al. 2004). Quantifying recovery rates is often difficult (Van Nes and Scheffer, 2007), primarily because of the limited number of long-term studies and because recovery of some state variables, such as abundance or species richness, may occur despite persistent alterations in community composition (Berumen and Pratchett, 2006). Another factor that limits our ability to quantify recovery

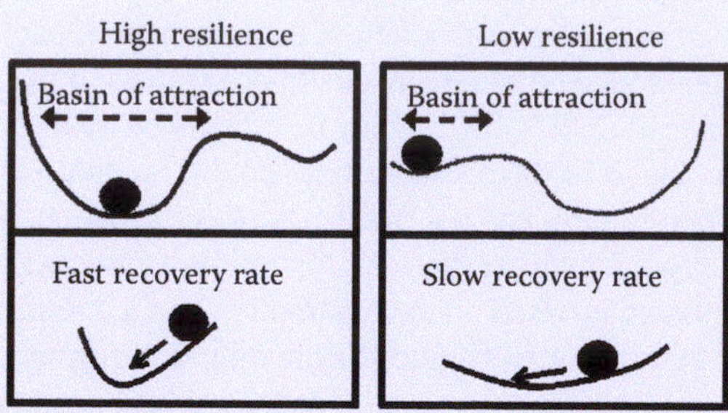

Figure 11.6 The relationship between resilience and recovery rate in communities. Size of the basin of attraction is a measure of the amount of disturbance a community can withstand before shifting to an alternative stable state. Communities recovering from previous exposure to contaminants generally will have a shallower basin of attraction compared to reference communities. (Modified from Van Nes, E. H. and Scheffer, M., *Am. Nat.*, 169, 738–747, 2007.)

has been the focus on single disturbance events (Paine et al. 1998). Recurrent disturbances in systems recovering from anthropogenic perturbations often have long lasting and cryptic effects.

Theoretically, the loss of ecological resilience caused by long-term exposure to a stressor decreases the size of attraction basins (Figure 11.6), thereby increasing the likelihood of a shift to an alternative stable state (Bellwood et al. 2004). At the population level, acclimating or adapting to one set of environmental stressors has a fitness cost and will likely increase susceptibility of organisms to other environmental stressors (Wilson, 1988; Clements, 1999; Gallently et al. 2007; Tanaka and Tatsuta, 2013). Levinton et al. (2003) attributed the rapid loss of tolerance in oligochaetes following cleanup at a metal-contaminated site to the cost of fitness associated with adaptation to cadmium. Reduced genetic diversity resulting from low population density, population bottlenecks and inbreeding depression is common in contaminated environments (Bickham et al. 2000) and has important long-term consequences (Mulvey and Diamond, 1991; Heagler et al. 1993). Although most research on the cost of tolerance has focused on population-level consequences, communities from contaminated environments may also be at greater risk from novel stressors (Clements, 1999). In this respect, loss of species diversity is analogous to reduced genetic diversity, with similar implications for ecological resilience. Just as genetic structure reflects the unique history of a population over evolutionary time, communities are a reflection of their unique history and etiology (Landis et al. 1996).

QUANTIFYING RESILIENCE AND RECOVERY IN A METAL-CONTAMINATED STREAM

Pollution is considered one of the most powerful agents of selection in aquatic ecosystems (Gallently et al. 2007), with tremendous potential to restructure populations and communities (Clements et al. 2000). Over the past 17 years our research group has measured recovery of the Arkansas River, a metal-polluted stream in the Southern Rocky Mountain ecoregion of Colorado (Clements, 2004). Mining operations have had a major impact on this watershed since the mid-1800s when gold was discovered near Leadville, Colorado. Concentrations of heavy metals (copper, cadmium, and zinc) are greatly elevated downstream from Leadville and often exceed acutely toxic levels. Since 1989 we have measured physicochemical characteristics, habitat quality, heavy metal concentrations, and macroinvertebrate community structure seasonally (spring and fall) at several locations upstream and downstream from sources of metal contamination. Three years after we began this research, state and federal natural resource agencies initiated a large-scale restoration program to improve water quality and reestablish trout populations in the upper Arkansas River basin. Because data were collected before and after remediation, this long-term research provides a unique opportunity to quantify ecological resilience in a system recovering from chronic pollution. Since 1994, we have observed recovery at some sites, while others have shown only modest changes in community composition, despite significant improvements in water quality (Figure 11.7). We have also observed considerable variation in the rate at which variables recovered (e.g., species richness vs. abundance of grazers).

In addition to evaluating long-term recovery of the Arkansas River, we conducted a series of stream mesocosm experiments to quantify ecological resilience. The goal of these experiments was to test the hypothesis that communities from chronically polluted locations are more susceptible to novel stressors compared to those from unpolluted locations. The physiological, morphological, and genetic mechanisms that allow organisms to tolerate contaminant exposure are especially important in benthic macroinvertebrate communities because

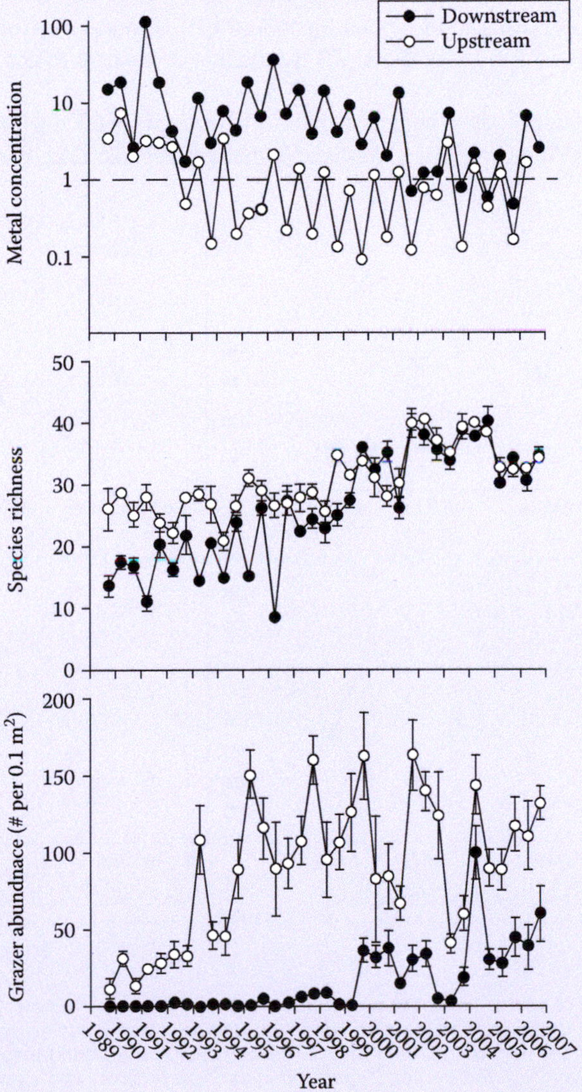

Figure 11.7 Long-term (1989–2006) changes in heavy metals, species richness and abundance of metal-sensitive grazers in the Arkansas River, CO. Horizontal line in upper panel is the U.S. EPA estimated safe metal concentration for Colorado streams.

of the relatively limited dispersal ability of most species. In these experiments, communities from reference and metal-polluted sites were exposed to a variety of novel stressors, including acidification, UVB radiation, and stonefly predation (Figure 11.8). As predicted, organisms from metal-polluted sites were more tolerant of heavy metals compared to naïve organisms (Figure 11.8a). However, in all instances, communities from metal-polluted sites were more sensitive to novel stressors (Figure 11.8b–d). We repeated these experiments and quantified sensitivity of reference and metal-polluted communities to UVB radiation, focusing on structural (abundance of EPT*) and functional (community metabolism) measures (Kashian et al. 2007; Clements et al. 2008). Results showed that regardless of site history, benthic communities were adversely affected by metals (Figure 11.9); however, the magnitude of these effects was much greater in communities from the reference site compared to the metal-polluted site. In contrast to the effects of metals, significant responses to enhanced UVB were limited to the metal-polluted site. Community metabolism and EPT abundance from the metal-polluted site were significantly different in the UVB treatments compared to the no UVB treatments ($p = .0002$).

Results of these experiments suggest a trade-off between metal tolerance and sensitivity to other stressors in the Arkansas River. We believe that greater tolerance of benthic communities

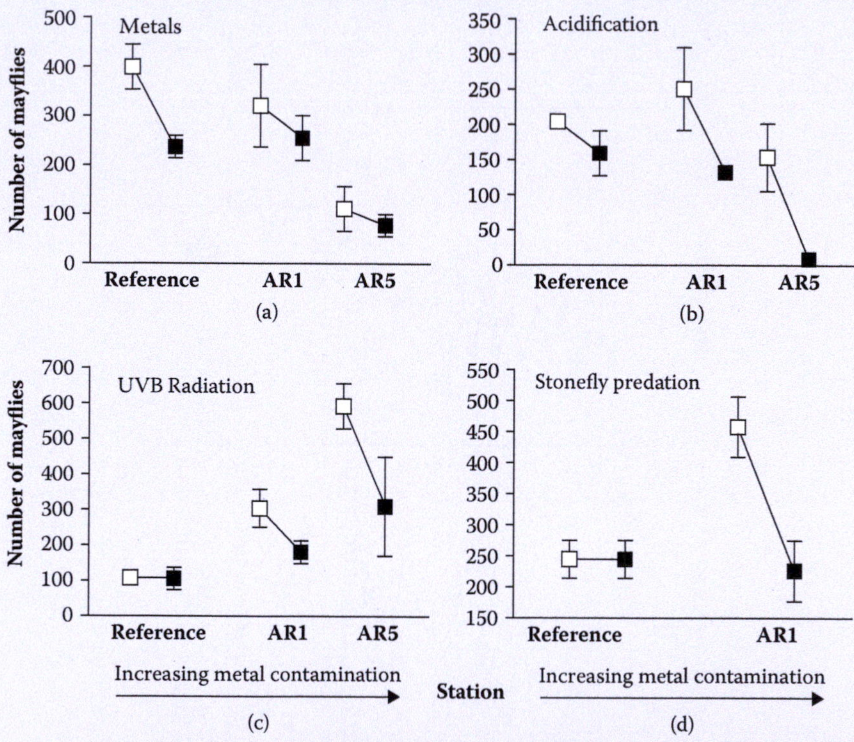

Figure 11.8 Results of stream mesocosm experiments showing effects of heavy metals (a), acidification (b), UVB radiation (c), and stonefly predation (d) on mayflies collected along a gradient of metal contamination. Results in each panel show abundance of mayflies in control (open symbols) and treated (closed symbols) mesocosms. The slope of each line reflects the relative impact of each perturbation.

* EPT refers to the insect taxa, Ephemeroptera (mayflies), Plecoptera (stoneflies), and Trichoptera (caddisflies).

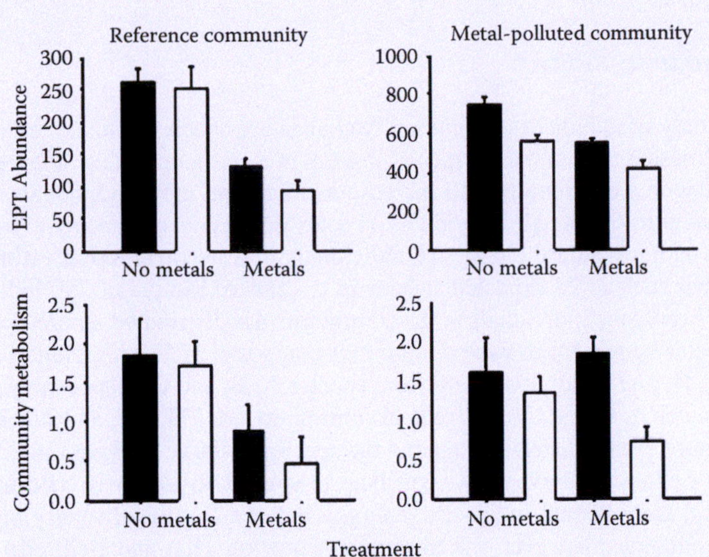

Figure 11.9 Individual and combined effects of metals and UVB on the total abundance of EPT organisms and community metabolism in stream mesocosms. Experiments were conducted using benthic communities collected from a reference site (left) and a metal-impacted site (right) in the Arkansas River watershed.

to metals has resulted from both population-level responses (e.g., acclimation and adaptation) and shifts in community composition (Clements, 1999; Courtney and Clements, 2000; Kashian et al. 2007). The underlying mechanisms responsible for the greater sensitivity of Arkansas River communities to acidification, UVB, and predation are uncertain, but these findings are consistent with the hypothesis that adaptation or acclimation to one set of environmental stressors may increase susceptibility to novel stressors (Wilson, 1988; Clements, 1999). As a first step at elucidating potential mechanisms, we measured genetic variability of microsatellite DNA in mayflies (*Baetis bicaudatus*) collected upstream and downstream from sources of metal contamination. Population genetics studies focused on *Baetis* because of its abundance and widespread distribution in Colorado streams. Results showed that populations from the two upstream sites were genetically distinct from all other locations and that heterozygosity was reduced at downstream stations. It is likely that mayflies in the Arkansas River experienced a population bottleneck and that metal contamination acted as a barrier that reduced gene flow between populations. These results highlight the importance of previous exposure to contaminants and demonstrate that community composition can significantly influence responses to anthropogenic disturbance. Greater susceptibility of metal-impacted communities to novel stressors may have implications for other Colorado streams because of the large number of watersheds in the region that are contaminated by mining pollution (Clements et al. 2000). Finally, these findings support the hypothesis that responses to multiple perturbations are often not additive and that superimposing stressors such as enhanced UVB or acidification on disturbed communities may result in ecological surprises (Paine et al. 1998). Susceptibility of communities to novel stressors is a reflection of exposure history that can be employed to assess recovery in aquatic ecosystems (Clements et al. 2010). Finally, understanding exposure history of populations and communities and the cost of tolerance are particularly important in ecosystems that are simultaneously experiencing effects of global change (Moe et al. 2013).

11.3.2 Structure

11.3.2.1 Community Indices

The most commonly used indices of community change are species richness, evenness, and diversity. **Species richness** is the number of species present in a community (e.g., Figure 11.10). Because the tally of species in a community will increase as more and more individuals are sampled and it can be impractical to sample all individuals, species richness is commonly expressed relative to that of a sample with a standard number of individuals in it. A **rarefaction estimate of richness** produces a number such as 25 expected species in a standard sample of 250 individuals. **Species evenness** is how evenly the individuals in the community are distributed among species. For example, let three species be present in two communities composed of 500 individuals each. In the first community, 450, 41, and 9 individuals are from species A, B, and C, respectively. The numbers of individuals in species A, B, and C in the second community are 134, 138, and 228, respectively. The individuals are more evenly distributed among the species in the second community.

Both species richness and evenness contribute to **species diversity** (sometimes referred to as heterogeneity) and are reflected in species diversity indices. Species diversity may be quantified with several formulations; however, the **Shannon** (Equation 11.1) and **Brillouin** (Equation 11.2) **indices** are the most common.

$$H' = -\sum_{i=1}^{S} p_i \ln p_i \tag{11.1}$$

$$H = \frac{1}{N} \ln \frac{N!}{\prod_{i=1}^{S} n_i!} \tag{11.2}$$

where S = the total number of species and p_i = the proportion of all individuals that are species i as estimated by the number of individuals of species i (n_i) divided by the total number of individuals (N) in the S species.

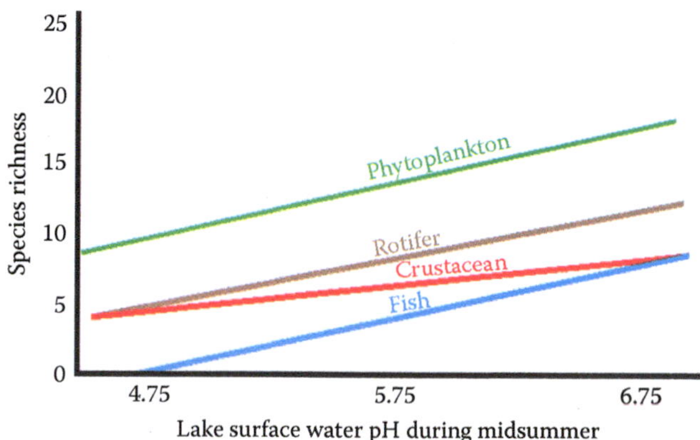

Figure 11.10 Species richness measured for 30 Adirondack lakes (New York) impacted to different degrees by acid precipitation. The lines show the general trends for phytoplankton, zooplankton rotifer, crustacean, and fish species from these lakes (lines from selected panels of Figure 1 of Nierzwicki-Bauer et al. [2010]). Clearly, lowered pH results in lowered species richness including the absence of fish in several low pH lakes.

Both give similar estimates, but Equation 11.2 gives estimates lower than Equation 11.1. This difference arises because the Shannon index is a diversity estimate *for the community* from which the sample was taken but the Brillouin index is a diversity estimate *for the sample*. The diversity in the sample will be lower than the diversity predicted for the entire community from which the sample was taken. Both of these indices are associated with diversity of the particular community being sampled, so estimation of diversities characteristics in the context of a metacommunity requires some additional computations (Magurran, 1988; Mouquet and Loreau, 2003). Diversity within a community is called local or α diversity, and that measuring diversity among communities is called β diversity. Together, α and β diversities contribute to the total (γ) diversity for a region.*

Species evenness can be estimated for the community (Equation 11.3, **Pielou's J′**) or for the sample (Equation 11.4, **Pielou's J**). The ln S and H_{MAX} in these equations are the maxima for Shannon (H') and Brillouin (H) indices, respectively. Consequently, these evenness indices are the estimated species diversity divided by the maximum possible species diversity for that community or sample.

$$J' = \frac{H'}{\ln S} \tag{11.3}$$

$$J = \frac{H}{H_{MAX}} \tag{11.4}$$

where

$$H_{MAX} = \frac{1}{N} \ln\left[\frac{N!}{[(N/S)!]^{S-r}[(\{N/S\}+I)!]^{r}}\right]$$

with [N/S] being the integer part of the quotient, N/S, and r being $N - S[N/S]$.

Often, but not always, values for species indices decline as a consequence of pollution. Species diversity dropped in periphyton communities below heavy metal mine discharges (Austin and Deniseger, 1985), in periphyton communities in the presence of high zinc concentrations (Williams and Mount, 1965), and in stream macroinvertebrate communities below coal mine drainage (Chadwick and Canton, 1983). Both diversity and species richness dropped for lake algal communities exposed to mining wastes with a few, tolerant species becoming very abundant (Austin et al. 1985). Ford (1989) indicates that richness is a relatively good measure of effect to plankton and benthic communities, although richness may increase slightly at low levels of pollution.

Often a statistically significant difference in species richness or species diversity is used to suggest an adverse impact of toxicants on communities. Aside from the problem of equating statistical and biological significance, this approach is compromised by our lack of knowledge regarding functional redundancy within communities. **Functional redundancy** involves an apparently unaltered maintenance of community functioning despite changes in structure. Species might drop out and be replaced, yet the community will still appear to function normally.

* Species diversity and biodiversity are not identical terms (Hamilton, 2005) although they are commonly used as synonyms. According to The International Convention on Biological Diversity (2003), "Biological diversity [**biodiversity**] means the variation among living organisms from all sources including, *inter alia*, terrestrial, marine and other aquatic ecosystems and the ecological complexes of which they are part; this includes diversity within species, between species, and of ecosystems."

The rivet popper and redundancy hypotheses are two unresolved hypotheses germane to toxicant impact on community functioning. The **rivet popper hypothesis** suggests that species in a community are like rivets that hold an airplane together and contribute to its proper functioning (Ehrlich and Ehrlich, 1981). Each loss of a rivet weakens the structure by a small but noticeable amount. The loss of too many rivets eventually leads to a catastrophic failure in function. In contrast, the **redundancy hypothesis** holds that many species are redundant and the loss of some species will not influence the community function as long as crucial (keystone and dominant) species are maintained (Walker, 1991). There are often guilds* of similarly functioning species to provide consistency of function if one or a few member species are lost. Pratt and Cairns (1996) emphasize the importance to ecotoxicology of determining whether the rivet popper or redundancy hypothesis best describes real biological communities. The answer is needed to decide how much toxicant-induced change in a community is required to degrade its functioning. Currently, there is good evidence to support the rivet popper hypothesis (Baskin, 1994). Given our present lack of understanding, it seems prudent to assume that the conservative rivet popper hypothesis is the best working model.

There are two hypotheses that provide plausible mechanisms for the influence of species diversity on system stability, the insurance hypothesis and portfolio effect (Free and Barton, 2007). The **insurance hypothesis** suggests that the presence of species that do not contribute much to stability under basal conditions might nonetheless increase the chances of the system remaining stable when conditions shift away from basal conditions because those species are now better adapted to the new conditions. The analogy here is that you might never need an insurance policy but, if you ever do, it becomes a crucial resource for coping with an adverse event. The **portfolio effect** assumes that stability results from the net activities of all species and the more contributing species there are in the system, the lower the chances that the system will suddenly change when conditions that might disfavor some of the contributing species change. The financial analogy is having a diverse portfolio to ensure a steady net return on your investment.

Species abundance curves are also used to describe community shifts as a consequence of toxicant exposure (Figure 11.11). These curves are based on the **Law of Frequencies**, which states that there is a relationship between the numbers of species and the number of individuals in a community (Fisher et al. 1943). The numbers of species falling into different abundance classes are plotted against abundance class. In the classic approach of Preston (1948), abundance classes are defined as doublings in abundances, e.g., 2, 4, 8, 16, 32, and so forth individuals present for a species. These $\log_2$ classes (e.g., 1, 2–3, 4–7, 8–15, 16–31, … individuals) are called **octaves**. A plot similar of these data (Figure 11.11) describes a log normal distribution. This **log normal model** for species abundance is thought to reflect a community structure in which several factors influence species interactions and the resulting allocation of resources.

Patrick (1973) noted that this log normal curve shifts in a predictable way for diatom communities exposed to organic pollution. The mode drops down and the right tail extends outward to include more octaves with high numbers of individuals. There is a shift toward fewer species of intermediate or rare abundances and a few very dominant species. Herricks and Cairns (1982) suggested that this shift results from a rise to dominance of opportunistic species and a disruption of equilibrium. This shift suggested to May (1976a) and Odum (1985) that the disordering effect of pollution has forced a reversion to an earlier successional stage. The diverse processes allowing better species packing in a mature community (K-strategy) become less important in shaping community structure than those

* An ecological **guild** is a "group of functionally similar species whose members interact strongly with one another but weakly with the remainder of the community" (Smith, 1986).

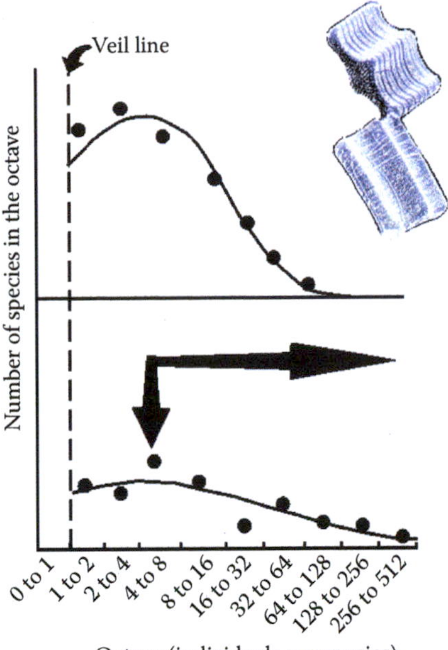

Figure 11.11 Species abundance curves (log normal model) before (top panel) and after (bottom panel) toxicant exposure. Note the transition from a common log normal distribution to a curve with a lowered mode and extended right tail. May (1976a) suggests that such a transition may reflect a reversion to an earlier successional stage when interspecies interactions were less important in shaping community species structure (i.e., away from the importance of species interactions toward an r-selection strategy).

associated with earlier successional stages (classic r-strategy).* Based on this premise, Bongers (1990) proposed a community **maturity index** for pollution based on the proportions of species in a soil nematode community that fell into various categories ranging from colonizers (r-strategists) to persisters (K-strategists). Regardless, this shift in the log normal species abundance curve is useful as an indicator of pollutant effect (Gerhart et al. 1977; Gray, 1979).

Finally, more comprehensive indices may be applied to detect changes in the biota inhabiting notionally impacted sites. Currently, one of the most widely applied for aquatic systems is the **index of biotic integrity (IBI)**. Originally, this index combined 12 qualities of fish assemblages of warm-water, low-gradient streams to determine the degree of stream degradation. These qualities included information on species richness and composition, trophic characteristics, and abundance and condition (Table 11.1). Numerical scores are generated for each quality and summed to produce the IBI for a site. These IBI scores are compared to those expected in the particular area for an undisturbed system. This specialized index has been successfully modified for a variety of aquatic habitats (e.g., Steedman, 1988) and enjoys widespread application today.

* Although not able to fully explain individual success, these two strategies do provide sufficient information here. An **r-strategy** (r is the intrinsic rate of increase) or opportunistic strategy might be taken by species coming into an uninhabited/unexploited habitat such as a newly plowed field. Selection favors species that establish themselves quickly, grow quickly to exploit as many resources as possible, and produce many offspring quickly. A **K-strategy** (K is the carrying capacity) or equilibrium strategy involves important interactions among species that allow coexistence of many species in the community. Equilibrium species are more effective competitors than opportunistic species. Many factors including interactions among species determine the structure of such a mature community whereas the early successional community structure may be determined more by **niche preemption**, a rapid use and preemption of resources by any species that exploits them before another can. In actuality, the r- and K- strategies are extremes in a spectrum of possible strategies.

Table 11.1 Qualities (Metrics) Included in the Original Index of Biological Integrity

Category		Specific Quality
Species richness and composition	1	Total number of species
	2	Number of darter species
	3	Number of sunfish species
	4	Number of sucker species
	5	Number of intolerant species
	6	Proportion of all individuals that are green sunfish, a pollution-tolerant species.
Trophic composition	7	Proportion of all individuals that are omnivores
	8	Proportion of all individuals that are insectivorous cyrinids
	9	Proportion of all individuals that are piscivores
Abundance and condition	10	Total number of individuals in the sample
	11	Proportion of total that are hybrids
	12	Total number of individuals with signs of disease or some abnormality

Source: Modified slightly from Appendix Table 1 in Karr, J. R., et al. *Assessing Biological Integrity in Running Waters. A Method and Its Rationale*. Champaign, IL, Illinois Natural History Survey Special Publication 5, 1986.

VIGNETTE 11.2 Biological Integrity and Ecological Health†

James R. Karr
University of Washington

Six decades ago, Aldo Leopold discovered that important lessons often come when least expected. "In those days," he wrote, "we had never heard of passing up a chance to kill a wolf" (Leopold, 1949). Fewer wolves, the young Leopold thought, would mean more deer—a hunter's paradise. After pumping lead into a pack of big pups led by a female, Leopold and his friends "reached the old wolf in time to watch a fierce green fire dying in her eyes. I realized then, and have known ever since, that there was something new to me in those eyes."

Leopold spent decades afterward protecting and putting words to the fire he saw in the old wolf's eyes—a fire peculiar to living things, including living landscapes. His writings speak often of "land health" as the "capacity of the land for self-renewal;" they also speak of "integrity:" "A thing is right when it tends to preserve the integrity, stability, and beauty of the biotic community. It is wrong when it tends otherwise."

Since Leopold's time, the terms *health* and *integrity* have become lightning rods, especially among scientists. Some argue that such value-laden words should not be applied to multi-species assemblages, ecosystems, or landscapes; others hold that talking about ecological health or biological integrity is beyond the purview of science. Yet the words are particularly useful in policymaking arenas precisely because they are familiar and imply values worth protecting. It seems a natural intuitive leap from "my health" or the nation's "economic health" to "ecological health" or "land health." And, as a goal of policy or law, sustaining a place's health and integrity has greater direct appeal than protecting abstractions like "system dynamics" or "ecosystem functions."

Like people, ecosystems or landscapes can be more or less "ill". An ill person may be suffering from a cold or dying of cancer; an ill landscape may be degraded by loss of a few sensitive species or all of its vegetation. An ill river may have game fish populations depleted

† Current address: 102 Galaxy View Court, Sequim, Washington, United States.

by overfishing or have no fish at all; or only a few of the river's most tolerant invertebrates may remain after severe chemical pollution.

The healthiest places are those that have undergone little or no disturbance at human hands. These areas support the full range of biological parts and processes characteristic of the region; that is, they show a full complement of plants, animals, and microbes and their genetic diversity as well as a full array of ecological processes such as nutrient cycling, birth, death, competition, and mutualism. Because such places support a thriving living system, they retain the capacity to regenerate, reproduce, sustain, adapt, develop, and evolve; they retain the full legacy of wild nature, or in Leopold's words, they still have "all the parts."

Complete, unimpaired living systems possess biological integrity. They constitute one end in a spectrum of biological condition and provide a benchmark against which other sites can be evaluated. A "normal," or benchmark, body temperature of 37°C (98.6°F) provides a similar standard for humans.

Defining biological integrity, however, is only the first step toward using the concept in science, policymaking, or law. For credibility in any of these fields, practitioners need tools for translating the subjective concept into something objective; they need tools both to quantify and to describe. Fortunately, the toolbox has been expanded in recent decades, enabling practitioners to evaluate sites and rank them along a gradient of biological condition according to how far they diverge from integrity.

Links between biology and human impacts came to the forefront more than a century ago, when pollution, particularly from raw sewage, was found to harm living systems. Through much of the twentieth century, however, efforts to track the health of water bodies focused instead on the presence of chemical contaminants; the assumption was that chemically clean water was good enough to protect river health, that physical and chemical measures were effective surrogates of biological condition. This assumption proved wrong.

We now know that human influences on living systems fall into five major classes: changes in energy sources, chemical pollution, modification of seasonal flows, physical habitat alteration, and shifts in biotic interactions (Figure 11.12). Given the choice of measuring all such influences or of measuring the condition of the biota—which includes the

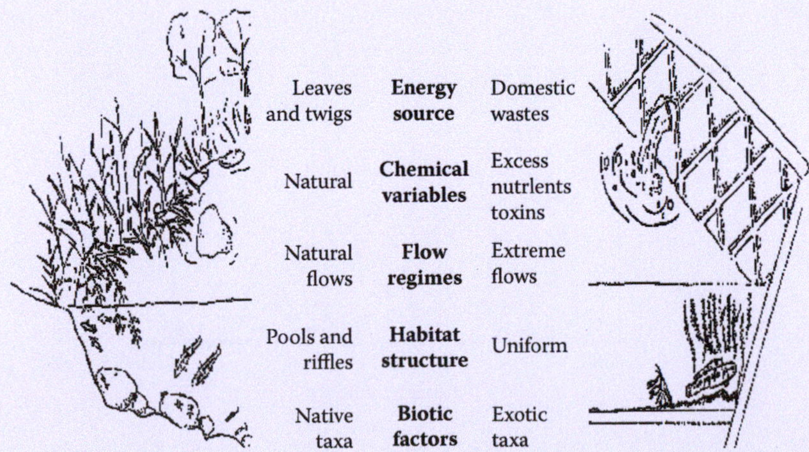

Figure 11.12 Five principle features, with examples, that determine river health. Left, a natural river; right, a modified river. (From Karr, J. R. and Rossano, E. M., *Ecol. Civil Eng.*, 4, 3–18, 2001.)

prime witnesses, and victims of environmental change, many agencies and institutions have shifted to direct measurement of biological condition. Biological monitoring and assessment, or biomonitoring, detects and evaluates human-caused biotic changes apart from those occurring naturally; the techniques have gained widespread acceptance as part of the water and land manager's toolkits.

Two major approaches to river biomonitoring have emerged: the IBI and the river invertebrate prediction and classification system (RIVPACS). IBI and other multimetric indexes comprise of biological metrics that count, for example, the number of kinds of organisms present at a site (taxa richness, that is, biodiversity) or the relative abundance of trophic groups such as predators. RIVPACS analyses compare the number of taxa found at a test site with the taxa richness at an undisturbed site, as predicted by multivariate statistical models. Common themes in these two approaches include (1) focus on biological endpoints to define river health; (2) use of natural or "undisturbed sites" as a benchmark; (3) standardized sampling, laboratory, and analytical methods, including statistics; (4) numerical ranking of sites according to their condition; and (5) a scientifically rigorous foundation for water policy. Such measures provide better information about the biological dimensions of environmental quality than do physical or chemical measures.

Index of biological integrity is an especially powerful biomonitoring tool for several reasons (Karr, 2006). First, like the index of leading economic indicators, IBI bases its conclusions about river health on an ensemble of biological indicators (metrics), each measuring a different aspect of the biotic community. Second, the metrics in the index reflect tested and predictable responses to human influences; each metric has its own "dose-response" curve associated with human land uses or other impacts (Figure 11.13). Third, metrics are chosen to reflect the effects of diverse human actions, such as logging, urbanization, or agriculture. Adaptations of IBI are now available for various organisms (fishes, insects, birds, vascular plants, and algae) and for diverse environments (rivers, wetlands, coastlines, and land areas). Multimetric biomonitoring programs modeled on the framework of IBI are now employed by scientists and water managers in more than 65 countries and on all continents but Antarctica.

In seven Japanese watersheds, for example, IBI's multiple biological metrics revealed a much more refined picture of river health than the single parameter of biochemical oxygen demand (BOD) (Figure 11.14). Minimally impaired biological condition (high IBI values) occurred only at sites where BOD was low (<1.75 ppm), but not all sites with low BOD had high IBIs.

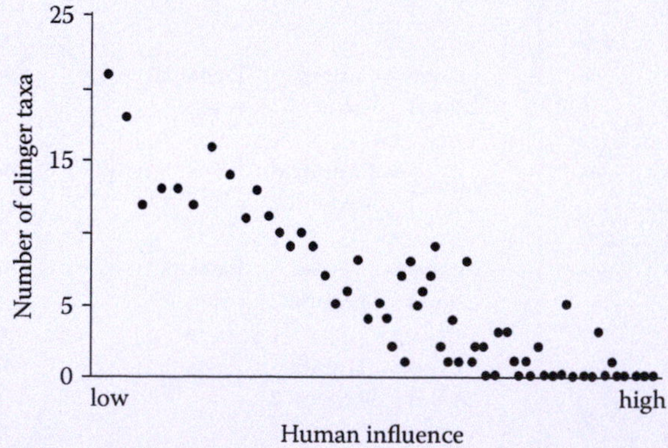

Figure 11.13 Dose-response curve for taxa richness of clingers—benthic invertebrates that cling to rocks enabling them to live in the interstitial spaces between rocks—in standard samples from 65 Japanese streams ranked according to intensity of human influence. (From Karr, J. R. and Chu, E. W., *Restoring Life in Running Waters: Better Biological Monitoring*, Island Press, Washington, DC, 1999.)

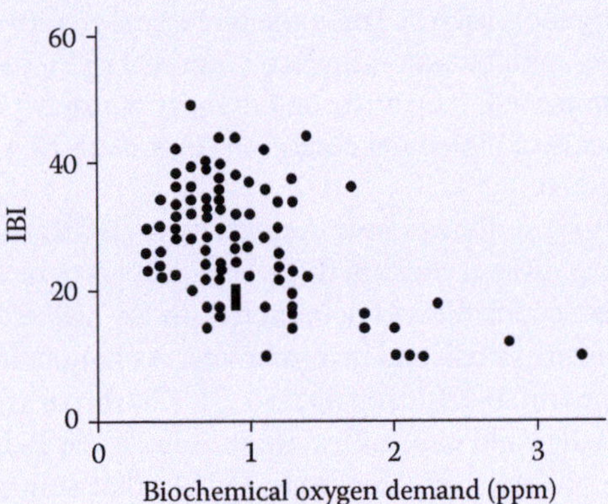

Figure 11.14 Index of biological integrity (IBI) compared with biochemical oxygen demand (BOD) for 100 sites from nine watersheds in Chugoku district, Japan. (From Karr, J. R. and Rossano, E. M., *Ecol. Civil Eng.*, 4, 3–18, 2001.)

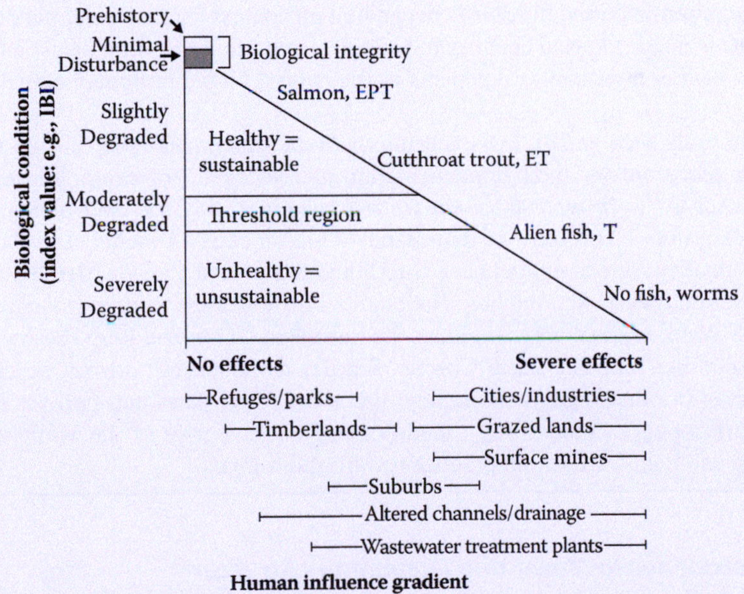

Figure 11.15 Relationship between biological condition and a hypothetical, multimetric measure of human activity, with examples. Different human activities result in biological changes such as different dominant organisms along a descending slope of biological condition. (Modified after Karr in Pimentel, D. et al. *Ecological Integrity: Integrating Environment, Conservation, and Health*, Washington, DC, Island Press, 2000.)

Management or policy decisions based solely on BOD would fail to recognize the wider biotic degradation at a substantial number of sites in the seven watersheds. In short, IBI describes a river's ecological health with greater relevance, precision, and clarity than does water chemistry, thus offering a better guide for river protection or restoration.

Built into biomonitoring tools such as IBI is the biological reality that the condition of living systems varies continuously with human influence (Figure 11.15). Instead of designating water bodies "impaired" or "unimpaired," scientists and managers can express ecological health with greater precision along a scale of biological condition (Figure 11.15, *y* axis) in relation to human influence (Figure 11.15, *x* axis).

In the U.S. Pacific Northwest, for example, benthic IBIs (B-IBI) have been calculated on the basis of 10 metrics, including diversity within three key insect orders: mayflies (Ephemeroptera), stoneflies (Plecoptera), and caddisflies (Trichoptera); index values range up to 50. Healthy streams support a high diversity of fish and invertebrates. As human influence increases, salmon and most stoneflies disappear if B-IBI drops below 35. Cutthroat trout largely disappear, and only the most tolerant mayflies and caddisflies are present when B-IBI drops below 20. These shifts in the biota and the progressively declining B-IBIs that summarize those shifts quantitatively are associated with a variety of human land uses, also varying continuously in their impacts, from protected areas in parks and refuges through lightly or heavily logged forestlands and farms through suburban and urban development.

Watershed managers can also overlay legal and regulatory categories—such as the U.S. Clean Water Act's "designated uses"—on the sliding scales of biological condition and human influence (see Figure 11.15). Ultimately, society decides whether sites or regions are impaired or unimpaired, acceptable or unacceptable, within such legal or regulatory limits. These limits may have different unimpaired-impaired thresholds depending on context (e.g., national park or large city). Managers must use care to avoid declines in biological condition that drop below a tipping point, beyond which neither important components of the natural biota nor human activity can persist in a place.

Backed by tools such as IBI, laws and international agreements that invoke integrity and health have a scorecard for their implementation and success. For example, the goal of the Clean Water Act is "to restore and maintain the chemical, physical and biological integrity of the Nation's waters." The spirit of that mandate has since been included in the 1987 Great Lakes Water Quality Agreement between the United States and Canada, the 1988 amendment to Canada's National Park Act, the U.S. National Wildlife Refuge System Improvement Act of 1997, the 1998 Water Framework that guides the European Union, the 1999 Freshwater Strategy for British Columbia, and The Earth Charter. Similar mandates are present in legislative and policy initiatives in other regions throughout the world. Such laws and policies lay out goals for protecting Leopold's "green fire," "land health," or "integrity of the biotic community;" biomonitoring can help drive actual practice toward those goals.

11.3.2.2 Approaches to Measuring Community Structure

Changes in community structure are studied with laboratory, microcosm, mesocosm, and field approaches. These range from straightforward experiments such as those described above for simple species interactions to those involving whole ecosystem exposures such as that shown in Figure 11.16. Laboratory experiments have many advantages, including the ability to randomly assign treatments, the ability to achieve adequate treatment replication, more control of potentially confounding factors, and more control over exposure dosing. Because of practical constraints, laboratory experiments tend to lose some ecological realism, generality, and ability to address processes occurring at larger temporal or spatial scales (Newman and Clements, 2008). Laboratory studies can include two or more species. Laboratory systems designed to simulate some component of an ecosystem such as multiple species assemblages are called **microcosms**. Pontasch et al. (1989) examined changes in stream macroinvertebrate assemblages in response to an industrial discharge

Figure 11.16 The Biology Gamma Forest at the Brookhaven National Laboratory (Long Island, New York) as it appeared in 1964. This eastern deciduous forest, which was dominated by white oak, scarlet oak, and pitch pine, was exposed to 9500 Ci of ^{137}Cs for approximately 6 months, beginning in 1961 (Woodwell, 1962, 1963). The radiation source was drawn up remotely from inside an underground pipe to expose the woodland. Exposure was many thousand roentgens at this γ source (center of barren spot) and decreased inversely with distance from the source. Zones composed of species of different tolerances ringed the source. Pitch pine (*Pinus rigida*) was the most sensitive with death occurring at 20 r day^{-1}. At the other extreme, sedge (*Carex pensylvanica*) was the most tolerant, surviving 350 r day^{-1}. (Courtesy of Brookhaven National Laboratory.)

by exposing assemblages in laboratory stream microcosms. Niederlehner et al. (1985) examined cadmium's influence on species richness of protozoan communities by exposing naturally colonized substrates to cadmium in laboratory microcosms. Although referred to as mesocosm studies by the authors, 85 L aquaria (microcosms?) were used by Le Jeune et al. (2007) to study copper sulfate effects on the microbial loop by which nutrients are taken into the food web and recycled. Terrestrial microcosms can involve plant growth chambers or soil columns (Gillett, 1989).

Between field and laboratory studies are those involving **mesocosms**, relatively large experimental systems also designed to simulate some key component of an ecosystem. Mesocosms are delimited and enclosed to a lesser extent than are microcosms. They are normally used outdoors or, in some manner, incorporated intimately with the ecosystem that they are designed to reflect. They differ from microcosms by being larger, being located outdoors as a rule, and having a lower degree of control by the researcher (Gillett, 1989). Although mesocosms vary considerably in their design (e.g., Figure 11.17), mesocosm studies all have the common goal of obtaining more realism than achievable with microcosms and more tractability than afforded by field surveys. Liber et al. (1992) conducted a mesocosm-based study to examine natural zooplankton community response to 2,3,4,6-tetrachlorophenol with *in situ* plastic bags extending upwards from the sediments to the surface of a freshwater body. The bags allowed treatments of different concentrations of toxicant and replication within treatments. Goldsborough and Robinson (1986) used similar *in situ* marsh enclosures to study periphyton assemblage response to the triazine herbicides, simazine and terbutryn. Roussel et al. (2007a,b) used

Figure 11.17 Two types of mesocosms used to study fate and effects of contaminants. The top panel shows the indoor mesocosms of the Procter & Gamble Company's experimental streams facility (ESF). This system has the great advantage of more control over conditions (e.g., light and temperature) than normally afforded by outside mesocosms. Eight 12 m long channels allow replication of treatments and production of exposure concentration gradients. The top (head) section of each stream is paved with clay tiles for colonization by algae and microorganisms. Trays of gravel and sand are placed downstream of the tiles and afford substrate for invertebrates. (Courtesy Mr. John Bowling of Procter & Gamble, Co.) The bottom panel shows several outdoor, pond mesocosms used in similar fashion for examining pollutant effects. (Courtesy of Dr. Thomas La Point, North Texas University.)

copper-spiked pond mesocosms like those shown in Figure 11.17 to study primary producer and fish (*Gasterosteus aculeatus*) individual and population effects. Flowing systems may also be studied with mesocosms as evidenced by the work of McCormick et al. (1991) who studied diatom and protozoan assemblages in experimental stream channels dosed with the surfactant, dodecyl trimethyl ammonium chloride. A more recent study spiked the herbicide metazachlor to indoor stream mesocosms (106 m long and 1 m wide) like those shown in Figure 11.17 (Mohr et al. 2007).

Field studies may also vary from surveys of contaminated systems to whole or partial ecosystem manipulations. As experimental manipulations of natural systems afford stronger inference than surveys, a variety of studies have attempted such large scale and expensive manipulations (e.g., Figure 11.18). To examine effects of radiation exposure, terrestrial systems in several geographical regions were irradiated and changes in associated communities studied. The communities included old fields in South Carolina (Monk, 1966), woodlands in Georgia (Schnell, 1964), and forests and

Figure 11.18 Whole lake or split lake studies such as shown here afford *in situ* but expensive information on responses of aquatic communities to anthropogenic materials. Although not associated with a conventional toxicant, this particular split lake study of nutrient enrichment (Canadian experimental Lake 226: P, N, and C added to the bottom section and only N and C added to the top section of the lake) clearly shows the phytoplankton community response to an ecotoxicant. (Courtesy of Dr. Ken Mills.)

old fields in New York (Woodwell, 1962, 1963). Perhaps the most noteworthy whole lake studies were conducted in Canada's Experimental Lakes Area (ELA) that, since the 1960s, explored nutrient loading, acidification, UV radiation, radionuclides, cadmium pollution, organic pollutants, endocrine disruptors (EE2), and mercury cycling or effects (Blanchfield et al. 2009). Illustrative is one study conducted in the ELA concerned the impact of acidification on lake communities in which acid was added to an entire experimental lake (Schindler, 1996). Sensitive species were identified as they disappeared from the lake. Functional redundancy was suggested by a shift in lake trout (*Salvelinus namaycush*) predation. As the pH-sensitive fathead minnow (*Pimephales promelas*) population declined, lake trout shifted their predation effort to the more pH-tolerant pearl dace (*Semotilus margarita*). Community shifts were also examined as a consequence of copper spiking of streams in Ohio (Winner et al. 1980) and California (Leland and Carter, 1984). High concentrations of copper shifted the insect assemblage away from caddisflies toward more tolerant midges. It also reduced diatom species richness.

More often used than field manipulations are field surveys of impacted locations that provide less structured yet much less expensive observation of the consequences of toxicant introduction to communities. Such **biomonitoring**,* the widely applied monitoring with selected sampling protocols of a subset of an entire community with the goal of assessing community condition (Herricks and Cairns, 1982), can produce a simple listing of species present or a much more complex analysis of data. Herricks and Cairns (1982) suggest three general types of biomonitoring efforts. The first simply describes the biota, perhaps summarizing results as a species-abundance list. The second may involve the formulation of a hypothesis that is then tested with field observations. The last type combines these two approaches to formally test conclusions (hypotheses) derived from the descriptive phase of the biomonitoring effort. Clearly inferential strength is highest for this last type of biomonitoring.

11.3.3 Function

Changes in community functioning are used less often by ecotoxicologists than are structural changes because it is generally believed that feedback loops and functional redundancies make community functions less sensitive to toxicants than community structure (Odum, 1985; Forbes and Forbes, 1994). Nonetheless, important community functions can be modified by toxicants. Certainly, modified functioning is implied by any change in functional groups or guilds such as the shift in macroinvertebrate shredder and collector groups measured around coal mine drainage (Chadwick and Canton, 1983). Blanck (1985) also suggested that natural periphyton photosynthetic activity, measured as $^{14}CO_2$ incorporation, could be used as an ecotoxicological test. Giesy (1978) measured a significant drop in leaf litter decomposition rates at elevated cadmium concentrations, suggesting another important function influenced by toxicants. Cairns and coworkers (e.g., Niederlehner et al. 1985; Cairns et al. 1986; McCormick et al. 1991) demonstrated a clear concentration–response relationship for colonization by protozoa of artificial substrates. They fit colonization data under various toxicant concentrations using the **MacArthur–Wilson model** of island colonization,

$$S_t = S_{EQ}(1 - e^{-Gt}) \tag{11.5}$$

where S_t = the number of species present at time t, S_{EQ} = the equilibrium number of species for the island, and G = the rate constant for colonization of the island. A sensitive assay was developed and demonstrated with a series of toxicants (Figure 11.19).

Change in community tolerance has also been proposed as a measure of functional change with pollutant exposure. It is measured as **pollution-induced community tolerance (PICT)**, a net increase in tolerance to pollution resulting from species composition shifts in the community, acclimation of individuals, and genetic changes in populations in the community. Procedurally, a previously exposed community and an unexposed community might be challenged with a toxicant and the difference in responses of the two communities used to reflect adaptation. Kaufman (1982) reported that a periphyton community adapted to copper displayed a smaller decrease in ATP and chlorophyll upon repeated exposure to high copper concentrations relative to a periphyton community with no previous exposure. Similar results were obtained for 4,5,6-trichloroguaiacol-adapted periphyton communities exposed for a second time to this toxicant (Molander et al. 1990).

* Qualifiers are frequently made for the term, biomonitoring. As an example, Hopkin (1993) defines the monitoring of community changes along a gradient or among sites differing in levels of pollution as **Type 1 biomonitoring**. **Type 2 biomonitoring** involves the measurement of bioaccumulation in organisms among sites notionally varying in the level of contamination. **Type 3 biomonitoring** attempts to define the effects on organisms using tools such as biochemical markers in sentinel species or some measure of diminished fitness of individuals. **Type 4 biomonitoring** involves the detection of genetically based resistance in populations of contaminated areas.

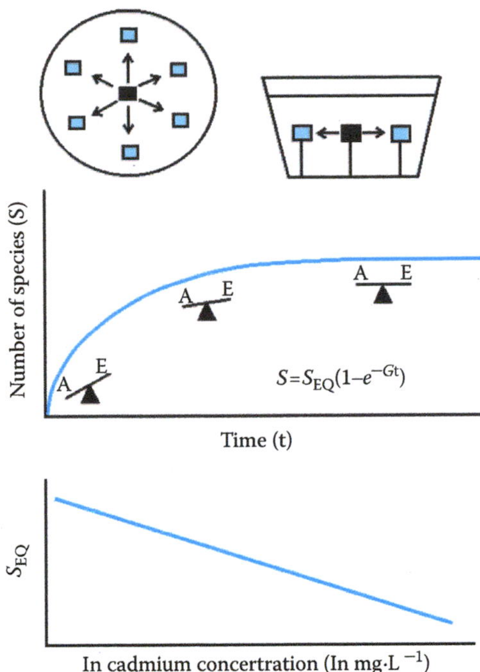

Figure 11.19 The protozoan community colonization assay developed by Cairns and coworkers (e.g., Niederlehner et al. 1985; Cairns et al. 1986). A polyurethane foam substrate (black square in top diagram) is allowed to accumulate species in a natural stream and then brought to the laboratory to serve as a source or epicenter for colonization of other, uncolonized foam substrates (blue squares in top diagram). The dynamics of species colonization on foam substrates is measured using the MacArthur–Wilson model for island colonization (center plot). This is done with substrates submersed in containers filled with different concentrations of toxicant (bottom plot) to determine the effect of toxicants on the process. (Modified from Figures 1, 2, and 4 in Cairns et al. *Environ. Monit. Assess.*, 6, 207–220, 1986.)

11.4 ECOSYSTEM QUALITIES

Many of the changes predicted by Odum (1985) to occur in ecosystems impacted by toxicants (Table 11.2) have already been discussed in the context of communities. However, a few remain to be discussed more fully. Particularly relevant are those related directly to nutrient cycling and energy flow.

That energy flow in agricultural ecosystems is changed with pesticide application is self-evident from the goal of increasing crop biomass. Species diversity, food chain length, predator–prey ratios, dominance, and other structural qualities of arthropod communities also shift in ecosystems to which pesticides are applied (e.g., Pimentel and Edwards, 1982; Deb, 2009). Associated with such changes is a change in energy flow via trophic exchange. Diminished decomposition rates in soils and recycling of minerals by soil organisms in agricultural systems and adjacent natural ecosystems are another category of less obvious effects. Pimentel (Pimentel and Edwards, 1982; Pimentel and Levitan, 1986; Pimentel et al. 1991) reports that application of pesticides can shift elemental cycling in fields and composition of crops such as corn and beans.

Changes in nonagricultural system's energy flow and material cycling have been demonstrated over a wide range of scales. Acidification of streams in the Great Smoky Mountains diminished leaf decomposition, bacterial production, and microbial respiration rate (Mulholland et al. 1987). Odum (1985) indicates that enhanced loss of calcium from forested watersheds is a good indicator of functional damage to the watershed. Smaller scale microcosm and mesocosms studies also

Table 11.2 Odum's Predicted Changes in Ecosystems Experiencing Toxicant Stress

Component/Quality	Predicted Change
Energetics	Increased community respiration
	Imbalance production/respiration, i.e., P/R < 1 or P/R > 1
	Increased maintenance: biomass, i.e., increase in production/biomass and respiration/biomass
	Increased importance of energy from outside the ecosystem
Nutrients	Increased export of primary production
	Increased turnover of nutrients
	Decreased cycling of nutrients
	Increased loss of nutrients as a result of the two above changes
Community structure	Increased proportion of species that are r-strategists
	Decreased size of organisms
	Decreased life span of organisms
	Shortened food chains
	Decreased species diversity with increased species dominance
Ecosystem	Decreased internal cycling and increased importance of input and output from outside sources
	Regression to an earlier successional stage
	Functions (e.g., community metabolism) changed less than structural components (e.g., species richness)
	Decreased positive (e.g., mutualism) and increased negative (e.g., disease or parasitism) interactions

Source: Modified from Odum, E. P., *BioScience*, 35, 419–422, 1985.

have been used to examine energy flow and material cycling under the influence of toxicants. As a laboratory-based example, Bundschuh et al. (2011) noted decreased freshwater amphipod processing of dead leaf material in the presence of mercury-contaminated sediment. Copper addition to microcosms reduced primary production and dissolved organic carbon production (Hedtke, 1984). Spiking of experimental streams with triphenyl phosphate (Fairchild et al. 1987) did not lower leaf decomposition rates relative to control streams. Also, rooted flora increased and net nutrient retention increased with treatment. Clearly, these important functions of ecosystems are influenced by toxicants. However, there will be exceptions (e.g., Fairchild et al. 1987) to general predictions, such as those of Odum's, due to complex changes in the community structure.

As mentioned in Chapter 2 regarding cultural eutrophication, an excess of nutrients predictably result in shifts in dominant algal species and abundances to produce a dysfunctional aquatic ecosystem. For example, a conventional Vollenweider model (Vollenweider, 1975; Tapp, 1978; Reynolds, 1992) was used to predict dysfunction before construction of the cooling water reservoir depicted in Figure 2.4. Newman (1986) estimated from the phosphorus loading to this proposed one-pass cooling water reservoir and compared it to that of a nearby cooling lake that recirculated cooling water instead of drawing new water continuously from the nutrient-rich Savannah River. The loadings for these two reservoirs were 8.57 and 0.65 g of phosphorus per square meter of lake surface, respectively. A Vollenweider model uses such loadings to estimate expected phytoplankton biomass after accounting for the influence of lake morphology and hydrology. To account for lake morphology and hydrology, the mean lakes depth is divided by the estimated mean water residence time. As shown in Figure 11.20, the recirculating cooling lake (Par Pond) was correctly characterized as

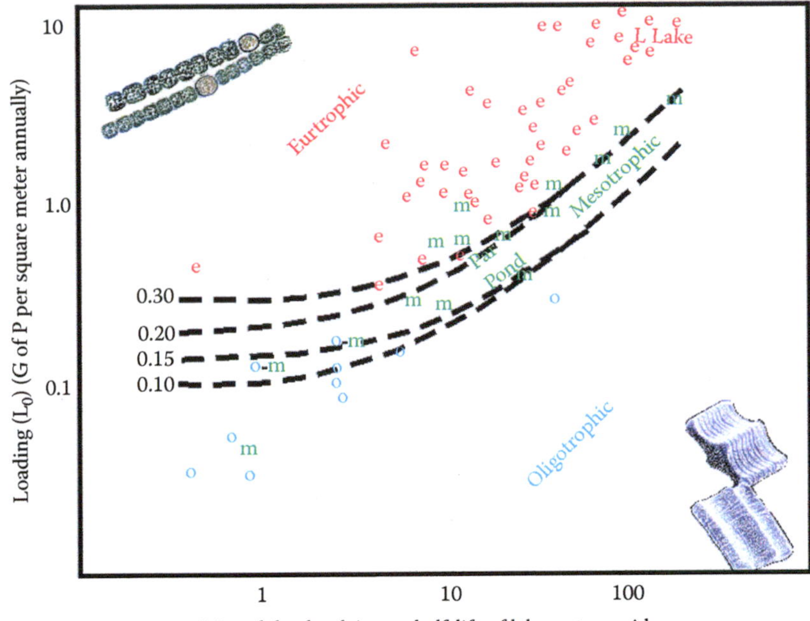

Figure 11.20 A simple Vollenweider plot of a series of European and U.S. water bodies using phosphorus loading and hydrology metrics. Lentic systems that were eutrophic, mesotrophic, or oligotrophic were coded as such with colored e, m, and o, respectively. Those transitional between mesotrophic and oligotrophic are depicted with an o-m. The position for Par Pond (recirculating cooling lake) and then-planned L Lake (once-through cooling lake) are also shown. Lines depicting the general trends for lakes with phosphorus loadings of 0.10, 0.15, 0.20, and 0.3 are also shown to suggest zones of transition between oligotrophic, mesotrophic and eutrophic water bodies.

mesotrophic* and had modest risk of dense algal blooms and resulting fish kills due to low nighttime oxygen concentrations. In contrast, the proposed once-through cooling water lake (L Lake) was predicted to be highly eutrophic, and when built, did experience dense algal blooms (Figure 2.5) and frequent fish kills during low oxygen conditions.

Cooling water discharges can influence function in other ways as illustrated by thermal discharge into a wetland near L Lake and Par Pond. Pen Branch Creek (South Carolina) received thermal discharge from a nuclear reactor for several decades (Figure 2.35). The hot, nutrient-rich discharge killed the cypress and tupelo trees of the downstream river swamp, opening up the tree canopy (lower photograph of Figure 2.35). The increased availability of light and warm, nutrient-rich waters allowed dense blue-green algal mats to form and influenced water quality (Figure 11.21). Dissolved oxygen concentrations fell below 4 $mg \cdot L^{-1}$ during nights and rose above saturation concentrations during the day. These rapid changes in the balance between productivity and respiration also resulted in wide diurnal fluctuations in water pH. Clearly, the complex changes brought about by the thermal pollution caused a substantial change in the wetland community functioning.

* Respectively, oligotrophic, mesotrophic, and eutrophic refer to the "scant/few food/feeding," "middle food/feeding," or "rich/good food/feeding" states of available nutrients. An **oligotrophic** water body has low productivity due to low nutrient levels and will have a characteristic diverse ecological community including diatom and green algal (such as desmids) species as the dominant phytoplankton. They have clear waters and no dense algal blooms to cause low oxygen episodes in surface waters. Turbid waters of highly productive **eutrophic** systems result from dense algal growth. Eutrophic systems are characterized by periodic low oxygen episodes that can cause fish kills. Eutrophic systems have low species diversity with blue-green algae such as the *Anabeana* in Figure 2.5 or *Microcystis* species as frequent dominants. **Mesotrophic** systems are intermediate between oligotrophic and eutrophic systems.

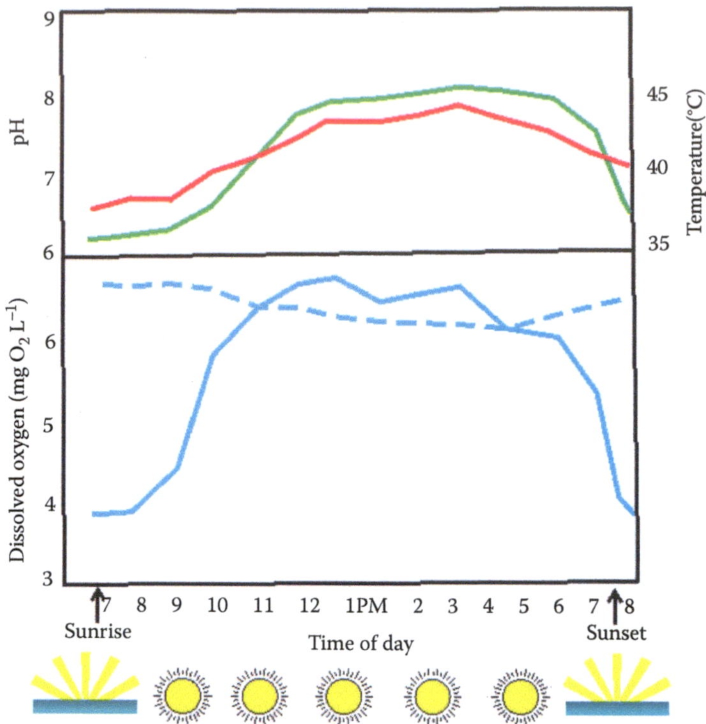

Figure 11.21 Daily shifts in water quality measured in a region of river swamp where the tree canopy had been removed due to past discharge of hot cooling waters (see lower photograph in Figure 2.35). The aquatic community shifted substantially with the destruction of the cypress and tupelo trees, and opening up of the canopy. On the day depicted here, warm nutrient-rich cooling water flowed over thick mats of blue-green algae (cyanobacteria) in this stream delta region of the swamp. The water temperature (red line) increased slightly and then decreased from dawn to dusk. Intense microfloral respiration had drawn dissolved oxygen concentrations (solid blue line) down to approximately 4 $mg{\cdot}L^{-1}$ until dawn when microfloral photosynthesis rapidly raised it above dissolved oxygen saturation (dashed blue line). Small bubbles of oxygen appeared in the algal mat at the most productive time of day. Strong shifts in microfloral respiration and photosynthesis created wide shifts in water pH (green line). (Figure drawn from data in Newman (1986) unpublished report.)

11.5 SUMMARY

Beginning with a brief discussion of ecological communities, metacommunities, species assemblages, and niche, this chapter outlines general laboratory, mesocosm, and field approaches to determining the influence of toxicants on communities or species assemblages. General methods of estimating community effects include the use of indicator species (sensitive and tolerant species), community level NOEC estimation, and biomonitoring. Simple species interactions, e.g., predator–prey and interspecies competition, were shown to be susceptible to toxicants. The reduction in fitness of individuals participating in such simple interactions was placed into the context of the principle of allocation. Assuming an equilibrium model for communities, qualities contributing to community vulnerability to toxicant effect were detailed, including community elasticity, inertia, and resilience. The question of whether a community can be rendered accurately to an equilibrium context was then brought up and contrasted with the community conditioning hypothesis. Measures of community structure and function were then discussed relative to toxicant effects. Such changes were placed into the context of community successional regression, functional redundancy theory, and the law of frequencies. Changes in ecosystem energy flow and material cycling, although also implied in discussions of community shifts, were then described briefly. Some of these ecosystem changes will be discussed again in a wider geographical context in Chapter 12.

SUGGESTED READINGS

Atchison, G. J., M. B. Sandheinrich, and M. D. Bryan, Effects of environmental stressors on interspecific interactions of aquatic animals, in *Ecotoxicology: A Hierarchical Treatment*, Newman, M. C. and CH. Jagoe, Eds., CRC Press, Boca Raton, FL, 1996.

Gillett, J. W., The role of terrestrial microcosms and mesocosms in ecotoxicological research, in *Ecotoxicology: Problems and Approaches*, Levin, S. A., M. A. Harwell, J. R. Kelly, and K. D. Kimball, Eds., Springer-Verlag, New York, 1989.

Graney, R. L., J. P. Giesy, and J. R. Clark. Field studies, in *Fundamentals of Aquatic Toxicology*, 2nd Edition, Rand, G. M., Ed., Taylor & Francis, Washington, DC, 1995.

Hopkin, S. P., *In situ* biological monitoring of pollution in terrestrial and aquatic ecosystems, in *Handbook of Ecotoxicology*, Calow, P., Ed., Blackwell Scientific Publications, London, United Kingdom, 1993.

Newman, M. C., *Quantitative Ecotoxicology,* 2nd Edition, Taylor & Francis/CRC Press, Boca Raton, FL, 2013.

Newman, M. C. and W. H. Clements, *Ecotoxicology: A Comprehensive Treatment*, Taylor & Francis/CRC Press, Boca Raton, FL, 2008.

Odum, E. P., Trends expected in stressed ecosystems. *BioScience,* 35, 419–422, 1985.

Woodwell, G. M., The ecological effects of radiation. *Sci. Am.*, 208, 2–11, 1963.

CHAPTER 12

Landscape to Global Effects

Even though the pattern of our relationship to the environment has undergone a profound transformation, most people still do not see the new pattern... The sights and sounds of this change are spread over an area too large for us to hold in our field of awareness.

Gore (1992)

12.1 GENERAL

"Is it bigger than a bread box?" This is the traditional opening* to a familiar guessing game in which an object is eventually identified from answers to a series of questions. It reflects our tendency to categorize things by size or spatial scale. This tendency extends to topics traditionally classified as within or outside the purview of ecotoxicology. Customarily, but not always correctly, ecotoxicology focuses on scales up to the traditional ecosystem, such as the fate and effects of pollutants in a field, forest, lake, or stream. Fortunately, an increasing number of studies have begun extending beyond this framework.

Divergent answers would result if one asked established ecotoxicologists to decide whether a topic such as the current pollination crisis, global warming, widespread forest decline in central Europe, ocean acidification, or global distillation were in the purview of ecotoxicology. Some would feel that, if the context of the problem were bigger than a traditional ecosystem, then it would be better handled in biogeochemistry, landscape or conservation ecology, soil sciences, atmospheric chemistry, or remote sensing technology. There would be a contrastingly uniform affirmation if the question involved PCB bioaccumulation in trout of a lake or a pollution-induced decrease in arthropod species diversity in forest litter. One obvious reason for this bias is that much in ecotoxicology was borrowed in the 1970s from the sciences of toxicology and ecology. The emphasis in toxicology was and remains appropriately on the individual. Until a few decades ago, the dominant context of ecology was the ecosystem or lower levels of biological organization.

In Chapters 9 and 10, we suggested that the single species bias in much of ecotoxicology grew out of the early transplanting of ideas and approaches from mammalian toxicology. Although still present in ecotoxicology, single species effects are generally accepted as important, but not inclusive, features deserving consideration relative to many pressing problems. Higher level effects can be as or more important than those at the level of the individual. Similarly, the conventional

* For the young reader who does not know what a bread box is, the question might be changed to "Is it bigger than an Xbox 360?"

ecosystem* bias is opined here to be inadequate in many cases too. This opinion is reinforced by Cairns (1993), Catallo (1993), and Holl and Cairns (1995), who argued decades ago that a landscape context for ecotoxicology is also needed. A reasonable argument could be made to extend the context beyond the landscape scale. As examples from the **cryosphere** (that part of the earth that is perennially frozen), the concentrations of persistent organic pollutants (POPs) measured in Adelie penguins (*Pygoscelis adeliae*) nesting at the southern pole (Figure 12.1) or changes in polar bear (*Ursus maritimus*) conditions at the northern pole (Figure 12.2) are inexplicable unless a global vantage is taken during inquiry. The search for remedies is also ineffective without a global vantage. The traditional, but now too confining, bias toward the ecosystem, single community, or lower levels is designated the **ecosystem incongruity** here.

Supporting examples are easy to find. A landscape example involves copper mining and smelting in Copperhill, Tennessee, (Figure 12.3, top panel) and also Mount Lyell, Tasmania (Figure 12.3, bottom panel). By killing vegetation and stripping nutrients from the soil, acidic fumes from smelting transformed forested landscapes to desert-like surroundings. A larger scale example is pollution from the Kuwait oil fires (Figure 1.6), an event influencing significant land (desert and urban) and marine components of a country. An even more encompassing example is the Trans-Alaska Oil Pipeline (Figure 12.4). It extends south from Prudhoe Bay (Arctic Ocean) up the North Slope over the Brooks Range to cross the Arctic Circle, Yukon River, and the Alaska Range to end at the Valdez marine terminal on Prince William Sound (Pacific Ocean). For this one project, risk of damage exists for tundra, taiga, boreal forest, river, mountain, lake, fjord, and intertidal ecosystems. One accident associated with only one segment of this project, the 1989 *Exxon Valdez* spill, spread oil out into parts of Cook Inlet and Alaska Sound, and covered 30,000 km^2 of Alaskan waters. A large hypothetical spill onto the tundra could conceivably have an impact beyond that ecosystem because many bird species spend part of their time there and migrate to Asia (e.g., the wheatear, *Oenanthe oenanthe* nesting in rocky fields of the tundra), North America (e.g., sandhill crane, *Grus*

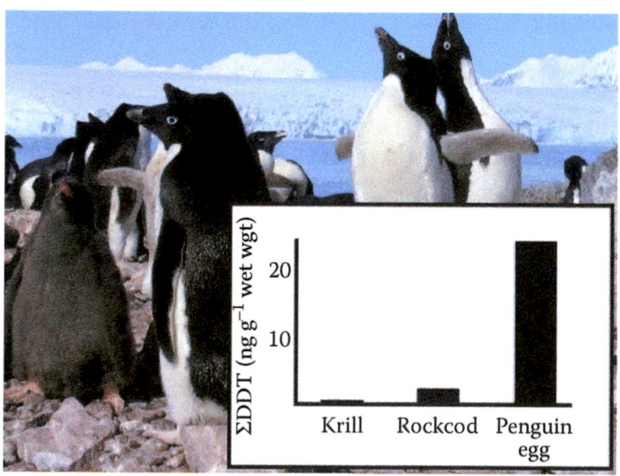

Figure 12.1 The accumulation of ΣDDT in Adelie penguin (*Pygoscelis adeliae*) eggs in the Antarctic. Diverse persistent organic pollutants (POPs) are found to accumulate to high levels in these and other polar species inhabiting regions distant from the use of the POPs, e.g., Bidleman et al. (1993), Kumar et al. (2002), Goerke et al. (2004), Vorkamp et al. (2004), and Corsolini et al. (2007). (Data from Table 1 of Corsolini, S. et al. *Environ. Pollut.*, 140, 371–382, 2006 and photograph courtesy of Heidi Geisz [College of William & Mary, VIMS]).

* Note that the word, biosphere, was applied in Chapter 1 instead of the usual ecosystem in the definition of ecotoxicology. The intent in doing so was to unfetter discussion from the conventional ecosystem context and allow free consideration of landscape, regional, continental, and global scales.

Figure 12.2 Pack ice in Ungava Bay of the Arctic Ocean (northeast Canada) as seen on July 1, 2007 during a historically thin summer polar ice condition. Pack ice melt is normally slow until mid-July when it accelerates in this part of the Arctic Ocean. Feeding by polar bears on seals is strongly influenced by pack ice dynamics. Scientist and Native American Inuit groups have begun to focus on potential impacts of global warming trends on the persistence of polar bear populations (Derocher et al. 2004; Wiig, 2005; Stirling and Parkinson, 2006). Some predictions indicate the potential for substantial thinning (Bluhm and Gradinger, 2008; Walsh, 2008) or even complete disappearance of summer ice (Kerr, 2007; Serreze et al. 2007; Stroeve et al. 2007). Like the penguins shown in Figure 12.1, polar bears can accumulate high concentrations of POPs (Kumar et al. 2002), which suggests a complex situation that exists for polar bears at this time. (Both photographs taken by M. C. Newman.)

canadensis breeding in tundra marshes), and South America (e.g., golden plover, *Pluvialis dominica* nesting on tundra hillsides). Birds that are members of metapopulations spread across several continents could be harmed by an oil spill in Alaska. Clearly, any preoccupation with a classic ecosystem or local ecological community, rather than a landscape, metacommunity, or larger context would result in an insufficient description of potential consequences of the Trans-Alaska Oil Pipeline. Our "stress signature" on the earth now extends up to a global context (e.g., Figure 12.5), touching even remote regions of the world, e.g., the movement of airborne soot (black carbon) to the extreme heights of the Himalayas (Qiu, 2013).

There is a second and equally good reason why the ecosystem bias is no longer sensible. The ecosystem focus draws attention away from important qualities of landscapes that are a matrix of "ecosystems." Unique properties emerge in this landscape* context. The source–sink framework for metapopulation and metacommunities dynamics discussed in Chapters 10 and 11 are obvious examples. Maurer and Holt (1996) modeled effects of pesticides on mobile wildlife in a complex landscape and found that inclusion of source–sink dynamics was crucial for predicting impact on populations. Another classic example involves **ecotones**, areas of transition between two or more kinds of communities (Odum, 1971). Ecotones often have species assemblages with high species richness and high abundance of individuals relative to those of the adjacent communities. There are several reasons underlying such an **edge effect**. Species from contiguous habitats are present in the ecotone, increasing species richness at this edge relative to the two adjacent habitats. Some species can exploit both habitats in different ways, increasing their abundances at the ecotone. A species may nest in the forest but forage on grains in an adjacent field. Finally, unique species adapted to the ecotone, e.g., estuarine species, add to species richness. Consequently, pesticide application to agricultural fields may

* **Landscape** is an ambiguous term used in many contexts. Here, it is used to simply denote the sum total aspect of a geographical area (Monkhouse, 1965).

Figure 12.3 In the top photograph, landscape modification by smelting and mining activities in Copperhill, Tennessee, is shown. Copperhill is situated in the Blue Ridge Mountains at the convergence of northern Georgia, western North Carolina, and southern Tennessee. The Ducktown Mining District began smelting circa 1854, and rapidly developed during the next four decades. Sulfuric acid and sulfur dioxide releases were greatly reduced after 1910. Tree growth, as measured from growth rings, was slowed from 1863 to 1912 in the nearby Great Smoky Mountains National Park (88 km upwind) because of the emissions from smelting (Baes and McLaughlin, 1984). This photograph was taken more than 70 years after emission reductions occurred (1982) and shows a desert-like landscape instead of the typical, forested landscape. In the bottom photograph, a similar situation is shown for the Mount Lyell copper mines and smelters near Queenstown, Tasmania. The Mount Lyell Mining and Railway Company began operations in 1893. Smelting ended in 1969, 39 years before this photograph was taken. (Photographs by M. C. Newman)

not have the same predicted ecotoxicological consequences for areas with an extensive network of hedgerows or woodland patches among the fields compared with those without. Such differences due to ecotones become important as the trend toward large agroindustrial farming and away from small farms continues in many parts of the world. Another important class of ecotone, estuaries at river mouths, is extremely vulnerable to contaminants from upriver sources and from port cities along their shores. Any unwarranted preoccupation with the traditional ecosystem or community context tends to draw attention away from the unique qualities of ecotones and other important features.

The third and final reason why we should extend our spatial context for ecotoxicology is simply that we now have the tools and data to do so. Affordable computer costs and an increase in computational power allow diverse data sets, including inexpensive high altitude and satellite data, to be integrated into a coherent and informative form by researchers and managers (see Figures 12.6

Figure 12.4 The Trans-Alaska Oil Pipeline as it passes across the taiga, a transitional community between the tundra and boreal forest communities. (Photograph by M. C. Newman)

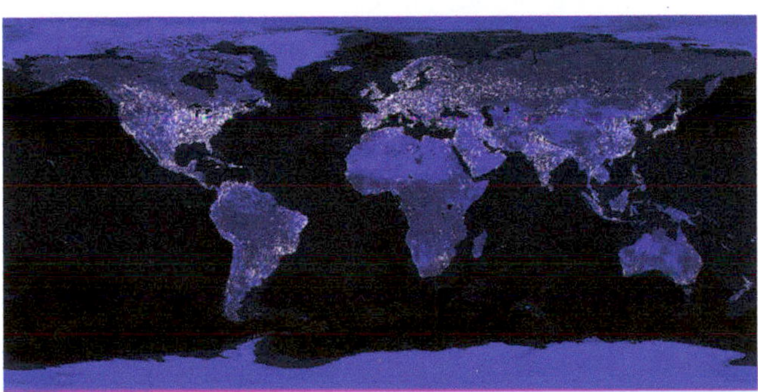

Figure 12.5 The global pattern of night lights visualized by compositing two hundred Defense Meteorological Satellite Program (DMSP) images (http://antwrp.gsfc.nasa.gov/apod/earth.html). (Courtesy of NASA, C. Mayhew and R. Simmon (NASA/GSFC), NOAA/NGDC, DMSP Digital Archive)

through 12.9 as examples.) Computerized **geographic information systems** (GIS) now handle these data at a reasonable cost, allowing one to archive, organize, integrate, analyze, and display many kinds of spatial information with a common coordinate system (Avery and Berlin, 1985). Data of different types such as land use, vegetation, rates of pesticide application, soil type, weather, and air or water quality can be merged and compared statistically to provide invaluable insights for effective stewardship of resources and environmental regulation. More recent examples include the global mapping of NO_2 emissions (Ghude et al. 2009) and also the estimated change in ice-sheet mass linked to global climate change (Shepherd et al. 2012). Many books, such as those by Chuvieco and Huete (2009), Michener et al. (1994), Korte (2000), and Wang and Weng (2013), now detail methods for doing so, and a wide range of affordable imagery and maps are available.

Some of this imagery is produced by remote sensing. **Remote sensing** technologies allow the acquisition and analysis of data without requiring physical contact with the land or water surface being studied. Most determine qualities or characteristics of areas of interest based on measurements of visible light, infrared radiation, or radio energy coming from them (Sabins, 1987). For example,

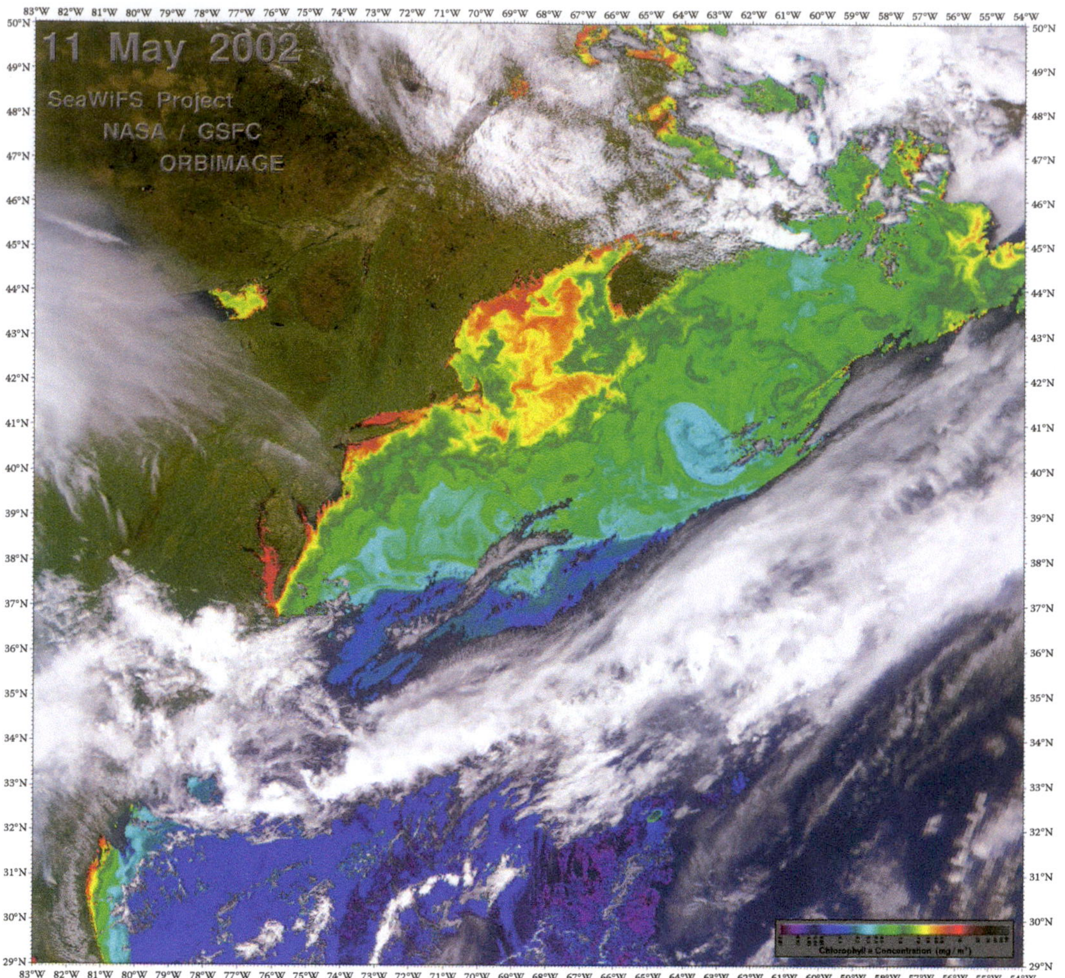

Figure 12.6 Chlorophyll (mg·M^{-3}) off the heavily developed east coast of the United States showing clearly high coastal productivity (e.g., red and orange reflect areas of approximately 10–50 and 2–10 mg·M^{-3}) with productivity being lowest offshore (blue areas have approximately 0.01–0.05 mg·M^{-3} of chlorophyll). Intermediate productivity areas of yellow (yellow areas are approximately 0.5–2 mg·M^{-3} of chlorophyll) to green are obvious as regions of mixing between the nutrient enriched coastal waters and the waters of the Gulf Stream. A low chlorophyll gyre is seen in this image as a mass of Gulf Stream water spun off into the more productive waters. Although increased nutrient-driven productivity is shown here only for a small area of the ocean, increased nutrient inputs to the world's oceans have produced a global crisis in which large expanses of the oceans called "dead zones" are experiencing chronic hypoxia (Diaz, 2001; Diaz and Rosenberg, 2008). (Courtesy of NASA SeaWiF Project, NASA/Goddard Space Flight Center and ORBIMAGE.)

infrared spectral characteristics may be used to define vegetation community types over a wide area. As another example, data from sensitive radiation sensors mounted in aircraft are used to map γ irradiation over large areas of U.S. Department of Energy nuclear facilities where past releases occurred. Oil slicks on sea surfaces are detected and tracked by their higher radiance of ultraviolet and blue light (Sabins, 1987).

This type of spatial information has quickly been incorporated into environmental regulation and management activities. As only one of many examples, the U.S. EPA now has placed U.S. vegetation types, a Toxic Release Inventory (TRI), air pollution, areas of air quality nonattainment, and Superfund sites into a GIS format (Reichhardt, 1996).

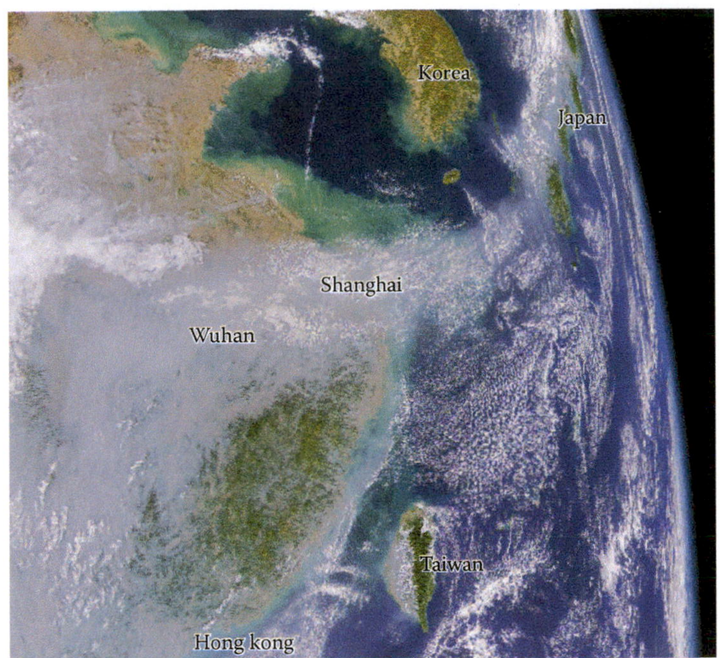

Figure 12.7 A mass of haze is seen in this SEAWiF image that originates in eastern China, passes over the large cities of Wuhan and Shanghai, and then between South Korea and Japan. The rapid industrial development of China that depends heavily on coal burning produces such a visible bank of aerosol haze as well as sulfur dioxide that is not confined by national borders. (Courtesy of NASA SeaWiF Project, NASA/Goddard Space Flight Center and ORBIMAGE.)

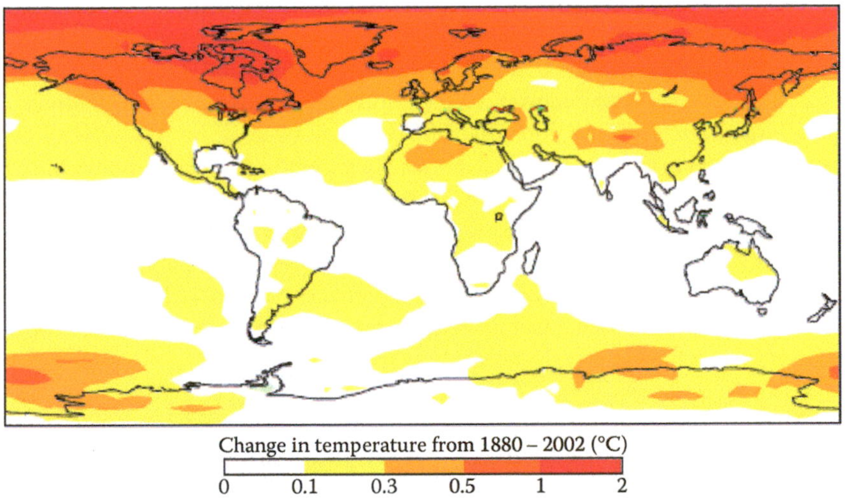

Figure 12.8 NASA estimation of the amount of the total increase in temperature that can be attributed to increases in soot generation globally. The period over which the change is calculated is 1880 to 2002. (Courtesy of NASA Goddard Institute of Space Studies.)

In a departure from the approach used in earlier chapters, this chapter will be largely based on examples and imagery. Each example or image will be selected to represent an ecotoxicological issue at a particular spatial scale, i.e., landscape, regional, continental, hemispheric, or global scale. The general trend should become obvious from these examples that a contaminant's

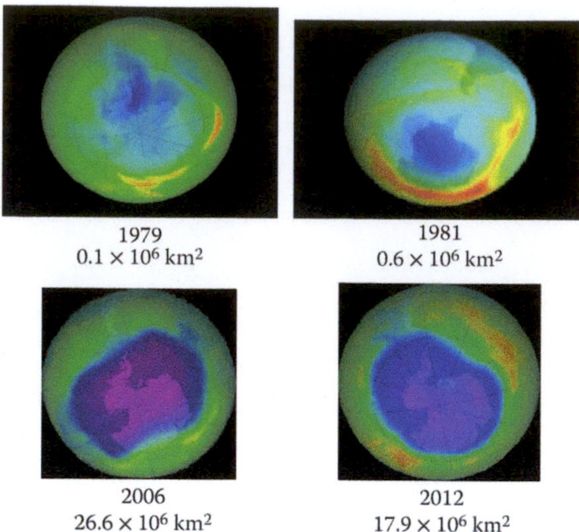

Figure 12.9 The Antarctic ozone hole as generated from data taken from August/September 1979, 1981, 2006, and 2012, showing the emergence of a hole extending outward to cover approximately 20 million square kilometers of the earth's surface. The hole reached its maximum size (26.6 million square kilometers) in 2006 and seems to have stabilized, notionally because of the Montreal Protocol. (Courtesy of NASA/Goddard Space Flight Scientific Visualization Studio. See http://www.nasa.gov/earth/lookingatearth/ozone_record.html and other related NASA webpages)

potential for dispersal and its spatial scale for concern increase with the degree to which it is associated with the more mobile components of the environment (atmosphere mobility > hydrosphere mobility > pedosphere* mobility > lithosphere mobility). For example, contaminants associated primarily with the atmosphere, such as those giving rise to acid precipitation, will have effects over wide expanses. Some exceptions to this trend occur if large amounts of a material (e.g., a pesticide) are applied over a wide region† committed predominantly to one human activity (e.g., a large agricultural region of North America), or if the human activity giving rise to the contamination is occurring over extensive areas (e.g., lead contamination of North American soils resulting from past widespread use of leaded gasoline). An unfortunate corollary is that cause and effect relationships are clearest locally, but become increasingly more challenging to assign with distance from a source (Cairns and Pratt, 1990). Consequently, some of the widest spread problems of global concern such as ozone depletion or global climate change were quite difficult to assign a cause and then enact an effective remedy. However, this impediment is being reduced each year as technologies advance.

12.2 LANDSCAPES AND REGIONS

Often landscape studies are framed around some physical feature such as a watershed. As a typical example, Richards et al. (1993) used GIS methods to categorize land use in a Michigan catchment and linked land use with macroinvertebrate community composition of associated

* The **pedosphere** is that part of the earth made up of soils and where important soil processes are occurring (Ugolini and Spaltenstein, 1992).

† As used here, a geographic **region** is an "area of the earth's surface differentiated by its specific characteristics" (Monkhouse, 1965).

water bodies. There was a direct linkage between agricultural activity and stream substrate quality. In turn, substrate quality influenced the abundances of Ephemeropteran, Plecopteran, and Trichopteran insect taxa in benthic communities. Recommendations for modifying land use and predictions of change under various restoration scenarios were generated from this study. Richards and Host (1994) successfully applied this method again to Minnesota catchments along the shores of Lake Superior. A similar approach was taken to categorize and then project future problem areas from nonpoint pollution in the St. Johns River Basin in Florida (Adamus and Bergman, 1995). Their analysis integrated information on contaminant amounts and concentrations in surface runoff, sites of stormwater treatment and efficiency of that treatment, projected changes in land use, soil types, rainfall, hydrology, and current water quality. The inset of Figure 12.10 shows the predicted sites of significant pollutant generation along the St. Johns River.

Also illustrated in Figure 12.10 are features important in predicting contaminant impact at a larger scale, that is, at the scale of a large U.S. state. The dominant vegetation changes considerably in the lower subtropical half of this state and determines the specific communities at risk and the *milieu* in which the contaminant effect might or might not be expressed. The scale of an entire state might also be required to encompass metacommunity dynamics. Because U.S. laws and regulations are applied by states, the arena for dealing with contaminants may be defined by state boundaries. For example, state fish consumption advisories and bans are determined by contaminant concentrations in game species. States in the United States establish their own, occasionally divergent, criteria

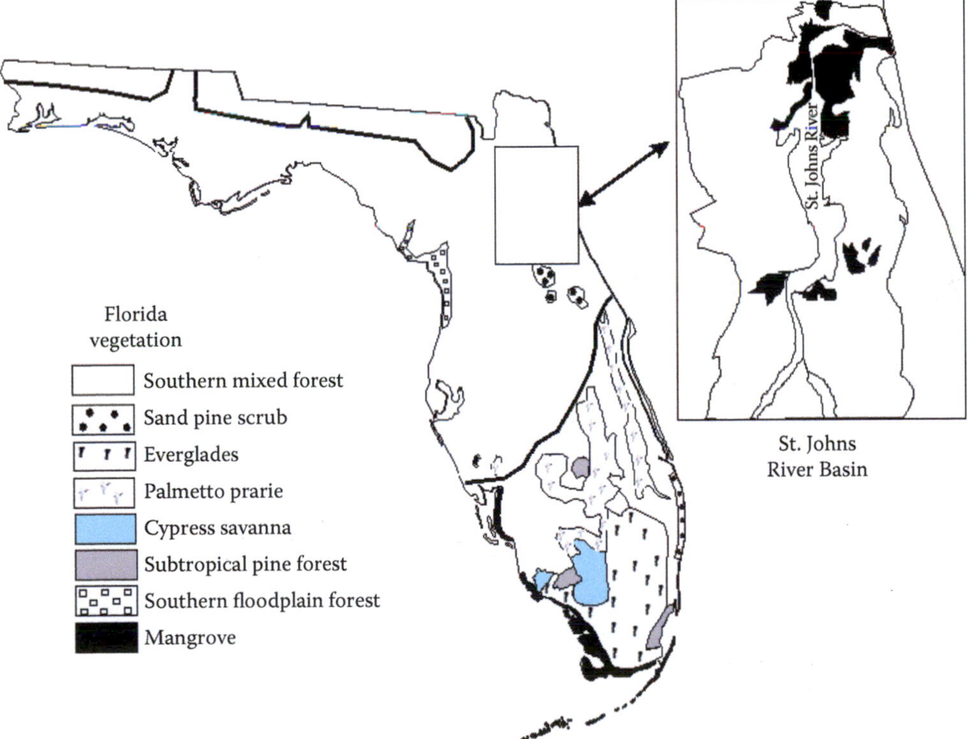

Figure 12.10 Three scales (river basin, vegetation type, and Type II ecoregion) of consideration for Florida. (Modified by combining spatial data from Figure 5 of Adamus, C.L. and M.J. Bergman, *Water Resour. Bull.*, 31, 647–655, 1995; an ecoregion map from Omernik, J.M., *Ann. Assoc. Am. Geogr.*, 77, 118–125, 1987; and a U.S. Geological Survey vegetation map [Sheet 90].)

using the U.S. Food and Drug Administration (FDA) action levels or U.S. EPA risk-based methods (Cunningham et al. 1994).

Often transcending U.S. state borders are **ecoregions** ("mapped classification[s] of ecosystem regions of the U.S. ... generally considered to be regions of relative homogeneity in ecological systems or in relationships between organisms and their environment"* [Omernik, 1987]). Inherent in their use is the working principle that contaminant effects vary more in important ways among ecoregions than within ecoregions, e.g., ecoregions with high carbonate soils will be less sensitive to acid precipitation effects than those with mineral chemistries reflecting underlying granite or sandstone mineralogy (Glass et al. 1982). Ecoregions are used to manage aquatic and terrestrial resources of the United States based on land use, land surface features, vegetation, and soil types. In Figure 12.10, the dark lines crossing central Florida and those dipping down along the northern border of the panhandle define the edges of the three Type II ecoregions in Florida. The southeastern Coastal Plain ecoregion is the northernmost: the southern Coastal Plain (approximately the upper half of Florida) and southern Florida Coastal Plain (approximately the lower half of Florida) ecoregions occupy most of the State.

Hughes and Larsen (1988) applied the ecoregion classification to the formulation of a surface water protection strategy for the contiguous United States. Their aim was to develop a more accurate and appropriate framework for water quality criteria based on ecoregions, rather than the entire country. Assuming correctly that water bodies within ecoregions were more similar to each other than to water bodies of other ecoregions, such diverse qualities as the index of biological integrity for fish assemblages, phosphorus concentrations, and dissolved oxygen concentrations were successfully classified for various ecoregions.

12.3 CONTINENTS AND HEMISPHERES

Many problems such as acid rain† fit somewhere between the spatial scales of ecoregions and continents (Figure 12.11). Continental networks of precipitation monitoring have documented the spatial scale of the acid precipitation problem and concordance of precipitation pH with sources of acid-generating contaminants (Barrie and Hales, 1984). Industrialized areas emit sulfur and nitrogen oxides that combine with atmospheric water to form H^+, SO_4^{-2}, NO_3^-. Much of the sulfur dioxide produced by North American sources involves the burning of coal that can contain 1.5% to 5% sulfur, and the roasting of lead, nickel, and zinc sulfides to produce metals. Another source is the burning of oil that can contain 2.5% to 3.5% sulfur (Bridgman, 1994). In contrast to sulfur dioxide sources that tend to be industrial or commercial sources with tall smoke stacks, nitrogen oxides sources are primarily near-ground sources such as automobiles (Bridgman, 1994). This combined with the fact that the acid-producing reactions for sulfur dioxide are slower than those for nitrogen oxides give explanation for the general observation that N-related pH problems tend to be more localized than S-related pH problems.

Equations 12.1 through 12.6 summarize the general reactions leading to precipitation with high H^+ concentrations (Equations 12.2 through 12.3 for sulfur dioxide and Equations 12.4 through 12.6 for nitrogen oxides [Bridgman, 1990; Bunce, 1991]). Some occur in the gaseous phase, whereas others in aqueous phases such as cloud and mist droplets. An oxidation occurs in the first step

* It is apparent from Figure 12.10 that some ecoregions such as the southern Florida Coastal Plain ecoregion are more heterogeneous than others. It contains the Everglades, palmetto prairie, subtropical pine forest, sand pine scrub, cypress savanna, and mangrove swamps.

† In equilibrium with gaseous carbon dioxide (Equation 12.1), liquid water in the atmosphere is predicted to have a pH of 5.7. **Acid precipitation**, including rain, fog (Hileman, 1983), snow, or other forms of precipitation, is defined as that with a pH below 5.7.

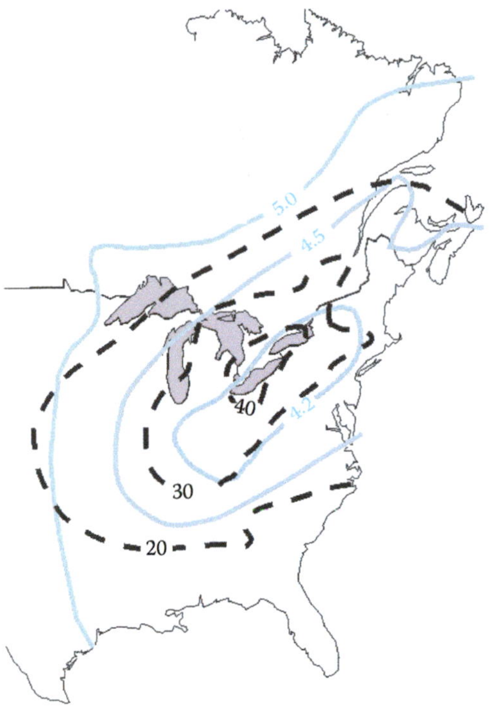

Figure 12.11 The pH (solid blue lines and numbers) and sulfate ion concentration (μm·L^{-1}) (dashed black lines and open numbers) in precipitation measured in 1980. In general, pH and sulfate contours coincide with the spatial pattern of sulfur dioxide emissions reported by a joint U.S./Canadian working group. (Modified from Figures 2 and 3 of Barrie, L.A. and J.M. Hales, *Tellus*, 36B, 333–335, 1984.)

of Equation 12.3 and in Equation 12.4. Although not explicitly indicated as such, the first step in Equation 12.6 is a catalyzed reaction.

$$CO_2 + H_2O \longleftrightarrow H_2CO_3 \longleftrightarrow H^+ + HCO_3^- \quad (12.1)$$

$$SO_2 + H_2O \leftrightarrow SO_2 \cdot H_2O \leftrightarrow H^+ + HSO_3^- \quad (12.2)$$

$$SO_2 \rightarrow SO_3 + H_2O \rightarrow H_2SO_4 \leftrightarrow 2H^+ + SO_4^{-2} \quad (12.3)$$

$$NO \rightarrow NO_2 \quad (12.4)$$

$$2NO_2 + H_2O \rightarrow HNO_2 + HNO_3 \rightarrow 2H^+ + NO_3^- + NO_2 \quad (12.5)$$

$$NO_2 + OH \rightarrow HNO_3 \leftrightarrow H^+ + NO_3^- \quad (12.6)$$

These gases can disperse to hundreds to thousands of kilometers from their sources (Cowling and Linthurst, 1981), causing widespread problems in parts of North America, northern Europe (Likens, 1976), and China (Bridgman, 1994; Zeng et al. 2008).*

The impact of low pH precipitation is not solely a function of proximity to and magnitude of a source (Ravera, 1986). Different regions are inherently more sensitive than others. Soil type and underlying mineralogy influence the capacity to buffer pH changes, and consequently, influences sensitivity to the effects of acid precipitation. An area with an underlying geology of granite, granitic gneisses, or quartz sandstones will have very poor buffering capacity and be sensitive to low pH precipitation. Those areas with sandstone or shale mineralogies are poorly to moderately buffered, and those with limestone or dolomitic geologies will have high buffering capacity and be insensitive to acid precipitation (Glass et al. 1982). Bedrock geology maps can be combined with maps of the distribution of acid precipitation to predict areas of high or low concern. Glass et al. (1982) related bedrock geology, sources of acid precipitation, and stream alkalinity† to define pH-sensitivity classes of surface water bodies for New York State. Schindler (1988) examined the acid-neutralizing capacity of North American lakes and documented a gradual decrease in the northeastern United States (New England and New York), northeastern Canada, and areas of Canada above the Great Lakes.

Effects of acid precipitation to aquatic biota may be sudden or gradual. Releases during seasonal thaws of pollutants accumulated in snowpack may cause high mortality to or diminished spawning success of downstream fish (Cowling and Linthurst, 1981; Bridgman, 1994). Schindler (1988) suggests that, in general, autumn-spawning fishes will be more sensitive than spring-spawning fishes to sudden changes because their pH-sensitive hatchlings tend to be in shallow, nearshore waters when the spring thaw produces pulses of low pH and high aluminum water. Slow deterioration of aquatic systems involves lowered buffering capacity as acid in precipitation "titrates" the entire system downward toward damagingly low pH conditions. Slow deterioration can involve the shift in equilibrium for various biogeochemical processes until a dysfunctional condition emerges. For example, acidic conditions can increase dissolved aluminum flux into overlying waters from solid forms in sediments until toxic concentrations are reached. Low pH conditions can increase leaching of aluminum and other metals from soils and minerals of a watershed, having toxic consequences to aquatic biota. Calcium and magnesium leaching can also increase in a watershed as a consequence of acid precipitation (Smith, 1981; Schindler, 1988). Cronan and Schofield (1979) showed that atmospheric inputs of sulfuric and nitric acid to the pH-sensitive aquatic systems of the Adirondack Mountain region of New York resulted in high dissolved aluminum concentrations in surface waters. The geology of this sensitive area is dominated by granitic gneisses, resulting in poorly buffered waters. They (Cronan and Schofield, 1979) expressed concerns regarding aluminum toxicity there and in other areas of the United States and Europe that have silicate bedrock.

Regardless of the exact mechanism of demise during the decline in freshwater systems, it is clear that aquatic systems distributed over wide swaths of continents are being damaged by acid precipitation. Baker et al. (1991) estimated that the atmospheric input of acid anions represents the dominant anion flux into 75% of 1,180 acid-sensitive lakes and 47% of 4,670 acid-sensitive streams surveyed in the United States. In a survey of 5,000 lakes of southern Norway, 1,750 had lost fish species and another 900 were seriously impacted due to acid precipitation (Bridgman, 1994). In

* Although the discussion here revolves around wet precipitation, fluxes of both dry and wet material can contribute pollutants to sites of effect. **Wet deposition** includes pollutants formed in the liquid media of the precipitation and that incorporated into the precipitation during rain out. **Dry deposition** is the flux of particles and gases such as SO_2, HNO_3 and NH_3 to surfaces (Stumm et al. 1987).

† **Alkalinity** is the capacity of natural water to neutralize acid and is measured by titration of a water sample with a dilute acid to a specific pH endpoint. Most often in freshwaters, it is a function of carbonate (CO_3^{2-}), bicarbonate (HCO_3^-), and hydroxide (OH^-) concentrations, i.e., the carbon dioxide-bicarbonate-carbonate buffering of the water. However, dissolved organic compounds, borates, phosphates, and silicates can also contribute to alkalinity.

southern Ontario, 56% of surveyed lakes had reduced fish populations and an extraordinary 24% had no fish at all due to acid precipitation (Bridgman, 1994).

Other effects of acid precipitation on terrestrial components of the biosphere are also significant and widespread. At low levels, the nitrogen and sulfur added to a forest by acid precipitation can enhance growth, that is, the **fertilization effect** (Bridgman, 1994). However, acid precipitation can also increase nutrient leaching from foliage and forest soils, and accelerate mineral weathering (Smith, 1981). Acid leaching of calcium, magnesium, potassium, and sodium from decomposing forest litter, and calcium, potassium, and magnesium from soils has been demonstrated (Smith, 1981). Leaching of magnesium and potassium from soils may produce a deficiency of these essential plant nutrients. Release of aluminum from solid phases in soils can result in direct toxicity to vegetation (Cowling and Linthurst, 1981; Smith, 1981). Acid precipitation may cause necrotic lesions on foliage, increased plant susceptibility to disease, increased rate of wax erosion from foliage surfaces, and lower nitrogen fixation by legumes via the inhibition of root nodule formation (Cowling and Linthurst, 1981). The combination of all of these effects of acid precipitation on forests is a major explanation forwarded for the widespread forest damage in large tracts of North America (Nihlgård, 1985) and the **waldsterben**, that is, "the widespread and substantial decline in growth and the change in behavior of many softwood and hardwood forest ecosystems in central Europe" (Schütt and Cowling, 1985).

Dubbed by Service (2012) as "the other CO_2 problem," increasing carbon dioxide concentrations in our atmosphere has produced a gradual acidification of the world's oceans. As illustrated in Equation 12.1, an increase in dissolved carbon dioxide in the oceans shifts equilibria toward higher concentrations of H^+. The ocean's acidity is projected to increase by 150% by 2100, causing widespread harm to marine ecosystems. The biogenic deposition of calcareous structures such as coral skeleton and organism shells are predicted to shift such that massive damage could occur (Hoegh-Guldgerg et al. 2007; Pandolfi et al. 2011). Before the Industrial Revolution, 98% of the world's coral reefs grew in waters with chemistries more than 3.5 times above the saturation point of aragonite, the skeletal material of corals; however, projected future exceedance of 450 ppm of atmospheric carbon dioxide concentration will reduce that percentage to only 8% of all existing world's reefs (Cao and Caldeira, 2008). At that point, the world's coral reefs will deteriorate, being unable to sustain themselves.

Atmospheric dispersal of pollutants in aerosols or particulates can also encompass wide areas (Figures 12.7, 12.8, and 12.12). Worldwide increase in soot from burning has contributed to global warming. The global increase in burning also elevates carbon monoxide concentrations and most certainly PAH deposition over very wide areas. Dry deposition also plays an important role in the acid deposition-related problems just discussed. Much of the widespread metal deposition associated with the Copperhill smelting shown in Figure 12.3 was associated with particulate transport. Elevated iron concentrations were measured in tree rings 88 km far from the Copperhill source (Baes and McLaughlin, 1984): particulate-associated iron moved long distances from its source at ore processing. Generally, widespread deposition of particulate-associated cadmium, manganese, lead, and zinc in forests of Tennessee has been documented: one-third or more of the annual flux of cadmium, lead, and zinc to a Tennessee Valley forest was from atmospheric deposition (Lindberg et al. 1982). Hirao and Patterson (1974) found that most of the lead in sedge (*Carex scopulorum*) and voles (*Microtus montanus*) inhabiting a remote High Sierra valley in California came long distances from automotive and industrial sources in Los Angeles and San Francisco. Much of the lead (12 kg vs. 1 kg from other sources) entered the valley associated with snow at that time. And still wider transport of lead has been documented. Analysis of Antarctic ice cores shows an increase in lead flux to the Antarctic after the worldwide onset of industrialization (Boutron and Patterson, 1983). A similar trend has been noted for mercury (Schuster et al. 2002; Slemr et al. 2003). To the earth's other pole, the flux of particulate-associated contaminants is significant from Eurasia (Pacyna, 1995). In air above remote parts of the northern Atlantic Ocean, levels of many metals are elevated

Figure 12.12 Intense air pollution in Beijing, People's Republic of China. The upper photograph of Beijing's Forbidden City was taken at 1:00 PM in the afternoon on June 11, 2010 from Beihai Gongyuan. These ancient buildings sit in a haze generated by automobile exhaust. If the air pollution was not present, the modern Beijing cityscape would be visible, filling the entire background of the photograph. The lower, street-level photograph showing the haze and automobile sources was taken 2 hours later at a busy Beijing intersection. (Photographs by M. C. Newman.)

far above background concentrations (Figure 12.13) (Duce and Zoller, 1975). Particulate-associated radionuclides released into the atmosphere during the Chernobyl reactor meltdown were distributed over most of the Northern Hemisphere and were predicted to result in elevated cancer deaths throughout Europe and portions of the former Soviet Union (Barnaby, 1986; Anspaugh et al. 1988).

Another phenomenon with a scale encompassing an entire continent is ozone (O_3) depletion by chlorofluorocarbons (CFCs) (Figure 12.14). The introduction of CFCs* such as CFC12 (dichlorofluoromethane or CF_2Cl_2) and CFC11 (trichlorofluoromethane or $CFCl_3$) has shifted the balance among reactions taking place in the stratosphere to disfavor the maintenance of normal levels of ozone.

* The CFC structures can be derived easily from their names. Ninety is added to the number in the CFC's name, e.g., CFC11 produces 90+11 or 101. This number codes for one carbon atom (*hundreds* digit), zero hydrogen atoms (*tens* digit), and one fluorine atom (*units* digit). Since the carbon atom forms four covalent bonds and only one is occupied (by a fluorine atom), the other three bonds must be with three chlorine atoms. So, CFC11 is $CFCl_3$.

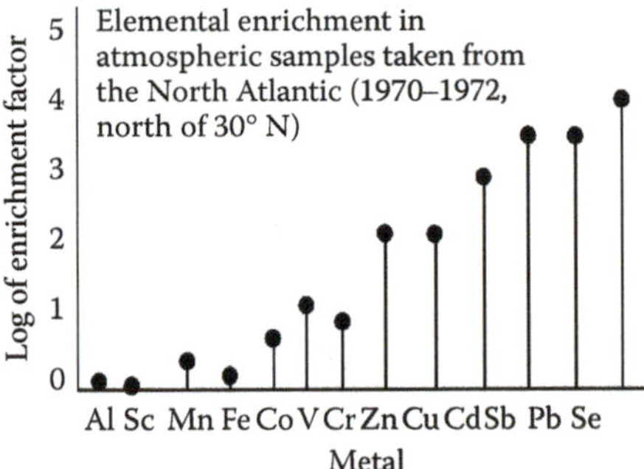

Figure 12.13 Metal particulate dispersal over the remote north Atlantic Ocean as evidenced by increased enrichment factors. The **enrichment factor** (EF_{crust}) for an element is its concentration (X) measured in air samples divided by that expected in the earth's crust: $EF_{crust} = [X/Al]_{air}/[X/Al]_{crust}$. Both air and crustal concentrations are normalized to aluminum concentrations. Aluminum is a ubiquitous element comprising about 8% of crustal material. Increases in EF_{crust} above 10^0 (= 1) imply enrichment from anthropogenic sources. (Modified from Figure 1 of Duce, R.A. and W.H. Zoller, *Science*, 187, 59–61, 1975. With permission.)

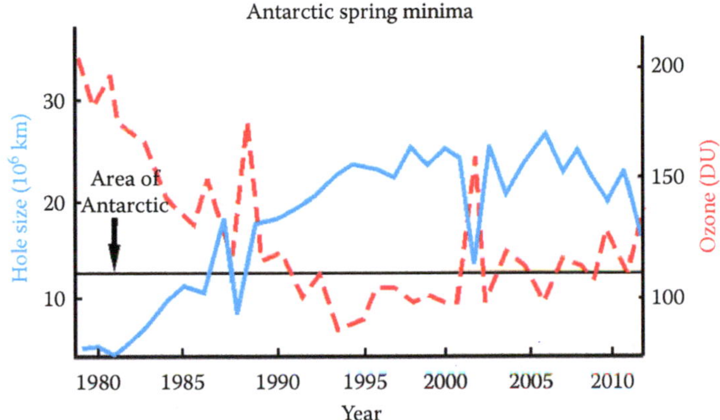

Figure 12.14 The increase in and then leveling off of the size of the Antarctic ozone hole (blue line) and also decrease in spring minima ozone concentration (red line). Ozone concentrations are expressed in **Dobson units (DU)**. One DU is the equivalent of 0.001 mm thickness of pure ozone at 1 atmosphere. To give some scale to this unit, if all of the ozone in the atmosphere were to be brought down to form a pure layer of ozone at sea level, it would be 3 mm thick (Bunce, 1991). (Data from NASA, http://ozonewatch.gsfc.nasa.gov/meteorology/ytd_data.txt.)

Ozone is formed by the **Chapman mechanism** (Equations 12.7 through 12.9) (Bunce, 1991). Energy from sunlight of wavelengths <240 nm and <325 nm is required for Equations 12.7 and 12.8, respectively. A catalyst is required for Equation 12.9:

$$O_2 \rightarrow 2O \tag{12.7}$$

$$O + O_3 \rightarrow 2O_2 \tag{12.8}$$

$$O + O_2 \rightarrow O_3 \rightarrow O_2 + O \tag{12.9}$$

Chloride also becomes involved in the ozone-generating and ozone-depleting reactions in the stratosphere (Equations 12.10 through 12.14) (Zurer, 1988). Equations 12.12 and 12.14 require a catalyst and Equation 12.13 requires light energy.

$$Cl + O_3 \rightarrow ClO + O_2 \tag{12.10}$$

$$ClO + ClO \rightarrow Cl_2O_2 \tag{12.11}$$

$$ClO + O \rightarrow Cl + O_2 \tag{12.12}$$

$$Cl_2O_2 \rightarrow Cl + ClOO \tag{12.13}$$

$$ClOO \rightarrow Cl + O_2 \tag{12.14}$$

Nitrogen species can shift these reactions such that some of the Cl is bound up in nitrogen compounds and unavailable to react with ozone, e.g., $ClO + NO_2 \rightarrow ClONO_2$ (Zurer, 1987; Kerr, 1988a). However, excess Cl from the breakdown of CFCs can overwhelm this sequestering process with a net effect of decreasing ozone concentrations. Molecular chlorine (Cl_2) and hypochlorous acid (HOCl) generated by Reactions 12.15 and 12.16 are readily converted to free radicals that destroy ozone (Zurer, 1987).

$$H_2O + ClONO_2 \rightarrow HNO_3 + HOCl \tag{12.15}$$

$$HCl + ClONO_2 \rightarrow HNO_3 + Cl_2 \tag{12.16}$$

Ozone destruction as a consequence of CFC accumulation in the stratosphere and circulation patterns above the Antarctic produced an alarming **ozone hole** in recent years. Meteorological conditions are less favorable for the formation of a similar hole above the Arctic because the polar vortex does not last as long. Nonetheless, elevated chlorine monoxide (ClO) concentrations and ozone thinning have been reported there too (Zurer, 1989; Kerr, 1992).

The concern for the destruction of vast parts of the ozone layer was heightened by the realization that the expected lifetime of CFCs in the atmosphere is quite long, i.e., 70 years for CFC11 and 110 years for CFC12 (Thompson, 1992). Any remedial action will take considerable time to reverse the damage caused by CFC releases. Fortunately, the **Montreal Protocol**, an international treaty to limit and eventually eliminate the use of CFCs, was endorsed by 70 countries and then signed into law by 1987 (Crawford, 1987; Bunce, 1991). Although it will take a long time to reduce CFCs in the stratosphere, there are already some encouraging indications that chlorine levels might have leveled off in the stratosphere (Kerr, 1996) (Figure 12.9).

Our ability to predict ozone depletion's effects to humans and ecological systems remains inadequate. Ozone absorbs UV light with wavelengths of 290 to 330 nm or roughly the UVB range (280–320 nm) as it enters the earth's atmosphere. This UVB can cause skin cancer and speculations are that the incidence of skin cancers could increase as stratospheric ozone levels drop. However, Bunce (1991) provides a tempering comparison that "for people living in the middle latitudes, the increased risk of skin cancer due to each 1% decrease in ozone levels is equivalent to that posed by moving 20 km closer to the equator." Effects to ecological entities remain equivocal at this time but some scientists have speculated that ozone depletion might be linked to the global decline in amphibian species. Increased UVB penetration into the surface waters of the oceans is also speculated to decrease marine phytoplankton photosynthesis (Baird, 1995). Boreal lake phytoplankton community abundance and composition has been related to UVB radiation (Xenopoulos et al. 2009). A final, large-scale impact of polar ozone depletion was discovered only recently. The ozone hole enhanced tropospheric warming, and in doing so, shifted the westerlies around the Antarctic continent. The changes in these winds shifted the Antarctic Circumpolar Current so as to enhance the draw of deep waters up into the ocean surface waters (Kerr, 2013).

12.4 BIOSPHERE

12.4.1 General

Some human activities can stretch out to involve entire hemispheres or even the planet. As depicted in Figure 12.14, the Antarctic ozone hole extended in the late 1980s beyond the boundaries of that continent. As a recent example requiring a global vantage, Halpern et al. (2008) analyzed the total human impact on the earth's oceans, emphasizing that only a management scheme framed on a large scale can provide adequate solutions (see also Figure 12.15). Two additional phenomena that are global in nature will be discussed in this section, global distillation of persistent organic pollutants and global warming (climate change).

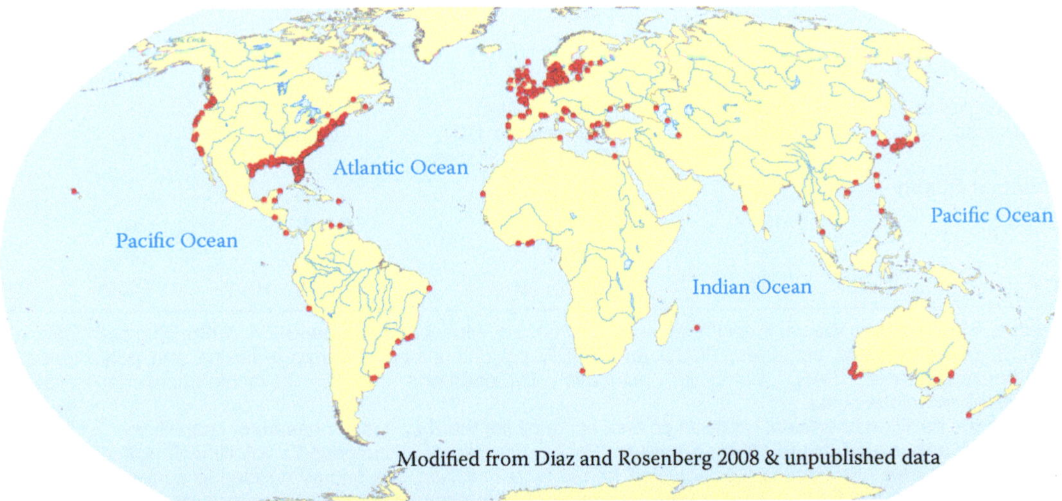

Figure 12.15 The global distribution of cultural eutrophication-associated anoxic dead zones reported in the literature by the end of 2008. This figure is an updated rendering of similar data shown in Figure 1 of Diaz and Rosenberg (2008) that correlated the spatial distribution of dead zones with that of human influence (i.e., the human "footprint") across the globe. (Courtesy of Robert Diaz, Virginia Institute of Marine Science, College of William & Mary, Virginia.)

12.4.2 Global Movement of Persistent Organic Pollutants

As first proposed by Goldberg (1975), many persistent organic pollutants (POPs)* are subject to extensive movement and redistribution on a global scale (Table 12.1). A POP will vaporize and move in the atmosphere until it reaches a characteristic temperature at which it condenses. Depending on several qualities of a POP, it becomes associated with a less mobile solid or liquid phase. The extent to such movement of POPs from their sites of release to cooler latitudes depends on each POP's rate of degradation, vapor pressure, water solubility, and lipophility (Wania and Mackay, 1993; Loganathan and Kannan, 1994; Simonich and Hites, 1995; Kalantzi et al. 2001; Fernández and Grimalt, 2003).

According to the **cold condensation theory**, POPs in the air will condense onto soil, water, and biota at cool temperatures, and consequently, the ratios for POP concentrations in the air and condensed phases decrease from warmer to cooler climates (Wania and Mackay, 1995). This leads to **global distillation** in which POPs migrate from warm regions of release to cold regions of condensation (Figure 12.16). This distillation can involve seasonal cycling of temperatures so that movement toward the higher latitudes occurs in annual pulses or jumps (**grasshopper effect**) (Wania and Mackay, 1996). Because POPs differ in their individual rates of degradation, vapor pressures, water solubilities, and lipid solubilities, a **global fractionation** occurs in which some POPs move more rapidly and further than others toward polar regions. Some, because of their temperatures of

Table 12.1 Persistent Organic Pollutant Mobility in a Global Context

Pollutant Classes (Subclassified by Number of Cl Atoms, Rings, or by Pesticide Type)	Relative Mobility Class: Rapidly Deposited and Retained near Source	Preferential Deposition and Accumulation in Mid-latitudes	Preferential Deposition and Accumulation in Polar Latitudes	Worldwide Dispersion and Deposition
Chlorobenzenes			5 to 6 Cl	0 to 4 Cl
PCBs[a]	8 to 9 Cl	4 to 8 Cl	1 to 4 Cl	0 to 1 Cl
PCDDs[a] and PCDFs[a]	4 to 8 Cl	2 to 4 Cl	0 to 1 Cl	
PAHs[a]	>4 Rings	4 Rings	3 Rings	2 Rings
Organochlorine Pesticides	Mirex	Polychlorinated Camphenes, DDT, DDE, Chlorodanes	HCB[a], HCCHs[a], Dieldrin	
Pollutant Quality				
Log K_{OA}[b]	>10	8 to 10	6 to 8	<6
P_L[c]	<−4	−4 to −2	−2 to 0	>0
T_C[d]	>+30	−10 to +30	−50 to−10	<−50

Source: Modified from Wania, F. and D. Mackay, *Environ. Sci. Technol.*, 30, 390A–396A, 1996. With permission.

[a] PCB, polychlorinated biphenyls; PCDD and PCDF, polychlorinated di-benzo-*p*-dioxins and polychlorinated di-benzo-*p*-furans; PAH, polycyclic aromatic hydrocarbons; HCB, hexachlorobenzene; HCCHs, hexachlorocyclohexanes.

[b] K_{OA} is the partition coefficient between octanol and air. Like the K_{OW}, it is a measure of lipophilicity.

[c] P_L is the **subcooled liquid vapor pressure** (Pa), a measure of a compound's volatility. Specifically, it is the liquid vapor pressure corrected or adjusted for the heat of fusion, the energy needed to convert a mole of a compound from a solid to a liquid phase. Its use allows the expression of *liquid* vapor pressures at a specific temperature for organic compounds with widely varying melting temperatures.

[d] T_C is the **temperature of condensation**, the temperature (°C) at which the compound condenses or partitions from the gaseous to the nongaseous phase.

* As defined in Chapter 2, **Persistent organic pollutants (POPs)** are those organic pollutants that are long-lived in the environment. Many tend to increase in concentration as they move through food webs (Wania and Mackay, 1996). According to Wania and Mackay (1996), they are also called **bioccumulative chemicals of concern (BCCs)** and **persistent toxicants that bioaccumulate (PTBs)**.

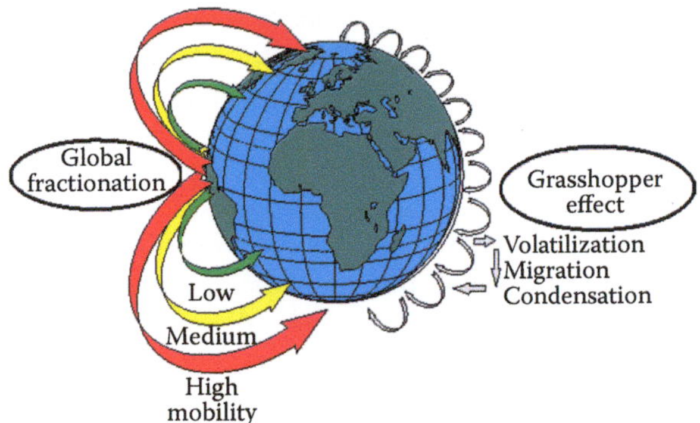

Figure 12.16 The movement of persistent organic pollutants on a global scale. (Modified from Figure 1 of Wania, F. and D. Mackay, *Environ. Sci. Technol.*, 30, 390A–396A, 1996. With permission.)

condensation, will be unable to move beyond a certain point toward cooler latitudes. The net result is a redistribution of different POPs from the equator or site of origin toward the cold poles. The lipophilicity (K_{OW}) of POPs also influences POP movement via the **retention effect**. Those POPs with high lipophility tend to be held more firmly than less lipophilic POPs in solid phases such as soil and vegetation. Consequently, they spend less time in the atmosphere and are less available for transport in that medium (Wania and Mackay, 1995). All else being equal, more lipophilic POPs move slower toward higher latitudes than less lipophilic POPs.

Actual global distributions of POPs conform to the global distillation and fractionation scheme outlined above. In a survey of tree bark from sites around the world (Simonich and Hites, 1995), global distillation was apparent for the volatile hexachlorobenzene but not apparent for the less volatile endosulfan and DDT. Clear movement of volatile organic compounds into Arctic systems has also been documented (Chernyak et al. 1995; Muir et al. 1995). Mackay and Wania (1995) produced a model that accurately predicted POPs global distributions based on these processes.

12.4.3 Global Warming

Global warming is thought to result primarily from the increased atmospheric carbon dioxide (CO_2) concentrations generated by fossil fuel burning and the worldwide destruction of forests (Woodwell, 1978; Khalil and Rasmussen, 1984). Much of the concern centers around atmospheric carbon dioxide, methane, and nitrous oxide that have rapidly increased since preindustrial times (Hileman, 1989). These gases, along with water vapor, are relatively transparent to light entering the earth's atmosphere but absorb long-wave, infrared radiation radiating from the earth's surface toward space. Because the net energy balance of sunlight influx and infrared radiation efflux from the earth's surface determines the steady state temperature of the earth (the **greenhouse effect**), increases in these greenhouse gases* produce an increase in global temperatures. Although there were still some residual disagreements in the early 2000s about whether this is occurring (e.g., Hileman, 1984; White, 1990; Thompson, 1992; Lindzen, 1994; Maswood, 1995), nearly all such rational disagreements have been resolved at this time in the scientific arena: Global warming is occurring and will have widespread effects on human and nonhuman species. Abrupt changes in economic and agricultural activities will very likely occur, sea level will change, and tropical diseases will extend into other regions of the world (Kerr, 1988b). Relative to changes that occur between ice ages, species ranges

* **Greenhouse gases** include water vapor, carbon dioxide, methane, nitrous oxide (N_2O), CFCs, methylchloroform, carbon tetrachloride, and the fire retardant—halon. Ozone in the *troposphere* may also act as a greenhouse gas (Hileman, 1989). All greenhouse gases are not equal relative to infrared radiation absorption. For example, one molecule of CFC11 absorbs 10,000 more infrared radiation as one molecule of carbon dioxide (Anonymous, 1985).

would have to respond faster to global warming. Some will fail and their extinctions will decrease global biodiversity (Roberts, 1988). Plant species that normally migrate during ice age-interglacial cycles would be severely challenged by the extremely rapid rate of temperature change. Our current landscape is highly fragmented because of human activity. Such plant species in fragmented habitats would be "man-locked" (Roberts, 1988), unable to migrate, and at a high risk of extinction.

Before leaving this issue of increased gaseous emissions and their consequences, it is important to note that these elevated levels of atmospheric gases carry other serious consequences. An important example already discussed is **ocean acidification**, the decrease in ocean pH as a consequence of increased atmospheric carbon dioxide levels and general ocean warming (Hoegh-Guldberg et al. 2007). Increases in atmospheric carbon dioxide and ocean temperatures shift equilibria between dissolved carbon dioxide, bicarbonate, carbonate, and solid carbonate phases. This acidification is currently being assessed relative to coral (and other species) calcification in which it might shift equilibria away from a state conducive to hard coral formation.

12.5 SUMMARY

In this chapter, the point was made that a context larger than the ecosystem is required to fully grasp many crucial problems facing us today. Discussion relied heavily on imagery and examples of important issues emerging at increasingly broader spatial scales. Unfortunately, as the scale of problems becomes wider, the potential scope of associated harm increases and our ability to quickly assign a cause–effect relationship decreases. This makes environmental assessment and management of such problems exceedingly difficult and prone to divergent conclusions. However, politicians, environmental policymakers, and environmental decision makers have begun effective action to meet many of these challenges. Sound approaches are now emerging based on watershed (e.g., Diamond and Serveiss, 2001), regional (Smith et al. 2003; Jansen et al. 2005), and transnational (Mora, 1997; Farrell and Keating, 2002) scales. The global loss of biodiversity has been defined (Cardinale et al. 2012) and dangers to humankind and ecosystems are being actively addressed (McCann, 2000; Loreau, 2010). This includes the global loss of terrestrial pollinators and impacts on dependent plants species (Thompson, 2001; Devine and Furlong, 2007; Potts et al. 2010; Dicks, 2013; Tylianakis, 2013), general global downgrading of trophic webs (Estes et al. 2011), global mapping of human impacts on oceans (Halpen et al. 2008), and carbon emissions (Hertwich and Peters, 2009).

SUGGESTED READINGS

Anspaugh, L. R., R. J. Catlin, and M. Goldman, The global impact of the Chernobyl reactor accident. *Science,* 242, 1513–1519, 1988.

Cowling, E. B., Acid precipitation in historical perspective. *Environ. Sci. Technol.,* 16, 110A–123A, 1982.

Diaz, R. J. and R. Rosenberg, Spreading dead zones and consequences for marine ecosystems. *Science,* 321, 926–929, 2008.

Halpern, B. S., S. Walbridge, K. A. Selkoe, C. V. Kappel, F. Micheli, C. D'Agrosa, J. F. Bruno et al. A global map of human impact on marine ecosystems. *Science,* 319, 948–952, 2008.

Houghton, R. A., The global effects of tropical deforestation. *Environ. Sci. Technol.,* 24, 414–422, 1990.

Omernik, J. M., Ecoregions of the conterminous United States. *Ann. Assoc. Am. Geogr.,* 77, 118–125, 1987.

Wania, F. and D. Mackay, Tracking the distribution of persistent organic pollutants. *Environ. Sci. Technol.,* 30, 390A–396A, 1996.

Zeng, N., Y. Ding, J. Pan, H. Wang, and J. Gregg, Climate change—The Chinese challenge. *Science*, 319, 730–731, 2008.

Zurer, P. S., Arctic ozone loss: Fact-finding mission concludes outlook is bleak. *Chem. Eng. News,* 67, 29–31, 1989.

CHAPTER 13

Risk Assessment of Contaminants

13.1 OVERVIEW

13.1.1 Real and Perceived Risk

... many likelihoods informed me of this before, which hung so tottering in the balance that I could neither believe nor misdoubt ...

Shakespeare, *All's Well That Ends Well,* **Act 1: Sc III**

There is a hackneyed expression that a coward dies a thousand deaths. Perhaps, the modern equivalent is that an ill-informed citizen is poisoned by a thousand pollutants. Any dire misperception of a world full of deadly pollutants would be exaggerated for most people and ecological entities. Of course, there is also the counterpoint misconception that what you do not know will not hurt you. This belief is also demonstrably exaggerated for environmental toxicants.

Often perceived risk is not the same as actual or objective risk.* It is the responsibility of the risk assessors to define actual risk to bring perceptions and beliefs into line with the true state of things so that the best possible decision can be made to prevent harm.

13.1.2 Logic of Risk Assessment

Inferences are movements of thought within the sphere of belief.

Josephson and Josephson (1996)

This chapter brings our discussion squarely into the realm of technology as applied to environmental problem solving. Techniques estimating both human and nonhuman risk are described together using the reasonable approach developed for the U.S. Comprehensive Environmental Response, Compensation, and Liability Act (CERCLA) assessment process. However, many germane particulars of the approach have been discussed already. Chief among them are the logic of scientific enquiry (Chapter 1), bioaccumulation (Chapters 3 and 4), trophic transfer of contaminants (Chapter 5), biomarkers (Chapter 6), indicators of effects to individuals (Chapters 6 through 9), the no observed effect level/lowest observed effect level (NOEL/LOEL) approach to sublethal and chronic lethal effects (Chapter 8), models of toxic response including survival time models (Chapter 9), life tables (Chapter 10), Hill's aspects of disease association (Chapter 10), community effects (Chapter 11), and approaches to large-scale processes (Chapter 12).

* Please see Adams (1995) for enriching discussion of risk perception.

The logic of scientific enquiry—indeed, the logic of any effort to enhance belief about physical phenomena—is frequently more complicated than laid out in Chapter 1. For example, the straightforward "reject or accept" context for testing the mettle of a working hypothesis is a logical luxury not always available to the ecotoxicologist (Newman et al. 2007; Newman, 2008). Often, an ecotoxicologist assessing the risk from contamination bolsters belief using less discerning tools such as Hill's aspects of disease association. Information is gathered until a balanced, qualitative judgment can be made based on a **weight of evidence** (preponderance of evidence) approach.* Note that this vague term, weight of evidence, might mean to some that a *reasonable person* reviewing the available information *could* agree that the conclusion was plausible (Apple et al. 1986). A heavy burden is been placed on users of such judgments, whose basis could range from opiniatretry (i.e., mere opinion) (Locke, 1690) to very sound reasoning. At the other extreme, the statistical use of the term weight of evidence implies to Kotz and Johnson (1988), a quantitative or semiquantitative estimate of the degree to which the evidence supports or undermines the conclusion. Of these two definitions, the first definition seems to most accurately describe most environmental risk assessment activities; however, quantitative methods are also available (Newman and Clements, 2008; Newman and Evans, 2002; Newman et al. 2007; Newman, 2013). Evidence describing clear, consistent, and plausible toxic effects would be judged as having considerable weight. Regardless of its shortcomings, the qualitative weight of evidence approach allows movement toward belief using incomplete knowledge. Although this approach is often labeled as less scientific (i.e., lacking logical rigor) than that described in Chapter 1, the distinction is one of degree. Indeed, a formal application of this qualitative approach, the Delphi method, can produce very reliable conclusions from expert opinions (Cooke, 1991).†

The weight of evidence approach, as applied in risk assessment, has major elements of abductive inference and sometimes formal probabilistic induction. (Both abductive inference and probabilistic induction permeate traditional and modern scientific methods too.) **Abductive inference** is simply inference to the best explanation. It uses information gathered about a phenomenon or situation to produce the hypothesis that best explains the available data. Josephson and Josephson (1996) give the following example of abductive inference:

1. D is a collection of data.
2. H explains D or, if true, H would explain D.
3. No other explanation (hypothesis) explains D as well as H.

   ∴ H is *probably* true.

This linkage of the weight of evidence approach in risk assessment to abductive logic is not a trivial point. Abductive inference can be formalized in computer programs (Josephson and Josephson, 1996) that could easily be adopted for risk assessments. **Probabilistic induction** uses probabilities associated with competing theories or explanations to calculate which is most probably true. Credibilities are assigned to competing explanations based on their associated probabilities (Howson and Urbach, 1989). Instead of the quantal (fail-to-reject vs. reject) falsification of a working hypothesis as described in Chapter 1, such probabilistic induction considers a hypothesis lacking

* This common strategy or heuristic for belief adjustment is discussed as the "bucket theory of the mind" in philosophy (Musgrave, 1993) or qualitative global introspection in causal assessment (Arimone et al. 2005; Lane et al. 1987; Newman and Evans, 2002).

† Created by the RAND Corporation to improve accuracy of expert opinion elicitations, the **Delphi method** develops a set of issues, gathers responses from a set of experts who provide independent answers. The compiled answers and statistics are sent back to all of the experts who can then argue their points and/or revise their judgments. The summarized responses are compiled and sent out for additional rounds until a final set of judgments are obtained for use by decision makers (Cooke, 1991).

credibility if it becomes sufficiently improbable.* Like the traditional conclusion to designate an accepted explanation as the best approximation of reality available at the moment, this conclusion may be reconsidered later if conflicting facts emerge. In reality, if not current tradition, these approaches are no less valuable in fostering the growth of knowledge than the classic methods described earlier. Indeed, the enhanced status of a hypothesis surviving repeated and rigorous testing is a straightforward permutation of abductive inference (Newman and Clements, 2008). It would also lead to probabilistic induction if done rigorously and with ample statistical power so as to generate probabilities. However, probabilistic induction can be difficult to implement in some instances because it requires quantified probabilities (Arimone et al. 2005). Regardless, both qualitative abductive inference and quantitative probabilistic induction can contribute to the weight of evidence approach as applied to risk assessments.

VIGNETTE 13.1 Why Risk Assessment?

Glenn W. Suter, II
U.S. Environmental Protection Agency

Risk assessment is a technical support for decision making under uncertainty. It is based on the realization, dating to the seventeenth century, that is, although the future is unpredictable, one can estimate the likelihood of alternative outcomes of an action. Risk assessment is used when a decision must be made that has uncertain outcomes due to varying conditions or uncertainty concerning the nature of the situation. It implies alternatives with qualitatively or quantitatively different possible outcomes. Hence, risk assessment estimates the absolute or relative probabilities of prescribed negative outcomes of alternative choices. It has been plausibly argued that basing decisions on estimates of risk rather than auguries, prayers, astrology, or intuition is the defining feature of modern culture (Bernstein, 1996). Risk assessment is applied to many activities including insurance, engineering, forest fire management, investment, and as discussed in this book, the management of chemicals in the environment.

The principal alternative to risk assessment is rule-based decision making. For example, offshore disposal of sewage sludge was banned in the United States, not because of risks it posed, but because of popular concern. No analysis indicated that deep offshore disposal of sludge has significant ecological or health risks, or that the risks from land disposal or incineration of sludge are less. The precautionary principle† is often cited as an alternative to risk assessment. However, this principle simply requires that an action be demonstrated to be safe with reasonable confidence before an action is approved. It is a management principle that is applied after a technical analysis, not an alternative to analysis.

Ecological risk assessment is concerned with assessing risks to nonhuman organisms, populations, or ecosystems. It consists of a problem formulation, an analysis of exposure and of the relationship of exposure to response, and a characterization of risks. It is connected to the risk management process at the beginning when the problem to be assessed and the goals of the assessment are defined, and at the end when the results are communicated. Individuals and organizations, termed stakeholders, which have an interest in the risk management decision, may also be involved in helping to define the problem and goals and in communicating or evaluating the

* Obviously, probabilistic induction is one important logical extension of conventional statistical analyses used in many risk assessments.

† The **precautionary principle** is based on the conservative policy that, even in the absence of any clear evidence and in the presence of high scientific uncertainty, action should be taken if there is any reason to think that harm might be caused.

results. The nature of interactions among risk assessors, risk managers, and stakeholders depends on the cultural and legal context.

The case of Bt corn and monarch butterflies may serve to illustrate the risk assessment process. Genetically modified corn, containing the *Bacillus thuringiensis* (Bt) delta-endotoxin, was developed to combat corn borers and corn earworms. Because the corn pollen in some transformation events contained expressed endotoxin, it constituted a potential hazard to lepidopteran larvae in or adjacent to a corn field. This issue was raised by a study that demonstrated that experimental exposures to Bt corn pollen increased mortality and arrested the development of monarch butterfly (*Danaus plexippus*) larvae. The extreme precautionary approach, advocated by some environmental groups, would ban the genetically modified corn based on this hazard. However, a risk-based regulatory approach would determine whether there was a significant probability of adverse effects, and a comparative risk approach would determine whether that risk was greater than the risks from alternative methods of controlling corn borers and earworms. The following summary is based on the U.S. Environmental Protection Agency's (EPA's) assessment of plant-incorporated Bt endotoxin that also considered many risks other than those to monarch butterflies (EPA, 2001).

The goal of the assessment is to determine what restrictions on the use of Bt corn, if any, are required to protect monarch butterflies. The problem formulation could identify the Bt pollen as the stressor of concern, the U.S. Corn Belt as the environment of concern, and the abundance of monarch butterflies as the assessment endpoint.

The analysis of exposure determined the distribution and abundance of the butterflies and the distribution of the larval food (milkweed—*Asclepias* spp.) with respect to cornfields. It then estimated the density of corn pollen or endotoxin protein on milkweed leaves in cornfields and outside fields as a function of distance from cornfields and the rate of degradation of the endotoxin in pollen. The exposure estimate could be refined by determining the frequency of oviposition in cornfields relative to other habitats and the larval feeding pattern in terms of leaf height, leaf age, leaf surface, and so on, relative to the distribution of pollen on milkweed plants. Monarch butterflies were found to oviposit on milkweed in cornfields and there was 0%–75% overlap of their presence with the period of pollen shedding.

The analysis of effects estimated the response of larvae to pollen consumption. Available studies examined effects of pollen from various modified corn varieties in the laboratory and field and of the endotoxin proteins. Responses considered included death, growth, maturation, emergence, and wing length.

The risk characterization concluded that significant effects are unlikely. Mean observed pollen densities were 170 cm^{-2} in corn fields and 63 cm^{-2} at the field edge. The maximum observed density was 900 cm^{-2} and the estimated worst case was 1400 cm^{-2}. Experimental concentrations, at least as high as 4000 cm^{-2}, caused no observed significant effects. Conservatively, 1/100,000 monarch butterflies were estimated to be exposed to sublethal levels of the pollen. This estimate is based on assumptions that 50% of monarchs use the Corn Belt, 18% of that habitat is cornfields, 25% of that is Bt corn, there is a 50% overlap of pollen shedding and larval occurrence, and 0.1% of those larvae experience sublethal toxic effects. A field study found that Bt pollen had no detectable effects on monarch larvae while larvae exposed to conventional pesticide sprays or drift were killed (Stanley-Horn et al. 2001). Hence, the risk of Bt pollen to monarch butterflies is estimated to be low in absolute terms. A subsequent assessment, based on test data from longer exposures, estimated that, at the current 37% adoption rate, monarch larvae experience a 0.3% increase in mortality (Vaituzis, 2004). Although a formal comparative assessment was not performed, the risk appears to be much lower than that associated with the most likely alternative. However, the U.S. EPA will continue to monitor the results of any additional studies addressing this issue.

This case demonstrates that risk is a function of the magnitude and likelihood of exposure, and the quantitative relationship of effects to exposure, and that the existence of a hazard does not provide an adequate basis for decision making. The analysis is relatively simple because the data are relatively abundant, so little modeling is required, and because the results are far from suggesting a potentially significant risk. If the assessment had suggested that biologically significant larval mortality could occur in corn fields, a more complex assessment might be required. Such an assessment might model the demographics of monarch butterflies and the distribution of exposure levels given weather, the behavior of the butterflies and the abundance of milkweed in corn fields and elsewhere. It might also formally assess risks from conventional pesticides and from Bt sprays for comparison to risks from Bt corn pollen. In any case, it should provide a scientific basis for environmental management decisions in an uncertain world.

13.1.3 Expressions of Risk

In this chapter, the preoccupation will be on the expression of **risk** as a probability of some adverse consequence occurring to an exposed human or ecological entity. For example, one might estimate a less than 1 in 100,000 chance of dying due to exposure to a particular toxicant. Often, risk is defined more precisely as "the product of the probability and frequency of effect (e.g., probability of an accident × the number of expected mortalities)" (Suter, 1993). The concept, as applied to environmental risk assessment includes the probability of an event occurring that *could* lead to an adverse effect and the probability of an adverse effect given that the event *did occur*. This will be the context for the term in most of this chapter. However, other expressions of risk have already been discussed that are equally valuable. In an epidemiological context (Chapter 10), risk was expressed in units such as a relative risk ratio. In discussions of time-to-event models (Chapter 9), risks were expressed as relative risks and modeled as proportional hazards. Although not used as often as probabilities in environmental risk assessments, relative risks are often used to explain risk to the general public. For example, people living in U.S. states with very low selenium levels in soils and waters have a higher risk (expressed as a percentage above the national average) of dying from heart disease than those living in States with normal levels of selenium (Anonymous, 1976). Survival functions (Chapter 9) and tabulations of age-specific life expectancies (Chapter 10) are also useful expressions of fatal risk that can easily be incorporated into predictive models.

In communicating with the public, the conventional expression of risk as a probability (e.g., 10^{-5} or 1 chance in 100,000 of the adverse outcome occurring) can be less intuitive than its expression as a change in life expectancy. Consequently, expression of risk in terms of life expectancy is worth exploring for a moment before proceeding to the more conventional context of risk. Also, this expression of risk is very closely tied to and extends the computations discussed in Chapters 9 and 10. The **loss of life expectancy** (LLE) is estimated as the simple difference between the life expectancy with (E_x) or without (E) the risk factor being present (Cohen and Lee, 1979).

$$\Delta E = E_x - E \tag{13.1}$$

Loss of life expectancy is expressed in days, months, or years depending on the magnitude of the risk factor's effect. For example, the LLE for the average American (age 0–55 years) due to cancer is 0.34 (males) and 0.32 (females) years (Cohen and Lee, 1979). A uranium miner (1970–1972 statistics) has a mortality rate of 232×10^{-5} year^{-1}, or a LLE of 1160 days relative to that of the average person. Cohen (1981) gives an example of a gain in life expectancy (negative LLE) that is a particularly fascinating statistic to the author. There is a gain of 500 days if one's occupation were that of a university teacher. There is an intuitive statistic with which I can live!

13.1.4 Risk Assessment

Assessment of contaminant-associated risk is mandated in key U.S. federal laws, including RCRA and CERCLA as amended by SARA (see Appendix 3). It is also implied by the use of the term "unreasonable risks" in FIFRA and TSCA (Suter, 1993). For this reason, the EPA and other regulatory agencies have developed numerous documents and regulations ensuring that the intent of such laws is met relative to protecting human health and the environment. For example, assessment of risk to humans is detailed in guidelines for Superfund sites (EPA, 1989d, 1996; http://www.epa.gov/raf/frameworkhhre.htm) and to ecological entities in several EPA guidance documents (EPA, 1989c, 1998). The remainder of this chapter is a condensed version of these guideline documents.

As defined in such regulations, **risk assessment** is the process by which one estimates the probability of some adverse effect(s) of an existing or planned exposure to either human or ecological entities. Two general categories of risk assessments exist: (1) retroactive and (2) predictive. A **retroactive risk assessment** deals with an existing condition such as an existing contaminated seepage basin and the **predictive risk assessment** deals with a planned or proposed condition such as a planned discharge of a waste.* In some situations, the predictive risk assessment might also deal with a future consequence of an existing situation. For example, a predictive assessment might estimate the consequences of a contaminated plume of ground water that will outcrop soon to a stream or reach nearby drinking water wells. In some situations, a risk assessment might be framed as a **comparative risk assessment**. For example, the risks of two or more alternate actions or processes might be assessed and compared before deciding which to implement.

A risk assessment is carried out by a **risk assessor**, a person or group of people "who actually organizes and analyses site data, develops exposure and risk calculations, and prepares a risk assessment report" (EPA, 1989d). The assessment is then provided to a **risk manager**, "the individual or group who serves as primary decision-maker for a site" (EPA, 1989d). This distinction between the roles of the risk assessor and manager is important: The assessor does not make any decisions regarding the action to be taken as a consequence of the assessment although potential remedial actions may be detailed in the assessor's report. A decision is the responsibility of the risk manager who must also weigh all costs, benefits, and risks. As a human risk example, an extremely low and debatable risk of cancer associated with consuming small amounts of residual pesticide in food may or may not be deemed acceptably low relative to the widespread economic and nutritional risk associated with abandoning the pesticide's use in agriculture. A common ecological example involves the certain risk of damage through the destruction of habitat during removal of contaminated soil from an area that is predicted to have a very small, but measurable, toxic risk to endemic species. Is it wise, and in the spirit of federal law, to destroy an invaluable and fragile habitat to remove contaminated soil that presents a finite, but very low, risk?

Permutations of the **National Academy of Science (NAS) paradigm** (National Research Council, 1983) (Figure 13.1) are used for both human and ecological risk assessments. As already mentioned in Vignette 13.1, there are four components to this paradigm: (1) hazard identification, (2) exposure assessment, (3) dose–response assessment, and (4) risk characterization. In the first, relevant data on

* Formally, a risk assessment is different from a hazard assessment. A **hazard assessment** compares the expected environmental concentration (EEC) to some estimated threshold effect (ETT) with the intent of deciding if (1) a situation is safe, (2) a situation is not safe, or (3) there is not enough information to decide. Often, a **hazard quotient** (HQ = EEC/ETT) is used as a crude indicator of hazard. (This use of the EEC divided by some endpoint value for adverse effect is called the **quotient method**.) A risk assessment is like a hazard assessment except that it has as its goal the generation of a quantitative estimate (probability) of some adverse effect occurring.

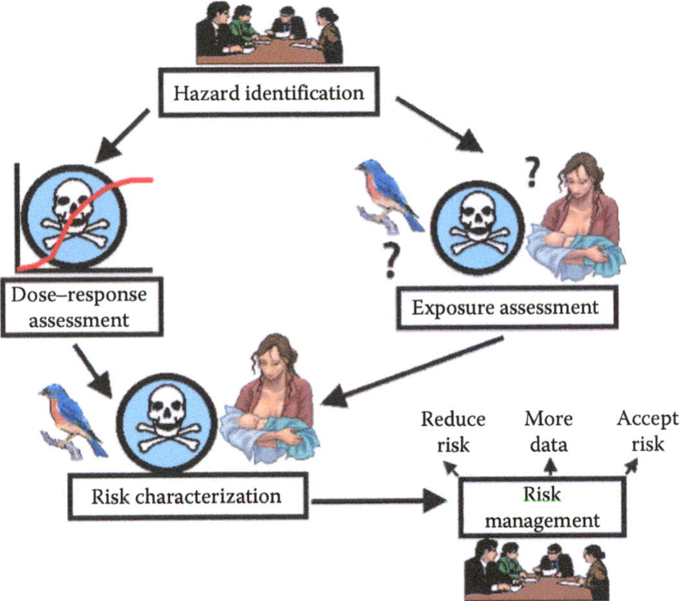

Figure 13.1 The National Academy of Science paradigm for risk assessment. The risk assessor identifies the hazard, assesses the potential for exposure (or current exposure) and the dose–response relationship for the relevant entities being protected and the relevant toxicants. This information is integrated to characterize the risk. At that point, the assessment is provided to the risk manager who then decides to accept or reduce the associated risk or, if insufficient information is available to make a good decision, to seek out more information.

the situation are gathered and chemicals of potential concern highlighted. **Exposure** (contact with the contaminant) assessment estimates the magnitude of releases, identifies possible pathways of exposure, and estimates potential exposure. Dose–response assessment gathers together relevant toxicological data relating exposure to relevant effects. Risk characterization then integrates this information to assess the potential or existing risk of an adverse effect. Also included in this fourth component is a statement of the uncertainty involved in the risk estimate and the general quality of data used in the assessment. These components will be described to particulars for human and ecological assessments.

13.2 HUMAN RISK ASSESSMENT

13.2.1 General

> *... scientific activity is not the indiscriminate amassing of truths; science is selective and seeks the truths that count for most, either in point of intrinsic interest or as instruments for coping with the world.*
>
> **Quine (1982)**

Obviously, human risk assessment has as its goal the estimation of the probability or likelihood of an adverse effect to humans occurring as a result of a defined exposure. For U.S. Superfund sites, it is applied as one part of a general **remedial investigation and feasibility study (RI/FS)*** that has the larger goal of implementing "remedies that reduce, control, or eliminate risks to human health

* The **remedial investigation (RI)** has three parts: characterization of the type and degree of the contamination, human risk assessment, and ecological risk assessment (EPA, 1994). The **feasibility study (FS)** explores the various options for remediation. The chapter will focus on the last two parts of the RI.

and the environment" or, more specifically, the accumulation of "information sufficient to support an informed risk management decision regarding which remedy appears to be most appropriate for a given site" (EPA, 1989d). The human risk assessment part of the RI/FS has three components: (1) a baseline risk assessment, (2) the refinement of the initial remediation goals, and (3) the evaluation of the risk associated with candidate remedial activities. The first stage involves the application of the NAS risk assessment paradigm to the specific site with the intention of determining if the contaminants at the waste site present, or could present in the future, a risk if left alone, i.e., the **"no action" alternative** to remediation of the site. In the assessment, consideration should include possible exposure from one or several routes, and effect to one or several subpopulations of humans. Sections 13.2.2 through 13.2.5) give as an illustration the details for the baseline human risk assessment based on the NAS paradigm.

13.2.2 Hazard Identification (Data Collection and Data Evaluation)

From among the many chemicals present at the site, a tentative list of those of most concern (**chemicals of potential concern**) is made in this phase of the risk assessment. The concentrations of these chemicals are compared with background concentrations. Concentrations are measured in all relevant media, such as, air, surface water, ground water, soils, sediments, game species, and edible fish. In contrast to ecological risk assessments, contaminant concentrations in biological media most relevant in human risk assessment are edible parts, that is, the muscle fillet from a fish rather than the entire fish. Concentrations in all relevant media from the waste site are compared to background levels.

The various contaminant sources are described. Qualities of the environmental setting are also compiled, especially those qualities "that may affect the fate, transport, and persistence of the contaminants" (EPA, 1989d). (As part of the entire RI/FS, a preliminary series of tests might also be done to assess various remediation alternatives at this stage.) The quality of the data needed for the assessment should also be defined, e.g., types of effects data or required detection limits for the chemicals of potential concern. Also, preliminary exposure scenarios are formulated at this early stage. Usually, risk assessors establish a formal scoping process to identify what data must be collected at this point.

After the data have been collected, they are used to do the following: (1) determine the analytical data quality and adequacy of sampling, (2) compare site concentrations to background concentrations, (3) identify chemicals of potential concern from the initial list of compounds, and (4) determine if the compiled data set is adequate to proceed. A rough screening is done to highlight contaminants for further consideration. According to the EPA (1989d), a chemical concentration may be multiplied by a toxicity value to generate an associated **risk factor**.

$$R_{ij} = C_{ij} T_{ij} \tag{13.2}$$

where R_{ij} = the risk factor for the chemical i in association with media j, C_{ij} = the concentration of i in media j, and T_{ij} = a toxicity value for i in j. It is further suggested that the total risk score for a mixture of contaminants in a medium can be estimated as the sum of the individual R_{ij} values: $R_j = R_{1j} + R_{2j} + R_{3j} + \ldots R_{ij}$. The magnitude of the risk factor dictates whether the chemical is retained for further consideration. Obviously, the main virtue of this approach is expediency, not accurate inclusion of all toxicological factors involved in the expression of an adverse effect. It is felt that this crude estimate of risk is adequate, assuming effect additivity and a Selyean stress context for mixture effects.

Determining a **toxicity value** can involve one of the two methods. One method uses the slope of a published effect–dose relationship: R_{ij} = Slope × C_{ij}. The other uses a RfD value in a similar manner. Details of these approaches will be described in Section 13.2.4

Note that, although the methods below will focus on long exposure periods, short-duration exposures such as those associated with the consumption of tainted drinking water can be assessed using 1-day, 10-day, or other short-duration health advisories. The U.S. EPA Office of Drinking Water has developed such health advisories based on the ingestion of individual chemicals. These **health advisory concentrations** identify concentrations below which no impact to human health is anticipated for the specified exposure duration. They are usually derived by a process like that described below for RfD values. Often, they are applied by public officials in dealing with spills, short-term exposures, or similar situations although longer term advisory concentrations are also available.

13.2.3 Exposure Assessment

The goal of human exposure assessment is to determine or estimate the route, magnitude, frequency, and duration of exposure (EPA, 1989d). In such an assessment, exposures are estimated for specific subpopulations (e.g., hypersensitive individuals, elders, children, or remediation workers) by specific routes (e.g., dust inhalation during remediation, drinking water, or fish consumption). **Exposure pathways** (the avenues by which an individual is exposed to a contaminant including the source and route to contact) can include inhalation, ingestion from various media, and dermal contact and absorption.

The **reasonable maximum exposure (RME)** is calculated for chemicals of potential concern. This conservative estimate of exposure is computed differently depending on the route of exposure. Exposure concentrations are often taken from among the highest measured, e.g., the concentration at the 95% upper confidence limit instead of the mean concentration. The intake is then estimated using an equation such as the general Equation 13.3 (EPA, 1989d):

$$I = C\frac{(CR)(\text{EFD})}{\text{BW}}\frac{1}{\text{AT}} \tag{13.3}$$

where I = the intake (e.g., [mg of contaminant][kg of body mass]$^{-1}$[day]$^{-1}$), C = contaminant concentration (e.g., mg·L^{-1}), CR = contact rate (e.g., L·day^{-1}), EFD = an estimate of frequency and duration of exposure composed of exposure frequency (EF) (e.g., days year^{-1}) and exposure duration (ED) (e.g., years), BW = body weight or mass (kg), and AT = the time over which exposure is averaged. The AT may be estimated as ED × 365 days year^{-1} for a noncarcinogen or 70 years × 365 days year^{-1} for a carcinogen. The exact formulation of this equation changes depending on the exposure route, but the general approach remains the same, allowing estimation of the appropriate RMEs. Appendix 10 is a listing of those formulae as supplied in EPA guidelines (EPA, 1989d). EPA documents and other sources are drawn on for specific variable values used in such calculations.

13.2.4 Dose–Response Assessment

The goal of the dose–response assessment is to gather all information useful in establishing a relationship between the extent of contamination and the likelihood or magnitude of an adverse effect. A wide range of information is drawn upon for human risk assessments including human epidemiological data, data derived from study of nonhuman animals, and predictive models such as toxicokinetics models or QSARs. General mechanistic information is also sought in making judgments. The most valuable data are those from long-term studies involving humans, e.g., Japanese atomic bomb studies for radiation effects. But such studies are often less structured than nonhuman animal studies. There is an obvious reason for this. Most human information comes from effects observed after accidents, often occupational accidents with very high exposures. From these unstructured experiences, it is difficult to isolate dose effects from uncontrolled covariates (e.g., age, sex, or other risk factors) and to extrapolate downward to the lower, chronic exposure scenarios normally associated with environmental contaminants. The use of nonhuman animal studies carries the uncertainty of extrapolation to humans. Most animal to human extrapolations involve allometric modification of effects, e.g., adjusting effect by weight$^{2/3}$ (or weight$^{3/4}$) if milligram-per-day intakes are used. If intake is expressed as mg·kg^{-1} day^{-1}, the allometric adjustment for the difference in weights between the study animal and the humans might be weight$^{1-2/3}$ or weight$^{1/3}$ (see Chapter 9 for further explanation). Sometimes, the allometric physiologically based pharmacokinetic (PBPK) models discussed in previous chapters are applied to reduce inaccuracies in extrapolation. In general, results with high consistency among animal species provide increased confidence in extrapolating to humans (EPA, 1989d). As you will see, various uncertainty factors (UFs) are used to compensate for the uncertainty associated with such problems. For carcinogens, a weight of evidence classification has also been established to aid the risk manager in understanding the degree of uncertainty involved in each risk calculation.

For noncarcinogenic effects, the **reference dose (RfD)** is most often applied to risk estimation. It is the best available estimate of the daily exposure for humans, including the most sensitive subpopulation, that will result in no significant risk of an adverse health effect if not exceeded. It is assumed to be accurate to only within an order of magnitude, i.e., 10-fold of the true value. There are different types of RfDs. In most assessments, a **chronic RfD** is assumed unless indicated otherwise. It is the RfD associated with exposures spanning an individual's lifetime. There are also **subchronic RfDs (RfD_s)** that are derived from short-term exposure data and **developmental RfDs (RfD_{dt})** that focus on developmental consequences of a single, maternal exposure during development. Which RfD is most appropriate depends on the exposure scenario of concern. Calculations can be done for several scenarios if warranted. For example, a chronic scenario might assess the risk of leaving the site as it exists, and an acute scenario might assess risk to workers during remediation.

The RfDs incorporate toxicity data (i.e., no observed adverse effect level [NOAEL]) and qualitative UFs associated with these data. First, all data sets for the toxicant in the appropriate media are compiled. In the absence of data for the appropriate media or exposure route, data from other forms or routes might be used after some adjustment. Unless there is sound evidence indicating otherwise, the conservative assumption is made that humans are as sensitive as the most sensitive species tested. Under this assumption, the human or nonhuman animal study with the lowest observed adverse effect level (LOAEL) is designated the **critical study**. The associated effect is called the **critical toxic effect**. Notice that the LOAEL comes from a suite of LOAELs derived by methods discussed in Chapter 8 and that there are compromises associated with the application of LO(A)ELs and NO(A)ELs. Once the LOAEL is found, the NOAEL or highest toxicant level tested that had no adverse effect is identified for that critical study/effect. This NOAEL is the measure of toxicity used to derive the RfD. The LOAEL can be used after adjustment if an NOAEL is not available.

UFs and a modifying factor (MF) are now applied to the NOAEL to compensate in a conservative direction for uncertainty or unaccounted factors. **Uncertainty factors** are often, but not always, factors of 10 that lower the NOAEL to compensate for various sources of uncertainty including variation in sensitivity within the human population (UF_H), extrapolation from other species to humans (UF_A), use of subchronic rather than chronic or lifetime exposure data (UF_S), and use of a LOAEL instead of the NOAEL (UF_L). UFs can take other values, e.g., EPA's data base (IRIS) lists those for hexavalent chromium as $UF_H = 10$, $UF_A = 10$ (data from rats), and $UF_S = 5$ (data involved a chronic, but less than lifetime, exposure). Expert opinion may be applied to modifying a NOAEL even further. Expert opinion is incorporated through a **modifying factor** that ranges from 1 to 10. The RfD is then generated using the NOAEL (or LOAEL), UFs, and MF. Obviously, the default values for the UFs and MF are 1 if the associated uncertainties or modifying circumstances are insignificant. Commonly, the UF_H is set at 10 to ensure that the most sensitive humans are protected. The UF_L is 10 if LOAEL is used instead of NOAEL in Equation 13.4; otherwise, it is 1:

$$\text{RfD} = \frac{\text{NOAEL}}{(\text{UF}_\text{H})(\text{UF}_\text{A})(\text{UF}_\text{S})(\text{UF}_\text{L})(\text{MF})}. \tag{13.4}$$

Clearly, the approach just described for noncarcinogenic effects is based on a threshold model or notion of toxicant effect. Below a certain level, the human individual is protected from any adverse effect. That protective level is assumed to be above the RfD but below the LOAEL.

In contrast, a nonthreshold model is assumed in dealing with the risk of carcinogenic effects. A **slope factor (SF)** (risk or probability of occurrence per unit of dose or intake) for the risk–dose model is then used to estimate the probability of a cancer under a particular exposure scenario. Although based primarily on intake, models may be expressed in terms of concentration, dose, or intake. The SF is defined by EPA as "a plausible upper-bound estimate of the probability of a response per unit intake of a chemical over a lifetime" (EPA, 1989d). Usually, the upper 95% confidence limit for the estimated slope of the risk–dose or risk–intake curve is used as the SF. The exposure dose or intake is multiplied by the SF to estimate risk:

$$\text{Risk} = (\text{CDI})(\text{SF}) \tag{13.5}$$

where risk = the probability of developing cancer and CDI = the chronic daily intake averaged over a lifetime (70 years) (mg kg^{-1} day^{-1}). Estimation of daily intakes is illustrated generically in Equation 13.3 and specifically for various sources in Appendix 10. All risk estimations are accompanied by a qualitative **EPA weight of evidence classification** because the strengths of evidence for specific chemicals being human carcinogens vary widely. This classification informs the risk manager about the strength of the evidence supporting the risk calculation.

The EPA has compiled RfDs, SFs, drinking water health advisories (1-day, 10-day, longer term, and life time advisories), and important associated information into a large data base, **Integrated Risk Information System (IRIS)**. At this time, IRIS is accessible through the U.S. Government's Web site (http://cfpub.epa.gov/ncea/iris/index.cfm). As an example of how easily these data may be found, the author spent less than 10 minutes logging onto IRIS to retrieve the following information for cadmium. The cadmium RfD is 0.0005 mg·kg^{-1} day^{-1} based on proteinuria* (the critical effect) in humans after imbibing cadmium in drinking water. The NOAELs were 0.005 mg·kg^{-1} day^{-1} for water and 0.01 mg·kg^{-1} day^{-1} for food. The UF_H is 10 and MF is 1. Confidence in these data was judged to be high because of the extensive human and animal data sets, and sound PBPK models available to the assessor.†

13.2.5 Risk Characterization

The data collected in the previous steps are now combined in this last step to generate a statement of risk. To this end, the exposure information including intake rates and dose–response data are used in the calculations below. The final statement of risk may be qualitative or quantitative. Regardless, it must include an explanation, specific details, and any important qualifiers.

A hazard quotient (Equation 13.6) is estimated for each noncarcinogenic effect using the exposure level or intake (E) and associated RfD. If the quotient does not exceed 1, the human population is assumed to be safe. Hazard quotients for chronic, subchronic and shorter term exposure may need to be estimated also depending on the exposure scenario of concern:

* **Proteinuria** is the presence of protein in the urine. The implication is kidney damage caused by cadmium that has accumulated in the renal cortex.

† Here, again, is another opportunity for confusion if the distinct goals of scientific, technical, and practical ecotoxicologists are not kept in mind and respected. Data from very sound scientific studies may be judged of "low" confidence relative to the very specific application in estimating a RfD. On the other hand, an RfD is meaningless in a scientific sense. What might be good for one purpose might be inadequate for another.

$$\text{Hazard quotient} = \frac{E}{\text{RfD}}. \tag{13.6}$$

However, the quotient estimated above is not useful if there is more than one chemical of potential concern. A hazard quotient can be calculated for situations involving several (x) chemicals of potential concern:

$$\text{Hazard index}_{\text{Total}} = \sum_{i=1}^{x} \frac{E_i}{\text{RfD}_i} \tag{13.7}$$

where E_i = exposure levels (or intakes) and RfD_i = reference doses for the i chemicals expressed in similar units and covering the same exposure durations as E_i. With temperance, summations could be done for contaminants in different media. Again, a quotient less than 1 suggests protection of the human population. For chronic exposures, a chronic hazard index can be estimated by Equation 13.7 if chronic RfD_is are used and CDI_i is used as E_i. Similarly, a subchronic hazard index is generated with a subchronic EfD_i and subchronic daily intake rates (SDI_is).

Risks of carcinogenic effect are estimated with Equation 13.5 assuming a multistage model of carcinogenicity. This linearized, multistage model is appropriate only at low doses. Consequently, it should be replaced by the first-order or **one-hit risk model** if estimated risk is 0.01 or higher:

$$\text{Risk} = 1 - e^{-(\text{CDI})(\text{SF})} \tag{13.8}$$

Of course, other plausible models such as a gamma multiple hit or Weibull models exist as discussed regarding the dynamics of carcinogenesis in Chapters 7 and 9, and expressed mathematically in Chapter 9.

If several carcinogens are considered together, a total cancer risk is approximated as the simple sum of the individual risks assuming independence of effects. This might not be valid as suggested from our previous discussions of carcinogenesis. Again, risk can also be summed across different media with an understanding of the limits of such a calculation.

13.2.6 Summary

I prefer the errors of enthusiasm to the indifference of wisdom.

Anatole France (Quoted in Casti [1989])

The techniques described earlier for assessing risk from noncarcinogenic and carcinogenic contaminants (Figure 13.2) can be criticized easily for inconsistency with scientific knowledge. However, as detailed in Chapter 1, the goal of this crucial process is not scientific and should not be judged from that context alone. The goal is to protect human health. Proponents are acutely aware of the many approximations and compromises in the approach. They conditionally accept those errors to fulfill their immediate goal. Indeed, they incorporate UFs and conservative SFs so as to ensure that the results will be biased toward caution and away from any possible harm to humans. These modifications are not made to more accurately define the threshold levels for the contaminant effects. Unfortunately, this dictates that more time and money than necessary be spent in remediation. It also assures that more valuable habitat than necessary will be damaged or destroyed during unnecessary or excessive remediation activities.

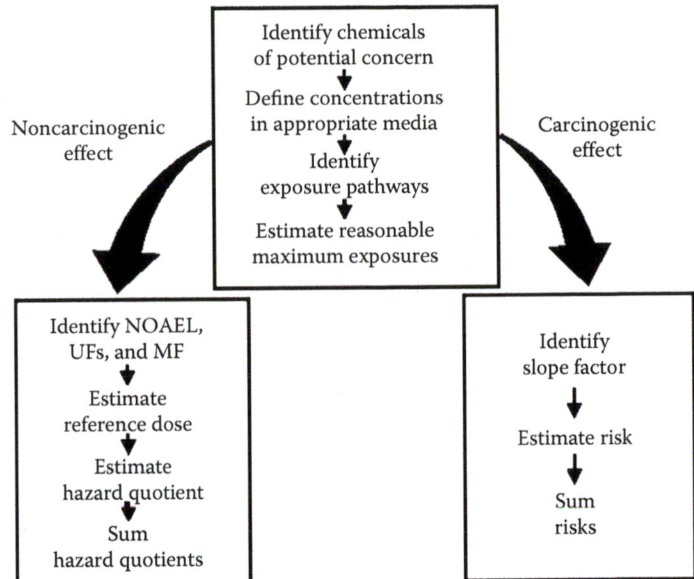

Figure 13.2 The general sequence of steps toward assessment the hazard/risk associated with noncarcinogenic and carcinogenic effects of contaminants. The final step of summing hazard quotients or risks for the individual chemicals or media may not be required or may not be appropriate (see text for more details).

13.3 ECOLOGICAL RISK ASSESSMENT

"If seven maids with seven mops
Swept it for half a year,
Do you suppose," the Walrus said,
"That they could get it clear?"
"I doubt it," said the Carpenter,
And shed a bitter tear.

Carroll (1872)

13.3.1 General

The first two parts of a remedial investigation (characterization of the contamination and human risk assessment) have been described briefly to this point. Now, let us consider the third part of a remedial investigation, the ecological risk assessment. Like human risk assessment, the goal of ecological risk assessment is the estimation of the likelihood* of a specified adverse effect or ecological event due to a defined exposure to a stressor.† Relevant effects can range from the suborganismal to the landscape scale. Unlike human risk assessment, ecological risk assessments must consider many species with diverse niches and phylogenies. It should even consider ecological entities, e.g., communities or metacommunities, composed of many species occupying a

* Likelihood is interpreted even more loosely in ecological assessments than in human assessments (Norton et al. 1992). "Descriptions of risk may range from qualitative judgments to quantitative probabilities. While risk assessments may include quantitative risk estimates, the present state of the science often may not support such quantitation" (EPA, 1996). The term may not always imply the generation of a probability with an associated statistical statement of confidence.

† Stressor is defined in ecological risk assessments as "any chemical, physical, or biological entity that can induce adverse effects on ecological components, that is, individuals, populations, communities, or ecosystems" (Norton et al. 1992). Obviously, there is considerable latitude in this definition and, as discussed before, many effects at higher levels will more often be ambiguous than clearly adverse.

heterogeneous landscape. Also, in contrast to human risk assessment in which extrapolation to one species (humans) is often done from many species (e.g., mouse, rat, or dog toxicity data), ecological risk assessment extrapolates from one or a few species to many.

Ecological assessments may be retroactive or predictive. The predictive assessment generally adheres to the NAS paradigm (Figure 13.1), but retroactive assessment relays less on this paradigm and more on surveys of contamination and ecological impact, models of fate and effects, and epidemiological data.

The ecological risk assessment process is organized slightly differently from the NAS paradigm, although the overall logic remains the same (Figure 13.3). The first step is problem formulation, a process involving both the risk assessor and the risk manager. Next is the analysis step. The analysis step has two components similar to the NAS paradigm's exposure and dose–response assessments. It also has parts of the hazard identification component of the NAS paradigm. In the analysis and risk characterization stages, there might be reexamination of various actions or decisions as new information arises. In the last step, risk characterization, the information generated in the analysis step and the context developed in the problem formulation step come together. After the risk characterization step, the risk assessors and managers review the results relative to the original needs set out during problem formulation and any needs that might have emerged during the process.

13.3.2 Problem Formulation

Problem formulation includes the initial planning and scoping that establishes the framework around which the assessment is done (Norton et al. 1992). It includes the selection of assessment endpoints, a conceptual model, and a plan of analysis. The **assessment endpoint** is the valued ecological entity to be protected (e.g., bald eagles nesting by a contaminated lake) and the precise

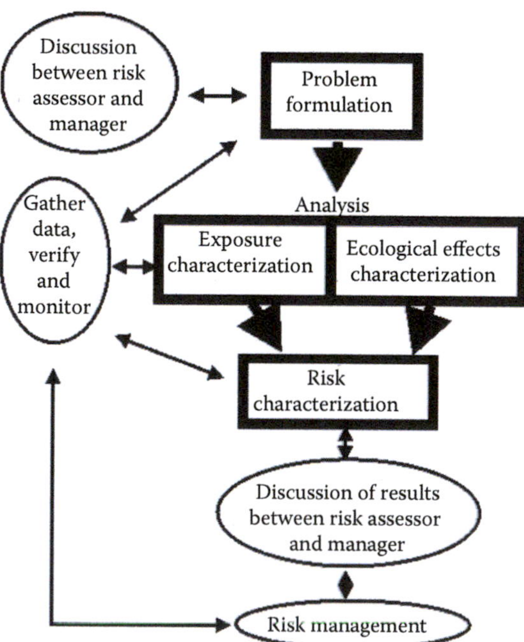

Figure 13.3 The general framework of an ecological risk assessment. The boxes and thick arrows reflect the risk assessment proper and is similar to the National Academy of Science paradigm (Figure 13.1). The narrow, two-way arrows reflect interactions between the risk assessor and the risk manager during the process. They also reflect the continual accrual and evaluation of information during the entire process. (Modified from Figure 2 in EPA, *Review Draft*, p. 564, 1994.)

quality to be measured for this entity (e.g., adult survival and nesting success of bald eagles).* Including the measured quality in the definition ensures that the assessor specifies clearly an effect or improvement after remediation that can be quantified. For example, if one incorrectly decided that the assessment endpoint was the vague "functional integrity of the stream ecosystem," it would be very difficult to know what exactly should be measured to establish whether an adverse effect is present or to document any improvement after remediation.

Sometimes, the specific quality of the assessment endpoint cannot be measured in the valued entity and it is measured in a surrogate instead. For example, risk assessment for an endangered species may require unacceptable destructive sampling to determine egg viability or body burden of a contaminant. Instead, these qualities may be measured for a closely related or ecologically similar species, and associated results applied to predicting risk to the endangered species.

The EPA (1996) suggests three qualities for a good assessment endpoint: (1) it should be ecologically relevant to the ecosystem being assessed, (2) it should be susceptible to the stressor, and (3) it is desirable, but not necessary, that it be valued by society. The last quality increases the value of the assessment to the risk manager. Suter (EPA, 1989c) further suggests that an ideal assessment endpoint should have an unambiguous operational definition and that it should be readily measurable or predictable from measurements. He gives the example of "balanced indigenous populations" as an ambiguous, and therefore compromised, assessment endpoint. Assessment endpoints are often, but not always, legally protected entities (e.g., survival and reproduction of an endangered or threatened species) or economic entities (e.g., successful reproduction of a salmon species). They can also be important ecological qualities, e.g., species diversity or some reflection of biodiversity. Suter (EPA, 1989c) tabulated (Table 13.1) some examples of assessment and measurement endpoints that may applied to various levels of ecological organization.

The **conceptual model** links the assessment endpoint and the stressor of concern. It evaluates possible exposure pathways, effects, and ecological receptors. Conceptual models include hypotheses of risk and a diagram of the conceptual model. The **risk hypotheses** are clear statements of postulated or predicted effects of the stressor on the assessment endpoint. The **conceptual model**

* In many publications (e.g., EPA, 1989c; Norton et al. 1992), the distinction is made that the **assessment endpoint** or **receptor** is the ecological entity or value to be protected (e.g., a population of the endangered bald eagle nesting by a contaminated lake) and the **measurement endpoint** is a measurable response to the stressor (e.g., number of fledglings produced per nest each year) that is related to the valued qualities of the assessment endpoint (e.g., reproductive success of the bald eagles). Some logical or quantitative model must link the two endpoints.

Table 13.1 Examples of Endpoints for Ecological Risk Assessments

Level of Ecological Organization	Assessment Endpoint	Measurement Endpoint
Population	Extinction	Occurrence
	Abundance	Abundance
	Yield or production	Reproductive performance
	Age or size class structure	Age/size class structure
	Mass mortality	Frequency of mass mortality
Community	Market sport value	Number of species
	Recreational value	Species diversity/richness
	Change to a less useful or appealing state	Index of biological integrity
Ecosystem	Productive capacity	Biomass
		Productivity
		Nutrient dynamics

Source: From EPA, *Ecological Assessment of Hazardous Waste Sites*, National Technical Information Service, Springfield, VA, 1989c.

diagram (Figure 13.4) shows the pathways of exposure and illustrates areas of uncertainty or concern. It is a visual aid for communicating to the risk manager the model from which the risk hypotheses emerge.

In the final step of problem formulation, the risk hypotheses are examined carefully and a plan of analysis is produced. "Here, risk hypotheses are evaluated to determine how they will be assessed using available and new data" (EPA, 1996). An **analysis plan** defines the format and design of the assessment, explicitly states the required data, and describes the methods and design for data analysis. It describes what will or will not be analyzed. Measurement endpoints, those qualities that will be measured to assess effect to the assessment endpoint, are also stated in the analysis plan. A measurement endpoint may involve measurements derived directly from the valued ecological entity or from its surrogate.

13.3.3 Analysis

Again, the analysis step has two components (exposure characterization and ecological effects characterization) that are very similar to the exposure and dose–response assessments of the NAS paradigm. The exposure and ecological effects characterizations are done in tandem with considerable exchange of information occurring between the two components.

Some aspects of hazard assessment of the NAS paradigm are inserted as part of the analysis step of an ecological risk assessment. The gathering of relevant data and identification of chemicals of potential concern done in the hazard assessment step of the NAS paradigm are also done in the

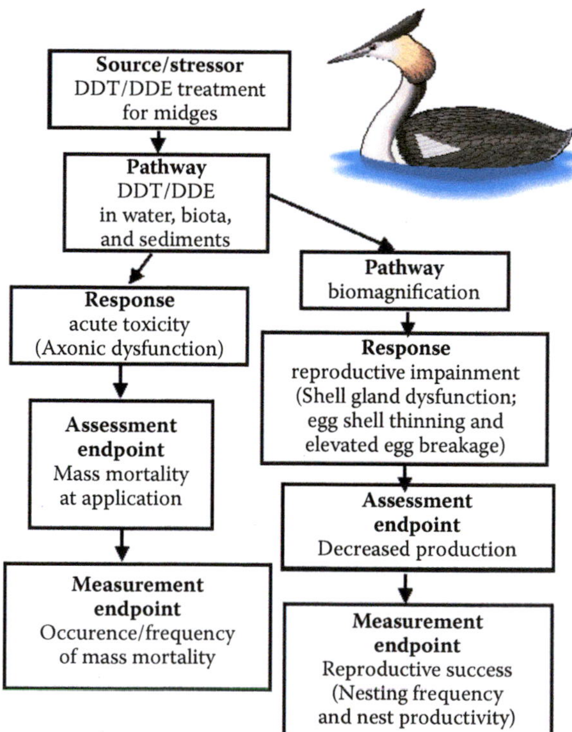

Figure 13.4 A conceptual model diagram for the pesticide spraying of nonbiting midges as described in Chapter 1 and depicted in Figure 1.2. Here, the assessment and measurement endpoints are separated for clarity. They could have been combined as the assessment endpoint. The valued ecological entity is the Western Grebe population.

analysis step of the ecological risk assessment. However, some of the initial data gathering also occurred during the problem formulation step of the ecological risk assessment.

13.3.3.1 Exposure Characterization

Exposure characterization describes the characteristics of any contact between the contaminant and the ecological entity of concern. It summarizes this information in an exposure profile. Temporal and spatial patterns in contaminant distribution are defined in addition to the amount of contaminant present. The source of the contaminant, any potential costressors, transport pathways (e.g., outcropping of contaminated groundwater to the stream, or ingestion of prey after biomagnification), and type of contact are defined. In this characterization of exposure, quantitative methods described earlier for human exposure are applied. Concentration, duration, and frequency of exposure must be considered, including consideration of factors such as seasonal cycles and home ranges of species. For example, the contaminated region may be used only 10% of the time by a free ranging species. Or a contaminated food may be ingested only during one season. Estimates of exposure duration and frequency should take such factors into consideration. The final **exposure profile** "quantifies the magnitude, and spatial and temporal pattern of exposure for the scenarios developed during problem formulation" (Norton et al. 1992).

13.3.3.2 Ecological Effects Characterization

Ecological effects characterization "describes the effects that are elicited by a stressor, links these effects with the assessment endpoints, and evaluates how effects change with varying stressor levels" (EPA, 1996). It also specifies the strength of evidence associated with the effects characterization, and level of confidence in the causal linkage between the contaminant and the effect (Norton et al. 1992). Information generated with many of the methods described previously are brought together to develop a stressor–response profile for the valued ecological entity.* Which methods are applied would depend on the nature of the exposure and the ecological entity of concern. Unfortunately, most dose–response information are generated for effects to individuals, yet those most needed here are dose–response information for populations, metapopulations, communities, metacommunities, ecosystems, and metaecosystems in landscapes. The unfortunate consequence of this imbalance between available information and need is compromised ecological risk assessments.

13.3.4 Risk Characterization

Risk characterization draws together the information from previous steps to produce a statement of the likelihood of an adverse effect to the assessment endpoint (EPA, 1998). Risk may be expressed in several ways including a simple qualitative judgment or hazard quotient. It could involve a richer interpretation including description of the influences of concentration and temporal variations on estimates of effect. It could use complex models that also generate some estimate of confidence in the risk predictions.

The final statement of risk must include details about the adequacy of the data going into the judgment, uncertainty involved in the conceptual mode or calculations, and weight of evidence for each causal relationship. If a surrogate were used in place of the assessment endpoint, confidence in

* Data bases like IRIS are being developed for ecological effects. For example, the Oak Ridge National Laboratory has data bases for aquatic biota, wildlife, terrestrial plants, sediments, and soil invertebrates/microbial processes (http://www.esd.ornl.gov/programs/ecorisk/ecorisk.html). Details are provided in extensive documentation obtained from this same web site. Also, EPA has an ECOTOX data base of ecological effects information (http://cfpub.epa.gov/ecotox/).

the associated extrapolation should be addressed. For example, the confidence in the extrapolation would be high if there were very little variation among raptors in the dose–effect relationship and a surrogate raptor with a very similar niche were used for bald eagles. Some statement about the significance of the adverse effect should also be made so that the risk manager may better judge the seriousness of risk consequences.

13.3.5 Summary

The expedient compromises and conservative biases already discussed for human risk assessments are also present in ecological risk assessments. More are added because the manifestation of effects could involve several species or levels of ecological organization. There is a tendency to address the level at which most information exists, i.e., the individual level. Understandably, implications are most often to the level of population or community viability.

13.4 RADIATION RISK ASSESSMENT

13.4.1 Characteristics of Types of Radiations and Their Effects

The qualities of relevant radiations were described briefly in Chapter 2 so only a general reminder is required here to move onto discussions of radiation risks. Recollect that *x*- and γ-rays are photons of electromagnetic radiation but the α (helium nuclei with a +2 charge) and β (electrons or positrons) radiations are particles. Characteristics such as mass, energy, and source of a radiation greatly influence the risk associated with exposure. For example, the mass and charge of a β particle dictates that it has minimum tissue penetration. In contrast, a γ photon will penetrate much deeper into tissue. Harm from radiation is a function of the ability of the photon or particle to generate ion pairs as it interacts with tissue. Because most of the molecules in tissues are water, most ionization and free radical generation involves water in the presence of oxygen to produce H_2O^-, OH^-, H^+, $\cdot H$, $\cdot OH$, $HO_2\cdot$, HO_2^-, and H_2O_2 (Cockerham et al. 2008). Not only do photons and particles differ in the depth into which they will penetrate tissues but also differ in their ability to transfer energy into that tissue such that ion pairs are generated. An α particle has only very shallow penetration but, because of its relatively high mass and charge, it can dissipate considerable energy in its short ionization track. As discussed in Chapter 2, the α particle has a high linear energy transfer (LET). A high-energy *x*-ray photon passes deeply into tissue, but relative to the α particle, transfers small amounts of energy to the tissues along its ionization track, i.e., it has a lower LET than the α particle.

Also important in determining risk from radiation exposure are the nature and source of the radioactive material. A radionculide that is an analog of an essential element will move into an organism differently than one that is not an essential element analog. How a radionuclide moves through geochemical cycles is also important as in the case of ^{137}Cs, which moves like potassium in geochemical cycles. As previously discussed, the geochemistry and gaseous dynamics of radon strongly influence human exposure.

Damage from ionizing radiation includes somatic and reproductive effects. However, most attention during risk assessments is focused on cancers after exposure.

13.4.2 Expressing Radiation Dose and Effect

As mentioned in Chapters 1 and 2, standard units used in the literature to express radiation dose and effect include older units and the newer International System units. Curies and the newer becquerels (Bq = 1 disintegration s^{-1}; 1 Ci = 2.7×10^{-11} Bq) are units of radioactivity. Measures of radioactivity are not directly useful expressions of dose and some measure of a potentially harmful dose is required instead. Such an absorbed dose is expressed as the amount of energy deposited in tissue, joules (J) per kilogram of tissue. The new standard unit of absorbed dose is the Gray (1 Gy = 1 J·kg^{-1}) and the older unit was the rad (erg·g^{-1}), which is equivalent to 0.01 Gy. The Gray is still not the most directly useful dose metric in risk assessments and more calculations are required to provide a more appropriate expression.

The dose rate (Gy·day^{-1}) can be estimated by assuming that the entire dose is delivered solely to the tissues. A dose rate is generated by knowing the radionuclide decay rate (disintegrations·day^{-1}) and the amount of energy effectively transferred to tissue (**specific effective energy**, MeV·disintegration^{-1}). The last two pieces of information are taken from tables published by organizations like the U.S. Department of Energy (e.g., Kocher, 1981) or the International Commission on Radiological Protection (ICRP) (www.icrp.org).

Expression of dose needs to accommodate an additional feature because the various types of radiation differ in their ability to cause damage. To create a common scale for different radiation types, the dose rate is multiplied by a **quality factor** (Q) or **radiation weighting factor** to get an estimate of a **dose equivalent**. The Q is the biological damage from the particular radiation of interest divided by that expected of a reference radiation type (Wang et al. 1975): It is a simple conversion factor from a table in which *x*- and γ-rays are used as the reference ($Q = 1$). This dose metric incorporates the amount of energy transferred to tissue and the amount of expected damage from the energy dissipated in the tissue. It allows easy comparison among different kinds of radiation. As an example, Table 16.3 in Wang et al. (1975) or Table 25.2 of Harley (2008) give Q values of 1 for γ- and *x*-rays and β particles, but a Q value of 20 for α particles and (nonthermal) neutrons. As mentioned in Chapter 1, this dose equivalent can be expressed with either, the older rem (roentgen equivalent man) or the newer sievert (Sv) units. Figure 13.5 illustrates the expression of annual human exposure as dose equivalents.

Which tissue is exposed also influences the risk associated with dose, so an additional calculation is often done to express the doses to different organs relative to that which would manifest if the exposure had been a uniform one over the entire body. This final dose expression modification produces what was once called an **effective dose equivalent** but is now called an **effective dose** (ED). With the ED, the dose equivalents for different parts of the body are combined using published

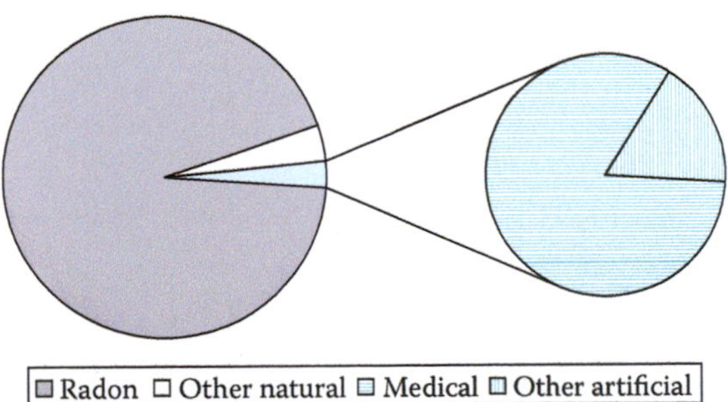

Figure 13.5 The estimated annual dose equivalent of ionizing radiations received by humans. (Data from Table 18.2 of Cockerham et al. [2008].) Most of the average annual dose equivalent is from radon (24.0 mSv) with a modest amount of additional natural dose from cosmic rays (0.27 mSv), terrestrial sources (0.28 mSv), and internal sources (0.39 mSv) such as radiopotassium in the body. The dose equivalent from cosmic sources depends on elevation and the value given here is for 0–150 m above sea level (Cockerham et al. 2008). Annual effective dose from artificial sources (0.64 mSv) is composed primarily of medical diagnosis *x*-rays (0.39 mSv), consumer products (0.10 mSv), and nuclear medicine (0.14 mSv).

organ weighting factors (see Publication 60 of the ICRP [1991]) to express the dose in terms of a uniform exposure to the entire body with an equivalent risk (Figure 13.6). It allows comparisons of partial exposures involving different portions of the body, e.g., radon exposure through inhalation and whole body exposure from natural ^{40}K. For a radionuclide that has been assimilated into tissue, this dose can be used with the amount of time that an effect might be expected to manifest (e.g., a 70-year life span), biological elimination rates, and decay rates to predict a **committed effective dose**. The committed effective dose is then used in a dose–effect model to predict risk from the exposure (Hinton, 2003; Harley, 2008). These metrics are summarized in Figure 13.7.

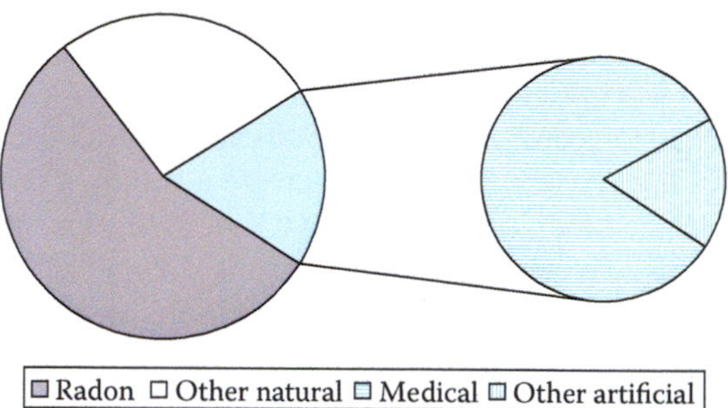

Figure 13.6 The expression of the doses shown in Figure 13.5 as effective dose equivalents. (Data from Table 18.2 of Cockerham et al. [2008].) Radon = 2.0 mSv, cosmic rays = 0.27 mSv, terrestrial sources = 0.28 mSv, natural internal = 0.39 mSv, medical *x*-rays = 0.39 mSv, nuclear medicines = 0.14 mSv, and consumer products = 0.10 mSv. Notice that radon (and its daughters) has an effective dose equivalent that accounts for a large portion of the human annual radiation dose. Radon's effective dose equivalent is different from its dose equivalent because its influence is heavily focused on pulmonary exposure.

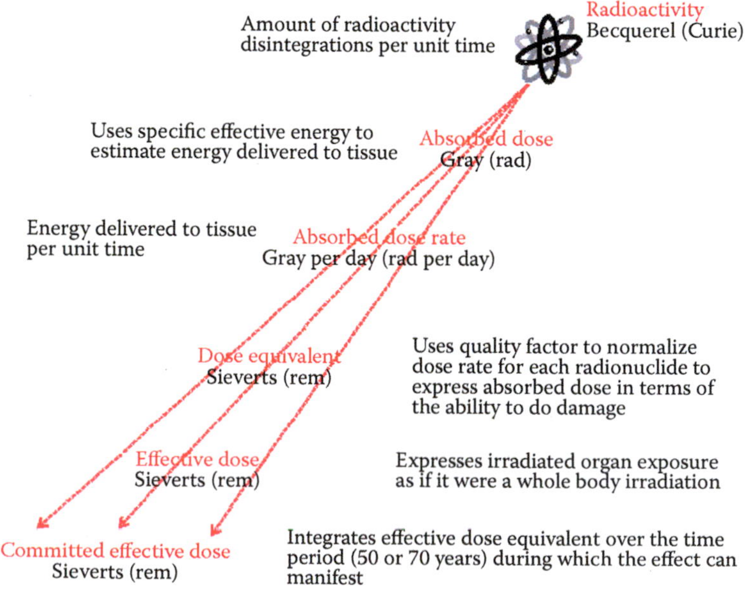

Figure 13.7 Summary of radiation dose and effect metrics. International System units are placed under each metric. Older, but still used, metrics are given in parentheses after the International System units (see text for details).

13.4.3 Models of Radiation Effect

Selection of a model for estimating risk from radiation dose begins with the determination of whether the effect is stochastic or nonstochastic. **Nonstochastic effects** are those that are directly relatable to a dose above a certain threshold such as the skin erythema from *x*-ray exposure or the acute radiation syndrome. However, most radiation risk assessments focus on **stochastic effects** such as cancers in which it is the probability of an adverse effect that increases with dose. Stochastic effects are modeled with no threshold in most radiation risk assessments. Models similar to those described earlier are used to link dose to probability of a lethal cancer. More complex models incorporating different environmental compartments or toxicokinetics are required in some cases. Dose–effect relationships for humans are derived using data from industrial accidents or high-exposure situations, post–nuclear weapons denotation surveys in Japan, medical situations, or nuclear reactor accidents such as the Chernobyl accident. Descriptions of such studies are provided by Cockerham and Cockerham (1994), Cockerham et al. (2008), and Harley (2008).

Human risk assessments for radiation have been developed to a relatively sophisticated level; however, those to nonhuman species are often done assuming that application of human data will provide a good estimate of risk to other species (Cockerham and Cockerham, 1994; Hinton, 2003).

VIGNETTE 13.2 Radiation and Ecological Risk

Eric L. Peters
Chicago State University

Of all pollutants, probably none have so colorful a history of cultural representation as do radiation and radionuclides. Many persons (including myself) were reared on films that presented the seemingly logical, inevitable, and horrific results of exposure to radiation and nuclear fallout: giant insects and other arthropods would stalk the land, monstrous octopods and reptiles would rise from the depths of the oceans and smash cities flat, and horrific mutations of body and mind would threaten all "normal" human life. Interestingly, a common theme of many of these depictions was that radiation might greatly *enhance* some organisms' size, strength, or other abilities to generate destruction. Although admittedly silly, these films also reflected a widespread and nebulous perception of the true risks of radiation (and the use of nuclear reactions by humans in particular) that persists to this day.

The Earth is commonly considered to have formed at least 4.6 billion years ago, and we are only now fully appreciating how radiation has influenced the evolution of life on the earth since its origin (perhaps 3.8 billion years ago). Recently, it has been discovered that heat generated within the Earth, which is produced by the decay of radioactive elements, provides the geothermal heat that supports chemoautotophic Archaea. These bacteria are apparently very widely distributed within the earth's crust and across the vast expanses of the hydrothermal rift vents. It is likely that this heat prevented the extinction of all life during the Snowball Earth period (the repetitive multimillion-year bouts of freezing, thawing, and heating of the oceans) that immediately preceded the evolution of animal life. Furthermore, all organisms contain both primordial (e.g., ^{40}K) and cosmogenic radionuclides (e.g., ^{14}C), and these combine with emitted radiations from terrestrial radionuclides in the earth and solar/cosmic radiations. When these radiations alter the DNA of sex cells (or, more likely, the germ cells that produce them), the resulting genetic mutations provide much of the raw material that underlies evolution and subsequent natural selection.

The potential of ionizing radiations to influence the heredity of organisms was recognized soon after their discovery. By the earliest decades of the twentieth century, Thomas H. Morgan and his students had employed *x*-rays to generate mutations in fruit flies (*Drosophila*) and used these mutant strains to produce the first maps of gene locations on chromosomes. These efforts were especially impressive, considering that another three decades would follow before DNA was recognized as the carrier of genetic information and its molecular structure determined. In the early twentieth century, radium needles were being used to successfully treat breast cancer (even as employees of watchmaking companies were developing cancer from swallowing radium paint when they wetted their brushes in their mouths). Over the past century, radiation effects on humans have become better characterized, to the point where radiation can be used in a wide variety of diagnostic and therapeutic contexts with fewer side effects.

Accomplishing this has required an enormous and still ongoing effort, and assessing the actual risks associated with exposure to radiation is not at all perfect. Notably, no effects of radiation on humans have ever been observed at acute exposures <100 mSv (10 rem). Natural sources of exposure average about 3 mSv year in the United States, but can be much higher in areas with a large amount of terrestrial radionuclides (e.g., Kerala, India). Routine medical diagnostic procedures may add up to another 3 mSv year^{-1}. Up to 20 mSv year^{-1} is permitted for radiation workers (members of the general public are limited to 1 mSv from reactor releases). Up to 50 mSv in a single year is permitted for emergency workers (although workers attempting to control the damage to the Fukushima-Daiichi reactors were allowed to volunteer for single exposures of up to 250 mSv), but most radiation workers average no more than 2 mSv year^{-1} from occupational exposure. To put this in perspective, a single cross-country airplane flight increases radiation exposure by about 25 μSv (0.025 mSv). Some airline flight crews are therefore exposed to radiation at levels (5–9 mSv year^{-1}) that exceed what is usual for workers at nuclear reactors. Yet, flight crews and frequent flyers are not monitored for radiation exposure, nor are they limited in their exposure time due to radiation concerns. To attempt to do so would vastly complicate and increase the costs of air travel and provide negligible benefits to the persons affected (relative to the other risks of flying).

Astronauts on the International Space Station often remain there for months and are exposed to even higher amounts of radiation during that time (doses to blood-forming organs are estimated to range from 0.4 to 1.1 mSv day^{-1} (Cucinotta et al. 2001). The Apollo astronauts (the only humans to have traveled beyond the protection of the Earth's magnetic field) experienced radiation exposures that exceeded the upper end of this range. Their command modules were better shielded, but their landers were made largely of aluminized Mylar plastic not much thicker than an inflatable swimming pool. As Alan Bean, one of the astronauts on Apollo 12 (and the fourth man to walk on the moon) observed, "When you're an astronaut, you buy into a lot of risk … [and] … if you can't buy into it, don't be an astronaut" (Interview, *When We Left Earth*, The Discovery Channel, 2008).

Even where the absolute best possible information is available, the fact that public perceptions of the likely risks of radiation exposure remain incomplete and not well aligned between costs and benefits is not entirely attributable to insufficient educational presentations or their own learning efforts. Even measuring radiation exposure and dose (much less assessing the resulting risks) is enormously complicated. This is true even for humans under well-understood circumstances (e.g., in a therapy or diagnostic facility). In the context of an environmental release, hundreds of additional factors need to be considered. These include, but certainly not limited to, the following:

Physical/chemical processes

- Half-life
- Physical and chemical forms of the radionuclide
- Chemical form of the radionuclide
- Dispersion by wind (e.g., wind speed)
- Wet and dry deposition
- Electrostatic attraction between radionuclides and surfaces
- Sorption and adsorption
- Sedimentation rates
- Resuspension
- Dispersion by water (e.g., mixing currents)
- Soil composition and chemistry
- Foliage type and physiognomy
- Vertical and horizontal structure

Biological/ecological processes

- Biochemical/physiological properties of radionuclides (e.g., biological availability)
- Radiation sensitivity
- Inhalation
- Absorption and bioaccumulation
- Biomagnification
- Diet and drinking
- Assimilation
- Excretion, exfoliation, and depuration
- Metabolic rate
- Radiation damage/repair systems
- Longevity
- Successional state; community stability

The impacts of many of these factors remain incompletely understood even for human exposures and doses, and humans are the only species for which much information is even available. Furthermore, exposures to humans and other species can decline faster than the decay rate of the isotope (unless the system is disturbed), as studies of "ecological half-life" (e.g., Paller et al. 1999) have shown. We thus have large information gaps when it comes to evaluating ecological risks of radiation and radionuclide releases. Some assumptions may be conservative (for example, it is likely that exposure to an acute dose produces a greater effect than a longer term exposure to the same total dose, but both are generally assumed to be equivalent in effect). Other assumptions may not be: It is often assumed that, if the environmental releases remain within safety limits for humans, then other species will be sufficiently protected. Some species, however, are more (or less) sensitive to radiation than are humans. For example, coniferous trees near the Chornobyl* Nuclear Power Plant were quickly killed when radiation destroyed their sensitive meristems, but many other plant and animal species survived (and even thrived). Many bacteria and protists, as well as many animals, e.g., insects, mollusks, and turtles, are more resistant to radiation than are humans and other mammals.

Although thousands of papers have been published on radionuclide levels in biota, little systematic information exists on the long-term environmental effects of radiation on nonhuman species.

* The Ukrainian spelling of Chernobyl is used in this vignette.

Probably the best data from a quantitative standpoint are the studies conducted in the 1960s and 1970s that used large sealed radiation sources to chronically irradiate small areas with γ radiation over a period of few years. These studies were conducted in different biomes, including: a temperate forest, Long Island, New York (see Figure 11.16), a short grass prairie (Colorado), and a subtropical rain forest (Puerto Rico). The general results were that chronic exposures to levels of radiation could selectively eliminate some species and could drive the communities to an earlier stage of ecological succession, but only when those locations were exposed for years to extremely high doses (1 cGy or more per day).

Environmental releases of radionuclides can and do occur, but the releases to date have differed so broadly in their magnitude, geographical dispersal, and composition, that quantitative evaluations and comparisons of their potential effects are extraordinarily difficult. Aside from nuclear weapons testing, the best-known instances so far have been:

- The fire at the Windscale weapons production facility (Sellafield, United Kingdom) in 1957.
- The explosions of a waste tank (near Kyshtym, Russia) in 1957.
- The partial core meltdown of a reactor at the Three Mile Island (TMI) Nuclear Generating Station (Pennsylvania) in 1979.
- The destruction of the Chornobyl Nuclear Power Plant reactor No. 4 (Pripyat, Ukraine) in 1986.
- The explosions and meltdown of three reactors at the Fukushima Daiichi Nuclear Power Plant (Fukushima Prefecture, Japan) in 2011.

All of these events were avoidable, but they differed greatly in their magnitude. The partial core meltdown of the Three Mile Island (TMI) nuclear reactor in the United States is frequently cited in the same breath as the catastrophic explosion of the reactor at the Chornobyl nuclear power plant near the Ukraine–Belarus border. Yet, not a single injury has been attributed to the TMI accident, whereas the release of radiation from the Chornobyl reactor was tens of millions of times greater, killed at least 35 persons, and has produced a significant increase in cancers and other health problems. Fortunately, there are very few cases of widespread environmental contamination have produced doses anywhere near that scale. Hundreds of nuclear weapons were tested in the atmosphere and in the oceans from the 1940s until the 1980s, and their fallout has dispersed all over the world. Most of this fallout consisted of radionuclides with very short half-lives that most affected the areas closest to the detonations (although the crew of the tuna boat *Daigo Fukuryū Maru* [Lucky Dragon 5] was exposed to the fallout from the Castle Bravo thermonuclear test, conducted on Bikini atoll in 1954: one crew member died and several experienced radiation sickness). Estimates of doses to "downwinders" in the United States, who were exposed to fallout from nuclear weapons tests in Nevada, were initiated years after such tests had ceased. Leaks from weapons production and commercial power reactors have occurred that have contaminated rivers, lakes, and oceans worldwide, again to widely varying degrees. Prominent examples in the United States are the releases from the Hanford facility into the Columbia River (Washington State) and from the Savannah River Site (South Carolina). These, however, were many orders of magnitude less than releases from the Mayak nuclear weapons facilities into the Techa River and Lake Karachay in Russia, whose inhabitants were exposed to high levels of radiation over many years.

Such high exposures are rare, and the long-term effects of low exposures are difficult to quantify. The latency periods for radiation-induced cancers in humans are often decades long, and are difficult to impossible to distinguish from the background cancer levels of a population (the elevation of thyroid cancers in children after the Chernobyl accident is a rare exception). Evaluating exposures and translating them into risks requires many decades, and previous estimates may need revision. The lack of planning in assessing the potential effects of these chronic doses means that human standards for evaluating exposure effects rely heavily on short-term exposures (e.g., the survivors of the atomic explosions at Hiroshima and Nagasaki). As the survivors, who were children at the time, began to reach their 70s and 80s, their risks of developing cancers from their radiation exposures began to be seen as greater than previously thought. The collapse of the Soviet Union essentially ended meaningful monitoring of the citizens exposed to radionuclides from the Chernobyl reactor, and the widespread devastation wrought by the earthquake and tsunami in Japan continues to compromise the ability to completely evaluate the magnitude and long-term effects of their radionuclide releases from this facility.

It may be more useful to compare risks from radiation exposure in the context of other (more likely) ones. For example, even omitting all other pollutants, more radionuclides are emitted by many coal-burning power plants in a single year than *should* be released from a nuclear reactor into the environment during its entire operating lifespan. Combusting wood has a similar effect in concentrating radionuclides in the remaining ash. Imported herbs and spices are routinely irradiated to kill pests and pathogens, yet many persons oppose the idea of sterilizing food by radiation because it produces a decrease in vitamin content and an increase in the generation of "radiolytic" compounds (cooking food, especially at high temperatures, also generates carcinogenic compounds and to a much greater degree). Smokers inhale radioactive polonium, which concentrates in tobacco and contributes to their cancer risk, although this added risk is certainly less than the risks to themselves and others from simultaneously inhaling the other constituents of tobacco smoke (e.g., PAHs and cyanide: three puffs on a cigarette is estimated to decrease one's life span by the same amount as the 10 μSv radiation exposure from a transcontinental plane flight).

Our present consumption rates of fossil fuels are unsustainable, and there is increasing evidence that our dependence on them for energy is degrading the earth's ecosystems and threatens our civilization, if not our species' existence. As a result, after several decades of dormancy, nuclear energy is again being advanced as a more "green" alternative (although natural gas as a bridging fuel to sustainable technologies that have yet to be realized is promoted even more aggressively). We should recognize that almost all existing high-level nuclear wastes remain within the facilities that generated them. None have been stabilized for long-term storage to any significant degree, much less transported to and deposited within an identified long-term storage site. That this remains true has been possible only because the mass of these wastes produced so far can be measured in the range of megatons and that their total volume is small enough to fit within a large sports stadium. Compared with the enormous bulk of carcinogenic ash and the release of atmospheric pollutants from burning coal; however, these would seem to represent a known, relatively well-supervised, and manageable amount. In contrast, releases of mercury from coal have eclipsed those from other human activities and now collectively exceed natural releases. Inputs of additional CO_2 into the atmosphere from fossil fuels are now nearly one-third of the 100-billion-ton carbon exchange for the entire planet each year.

My point is not to assert that radiation or radioactive materials cannot or should not be taken seriously as potential risks. A very important issue (which radiation typifies perhaps better than most) is the *voluntary* assumption of risk based on informed choice, as with the astronauts above. This might be best considered as a continuum: to voluntarily assume any risk requires an educational effort by the persons conducting the risky activity to provide meaningful information as to the potential consequences, together with a learning effort by the persons who may be affected to acquire enough information to make a sufficiently informed decision. Achieving these goals can (but need not) be frustrating for all involved, yet the consequence of not trying is the likelihood of creating misunderstanding and mistrust, even in the face of sincere efforts to communicate both risks and benefits. As in any educational situation, the relative contributions of the educational and learning efforts need to be balanced. Complex information about risks needs to be synthesized and explained in a transparent and honest way that invites questioning and is responsive to the individual audience. The recipients of the information likewise need to actively engage in the learning process. At one extreme, many persons may not want to be informed about risks, and many may look to some external entity (e.g., the government) to solve the problem. At the other extreme, some persons may be suspicious of external decisions and view them as paternalistic at best and evidence of conspiracies at worst.

It is notable that the worst nuclear releases have tended to occur in situations where the public's ability to assume the risk was essentially nonexistent or was hijacked by their governments. In the case of facilities in the former Soviet Union, no consideration was made for public input at all (citizens of the Former Soviet Union were exposed to decades of radiation from the military nuclear weapons complexes; persons in the area surrounding the destroyed Chernobyl reactor were not evacuated until 2 days afterward). Decisions made about the construction and operation of the Fukushima Daiichi power plant would seem to be largely responsible for their failure, but many of these were made without consulting the public (although the government was consulted, sufficient oversight seems to have been absent both before and after the earthquake). Construction of the plant at the location chosen (at the juncture of three crustal plates) seems amazingly foolish in retrospect, yet a sister plant with four reactors about 11 km away was undamaged by the 9.0 magnitude earthquake. A previous earthquake in the 1970s damaged another facility and demonstrated that upgrades were need. On the basis of previous studies, the seawall in front of the Fukushima Daiichi plant was not high enough, and it was recommended (just 4 days before the earthquake) that it should be raised in case of a tsunami. In the decades preceding, however, both the emergency diesel and the backup battery-powered generators were situated where they would be immediately ruined by salt water should a tsunami breach the seawall. Both the reactor fuel and the stored fuel rods required a constant flow of water to prevent them from boiling dry, rather than being able to be stabilized in place under passive cooling during the weeks that it would require to restore pumping systems in the event of a disaster (this has still not been accomplished at the time of this writing: 3 years later). These were all easily predictable situations that, given the decades-long operation of the plant, might have been greatly mitigated, even if they had been addressed retroactively.

If we continue to operate nuclear power reactors, how might we create facilities that will isolate large quantities of highly toxic wastes that are likely to remain dangerous for periods that are hundreds of times longer than any human-built structure has persisted (a few thousand years at best so far) and that might still exist long after our species has become extinct? How could future generations be educated about the risks of these wastes long after we have given our common assent to producing and storing them? Given our difficulty of preserving records (hundreds of thousands of printed works and movies have already been lost) and the accelerating rate of our cultural evolution, how can we inform members of our species living hundreds or thousands of generations from now about the danger of these pollutants so that they could protect themselves and their environment? Do we attempt to erect monuments that would last over many millennia that would warn future descendants? Should we establish a "nuclear priesthood" that would carry these dangers forward until humans either no longer exist or the problem is otherwise solved? These are not trivial questions, and a considerable amount of thought has been devoted to this issue with respect to nuclear wastes.

The worst nuclear releases to date has never produced a glowing moonscape devoid of life. South Pacific Islands destroyed by nuclear blasts just a few decades ago teem with marine life today. The area around Chernobyl is rich in wildlife, which thrives free of human interference. Even the sarcophagus constructed to contain the core leaks air, water, and wildlife (swallows fly in and out at will), while the world waits for it to collapse. All evidence to date indicates that life on Earth will likely survive our adventures with nuclear energy, a great deal easier than it will our burning fossil fuels.

13.5 CONCLUSION

Environmental risk assessment terminology and technology change quickly. Indeed, terminology and emphasis have changed from 1989 to now as evidenced in the EPA and ICRP documents used to frame this chapter. It is anticipated that this treatment will require careful revision with every new edition. Regardless, the basic risk assessment paradigm is intelligent and insightful: It will probably remain intact into the near future.

What is painfully needed at this time is a sound data set for effects at all levels of biological organization. Also, basic ecotoxicological tests produce data that are inadequate to the task. As an important example, temporal dynamics are given minimal consideration in most ecotoxicological tests. As we discussed in Chapter 9, only convention in the field inhibits the construction of more complete models incorporating time effectively. Neglected methods for expressing many toxicant-related effects in terms of risk probabilities are also discussed in Chapter 9. Finally, unjustified conceptual shortcuts are made in ecological risk assessments such as assuming that the most sensitive life stage is the critical life stage for a population or that effects are generally (concentration) additive. As we already discussed in earlier chapters, these are dubious general assumptions. Hopefully, these and other shortcomings in risk assessments will be resolved soon.

SUGGESTED READINGS

Calow, P. P., *Handbook of Environmental Risk Assessment and Management*, Blackwell Science, Oxford, United Kingdom, 1998.

EPA, *Risk Assessment Guidance for Superfund. Volume I. Human Health Evaluation Manual (Part A). Interim Final.* EPA 540/1-89/002. National Technical Information Service, Springfield, VA, 1989.

EPA, *Proposed Guidelines for Ecological Risk Assessment,* EPA/630/R-95/002B. Risk Assessment Forum, U. S. Environmental Protection Agency, Washington, DC, 1996, p. 564.

Harley, N. H., Chapter 25. Health effects of radiation and radioactive materials, in *Casarett & Doull's Toxicology: The Basic Science of Poisons,* 7th Edition, Klaassen, C. D., Ed., McGraw Hill, New York, 2008.

Neely, W. B., *Introduction to Chemical Exposure and Risk Assessment.* CRC Press, Boca Raton, FL, 1994.

Newman, M. C., *Quantitative Ecotoxicology,* 2nd Edition, Taylor & Francis/CRC Press, Boca Raton, FL, 2013.

Newman, M. C. and W. H. Clements, *Ecotoxicology: A Comprehensive Treatment*, Taylor & Francis/CRC Press, Boca Raton, FL, 2008.

Suter, II., G. W., *Ecological Risk Assessment,* 2nd Edition. Taylor & Francis/CRC Press, Boca Raton, FL, 2007.

CHAPTER 14

Conclusions

14.1 OVERVIEW

We have discussed the diverse goals of ecotoxicology and the many disciplines contributing to our understanding of contaminant fate and effects. We discussed contaminant sources, cycling, and accumulation in components of the biosphere, including transfers into individuals and through trophic webs. Then, effects from the suborganismal to the biosphere levels were discussed in a series of chapters that took up most of the book. Finally, the technical issues of estimating chemical and radiological risks to humans and the environment were explored. Legal aspects of ecotoxicants were discussed in the appendices for the interested reader. Now, all that is left to do is to remind the reader of the importance and context of ecotoxicology. That is the goal of this very brief chapter.

14.2 PRACTICAL IMPORTANCE OF ECOTOXICOLOGY

Not one would mind, neither bird nor tree
If mankind perished utterly;

Teasdale (1937)

In looking back at the enormous human suffering of World War I, Sarah Teasdale wrote the poem from which the above excerpt was taken. She correctly observed that, although the suffering was of such enormity that it will stand out in history for centuries, it was trivial in the larger context of nature. The biosphere can do just fine without our help. As Stephen Jay Gould (quoted in Wheeler, 1996) expressed it, "Humans are simply a dot on one side of the curve of biological complexity … a dot that could easily disappear."

Does this mean that we are free to contaminate the biosphere with abandon? This is certainly the case in the context of ensuring that life continues on the earth. Our influence on the permanence of life on the earth is minuscule. One meteorite striking the earth at the end of the Cretaceous period had more influence on the trajectory of life on the earth than we will ever have—even in the case of global climate change. In a few billion years, life will end when the sun expands to engulf our planet: our efforts cannot change the physics of star aging. However, this is certainly not true in the pragmatic context of ensuring that the biosphere remains compatible with and appealing to human life. It would be a catastrophic mistake to continue to contaminate the biosphere. In the case of climate change, allowing things to continue without conscientious efforts to mitigate would result in a profound change in the biosphere relative to our needs and esthetics. And the longer mitigation is delayed, the worse the consequences will be. Stocker (2013) calculates that a scheme reducing carbon dioxide emissions by 6.5% now would result in only a 1.5° warming, whereas one with 1.0% reduction beginning in 2040 would result in a 6° warming. Less visible, but equally unsettling, trends in ecological systems reveal a global decrease in plant pollinators (Tylianakis, 2013) of wild and crop plants, a decrease in overall biodiversity (Cardinale et al., 2012), and a trophic downgrading of the planet's food webs (Estes et al., 2011).

Presently, we are having enormous impact on the biosphere relative to its ability to provide an appealing home for us and for valued species. Many issues have emerged that cannot be put off until the global economy improves or nations become comfortable with working together to solve common problems. The science of ecotoxicology is now a part of our knowledge needed to protect the quality of our lives. It is extremely important to understand that ecotoxicological knowledge and activities are no more or less important than our industrial knowledge and activities. Few people would willingly abandon the right to clean water and an aesthetically pleasing environment. Similarly, few people would deny themselves or their children the fruits of our industry—an industry that produces waste. The key is to work diligently toward an informed and insightful balance between these two facets of our lives. And to do so before it is too late.

> *The main threats to sustained human existence now come from people, not from nature. Ecological shocks that irreversibly degrade the biosphere could be triggered by the unsustainable demands of a growing world population ... political tensions will probably stem from scarcity of resources, aggravated by climate change.*
>
> **Rees (2013)**

In the Preface of this book, I presented a counterargument to the well-known phrase from Ecclesiastes that: "Nothing is new under the sun. Even that thing of which they say, 'See, this is new!' has already existed in the ages that preceded us." Referring to these same words, McNeill (2000) stressed that a new, fundamental change has occurred in our relationship to the environment:

> *There is something new under the sun ... the place of humankind within the natural world is not what it was. In this respect at least, modern times are different, and we would do well to remember that.*

14.3 SCIENTIFIC IMPORTANCE OF ECOTOXICOLOGY

It may not seem to be true if you are reading this chapter as part of a college course that, at long last, is finally drawing to an end, but there is also intrinsic value to the material described so far. There is enormous value in simply learning more about your world. With ecotoxicology, one has the added value of the knowledge having obvious and immediate utility.

Our world is filled with fascinating concepts and more appear daily. In the 2 years that passed in writing the first edition of this book, many new discoveries were made about our world. The conclusion from the Viking lander missions that life did not exist on Mars was thrown open to question based on evidence from a Martian meteorite found on the Antarctic ice. As the second edition was being revised, NASA detected large amounts of water on Mars. The scientific debate about life on Mars continued, e.g., Formisano et al. (2004). As this fourth edition was being revised, the Mars rover, Curiosity, had drilled into Martian rock to confirm that ancient conditions there had been conducive to life. Although only a handful of exoplanets (planets outside our solar system) were known when the first edition of this book was written, the count has increased quickly to 861 as this chapter is being rewritten. More are being found almost daily, and perhaps more importantly, methods are being developed to ascertain their qualities such as their capability of harboring life. While the first edition was being written (1997), the first mammalian clone, a sheep named Molly, was produced from a mature cell of another individual. As the second edition went to press (2003), the field of cloning had shot forward at such a rate that all countries of the world were intently focused on the legal and moral issues surrounding human cloning. As the third edition was being written, genes had been inserted into mouse skin cells to produce pluripotent* stem cells (Vogel and Holden, 2007) and primate stem cells were being made from adult cells (Wilmut and Taylor, 2007). Now wealthy clients who contract with specialty biotechnology companies to clone their dead pet, must be counseled that cloning is a form of reproduction, not resurrection! A new form of matter (Bose–Einstein condensate) was generated in 1995 and, when writing of the third edition began in 2006, was used to literally quantum teleport cesium atoms between two locations (Sherson et al., 2006). The existence of the previously theoretical Higgs boson was just confirmed at CERN's Large Hadron Collider. What were once the props of science fiction writers are now the realities of our world. New facts enhance our appreciation of the world and facilitate wise decisions. Although I admit to more than a small bias here, those associated with ecotoxicology are certainly among the most interesting and useful.

* Pluripotent implies capable of going on to become many types of cells during development. It would not be too difficult to imagine this as an important step in a progression toward being able to produce complete individuals from such genetic manipulation of adult skin cells.

APPENDIX 1

International System (SI) of Units Prefixes

Factor	Unit Prefix (Symbol)	Factor	Unit Prefix (Symbol)
10^{15}	peta (P)	10^{-1}	deci (d)
10^{12}	tera (T)	10^{-2}	centi (c)
10^{9}	giga (G)	10^{-3}	milli (m)
10^{6}	mega (M)	10^{3}	kilo (k)
10^{-6}	micro (μ)	10^{-9}	nano (n)
10^{2}	hecto (h)	10^{-12}	pico (p)
10^{1}	deka (da)	10^{-15}	femto (f)

APPENDIX 2

Miscellaneous Conversion Factors*

1 acre (a) = 4.046873 × 10^3 meter^2 (m^2)
1 atmosphere (atm) = 1.013250 × 10^5 pascal (Pa)
1 bar (bar) = 1.000000 × 10^5 pascal (Pa)
1 barrel of oil (bbl) = 4.2 × 10^1 gallons (gal) = 1.589873 × 10^{-1} meter^3 (m^3)
1 becquerel (Bq) = 1 disintegration per minute (dpm) = 27.027 × 10^{-12} curie (Ci)
1 British thermal unit (Btu) = 1.055 × 10^3 joule (J)
1 calorie (Cal) = 4.186800 joule (J)
1 cubic foot (ft^3) = 2.831685 × 10^{-2} cubic meter (m^3)
1 curie (Ci) = 2.2 × 10^6 disintegrations per minute (dpm) = 3.700 × 10^{10} becquerel (Bq)
Dalton (Da) = an arbitrary atomic mass unit that is 1/12 the mass of a carbon atom with mass number of 12
degrees Celsius (°C) = degree Kelvin (K) – 273.16
degrees Celsius (°C) = (5/9) (degree Fahrenheit – 32) (F)
1 electron volt (eV) = 1.602 × 10^{-12} erg = 1.602 × 10^{-19} joule (J)
1 foot (ft) = 3.048000 × 10^{-1} meter (m)
1 gallon (gal) = 3.785412 × 10^{-3} cubic meter (m^3)
1 Gray (Gy) = 1 Joule of energy deposited per kilogram (kg) of tissue
1 liter (L) of crude oil ≈ 868.63 gram (g) based on the density of Kuwait crude oil at 20°C
1 micron (μ) = 1.000000 × 10^{-6} meter (m)
1 mile (mi) = 1.6093 × 10^3 meter (m)
1 millibar (mbar) = 1.000000 × 10^2 Pascal (Pa)
1 millimeter of mercury = 1.33322 × 10^2 Pascal (Pa)
1 pound (lb) = 4.535924 × 10^{-1} kilogram (kg)
1 rad (absorbed radiation dose) = 1.000000 × 10^{-2} gray (Gy)
1 rem (radiation dose equivalent) = 1.000000 × 10^{-2} sievert (Sv)
1 Roentgen (R) = 2.58 × 10^{-4} coulomb per kg of air
1 seivert (Sv) = 1.0 × 10^5 mrem
1 (short) ton (tn) = 2.000 × 10^3 pound (lb) = 9.071847 × 10^2 kilogram (kg) = 0.907 tonnes (t)
1 torr (mm Hg) = 1.33322 × 10^2 Pascal (Pa)
1 yard (yd) = 9.144000 × 10^{-1} meter (m)

* Extracted from *CRC Handbook of Chemistry and Physics* (Lide, 1992) and miscellaneous sources.

APPENDIX 3

Summary of U.S. Laws and Regulations

OVERVIEW

Laws and regulations are important practical aspects of ecotoxicology that deserve discussion.* Discussion is focused on U.S. federal law in this appendix. Guest authors of Appendices 4 through 8 detail the important topics of European Union, Chinese, Australian, South African, and Indian legislation. Additional information pertinent to Canada and the **European Community (EC)** law is also provided in Foster (1985) for water, Elsom (1987) for air, and Frankel (1995) for marine resources. European community activities and the **Organization for Economic Cooperation and Development (OECD)** policies are described in Blok and Balk (1995), Grandy (1995), and Crane et al. (2002).

The evolution of the U.S. environmental movement, including the eventual establishment of the federal environmental legislation outlined here, is carefully examined in the excellent book, *First along the River*, by Benjamin Kline (2011). He defines the early 1800s as the time when the U.S. frontiers were closing and general misuse of land and natural resources become obvious to the public. The nation moved into an industrialization phase ca. the 1880s, bringing additional issues of workplace, urban, and eventually general environmental pollution to the public's attention. Conservation emerged as an important political theme in the first three decades of the twentieth century and an environmental movement spread across the country in the 1940s and 1960s. Environmental laws passed before the 1970s were brought together to form what was to become the central environmental legislation of the U.S. as sketched out below.

CREATION AND IMPLEMENTATION

Perhaps the best way to begin is with a description of how a piece of U.S. legislation is implemented. Once the U.S. law is established, a federal agency such as the Environmental Protection Agency (EPA) is charged with developing the associated regulations that are intended to enforce the particulars of the law. When EPA is so charged, its proposed rules are published initially in the *Federal Register* (*FR*) and remain open for public comment for a specified period. After that time has elapsed, EPA repeats the internal process of rule development with consideration of any public comment and other input. Final rules are then established and published in the *FR*. These regulations may be open to interpretation by affected parties but redress occurs through the courts after this point in the process. During the next revision of the *Code of Federal Regulations* (*CFR*), the regulations are incorporated into this compilation of federal code.

* Only a few sets of national environmental laws are provided here; however, Wikipedia has the following web pages that provide summaries for laws of other countries, http://en.wikipedia.org/wiki/List_of_environmental_laws_by_country. See also http://en.wikipedia.org/wiki/Environmental_law.

The acts and laws established by the U.S. Congress are identified using references such as 42 USC §§6901 to 6992k or PL 91-190 for the National Environmental Policy Act (NEPA). USC and PL are abbreviations for U.S. Code and Public Law, respectively. As mentioned, a law* requires a federal agency such as the EPA to develop **regulations** that provide specific details insuring that the intent of the law is met. According to Mackenthun and Bregman (1992), such regulations "specify conditions and requirements to be met, ... provide a schedule for compliance, and ... record any exceptions to the regulated community." Within the text of these materials, legal definitions are also provided for terms such as contaminant (40 CFR Sections 230.3, 300.6, 310.11, etc.), point source (40 CFR Sections 233.3 and 260.10), or pollutant (40 CFR Section 122.2) that may vary among laws and deviate from scientific definitions or common use. Therefore, it is important to understand the legal context of such terms. Legal dictionaries of environment-related terms (e.g., King, 1993) should be consulted to avoid confusion when working under a specific set of regulations. Regulations from EPA are published in the *CFR* under Title 40 (Protection of Environment) and updated as needed in the *FR*. The daily issuances of the *FR* augment the *CFR*, and consequently, these sources together define current federal regulations. These specific codifications of environmental regulation are referenced with numbers such as 40 CFR Parts 1500 to 1517. The numbers 1500 to 1517 identify specific parts of Title 40 of *CFR*; here, a section providing details on National Environmental Policy Act.

With this background established, specific laws and associated regulations can now be outlined. Each is presented with relevant references noted from the *CFR* and *FR*.

NATIONAL ENVIRONMENTAL POLICY ACT

The purposes of this Act are: To declare a national policy which will encourage productive and enjoyable harmony between man and his environment; to promote efforts which will prevent or eliminate damage to the environment and biosphere and stimulate the health and welfare of man; to enrich the understanding of the ecological systems and natural resources important to the Nation; and to establish a Council on Environmental Quality.

(42 USC 4321)

The **National Environmental Policy Act** or NEPA (1969, PL 91-190; 42 USC §§6901-6992k; see 40 CFR 1500-1517) was signed into law on January 1, 1970 by President Richard Nixon. Combined with the Environmental Quality Improvement Act (EQIA, 1970), NEPA formed the cornerstone for U.S. environmental policy (Foster, 1985). NEPA established a federal commitment to judge the government's actions relative to environmental impacts and to take action fostering "harmony" between human activities and nature. It established the **Council on Environmental Quality** to aid and advise the President during preparation of the annual **Environmental Quality Report**. This council assesses federal programs relative to NEPA and researches various topics on environmental quality.

NEPA requires federal agencies to prepare an **environmental impact statement** (EIS) for any major federal action that could have an adverse environmental impact. In addition to direct actions, federal actions include issuance of federal permits, leases or licenses, and granting, contracting, or loaning of federal monies (McGregor, 1994). NEPA does not cover nonfederal actions and there are no defined rights of individuals to live without environmental damage included in NEPA (Freedman, 1987). The EIS outlines possible adverse impacts, details alternatives to the action

* Although we are discussing only U.S. federal law here, **environmental law** is much more inclusive. McGregor (1994) defines environmental law as "a body of federal, state, and local legislation, in the form of statutes, bylaws, ordinances, and regulations, plus court-made principles known as common law." Of course, international treaties and conventions also are becoming more important to U.S. environmental law as well.

including the "no-action" alternative, and describes irreversible impacts to resources (McGregor, 1994). Before an EIS, a shorter and preliminary **environmental assessment** (EA) may be used to determine if a full EIS is necessary. A statement of no significant effect ("**finding of no significant impact**" or FONSI) is made if no EIS is thought to be needed. Otherwise, a full EIS is developed. This process of incorporating environmental impact into the decision making of the federal government is open to public comment, and involves relevant state, local, and federal agencies with the EPA overseeing much of the process.

CLEAN AIR ACT

The purposes of [the Clean Air Act] *are —*

(1) to protect and enhance the quality of the Nation's air resources so as to promote the public health and welfare and the productive capacity of its population;
(2) to initiate and accelerate a national research and development program to achieve the prevention and control of air pollution;
(3) to provide technical and financial assistance to State and local governments in connection with the development and execution of their air pollution prevention and control programs; and
(4) to encourage and assist the development and operation of regional air pollution prevention and control programs.

...

A primary goal of this Act is to encourage or otherwise promote reasonable Federal, State, and local governmental actions, consistent with the provisions of this Act, for pollution prevention.

(42 USC 7401)

The **Clean Air Act** or CAA (1977, PL 9595, 42 USC §§7401-7671q; see 40 CFR50-99; also 1990 Clean Air Act Amendments [CAAA], PL 101-549, 104 Stat. 2399) was proceeded by the 1970 Clean Air Act that developed air emission standards for States and the federal government.* Overall, the CAA of 1977 was designed to regulate air pollution from stationary (e.g., factories) and mobile (e.g., automobiles) sources so that human health and the environment were protected. It is overseen and administered by the EPA, but responsibility for control and prevention of air pollution is passed to state and local governments. This act sets maximum allowable levels of pollution and dictates emission levels designed to achieve these limits. It establishes national emission standards for pollutants. Stationary sources are controlled by a state permit system. As detailed in the CAA, mobile sources are regulated through processes such as motor vehicle emissions standards or vehicle inspection programs. States are required to develop **State Implementation Plans** (SIPs) outlining steps toward meeting and maintaining **national ambient air quality standards** (NAAQS). Section 109 of this act requires the EPA to establish and periodically revise NAAQS. The act also includes details for prevention of significant deterioration (PSD) in geographical areas that have attained the NAAQS.

Importantly, the 1990 CAAA shifted the control of hazardous emissions to a technology-based context from a risk-based one. The federal government was given more responsibility and authority to meet the intent of these modifications. The EPA now has a mandate to set maximum achievable control technology (MACT) emission standards according to the modification of CAA Section 112 in 1990.

* As mentioned in Chapter 1, air pollution issues in urban regions were among the first environmental pollution addressed legally in the U.S. The 1970 CAA, 1977 CAA and CAAA followed the 1955 Air Pollution Control, 1963 Clean Air, and 1967 Air Quality Acts. Industrialized cities had created air pollution boards such as that established in 1947 by St. Louis.

FEDERAL INSECTICIDE, FUNGICIDE, AND RODENTICIDE ACT

[FIFRA's aim is] *To regulate the marketing of economic poisons and devices, and for other purposes.*

(40 CFR 150-189)

The **Federal Insecticide, Fungicide, and Rodenticide Act** or FIFRA (1972, 7 USC 136, PL 95396; see also 40 CFR 150-189) controls the production, labeling, shipping, sale, and use of **pesticides** ("substances used to prevent, destroy, repel, or mitigate any pest") (Mackenthun and Bregman, 1992). Before the 1972 act, a 1910 pesticide law existed but it addressed pesticide consumer fraud issues, not primarily environmental or health safety. The FIFRA (1947) involved pesticide registration and labeling issues and was the predecessor of the current FIFRA that was established in 1972 and modified since then with important amendments such as PL 104-170, 1996 for food quality issues.

Pesticides may be individual substances or mixtures including plant regulators, insecticides, fungicides, rodenticides, desiccants, and defoliants (Freedman, 1987). FIFRA defines certification of restricted-use pesticide applicators, defines the conditions for canceling the registration for a pesticide, and establishes pesticide disposal requirements (Mackenthun and Bregman, 1992). FIFRA requires submittal of materials for pesticide registration that includes the specific labeling and directions for use, and results of product tests. Specifics of ecological effects testing and environmental fate are given in 40 CFR 158.145 and 40 CFR 158.130, respectively (Touart, 1995). Submitted information must demonstrate that the product will work as intended without unacceptable human or ecological risk. Registration is renewed periodically and, if a pesticide is found after registration to cause unreasonable risk, the EPA can cancel its registration under Section 6 of FIFRA (see 40 CFR 154).

MARINE PROTECTION, RESEARCH, AND SANTUARIES ACT

The Congress declares that it is the policy of the United States to regulate the dumping of all types of materials into ocean waters and to prevent or strictly limit the dumping into ocean waters of any material which would adversely affect human health, welfare, or amenities, or the marine environment, ecological systems, or economic potentialities.... It is the purpose of this Act to regulate (1) the transportation by any person of material from the United States and, in the case of United States vessels, aircraft, or agencies, the transportation of material from a location outside the United States, when in either case the transportation is for the purpose of dumping the material into ocean waters, and (2) the dumping of material transported by any person from a location outside the United States, if the dumping occurs in the territorial sea or the contiguous zone of the United States.

(40 CFR 220-233)

The **Marine Protection, Research, and Sanctuaries Act** or MPRSA (1972, 33 USC 1401; see 40 CFR 220-233) recognizes that the practice of ocean dumping is unsound, states the U.S. policy regarding dumping, and establishes limitations and prohibitions on dumping. It prohibits or limits dumping of various materials from the United States, or by U.S. vehicles (Freedman, 1987). The EPA limits dumping of wastes except dredge spoils through a permit system. The Army Corp of Engineers uses EPA criteria and regulates dredge spoils dumping (Mackenthun and Bregman, 1992). The Ocean Dumping Ban Act (1988) amended the MPRSA and put a time limit on dumping of materials such as sewage sludge, industrial waste, and medical waste (Mackenthun and Bregman, 1992). Also prohibited is the dumping of radioactive waste, and radioactive, chemical or biological weapons material (Freedman, 1987).

CLEAN WATER ACT

The objective of this Act [Clean Water Act] *is to restore and maintain the chemical, physical, and biological integrity of the Nation's waters.*

(40 CFR 100)

The **Clean Water Act** or CWA (1977, PL 95-217, 33 USC §§1251 to 1387; see 40 CFR Sections 100 to 140, 400 to 699; see also PL 100-4, Feb. 4, 1987 and PL 107–303, November 27, 2002 amendments) limits the discharge of pollutants (primarily industrial and municipal discharges) into navigable waters. This is accomplished by imposing limits on discharges using either, a state or EPA permitting system (**National Pollutant Discharge Elimination System** or NPDES). State limits are developed under the guidance of the EPA. The immediate goal of this legislation was to improve water quality so as to insure that waters are swimmable and fishable (McGregor, 1994). The overall goal is to "restore and maintain the chemical, physical, and biological integrity" of U.S. water bodies.

SAFE DRINKING WATER ACT

[This Act that pertains to public water systems, not private sources,] *specifies contaminants which, in the judgment of the Administrator, may have any adverse effect on the health of persons; ...*

(B) specifies for each such contaminant either—

(i) a maximum contaminant level, if, in the judgment of the Administrator, it is economically and technologically feasible to ascertain the level of such contaminant in water in public water systems, or

(ii) if, in the judgment of the Administrator, it is not economically or technologically feasible to so ascertain the level of such contaminant, each treatment technique known to the Administrator which leads to a reduction in the level of such contaminant sufficient to satisfy the requirements of section 1412; and

(C) contains criteria and procedures to assure a supply of drinking water which dependably complies with such maximum contaminant levels; including accepted methods for quality control and testing procedures to insure compliance with such levels and to insure proper operation and maintenance of the system, ...

(40 CFR 141)

The **Safe Drinking Water Act** or SDWA (1974; 42 USC §§300f to 300j-26; see 40 CFR 141-149; amended in 1986 and 1996 (PL 104-182)) protects the public water supply (systems supplying more than 25 people or possessing more than 15 service outlets) by setting quality standards, and protecting source aquifers from contamination by establishing permitting requirements for underground disposal of waste. The EPA has oversight but individual states are assigned responsibility for enforcement. The EPA establishes primary and secondary "at the tap" standards for public water that the states can either use without modification or use to establish even more stringent standards (Mackenthun and Bregman, 1992). Primary standards ("maximum contaminant levels" or MCLs) are those for a list of specific contaminants that adversely affect health; secondary standards are EPA recommendations for qualities that protect the public welfare (e.g., water taste, color, or smell) (McGregor, 1994). The present MCL list includes qualities under the categories of volatile organic compounds, organic compounds, inorganic chemicals including radionuclides, microbiology, and turbidity (Mackenthun and Bregman, 1992). This list of substances of concern is reviewed and revised every 3 years by the EPA. In the 1986 amendment, a ban that specified limits was added to restrict lead use in public water system components such as solder and fixtures.

Disposal of wastes into potential underground drinking water sources is also closely controlled under the SDWA through the use of a permit system or designation of aquifers as "sole source aquifers" to be protected. Federal activities are restricted in areas with sole source aquifers. The EPA also sets strict regulations for underground waste injection activities under the SDWA.

TOXIC SUBSTANCE CONTRAL ACT

It is the policy of the United States that—

(1) adequate data should be developed with respect to the effect of chemical substances and mixtures on health and the environment and that the development of such data should be the responsibility of those who manufacture and those who process such chemical substances and mixtures;

(2) adequate authority should exist to regulate chemical substances and mixtures which present an unreasonable risk of injury to health or the environment, and to take action with respect to chemical substances and mixtures which are imminent hazards; and

(3) authority over chemical substances and mixtures should be exercised in such a manner as not to impede unduly or create unnecessary economic barriers to technological innovation while fulfilling the primary purpose of this Act to assure that such innovation and commerce in such chemical substances and mixtures do not present an unreasonable risk of injury to health or the environment.

(40 CFR 700-799)

The **Toxic Substances Control Act** or TSCA (1976, 15 USC 2601; see 40 CFR 700-799) regulates, through the EPA, the manufacture, processing, transport, use, import, and disposal of chemicals or chemical mixtures that may pose an unreasonable risk to human health or the environment. Consideration is given to the quality and quantity of substances being produced. It does not cover items already regulated by other laws such as cosmetics, drugs, firearms and ammunition, food additives, pesticides, and nuclear materials (Mackenthun and Bregman, 1992). However, it does include products of biotechnology such as genetically altered microbes intended for release into the environment. It specifically regulates some existing chemicals not already regulated by other laws, such as polychlorinated biphenyls. The EPA requires producers to give 90-day notification of intent to manufacture or import, or to put a chemical product to new use. It requires testing to clarify the potential for unreasonable risk and a statement of the intended use of the material. The EPA may then permit, deny permission within 90 days of notice, or delay permit issuance until further study has been completed. The EPA may inspect facilities or investigate any actions that may violate this law. The law provides for criminal liability under specific conditions, that is, knowing and willful violation (McGregor, 1994). There are also record keeping requirements and an obligation to notify the EPA if new information arises relative to the risk associated with the chemical or mixture (Freedman, 1987).

RESOURCE CONSERVATION AND RECOVERY ACT

The Congress hereby declares it to be the national policy of the United States that, wherever feasible, the generation of hazardous waste is to be reduced or eliminated as expeditiously as possible. Waste that is nevertheless generated should be treated, stored, or disposed of so as to minimize the present and future threat to human health and the environment.

(42 USC 6902)

The **Resource Conservation and Recovery Act** or RCRA (1976, 42 USC 6901-6987; see 40 CFR 240-299 and 1984 Hazardous and Solid Waste Amendments, PL 98-616) regulates production, storage, treatment, and final disposal of hazardous wastes (McGregor, 1994), i.e., "cradle-to-grave" regulation. The EPA is required to develop standards for generation, transport, storage, and treatment of hazardous waste, and to establish a permit system for storage, treatment, and disposal (Mackenthun and Bregman, 1992). Disposal requirements for small-quantity generators, such as research laboratories are also defined by this law. RCRA requires accurate record keeping. With the 1984 amendment, clear legal penalties were specified for knowing violation of a permit, illegal transport or treatment of waste, or falsifying or altering records. Disposal of waste into underground formations was also restricted or prohibited by the 1984 Amendment.

COMPREHENSIVE ENVIRONMENTAL RESPONSE, COMPENSATION, AND LIABILITY ACT

> [CERCLA is intended to] *provide for liability, compensation, cleanup, and emergency response for hazardous substances released into the environment and the cleanup of inactive hazardous waste disposal sites.*
>
> **(26 USC 4611-4612)**

The **Comprehensive Environmental Response, Compensation, and Liability Act** or CERCLA or Superfund Act (1980, 42 USC 9601-9615, 9631-9633, 9651-9657, 26 USC 4611-4612, 4661-4662, 4681-4682; see 40 CFR 300-374, amended in 1986 with the Superfund Amendments and Reauthorization Act or SARA) gives EPA the authority to respond to hazardous releases, cleanup or require cleanup of hazardous waste sites, and assign liability for released hazardous waste (Mackenthun and Bregman, 1992). Although costs are first sought from responsible parties, CERCLA also established two funds (Hazardous Substance Response Trust Fund and Post-Closure Liability Trust Fund) with various tax sources (oil, chemical, and hazardous waste activities) to cover some cleanup costs (Freedman, 1987). CERCLA also establishes a National Contingency Plan to identify sites for cleanup, and define the required cleanup activities (Freedman, 1987). At this time, the EPA has compiled a listing (National Priority List or NPL) of sites for cleanup under CERCLA.*

* Details about the current NPL sites listed by the EPA are provided on the web location, http://www.epa.gov/superfund/sites/npl/.

APPENDIX 4

Summary of European Union Laws and Regulations*

Mark Crane and Albania Grosso

The **European Union (EU)**, established in 1957 by the Treaty of Rome, differs from the U.S. federal government in that EU Member States are sovereign nations that have surrendered only some law making and enforcing powers. The Treaty of Rome was a tool to establish a common market, expand economic activity, promote living standards, and, perhaps most importantly, encourage political stability in Western Europe after the ravages of two World Wars. The **European Community** formed by the Treaty of Rome unified the European Economic Community, the European Atomic Energy Community, and the European Coal and Steel Community. This has expanded in both membership and aims over the decades since its formation. The postwar organization, set up primarily for economic and security reasons, has now evolved into the EU, with 28 Member States in 2013 (in order of population size): Germany, United Kingdom, France, Italy, Spain, Poland, Romania, the Netherlands, Belgium, Greece, Portugal, Czech Republic, Hungary, Sweden, Austria, Bulgaria, Denmark, Finland, Slovakia, Ireland, Croatia, Lithuania, Slovenia, Latvia, Estonia, Cyprus, Luxembourg, and Malta. Candidate countries for membership of the EU, who are in the process of "transposing" (or integrating) EU legislation into national law, are Iceland, Montenegro, Serbia, the former Yugoslav Republic of Macedonia, and Turkey. There are also three further "potential candidates" who do not yet fulfill the requirements for EU membership: Albania, Bosnia and Herzegovina, and Kosovo.

There are four major EU institutions responsible for legislation under the Treaty of Rome: the European Commission, European Parliament, Council of Ministers, and European Court. The Commission is the supreme EU executive, comprising independent members appointed by individual Member States. Members of the Commission are charged with operating in the interests of the community as a whole, and not as national representatives. Each Commissioner has responsibility for an area of community policy, which includes a Commissioner for the Environment, and their main function is to propose EU legislation. The Commission civil servants are divided among different Directorates that report to the Council of Ministers.

Legislation generally emerges from the Commission in the form of **Directives** or **Regulations**. These are usually enforceable across all Member States and are the main basis for statutory controls in EU environmental legislation. **Directives** empower the Commission to define objectives, standards, and procedures, but allow Member States some flexibility in implementation, so they can use their own national legislative processes to "transcribe" the Directive and develop their own regulations. These national legislative processes can differ fundamentally among the different

* For further information, see also the Wikipedia page, http://en.wikipedia.org/wiki/Environmental_policy_of_the_European_Union or the European Commission page, http://europa.eu/legislation_summarie/environment/.

EU states. For example, in the United Kingdom and Ireland case law forms the legal base, while in much of continental Europe legal principles derive from the Napoleonic Code. A **Regulation** is a direct legislative act of the European Union that becomes immediately enforceable as law in all Member States simultaneously.

The European Parliament is a consultative and advisory body, and its function is to assess proposals for legislation, by commenting on the Commission's proposals. The Council of Ministers comprises government ministers from each Member State, and is the primary decision-making body in the EU, with a main function to consider proposals from the Commission. The main function of the European Court is to interpret and apply all community law. A **European Environment Agency** exists, but this is currently a small body responsible largely for gathering data on environmental indicators from Member States so that general trends in Europe can be reported upon. The practical aspects of environmental policy implementation, monitoring, and enforcement fall largely to individual agencies within the Member States.

Environmental legislation adopted by the EU over the past two decades can be divided into three broad categories as follows:

- Directives and Regulations that try to limit or prohibit discharges of dangerous substances into the environment by industrial plants
- Directives and Regulations setting environmental quality objectives for various uses of the environment (e.g., use of potable water, protection of shellfisheries, or protection of the wider aquatic environment)
- Geographically specific regulations on pollution to help protect areas such as the North, Baltic, and Mediterranean Seas

There are several European Directives and Regulations that are important for those with an interest in ecotoxicology.

The **REACH (Registration, Evaluation, Authorisation and Restriction of Chemicals)** Regulations came into force in June 2007 and required EU manufacturers or importers of most chemical substances on their own or in preparations to register the substance if they manufacture or import it in quantities of 1 t or more per year. Deadlines for registration depend on whether a substance was already on the market and could be "preregistered" as a "phase-in substance" in 2008, or is a new, nonphase-in substance. Nonphase-in substances must be registered under REACH before they can be marketed in the EU. In contrast, phase-in substances can remain on the market subject to the provision of data to the European Chemicals Agency (ECHA) by deadlines in 2010, 2013, and 2018.

REACH provides a timetable for substance registration based on the inherent hazard and production volume of each substance. The first deadline was for substances manufactured or imported at greater than or equal to 1000 t/year, or which were considered to be carcinogenic, mutagenic, or reprotoxic (CMR) at volumes of greater than or equal to 1 t/year, or were classified as R50/R53 at volumes of >100 t/year. The deadline for the registration of these substances was by December 1, 2010. During this first phase of registration 3801 substances were registered as either full registrations, on-site isolated intermediates, or transported isolated intermediates. A second phase of registration was completed by June 1, 2013 for substances manufactured or imported by a company at greater than or equal to 100–1000 t/year. The final registration phase for phase-in substances is scheduled for completion by June 1, 2018 for substances manufactured or imported by a company at greater than or equal to 1–100 t/year.

The four basic registration steps identified in the REACH regulations are as follows:

- Step 1—Gather and share existing information
- Step 2—Consider information needs
- Step 3—Identify information gaps
- Step 4—Generate new data/propose testing strategy

Each manufacturer, importer, or "only representative" (of manufacturers based outside the EU) is individually obliged to submit a separate registration for each of his substances. However, registrants for the same substance must jointly submit information on its hazardous properties, classification and labeling, future testing proposals, and safe uses via a lead registrant who is selected through discussion amongst all registrants. The Substance Information Exchange Forum (SIEF) is the main mechanism for sharing data for phase-in substances after preregistration. The "inquiry process" is the mechanism used for nonphase-in substances.

Manufacturers or importers of substances above 1 t/year must submit a technical dossier electronically to ECHA. Submission must be via IUCLID 5 software and must contain information on the following:

- Manufacturer/importer identity
- Substance identity, manufacture, and use
- Substance classification and labeling
- Safe use guidance
- Substance intrinsic properties in the form of robust summaries
- Whether the above information has been reviewed by an assessor
- Proposals for further testing, if relevant
- Main use categories, type of uses, and significant routes of exposure (for substances registered in quantities of 1–10 t/year)

Manufacturers or importers of substances above 10 t/year must submit a Chemical Safety Report (CSR) to ECHA, which documents a Chemical Safety Assessment (CSA) for the substance. If the substance is classified as Dangerous; Persistent, Bioaccumulative and Toxic (PBT); or very Persistent and very Bioaccumulative (vPvB), an exposure assessment needs to be performed, using exposure scenarios. ECHA has been working on standardizing the approach and format of the exposure scenarios included in the submitted CSRs, for example, by developing the Chemical Safety Assessment and Reporting (CHESAR) tool to enable registrants to carry out their safety assessments in a structured, harmonized, and efficient way. The tool uses substance-specific data directly imported from IUCLID and reports exposure and risk calculation results, which can then be imported directly back into IUCLID for submission to ECHA.

Annex VI of REACH is applicable for all substances. The required information is as follows:

- **General information on the registrant** (e.g., name, address, contact person, and production sites)
- **Substance identity** (e.g., name, EINECS/CAS number, formula, and impurities)
- **Manufacture and use** (e.g., overall manufacture, form supplied to downstream users, waste quantities, and unsafe uses)
- **Classification and labeling** (e.g., hazard classification and specific concentration limits)
- **Guidance on safe use** (e.g., Safety Data Sheet information)
- **Exposure** (use category, use specification, significant routes of human and environmental exposure, exposure pattern)

In addition to the Annex VI information requirements, hazard data are required for substances, depending on the manufacture or imported volume of the substance (tonnage band).

With expert argument it is possible to adapt or avoid the tests identified in the table above if testing does not appear to be scientifically necessary (e.g., it is possible to use alternative nontesting approaches such as quantitative structure–activity relationship and read-across, or weight of evidence from existing studies), or if testing is not technically possible, or an exposure assessment for the specific substances suggests that particular tests for a hazard are not required.

Registrants needed to notify ECHA electronically by December 1, 2010, about classification and labeling of a substance alone or in a preparation that is placed on the market and for which a registration had not already been submitted. This had to be done if the substance met criteria for classification as dangerous under Directive 67/548/EEC (the Directive on Classification, Packaging,

REACH Information Requirements

<table>
<tr><th colspan="4">Tonnage Band</th><th>Information Requirements</th></tr>
<tr><td rowspan="4">>1000 t/year</td><td rowspan="3">100–1000 t/year</td><td rowspan="2">10–100 t/year</td><td>1–10 t/year[a]</td><td>REACH Annex VII
Physicochemical properties (e.g., melting and boiling point, solubility, octanol–water partitioning, and explosivity)
Toxicology (skin and eye irritation, skin sensitization, mutagenicity, acute toxicity)
Ecotoxicology (acute toxicity to invertebrates, algal growth inhibition, ready biodegradability)</td></tr>
<tr><td></td><td>REACH Annex VIII
Toxicology (repeated dose toxicity, reproductive toxicity, toxicokinetics)
Ecotoxicology (acute fish toxicity, sludge respiration, hydrolysis, sorption)</td></tr>
<tr><td></td><td></td><td>REACH Annex IX
Physicochemical properties (stability in organic solvents, dissociation constant, viscosity)
Toxicology (further repeated dose and reproductive toxicity tests)
Ecotoxicology (long-term invertebrate and fish tests; degradation in water, soil and sediment; fish bioaccumulation; tests with terrestrial invertebrates, plants and microbes)</td></tr>
<tr><td></td><td></td><td></td><td>REACH Annex X
Toxicology (further reproduction studies, carcinogenicity study)
Ecotoxicology (further degradation, fate and behavior studies; long-term toxicity to terrestrial invertebrates, plants, and to aquatic sediment organisms; long-term toxicity to birds)</td></tr>
</table>

[a] *REACH Annex III on identifying CMRs and dangerous substances also applies to substances in this tonnage range*

and Labelling of Dangerous Substances) or exceeded concentration limits in Directive 1999/45/EC if present in a preparation. ECHA collated all information received into a Classification & Labelling Inventory, which identifies whether registrants and notifiers differ in their labeling or classification for the same substance. If this occurs, the registrants and notifiers are required to make every effort to come to an agreed classification.

Along with REACH, the **Water Framework Directive** (WFD) (2000/60/EC) is one of the most important pieces of European environmental legislation of relevance to ecotoxicologists, requiring all inland and coastal waters to achieve "good status" by 2015. It does this by establishing a river basin district structure within which demanding environmental objectives are set, including ecological targets for surface waters and the use of Environmental Quality Standards (EQS; similar to Water Quality Criteria in the United States as described in Appendix 3) for individual chemical pollutants.

Article 16 of the Directive describes broadly how and by when EQS for pollutants should be developed:

1. Pollutants presenting a significant risk to or via water should be identified by the European Commission and classified as priority substances (PS), with the most hazardous of these classed as priority hazardous substances (PHS). For PS and PHS, measures should aim at progressive reduction and cessation of discharges, respectively, by 2025. Priority substance lists should be reviewed at 4-year intervals.
2. Quality standards should be submitted by the European Commission, which are applicable to concentrations of PS in surface water, sediments, or biota.

In 2008, the European Commission published an EQS "Daughter" Directive to the WFD to deal with the control of priority substances, which has now itself been replaced by an updated EQS Directive (2013/39/EU). In the new EQS Directive, EQS are set for surface waters or biota (usually fish) for 48 priority substances, 21 of which are identified as PHS. Annual Average EQS for surface waters are set for all of these substances except hexachlorobenzene, hexachlorobutadiene, and mercury for which only maximum allowable concentrations and biota EQS have been set. Only biota EQS are set for dioxins and dioxin-like compounds. Sediment EQS have not been set because of the technical problems associated with these, but trends in sediment and biota concentrations of PS and PHS should be monitored by Member States. There is now comprehensive guidance on how to develop and monitor an EQS in different matrices, which helps to ensure greater consistency between Member States, and greater consistency between the development of EQS and the development of PNECs for the same substances across different European Regulations and Directives.

The historical use of aquatic systems as repositories for human wastes goes some way toward explaining why so much emphasis has been placed on aquatic pollution prevention in EU law. However, air and land pollution have not been ignored. Framework Directives exist for **Waste Disposal** (84/360/EEC and 2008/98/EC) and **Large Combustion Sites** (88/609/EEC), or waste incinerators. Limits for air pollution have been set for lead, oxides of nitrogen, sulfur dioxide, and particulates. These guidelines are intended to protect both human health and to serve as long-term safeguards for the environment. There are also atmospheric limit values for volatile organic compounds (as total organic carbon), dioxins and furans, and for inorganic elements and compounds such as chlorine, fluorine, ammonia, and metals.

Other important Directives with relevance for ecotoxicologists include the **Plant Protection Products Directive** (Directive 91/414/EEC), which came into force in 1991. The purpose of this Directive was to harmonize arrangements for the marketing of plant pesticides within the European Union. Many of the data requirements for this Directive are similar to those required for new substances, e.g., toxicity profiles for fish and algae, but higher tier testing, including the use of microcosms, is more likely because pesticides are toxic substances that are deliberately spread in the environment at lethal concentrations. A risk assessment of product impacts on humans and wildlife must be carried out for both new and existing pesticides. The European Food Safety Authority (EFSA), based in Parma, Italy, is the European agency with responsibility for pesticide regulation. In June 2011, the Plant Protection Products Directive was replaced by the **Plant Protection Products Regulation** (EC 1107/2009) in an effort to provide greater uniformity across Member States. The main elements of this are as follows:

- Active substances, safeners,* and synergists are approved at EU level, following assessment against a set of agreed criteria, which cover both the intrinsic properties of active substances, safeners, and synergists (i.e., an assessment of their hazard), and the risks arising from the use of plant protection products that contain them.
- Active substances that have been shown to be without unacceptable risk to people or the environment are added to the list of approved active substances contained in Commission Implementing Regulation (EU) No. 540/2011. Inclusion is for a maximum period of 15 years but is renewable, and can be subject to conditions and be reviewed at any time.
- Substances that demonstrate a less favorable toxicological profile but still satisfy the criteria for approval may be approved as candidates for substitution.
- Existing active substances that have already undergone a review under the earlier Directive (91/414/EEC) are considered to be approved.

* Safeners are substances mixed with herbicides to make the crop species safer relative to harm from the herbicide. They decrease the harm to the crop species relative to that done to weed species.

- Products containing approved active substances are authorized by Member States working within three "zones," which cover the different climates and landscapes from northern to southern Europe. Member States in each zone share assessments and recognize each other's authorizations.
- Member States can only authorize the marketing and use of plant protection products after an active substance has been added to the list of approved active substances.
- Products containing active substances approved as candidates for substitution are subject to comparative assessment. Such products are withdrawn if that assessment identifies alternative products or methods of control, which are significantly safer and can be used without significant drawbacks.
- The Regulation also applies to safeners and synergists in plant protection products. New safeners and synergists will need to undergo the same procedures and satisfy the same requirements as for active substances. Existing safeners and synergists will undergo a review program to be established by the adoption of a Regulation by 14 December 2014 (Article 26).

There are also Directives or Regulations covering biocides, cosmetics, and veterinary and human medicines, which contain many similar provisions for environmental protection to the Plant Protection Products Regulation, including hazard, exposure, and risk assessments. The **Biocides Regulation** (EC 528/2012) replaced an earlier Directive and was applied from September 1, 2013. ECHA is the European agency, based in Helsinki, Finland, responsible for the regulation of biocides (as well as all chemicals falling under the REACH regulation, as discussed above). The **Cosmetics Regulation** (EC 1223/2009), implemented in July 2013, also replaces an earlier Directive; environmental concerns associated with cosmetics ingredients are considered to be covered by the REACH regulation. Veterinary medicines are regulated by Directive 2009/9/EC and human medicines by Regulation EC 726/2004. The European Medicines Agency, based in London, England, is the European agency responsible for regulating both human and veterinary medicines.

Finally, several Directives on nature conservation have implications for national agencies in Member States with responsibilities for regulating the environmental impacts of chemicals in either aquatic or terrestrial environments. The **Habitats Directive** requires Member States to take measures to maintain or restore natural habitats and wild species that are of nature conservation value. Special Areas of Conservation in each Member State have been agreed with the Commission, along with necessary measures to protect them. The **Birds Directive** has a similar nature conservation objective, but with an obvious focus on bird species, their eggs, nests, and wider habitats. The Directive on the **Assessment of the Effects of Certain Public and Private Projects on the Environment** requires an environmental assessment to be carried out before a decision is taken on whether development consent should be granted for certain types of project, which are likely to have significant environmental effects. The **Environmental Liability Directive** requires polluters to pay for and mitigate any environmental damage they cause, although the likely impact of this relatively new Directive remains unclear. Clearly, consideration of the potential effects of toxic chemicals falls under each of these Directives when these are relevant stressors.

In summary, the EU has developed a plethora of Directives and Regulations that seek to protect the environment. These rather piecemeal legal instruments are gradually being drawn together, often in the form of Regulations that apply in the same way across all of the EU, to provide a coherent framework for controlling the marketing and use of chemicals in the EU. This had not prevented individual Member States from implementing their own antipollution legislation to meet local environmental or political concerns, but the general move is away from individual Member State initiatives and toward environmental regulation through harmonized procedures throughout the EU.

APPENDIX 5

Summary of Modern Environmental Laws and Regulations of China*

Taiping Wang

The modern environment laws and regulations of China are those enacted after the formation of the People's Republic of China (PRC). They can roughly be divided into those emerging in three chronological stages: (1) the early stage of New China, (2) the pioneering stage, and (3) the stage since the reform and opening-up policy occurred. This appendix outlines the environmental laws and regulations, and their characteristics that emerged during these stages, emphasizing those that emerged after the reform and opening-up, i.e., the contemporary Chinese environmental laws.

LAWS OF THE EARLY STAGE OF NEW CHINA

From the foundation of the PRC in 1949 to the end of the 1960s, the immature Chinese environmental laws had many shortcomings and developed slowly. Without the support of precise and systematic theory, they tended to be infeasible or ineffective. They emphasized natural resource protection, and rarely considered pollution prevention and control. The main environmental laws and regulations established during this period included the Prescription of the State Council (SC) on the Active Protection and Rational Utilization of Wildlife Resources, the Trial Ordinance for Protection of Mineral Resources, the Ordinance for Protection of Reproduction of Fishery Resources, and the Provisional Compendium of Water and Soil Conservation.

LAWS OF THE PIONEERING STAGE

The pioneering stage of Chinese environmental laws extended from the 1960s to the end of the 1970s. The special modes of production and policy during the Cultural Revolution had accelerated impacts of humans on the environment and increased the severity of ecological damage and pollution. In 1972, the United Nations Conference on the Human Environment was held in Stockholm, Sweden. It was the first worldwide international conference on environmental issues in human history and played an important role in promoting the protection and improvement of the human environment. From then on, environmental protection was placed on the policy agendas of many countries and developed rapidly. Under the influence of this meeting and the growing worldwide focus on environmental protection, the SC enacted in 1973 the **Provisions Concerning Environmental Protection and Amelioration**, China's first comprehensive administrative law of environmental protection. With the principle of "overall planning, all-round consideration, and

* For more information, see websites such as The Ministry of Environmental Protection, PRC's http://english.sepa.gov.cn/ or that for the US EPA–China Environmental Law Initiative, http://www.epa.gov/ogc/china/initiative_home.htm.

comprehensive arrangement" to promote production and protect the environment, it brought forward guidelines, declaring key themes of "overall planning, reasonable distribution, and comprehensive utilization," "to turn harmfulness into benefit," "to rely on the multitude," and "everybody [to] start work to protect the environment and benefit the people." It specified such issues as reasonable industrial distribution, old city environment improvement, soil and plant protection, freshwater and marine area regulation, forestation, environmental monitoring, environmental scientific research, propagandizing and education, and the investment in and facilitation of environmental protection. In 1974, the SC promulgated the **Provisional Rules of Preventing Coastal Water Pollution**. It was the first formal law for coastal water pollution prevention. The Chinese government also promulgated a series of new environmental standards, such as Standards for Drinking Water Quality and Trial Standards of Industrial "Three Wastes" Discharge. Generally speaking, the environmental laws during this period mainly focused on pollution prevention with nature conservation being a secondary consideration. The associated planning, feasibility, and efficiency were poor although there were some improvements on integrity and generality relative to the laws of the early stage of New China. They did not have a strong backbone of scientific theory.

LAWS AFTER THE REFORM AND OPENING-UP POLICY

From the Third Plenum of the 11th Communist Party of China Central Committee in 1978 to the present, China has transformed from a socialistic, planned economy to a socialistic, market economy. The environmental laws have made great progress and continued to be consummated. The first time that environmental protection was incorporated into China's fundamental law was 1978. The revised Constitution of the PRC prescribed a State fundamental function that "protects the environment and natural resources; prevents and controls pollution and other public hazards." Approved by the 11th Session of the 5th National People's Congress (NPC), the **Environmental Protection Law of the PRC (for Trial Implementation)** specifically prescribed the "polluter-pays principle," and the establishment of environmental monitoring, environmental impact assessment, and discharge levy systems. The law summarized China's experiences with environmental protection and referred to foreign environmental laws. It incorporated some comprehensive and principled provisions such as the objectives, targets, guidelines, fundamental principles, and systems of environmental protection; the basic requirements and measures of natural environment protection; pollution and other public hazards prevention and control; scientific research, propaganda, and education; and encouragement and punishment. As the predecessor of the **1989 Environmental Protection Law of the PRC**, it indicated that China's environmental protection efforts had evolved to become a core legal theme. The Chinese environmental legal system began to take form.

In 1992, the United Nations Conference on Environment and Development held in Rio de Janeiro, Brazil, established Agenda 21, the program for future environmental and developmental action. It brought global environmental protection to the new stage of sustainable development, which requires protecting environment and ecology, economizing resources, and simultaneously controlling population growth while fostering economic development. Relative to production and consumption, it requires enhancement of production while reducing input, and maximizing utilization while minimizing discharge. In order to resolve the problems resulting from the continuous population growth and rapid economic development in China and to keep pace with the worldwide trend in environmental protection, the SC ratified the **Ten Major Countermeasures of China's Environment and Development** in August 1992. It pointed out that a strategy of sustainable development is the most reasonable mode of accelerating economic development while solving environmental problems. The Environment and Resources Protection Committee of NPC came into existence in March of 1993. Instead of relevant SC departments developing and presenting draft laws in coordination with the SC Bureau of Legislative Affairs, this new Committee could prepare draft laws and present them directly to the NPC or its Standing Committee for approval. The process was simplified and efficiency increased. It reflected the emergence of a new legislative mechanism for environmental and resource conservation. In March 1994, China's Agenda 21 was approved by the SC. It put forward the general stratagem and action plan for sustainable development. Environmental legislation was listed as one of the priority-program plans.

Contemporary Chinese environmental law is mainly composed of eight levels, each of which has a distinct role:

1. The Constitution of the PRC. This category is the basis of the entire environmental legislation system and has ultimate legal authority.
2. International treaties regarding environmental protection to which China has agreed. If the treaty "contains provisions differing from those contained in the laws of the PRC, the provisions of the international treaty shall apply, unless the provisions are ones on which the PRC has announced reservations" (Environmental Protection Law of the PRC). Therefore, these treaties have priority to other national laws.
3. Laws of environmental and resource protection (i.e., those constituted by the NPC and its Standing Committee about reasonable environment and resources exploitation, utilization, protection, and amelioration) have legal authority that is subordinate to the Constitution.
4. Administrative regulations of environmental and resource protection (i.e., those constituted by the SC) that must be consistent with the Constitution and laws.
5. Department rules and standards of environmental and resource protection (i.e., those constituted by the departments and ministries of the SC and other authorized national administrative departments), which must not conflict with the Constitution, laws, and administrative regulations.
6. Local regulations of environmental and resource protection (i.e., those constituted by the local People's Congress and its Standing Committee of provinces, autonomous regions, municipalities, and authorized others) that cannot be in conflict with the Constitution, laws, and administrative regulations.
7. Local administrative rules of environmental and resource protection (i.e., those constituted by the local people's government of provinces, autonomous regions, municipalities, and authorized others) that should conform to national laws, regulations, and local regulations.
8. Other environmental standardization documents (except for the aforementioned, those constituted according to the Constitution, laws, and regulations, by local People's Congress and its Standing Committee, or local people's government at or above the county level).

The major pollution prevention and control laws introduced during this period were the following. The **Marine Environment Protection Law of the PRC** was approved in 1982 and amended in 1999. It prescribed marine environmental stewardship, ecological protection, and the prevention and control of pollution damage to the marine environment by land-derived pollutants, marine and coastal construction projects, waste dumping, vessels, and relevant operations. The **Law of the PRC on Prevention and Control of Water Pollution** was put in force in 1984 and amended in 1996. This law is intended to "prevent and control water pollution, protect and improve the environment, safeguard human health, ensure effective water resources utilization, and promote socialist modernization drive progress." It prescribed in detail the standards for water environmental quality and water pollutant discharge, the prevention and control of water pollution, and its supervision and management. It is a relatively comprehensive law for inland water pollution prevention and control. The amended **Law of the PRC on the Prevention and Control of Atmospheric Pollution** was approved in 2000 and put into force on September 1 of the same year. Its purpose is to "prevent and control atmospheric pollution, protect and improve the human and ecological environments, safeguard human health, and promote the sustainable development of both economy and society." The law prescribes in detail the prevention and control of the atmospheric pollution caused by coal burning, by motor vehicle and vessel emissions, and by waste gas and particulates. The supervision and management of atmospheric pollution prevention and control were detailed as well. This comprehensive law is the basis for state and local legislation on atmospheric pollution. The **Law of**

the PRC on Prevention and Control of Environmental Pollution Caused by Solid Waste was approved in 1995, amended in 2004, and put into force on April 1, 2005. It prescribes the supervision and management for the prevention and control of solid wastes pollution, which includes industrial solid waste and domestic refuse. It has a special provision regarding the prevention and control of hazardous waste pollution. All of the aforementioned laws specify the legal liabilities associated with their violation.

Relative to natural resource conservation, the **Regulations of the PRC on Nature Reserves** were put into effective on December 1, 1994, with the intent of strengthening the creation and management of nature reserves, and protecting the natural environment and resources. It prescribes the creation and management of nature reserves, and establishes legal liability in case of violation. It mandates that the local administrative management department prescribe specific rules. Promulgated in 1997 and effective as of January 1, 1998, the **Law of the PRC on Conserving Energy** was established to promote energy conservation by all sectors of society, increase energy use efficiency, ensure national economic and social development, and meet the people's everyday needs. It mainly prescribed energy conservation management, reasonable use of energy, development of energy conservation technology, and the legal liability for transgressions. The **Water Law of the PRC** was put into force in October 1, 2002, after the amendment and approval that same year. Its purpose is to foster rational economic development, utilization, and protection of water resources, for the prevention and control of water disasters, and for water resource sustainable use. Prescribed by this law were specifics for the protection of water, water areas, and water projects, allocation and efficient use, management of disputes concerning water, law enforcement supervision, and the legal liability for violations. The development, utilization, protection, and management of seawater are stipulated separately.

To summarize, the federal environmental laws during this period have the following characteristics. First, the underlying theme for all of the laws is sustainable development. The laws involve whole-process control and the management of related entities such as wastes, products, resources, and environment. New legal systems are promoted such as comprehensive environmental impact assessment, discharge permitting, and clean production systems. Second, the implementation and effectiveness of the laws are strengthened by advances in science and technology, consummate legal theory, and solid Constitutional and political foundations. Third, they are gradually integrating with foreign and international laws, and foster international cooperation. The laws use for reference the advanced foreign laws and regulations as incorporated in the Vienna and Basel Conventions. While the federal environmental laws are becoming more robust, local environmental legislations have also made great advances and are becoming increasingly local and specific. Their technical underpinnings are improving. These local laws have been essential components of federal environmental laws and are even more effective than the latter.

UPDATE ADDED FOR 4TH EDITION (M. C. NEWMAN)

To keep pace with China's rapid development, Chinese environmental law has changed rapidly in the last few years. He et al. (2012, 2013 including supplementary materials) describe the current activities of the NPC in the revising of the 1989 Environmental Protection Law of the PRC. They (He et al. 2013) comment that the 2012 draft revisions reflect "the power of incrementalists: Only the most urgent, feasible, and commonly agreed-upon improvements that require little change of other existing environmental laws have been included." They also include a detailed comparison of the 1989 law with the 2012 proposed revisions. The 2012 revisions failed to be approved and renewed efforts are being made to draft a successful revision. Given the increasingly important role of China in global issues, including those environmental, the outcome of this legislative process will have worldwide repercussions.

APPENDIX 6

Regulation and Management of Chemicals in Australia: A 2013 Update*

Michael StJ. Warne

Australia has relatively little legislation relating to managing chemicals, pollution, and ecotoxicology at a national level (i.e., federal legislation) compared to that at the state and territory level. This has arisen because originally Australia existed as six separately governed colonies of England. In 1901, these colonies decided to unite to form a new nation—Australia. A constitution for Australia was written, but because the colonies (now termed "states" and "territories") did not wish to cede all their power to the future federal government, powers were split between the federal and state/territory governments. A third level of government, local government, was also established. All of the federal governments' powers are defined in and conferred by Section 51 and 52 of the constitution. All other powers not conferred in the constitution lie with the six state and two territory governments. The constitution confers federal control to such issues as trade, taxation, defense, and immigration. However, control of chemical management, environmental health, and occupational health and safety is the responsibility of the states and territories. In terms of environmental issues, the vast majority of the powers reside with the states and territories and this is reflected in environmental legislation at the state and territory level (e.g., the **State Environment Protection Policy [Waters of Victoria]**, the **Protection of the Environment Ordinance [POEO] Act of NSW**, and the **Environmental Protection Act [EPP Water] 2009** of Queensland).

It was realized that there would be numerous advantages to having a single national approach to environmental issues rather than each state and territory developing their own unique legislation and approach. Thus, in 1934, the Australian Agriculture Council was established, which consisted of a ministerial representative from each federal, state, and territory government. A standing committee consisting of heads of Departments of Agriculture from federal, state, and territory governments was also established to provide technical advice to the Council.

This system of establishing Australian councils proved highly successful, having a number of advantages including minimizing duplication, costs, and regulatory burden placed on industry. This approach has since been adopted for many other issues of national concern. Other councils have been established to address issues such as environment and conservation; natural resource management; water allocation; forestry, fisheries, and aquaculture; soil conservation; and water resources. In fact, this concept led to the establishment of the Council of Australian Governments (COAG), where the leaders of federal, state, and territory governments meet to discuss issues of national importance. This is the peak intergovernmental forum in Australia.

In February 2011, COAG adopted a new structure, so that there are three types of councils: standing councils, select councils, and legislative and governance fora. All councils make decisions

* For more general discussion, see the Australian Government Department of the Environment webpages: http://www.environment.gov.au/cleaner-environment/index.html.

by consensus. Whether the decisions or recommendations of the councils are implemented depends on the willingness of the various jurisdictions to implement them. However, the decisions of some councils must be enacted by all jurisdictions. Examples of this are the National Environment Protection Measures (NEPMs) that are legal documents.

Australia and New Zealand despite being separate countries have close ties and New Zealand has either formally or informally joined many of these councils. For example, New Zealand is a member of the Standing Councils for Environment and Water (SCEW), Energy and Resources (SCER), and Health (SCoH). However, New Zealand is its own sovereign nation and chooses which of these developments from the councils to adopt and how it will adopt them. In fact, in many instances, Australia and New Zealand have different legislative approaches, for example, the legislation and approaches to contaminated sites.

Despite the bulk of the environmental legislation in Australia being at the state level, this appendix will not discuss any such legislation as this would be confusing, repetitious, and not overly productive. Rather, it will focus on what national legislation there is and also on the non-legislative measures that exist at the national level that relate to the management of chemicals and environmental pollution.

The Australian Constitution gives the federal government powers relating to trade, and under these, it has implemented a number of acts that control the import and manufacture of chemicals or "chemicals goods." It has passed legislation that covers the four main ways in which chemicals are used in Australia. These are industry, agriculture, human medicines, and food production. The focus of each of these regulations is governed to a large degree by the end use of the chemicals.

The **Industrial Chemicals (Notification and Assessment) Act (1989)** controls the import and manufacture of industrial chemicals. The agency responsible for administering the act is the National Industrial Chemicals Notification and Assessment Scheme (NICNAS) that was established in 1990. It only assesses chemicals and determines whether they can be imported or manufactured; it does not register chemicals. Its principle focus is on new chemicals although it does have an existing chemicals assessment program. Chemicals that have been assessed and their use permitted in Australia, or chemicals that were in use prior to the establishment of NICNAS are listed in the Australian Inventory of Chemical Substances (AICS). Any chemical not on the AICS is considered a new chemical. As NICNAS was only established in 1990, the vast majority of over 40,000 chemicals on the AICS have never been subject to NICNAS review. In 2012, NICNAS adopted and implemented a new innovative Inventory Multi-tiered Assessment and Prioritisation (IMAP) framework to accelerate the assessment and thus address this backlog of unassessed chemicals. The Targeted Assessment Program is the means by which industrial chemicals can be reassessed and an order of assessment has been established with priority existing chemicals having the highest priority. In addition to assessing industrial chemicals, NICNAS makes recommendations to appropriate federal, state, and territory entities on worker, public, and environmental health issues associated with each assessed chemical. There are penalties associated with noncompliance with the act and NICNAS can require proponents to provide information necessary to conduct the assessment. NICNAS operates using a cost-recovery model.

The **Agricultural and Veterinary Chemicals Administration Act (1992)**, the **Agricultural and Veterinary Chemicals (Code) Act (1994)**, and the **Agricultural and Veterinary Chemicals Legislation Amendment Bill (2012)** control the import, manufacture, and use of agricultural and veterinary products. These acts are administered by the Australian Pesticides and Veterinary Medicines Authority (APVMA, formerly called the National Registration Authority). Before any agricultural and veterinary product can be imported or sold within Australia, it must be registered with the APVMA. The registration process considers potential effects on the public and workers, crops, animals, environmental health, trade, and efficacy of the product. A product may or may not be registered depending on the risk it poses or it may have its uses restricted. The AVPMA has the ability to reassess the registration of any product. This can and has led to the deregistration of products or changes to their permitted uses. For example, there have been recent changes to the permitted uses of the herbicide diuron (http://www.apvma.gov.au/publications/gazette/2011/24/gazette_2011-12-06_page_36.pdf). The APVMA controls all aspects of registered products up to the point of sale at which point state/territory governments take control. There are penalties associated with noncompliance with these acts.

The **Therapeutic Goods Act (1989)**, which is administered by the Therapeutic Goods Administration (TGA), covers the import, manufacture, and registration of prescription and nonprescription medicines for human consumption. All products that fall under the act must be registered in the Australian Register of Therapeutic Goods (ARTG) to be used in Australia.

The **Food Standards Australia New Zealand Act (1991)** and the **Australia New Zealand Food Standards Code** are administered by Food Standards Australia New Zealand (FSANZ) that was formerly called the Australia and New Zealand Food Authority. FSANZ is responsible for establishing maximum levels (MLs) of contaminants and natural toxins in foodstuffs (FSANZ, 2002). Consistent with the international codex system for developing food safety standards, FSANZ only develops MLs for foodstuffs that are major components of the normal dietary intake. The main means by which compliance with the MLs is assessed is through the Australian Total Dietary Survey that is conducted approximately every 2 years (e.g., 2001, 2003, 2005, 2008, 2011, 2014) (www.foodstandards.gov.au/science/monitoring/pages/australiantotaldiets1914.aspx) and the corresponding New Zealand survey that is conducted every 5 years (e.g., 2003/2004 and 2009) (http://foodsafety.govt.nz/policy-law/food-monitoring-programmes/total-diet-study/documents.htm). The acts covering the TGA and the FSANZ do not require any consideration of potential environmental impacts—only those to humans.

Australia has developed limits (which have a variety of names) for chemicals in various environmental compartments including air, water, soil, and sediment.

The limits for air are termed standards, and these are stated in the **Ambient Air Quality National Environment Protection Measure (NEPM)** (NEPC, 2003). The standards are for carbon monoxide, carbon dioxide, ozone, sulphur dioxide, lead, PM10, and PM2.5 particles.* In addition, there is an **Air Toxics NEPM** (2004) that established procedures to collect data in order to develop standards for toxicants by 2012. However, all the air standards only consider human health issues and so they will not be considered here further. In contrast, the limits for other environmental compartments take a much broader approach and consider human and environmental issues. Copies of all Australia's NEPMs are available from http://www.scew.gov.au/nepms.

Australia's waterways are managed through the **National Water Quality Management Strategy (NWQMS)** that was established in 1992. The guiding principle of the NWQMS is ecologically sustainable development and thus its principle objective is to "achieve sustainable use of the nation's water resources by protecting and enhancing their quality while maintaining economic and social development." Underpinning the policy is a series of guidelines that cover the following:

- Policies and processes to achieve water quality
- Effluent and sewerage system management
- Urban stormwater and recycled water
- Fresh and marine water quality (including sediment quality)
- Monitoring and reporting
- Groundwater protection
- Water recycling and reuse
- Drinking water

Copies of the NWQMS and all the associated guidelines are available from http://www.environment.gov.au/topics/water/water-quality/national-water-quality-management-strategy/. As their name implies, the water quality guidelines (WQGs) have no legal standing. However, they

* The designations PM10 and PM2.5 refer to categories of particulate matter with diameters less than or equal to 10 or 2.5 μm, respectively.

are almost invariably used by national, state, and territory jurisdictions in controlling chemicals in aquatic environments. The WQGs are the best scientific estimates of the concentrations that are considered safe. As such, they do not consider social, religious, or economic issues. These issues and the desired uses of the water body are considered by stakeholders in developing water quality objectives (WQOs)—which become the main means of managing water bodies. At least in terms of the WQGs for chemicals for ecosystem protection, WQGs have invariably become the WQOs without any modification.

The actual numerical or descriptive limits in the WQGs are termed Trigger Values (TVs); if the measured concentration in a water body exceeds the TV or lies outside the stated range of TVs, further action is triggered. This action can take the form of further investigation, development and implementation of management strategies, or remediation (ANZECC/ARMCANZ, 2000). The WQGs actively promote further investigation such as site-specific assessment, and decision trees were supplied to provide guidance on how these should be conducted. There are TVs for over 250 chemicals.

Documents 4 (The Australian and New Zealand Guidelines for Fresh and Marine Water Quality) and 7 (The Australian and New Zealand Guidelines for Water Quality Monitoring and Reporting) of the NWQMS are currently being reviewed and revised. Products resulting from this review and revision (including new sediment quality guidelines, a new BurrliOZ software for calculating TVs, and an improved method for deriving TVs) should begin being released in late 2014. The revision of these documents is evolutionary rather than revolutionary, in that they are building onto and improving the existing system rather than creating entirely new systems.

Australia only has the equivalent of soil quality guidelines for contaminated sites. These guidelines are termed Ecological, Human, and Groundwater Investigation Levels (EILs, HILs, and GILs, respectively) and are found in the **National Environment Protection (Assessment of Site Contamination)** Measure (NEPC, 1999 and the 2013 amendment). These investigation levels apply only to contaminated sites and act in a similar manner to the aquatic TVs—in that, if they are exceeded, further investigation or management action is triggered. They are not clean up criteria nor are they concentrations that can be polluted up to. A review of the NEPM began in 2005 and the revised NEPM was released in 2013 (http://www.comlaw.gov.au/Details/F2013C00288). In the 2013 amendment to the NEPM, most of the limitations identified in the review have been addressed and the number of HILs and EILs has been increased. In addition, a new framework for deriving EILs was developed and this preferentially derives soil-specific EILs (that vary according to the soil physicochemical properties found to control the toxicity and bioavailability of the contaminant) for both fresh and aged (i.e., contamination has been in the soil for greater than 2 years) contamination. No new GILs were derived as part of the revised NEPM, rather they still refer to the Australian WQGs (the Australian and New Zealand Guidelines for Fresh and Marine Water Quality) (ANZECC and ARMCANZ, 2000) for protecting ecosystems, the Australian drinking water guidelines 2011 (NHMRC & NRMMC, 2011) for protecting humans, and the Australian Recreational Water Quality guidelines (NHMRC, 2008) for indirect human exposure.

In 1998, the **National Environment Protection Council** established Australia's first NEPM—the **National Pollutant Inventory (NPI) NEPM** (http://www.npi.gov.au/index.html). Each state and territory government is responsible for implementing the NPI NEPM. The NPI consists of a database containing data on 93 chemicals that are released into the Australian environment from over 4300 industry entities in quantities above an annual threshold (http://www.npi.gov.au/npi-data/search-npi-data). The NPI database is maintained by the Department of the Environment (established in September 2013). The NPI provides information on the amounts and types of chemicals emitted from various sources at the postcode level. It also provides information on each of the 93 chemicals about their human and environmental health impacts, common uses, hazard rating, and physicochemical properties. This information is freely available to members of the government, industry, and the public. The goal of the NPI is to provide information that can assist with reducing the impacts associated with the release of pollutants. In 2008, the NPI NEPM was varied so that greenhouse gases and energy reporting are no longer required. These data are now covered by the **National Greenhouse and Energy Reporting Act 2007**, http:www.comlaw.gov.au/Details/C2012C00373.

The Australian government enacted the **Environment Protection (Sea Dumping) Act 1981** in 1984. This arose from international obligations associated with Australia being a signatory of the Prevention of Marine Pollution by Dumping of Wastes and Other Matter, 1972 (the London Convention) and the 1996 Protocol to the London Convention and the United Nations Convention on the Law of the Sea 1982. As part of the Act, Australia developed a set of Ocean Dumping Guidelines (Environment Australia, 2002), which provides a national framework for assessing the potential environmental impacts associated with the confined and unconfined disposal of dredging spoil in the sea. The guidelines use the sediment quality guideline (SQG) trigger values developed as part of the Australian and New Zealand WQGs (ANZECC/ARMCANZ, 2000) and recommend a process that compares measured concentrations with the trigger values. Exceeding the trigger values leads to site-specific assessment that considers background concentrations, elutriate concentrations, simultaneously extractable metals and acid volatile sulfates, and finally direct toxicity assessment. Overall, these guidelines have the same basis and philosophy as the Australian and New Zealand SQGs.

Australia also has limits for contaminants in various products including biosolids (treated sewage sludge) and composts (that use the biosolids guidelines).

A key feature of current management of chemicals in Australia is the move away from regulation toward cooperative arrangements between industry and the Australian, state, and territory governments, and the implementation of best management practice. An excellent example of this is the **Reef Water Quality Protection Plan 2013** (State of Queensland, 2013a).

Previous work, particularly the scientific consensus statement on water quality in the Great Barrier Reef (Department of Premier and Cabinet, 2008), identified that the major anthropogenic threats to the long-term sustainability of the Great Barrier Reef (GBR) that could be managed and reduced were diffuse agricultural pollutants, i.e., total suspended sediment (TSS), nutrients (nitrogen and phosphorus), and pesticides (particularly photosystem II inhibiting [PSII] herbicides). These findings were subsequently verified by the 2013 scientific consensus statement (Brodie et al. 2013). As a result, the Reef Plan 2013 revised the targets to improve land management practices and to reduce the loads (total amounts) of these pollutants being discharged from rivers to the GBR (Table A6.1). In addition, Reef Plan has the long-term goal to improve water quality such that by 2020 the water entering the GBR lagoon has no detrimental impact on the health and resilience of the GBR.

Table A6.1 Land Management Practice and Load Reduction Targets that were to be Achieved by 2018 According to the Reef Water Quality Protection Plan 2013

Land Management Targets	Loads Reduction Targets
90% of sugar cane, horticulture, cropping, and grazing lands are managed using best management practices in priority areas	At least a 50% reduction in anthropogenic end-of-catchment dissolved inorganic nitrogen loads in priority areas
A minimum of 70% late season groundcover on grazing lands	At least a 20% reduction in anthropogenic end-of-catchment loads of sediment and particulate nutrients in priority areas
There is no net loss of the extent, and an improvement in the ecological processes and environmental values, of natural wetlands	At least a 60% reduction in end-of-catchment loads of pesticides in priority areas
The extent of riparian vegetation is increased	

Since the implementation of Reef Plan 2009, progress toward meeting the targets and long-term goal has been tracked by the Paddock to Reef Integrated Monitoring, Modeling, and Reporting Program (Paddock to Reef Program) (http://www.reefplan.qld.gov.au/measuring-success/assets/paddock-to-reef.pdf). This program integrates information on management practices, catchment indicators, catchment and marine water quality, and the ecological health of the GBR. The program design incorporates monitoring and modeling activities across the paddock, catchment, and marine scales. The paddock monitoring and modeling component aims to understand and quantify the main mechanisms of the off-site migration of the pollutants and to determine the effectiveness of various management practices. The catchment monitoring and modeling component assesses water quality at the catchment scale to determine priority sources of pollutants and track changes in water quality over time. The catchment modeling is underpinned by the Great Barrier Reef Catchment Loads Monitoring Program (GBRCLMP) that measures the loads of TSS and nutrients at 25 sites (consisting of 11 catchments and 14 subcatchments) and pesticides at 11 sites (9 catchments and 2 subcatchments) of the 35 catchments that discharge to the GBR (http://www.reefplan.qld.gov.au/measuring-success/paddock-to-reef/catchment-loads-monitoring.aspx). The Marine Monitoring Program quantifies the same pollutants in the GBR as well as measuring the health of various key ecosystems of the GBR.

The magnitude of the pollutant loads depends on the climatic conditions, with wet "La Nina" years having considerably larger loads in Eastern Australia than the drier "El Nino" years. Source Catchment Modeling is used in the Paddock to Reef Program to remove annual variability in weather and generate loads under a normalized set of climatic conditions. Thus, progress to meeting the targets and the long-term goal of the Reef Plan can be determined. Results are released annually in a Reef Report Card (http://www.reefplan.qld.gov.au/measuring-success/report-cards.aspx). The monitoring and modeling provide a feedback loop to inform government, industry, and the community on the success of the land management practice changes that have been implemented and whether further improvements or greater levels of adoption are needed.

APPENDIX 7

Summary of Indian Environmental Laws and Regulations*

S. Bijoy Nandan

BACKGROUND

The modern environmental laws and regulations of India were formulated soon after the United Nations Conference on the Human Environment held at Stockholm in 1972. Two international conferences on environment and development—one at Stockholm in 1972 and another at Rio de Janerio in 1992—have influenced environmental policies in most countries, including India. The constitution of India commits itself to protect our environment through Article 48-A of the Directive Principles of State Policy, which reads, "The state shall endeavor to protect and improve the environment and to safeguard the forests and wildlife of the country." Later, Article 51-A (g) was added through the 42nd Amendment to the constitution in 1976, that enshrines a fundamental duty of the citizen to environmental protection. The 73rd Constitutional Amendment Act of 1992 on the revitalization of **panchayati raj**† in the Indian political system adds to the X1 schedule to the constitution. It has eight entries that are linked to environmental protection and conservation. Again, Entry 8 of the X11th Schedule adds to the Constitution by the 74th Amendment Act, 1992 for constituting urban local (municipal) bodies and the function of "protection of environment and protection of ecological effects" has been assigned to urban local bodies.

Gradually, from the Directive Principles to the Constitutional Amendments, increasing importance is given to environmental protection and promotion. Initially, only forests and wildlife were specifically mentioned in the Directive Principles. Now included are such important aspects of environmental protection as water management, watershed development, drinking water, nonconventional energy sources and environmentally significant community assets. These constitutional provisions have been translated into the "regulatory" laws that aim at environmental protection, exemplified in the umbrella Environment (Protection) Act of 1986. The whole legal framework provides for more specific laws. Important among them are the following:

- Water (Prevention and Control of Pollution) Act, 1974
- Air (Prevention and Control of Pollution) Act, 1981

* For more information see the webpages at the following sites: http://en.wikipedia.org/wiki/Category:Environmental_law_in_India and http://www.ceeraindia.org/documents/lib_tabofcon_160300.htm.

† A political system implemented widely in south Asia that provides a degree of local self-rule by assemblies (sabhas) made up of respected and informed community members. The role and importance of this system of decentralized administrative and judicial activities has shifted throughout Indian history. Most recently, the 73rd Constitutional Amendment established village, intermediate, and district panchayats with a village assembly (Gram Sabha) governing locally. Nearly all of India's urban and most praiseworthy rural citizenship is now represented in this tiered panchayat system of elected members.

- Environment Protection Act, 1986
- Wildlife Protection Act, 1972
- Coastal Ecology-Declaration of Coastal Stretches as Coastal Regulation Zone Notification, 1991
- Public Liability Insurance-Public Liability Insurance Act, 1991
- Tribunals/Authorities—The National Environment Tribunal Act, 1995; National Environment Appellate Authority Act, 1997

The first two of these specifically provide regulation of water and air pollution problems and create institutions such as central and state pollution control boards for implementing the provisions of the acts. The three acts pertinent to ecotoxicology are described in the following sections.

WATER (PREVENTION AND CONTROL OF POLLUTION) ACT, 1974 (AMENDMENT, 1988)

The Water (Prevention and Control of Pollution) Act of 1974 came into being at a time when the country had already prepared itself for rapid industrialization and urbanization. The need was keenly felt for the treatment of domestic and industrial effluents before discharge into rivers and streams. Pollution of streams, rivers, and other watercourses reduced the availability of potable water. In addition, it caused deterioration in the quality of vegetation and other living creatures in water. Destruction of fish populations had far reaching consequences on the economy. The Water Act was enacted in this context. Water being a state concern, Indian parliament passed the law after requests from some of the constituent states in the country.

The **Water (Prevention and Control of Pollution) Act**, 1974, was enacted to prevent and control water pollution, and to maintain or restore the wholesomeness of water in the country. The act was amended in 1988. The Water Act, defines the pollution of water as follows,

> *Pollution means such contamination of water of such alteration of the physical, chemical or biological properties of water or such discharge of any sewage or trade effluent or of any other liquid, gaseous or solid substance into water (whether directly or indirectly) as may, or is likely to create a nuisance or render such water harmful or injurious to public health or safety, or to domestic, commercial, industrial, agricultural or other legitimate uses, or to the life and health of animals or plants, or of aquatic organisms.*

Some important provisions of the 1974 Water Act and its 1988 Amendment were the following:

- In order to achieve its objective, pollution control boards at the central and state levels were created to establish and enforce standards for industries discharging pollutants into bodies of water. The central pollution control boards (CPCB) and state pollution control boards (SPCB) were constituted under Sections 3 and 4 of the act, respectively. The act was amended in 1978 and 1988 to clarify certain ambiguities and to vest more powers in pollution control boards.
- Functions and Powers of CPCB and SPCB.

The Water Act provides for the constitution of the central and state pollution control boards empowered to carry out a variety of functions to promote cleanliness of streams and wells, and to prevent and control pollution of water. The central board advises the central government, coordinates state board activities, provides them with technical assistance and guidance, organizes training of personnel for pollution control, collects and compiles technical and statistical data, lays down water quality standards, and executes nationwide pollution control programs. It may step into the shoes of the state board when the central government concludes that the state board has defaulted in complying with any direction issued by the central board, and that a grave emergency had arisen so as to warrant immediate action. State boards carry out programs identical to those of the central board within the territory of the state. They plan comprehensive programs, advice state governments, collect and disseminate information, collaborate with the central board in organizing training programs, inspect sewage and trade effluent treatment plants, lay down standards for water quality, evolve methods of treatment and disposal of sewage and trade effluents, formulate modes of utilization of sewage and trade effluents for agriculture, and advise the state governments about the location of industries. The most significant power of the state board is the one to make, vary, or revoke an order for the prevention or control of discharge. In exercising this power, the state board can request any person concerned to construct new, or modify existing, systems for disposal and to adopt remedial measures necessary for the control of water pollution.

While the central government can give directions to the central board, the latter in turn can issue directions to the state board. Both the central board and state board will be bound by the directions issued by the central government and state government, respectively. Consequent to the enactment of the Environment (Protection) Act (EPA), an amendment to the Water Act conferred more potent and meaningful powers on the boards. It laid down that a board may, in exercise of its powers and performance of its functions, issue any direction in writing to any person, officer, or authority, and such person, officer, or authority shall be bound to comply with such direction. This means that a state board as well as the central board can issue directions to an industry to stop functioning. As explanation, the provision declares that this includes the power to direct closure, prohibition, or regulation of any industry, operations, or process; or stoppage or regulation of supply of electricity, water, or any other services.

The Water Act confers on any person empowered by the board, the power to enter any place for performing the functions of the board and determine compliance with any order or direction. However, a plain reading of the statutory provision shows that the seizure of any plant requires evidence for prosecution. The identical provision in the EPA, on the other hand, confers a wider power. It can be used for prevention and mitigation of environmental pollution.

AIR (PREVENTION AND CONTROL OF POLLUTION) ACT, 1981 (AMENDMENT, 1987)

The objective of this act is to prevent, control, and abate air pollution; to establish boards to carry out the aforesaid purposes; and to confer and assign related power and functions to such boards powers. The central government used Article 253 to enact this law and made it applicable throughout India. The **Air (Prevention and Control of Pollution) Act**, last amended in 2003, defines air pollutant as "any solid, liquid or gaseous substance (including noise) present in the atmosphere in such concentration as may be or tend to be injurious to human beings or other living creatures or plants or property or environment."

The CPCBs and SPCBs created under the Water Act carry out the functions of the boards envisaged under the Air Act also. In the functions of a board, the two laws make identical provisions about criminal and administrative sanctions, powers of the government vis-à-vis the actions of the board, and in rule-making powers of the central and state governments. The distinguishing characteristics of the Air Act can be seen in the provision for declaration of air pollution control zones, inclusion of noise within the definition of air pollution, and control of pollution from motor vehicles. On consultation with boards, the state government can establish air pollution control areas within the state. The state government can prohibit the use of any fuel other than approved fuel in that area. Restrictions are imposed on the establishment and operation of any industrial plant within the area by an administrative consent system. The states are required to prescribe emission standards for industry and automobiles after consulting the central board and taking into consideration its ambient air quality standards. The board considers applications for grant of consent before allowing such establishment of operations.

ENVIRONMENT (PROTECTION) ACT, 1986

The 1986 **Environment (Protection) Act (EPA)** was enacted in the aftermath of the Bhopal gas tragedy in 1984 that claimed more than 3000 lives. The Statement of Objects and Reasons of this act refers to the decisions from the June 1972 Stockholm Conference and expresses concern about the decline in environmental quality, increasing pollution, loss of vegetative cover and biological diversity, excessive concentrations of harmful chemicals in the ambient atmosphere, growing risks of environmental accidents, and threats to ecosystems. The EPA was enacted under the provisions of Article 253 of the Constitution of India. It has 26 Sections and is divided into four chapters relating to (1) Preliminary, (2) General powers of the central government, (3) Prevention, control, and abatement of environmental pollution, and (4) Miscellaneous. The act was last amended in 1991.

According to this act, the environment includes "water, air, and land and the interrelationship which exists among and between water, air and land, and human beings, other living creatures, plants, microorganisms and property." Relative to environmental pollutants, it defines a hazardous substance as "any substance or preparation which, by reasons of its chemical or physiochemical properties or handling, is liable to cause harm to human beings, other living creatures, plants, micro-organisms, property or the environment."

The 42nd Constitution Amendment Act of 1976 inserted specific provisions for environmental protection in the form of Directive Principles of State Policy and Fundamental Duties. The State's responsibility with regard to environmental protection has been laid down under Article 48-A of Indian Constitution, which reads as follows: "The State shall endeavour to protect and improve the environment and to safeguard the forests and wildlife of the country." Environmental protection is a fundamental duty of every citizen of this country under Article 51-A (g) of the Indian Constitution, which reads as follows: "It shall be the duty of every citizen of India to protect and improve the natural environment including forests, lakes, rivers, and wildlife and to have compassion for living creatures."

The Act empowers the central government to take all appropriate measures to prevent and control pollution, and to establish effective machinery for the purpose of protecting and improving the quality of the environment and protecting, controlling, and abating environmental pollution. These measures include a wide range of activities such as coordination of actions by state governments, officers, and other authorities, planning and execution of nationwide programs, standard fixing, restriction of areas for industries and operations, establishing procedures and safeguards, inspection of any premises, and dissemination of environmental information. The central government can create authorities, appoint officers, issue binding directions, and delegate its power to issue directions including the power to direct closure, prohibition, or regulation of any industry, operation, or process; or stoppage or regulation of the supply of electricity, water, or any other service. Even in a case where there is a fear of an accident as a result of discharge of environmental pollutants, the person in charge of the place has a statutory duty to take preventive measures as well as to inform the authorities of the likelihood of an accident. This act also empowers and authorizes the central government to issue directions for the operation or process, prohibition, closure, or regulation of any industry. The central government is also authorized to stop or regulate the supply of electricity, water, or any other service directly without obtaining the order of the Court in this regard. The central government is also empowered to enter and inspect any place through any person or through any agency authorized by the central government. The act debars the Civil Courts from having any jurisdiction to entertain any suit or proceeding in respect of an action, direction, or order issued by the central government or other statutory authority under this act. Under the act, there will be supremacy of provision. In other words, the provisions of this act and the rules or orders made under this act shall have effect and supremacy over anything inconsistent contained in any enactment other than this act. Any equipment, industrial plants, or other objects can be seized for the purpose of taking measures to prevent or mitigate environmental pollution. This power lying hidden as a tag to "the entry and inspection" clause is wider in operation than such clauses in the Water Act and Air Act. Failure to comply with, or contravention of the provisions of, the EPA, or the rules made, or orders or directions issued, are punishable with imprisonment that may extend to five years, fine up to one lakh* rupees, or both. Any person can go in for a prosecution after issuing 60-day notice.

It is worth mentioning the names of a few important rules that have recently been specified under the Environment Act for the management and control of hazardous substance, which include hazardous chemicals, waste, and microorganisms.

1. Hazardous waste (management and handling) rules of 1989: objective is to control generation, collection, treatment, import, storage, and handling of hazardous waste.
2. The manufacture, storage, and import of hazardous chemical rules of 1989: defines the terms used in this context, and sets up an authority to annually inspect the industrial activity connected with hazardous chemical and storage facilities.
3. The manufacture, use, import, export, and storage of hazardous microorganism/genetically engineered organisms or cells rules of 1989: these were introduced to protect the environment nature and health given the increasing application of gene technology and microorganisms.
4. Biomedical waste (management and handling) rule of 1998: it is legally binding on the health care institutions to streamline the process of proper handling of hospital waste such as its segregation, disposal, collection, and treatment.

IMPLEMENTATION OF ENVIRONMENTAL PROTECTION LAWS

The central agency for implementing environmental protection legislation in India is the Ministry of Environment and Forests (MoEF). Besides giving direction to the Central Pollution Control Board (CPCE) about pollution prevention and control, the MoEF is also responsible for designing and implementing a wide range of programs relating to environmental protection. It states that "the focus of various programs of the Ministry and its associated organizations, aimed at prevention and control of pollution is on issues such as promotion of clean and low waste technologies, waste minimization, reuse or recycling, improvement of water quality, environmental audit, natural resource accounting, development of mass based standards, institutional and human resource development etc. The whole issue of pollution prevention and control is dealt with a combination of command and control methods as well as voluntary regulations, fiscal measures, promotion of awareness, involvement of public, etc." On the basis of the environmental laws and directions given by the Supreme Court, the central government has created many authorities for designing, implementing, and monitoring its environmental programs. At the state level, most states have set up Departments of Environments and State Pollution Control Boards (SPCBs).

The CPCB and the SPCBs are responsible for implementing legislations relating to pollution prevention and control. Pollution arises both from point sources, like factories and nonpoint sources, like automobiles. Source-specific effluent and emission standards have been fixed for polluting point sources. Because monitoring pollution generation is difficult for nonpoint sources, indirect measures of pollution prevention control are being adopted such as catalytic converters in automobile engine for new cars, lead-free petrol, fuel with low sulfur content, and periodic inspection of vehicles. In addition, ambient standards for air and water have been laid down and are regularly monitored by the CPCB with the support of the SPCBs.

* In south Asian counting systems, a lakh is 100,000 so the fine referred to here is 100,000 rupees or, at today's exchange rate, approximately $1,700.

SUGGESTED READINGS

Chakraborti, N. K., *Environmental Protection and the Law*. Ashish Publishing House, New Delhi, India, 1994.

Desai, B., *Environmental Laws of India: Basic Documents*. Lancers Books, New Delhi, India, 1994.

Dhanai, R., *Global Environmental Policies*. Cyber Tech Publications, New Delhi, India, 2010.

Dhar, D. N., Kumar, S., and Vaish, T., *Environment and Ecology*. Vayu Education of India, New Delhi, India, 2011.

Divan, S. and Rozencranz, A., *Environmental Law and Policy in India: Cases, Materials and Statutes*. 2nd Edition, Oxford University Press, New Delhi, India, 2001.

Leelakrishnan, P. *Environmental Law in India*. Butterworths, New Delhi, India, 1999.

Madhvan, K. N., *Environmental Law Pollution and Management*. A.K. Publications, New Delhi, India, 2009.

Maheshwara Swami, N., *Law Relating to Environmental Pollution and Protection*. Asia Law House, New Delhi, India, 1998.

Nandimath, O. V., *Handbook of Environmental Decision Making In India: An EIA Model*. Oxford University Press, New Delhi, India, 2008.

Naseem, M., *Environmental Law in India*. Kluwer Law International Publishers, The Netherlands, 2011.

Rajput, V., *Environmental Science*. Vayu Education of India, New Delhi, India, 2011.

Rodgers, W., *Environmental Law in Indian Country*. Thomson West Publishers, New York, 2005.

Sahasranaman, P. B., *Handbook of Environmental Law of India*. Oxford University Press, New Delhi, India, 2009.

Shanthakumar, S., *Introduction to Environmental Law*. 2nd Edition, LexisNexis Butterworths, New Delhi, India, 2007.

Singh, R. and Misra, S., *Environmental Law in India, Issues and Resources*. Concept Publishing Company, New Delhi, India, 1996.

Tiwari, A. K., *Environmental Laws in India*. Eastern Book Corporation, New Delhi, India, 2006.

APPENDIX 8

Regulation and Management of Hazardous Chemical Substances in South Africa*

Theunis Meyer and Claudine Roos

INTRODUCTION

Hazardous substances can potentially cause harm to human health and safety, as well as the environment throughout its entire life cycle, which presents significant risks during incidents and accidents. The importance of applying sound management practices during all phases of the chemical life cycle can, therefore, not be overemphasized. Depending on the type and degree of their hazardous properties, hazardous substances require different levels of control in their manufacturing, transportation, storage, use, and disposal to manage the potential risks appropriately.

This appendix provides an overview of the legal framework for the management of hazardous substances in South Africa, specifically focusing on provisions for public protection, worker and workplace protection, environmental protection under normal operational conditions, as well as environmental emergencies and disasters. It also gives an overview of the requirements for the transportation, storage, handling, use, and disposal of hazardous substances.

OVERARCHING LEGAL FRAMEWORK

International Legal Instruments

South Africa is a party to a range of international legal instruments, in the form of conventions, treaties, and protocols that regulate the control and management of hazardous substances, as part of global efforts to reduce the effects of hazardous substances on human health and safety, and the environment. As a signatory, such instruments require South Africa to ensure the capacity to uphold the provisions to which the country has committed itself.

Constitutional Framework

Section 24 of the Constitution of the Republic of South Africa 108 of 1996 guarantees everyone the right to an environment that is not harmful to their health or well-being as well as the right to have the environment protected, for the benefit of present and future generations. This provision in the Bill of Rights forms the cornerstone for all environmentally related legislation, including legislation to protect human health and well-being.

* The Wikipedia web pages at http://en.wikipedia.org/wiki/South_African_environmental_law provide additional details.

Municipal Bylaws Regulating Hazardous Substances

The control of hazardous substances has historically been achieved at the local government level through the use of bylaws. This trend is still being pursued and many local and district municipalities have published bylaws regulating the storage, transportation, and supply of hazardous substances and dangerous goods since the political transition in 1994.* These bylaws are primarily focused on the fire hazards posed by the substances and regulate aspects such as general requirements for the control of hazardous substances and dangerous goods, requirements for storerooms for hazardous substances and dangerous goods, and supply and delivery of flammable substances.

SPECIFIC LEGAL PROVISIONS FOR THE CONTROL AND MANAGEMENT OF HAZARDOUS SUBSTANCES AND DANGEROUS GOODS

A summary of the legislation applicable to the management and control of hazardous substances and dangerous goods in South Africa is provided in the table below. The table briefly provides the name of the applicable piece of legislation, its purpose, as well as the hazardous chemical-related aspects that each act aims to regulate.

Provisions for Public Protection

The two key acts dealing with hazardous substances management and control for the purposes of *public protection* are as follows

- National Health Act 61 of 2003
- Hazardous Substances Act 15 of 1973

Provisions for Worker and Workplace Protection

The key requirements for the management and control of hazardous substances for the *protection of employees within their workplaces,* primarily related to the storage and use of such substances, are stipulated in the following:

- Occupational Health and Safety Act 85 of 1993 (OHSA) with regulations for the control of hazardous chemical substances, asbestos, lead, and hazardous biological agents, as well as installations where more than a prescribed quantity of any substance is or may be kept, whether permanently or temporarily; or any substance is produced, processed, used, handled, or stored in such a form and quantity that it has the potential to cause a major incident (major hazardous installations).
- Mine Health and Safety Act 29 of 1996.

Provisions for General Environmental Protection

The **duty of care principle**† is translated into specific requirements to act responsibly in several pieces of legislation that applies generically to everybody in South Africa. This is especially important to the management and control of hazardous substances that may cause serious environmental pollution and even threaten human life.

* During the mentioned South African transition, its Parliament drafted the fifth Constitution (1994) that was adopted under President Nelson Mandela (1996).

† According to the **duty of care principle**, one (e.g., an employer) has a legal obligation to refrain from any conduct that places others (e.g., employees) at unreasonable risk of harm. One is obligated to use reasonable care and foresight, or when failing to do so, becomes legally liable. Although presented here in the context of South African law, it is generally relevant as in the case of the tort laws of U.S. states.

The key requirements for general environmental protection are contained in the following:

- National Environmental Management Act 107 of 1998
- National Water Act 36 of 1998
- National Environmental Management Air Quality Act 39 of 2004
- Environment Conservation Act 73 of 1989
- National Environmental Management Waste Act 59 of 2008
- Regulations promulgated in terms of the Mineral and Petroleum Resources Development Act 28 of 2002

Provisions for Environmental Emergencies and Disaster Management

Due to the harmful nature (flammable, toxic, etc.) of hazardous substances, they have the potential to cause serious danger to the public or potentially serious pollution of or detriment to the environment when a sudden, unexpected release occurs. Various acts deal with the management and control of hazardous substances during *environmental emergencies*. These include the following:

- National Environmental Management Act 107 of 1998
- National Water Act 36 of 1998
- Major Hazardous Installation regulations (published in terms of the OHSA)
- Fire Brigade Services Act 99 of 1987
- Disaster Management Act 57 of 2002

Provisions for the Transportation of Hazardous Substances

The transportation of hazardous substances by road is primarily regulated by the following:

- National Road Traffic Act 93 of 1996 (together with a number of incorporated SANS standards)
- Major Hazardous Installation regulations (published in terms of the OHSA)
- National Environmental Management Waste Act 59 of 2008

Provisions for the Storage and Use of Hazardous Substances

In addition to the requirements for the management and control of hazardous substances in workplaces, general environmental protection, and environmental emergencies, the storage of hazardous substances is also regulated by the National Building Regulations and Building Standards Act 103 of 1977.

Provisions for Hazardous Waste Management

The bulk of the waste generated by the manufacturing, transport, storage, and warehousing, as well as use of hazardous substances is hazardous as well. The legislation dealing with the management and control of hazardous substances in workplaces, general environmental protection, and environmental emergencies also regulate hazardous waste management.

CONCLUSION

The management of hazardous substances and dangerous goods in South Africa is not a simple task as it is regulated by a wide range of laws and regulations that each focus on different aspects of hazardous substance management and control as described earlier. Due to the fact that the detailed requirements are captured in bylaws, regulations, national, and international standards, as well as Codes of Practice, these should be recognized as part of the legislative framework for regulating hazardous substances.

Legislation	Purpose of Legislation	Aspects Regulated Related to Hazardous Substance Management and Control
National Environmental Management Act 107 of 1998 (NEMA)	The NEMA seeks to protect the environment through (among other mechanisms) facilitating cooperative environmental governance. The Act imposes a duty of care on every person who causes, has caused, or may cause significant pollution or degradation of the environment.	1. Principles for sustainable development 2. Implementation of a general duty of care and reasonable measures to prevent pollution and degradation of the environment 3. Control of environmental emergency incidents (reporting and response) 4. Authorization of the following hazardous substances (dangerous goods) related activities through the Environmental Impact Assessment (EIA) process a. construction, expansion, or decommissioning of facilities for the storage and handling of dangerous goods b. the bulk transportation of dangerous goods.
National Water Act 36 of 1998 (NWA)	The NWA has stewardship over water as a scarce and unevenly distributed national resource in South Africa. The Act recognizes the need for the integrated management of all aspects of water resources, the necessity to protect the quality of the nation's water resources in the interests of all water users and ultimately aims to achieve the sustainable use of water for the benefit of all users.	1. Principles for the protection, use, development, conservation, management, and control of water resources. 2. Implementation of the general duty of care and reasonable measures to prevent pollution and degradation of water resources. 3. Authorization of hazardous substance related water uses (i.e., disposal of hazardous waste, irrigation of land with water containing waste, discharge of water containing waste). 4. Control of environmental emergency incidents (reporting and response). 5. Regulation of dams with a safety risk.
National Environmental Management Waste Act 59 of 2008 (NEMWA)	The NEMWA provides for, among others, the regulation of waste management to protect health and the environment by providing reasonable measures for the prevention of pollution and ecological degradation and for securing ecologically sustainable development; institutional arrangements and planning matters; national norms and standards for regulating the management of waste; specific waste management measures; the licensing and control of waste management activities; the remediation of contaminated land; and compliance and enforcement.	1. General duty with regards to waste management 2. Implementation of the waste management hierarchy (reduction, reuse, recycling, and recovery) 3. Provisions for a. the storage of waste b. preventing the unauthorized use of waste c. authorization of waste management activities that might have a detrimental impact on the environment through waste management licensing and EIA process d. the remediation of contaminated land e. regulating the classification and disposal of hazardous waste at authorized landfill sites f. regulating the duties of persons transporting waste

Legislation	Purpose of Legislation	Aspects Regulated Related to Hazardous Substance Management and Control
National Environmental Management Air Quality Act 39 of 2004 (NEMAQA)	The NEMAQA was promulgated to protect the environment by providing reasonable measures for the prevention from pollution and ecological degradation and for securing ecologically sustainable development, while promoting justifiable economic and social development and to provide for national norms and standards regulating air quality monitoring, management, and control. It also allows for specific air quality management measures, including the declaration and management of priority areas, controlled emitters, and controlled fuels, as well as the licensing of listed activities that have or may have a significant detrimental effect on the environment, including health, social conditions, economic conditions, ecological conditions, or cultural heritage.	1. Regulation of the storage, use, application and disposal of hazardous substances that could result in atmospheric emissions. 2. Authorization of certain activities that may result in atmospheric emissions through the atmospheric emission licensing and EIA process. 3. Regulation of the incineration and thermal treatment of hazardous waste.
Environmental Conservation Act 73 of 1989	Although the Environmental Conservation Act (ECA) 73 of 1989 has been partially repealed and replaced by the NEMA, the *regulations for the prohibition of the use, manufacturing, import and export of asbestos and asbestos containing materials*; published in terms of the Environmental Conservation Act are still applicable to the management of asbestos and asbestos containing materials.	1. Regulation of the use, manufacturing, import, and export of asbestos and asbestos containing materials. 2. Regulation of the disposal of asbestos and asbestos containing materials.
Provincial nature conservation ordinances and acts	A number of nature conservation ordinances and provincial nature conservation acts regulate water pollution.	Regulation of the pollution of waters, prohibition of dumping or discharge of substances that may be harmful to the environment.
National Health Act 61 of 2003 (NHA)	The NHA seeks to provide a framework for a structured uniform national health system in South Africa, taking into account the obligations imposed by the Constitution and other laws on the national, provincial, and local governments with regard to health services. It also aims to protect, respect, promote, and fulfill the right of the people of South Africa to an environment that is not harmful to their health or well-being.	1. Public health and public protection against the potential hazards posed by hazardous substances. 2. Provincial health services include environmental pollution control services. 3. Municipal health services includes, among others water quality monitoring, waste management, health surveillance of premises, environmental pollution control, and chemical safety, but excludes control of hazardous substances.

(Continued)

Legislation	Purpose of Legislation	Aspects Regulated Related to Hazardous Substance Management and Control
Hazardous Substances Act 15 of 1973 (HSA)	The HSA provides for the control of substances that may cause injury or ill-health to or death of human beings by reason of their toxic, corrosive, irritant, strongly sensitizing or flammable nature, or the generation of pressure thereby in certain circumstances, and for the control of certain electronic products.	1. Public health and public protection against the potential hazards posed by hazardous substances. 2. Provides for the classification of hazardous substances or products into four hazard groups, primarily in relation to the degree of danger these pose to human health. 3. Regulates the prohibition and control of the importation, manufacture, sale, use, operation, application, modification, disposal or dumping of hazardous substances, and certain electronic products.
Occupational Health and Safety Act 85 of 1993 (OHSA) and regulations (Lead Regulations, Asbestos Regulations, Hazardous Chemical Substances Regulations, Hazardous Biological Agent Regulations, Major Hazard Installation Regulations)	The OHSA is one of the most explicit and applicable acts for worker and workplace protection as far as hazardous substances control are concerned. The Act provides for the health and safety of persons at work, the health and safety of persons in connection with the use of plant and machinery, as well as the protection of persons other than persons at work against hazards to health and safety arising out of or in connection with the activities of persons at work. It aims to protect employees and other persons against any dangers (hazards and risks) to their health and safety and emphasizes the general duties of employers, employees, and manufacturers, all of which are required to implement preventive and protective measures as far as reasonably practicable. Explicit regulations have been published under the OHSA for the control of hazardous substances. Although these requirements are primarily aimed at protecting people in the workplace, they also serve to protect the environment.	1. Worker and workplace protection against the possible hazards posed by the handling, storage, and use of hazardous substances. 2. Regulation and control of the storage and transportation of hazardous substances. 3. Prevention of certain hazardous substances (i.e., lead, asbestos, and hazardous biological agents) from being released into the environment or water systems. 4. Regulation of the disposal of hazardous substances. 5. Regulation of installations where more than a prescribed quantity of any substance is or may be kept, whether permanently or temporarily; or any substance is produced, processed, used, handled or stored in such a form and quantity that it has the potential to cause a major incident through notification of such installations, risk assessments and on-site emergency plans.
Mine Health and Safety Act 29 of 1996 (MHSA)	The MHSA provides for the protection of the health and safety of employees and other persons at mines. The Act aims to regulate employers' and employees' duties to identify hazards and eliminate, control and minimize the risk to health and safety.	Worker and workplace protection against the possible hazards posed by the handling, storage, and use of hazardous substances at mines.
Fire Brigade Services Act 99 of 1987 (FBSA)	This Act provides for the establishment, maintenance, employment, coordination, and standardization of fire brigade services, which include services for preventing the outbreak or spread of fire and reacting to fire, to protect life or property against a fire or other threatening danger.	Fire prevention and control related to hazardous substances.

Legislation	Purpose of Legislation	Aspects Regulated Related to Hazardous Substance Management and Control
Disaster Management Act 57 of 2002 (DMA)	The Act provides for an integrated and coordinated disaster management policy that focuses on preventing or reducing the risk of disasters, mitigating the severity of disasters, emergency preparedness, rapid and effective response to disasters and post-disaster recovery; as well as the establishment of disaster management centers to deal with disasters.	Management of and response to local, provincial, and national disaster incidents involving hazardous substances.
National Road Traffic Act 93 of 1996 (NRT) and National Road Traffic Regulations	The NRTA and its regulations dictate that no person shall offer dangerous goods for transportation in or on a vehicle, or accept dangerous goods after transportation, except as prescribed by the Act, and prohibit the transportation of dangerous goods under certain conditions. Numerous national standards have been developed for the management and control of hazardous substances, some of which have been incorporated into the NRT regulations.	1. Regulation of the transportation of dangerous goods on national roads, through controlling the a. construction and maintenance of vehicles b. fitness and behavior of drivers c. availability of appropriate information d. emergency response arrangements e. behavior of consignors, consignees, and operators
National Building Regulations and Building Standards Act 103 of 1977 (NBRBSA) and National Building Regulations (SANS 10400)	The NBRBSA and the National Building Regulations is a set of functional regulations that are used as a point of reference for the design and construction of buildings, to ensure the safety and health of people and animals that live in or around buildings.	Approval of the design and construction of buildings for the storage and use of hazardous substances.
Municipal bylaws	Bylaws regulating the storage, transportation, and supply of hazardous substances and dangerous goods to promote the achievement of a fire-safe environment for the benefit of all persons within the area of jurisdiction of a Municipality and provide for procedures, methods, and practices to regulate fire safety.	1. Fire hazards and fire protection of buildings 2. Storage and use of flammable and hazardous substances 3. Transportation of dangerous goods and hazardous substances 4. Housekeeping and public safety

APPENDIX 9

Derivation of Units for Simple Bioaccumulation Models

Peter Landrum*

Assume a closed system containing water and fish in which both the amounts in the fish and water are changing over time. The total amount in the system (*A*) is $Q_w + Q_f$ where the subscripts w and f refer to water and fish, respectively, and *Q* denotes the amount (mass) in the compartment. There is no direct reference to the size of the fish or water compartments. The amount in the fish can be described by the following equation:

$$\frac{dQ_f}{dt} = k_1 Q_w - k_2 Q_f \tag{A9.1}$$

where k_1 = the rate constant for the fractional reduction in the amount in the water (fraction of mass t^{-1}) and k_2 = the rate constant for the fractional reduction in the amount in the fish (fraction of mass t^{-1}). These are conditional rate constants because they depend on the specific experimental conditions, i.e., the sizes of the compartments. For example, assume two systems differing only in the relative sizes of the associated fish and water compartments. If one system had fish and water compartments of identical size, the rate constants would be different from another system in which the fish compartment was much smaller than the water compartment. Although the flux of material into the fish is the same for the two systems, the fractional reduction in the amount in the waters for the two systems (i.e., the rate constants) would be different. A larger fraction of the amount in the water of the system with equal compartment sizes for the fish and water would be removed per unit time than from the system with much larger water compartment. The integrated form of Equation A9.1 is Equation A9.2:

$$Q_a = \frac{k_1 A (1 - e^{-\{k_1 + k_2\} t})}{k_1 + k_2} \tag{A9.2}$$

The unit analysis for this is the following:

$$Q_a(\text{ng}) = \frac{k_1(\text{hour}^{-1})\, A(\text{ng})\, [1 - e^{-(k_1\{\text{hour}^{-1}\} + k_2\{\text{hour}^{-1}\})\, t(\text{hour})}]}{k_1(\text{hour}^{-1}) + k_2(\text{hour}^{-1})} \tag{A9.3}$$

or

$$\text{ng} = \text{ng}$$

* Provided by Peter Landrum of the NOAA-Great Lakes Research Laboratories. This issue of units is discussed in Chapter 3.

Now, the information is referenced to the size of the fish and the concentration in the water with the formulation of the equation for the change of concentrations:

$$C_f = \frac{k_u C_w}{k_e}(1 - e^{-\{k_e t\}}) \tag{A9.4}$$

Solving the units,

$$C_f\left(\frac{\text{ng}}{\text{g}}\right) = \frac{k_u\,(\text{mL/g hour})\,C_w\,(\text{ng/mL})}{k_e\,(\text{hour}^{-1})}[1 - e^{-(k_e\{\text{hour}^{-1}\}t\{\text{hour}\})}] \tag{A9.5}$$

or

$$\frac{\text{ng}}{\text{g}} = \frac{\text{ng}}{\text{g}}$$

It is clear from this analysis of units that the unit for k_u is mL (g^{-1}·$hour^{-1}$) and that of k_e is $hour^{-1}$. The k_u is the clearance rate of the water by the fish of a specific size.

Demonstration of the associated units for k_u can also be done with the relationships assuming steady state:

$$\frac{dQ_f}{dt} = k_1 Q_w - k_2 Q_f \tag{A9.6}$$

At steady state, $k_1Q_w = k_2Q_f$ or $Q_f/Q_w = k_1/k_2$ for this relationship describing changes in mass. To get to a bioconcentration factor (BCF = concentration in fish expressed as mass per unit of mass of fish divided by concentration in water expressed as mass per unit of volume of water), the relationship just derived between masses and constants is multiplied by volume/mass:

$$\left(\frac{\text{volume}}{\text{mass}}\right)\left(\frac{Q_f}{Q_w}\right) = \left(\frac{\text{volume}}{\text{mass}}\right)\left(\frac{k_1}{k_2}\right) = \frac{Q_f/\text{mass}}{Q_w/\text{volume}} = \frac{C_f}{C_w} \tag{A9.7}$$

Rearranging Equation A9.7 and linking the rearranged equation to the relationship between constants for the concentration-based formulation (i.e., k_u/k_e = BCF),

$$\frac{k_1\,(\text{volume/mass})}{k_2} = \frac{C_f}{C_w} = \frac{k_u}{k_e} \tag{A9.8}$$

Because both k_2 and k_e describe the fractional change in chemical in the fish per unit time, they are equivalent. But, the relationship between k_1 and k_u is k_1 (volume/mass) = k_u. The unit of k_u for the model of change in concentration over time is mL (g^{-1}·$hour^{-1}$), not $hour^{-1}$.

APPENDIX 10

Equations for the Estimation of Exposure

Equations and example values for variables were extracted directly from EPA guidelines (EPA, 1989d). Each is a derivation of the general equation provided in Chapter 13 (Equation 13.3). Note that units may change for variables among the equations.

1. Imbibed via drinking water or beverages made with drinking water:

$$\text{Intake (mg per kg-day)} = \frac{(\text{CW})(\text{IR})(\text{EF})(\text{ED})}{(\text{BW})(\text{AT})} \tag{A10.1}$$

where:

CW = concentration in the drinking water ($mg \cdot L^{-1}$)
IR = imbibing or ingestion rate ($L \cdot day^{-1}$), e.g., 2 $L \cdot day^{-1}$ for an adult
EF = exposure frequency ($days \cdot year^{-1}$), e.g., 365 $days \cdot year^{-1}$ for a resident
ED = exposure duration (years), e.g., 70 years for a lifetime
BW = body weight or mass (kg), e.g., 70 kg for an adult
AT = averaging time (days), e.g., ED × 365 days for a noncarcinogen or 70 years $lifetime^{-1}$ × 365 $days \cdot year^{-1}$ for a carcinogen

2. Imbibed from surface water while swimming:

$$\text{Intake (mg per kg-day)} = \frac{(\text{CW})(\text{CR})(\text{ET})(\text{EF})(\text{ED})}{(\text{BW})(\text{AT})} \tag{A10.2}$$

where:

CW = concentration in the drinking water ($mg \cdot L^{-1}$)
CR = contact rate ($L \cdot day^{-1}$), e.g., 0.050 $L \cdot day^{-1}$
ET = exposure time ($h \cdot event^{-1}$)
EF = exposure frequency ($days \cdot year^{-1}$), e.g., 7 $days \cdot year^{-1}$ as an average
ED = exposure duration (years), e.g., 70 years for a lifetime
BW = body weight or mass (kg), e.g., 70 kg for an adult
AT = averaging time (days), e.g., ED × 365 days for a noncarcinogen or 70 years $lifetime^{-1}$ × 365 $days \cdot year^{-1}$ for a carcinogen

3. Absorbed during dermal contact with water:

$$\text{Absorbed Dose (mg per kg-day)} = \frac{(\text{CW})(\text{SA})(\text{PC})(\text{ET})(\text{EF})(\text{ED})(\text{CF})}{(\text{BW})(\text{AT})} \tag{A10.3}$$

where:

CW = concentration in the drinking water ($mg \cdot L^{-1}$)
SA = surface area of skin in contact (cm^{-1}), which is dependent on sex, age, and contact type
PC = dermal permeability constant ($cm \cdot h^{-1}$), which is highly dependent on the specific chemical

ET = exposure time ($h \cdot day^{-1}$)
EF = exposure frequency ($days \cdot year^{-1}$), e.g., 7 $days \cdot year^{-1}$ as an average
ED = exposure duration (years), e.g., 70 years for a lifetime
CF = volumetric conversion factor for water ($1\ L \cdot [1000\ cm^3]^{-1}$)
BW = body weight or mass (kg), e.g., 70 kg for an adult
AT = averaging time (days), e.g., ED × 365 days for a noncarcinogen or 70 years $lifetime^{-1}$ × 365 $days \cdot year^{-1}$ for a carcinogen

4. Inhaled as a vapor in the air:

$$\text{Intake (mg per kg-day)} = \frac{(CA)(IR)(ET)(EF)(ED)}{(BW)(AT)} \quad (A10.4)$$

where:
CA = concentration in the air ($mg \cdot m^{-3}$)
IR = inhalation rate ($m^3 \cdot h^{-1}$), e.g., 30 $m^3 \cdot h^{-1}$ for an adult
ET = exposure time ($h \cdot day^{-1}$)
EF = exposure frequency ($days \cdot year^{-1}$)
ED = exposure duration (years), e.g., 70 years for a lifetime
BW = body weight or mass (kg), e.g., 70 kg for an adult
AT = averaging time (days), e.g., ED × 365 days for a noncarcinogen or 70 years $lifetime^{-1}$ × 365 $days \cdot year^{-1}$ for a carcinogen

5. Ingested in fish or shellfish from the contaminated area:

$$\text{Intake (mg per kg-day)} = \frac{(CF)(IR)(FI)(EF)(ED)}{(BW)(AT)} \quad (A10.5)$$

where:
CF = concentration in the food ($mg \cdot kg^{-1}$),
IR = ingestion rate ($kg \cdot meal^{-1}$), e.g., 0.284 $kg \cdot meal^{-1}$ (upper 95 confidence level for fish) consumption
FI = fraction ingested from the contaminated source (no units to this fraction)
EF = exposure frequency ($meals \cdot year^{-1}$)
ED = exposure duration (years), e.g., 70 years for a lifetime
BW = body weight or mass (kg), e.g., 70 kg for an adult
AT = averaging time (days), e.g., ED × 365 days for a noncarcinogen or 70 years $lifetime^{-1}$ × 365 $days \cdot year^{-1}$ for a carcinogen

6. Ingested in vegetable matter from the contaminated area:

$$\text{Intake (mg per kg-day)} = \frac{(CF)(IR)(FI)(EF)(ED)}{(BW)(AT)} \quad (A10.6)$$

where:
CF = concentration in the food ($mg \cdot kg^{-1}$)
IR = ingestion rate ($kg \cdot meal^{-1}$)
FI = fraction ingested from the contaminated source (no units to this fraction)
EF = exposure frequency ($meals \cdot year^{-1}$)
ED = exposure duration (years), e.g., 70 years for a lifetime
BW = body weight or mass (kg), e.g., 70 kg for an adult
AT = averaging time (days), e.g., ED × 365 days for a noncarcinogen or 70 years $lifetime^{-1}$ × 365 $days \cdot year^{-1}$ for a carcinogen

Study Questions

PREFACE AND CHAPTER 1

1. Define the terms highlighted in the text and check your definitions against those provided in the glossary.
   Anthropocene epoch
   Bad Old Days incongruity
   Becquerel
   Biomarkers
   Biomineralization
   Biomonitoring
   Boomerang paradigm
   Contaminant
   Corporatism
   Criteria (Legal)
   Dilution paradigm
   Ecological overshoot
   Ecoregions
   Ecotoxicology
   Gaia hypothesis
   Global distillation
   Green Revolution
   Industrial ecology
   Innovative science
   Itai-itai disease
   Laggard systems
   Method of multiple working hypotheses
   Minamata disease
   Natural capital
   Normal science
   Oklo reactors
   Paradigm
   Persistent, bioaccumulative toxicants
   Persistent organic pollutant
   Pharmaceuticals and personal care products
   Planetary boundary values
   Pollutant
   Precautionary principle
   Precipitate explanation
   Progressive Era
   Rare earth elements
   Rem
   Standards (Legal)
   Stress

Stressor
Township and Village Enterprise sector
Working hypothesis
Xenobiotic

2. Describe how the resource conservation, food safety, and industrial hygiene movements helped set the stage for the modern environmental ethic.
3. Briefly describe the transition from the Dilution Paradigm to the Boomerang Paradigm. Provide a definition of each paradigm in your answer with illustrative examples. Give a few examples of aspects of your daily life that might be indicative of this shift to the boomerang paradigm.
4. Briefly describe the Stockholm Convention and its goals.
5. Who was Alice Hamilton and what was her contribution to protecting human health?
6. How does one maintain conceptual coherency in dealing with a hierarchical field of study such as ecotoxicology?
7. Define the scientific, technical, and practical goals of ecotoxicologists.
8. Explain the evolution of scientific inquiry from the ruling theory, to the working hypothesis, to the multiple working hypotheses stages. Is the multiple working hypotheses approach prevalent in ecotoxicology today?
9. Compare and contrast normal and innovative science. What are their relative values in young versus mature sciences?
10. What are the abnormal practices called precipitate explanation, the tyranny of the particular, and *idola quantitatus*? Are they a problem in ecotoxicology today?
11. What qualities are valued in scientific, technological, and practical ecotoxicology?
12. Describe natural capitalism and its goals.
13. If polycyclic aromatic hydrocarbons are naturally occurring organic compounds, why are they discussed as environmental contaminants and pollutants in the book?

CHAPTER 2

1. Define the terms highlighted in the chapter and check your definitions against those provided in the glossary.

Acid mine (rock) drainage
Activation
Biosolids
Becquerel
Calcium sink
Chromophores
Complexation field diagram
Congener
Contaminant
Cultural (accelerated) eutrophication
Curie
Dense, nonaqueous phase liquids
Ecotoxicant
Essential element analog
Genetically modified organism (GMO)
GMO biosafety
Gray
Linear energy transfer (LET)
Methemoglobinemia
Na^+/I^- symporter (NIS) protein

Natural plant product pesticide
Perfluoroalkyl acids (PFAs)
Peroxisomes
Peroxisome proliferation
Pharmaceuticals and personal care products
Photomodification
Pollutant
Pyrethrins
Pyrethroids
Quality Factor (Q)
Rad
Radiation
Rem
Rotenone
Thermal pollution
Toxin
Vitellogenin
UVA, UVB, and UVC

2. Also provide a succinct explanation of the following background chemistry concepts and definitions.
   Aliphatic compound
   Aromatic compound
   Binding trends
   Bonds (Chemical)
   Electronegativity
   Electronegativity coefficient
   Hard Soft Acid Base theory
   Henry's Law
   Ionization
   Isotope
   Ligand
   Lipid solubility
   Metal
   Nuclide
   Oxyanion and oxycation
   Partitioning
   Polar bond
   Radioactivity decay series
3. If metals are naturally occurring elements, how do they become environmental contaminants and pollutants?
4. What are the major inorganic gases of concern? Explain why each is an issue requiring attention.
5. What is cultural eutrophication and how might it relate to the ocean dead zones of depleted oxygen described later in Chapter 12? What are some of the management strategies that are being used to cope with cultural eutrophication?
6. What human activities introduce chlorofluorocarbons into the environment and what are the adverse impacts of their release?
7. What are the commercial uses for dibenzodioxin and dibenzofuran compounds?
8. Both organochlorine and pyrethroid insecticides are synthetic organic molecules. Which of these two insecticide groups is less prone to microbial degradation? Why?
9. Please give the major classes of pesticides and compare them relative to mammalian toxicity, persistence in the environment, and their capacity to accumulate in organisms.
10. Describe why polychlorinated biphenyls are found as mixtures. Do you think that the relative amounts of each polychlorinated biphenyl congener might be different in the source (for

example, sediments) and the tissues of an organism that accumulate them from that source? Why or why not?

11. You decide to begin taking omega-3 supplements after hearing positive comments about how they promote cardiovascular health. Most omega-3 supplements are fish oil preparations. You pick up a bottle of 1500 mg odorless Norwegian Salmon oil capsules. You know from class that some pollutants accumulate in lipids (oils and fatty acids here) of fish, so you read the label that says, "This product is regularly tested (using AOAC international protocols) for freshness, potency and purity by an independent, FDA-registered laboratory and has been determined to be fresh, fully-potent and free of detectable levels of mercury, cadmium, lead, PCB's and 28 other contaminants." Using your knowledge of lipophilic and hydrophilic contaminants, explain whether you would have expected any of these (mercury, cadmium, lead, PCBs) to accumulate to high concentrations in fish oils. Might one expect some but not others of these to accumulate to high levels in oils (lipids)?
12. To what general class of contaminants does the following organic compound belong? Give details about this class of compounds including their origins, general fate in the environment, potential effects, and any regulatory actions taken to minimize their effects to human health and/or the environment. 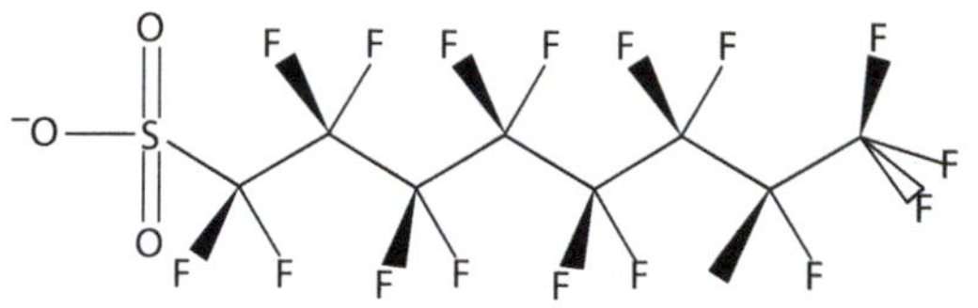
13. Describe the history, context, and consequences of pesticide use in Central America.

CHAPTER 3 (AND APPENDIX 8)

1. Define the terms highlighted in the chapter and check your definitions against those provided in the glossary.
   Absorption
   Accumulation factor
   Activation
   Active transport
   Apparent volume of distribution
   Assimilation efficiency
   Backstripping procedure
   Bioaccumulation
   Bioaccumulation factor
   Bioconcentration factor
   Biological half-life
   Biota-sediment accumulation factor
   Biotransformation
   Body burden
   Carrier protein
   Channel protein
   Class A, B, and intermediate metal cations
   Clearance volume–based model
   Depuration
   Diffusion
   Distribution phase
   Clearance

Effective half-life
Elimination
Endocytosis
Enterohepatic circulation
Exchange diffusion
Exocytosis
Facilitated diffusion
Freundlich isotherm equation
Fugacity
Gastrointestinal excretion
Glutathione S-conjugate export pump
Growth dilution
K_{ow}
Langmuir isotherm equation
Mean residence time
Membrane transport protein
Metallothionein
Monooxygenase
Multixenobiotic resistance (MXR)
Organic anion transporters (OATS)
P-glycoprotein (P-gp) pump
Pharmacokinetics (Toxicokinetics)
Phase I reaction
Phase II reaction
Physiologically based pharmacokinetics model
Porin
Rate constant–based model
Sorption
Uptake

2. Distinguish between steady state and chemical equilibrium. Which term is most accurate in describing bioaccumulation? Why?
3. What contaminants are most likely to pass into an organism via the lipid route? Which are least likely?
4. What parameter would you use to quantify the difference in adsorption affinities for a series of contaminants to a surface?
5. Would ionized or nonionized ammonia pass through the cell membrane fastest? Why?
6. How could you determine if passage of a contaminant across a biological membrane takes place by active or passive diffusion?
7. What is the reaction order of the following equation, $dC/dt = kC$? Is this reaction order common or uncommon for elimination kinetics?
8. Describe the role of biomineralization in the bioaccumulation of metals and metalloids.
9. Describe Phase I and II reactions in the biotransformation of organic contaminants.
10. What is enterohepatic circulation and why might it be important relative to realized effect of toxicant exposure?
11. Distinguish among elimination, depuration, and clearance.
12. Name and describe the three phases of renal elimination?
13. Describe the differences and similarities of rate constant–based, clearance volume–based, and fugacity-based models.
14. The rate constant for a single component, rate constant–based model is 0.1 $hour^{-1}$. What are the biological half-life and mean residence time predicted for this model?
15. Describe the backstripping method. Does it have weaknesses?
16. Explain the clearance volume concept.
17. Describe the models given in Equations 3.20 through 3.22. What are the meanings of the associated variables and constants?

18. The units of k_u are mL ($g^{-1} \cdot h^{-1}$) but the unit of k_e is hour^{-1}. Why?
19. You collect data for the elimination of contaminant X from bobwhite quail. When plotted as the natural log of concentration ($mg \cdot g^{-1}$) versus time (day), the data produce a straight line. You do linear regression on these data, producing the following model:
    Ln Concentration = 3.50 – 0.29*Time.
    a. What is the elimination rate constant for contaminant X from quail?
    b. How long would it take for 50% of contaminant X to be eliminated from a quail?
    c. What was the concentration of contaminant X in the study quail at time = 0?
20. Describe the role of biomineralization in the bioaccumulation of metals and metalloids.

CHAPTER 4

1. Define the terms highlighted in the chapter and check your definitions against those provided in the glossary.
   Absolute bioavailability
   Acid volatile sulfides
   Absorption rate constant
   Aerosol
   Allometry
   Assimilation efficiency
   Bioavailability
   Biological ligand model
   Biologically determinant element
   Biologically effective dose
   Biologically indeterminant element
   Biomimetic approach
   Chelates
   Cysteine-rich intestinal protein (CRIP)
   Essential element
   Exponential relationship
   Flory–Huggins Theory
   Free ion activity model
   Gastric emptying rate
   Henderson–Hasselbalch relationship
   Ligand
   Linear solvation energy relationships
   Mean absorption time
   Mean residence time
   Monodentate ligand
   Multidentate ligand
   pH-partition hypothesis
   Polychlorinated diphenyl ethers (PCDE)
   Power relationship
   Quantitative Structure-activity relationship (QSAR)
   Relative bioavailability
   Scaling
   Structure–activity relationship (SAR)
   Type A and B organisms
2. Discuss the definition of bioavailability relative to ecotoxicology and pharmacology.
3. Describe the various approaches to estimating bioavailability.
4. How could you estimate the relative bioavailability of a compound in imbibed water versus that in ingested plant matter?

5. Discuss the FIAM and exceptions to this model.
6. What factors influence bioavailability of contaminants in food?
7. What measures can be used to reflect the bioavailability of metals from oxic and anoxic sediments?
8. Describe the QSAR approach. What is the basis for the extensive use of K_{ow} in these models?
9. Explain the pH-partition hypothesis and relate the Henderson–Hasselbalch relationship to the bioavailability of a monobasic acid using this hypothesis.
10. How would you normalize for animal size effects on bioaccumulation?
11. A series of clam populations are sampled and analyzed for tissue cadmium concentrations. Cadmium concentrations are found to be directly dependent on ambient cadmium concentrations. In this case, is cadmium acting as a biologically determinant or indeterminant element? Explain your answer.
12. You get a job at a prestigious California aquarium as an environmental toxicologist and are greeted during your first week with the following issue. Sea otters in one harbor have high polychlorinated biphenyl ethers (PCBE) tissue concentrations but low tissue concentrations in another harbor. The person, who was not promoted into the job you just got, asks loudly in your first staff meeting if you had sufficient knowledge of the harbors and ecotoxicology to find the reason for this difference. Before you can respond, he asserts that, based on his extensive understanding of the harbors, the otters with high tissue concentrations feed primarily on sea urchins with an average of 5 μg/g of PCBE in them whereas the otters with lower tissue concentrations feed primarily on abalone with an average of 10 μg/g PCBE. With everyone at the meeting looking at you (their new boss) for your directions on how to determine what was the reason for the difference in otter PCBE tissue concentrations, you begin to describe a study you need done to ascertain the reason. Use what you learned in class to describe the otter feeding experiment you would have them do and the calculations you would do to determine the reason for the difference.

CHAPTER 5

1. Define the terms highlighted in the chapter and check your definitions against those provided in the glossary.
   Amplification, concentration, and fugacity
   Bioaccumulation factor (BF)
   Bioamplification
   Biomagnification
   Biomagnification factor (B)
   Biomagnification power (b)
   Biominification
   Bioreduction
   C_3 and C_4 plants
   Calcium sink
   Concentration factor (CF)
   Congener
   Discrimination factor
   Discrimination ratio
   Elemental analog
   Fatty acids
   Isotopic discrimination
   Periphyton
   Specific activity concept
   Trophic dilution
   Trophic enrichment

Trophic level
Twin trace technique

2. Describe how isotopic ratios of nitrogen, sulfur, and carbon might be used to define trophic structure of a lake community. What are the advantages and disadvantages of each of these elements for such use?
3. Describe how you would estimate biomagnification using body mass–weighted mean concentrations.
4. What elements might biomagnify? Why would these specific elements biomagnify?
5. What is "liquid" digestion by zooplankton and how does it relate to metal bioavailability from algae?
6. Would fish-eating sea birds accumulate more mercury than seabirds consuming other prey? Why?
7. How do birds eliminate mercury taken up in their diets?
8. Explain how measuring fatty acids might provide information about food sources of a species of marine finfish.
9. How might calcium sinks influence transfer of calcium analogs?
10. What qualities of organic compounds foster biomagnification?
11. Describe and contrast POP food web transfer when air-breathing versus water-breathing organisms dominate.
12. You are approached by a community group concerned about the potential biomagnification of an organic compound that has a log K_{ow} of 5.7 yet is metabolized very slowly in organisms. The compound is seeping from ground water into a lake that the public uses for fishing. Does it have the potential to biomagnify? Describe the approach you would use to determine if biomagnification was occurring and to quantify the possible biomagnification in the lake.

CHAPTER 6

1. Define the terms highlighted in the chapter and check your definitions against those provided in the glossary.

Acetylation
Adduct
AHH
Alkoxyradicals
Amidases
Amino acid conjugation
δ-aminolevulinic acid dehydratase (ALAD)
Aneuploidy
Carcinogenic
Catalase (CAT)
Cellular stress response
Chaperon
Clastogenic
Conjugation
Cytochrome P-450 monooxygenase
Epoxide hydrolases
EROD
Esterases
Flavin monooxygenase (FMO)
Free radical
Genotoxicity
Glucose-regulated proteins
Glucuronic acid
Glucuronidation

Glutathione
Glutathione peroxidase
Glutathione S-transferase
Haber–Weiss reaction
Haber–Weiss catalyzed reaction
Heat shock proteins (hsp)
Isoforms (viz. isozyme or isoenzyme)
Lipid peroxidation
Malondialdehyde
Metal transcription factor (MTF-1)
Metallothionein
Metallothionein-like protein
Micronuclei
Microsomes
Mixed function oxygenase
Mutagenic
NADPH cytochrome P-450 reductase
Oxidative stress
Oxyradical
Peroxisomes
Peroxyradicals
Porphyrins
Procarcinogen
Promoter element
Proteotoxicity
Radiosensitizer
Redox cycling
Sentinel species
Single-cell gel electrophoresis
Spillover hypothesis
Stress protein
Stress protein fingerprinting
Sulfotransferase
Superoxide dismutase (SOD)
Tail moment
Teratogenic
Transcription factor
Transgenic organism
Uridine diphospho glucuronosyltransferases

2. What are the qualities of the ideal molecular biomarker?
3. Describe in detail the components of the cytochrome P-450 monooxygenase system.
4. What biomarkers are associated with Phase II reactions?
5. Briefly summarize the qualities of metallothioneins.
6. What are the different functions of metallothioneins?
7. What is the spillover hypothesis? Are there any situations that you can identify where this concept would not be valid, e.g., mixtures of metals, and so forth? Why?
8. Describe proteotoxicity and the roles that stress proteins play in lessening the effects of proteotoxic agents.
9. Describe the types of molecular damage that can be produced by oxyradicals. How does the cell lessen the damage of oxyradicals?
10. Hydrogen peroxide is not an oxyradical, yet it contributes to oxidative stress. How?
11. How might you measure lipid peroxidation in a field population of a species thought to be experiencing oxidative stress?
12. How would you measure DNA damage?

13. A group of subsistence fishermen is concerned that they are subject to high mercury exposure from their diet. They hire you to determine if this is a legitimate concern. Design a biomarker study to assess the potential exposure.

CHAPTER 7

1. Define the terms highlighted in the chapter and check your definitions against those provided in the glossary.
   Acidophilic component
   Adenocarcinoma
   Adenoma
   Age pigment
   Apoptosis
   Apoptotic bodies
   Basophilic component
   Benign neoplasia
   Calculi
   Cancer progression
   Caseous necrosis
   Caspase enzymes
   Carcinoma
   Cardinal signs of inflammation
   Ceroid pigment
   Chloragosomes
   Chloride cell
   Cholestasis
   Chromatid
   Chromatid aberration
   Chromosomal aberration
   Coagulation necrosis
   Compensatory hyperplasia
   Cuprosomes
   Cytotoxicity
   Excessive hyperplasia
   Fat necrosis
   Gangrenous necrosis
   Genetic risk
   Granulation tissue
   Histopathology
   Homolog (metaphase)
   Hormonal hyperplasia
   Hormonal oncogenesis
   Hyperplasia
   Hypertrophy
   Inflammation
   Initiator
   Ischemia
   Karyolysis
   Latent (or latency) period
   Lesion
   Linear no-threshold theory

Lipofuscin (ceroid)
Liquefactive (cytolytic) necrosis
Macroexplanation
Malignant neoplasia
Metastasis
Mutagen
Necrosis
Neoplasia
Neoplastic hyperplasia
Oncogene
Pathologic hyperplasia
Pericytes
Physiologic hyperplasia
Primary lamellae
Problem of ecological inference
Promoter
Proto-oncogene
Ptolemaic incongruity
Pyknosis (pycnosis)
Reductionism (microexplanation)
Repair fidelity (of DNA)
Residual bodies
Sarcoma
Secondary lamellae
Sister chromatid
Sister chromatid exchange
Somatic death
Somatic risk
Suppressor gene
Target organ
Tetrad
Threshold theory
Thrombosis
Zenker's necrosis

2. Define and contrast the holistic and reductionist approaches to hierarchical subjects such as ecotoxicology. Which is the correct or best approach to ecotoxicology?
3. If ecologists wish to understand the influence of ecotoxicants on ecological entities, why are cellular, tissue, and organ biomarkers valuable in ecotoxicology?
4. Describe some of the characteristics of necrosis and compare these characteristics to those of apoptosis.
5. Describe the process of inflammation including its four cardinal signs. Are all relevant to all animal species?
6. If you suspected that a genotoxic agent was affecting mice inhabiting an area, what biomarkers would you use to test your suspicions? In your answer, explain why you would use each biomarker.
7. What are hyperplasia and hypertrophy? Which might manifest in gills of a fish that was exposed chronically to a dissolved pesticide that disrupts passage of ions across cell membranes?
8. An individual was exposed to high concentrations of a chemical with the properties of an initiator. Will that individual develop cancer? Explain your answer.
9. You are asked to determine if trout in a lake modified by acid precipitation (low pH and high aluminum concentrations) have been affected at the cellular, tissue, or organ level. What would you examine to answer this question? What changes might you expect? Link these biomarkers to consequences to individuals and to the trout population.

CHAPTER 8

1. Define the terms highlighted in the chapter and check your definitions against those provided in the glossary.
   Acetylcholinesterase inhibitors
   Adenylate energy charge
   Amelia
   Analysis of variance (ANOVA)
   Androgen receptor antagonist
   Antisymmetry
   Arcsine square root transformation
   Autophagy
   Bartlett's test
   Behavioral teratology
   Behavioral toxicology
   Biphasic dose-effect model
   Bonferroni adjustment
   Canalization
   Chlorosis
   Coughs or gill purges
   Developmental stability
   Developmental toxicity
   Directional asymmetry
   Dunnett's test
   Dunn–Šidák adjustment
   Ecological mortality
   Endocrine disruptor (modifier)
   Estrogenic chemicals
   FETAX
   Fluctuating asymmetry (FA)
   General adaptation syndrome (GAS)
   Genomic imprinting
   Homeopathic medicine
   Hormesis
   Immaterial significance incongruity
   Imposex
   Inverse probability error
   Karnofsky's law
   Law of similars
   Lordosis
   Lowest observed effect concentration (LOEC)
   Male-mediated toxicity
   Maulstick incongruity
   Maximum acceptable toxicant concentration (MATC)
   Metabolic scope
   Negative predictive value
   Neurathian bootstrap incongruity
   Nil hypothesis error
   No observed effect concentration (NOEC)
   Pericardial edema
   Phenotypic plasticity
   Phocomelia
   Positive predictive value
   Safe concentration

Scoliosis
Scope of activity
Scope of growth
Selyean stress
Shapiro–Wilk's test
Steel's many-one rank test
Sublethal effects
Teratogen
Teratogenic index
Teratology
t-test
Wilcoxon rank sum test
Williams's test

2. What sublethal effects are most often studied by ecotoxicologists?
3. What is Selyean stress? Which of the following responses is not associated with Selyean stress: increased adrenal size, induction of EROD activity, kidney damage by Cd, elevated blood pressure because of an emotional shock, or depletion of secretory granules in cells of the adrenal cortex?
4. A data set for growth retardation of an endemic, endangered fish suggests that concentrations of zinc in the range measured in your discharge consistently stimulates growth but local officials responsible for regulating your discharge insist on a linear no-threshold extrapolation of effect down to no more than a 5% reduction in growth (i.e., extend the dose-effect line downward from high effect concentrations to a concentration that is predicted to have only a 5% reduction in growth based on a linear no-threshold model). Extrapolation predicts 5% inhibition, but your data set suggests 20% enhancement of growth. Give your reasoning in deciding to conform to or contest the regulators' decision.
5. You are told that voles near a contaminated site have necrotic lesions in their livers. Discuss one reason why this observation *might* suggest that abnormal development will also be occurring in the young of this rodent.
6. Hatchlings of an endangered and long-lived sea turtle are closely monitored by state fish and wildlife technicians. Data from 20 years ago describe a sex ratio of 50:50 male:female of hatchlings for many populations of these turtles. Now, almost all turtles are females. Describe what you would do to determine the cause of this apparent shift and what you might do to counterbalance the shift. Should you intercede?
7. Periodic episodes of meadow lark mortality are noted near an agricultural field. The field was used for many years to grow cotton and, consequently, has a history of DDT and lead arsenate application. At present, carbamate pesticides are periodically applied to the fields. How would you determine which, if any, of these potential toxicants is producing the mortality?
8. Why are behavioral abnormalities used less often than other sublethal effects in the assessment of contaminant effects to individuals?
9. Tabulate the advantages and disadvantages of the post-ANOVA methods described in this chapter.
10. Compare and contrast the hypothesis testing and regression methods for analyzing sublethal effects data.
11. Given the following table of NOEC and LOEC values for toxicant X, estimate the concentration ("safe concentration") that will protect an endangered siren (amphibian) living below a continuous discharge containing toxicant X.

		Concentration ($\mu g \cdot L^{-1}$)		
Species	**Effect**	**NOEC**	**LOEC**	**Exposure Duration**
Fathead minnow	Growth	1	10	14 days
Bluegill sunfish	Egg hatch success	5	46	7 days
Leopard frog	Tadpole growth	0.09	0.60	7 days
FETAX assay	Heart development	0.17	1.7	4 days
Daphnia magna	Fecundity	0.001	0.008	7 days

12. Explain how the use of positive and negative predictive values might improve inferences from toxicity tests that might otherwise use NOEC and LOEC values to make conclusions.
13. Explain the Neurathian bootstrap incongruity and provide your thoughts about whether it is an issue of concern when assessing sublethal effects using conventional ANOVA-based methods.
14. Shortly after being hired by an environmental NGO, you are contacted about a potentially unresolved problem with PAH pollution in a water body near an abandoned creosote treatment facility. You know that a local university has faculty members with expertise in environmental physiology, histopathology, and biochemical toxicology. You must decide whether legal action should be taken to force a further clean up and need sound information to determine if there is an adverse effect to fish in the water body. Funds are only available for a modest study so you can only hire one of the university experts. Describe the study that you would fund, providing reasons you picked the biomarkers to be measured and what results you would expect if there were an adverse impact.

CHAPTER 9

1. Define the terms highlighted in the chapter and check your definitions against those provided in the glossary.
   Accelerated failure time model
   Acclimation
   Acclimatization
   Acute lethality
   Additive index
   Additivity
   Agonist
   Antagonism
   Antagonist
   Binomial method
   Chemical antagonism
   Chi-square (χ^2) ratio
   Chronic lethality
   Complete median lethal concentration
   Concentration additivity
   Concentration–response approach
   Cox proportional hazard model
   Critical life stage testing
   Dispositional antagonism
   Early life stage testing
   Elutriate test
   Flow-through test
   Functional antagonism
   Gavage
   Hardness
   Hit-and-run poisons
   Incipient median lethal concentration
   Independent joint action
   Induction antagonism
   Individual effective dose
   Individual tolerance
   Isobole approach
   Life-cycle study
   Litchfield method

Litchfield–Wilcoxon method
Logit
Maximum likelihood estimation (MLE)
Median effective concentration (EC50)
Median effective time (ET50)
Median lethal concentration (LC50)
Median lethal dose (LD50)
Median lethal time (LT50)
Median time-to-death (MTTD)
Metameter
Microtox® assay
Minimal time to response
Moving average method
Normal equivalent deviation (NED)
Normit
Partial kills
Photoinduced toxicity
Photosensitivity
Potentiation
Probit
Product-limit (Kaplan–Meier) method
Proportional diluter
Proportional hazard method
Receptor antagonism
Relative risk
Similar joint action
Spearman–Karber method
Spiked bioassay (SB) approach
Static renewal test
Static toxicity test
Summation rule
Synergism
Time–response approach
Toxic equivalency factor (TEF)
Toxic equivalent (TEQ)
Toxic unit (TU)
Up-and-down experimental design
Weakest link incongruity
Weibull metameter

2. The most critical (sensitive) life stage of a particular invertebrate is its first instar that has an incipient LC50 of 10 $\mu g \cdot L^{-1}$ and an NOEC of 5 $\mu g \cdot L^{-1}$ for growth during toxicant X exposure. Can a viable population of this invertebrate be maintained below an effluent with a maximum concentration of 5 $\mu g \cdot L^{-1}$ for this toxicant?
3. Compare the advantages and disadvantages of static, static-renewal, and flow-through (continuous and intermittent) tests.
4. What are the relative advantages of the dose– or concentration–response versus the survival time approach to measuring toxicity?
5. In what situations would it be essential to estimate a complete median lethal concentration instead of the conventional median lethal concentration?
6. A channel in a heavily industrialized estuary must be deepened. How would you assess the potential for acute and chronic toxicity to endemic species resulting from the dredging activities?
7. Defend the concept of individual effective dose.
8. You are required to accurately estimate the LC50, LC10, and LC5 for a toxicant. Describe the methods you would use to calculate these numbers. Give the reasons for selecting one approach over another.

9. Would normit and probit analyses (MLE method) of a data set produce the same LC5 estimate? Would they produce the same LC50 estimate? Give reasons for each of your responses.
10. Describe how you would estimate toxic incipiency.
11. What quantitative models would you use to predict the joint action of two toxicants with identical modes of lethal action? Which model would you use for the joint action of two toxicants with completely different modes of lethal action? How would you determine if their joint effects were additive or nonadditive?
12. What is the difference between concentration additivity and effect additivity?
13. Could a pair of chemicals that are functional antagonists also be receptor antagonists? Could they also be dispositional antagonists?
14. What is the difference between proportional hazard and accelerated failure time models? Are there any differences in the data collected to fit these two models?
15. Describe the Bliss method of accounting for the effect of individual size on toxic effect.
16. What is the difference between photoinduced toxicity and photosensitivity?
17. What is a split probit? What factors would result in a split probit plot?
18. What is the advantage of the up-and-down experimental design?

CHAPTER 10

1. Define the terms highlighted in the chapter and check your definitions against those provided in the glossary.
   Age-specific birth rate (m_x)
   Age-specific death rate (q_x)
   Age-specific number of individuals dying (d_x)
   Allozyme
   Beach head effect
   Carrying capacity (K)
   Concept of strategy
   Core-satellite hypothesis
   Cross-resistance or co-tolerance
   Disposable soma theory of ageing
   Doubling time
   Dynamic Energy Budget (DEB) Approach
   Ecological epidemiology
   Ecotoxicogenomics
   Effect at distance hypothesis
   Effective population size
   Environmental epidemiology
   Epidemiology
   Etiological agent
   Euler–Lotka equation
   Expected life span (e_x)
   Fecundity selection
   Finite rate of increase (λ)
   Gametic selection
   Genetic bottleneck
   Genetic drift
   Genetic hitchhiking
   Hardy–Weinberg equilibrium
   Heterosis
   Incidence
   Incidence rate

Industrial melanism
Intrinsic rate of increase (r)
Iteroparous species
Keystone habitat
Lefkovitch matrix
Leslie matrix
Liebig's law of the minimum
Limited lifespan paradigm
l_x
l_x table
$l_x m_x$ life table
Malthusian theory
Mean generation time (T_c)
Meiotic drive
Metapopulation
Minimum amount of suitable habitat
Minimum viable metapopulation
Minimum viable population (MVP)
Monogenic control
Multiple heterosis
Multiplicative growth factor per generation (λ)
Natural selection
Net reproductive rate (R_0)
Nine aspects of disease association
Occupancy model
Odds ratio
Optimal stress response
Polygenic control
Population
Prevalence (P)
Principle of allocation
Propagule rain effect
Rate ratio
Rate of living theory of ageing
Relative risk (RR)
Reproductive value (V_A)
Rescue effect
Resistance
Ricker model
Risk factor
Risk ratio (RR)
Roentgen (R)
Selection component
Semelparous species
Sexual selection
Shelford's law of tolerance
Stable population
Stress theory of ageing
Structured metapopulation model
Tolerance
Viability selection
Wahlund effect

2. Review the earlier chapters and identify 10 specific risk factors. Five should be qualities of individuals and five should be etiological factors.

3. Liver tumors in flounder are surveyed in a contaminated bay. Calculate the incidence rate as cases per 1000 flounder years of exposure. Assume the tumors kill each diseased individual within the sampling year. What was the mean prevalence rate over the 5 years of sampling?

Year	Number of Flounder Sampled	Number of Diseased Flounder
1	1583	112
2	2032	100
3	1911	216
4	2118	96
5	1625	97

4. The above population is compared to a reference population ("control") that has the qualities tabulated below. Use the mean incidence rate for both populations to calculate the rate ratio. Assuming that the contamination in the bay was the "treatment," calculate the odds ratio for each of the 5 years of the survey.

Year	Number of Flounder Sampled	Number of Diseased Flounder
1	3008	3
2	5025	11
3	4091	7
4	3557	6
5	4253	17

5. Describe the nine aspects of disease association. Include an example of each.
6. Explain the predisposition of ecotoxicologists in using laboratory sublethal and lethal test results to predict the viability or distribution of populations about a contaminated field site.
7. You would like to study the impact of a contaminant on a songbird species nesting in a contaminated location. Explain why you might select either, a Leslie or Lefkovitch matrix approach in this study.
8. In 1 year, a population increased from 100 to 150 individuals. Calculate λ and r for this population for that time. Assuming no change in growth dynamics, calculate the population size in 20 years of growth beginning with 100 individuals.
9. How could knowledge of V_a for the age classes of a population aid in estimating the viability of a population exposed to a toxicant over long periods?
10. What is a metapopulation? Describe how toxicants could affect a metapopulation. Include in your answer the concepts associated with the "rescue effect," "propagule rain," and "effect-at-distance."
11. Explain how metapopulation processes might influence the outcome of chronic exposure of a metapopulation to an environmental toxicant?
12. State with reasons why individual- or population-based effect metrics would be more sensitive to the effects of an environmental toxicant.
13. Using the principle of allocation and concept of strategy, relate how life histories of individuals in a population may shift in response to a toxicant.
14. Contrast how a stressor might influence longevity of an individual under the disposable soma and stress theories of ageing.
15. How might a toxicant influence the genetics of a population?
16. What factors influence the rate of tolerance acquisition in a population?
17. Describe the concept of selection components.
18. A disease has appeared in a population of endangered voles that live near a leather tanning factory that contaminated the surrounding soils with chromium. The town leaders argue that the disease is characteristic of the voles and the chromium-contaminated soil has nothing to do with it. You decide to study the voles by sampling from the contaminated site and a reference site. You collect a total of 750 voles with the disease. 500 of these diseased voles were collected from the area with the contaminated soil and 250 voles with the disease were collected from the area with clean soils. You

also collected 500 healthy voles of which 60 came from chromium-contaminated soil area and 440 came from clean soil area. Drawing on the techniques and materials discussed in class, use these data to decide whether the town leaders' collective opinion is sound. What is your decision? How much confidence is warranted for your decision (e.g., very confident, more likely correct than incorrect, and unsure)? What valuable piece of information would you request if you could ask/gather only one more piece of information?

CHAPTER 11

1. Define the terms highlighted in the chapter and check your definitions against those provided in the glossary.
   Biodiversity
   Biomonitoring
   Brillouin index
   Community
   Community conditioning hypothesis
   Ecosystem
   Elasticity
   Emergent properties
   Eutrophic
   Exploitation interference
   Functional redundancy
   Functional response
   Fundamental niche
   Guild
   Hutchinsonian niche
   Index of biological integrity (IBI)
   Inertia
   Info-disruption
   Insurance hypothesis
   Interference competition
   Interspecies competition
   Keystone species
   K-strategy
   Law of frequencies
   Lentic
   Log normal model
   Lotic
   MacArthur–Wilson model
   Maturity index
   Mesocosm
   Mesotrophic
   Metacommunity
   Microcosm
   Most sensitive species approach
   Niche preemption
   Numerical response
   Octave
   Oligotrophic
   Optimal foraging theory
   Pielou's J
   Pollution-induced community tolerance (PICT)

Pound-of-flesh incongruity
Rarefaction estimate of richness
Realized niche
Recovery rate
Redundancy hypothesis
Resilience
Rivet popper hypothesis
r-strategy
Saprobien spectrum
Shannon index
Species assemblage
Species diversity
Species evenness
Species richness
Taxocene
Type 1 biomonitoring
Type 2 biomonitoring
Type 3 biomonitoring
Type 4 biomonitoring
Vulnerability

2. What are the differences among the terms: community, species assemblage, guild, and metacommunity?
3. Describe the community-level NOEC concept, including its advantages and disadvantages.
4. How might optimal foraging theory help us to understand the effects of toxicants on populations in a community?
5. Describe the factors contributing to the ability of a community to successfully avoid irreversible damage by a pollutant?
6. What is the major difference between the Shannon and Brillouin indices of species diversity? Which one will have the largest value for any sample?
7. Describe the rivet popper hypothesis versus redundancy hypothesis debate. What is the significance of this debate relative to assessing an adverse effect of pollutants to ecological communities? How might the application of these hypotheses change for predictions of toxicant effects in a metacommunity context?
8. In general, how would the log normal, species abundance curve change with the introduction of a toxicant to an ecosystem? What is (are) the possible mechanism(s) for this shift in the curve?
9. What are the differences between microcosms and mesocosms? What are the relative advantages and disadvantages of these tools?
10. Assuming that the MacArthur–Wilson model accurately reflects colonization, predict the number of species inhabiting a beach devastated by an oil spill if the number of species originally on the beach was 23 (S_{EQ} = 23?), G = 0.10 (units of years), and 10 years have passed since the devastation occurred. How long would it take for 95% of the species to reestablish themselves? Would the community (species assemblage) be back to normal at that point?
11. What are the differences among oligotrophic, mesotrophic, and eutrophic water bodies?
12. Toxicants can have more impact than expected if they change keystone habitats, corridors between habitats, or keystone species. Explain the reason for each of these three entities (i.e., keystone habitat, habitat corridors, and keystone species).
13. How can a toxicant's effects on species interactions result in the disappearance of species even though the toxicant is present at concentrations that do not directly harm the individuals making up each species population?
14. Why do ecotoxicologists tend to look for toxicant-induced changes in community structure rather than function? Describe some changes in community function that have been described for communities exposed to toxicants.
15. Explain what a trophic cascade is and use one of the examples from the book to describe it. Explain how pollution might influence an ecological community by changing the size of the population of a predator species that was important in top-down regulation.

CHAPTER 12

1. Define the terms highlighted in the chapter and check your definitions against those provided in the glossary.
   Acid precipitation
   Alkalinity
   Bioaccumulative chemicals of concern (BCCs)
   Chapman mechanism
   Cold condensation theory
   Cryosphere
   Dobson unit
   Dry deposition
   Ecoregion
   Ecosystem incongruity
   Ecotone
   Edge effect
   Enrichment factor
   Fertilization effect
   Geographic information system (GIS)
   Global distillation
   Global fractionation
   Global warming
   Grasshopper effect
   Greenhouse effect
   Greenhouse gases
   K_{OA}
   Landscape
   Montreal Protocol
   Ocean acidification
   Ozone hole
   Pedosphere
   Persistent organic pollutants (POPs)
   Persistent toxicant that bioaccumulates (PTBs)
   Region (geographic)
   Remote sensing
   Retention effect
   Subcooled liquid vapor pressure
   Temperature of condensation
   Waldsterben
   Wet deposition
2. Give three reasons why the traditional ecosystem context for ecotoxicology should be expanded to a larger scale to address ecotoxicological problems facing us today.
3. What factors would influence the sensitivity of a watershed to acid precipitation? What effects might you expect in aquatic and terrestrial biota in a watershed with a granite bedrock geology?
4. How does the increased use of CFCs result in decreased ozone levels in the stratosphere above the Antarctic? Why is the problem less noticeable above the Arctic?
5. Describe the processes by which POPs are transported and differentially distributed from the equator to the earth's poles.
6. The process of mountain cold trapping was discussed in Chapter 2 as one occurring for persistent organic compounds along mountain slopes and the global distillation process is discussed in this chapter for the same compounds but at a global scale. Compare and contrast mountain cold trapping and global distillation. How might global climate change interact with these processes to modify persistent organic compounds' fate and transport?

7. Persistent pesticides have been found to move from valley areas where they are used into the mountains of Costa Rica. How high each pesticide travels up the mountain range seems to be related to its volatility and K_{ow}. Recollecting the class materials describing global distillation, explain this mountain cold trapping phenomena. Begin by explaining the global distillation process and then relate it to the mountain cold trapping.

CHAPTER 13

1. Define the terms highlighted in the chapter and check your definitions against those provided in the glossary.
Abductive inference
Analysis plan
Assessment endpoint
Chemicals of potential concern
Chronic RfD
Committed effective dose
Comparative risk assessment
Conceptual model
Conceptual model diagram
Critical study
Critical toxic effect
Delphi method
Developmental RfD
Dose equivalent
Effective dose (ED)
Effective dose equivalent
EPA weight of evidence classification
Exposure
Exposure characterization
Exposure pathways
Exposure profile
Feasibility study (FS)
Hazard assessment
Hazard quotient
Health advisory concentrations
Integrated Risk Information System (IRIS)
Loss of life expectancy
Measurement endpoint
Modifying factor
NAS paradigm
"No action" alternative
Nonstochastic effects
One-hit risk model
Precautionary principle
Predictive risk assessment
Probabilistic induction
Problem formulation
Proteinuria
Quotient method
Reasonable maximum exposure (RME)
Receptor
Reference dose (RfD)

Remedial investigation (RI)
Remedial investigation and feasibility study (RI/FS)
Retroactive risk assessment
Risk
Risk assessment
Risk assessor
Risk characterization
Risk factor
Risk hypotheses
Risk manager
Slope factor
Specific effective energy
Stochastic effects
Stressor
Subchronic RfD
Toxicity value
Uncertainty factor
Weight of evidence

2. What is the major difference between the risk assessment principle and the precautionary principle for making management decisions?
3. What are the four components of the NAS paradigm for risk assessment? What are their specific goals?
4. What is the major distinction between a hazard assessment and a risk assessment? As described in this chapter, is the assessment done for noncarcinogenic effects a hazard or risk assessment? Explain your answer.
5. Outline the steps in a human risk assessment. Check your outline against that shown in Figure 13.1.
6. Groundwater below a closed industrial building contains high concentrations of trichloroethylene (TCE) and the contaminated water is predicted to outcrop to a stream within the next 2 years. You are asked to do an assessment of the situation. Are you doing a predictive or retroactive risk assessment? Give the reasons for your answer.
7. What is the function of uncertainty factors? Is it the same as that of the slope factor? What is the function of the modifying factor?
8. Get onto IRIS to find the RfD and associated information for hexavalent chromium, i.e., chromium (IV). If the concentration in drinking water taken from a representative well near a hazardous waste site was 6 $mg{\cdot}L^{-1}$, what would the hazard quotient be? Provide the qualifiers needed by the risk manager to use this calculated value intelligently.
9. Contrast and give examples of assessment and measurement endpoints. What are the qualities of good assessment and measurement endpoints? Give some good and bad examples of both endpoints.
10. Describe the steps and products of an ecological risk assessment.
11. Why are α particles of little concern from an external irradiation standpoint, but of major concern if inhaled?
12. Describe how you would derive a reference dose (RfD) from NOEC and LOEC data for a (noncarcinogenic) toxicant. Include some explanation of the associated use of uncertainty and modifying factors.
13. Describe how one estimates risk associated with carcinogens, including any quantitative methods for estimating risk and qualitative classifications required in the statement of risk.
14. How does ecological risk assessment differ from human risk assessments? What is same in both assessment processes?
15. The book describes the concepts of radioactivity, absorbed dose, absorbed dose rate, dose equivalent, effective dose equivalent, and committed effective dose. Generally describe these terms. Do not worry about providing units unless you happen to need them to clarify your answers.

ENVIRONMENTAL LAW APPENDICES

1. Define the terms highlighted in the appendices and check your description against those provided in the glossary. (The relevant appendix is noted in brackets, e.g., [*U.S. Law Appendix*]).
   Agricultural and Veterinary Chemicals Administration Act (Australian Law Appendix)
   Agricultural and Veterinary Chemicals (Code) Act (Australian Law Appendix)
   Air (Prevention and Control of Pollution) Act (Indian Law Appendix)
   Air Toxics NEPM (Australian Law Appendix)
   Ambient Air Quality National Environment Protection Measure (NEPM) (Australian Law Appendix)
   Assessment of Effects of Certain Public and Private Projects on the Environment (European Union Law Appendix)
   Australia New Zealand Food Standards Code (Australian Law Appendix)
   Australian Councils (Australian Law Appendix)
   Biocides Regulation (European Union Law Appendix)
   Birds Directive (European Union Law Appendix)
   Clean Air Act (U.S. Law Appendix)
   Clean Water Act (U.S. Law Appendix)
   Comprehensive Environmental Response, Compensation, and Liability Act (CERCLA) (U.S. Law Appendix)
   Cosmetics Regulation (European Union Law Appendix)
   Council of Environmental Quality (U.S. Law Appendix)
   Directives (European Union Law Appendix)
   Duty of Care Principle (South African Law Appendix)
   Environment (Protection) Act (EPA) (Indian Law Appendix)
   Environmental assessment (U.S. Law Appendix)
   Environmental Conservation Act 73 (South African Law Appendix)
   Environmental impact statement (U.S. Law Appendix)
   Environmental Liability Directive (European Union Law Appendix)
   Environmental law (U.S. Law Appendix)
   Environmental Protection Law of the PRC (for Trial Implementation) (Chinese Law Appendix)
   Environmental Protection Law of the PRC 1989 (Chinese Law Appendix)
   Environmental Quality Report (U.S. Law Appendix)
   Environment Protection (Sea Dumping) Act (Australian Law Appendix)
   European Community (EC) (U.S. and European Union Law Appendices)
   European Environment Agency (European Union Law Appendix)
   European Union (EU) (European Union Law Appendix)
   Federal Insecticide, Fungicide, and Rodenticide Act (U.S. Law Appendix)
   Finding of no significant impact (U.S. Law Appendix)
   Food Standards Australia New Zealand Act (Australian Law Appendix)
   Habitats Directive (European Union Law Appendix)
   Hazardous Substances Act 15 (South African Law Appendix)
   Industrial Chemicals (Notification and Assessment) Act (Australian Law Appendix)
   Large Combustion Sites Directive (European Union Law Appendix)
   Law of the PRC on Conserving Energy (Chinese Law Appendix)
   Law of the PRC on the Prevention and Control of Atmospheric Pollution (Chinese Law Appendix)
   Law of the PRC on Prevention and Control of Environmental Pollution Caused by Solid Waste (Chinese Law Appendix)
   Law of the PRC on Prevention and Control of Water Pollution (Chinese Law Appendix)
   Marine Environment Protection Law of the PRC (Chinese Law Appendix)
   Marine Protection, Research, and Sanctuaries Act (U.S. Law Appendix)
   National ambient air quality standards (NAAQ) (U.S. Law Appendix)
   National Environmental Policy Act (U.S. Law Appendix)

National Environment Protection (Assessment of Site Contamination) Measure (Australian Law Appendix)
National Environment Protection Council (Australian Law Appendix)
National Environmental Management Act 107 (South African Law Appendix)
National Environmental Management Air Quality Act 39 (South African Law Appendix)
National Environmental Waste Management Act 59 (South African Law Appendix)
National Pollutant Discharge Elimination System (NPDES) (U.S. Law Appendix)
National Pollutant Inventory (NPI) NEPM (Australian Law Appendix)
National Water Act 36 (South African Law Appendix)
National Water Quality Management Strategy
Occupational Health and Safety Act 85 (South African Law Appendix)
Organization for Economic Cooperation and Development (U.S. Law Appendix)
Panchayati raj (Indian Law Appendix)
Pesticides, legal definition (U.S. Law Appendix)
Plant Protection Products Directive (European Union Law Appendix)
Provisional Rules of Preventing Coastal Water Pollution (Chinese Law Appendix)
Provisions Concerning Environmental Protection and Amelioration (Chinese Law Appendix)
Registration, Evaluation, Authorization, and Restriction of Chemicals (REACH) (European Union Law Appendix)
Regulation (U.S. Law Appendix)
Regulations of the PRC on Nature Reserves (Chinese Law Appendix)
Resource Conservation and Recovery Act (RCRA) (U.S. Law Appendix)
Safe Drinking Water Act (SDWA) (U.S. Law Appendix)
State implementation plans (SIPs) (U.S. Law Appendix)
Ten Major Countermeasures of China's Environment and Development (Chinese Law Appendix)
Therapeutic Goods Act (Australian Law Appendix)
Toxic Substances Control Act (TSCA) (U.S. Law Appendix)
Waste Disposal Directive (European Union Law Appendix)
Water Framework Directive (WFD) (European Union Law Appendix)
Water Law of the PRC (Chinese Law Appendix)
Water (Prevention and Control of Pollution) Act (Indian Law Appendix)

2. The environmental law appendices provide summaries of environmental laws and regulations for the United States, European Union, the People's Republic of China, Australia, India, Central America, and South Africa. Briefly contrast the manner in which these legal tools are used in each of these cases.
3. Sketch out five timelines for the various pieces of legislation enacted by the United States, European Union, China, Australia, India, and South Africa. If you can, try to explain the differences and similarities in these timelines.
4. Assume that you are responsible for planning an industrial activity (lead battery recycling) within each of the following four regulated locations, California, Portugal, Hunan Province, and New South Wales. Rank the relative level of effort you might have to expend in each of these cases to conform to the associated environmental legislation. Explain your ranking.
5. You are involved with an international lubricant-producing corporation that is based in Ohio. Will the REACH legislation influence your business in any way? Please explain your answer.

Glossary

Abductive Inference: Inference to the best, i.e., the most probable, explanation. It uses information gathered about a phenomenon or situation to produce a hypothesis that is the most probable explanation for the data. (Chapter 13)

Absolute Bioavailability: The bioavailability of a dose (D) estimated from the area under the curve (AUC) for any route or formulation of the compound divided by the AUC after direct injection of the same dose (D) into the bloodstream. (Chapter 4)

Absorption Rate Constant (k_a): A first-order rate constant for absorption calculated as MAT = k_a^{-1}, where MAT is the mean absorption time. (Chapter 4)

Acaricide: A chemical agent used intentionally to kill mites or ticks. (Chapter 2)

Accelerated Failure Time Model: A survival time model in which the time-to-death (ln TTD_i) of a particular type/class of individual (e.g., smoker) is changed ("accelerated") as some function of a covariate (e.g., classification relative to smoking habits). (Chapter 9)

Acclimation: The modification of biological functions, especially physiological, or structures to maintain or minimize deviations from homeostasis despite change in some environmental quality, such as temperature, salinity, light, radiation, or toxicant concentration. It is an expression of phenotypic plasticity of individuals in response to a sublethal change in some environmental factor. (Chapter 9)

Acclimatization: Like acclimation, acclimatization is the modification of biological functions, especially physiological, or structures to maintain or minimize deviations from homeostasis despite change in some environmental quality. However, these shifts are taking place under natural conditions, not under controlled laboratory conditions as is often the case with studies of acclimation. (Chapter 9)

Accumulation Factor (AF): The ratio of nonpolar organic compound concentration in the organism to that of sediments ([organism]/[sediment]) with the organism's concentration normalized to gram of lipid and sediment concentration normalized to gram of organic carbon. (Chapter 3)

Acetylation: In this book, the formation of amides in a xenobiotic by *N*-acetyltransferase. (Chapter 6)

Acetylcholinesterase Inhibitors: Compounds such as many organophosphate and carbamate insecticides that inhibit the normal functioning of the enzyme, acetylcholinesterase, which hydrolyzes the neurotransmitter, acetylcholine. (Chapter 8)

Acid Mine Drainage: This condition occurs when exposure of sulfide ores in metal mines to oxygen and chemoautotrophic bacteria generates low pH drainage with very high sulfate concentrations. This also occurs with pyrite in coal mines. (Chapter 2)

Acid Precipitation: Precipitation with a pH below 5.7 including rain, fog, snow, or other forms of precipitation. (Chapter 12)

Acid Rock Drainage: This condition occurs when exposure of sulfide ores from metal mines to oxygen and chemoautotrophic bacteria produces low pH drainage with very high sulfate concentrations. This also occurs with pyrite in coal piles. (Chapter 2)

Acid Volatile Sulfides (AVS): Sediment-associated sulfides extracted with 1 N HCl which are assumed to be primarily iron sulfides, especially metastable mackinawite and greigite. In some cases, pyrite can also become the dominant component of AVS (Morse and Rickard, 2004). (Chapter 4)

Acidophilic Component: A cell component such as the general cytoplasm that is readily stained by an acidic dye. (Chapter 7)

Activation: One possible consequence of biotransformation in which the effect of an active compound is worsened or an inactive compound is converted to one with an adverse bioactivity. (Chapter 3)

Active Transport: Movement of a substance up an electrochemical gradient that requires a carrier molecule and energy. (Chapter 3)

Acute Lethality: Death following a short and often intense exposure. The duration of an acute exposure in a conventional toxicity testing is generally 96 (4 days) or fewer hours of exposure. (Chapter 9)

Additive Index: An index for quantifying the joint action of toxicants in a mixture. (Chapter 9)

Additivity: The condition of additivity exists for two or more toxicants in mixture if the mixture effect level was simply that expected by summing the expected individual toxicant effects. (Chapter 9)

Adduct: Modification of the DNA molecule produced when a xenobiotic or a metabolite binds covalently to a base (most often) or another portion of the DNA molecule. (Chapter 6)

Adenocarcinoma: An adenoma that becomes malignant. (Chapter 7)

Adenoma: A benign epithelial tumor of glandular tissue. (Chapter 7)

Adenylate Energy Charge (AEC): An index thought to reflect the balance of energy transfer between catabolic and anabolic processes. AEC = (ATP + 1/2 ADP)/(ATP + ADP + AMP), where ATP, ADP, and AMP = concentrations of adenosine tri-, di-, and monophosphate, respectively. (Chapter 8)

Adsorption: The accumulation of a substance at the common boundary of two phases, most often adsorption from a solution onto the solid surface. Adsorption of contaminants can result from processes of ion exchange, weak bonding to surfaces such as that associated with van der Waal forces, or even, molecular orientation of large, dipolar organic compounds relative to solid (less polar) and water (more polar) phases. (Chapter 3)

Aerosol: A collection of liquid or solid particles suspended in a gas such as air. (Chapter 4)

Age Pigment: See **Ceroid Pigment** or **Ceroid**. (Chapter 7)

Age-Specific Birth Rate: The mean number of females born to a female of an age class *x*. (Chapter 10)

Age-Specific Death Rate: The proportion dying or probability of dying as tabulated for a life table interval or age class (*x*). It is the number of individuals of that age class dying divided by the number of individuals in that age class (and available to possibly die). For some application such as human demography, this rate might be multiplied by a standard number of individuals; e.g., 153 newborns dying of 100,000 newborns. (Chapter 10)

Age-Specific Number of Individuals Dying: The number of individuals dying in a life table interval (*x*). It is estimated as a simple difference, $d_x = l_x - l_{x+1}$. (Chapter 10)

Agonist: If chemical antagonism is framed in the special context of antidotes, the chemical whose effect is thought to be reduced is the agonist and the chemical that reduces that effect is the antagonist or antidote. (Chapter 9)

Agricultural and Veterinary Chemicals Administration Act (1992) and Agricultural and Veterinary Chemicals (Code) Act (1994): Australian laws that control the import, manufacture, and use of agricultural and veterinary products. (Appendix 6)

AHH (Aryl Hydrocarbon Hydroxylase): Activity units of benzo[a]pyrene hydroxylation by AHH used to reflect cytochrome P-450 monooxygenase activity. (Chapter 6)

Air (Prevention and Control of Pollution) Act: This Indian legislation was enacted in 1981 to prevent, control, and abate air pollution; to establish boards to do this; and to confer and assign related power and functions to such boards' powers. (Appendix 7)

Air Toxics National Environment Protection Measure (NEPM): The 2004 Australian legislation establishing procedures for collection of data for the purpose of developing standards for toxicants. (Appendix 6)

Aliphatic Compounds: These nonaromatic organic compounds (i.e., containing no six-carbon rings sharing resonating electrons) are extraordinarily diverse. They can incorporate single, double, or triple bonds, and branches in their structures. (Chapter 2)

Alkalinity: The capacity of natural water to neutralize acid as measured by titration of a water sample with a dilute acid to a specific pH endpoint. Most often, alkalinity is a function of carbonate (CO_3^{2-}), bicarbonate (HCO_3^-), and hydroxide (OH^-) concentrations, i.e., the carbonate–bicarbonate buffering system of the water. However, dissolved organic compounds, borates, phosphates, and silicates can also contribute to alkalinity. (Chapter 12)

Alkoxyradicals: Oxyradicals of organic compounds (R) of the form $RO^\bullet$. (Chapter 6)

Allelochemical: Toxic chemicals produced by plants as a defense mechanism against grazers or to inhibit competing plant species. (Chapter 6)

Allometry: The study of size and its consequences. (Chapter 4)

Allozyme: An allelic variant of an enzyme coded for by a particular locus. (Chapter 10)

Alpha (α) Particles: Fragments of the nucleus ejected from a radioactive atom to reduce excess energy and gain stability. They are relatively large in mass (7345 times larger than a β particle), consisting of 2 neutrons and 2 protons, and carrying a +2 charge. (Chapter 2)

Ambient Air Quality National Environment Protection Measure (NEPM): Australian law that sets standards for air pollutants including carbon monoxide, carbon dioxide, ozone, sulfur dioxide, lead, PM10, and PM2.5 particles. (Appendix 6)

Amelia: A developmental abnormality in which the individual is born without limbs. (Chapter 8)

Amidase: An enzyme that catalyzes the hydrolysis of the carbon–nitrogen bond of an amide. (Chapter 6)

Amiloride-Blockable Sodium Channel: A porin responsible for sodium readsorption within distal nephrons of the vertebrate kidney. Sodium in the tubule lumen diffuses passively via these sodium channels on the apical surface of nephron cells. (Chapter 3)

Amino Acid Conjugation: The conjugation of a xenobiotic possessing a carboxylic acid group with an endogenous amino acid. (Chapter 6)

δ-Aminolevulinic Acid Dehydratase (δ-ALAD or ALAD): An enzyme catalyzing the conversion of δ-aminolevulinic acid to porphobilnogen during heme synthesis. (Chapter 6)

Analog (Elemental): An element that behaves like, but not necessarily identical to, another element in biological processes; e.g., cesium is an analog of potassium, and strontium is an analog of calcium. (Chapters 2 and 5)

Analysis of Variance (ANOVA): One-way ANOVA breaks down the total variance (total sum of squares) in a data set into the variance among and within treatments, e.g., among the different concentration treatments of a bioassay and within replicates for each concentration treatment. The variance within treatments (mean sum of $\text{squares}_{\text{within}}$) is assumed to reflect the sampling or error variance, and that among treatments (sum of $\text{squares}_{\text{among}}$) is thought to estimate the error variance plus any additional variance that might be associated with the treatment effect. ANOVA is used often to test the null hypothesis of equal means among treatments in sublethal or chronic lethal assays. (Chapter 8)

Analysis Plan (of an Ecological Risk Assessment): A plan that defines the exact format and design of the assessment, explicitly states the data needed, and describes the methods and design for analyzing these data. (Chapter 13)

Androgen Receptor Antagonist: A xenobiotic that acts by blocking androgen receptor–mediated processes. (Chapter 8)

Aneuploidy: The deviation by loss or addition of chromosomes from the usual (e.g., $2N$) number of chromosomes. (Chapter 6)

Antagonism: Toxicant antagonism occurs if the actual effect level of the mixture was lower than the sum of the predicted effects for the individual toxicants in the mixture. (Chapter 9)

Antagonist: If chemical antagonism is framed in the special context of antidotes, the chemical whose effect is thought to be reduced is the agonist and the chemical that reduces that effect is the antagonist or antidote. (Chapter 9)

Anthropocene Epoch: A proposed new epoch of the earth dominated by the activities of humans that would start with the industrial revolution of the late eighteenth century and extend into present times. According to Steffen et al. (2011), four profound changes have resulted from human activities during that period: (1) climate change, (2) significant alteration of elemental cycles such as those of nitrogen, phosphorus, and sulfur, (3) substantial modification of the terrestrial water cycle, and (4) precipitation of the sixth mass extinction event in the history of life. (Chapter 1)

Antidote: See **Anatgonist**. (Chapter 9)

Antiporter: A membrane-associated transporter that moves one (or more) molecule/ion across a membrane in exchange for another molecule/ion. Contrast to **Symporter**. (Chapter 3)

Antirescue Effect: Increased risk of extinction of a metapopulation patch resulting from increased likelihood of immigration. (Chapter 10)

Antisense Oligonucleotides: Oligonucleotides that are complementary to messenger RNA and bind to them. (Chapter 6)

Antisymmetry: Antisymmetry occurs when structure is greater on one side than on the other, but it is not predetermined on which side the structure will be larger. (Contrast with **Directional Asymmetry** in which the presence and direction of the asymmetry are predetermined.) (Chapter 8)

Apoptosis: A type of programmed cell death that might occur if a cell is damaged or otherwise compromised. A programmed sequence of biochemical steps begin that is initiated that result in the death and removal of the cell from the tissue. (Chapter 7)

Apoptotic Bodies: Apoptotic cells can break into membrane-bound fragments called apoptotic bodies. (Chapter 7)

Apparent Volume of Distribution (V_d): A mathematical volume used in clearance volume–based models. It is expressed in units of volume of a reference compartment, often the blood or plasma compartment. (Chapter 3)

Arcsine Square Root Transformation: A common and useful transformation of effects data that often allows one to meet the assumption of homogeneous variances for proportions of exposed individuals responding,

$$\text{Transform } P = \arcsin\sqrt{P}$$

where P = the measured effect expressed as the proportion of the exposed organisms. (Chapter 8)

Aromatic Compounds: These compounds have one (or more) six-carbon ring(s) (C_6H_6) in which high stability is imparted because carbons bonds of the ring are intermediate between single and double bonds. The involved electrons are in resonance: they move among the six carbons of a ring to produce this state. (Chapter 2)

Aryl Hydrocarbon Receptor (AHR): A transcription factor involved in the induction of CYP1A. (Chapter 6)

Assessment Endpoint (in Ecological Risk Assessment): A valued ecological entity that is to be protected and the precise quality to be measured for this entity. Earlier, only the first part of this definition (the valued entity) was considered the assessment endpoint. See **Measurement Endpoint** also. (Chapter 13)

Assessment of the Effects of Certain Public and Private Projects on the Environment Directive: This European Union directive requires an environmental assessment to be carried out before a decision is made about whether development consent should be granted for certain types of projects, which are likely to have significant environmental effects. (Appendix 4)

Assimilation Efficiency (AE): A measure of the proportion of an ingested chemical that is assimilated into (initially) the alimentary epithelium of the feeding animal; the amount absorbed per amount ingested in food. (Chapters 3 and 4)

***Aufwuchs*:** See **Periphyton**. (Chapter 5)

Australian Councils: Australian environmental laws reside primarily in states and territories, but some issues require a broader and more coordinated effort. Councils of representatives from each federal, state, and territory government have proven to be highly successful for this purpose, having a number of advantages including minimizing duplication, costs, and regulatory burden placed on industry. (Appendix 6)

Autophagy: The destruction in lysosomes of organelles or damaged proteins by the cell so as to eliminate damaged cell components. (Chapter 8)

Backstripping or Backprojection Procedure: A method used to extract parameter estimates for elimination involving a multiexponential process. The procedure may be implemented graphically or mathematically with a computer program. (Chapter 3)

Bad Old Days Incongruity: The dubious working assumption that the worst environmental issues are in our past and present issues are being handled adequately with existing legislation and technology. (Chapter 1)

Bartlett's Test: A statistical test used to test the assumption of homogeneity of variances. In this book, it was used to assess this assumption prior to one-way analysis of variance of sublethal and chronic lethal effects data. (Chapter 8)

Basophilic Component: Component of a cell such as the nucleus that is readily stained by a basic dye. (Chapter 7)

Beach Head Effect: In a habitat mosaic, a beach head patch allows the invasive population to build to sufficient numbers so that it now can act as a source population for other invasions to other regions of the landscape. (Chapter 10)

Becquerel (Bq): The official unit of radioactivity that has replaced the curie. One curie is 3.7×10^{10} Bq. (Chapters 1 and 2)

Behavioral Teratology: The study of behavioral abnormalities in otherwise normal appearing individuals after exposure as an embryo to an agent. (Chapter 8)

Behavioral Toxicology: The science of abnormal behaviors produced by exposure to a chemical or a physical agent. (Chapter 8)

Benign Neoplasia: Neoplasia that tend to remain differentiated in their morphology and grow slowly such that they are not as invasive of neighboring tissues as other neoplasia. (Contrast with **Malignant Neoplasia**.) (Chapter 7)

Beta (β) Particles: Electrons or positrons (positive electron) ejected from the atom during radioactive decay. (Chapter 2)

Binding Trends (of Metals): The ionic bond stability of a metal with a hard ligand is estimated with the metal ion charge (Z) and radius (r): polarizing power = Z^2/r. The covalent bond strength is quantified by $\Delta\beta$. It is calculated with the stability constants (β_{MX}) of the metal fluoride and chloride complexes: $\Delta\beta = \log \beta_{MF} - \log \beta_{MCl}$. (Chapter 2)

Binomial Method: A method of estimating LC50, EC50, or LD50 if there are no partial kills. (Chapter 9)

Bioaccumulation: The net accumulation of a contaminant in (and in some occasional instances on) an organism from all sources including water, air, and solid phases of the environment. Solid phases include food sources. (Chapter 3)

Bioaccumulation Factor (BAF): The ratio of the contaminant concentration in the organism to that in the sediment or some other source, i.e., [organism]/[sediment] or [organism]/[food]. Often, this term is used in a broader sense if the exact source is not known or poorly defined as in a field survey (see **Bioaccumulation Factor [BF]**). For the purposes of avoiding confusion, BSAF (see **Biota-Sediment Accumulation Factor**) is used in this book to define sediment-associated bioaccumulation and BAF is used in a more general context. (Chapter 3)

Bioaccumulation Factor (BF): Like bioaccumulation factor above, the BF is the ratio of contaminant concentration in the organism to that in potential sources. The BF is based on the assumption that water and food (including sediments in some cases) may contribute to differences in concentrations measured for individuals at different trophic levels. (Chapter 5)

Bioamplification: See **Biomagnification**, **Concentration Amplification**, and **Fugacity Amplification**. (Chapter 5)

Bioavailability: The extent to which a contaminant in a source is free for uptake. In many definitions, especially those associated with pharmacology or mammalian toxicology, bioavailability implies the degree to which the contaminant is free to be taken up by the organism *and to cause an effect at the site of action.* (Chapter 4)

Bioccumulative Chemicals of Concern (BCCs): See **Persistent Organic Pollutants**. (Chapter 12)

Biocides Regulation: A recent European Union regulation (European Commission 528/2012) that replaced an earlier directive and came into effect on September 2013. It addresses the marketing and use of biocides. (Appendix 4)

Bioconcentration: The net accumulation in (and, in some cases, on) an organism of a contaminant from water only. (Chapter 3)

Bioconcentration Factor (BCF): The ratio of concentrations of contaminant in the organism and dissolved in water (the presumed or explicit source) ([organism]/[water]). (Chapter 3)

Biodiversity (Biological Diversity): According to The International Convention on Biological Diversity (2003), "Biological diversity [biodiversity] means the variation among living organisms from all sources including, *inter alia*, terrestrial, marine and other aquatic ecosystems and the ecological complexes of which they are part; this includes diversity within species, between species, and of ecosystems." (Chapter 11)

Biogenic PAHs: Polycyclic aromatic hydrocarbons (PAHs) produced naturally by fungi, plants, and bacteria. (Contrast with **Petrogenic PAHs** and also **Pyrogenic PAHs**.) (Chapter 2)

Biological Half-life ($t_{\frac{1}{2}}$): The time required for the amount or measured concentration of a contaminant in a compartment to decrease by 50%. (Chapter 3)

Biological Ligand Model: An approach that integrates chemical equilibrium predictions and potential metal binding with ligands associated with external phases and biological surfaces to model the potential for metal effects. (Chapter 4)

Biologically Determinant: A quality of an element such that its concentration in organisms remains relatively constant over a wide range of environmental concentrations. Many essential elements are biologically determinant due to their metabolic regulation. (Contrast with **Biologically Indeterminant**.) (Chapter 4)

Biologically Effective Dose: The dose of a chemical that is free to be taken up and to cause an effect at the site of action. (Chapter 4)

Biologically Indeterminant: A quality of an element such that its concentration in organisms is directly proportional to environmental concentrations. (Contrast with **Biologically Determinant**.) (Chapter 4)

Biomagnification: An increase in concentration from one trophic level (e.g., prey) to the next (e.g., predator) due to accumulation of contaminant from food. (Chapter 5)

Biomagnification Factor (*B*): The contaminant concentration at trophic level n (C_n) divided by that at the next lowest trophic level (C_{n-1}) (e.g., Bruggeman et al. 1981; Laskowski, 1991). This factor may be estimated with individual organisms of known or assumed trophic status. (Chapter 5)

Biomagnification Power (*b*): The exponent of the exponential relationship: concentration in an organism $= a_e^{b(\delta^{15}\mathrm{N})}$. It quantifies the proportional increase or decrease in concentration along a trophic food web. The δ^{15}N reflects the trophic status of the individual. (Chapter 5)

Biomarker: A cellular, tissue, body fluid, physiological, or biochemical change in extant individuals, which is used quantitatively during biomonitoring to either imply presence of significant pollutant or as an early warning system for imminent effects. A biomarker is also called a bioindicator by some authors. (Chapter 1)

Biomimetic Approach: A means of estimating bioavailability of contaminants in which the contaminant-containing phase is placed into a synthetic digestive juice solution and the amount of contaminant released is measured. (Chapter 4)

Biomineralization: Biologically mediated deposition of minerals. (Preface)

Biominification: See **Trophic Dilution**. (Chapter 5)

Biomonitoring: A widely applied practice of monitoring of a subset of an entire community with the goal of assessing community condition. (Chapters 1 and 11)

Biomonitoring (Type 1): The monitoring of community changes along a gradient or among sites differing in levels of pollution. (Chapter 11)

Biomonitoring (Type 2): The measurement of bioaccumulation in organisms among sites notionally varying in the level of contamination. (Chapter 11)

Biomonitoring (Type 3): The measurement of effects on organisms using tools such as biochemical markers in sentinel species or some measure of diminished fitness of individuals. (Chapter 11)

Biomonitoring (Type 4): The measurement of genetically based resistance in populations of contaminated areas. (Chapter 11)

Bioreduction: Trophic dilution that arises from the rendering of metals in the prey biomass less available to consumers as might occur with a metal being incorporated into granules. (Chapter 5)

Biosolid: In this book, stabilized sewage sludge is defined as biosolids. (Chapter 2)

Biota-Sediment Accumulation Factor (BSAF): The specific term used for a bioaccumulation factor if it is clear that the factor relates accumulated contaminant to that in sediments. Usually normalized as kilogram of sediment carbon per kilogram of lipid in the organism. See also **Bioaccumulation Factor (BAF).** (Chapter 3)

Biotic Ligand Model (BLM): A conceptual model that "focuses on activities of metal–ligand complexes and metal–ligand complexes and the metal–biotic ligand complexes formed at crucial sites on organism surfaces such as gill surfaces." The BLM is used to imply or predict relationships between dissolved metal concentration and bioavailability or effect (Newman and Clements, 2008). (Chapter 4)

Biotransformation: The biologically mediated transformation of one chemical compound to another. (Chapter 3)

Biphasic Dose–Effect Model: A model of dose–effect that, due to hormesis, is shaped like the threshold model in Figure 7.8, but the curve actually dips down from the control level before increasing with dose. The individuals at these low, subinhibitory concentrations are performing better than individuals exposed to lower or higher concentrations. (Chapter 8)

Birds Directive: An European Union directive with a similar nature as the Habitats Directive in that it has a conservation objective. However, it focuses on bird species, their eggs, nests, and wider habitats. (Appendix 4)

Body Burden: The total mass or amount of contaminant in (and in some cases, on) an individual. (Chapter 3)

Bonds (Chemical): Bonds can be classified generally as covalent or ionic depending on how equitably electrons are shared by the interacting atoms. An electron pair can become so predominately associated with one of the atoms that only electrostatic forces hold the charged ions together after transfer of the electron. The resulting bond is ionic. An atom that donates an electron pair during interaction is called a ligand or donor atom, whereas the atom accepting the electron pair is called the receptor atom. In contrast with the ionic bond, the electron pairs are shared more equitably, but possibly to differing degrees, with covalent bonds. (Chapter 2)

Boomerang Paradigm: What you throw away can come back to hurt you. (Chapter 1)

Borderline Metal Cations: Metal intermediate between Class A and B metals. These metals have one to nine outer orbital electrons. (Chapter 3)

Boundary Values (Planetary): Limits to which humanity can dissipate or use global resources while still maintaining its "freedom to pursue long-term social and economic development" (Rockström et al. 2009b). (Chapter 1)

Brillouin Diversity Index: A measure of the species diversity of a sample taken from a community. (Compare to **Shannon Diversity Index**.) (Chapter 11)

Brunson (Benson)–Roscoe Law of Reciprocity: This general photochemical law refers to the inverse relationship between the intensity and duration of light on the production of reactive oxygen species. The cumulative irradiance estimated with light intensity and exposure duration (intensity × time = cumulative irradiance) determines the amount of product such as reactive oxygen species generated by polycyclic aromatic hydrocarbons. (Chapter 9)

C_3 and C_4 Plants: This classification of plants is based on the number of carbon atoms of the first compound into which the CO_2 is initially incorporated by a plant during photosynthesis. C_4 plants combine CO_2 to form oxaloacetate and C_3 plants combine CO_2 to form glycerate 3-phosphate. The C_3 plants tend to dominate ecosystems and to have a higher $\delta\ ^{13}C$ than C_4 plants. (Chapter 5)

Calcium Sinks: Physical sinks such as arthropod cuticles or bone that renders calcium or its analogs less bioavailable during trophic interactions, and consequently, provide a mechanism for trophic dilution. (Chapters 2 and 5)

Calculi: Mercuric selenide granules found in tissues of toothed whales that sequester mercury. (Chapter 7)

Calmodulin: A membrane-associated messenger protein that functions to transport calcium. It is involved in the inflammation process and also apoptosis. (Chapter 3)

Canalization: The ability to produce consistent phenotype under different conditions. Contrast with **Phenotypic Plasticity.** (Chapter 8)

Cancer Progression: The change in the biological attributes of neoplastic cells over time that leads to malignancy. (Chapter 7)

Carcinogenic: Capable of causing cancer. (Chapter 6)

Carcinoma: A cancer involving epithelial cells that tends to be malignant, i.e., spread over or into the organ or nearby organs, and to metastasize to more distant organs. (Chapter 7)

Cardinal Signs of Inflammation: Heat, redness, swelling, and pain although heat is not relevant in the case of poikilotherms. (Chapter 7)

Carrier Proteins: Cell membrane–associated proteins that act as carriers to transfer hydrophilic contaminants across the membrane. (Chapter 3)

Carrying Capacity (K): The maximum population size expressed as total number of individuals, biomass, or density that a particular environment is capable of sustaining. (Chapter 10)

Caseous Necrosis: Necrosis (cell death) in which cells disintegrate and form a mass of fat and protein. (Chapter 7)

Caspase Enzymes: A class of proteases involved in apoptosis. (Chapter 7)

Catalase (CAT): An enzyme catalyzing the reaction, $2H_2O_2 \rightarrow 2H_2O + O_2$. It is involved in reducing oxidative stress. (Chapter 6)

Catalyzed Haber–Weiss Reaction: A greatly accelerated Haber–Weiss reaction ($O_{2-} \rightarrow H_2O_2 \rightarrow {}^{\bullet}OH + OH$) catalyzed by metal chelates. (Chapter 6)

Cellular Stress Response: An "orchestrated induction of key proteins that form the basis for the cell's protein repair and recycling system" (Sanders and Dyer, 1994). (Chapter 6)

Ceroid: A degradation product of lipid peroxidation of the same composition as lipofuscin but forming under different conditions. (Chapter 7)

Ceroid Pigment: Ceroid in tissues tends to have a yellow to light brown color. As a common example, this colored material is evident in human skin age spots. See **Ceroid**. (Chapter 7)

Channel Proteins: Cell membrane–associated transport proteins that form channels to allow solute passage through the membrane. (Chapter 3)

Chaperon: Stress proteins that associate with and direct the proper folding and coming together of proteins. They also protect proteins from denaturing and aggregating and enhance refolding to a functional conformation. (Chapter 6)

Chapman Mechanism: The series of reactions by which ozone is formed in the stratosphere (Reactions 12.7 through 12.9). (Chapter 12)

Chelate: A multidentate ligand. (Chapter 4)

Chemical Antagonism: Antagonism resulting when two toxicants react with one another to produce a less toxic product. (Chapter 9)

Chemicals of Potential Concern: During human risk assessments, a list of those chemicals of most concern (chemicals of potential concern) is made from a list of all chemicals present at the site. (Chapter 13)

χ^2 ("Chi" Square) Ratio: In this book, the ratio of χ^2 values for two candidate models is used to select the model that best fits the dose/concentration–effect data set. (Chapter 9)

Chloride Cells: Specialized cells on the gill found predominately on the primary lamellae, but also on the secondary lamellae. They function in ion regulation. (Chapter 7)

Chlorosis: The blanching of green color due to the lack of production or destruction of chlorophyll. (Chapter 8)

Cholestasis: Physical blockage of bile secretion as that manifesting with various disease conditions including those associated with toxicant exposure. (Chapter 7)

Chromatid: Before cell division, the DNA in each chromosome is duplicated to produce two chromatids. As the chromosome condenses, the chromosome appears as a pair of chromatids connected to a common centromere. (Chapter 7)

Chromatid Aberration: An aberration that occurs if only one strand (chromatid) is broken in the chromosome. (Chapter 7)

Chromosomal Aberration: Damage to chromosomes including breakage and loss of segments of DNA, addition of segments of DNA, or chromosomal rearrangements. Chromosomal breaks involve double strand breaks. (Chapter 7)

Chronic Lethality: Death resulting from prolonged exposure. By recent convention, a chronic test should be at least 10% of the duration of the species life span. (Chapter 9)

Chronic Reference Dose (Chronic RfD): The reference dose for chronic exposure. (Chapter 13)

Class A Metal Cation: Metal that has an inert gas electron configuration, high electronegativity, and a hard (difficult to polarize) outer sphere. (Chapter 3)

Class B Metal Cation: Metal with filled d orbital of 10–12 electrons and low electronegativity. It has a soft (easily polarizable) sphere readily deformed by adjacent ions; i.e., it easily forms covalent bonds with donor atoms such as sulfur. (Chapter 3)

Clastogenic: Capable of causing chromosome damage in living cells. (Chapter 6)

Clean Air Act (CAA): A U.S. federal act designed to regulate air pollution with the goal of protecting human health and the environment. (Appendix 3)

Clearance: As used in modeling, the rate of substance movement among compartments normalized to concentration. Clearance has units of flow, volume time^{-1}. (Chapter 3)

Clearance Volume–Based Model: A model of substance uptake, elimination, or bioaccumulation based on the distribution of the substance in, and clearance (see **Clearance** above) from and among compartments of different volumes. (Chapter 3)

Cloragosomes: Granules in earthworms with similar composition and function to the calcium phosphate granules of arthropods and molluscs. (Chapter 7)

Coagulation Necrosis: Necrosis (cell death) characterized by extensive cytoplasmic protein coagulation, which makes the cell appear opaque. The cell outline and arrangement in the tissue remain for some time after cell death. (Chapter 7)

Cold Condensation Theory: Persistent organic pollutants (POPs) in the air will condense onto soil, water, and biota at cool temperatures: consequently, the ratios for POP concentrations in the air and on condensed phases decrease as one moves from warmer to cooler climates. (Chapter 12)

Comet Electrophoresis: See **Single-Cell Gel Electrophoresis**. (Chapter 6)

Committed Effective Dose (CED): The dose unit to which risk factors are multiplied to estimate the probability of an individual experiencing a deleterious effect from exposure to radiation. CED accounts for the relative biological effectiveness of different types of radiation, characteristics of the radiation such as energy and half-life of the radioactivity, human physiology, and integrates dose from ingested radioactivity over a 50-year period. (Chapter 13)

Community: "An assemblage of populations living in a prescribed area or physical habitat: it is an organized unit to the extent that it has characteristics additional to its individual and population components … [it is] the living part of the ecosystem" (Odum, 1971). The community is made up of species that interact to form an organized unit (Magurran, 1988) although some species may interact only loosely. (Chapter 11)

Community Conditioning Hypothesis: Communities retain information about occurrences in their past and will not return to their original state after perturbation. (Chapter 11)

Comparative Risk Assessment: A risk assessment in which the risk from alternate actions, situations, or conditions might be compared. For example, a spill of a new heating oil substitute in a large harbor might be compared to that of conventional bunker oil. (Chapter 13)

Compensatory Hyperplasia: An excessive amount of hyperplasia occurring in response to injury or irritation that compensates for lost or damaged tissue; e.g., the enlargement of a remaining kidney after removal of a diseased, second kidney. (Chapter 7)

Complete Median Lethal Concentration: A median lethal concentration (complete LC50) that can be estimated from toxicity tests that includes the latent (postexposure) mortality occurring after cessation of exposure ends as well as the mortality occurring during the exposure period. (Chapter 9)

Complexation Field Diagram: A diagram based largely on hard soft acid base (HSAB) theory that describes how the stability of bonds changes among dissolved metals in freshwater and seawater systems. (Chapter 2)

Comprehensive Environmental Response, Compensation, and Liability Act (CERCLA): A U.S. federal act that gives the EPA the ability to respond to hazardous releases, cleanup or require cleanup of waste sites, and to identify liability for released hazardous waste. (Appendix 3)

Concentration Additivity: Concentration additivity occurs if concentrations of toxicants (adjusted for relative potency) can be added together to predict effects of toxicant mixtures under the assumption of additivity. (Chapter 9)

Concentration Amplification: Biomagnification of a persistent hydrophic compound involves concentration and fugacity amplification. Solvent switching and solvent depletion result in amplification. Solvent switching takes place when a contaminant partitions preferentially into one solvent or phase (e.g., lipids in an organism) relative to another (e.g., its food). There is an increase in concentration in the organism (i.e., concentration amplification), but the final media–organism fugacity does not change with solvent switching. When the organism is eaten, solvent depletion occurs with the digestion of the lipids containing the contaminant. The contaminant then partitions into the lipids of the consumer, creating concentration and fugacity amplification of the consumer. The dynamics are such that the contaminant D_{in} is greater than D_{out} for members of the food web. (Chapter 5)

Concentration Factor (CF): The quantitative expression of the change in concentration at different trophic levels relative to the concentration in the ultimate or lowest defined source, e.g., relative to the water concentration, CF = C_n/C_{water}. The change in concentration is expressed as a multiple of the source concentration. (Chapter 5)

Concentration–Response Approach: A test design in which a series of toxicant concentrations is delivered and the response is noted after a set duration of exposure, or in some cases, after a few time durations. (Chapter 9)

Concept of Strategy: See **Principle of Allocation.** (Chapter 10)

Conceptual Model (in Ecological Risk Assessment): The model that links and interrelates assessment endpoint(s) and a stressor. It includes evaluation of potential exposure pathways, effects, and ecological receptors. Conceptual models include hypotheses of risk and a diagram of the conceptual model. (Chapter 13)

Conceptual Model Diagram (in Ecological Risk Assessment): This diagram is part of the conceptual model described above. It is a diagram showing the pathways of exposure and illustrating areas of uncertainty or concern. It serves as a visual aid for communicating to the risk manager the model from which the risk hypotheses emerged. (Chapter 13)

Congener: In this book, "[a term used to] point up the relationship among members of a chemical family such as the PCBs" (Bunce, 1991). This term is relevant to mixtures of similar compounds such as polychlorinated biphenyls (PCBs) that differ along a common theme. For example, the PCB congeners in Aroclor 1221 are all of the biphenyls in the mixture with different numbers and positions of the substituted chlorine atoms. (Chapters 2 and 5)

Conjugation: In Phase II reactions, "the addition to foreign compounds of endogenous groups which are generally polar and readily available *in vivo*" (Timbrell 2000). (Chapter 6).

Constitutive Androstane Receptor (CAR): A nuclear receptor activated during the induction of the CYP2 gene. (Chapter 6)

Contaminant: "a substance released by man's activities." (Moriarty, 1983) (Chapters 1 and 2)

Core–Satellite Hypothesis: The assemblage of species metapopulations making up an ecological community will display a bimodal distribution of patch occupancy frequencies: most species metapopulations will either occupy many or very few habitat patches with few intermediate species metapopulations. (Chapter 10)

Corporatism: "the belief that all parts of society [are] necessary to its harmonious functioning, and that therefore all parts should cooperate to see to the welfare of each part." (Clark 1997) (Chapter 1)

Cosmetics Regulation: A recently implemented European Union regulation (July 2013) that addresses environmental concerns associated with ingredients of cosmetics. (Appendix 4)

Cosmic Rays: Radiation from outside the earth that includes electron-free nuclei of iron atoms, protons emitted from the sun, and subatomic muons and neutrinos. (Chapter 2)

Cotolerance: See **Cross-resistance**. (Chapter 10)

Cough (Gill Purge): An abrupt, periodic reversal of water flow over the gills that dislodges and eliminates excess mucus from the gill surfaces. (Chapter 8)

Council on Environmental Quality: A council established by National Environmental Policy Act to aid and advise the president during preparation of the annual Environmental Quality Report. (Appendix 3)

Cox Proportional Hazard Model: A semiparametric method allowing the examination of proportional hazards that does not require the assumption of any specific model for the underlying baseline hazard. (Chapter 9)

Criteria: Estimated concentrations of toxicants based on current scientific information that, if not exceeded, are believed to protect organisms or a defined use of a water body. (Chapter 1)

Critical Life Stage Testing: Toxicity testing focused on the life stage of a species thought to be most sensitive to the toxicant such as newly hatched individuals. (Chapter 9)

Critical Study: During human risk assessment, the human or nonhuman animal study with the lowest-observed-adverse-effect-level (LOAEL). (Chapter 13)

Critical Toxic Effect: In human risk assessment, the effect associated with the critical study. (Chapter 13)

Cross-resistance: The condition in which enhanced tolerance to one toxicant enhances tolerance to another. (Chapter 10)

Cryosphere: That portion of the earth which is composed of frozen water such as the frozen poles, glaciers, permafrost, and ocean ice sheets. (Chapter 12)

Cultural (Accelerated) Eutrophication: The greatly accelerated aging of water bodies due to the addition of excess nitrogen and phosphorus nutrients to aquatic systems by human activities. This abnormal condition changes the structure and functioning of the associated water body's ecological community. This is often characterized by dense algal blooms and episodic anoxic or near anoxic events. (Chapter 2)

Cuprosomes: Sulfur-rich copper storage granules found in soil isopods. (Chapter 7)

Curie (Ci): A measure of radioactivity equivalent to 2.2×10^6 dpm (disintegrations per minute). (Chapters 1 and 2)

Cysteine-Rich Intestinal Protein (CRIP): A protein that serves in zinc uptake by cells in the intestine wall. (Chapter 4)

Cytochrome P-450 Monooxygenase: A 45–60 kDa hemoprotein associated with membranes, especially those of the endoplasmic reticulum. It is active in Phase I biotransformation of not only organic compounds but also fatty acids, cholesterol, and steroid hormones. See also **Monooxygenase**. (Chapter 6)

Cytolytic Necrosis: See **Liquefactive Necrosis**. (Chapter 7)

Cytotoxicity: Toxicity causing cell death. (Chapter 7)

Decay Series: A radionuclide transforms and begins a sequence or cascade of element or isotope production until the most energetically stable radionuclide(s) is (are) formed. See also **Radioactive Decay Series**. (Chapter 2)

Delphi Method: A method for eliciting expert opinions that develops a set of issues and then gathers responses from experts who provide independent answers regarding these issues. Next, the compiled answers and statistics are sent back to all of the experts who can argue their points and possibly revise their judgments. The summarized responses are compiled and sent out for additional iterations until a final set of judgments is obtained for use by decision makers (Cooke, 1991). (Chapter 13)

Dense, Nonaqueous Phase Liquids (DNAPLs): An important class of contaminants denser than and immiscible in water. They are often associated with groundwater and drinking water contamination. (Chapter 2)

Depuration: The loss of contaminant from an organism that is measured after the organism has been placed into a clean environment and allowed to eliminate the contaminant. (Chapter 3)

Developmental Reference Dose (RfD_{dt}): A reference dose determined for developmental consequences of a single, maternal exposure during development. (Chapter 13)

Developmental Stability: The capacity of an organism to develop into a consistent phenotype in an environment. (Chapter 8)

Developmental Toxicity: A broad area of toxic effect that considers altered growth and functional deficiencies in addition to classic teratogenic effects. (Chapter 8)

Diene: An alkene with two double bonds in its structure. (Chapter 2)

Diffusion: The movement of a contaminant down an electrochemical gradient that requires no energy. (Chapter 3)

Dilution Paradigm: The solution to pollution is dilution. (Chapter 1)

Directional Asymmetry: The deviation for a population from a mean of zero for the difference between a trait measured from the right and left sides of bilaterally symmetrical individuals from that population. For example, measurement of the difference in weights of left and right arms of right-handed humans would display directional asymmetry because most such humans will have larger right arms. With directional asymmetry, both presence and direction of asymmetry are predetermined. (Contrast with **Antisymmetry** in which the direction of asymmetry is not predetermined). (Chapter 8)

Directive: Legislation, including environmental legislation, generally emerges from the European Commission in the form of directives. (Appendix 4)

Discrimination Ratio (or Factor): A ratio measuring the degree of isotopic discrimination (see definition below) with a ratio of 1 indicating no discrimination. In the context of discrimination between elemental analogs such as cesium and potassium in a trophic exchange, a discrimination factor or ratio is expressed as $[Cs]_{food}/[K]_{food}$ divided by $[Cs]_{body}/[K]_{body}$. (Chapter 5)

Disposable Soma Theory of Ageing: Ageing is a consequence of the gradual accumulation of cellular damage via random molecular defects. (Chapter 10)

Dispositional Antagonism: Antagonism involving toxicant mixture effects on the uptake, movement within the organism, deposition at specific sites, and elimination of the toxicants. The presence of the two toxicants together shifts one or more of these processes to lower the impact of the toxicants at the site(s) of action or target organ(s). (Chapter 9)

Distribution Phase: A common initial phase in a plasma or blood concentration–time curve that reflects an initial period of distribution of the introduced chemical among compartments. (Chapter 3)

Dobson Units (DU): A measure of atmospheric ozone levels that is equivalent of 0.001 mm thickness of pure ozone at 1 atm. (Chapter 12)

Dose Equivalent: A measure of radiation dose that takes into account the different abilities of different kinds of radiation to cause biological damage. See **Effective Dose Equivalent**. (Chapter 13)

Doubling Time: The doubling time of a population is the estimated time required for the population to double its present size. It (t_d) is estimated from the intrinsic rate of increase (r) as $(\ln 2)/r$. (Chapter 10)

Dry Deposition: The flux or deposition of particles and gases such as SO_2, HNO_3, and NH_3 onto surfaces. (Chapter 12)

Dunnett's Test: A parametric, post-ANOVA test used often in the analysis of sublethal and chronic lethal effects data. (Chapter 8)

Duty of Care Principle: One has a legal obligation to refrain from any conduct that places unreasonable risk on another. One is obligated to practice reasonable care and foresight, or when failing to do so, to be legally culpable for any consequent harm. (Appendix 8)

Dynamic Energy Budget (DEB) Approach: A theory-rich approach using energy budgeting for individuals as the central theme around which survival, growth, and reproduction under the influence of toxicants are modeled. Standard toxicity test data may be incorporated directly into the DEB approach. (Chapter 10)

Early Life Stage (ELS) Test: A critical life stage test using early life stages such as embryos or larvae based on the observation or assumption that the early life stage is the most sensitive in the species' life cycle. (Chapter 9)

Ecological Epidemiology: The name given to epidemiological methods applied to determining the cause, incidence, prevalence, and distribution of adverse effects to nonhuman species inhabiting contaminated sites. It is frequently associated with retrospective ecological risk assessment. (Chapter 10)

Ecological Mortality (or Death): The toxicant-related diminution of fitness of an individual functioning within an ecosystem that is of a magnitude sufficient to be equivalent to somatic death. (Chapter 8)

Ecological Overshoot: Using resources faster than they can be regenerated. (Chapter 1)

Ecoregion: A relatively homogeneous region in an ecosystem, or association between organisms and their environment. Ecoregions are usually defined spatially with maps. (Chapters 1 and 12)

Ecosystem: The functional unit of ecology including the biotic community and its abiotic environment functioning together as a unit to direct the flow of energy and cycling of materials. (Chapter 11)

Ecosystem Incongruity: The traditional, but now too confining, bias toward the ecosystem or lower levels in ecotoxicology. (Chapter 12)

Ecotone: Area of transition between two or more community types. (Chapter 12)

Ecotoxicant: Any agent that has an adverse effect on any biological entity within the biochemical to biospheric scale. Adverse effects include conventional lethal and sublethal effects of toxicants. They also include unconventional, adverse modifications of essential habitat, intra- or interspecies interactions, metapopulation or metacommunity dynamics, material cycling, or energy flow. (Preface and Chapter 2)

Ecotoxicogenomics: The study of toxicant effects on organism genomes and related changes in biological functions. (Chapter 10)

Ecotoxicology: The science of contaminants in the biosphere and their effects on constituents of the biosphere, including humans. (Chapter 1)

Edge Effect: Ecotones often have species assemblages with high species richness and high abundance of individuals relative to those of the adjacent communities. (Chapter 12)

Effect at Distance Hypothesis: In metapopulations, the possibility of an individual being exposed in one patch but, after migration to another, having the effect of toxicant exposure expressed in another patch. (Chapter 10)

Effective Dose (Biologically): A term used in pharmacology to define the amount of drug entering the blood and available to have a pharmacological effect. It is used in the context of drug bioavailability from different routes of administration. (Chapter 4)

Effective Dose (ED, Radiation): Most recent term for the **Effective Dose Equivalent** (see below). (Chapter 13)

Effective Dose Equivalent (Radiation): Recognizing that biological effects from a uniform irradiation of the whole body are different than effects from a similar dose concentrated in specific tissues, effective dose equivalent weights the radiation dose to different organs or tissues. The fractional contribution of organs and tissues to the total risk is normalized to when the entire body is uniformly irradiated. (Chapter 13)

Effective Half-life (k_{eff}): An estimated half-life in a compartment model that has numerous elimination mechanisms, each with an associated k_i. It is equal to (ln 2)/Σk_i. (Chapter 3)

Effective Population Size: The number of individuals in a population contributing genes to the next generation. (Chapter 10)

Elasticity (Community): The ability of a community to return to its prestressed condition. (Chapter 11)

Electronegativity: The atom's attraction for electrons during reaction with another atom. It dictates many qualities of an atom during chemical reaction such as bond ionic/covalent nature, strength, or length. (Chapter 2)

Electronegativity Coefficient: There are several coefficients used to quantify the degree of electronegativity (e.g., Mulliken, Allred–Rochow, and Pauling). Their values reflect the atom's tendency (in a molecule) to attract electrons during chemical reaction. The relative electronegativities of two atoms determine important bond qualities such as the degree to which the bond displays ionic versus covalent characteristics. (Chapter 2)

Elimination: The loss or metabolism of a contaminant resulting in a decrease in the amount of contaminant within an organism. (Chapter 3)

Elutriate Test: A test in which a nonbenthic species such as *Daphnia magna* is exposed to an elutriate produced by mixing the test sediment with water and then centrifuging the mixture. (Chapter 9)

Emergent Properties: Properties emerging in hierarchical systems such as ecological communities or ecosystems that cannot be predicted solely from our limited understanding of the system's parts or components. (Chapter 11)

Endocrine Disruptor: Environmental pollutants that interfere with the normal functioning of human and animal endocrine systems. (Chapter 8)

Endocrine System: A system composed of several tissues that broadly includes any tissue or cells that releases a chemical messenger (hormone) that signals or induces a physiological response in some target tissue. (Chapter 2)

Endocytosis: Uptake of solids (phagocytosis) or liquids (pinocytosis) by cells through a process of engulfing the material and enclosure in a cellular vacuole. (Chapter 3)

Enrichment Factor (EF_{crust}): A measure of anthropogenic enrichment of an element above natural levels. It is an element's concentration (X) measured in air samples divided by that expected in the earth's crust: $EF_{crust} = [X/Al]_{air}/[X/Al]_{crust}$. Both air and crustal concentrations are normalized to Al concentrations. (Chapter 12)

Enterohepatic Circulation: Recirculation of toxicant back to the liver after passage into the intestine in bile and then reabsorption in the intestine. (Chapter 3)

Environment (Protection) Act (EPA): The Statement of Objects and Reasons from this 1986 Indian law refers to the decisions from the June 1972 Stockholm Conference and expresses concern about the decline in environmental quality, increasing pollution, loss of vegetative cover and biological diversity, excessive concentrations of harmful chemicals in the ambient atmosphere, growing risks of environmental accidents, and threats to ecosystems. (Appendix 7)

Environment Protection (Sea Dumping) Act: Enacted in 1984, this Australian law arose from international obligations associated with Australia being a signatory of the Prevention of Marine Pollution by Dumping of Wastes and Other Matter, 1972 and the 1996 Protocol to the London Convention, and the United Nations Convention on the Law of the Sea 1982. As part of the act, Australia developed a set of Ocean Dumping Guidelines that provide a national framework for assessing the potential environmental impacts associated with the confined and unconfined disposal of dredging spoil in the sea. (Appendix 6)

Environmental Assessment (EA): A short, preliminary assessment of potential environmental damage used to determine if a full environmental impact statement is required. (Appendix 3)

Environmental Conservation Act 73 (ECA) of 1989: Although the ECA of South Africa has been partially repealed and replaced by the 1998 National Environmental Management Act, the regulations covering asbestos in the ECA are still applicable to the management of asbestos and asbestos-containing materials. (Appendix 8)

Environmental Epidemiology: A subdiscipline of human epidemiology concerned with diseases caused by chemical or physical agents in the environment. (Chapter 10)

Environmental Impact Statement (EIS): A document required by the U.S. National Environmental Policy Act that outlines possible impacts, describes impacts to resources, and details alternatives for any major federal action. (Appendix 3)

Environmental Law: "a body of federal, state, and local legislation, in the form of statutes, bylaws, ordinances, and regulations, plus court-made principles known as common law" (McGregor, 1994) (Appendix 3)

Environmental Liability Directive: An European Union directive that requires polluters to pay for and mitigate any environmental damage they cause. (Appendix 4)

Environmental Protection Law of the PRC: A 1989 law with which China's environmental protection efforts evolved to become a core legal theme. (Appendix 5)

Environmental Protection Law of the PRC (for Trial Implementation): This Chinese law specifically prescribes the "polluter-pays principle," and the establishment of environmental monitoring, environmental impact assessment, and discharge levy systems. (Appendix 5)

Environmental Quality Report: An annual report from the U.S. president's office discussing current environmental conditions and trends. Under the National Environmental Policy Act, the Council on Environmental Quality aids and advises the president in the preparation of this report. (Appendix 3)

EPA Weight of Evidence Classification: A classification used in human risk assessment involving carcinogenic effects that communicates to the risk manager the strength of the evidence supporting the risk calculation, e.g., "Human Carcinogen," "Possible Human Carcinogen," or "Evidence of Non-carcinogenicity for Humans." There are currently five such classes used by EPA. (Chapter 13)

Epidemiology: The science concerned with the cause, incidence, prevalence, and distribution of infectious and noninfectious diseases in populations. (Chapter 10)

Epizootic: Outbreak of disease in a population or in a large number of individuals of a (nonhuman) species. (Chapter 10)

Epoxide Hydrolase: Phase I enzyme that adds water to an epoxide to produce a dihydrodiol. (Chapter 6)

EROD (Ethoxyresorufin O-deethylase): Units of activity for O-deethylation of ethoxyresorufin by ethoxyresorulin O-deethylase. Used to reflect cytochrome P-450 monooxygenase activity. (Chapter 6)

Essential Element: An element essential for the normal functioning of a living organism. The following are the essential elements: H, Na, K, Mg, Ca, V, Cr, Mo, Mn, Fe, Co, Ni, Cu, Zn, B, C, Si, N, P, O, S, Se, F, Cl, and I. It is presently unclear if Sn, As, and Br are also essential. (Chapter 4)

Esterase: An enzyme catalyzing ester hydrolysis into an acid and alcohol. (Chapter 6)

Estrogenic Chemicals: Contaminants possessing biological activities like estrogen that causes changes in the sexual characteristics of individuals. (Chapter 8)

Etiological Agent: An agent responsible for causing, initiating, or promoting a disease. (Chapter 10)

Euler–Lotka Equation: An equation used to estimate the intrinsic rate of increase from life table data. (Chapter 10)

European Community (EC): An organization formed in 1957 to foster a stronger union among European countries. With the Maastricht agreement, it became the European Union (EU) in 1991 which is composed of (listed by population size) Germany, United Kingdom, France, Italy, Spain, Poland, Romania, the Netherlands, Belgium, Greece, Portugal, Czech Republic, Hungary, Sweden, Austria, Bulgaria, Denmark, Finland, Slovakia, Ireland, Croatia, Lithuania, Slovenia, Latvia, Estonia, Cyprus, Luxembourg, and Malta. Countries principally from the old Soviet Bloc that recently joined the EU include Bulgaria, Cyprus, Czech Republic, Estonia, Hungary, Latvia, Lithuania, Malta, Romania, Slovakia, and Slovenia. The EC has its own budget and the power to make and enforce laws including environmental laws. (Appendices 3 and 4)

European Environment Agency: An European Union agency responsible largely for gathering data on environmental indicators from member states so that general trends in Europe can be documented and communicated. (Appendix 4)

European Union (EU): Established in 1957 by the Treaty of Rome differs from the U.S. federal government in that EU member states are sovereign nations that have surrendered only some law making and enforcing powers. See **European Community**. (Appendix 4)

Eutrophic Systems: Eutrophic aquatic systems are highly productive, nutrient-rich systems characterized by turbid waters resulting from dense algal growth. Eutrophic systems are characterized by periodic low oxygen episodes that can cause fish kills. Eutrophic systems have low species diversity with blue-green algae such as the *Anabeana* in Figure 2.4

or *Microcyctis* sp. as frequent dominants. Contrast with **Oligotrophic Systems** and **Mesotrophic Systems**. (Chapter 11)

Excessive Hyperplasia: A type of pathologic hyperplasia that can occur during response to injury or irritation. It is distinct in that it is an excessive hyperplasia such as seen with hyperthyroidism. It involves inappropriate response to growth factors or hormonal stimuli, but not a hereditary change in cells. Compare to **Neoplastic Hyperplasia**. (Chapter 7)

Exchange Diffusion: Diffusion across a membrane by means of a carrier molecule that requires no energy and involves the exchange of two ions across the membrane. (Chapter 3)

Expected Life Span: The calculated life span for individuals of age class x of a life table. It can be estimated as T_x/l_x. (Chapter 10)

Exploitation Competition: Interspecies competition in which species compete for some limiting resource such as food. (Chapter 11)

Exponential Relationship: A mathematical relationship in which the Y variable is related to some constant raised to the X variable, i.e., $Y = a10^{bX}$. Compare to **Power Relationship**. (Chapter 4)

Exposure: In risk assessment, contact with the contaminant or stressor. (Chapter 13)

Exposure Characterization (in Ecological Risk Assessment): A description of the presence and characteristics of contact between the contaminant and the ecological entity of concern, and a summary of this information in an exposure profile. (Chapter 13)

Exposure Pathways: The avenues by which an individual is exposed to a contaminant including the source and route to contact. (Chapter 13)

Exposure Profile (in Ecological Risk Assessment): A profile that "quantifies the magnitude and spatial and temporal pattern of exposure for the scenarios developed during problem formulation" (Norton et al. 1992). (Chapter 13)

Exocytosis: Fusion of intracellular vesicles with the cell membrane followed by emptying the vesicle contents to the cell exterior. (Chapter 3)

Facilitated Diffusion: Diffusion down a gradient not requiring energy, but occurring at a rate faster than expected by simple diffusion alone. (Chapter 3)

Fat Necrosis: Necrosis (cell death) that involves deposits of saponified fats in dead fat cells. (Chapter 7)

Fatty Acids: These biomolecules are chains with even numbers of carbons, commonly 14 and 24 carbons. There is a methyl (CH_3–) group on one end and a carboxyl (–COOH) group on the other. If the fatty acid has no, 1, ≥2, ≥4 double bonds, they are categorized as saturated fatty acids (SFA), monosaturated fatty acids (MUFA), polyunsaturated fatty acids (PUFA), or highly unsaturated fatty acids (LCFA), respectively. Fatty acids with ≥24 carbons are labeled as long-chain fatty acids. A fatty acid might be identified as $A{:}B\text{n} - x$, where A = the number of carbon atoms, B = number of double bonds, and x = the position of the first double bond relative to the terminal methyl group (Kelly and Scheibling, 2012). Under this system, double bonds in the molecule are assumed to be separated by a $-CH_2^-$ group (i.e., "methylene separated") as is most frequently the case (Budge et al. 2006). (Chapter 5)

Feasibility Study (FS): Part of a Remedial Investigation and Feasibility Study that explores various options for remediation. (Chapter 13)

Fecundity Selection: A component of the life cycle of an individual in which natural selection can occur, involving the production of more offspring by matings of certain genotype pairs than are produced by other genotype pairs. (Chapter 10)

Federal Insecticide, Fungicide, and Rodenticide Act (FIFRA): A U.S. federal law designed to control the production, distribution, labeling, sale, and use of pesticides. (Appendix 3)

Fertilization Effect: The nitrogen and sulfur added at low levels to a forest in acid precipitation can enhance growth. (Chapter 12)

FETAX (Frog Embryo Teratogenesis Assay—*Xenopus*): A teratogenesis assay using embryos of the frog, *Xenopus laevis*. (Chapter 8)

Filaments (Gill): See **Primary Lamellae**. (Chapter 7)

Finding of No Significant Impact (FONSI): A statement of no significant impact of a major federal action concluded after an environmental assessment. (Appendix 3)

Finite Rate of Increase (λ): The rate of increase of population size measured over set intervals such as between age classes of a life table or generations of a population with nonoverlapping generations, e.g., an annual plant. (Chapter 10)

Flavin (-Containing) Monooxygenase (FMO): A flavin-containing, Phase I monooxygenase found in microsomes that is capable of oxidizing nucleophilic (electron-deficient) nitrogen, sulfur, and phosphorus groups in xenobiotic compounds. (Chapter 6)

Flory–Huggins Theory: A quantitative theory relating solubility (partition coefficient) of compounds in dilute solutions to solvent molecular size (volume). In environmental toxicology, it has been used to explain the nonideal behavior of the K_{ow} in reflecting partitioning between water and lipids for very lipophilic compounds. (Chapter 4)

Flow-Through Test: An aquatic toxicity test that has a constant (continuous flow-through test) or nearly constant (intermittent flow-through test) flow of the toxicant solutions through the exposure tanks. (Chapter 9)

Fluctuating Asymmetry: Deviation from perfect bilateral symmetry for a population that is thought to reflect developmental instability. A quality is measured from the right and left sides of a bilaterally symmetrical species and the unsigned difference (d = |Right − Left|) is calculated. The variance in d for the population is a measure of fluctuating asymmetry. (Chapter 8)

Food Standards Australia New Zealand Act (1991) and Australia New Zealand Food Standards Code: These laws are administered by Food Standards Australia New Zealand (FSANZ), which is responsible for establishing maximum levels of contaminants and natural toxins in foodstuffs. (Appendix 6)

Free-Ion Activity Model (FIAM): "The universal importance of free metal ion activities in determining the uptake, nutrition and toxicity of all cationic trace metals." (Campbell and Tessier, 1996) (Chapter 4)

Free Radical: A molecule having an unshared electron. (That electron is usually designated by a dot,·.) Free radicals are extremely reactive. (Chapter 6)

Freundlich Isotherm Equation: An empirical relationship quantifying adsorption. (Chapter 3)

Fugacity: The escaping tendency of a substance in a phase. Fugacity (*f*) is expressed in pressure (Pa). (Chapter 3)

Fugacity Amplification: Biomagnification of a persistent hydrophobic compound involves concentration and fugacity amplification. Solvent switching and solvent depletion result in such amplification. Solvent switching takes place when a contaminant partitions preferentially into one solvent or phase (e.g., lipids in an organism) relative to another (e.g., its food). There is an increase in concentration in the organism but the final media–organism fugacity does not change with solvent switching. When the organism is eaten, solvent depletion occurs with the digestion of the lipids containing the contaminant. The contaminant then partitions into the lipids of the consumer, creating concentration and fugacity amplification of the consumer. The dynamics are such that the contaminant D_{in} is greater than D_{out} for members of the food web. (Chapter 5)

Fullerene: A large molecule composed of 60 or more carbon atoms with the form of a hollow sphere, ellipsoid, or tube. (Chapter 2)

Functional Antagonism: Antagonism resulting from two chemicals eliciting opposite physiological effects and, as a consequence, counterbalancing each other's effect. (Chapter 9)

Functional Redundancy: An apparently unaltered maintenance of community functioning despite changes in community structure. (Chapter 11)

Functional Response: A change in some predator function, such as prey consumption rate, in response to changes in prey density. (Chapter 11)

Fundamental Niche: A species has a certain (Hutchinsonian) niche in which it could exist and function based on its physiological and other limits. In contrast to the **Realized Niche**, this fundamental niche includes all the possible niche volume. (Chapter 11)

Gaia Hypothesis: A hypothesis forwarded by James Lovelock that the Earth's temperature, albedo, and surface chemistry are homeostatically regulated by the sum of all the biota of the earth. "… organisms and their material environment evolve as a single coupled system, from which emerges the sustained self-regulation of climate and chemistry at a habitable state for whatever is the current biota." (Lovelock 2003) (Preface)

Gametic Selection: Natural selection involving differential success of gametes produced by heterozygotes. (Chapter 10)

Gamma (γ) Rays: Electromagnetic photons emitted from the nucleus. (Chapter 2)

Gangrenous Necrosis: Combination of coagulation and liquefactive necrosis often resulting from puncture and subsequent infection. (Chapter 7)

Gastric Emptying Rate: The rate at which the contents of the stomach are emptied into the small intestine. (Chapter 4)

Gastrointestinal Excretion: Excretion through the intestinal mucosa by active or passive processes. This may involve loss by normal cell sloughing of the intestine wall. Metals such as cadmium and mercury can undergo significant levels of gastrointestinal excretion. (Chapter 3)

Gavage: To use a small tube inserted into the throat to feed or dose an organism. (Chapter 9)

General Adaptation Syndrome (GAS): The specific syndrome associated with Selyean stress composed of three phases: the alarm reaction, adaptation or resistance, and exhaustion phases. The goal in all phases of the GAS is to regain or resist deviation from homeostasis. (Chapter 8)

Genetic Bottleneck: A bottleneck in the transmission of genes from one generation to the next occurs if too few individuals are available to ensure any particular allele makes it into future generations. (Chapter 10)

Genetic Drift: The random change in allele frequencies in a population. It is accelerated by low **Effective Population Sizes**. (Chapter 10)

Genetic Hitchhiking: Used in this book to describe the situation in which a scored locus is acting only as a marker for a closely linked gene that is actually responsible for the difference in tolerance among genotypes. More generally, it is the condition "in which a given allele changes in frequency as a result of linkage or gametic phase disequilibrium with another selected locus" (Endler, 1986). (Chapter 10)

Genetic Risk: The risk to the progeny of the exposed individual of an adverse effect associated with heritable genetic damage, e.g., damage to germ cells leading to a nonviable fetus or an offspring with a birth defect. (Chapter 7)

Genetically Modified Organism (GMO) Biosafety: "… the effects of [genetically modified] crops and their products on human health and the environment." (Lu and Snow, 2005) (Chapter 2)

Genetically Modified Organisms (GMO): Organisms that have had their genome modified by means of genetic engineering. (Chapter 2)

Genomic Imprinting: Imprinting occurring during gametogenesis provides the mechanism by which an individual only inherits one functional copy of specific genes from the two parents, so that either the paternal or the maternal gene is expressed, but not both (Villar et al. 1995). Differences in maternal and paternal gene expression are established during gametogenesis by this process, which involves cytosine methylation (Rogers and Kavlock, 2008). (Chapter 8)

Genotoxicity: Damage by a physical or chemical agent to genetic materials, e.g., chromosomes or DNA. (Chapter 6)

Geographic Information Systems (GIS): Computerized systems to handle spatial data at a reasonable cost. Most allow one to archive, organize, integrate, statistically analyze, and display many kinds of spatial information using a common coordinate system. (Chapter 12)

Global Distillation: A process by which persistent and relatively volatile organochlorine compounds are distilled from warmer regions of use to cooler regions of the globe. (Preface and Chapter 12)

Global Fractionation: Because persistent organic pollutants (POPs) differ in their individual rates of degradation, vapor pressures, and lipophilities, a fractionation occurs in which some POPs move more rapidly than others toward the polar regions. The net result is a redistribution of the different POPs from the Equator or site of origin toward the cold polar regions of the Earth. (Chapter 12)

Global Warming: A general warming of the earth thought to result from the increased atmospheric carbon dioxide (CO_2) concentrations from fossil fuel burning, release of other greenhouse gases, and the worldwide destruction of forests. (Chapter 12)

Glucocorticoids: These compounds function in glucose metabolism, the Selyean stress response, and immune response. (Chapter 2)

Glucose-Regulated Proteins (grps): Proteins that, under low glucose or oxygen conditions, are part of the cellular stress response. The grps are structurally similar to heat shock proteins, are present at basal levels in unstressed cells, and are induced in glucose- or oxygen-deficient cells exposed to toxicants, which modify calcium metabolism, e.g., lead (Sanders, 1990). (Chapter 6)

Glucuronic Acid: A carbohydrate that can be conjugated to xenobiotics by uridine 5′-diphosphoglucuronosyltransferase. (Chapter 6)

Glucuronidation: The transfer of glucuronic acid from uridine diphosphate glucuronic acid to electrophilic xenobiotics. (Chapter 6)

Glutathione (GSH): A tripeptide composed of cysteine, glutamate, and glycine, specifically γ-glutamyl-L-cysteinyl-glycine. (Chapter 6)

Glutathione Peroxidase: An enzyme catalyzing the reaction, 2 Reduced Glutathione (GSH) + $H_2O_2 \rightarrow$ Oxidized Glutathione (GSSG) + H_2O. It is involved in reducing oxidative stress. (Chapter 6)

Glutathione S-Conjugate Export Pump: A Phase II, ATP-dependent pump that removes glutathione conjugates from the cell. (Chapter 3)

Glutathione S-Transferase (GST): A Phase II enzyme that conjugates glutathione with a xenobiotic or its metabolite. (Chapter 6)

Granulation Tissue: During the repair stage of the inflammation process, small blood vessels begin to form and connective tissue begins to grow in a mass called the granulation tissue. (Chapter 7)

Grasshopper Effect: Global distillation of POPs can involve seasonal cycling of temperatures such that movement toward the higher latitudes occurs in annual pulses. This is called the grasshopper effect. (Chapter 12)

Gray: A unit of radiation dose that has replaced the rad. A rad is 0.01 Grays. See also **Rad**. (Chapter 2)

Green Revolution: A concerted program in the 1940s to 1970s that brought agricultural practices developed in industrialized countries to developing countries including high-yield crop strains, improved tillage and irrigation practices, chemical fertilizers, and pesticides. (Chapter 1)

Greenhouse Effect: Greenhouse gases are relatively transparent to light but absorb long-wave, infrared radiation that is radiating back from the Earth's surface. The net balance for sunlight influx, infrared radiation absorption by greenhouse gases, and infrared efflux from the earth's surface determines the steady state temperature of the Earth. The net warming of the Earth is called the greenhouse effect. (Chapter 12)

Greenhouse Gases: Atmospheric gases that are relatively transparent to sunlight entering the atmosphere but absorb infrared radiation being generated at the Earth's surface. They include water vapor, carbon dioxide, methane, nitrous oxide, chlorofluorocarbons, methylchloroform, carbon tetrachloride, and the fire retardant, halon. Ozone *in the troposphere* may also act as a greenhouse gas. (Chapter 12)

Growth Dilution: The decrease in contaminant concentration in a growing organism because the amount of tissue in which the contaminant is distributed is increasing. (Chapter 3)

Guild (Ecological): A "group of functionally similar species whose members interact strongly with one another but weakly with the remainder of the community" (Smith, 1986). (Chapter 11)

Gynaecomastia: The abnormal enlargement of breasts of boys or men. (Chapter 2)

Haber–Weiss Reaction: The reaction, $O_2^{\bullet} \rightarrow H_2O_2 \rightarrow {}^{\bullet}OH + OH^-$. (Chapter 6)

Habitats Directive: An European Union directive that requires member states to take measures to maintain or restore natural habitats and wild species that are of nature conservation value. (Appendix 4)

Hard Soft Acid Base (HSAB) Theory: In this book, HSAB theory provides a general scheme for quantifying metal-binding tendencies. The hard-soft label has to do with the propensity of the metal's outer electron shell to deform (polarize) during interactions with ligands. Acid–base refers to the Lewis acid or base context for predicting the nature and stability of the metal interaction with ligands. (Chapter 2)

Hardness (of Water): The combined concentrations of dissolved calcium and magnesium. (Chapter 9)

Hardy–Weinberg Equilibrium: The frequency of genotypes will remain constant through time if the following conditions are met: (1) the population is large ("infinite") and composed of randomly mating, diploid organisms with overlapping generations, (2) no selection is occurring, and (3) mutation and migration are negligible. (Chapter 10)

Hazard Assessment: An assessment that compares the expected environmental concentration (EEC) to some estimated threshold effect (ETT) with the intent of deciding if (1) a situation is safe, (2) a situation is not safe, or (3) there is not enough information to decide. (Chapter 13)

Hazard Quotient (HQ): A crude indicator of hazard calculated as the expected environmental concentration (EEC) divided by some estimated threshold effect concentration (ETT): HQ = EEC/ETT. (Chapter 13)

Hazardous Substance Act 15 (HSA) of 1973: The HSA of South Africa provides for the control of substances that may cause injury or ill-health to or death of human beings by reason of their toxic, corrosive, irritant, strongly sensitizing, or flammable nature or the generation of pressure thereby in certain circumstances, and for the control of certain electronic products. (Appendix 8)

Health Advisory Concentrations: Concentrations below which no impact on human health is anticipated for the specified duration of exposure. They often are applied by public officials in dealing with spills, short-term exposures, or similar situations although longer term advisory concentrations are also commonly applied; e.g., those used to establish a fish consumption advisory for a contaminated region of a river. (Chapter 13)

Heat Shock Proteins (hsp): Stress proteins induced by an abrupt shift in temperature that function to reduce associated protein damage in cells. (Chapter 6)

Heavy Metal: This term originated from early studies of the harmful effects of metallic elements such as mercury, lead, and cadmium, which all have very high specific gravities. Although this term is properly applied to metals with specific gravities of five or higher, it is sometimes applied to other metallic elements of environmental concern regardless of their density. This term is not used as a rule in this book because it is not as useful as the Class A, intermediate, and B metal classification for describing trends in binding, bioaccumulation, or effects. (Chapter 2)

Henderson–Hasselbalch Equation: The relationship between pH and the ratio of conjugate base (B^-) to acid (BH), $pH = pK_a + \log([B^-]/[BH])$ (Piszkiewicz, 1977). (Chapter 4)

Henry's Law: A compound's concentration in the aqueous phase (C_c) will be proportional to its partial pressure in the gas phase (P_c), i.e., $K = C_c/P_c$. (Chapter 2)

Heterosis: The superior performance of heterozygotes (relative to homozygotes). (Chapter 10)

Histopathology: Change in cells and tissues associated with a communicable or noncommunicable disease. (Chapter 7)

Hit-and-Run Poisons: Poisons with delayed lethal effects after acute exposure. (Chapter 9)

Holism: In contrast to reductionism, understanding by focusing on the description of phenomena at higher levels of organization rather than applying knowledge of the nature of the parts of the phenomena of interest. (Chapter 7)

Homeopathic Medicine: A branch of medicine, founded by Samuel Hahnemann, that is based on the law of similars, i.e., a drug that induces symptoms similar to those of the disease that will aid the body in defending itself by stimulating its natural responses. (Chapter 8)

Homolog: A pair of homologous chromosomes. (Chapter 7)

Hormesis: A stimulatory effect exhibited with exposure to low, subinhibitory levels of some toxicants or physical agents. Calabrese (2008 a,b) recently defined hormesis generally as "a biphasic dose–response phenomenon characterized by a low-dose stimulation and high-dose inhibition." Typically, hormesis is not a toxicant-specific response. (Chapter 8)

Hormetic Dose Response: A biphasic dose response characterized by a low-dose stimulation and a high-dose inhibition. (Chapter 8)

Hormonal Hyperplasia: One of two forms of physiologic hyperplasia that occurs when there are hormone-associated changes in organ or tissue function, such as an increase in cells of the human breast or uterus associated with childbearing. (Chapter 7)

Hormonal Oncogenesis: Tumor production resulting from high levels of hormones (a promoter) with associated hyperplasia. (Chapter 7)

Hutchinsonian Niche: A niche is "… the certain biological activity space in which an organism exists in a particular habitat. This space is influenced by the physiological and behavioral limits of a species and by effects of environmental parameters (physical and biotic, such as temperature and predation) acting on it." (Wetzel, 1982). (Chapter 11)

Hyperplasia: An increase in the number of cells in a tissue or organ via mitosis that results in an increase in tissue or organ size. (Chapter 7)

Hypertrophy: An increase in cell size (and function) resulting from an increase in the mass of cellular structural components often as a compensatory response. (Chapter 7)

Immaterial Significance Incongruity: The inverse probability error, nil hypothesis error, and other inferentially flawed characteristics of current hypothesis testing practice create a fundamental inferential incongruity that is called the insignificant significance incongruity in this book. It refers to the preoccupation with statistically significant effects (from a nil hypothesis) as opposed to biologically material differences. (Chapter 8)

Imposex: The development (imposition) of male characteristics such as a penis or vas deferens in females. (Chapter 8)

Incidence: The number of new individuals having a disease in a certain time interval. (Chapter 10)

Incidence Rate: Incidence rate of a disease for a nonfatal condition is calculated as the number of individuals with the disease divided by the total time that the population had been exposed. It is expressed in units of individuals or cases per unit of exposure time, e.g., 10 new cases this year. (Chapter 10)

Incipient Median Lethal Concentration: The concentration below which 50% of individuals will live indefinitely relative to the lethal effects of the toxicant. (Chapter 9)

Independent Joint Action: Independent action of toxicants occurs if each toxicant produces an effect independent of the other and by a different mode of action. (Chapter 9)

Index of Biotic Integrity (IBI): A composite index combining 12 qualities of fish communities of warm water, low-gradient streams to determine the level of stream degradation. This index has been modified for diverse communities/taxocenes and widely used in the United States. (Chapter 11)

Individual Effective Dose (IED) Concept: A concept forming the basis for most dose–response models, which holds that there is a unique smallest dose needed to kill any particular individual. The IED is a characteristic of an individual. (Chapter 9)

Individual Tolerance Concept: See **Individual Effective Dose Concept**. (Chapter 9)

Induction Antagonism: The administration of one agent that induces enzymes before exposure to a second agent, resulting in lowered toxic effect of the second chemical. (Chapter 9)

Industrial Chemicals (Notification and Assessment) Act (1989): The piece of Australian legislation that controls the import and manufacture of industrial chemicals. (Appendix 6)

Industrial Ecology: The study of the flows of materials and energy in the industrial environment and the effects of these flows on natural systems. (Chapter 1)

Industrial Melanism: The gradual increase to predominance of melanic forms in industrialized regions. (Chapter 10)

Inertia (Community): A community's ability to resist change. (Chapter 11)

Inflammation: A response to cell injury or death that attempts to isolate and destroy the offending agent and any damaged cells. (Chapter 7)

Info-disruption: Pollutant disruption of the ability of an organism in a community (receiver) to sense and respond accordingly to cues such as those that reflect the presence of a natural enemy or predator (signaler). (Chapter 11)

Initiator: An agent producing cancer by converting normal cells to latent tumor cells. (Chapter 7)

Innovative Science: An activity within any scientific discipline that questions existing paradigms and/or formulates new paradigms. (Chapter 1)

Insurance Hypothesis: A hypothesis to explain the effect of species diversity on systems stability that suggests species that do not contribute to stability under basal conditions might increase the probability of the system remaining stable when conditions shift away from basal conditions because those species might be better adapted to the new conditions. (Chapter 11)

Integrated Risk Information System (IRIS): The Environmental Protection Agency's large database containing reference doses, slope factors, drinking water health advisories (one-day, 10-day, longer term, and lifetime advisories), and associated information. (Chapter 13)

Interference Competition: Interspecies competition in which one species interferes with another as might occur with territoriality or aggressive behavior. (Chapter 11)

Intermediate Metal Cations: See **Borderline Metal Cations**. (Chapter 3)

Interspecies Competition: The interference with or inhibition of one species by another. (Chapter 11)

Intrinsic (or Malthusian) Rate of Increase: The rate of increase in size of a population growing under no constraints. (Chapter 10)

Inverse Probability Error: The term used in this book for a common inference error from probabilities. A low probability of a hypothesis such as a null hypothesis is incorrectly used to infer that some alternate hypothesis is true. The false assertion that improbability of a primary hypothesis as inferred with a test statistic's p value means that the alternate hypothesis is probable. (Newman, 2008). (Chapter 8)

Ionization (Chemical): Relative to discussions in this book, the ionizable group of an organic compound might be completely, partially, or sparingly ionized depending on the solution pH and the group's dissociation constant, pK_a. (Chapter 2)

Ischemia: Localized inadequacy of blood supply to or anemia of tissue resulting from an obstruction of blood flow such as that associated with a wound. (Chapter 7)

Isobole Approach: An approach used to visualize or quantify joint action of chemical mixtures. (Chapter 9)

Isoenzymes: Different forms of the same enzyme, which are coded for by different gene loci. See also **Isozyme**. (Chapter 6)

Isoforms: Different forms differing in their substrate specificity such as cytochrome P-450 monooxygenase isoforms. (Chapter 6)

Isotopes: Nuclides with the same number of protons but different numbers of neutrons are called isotopes. The number of protons determines the chemical identity of an atom. For example, all atoms with 82 protons are lead; however, lead can have 122, 124, 125, or 126 neutrons. Thus ^{204}Pb (122 neutrons + 82 protons), ^{206}Pb, ^{207}Pb, and ^{208}Pb are all isotopes of lead. (Chapter 2)

Isotopic Discrimination: The differential behavior of isotopes occurring if the rate or extent of participation in some biological or chemical process depends observably on the mass of the isotope. Also called the isotope effect. (Chapter 5)

Isozyme: Multiple forms of an enzyme. (Chapter 6)

Itai-itai Disease: An epidemic of human cadmium poisoning (1940–1960) linked to water contaminated with mine wastes used to irrigate rice fields. Itai-itai literally means "ouch-ouch" and reflects the extreme joint pain of victims. (Chapter 1)

Iteroparous Species: A species that reproduces more than once. (Chapter 10)

Kaplan–Meier Method: See **Product-Limit Method**. (Chapter 9)

Karnofsky's Law: Any agent will be teratogen if it is present at concentrations or intensities producing cell toxicity. (Chapter 8)

Karyolysis: The disintegration of the cell nucleus with necrosis (cell death). (Chapter 7)

Keystone Habitat: A high-quality habitat patch essential to maintaining the vitality of the metapopulation. (Chapter 10)

Keystone Species: A species that influences the ecological community by its activity or role, not its numerical dominance. (Chapter 11)

K_{OA}: The partition coefficient for a compound between n-octanol and air phases. Like the K_{ow}, it or its log-transformed value is a measure of lipophilicity. (Chapter 12)

K_{ow}: The partition coefficient for a compound between n-octanol and water phases, i.e., concentration in octanol/concentration in water at equilibrium. It or its log-transformed value is used to reflect lipophilicity of compounds. (Chapter 3)

K-Strategy: An equilibrium strategy for species involving effective interactions with each other in the community, allowing coexistence of many species. Equilibrium species are more effective competitors than opportunistic species. Contrast to **r-Strategy**. (Chapter 11)

Laggard System: Systems with adverse effects that only slowly emerge to a critical state after the conditions are established for their emergence. (Chapter 1)

Landscape: The sum total aspect of any geographical area. (Chapter 12)

Langmuir Isotherm Equation: Theoretically derived relationship for quantifying adsorption. (Chapter 3)

Latent (Latency) Period: The time or lag between exposure to the carcinogenic agent and the appearance of cancer. (Chapter 7)

Law (United States): "a body of federal, state, and local legislation, in the form of statutes, bylaws, ordinances, and regulations, plus court-made principles known as common law." (McGregor, 1994) (Appendix 3)

Law of Frequencies: There exists a relationship between the numbers of species and the number of individuals in a community. (Chapter 11)

Law of Similars: A foundation premise of homeopathic medicine that a drug which induces symptoms similar to those of the disease will aid the body in defending itself by stimulating the body's natural responses. (Chapter 8)

Law of the PRC on Conserving Energy: Enacted in 1998, this Chinese law was established to promote energy conservation by all sectors of society, increase energy use efficiency, ensure national economic and social development, and meet citizen's everyday needs. (Appendix 5)

Law of the PRC on Prevention and Control of Environmental Pollution Caused by Solid Waste: This law was approved in 1995, amended in 2004, and put into force in 2005. It prescribes the supervision and management for the prevention and control of solid waste pollution, which includes industrial solid waste and domestic refuse. (Appendix 5)

Law of the PRC on Prevention and Control of Water Pollution: A law put in force in 1984 and amended in 1996 that is intended to "prevent and control water pollution, protect and improve the environment, safeguard human health, ensure effective water resource utilization, and promote socialist modernization drive progress." (Appendix 5)

Law of the PRC on Prevention and Control of Atmospheric Pollution: This amended law was approved in 2000 and put into force on September 1 of the same year. Its purpose is to "prevent and control atmospheric pollution, protect and improve the human and ecological environments, safeguard human health, and promote the sustainable development of both economy and society." (Appendix 5)

Lefkovitch Matrix: A matrix used to analyze stage-specific demographic data. (Chapter 10)

Lentic Waters: Lentic aquatic systems are standing waters such as lakes and ponds. Contrast with **Lotic Waters**. (Chapter 11)

Lesion: Alterations in cells, tissues, or organs indicating exposure or damage. (Chapter 7)

Leslie Matrix: A matrix used to analyze age-specific demographic data that is constructed with the probability (P_x) of a female alive in period x_i to x_{i+1} being alive in period x_{i+1} to x_{i+2} in the matrix subdiagonal and the number of daughters (F_x) born in the interval t to $t + 1$ per female of age x to $x + 1$ in the top row of the matrix. (Chapter 10)

Liebig's Law of the Minimum: A population's size (number of individuals or biomass) is limited by some essential factors in the environment that is scarce relative to the amount of other essential factors, e.g., phosphorus limited algal growth in a lake. (Chapter 10)

Life-Cycle Studies: Comprehensive studies to determine the impact of a substance or mixture on the survival, growth, reproduction, development, or other important qualities at all stages of a species' life cycle. (Chapter 9)

Ligand: An anion or molecule that forms a coordination compound or complex with metals. (Chapters 2 and 4)

Limited Lifespan Paradigm: An inherent quality of an individual is their genetically defined maximum lifespan. (Chapter 10)

Linear Energy Transfer: The average energy released by ionizing radiation per unit path length through a medium, usually expressed in thousands of electron volts per micron of path length. For example, the linear energy transfer ($keV\mu^{-1}$) in water for a 1.2 MeV γ-ray emitted from ^{60}Co, a 0.6 keV α-particle from tritium, and a 5.3 MeV α-particle from polonium are 0.3, 5.5, and 110.0, respectively. The probability of damage increases with the amount of energy released. (Chapter 2)

Linear Solvation Energy Relationship (LSER): A class of quantitative structure–activity relationships based on molecular volume, ability to form hydrogen bonds, and polarity or ability to become polarized. (Chapter 4)

Linear-No Threshold Theory: This theory, relating incidence or risk of cancer to a dose, is based on several radiation-induced cancer studies suggesting no threshold dose. It assumes that any lack of cancers below a certain dose reflects our inability to measure low incidences at these exposure levels, not a threshold of effect. (Chapter 7)

Lipid Peroxidation: The oxidation of polyunsaturated lipids in membranes resulting in cell damage during xenobiotic exposures. (Chapter 6)

Lipid Solubility: The lipophilicity of a compound is quantified most often using a coefficient (K_{ow} or P) for its partitioning between a lipid (n-octanol) and water phase. (Chapter 2)

Lipinski's Rule of 5: General rules of thumb developed for drug screening that indicate poor potential for absorption or permeation after ingestion of a compound. The rules are the following: (1) more than 5 hydrogen bond donors, that is, the sum of NH and OH groups exceeds 5; (2) more than 10 hydrogen bond acceptors, that is, the sum of N and O atoms exceeds 10; (3) a molecular weight greater than 500 Da; and (4) a calculated log K_{ow} value above 5. Poor bioavailability is anticipated if more than one of these rules is violated. These rules are clearly also pertinent to chemicals other than drugs such as contaminants. (Chapter 4)

Lipofuscin (Age Pigment): A degradation product of lipid oxidation that accumulates in cell vacuoles with age or exposure to some toxicants such as copper. (Chapter 7)

Liquefactive (Cytolytic) Necrosis: Necrosis characterized by a rapid breakdown of the cell as a consequence of the release of cellular enzymes. (Chapter 7)

Litchfield Method: A simple, semigraphical method for analyzing survival time data and estimating LT50 values. (Chapter 9)

Litchfield–Wilcoxon Method: A semigraphical method for estimating an LC50, EC50, or LD50 value. Although very easy to perform, it is the most subjective method for such estimations because it involves fitting a line to data by eye. (Chapter 9)

Log Normal Model: A model fit to species-abundance curves that is thought to reflect a community structure in which several factors influence species interactions and subsequent allocation of resources. (Chapter 11)

Logit: A metameter used for dose- or concentration–response data under the assumption of a log logistic model. Although it has the form logit (P) = ln $[P/(1 - P)]$, its common transform ([logit as just calculated]/2 + 5) is often used because the associated values are very close to those of the probit metameter. (Chapter 9)

Lorax Incongruity: The delusion of selfless motivation in environmental stewardship or advocacy. The Lorax is a character in a popular children's book by Dr. Suess who "speaks for the trees, for the trees have no tongues." (Preface)

Lordosis: The extreme and abnormal forward curvature of the spine. (Chapter 8)

Loss of Life Expectancy (LLE): A calculated estimate of loss in lifetime associated with a risk factor. It is estimated as the simple difference between life expectancy without the risk factor and life expectancy with the risk factor. (Chapter 13)

Lotic Waters: Lotic aquatic systems are flowing water bodies such as streams and rivers. Contrast with **Lentic Waters**. (Chapter 11)

Lowest Observed Effect Concentration (or Level) (LOEC or LOEL): The lowest concentration in a test with a statistically significant difference in mean response from the control mean response. (Chapter 8)

l_x: From a life table, the number of individuals in a cohort alive at age class, x. (Chapter 10)

l_x Life Table (Schedule): A life table that summarizes mortality data for populations. (Chapter 10)

$l_x m_x$ Life Table: A life table that summarizes both mortality and natality data for populations. (Chapter 10)

MacArthur–Wilson Model of Island Colonization: The model, $S_t = S_{EQ}(1 - e^{-Gt})$, where S_t = the number of species present at time t, S_{EQ} = the equilibrium number of species for the island, and G = the rate constant for colonization of the island. (Chapter 11)

Macroexplanation: An approach to understanding something in which the properties of the whole are used to explain the nature or behavior of the parts. (Chapter 7)

Male-Mediated Toxicity: Disease or birth defects produced as a consequence of a father's exposure to a physical or chemical agent. (Chapter 8)

Malignant Neoplasia: Neoplasia in which cells take on undifferentiated forms, tend to grow rapidly, and invade other tissues. Malignant neoplasia are more life threatening than benign neoplasia. Also pieces of the original malignant cancerous growth can dislodge, move to another tissue via the circulatory or lymphatic system, and establish other foci of cancerous growth. (Chapter 7)

Malondialdehyde: A breakdown product of lipid peroxidation used as an indicator of oxidative damage. (Chapter 6)

Malthusian Theory: A series of assumptions and observations regarding limitations on human populations as developed by Thomas R. Malthus (1766–1834). (Chapter 10)

Marine Environment Protection Law of the PRC: This law was approved in 1982 and amended in 1999. It prescribed marine environmental stewardship, ecological protection, and the prevention and control of pollution damage to marine environment by land-derived pollutants, marine and coastal construction projects, waste dumping, vessels, and relevant operations. (Appendix 5)

Marine Protection, Research, and Sanctuaries Act (MPRSA): A U.S. federal act that recognizes the unsoundness of ocean dumping, states the U.S. policy regarding dumping, and establishes limitations and prohibition on dumping. (Appendix 3)

Maturity Index: An index for pollution based on the proportions of species in a soil nematode community that fell into various categories ranging from colonizers (r-strategists) to persisters (K-strategists). (Chapter 11)

Maulstick Incongruity: The incongruous assignment of ecological or biological significance of a contaminant's effect based primarily on statements of statistical significance. (Chapter 8)

Maximum Acceptable Toxicant Concentration (MATC): " ... an undetermined concentration within the interval bounded by the NOEC and LOEC that is presumed safe by virtue of the fact that no statistically significant adverse effect was observed." (Weber et al. 1989). (Chapter 8)

Maximum Likelihood Estimation (MLE): In this book, a parametric method used to fit dose- or concentration–effect data to the log normal, log logistic, or other models. Probit and logit approaches are most often applied with MLE methods. (Chapter 9)

Mean Absorption Time (MAT): The mean time required for absorption of a drug or contaminant calculated as the difference in mean residence time (MRT) of the material introduced by the (noninstantaneous) route of interest and the MRT for the same material injected intravenously. (Chapter 4)

Mean Generation Time (T_c): The predicted generation time for a population estimated from life tables as the sum of the xl_xm_x column divided by R_0. (Chapter 10)

Mean Residence Time (τ or MRT): An estimated mean time that a chemical (e.g., molecule, ion, or atom) remains in a compartment. (Chapters 3 and 4)

Measurement Endpoint (in Ecological Risk Assessment): That measurable response to the stressor (e.g., fledglings produced per nest each year) that is related to the valued qualities of the assessment endpoint (e.g., reproductive success of bald eagles). (Chapter 13)

Median Effective Concentration (EC50): For sublethal or ambiguously lethal effects, the concentration predicted to affect 50% of exposed individuals by a predetermined time, such as, 96 hours. (Chapter 9)

Median Effective Time (ET50): For sublethal or ambiguously lethal effects, the predicted time until 50% of the exposed individuals respond. (Chapter 9)

Median Lethal Concentration (LC50): The concentration predicted to result in death for 50% of exposed individuals by a predetermined time, e.g., 96 hours. (Chapter 9)

Median Lethal Dose (LD50): The dose predicted to result in death for 50% of the exposed individuals by a predetermined time, e.g., 96 hours. (Chapter 9)

Median Lethal Time (LT50): The time predicted to result in death for 50% of the exposed individuals. (Chapter 9)

Median Time-to-Death (MTTD): Like the LT50, the time predicted to result in death for 50% of the exposed individuals. (Chapter 9)

Meiotic Drive: Natural selection occurring at a component of the life cycle of an individual that involves the differential production of gametes by different heterozygous genotypes. (Chapter 10)

Membrane Transport Proteins: Cell membrane–associated proteins involved in transport of solutes. (Chapter 3)

Meristic Character: A characteristic of an individual for which the phenotype is expressed in discrete, integral terms such as number of bristles or number of nodules. (Chapter 8)

Mesocosms: Relatively large experimental systems designed to simulate some feature or quality of an ecosystem. Mesocosms are delimited and enclosed to a lesser extent than are microcosms. They are normally used outdoors, or in some manner, incorporated intimately with the ecosystem that they are designed to reflect. (Chapter 11)

Mesotrophic Systems: Aquatic systems intermediate between oligotrophic and eutrophic systems. Compare to **Eutrophic Systems** and **Mesotrophic Systems**. (Chapter 11)

Metabolic Scope: See **Scope of Activity**. (Chapter 8)

Metacommunity: "a set of local communities that are linked by dispersal of multiple potentially interacting species." (Leibold et al. 2004) (Chapter 11)

Metal: Metals are elements known for their lustrous appearance, malleability, ductility, and conductivity. With the exception of hydrogen, they make up the left two-thirds of the periodic table. (Chapter 2)

Metal Fume Fever: A disease caused by breathing metal-rich fumes, often containing zinc oxide particles, that are produced during smelting, metal casting, welding, and similar activities. (Chapter 2)

Metalloid: Metalloids are intermediate in properties between the metallic and nonmetallic elements, and line up between them in the periodic table. They have a less lustrous appearance than metals, are semiconductors, and include elements such as silicon, arsenic, and antimony. (Chapter 2)

Metallothionein: "polypeptides resembling equine metallothionein in several of their features." (Kojima et al. 1999). They are typically small proteins, ranging from 2 to 7 kDA, but by recent convention now include phytochelatins, which are polypeptides. Metallothioneins function in the uptake, internal compartmentalization, sequestration, and excretion of essential and nonessential metals (Chapters 3 and 6)

Metallothionein-Like Proteins: Poorly characterized, metal-binding proteins or proteins not conforming precisely to the classic properties of metallothioneins. (Chapter 6)

Metameter: A measurement or a transformation of a measurement used in the analysis of biological tests, e.g., the probit metameter of the proportion of individuals responding. (Chapter 9)

Metapopulation: "a set of local populations which interact via dispersing individuals among local populations; though not all local populations in metapopulations interact directly with every other local population." (Hanski, 1996). (Chapter 10)

Metastasis: The process in which pieces of a cancerous growth dislodge and move to other tissues via the circulatory or lymphatic system to establish other loci of cancerous growth. This process leads to the spread of a cancer from the site of origin to other sites in the body. (Chapter 7)

Methemoglobinemia: Referred to as blue-baby syndrome in humans because of the initial skin color of afflicted babies. It is caused by the reaction of nitrite (and some drugs) with hemoglobin (oxidation of ferrous iron to ferric iron) to produce methemoglobin, which is incapable of the normal transport of molecular oxygen in the blood of the newborn. It can be caused directly by nitrite in drinking water or by the conversion of nitrate to nitrite in the baby's anaerobic stomach. It can also be caused in ruminants by the consumption of plants with high nitrate content. (Chapter 2)

Method of Multiple Working Hypotheses: A method proposed by Chamberlin (1897) to reduce precipitate explanation by considering all plausible hypotheses simultaneously in testing so that equal amounts of effort and attention are provided to each. (Chapter 1)

Microcosm: Laboratory study systems designed to simulate some component of an ecosystem such as multiple species assemblages. (Chapter 11)

Microexplanation: See **Reductionism**. (Chapter 7)

Micronuclei: Membrane-bound masses of chromatin separate from the nucleus proper. McBee (Vignette 7.1) describes them as "cytoplasmic nuclear bodies that are formed when whole or fragmented chromosomes are not incorporated into the nuclei of daughter cells or when small fragments of chromatin are retained from polychromatic erythrocytes after expulsion of the nucleus in the process of erythrocyte maturation in mammals." (Chapter 6)

Microsome: A cell fraction composed of membrane vesicles. (Chapter 6)

Microtox® Assay: A rapid, bacterial assay in which a decrease in bioluminescence is thought to reflect toxic action. (Chapter 9)

Minamata Disease: A human poisoning epidemic in Minamata and then in Niigata, Japan, resulting from organic mercury release from industrial sources and consequent contamination of seafood. The first case was reported in 1953 and almost 1000 victims were identified by 1975. (Chapter 1)

Minimal Time to Response: The minimum time required for a toxicant's effect to manifest. Regardless of the toxicant concentration, the effect cannot occur any faster. (Chapter 9)

Minimum Amount of Suitable Habitat (MASH): The minimum density of habitat patches needed to ensure metapopulation persistence. (Chapter 10)

Minimum Viable Metapopulation (MVM): The minimum number of interacting subpopulations needed to ensure long-term persistence of a metapopulation, e.g., a 95% chance of metapopulation persistence for 100 years. (Chapter 10)

Minimum Viable Population (MVP): A size below which extinction of the population in a habitat patch will occur. (Chapter 10)

Mixed Function Oxidase (MFO): An old term that is generally being displaced by more specific terms. It refers to the P-450 complex composed of cytochrome P-450, NADPH-cytochrome P-450 reductase, NADPH, and O_2. See also **Cytochrome P-450 Monooxygenases** and **Monooxygenase**. (Chapter 6)

Modifying Factor: A factor based on expert opinion that is used to adjust the NOAEL (no observed adverse effect level) (or LOAEL [low observed adverse effect level]) downward during risk assessment. (Chapter 13)

Monodentate Ligand: A ligand that shares one pair of electrons with a cation. (Chapter 4)

Monogenic Control: Control of some quality by a single gene. (Chapter 10)

Monooxygenase: One of a general class of enzymes involved in Phase I reactions with xenobiotics. Their action involves the addition of an oxygen atom (from O_2) to the xenobiotic and reduction of the remaining oxygen atom to produce water. Although the term is slowly being displaced by more specific terms, these enzymes were also called mixed function oxidases and abbreviated as MFO. (Chapters 3 and 6)

Montreal Protocol: An international treaty to limit and eventually eliminate the use of chlorofluorocarbons that was signed into law in 1988. (Chapter 12)

Most Sensitive Species Approach: An ecotoxicological approach in which results for the most sensitive of all tested species are used as an indicator of that concentration most likely to protect the entire community. (Chapter 11)

Mountain Cold Trapping: The enrichment of certain persistent organic compounds in various media (e.g., soil, water, snow, or tissues) with increasing elevation up mountain slopes. The mechanism for cold trapping, as implied in its name, is compound transition out of the atmospheric gaseous phase and into solid phases such as particulates, water droplets (rain and fog), snow, and components of the land surface at the colder temperatures that prevail at higher elevations. The increased amount of water moving from gaseous to liquid or solid phases with increased altitude also contributes to the movement of persistent organic compounds out of the atmosphere and onto the land and biota at higher elevations. (Chapter 2)

Moving Average Method: Here, a method of estimating LC50, EC50, or LD50 values. It may be implemented with straightforward equations if the toxicant concentrations are set in a geometric series, and there are equal numbers of individuals exposed in each treatment. (Chapter 9)

MTF-1 (Metal Transcription Factor-1): A zinc-responsive transcription factor responsible for regulating the expression of major metallothionein genes in varied species. (Chapter 6)

Multidentate Ligand: A ligand that shares more than one pair of electrons with the cation. They are also called chelates. (Chapter 4)

Multiple Heterosis: The generally higher fitness of an individual as a result of the combined effect of heterozygote superiority (heterosis) at each of a series of loci. (Chapter 10)

Multiplicative Growth Factor per Generation: See **Finite Rate of Increase**. (Chapter 10)

Multixenobiotic Resistance (MXR): Increased resistance to xenobiotic exposure due to an increase in MXR activity. (Chapter 3)

Mutagen: A physical or chemical entity capable of producing mutations. (Chapter 7)

Mutagenic: Capable of causing mutations. (Chapter 6)

Na^+/I^- Symporter (NIS) Protein: This protein facilitates I^- uptake by thyroid and breast cells. (Chapter 2)

$Na^+/K^+/2Cl^-$ Symporter: An ion regulation–associated symporter active in the loop of Henle. (Chapter 3)

NADPH Cytochrome P-450 Reductase: A flavoprotein that transfers electrons to the cytochrome P-450 isoform assemblage. (Chapter 6)

Nanomaterial: A natural or manmade material having at least one dimension of 100 nm or less. (Chapter 2)

Nanoparticle (NP): Natural or manmade particles with minimally two dimensions between 1 and 100 nm. Although nanoparticles can be natural in origin as in the case of asbestos fibers, the term is now used primarily for manufactured particles. (Chapter 2)

NAS Paradigm: A paradigm used for both human and ecological risk assessments. There are four components to this paradigm: hazard identification, exposure assessment, dose–response assessment, and risk characterization. (Chapter 13)

National Ambient Air Quality Standards (NAAQS): Standards of air quality that Section 109 of the U.S. Clean Air Act requires the Environmental Protection Agency to establish and periodically revise. (Appendix 3)

National Environment Protection (Assessment of Site Contamination) Measure: This 1999 Australian law contains the only limits relating to contaminants in terrestrial ecosystems, these being the Ecological, Human, and Groundwater Investigation Levels. (Appendix 6)

National Environment Protection Council (NEPC): An Australian council established in 1995 to develop and implement nationally consistent environmental policies, known as National Environment Protection Measures. (Appendix 6)

National Environmental Management Act (NEMA) 107 of 1998: NEMA of South Africa seeks to protect the environment through (among other mechanisms) facilitating cooperative environmental governance. The Act imposes a duty of care on every person who causes, has caused, or may cause significant pollution or degradation of the environment. (Appendix 8)

National Environmental Management Air Quality Act 39 (NEMAQA) of 2004: The NEMAQA of South Africa was promulgated to protect the environment by providing reasonable measures for the prevention of pollution and ecological degradation and for securing ecologically sustainable development, while promoting justifiable economic and social development and provide for national norms and standards regulating air quality monitoring, management, and control. It also allows for specific air quality management measures, including the declaration and management of priority areas, controlled emitters, and controlled fuels, as well as the licensing of listed activities that have or may have a significant detrimental effect on the environment, including health, social conditions, economic conditions, ecological conditions, or cultural heritage. (Appendix 8)

National Environmental Management Waste Act 59 (NEMWA) of 2008: The NEMWA of South Africa provides for the regulation of waste management to protect health and the environment by providing reasonable measures for the prevention of pollution and ecological degradation and for securing ecologically sustainable development, institutional arrangements and planning matters, national norms and standards for regulating the management of waste, specific waste management measures, the licensing and control of waste management activities, the remediation of contaminated land, and compliance and enforcement. (Appendix 8)

National Environmental Policy Act (NEPA): A U.S. federal law that, combined with the Environmental Quality Improvement Act, forms the cornerstone for U.S. environmental improvement policy. NEPA established the federal government's commitment to judge their actions relative to environmental impacts and to action that fostered harmony between human activities and nature. It established the council on Environmental Quality and the requirement of an environmental impact statement. (Appendix 3)

National Health Act 61 (NHA) of 2003: The South African NHA aims to provide a framework for a structured uniform national health system, taking into account the obligations imposed by the Constitution and other laws on the national, provincial, and local governments with regard to health services. It also aims to protect, respect, promote, and fulfill the right of the people of South Africa to an environment that is not harmful to their health or well-being. (Appendix 8)

National Pollutant Discharge Elimination System (NPDES): A U.S. state or Environmental Protection Agency-permitting system mandated by the Clean Water Act that imposes limits on discharges. (Appendix 3)

National Pollutant Inventory (NPI): Established in 1998, the NPI consists of a database containing data on 93 chemicals that are released into the Australian environment in quantities above a certain annual threshold. The NPI provides information about each of these chemicals with the goal of providing information that can assist with reducing the impacts associated with the release of pollutants. (Appendix 6)

National Water Act 36 of 1998: The NWA of South Africa has stewardship over water as a scarce and unevenly distributed national resource in South Africa. The act recognizes the need for the integrated management of all aspects of water resources, the necessity to protect the quality of the nation's water resources in the interests of all water users, and ultimately aims to achieve the sustainable use of water for the benefit of all users. (Appendix 8)

National Water Quality Management Strategy (NWQMS): This 1992 law focuses on Australian waterways. The guiding principle of the NWQMS is ecologically sustainable development and thus its principle objective is to "achieve sustainable use of the nation's water resources by protecting and enhancing their quality while maintaining economic and social development." (Appendix 6)

Natural Capital: Resources, living systems, and ecosystem services. (Chapter 1)

Natural Plant Product Pesticide: A pesticide that is produced naturally such as a pyrethrin or rotenone. (Chapter 2)

Natural Selection: The process by which genes from the most fit individuals are overrepresented in the next generation. (Chapter 10)

Necrosis: Cell death resulting from disease or injury. (Chapter 7)

Negative Predictive Value: The probability of the null hypothesis being true given a nonsignificant hypothesis test result. (Chapter 8)

Neonicotinoid Insecticide: A pesticide produced by modification of the structure of the naturally produced insecticide, nicotine. (Chapter 2)

Neoplasia: "Hyperplasia which is caused, at least in part, by an intrinsic heritable abnormality in the involved cells." (La Via and Hill, 1971) (Chapter 7)

Neoplastic Hyperplasia: Hyperplasia resulting from a hereditary change in the cell such that it no longer responds properly to chemical signals that normally control cell growth. Such cells can result in cancerous growth. (Chapter 7)

Net Reproductive Rate (R_0): The expected number of females to be produced during the lifetime of a newborn female as estimated with a life table. (Chapter 10)

Neurathian Bootstrap Incongruity: The moniker given in this book to our continued retention of the NOEC/LOEC (no observed effect concentration/lowest observed effect concentration) approach despite its fundamental shortcomings. It is based on the general description of how sciences progress as introduced by Otto Neurath who used the analogy of a ship at sea to describe how difficult it is to change an established scientific approach. Quine (1960) describes what has become known as the Neurathian bootstrap, "Neurath has likened science to a boat which, if we are to rebuild it, we must rebuild it plank by plank while staying afloat in it … Our boat stays afloat because at each alteration we keep the bulk of it intact as a going concern." (Chapter 8)

Niche Preemption: A rapid use and preemption of resources by a species that exploits them to the exclusion or severe disadvantage of another species. (Chapter 11)

Nil Hypothesis Error: A common habit in hypothesis testing in which, without contemplation of other potentially more meaningful null hypotheses, one tests for any difference without regard for whether that difference might be biologically meaningful. (Chapter 8)

Nine Aspects of Disease Association: Hill (1965) defined nine aspects of evidence fostering the accuracy of linkage between a risk factor and disease: strength of association, consistency of association, specificity of association, temporal association, biological gradient (dose–response) in the association, biological plausibility, coherence of the association, experimental support of association, and analogy. (Chapter 10)

No Action Alternative (to remediation of the site): A scenario in which one assesses if the contaminants at the waste site presents, or will present in the future, a risk if left alone. (Chapter 13)

No Observed Effect Concentration (or Level) (NOEC or NOEL): The highest concentration in a test for which there was no statistically significant difference in response from that of the control. (Chapter 8)

Nonstochastic Effect: In contrast to stochastic health effects of radiation, nonstochastic health effects are those dependent on the magnitude of the dose in excess of a threshold. Some nonstochastic health effects of radiation include acute radiation syndrome, opacification of the eye lens, erythema of the skin, and temporary impairment of fertility. (Chapter 13)

Normal Equivalent Deviate (NED): As used in this book, the proportion dying in a toxicity test expressed in terms of standard deviations from the mean of a normal curve. (Chapter 9)

Normal Science: A major activity of any scientific discipline that stays inside the framework of established paradigms, increasing the amount and accuracy of knowledge within that framework. (Chapter 1)

Normit: The metameter equal to the **Normal Equivalent Deviation**. The resulting analysis of dose- or concentration–effect data with the normit metameter is often called normit analysis and is essentially equivalent to probit analysis. (Chapter 9)

Nuclide: An atomic nucleus with a specified number of neutrons and protons (and amount of energy). (Chapter 2)

Numerical Response: A change in predator or grazer number through increased reproductive output, decreased mortality, or increased immigration in response to changes in prey or food densities. (Chapter 11)

Occupancy Model: Metapopulation models that focus on whether or not patches are occupied. Compare to **Structured Metapopulation Model.** (Chapter 10)

Occupational Health and Safety Act 85 (OHSA) of 1993: The OHSA of South Africa is one of the most explicit and applicable acts for worker and workplace protection as far as hazardous substances control are concerned. OHSA provides for the health and safety of persons at work, the health and safety of persons in connection with the use of plant and machinery, and the protection of persons other than persons at work against hazards to health and safety arising out of or in connection with the activities of persons at work. It aims to protect employees and other persons against any dangers (hazards and risks) to their health and safety and emphasizes the general duties of employers, employees, and manufacturers, all of which are required to implement preventive and protective measures as far as reasonably practicable. (Appendix 8)

Ocean Acidification: The decrease in ocean pH as a consequence of increased atmospheric carbon dioxide levels and ocean warming. (Chapter 12)

Octaves: Log_2 classes (e.g., 1–2, 2–4, 4–8, 8–16, 16–32, … individuals) used in species abundance curves and representing doublings of the numbers of individuals in a species. (Chapter 11)

Odds Ratio: A measure of relative risk in case–control studies in epidemiology. The number of disease cases that were (a) or were not (b) exposed, and the number of controls (no disease cases) that were (c) and were not (d) exposed to the risk factor such as an etiological agent are used to estimate the odds ratio: odds ratio = (a/b)/(c/d) or (ad)/(bc). (Chapter 10)

Oklo Natural Reactors: Naturally occurring nuclear reactors arising through biogeochemical processes approximately 1.8 billion years ago in Oklo (Gabon, Africa). (Preface)

Oligotrophic System: An oligotrophic water body has low productivity due to low nutrient levels and will have a characteristically diverse ecological community including diatom and specific green algal (such as desmids) species as the dominant phytoplankton. They have high water clarity and no dense algal blooms to cause low oxygen episodes in their surface waters. Contrast with **Eutrophic System** and **Mesotrophic System**. (Chapter 11)

Omnivory: As used in this book, omnivory refers to feeding upon many species that occupy different trophic positions. (Chapter 5)

Oncogene: An altered gene involved in cancer. Cancer results from the mutation of a proto-oncogene that was involved in the normal growth and differentiation of cells into this altered gene. (Chapter 7)

One-Hit Risk Model (for carcinogenic effect): In risk assessment, this model (Equation 13.8) is used to predict risk if the estimated risk is 0.01 or higher. (Chapter 13)

Optimal Foraging Theory: The theory that the ideal forager will obtain a maximum net rate of energy gain by optimally allocating its time and energy among the various components of foraging. (Chapter 11)

Optimal Stress Response: The optimal stress response involves a shift in the balance in energy allocation between somatic growth rate and longevity (survival) to optimize Darwinian fitness under stressful conditions. (Chapter 10)

Organic Anion Transporters (OATs): The OATs are involved in the removal from the cell of contaminants or their metabolites that are organic anions. (Chapter 3)

Organization for Economic Cooperation and Development (OECD): An organization founded in 1960 with the purpose of enhancing economic growth, living standards and financial stability, and development of and holding forums on associated policies. Member countries are Australia, Austria, Belgium, Canada, Denmark, Finland, France, Germany, Greece, Iceland, Ireland, Italy, Japan, Luxembourg, Mexico, Netherlands, New Zealand, Norway, Portugal, Spain, Sweden, Switzerland, Turkey, United Kingdom, and United States. (Appendix 3)

Oxidative Stress: The damage to biomolecules from free radicals. (Chapter 6)

Oxyanion/Oxycation: Many elements combine with oxygen to form oxygen-containing, negative (oxyanion or oxoanion) or positive (oxycation or oxocation) ions. (Chapter 2)

Oxyradical: A free radical involving an unshared electron of oxygen, e.g., $RO^{\bullet}$. (Chapter 6)

Ozone Hole: A hole or extreme thinning of ozone above the Antarctic due to the combined effects of circulation patterns above the Antarctic and ozone destruction as a consequence of chlorofluorocarbon accumulation in the stratosphere. (Chapter 12)

Panchayati Raj: A political system implemented widely in south Asia that provides a degree of local self-rule by assemblies (*sabhas*) of respected and informed community members. The role and importance of this system of decentralized administrative and judicial activities has shifted throughout Indian history. Most recently, the 73rd Constitutional Amendment of India established village, intermediate, and district *panchayats* with a village assembly (*Gram Sabha*) governing locally. Nearly all of India's urban, and most praiseworthy, rural citizenship is now represented in this tiered *panchayat* system of elected members. (Appendix 7)

Paradigms: Generally accepted concepts that, in a healthy science, have withstood rigorous testing, and as a consequence, are given enhanced status as explanations of fact and observation. (Chapter 1)

Partial Kill: A treatment in a toxicity test in which some, but not all, exposed individuals are killed. (Chapter 9)

Partition Coefficient: The quantitative expression for the concentration of a contaminant in one phase relative to the concentration in another phase, e.g., K_d or $K_p = [X_{\text{Phase b}}]/[X_{\text{Phase a}}]$. (Chapter 2)

Partitioning: Depending on its chemical properties, a charged or uncharged chemical substance will preferentially associate with (partition between) different environmental phases. (Chapter 2)

Pathologic Hyperplasia: Hyperplasia including excessive and neoplastic hyperplasia that create a pathological condition. Pathologic neoplasia can be contrasted with **Physiologic Hyperplasia**. (Chapter 7)

Pedosphere: That part of the earth made up of soils and where important soil processes are occurring. (Chapter 12)

Perfluoroalkyl Acids (PFAs): Synthetic, perfluorinated, straight- or branched-chain organic acids with a variety of reactive moieties, including among others, carboxylate or sulfonate groups. (Chapter 2)

Pericardial Edema (Effusion): The build up of fluids in the pericardium (the sac surrounding the heart and roots of the major blood vessels). (Chapter 8)

Pericytes: A relatively undifferentiated type of cell associated with capillaries or small blood vessels. (Chapter 7)

Periphyton: Periphyton or *aufwuchs* is the term used for the microfloral community growing on submerged surfaces. In this book, bioaccumulation in procedurally defined periphyton is also discussed. In contrast to periphyton which is a type of ecological community, procedurally defined periphyton is the material covering submerged surfaces in lentic and lotic systems which can contain abiotic material such as detritus and silt as well as biota. (Chapter 5)

Peroxisome: Membrane-bound organelles in the cytoplasm (endoplasmic reticulum) that can generate but also consume hydrogen peroxide. They are associated with oxidative reactions. (Chapter 6)

Peroxisome Proliferation: An increase in peroxisomes caused by some compounds that elevate oxidative stress and associated cell damage and death. (Chapter 2)

Peroxyradicals: Oxyradicals of organic compounds (R) of the form, $ROO^{\bullet}$. (Chapter 6)

Persistent Organic Pollutants (POPs): Organic pollutants that are long lived in the environment. Many tend to increase in concentration as they move through trophic webs. Often, **Persistent, Bioaccumulative Toxicants** and Persistent, Bioaccumulative Chemicals (PBTs) are the explicit terms used for POPs that tend to biomagnify in trophic webs. (Chapters 1 and 12)

Persistent Toxicants that Bioaccumulate (PTBs): See **Persistent Organic Pollutants**. (Chapter 12)

Persistent, Bioaccumulative Toxicants: See **Persistent Organic Pollutants**. (Chapter 1)

Pesticide Safety Factor: The quotient of $LD50_{mammal}/LD50_{insect}$ that suggests safety of use relative to nontarget mammalian species. (Chapter 2)

Petrogenic PAH: Polycyclic aromatic hydrocarbons (PAHs) from diverse petrochemical sources such as oil spills. They form slowly under low temperature conditions of petrogenesis. Contrast with **Pyrogenic PAH** and also **Biogenic PAH**. (Chapter 2)

P-glycoprotein (P-pg) Pump: An energy-requiring pump that eliminates metabolites in the cell. It is the first line of defense as contaminants that could potentially gain entry to the cell and are actively pumped out of the cell immediately as they begin entry. (Chapter 3)

Pharmaceuticals and Personal Care Products (PPCPs): A collective term used for the emerging class of contaminants that are initially used as pharmaceuticals (e.g., synthetic estrogens) or personal care products (e.g., nitro musks in soaps). (Chapters 1 and 2)

Pharmacokinetics: The study and predictive modeling of the internal kinetics of drugs and other chemicals. (Chapter 3)

Phase I Reactions: Reactions in the biotransformation of organic contaminants in which reactive groups are added or made more available. Although oxidation reactions are the most important Phase I reactions, hydrolysis and reduction reactions are also significant. (Chapter 3)

Phase II Reactions: Reactions in the biotransformation of organic contaminants in which conjugates are formed that inactivate the compound and foster elimination. (Chapter 3)

Phenotypic Plasticity: An ability to produce different or a range of phenotypes under different conditions. Contrast with **Canalization**, which is its converse. (Chapter 8)

Phocomelia: A developmental abnormality in which the individual is born with extremely short limbs because the long bones have failed to develop properly. The term is derived from the Greek words *phoke* and *melos* that means seal and fin, respectively (Taussig, 1962). This refers to the appearance of the extremely short limbs of afflicted individuals. (Chapter 8)

Photo-Induced Toxicity: Toxicity of a chemical in the presence of light due to the production of toxic, photolysis products (**photomodification**) or **photosensitization**. Photo-induced toxicity is particularly relevant to polycyclic aromatic hydrocarbon toxicity of nonhuman species. (Chapter 9)

Photomodification: Ultraviolet modification of a compound such as a polycyclic aromatic hydrocarbon that might involve photooxidation or photolysis. (Chapter 2)

Photosensitivity: Sensitivity of cutaneous tissues to the effects of light evoked by a chemical. See also **Photo-Induced Toxicity** and **Photosensitization**. (Chapter 9)

Photosensitization: Production of a reactive oxygen species by a compound's interaction with light. (Chapter 9)

pH–Partition Hypothesis: Bioavailability is determined by the diffusion of the unionized form through the gastrointestinal lumen as determined by pK_a and pH. (Chapter 4)

Physiologic Hyperplasia: A type of nonpathological hyperplasia in response to a variety of normal stimuli such as that involved in the tissue repair process. (Chapter 7)

Physiologically Based Pharmacokinetics (PBPK) Model: A pharmacokinetics model that includes physiological and anatomical features in describing internal kinetics. (Chapter 3)

Phytochelatin: A class of polypeptides in plants that are induced by and bind to metals. They may function in the regulation and detoxification of metals by plants. Genes for phytochelatins have now been isolated in several invertebrate species. (Chapters 3 and 6)

Pielou's J: A common measure of species evenness *for a sample from a community.* (Chapter 11)

Pielou's J': A common measure of species evenness *for a community.* (Chapter 11)

Plant Protection Products Directive: This European Union (EU) directive came into force in 1991 with the purpose of harmonizing arrangements for the marketing of plant pesticides within the EU. (Appendix 4)

Plant Protection Products Regulation: Replaced the European Union Plant Protection Products Directive in 2011 to provide greater uniformity across member states. (Appendix 4)

Polar Bond: A covalent bond can be a polar bond if the electron cloud associated with the bound atoms is denser around one atom than the other, e.g., the O–H bond is a polar bond. (Chapter 2)

Pollutant: "A substance that occurs in the environment at least in part as a result of man's activities, and which has a deleterious effect on living organisms." (Moriarty, 1983) (Chapters 1 and 2)

Pollution-Induced Community Tolerance (PICT): An increase in tolerance to pollution resulting from species composition shifts in the community, acclimation of individuals, and genetic changes in populations in the community. (Chapter 11)

Polychlorinated Diphenyl Ethers (PCDEs): Compounds similar in structure to the polybrominated diphenyl ethers described in Chapter 2 except they are polychlorinated, not polybrominated. (Chapter 4)

Polygenic Control: Control of some quality by several genes. (Chapter 10)

Population: A group of individuals occupying a defined space at a particular time. (Chapter 10)

Porins: Pores in the cell membrane that are nonspecific relative to ions passing through them. (Chapter 3)

Porphyrins: Molecules having a tetrapyrrole ring (four simple, heterocyclic nitrogen ring compounds bound together to form a ring) that are produced as intermediates during heme synthesis in animals. (Chapter 6)

Portfolio Effect: This explanation for the influence of species diversity on system stability assumes that stability is a result of the net action of all species activities and the presence of many species contributing to this net state decreases the chance that the system will experience sudden changes when conditions shift to disfavor some contributing species. (Chapter 11)

Positive Predictive Value: The probability that the alternate hypothesis is true given a significant hypothesis test result. (Chapter 8)

Positron: A particle with the same mass of an electron but possessing a positive charge. Also called a positive electron. (Chapter 2)

Potentiation: Enhanced toxicity of a chemical in the presence of a second chemical that is not itself toxic at its concentration in the mixture. (Chapter 9)

Pound-of-Flesh Incongruity: In early and some current applications of the species sensitivity distribution approach, the neglect of the ecological qualities of sensitive species and focus on number is so pervasive that it deserves a name, the pound-of-flesh incongruity. Settling on a 5% of species cut-off is like Antonio's naïve agreement with Shylock of a "pound of flesh" bond to secure a loan in Shakespeare's *Merchant of Venice*. It matters which pound of flesh is taken if the debt comes due. (Chapter 11)

Power Relationship: A mathematical relationship in which the Y variable is related to the X variable raised to some power. For example, $Y = aX^b$. Compare to **Exponential Relationship**. (Chapter 4)

Precautionary Principle: The conservative policy that, even in the absence of any clear evidence and in the presence of high scientific uncertainty, action should be taken if there is any reason to think that harm might be caused. (Chapters 1 and 13)

Precipitate Explanation: The unreliable scientific practice of uncritical or untested acceptance of an explanation based on some ruling theory. (Chapter 1)

Predictive Risk Assessment: A risk assessment dealing with a planned or proposed future condition. (Chapter 13)

Pregnane-X-Receptor (PXR): A nuclear receptor activated during the induction of the CYP3 gene. (Chapter 6)

Prevalence: The total number or proportion of individuals with the disease at a particular time. The incidence rate of a disease multiplied by the amount of time that individuals were at risk. (Chapter 10)

Primary Lamellae (Filaments): Gill structures extending outward at right angles from the branchial arches. (Chapter 7).

Principle of Allocation: There exists a cost or trade-off to every allocation of energy resources. Energy spent by an individual organism on one function, process, or structure cannot be spent on another. Optimal allocation of resources enhances Darwinian fitness. (Chapter 10)

Probabilistic (Bayesian) Induction: A type of induction that uses probabilities associated with competing theories or explanations to decide which is most probable based on available evidence. Certainties or credibilities are then assigned to competing explanations based on their associated probabilities. Instead of the quantal (accept or reject) falsification of a working hypothesis, Bayesian induction considers a hypothesis falsified if it were sufficiently improbable. (Chapter 13)

Probit: A metameter produced by adding 5 to the **Normal Equivalent Deviation**. Forming the basis of probit analysis, it was first proposed in the 1930s to avoid negative numbers and is still used in lieu of the **Normit** based on convention. (Chapter 9)

Problem Formulation (of ecological risk assessment): The planning and scoping phase that establishes the framework around which the risk assessment is done. (Chapter 13)

Problem of Ecological Inference: The central problem with macroexplanation in which difficult inferences about parts (e.g., individuals) are made from aggregated properties of the whole (e.g., population made up of individuals). (Chapter 7)

Procarcinogen: A compound that is converted to a carcinogen. (Chapter 6)

Product-Limit (Kaplan–Meier) Method: A nonparametric method for analyzing time-to-death or survival time data that does not require a specific model for the survival curve. (Chapter 9)

Progressive Era: Occurring during the U.S. industrial revolution and urbanization (circa late 1890s–1917), the goal of Progressive Era was to transform democracy to a more just political process by replacing established traditions with modern ones, including innovations suggested by scientific and technological scrutiny. A scientific lens was focused on a wide range of social and political issues including racial inequalities, women's voting rights, limits of capitalism, labor rights and safety, and immigration. (Chapter 1)

Promoter: An agent producing cancer by enhancing the growth of mutated cells. (Chapter 7)

Promoter Elements: Regulatory sequences upstream of the coding region of a gene. (Chapter 6)

Propagule Rain Effect: Relative to metapopulation dynamics, the presence of a seed bank or many dormant stage individuals for a species that continually introduces individuals to the patch regardless of the density of occupancy in the surrounding patches. This propagule rain increases the likelihood of population reappearance and decreases the likelihood of extinction in a patch. (Chapter 10)

Proportional Diluter: A special apparatus used in flow-through toxicity tests to mix and deliver a series of dilutions of a toxic solution to exposure tanks. (Chapter 9)

Proportional Hazard Model: A survival or time-to-death model that relates the hazard (proneness to die or risk of dying at any time, *t*) of one group (e.g., smokers) quantitatively to that of another group (e.g., reference group of nonsmokers). (Chapter 9)

Protein Corona: This term is applied to the assemblage of biological macromolecules that form around a nanoparticle within organisms. Some corona components might change as the nanoparticle moves from one component (e.g., plasma of the circulatory system) to another (e.g., cytoplasm of a cell). (Chapter 2)

Proteinuria: The presence of protein in the urine. (Chapter 13)

Proteotoxicity: A toxic or adverse effect of a chemical or physical agent with an underlying mechanism of protein damage. (Chapter 6)

Proto-oncogene: A gene involved in some way with the normal growth (enhancement) and differentiation of cells which, on mutation, becomes an **Oncogene**. (Chapter 7)

Provisional Rules of Preventing Coastal Water Pollution: China's first formal law for coastal water pollution prevention enacted in 1974. (Appendix 5)

Provisions Concerning Environmental Protection and Amelioration: China's first comprehensive administrative law of environmental protection enacted in 1973. (Appendix 5)

Ptolemaic Incongruity: The false paradigm that any particular level of biological organization holds a more central or important role than another in the science of ecotoxicology. (Chapter 7)

Pycnosis (Pyknosis): One of the more obvious indications of necrosis in which the distribution of chromatin in the nucleus changes. The chromatin condenses into a strongly staining, basophilic mass. (Chapter 7)

Pyrethrin: A natural compound pesticide similar to pyrethrum, which is produced by the plants *Chrysanthemum cineum* and *C. cinerariaefolium*. (Chapter 2)

Pyrethroid: A synthetic analog or derivative of a **pyrethrin**. (Chapter 2)

Pyrogenic PAH: PAHs produced during incomplete high-temperature combustion. Contrast with **Petrogenic PAH** and also **Biogenic PAH**. (Chapter 2)

Quality Factor (Q): The quality factor, or radiation-weighting factor, adjusts for the different capacities of various forms of radiation to damage tissues. (Chapter 2)

Quantitative Structure–Activity Relationship (QSAR): A quantitative, often statistical, relationship between a molecular quality or set of qualities and some activity, such as bioavailability or toxicity. (Chapter 4)

Quantum Dot: As discussed in this book, quantum dots are semiconductor nanocrystals with electronic properties dictated by their crystalline structure and size. They are commonly (colloidal semiconductor) nanocrystals of cadmium selenide, cadmium telluride, cadmium sulfide, or zinc selenide with diameters of approximately 2–10 nm. (Chapter 2)

Quotient Method: The use of the hazard quotient as a crude indicator of hazard. (Chapter 13)

Rad: An expression of radiation dose that is 100 ergs of ionizing radiation energy deposed in a gram of material. See **Gray** as the new unit replacing the Rad. (Chapter 2)

Radiation Weighting Factor: See **Quality Factor**. (Chapter 13)

Radioactive Decay Series: There are three natural and one artificial series of radionuclides in which one (parent) radionuclide decays to another (progeny) in a series of steps until the most stable element is formed. The series include the uranium–radium, thorium, actinium, and neptunium series. (Chapter 2)

Radiosensitizers: Chemicals, such as derivatives of nitroimidazoles, used during radiation treatment to enhance the production of free radicals which then kill cancer cells. (Chapter 6)

Rare Earth Elements: These include La, Ce, Pr, Nd, Pm, Sm, Eu, Gd, Tb, Dy, Ho, Er, Tm, Yb, Lu, and Y. They are the lanthanides with atomic numbers 57 to 71 plus scandium (atomic number 21) and yttrium (atomic number 39). (Chapter 1)

Rarefaction Estimate of Richness: An estimate of species richness calculated for a sample with a standard number of individuals in it. (Chapter 11)

Rate Constant-based Model: A model that uses rate constants to quantify the rate of change in concentration or amount of toxicant. (Chapter 3)

Rate of Living Theory of Ageing: The total metabolic expenditure of a genotype is generally fixed, and as a consequence, longevity depends on the rate of energy expenditure. (Chapter 10)

Rate Ratio: The ratio of disease incidence rates for two populations. Rate ratio = I_A/I_0, where I_A is the incidence rate in population A, and I_0 is the incidence rate in the reference or control population. (Chapter 10)

REACH (Registration, Evaluation, Authorisation and Restriction of Chemicals) Regulations: These European Union (EU) regulations came into force in June 2007 and require EU manufacturers or importers of most chemical substances alone or in preparations to register the substance if they manufacture or import it in quantities of 1 ton or more annually. (Appendix 4)

Realized Niche: That part of a species' **Fundamental Niche** that it actually occupies. (Chapter 11)

Reasonable Maximum Exposure (RME): Exposure calculated for a chemical of potential concern during a risk assessment. It is conservative estimate of exposure that is computed differently depending on the route of exposure. (Chapter 13)

Receptor Antagonism: Antagonism that involves the binding of the toxicants to the same receptor and one toxicant blocking the other from fully expressing its toxicity. (Chapter 9)

Recovery Rate: The rate at which a community returns to equilibrium following disturbance. (Chapter 11)

Redox Cycling: In the context of contaminant (e.g., quinones, aromatic nitro compounds, aromatic hydroxylamines, bipyridyls, and some chelated metals) involvement in the generation of oxyradicals, redox cycling occurs if contaminants are reduced to free radicals and then participate in redox reactions to produce the superoxide radical from molecular oxygen. The contaminant returns to its original form at the end of the redox reactions and is available to recycle many times through this process and produce more oxyradicals. (Chapter 6)

Reductionism: An approach that attempts to explain the behavior of something based on an understanding of its parts. (Chapter 7)

Redundancy Hypothesis: The hypothesis that many species are redundant and their disappearance from a community will not influence the community function as long as crucial (e.g., keystone and dominant) species populations are maintained. (Chapter 11)

Reference Dose (RfD) (for noncarcinogenic effects): The best estimate of the daily exposure for humans, including the most sensitive subpopulation, which will result in no significant risk of an adverse health effect if it is not exceeded. (Chapter 13)

Region (Geographical): An "area of the earth's surface differentiated by its specific characteristics" (Monkhouse, 1965). (Chapter 12)

Regulation (Legal): In the United States, specific regulations are produced by federal agencies to insure that the intent of a law is met. Environmental regulations "specify conditions and requirements to be met, … provide a schedule for compliance, and … record any exceptions to the regulated community" (Mackenthun and Bregman, 1992). As described in Appendix 4 relative to the European Union (EU), an EU regulation is a direct legislative act that becomes immediately enforceable as law in all member states simultaneously. (Appendix 4)

Regulations of the PRC on Nature Reserves: The Chinese regulations were put into effective on December 1, 1994 with the intent of strengthening the creation and management of nature reserves and protecting the natural environment and resources. (Appendix 5)

Relative Bioavailability: The **Bioavailability** estimated for a dose administered by any route or formulation relative to a dose administered through a reference (or alternate) route or in a reference formulation. (Chapter 4)

Relative Risk: In survival time analysis, the risk of one group expressed as a multiple of that of another. Relative risk is often estimated with a hazard model. In epidemiology, it is the ratio of occurrences of the disease in two populations. (Chapters 9 and 10)

Rem (Roentgen Equivalent Man): A measure of radiation dose that accounts for differences in radiation dose potential to cause biological damage. (Chapter 2)

Remedial Investigation (RI): Part of a **Remedial Investigation and Feasibility Study** that has the following three parts: characterization of the type and degree of the contamination, human risk assessment, and ecological risk assessment (EPA, 1994). (Chapter 13)

Remedial Investigation and Feasibility Study (RI/FS): For a U.S. Superfund site, a study that has as its goal the implementation of "remedies that reduce, control, or eliminate risks to human health and the environment" or, more specifically, the accumulation of "information sufficient to support an informed risk management decision regarding which remedy appears to be most appropriate for a given site" (EPA, 1989d). (Chapter 13)

Remote Sensing: Technologies that allow the acquisition and analysis of data without requiring physical contact with the land or water surface being studied. Most determine qualities or characteristics of areas of interest based on measurements of visible light, infrared radiation, or radio energy coming from them. (Chapter 12)

Repair Fidelity (of DNA): The accuracy in repairing and returning the DNA to its original state after damage. (Chapter 7)

Reproductive Value (V_A): The expected contribution of offspring during the life of an individual of an age class x in a life table. (Chapter 10)

Rescue Effect: The increased probability of a vacated patch reoccupation in a metapopulation as the proportion of nearby, occupied patches increases. (Chapter 10)

Residual Body: A cell vacuole containing lipofuscin, a degradation product of lipid oxidation. (Chapter 7)

Resilience (Community): The number of times a community can return to its normal state after perturbation or the level of perturbation a community can withstand before it is pushed to an alternative stable state. (Chapter 11)

Resistance: The term, resistance, is often reserved for the enhanced ability of an individual to cope with a factor due to genetic adaptation. The term, tolerance, is often reserved for enhanced abilities associated with physiological acclimation. Tolerance is used in this book for both acclimation and genetic adaptation. (Chapter 10)

Resource Conservation and Recovery Act (RCRA): A U.S. federal law that regulates production, treatment, storage, and disposal of hazardous wastes. (Appendix 3)

Respiratory Lamellae: See **Secondary Lamellae**. (Chapter 7)

Retention Effect: The K_{ow} values of persistent organic pollutants (POPs) influence their global movement toward higher latitudes. The POPs with high lipophility tend to be held more firmly in solid phases such as soil and vegetation than less-lipophilic POPs. Consequently, they spend less time in the atmosphere and are less available for transport in that medium. This decreased global movement due to lipophility is referred to as the retention effect. (Chapter 12)

Retroactive Risk Assessment: A risk assessment dealing with an existing condition. (Chapter 13)

Ricker Model: A difference equation model (Equation 10.10) for growth of populations with non-overlapping generations or experimental designs with discrete intervals of population growth. (Chapter 10)

Risk: As used in risk assessments, the probability (or likelihood) of some adverse consequence occurring to an exposed human or to an exposed ecological entity. (Chapter 13)

Risk Assessment: The process by which one estimates the probability or likelihood of some adverse effect(s) of a present or planned release to either human or ecological entities. (Chapter 13)

Risk Assessor: A person (or more often, group of people) "who actually organizes and analyses site data, develops exposure and risk calculations, and prepares a risk assessment report" (EPA, 1989d). (Chapter 13)

Risk Characterization (in Ecological Risk Assessment): The last step of the ecological risk assessment that draws together the information generated from previous steps to produce a statement of the likelihood of an adverse effect to the assessment endpoint. (Chapter 13)

Risk Factor: Any quality of an individual (e.g., age) or an etiological factor (e.g., chronically exposed to high levels of the toxicant) that modifies individual's risk of developing the disease or pathological condition in question (Chapter 10). Specific to chemical risk assessment (Chapter 13), a chemical concentration multiplied by a toxicity value estimates the risk factor. Relative to radiation's effects, a risk factor gives the probability of a deleterious effect for each millisievert of dose received. (Chapter 13)

Risk Hypotheses (in Ecological Risk Assessment): As part of the conceptual model, risk hypotheses are clear statements of postulated or predicted effects of the contaminant on the assessment endpoint. (Chapter 13)

Risk Manager: "The individual or group who serves as primary decision-maker for a [waste] site." (EPA, 1989d) (Chapter 13)

Rivet Popper Hypothesis: The hypothesis that species in a community are like rivets that hold an airplane together and contribute to its proper functioning. The loss of each rivet weakens the structure. (Chapter 11)

Roentgen (R): A measure of the amount of energy deposited in some material by a certain amount of radiation. It is expressed relative to energy dissipation in 1 cc of dry air. Use of R to express dose allows one to normalize for the different amounts of energy that are deposited in materials such as tissues by different types of radiation. (Chapter 10)

Roentgen Equivalent Man (Rem): A measure of radiation that takes into account the differences in potential biological effects of different types of radiation. It relates the dose received to potential biological damage. (Chapter 1)

Rotenone: A natural plant product pesticide derived from the plant genus, *Derris* sp. (Chapter 2)

r-Strategy: An opportunistic strategy favoring species that establish themselves quickly, grow quickly to exploit as many resources as possible, and produce many offspring. Contrast to **K-Strategy**. (Chapter 11)

Rules of Practical Causal Inference: Fox's (1991) rules of practical causal inference are used in ecotoxicology to infer causality for toxicant exposure/effect scenarios. (Chapter 10)

Safe Concentration: "The highest concentration of toxicant that will permit normal propagation of fish and other aquatic life in receiving waters. The concept of a 'safe concentration' is a biological concept, whereas the 'no observed effect concentration' is a statistically defined concentration." (Chapter 8)

Safe Drinking Water Act (SDWA): A U.S. federal law that protects the water supply of the public (systems supplying more than 25 people or with more than 15 service outlets) by setting water quality standards and also protects source aquifers from contamination by establishing permitting requirements for underground disposal of waste. (Appendix 3)

Safener: A substance mixed with herbicides to make the crop species safer relative to harm from the herbicide. They decrease the harm to the crop species relative to that done to the targeted weed species. (Appendix 4)

Saprobien Spectrum: The characteristic change in community composition at different distances below the discharge of putrescible organic waste to a river or stream. (Chapter 11)

Sarcoma: A malignant cancer involving mesodermal cells such as those of muscle, cartilage, blood vessels, fat, or bone. (Chapter 7)

Scaling: Handling or transformation of allometric data to produce a quantitative relationship between organism (or species) size and some characteristics such as metabolic rate, gill surface area, lung ventilation rate, or biochemical activity. (Chapter 4)

Scoliosis: Lateral curvature of the spine. (Chapter 8)

Scope of Activity: The difference between the rates of oxygen consumption under maximum and minimal activity levels. It reflects the respiratory capacity available for the diverse demands on and activities of an organism. (Chapter 8)

Scope of Growth: An index (P = production) calculated as the amount of energy taken into the organism in its food (A) minus the energy used for respiration (R) and excretion (U): $P = A - R - U$. It is the amount of energy available for growth or production of young. (Chapter 8)

Secondary Lamellae (Respiratory Lamellae): These gill structures are parallel rows of projections on the dorsal and ventral sides of each **Primary Lamellae**. They are the primary sites of gas exchange of the gills. (Chapter 7)

Selection Components: Components of the life cycle of an individual on which natural selection can act. They are **Viability Selection**, **Sexual Selection**, **Meiotic Drive**, **Gametic Selection**, and **Fecundity Selection**. (Chapter 10)

Selyean Stress: A nonspecific response of the body if extraordinary demands are made of it. " … the state manifested by a specific syndrome which consists of all the nonspecifically induced changes within a biological system." (Selye, 1956) (Chapter 8)

Semelparous Species: A species that reproduces once. (Chapter 10)

Sentinel Species: A feral, caged, or endemic species used in measuring and indicating the level of contaminant or effect during a biomonitoring exercise. The proverbial canary in the coal mine is an example of a sentinel species. (Chapter 6)

Sexual Selection: A component of the life cycle of an individual in which natural selection can occur, involving differential mating success of individuals. (Chapter 10)

Shannon Diversity Index: A common measure of species diversity of a community. Compare with **Brillouin Diversity Index**. (Chapter 11)

Shapiro–Wilk's Test: A statistical test of the null hypothesis that data are normally distributed. Discussed in this book relative to testing the assumption of normality before performing one-way ANOVA on sublethal or chronic lethal effects data. (Chapter 8)

Shelford's Law of Tolerance: A species' tolerance(s) along an environmental gradient (or series of environmental gradients) determines its population distribution and size in the environment. (Chapter 10)

Similar Joint Action: Toxicants in mixture act by the same mode of action and "one component can be substituted at a constant proportion for the other … toxicity of a mixture is predictable directly from that of the constituents if their relative proportions are known" (Finney, 1947). (Chapter 9)

Single-Cell Gel Electrophoresis: Also called **Comet Electrophoresis**. An electrophoretic technique related to the alkaline unwinding assay in which the pattern of DNA migration from cells embedded in a gel media is used as a measure of DNA strand breakage. (Chapter 6)

Sister Chromatid: At the metaphase plate, chromosomes are composed of two chromatids called sister chromatids. (Chapter 7)

Sister Chromatid Exchange (SCE): The exchange of DNA between sister chromatids as a consequence of DNA breakage followed by reunion and crossing over of DNA segments of the chromatids. (Chapter 7)

Slope Factor (SF): In human risk estimation, it is the slope (risk or probability of occurrence per unit of dose or intake) for the risk–dose model used to estimate the probability of a cancer at a specified exposure. (Chapter 13)

Somatic Death: Death of an individual organism. (Chapter 7)

Somatic Risk: The risk of an adverse effect to the exposed individual associated with genetic damage to somatic cells, e.g., damage leading to cancer. (Chapter 7)

Sorption: This term will be used instead of adsorption if the specific mechanism by which a compound in solution becomes associated with a solid surface is unknown or poorly defined. (Chapter 3)

Spearman–Karber Method: A nonparametric method to estimate LC50, EC50, or LD50 values when it is difficult or unnecessary to assume a specific model for the dose/concentration–effect data. (Chapter 9)

Species Assemblage: Any operationally defined grouping of species studied in a community. See also **Taxocene**. (Chapter 11)

Species Diversity (= Heterogeneity): The heterogeneity or diversity of the community that considers both species richness and evenness. (Chapter 11)

Species Evenness: The degree to which the individuals in the community are evenly or uniformly distributed among species. (Chapter 11)

Species Richness: The number of species present in the community. (Chapter 11)

Specific Activity Concept (for radiotracer use): The radionuclide used to trace or quantify the movement of a stable nuclide (e.g., ^{14}C for stable C) is assumed to behave identically in chemical and biological processes as its nonradioactive analog (e.g., stable C). (Chapter 5)

Specific Effective Energy: The amount of energy transferred to tissue during irradiation. (Chapter 13)

Spiked Bioassay Approach (SB): A sediment toxicity test method used to generate a concentration–response model for or test hypotheses about effects to individuals placed in sediments spiked with different amounts of toxicant. (Chapter 9)

Spillover Hypothesis: Based on the assumption that binding by metallothionein sequesters toxic metals away from sites of action, this hypothesis states that toxic effects will begin to be seen after exceeding the capacity of the metallothionein present at any time to bind metals. The unbound metals then "spill over" to interact at sites of adverse action. (Chapter 6)

Split Probit: A distinct change in slope of a probit plot (log of concentration or dose vs. the probit of the proportion responding). (Chapter 9)

Stable Population: If conditions do not change with time, a population with a particular *r* will eventually establish a stable distribution of individuals among the various age or stage classes. Such a population is called a stable population. (Chapter 10)

Standard: U.S. legal limits (concentration or intensity) permitted for a specific water body, based on criteria and the specified use of a water body. (Chapter 1)

State Implementation Plan (SIP): U.S. state plans required by the Clean Air Act that outline steps toward meeting and maintaining national air quality standards. (Appendix 3)

Static Toxicity Test: A type of aquatic toxicity test in which the exposure water is not changed during the test. (Chapter 9)

Static-Renewal Test: A modified **Static Toxicity Test** in which solutions are completely or partially replaced with new solutions at set periods during exposures, or organisms are periodically transferred to new solutions. (Chapter 9)

Steel's Many-One Rank Test: A nonparametric, post-ANOVA test often used in the analysis of sublethal and chronic lethal effects data. (Chapter 8)

Stochastic Effect: Effects for which it is the probability of an adverse effect that increases with exposure dose or concentration. Cancer and genetic disorders initiated by irradiation are probabilistic. The associated risk of incurring cancer or adverse genetic effects is proportional, without a threshold, to the exposure dose or concentration in the relevant tissue. (Chapter 13)

Stress: "At any level of ecological organization, a response to or effect of a recent, disorganizing or detrimental factor" (Newman, 1995). See also **Seylean Stress** for the definition of stress to an individual organism. (Chapter 1)

Stress Protein Fingerprinting: The use of the patterns of stress protein induction seen in the field to suggest the particular toxicant inducing the response. Patterns from organisms sampled from the field can be compared to those obtained with single candidate toxicants in the laboratory. (Chapter 6)

Stress Proteins: A class of proteins involved in lessening the damage to proteins (see **Cellular Stress Response**) associated with a variety of stressors including heat, anoxia, ultraviolet radiation, arsenate, metals, and some xenobiotics. This term is also used by several ecotoxicologists (e.g., Sanders and Dyer, 1994) in a more generic context to mean a protein induced in response to a stressor. Such a definition would include proteins such as metallothioneins. See also **Heat Shock Proteins**. (Chapter 6)

Stress Theory of Ageing: Stress shortens longevity by accelerating energy expenditure (see **Rate of Living Theory of Ageing**). Selection takes place for resistance to stress, and as an epiphenomenon, individuals resistant to stress will predominate in extreme age classes of a population. The diminution of homeostasis under stress with age should be slowest in individuals with highest longevity. (Chapter 10)

Stressor: That which produces stress (Chapter 2). In ecological risk assessment, "any chemical, physical, or biological entity that can induce adverse effects on *ecological components*, that is, individuals, populations, communities, or ecosystems" (Norton et al. 1992). (Chapter 13)

Structure–Activity Relationship (SAR): A relationship between molecular qualities and some activity such as bioavailability or toxicity. (Chapter 4)

Structured Metapopulation Model: Unlike **Occupancy Models**, structured metapopulation models focus on the distribution of population sizes among habitat patches. (Chapter 10)

Subchronic Reference Dose (RfD_s): A reference dose derived from short-term exposure data. (Chapter 13)

Subcooled Liquid Vapor Pressure (P_L): The liquid vapor pressure corrected or adjusted for the heat of fusion, the energy needed to convert a mole of a compound from a solid to a liquid phase. Its use allows the expression of *liquid* vapor pressures at a specific temperature for organic compounds with widely varying melting temperatures. (Chapter 12)

Sublethal Effects: Effects seen at concentrations below those producing direct somatic death, e.g., slowed growth of an individual or diminished reproduction. (Chapter 8)

Sulfotransferase: A Phase II enzyme that conjugates sulfates to xenobiotics or their metabolites. (Chapter 6)

Summation Rule: Concentrations of toxicants can be added together to predict effects under the assumption of additivity. Although applied frequently in different ecotoxicological methods, this rule is not generally valid. (Chapter 9)

Superoxide Dismutase (SOD): An enzyme catalyzing the reaction, $2\ O^{-}_2 + 2H^+ \rightarrow H_2O_2 + O_2$. It functions to reduce oxidative stress. (Chapter 6)

Suppressor Gene: A gene that functions normally to suppress cell growth and may inhibit abnormal growth. (Chapter 7)

Symporter (or Cotransporter): A membrane-associated transporter that moves two or more molecules/ions together in the same direction across a membrane. Contrast with **Antiporter**. (Chapter 3)

Synergism: Toxicant synergism occurs if the joint effect of a mixture was greater than that predicted by summing the predicted, separate effects for the individual toxicants in the mixture. (Chapter 9)

Tail Moment: In **Single-Cell Gel Electrophoresis**, the tail length times the fraction of the DNA that is in the tail as opposed to the comet head is used as a measure of DNA strand breakage. (Chapter 6)

Target Organ: The specific or characteristic organ in which lesions from a toxicant occur or are expected to occur based on toxicant transport to, accumulation in, or activation by that organ. (Chapter 7)

Taxocene: A taxonomically defined subset of the entire ecological community. (Chapter 11)

Temperature of Condensation (T_C): The temperature (°C) at which the compound condenses or partitions from the gaseous to the nongaseous phase. (Chapter 12)

Ten Major Countermeasures of China's Environment and Development: Enacted in 1992, this Chinese law pointed out that a strategy of sustainable development is the most reasonable mode of accelerating economic development while solving environmental problems. (Appendix 5)

Teratogen: A chemical or physical agent capable of causing a developmental malformation. (Chapter 8)

Teratogenic: Capable of causing developmental malformations. (Chapter 6)

Teratogenic Index (TI): The mortality of eggs expressed as an LC50 divided by the TC50 (EC50 for production of abnormal embryos). The TI value is thought to reflect the developmental hazard of a contaminant. (Chapter 8)

Teratology: The science of fetal and embryonic abnormal development of anatomical structures. (Chapter 8)

Terpene: A terpene is made up of isoprene units, i.e., $H_2C{=}C(CH_3){-}CH{=}CH_2$ that are often inserted head to tail in a compound. (Chapter 2)

Tetrad: Two homologous chromosomes composed of two chromatids, which come together at the metaphase plate to form a tetrad. (Chapter 7)

Therapeutic Goods Act (1989): The Australian legislation that covers the import, manufacture, and registration of prescription and nonprescription medicines for human consumption. (Appendix 6)

Thermal Pollution: Any shift in environmental thermal conditions of sufficient magnitude to adversely impact human or ecological well-being. (Chapter 2)

Threshold Theory: This theory assumes no response (dose-related incidence or risk of cancer) below a certain low dose. Above the threshold, the slope of the response versus dose curve increases rapidly such that the dose–response curve takes on the appearance of a hockey stick. (Chapter 7)

Thrombosis: A clot formed in the circulatory system that could cause infarction of local tissue. (Chapter 7)

Tolerance: This term is often reserved for the enhanced ability to cope with a factor due to physiological acclimation. **Resistance** is used if the enhanced abilities are associated with genetic adaptation. Tolerance is used in this book for both acclimation and genetic adaptation. (Chapter 10)

Township and Village Enterprise (TVE) sector: A sector composed of 20 million small factories throughout the Chinese countryside that are contributing significantly to China's current economic upsurge. (Chapter 1)

Toxaphene: A pesticide produced by adding chlorine atoms to the camphene molecule. (Chapter 2)

Toxic Equivalence (TEQ): The combined toxicity of dioxins, dibenzofurans, or dioxin-like polychlorinated biphenyls (i.e., compounds that have very similar toxic modes of action involving binding to the aryl hydrocarbon receptor) expressed in units of toxicity of 2,3,7,8-tetrachlorodibenzo-p-dioxin (TCDD). This summing to compound effects to generate a TEQ is done using the **Toxic Equivalency Factors** defined above. (Chapter 9)

Toxic Equivalency Factor (TEF): An empirically derived factor that scales the toxicity of a dioxin, dibenzofuran, or dioxin-like polychlorinated biphenyl (i.e., compounds that have toxic modes of action involving binding to the aryl hydrocarbon receptor) to that of TCDD. (Chapter 9)

Toxic Substances Control Act (TSCA): A U.S. law that regulates, through the Environmental Protection Agency, the manufacture, processing, transport, use, import, and disposal of chemicals or chemical mixtures that might pose an unreasonable risk to health or environment. (Appendix 3)

Toxic Unit (TU): Amount or concentration of a toxicant expressed in units of lethality such as LD50 or LC50. For example, if toxic units are based on the LC50, a chemical with an LC50 of 20 mg L^{-1} would be present at 0.5 TU in a 10 mg L^{-1} solution. (Chapter 9)

Toxicity Value: In risk assessment, it is a factor used to estimate a risk factor. Its estimation can involve one or two methods. One can use the slope of a published effect–dose relationship to estimate the risk factor: $R = \text{Slope} \cdot C$, where C = toxicant concentration. The other is to use a **Reference Dose (RfD)** value in a similar manner. (Chapter 13)

Toxicokinetics: The study and predictive modeling of the internal kinetics of toxicants. (Chapter 3)

Toxin: A toxicant of animal, plant, fungal, or microbial origin such as toxicants in animal venom or belladonna alkaloids in some plants. (Chapter 2)

Transcription Factor: Protein involved in the regulation of gene expression that binds with specific promoter elements of a gene. (Chapter 6)

Transgenic Organism: An organism modified by genetic engineering. (Chapter 6)

Trophic Cascade: Strong, common features of communities in which the presence of a predator depresses prey density (or some prey function/behavior) such that the next lower trophic levels/positions are released from the negative influence of the prey. Trophic cascades often become apparent when predator densities are increased or decreased. (Chapter 11)

Trophic Dilution: The decrease in contaminant concentration as trophic level increases. Trophic dilution results from a net balance of ingestion rate, uptake from food, internal transformation, and elimination processes favoring loss of contaminant that enters the organism via food. (Chapter 5)

Trophic Enrichment: See **Biomagnification**. (Chapter 5)

Trophic Level: A conceptual level in a trophic web, that is, primary producer, primary consumer, secondary consumer, and tertiary consumer. In the presence of omnivory, discrete trophic levels become an unrealistic depiction of trophic position of species. (Chapter 5)

***t* Test with a Bonferroni Adjustment:** A parametric, post-ANOVA test used often in the analysis of sublethal and chronic lethal effects data. (Chapter 8)

***t* Test with a Dunn-Šidák Adjustment:** A parametric, post-ANOVA test used rarely in the analysis of sublethal and chronic lethal effects data that has slightly better statistical power than the ***t* test with a Bonferroni Adjustment**. (Chapter 8)

Twin Tracer Technique: A technique that introduces together a radiotracer of the substance being assimilated and an inert tracer which will not be assimilated and to which the substance of interest's assimilation will be gauged. (Chapter 5)

Type A Organism: According to the scheme of Campbell et al. (1988), an organism in contact with the sediments is unable to ingest particulates. The implication is bioavailability from interstitial water but not sediment-associated particulates. Examples include rooted macrophytes and benthic algae. (Chapter 4)

Type B Organism: According to the scheme of Campbell et al. (1988), an organism in contact with the sediments that is capable of ingesting particulates. The implication is bioavailability from interstitial water, and sediment-associated particulates are potentially important. Some examples include detritivores and suspension feeders. (Chapter 4)

Uncertainty Factors: In risk assessment, factors to adjust the NOAEL (no observed adverse effect level) (or LOAEL [low observed adverse effect level]) downward to compensate in a conservative direction for uncertainty. (Chapter 13)

Uniporter: A protein integrated into the membrane that facilitates transport of a single molecule/ion across the membrane without any exchange as would occur with an antiporter. Contrast to **Antiporter** and **Symporter**. (Chapter 3)

Uptake: The movement of a contaminant into an organism. (Chapter 3)

Up-and-Down Technique: A method of calculating LD50 that uses the responses (died/lived) of a series of dosed individuals. This method has the advantage of using fewer individuals than the traditional experimental design for the estimation of LD50 or LC50. (Chapter 9)

Urban Heat Island Effect: The temperature in cities is higher than that of the surrounding rural/suburban region because the city structures and materials used to build the city more effectively captures heat than those of the nearby countryside. (Chapter 2)

Uridinediphospho glucuronosyltransferase (UDP-glucouronosyltransferase, UDP-GT): A Phase II enzyme that transfers glucuronic acid from uridine diphosphate glucuronic acid to electrophilic xenobiotics or xenobiotic metabolites. It also binds covalently with electrophilic compounds such as polycyclic aromatic hydrocarbons. (Chapter 6)

UVA: Ultraviolet radiation with wavelengths in the range of 320 to 400 nm. (Chapter 2)

UVB: Ultraviolet radiation with wavelengths in the range of 290 to 320 nm. (Chapter 2)

UVC: Ultraviolet radiation with wavelengths less than 290 nm. (Chapter 2)

Valinomycin: A membrane-associated carrier protein (uniporter) that transports K^+ across membranes. (Chapter 3)

Viability Selection: Natural selection within the life cycle of an individual that occurs by the differential survival of individuals. The period in which it can occur begins with the formation of the zygote and continues throughout the life of the individual. (Chapter 10)

Vital Rates: Rates at which important life-cycle processes such as birth, migration, and death occur for individuals in populations. (Chapter 10)

Vitellogenin: A protein normally synthesized by females during egg production. (Chapter 2)

Vulnerability (Community): Susceptibility to irreversible damage by toxicants. (Chapter 11)

Wahlund Effect: There will be a net deficit of heterozygotes when two populations, each in Hardy–Weinberg equilibrium but with different allele frequencies, are mixed and the genotype frequencies quantified in a combined population sample. (Chapter 10)

Waldsterben: "The widespread and substantial decline in growth and the change in behavior of many softwood and hardwood forest ecosystems in central Europe." (Schütt and Cowling, 1985) (Chapter 12)

Water (Prevention and Control of Pollution) Act: Enacted in 1974, this Indian legislation was enacted to prevent and control water pollution, and to maintain or restore the wholesomeness of water in the country. (Appendix 7)

Water Framework Directive: One of the most important pieces of European environmental legislation in recent years, requiring all inland and coastal waters to achieve "good status" by 2015. It will do this by establishing a river basin district structure within which demanding environmental objectives will be set, including ecological targets for surface waters and the use of environmental quality standards for individual chemical pollutants. (Appendix 4)

Water Law of the PRC: This law was put into force in October 1, 2002, after the amendment and approval that same year. Its purpose is to foster rational economic development, utilization, and protection of water resources, for the prevention and control of water disasters, and for water resource sustainable use. (Appendix 5)

Weakest Link Incongruity: An incongruous extension of the critical life stage concept to suggest that protection of the most sensitive stage will ensure protection of all life stages. The dubious extension is made in which one assumes that exposure of field populations to concentrations identified in testing as causing significant mortality at a critical life stage will result in significant impact on the field population. This might or might not be true. (Chapter 9)

Weibull Metameter: A metameter used occasionally in dose- or concentration–effect data analysis. It has the form, $U(P) = \ln(-\ln(1 - P))$, where P is the proportion dead or responding. (Chapter 9)

Weight of Evidence: In risk assessment, this phrase appears to refer to whether a *reasonable person* reviewing the available information *could* agree that the conclusion was plausible. The more the evidence supports the conclusion, the stronger the weight of evidence. It could mean a quantitative, semiquantitative, or qualitative estimate of the degree to which the evidence supports or undermines the conclusion. (Chapter 13)

Wet deposition: Deposition in the precipitation of contaminants that were formed in the liquid media of the precipitation and that were incorporated into the precipitation during rain out. (Chapter 12)

Wilcoxon Rank Sum Test with Bonferroni's Adjustment: A nonparametric, post-ANOVA test often used in the analysis of sublethal and chronic lethal effects data. (Chapter 8)

Williams's Test: A parametric test that is more powerful than many post-ANOVA tests used to analyze sublethal and chronic lethal data. It assumes a monotonic trend (consistent increase or decrease) that might occur with increasing concentration. (Chapter 8)

Working Hypothesis: A hypothesis that is used to determine fact during scientific inquiry. It is not assumed to be true and only serves to test facts. (Chapter 1)

Xenobiotic: A "foreign chemical or material not produced in nature and not normally considered a constitutive component of a specified biological system. [It is] usually applied to manufactured chemicals" (Rand and Petrocelli, 1985). (Chapter 1)

X-rays: Electromagnetic photons emitted from the inner orbital shells of electrons that surround the nucleus. (Chapter 2)

Zenker's Necrosis: Necrosis (cell death) that occurs in skeletal muscle, which is similar to coagulation necrosis. (Chapter 7)

References

Abarca, L. and C. Ruepert, Plaguicidas encontrados en el Valle de la Estrella: Estudio preliminar. *Tecnol. Marcha*, 12, 31–38, 1992.

Abrahamson, A., C. Andersson, M.E. Jönsson, O. Fogelberg, J. Örberg, B. Brunström, and I. Brandt, Gill EROD in monitoring CYP1A inducers in fish: A study in rainbow trout (*Oncorhychus mykiss*) caged in Stockholm and Uppsala waters. *Aquat. Toxicol.*, 85, 1–8, 2007.

Ackerman, J.T., J.Y. Takekawa, C.A. Eagles-Smith, and S.A. Iverso, Mercury contamination and effects on survival of American avocet and black-necked stilt chicks in San Francisco Bay. *Ecotoxicology*, 17, 103–116, 2008.

Adams, J., *Risk*, UCL Press, London, United Kingdom, 1995.

Adams, R., Surviving "radium girl" hits 100. *Waterbury Republican-American*, Waterbury, CT, 1B–2B, 2005.

Adams, S.M., Biological indicators of stress in fish. *Am. Fish. Soc. Symp.*, 8, 1–8, 1990.

Adams, S.M., W.D. Cumby, M.S. Greeley, Jr., M.G. Ryon, and E.M. Schilling, Relationships between physiological and fish population responses in a contaminated stream. *Environ. Toxicol. Chem.*, 11, 1549–1557, 1992.

Adamus, C.L. and M.J. Bergman, Estimating nonpoint source pollution loads with a GIS screening model. *Water Resour. Bull.*, 31, 647–655, 1995.

Adolph, E.F., Quantitative relations in the physiological constitutions of mammals. *Science*, 109, 579–585, 1949.

Agard, D.A., To fold or not to fold. *Science*, 260, 1903–1904, 1993.

Agius, C. and R.J. Roberts, Melano-macrophage centres and their role in fish pathology. *J. Fish Dis.*, 26, 499–509, 2003.

Aguilar, A. and A. Borrell, Reproductive transfer and variation of body load of organochlorine pollutants with age in fin whales (*Balaenoptera physalus*). *Arch. Environ. Contam. Toxicol.*, 27, 546–554, 1994.

Ahlbom, A., *Biostatistics for Epidemiologists*, Lewis Publishers, Boca Raton, FL, 1993.

Ahrland, S., Thermodynamics of complex formation between hard and soft acceptors and donors. *Struct. Bond.*, 5, 118–149, 1968.

Akçakaya, H.R., J.D. Stark, and T.S. Bridges, Eds., *Demographic Toxicity: Methods in Ecological Risk Assessment*, Oxford University Press, Oxford, 2008.

Akhil, P.S.and C.H. Sujatha, Prevalence of organochlorine pesticide residues in groundwaters of Kasargod District, Kerala, India. *Toxicol. Environ. Chem.*, 94, 1718–1725, 2012.

Alados, C.L., J. Escos, and J.M. Emlen, Developmental stability as an indicator of environmental stress in the Pacific hake (*Merluccius productus*). *Fish. Bull. NOAA*, 91, 587–593, 1993.

Alados, C.L., T. Navarro, J. Esćos, B. Cabezudo, and J.M. Emlen, Translational and fluctuating asymmetry as tools to detect stress in stress-adapted and nonadapted plants. *Int. J. Plant Sci.*, 162, 607–616, 2001.

Alados, C.L., M.L. Giner, and Y. Pueyo, An assessment of the differential sensitivity of four summer-deciduous chamaephytes to grazing and plant interactions using translational asymmetry. *Ecol. Indicators*, 6, 554–566, 2006.

Albers, P.H., Petroleum and individual polycyclic aromatic hydrocarbons, in *Handbook of Ecotoxicology*, D.J. Hoffman, B.A. Rattner, G.A. Burton, Jr., and J. Vairns, Jr., Eds., Lewis Publishers, Boca Raton, FL, 2003.

Alberts, B., D. Bray, J. Lewis, M. Raff, K. Roberts, and J.D. Watson, *Molecular Biology of the Cell*, Garland Publishing, New York, 1983.

Alegria, H.A., T.F. Bidleman, and T.J. Shaw, Organochlorine pesticides in ambient air of Belize, Central America. *Environ. Sci. Technol.*, 34, 1953–1958, 2000.

Alho, C.J.R. and L.M. Vieira, Fish and wildlife resources in the Pantanal wetlands of Brazil and potential disturbances from the release of environmental contaminants. *Environ. Toxicol. Chem.*, 16, 71–74, 1997.

Alimba, C.G., A.A. Bakare, and C.A. Latunji, Municipal landfill leachates induced chromosome aberrations in rat bone marrow cells. *African J. Biotechnol.*, 5, 2053–2057, 2006.

Allen, H.E., R.H. Hall, and T.D. Brisbin, Metal speciation. Effects on aquatic toxicity. *Environ. Sci. Technol.*, 14, 441–442, 1980.

Aloj Totaro, E., F.A. Pisani, and P. Glees, The role of copper level in the formation of neuronal lipofuscin in the spinal ganglia of *Torpedo. Mar. Environ. Res.*, 15, 153–163, 1985.

Aloj Totaro, E., F.A. Pisanti, P. Glees, and A. Continillo, The effect of copper pollution on mitochondrial degeneration. *Mar. Environ. Res.*, 18, 245–253, 1986.

Al-Sabti, K., Frequency of chromosomal aberrations in the rainbow trout, *Salmo gairdneri* Rich., exposed to five pollutants. *J. Fish Biol.*, 26, 13–19, 1985.

Al-Sabti, K. and J. Hardig, Micronucleus test in fish for monitoring the genotoxic effects of industrial waste products in the Baltic Sea, Sweden. *Comp. Biochem. Physiol.*, 97C, 179–182, 1990.

Al-Sabti, K. and B. Kurelec, Chromosomal aberrations in onion (*Allium cepa*) induced by water chlorination by-products. *Bull. Environ. Contam. Toxicol.*, 34, 80–88, 1985a.

Al-Sabti, K. and B. Kurelec, Induction of chromosomal aberrations in the mussel *Mytilus galloprovincialis* Watch. *Bull. Environ. Contam. Toxicol.*, 35, 660–665, 1985b.

Alzieu, C., Biological effects of tributyltin in marine organisms, in *Tributyltin: Case Study of an Environmental Contaminant*, de Mora, S.J., Ed., Cambridge University Press, Cambridge, United Kingdom, 1996.

Alzieu, C., Impact of tributyltin on marine invertebrates. *Ecotoxicology*, 9, 71–76, 2000.

Amaraneni, S.R. and R.R. Pillale, Concentrations of pesticide residues in tissues of fish from Kolleru Lake in India. *Environ.Toxicol.*, 16, 550–556, 2001.

Amarasekare, P., Competitive coexistence in spatially structured environments: A synthesis. *Eco. Lett.*, 6, 1109–1122, 2003.

Amarasekare, P. and R.M. Nisbet, Spatial heterogeneity, source-sink dynamics, and the local coexistence of competing species. *Am. Nat.*, 158, 572–584, 2001.

Ambo-Rappe, R., D.L. Lajus, and M.J. Schreider, Translational fluctuating asymmetry and leaf dimension in seagrass, *Zostera capricorni* Aschers in a gradient of heavy metals. *Environ. Bioindic.*, 2, 99–116, 2007.

Ambo-Rappe, R., D.L. Lajus, and M.J. Schreider, Increased heavy metal and nutrient contamination does not increase fluctuating asymmetry in the seagrass, *Halophila ovalis. Ecol. Indic.*, 8, 100–103, 2008.

Ambo-Rappe, R., D.L. Lajus, and M.J. Schreider, Heavy metal impact on growth and leaf asymmetry of seagrass, *Halophila ovalis. J. Environ. Chem. Ecotoxicol.*, 3, 149–159, 2011.

Ambrose, P., Osprey revival from DDT complete in Chesapeake Bay. *Mar. Pollut. Bull.*, 42, 338, 2001.

American Public Health Association, *Standard Methods for the Examination of Water and Wastewater,* 15th Edition, American Public Health Association, Washington, DC, 1981.

American Society for Testing and Materials, Standard guide for conducting the frog embryo teratogenesis assay: *Xenopus* (FETAX), in *Annual Book of ASTM Standards*, American Society for Testing and Materials, Philadelphia, PA, 1993.

Ames, L.J., J.D. Felley, and M.E. Smith, Amounts of asymmetry in centrarchid fish inhabiting heated and nonheated reservoirs. *Trans. Am. Fish. Soc.*, 108, 485–489, 1979.

Amiard, J.C., C. Amiard-Triquet, S. Barka, J. Pellerin, and P.S. Rainbow, Metallothioneins in aquatic invertebrates: Their role in metal detoxification and their use as biomarkers. *Aquat. Toxicol.*, 76, 160–202, 2006.

Amiard-Triquet, C., D. Pain, G. Mauvais, and L. Pinault, Lead poisoning in waterfowl: Field and experimental data, in *Impact of Heavy Metals on the Environment*, Vernet, J.P., Ed., Elsevier Science Publishers B.V., Amsterdam, The Netherlands, 1992.

AMPS, Contributions of the Expert Group on Analysis and Monitoring of Priority Substances to the Water Framework Directive Expert Advisory Forum on Priority Substances and Pollution Control. Brussels (BE): European Commission. EUR 21587 EN, 2005.

Anderson, D., Male-mediated developmental toxicity. *Toxicol. Appl. Pharmacol.*, 207, 506–513, 2005.

Anderson, D.P., Immunological indicators: Effects of environmental stress on immune protection and disease outbreaks. *Am. Fish. Soc. Symp.*, 8, 38–50, 1990.

Anderson, E.V., Phasing lead out of gasoline: Hard knocks for lead alkyls producers. *Chem. Eng. News.*, Feb. 6, 12–16, 1978.

Anderson, H.R., Air pollution and mortality: A history. *Atmos. Environ.*, 43, 142–152, 2009.

Anderson, P.D. and L.J. Weber, Toxic response as a quantitative function of body size. *Toxicol. Appl. Pharmacol.*, 33, 471–483, 1975.

Andrew, R.W., K.E. Biesinger, and G.E. Glass, Effects of inorganic complexing on the toxicity of copper in *Daphnia magna. Water Res.*, 11, 309–315, 1977.

Andrews, G.K., Regulation of metallothionein gene expression by oxidative stress and metal ions. *Biochem. Pharmacol.*, 59, 95–104, 2000.

Angeler, D.G. and M. Alvarez-Cobelas, Island biogeography and landscape structure: Integrating ecological concepts in a landscape perspective of anthropogenic impacts in temporary wetlands. *Environ. Pollut.*, 138, 420–424, 2005.

Ankley, G.T., B.W. Brooks, D.B. Huggett, and J.P. Sumpter, Repeating history: Pharmaceuticals in the environment. *Environ. Sci. Technol.*, 41, 8211–8217, 2007.

Ankley, G.T., K.M. Jensen, E.A. Makynen, M.D. Kahl, J.J. Korte, M.W. Hornung, T.A. Henry et al, Effects of the androgenic growth promoter 17 β-trenbolone on fecundity and reproductive endocrinology of the fathead minnow. *Environ. Toxicol. Chem.*, 22, 1350–1360, 2003.

Ankley, G.T., G.L. Phipps, E.N. Leonard, D.A. Benoit, V.R. Mattson, P.A. Kosian, A.M. Cotter, J.R. Dierkes, D.J. Hansen, and J.D. Mahony, Acid-volatile sulfide as a factor mediating cadmium and nickel bioavailability in contaminated sediments. *Environ. Toxicol. Chem.*, 10, 1299–1307, 1991.

Anonymous, Heart disease, cancer linked to trace metals. *Chem. Eng. News*, May 3, 24–27, 1976.

Anonymous, Hooker settle on Hyde Park dump cleanup. *Chem. Eng. News*, Jan. 26, 10, 1981.

Anonymous, Most of Love Canal habitable, EPA says. *Chem. Eng. News*, July 19, 6, 1982.

Anonymous, Lead use in gasoline: EPA proposes 91% cut by 1986. *Chem. Eng. News*, Aug. 6, 4, 1984a.

Anonymous, India's chemical tragedy: Death toll at Bhopal still rising. *Chem. Eng. News*, Dec. 10, 6–7, 1984b.

Anonymous, Global climate warming: Trace gases other than CO_2 play role. *Chem. Eng. News*, May 6, 6–7, 1985.

Anonymous, Appendix II: Method 1311 toxicity characteristic leaching procedure. *Federal Register*, 55, 11863–11875, 1990.

Anspaugh, L.R., R.J. Catlin, and M. Goldman, The global impact of the Chernobyl reactor accident. *Science*, 242, 1513–1519, 1988.

Anthony, R.G., A.K. Miles, M.A. Ricca, and J.A. Estes, Environmental contaminants in bald eagle eggs from the Aleutian archipelago. *Environ. Toxicol. Chem.*, 26, 1843–1855, 2007.

Antizar-Ladislao, B., Environmental levels, toxicity and human exposure to tributyltin (TBT)-contaminated marine environment: A review. *Environ. Int.*, 34, 292–308, 2008

ANZECC/ARMCANZ, *National Water Quality Management Strategy, Australian and New Zealand Guidelines for Fresh and Marine Water Quality*. Australian and New Zealand Environment and Conservation Council and Agriculture and Resource Management Council of Australia and New Zealand, Canberra, Australia, 2000.

Apple, G.J., W.G. Hunter, and S. Bisgaard, Scientific data and environmental regulation, in *Statistics and the Law*, DeGroot, M.H., S.E. Fienberg, and J.B. Kadane, Eds., Wiley, New York, 1986.

Arbeláez, M.P. and S. Henao, Situación epidemiológica de las intoxicaciones agudas por plaguicidas en el Istmo Centroamericano. Organización Panamericana de la Salud (OPS), San José, Costa Rica, p. 59 2002.

Arbeli, Z. and C. Fuentes, Accelerated biodegradation of pesticides: An overview of the phenomenon, its basis and possible solutions; and a discussion on the tropical dimension. *Crop Prot.*, 26, 1733–1746, 2007.

Arcos, J.M., X. Ruiz, S. Bearhop, and R.W. Furness, Mercury levels in seabirds and their fish prey at the Ebro Delta (NW Mediterranean): The role of trawler discards as a source of contamination. *Mar. Ecol.Prog. Ser.*, 232, 281–290, 2002.

Ares, J., Time and space issues in ecotoxicology: Population models, landscape pattern analysis, and long-range environmental chemistry. *Environ. Toxicol. Chem.*, 22, 945–957, 2003.

Arimone, Y., B. Bégaud, G. Miremont-Salamé, A. Fourrier-Réglat, N. Moore, M. Molimard, and F. Haramburu, Agreement of expert judgment in causality assessment of adverse drug reactions. *Eur. J. Clin. Pharmacol.*, 61, 169–173, 2005.

Armitage, P. and I. Allen, Methods of estimating the LD50 in quantal response data. *J. Hyg. Cambridge*, 48, 298–322, 1950.

Arnot, J.A. and F.A.P.C. Gobas, A food web bioaccumulation model for organic chemicals in aquatic ecosystems. *Environ. Toxicol. Chem.*, 23, 2343–2355, 2004.

Arnot, J.A., D. Mackay, T.F. Parkerton, and M. Bonnell, A database of fish biotransformation rates for organic chemicals. *Environ. Toxicol. Chem.*, 27, 2263–2270, 2008.

Arnot, J.A., W. Meylan, J. Tunkel, P.H. Howard, D. Mackay, M. Bonnell, and R.S. Boethling, A quantitative structure-activity relationship for predicting metabolic biotransformation rates for organic chemicals in fish. *Environ. Toxicol. Chem.*, 28, 1168–1177, 2009.

Atchison, G.J., M.G. Henry, and M.B. Sandheinrich, Effects of metals on fish: A review. *Environ. Biol. Fish.*,18, 11–25, 1987.

Atchison, G.J., M.B. Sandheinrich, and M.D. Bryan, Effects of environmental stressors on interspecific interactions of aquatic animals, in *Ecotoxicology: A Hierarchical Treatment*, Newman, M.C. and C.H. Jagoe, Eds., CRC Press, Boca Raton, FL, 1996.

Atlas, R.M. and T.C. Hazen. Oil biodegradation and bioremediation: A tale of the two worst spills in U.S. history. *Environ. Sci. Technol.*, 45, 6709–6715, 2011.

Au, D.W.T., The application of histo-cytopathological biomarkers in marine pollution monitoring: A review. *Mar. Pollut. Bull.*, 48, 817–834, 2004.

Auletta, C.S., Acute, subchronic, and chronic toxicology, in *Handbook of Toxicology,* 2nd Edition, Derelanko, M.J. and M.A. Hollinger, Eds., CRC Press, Boca Raton, FL, 2002.

Aunan, K., J. Fang, T. Hu, H.M. Seip, and H. Vennemo, Climate change and air quality: Measures with co-benefits in China. *Environ. Sci. Technol.*, 40, 4822–4829, 2006.

Auslander, M., Y. Yudkovski, V. Chalifa-Caspi, B. Herut, R. Ophir, R. Reinhardt, P.M. Neumann, and M. Tom, Pollution-affected fish hepatic transcriptome and its expression patterns on exposure to cadmium. *Mar. Biotechnol.*, 10, 250–261, 2008.

Aust, A.E., Mutations and cancer, in *Genetic Toxicology*, Li, A.P. and R.H. Heflich, Eds., CRC Press, Boca Raton, FL, 1991.

Austin, A. and J. Deniseger, Periphyton community changes along a heavy metals gradient in a long narrow lake. *Environ. Exp. Bot.*, 25, 41–52, 1985.

Austin, A., J. Deniseger, and M.J.R. Clark, Lake algal populations and physico-chemical changes after 14 years input of metallic wastes. *Water Res.*, 19, 299–308, 1985.

Avery, T.E. and G.L. Berlin, *Interpretation of Aerial Photographs*, Macmillan, New York, 1985.

Azenha, M., M.T. Vasconcelos, and J.P.S. Cabral, Organic ligands reduce copper toxicity in *Pseudomonas syringae. Environ. Toxicol. Chem.*, 14, 369–373, 1995.

Babukutty, Y. and J. Chacko, Chemical partitioning and bioavailability of lead and nickel in an estuarine system. *Environ. Toxicol. Chem.*, 14, 427–434, 1995.

Bacci, E., *Ecotoxicology of Organic Contaminants*, CRC/Lewis Publishers, Boca Raton, FL, 1996.

Bacon, F., *Novum Organum*, Wiley, New York, 1620, Reprinted 1944.

Baes, C.F. Jr. and S.B. McLaughlin, Trace elements in tree rings: Evidence of recent and historical air pollution. *Science*, 224, 494–497, 1984.

Baird, C., *Environmental Chemistry*, W.H. Freeman and Co., New York, 1995.

Bakan, D., The test of significance in psychological research. *Psychol. Bull.*, 66, 423–437, 1966.

Baker, A.J.M. and P.L. Walker, Physiological responses of plants to heavy metals and the quantification of tolerance and toxicity. *Chem. Spec. Bioavailab.*, 1, 7–18, 1989.

Baker, C.E. and P.B. Dunaway, Retention of ^{134}Cs as an index to metabolism in the cotton rat *(Sigmodon hispidus). Health Phys.*, 16, 227–230, 1969.

Baker, L.A., A.T. Herlihy, P.R. Kaufmann, and J.E. Eilers, Acidic lakes and streams in the United States: The role of acidic deposition. *Science*, 252, 1151–1154, 1991.

Baker, R., B. Lavie, and E. Nevo, Natural selection for resistance to mercury pollution. *Experientia*, 41, 697–699, 1985.

Baldwin, W.S., P.B. Marko, and D.R. Nelson. The cytochrome P450 (CYP) gene superfamily in *Daphnia pulex. BMC genomics*, 10, 169, 2009.

Balmford, A., Pollution, politics, and vultures. *Science*, 339, 653–654, 2013.

Banni, M., F. Dondero, J. Jebali, H. Guerbej, H. Boussetta, and A. Viarengo, Assessment of heavy metal contamination using real-time PCR analysis of mussel metallothionein mt10 and mt20 expression: A validation along the Tunisian coast. *Biomarkers*, 12, 369–383, 2007.

Bantle, J.A., FETAX: A developmental toxicity assay using frog embryos, in *Fundamentals of Aquatic Toxicology: Effects, Environmental Fate, and Risk Assessment,* 2nd Edition, Rand, G.M., Ed., Taylor & Francis, Washington, DC, 1995.

Bantle, J.A. and T.D. Sabourin, Standard guide for conducting the frog embryo teratogenesis assay*: Xenopus* (FETAX). American Society of Testing and Materials, *Am. Soc. Test. Mat. Spec. Pub.* E1439–91, 1–11, 1991.

Barber, M.C., A review and comparison of models for predicting dynamic chemical bioconcentration in fish. *Environ. Toxicol. Chem.*, 22, 1963–1992, 2003.

Barber, M.C., Dietary uptake models used for modeling the bioaccumulation of organic contaminants in fish. *Environ. Toxicol. Chem.*, 27, 755–777, 2008.

Barber, M.C., L.A. Suarez, and R.R. Lassiter, Modeling bioconcentration of nonpolar organic pollutants by fish. *Environ. Toxicol. Chem.*, 7, 545–558, 1988.

Bard, S.M., Multixenobiotic resistance as a cellular defense mechanism in aquatic organisms. *Aquat. Toxicol.*, 48, 357–389, 2000.

Barnaby, F., Chernobyl: The consequences in Europe. *Ambio*, 15, 332–334, 1986.

Barnes, J.M., Poisons that hit and run. *New Sci.*, 38, 619, 1968.

Barnthouse, L.W., G.W. Suter, II, A.E. Rosen, and J.J. Beauchamp, Estimating responses of fish populations to toxic contaminants. *Environ. Toxicol. Chem.*, 6, 811–824, 1987.

Barnthouse, L.W., W.R. Munns, Jr., and M.T. Sorensen, Eds., *Population-Level Ecological Risk Assessment*, Taylor & Francis/CRC, Boca Raton, FL, 2008.

Barrett, J., *Atomic Structure and Periodicity*, Royal Society of Chemistry Book Publ., Cambridge, UK, 2002.

Barrie, L.A. and J.M. Hales, The spatial distributions of precipitation acidity and major ion wet deposition in North America during 1980. *Tellus*, 36B, 333–335, 1984.

Barron, M.G., Bioaccumulation and bioconcentration in aquatic organisms, in *Handbook of Ecotoxicology*, D.J. Hoffman, B.A. Rattner, G.A., Burton, Jr., and J. Cairns, Jr., Eds., CRC Press, Boca Raton, FL, 1995.

Barron, M.G., G.R. Stehly, and W.L. Hayton, Pharmacokinetic modeling in aquatic animals. I. Models and concepts. *Aquat. Toxicol.*, 18, 61–86, 1990.

Bartholomew, G.A., The roles of physiology and behaviour in the maintenance of homeostasis in the desert environment, in *Symposia of the Society for Experimental Biology, No. 18*, Academic Press, New York, 1964.

Baskin, Y., Ecologists dare to ask: How much does diversity matter? *Science*, 264, 202–203, 1994.

Batley, G.E., S.C. Apte, and J.L. Stauber, Speciation and bioavailability of trace metals in water: Progress since 1982. *Aust. J. Chem.*, 57, 903–919, 2004.

Battaglia, B., P.M. Bisol, V.U. Fossato, and E. Rodino, Studies on the genetic effects of pollution in the sea. *Rapp. P. V. Reun. Cons. Int. Explor. Mer.*, 179, 267–274, 1980.

Baumann, P.C., M.J. Mac, S.B. Smith, and J.C. Harshbarger, Tumor frequencies in walleye (*Stizostedion vitreum)* and brown bullhead (*Ictalurus nebulosus)* and sediment contaminants in tributaries of the Laurentian Great Lakes. *Can. J. Fish. Aquat. Sci.*, 48, 1804–1810, 1991.

Baumann, P.C., PAH, metabolites, and neoplasia in feral fish populations, in *Metabolism of Polycyclic Aromatic Hydrocarbons in the Aquatic Environment*, Varanasi, U., Ed., CRC Press, Boca Raton, FL, 1989.

Baumann, P.C., M.J. Mac, S.B. Smith, and J.C. Harshbarger, Tumor frequencies in walleye (*Stizostedion vitreum)* and brown bullhead (*Ictalurus nebulosus)* and sediment contaminants in tributaries of the Laurentian Great Lakes. *Can. J. Fish. Aquat. Sci.*, 48, 1804–1810, 1991.

Beach, L.R. and R.D. Palmiter, Amplification of the metallothionein-I gene in cadmium-resistant mouse cells, *Proc. Natl. Acad. Sci USA*, 78, 2110–2114, 1981.

Beach, S.A., J.L. Newsted, K. Coady, and J.P. Giesy, Ecotoxicological evaluation of perfluorooctanesulfonate (PFOS). *Rev. Environ. Contam. Toxicol.*, 186, 133–174, 2006.

Bearhop, S., R.A. Phillips, D.R. Thompson, S. Waldron, and R.W. Furness, Variability in mercury concentrations of great skuas *Catharacta skua*: The influence of colony, diet and trophic status inferred from stable isotope signatures. *Mar. Ecol.Prog. Ser.*, 193, 261–268, 2000.

Beaty, B.J., R.S. Mackie, K.S. Mattingly, J.O. Carlson, and A. Rayms-Keller, The midgut epithelium of aquatic arthropods: A critical target organ in environmental toxicology. *Environ. Health Perspect.*, 110, 911–914, 2002.

Bechmann, R.K., Use of life tables and LC50 tests to evaluate chronic and acute toxicity effects of copper on the marine copepod *Tisbe furcata* (Baird). *Environ. Toxicol. Chem.*, 13, 1509–1517, 1994.

Beck, B.D., E.J. Calabrese, T.M. Slayton, and R. Rudel, The use of toxicology in the regulatory process, in *Principles and Methods of Toxicology,* 5th Edition, Hayes, A.W., Ed., CRC/Taylor & Francis Press, Boca Raton, FL, 2008.

Becker, P.H., J. Gonzalez-Solis, B. Behrends, and J. Croxall, Feather mercury levels in seabirds at South Georgia: Influence of trophic position, sex and age. *Mar. Ecol. Prog. Ser.*, 243, 261–269, 2002.

Becker, P.H., D. Henning, and R.W. Furness, Differences in mercury contamination and elimination during feather development in gull and tern broods. *Arch. Environ. Contam. Toxicol.*, 27, 162–167, 1994.

Beeby, A., Toxic metal uptake and essential metal regulation in terrestrial invertebrates: A review, in *Metal Ecotoxicology: Concepts and Applications*, M.C. Newman and A.W. McIntosh, Eds., Lewis Publishers, Chelsea, MI, 1991.

Beitinger, T.L., Behavioral reactions for the assessment of stress in fishes. *J. Great Lakes Res.*, 16, 495–528, 1990.

Belden, J., S. McMurry, L. Smith, and P. Reilley, Acute toxicity of fungicide formulations to amphibians at environmental relevant concentrations. *Environ. Toxicol. Chem.*, 29, 2477–2480, 2010.

Bell, M., and W. Hoar, Some effects of ultraviolet radiation on sockeye salmon eggs and alevins. *Can. J. Res.*, 28, 35–43, 1950.

Bellwood, D.R., T.P. Hughes, C. Folke, and M. Nystrom, Confronting the coral reef crisis. *Nature*, 429, 827–833, 2004.

Benguira, S. and A. Hontela, Adrenocorticotropin- and cyclic adenosine 3',5'-monophsphate-stimulated cortisol secretion in interregnal tissues of rainbow trout exposed in vitro to DDT compounds. *Environ. Toxicol. Chem.*, 19, 842–847, 2000.

Bennedetto, A.V., The psoralens: An historical perspective. *Cutis*, 20, 469–471, 1977.

Bennett, R.F., Industrial manufacture and applications of tributyltin compounds, in *Tributyltin: Case Study of an Environmental Contaminant*, de Mora, S.J., Ed., Cambridge University Press, Cambridge, United Kingdom, 1996.

Benson, A.A. and R.E. Summons, Arsenic accumulation in Great Barrier Reef invertebrates. *Science*, 211, 482–483, 1981.

Bercovitz, K. and D. Laufer, Lead release from human trabecular bone, in *Impact of Heavy Metals on the Environment*, Vernet, J.P., Ed., Elsevier Science Publishers B.V., Amsterdam, The Netherlands, 1992.

Berenbaum, M.C., The expected effect of a combination of agents: The general solution. *J. Theor. Biol.*, 114, 413–431, 1985.

Bergeron, J.M., D. Crews, and J.A. McLachlan, PCBs as environmental estrogens: Turtle sex determination as a biomarker of environmental contamination. *Environ. Health Perspect.*, 102, 780–781, 1994.

Berglind, R., Combined and separate effects of cadmium, lead and zinc in ALA-D activity, growth and hemoglobin content in *Daphnia magna*. *Environ. Toxicol. Chem.*, 5, 989–995, 1986.

Beringer, J.E., Releasing genetically modified organisms: Will any harm outweigh any advantage? *J. Appl. Ecol.*, 37, 207–214, 2000.

Berkson, J., Why I prefer logits to probits. *Biometrics*, 7, 327–339, 1951.

Berkson, J., Maximum likelihood and minimum χ^2 estimates of the logistic function. *J. Am. Stat. Assoc.*, 50, 130–162, 1955.

Bernstein, P.L., *Against the Gods: The Remarkable Story of Risk*, Wiley, New York, 1996.

Berumen, M.L. and M.S. Pratchett, Recovery without resilience: Persistent disturbance and long-term shifts in the structure of fish and coral communities at Tiahura reef, Moorea. *Coral Reefs*, 25, 647–653, 2006.

Betts, K.S., Formulating green flame retardants. *Environ. Sci. Technol.*, 41, 7201–7202, 2007.

Beyer, W.N., A reexamination of biomagnification of metals in terrestrial food chains. *Environ. Toxicol. Chem.*, 5, 863–864, 1986.

Bezel, V.S. and V.N. Bolshakov, Population ecotoxicology of mammals, in *Bioindications of Chemical and Radioactive Pollution*, Krivolutsky, D.A., Ed., CRC Press, Boca Raton, FL, 1990.

Bhattacharya, A. and S. Bhattacharya, Induction of oxidative stress by arsenic in *Clarias batrachus*: Involvement of peroxisomes. *Ecotoxicol. Environ. Saf.*, 66, 178–187, 2007.

Bickham, J.W., B.G. Hanks, M.J. Smolen, T. Lamb, and J.W. Gibbons, Flow cytometric analysis of the effects of low-level radiation exposure on natural populations of slider turtles *(Pseudemys scripta)*, *Arch. Environ. Contam. Toxicol.*, 17, 837–841, 1988.

Bickham, J.W., S. Sandhu, P.D.N. Hebert, L. Chikhi, and R. Athwal, Effects of chemical contaminants on genetic diversity in natural populations: Implications for biomonitoring and ecotoxicology. *Mutat. Res. Rev. Mutat.*, 463, 33–51, 2000.

Bidleman, T.F., M.D. Walla, R. Roura, E. Carr, and S. Schmidt, Organochlorine pesticides in the atmosphere of the Southern Ocean and Antarctica, January-March, 1990. *Mar. Pollut. Bull.*, 26, 258–262, 1993.

Biesinger, K.E. and G.M. Christensen, Effects of various metals on survival, growth, reproduction, and metabolism of *Daphnia magna*. *J. Fish. Res. Board Can.*, 29, 1691–1700, 1972.

Biesmeijer, J.C., S.P.M. Robeerts, M. Reemer, R. Ohlemhller, M. Edwards, T. Peeters, A.P. Schaffers et al, Parallel declines in pollinators and insect-pollinated plants in Britain and the Netherlands. *Science*, 313, 351–354, 2006.

Biggins, P.D.E. and R.M. Harrison, Chemical speciation of lead compounds in street dusts. *Environ. Sci. Technol.*, 14, 336–339, 1980.

Billiard, S.M., J.N. Meyer, D.M. Wassenberg, P.V. Hodson, and R.T. Di Giulio, Non-additive effects of PAHs on early vertebrate development: Mechanisms and implications for risk assessment. *Toxicol. Sci.*, 2007.

Binz, P.A. and J.H.R. Kägi, Metallothionein: Molecular evolution and classification, in *Metallothionein IV*, Klaassen, C., Ed., Birkhäuser-Verlag, Basel, Switzerland, 1999.

Birge, W.J., R.D. Hoyt, J.A. Black, M.D. Kercher, and W.A. Robison, Effects of chemical stresses on behavior of larval and juvenile fishes and amphibians, in *Water Quality and the Early Life Stages of Fishes*, Fuiman, L.A., Ed., American Fisheries Society, Bethesda, MD, 1993.

Birnbaum, L.S. and D.F. Staskal, Brominated flame retardants: Cause for concern? *Environ. Health Perspect.*, 112, 9–17, 2004.

Bishop, J.A., L.M. Cook, and J. Muggleton, The response of two species of moths to industrialization in northwest England I: Polymorphisms for melanism. *Philos. Trans. R. Soc. B*, 281, 491–515, 1978.

Bishop, J.S., An experimental study of the cline of industrial melanism in *Biston betularia* (L.)(Lepidoptera) between urban Liverpool and rural North Wales. *J. Animal Ecology*, 41, 209–243, 1972.

Bishop, W.E. and A.W. McIntosh, Acute lethality and effects of sublethal cadmium exposure on ventilation frequency and cough rate of bluegill (*Lepomis macrochirus*). *Arch. Environ. Contam. Toxicol.*, 10, 519–530, 1981.

Bjorksten, T.A., K. Fowler, and A. Pomiankowski, What does sexual trait FA tell us about stress? *Trends Ecol. Evol.*, 15, 163–166, 2000a.

Bjorksten, T.A., K. Fowler, and A. Poiankowski, What does sexual trait FA tell us about stress? *Trends Ecol. Evol.*, 15, 163–166, 2000b.

Bjorksten, T.A., K. Fowler, and A. Pomiankowski, Untitled. *Trends Ecol. Evol.*, 15, 331, 2006b.

Blanchfield, P.J., M.J. Paterson, J.A. Shearer, and D.W. Schindler, Johnson and Vallentyne's legacy: 40 years of aquatic research at the Experimental Lakes Area. *Can. J. Fish. Aquat. Sci.*, 66, 1831–1836, 2009.

Blanck, H., A simple, community level, ecotoxicological test system using samples of periphyton. *Hydrobiologia*, 124, 251–261, 1985.

Blanco, G., J.A. Sanchez, E. Vazques, E. Garcia, and J. Rubio, Superior developmental stability of heterozygotes at enzyme loci in *Salmo salar* L. *Aquaculture*, 84, 199–209, 1984.

Blaylock, B.G., Radionuclide data bases available for bioaccumulation factors for freshwater biota. *Nucl. Safety*, 23, 427–438, 1982.

Blindauer, C.A., Metallothioneins with unusual residues: Histidines as modulators of zinc affinity and reactivity. *J. Inorg. Biochem.*, 102, 507–521, 2008.

Bliss, C.I., The calculation of the dosage-mortality curve. *Ann. Appl. Biol.*, 22, 134–307, 1935.

Bliss, C.I., The size factor in the action of arsenic upon silkworm larvae. *J. Exp. Biol.*, 13, 95–110, 1936.

Blok, J. and F. Balk, Environmental regulation in the European Community, in *Fundamentals of Aquatic Toxicology: Effects, Environmental Fate, and Risk Assessment,* 2nd Edition, Rand, G.M., Ed., Taylor & Francis, Washington, DC, 1995.

Bluhm, B.A. and R. Gradinger, Regional variability in food availability for Arctic marine mammals. *Ecol. Appl.*, 18, S77–S96, 2008.

Blum, A., The fire retardant dilemma. *Science*, 318, 194, 2007.

Blum, D.J.W. and R.E. Speece, Determining chemical toxicity to aquatic species. *Environ. Sci. Technol.*, 24, 284–293. 1990.

Blus, L.J., Organochlorine pesticides, in *Handbook of Ecotoxicology*, Hoffman, D.J., B.A. Rattner, G.A. Burton, Jr., and J. Cairns, Jr., Eds., Lewis Publishers, Boca Raton, FL, 2003.

Boethling, R.S. and D. Mackay, *Handbook of Property Estimation Methods for Chemicals: Environmental and Health Sciences*, Lewis Publishers, Boca Raton, FL, 2000.

Bolognesi, C., R. Rabboni, and P. Roggieri, Genotoxicity biomarkers in *M. galloprovincialis* as indicators of marine pollutants. *Comp. Biochem. Physiol.*, 113, 319–323, 1996.

Bongers, A.B.J., M.Z. Ben-Ayed, Z.B. Doulabi, J. Komen, and C.J.J. Richter, Origin of variation in isogenic, gynogenetic, and androgenetic strains of common carp, *Cyprinus carpio*. *J. Exp. Zool.*, 277, 72–79, 1997.

Bongers, T., The maturity index: An ecological measure of environmental disturbance based on nematode species composition. *Oecologia*, 83, 14–19, 1990.

Bonneris, E., A. Giguère, O. Perceval, T. Buronfosse, S. Masson, L. Hare, and P.G.C. Campbell, Sub-cellular partitioning of metals (Cd, Cu, Zn) in the gills of a freshwater bivalve, *Pyganodon grandis*: Role of calcium concretions in metal sequestration. *Aquat. Toxicol.*, 71, 319–334, 2005.

Bonnomet, V. and C. Alvarez, Implementation of requirements on priority substances within the context of the Water Framework Directive. Methodology for setting EQS: Identifying gaps and further developments. Background document. Limoges (FR): International Office for Water (INERIS). Report ENV.D.2/ATA/2004/0103, p. 49, 2006.

Booth, W., Postmortem on Three Mile Island. *Science*, 238, 1342–1345, 1987.

Borgmann, U., Metal speciation and toxicity of free ions to aquatic biota, in *Aquatic Toxicology*, Nriagu, J.O., Ed., Wiley, New York, 1983.

Bornschein, R.L. and S.R. Kuang, Behavioral effects of heavy metal exposure, in *Biological Effects of Heavy Metals*, Foulkes, E.C., Ed., CRC Press, Boca Raton, FL, 1990.

Bornstein, R.D., Observations of the urban heat island effect in New York City. *J. Appl. Meteorol.*, 7, 575–582, 1968.

Borovec, J., Changes in incidence of carcinoma *in situ* after the Chernobyl disaster in Central Europe. *Arch. Environ. Contam. Toxicol.*, 29, 266–269, 1995.

Borrell, A., G. Cantos, T. Pastor, and A. Aguilar, Levels of organochlorine compounds in spotted dolphins from the Coiba archipelago, Panama. *Chemosphere*, 54, 669–677, 2004.

Bortone, S.A., W.P. Davis, and C.M. Bundrick, Morphological and behavioral characters in mosquitofish as potential bioindication of exposure to Kraft mill effluent. *Bull. Environ. Contam. Toxicol.*, 43, 370–377, 1989.

Boudreau, M., S.C. Courtenay, D.L. MacLatchy, C.H. Bérubé, J.L. Parrott, and G.J. Van Der Kraak, Utility of morphological abnormalities during early-life development of the estuarine mummichog, *Fundulus heteroclitus*, as an indicator of estrogenic and antiestrogenic endocrine disruption. *Environ. Toxicol. Chem.*, 23, 415–425, 2004.

Bouquegneau, J.M., Evidence for the protective effect of metallothioneins against inorganic mercury injuries to fish. *Bull. Environ. Contam. Toxicol.*, 23, 218–219, 1979.

Boutron, C.F. and C.C. Patterson, The occurrence of lead in Antarctic recent snow, firn deposited over the last two centuries and prehistoric ice. *Geochim. Cosmochim. Acta*, 47, 1355–1368, 1983.

Bowling, J.W., G.J. Leversee, P.F. Landrum, and J.P. Giesy, Acute mortality of anthracene-contaminated fish exposed to sunlight. *Aquat. Toxicol.*, 3, 79–90, 1983.

Boxall, A.B.A., L.A. Fogg, P.A. Blackwell, P. Kay, P. Pemberton, and A. Croxford, Veterinary medicines in the environment. *Rev. Environ. Contam. Toxicol.*, 180, 1–91, 2004.

Boxall, A.B.A., D.W. Kolpin, B. Halling-Sorensen, and J. Tools, Are veterinary medicines causing environmental risks? *Environ. Sci. Technol.*, 287A–294A, 2003.

Boyd, I.L., The art of ecological modeling. *Science*, 337, 306–307, 2012.

Boyden, C.R., Trace element content and body size in molluscs. *Nature*, 251, 311–314, 1974.

Boyden, C.R., Effect of size upon metal content of shellfish. *J. Mar. Biol. Assoc. U.K.* 57, 675–714, 1977.

Bradley, R.W. and J.B. Sprague, The influence of pH, water hardness, and alkalinity on the acute lethality of zinc to rainbow trout (*Salmo gairdneri*). *Can. J. Fish. Aquat. Sci.*, 42, 731–736, 1985.

Branches, F.J.P., T.B. Erickson, S.E. Aks, and D.O. Hryhorczuk, The price of gold: Mercury exposure in the Amazonian rain forest. *J. Toxicol. Clin. Toxicol.*, 31, 295–306, 1993.

Branson, D.R., G.E. Blau, H.S. Alexander, and W.B. Neely, Bioaccumulation of 2,2,4,4–tetrachlorobiphenyl in rainbow trout as measured by an accelerated test. *Trans. Am. Fish. Soc.*, 104, 785–792, 1975.

Braun, W., M. Vasak, A.H. Robbins, C.D. Stout, G. Wagner, J.H.R. Kägi, and K. Wuthrich, Comparison of the NMR solution structure and the x-ray crystal structure of rat metallothionein-2, *Proc. Natl. Acad. Sci. U.S.A.*, 89, 10124–10128, 1992.

Braune, B.M., Temporal trends of organochlorines and mercury in seabird eggs from the Canadian Arctic, 1975–2003. *Environ. Pollut.*, 148, 599–613, 2007.

Braune, B.M., P.M. Outridge, A.T. Fisk, D.C.G. Muir, P.A. Helm, K. Hobbs, P.F. Hoekstra et al, Persistent organic pollutants and mercury in marine biota of the Canadian Arctic: An overview of spatial and temporal trends. *Sci. Total Environ.*, 351, 4–56, 2005.

Bravo, V., E. de la Cruz, G. Herrera, and F. Ramírez, Uso de plaguicidas en cultivos agrícolas como herramienta para el monitoreo de peligros en salud. *Revista UNICIENCIA*, 27, 351–76, 2013.

Bravo, V., E. de la Cruz, F. Ramírez, S. Berrocal, G. Herrera, and C. Wesseling, Plaguicidas importados y sus peligros para la salud. Costa Rica, 1980–2009. Submitted (Revista de Salud Pública del Ministerio de Salud, Costa Rica).

Bravo, V., S. Berrocal, F. Ramirez, E. de la Cruz, N. Canto, A. Tatis, W. Mejía, and T. Rodríguez, Importación de plaguicidas y peligros en salud, América Central, período 2005–2009. Submitted (*Rev. Panam. Salud Públ.*).

Brecken-Folse, J.A., F.L. Mayer, L.E. Pedigo, and L.L. Marking, Acute toxicity of 4-nitrophenol, 2,4-dinitrophenol, terbufos and trichlorfon to grass shrimp (*Palaemonetes* spp.) and sheepshead minnows (*Cyprinodon variegatus*) as affected by salinity and temperature. *Environ. Sci. Technol.*, 13, 67–77, 1994.

Brendler-Schwaab, S., A. Hartmann, S. Pfuhler, and G. Speit, The in vivo comet assay: Use and status in genotoxicity testing. *Mutagenesis*, 20, 245–254, 2005.

Brezonik, P.L., S.O. King, and C.E. Mach, The influence of water chemistry on trace metal bioavailability and toxicity to aquatic organisms, in *Metal Ecotoxicology: Concepts and Applications*, Newman, M.C. and A.W. McIntosh, Eds., Lewis Publishers, Chelsea, MI, 1991.

Bricelj, V.M., A.E. Bass, and G.R. Lopez, Absorption and gut passage time of microalgae in a suspension feeder: An evaluation of the ^{51}Cr:^{14}C twin tracer technique. *Mar. Ecol. Prog. Ser.*, 17, 57–63, 1984.

Bridges, T.S., A.R. Akçakaya, and B. Bunch, *Leptocheirus plumulosus* in the Upper Chesapeake Bay: Sediment toxicity effects at the metapopulation level, in *Demographic Toxicity: Methods in Ecological Risk Assessment*, H.R. Akçakaya, J.D. Stark and T.S. Bridges, Eds., Oxford University Press, Oxford, 2008.

Bridgman, H., *Global Air Pollution: Problems for the 1990s*. Wiley, New York, 1994.

Broad, W.J., Sir Isaac Newton: Mad as a hatter. *Science*, 213, 1341–1344, 1981.

Broderius, S.J., L.L. Smith, Jr., and D.T. Lind, Relative toxicity of free cyanide and dissolved sulfide forms to the fathead minnow (*Pimephales promelas*). *J. Fish. Res. Board Can.*, 34, 2323–2332, 1977.

Brodie, J., J. Waterhouse, B. Schaffelke, F. Kroon, P. Thorburn, J. Rolfe, J. Johnson et al. 2013 Scientific Consensus Statement: Land use impacts on Great Barrier Reef water quality and ecosystem protection. State of Queensland, Brisbane, Queensland. Available from http://www.reefplan.qld.gov.au/about/assets/scientific-consensus-statement-2013.pdf, accessed 7/10/2013.

Broman, D., C. Näf, C. Rolff, Y. Zebühr, B. Fry, and J. Hobbie, Using ratios of stable nitrogen to estimate bioaccumulation and flux of polychlorinated dibenzo-p-dioxins (PCDDs) and dibenzofurans (PCDFs) in two food chains from the northern Baltic. *Environ. Toxicol. Chem.*, 11, 331–345, 1992.

Brouwer, A., A.J. Murk, and J.H. Koeman, Biochemical and physiological approaches in ecotoxicology. *Funct. Ecol.*, 4, 75–281, 1990.

Brown, B.E., Lead detoxification by a copper-tolerant isopod. *Nature*, 276, 388–390, 1978.

Brown, S.B., B.A. Adams, D.G. Cyr, and J.G. Eales, Contaminant effects on the teleost fish thyroid. *Environ. Toxicol. Chem.*, 23, 1680–1701, 2004a.

Brown, S.B., R.E. Evans, L. Vandenbyllardt, K.W. Finnon, V.P. Palace, A.S. Kane, A.Y. Yarechewski, and D.C.G. Muir, Altered thyroid status in lake trout (*Salvelinus namaycush*) exposed to co-planar 3,3',4,4',5-pentachlrobiphenyl. *Aquat. Toxicol.*, 67, 75–85, 2004b.

Brown, T.A. and A. Shrift, Selenium: Toxicity and tolerance in higher plants. *Biol. Rev.*, 57, 59–84, 1982.

Brown, T.N., J.A. Arnot, and F. Wania. Iterative fragment selection: A group contribution approach to predicting fish biotransformation half-lives. *Environ. Sci. Technol.*, 46, 8253–8260, 2012.

Bruce, R.D., An up-and-down procedure for acute toxicity testing. *Fundam. Appl. Toxicol.* (now *Toxicol. Sci.*), 5, 151–157, 1985.

Bruce, R.D., A confirmatory study of the up-and-down method for acute oral toxicity testing. *Fundam. Appl. Toxicol.* (now *Toxicol. Sci.*), 8, 97–100, 1987.

Bruggeman, W.A., L.B.J.M. Martron, D. Kooiman, and O. Hutzinger, Accumulation and elimination kinetics of di-, tri- and tetra chlorobiphenyls by goldfish after dietary and aqueous exposure. *Chemosphere*, 10, 811–832, 1981.

Brühl, C.A., T. Schmidt, S. Pieper, and A. Alscher, Terrestrial pesticide exposure of amphibians: An underestimated cause of global decline? *Scientific Reports*, 3, 1135, 2013.

Brulle, F., C. Cocquerele, A.N. Wamalah, A.J. Morgan, P. Kille, A. Leprêtre, and F. Vandenbulcke, cDNA cloning and expression analysis of *Eisenia fetida* (Annelida: Oligochaeta) phytochelatin synthase under cadmium exposure. *Ecotoxicol. Environ. Saf.*, 71, 47–55, 2008.

Brumley, C.M., V.S. Haritos, J.T. Ahokas, and D.A. Holdway, Validation of biomarkers of marine pollution exposure in sand flathead using Aroclor 1254. *Aquat. Toxicol.*, 31, 249–262, 1995.

Bryan, G.W. and P.E. Gibbs, Impact of low concentrations of tributyltin (TBT) on marine organisms: A review, in *Metal Ecotoxicology: Concepts and Applications*, Newman, M.C. and A.W. McIntosh, Eds., Lewis Publishers, Chelsea, MI, 1991.

Bryan, G.W., P.E. Gibbs, G.R. Burt, and L.G. Hummerstone, The effects of tributyltin (TBT) accumulation on adult dogwhelks, *Nucella lapillus*: Long-term field and laboratory experiments. *J. Mar. Biol. Assoc. U.K.*, 67, 525–544, 1987.

Bucheli, T.D. and K. Fent, Induction of cytochrome P450 as a biomarker of environmental contamination in aquatic ecosystems. *Crit. Rev. Environ. Sci. Technol.*, 25, 201–268, 1995.

Budge, S.M., S.J. Iverson, and H.N. Koopman, Studying trophic ecology in marine ecosystems using fatty acids: A primer on analysis and interpretation. *Mar. Mamm. Sci.*, 22, 759–801, 2006.

Bueno, A.M.S., J.M.S. Agostini, K. Gaidzinski, J. Moreira, and I. Brognoli, Frequencies of chromosomal aberrations in rodents collected in the coal-field and tobacco culture region of Criciúma, South Brazil. *J. Toxicol. Environ. Health*, 36, 91–102, 1992.

Buesen, R., M. Mock, A. Seidel, J. Jacob, and A. Lampen, Interaction between metabolism and transport of benzo[a]pyrene and its metabolites in enterocytes. *Toxicol. Appl. Pharmacol.*, 183, 168–178, 2002.

Buffle, J. and M.L. Tercier-Waeber, Voltammetric environmental trace-metal analysis and speciation: From laboratory to in situ measurements. *Trends Analyt. Chem.*, 24, 172–191, 2005.

Buikema, A.L., B.R. Niederlehner, and J. Cairns, Jr., Biological monitoring Part IV: Toxicity testing. *Water Res.*, 16, 239–262, 1982.

Bunce, N., *Environmental Chemistry*, Wuerz Publishing, Winnipeg, Canada, 1991.

Bundschuh, M., M.C. Newman, J.P. Zubrod, F. Seitz, R.R. Rosenfeldt, and R. Schulz, Misuse of null hypothesis significance testing: Would estimation of positive and negative predictive values improve certainty of chemical risk assessment? *Environ. Sci. Pollut. Res.*, 20, 7341–7347, 2013.

Bundschuh, M., J.P. Zubrod, F. Seitz, M.C. Newman, and R. Schulz, Mercury-contaminated sediments affect amphipod feeding. *Arch. Environ. Contam. Toxicol.*, 60, 437–443, 2011.

Bureau of Radiological Health. *Radiological Health Handbook.* Compiled by Bureau of Radiological Health. U.S. Department of Health, Education, and Welfare, Food and Drug Administration, Rockville, MD, p. 458, 1970.

Burger, J., M.H. Lavery, and M. Gochfeld, Temporal changes in lead levels in common tern feathers in New York and relationship of field levels to adverse effects in the laboratory. *Environ. Toxicol. Chem.*, 13, 581–586, 1994.

Burgess, N.M. and M.W. Meyer, Methylmercury exposure associated with reduced productivity in common loons. *Ecotoxicology*, 17, 83–91, 2008.

Burgess, R.M., M.J. Ahrens, and C.W. Hickey, Geochemistry of PAHs in aquatic environments: Source, persistence and distribution, in *PAHs: An Ecotoxicological Perspective*, Douben, P.E.T., Ed., Wiley, New York, 2003.

Burke L., K. Reytar, M. Spalding, and A. Perry, *Reefs at Risk Revisited.* World Resources Institute (WRI), Washington, DC, 2011.

Burkhard, L.P., J.A. Arnot, M.R. Embry, K.J. Farley, R.A. Hoke, M. Kitano, H.A. Leslie et al, Comparing laboratory and field measured bioaccumulation endpoints. *Integr. Environ. Assess. Manag.*, 8, 17–31, 2012.

Burkhardt, W., Zur frage der photosensibilisierden wirkung des teers. *Schweiz. Mediz. Wochen.*, 4, 82, 1939.

Bury, N.R., M. Grosell, A.K. Grover, and C.M. Wood, ATP-dependent silver transport across the basolateral membrane of rainbow trout gills. *Toxicol. Appl. Pharmacol.*, 159, 1–8, 1999.

Bustnes, J.O., S.A. Hanssen, I. Folstad, K.E. Erikstad, D. Hasselquist, and J.U. Skaare, Immune function and organochlorine pollutants in Arctic breeding glaucous gulls. *Arch. Environ. Contam. Toxicol.*, 47, 530–541, 2004.

Butcher, S.S., R.J. Charlson, G.H. Orians, and G.V. Wolfe, Eds., *Global Biogeochemical Cycles*, Academic Press, London, United Kingdom 1992.

Butler, D., Global observation projects gets green light. *Nature*, 433, 789, 2005.

Butler, R.A. and G. Roesijadi, Disruption of metallothionein expression with antisense oligonucleotides abolishes protection against cadmium cytotoxicity in molluscan hemocytes. *Toxicol. Sci.*, 59, 101–107, 2001.

Cabana, G. and J.B. Rasmussen, Modelling food chain structure and contaminant bioaccumulation using stable nitrogen isotopes. *Nature*, 372, 255–257, 1994.

Cabana, G., A. Tremblay, J. Kalff, and J.B. Ramussen, Pelagic food chain structure in Ontario lakes: A determinant of mercury in lake trout (*Salvelinus namaycush*). *Can. J. Fish. Aquat. Sci.*, 51, 381–389, 1994.

Cade, T.J., Exposure of California condors to lead from spent ammunition. *J. Wildl. Manage.* 71, 2125–2133, 2007.

Cade, T.J., J.L. Liner, C.M. White, D.G. Rosen, and L.G. Swartz, DDE residues and eggshell changes in Alaskan falcons and hawks. *Science*, 172, 955–957, 1971.

Cairns, J. Jr., Thermal pollution: A cause for concern, *J. Water Pollut. Control Fed.*, 43, 55–66, 1971.

Cairns, J. Jr., Heated waste-water effects on aquatic ecosystems, in *Thermal Ecology II*, Esch, G.W. and R.W. McFarlane, Eds., National Technical Information Center, Springfield, VA, 1976.

Cairns, J. Jr., The myth of the most sensitive species. *Bioscience*, 36, 670–672, 1986.

Cairns, J. Jr., Will there ever be a field of landscape toxicology? *Environ. Toxicol. Chem.*, 12, 609–610, 1993.

Cairns, J. Jr., Ecological tipping points: A major challenge for experimental sciences. *Asian J. Exp. Sci.*, 18, 1–16, 2004.

Cairns, Jr., J. and D.I. Mount, Aquatic toxicology. *Environ. Sci. Technol.*, 24, 154–161, 1990.

Cairns, Jr., J. and J.R. Pratt, Biotic impoverishment: Effects of anthropogenic stress, in *The Earth in Transition: Patterns and Processes of Biotic Impoverishment*, G.M. Woodwell, Ed., Cambridge University Press, Cambridge, 1990.

Cairns, J. Jr., J.R. Pratt, B.R. Niederlehner, and P.V. McCormick, A simple cost-effective multispecies toxicity test using organisms with a cosmopolitan distribution. *Environ. Monit. Assess.*, 6, 207–220, 1986.

Calabrese, E.J., Hormetic dose-response relationships in immunology: Occurrence, quantitative features of the dose-response, mechanistic foundations and clinical implications. *Crit. Rev. Toxicol.*, 35, 89–306, 2005a.

Calabrese, E.J., Cancer biology and hormesis: Human tumor cell lines commonly display hormetic (biphasic) dose responses. *Crit. Rev. Toxicol.*, 35, 463–582, 2005b.

Calabrese, E.J., Historical blunders: How toxicology got the dose-response relationship half right. *Cell. Mol. Biol.*, 51, 643–654, 2005c.

Calabrese, E.J., Factors affecting the historical rejection of hormesis as a fundamental dose response model in toxicology and the broader biomedical sciences. *Toxicol. Appl. Pharmacol.*, 206, 365–366, 2005d.

Calabrese, E.J., Converging concepts: Adaptive response, preconditioning, and the Yerkes-Dodson law are manifestations of hormesis. *Ageing Res. Rev.*, 7, 8–20, 2008a.

Calabrese, E.J., Hormesis: Why it is important to toxicology and toxicologists. *Environ. Toxicol. Chem.*, 27, 1451–1474, 2008b.

Calabrese, E.J., Hormesis is central to toxicology, pharmacology and risk assessment. *Hum. Exp. Toxicol.*, 29, 249–261, 2010.

Calabrese, E.J., Toxicology rewrites its history and rethinks its future: Giving equal focus to both harmful and beneficial effects. *Environ. Toxicol. Chem.*, 30, 2658–2673, 2011.

Calabrese, E.J., Origin of the linearity no threshold (LNT) dose response concept. *Arch. Toxicol.*, 87, 1621–1633, 2013a.

Calabrese, E.J., Hormetic mechanisms. *Crit. Rev. Toxicol.*, 43, 580–606, 2013b.

Calabrese, E.J., K.A. Bachmann, A.J. Bailer, P.M. Bolger, J. Borak, L. Cai, N. Cedergreen et al, Biological stress response terminology: Integrating the concepts of adaptive response and preconditioning stress within a hormetic dose-response framework. *Toxicol. Appl. Pharmacol.*, 222, 122–128, 2007.

Calabrese, E.J. and L.A. Baldwin, The frequency of U-shaped dose-responses in the toxicological literature. *Toxicol. Sci.*, 62, 330–338, 2001.

Calabrese, E.J. and L.A. Baldwin, The hormetic dose response model is more common than the threshold model in toxicology. *Toxicol. Sci.*, 71, 246–250, 2003.

Calabrese, E.J. and R. Blain, The occurrence of hormetic dose responses in the toxicological literature, the hormesis database: An overview. *Toxicol. Appl. Pharmacol.*, 202, 289–301, 2005.

Calabrese, E.J., and R. Blain, Hormesis and plant biology. *Environ. Pollut.* 157, 42–48, 2009.

Calabrese, E.J., and R. Blain, The hormesis database: The occurrence of hormetic dose responses in the toxicological literature. *Regul. Toxicol. Pharmacol.*, 61, 73–81, 2011.

Calabrese, E.J., I. Iavicoli, and V. Calabrese, Hormesis: Its impact on medicine and health. *Hum. Exp. Toxicol.*, 32, 120–152, 2013.

Calabrese, E.J., M.E. McCarthy, and E. Kenyon, The occurrence of chemically induced hormesis. *Health Phys.*, 52, 531–541, 1987.

Calabrese, E.J., J.W. Staudenmayer, E.J. Stanek, and G.R. Hoffmann, Hormesis outperforms threshold model in NCI anti-tumor drug screening data. *Toxicol. Sci.*, 94, 368–378, 2006.

Calero, S., I. Fomsgaard, M. Lacayo, V. Martinez, and R. Rugama, Toxaphene and other organochlorine pesticides in fish and sediment from Lake Xolotlan, Nicaragua. *Int. J. Environ. Anal. Chem.*, 53, 297–305, 1993.

Calow, P., R.M. Sibly, and V.E. Forbes, Risk assessment on the basis of simplified population dynamics' scenarios. *Environ. Toxicol. Chem.*, 16, 1983–1989, 1997.

Camargo, J.A., Contribution of Spanish-American silver mines (1570–1820) to the present high mercury concentrations in the global environment: A review. *Chemosphere*, 48, 51–57, 2002.

Camill, R., C.M. Reddy, D.R. Yoerger, B.A.S. Van Mooy, M.V. Jakuba, J.C. Kinsey, C.P. McIntyre, S.P. Sylva, and J.V. Maloney, Tracking hydrocarbon plume transport and biodegradation at *Deepwater Horizon*. *Science*, 330, 201–204, 2010.

Camner, P., T.W. Clarkson, and G.F. Nordberg, Route of exposure, dose and metabolism of metals, in *Handbook on the Toxicology of Metals*, Friberg, L., G.F. Nordberg, and V.B. Vouk, Eds., Elsevier/North-Holland Biomedical Press, Amsterdam, The Netherlands, 1979.

Campbell, F.L., Relative susceptibility to arsenic in successive instars of the silkworm. *J. Gen. Physiol.*, 9, 727–733, 1926.

Campbell, P.G.C., Interactions between trace metals and organisms: Critique of the free-ion activity model, in *Metal Speciation and Bioavailability in Aquatic Systems*, Tessier, A. and D.Turner, Eds., Wiley, Chichester, United Kingdom, 1995.

Campbell, P.G.C., P.M. Chapman, and B.A. Hale, Risk assessment of metals in the environment. *Issues Environ. Sci. Technol.*, 21, 102–131, 2006.

Campbell, W.B, J.M. Emlen, and W.K. Hershberger, Thermally induced chronic developmental stress in coho salmon: Integrating measures of mortality, early growth, and developmental instability. *Oikos*, 81, 398–410, 1998.

Campbell, L.M., A.T. Fisk, X. Wang, G. Köck, and D.C.G. Muir, Evidence for biomagnification of rubidium in freshwater and marine food webs. *Can. J. Fish. Aquat. Sci.*, 62, 1161–1167, 2005.

Campbell, P.G.C. and L. Hare, Metal detoxification in freshwater animals. Roles of metallothioneins, in *Metallothioneins and Related Chelators*, Sigel, A., H.Sigel, and R.K.O.Sigel, Eds., Royal Society of Chemistry, Cambridge, United Kingdom, 2009.

Campbell, P.G.C., L.D. Kraemer, A. Giguère, L. Hare, and A. Hontela, Subcellular distribution of cadmium and nickel in chronically exposed wild fish: Inferences regarding metal detoxification strategies and implications for setting water quality guidelines for dissolved metals. *Hum. Ecol. Risk Assess.*, 14, 290–316, 2008.

Campbell, P.G.C., A.G. Lewis, P.M. Chapman, A.A. Crowder, W.K. Fletcher, B. Imber, S.N. Luoma, P.M. Stokes, and M. Winfrey, *Biologically Available Metals in Sediments*, NRCC No. 27694, NRCC/CNRC Publications, Ottawa, Canada, 1988.

Campbell, P.G.C. and A. Tessier, Ecotoxicology of metals in the aquatic environment: Geochemical aspects, in *Ecotoxicology: A Hierarchical Treatment*, Newman, M.C. and C.H. Jagoe, Eds., CRC Press, Boca Raton, FL, 1996.

Campfens, J. and D. Mackay, Fugacity-based model of PCB bioaccumulation in complex food webs. *Environ. Sci. Technol.*, 31, 577–583, 1997.

Campoy, C., M. Jimenez, M.F. Olea-Serrano, M. Moreno-Frias, F. Cañabate, N. Olea, R. Bayés, and J.A. Molina-Font, Analysis of organochlorine pesticides in human milk: Preliminary results, *Early Hum. Dev.* 65 Suppl., S183–190, 2001.

Cañas, J.E. and T.A. Anderson, Organochlorine contaminants in eggs: The influence of contaminated nest material. *Chemosphere*, 47, 585–589, 2002.

Cao, L., and K. Caldeira, Atmospheric CO_2 stabilization and ocean acidification. *Geophys. Res. Lett.*, 35, L19609, 2008.

Carballa, M., F. Omil, J.M. Lema, M. Llompart, C. García-Jares, I. Rodríguez, M. Gómez, and T. Ternes, Behavior of pharmaceuticals, cosmetics and hormones in a sewage treatment plant. *Water Res.*, 38, 2918–2926, 2004.

Cardinale, B.J., J.E. Duffy, A. Gonzalez, D.U. Hooper, C. Perrings, P. Venail, A. Narwani et al, Biodiversity loss and its impact on humanity. *Nature*, 486, 59–67, 2012.

Carey, C., W.R. Heyer, J. Wilkinson, R.A. Alford, J.W. Arntzen, T. Halliday, L. Hungerford et al, Amphibian declines and environmental change: Use of remote-sensing data to identify environmental correlates. *Conserv. Biol.*, 15, 903–913, 2001.

Carlson, A.R., G.L. Phipps, V.R. Mattson, P.A. Kosian, and A.M. Cotter, The role of acid-volatile sulfide in determining cadmium bioavailability and toxicity in freshwater sediments. *Environ. Toxicol. Chem.*, 10, 1309–1319, 1991.

Carlton, W.H., L.R. Bauer, A.G. Evans, L.A. Geary, C.E. Murphy Jr., J.E. Pinder, and R.N. Strom, Cesium in the Savannah River Site Environment. Westinghouse Savannah River Co., Savannah River Site, Aiken, SC, 1992.

Carney, S.A., R.E. Peterson, and W. Heideman, 2,3,7,8-Tetrachlorodibenzo-p-dioxin activation of the aryl hydrocarbon receptor/aryl hydrocarbon receptor nuclear translocator pathway causes developmental toxicity through a CYP1A-independent mechanism in zebrafish. *Mol. Pharmacol.*, 66, 512–521, 2004.

Carrol, J.J,. S.J. Ellis, and W.S. Oliver, Influences of hardness constituents on the acute toxicity of cadmium to brook trout (*Salvelinus fontinalis*). *Bull. Environ. Contam. Toxicol.*, 22, 575–581, 1979.

Carroll, L., *Through the Looking-Glass*, in *The Best of Lewis Carroll*, Castle, a division of Book Sales, Inc., Secaucus, NJ, 1872.

Carson, R., *Silent Spring*, Houghton-Mifflin Co., Boston, MA, 1962.

Carter, W.H., C. Gennings, J.G. Staniwallis, E.D. Campbell, and K.L. White, A statistical approach to the construction and analysis of isobolograms. *J. Am. Coll. Toxicol.*, 7, 963–973, 1988.

Carvalho, F.P., J.P. Villeneuve, C. Cattini, I. Tolosa, S. Montenegro-Guillen, M. Lacayo, and A. Cruz, Ecological risk assessment of pesticide residues in coastal lagoons of Nicaragua. *J. Environ. Monit.*, 4, 778–787, 2002.

Carvolho, F.P., F. Gonzalez Faria, J.P. Villeneuve, C. Cattini, M. Hernandez-Garza, L.D. Mee, and S.W. Fowler, Distribution, fate and effects of pesticide residues in tropical coastal lagoons of north western Mexico, *Environ. Technol.*, 23, 1257–1270, 2002.

Casida, J.E. and G.B. Quistad, Golden age of insecticide research: past, present, or future? *Ann. Rev. Entom.*, 43, 1–16, 1998.

Casti, J.L., *Paradigms Lost: Tackling the Unanswered Mysteries of Modern Science*, Avon Books, New York, 1989.

Castilho, J.A.A., N. Fenzl, S.M. Guillen, and F.S. Nascimento, Organochlorine and organophosphorus pesticide residues in the Atoya river basin, Chinandega, Nicaragua. *Environ. Pollut.*, 110, 523–533, 2000.

Castillo, L.E., E. De La Cruz, and C. Ruepert, Ecotoxicology and pesticides in tropical aquatic ecosystems of Central America. *Environ. Toxicol. Chem.*, 16, 41–51, 1997.

Castillo, L.E., E. Martinez, C. Ruepert, C. Savage, M. Gilek, M. Pinnock, and E. Solis, Water quality and macroinvertebrate community response following pesticide applications in a banana plantation, Limon, Costa Rica. *Sci. Total Environ.*, 367, 418–432, 2006.

Castillo, L.E. and C. Ruepert, Pesticide impacts of banana and pineapple plantations and adjacent conservation area in Costa Rica. *SETAC Globe*, 6, 26–27, 2005.

Castillo, L.E., C. Ruepert, F. Ramírez, B. Van Wendel de Joode, V. Bravo, and E. De la Cruz. Plaguicidas y otros contaminantes. Décimo Octavo Informe Estado de La Nación en Desarrollo Humano Sostenible. San José, Costa Rica, 32p, 2012. Available at http://www.estadonacion.or.cr.

Castillo, L.E., C. Ruepert, and E. Solis, Pesticide residues in the aquatic environment of banana plantation areas in the North Atlantic zone of Costa Rica. *Environ. Toxicol. Chem.*, 19, 1942–1950, 2000.

Castro, L.F.C., D. Lima, A. Machado, C. Melo, Y. Hiromori, J. Nishikawa, T. Nakanishi, M.A. Reis-Henriques, and M.M. Santos, Imposex induction is mediated through the retinoid X receptor signaling pathway in the neogastropod *Nucella lapillus*. *Aquat. Toxicol.*, 85, 57–66, 2007.

Caswell, H., Demography meets ecotoxicology: Untangling the population level effects of toxic substances, in *Ecotoxicology: A Hierarchical Treatment*, Newman, M.C. and C.H. Jagoe, Eds., CRC/Lewis Publishers, Boca Raton, FL, 1996.

Caswell, H., *Matrix Population Models: Construction, Analysis, and Interpretation,* 2nd Edition, Sinauer Associates, Inc., Sunderland, MA, 2001.

Catallo, W.J., Ecotoxicology and wetland ecosystems: Current understanding and future needs. *Environ. Toxicol. Chem.*, 12, 2209–2224, 1993.

CDC (Center for Disease Control), *Pocket Guide to Chemical Hazards*, Center for Disease Control, Atlanta, GA, 1997.

Cengiz, E.I. and E. Ünlü, Histopathology of gills in mosquitofish, *Gambusia affinis*, after long-term exposure to sublethal concentrations of malathion, *J. Environ. Sci. Health*, B38, 581–589, 2003.

Centre for Science and Environment (CSE), Report on the contamination of endosulfan in the villagers of Kasargod, Kerala, India. *Down to Earth*, 9 *(19)*, 2001.

Cerejiera,M.J., P. Viana, S. Batista, T. Pereira, E. Silva, M.J. Valério, A. Silva, M. Ferreira, and A.M. Silva-Fernandes, Pesticides in Portuguese surface and ground water, *Water Res.*, 37, 1055–1063, 2003.

Chabicovsky, M., W. Klepal, and R. Dallinger, Mechanisms of cadmium toxicity in terrestrial pulmonates: Programmed cell death and metallothionein overload. *Environ. Toxicol. Chem.*, 23, 648–655, 2004.

Chadwick, J.W. and S.P. Canton, Coal mine drainage on a lotic ecosystem in northwest Colorado, U.S.A. *Hydrobiologia*, 107, 25–33, 1983.

Chamberlin, T.C., The method of multiple working hypotheses. *J. Geol.*, 5, 837–848, 1897.

Chan, K.M., K.M.Y. Leung, K.C. Cheung, M.H. Wong, and J.W.Qiu, Seasonal changes in imposex and tissue burden of butyltin compounds in *Thais clavigera* populations along the coastal area of Mirs Bay, China. *Mar. Pollut. Bull.*, 57, 645–651, 2008.

Chandler, G.T., T.L. Cary, A.C. Bejarano, J. Pender, and J.L. Ferry, Population consequences of fipronil and degradates to copepods at field concentrations: An integration of life cycle testing with Leslie matrix population modeling. *Environ. Sci. Technol.*, 38, 6407–6414, 2004.

Chao, T.C., S.M. Maxwell, and S.Y. Wong, An outbreak of aflatoxicosis and boric acid poisoning in Malaysia: A clinicopathological study. *J. Pathol.*, 164, 225–233, 2005.

Chapman, G.A., Sea urchin sperm cell test, in *Fundamentals of Aquatic Toxicology: Effects, Environmental Fate, and Risk Assessment,* 2nd Edition, Rand, G.M., Ed., Taylor & Francis, Washington, DC, 1995.

Chapman, P.M., Emerging substances: Emerging problems? *Environ. Toxicol. Chem.*, 25, 1445–1447, 2006.

Chapman, P.M., R.S. Caldwell, and P.F. Chapman, A warning: NOECs are inappropriate for regulatory use. *Environ. Toxicol. Chem.*, 15, 77–79, 1996.

Chase, J.M., Community assembly: When should history matter? *Oecologia*, 136, 489–498, 2003.

Chattopadhyay, J., Effect of toxic substances on a two-species competitive system. *Ecol. Modell.*, 84, 287–289, 1996.

Chaumot, A., S. Charles, P. Flammarion, and P. Auger, Do migratory or demographic disruptions rule the population impact of pollution in spatial networks? *Theor. Popul. Biol.*, 64, 473–480, 2003.

Chaves, A., D. Shea, and W.G. Cope, Environmental fate of chlorothalonil in a Costa Rican banana plantation. *Chemosphere*, 69, 1166–1174, 2007.

Chen, Z. and L.M. Mayer, Assessment of sedimentary Cu availability: A comparison of biomimetic and AVS approaches. *Environ. Sci. Technol.*, 33, 650–652, 1999.

Chen, D.G. and J.G. Pounds, A nonlinear isobologram model with Box-Cox transformation to both sides of chemical mixtures. *Environ. Health Perspect.*, 106, 1367–1371, 1998.

Cheng, S.Y. and J.C. Chen, Study on the oxyhemocyanin, deoxyhemocyanin, oxygen affinity and acid-base balance of *Marsupenaeus japonicus* following exposure to combined elevated nitrite and nitrate. *Aquat. Toxicol.*, 61, 181–193, 2002.

Cherian, M.G. and H.M. Chan, Biological functions of metallothionein: A review, in *Metallothionein III: Biological Roles and Medical Implications*, Suzuki, K.T., N. Imura and M. Kimura, Eds., Birkhäuser Verlag, Basel, Switzerland, 1993.

Chernyak, S.M., L.L. McConnell, and C.P. Rice, Fate of some chlorinated hydrocarbons in arctic and far eastern ecosystems in the Russian Federation. *Sci. Total Environ.*, 160/161, 75–85, 1995.

Cherry, D.S. and J. Cairns, Jr., Biological monitoring Part V: Preference and avoidance studies. *Water Res.*, 16, 263–301, 1982.

Chesman, B.S., S. O'Hara, G.R. Burt, and W.J. Langston, Hepatic metallothionein and total oxyradical scavenging capacity in Atlantic cod *Gadus morhua* caged in open ocean contaminant gradients. *Aquat. Toxicol.*, 84, 310–320, 2007.

Chew, R.D. and M.A. Hamilton, Toxicity curve estimation: Fitting a compartment model to median survival times. *Trans. Am. Fish. Soc.*, 114, 403–412, 1985.

Chilcutt, C.F. and B.E. Tabashnik, Contamination of refuges by *Bacillus thuringiensis* toxin genes from transgenic maize. *Proc. Natl. Acad. Sci. U.S.A.*, 101, 7526–7529, 2004.

Chiou, C.T., Partition coefficients of organic compounds in lipid-water systems and correlations with fish bioconcentration factors. *Environ. Sci. Technol.*, 19, 57–62, 1985.

Chitra, K.C., C. Latchoumycandane, and P.P. Mathur, Chronic effects of endosulfan on the testicular functions of rat. *Asian J. Androl.*, 1, 203–206, 1999.

Cho, E., A.J. Bailer, and J.T. Oris, The effect of methyl tert-butyl ether on the bioconcentration and photo-induced toxicity of fluoranthene in fathead minnow larvae (*Pimephales promelas*). *Environ. Sci. Technol.*, 37, 1306–1310, 2003.

Choi, J. and J.T. Oris, Anthracene photoinduced toxicity to PLHC-1 cell line (*Poeciliopsis lucida*) and the role of lipid peroxidation in toxicity. *Environ. Toxicol. Chem.*, 19, 2699–2706, 2000.

Choppin, G.R. and J. Rydberg, *Nuclear Chemistry: Theory and Applications*, Pergamon Press, Oxford, 1980.

Chowdhari, D.K., A. Nazir, and D.K. Saxena, Effect of three chlorinated pesticides on hrsw stress gene in transgenic, *Drosophila melanogaster, J. Biochem. Mol. Toxicol.*, 15, 173–186, 2001.

Christensen, E.R., Dose-response functions in aquatic toxicity testing and the Weibull model. *Water Res.*, 18, 213–221, 1984.

Christensen, E.R and N. Nyholm, Ecotoxicological assays with algae: Weibull dose-response curves. *Environ. Sci. Technol.*, 18, 713–718, 1984.

Chrousos, G.P., Stressors, stress, and neuroendocrine integration of the adaptive response. *Ann. N.Y. Acad. Sci.*, 851, 311–335, 1998.

Chuvieco, E. and A. Huete, *Fundamentals of Satellite Remote Sensing*, Taylor & Francis/CRC Press, Boca Raton, FL, 2009.

Clark, A.J., The historical aspects of quackery, Part 2. *Br. Med. J.*, 1927, 960, 1927.

Clark, A.J., *Handbook of Experimental Pharmacology*, Verlag Von Julius Springer, Berlin, Germany, 1937.

Clark, C. *Radium Girls: Women and Industrial Health Reform, 1910–1935*. The University of North Carolina, Chapel Hill, NC, 1997.

Clark, J.P., F.A.P.C. Gobas, and D. Mackay, Model of organic chemical uptake and clearance from fish from food and water. *Environ. Sci. Technol.*, 24, 1203–1213, 1990.

Clark, J.B. and J.C. Harshbarger, Epizootiology of neoplasms in bony fish from North America. *Sci. Total Environ.*, 94, 1–32, 1990.

Clark, K.E. and D. Mackay, Dietary uptake and biomagnification of four chlorinated hydrocarbons by guppies. *Environ. Toxicol. Chem.*, 10, 1205–1217, 1991.

Clarke, C.A., G.S. Mani, and G. Wynne, Evolution in reverse: Clean air and the peppered moth. *Biol. J. Linn. Soc.*, 26, 189–199, 1985.

Clayton, J.R. Jr., S.P. Pavlou, and N.F. Breitner, Polychlorinated biphenyls in coastal marine zooplankton: Bioaccumulation by equilibrium partitioning. *Environ. Sci. Technol.*, 11, 676–682, 1977.

Clemens, S., J.I. Schroeder, and T. Degenkolb, *Caenorhabditis elegans* expresses a functional phytochelatin synthase. *Eur. J. Biochem.*, 268, 3640–3643, 2001.

Clements, W.H., Metal tolerance and predator-prey interactions in benthic macroinvertebrate stream communities. *Ecol. Appl.*, 9, 1073–1084, 1999.

Clements, W.H., Small-scale experiments support causal relationships between metal contamination and macroinvertebrate community responses. *Ecol. Appl.*, 14, 954–967, 2004.

Clements, W.H., M.L. Brooks, D.R. Kashian, and R.E. Zuellig, Changes in dissolved organic material determine exposure of stream benthic communities to UV-B radiation and heavy metals: Implications for climate change. *Glob. Change Biol.*, 14, 2201–2214, 2008.

Clements, W.H., D.M. Carlisle, J.M. Lazorchak, and P.C. Johnson, Heavy metals structure benthic communities in Colorado mountain streams. *Ecol. Appl.*, 10, 626–638, 2000.

Clements, W.H., C.W. Hickey, and K.A. Kidd. How do aquatic communities respond to contaminants? It depends on the ecological context. *Environ. Toxicol. Chem.*, 31, 1932–1940, 2012.

Clements, W.H. and P.M. Kiffney, Assessing contaminant effects at higher levels of biological organization. *Environ. Toxicol. Chem.*, 13, 357–359, 1994.

Clements, W.H., N.K.M. Vieira, and S.E. Church, Quantifying restoration success and recovery in a metal-polluted stream: A 17-year assessment of physicochemical and biological responses. *J. Appl. Biol.*, 47, 899–910, 2010.

Clotfelter, E.D., A.M. Bell, and K.R. Levering, The role of animal behaviour in the study of endocrine-disrupting chemicals. *Anim. Behav.*, 68, 665–676, 2004.

Cockerham, L.G. and M.B. Cockerham, Environmental ionizing radiation, in *Basic Environmental Toxicology*, Cockerham, L.G. and B.S. Shane, Eds., CRC Press, Boca Raton, FL, 1994.

Cockerham, L.G. and B.S. Shane, *Basic Environmental Toxicology*, CRC Press, Boca Raton, FL, 1994.

Cockerham, L.G., T.L. Walden, Jr., C.E. Dallas, G.A. Mickley, Jr. and M.A. Landauer, Ionizing radiation, in *Principles and Methods of Toxicology,* 5th Edition, Hayes, A.W., Ed., CRC/Taylor & Francis Press, Boca Raton, FL, 2008.

Cohen, B.L., Perspective on occupational mortality risks. *Health Phys.*, 40, 703–724, 1981.

Cohen, B.L., A test of the linear-no threshold theory of radiation carcinogenesis. *Environ. Res.*, 53, 193–220, 1990.

Cohen, J., The earth is round (p<.05). *Am. Psychol.*, 49, 997–1003, 1994.

Cohen, B.L. and I.S. Lee, A catalog of risks. *Health Phys.*, 36, 707–722, 1979.

Colborn, T., D. Dumanoski, and J. Peterson Myers, *Our Stolen Future*, Penguin Books USA, New York, 1996.

Colborn, T., Clues from wildlife to create an assay for thyroid system disruption. *Environ. Health Perspect.*, 110, 363–367, 2002.

Comins, H.N., The development of insecticide resistance in the presence of migration. *J. Theor. Biol.*, 64, 177–197, 1977.

Connell, D.W., *Bioaccumulation of Xenobiotic Compounds*, CRC Press, Boca Raton, FL, 1990.

Connell, D.W. and D.W. Hawker, Use of polynomial expressions to describe the bioconcentration of hydrophobic chemicals by fish. *Ecotoxicol. Environ. Saf.*, 16, 242–257, 1988.

Connolly, J.P. and C.J. Pedersen, A thermodynamic-based evaluation of organic chemical accumulation in aquatic organisms. *Environ. Sci. Technol.*, 22, 99–103, 1988.

Considine, D.M., *Van Nostrand's Scientific Encyclopedia,* 8th Edition, Van Nostrand Reinhold, New York, 1995.

Cook, R.R. and E.J. Calabrese, The importance of hormesis to public health. *Environ. Health Perspect.*, 114, 1631–1635, 2006.

Cook, L.M. and B.S. Grant, Frequency of insularia during the decline in melanics in the peppered moth *Biston betularia* in Britain. *Heredity*, 85, 580–585, 2000.

Cook, L.M., B.S. Grant, I.J. Saccheri, and J. Mallet, Selective bird predation on the peppered moth: The last experiment of Michael Majerus. *Biol. Lett.*, 8, 609–612, 2012.

Cooke, A.S., Shell thinning in avian eggs by environmental pollutants. *Environ. Pollut.*, 4, 85–152, 1973.

Cooke, A.S., Egg shell characteristics of gannets *Sula bassana*, shags *Phalacrocorax aristotelis* and great black-backed gulls *Larus marinus* exposed to DDE and other environmental pollutants. *Environ. Pollut.*, 19, 47–65, 1979.

Cooke, R.M., *Experts in Uncertainty: Opinion and Subjective Probability in Science*, Oxford University Press, Oxford, 1991.

Cooney, R.V. and A.A. Benson, Arsenic metabolism in *Homarus americanus. Chemosphere*, 9, 335–341, 1980.

Cooper, E.L., *Comparative Immunology*, Prentice-Hall, Englewood Cliffs, NJ, 1976.

Cooper, G.L., A.A. Bickford, B.R. Charlton, F.D. Galey, D.H. Willoughby, and M.A. Grobner, Copper poisoning in rabbits associated with acute intravascular hemolysis. *J. Vet. Diagn. Invest.*, 8, 394–396, 1996.

Cooper, S.P., K. Burau, A. Sweeney, T. Robinson, M.A. Smith, E Symanski, J.S. Colt, J. Laseter, and S. Har Zahn, Prenatal exposure to pesticides: A feasibility study among migrant and seasonal workers. *Am. J. Ind. Med.*, 40, 578–585, 2001.

Cordasco, E.M., S.L. Demeter, and C. Zenz, *Environmental Respiratory Diseases*, Van Nostrand Reinhold, New York, 1995.

Corn, M., Corporations viewed as environmental bad guys. *Chem. Eng. News*, May 3, 47–48, 1982.

Correa, M., Physiological effects of metal toxicity on the tropical freshwater shrimp *Macrobrachium carcinus* (Linneo, 1758). *Environ. Pollut.*, 45, 149–155, 1987.

Correa, M. and H.I. Garcia, Physiological responses of juvenile white mullet, *Mugil curema*, exposed to benzene. *Bull. Environ. Contam. Toxicol.*, 44, 428–434, 1990.

Correia, A.D., M.H. Costa, K.P. Ryan, and J.A. Nott, Studies on biomarkers of copper exposure and toxicity in the marine amphipod *Gammarus locusta* (Crustacea): I. copper-containing granules within the midgut gland. *J. Mar. Biol. Assoc. U.K.*, 82, 827–834, 2002.

Corsolini, S., N. Borghesi, A. Schiamone, and S. Focardi, Polybrominated diphenyl ethers, polychlorinated dibenzo-dioxins, -furans, and –biphenyls in three species of Antarctic penguins. *Environ. Sci. Pollut. Res.*, 14, 421–429, 2007.

Corsolini, S., A. Covaci, N. Ademollo, S. Focard, and P. Schepens, Occurrence of organochlorine pesticides (OCPs) and their enantiomeric signatures, and concentrations of polybrominated diphenyl ethers (PBDEs) in the Adelie penguin food web, Antarctica. *Environ. Pollut.*, 140, 371–382, 2006.

Cossa, D., E. Bourget, D. Pouliot, J. Piuze, and J.P. Chanut, Geographical and seasonal variations in the relationship between trace metal content and body weight in *Mytilus edulis*. *J. Mar. Biol. Assoc. U.K.*, 58, 7–14, 1980.

Couillard, Y., P.G.C. Campbell, and A. Tessier, Response of metallothionein concentrations in a freshwater bivalve (*Anodonta grandis*) along an environmental cadmium gradient. *Limnol. Oceanogr.*, 38, 299–313, 1993.

Courtney, L.A. and W.H. Clements, Sensitivity to acidic pH in benthic invertebrate assemblages with different histories of exposure to metals. *J. N. Am. Benthol. Soc.*, 19, 112–127, 2000.

Cowling, E.B., Acid precipitation in historical perspective. *Environ. Sci. Technol.*, 16, 110A–123A, 1982.

Cowling, E.B. and R.A. Linthurst, The acid precipitation phenomenon and its ecological consequences. *Bioscience*, 31, 649–654, 1981.

Cox, C., Nonyl phenol and related chemicals. *J. Pestic. Reform.*, 16, 15–20, 1996.

Craig, E.A., The heat shock response. *Crit. Rev. Biochem. Mol.*, 18, 239–280, 1985.

Craig, E.A., Chaperones: Helpers along the pathways to protein folding. *Science*, 260, 1902–1903, 1993.

Crane, M., Proposed development of sediment environmental quality standards under the European Water Framework Directive: A critique. *Toxicol. Lett.*,142, 195–206, 2003.

Crane, M., M.C. Newman, P.F. Chapman, and J. Fenlon, *Risk Assessment with Time to Event Models*, Lewis Publishers, Boca Raton, FL, 2002.

Crawford, M., Landmark ozone treaty negotiated. *Science*, 237, 1557–1558, 1987.

Crecelius, E.A., J.T. Hardy, C.I. Bobson, R.L. Schmidt, C.W. Apts, J.M. Gurtisen, and S.P. Joyce, Copper bioavailability to marine bivalves and shrimp: Relationship to cupric ion activity. *Mar. Environ. Res.*, 6, 13–26, 1982.

Creed, E.R., D.R. Lees, and J.G. Duckett, Biological method of estimating smoke and sulphur dioxide pollution. *Nature*, 244, 278–280, 1973.

Crist, R.H., K. Oberhoiser, D. Schwartz, J. Marzoff, D. Ryder, and D.R. Crist, Interactions of metals and protons with algae. *Environ. Sci. Technol.*, 22, 755–760, 1988.

Cristaldi, M.E., L. D'Arcangelo, L.A. Ieradi, D. Mascanzoni, T. Mattei, and I. Castelli van Axel, ^{137}Cs determination and mutagenicity tests in wild *Mus musculus domesticus* before and after the Chernobyl accident. *Environ. Pollut.*, 64, 1–9, 1990.

Crommentuijn, T., M. Polder, D. Sijm, J. De Bruijn, and E. Van de Plassche, Evaluation of the Dutch environmental risk limits for metals by application of the added risk approach. *Environ. Toxicol. Chem.*, 19, 1692–1701, 2000a.

Crommenttuijn, T., D. Sijm, J. de Bruijn, K. van Leeuwen, and E.V. de Plassche, Maximum permissible and negligible concentrations of some organic substances and pesticides. *J. Environ. Manage.*, 58, 297–312, 2000b.

Crowell, M. and C. McCay, The lethal dose of ultraviolet light for brook trout (*Salvelinus fontinalis*). *Science*, 72, 582–583, 1930.

Crutzen, P.J., The geology of mankind, *Nature*, 415, 23, 2002.

CSTEE (Scientific Committee on Toxicity, Ecotoxicity, and the Environment), Opinion of the CSTEE on the setting of environmental quality standards for the priority substances included in Annex X of Directive 2000/60/EC in accordance with Article 16 thereof. Adopted by the CSTEE during the 43rd Plenary Meeting of 28 May 2004.

Cucinotta, F.A.,W. Schimmerling, J.W. Wilson, L.E. Peterson, G.D. Badhwar, P.B. Saganti, and J.F. Dicello, Space radiation cancer risks and uncertainties for Mars missions. *Radiat. Res.*, 156, 682–688, 2001.

Culliton, B.J., Continuing confusion over Love Canal. *Science*, 209, 1002–1003, 1980.

Cunningham, P.A., S.L. Smith, J.P. Tippett, and A. Greene, A national fish consumption advisory data base: A step toward consistency. *Fisheries*, 19, 14–23, 1994.

Curtsinger, J.W., H.H. Fukui, D.R. Townsend, and J.W. Vaupel, Demography of genotypes: Failure of the limited life-span paradigm in *Drosophila melanogaster. Science*, 258, 461–463, 1992.

Cushing, C.E. and D.G. Watson, Cycling of zinc-65 in a simple food web, in *Proceedings of the Third National Symposium on Radioecology, Oak Ridge, Tennessee*, 1971. The Atomic Energy Commission, Oak Ridge National Laboratory, and the Ecological Society of America, Oak Ridge, TN, 1971.

Custodia, N., S.J. Won, A. Novillo, M. Wieland, C. Li, and I.P. Callard, *Caenorhabditis elegans* as an environmental monitor using DNA microarray analysis. *Ann. N. Y. Acad. Sci.*, 948, 32–42, 2001.

Cutshall, N., Turnover of zinc-65 in oysters. *Health Phys.*, 26, 327–331, 1974.

Daley, J.M., L.D. Corkum, and K.G. Drouillard, Aquatic to terrestrial transfer of sediment associated persistent organic pollutants is enhanced by bioamplification processes. *Environ. Toxicol. Chem.*, 30, 2167–2174, 2011.

Daley, J.M., T.A. Leadley, and K.G. Drouillard, Evidence for bioamplification of nine polychlorinated biphenyl (PCB) congeners in yellow perch (*Perca flavascens*) eggs during incubation. *Chemosphere*, 75, 1500–1505, 2009.

Dallinger, R., Y.J. Wang, B. Berger, E.A. Mackay, and J.H.R. Kagl, Spectroscopic characterization of metallothionein from the terrestrial snail, *Helix pomatia. Eur. J. Biochem.*, 268, 4126–4133, 2001.

Dalsgaard, J., M. St. John, G. Kattner, D. Müller-Navarra, and W. Hagen, Fatty acid trophic markers in the pelagic marine environment. *Adv. Mar. Biol.*, 46, 225–340, 2003.

Daly, G.L., Y.D. Lei, C. Teixeira, D.C.G. Muir, L.E. Castillo, and F. Wania, Accumulation of current-use pesticides in neotropical montane forests. *Environ. Sci. Technol.*, 41, 1118–1123, 2007a.

Daly, G.L., Y.D. Lei, C. Teixeira, D.C.G. Muir, C.G. Dereck, L.E. Castillo, L.M.M. Jantunen, and F. Wania, Organochlorine pesticides in the soils and atmosphere of Costa Rica. *Environ. Sci. Technol.*, 41, 1124–1130, 2007b.

Daniels, R.E. and J.D. Allan, Life table evaluation of chronic exposure to a pesticide. *Can. J. Fish. Aquat. Sci.*, 38, 485–494, 1981.

Daniels, P.J. and G.K. Andrews, Dynamics of the metal-dependent transcription factor complex in vivo at the mouse metallothionein-I promoter. *Nucleic Acids Res.*, 31, 6710–6721, 2003.

Danzmann, R.G., M.M. Feruson, F.W. Allendorf, and K.L. Knudsen, Heterozygosity and developmental rate in a strain of rainbow trout (*Salmo gairdneri*). *Evolution*, 40, 86–93, 1986.

Dauble, D.D. and T.M. Poston, Radionuclide concentrations in white sturgeons from the Hanford Reach of the Columbia River. *Trans. Am. Fish. Soc.*, 123, 565–573, 1994.

Dauer, L.T., P. Zanzonico, R.M. Tuttle, D.M. Quinn, and H.M. Strauss, The Japanese tsunami and resulting nuclear emergency at the Fukushima Daiichi power facility: Technical, radiologic, and response perspectives. *J. Nucl. Med.*, 52, 1423–1432, 2011.

Daughton, C.G., Environmental stewardship and drugs as pollutants. *Lancet*, 360, 1035–1036, 2002.

Daughton, C.G., Non-regulated water contaminants: Emerging research. *Environ. Impact Asses.*, 24, 711–732, 2004.

Davidson, B. and R.W. Bradshaw, Thermal pollution of water systems. *Environ. Sci. Technol.*, 1, 618–630, 1967.

Davidson, C., Declining downwind: Amphibian population declines in California and historical pesticide use. *Ecol. Appl.*, 14, 1892–1902, 2004.

Davies, P.H., J.P. Goettl, Jr., J.R. Sinley, and N.F. Smith, Acute and chronic toxicity of lead to rainbow trout *Salmo gairdneri*, in hard and soft water. *Water Res.*, 10, 199–206, 1976.

Davis, J.J. and R.F. Foster, Bioaccumulation of radioisotopes through aquatic food chains. *Ecology*, 39, 530–535, 1958.

Day, K. and N.K. Kaushik, An assessment of the chronic toxicity of the synthetic pyrethroid, Fenvalerate, to *Daphnia galeata mendotae*, using life tables. *Environ. Pollut.*, 44, 13–26, 1987.

Day, T.A. and P.J. Neale, Effects of UV-B radiation on terrestrial and aquatic primary producers. *Annu. Rev. Ecol. Syst.*, 33, 371–396, 2002.

Dean, M. and T. Annilo, Evolution of the ATP-binding cassette (ABC) transporter superfamily in vertebrates. *Annu. Rev. Genomics Hum. Genet.*, 6, 123–142, 2005.

Deb, D., Biodiversity and complexity of rice farm ecosystems: An empirical assessment. *Open Ecol. J.*, 2, 112–129, 2009.

Decourtye, A., J. Devillers, S. Cluzeau, M. Charreton, and M.H. Pham-Delégue, Effects of imidacloprid and deltamethrin on associative learning in honeybees under semi-field and laboratory conditions. *Ecotoxicol. Environ. Saf.*, 57, 410–419, 2004.

Deevey, E.S. Jr., Life tables for natural populations of animals. *Q. Rev. Biol.*, 22, 283–314, 1947.

Degitz, S.J., G.W. Holcombe, K.M. Flynn, P.A. Kosian, J.J. Korte, and J.E. Tietge, Progress towards development of an amphibian-based thyroid screening assay using *Xenopus laevis*. Organismal and thyroidal responses to the model compounds 6-propylthiouracil, methimazole, and thyroxine. *Toxicol. Sci.*, 87, 353–364, 2005.

De Guise, S., J. Maratea, and C. Perkins, Malathion immunotoxicity in the American lobster (*Homarus americanus*) upon experimental exposure. *Aquat. Toxicol.*, 66, 419–425, 2004.

Deines, A.M., V. Chen, and W.G. Landis, Modeling the risks of non-indigenous species introductions using a patch-dynamics approach incorporating contaminant effects as a disturbance. *Risk Anal.*, 6, 1637–51, 2005.

De Jager, M., F.J. Weissing, P.M.J. Herman, B.A. Nolet, and J. van de Koppel, Lévy walks evolve through interaction between movement and environmental complexity. *Science*, 332, 1551–1553, 2011.

De Jonge, M., F. Dreesen, J. De Paepe, R. Blust, and L. Bervoets, Do Acid Volatile Sulfides (AVS) influence the accumulation of sediment-bound metals to benthic invertebrates under natural field conditions? *Environ. Sci. Technol.*, 43, 4510–4516, 2009.

De Lacerda, L.D., W.C. Pfeiffer, A.T. Ott, and E.G. da Silveira, Mercury contamination in the Madeira River, Amazon: Hg inputs to the environment. *Biotropica.*, 21, 91–93, 1989.

De la Cruz, E. V. Bravo Durán, F. Ramírez and L.E.Castillo. Environmental hazards associated with pesticide import into Costa Rica, 1977–2009. *J Environ Biol,* 35, 35–42, 2014.

De la Cruz, V. Bravo-Durán, F. Ramírez, and L.E.Castillo, Environmental hazards associated with pesticide import into Costa Rica, 1977–2009. *J. Environ. Biol.*, 35, 43–55, 2014.

De la Cruz, E and L.E. Castillo, Pesticide use in Costa Rica and their impact on coastal ecosystems, in *Pesticide Residues in Coastal Tropical Ecosystems: Distribution, Fate and Effects*, M. Taylor, S. Kleine, F. Carvalho, D. Barcelo, and J. Everaarts, Eds., CRC/Taylor & Francis, Boca Raton, FL, 2003.

De la Cruz, C., C. Ruepert, C. Wesseling, P. Monge, F. Chaverri, L. Castillo, and V. Bravo, Los plaguicidas de uso agropecuario en Costa Rica: Impacto en la salud y el ambiente [The pesticides in agricultural use in Costa Rica: Impact on health and the environment]. Informe de consultoria para el Area de Servicio Agropecuario y Medio Ambiente de la Contraloria General de la Republica, Instituto Regional de Estudios en Sustancias Toxicas (IRET). Universidad Nacional, Heredia, Costa Rica, 2004.

De Mora, S.J., Tributyltin debate: Ocean transportation versus seafood harvesting, in *Tributyltin: Case Study of an Environmental Contaminant*, de Mora, S.J., Ed, Cambridge University Press, Cambridge, United Kingdom, 1996.

Dell'Omo, G., Ed., *Behavioural Ecotoxicology*, Wiley and Sons, New York, 2002.

Department of Premier and Cabinet. 2008. Scientific consensus statement on water quality in the Great Barrier Reef. Department of Premier and Cabinet, Brisbane, Australia. 6p. Available at http://www.reefplan.qld.gov.au/about/assets/scientific-consensus-statement-on-water-quality-in-the-gbr.pdf, accessed September 29, 2013.

Department of Premier and Cabinet. 2009. Reef water quality protection plan 2009. Department of Premier and Cabinet, Brisbane, Australia. 32p. Available at http://www.reefplan.qld.gov.au/about/assets/reefplan-2009.pdf, accessed September 29, 2013.

Derksen, J.G.M., G.B.J. Rijs, and R.H. Jongbloed, Diffuse pollution of surface water by pharmaceutical products. *Water Sci. Technol.*, 49, 213–221, 2004.

Derocher, A.E., N.J. Lunn, and I. Stirling, Polar bears in a warming climate. *Integr. Comp. Biol.*, 44, 163–176, 2004.

Derycke, S., F. Hendrickx, T. Backeljau, S. D'Hondt, L. Camphijn, M. Vinca, and T. Moens, Effects of sublethal abiotic stressors on population growth and genetic diversity of *Pellioditis marina* (Nematoda) from the Westerschelde estuary. *Aquat. Toxicol.*, 82, 110–119, 2007.

Devine, G.J. and M.J. Furlong, Insecticide use: Context and ecological consequences. *Agric. Human Values*, 24, 281–306, 2007.

DeWitt, J.C., D.S. Millsap, R.L. Yeager, SS. Heise, D.W. Sparks, and D.S. Henshel, External heart deformities in passerine birds exposed to environmental mixtures of polychlorinated biphenyls during development. *Environ. Toxicol. Chem.*, 25, 541–551, 2006.

Diamond, S.A., Photoactivated toxicity in aquatic environments, in *UV Effects in Aquatic Organisms and Ecosystems*, D. Hader and G. Jori, Eds., Wiley, New York, 2003.

Diamond, S.A., M.C. Newman, M. Mulvey, P.M. Dixon, and D. Martinson, Allozyme genotype and time to death of mosquitofish, *Gambusia affinis* (Baird and Girard), during acute exposure to inorganic mercury. *Environ. Toxicol. Chem.*, 8, 613–622, 1989.

Diamond, J.M., M.J. Parson, and D. Gruber, Rapid detection of sublethal toxicity using fish ventilatory behavior. *Environ. Toxicol. Chem.*, 9, 3–11, 1990.

Diamond, J.M. and V.B. Serveiss, Identifying sources of stress to native aquatic fauna using a watershed ecological risk assessment framework. *Environ. Sci. Technol.*, 35, 4711–4718, 2001.

Diaz, R.J., Overview of hypoxia around the world. *J. Environ. Qual.*, 30, 275–281, 2001.

Diaz, R.J. and R. Rosenberg, Marine benthic hypoxia: A review of its ecological effects and the behavioral responses in benthic macrofauna. *Oceanogr. Mar. Biol. Assoc. Annu. Rev.*, 33, 245–303, 1995.

Diaz, R.J. and R. Rosenberg, Spreading dead zones and consequences for marine ecosystems. *Science*, 321, 926–929, 2008.

Dicks, L., Bees, lies, and evidence-based policy. *Nature*, 494, 283, 2013.

Dickson, D., Details of 1957 British nuclear accident withheld to avoid endangering U.S. ties. *Science*, 239, 137, 1988.

Diepens, N., S. Pfennig, P. Van den Brink, J. Gunnarsson, C. Ruepert, and L.E.Castillo, Effect of pesticides used in banana and pineapple plantations on aquatic ecosystems in Costa Rica. *J. Environ. Biol.*, 35, 73–84, 2014.

Di Giulio, R.T., W.H. Benson, B.M. Sanders, and P.A. Van Veld, Biochemical mechanisms: Metabolism, adaptation, and toxicity, in *Fundamentals of Aquatic Toxicology: Effects, Environmental Fate, and Risk Assessment,* 2nd Edition, Rand, G.M., Ed., Taylor & Francis, Washington, DC, 1995.

Di Giulio, R.T., P.C. Washburn, R.J. Wenning, G.W. Winston, and C.S. Jewell, Biochemical responses in aquatic animals: A review of determinants of oxidative stress. *Environ. Toxicol. Chem.*, 8, 1103–1123, 1989.

Dillon, T.M. and M.P. Lynch, Physiological responses as determinants of stress in marine and estuarine organisms, in *Stress Effects on Natural Ecosystems*, Barrett, G.W. and R. Rosenberg, Eds., Wiley, Chichester, United Kingdom, 1981.

Di Toro, D.M., J.D. Mahony, D.J. Hansen, K.J. Scott, M.B. Hicks, S.M. Mayr, and M.S. Redmond, Toxicity of cadmium in sediments: The role of acid volatile sulfide. *Environ. Toxicol. Chem.*, 9, 1487–1502, 1990.

Di Toro, D.M., J.A. McGrath, D.J. Hansen, W.J. berry, P.R. Paquin, R. Mathew, K.B. Wu, and R.C. Santore, Predicting sediment metal toxicity using a sediment biotic ligand model: Methodology and initial application. partitioning. *Environ. Toxicol. Chem.*, 24, 2410–2427, 2005.

Di Toro, D.M., C.S. Zarba, D.J. Hansen, W.J. Berry, R.C. Swartz, C.E. Cowan, S.O. Pavlou, H.E. Allen, N.A. Thomas, and P.R. Paquin, Technical basis for establishing sediment quality criteria for nonionic organic chemicals using equilibrium partitioning. *Environ. Toxicol. Chem.*, 10, 1541–1583, 1991.

Dixon, W.J., The Up-and-Down method from small samples, *J. Am. Stat. Assoc.*, 60, 967–978, 1965.

Dixon, W.J., Staircase bioassay: The up-and-down method, *Neurosci. Biobehav. Rev.*, 15, 47–50, 1991.

Dixon, D.R. and K.R. Clarke, Sister chromatid exchange: A sensitive method for detecting damage caused by exposure to environmental mutagens in the chromosomes of adult *Mytilus edulis. Mar. Biol. Lett.*, 3, 163–172, 1982.

Dixon, W.J. and F.J. Massey, Jr., *Introduction to Statistical Analysis*. McGraw-Hill, New York, 1969.

Dixon, W.J. and A.M. Mood, A method for obtaining and analyzing sensitivity data, *J. Am. Stat. Assoc.*, 43, 109–126, 1948.

Dixon, P.M. and M.C. Newman, Analyzing toxicity data using statistical models of time-to-death: An introduction, in *Metal Ecotoxicology: Concepts and Applications*, Newman, M.C. and A.W. McIntosh, Eds., Lewis Publishers, Chelsea, MI, 1991.

Dixon, D.G. and J.B. Sprague, Acclimation to copper by rainbow trout (*Salmo gairdneri*): A modifying factor in toxicity. *Can. J. Fish. Aquat. Sci.*, 38, 880–888, 1981a.

Dixon, D.G. and J.B. Sprague, Acclimation-induced changes in toxicity of arsenic and cyanide in rainbow trout, *Salmo gairdneri* Richardson. *J. Fish Biol.*, 18, 579–589, 1981b.

Dmitriev, S.G., and V.M. Zakharov, Estimate of cytogenetic homeostasis in natural populations of some small murid rodents. *Russ. J. Devel. Bio.*, 32, 373–380, 2001.

Dodge, E.A. and T.L. Theis, Effect of chemical speciation on the uptake of copper by *Chironomus tentans. Environ. Sci. Technol.*, 13, 1287–1288, 1979.

Dohán, O., C. Portulano, C. Basquin, A. Reyna-Neyra, L.M. Amzel, and N. Carrasco, The Na^+/I^- symporter (NIS) mediates electroneutral active transport of the environmental pollutant perchlorate. *Proc. Natl. Acad. Sci. U.S.A.*, 104, 20250–20255, 2007.

Dohm, M.R., W.J. Mautz, J.A. Andrade, K.S. Gellert, L.J. Salas-Ferguson, N. Nicolaisen, and N. Fugie, Effects of ozone exposure on nonspecific phagocytic capacity of pulmonary macrophages from an amphibian, *Bufo marinus. Environ. Toxicol. Chem.*, 24, 205–210, 2005.

Dolphin, R., Lake County Mosquito Abatement District Gnat Research Program. Clear lake gnat (*Chaoborus astictopus*). *Proceedings of 27th Annual Conference of the California Mosquito Control Association*, 47–48, 1959.

Domouhtsidou, G.P. and V.K. Dimitriadis, Ultrastructural localization of heavy metals (Hg, Ag, Pb, and Cu) in gills and digestive gland of mussels, *Mytilus galloprovincialis* (L.). *Arch. Environ. Contam. Toxicol.*, 38, 472–478, 2000.

Donald, P.F., R.E. Green, and M.F. Heath, Agricultural intensification and the collapse of Europe's farmland bird populations. *Proc. R. Soc. Lond. B*, 268, 25–29, 2001.

Dondero, F., A. Dagnino, H. Jonsson, F. Capri, L. Gastaldi, and A. Viarengo, Assessing the occurrence of a stress syndrome in mussels (*Mytilus edulis*) using a combined biomarker/gene expression approach. *Aquat. Toxicol.*, 78, S13–S24, 2006.

Doniach, I. and J.C. Mottram, Sensitization of the skin of mice to light by carcinogenic agents. *Nature*, 140, 588, 1937.

Donkin, S.G. and D.B. Dusenbery, Using the *Caenorhabditis elegans* soil toxicity test to identify factors affecting toxicity of four metal ions in intact soil. *Water Air Soil Pollut.*, 86, 359–373, 1994.

Donnelly, K.C., C.S. Anderson, G.C. Barbee, and D.J. Manek, Soil toxicology, in *Basic Environmental Toxicology*, L.G. Cockerham, and B.S. Shane, Eds., CRC Press, Boca Raton, FL, 1994.

Dopp, E., C.M. Barker, D. Schiffmann, and C.L. Reinisch, Detection of micronuclei in hemocytes of Mya *arenaria* association with leukemia and induction with an alkylating agent. *Aquat. Toxicol.*, 34, 31–45, 1996.

Doust, J.L., M. Schmidt, and L.L. Doust, Biological assessment of aquatic pollution: A review, with emphasis on plants as biomonitors. *Biol. Rev.*, 69, 147–186, 1994.

Douwes, P., K. Mikkola, B. Petersen, and A. Vestergren, Melanism in *Biston betularia* from north-west Europe (Lepidoptera: Geometridae). *Ent. Scand.*, 7, 261–266, 1976.

Downs, T.D. and R.F. Frankowski, Influence of repair processes on dose-response models. *Drug Metab. Rev.*, 13, 839–852, 1982.

Doxon, D.R., A.M. Pruski, L.R.J. Dixon, and A.N. Jha, Marine invertebrate eco-genotoxicology: A methodological overview. *Mutagenesis*, 17, 495–507, 2002.

Driscoll, C.T., V. Blette, C. Yan, C.L. Schofield, R. Munson, and J. Holsapple, The role of dissolved organic carbon in the chemistry and bioavailability of mercury in remote Adirondack lakes. *Water Air Soil Pollut.*, 80, 499–508, 1995.

Drouillard, K.G., G. Paterson, J. Liu, and G.D. Haffner, Calibration of the gastrointestinal magnification model to predict maximum biomagnification potentials of polychlorinated biphenyls in a bird and fish. *Environ. Sci. Technol.*, 46, 10279–10286, 2012.

Drummond, R.A., G.F. Olson, and A.R. Batterman, Cough response and uptake of mercury by brook trout, *Salvelinus fontinalis*, exposed to mercuric compounds at different hydrogen-ion concentrations. *Trans. Am. Fish. Soc.*, 2, 244–249, 1974.

Drummond, R.A., C.L. Russom, D.L Geiger, and D.L. DeFoe, behavioral and morphological changes in fathead minnows, *Pimephales promelas*, as diagnostic endpoints for screening chemicals according to modes of action, in *Aquatic Toxicology. 9th Aquatic Toxicity Symposium.* American Society for Testing and Materials, STP 921, Philadelphia, PA, 1986.

Dubois, M. and L. Hare, Selenium assimilation and loss by an insect predator and its relationship to Se subcellulare partitioning in two prey types. *Environ. Pollut.*, 157, 772–777, 2009a.

Dubois, M. and L. Hare, Subcellular distribution of cadmium in two aquatic invertebrates: Change over time and relationship to Cd assimilation by a predatory insect. *Environ. Sci. Technol.*, 42, 5144–5149, 2009b.

Duce, R.A. and W.H. Zoller, Atmospheric trace metals at remote northern and southern hemisphere sites: Pollution or natural? *Science*, 187, 59–61, 1975.

Dudal, Y. and F. Gerard, Accounting for natural organic matter in aqueous chemical equilibrium models: A review of the theories and applications. *Earth-Sci. Rev.*, 66, 199–216, 2004.

Duffus, J.H., *Environmental Toxicology*, Wiley, New York, 1980.

Duffus, J.H., "Heavy metals"—a meaningless term? *Pure Appl. Chem.*, 74, 793–807, 2002.

Dumas, J. and L. Hare, The internal distribution of nickel and thallium in two freshwater invertebrates and its relevance to trophic transfer. *Environ. Sci. Technol.*, 42, 5144–5149, 2008.

Dunbar, C.E., Sunburn in fingerling trout. *Prog. Fish. Cult.*, 21, 74, 1959.

Dunson, W.A. and J. Travis, The role of abiotic factors in community organization. *Am. Nat.*, 138, 1067–1091, 1991.

Dworkin, M., Endogenous photosensitization in a carotenoidless mutant of *Rhodopseudomonas spheroides*. *J Gen. Physiol.*, 41, 1099–1112, 1958.

Dwyer, F.J., C.J. Schnitt, S.E. Finger, and P.M. Mehrle, Biochemical changes in longear sunfish, *Lepomis megalotis*, associated with lead, cadmium and zinc from mine tailings. *J. Fish Biol.*, 33, 307–317, 1988.

Dye, J.A., M. Venier, L. Zhu, C.R. Ward, and L.S. Birnbaum, Elevated PBDE levels in pet cats: Sentinels for humans? *Environ. Sci. Technol.*, 41, 6350–6356, 2007.

Eaton, D.C., A. Becchetti, H. Ma, and B.N. Ling, Renal sodium channels: Regulation and single channel properties. *Kidney Int.*, 48, 941–949, 1995.

Eberhardt, L.L., Relationship of cesium-137 half-life in humans to body weight. *Health Phys.*, 13, 88–90, 1967.

Echeverría-Sáenz, S., F. Mena, M. Pinnock, C. Ruepert, K. Solano, E. de la Cruz, B. Campos, J. Sánchez-Avila, S. Lacorte, and C. Barata, Environmental hazards of pesticides from pineapple crop production in the Río Jiménez watershed (Caribbean Coast, Costa Rica). *Sci. Total Environ.*, 440, 106–114, 2012.

Eckl, P.M. and D. Riegler, Levels of chromosomal damage in hepatocytes of wild rats living within the area of a waste disposal plant. *Sci. Total. Environ.*, 196, 41–149, 1997.

Ecobichon, D.J., Pesticide use in developing countries. *Toxicology*, 160, 27–33, 2001.

Edmonds, J.S. and K.A. Francesconi, Isolation and identification of arsenobetaine from the American lobster, *Homarus americanus. Chemosphere*, 10, 1041–1044, 1981.

Edwards, M., Lethal legacy: Pollution in the former U.S.S.R. *Nat. Geogr.*, 186, 70–99, 1994.

EFSA (European Food Safety Authority), Conclusion on the peer review of the pesticide risk assessment for bees for the active substance clothianidin. *EFSA J.*, 11, 3066, 58 p., 2013a, Available at www.efsa.europa.eu/efsajournal.

EFSA (European Food Safety Authority), Conclusion on the peer review of the pesticide risk assessment for bees for the active substance imidacloprid. *EFSA J.*, 11, 3068, 55 p., 2013b, Available at www.efsa.europa.eu/efsajournal.

EFSA (European Food Safety Authority), Conclusion on the peer review of the pesticide risk assessment for bees for the active substance thiamethoxam. *EFSA J.*, 11, 3067, 68 p., 2013c, Available at www.efsa.europa.eu/efsajournal.

Ehrlich, P.R. and A.H. Ehrlich, *Extinction, the Causes and Consequences of the Disappearance of Species*, Random House, New York, 1981.

Eichhorn, G.L., Active sites of biological macromolecules and their interaction with heavy metals, in *Ecological Toxicology: Effects of Heavy Metal and Organohalogen Compounds*, McIntyre, A.D. and C.F. Mills, Eds., Plenum Press, New York, 1975.

Eichhorn, G.L., J.J. Butzow, P. Clark and Y.A. Shin, Studies on metal ions and nucleic acids, in *Effects of Metals on Cells, Subcellular Elements, and Macromolecules*, Maniloff, J., J.R. Coleman, and M.W. Miller, Eds., Charles C. Thomas Publisher, Springfield, IL, 1970.

Eide, I. and H.G. Johnsen, Mixture design and multivariate analysis in mixture research. *Environ. Health Perspect.*, 106, 1373–1376, 1998.

Eisler, R., Radiation hazards to fish, wildlife, and invertebrates: A synoptic review, Biological Report 26, December 1994, U.S. Dept. of Interior, National Biological Service, Washington, DC, 1994.

Ejnik, J., A. Munoz, T. Gan, C.F. Shaw, and D.H. Petering, Interprotein metal ion exchange between cadmium-carbonic anhydrase and apo- or zinc-metallothionein. *J. Biol. Inorg. Chem.*, 4, 784–790, 1999.

El-Alawi, Y.S., J. McConkey, D.G. Dixon, and B.M. Greenberg, Measurement of short- and long-term toxicity of polycyclic aromatic hydrocarbons using luminescent bacteria. *Ecotoxicol. Environ. Saf.*, 51, 12–21, 2002.

Ellenton, J.A. and M.F. McPherson, Mutagenicity studies of herring gulls from different locations on the Great Lakes. I. Sister-chromatid exchange rates in herring gull embryos, *J. Toxicol. Environ. Health*, 12, 317–324, 1983.

Ellgehausen, H., J.A. Guth, and H.O. Essner, Factors determining the bioaccumulation potential of pesticides in the individual compartment of aquatic food chains. *Ecotoxicol. Environ. Saf.*, 4, 134–157, 1980.

Elliott, J.M., Tolerance and resistance to thermal stress in juvenile Atlantic salmon, *Salmo salar. Freshw. Biol.*, 25, 61–70, 1991.

Ellis, D., *Environments at Risk: Case Histories of Impact Assessment*, Springer-Verlag, Berlin, Germany, 1989.

Ellis, E.C., Anthropogenic transformations of the terrestrial biosphere. *Philos. Trans. R. Soc. A*, 369, 1010–1035, 2012.

Elsom, D., *Atmospheric Pollution: Causes, Effects and Control Policies*, Basil Blackwell, New York, 1987.

Ember, L.R., Environmental lead: Insidious health problem. *Chem. Eng. News*, June 23, 28–35, 1980.

Ember, L.R., EPA study backs cut in lead use in gas. *Chem. Eng. News*, April 9, 18, 1984.

Emlen, J.M. and K.R. Springman, Developing methods to assess and predict population level effects of environmental contaminants. *Intgr. Environ. Assess. Manag.*, 3, 157–165, 2007.

Endler, J.A., *Natural Selection in the Wild*, Princeton University Press, Princeton, NJ, 1986.

Environment Australia. *National Ocean Disposal Guidelines for Dredged Material*. Environment Australia, Canberra, Australia, 182 p., 2002.

EPA, *Water Quality Standards Handbook, December 1983*. National Technical Information Service, Springfield, VA, 66 p., 1983.

EPA, *Methods for Measuring the Acute Toxicity of Effluents to Freshwater and Marine Organisms, PB85-205383 March 1985*. National Technical Information Service, Springfield, VA, 216 p., 1985a.

EPA, *Ambient Water Quality Criteria for Cadmium: 1984, PB85-227031 January 1985*. Nation Technical Information Service, Springfield, VA, 127 p., 1985b.

EPA, *Ambient Water Quality Criteria for Copper: 1984, PB85-227023, January 1985*. National Technical Information Service, Springfield, VA, 142 p., 1985c.

EPA, *Ambient Water Quality Criteria for Lead: 1984, PB85-227437, January 1985*. National Technical Information Service, Springfield, VA, 81 p., 1985d.

EPA, *Guidelines for Deriving Numerical National Water Quality Criteria for The Protection of Aquatic Organisms and Their Uses, PB85-227049, January 1985*. National Technical Information Service, Springfield, VA, 98 p, 1985e.

EPA, *Ambient Water Quality Criteria for Zinc: 1987, PB87-153581, February 1987*. National Technical Information Service, Springfield, VA, 214 p, 1985f.

EPA, *The Enhanced Stream Water Quality Models QUAL2E and QUAL2E-UNCAS: Documentation and User Manual, EPA/600/3-87/007, May 1987*. National Technical Information Service, Springfield, VA, 189 p, 1987a.

EPA, Radiation protection guidance to federal agencies for occupational exposure. *Federal Register* 52, 2822, 1987b.

EPA, *Short-Term Methods for Estimating the Chronic Toxicity of Effluents and Receiving Waters to Marine and Estuarine Organisms, PB89-220503 May 1988*. National Technical Information Service, Springfield, VA, 415 p, 1988a.

EPA, *Ambient Water Quality Criteria for Aluminum: 1988, PB88-245998 August 1988*. National Technical Information Service, Springfield, VA, 47 p, 1988b.

EPA, *Short-Term Methods for Estimating the Chronic Toxicity of Effluents and Receiving Waters to Freshwater Organisms, EPA/600/4-89/001 March 1989*. National Technical Information Service, Springfield, VA, 249 p, 1989a.

EPA, *Short-term Methods for Estimating the Chronic Toxicity of Effluents and Surface Waters to Freshwater Organisms. Supplement, PB90-145764 September 1989*. National Technical Information Service, Springfield, VA, 262 p, 1989b.

EPA, *Ecological Assessment of Hazardous Waste Sites, EPA 600/3-89/013 March 1989*. National Technical Information Service, Springfield, VA, 260 p, 1989c.

EPA, *Risk Assessment Guidance for Superfund. Volume I. Human Health Evaluation Manual (Part A). Interim Final, EPA 540/1-89/002 December 1989*. National Technical Information Service, Springfield, VA, 290 p., 1989d.

EPA, *Summary Report on Issues in Ecological Risk Assessment, EPA/625/3-91/018 February 1991*. National Technical Information Service, Springfield, VA, 46 p, 1991a.

EPA, *MINTEQA2/PRODEFA2, A Geochemical Assessment Model for Environmental Systems: Version 3.0 User's Manual, EPA/600/3-91/021 March 1991*. National Technical Information Service, Springfield, VA, 106 p, 1991b.

EPA, *Ecological Risk Assessment Guidance for Superfund: Process for Designing and Conducting Ecological Risk Assessment, Review Draft, September* 26, National Technical Information Service, Springfield VA, 564 p, 1994.

EPA, *Proposed Guidelines for Ecological Risk Assessment, EPA/630/R-95/002B August 1996*. National Technical Information Service, Springfield, VA, 247 p, 1996.

EPA, *Biopesticides Registration Action Document:* Bacillus thuringiensis *Plant-Incorporated Protectants*, 2001. Available at http://www.epa.gov/pesticides/biopesticides/reds/brad_bt_pip2.htm.

EPA, *World Wildlife Fund Comments on Re-registration of Endosulfan Submitted to Public Information and Record Integrity Branch, Information Resources and Services Division (7502c)*, Office of Pesticide Program. Environmental Protection Agency, Washington, DC, 2001.

EPA, *An Examination of EPA Risk Assessment Principles and practices.* EPA/100/B/001), Environmental Protection Agency, Washington, DC, 192 p, 2004.

EPA, MINTEQA2, U.S. Environmental Protection Agency, Ecosystems Research Division, Center for Exposure Assessment Modeling (CEAM), Athens, GA, 2006.

Erickson, R.J. and J.M. McKim, A simple flow-limited model for exchange of organic chemicals at fish gills. *Environ. Toxicol. Chem.*, 9, 159–165, 1990.

Erickson, J.M., M. Rahire, and J.D. Rochaix, Herbicide resistance and cross-resistance: Changes at three distinct sites in the herbicide-binding protein. *Science*, 228, 204–207, 1985.

Ericson, G. and C. Larsson, DNA adducts in perch (*Perca fluviatilis*) living in coastal water polluted with bleached pulp mill effluents. *Ecotoxicol. Environ. Saf.*, 46, 167–173, 2000.

Esch, G.W. and T.C. Hazen, Stress and body condition in a population of largemouth bass: Implications for red-sore disease. *Trans. Am. Fish. Soc.*, 109, 532–536, 1980.

Escos, J.M., C.L. Alados, and J.M. Emlen, Developmental stability in the Pacific hake parasitized by the *Myxosporean* Kuoda spp. *Trans. Am. Fish. Soc.*, 124, 943–945, 1997.

Estes, J.A., J. Terborgh, J.S. Brashares, M.E. Power, J. Berger, W.J. Bond, S.R. Carpenter et al, Trophic downgrading of planet Earth. *Science*, 333, 301–306, 2011.

European Commission, Technical Guidance Document in support of Commission Directive 93/67/EEC on risk assessment for new notified substances, Commission Regulation (EC) No. 1488/94 on Risk Assessment for Existing Substances and Directive 98/8/EC of the European Parliament and of the Council concerning the placing of biocidal products on the market. Ispra (IT): EC, 2003.

European Commission, Proposal for a Directive of the European Parliament and of the Council on Environmental Quality Standards in the Field of Water Policy and Amending Directive 2000/60/EC. COM(2006) 397 final. Brussels, 2006.

Evans, D.H., The fish gill: Site of action and model for toxic effects of environmental pollutants. *Environ. Health Perspect.*, 71, 47–58, 1987.

Evans, H.J., Leukemia and radiation. *Nature*, 345, 16–17, 1990.

Evans, M.S., G.E. Noguchi, and C.P. Rice, The biomagnification of polychlorinated biphenyls, toxaphene, and DDT compounds in a Lake Michigan offshore food web. *Arch. Environ. Contam. Toxicol.*, 20, 87–93, 1991.

Evenson, R.E. and D. Gollin, Assessing the impact of the green revolution, 1960 to 2000. *Science*, 300, 758–762, 2003.

Evers, D.C., N.M. Burgess, L. Champoux, B. Hoskins, A. Majoe, W.M. Goodale, R.J. Taylor, R. Poppenga, and T. Daigle, Patterns and interpretation of mercury exposure in freshwater avian communities in northeastern North America. *Ecotoxicology*, 14, 193–221, 2005.

Evers, D.C., Y.J. Han, C.T. Driscoll, N.C. Kamman, M.W. Goodale, K.F. Lambert, T.M. Holsen, C.Y. Chen, T.A. Clair, and T. Butler, Biological mercury hotspots in the northeastern United States and southeastern Canada. *Bioscience*, 57, 29–43, 2007.

Evers, D.C., L.J. Savoy, C.R. DeSorbo, D.E. Yates, W. Hanson, K.M. Taylor, L.S. Siegel et al, Adverse effects from environmental mercury loads on breeding common loons. *Ecotoxicology*, 17, 69–81, 2008.

Evers, D.C., K.M. Taylor, A. Major, R.J. Taylor, R.H. Poppenga, and A.M. Scheuhammer, Common loon eggs as indicators of methylmercury availability in North America. *Ecotoxicology*, 12, 69–81, 2003.

Exeley, C., J.S. Chappell, and J.D. Birchall, A mechanism of acute aluminum toxicity in fish. *J. Theor. Biol.*, 151, 417–428, 1991.

Extoxnet (Extension Toxicology Network), A Pesticide Information Project of Cooperative Extension Offices of Cornell University, Michigan State University, Oregon State University and University of California at Davis, 1993. http://ace.ace.orst.edu/info/extoxnet/.

Extoxnet (Extension Toxicology Network), Pesticide information profile: Endosulfan, 1996. http://ace.ace.orst.edu/info/extoxnet/.

Fabacher, D.L. and H. Chambers, Rotenone tolerance in mosquitofish. *Environ. Pollut.*, 3, 139–141, 1972.

Fagerström, T., Body weight, metabolic rate, and trace substance turnover in animals. *Oecologia*, 29, 99–104, 1977.

Fagin, D., The learning curve, *Nature*, 490, 462–465, 2012.

Fairbairn, D.W., P.L. Olive, and K.L. O'Neill, The comet assay: A comprehensive review. *Mutat. Res.*, 339, 37–59, 1995.

Fairchild, J.F., T. Boyle, W.R. English, and C. Rabeni, Effects of sediment and contaminated sediment on structural and functional components of experimental stream ecosystems. *Water Air Soil Pollut.*, 36, 271–293, 1987.

Fan, W. and W.X. Wang, Extraction of spiked metals from contaminated coastal sediments: A comparison of different methods. *Environ. Toxicol. Chem.*, 22, 2659–2666, 2003.

Farrell, A.E. and T.J. Keating, Transboundary environmental assessment: Lessons from OTAG. *Environ. Sci. Technol.*, 36, 2537–2544, 2002.

Fatoki,O.S. and O.R. Awofolus, Levels of organochlorine pesticide residues in marine, surface, ground and drinking water from eastern Cape Province in South Africa. *J. Environ. Sci. Health B*, 39, 101–114, 2004.

Fenske, M., G. Maack, C. Schäfers, and H. Segner, An environmentally relevant concentration of estrogen induces arrest of male gonad development in zebrafish, *Danio rerio. Environ. Toxicol. Chem.*, 24, 1088–1098, 2005.

Fenster, L., K. Waller, G. Windham, T. Henneman, M. Anderson, P. Mendola, J.W. Overstreet, and S.H. Swan, Trihalomethane levels in home tap water and semen quality. *Epidemiology*, 14, 650–658, 2003.

Fernández, P. and J.O. Grimalt, On the global distribution of persistent organic pollutants. *Chimia*, 57, 514–521, 2003.

Ferson, S. and H.R. Akçakaya, *Modeling Fluctuations in Age-Structured Populations*, Exeter Software, Setauket, NY, 1991.

Fidler, F., G. Cumming, M. Burgman, and N. Thomason, Statistical reform in medicine, psychology and ecology. *J. Socio. Econo.*, 33, 615–630, 2004.

Fiering, S., E. Whitelaw, and D.I. Martin, To be or not to be active: The stochastic nature of enhancer action. *Bioessays*. 22, 381–387, 2000.

Findlay, G.M., Ultra-violet light and skin cancer. *Lancet*, 215, 1070–1073, 1928.

Finkelstein, M.E., K.A. Grasman, D.A. Croll, B.R. Tershy, B.S. Keitt, W.M. Jarman, and D.R. Smith, Contaminant-associated alteration of immune function in black-footed albatross (*Phoebastria nigripes*), a North Pacific predator. *Environ. Toxicol. Chem.*, 26, 1896–1903, 2007.

Finney, D.J., *Probit Analysis: A Statistical Treatment of the Sigmoid Response Curve*, Cambridge University Press, Cambridge, United Kingdom, 1947.

Finney, D.J., *Statistical Method in Biological Assay*, Charles Griffin & Co. Ltd., London, United Kingdom, 1971.

Finney, L.A. and T.V. O'Halloran, Transition metal speciation in the cell: Insights from the chemistry of metal ion receptors. *Science*, 300, 931–936, 2003.

Fiorucci, S., G. Rizzo, A. Donini, E. Distrutti, and L. Santucci, Targeting farnesoid X receptor for liver and metabolic disorders. *Trends Mol. Med.*, 13, 298–309, 2007.

Fisher, R.A., A.S. Corbet, and C.B. Williams, The relation between the number of species and the number of individuals in a random sample of an animal population. *J. Anim. Ecol.*, 12, 42–58, 1943.

Fisher, N.S., J.L. Teyssié, S. Krishnaswami, and M. Baskaran, Accumulation of Th, Pb, U, and Ra in marine phytoplankton and its geochemical significance. *Limnol. Oceanogr.*, 32, 131–142, 1987.

Fleeger, J.W., K.R. Carman, and R.M. Nisbet, Indirect effects of contaminants in aquatic ecosystems. *Sci. Total Environ.*, 317, 207–233, 2003.

Food and Agriculture Organization of the United Nations (FAO), FAOSTAT, 2012, faostat.fao.org.

Forbes, V.E. and P. Calow, Is the per capita rate of increase a good measure of population-level effects in ecotoxicology? *Environ. Toxicol. Chem.*, 18, 1544–1556, 1999.

Forbes, V.E. and P. Calow, Species sensitivity distributions revisited: A critical appraisal. *Hum. Ecol. Risk Assess.*, 8, 473–492, 2002.

Forbes, V.E., P. Calow, and R.M. Sibly, Are current species extrapolation models a good basis for ecological risk assessment? *Environ. Toxicol. Chem.*, 20, 442–447, 2001a.

Forbes, V.E., P. Calow, and R.M. Sibly, The extrapolation problem and how population modeling can help. *Environ. Toxicol. Chem.*, 27, 1987–1994, 2008.

Forbes, V.E. and T.L. Forbes, *Ecotoxicology in Theory and Practice*, Chapman & Hall, London, United Kingdom, 1994.

Forbes, V.E., M. Olsen, A. Palmqvist, and P. Calow, Environmentally sensitive life-cycle traits have low elasticity: Implications for theory and practice. *Ecol. Appl.*, 20, 1449–1455, 2010.

Forbes, V.E., R.M. Sibly and P. Calow, Determining toxicant impacts on density-limited populations: A critical review of theory, practice and results. *Ecol. Appl.*, 11, 1249–1257, 2001b.

Ford, J., The effects of chemical stress on aquatic species composition and community structure, in *Ecotoxicology: Problems and Approaches*, Levin, S.A., M.A. Harwell, J.R. Kelly, K.D. Kimball, Eds., Springer-Verlag, New York, 1989.

Foreman, R.T.T. and M. Godron, *Landscape Ecology*, Wiley, New York, 1986.

Formisano, V., S. Atreya, T. Encrenaz, N. Ignatiev, and M. Giuranna, Detection of methane in the atmosphere of Mars. *Science*, 306, 1758–1761, 2004.

Forni, A., Chromosomal effects of lead. A critical review, in *Reviews on Environmental Health, Volume. 3*, James, G.V., Ed., Freund Publishing House, Tel-Aviv, Israel, 1980.

Foster, R.B., Environmental legislation, in *Fundamentals of Aquatic Toxicology*, Rand, G.M. and S.R. Petrocelli, Eds., Hemisphere Publishing, Washington, DC, 1985.

Fournie, J.W. and W.E. Hawkins, Exocrine pancreatic carcinogenesis in the guppy *Poecilia reticulata. Dis. Aquat. Org.*, 52, 191–198, 2002.

Fournie, J.W. and W.K. Vogelbein, Exocrine pancreatic neoplasms in the mummichog (*Fundulus heteroclitus*) from a creosote contaminated site. *Toxicol. Pathol.*, 22, 237–247, 1994.

Fowler, B.A., C.E. Hildebrand, Y. Kojima, and M. Webb, Nomenclature of metallothionein, in *Metallothionein IV*, J.H.R. Kägi and Y. Kojima, Eds., Birkhäuser-Verlag, Basel, Switzerland, 1987.

Fox, G.A., Practical causal inference for ecoepidemiologists. *J. Toxicol. Environ. Health*, 33, 359–373, 1991.

Fox, G.A., S.W. Kennedy, R.J. Norstrom, and D.C. Wigfield, Porphyria in herring gulls: A biochemical response to chemical contamination of Great Lakes food chains. *Environ. Toxicol. Chem.*, 7, 831–839, 1988.

Fox, H.E., S.A. White, M.H.F. Kao, and R.D. Fernald, Stress and dominance in a social fish. *J. Neurosci.*, 17, 6463–64169, 1997.

Foy, C.D., R.L. Chaney, and M.C. White, The physiology of metal toxicity in plants. *Ann. Rev. Plant Physiol.*, 29, 511–566, 1978.

Fracácio, R., N.F. Verani, E.L.G. Espíndola, O. Rocha, O. Rigolin-Sá, and C.A. Andrade, Alterations on growth and gill morphology of *Danio rerio* (Pisces, Cyprinidae) exposed to the toxic sediments. *Braz. Arch. Biol. Technol.*, 46, 685–695, 2003.

Frampton, G.K., S. Jänsch, J.J. Scott-Fordmand, J. Römbke, and P.J. Van den Brink, Effects of pesticides on soil invertebrates in laboratory studies: A review and analysis using species sensitivity distributions. *Environ. Toxicol. Chem.*, 25, 2480–2489, 2006.

Frankel, E.G., *Ocean Environmental Management: A Primer on the Role of the Oceans and How to Maintain their Contributions to Life on Earth*, Prentice Hall PTR, Englewood Cliffs, NJ, 1995.

Franklin, M.R. and G.S. Yost, Biotransformation: A balance between bioactivation and detoxification, in *Principles of Toxicology: Environmental and Industrial Applications,* 2nd Edition, Williams, P.L., R.C. James, and S.M. Roberts, Eds., Wiley, New York, 2000.

Franson, J.C., L. Sileo, and N.J. Thomas, Causes of eagle deaths, in *Our Living Resources*, LaRoe, E.T., G.S. Farris, C.E. Puckett, P.D. Doran, and M.J. Mac, Eds., U.S. Department of the Interior: National Biological Service, Washington, DC, 1995.

Fransson-Steen, R., S. Flodstram, and L. Warngard, Insecticide, endosulfan and its two stereo isomers promote the growth of altered hepatic loci in rats. *Carcinogenesis*, 13, 2299–2303, 1992.

Fraüsto da Silva, J.J.R. and R.J.P. Williams, *The Biological Chemistry of the Elements: The Inorganic Chemistry of Life*, Oxford University Press, Oxford, 1991.

Free, A. and N.H. Barton, Do evolution and ecology need the Gaia hypothesis? *Trends Ecol. Evol.*, 22, 611–619, 2007.

Freedman, W., *Federal Statutes on Environmental Protection, Regulation in the Public Interest.* Quorum Books, New York, 1987.

French, N.R., Comparison of radioisotope assimilation by granivorous and herbivorous mammals, in *Radioecological Concentration Processes, Proceedings of an International Symposium Held in Stockholm, 1966*, Åberg, B. and F.P. Hungate, Eds., Pergamon Press, New York, 1967.

Fridovich, I., Superoxide radical: An endogenous toxicant. *Ann. Rev. Pharmacol.*, 23, 239–257, 1983.

Friesen, V.L., A.L. Smith, E. Gómez-Díaz, M. Bolton, R.W. Furness, J. González-Solís, and L.R. Monteiro, Sympatric speciation by allochrony in a seabird. *Proc. Natl. Acad. Sci. U.S.A.*, 104, 18589–18594, 2007.

Frodello, J.P., M. Roméo, and D. Viale, Distribution of mercury in the organs and tissues of five toothed-whale species of the Mediterranean. *Environ. Pollut.*, 108, 447–452, 2000.

Fromm, P.O., A review of some physiological and toxicological responses of freshwater fish to acid stress. *Environ. Biol. Fish.*, 5, 79–93, 1980.

Fromm, P.O. and J.R. Gillette, Effect of ambient ammonia on blood ammonia and nitrogen excretion of rainbow trout (*Salmo gairdneri*). *Comp. Biochem. Physiol.*, 26, 887–896, 1968.

Fry, B., Food web structure on Georges Bank from stable C, N, and S isotopic compositions. *Limnol. Oceanogr.*, 33, 1182–1190, 1988.

Fry, B., Stable isotope diagrams of freshwater food webs. *Ecology*, 72, 2293–2297, 1991.

Fry, D.M. and C.K. Toone, DDT-induced feminization of gull embryos. *Science*, 213, 922–924, 1981.

FSANZ, Food Standards Australia New Zealand, Website: www.foodstandards.gov.au, Canberra, Australia, 2002.

FSANZ, The 20th Australian Total Diet Survey. A total diet survey of pesticides residues and contaminants. Food Standards Australia New Zealand, Canberra, Australia, 62 p, 2003.

Fu, B.J., X.L. Zhuang, G.B. Jiang, J.B. Shi, and Y.H. Lü, Environmental problems and challenges in China. *Environ. Sci. Technol.*, 41, 7597–7602, 2007.

Furness, R.W., S.A. Lewis, and J.A. Mills, Mercury levels in the plumage of red-billed gulls *Larus novaehollandiae scopulinus* of known sex and age. *Environ. Pollut.*, 63, 33–39, 1990.

Furness, R.W., S.J. Muirhead, and M. Woodburn, Using bird feathers to measure mercury in the environment: Relationships between mercury content and moult. *Mar. Pollut. Bull.*, 17, 27–30, 1986.

Fürst, P., S. Hu, R. Hackett, and D. Hamer, Copper activates metallothionein gene transcription by altering the conformation of a specific DNA binding protein. *Cell*, 55, 705–717, 1988.

Gächter, R. and W. Geiger, MELIMEX, an experimental heavy metal pollution study: Behavior of heavy metals in an aquatic food chain. *Schweiz. Z. Hydrol.*, 41, 277–290, 1979.

Gad, S.C., Statistical analysis of behavioral toxicology data and studies. *Arch. Toxicol. Suppl.* 5, 256–266, 1982.

Gaddum, J.H., Bioassays and mathematics. *Pharacol. Rev.*, 5, 87–134, 1953.

Gagné, F., C. Blaise, J. Pellerin, E. Pelletier, and J. Starnd, Health status of *Mya arenaria* bivalves collected from contaminated sites in Canada (Saguenay Fjord) and Denmark (Odense Fjord) during their reproductive period. *Ecotoxicol. Environ. Saf.*, 64, 348–361, 2006.

Gallegos, A.F. and F.W. Whicker, Radiocesium retention by rainbow trout as affected by temperature and weight, in *Proceedings of the Third National Symposium on Radioecology*, Oak Ridge National Laboratories, Oak Ridge, TN, 1971.

Gallently, B.C., M.W. Blows, and D.J. Marshall, Genetic mechanisms of pollution resistance in a marine invertebrate. *Ecol. Appl.*, 17, 2290–2297, 2007.

Gallopin, G. and S. Öberg, Quality of life, in *An Agenda of Science for Environment and Development into the 21st Century*, Dooge, J.C.I., G.T. Goodman, J.W.M. la Rivière, J. Marton-Lefèvre, T. O'Riordan, and F. Praderie, Eds., Cambridge University Press, Cambridge, United Kingdom, 1992.

Galtsoff, P.S., *The American Oyster* Crassostrea virginica *Gmelin*, Fishery Bull. of the Fish and Wildlife Service Volume 64, U.S. Printing Office, Washington, DC, 1964.

Gao, J.M., J.Y. Hu, Y. Wan, W. An, L. An, and Z.G. Zheng, Butyltin compounds distribution in the coastal waters of Bohai Bay, People's Republic of China. *Bull. Environ. Contam. Toxicol.*, 72, 945–953, 2004.

Garcia, J.S., C.S. de Magalhaes, and M.A.Z. Arruda, Trends in metal-binding and metalloprotein analysis. *Talanta*, 69, 1–15, 2006.

Gardner, M.J., M.P. Snee, A.J. Hall, C.A. Powell, S. Downes, and J.D. Terrell, Results of case-control study of leukaemia and lymphoma among young people near Sellafield nuclear plant in West Cumbria. *Br. Med. J.*, 300, 423–434, 1990.

Garvey, J.S., Metallothonein: A potential biomonitor of exposure to environmental toxins, in *Biomarkers of Environmental Contamination*, McCarthy, J.F. and L.R. Shugart, Eds., Lewis Publishers, Boca Raton, FL, 1990.

Gauthier-Clerc, M., C. Le Bohec, N. Crini, M. Coeurdassier, P.M. Badot, P. Giraudoux, and Y. Le Maho, Mercury concentrations in king penguin feathers at Crozet Islands (subAntarctic): Temporal trend between 1966–1974 and 2000–2001. *Environ. Toxicol. Chem.*, 24, 125–128, 2005.

Gaylor, D.W., F.F. Kadlubar, and F.A. Beland, Application of biomarkers to risk assessment. *Environ. Health Perspect.*, 98, 139–141, 1992.

Geckler, J.R., W.B. Hornung, T.M. Neiheisel, Q.H. Pickering, E.L. Robinson, and C.E. Stephan, *Validity of Laboratory Tests for Predicting Copper Toxicity in Streams*, U.S. Environmental Protection Agency, Ecological Research Series EPA-600/3-76, Duluth, MN, 1976.

Geisel, T.S. and A.S. Geisel, *The Lorax*, Random House, New York, 1971.

Geist, J., I. Werner, K.J. Eder, and C.M. Leutenegger, Comparisons of tissue-specific transcription of stress response genes with whole animal endpoints of adverse effect in striped bass (*Morone saxatilis*) following treatment with copper and esfenvalerate. *Aquat. Toxicol.*, 85, 28–39, 2007.

Gendron, A.D., C.A. Bishop, R. Fortin, and A. Hontela, In vivo testing of the functional integrity of the corticosterone-producing axis in mudpuppy (Amphibia) exposed to chlorinated hydrocarbons in the wild. *Environ. Toxicol. Chem.*, 16, 1694–1706, 1997.

George, S.G., Enzymology and molecular biology of Phase II xenobiotic-conjugating enzymes in fish, in *Aquatic Toxicology: Molecular, Biochemical and Cellular Perspectives*, Malins, D.C. and G.K. Ostrander, Eds., CRC Press, Boca Raton, FL, 1994.

Geping Qu, Ed., *Reading Book on Environmental Protection*, Red Flag Press, Beijing, China, 1999.

Geret, F. and R.P. Cosson, Induction of specific isoforms of metallothionein in mussel tissues after exposure to cadmium or mercury. *Arch. Environ. Contam. Toxicol.*, 42, 36–42, 2002.

Gerhardt, A., M.K. Ingram, I.J. Kang, and S. Ulitzur, In situ on-line toxicity biomonitoring in water: Recent developments. *Environ. Toxicol. Chem.*, 25, 2223–2271, 2006.

Gerhart, D.Z., S.M. Anderson, and J. Richter, Toxicity bioassays with periphyton communities: Design of experimental streams. *Water Res.*, 11, 567–570, 1977.

Gevertz, A.K., A.J. Tucker, A.M. Bowling, C.E. Williamson, and J.T. Oris, Differential tolerance of native and non-native fish exposed to ultraviolet radiation and fluoranthene in Lake Tahoe (CA/NV). *Environ. Toxicol. Chem.*, 31, 1129–1135, 2012.

Geyer, H., D. Sheehan, D. Kotzias, D. Freitag, and F. Korte, Prediction of ecotoxicological behavior of chemicals: Relationship between physiochemical properties and bioaccumulation of organic compounds in the mussel. *Chemosphere*, 11, 1121–1134, 1982.

Ghude, S.D., R.J. Van der A, G. Beig, S. Fadnavis, and S.D. Polade, Satellite derived trends in NO_2 over the major global hotspot regions during the past decade and their inter-comparison. *Environ. Pollut.*, 157, 1873–1878, 2009.

Giattina, J.D. and R.R. Garton, A review of the preference-avoidance responses of fishes to aquatic contaminants. *Residue Rev.*, 87, 43–90, 1983.

Gibaldi, M., *Biopharmaceutics and Clinical Pharmacokinetics*, Lea and Febiger, Philadelphia, PA, 1991.

Gibaldi, M. and D. Perrier, *Pharmacokinetics,* 2nd Edition, Marcel Dekker, New York, 1982.

Gibbs, P.E. and G.W. Bryan, TBT-induced imposex in neogastropod snails: Masculinization to mass extinction, in *Tributyltin: Case Study of an Environmental Contaminant*, de Mora, S.J., Ed., Cambridge University Press, Cambridge, United Kingdom, 1996.

Gibbs, P.E., G.W. Bryan, P.L. Pascoe, and G.R. Burt, The use of the dog-whelk, *Nucella lapillus*, as an indicator of tributyltin (TBT) contamination. *J. Mar. Biol. Assoc. U.K.*, 67, 507–523, 1987.

Gibbs, P.E., P.L. Pascoe, and G.R. Burt, Sex change in the female dog-whelk, *Nucella lapillus*, induced by tributyltin from antifouling paints. *J. Mar. Biol. Assoc. U.K.*, 68, 715–731, 1988.

Gibbs, M.H., L.F. Wicker, and A.J. Stewart, A method for assessing sublethal effects of contaminants in soils to the earthworm, *Eisenia foetida*. *Environ. Toxicol. Chem.*, 15, 360–368, 1996.

Giesy, J.P., Cadmium inhibition of leaf decomposition in an aquatic microcosm. *Chemosphere*, 6, 467–475, 1978.

Giesy, J.P., S.R. Denzer, C.S. Duke, and G.W. Dickson, Phosphoadenylate concentrations and energy charge in two freshwater crustaceans: Responses to physical and chemical stressors. *Verh. Internat. Verein. Limnol.*, 21, 205–220, 1981.

Giesy, J.P. and R.A. Hoke, Freshwater sediment quality criteria: Toxicity bioassessment, in *Sediments: Chemistry and Toxicity of In-Place Pollutants*, Baudo, R., J.P. Giesy, and H. Muntau, Eds., Lewis Publishers, Chelsea, MI, 1990.

Giesy, J.P. and K. Kannan, Global distribution of perfluorooctane sulfonate in wildlife. *Environ. Sci. Technol.*, 35, 1339–1342, 2001.

Giesy, J.P. and K. Kannan, Perfluorochemical surfactants in the environment. *Environ. Sci. Technol.*, 35, 1339–1342, 2001.

Giesy, J.P. and K. Kannan, Global distribution of perfluorooctane sulfonate in wildlife. *Environ. Sci. Technol.*, 35, 1339–1342, 2002.

Giesy, J.P., J.L. Newsted, and J.T Oris, Photo-enhanced toxicity: Serendipity of a prepared mind and flexible program management. *Environ. Toxicol. Chem.*, 32, 969–971, 2013.

Gigerenzer, G., Mindless statistics. *J. Socio Econ.*, 33, 587–606, 2004.

Giguère, A., P.G.C. Campbell, L. Hare, and P. Couture, Sub-cellular partitioning of cadmium, copper, nickel and zinc in indigenous yellow perch (*Perca flavescens*) sampled along a polymetallic gradient. *Aquat. Toxicol.*, 77, 178–189, 2006.

Gilbert, F., A. Gonzalez, and I. Evans-Freke, Corridors maintain species richness in the fragmented landscapes of a microecosystem. *Proc. R. Soc. Lond. B Biol. Sci.*, 265, 577–582, 1998.

Gillespie, R.B., Allozyme frequency variation as an indicator of contaminant-induced impacts in aquatic populations, in *Techniques in Aquatic Toxicology*, Ostrander, G.K., Ed., CRC Press, Boca Raton, FL, 1996.

Gillespie, R.B. and S.I. Guttman, Effects of contaminants on the frequencies of allozymes in populations of the central stoneroller. *Environ. Toxicol. Chem.*, 8, 309–317, 1989.

Gillett, J.W., The role of terrestrial microcosms and mesocosms in ecotoxicological research, in *Ecotoxicology: Problems and Approaches*, Levin, S.A., M.A. Harwell, J.R. Kelly, and K.D. Kimball, Eds., Springer-Verlag, New York, 1989.

Glaser, J.A. and S.R. Matten, Sustainability of insect resistance management strategies for transgenic Bt corn. *Biotechnol. Adv.*, 22, 45–69, 2003.

Glass, N.R., D.E. Arnold, J.N. Galloway, G.R. Hendrey, J.J. Lee, W.W. McFee, S.A. Norton, C.F. Powers, D.L. Rambo, and C.L. Schofield, Effects of acidic precipitation. *Environ. Sci. Technol.*, 16, 163A–169A, 1982.

Glynn, P.W., L.S. Howard, E. Corcoran, and A. Freay, The occurrence and toxicity of herbicides in reef building corals. *Mar. Pollut. Bull.*, 15, 370–374 1984.

Gobas, F.A.P.C., A model for predicting the bioaccumulation of hydrophobic organic chemicals in aquatic food-webs: Application to Lake Ontario. *Ecol. Modell.*, 69, 1–17, 1993.

Gobas, F.A.P.C. and D. Mackay, Dynamics of hydrophobic organic chemical bioconcentration in fish. *Environ. Toxicol. Chem.*, 6, 495–504, 1987.

Gobas, F.A.P.C., D.C.G. Muir, and D. Mackay, Dynamics of dietary bioaccumulation and faecal elimination of hydrophobic organic chemicals in fish. *Chemosphere*, 17, 943–962, 1988.

Gobas, F.A.P.C., J. Wilcockson, R. W. Russell, and G. D. Haffner, Mechanism of biomagnification in fish under laboratory and field conditions. *Environ. Sci. Technol.*, 33, 133–141, 1999.

Goerke, H., K. Weber, H. Bornemann, S. Ramdohr and J. Plötz, Increasing levels and biomagnification of persistent organic pollutants (POPs) in Antarctic biota. *Mar. Pollut. Bull.*, 48, 295–302, 2004.

Goksøyr, A. and L. Förlin, The cytochrome P-450 system in fish, aquatic toxicology and environmental monitoring. *Aquat. Toxicol.*, 22, 287–312, 1992.

Goldberg, E.D., Synthetic organohalides in the sea. *Proc. R. Soc. Lond.* B, 189, 277–289, 1975.

Goldberg, E.D., The mussel watch concept. *Environ. Monit. Assess.*, 7, 91–103, 1986.

Goldsborough, L.G. and G.G.C. Robinson, Changes in periphytic algal community structure as a consequence of short herbicide exposures. *Hydrobiologia*, 139, 177–192, 1986.

Goldstein, R.S. and R.G. Schnellmann, Toxic responses of the kidney, in *Casarett & Doull's Toxicology: The Basic Science of Poisons,* 5th Edition, Klaassen, C.D., Ed., McGraw-Hill, New York, 1996.

Goldstone, J.V., Environmental sensing and response genes in cnidaria: The chemical defensome in the sea anemone *Nematostella vectensis. Cell Biol. Toxicol.*, 24, 483–502, 2008.

Goldstone, J.V., H.M. Goldstone, A.M. Morrison, A.Tarrant, S.E. Kern, B.R. Woodin, and J.J. Stegeman, Cytochrome P450 1 genes in early deuterostomes (tunicates and sea urchins) and vertebrates (chicken and frog): Origin and diversification of the CYP1 gene family. *Mol. Biol. Evol.*, 24, 2619–2631, 2007.

Goldstone, J.V., A. Hamdoun, B.J. Cole, M. Howard-Ashby, D.W. Nebert, M. Scally, M. Dean, D. Epel, M.E. Hahn, and J.J. Stegeman, The chemical defensome: Environmental sensing and response genes in the *Strongylocentrotus purpuratus* genome. *Dev. Biol.*, 300, 366–384, 2006.

Goldstone, J.V., A.G. McArthur, A. Kubota, J. Zanette, T. Parente, M.E. Jonsson, D.R. Nelson, and J.J. Stegeman, Identification and developmental expression of the full complement of cytochrome P450 genes in Zebrafish. *BMC genomics*, 11, 643, 2010.

Golley, F.B., *A History of the Ecosystem Concept in Ecology. More than the Sum of the Parts*, Yale University, New Haven, CT, 1993.

Gónara, B., L. Herrero, J.J. Ramos, J.R. Mateo, M.A. Fernández, J.F. García, and M.J. Gonzáez, Distribution of polybrominated diphenyl ethers in human umbilical cord serum, paternal serum, maternal serum, placentas, and breast milk from Madrid population, Spain. *Environ. Sci. Technol.*, 41, 6961–6968, 2007.

Gonzalez-Farias, F., X. Gisneros Estrada, Z.C. Fuentes Rui, G. Diaz Gonzalez, and V. Botello, Pesticides distribution in sediments of tropical coastal lagoon adjacent to an irrigation district in northwest Mexico. *Environ.Technol.*, 23, 1247–1256, 2002.

Gonzalez-Mendoza, D., A. Quiroz Moreno, and O. Zapata-Perez, Coordinated responses of phytochelatin synthase and metallothionein genes in black mangrove, *Avicennia germinans*, exposed to cadmium and copper. *Aquat. Toxicol.*, 83, 306–314, 2007.

Goodyear, C.P., A simple technique for detecting effects of toxicants or other stresses on a predator-prey interaction. *Trans. Am. Fish. Soc.*, 101, 367–370, 1972.

Gore, A., *Earth in the Balance. Ecology and the Human Spirit*, Penguin Books USA, New York, 1992.

Gore, A., *An Inconvenient Truth*, Rodale, New York, 2006.

Gorman, M., *Environmental Hazards. Marine Pollution*, ABC-CLIO, Santa Barbara, CA, 1993.

Gotelli, N.T., Metapopulation models: The rescue effect, the propagule rain, and the core-satellite hypothesis. *Am. Nat.*, 138, 768–776, 1991.

Gouin, T., F. Wania, C. Ruepert, and L. Castillo, Field testing passive air samplers for current-use pesticides in a tropical environment. *Environ. Sci. Technol.*, 42, 6625–6630, 2008.

Govt. of Kerala., The health hazards of aerial spraying of endosulfan in Kasargod District, Kerala. *Report of the Expert Committee, Government of Kerala*, Kerala, India, 2003.

Graedel, T.E. and B.R. Allenby, *Industrial Ecology*, Prentice-Hall, Upper Saddle River, NJ, 1995.

Graham, J.H., J.M. Emlen, and D.C. Freeman, Developmental stability and its applications in ecotoxicology. *Ecotoxicology*, 2, 175–184, 1993b.

Graham, J.H., D.C. Freeman, and J.M. Emlen, Developmental stability: A sensitive indicator of populations under stress, in *Environmental Toxicology and Risk Assessment, ASTM STP 1179*, Landis, W.G., J.S. Hughes and M.A. Lewis, Eds., American Society for Testing and Materials, Philadelphia, PA, 1993a.

Graham, J.H., S. Raz, H. Hel-Or, and E. Nevo, Fluctuating asymmetry: Methods, theory, and applications. *Symmetry*, 2, 466–540, 2010.

Graham, J.H., K.E. Roe, and T.B. West, Effects of lead and benzene on the developmental stability of *Drosophila melanogaster. Ecotoxicology*, 2, 185–195, 1993b.

Grandy, N.J., Role of the OECD in chemicals control and international harmonization of testing methods, in *Fundamentals of Aquatic Toxicology: Effects, Environmental Fate, and Risk Assessment,* 2nd Edition, Rand, G.M., Ed., Taylor & Francis, Washington, DC, 1995.

Graney, R.L. Jr., D.S. Cherry, and J. Cairns, Jr., The influence of substrate, pH, diet and temperature upon cadmium accumulation in the Asiatic clam (*Corbicula fluminea*) in laboratory artificial streams. *Water Res.*, 18, 833–842, 1984.

Graney, R.L., J.P. Giesy, and J.R. Clark, Field studies, in *Fundamentals of Aquatic Toxicology,* 2nd Edition, Rand, G.M., Ed., Taylor & Francis, Washington, DC, 1995.

Grant, A., Population consequences of chronic toxicity: Incorporating density dependence into the analysis of life table response experiments. *Ecol. Modell.*, 105, 325–335, 1998.

Grant, B.S., Allelic melanism in American and British peppered moths, *J. Hered.*, 95, 97–102, 2004.

Grant, B.S., A.D. Cook, D.F. Owen, and C.A. Clarke, Geographic and temporal variation in the incidence of melanism in peppered moth populations in America and Britain. *J. Hered.*, 89, 465–471, 1998.

Grant, B.S., D.F. Owen, and C.A. Clarke, Parallel rise and fall of melanic peppered moths in America and Britain. *J. Hered.*, 87, 351–357, 1996.

Grant, B.S. and L.L. Wiseman, Recent history of melanism in American peppered moths. *J. Hered.*, 93, 86–99, 2002.

Grant, P.B.C., M.B. Woudneh, and P.S. Ross, Pesticides in blood from spectacled caiman (*Caiman crocodilus*) downstream of banana plantations in Costa Rica. *Environ. Toxicol. Chem.*, 32, 2576–2583, 2013.

Gray, J.S., Pollution-induced changes in populations. *Philos. Trans. R. Soc. B*, 286, 545–561, 1979.

Gray, R.H., Fish behavior and environmental assessment. *Environ. Toxicol. Chem.*, 9, 53–67, 1990.

Green, A.S. and G.T. Chandler, Life-table evaluation of sediment-associated chlorpyrifos chronic toxicity to the benthic copepod, *Amphiascus tenuiremis. Arch. Environ. Contam. Toxicol.*, 31, 77–83, 1996.

Gregus, Z., Mechanisms of toxicity, in *Casarett & Doull's Toxicology: The Basic Science of Poisons,* 7th Edition, Klaassen, C.D., Ed., McGraw Hill, New York, 2008.

Griffin, J.L., Temperature tolerance of pathogenic and nonpathogenic free-living amoebas. *Science*, 178, 869–870, 1972.

Grill, E., E.L. Winnacker, and M.H. Zenk, Phytochelatins: The principal heavy-metal complexing peptides in higher plants. *Science*, 230, 674–676, 1985.

Grimm, V, R. Ashauer, V. Forbes, U. Hommen, T.G. Preuss, A. Schmidt, P.J. van den Brink, J. Wogram, and P. Thorbek, CREAM: A European project on mechanistic effect models for ecological risk assessment of chemicals. *Environ. Sci. Pollut. Res.*, 16, 614–617, 2009.

Groen, J.P., O. Tajima, V.J. Feron, and E.D. Schoen, Statistically designed experiments to screen chemical mixtures for possible interactions. *Environ. Health Perspect.*, 106, 1361–1365, 1998.

Grosch, D.S., *Biological Effects of Radiations*, Blaisdell Publishing Co., Waltham, MA, 1965.

Grünbaum, D., Why did you Lévy? *Science*, 332, 1514–1515, 2011.

Guengerich, F.P., Cytochrome P450 and chemical toxicology. *Chem. Res. Toxicol.*, 21, 70–83, 2008.

Guillette, L.J., J.W. Brock, A.A. Rooney, and A.R. Woodward, Serum concentrations of various environmental contaminants and their relationship to sex steroid concentrations and phallus size in juvenile American alligators. *Arch. Environ. Contam. Toxicol.*, 36, 447–455, 1999.

Guillette, L.J., D.B. Pickford, D.A. Crain, A.A. Rooney, and H.F. Percival, Reduction in penis size and plasma testosterone concentrations in juvenile alligators living in a contaminated environment. *Gen. Comp. Endocrinol.*, 101, 32–42, 1996.

Gunderson, M.P., D.S. Bermudez, T.A. Bryan, D.A. Crain, S. DeGala, T.M. Edwards, S.A.E. Kools, M.R. Milnes, and L.J. Guillette, Jr., Temporal and spatial variation in plasma thyroxine (T_4) concentrations in juvenile alligators collected from Lake Okeechobee and the northern Everglades, Florida, USA. *Environ. Toxicol. Chem.*, 21, 914–921, 2002.

Guo, W., J.S. Park, Y. Wang, S. Gardner, C. Baek, M. Petreas, and K. Hooper, High polybrominated diphenyl ether levels in California house cats: House dust a primary source? *Environ. Chem. Toxicol.*, 31, 301–306, 2012.

Guruge, K.S., L.W.Y. Yeung, N. Yamanaka, S. Miyazaki, P.K.S. Lam, J.P. Giesy, P.D. Jones, and N. Yamashita, Gene expression profiles in rat liver treated with pentadecafluorooctanoic acid (PFOA). *Toxicol. Sci.*, 89, 93–107, 2006.

Gustafsson, J.P. *Visual MINTEQ,* 3rd Edition, KTH Royal Institute of Technology, Department of Land and Water Resources Engineering, Stockholm, Sweden, 2012.

Haasch, M.L., R. Prince, P.J. Wejksnora, K.R. Cooper, and J.J. Lech, Caged and wild fish: Induction of hepatic cytochrome P-450 (CYP1A1) as an environmental monitor. *Environ. Toxicol. Chem.*, 12, 885–895, 1993.

Hagger, J.A., M.H. Depledge, J. Oehlmann, S. Jobling, and T.S. Galloway, Is there a causal association between genotoxicity and the imposex effect? *Environ. Health Perspect.*, 114 (S–1), 20–26, 2006.

Hahn, M.E., Aryl hydrocarbon receptors: Diversity and evolution. *Chem. Biol. Interact.*, 141, 131–160, 2002.

Hahn, M.E., B.L. Woodward, J.J. Stegeman, and S.W. Kennedy. Rapid assessment of induced cytochrome P4501A (CYP1A) protein and catalytic activity in fish hepatoma cells grown in multi-well plates: Responses to TCDD, TCDF, and two planar PCBs. *Environ. Toxicol. Chem.*, 15, 582–591, 1996.

Hale, R.C., J. Greaves, K. Gallagher, and G.G. Vadas, Novel chlorinated terphenyls in sediments and shellfish of an estuarine environment. *Environ. Science. Technol.*, 24, 1727–1731, 1990.

Hale, R.C., M.J. La Guardia, E.P. Harvey, M.O. Gaylor, T. Matteson Mainor, and W.H. Duff, Flame retardants: Persistent pollutants in land-applied sludges. *Nature*, 412, 140–141, 2001.

Hale, R.C., C.L. Smith, P.O. De Dur, E. Harvey, E.O. Bush, M.J. La Guardia, and G.G. Vadas, Nonylphenols in sediments and effluents associated with diverse wastewater outfalls. *Environ. Toxicol. Chem.*, 19, 946–952, 2000.

Hall, R.J., Impact of pesticides on bird populations, in *Silent Spring Revisited*, Marco, G.J., R.M. Hollingworth, and W. Durham, Eds., American Chemical Society, Washington, DC, 1987.

Halpern, B.S., S. Walbridge, K.A. Selkoe, C.V. Kappel, F. Micheli, C. D'Agrosa, J.F. Bruno et al, A global map of human impact on marine ecosystems. *Science*, 319, 948–952, 2008.

Ham, L., R. Quinn and D. Pascoe, Effects of cadmium on the predator-prey interaction between the turbellarian *Dendrocoelum lacteum* (Müller, 1774) and the isopod crustacean *Asellus aquaticus* (L.). *Arch. Environ. Contam. Toxicol.*, 29, 358–365, 1995.

Hamelink, J.L., P.F. Landrum, H.L. Bergman, and W.H. Benson, Eds., *Bioavailability: Physical, Chemical and Biological Interactions*, CRC Press, Boca Raton, FL, 1994.

Hamer, D.H., D.J. Thiele, and J.E. Lemontt, Function and autoregulation of yeast copper-thionein, *Science*, 228, 685–690, 1985.

Hamilton, A., Forty years in the poisonous trade. *Am. J. Ind. Med.*, 7, 3–18, 1985.

Hamilton, A.J., Species diversity of biodiversity? *J. Environ. Manage.*, 75, 89–92, 2005.

Hamilton, S.J. and P.M. Mehrle, Metallothionein in fish: Review of its importance in assessing stress from metal contaminants. *Trans. Am. Fish. Soc.*, 115, 596–609, 1986.

Hamilton, M.A., R.C. Russo, and R.V. Thurston, Trimmed Spearman-Karber method for estimating median lethal concentrations in toxicity bioassays. *Environ. Sci. Technol.*, 11, 714–719, 1977.

Handley-Goldstone, H.M., M.W. Grow, and J.J. Stegeman, Cardiovascular gene expression profiles of dioxin exposure in zebrafish embryos. *Toxicol. Sci.*, 85, 683–693, 2005.

Handschin, C. and U.A. Meyer, Induction of drug metabolism: The role of nuclear receptors. *Pharmacol. Rev.*, 55, 649–673, 2003.

Handy, R.D., T.S. Galloway, and M.H. DePledge, A proposal for the use of biomarkers for the assessment of chronic pollution and in regulatory toxicology. *Ecotoxicology*, 12, 331–343, 2003.

Handy, R.D., R.O. Owen, and E. Valsami-Jones, The ecotoxicology of nanoparticles and nanomaterials: Current status, knowledge gaps, challenges, and future needs. *Ecotoxicology*, 17, 315–325, 2008.

Hansen, L.G. and B.S. Shane, Xenobiotic metabolism, in *Basic Environmental Toxicology*, Cockerham, L.G., and B.S. Shane, Eds., CRC Press, Boca Raton, FL, 1994.

Hansen K.J., L.A. Clemen, M.E. Ellefson, and H.O. Johnson, Compound-specific, quantitative characterization of organic fluorochemicals in biological matrices. *Environ. Sci. Technol.*, 35, 766–770, 2001.

Hansen Jesse, L.C., and J.J. Obrycki, Field deposition of Bt transgenic corn pollen: Lethal effects on the monarch butterfly. *Oecologia*, 125, 241–248, 2000.

Hanski, I. and M. Gilpin, Metapopulation dynamics: Brief history and conceptual domain. *Biol. J. Linn. Soc.*, 42, 3–16, 1991.

Hanski, I. and M. Gyllenberg, Two general metapopulation models and the core-satellite species hypothesis. *Am. Nat.*, 142, 17–41, 1993.

Hanski, I., A. Moilanen, and M. Gyllenberg, Minimum viable metapopulation size. *Am. Nat.*, 147, 527–541, 1996.

Hanski, I. and O. Ovaskainen, Metapopulation theory for fragmented landscapes. *Theor. Popul. Biol.*, 64, 119–127, 2003.

Hanson, M.L., D.W. Graham, E. Babin, D. Azam, M.A. Coutellec, C.W. Knapp, L. LaGadic, and T. Caquet, Influence of isolation on the recovery of pond mesocosms from the application of an insecticide. 1. Study design and planktonic community responses. *Environ. Toxicol. Chem.*, 26, 1265–1279, 2007.

Haq, F., M. Mahoney, and J. Koropatnick, Signaling events for metallothionein induction. *Mutat. Res.*, 533, 211–226, 2003.

Hare, L.A. and A. Tessier, The aquatic insect *Chaoborus* as a biomonitor of trace metals in lakes. *Limnol. Oceanogr.*, 43, 1850–1859, 1998.

Harley, N.H., Health effects of radiation and radioactive materials, in *Casarett & Doull's Toxicology: The Basic Science of Poisons,* 7th Edition, Klaassen, C.D., Ed., McGraw Hill, New York, 2008.

Harrison, F.L. and I.M. Jones, An in vivo sister-chromatid exchange assay in the larvae of the mussel *Mytilus edulis*: Response to 3 mutagens. *Mutat. Res.*, 105, 235–242, 1982.

Hart, A., D. Balluff, R. Barfknecht, P. Chapman, T. Hawkes, G. Joermann, A. Leopold, and R. Luttik, *Avian Effects Assessment: A Framework For Contaminants Studies*, SETAC Press, Pensacola, FL, 2001.

Hatakeyama, S. and M. Yasuno, Effects of cadmium on the periodicity of parturition and brood size of *Moina macrocopa* (Cladocera). *Environ. Pollut.*, 26, 111–120, 1981.

Hausbeck, J.S., Analysis of trace metal contamination of coal stripmines and bioaccumulation of trace metals by *Peromyscus leucopus*, M.S. thesis, Oklahoma State University, Stillwater, OK, 1995.

Haux, C. and L. Förlin, Biochemical methods for detecting effects of contaminants on fish. *Ambio*, 17, 376–380, 1988.

Hawken, P., A. Lovins, and L.H. Lovins, *Natural Capitalism: Creating the Next Industrial Revolution*, Little, Brown and Co., New York, 1999.

Hawkins, W.E., W.W. Walker, R.M. Overstreet, J.S. Lytle, and T.F. Lytle, Carcinogenic effects of some polynuclear aromatic hydrocarbons on the Japanese medaka and guppy in waterborne exposures. *Sci. Total Environ.*, 94, 155–167, 1990.

Hayashi, T.I., M. Kamo, and Y. Tanaka, Population-level ecological effect assessment: Estimating the effect of toxic chemicals on density-dependent populations. *Ecol. Res.*, 24, 945–954, 2009.

Hays, K.A., An ecotoxicological study of white-footed mice (*Peromyscus leucopus*) from Tar Creek Superfund Site., Ph.D. Diss., Oklahoma State University, Stillwater, OK, 2010.

Hays, K.A., and K. McBee., Flow cytometric analysis of red-eared slider turtles (*Trachemys scripta*) from Tar Creek Superfund Site, *Ecotoxicology*, 16, 353–361, 2007.

Hayton, W.L., Pharmacokinetic parameters for interspecies scaling using allometric techniques. *Health Phys.*, 57 (Sup. 1), 159–164, 1989.

He, G., Y. Lu, A.P.J. Mol, and T. Beckers, Changes and challenges: China's environmental management in transition. *Environ. Dev.* 3, 25–38, 2012.

He, G., L. Zhang, A.P.J. Mol, Y. Lu, and J. Liu, Revising China's environmental law. *Science*, 341, 133, 2013.

He, G., L. Zhang, A.P.J. Mol, Y. Lu, and J. Liu, Supplementary materials for revising China's environmental law. *Science*, 341, 133, 2013.

Heading, R.C., J. Nimmo, L.F. Prescott, and P. Tothill, The dependence of paracetamol absorption on the rate of gastric emptying. *Br. J. Pharmacol.*, 47, 415–421, 1973.

Heagler, M.G., M.C. Newman, M. Mulvey, and P.M. Dixon, Allozyme genotype in mosquitofish, *Gambusia holbrooki*: Temporal stability, concentration effects and field verification. *Environ. Toxicol. Chem.*, 12, 385–395, 1993.

Hedtke, S.F., Structure and function of copper-stressed aquatic microcosms. *Aquat. Toxicol.*, 5, 227–244, 1984.

Hemmer, M.J., C.J. Bowman, B.L. Hemmer, S.D. Friedman, D. Marcovich, K.J. Kroll, and N.D. Denslow, Vitellogenin mRNA regulation and plasma clearance in male sheepshead minnows, (*Cyprinodon variegatus*) after cessation of exposure to 17β-estradiol and p-nonylphenol. *Aquat. Toxicol.*, 58, 99–112, 2002.

Hemmer, M.J., B.L. Hemmer, C.J. Bowman, K.J. Kroll, L.C. Folmar, D. Marcovich, M.D. Hoglund, and H.D. Denslow, Effects of p-nonylphenol, methoxychlor, and endosulfan on vitellogenin indication and expression in sheepshead minnow (*Cyprinodon variegatus*). *Environ. Toxicol. Chem.*, 20, 336–343, 2001.

Hempelmann, L.H., Risk of thyroid neoplasms after irradiation in childhood. Studies of populations exposed to radiation in childhood show a dose response over a wide dose range. *Science*, 160, 159–163, 1968.

Hendricks, J.D., T.R. Meyers, D.W. Shelton, J.L. Casteel, and G.S. Bailey, Hepatocarcino- genicity of benzo[a] pyrene to rainbow trout by dietary exposure and intraperitoneal injection. *J. Natl. Cancer Inst.*, 74, 839–851, 1985.

Hendricks, A.J. and A. Heikens, The power of size. 2. rate constants and equilibrium ratios for accumulation of inorganic substances related to species weight. *Environ. Toxicol. Chem.*, 20, 1421–1437, 2001.

Hendricks, A.J., A. van der Linde, G. Cornelissen, G., and D.T.H.M. Sijm, The power of size. 1. rate constants and equilibrium ratios for accumulation of organic substances related to octanol-water partition ratio and species weight. *Environ. Toxicol. Chem.*, 20, 1399–1420, 2001.

Hennig, H.F.K.O., Metal-binding proteins as metal pollution indicators. *Environ. Health Perspect.*, 65, 175–187, 1986.

Henny, C.J. and G.B. Herron, DDE, selenium, mercury, and white-faced ibis reproduction at Carson Lake, Nevada. *J. Wildl. Manage.*, 53, 1032–1045, 1989.

Henriques, W., R.D. Jeffers, T.E. Lacher, Jr., and R.J. Kendall, Agrochemical use in banana plantations in Latin America: Perspectives on ecological risk. *Environ. Toxicol. Chem.*, 16, 91–99, 1997.

Henry, M.G. and G.J. Atchison, Influence of social rank on the behavior of bluegill, *Lepomis macrochirus* Rafinesque, exposed to sublethal concentrations od cadmium and zinc. *J. Fish. Biol.*, 15, 309–315, 1979.

Henry, M.G. and G.J. Atchison, Metal effects on fish behavior: Advances in determining the ecological significance of responses, in *Metal Ecotoxicology. Concepts & Applications*, Newman, M.C. and A.W. McIntosh, Eds., Lewis Publishers, Chelsea, MI, 1991.

Henschler, D., The origin of hormesis: Historical background and driving forces. *Hum. Exp. Toxicol.*, 25, 347–351, 2006.

Herb, W.R., B. Janke, O. Mohseni, and H.G. Stefan, Thermal pollution of streams by runoff from paved surfaces. *Hydrol. Process*, 22, 987–999, 2008.

Herricks, E.E. and J. Cairns, Jr., Biological monitoring Part III: Receiving system methodology based on community structure. *Water Res.*, 16, 141–153, 1982.

Herrmann, M., Endosulfan preliminary dossier, Website: www.unece.org/env/popsxg/docs/2000–2003/dossierendosulfan-may03.pdf, 2003.

Hertwich, E.G. and G.P. Peters, Carbon footprint of nations: A global, trade-linked analysis. *Environ. Sci. Technol.*, 43, 6414–6420, 2009.

Hesslein, R.H., M.J. Capel, D.E. Fox, and K.A. Hallard, Stable isotopes of sulfur, carbon, and nitrogen as indicators of trophic level and fish migration in the Lower Mackenzie River Basin, Canada. *Can. J. Fish. Aquat. Sci.*, 48, 2258–2265, 1991.

Heusner, A.A., What does the power function reveal about structure and function in animals of different size. *Annu. Rev. Physiol.*, 49, 121–133, 1987.

Heylin, M., Bhopal. *Chem. Eng. News*, Feb. 11, 14–15, 1985.

Hickey, C.W., D.S. Roper, P.T. Holland, and T.M. Trower, Accumulation of organic contaminants in two sediment-dwelling shellfish with contrasting feeding modes: Deposit- (*Macomona liliana*) and filter-feeding (*Austrovenus stutchburyi*). *Arch. Environ. Contam. Toxicol.*, 29, 221–231, 1995.

Hickey, D.A. and T. McNeilly, Competition between metal tolerant and normal plant populations: A field experiment on normal soil. *Evolution*, 29, 458–464, 1975.

Hickie, B.E., P.S. Ross, R.W. MacDonald, and J.K.B. Ford, Killer whales (*Orcinus orca*) face protracted health risks associated with lifetime exposure to PCBs. *Environ. Sci. Technol.*, 41, 6613–6619, 2007.

Hickie, J.J. and D.W. Anderson, Chlorinated hydrocarbons and eggshell changes in raptorial and fish-eating birds. *Science*, 162, 271–273, 1968.

Hightower, L.E., Heat shock, stress proteins, chaperons, and proteotoxicity. *Cell*, 66, 191–197, 1991.

Hileman, B., Acid fog. *Environ. Sci. Technol.*, 17, 117A–120A, 1983.

Hileman, B., Recent reports on the greenhouse effect. *Environ. Sci. Technol.*, 18, 454–455, 1984.

Hileman, B., Global warming. *Chem. Eng. News*, March 13, 25–44, 1989.

Hilje, L., L.E. Castillo, L. Thrupp, and C. Wesseling, El Uso de los Plaguicidas en Costa Rica. Ed. Heliconia/UNED, San José, Costa Rica, 1987.

Hill, A.B., The environment and disease: Association or causation? *Proc. R. Soc. Med.*, 58, 295–300, 1965.

Hinton, D.E., Cells, cellular responses, and their markers in chronic toxicity of fishes, in *Aquatic Toxicology. Molecular, Biochemical and Cellular Perspectives*, Malins, D.C. and G.K. Ostrander, Eds., CRC Press, Boca Raton, FL, 1994.

Hinton, T.G., Risk from exposure to radiation, in *Fundamentals of Ecotoxicology*, Newman, M.C. and M.A. Unger, CRC/Lewis Publishers, Boca Raton, FL, 2003.

Hinton, D.E., R.C. Lantz, J.A. Hampton, P.R. McCuskey, and R.S. McCuskey, Normal versus abnormal structure: Considerations in morphologic responses of teleosts to pollutants. *Environ. Health Perspect.*, 71, 139–146, 1987.

Hinton, D.E. and D.J. Laurén, Integrative histopathological approaches to detecting effects of environmental stressors on fishes. *Amer. Fish. Soc. Symposium*, 8, 51–66, 1990a.

Hinton, D.E. and D.J. Laurén, Liver structural alterations accompanying chronic toxicity in fishes: Potential biomarkers of exposure, in *Biomarkers of Environmental Contamination*, McCarthy, J.F. and L.R. Shugart, Eds., Lewis Publishers, Boca Raton, FL, 1990b.

Hinton, T.G., F.W. Whicker, J.E. Pinder III and S.A. Ibrahim, Comparative kinetics of ^{47}Ca, ^{85}Sr and ^{226}Ra in the freshwater turtle, *Trachemys scripta. J. Environ. Radioact.*, 16, 25–47, 1992.

Hirao, Y. and C.C. Patterson, Lead aerosol pollution in the High Sierra overrides natural mechanisms which exclude lead from a food chain. *Science*, 184, 989–992, 1974.

Hjollund, N.H.I., J.P.E. Bonde, T. Kold Jensen, T. Brink Henriksen, A.M. Andersson, H.A., Kolstad, E. Ernst, A. Giwercman, N.E. Skakkebæk, and J. Olsen, Male-mediated spontaneous abortion among spouses of stainless steel welders. *Scand. J. Work Environ. Health*, 26, 187–192, 2000.

Hoak, R.D., The thermal pollution problem. *J. Water Pollut. Control Fed.*, 33, 1267–1276, 1961.

Hobson, J.F. and W.J. Birge, Acclimation-induced changes in toxicity and induction of metallothionein-like proteins in the fathead minnow following sublethal exposure to zinc. *Environ. Toxicol. Chem.*, 8, 157–169, 1989.

Hoegh-Guldberg, O., P.J. Mumby, A.J. Hooten, R.S. Steneck, P. Greenfield, E. Gomez, C.D. Harvell et al, Coral reefs under rapid climate change and ocean acidification. *Science*, 318, 1737–1742, 2007.

Hoekstra, J.A. and P.H. Van Ewijk, Alternatives for the no-observed-effect-level. *Environ. Toxicol. Chem.*, 12, 187–194, 1993.

Hoffman, D.J., B.A. Rattner, G.A. Burton, Jr., and J. Cairns, Jr., *Handbook of Ecotoxicology*, CRC Press, Boca Raton, FL, 1995.

Hoffmann, U. and H.K. Kroemer, The ABC transporters MDR1 and MRP2: Multiple functions in disposition of xenobiotics and drug resistance. *Drug Metab. Rev.*, 36, 669–701, 2004.

Hoffmann, A.A. and R.E. Woods, Associating environmental stress with developmental stability: Problems and patterns, in *Developmental Instability: Causes and Consequences*, M. Polak, Ed., Oxford University Press, New York, 2003.

Hoh, E, and R.A. Hite, Sources of toxaphene and other organochlorine pesticides in North America as determined by air measurements and potential source contribution function analysis, *Environ. Sci. Technol.*, 38, 4187–4194, 2004.

Holden, C., Report warns of looming pollination crisis in North America. *Science*, 314, 397, 2006.

Holl, K.D. and J. Cairns, Jr., Landscape indicators in ecotoxicology, in *Handbook of Ecotoxicology*, D.J. Hoffman, B.A. Rattner, G.A. Burton, Jr., and J. Cairns, Jr., Eds., Lewis Publishers, Boca Raton, FL, 1995.

Holling, C.S., Resilience and stability of ecological systems. *Annu. Rev. Ecol. Syst.*, 4, 1–24, 1973.

Holloway, G.J., R.M. Sibly, and S.R. Povey, Evolution in toxin-stressed environments. *Funct. Ecol.*, 4, 289–294, 1990.

Holyoak, M., Habitat subdivision causes changes in food web structure. *Eco. Lett.*, 3, 509–515, 2000.

Hontela, A., Interrenal dysfunction in fish from contaminated sites: *In vivo* and *in vitro* assessment. *Environ. Toxicol. Chem.*, 17, 44–48, 1998.

Hopkin, S.P., Ecophysiological strategies of terrestrial arthropods for surviving heavy metal pollution, in *Proceedings of the 3rd European Congress of Entomology, Amsterdam, 1986*. Nederlandse Entomologische Vereniging, Amsterdam, The Netherlands, 1986.

Hopkin, S.P., *Ecophysiology of Metals in Terrestrial Invertebrates*, Elsevier Applied Science, London, United Kingdom 1989.

Hopkin, S.P., Ecological implications of "95% protection levels" for metals in soil. *Oikos*, 66, 137–141, 1993.

Hopkin, S.P., C.A.C. Hames, and A. Dray, X-ray microanalytical mapping of the intracellular distribution of pollutant metals. *Microsc. Anal.*, November 1989, 1–4, 1989.

Hopkin, S.P. and J.A. Nott, Some observations on concentrically structured, intracellular granules in the hepatopancreas of the shore crab, *Carcinis maenas* (L.). *J. Mar. Biol. Assoc. U.K.*, 59, 867–877, 1979.

Hopkins, L.A., Mercury concentrations in tissues of ospreys from the Carolinas, USA. *J. Wildl. Manage.*, 71, 1819–1829, 2007.

Horiguchi, T., T. Nishikawa, Y. Ohta, H. Shiraishi, and M. Morita, Retinoid X receptor gene expression and protein content in tissues of the rock shell *Thais clavigera. Aquat. Toxicol.*, 84, 379–388, 2007.

Horiguchi, T., H. Shiraishi, M. Shimizu, and M. Morita, Imposex and organotin compounds in *Thais clavigera* and *T. bronni* in Japan. *J. Mar. Biol. Assoc. U.K.*, 74, 651–669, 1994.

Horiguchi, T., C. Hyeon-Seo, H. Shiraishi, Y. Shibata, M. Soma, M. Morita, and M. Shimizu, Field studies on imposex and organotin accumulation in the rock shell, *Thais clavigera*, from the Seto Inland Sea and the Sanriku region, Japan. *Sci. Total Environ.*, 214, 65–70, 1998.

Houghton, R.A., The global effects of tropical deforestation. *Environ. Sci. Technol.*, 34, 416–422, 1990.

Howard, L., *Of the Climate of London, Deduced from Meteorological Observations Made at Different Places in the Neighbourhood of the Metropolis. Volume 2.* W. Phillips, London, United Kingdom, 1820.

Howard, B., P.C.H. Mitchell, A. Ritchie, K. Simkiss, and M. Taylor, The composition of invertebrate granules from metal-accumulating cells of the common garden snail (*Helix aspersa*). *Biochem. J.*, 194, 507–511, 1981.

Howe, G.E., L.L. Marking, T.D. Bills, J.J. Rach, and F.L.Mayer, Jr., Effects of water temperature and pH on toxicity of terbufos, trichlorfon, 4-nitrophenol and 2,4-dinitrophenol to the amphipod, *Gammarus pseudolimnaeus* and rainbow trout, *Oncorhynchus mykiss. Environ. Toxicol. Chem.*, 13, 51–66, 1994.

Howell, W.M., D.A. Black, and S.A. Bortone, Abnormal expression of secondary sex characteristics in a population of mosquitofish, *Gambusia affinis holbrooki*: Evidence for environmentally-induced masculinization. *Copeia*, 1980, 676–681, 1980.

Howson, C. and P. Urbach, *Scientific Reasoning: The Bayesian Approach*, Open Court Publishing Co., La Salle, IL, 1989.

Hu, W.Y., P.D. Jones, W. DeCoen, L. King, P. Fraker, J.L. Newsted, and J.P. Giesy, Alternations in cell membrane properties caused by perfluorinated compounds. *Comp. Biochem. Physiol. C*, 135, 77–88, 2002a.

Hu, W.Y., P.D. Jones, B.L. Upham, J.E. Trosko, C. Lau, and J.P. Giesy, Inhibition of gap junctional intercellular communication by perfluorinated compounds in dolphin kidney and rat liver epithelial cell lines. *Toxicol. Sci.*, 68, 429–436, 2002b.

Huang, P.C., Metallothionein structure/function interface, in *Metallothionein III: Biological Roles and Medical Implications*, K.T. Suzuki, N. Imura, and M. Kimura, Eds., Birkhäuser Verlag, Basel, Switzerland, 1993.

Huang, R.P. and E.D. Adamson, Characterization of the DNA-binding properties of the early growth response-1 (Egr-1) transcription factor: Evidence for modulation by a redox mechanism. *DNA Cell Biol.*, 12, 265–273, 1993.

Hubert, W.A. and C.B. Alexander, Observer variation in counts of meristic traits affects fluctuating asymmetry. *N. Am. J. Fish. Manag.*,15, 156–158, 1995.

Huckabee, J.W., J.W. Elwood, and S.G. Hildebrand, Accumulation of mercury in freshwater biota, in *The Biogeochemistry of Mercury in the Environment*, Nriagu, J.O., Ed., Elsevier/North-Holland Biomedical Press, Amsterdam, The Netherlands, 1979.

Hughes, R.M. and D.P. Larsen, Ecoregions: An approach to surface water protection. *J. Water Pollut. Control Fed.*, 60, 486–493, 1988.

Hulett, L.D. Jr., A.J. Weinberger, K.J. Northcutt, and M. Ferguson, Chemical species in fly ash from coal-burning power plants. *Science*, 210, 1356–1358, 1980.

Humbert, S., M. Margni, R. Charles, O.M.T. Salazar, A.L. Quiros, and O. Jolliet, Toxicity assessment of the main pesticides used in Costa Rica. *Agric. Ecosyst. Environ.*, 118, 183–190, 2007.

Hunt, E.G. and A.I. Bischoff, Inimical effects on wildlife of periodic DDD applications to Clear Lake. *Calif. Fish Game*, 46, 91–106, 1960.

Hunt, G.L. and M.W. Hunt, Jr., Female-female pairing in western gulls (*Larus occidentalis*) in southern California. *Science*, 196, 1466–1467, 1977.

Hursh, J.B., M.R. Greenwood, T.W. Clarkson, J. Allen, and S. Demuth, The effect of ethanol on the fate of mercury vapor inhaled by man. *J. Pharmacol. Exp. Ther.*, 214, 520–527, 1980.

Husby, M.P., J.S. Hausbeck, and K. McBee, Chromosomal aberrancy in white-footed mice (*Peromyscus leucopus*) collected on abandoned coal strip mines, Oklahoma, USA. *Environ. Toxicol. Chem.*, 18, 919–925, 1999.

Husby, M.P. and K. McBee, Nuclear DNA content variation and double-strand DNA breakage in white-footed mice (*Peromyscus leucopus*) collected on abandoned coal strip mines, Oklahoma, USA. *Environ. Toxicol. Chem.*, 18, 926–931, 1999.

Hussain, I., M.Z. Khan, A. Khan, I. Javed, and M.K. Saleemi, Toxicological effects of diclofenac in four avian species. *Avian Pathol.*, 37, 315–321, 2008.

Huxley, J.S., Relative growth and form transformation. *Proc. R. Soc. Lond. B Bio. Sci.*, 137, 465–470, 1950.

Hylland, K., T. Nissen-Lie, P.G. Christensen, and M. Sandvik, Natural modulation of hepatic metallothionein and cytochrome P4501A in flounder, *Platichthys flesus* L, *Mar. Environ. Res.*, 46, 51–55, 1998.

Ibarluzea, J,J., M.F. Fernandez, L. Santa Marina, M.F. Olea-Serrano, A.M. Rivas, J.J. Aurrekoetxea, J. Expósito et al, Breast cancer risk and the combined effect of environmental estrogens, *Cancer Causes Control*, 15, 591–600, 2004.

Incardona, J.P., H.L. Day, T.K. Collier, and N.L. Scholz, Developmental toxicity of 4-ring polycyclic aromatic hydrocarbons in zebrafish is differentially dependent on AH receptor isoforms and hepatic cytochrome P4501A metabolism. *Toxicol. Appl. Pharmacol.*, 217, 308–321, 2006.

Incardona, J.P., T.L. Linbo, and N.L. Scholz, Cardiac toxicity of 5-ring polycyclic aromatic hydrocarbons is differentially dependent on the aryl hydrocarbon receptor 2 isoform during zebrafish development. *Toxicol. Appl. Pharmacol.*, 257, 242–249, 2011.

Instituto Regional de Estudios en Sustancias Toxicas, Pesticide Database. Instituto Regional de Estudios en Sustancias Toxicas, Universidad Nacional Costa Rica, 2003.

International Atomic Energy Agency, *Environmental Consequences of the Chernobyl Accident: Twenty Years of Experience*, IAEA, Vienna, Austria, 2006.

International Atomic Energy Agency, *Energy and Nuclear Power Estimates for the Period up to 2030.* IAEA, Vienna, Austria, 2007.

The International Convention on Biological Diversity. Convention on Biological Diversity: Article 2: Use of terms. 2003. http://www.biodiv.org/convention.

International Council on Mining and Metals, *Metals Environmental Risk Assessment Guidance (MERAG).* International Council on Mining and Metals, London, United Kingdom, 2007.

Ishitawa, T., The ATP-dependent glutathione S-conjugate export pump. *Trends Biochem. Sci.*, 18, 164–166, 1992.

Jaeschke, H., Toxic responses of the liver, in *Casarett & Doull's Toxicology: The Basic Science of Poisons,* 7th Edition, Klaassen, C.D., Ed., McGraw-Hill, New York, 2008.

Jager T., Bad habits die hard: The NOEC's persistence reflects poorly on ecotoxicology. *Environ. Toxicol. Chem.*, 31, 228–229, 2012.

Jagoe, C.H., Responses at the tissue level: Quantitative methods in histopathology applied to ecotoxicology, in *Ecotoxicology: A Hierarchical Treatment*, Newman, M.C. and C.H. Jagoe, Eds., CRC Press, Boca Raton, FL, 1996.

Jagoe, C.H., A. Faivre and M.C. Newman, Morphological and morphometric changes in the gills of mosquitofish (*Gambusia holbrooki*) after exposure to mercury (II). *Aquat. Toxicol.*, 34, 163–183, 1996.

Jagoe, R. and M.C. Newman, Bootstrap estimation of community NOEC values. *Ecotoxicology*, 6, 293–306, 1997.

Jakobsson, E. and L. Asplund, Polychlorinated naphthalenes (PCNs), in *The Handbook of Environmental Chemistry, Volume 3, Part K. New Types of Persistent Halogenated Compounds*, Paasivirta J., Ed., Springer-Verlag, Berlin, Germany, 2000.

Jamil,K., A.P. Shaik, M. Mahboob, and D. Krishna, Effect of organophosphorus and organochlorine pesticides (monochrotophos, chlorpyriphos, dimethoate and endosulfan) on human lymphocytes in vitro. *Drug Chem. Toxicol.*, 27, 133–144, 2004.

Janes, N. and R.C. Playle, Modeling silver binding to gills of rainbow trout (*Oncorhynchus mykiss*). *Environ. Toxicol. Chem.*, 14, 1847–1858, 1995.

Janicki, R.H. and W.B. Kinter, DDT: Disrupted osmoregulatory events in the intestine of the eel *Anguilla rostrata* adapted to seawater. *Science*, 173, 1146–1148, 1971.

Jansen, H.G.P., B.A.M. Bouman, R.A. Schipper, H. Hengsdijk, and A. Nieuwenhuyse, An interdisciplinary approach to regional land use analysis using GIS, with applications to the Atlantic Zone of Costa Rica. *Agric. Econ.*, 32, 87–104, 2005.

Jardine, T.D., K.A. Kidd, and A.T. Fisk, Applications, considerations, and sources of uncertainty when using stable isotope analysis in ecotoxicology. *Environ. Sci. Technol.*, 40, 7501–7511, 2006.

Jenny, M.J., A.H. Ringwood, K. Schey, G.W. Warr, and R.W. Chapman, Diversity of metallothioneins in the American oyster, *Crassostrea virginica*, revealed by transcriptomic and proteomic approaches. *Eur. J. Biochem.*, 271, 1702–1712, 2004.

Jeschke, P. and R. Nauen, Neonicotinoids: From zero to hero in insecticide chemistry, *Pest Manag. Sci.*, 64, 1084–1098, 2008.

Jeschke, P., R. Nauen, M. Schindler, and A. Elbert, Overview of the status and global strategy for neonicotinoids, *J. Agric. Food Chem.*, 59, 2897–2908, 2011.

Jia, G., H. Wang, L. Yan, X. Wang, R. Pei, T. Yan, Y. Zhao, and X. Guo, Cytotoxicity of carbon nanomaterials: Single-wall nanotube, multi-wall nanotube, and fullerene. *Environ. Sci. Technol.*, 39, 1378–1383, 2005.

Jiang, G.B., Q.F. Zhou, J.Y. Liu, and D.J. Wu, Occurrence of butyltin compounds in the waters of selected lakes, rivers and coastal environments from China. *Environ. Pollut.*, 115, 81–87, 2001.

Jobling, M., *Environmental Biology of Fishes*, Chapman & Hall, London, United Kingdom, 1995.

Jobling, S., D. Sheahan, J.A. Osborne, P. Matthiessen, and J.P. Sumpter, Inhibition of testicular growth in rainbow trout (*Oncorhychus mykiss*) exposed to estrogenic alkylphenolic chemicals. *Environ. Toxicol. Chem.*, 15, 194–202, 1996.

Johansson-Sjöbeck, M.L. and Å. Larsson, The effect of cadmium on the hematology and on the activity of δ-aminolevulinic acid dehydratase (ALA-D) in blood and hematopoietic tissues of the flounder, *Pleuronectes flesus* L. *Environ. Res.*, 17, 191–204, 1978.

Johansson-Sjöbeck, M.L. and Å. Larsson, Effects of inorganic lead on delta-aminolevulinic acid dehydratase activity and hemotological variables in the rainbow trout, *Salmo gairdnerii. Arch. Environ. Contam. Toxicol.*, 8, 419–431, 1979.

Johnson, A.R., Landscape ecotoxicology and assessment of risk at multiple scales. *Hum. Ecol. Risk Assess.*, 8, 127–146, 2002.

Johnson, D.R. and R.M. Hansen, Effects of range treatment with 2,4-D on rodent populations. *J. Wildl. Manag.*, 33, 125–132, 1969.

Jones, K.A. and T.J. Hara, Behavioral alterations in Arctic char (*Salvelinus alpinus*) briefly exposed to sublethal chlorine levels. *Can. J. Fish. Aquat. Sci.*, 45, 749–752, 1988.

Jones, T.A., M.M. Chumchal, R.W. Drenner, G.N. Timmins, and W.H. Nowlin, Bottom-up nutrient and top-down fish impacts on insect-mediated mercury flux from aquatic ecosystems. *Environ. Toxicol. Chem.*, 32, 1–7, 2013.

Jones, P.D., W.Y. Hu, W. de Coen, J.L. Newsted, and J.P. Giesy, Binding of perfluorinated fatty acids to serum proteins. *Environ. Toxicol. Chem.*, 22, 2639–2649, 2003.

Jones, N.J. and J.M. Parry, The detection of DNA adducts, DNA base changes and chromosome damage for the assessment of exposure to genotoxic pollutants. *Aquat. Toxicol.*, 22, 323–344, 1992.

Jones, M.M. and W.K. Vaughn, HSAB theory and acute metal ion toxicity and detoxification processes. *J. Inorg. Nucl. Chem.*, 40, 2081–2088, 1978.

Jones, O.A.H., N. Voulvoulis, and J.N. Lester, Potential impact of pharmaceuticals on environmental health. *B. World Health Organ.*, 81, 768–769, 2003.

Jönsson, M.E., B. Brunström, K. Ingebrigsten, and I. Brandt, Cell-specific CYP1A expression and benzo[a] pyrene adduct formation in gills of rainbow trout (*Oncorhychus mykiss*) following CYP1A induction in the laboratory and in the field. *Environ. Toxicol. Chem.*, 23, 874–882, 2004.

Jørgensen, S.E., *Modelling in Ecotoxicology*, Elsevier, New York, 1990.

Jørgensen, S.E. and B. Halling-Sorensen, Drugs in the environment. *Chemosphere*, 40, 691–699, 2000.

Jörundsdóttir, K., J. Svavarsson, and K.M.Y. Leung, Imposex levels in the dogwhelk *Nucella lapillus* (L.): Continuing improvement at high latitudes. *Mar. Pollut. Bull.*, 51, 744–749, 2005.

Josephson, J.R. and S.G. Josephson, *Abductive Inference: Computation, Philosophy, Technology*, Cambridge University Press, Cambridge, United Kingdom, 1996.

Joshi, S., Children of endosulfan, *Down to Earth*, 9, 19, 2001. http://www.downtoearth.org/node/15838.

Judy, J.D., J.M. Unrine, and P.M. Bertsch, Evidence for biomagnifications of gold nanoparticles within a terrestrial food chain. *Environ. Sci. Technol.*, 45, 776–781, 2011.

Judy, J.D., J.M. Unrine, W. Rao, and P.M. Bertsch, Bioaccumulation of gold nanoparticles by *Manduca sexta* through dietary uptake of surface contaminated plant tissue. *Environ. Sci. Technol.*, 46, 12672–12678, 2012.

Ju-Nam, Y. and J.R. Lead, Manufactured nanoparticles: An overview of their chemistry, interactions and potential environmental implications. *Sci. Total Environ.*, 400, 396–414, 2008.

Jung, T., N. Bader, and T. Grune, Lipofuscin. Formation, distribution, and metabolism. *Ann. N.Y. Acad. Sci.*, 97–111, 2007.

Jung, D., C.W. Matson, L.B. Collins, G. Laban, H.M. Stapleton, J.W. Bickham, J.A. Swenberg, and R.T. Di Giulio, Genotoxicity in Atlantic killifish (*Fundulus heteroclitus*) from a PAH-contaminated Superfund site on the Elizabeth River, Virginia. *Ecotoxicology*, 20, 1890–1899, 2011.

Jung, R.E. and C.H. Jagoe, Effects of low pH and aluminum on body size, swimming performance, and susceptibility to predation of green tree frog (*Hyla cinerea*) tadpoles. *Can. J. Zool.*, 73, 2171–2183, 1995.

Kac, M., Some mathematical models in science. *Science*, 166, 695–699, 1969.

Kagabu, S., Chloronicotinyl insecticides: Discovery, application and future perspective. *Rev. Toxicol.*, 1, 75–129, 1997.

Kägi, J.H.R., Overview of metallothionein, in *Methods in Enzymology Volume 205, Metallobiochemistry, Part B: Metallothionein and Related Molecules*, Riordan, J.F. and B.L. Vallee, Eds., Academic Press., San Diego, CA, 1991.

Kahn, B. and K.S. Turgeon, The bioaccumulation factor for phosphorus-32 in edible fish tissue. *Health Phys.*, 46, 321–333, 1984.

Kahn, J. and J. Yardley, As China roars, pollution reaches deadly extremes. *The New York Times*, August 26, 2007.

Kaiser, K.L., M.B. McKinnon, D.H. Stendahl, and W.B. Pett, Response threshold levels of selected organic compounds for rainbow trout (*Oncorhynchus mykiss*). *Environ. Toxicol. Chem.*, 14, 2107–2113, 1995.

Kalantzi, O.I., R.E. Alcock, P.A. Johnston, D. Santillo, R.L. Stringer, G.O. Thomas, and K.C. Jones, The global distribution of PCBs and organochlorine pesticides in butter. *Environ. Sci. Technol.*, 35, 1013–1018, 2001.

Kammenga, J.E., M. Busschers, N.M. Van Straalen, P.C. Jepson, and J. Bakker, Stress induced fitness is not determined by the most sensitive life-cycle trait. *Funct. Ecol.*, 10, 106–111, 1996.

Kammenga, J. and R. Laskowski, Eds., *Demography in Ecotoxicology*, Wiley, Chicester, United Kingdom, 2000.

Kammerbauer, J. and J. Moncada, Pesticide residue assessment in three selected agricultural production systems in the Choluteca River Basin of Honduras. *Environ. Pollut.*, 103, 171–181, 1998.

Kampa, M. and E. Castanas, Human health effects of air pollution. *Environ. Pollut.*, 151, 362–367, 2008.

Kamunde, C., C. Clayton, and C.M. Wood, Waterborne vs. dietary copper uptake in rainbow trout and the effects of previous waterborne copper exposure. *Am. J. Physiol.Regul. Integr.*, 283, R69–R78, 2002.

Kang, Y.J., Metallothionein redox cycle and function. *Exp. Biol. Med.*, 231, 1459–1467, 2006.

Kania, H.K. and J. O'Hara, Behavioral alterations in a simple predator-prey system due to sublethal exposure to mercury. *Trans. Am. Fish. Soc.*, 103, 134–136, 1974.

Kannan, K., S.P. Hansen, C.J. Franson, W.W. Bowerman, K.J. Hansen, P.D. Jones, and J.P. Giesy, Perfluorooctane sulfonate in fish-eating water birds including bald eagles and albatrosses. *Environ. Sci. Technol.*, 35, 3065–3070, 2001b.

Kannan, K., J. Koistinen, K. Beckmen, T. Evans, J. Gorzelany, K. Hansen, P.D. Jones, E. Helle, M. Nyman, and J.P. Giesy, Perfluorooctane sulfonate and related fluorinated organic chemicals in marine mammals. *Environ. Sci. Technol.*, 35, 1593–1598, 2001a.

Karin, M. and H.R. Herschman, Induction of metallothionein in HeLa cells by dexamethasone and zinc. *Eur. J. Biochem.*, 113, 267–272, 1981.

Karr, J.R., Seven foundations of biological monitoring and assessment. *Biol. Ambient.*, 20, 7–18, 2006.

Karr, J.R. and E.W. Chu, *Restoring Life in Running Waters: Better Biological Monitoring*, Island Press, Washington, DC, 1999.

Karr, J.R., K.D. Fausch, P.L. Angermeier, P.R. Yant, and I.J. Schlosser, *Assessing Biological Integrity in Running Waters. A Method and Its Rationale.* Illinois Natural History Survey Special Publication 5, Champaign, IL, September 1986.

Karr, J.R. and E.M. Rossano, Applying public health lessons to protect river health. *Ecol. Civ. Eng.*, 4, 3–18, 2001.

Kashian, D.R., R.E. Zuellig, K.A. Mitchell, and W.H. Clements, The cost of tolerance: Sensitivity of stream benthic communities to UV-B and metals. *Ecol. Appl.*, 17, 365–375, 2007.

Kasumyan, A.O., Effects of chemical pollutants on foraging behavior and sensitivity of fish to food stimuli. *J. Ichthyol.*, 41, 76–87, 2001.

Kaufman, L.H., Stream *aufwuchs* accumulation: Disturbance frequency and stress resistance and resilience. *Oecologia*, 52, 57–63, 1982.

Keiser, R.K., J.A. Amado, and R. Murillo, Pesticide levels in estuarine and marine fish and invertebrates from the Guatemala Pacific coast. *Bull. Mar. Sci.*, 23, 905–924, 1973.

Keklak, M.M., M.C. Newman, and M. Mulvey, Enhanced uranium tolerance of an exposed population of the Eastern moquitofish (*Gambusia holbrooki* Girard 1859). *Arch. Environ. Contam. Toxicol.*, 27, 20–24, 1994.

Kelce, W.R., C.R. Stone, S.C. Laws, L.E. Gray, J.A. Kemppainen, and E.M. Wilson, Persistent DDT metabolite p,p'-DDE is a potent androgen receptor antagonist. *Nature*, 375, 581–585, 1995.

Kellie-Smith, O. and P.M. Cox, Emergent dynamics of the climate-economy system in the Anthropocene, *Philos. Trans. R. Soc. A*, 369, 868–886, 2011.

Kelly, B.C., F.A.P.C. Gobas, and M.S. McLachlan, Intestinal absorption and biomagnification of organic contaminants in fish, wildlife, and humans. *Environ. Toxicol. Chem.*, 23, 2324–2336, 2004.

Kelly, B.C., M.G. Ikonomou, J.D. Blair, A.E. Morin, and F.A.P.C. Gobas, Food web-specific biomagnification of persistent organic pollutants. *Science*, 317, 236–239, 2007.

Kelly, J.R. and R.E. Scheibling, Fatty acids as dietary tracers in benthic food webs. *Mar. Ecol. Prog. Ser.*, 446, 1–22, 2012,

Kemper, R.A., J.R. Hayes, and M.S. Bogdanffy, Metabolism: A determinant of toxicity, in *Principles and Methods of Toxicology,* 5th Edition, Hayes, A.W., Ed., CRC/Taylor & Francis Press, Boca Raton, FL, 2008.

Kennish, M.J., *Ecology of Estuaries: Anthropogenic Effects*, CRC Press, Boca Raton, FL, 1992.

Kennish, M.J., *Practical Handbook of Estuarine and Marine Pollution*, CRC Press, Boca Raton, FL, 1997.

Kenow, K.P., K.A. Grasman, R.K. Hines, M.W. Meyer, A. Gendron-Fitzpatrick, M.G. Spalding, and B.R. Gray, Effects of methylmercury exposure on the immune function of juvenile common loons (*Gavia immer*). *Environ. Toxicol. Chem.*, 26, 1460–1469, 2007.

Kerr, R.A., Ozone hole bodes ill for the globe. *Science*, 241, 785–786, 1988a.

Kerr, R.A., Is there life after climate change? *Science*, 242, 1010–1013, 1988b.

Kerr, R.A., New assaults seen on Earth's ozone shield. *Science*, 255, 797–798, 1992.

Kerr, R.A., Ozone-destroying chlorine tops out. *Science*, 271, 32, 1996.

Kerr, R.A., How urgent is climate change? *Science*, 318, 1230–1231, 2007.

Kerr, R.A., The psst that pierced the sky is now churning the sea, *Science*, 339, 500, 2013.

Kerr, R., E. Kintisch, and E. Stokstad, Will Deepwater Horizon set a new standard for catastrophe? *Science*, 328, 674–675, 2010.

Kersting, K., Normalizing ecosystem strain: A system parameter for analysis of toxic stress in (micro-) ecosystems. *Ecol. Bull.*, 36, 150–153, 1984.

Kettlewell, H.B.D., Selection experiments on industrial melanism in the *Lepidoptera. Heredity*, 9, 323–342, 1955.

Kettlewell, B., *The Evolution of Melanism*, Clarendon Press, Oxford, 1973.

Key, B.D., R.D. Howell, and C.S. Criddle, Fluorinated organics in the biosphere. *Environ. Sci. Technol.*, 31, 2445–2454, 1997.

Khalil, A.M., Chromosome aberrations in blood lymphocytes from petroleum refinery workers. *Arch. Environ. Contam. Toxicol.*, 28, 236–239, 1995.

Khalil, M.A.K. and R.A. Rasmussen, Carbon monoxide in the earth's atmosphere: Increasing trend. *Science*, 224, 54–56, 1984.

Khan,P.K. and S.P. Sinha, Ameliorating effect of vitamin C on murine sperm toxicity induced by three pesticides (endosulfan, phosphamidon and mancozeb), *Mutagenesis*, 11, 1133–1136, 1996.

Khayat, A.I. and Z.A. Shaikh, Dose-effect relationship between ethyl alcohol pretreatment and retention and tissue distribution of mercury vapor in rats. *J. Pharmacol. Exp. Ther.*, 223, 649–653, 1982.

Khera, K.S., Teratogenic and genetic effects of mercury, in *The Biogeochemistry of Mercury in the Environment*, Nriagu J.O., Ed., Elsevier/North-Holland Biomedical Press, Amsterdam, The Netherlands, 1979.

Kidd, K.A., R.H. Hesslein, R.J.P. Fudge, and K.A. Hallard, The influence of trophic level as measured by $\delta\ ^{15}N$ on mercury concentrations in freshwater organisms. *Water Air Soil Pollut.*, 80, 1011–1015, 1995.

Kimura-Kuoda, J., Y. Komuta, Y. Kuroda, M. Hayashi, and H. Kawano, Nicotine-like effects of the neonicotinoid insecticides acetamiprid and imidacloprid on cerebellar neurons from neonatal rats. *PLoS ONE*, 7, e32432 (1–11), 2012.

King, J.J., *The Environmental Dictionary,* 2nd Edition, Executive Enterprises Publications Co., New York, 1993.

Kingtong, S., Y. Chitramvong, and T. Janvilisri, ATP-binding cassette multidrug transporters in Indian-rock oyster *Saccostrea forskali* and their role in the export of an environmental organic pollutant tributyltin. *Aquat. Toxicol.*, 85, 124–132, 2007.

Kintisch, E. and E. Stokstad, Ocean CO_2 studies look beyond coral. *Science*, 319, 1029, 2008.

Kissa, E., *Fluorinated Surfactants and Repellents.* 2nd Edition, Revised and Expanded, Marcel Dekker, New York, 2001.

Klaassen, C.D., M.O. Amdur, and J. Doull, Eds., *Casarett and Doull's Toxicology: The Basic Science of Poisons*, Macmillan, New York, 1987.

Klaassen, C.D., J. Liu, and S. Choudhuri, Metallothionein: An intracellular protein to protect against cadmium toxicity. *Annu. Rev. Pharmacol.*, 39, 267–294, 1999.

Klausmeier, C.A., Habitat destruction and extinction in competitive and multualistic metacommunities. *Eco. Lett.*, 4, 57–63, 2001.

Klaverkamp, J.F., M.D. Dutton, H.S. Majewski, R.V. Hunt, and L.J. Wesson, Evaluating the effectiveness of metal pollution controls in a smelter by using metallothionein and other biochemical responses in fish, in *Metal Ecotoxicology: Concepts and Applications*, Newman, M.C. and A.W. McIntosh, Eds., Lewis Publishers, Chelsea, MI, 1991.

Klein, W., S. Denzer, M. Herrchen, P. Lepper, M. Müller, R. Sehrt, A. Storm, and J. Volmer, Revised proposal for a list of priority substances in the context of the Water Framework Directive (COMMPS procedure). Final report 98/788/3040/ DEB/E1. Fraunhofer Institute, Schmallenberg, Denmark,1999.

Kleinow, K.M., M.J. Melancon, and J.J. Lech, Biotransformation and induction: Implications for toxicity, bioaccumulation and monitoring of environmental xenobiotics in fish. *Environ. Health Perspect.*, 71, 105–119, 1987.

Klemens, J.A., M.L. Wieland, V.J. Flanagin, J.A. Frick, and R.G. Harper, A cross-taxa survey of organochlorine pesticide contamination in a Costa Rican wildland. *Environ. Pollut.*, 122, 245–251, 2003.

Klerks, P.L. and J.S. Levinton, Rapid evolution of metal resistance in a benthic oligochaete inhabiting a metal-polluted site, *Biol. Bull.*, 176, 135–141, 1989.

Klerks, P.L. and J.S. Weis, Genetic adaptation to heavy metals in aquatic organisms: A review. *Environ. Pollut.*, 45, 173–205, 1987.

Klinck, J.S., W.W. Green, R.S. Mirza, S.R. Nadella, M. Chowdhury, C.M. Wood, and G.G. Pyle, Branchial cadmium and copper binding and intestinal cadmium uptake in wild yellow perch (*Perca flavescens*) from clean and metal-contaminated lakes. *Aquat. Toxicol.*, 84, 198–207, 2007.

Kline, B., *First along the River: A Brief History of the U.S. Environmental Movement,* 4th Edition. Rowman & Littlefield Publishers, Lanham, MD, 2011.

Kling, G.W., B. Fry, and W.J. O'Brien, Stable isotopes and planktonic trophic structure in Arctic lakes. *Ecology*, 73, 561–566, 1992.

Klingenberg, C.P. and G.S. McIntyre, Geometric morphometrics of developmental instability: Analyzing patterns of fluctuating asymmetry with Procrustes methods. *Evolution*, 52, 1363–1375, 1998.

Klopsteg, P.E., Environmental science. *Science*, 152, 595, 1966.

Klugh, A.B., The effect of the ultra-violet component of sunlight on certain marine organisms. *Can. J. Res.*, 1, 100–109, 1929.

Knapen, D., H. Reynders, L. Bervoets, E. Verheyen, and R. Blust, Metallothionein gene and protein expression as a biomarker for metal pollution in natural gudgeon populations. *Aquat. Toxicol.*, 82, 163–172, 2007.

Knight, T.M., M.W. McCoy, J.M. Chase, K.A. McCoy, and R.D. Holt, Trophic cascades across ecosystems. *Nature*, 437, 880–883, 2005.

Knoph, M.B. and Y.A. Olsen, Subacute toxicity of ammonia to Atlantic salmon (*Salmo salar* L.) in seawater: Effects on water and salt balance, plasma cortisol and plasma ammonia levels. *Aquat. Toxicol.*, 30, 295–310, 1994.

Kobayashi, M., K. Itoh, T. Suzuki, H. Osanai, K. Nishikawa, Y. Katoh, Y. Takagi, and M. Yamamoto, Identification of the interactive interface and phylogenic conservation of the Nrf2-Keap1 system. *Genes Cells*, 7(8), 807–820, 2002.

Koch, B.T., J.E. Garvey, J. You, and M.J. Lydy, Elevated organochlorines in the brain-hypothalamic-pituitary complex of intersexual shovelnose sturgeon. *Environ. Toxicol. Chem.*, 25, 1689–1697, 2006.

Kocher, D.C., *Radioactive Decay Data Tables*, U.S. Department of Energy, National Technical Information Service, Springfield, MD, 1981.

Kochevar, I.E., R. Armstrong, J. Einbinder, R. Walther, and L. Harber, Coal tar phototoxicity: Active compounds and action spectra. *Photochem. Photobiol.*, 36, 65–69, 1982.

Koehn, R.K. and B.L. Bayne, Towards a physiological and genetical understanding of the energetics of the stress response. *Biol. J. Linn. Soc.*, 37, 157–171, 1989.

Koehn, R.K. and P.M. Gaffney, Genetic heterozygosity and growth rate in *Mytilus edulis. Mar. Biol.*, 82, 1–7, 1984.

Kojima, Y., P.A. Binz, and J.H.R. Kägi, Nomenclature of metallothionein: Proposal for revision, in *Metallothionein IV*, Klaassen, C., Ed., Birkhäuser Verlag, Basel, Switzerland, 1999.

Kolaja, G.J. and D.E. Hinton, DDT-induced reduction in eggshell thickness, weight, and calcium is accompanied by calcium ATPase inhibition, in *Animals as Monitors of Environmental Pollutants*, National Academy of Sciences, Washington, DC, 1979.

Kolkwitz, R. and M. Marsson, Ökologie der pflanzlichen Saprobien. *Ber. Dtsch. Bot. Ges.*, 26, 505–519, 1908.

Kondo, S., *Health Effects of Low-Level Radiation*, Kinki University Press, Osaka, Japan, 1993.

Kondrashov, V., S.J. Rothenberg, D. Chettle, and J. Zerwekh, Evaluation of potentially significant increase of lead in the blood during long-term bed rest and space flight. *Physiol. Meas.*, 26, 1–12, 2005.

Kooijman, S.A.L.M., Parametric analyses of mortality rates in bioassays. *Water Res.*, 15, 107–119, 1981.

Kooijman, S.A.L.M., *Dynamic Energy Budgets in Biological Systems*, Cambridge University Press, Cambridge, United Kingdom, 1993.

Kooijman, S.A.L.M. and J.J.M. Bedaux, *The Analysis of Aquatic Toxicity Data*, VU University Press, Amsterdam, The Netherlands, 1996.

Kooijman, S.A.L.M., T. Sousa, L. Pecquerie, J. van der Meer, and T. Jager, From food-dependent statistics to metabolic parameters, a practical guide to the use of dynamic energy budget theory. *Biol. Rev.*, 83, 533–552, 2008.

Kopp, R.L., S.I. Guttman, and T.E. Wissing, Genetic indicators of environmental stress in central mudminow (*Umbra limi*) populations exposed to acid deposition in the Adirondack Mountains. *Environ. Toxicol. Chem.*, 11, 665–676, 1992.

Körner, C., Biosphere responses to CO_2 enrichment. *Ecol. Appl.*, 10, 1590–1619, 2000.

Koropatnick, J., Amplification of metallothionein-1 genes in mouse liver cells in situ: Extra copies are transcriptionally active, *Proc. Soc. Exp. Biol. Med.*, 188, 287–300, 1988.

Korte, G., *The GIS Book,* 5th Edition, Onword Press, Santa Fe, NM, 2000.

Koss, G., E. Schuler, B. Arndt, J. Seidel, S. Seubert, and A. Seubert, A comparative toxicological study of pike (*Esox lucius* L.) from the River Rhine and River Lahn. *Aquat. Toxicol.*, 8, 1–9, 1986.

Kotz, S. and N.L. Johnson, *Encyclopedia of Statistical Sciences, Volume 8*, Wiley, New York, 1988.

Kozhara, A.V., Phenotypic variance of bilateral characters as an indicator of genetic and environmental conditions in bream *Abramis brama* (L.) (Pisces, Cyprinidae) population. *J. Appl. Ichthyol.*, 10, 167–181, 1994.

Krahn, M.M., M.S. Myers, D.G. Burrows, and D.C. Malins, Determination of metabolites of xenobiotics in the bile of fish from polluted waterways. *Xenobiotica*, 16, 957–973, 1984.

Kramer, V.J. and M.C. Newman, Inhibition of glucosephosphate isomerase allozymes of the mosquitofish, *Gambusia holbrooki*, by mercury. *Environ. Toxicol. Chem.*, 13, 9–14, 1994.

Kramer, V.J., M.C. Newman, M. Mulvey, and G.R. Ultsch, Glycolysis and Krebs cycle metabolites in mosquitofish, *Gambusia holbrooki*, Girard 1859, exposed to mercuric chloride: Allozyme genotype effects. *Environ. Toxicol. Chem.*, 11, 357–364, 1992.

Krantz, D.H., The null hypothesis testing controversy in psychology. *J. Am. Stat. Assoc.*, 44, 1372–1381, 1999.

Krantzberg, G., Spatial and temporal variability in metal bioavailability and toxicity of sediment from Hamilton Harbour, Lake Ontario. *Environ. Toxicol. Chem.*, 13, 1685–1698, 1994.

Krasnov, A., H. Koskinen, C. Rexroad, S. Afanasyev, H. Molsa, and A. Oikari, Transcriptome responses to carbon tetrachloride and pyrene in the kidney and liver of juvenile rainbow trout (*Oncorhynchus mykiss*). *Aquat. Toxicol.*, 74, 70–81, 2005.

Krause, P.R., Effects of an oil production effluent on gametogenesis and gamete performance in the purple sea urchin (*Strongylocentrotus purpuratus* Stimpson). *Environ. Toxicol. Chem.*, 13, 1153–1161, 1994.

Kretschmann, A., R. Ashauer, J. Hollender, and B.I. Escher, Toxicokinetic and toxicodynamic model for diazinon toxicity: Mechanistic explanation of differences in the sensitivity of *Daphnia magna* and *Gammarus pulex. Environ. Toxicol. Chem.*, 31, 2014–2022, 2012.

Krueger, J., Null hypothesis significance testing. *Am. Psychol.*, 56, 16–26, 2001.

KSCSTE, (Kerala State Council for Science, Technology and Environment), Report on monitoring of endosulfan residues in the eleven panchayaths of Kasargod District, Kerala. Kerala State Council for Science, Technology and Environment, Thiruvananthapuram, Kerala, India, June 2011.

Kuhn, T.S., *The Structure of Scientific Revolutions*, The University of Chicago Press, Chicago, IL, 1970.

Kuhn, T.S., *The Essential Tension*, The University of Chicago Press, Chicago, IL, 1977.

Kumar, G., D.S. Tripathi, and S.K. Roy, Cytological effects on plants by an accidental leakage of methylisocyanate (MIC) gas. *Environ. Exper. Bot.*, 29, 261–271, 1989.

Kumar, K.S., K. Kannan, S. Corsolini, T. Evans, J.P. Giesy, J. Nakanishi, and S. Masunaga, Polychlorinated dibenzo-*p*-dioxins, dibenzofurans and polychlorinated biphenyls in polar bear, penguin and south polar skua. *Environ. Pollut.*, 119, 151–161, 2002.

Lacayo-Romero, M., J. Quillaguaman, B. van Bavel, and B. Mattiasson, A toxaphene-degrading bacterium related to *Enterobacter cloacae*, strain D1 isolated from aged contaminated soil in Nicaragua. *Syst. Appl. Microbiol.*, 28, 632–639, 2005.

Łagisz, M., P. Kramarz, R. Laskowski, and M. Tobor, Population parameters of the beetle *Pterostichus oblongopunctatus* F. from metal contaminated and reference areas. *Arch. Environ. Contam. Toxicol.*, 69, 243–249, 2002.

Lagus, D.L. and V.R. Alekseev, Components of morphological variation in baikalian endemial cyclopid *Acanthocyclops signifier* complex from different localities. *Hydrobiologia*, 417, 25–35, 2000.

Laity, J.H. and G.K. Andrews, Understanding the mechanisms of zinc-sensing by metal-response element binding transcription factor-1 (MTF-1). *Arch. Biochem. Biophys.*, 463, 201–210, 2007.

Lajus, D.L., Variation patterns of bilateral characters: Variation among characters and among populations in the White Sea herring, *Clupea pallasi marisalbi* (Berg) (Clupeidae, Teleostei). *Biol. J. Linn. Soc.*, 74, 237–253, 2001.

Lajus D.L., M. Ciostek, M. Draszanowska, and T. Sywula, Geographic and ontogenetic patterns of chaetotaxy variation in glacial relict *Saduria entomon* (Crustacea, Isopoda): Inter-population, inter-individual and intra-individual variations (fluctuating asymmetry). *Ann. Zool. Fennici*, 40, 411–419, 2003.

Lajus, D.L., J.H. Graham, and A.V. Kozhara,, Developmental instability and the stochastic component of total phenotypic variance, in *Developmental Instability: Causes and Consequences*, Polak M., Ed., Oxford University Press, New York, 2003.

Lamb, T., J.W. Bickham, J.W. Gibbons, M.J. Smolen, and S. McDowell, Genetic damage in a population of slider turtles (*Trachemys scripta*) inhabiting a radioactive reservoir. *Arch. Environ. Contam. Toxicol.*, 20, 138–142, 1991.

Lamb, T., J.W. Bickham, T.B. Lyne, and J.W. Gibbons, The slider turtle as an environmental sentinel: Multiple tissue assays using flow cytometric analysis. *Ecotoxicology*, 4, 5–13, 1995.

Lamborg, C.H., K.R. Rolfhus, W.F. Fitzgerald, and G. Kim, The atmospheric cycling and air-sea exchange of mercury species in the South and equatorial Atlantic Ocean. *Deep-Sea Res. Part 2*, 46, 957–977, 1999.

Lance, B.K., D.B. Irons, S.J. Kendall, and L.L. McDonald, An evaluation of marine bird population trends following the Exxon Valdez oil spill, Prince William Sound Alaska. *Mar. Pollut. Bull.*, 42, 298–309, 2001.

Landis, W.G., L.A. Lenart, and J.A. Spromberg, Patch dynamics of horizontal gene transfer with application to the ecological risk assessment of genetically engineered organisms. *Hum. Ecol. Risk Assess.*, 6, 875–99, 2000.

Landis, W.G., R.A. Matthews, and G.B. Matthews, The layered and historical nature of ecological systems and the risk assessment of pesticides. *Environ. Toxicol. Chem.*, 15, 432–440, 1996.

Landis, W.G. and M.H. Yu, *Introduction to Environmental Toxicology*, Lewis Publishers, Boca Raton, FL, 1995.

Landrigan, P.J., B. Sonawane, D. Mattison, M. McCally, and A. Garg, Chemical contaminants in breast milk and their impacts on children's health: An overview. *Environ. Health Perspect.*, 110, A313–A315, 2002.

Landrum, P.F., G.A. Karkley, and J. Kukkonen, Evaluation of organic contaminant exposure in aquatic organisms: The significance of bioconcentration and bioaccumulation, in *Ecotoxicology: A Hierarchical Treatment*, Newman. M.C. and C.H. Jagoe, Eds., CRC Press, Inc., Boca Raton, FL, 1996.

Landrum, P.F., H. Lee, II, and M.J. Lydy, Toxicokinetics in aquatic systems: Model comparisons and use in hazard assessment. *Environ. Toxicol. Chem.*, 11, 1709–1725, 1992.

Landrum, P.F., M. Leppänen, S.D. Robinson, D.C. Gossiaux, G.A. Burton, M. Greenberg, J.V.K. Kukkonen, B.J. Eadie, and M.B. Lansing, Effect of 3,4,3',4'-tetrachlorobiphenyl on the reworking behavior of *Lumbriculus variegatus* exposed to contaminated sediment. *Environ. Toxicol. Chem.*, 23, 178–186, 2004.

Landrum, P.F. and M.J. Lydy, personal communication during toxicokinetics short course, Nov. 3, 1991, Society of Environmental Toxicology and Chemistry 12th Annual Meeting, 1991.

Landrum, P.F. and J.A. Robbins, Bioavailability of sediment-associated contaminants to benthic invertebrates, in *Sediments: Chemistry and Toxicity of In-Place Pollutants*, Baudo, R., J. Giesy, and H. Muntau, Eds., Lewis Publishers, Chelsea, MI, 1990.

Lane, D.A., M.S. Kramer, T.A. Hutchinson, J.K. Jones, and C. Naranjo, The causality assessment of adverse drug reactions using a Bayesian approach. *Pharmaceut. Med.*, 2, 265–283, 1987.

Lange, B.W., *Drosophila melanogaster* metallothionein genes: Selection for duplications? Ph.D. Diss., Duke University, Durham, NC, 1989.

Langston, W.J. and G.R. Burt, Bioavailability and effects of sediment-bound TBT in deposit-feeding clams, *Scrobicularia plana. Mar. Environ. Res.*, 32, 61–77, 1991.

Lapp, R.E., Nevada test fallout and radioiodine in milk. *Science*, 137, 756–758, 1962.

Larson, R.A. and E.J. Weber, *Reaction Mechanisms in Environmental Organic Chemistry*, CRC Press, Boca Raton, FL, 1994.

Larson, R.J. and D.M. Woltering, Linear alkylbenzene sulfonate (LAS), in *Fundamentals of Aquatic Toxicology: Effects, Environmental Fate, and Risk Assessment,* 2nd Edition, Rand, G.M., Ed., Taylor & Francis, Washington, DC, 1995.

Larssen, T., E. Lydersen, D. Tang, Y. He, J. Gao, H. Liu, L. Duan et al, Acid rain in China. *Environ. Sci. Technol.*, 40, 418–425, 2006.

Larsson, Å., B.E. Bengtsson, and C. Haux, Disturbed ion balance in flounder, *Platichthys flesus* L. exposed to sublethal levels of cadmium. *Aquat. Toxicol.*, 1, 19–35, 1981.

Laskowski, R., Are the top carnivores endangered by heavy metal biomagnification? *Oikos*, 60, 387–390, 1991.

Laskowski, R., Why short-term bioassays are not meaningful: Effects of a pesticide (imidacloprid) and a metal (cadmium) on pea aphids (*Acyrthosiphon pisum* Harris). *Ecotoxicology*, 10, 177–183, 2001.

Laurén, D.J. and D.G. McDonald, Effects of copper on branchial ionoregulation in the rainbow trout, *Salmo gairdneri* Richardson. Modulation by water hardness and pH. *J. Comp. Physiol., B*, 155, 635–644, 1985.

Laurent, T., A.S. Ho, and G. Maroni, Recent evolutionary history of the metallothionein gene Mtn in *Drosophila. Genet. Res. (Camb.)*, 58, 203–210, 1991.

La Via, M.F. and R.B. Hill, Jr., *Principles of Pathobiology*, Oxford University Press, New York, 1971.

Lavie, B. and E. Nevo, Heavy metal selection of phosphoglucose isomerase allozymes in marine gastropods. *Mar. Biol.*, 71, 17–22, 1982.

Lavie, B. and E. Nevo, Genetic selection of homozygote allozyme genotypes in marine gastropods exposed to cadmium pollution. *Sci. Total Environ.*, 57, 91–98, 1986.

Lavoie, M., P.G.C. Campbell, and C. Fortin, Extending the biotic ligand model to account for positive and negative feedback interactions between cadmium and zinc in a freshwater alga. *Environ. Sci. Technol.*, 46, 12129–12136, 2012.

Law, B., Nitrogen deposition and forest carbon. *Nature*, 496, 307–308, 2013.

Laxen, D.P.H. and R.M. Harrison, The highway as a source of water pollution: An appraisal with the heavy metal lead. *Water Res.*, 11, 1–11, 1977.

Lazarow, P.B. and Y. Fujiki, Biogenesis of peroxisomes. *Annu. Rev. Cell Biol.*, 1, 489–530, 1985.

Leamy, L.J., Morphometric studies in inbred and hybrid house mice. V. Directional and fluctuating asymmetry. *Am. Nat.*, 123, 579–93, 1984.

Leamy, L.J. and C.P. Klingenberg, The genetics and evolution of fluctuating asymmetry. *Annu. Rev. Ecol. Evol. Syst.*, 36, 1–21, 2005.

Leary, R.F. and F.W. Allendorf, Fluctuating asymmetry as an indicator of stress: Implications for conservation biology. *Trends Ecol. Evol.*, 4, 214–217, 1989.

Leary, R.F, F.W. Allendorf, and K.L. Knudsen, Developmental stability and enzyme heterozygosity in rainbow trout. *Nature*, 301, 71–72, 1983.

Leary, R.F., F.W. Allendorf and K.L. Knudsen, Effects of rearing density on meristics and developmental stability of rainbow trout. *Copeia*, 1991, 44–49, 1991.

Leary, R.F., F.W. Allendorf and K.L. Knudsen, Genetic, environmental, and developmental causes of meristic variation in rainbow trout. *Acta Zool. Fennici.*, 191, 79–95, 1992.

Lease, H.M., J.A. Hansen, H.L. Bergman, and J.S. Meyer, Structural changes in gills of Lost River suckers exposed to elevated pH and ammonia concentrations. *Comp. Biochem. Physiol. C*, 134, 491–500, 2003.

Lech, J.J. and M.J. Vodicnik, Biotransformation, in *Fundamentals of Aquatic Toxicology*, Rand, G.M. and S.R. Petrocelli, Eds., Hemisphere Publishing Corp., Washington, DC, 1985.

Lechelt, M., W. Blohm, B. Kirschneit, M. Pfeiffer, E. Gresens, J. Liley, R. Holz, C. Lüering, and C. Moldaenke, Monitoring of surface water by ultra-sensitive *Daphnia* toximeter. *Environ. Toxicol.*, 15, 390–400, 2000.

Le Jeune, A.-H., M. Charpin, D. Sargos, J.-F. Lenain, V. Deluchat, N. Ngayila, M. Baudu, and C. Amblard. Planktonic microbial community responses to added copper. *Aquat. Toxicol.*, 83, 223–237, 2007.

Lee, G., Lawmakers move to check CFC phaseout. *Washington Post*, Sept. 21, A13, 1995.

Lee, J.M., 'Silent Spring' is now noisy summer. *The New York Times*, July 22, 1962.

Lee, B., S.B. Griscom, J. Lee, H.J. Choi, C. Koh, S.N. Luoma, and N.S. Fisher, Influences of dietary uptake and reactive sulfides on metal bioavailability from aquatic sediments. *Science*, 287, 282–284, 2000.

Lees, D.R. and E.R. Creed, The genetics of the insularia forms of the peppered moth, *Biston betularia. Heredity*, 39, 67–73, 1977.

Lefkovitch, L.P., The study of population growth in organisms grouped by stages. *Biometrics*, 21, 1–18, 1965.

Leggett, R.W., Predicting the retention of Cs in individuals. *Health Phys.*, 50, 747–759, 1986.

Leibold, M.A., M. Holyyoak, M. Mouquet, P. Amarasekare, J.M. Chase, M.F. Hoopes, R.D. Holt et al, The metacommunity concept: A framework for multi-scale community ecology. *Eco. Lett.*, 7, 601–613, 2004.

Leknes, I.L., Melano-macrophage centres and endocytic cells in kidney and spleen of pearl gourami and platyfish (Anabantidae, Poeciliidae: Teleostei), *Acta Histochem.*, 109, 164–168, 2007.

Leland, H.V. and J.L. Carter, Effects of copper on species composition of periphyton in a Sierra Nevada, California, stream. *Freshw. Biol.*, 14, p. 281–296, 1984.

Leland, H.V. and J.S. Kuwabara, Trace elements, in *Fundamentals of Aquatic Toxicology*, Rand, G.M. and S.R. Petrocelli, Eds., Hemisphere Publishing Corp., Washington, DC, 1985.

Lemly, A.D., Aquatic selenium pollution is a global environmental issue, *Ecotoxicol. Environ. Saf.*, 59, 44–56, 2004.

LeNoir, J.S., L.L. McConnell, G.M. Fellers, T.M. Cahill, and J.N. Seiber, Summertime transport of current-use pesticides from California's Central Valley to the Sierra Nevada Mountain Range, U.S.A. *Environ. Toxicol. Chem.*, 18, 2715–2722, 1999.

Lens, L., S. Van Dongen, S. Kark, and E. Matthysen, Fluctuating asymmetry as an indicator of fitness: Can we bridge the gap between studies? *Biol. Rev.*, 77, 27–38, 2002.

Lenz, W., The susceptible period for thalidomide malformations in man and monkey. *Ger. Med. Mon.*, 4, 197–198, 1968.

Lenz, W., Malformations caused by drugs in pregnancy. *Am. J. Dis. Child.*, 2, 99–106, 1996.

Leonard, S.S., J.J. Bower, and X. Shi, Metal-induced toxicity, carcinogenesis, mechanisms and cellular responses. *Mol. Cell. Biochem.*, 255, 3–10, 2004.

Leopold, A., *A Sand County Almanac.* Oxford University Press, Oxford, 1949. (Reprinted 1966, Sierra Club/ Ballantine Books, New York.)

Lepkowski, W., Bhopal. Indian city begins to heal but conflicts remain. *Chem. Eng. News*, Dec. 2, 18–32, 1985.

Lepper, P., Manual on the methodological framework to derive environmental quality standards for priority substances in accordance with Article 16 of the Water Framework Directive (2000/60/EC). Schmallenberg (DE): Fraunhofer-Institute Molecular Biology and Applied Ecology, 2005.

Leslie, P.H., On the use of matrices in certain population mathematics. *Biometrika*, 33, 183–212, 1945.

Leslie, P.H., Some further notes on the use of matrices in population mathematics. *Biometrika*, 35, 213–245, 1948.

Leung, B. and M. Forbes, Fluctuating asymmetry in relation to stress and fitness: Effect of trait type as revealed by meta-analysis. *Ecoscience*, 3, 400–413, 1996.

Leung, B., M.R. Forbes, and D. Houle, Fluctuating asymmetry as a bioindicator of stress: Comparing efficacy of analyses involving multiple traits. *Am. Nat.*, 155, 101–115, 2000.

Leung, B., L. Knopper, and P. Mineau, A critical assessment of the utility of fluctuating asymmetry as a biomarker of anthropogenic stress, in *Developmental Instability: Causes and Consequences*, Polak, M., Ed., Oxford University Press, New York, 2003.

Leung, K.M.Y., R.P.Y. Kwong, W.C. Ng, T. Horiguchi, J.W. Qiu, R. Yang, M. Song, G. Jiang, G.J. Zheng, and P.K.S. Lam, Ecological risk assessments of endocrine disrupting organotin compounds using marine neogastropods in Hong Kong. *Chemosphere*, 65, 922–938, 2006.

Leung, K.M.Y., J.R. Wheeler, D. Morritt, and M. Crane, Endocrine disruption in fishes and invertebrates: Issues for saltwater ecological risk assessment, in *Coastal and Estuarine Risk Assessment*, Newman, M.C., M.H. Roberts, Jr., and R.C. Hale, Eds., CRC Press, Boca Raton, FL, 2002.

Levin, S.A., M.A. Harwell, J.R. Kelly, and K.D. Kimball, *Ecotoxicology: Problems and Approaches*, Springer-Verlag, New York, 1989.

Levins, R., Some demographic and genetic consequences of environmental heterogeneity for biological control. *Bull. Entomol. Soc. Am.*, 15, 237–240, 1969.

Levinton, J.S., E. Suatoni, W. Wallace, R. Junkins, B. Kelaher, and B.J. Allen, Rapid loss of genetically based resistance to metals after the cleanup of a Superfund site. *Proc. Natl. Acad. Sci. U.S.A.*, 100, 9889–9891, 2003.

Lewis, G.N., The law of physico-chemical change. *Proc. Am. Acad. Arts Sci.*, 37, 49–69, 1901.

Lewis, S.A., P.H. Becker, and R.W. Furness, Mercury levels in eggs, internal tissues and feathers of herring gulls *Larus argentatus* from the German Wadden Sea coast. *Environ. Pollut.*, 80, 293–299, 1993.

Lewis, S.A. and R.W. Furness, Mercury accumulation and excretion in laboratory reared Black-headed gull *Larus ridibundus* chicks. *Arch. Environ. Contam. Toxicol.*, 21, 316–320, 1991.

Lewis, S.A. and R.W. Furness, The role of eggs in mercury excretion by quail *Coturnix coturnix* and the implications for monitoring mercury pollution by analysis of feathers. *Ecotoxicology*, 2, 55–64, 1993.

Lewis, S., R.H. Handy, B. Cordi, Z. Billinghurst, and M.H. DePledge, Stress proteins (HSP's): Methods of detection and their use as an environmental biomarker. *Ecotoxicology*, 8, 351–368, 1999.

Li, M.I., Ecological restoration of mineland with particular reference to the metalliferous mine wasteland in China: A review of research and practice. *Sci. Total Environ.*, 357, 38–53, 2006.

Li, A.P. and R.H. Heflich, *Genetic Toxicology*, CRC Press, Boca Raton, FL, 1991.

Li, Z., The Incidence of Imposex in Hong Kong and the value of *Thais clavigera* (Gastropoda: Muricidae) as a Bioindicator of TBT Pollution. PhD Diss., The University of Hong Kong, Hong Kong, China, 2000.

Li, X., M.A. Schuler, and M.R. Berenbaum, Molecular mechanisms of metabolic resistance to synthetic and natural xenobiotics. *Annu. Rev. Entomol.*, 52, 231–253, 2007.

Li, Q., H. Zhang, X. Liu, and J. Huang, Urban heat island effect on annual mean temperature during the last 50 years in China. *Theor. Appl. Climatol.* 79, 165–174, 2004.

Liber, K., N.K. Kaushik, K.R. Solomon, and J.H. Carey, Experimental designs for aquatic mesocosm studies: A comparison of the "ANOVA" and "Regression" design for assessing the impact of tetrachlorophenol on zooplankton populations in limnocorrals. *Environ. Toxicol. Chem.*, 11, 61–77, 1992.

Lide, D.R., *CRC Handbook of Chemistry and Physics,* 73rd Edition, CRC Press, Boca Raton, FL, 1992.

Likens, G.E., Acid precipitation. *Chem. Eng. News*, Nov. 22, 29–44, 1976.

Likens, G.E. and F.H. Borman, Acid rain: A serious environmental problem. *Science*, 184, 1176–1179, 1974.

Lillebø, A.I., D.F.R. Cleary, B. Marques, A. Reis, T. Lopes de Silva, and R. Calado, Ragworm fatty acid profiles reveal habitat and trophic interactions with halophytes and with mercury. *Mar. Pollut. Bull.*, 64, 2528–2534, 2012.

Lindberg, S.E., R.C. Harris, and R.R. Turner, Atmospheric deposition of metals to forest vegetation. *Science*, 215, 1609–1612, 1982.

Lindqvist, L. and M. Block, Excretion of cadmium and zinc during moulting in the grasshopper *Omocestus viridulus* (Orthoptera). *Environ. Toxicol. Chem.*, 13, 1669–1672, 1994.

Lindzen, R.S., On the scientific basis for global warming scenarios. *Environ. Pollut.*, 83, 125–134, 1994.

Linke-Gamenick, I., V.E. Forbes, and R.M. Sibly, Density-dependent effects of a toxicant on life-history traits and population dynamics of a capitellid polychaete. *Mar. Ecol. Prog. Ser.*, 184, 139–148, 1999.

Lipinski, C.A., F. Lombardo, B.W. Dominy, and P.J. Feeney, Experimental and computational approaches to estimate solubility and permeability in drug discovery and development settings. *Adv. Drug Deliver. Rev.*, 46, 3–26, 2001.

Lipman, Z., A dirty dilemma. A hazardous waste trade. *Harvard Int. Rev.*, 2, 67–71, 2002.

Lipnick, R.L., A perspective on quantitative structure-activity relationships in ecotoxicology. *Environ. Toxicol. Chem.*, 4, 255–257, 1985.

Lipnick, R.L., Structure-activity relationships, in *Fundamentals of Aquatic Toxicology: Effects, Environmental Fate, and Risk Assessment,* 2nd Edition, Rand, G.M., Ed., Taylor & Francis, Washington, DC, 1995.

Litchfield, J.T., A method for rapid graphic solution of time-per cent effect curves. *J. Pharmacol. Exp. Ther.*, 97, 399–408, 1949.

Litchfield, J.T. and F. Wilcoxon, A simplified method of evaluating dose-effect experiments. *J. Pharmacol. Exp. Ther.*, 96, 99–113, 1949.

Little, E.E., Behavioral toxicology: Stimulating challenges for a growing discipline. *Environ. Toxicol. Chem.*, 9, 1–2, 1990.

Little, E.E., J.F. Fairchild, and A.J. DeLonay, Behavioral methods for assessing impacts of contaminants on early life stage fishes, in *Water Quality and the Early Life Stages of Fishes*, Fuiman, L.A., Ed., American Fisheries Society, Bethesda, MD, 1993.

Little, E.E. and S.E. Finger, Swimming behavior as an indicator of sublethal toxicity in fish. *Environ. Toxicol. Chem.*, 9, 13–20, 1990.

Liu, Y., J. Binz, M.J. Numerick, S. Dennis, G. Luo, B. Desai, K.I. MacKensie, T.A. Mansfield, S.A. Kliewer, B. Goodwin, and S.A. Jones, Hepatoprotection by the farnesoid X receptor agonist GW4064 in rat models of intra- and extrahepatic cholestasis. *J. Clin. Invest.*, 112, 1678–1687, 2003.

Liu, J. and J. Diamond, China's environment in a globalizing world. *Nature*, 435, 1179–1186, 2005.

Liu, Y.P., J. Liu, M.B. Iszard, G.K. Andrews, R.D. Palmiter, and C.D. Klaassen, Transgenic mice that overexpress metallothionein-I are protected from cadmium lethality and hepatotoxicity, *Toxicol. Appl. Pharmacol.*, 135, 222–228, 1995.

Liu, C., K. Yu, X. Shi, J. Wang, P.K.S. Lam, R.S.S. Wu, and B. Zhuo, Induction of oxidative stress and apoptosis by PFOS and PFOA in primary cultured heptaocytes of freshwater tilapia (*Oreochromis niloticus*). *Aquat. Toxicol.*, 82, 135–143, 2007.

Lloyd, R. and D.W.M. Herbert, The influence of carbon dioxide on the toxicity of un-ionized ammonia to rainbow trout *(Salmo gairdneri* Richardson). *Ann. Appl. Biol.*, 48, 399–404, 1960.

Lloyd, R. and L.D. Orr, The diuretic response by rainbow trout to sub-lethal concentrations of ammonia. *Water Res.*, 3, 335–344, 1969.

Lobinski, R., J.S. Becker, H. Haraguchi, and B. Sarkar. Metallomics: Guidelines for terminology and critical evaluation of analytical chemistry approaches (IUPAC Technical Report). *Pure Appl. Chem.*, 82, 493–504, 2010.

Locke, J., *An Essay Concerning Human Understanding*, Collated by A.C. Foster in 1959. Dover Publications, New York, 1690.

Lofts, S. and E. Tipping. Assessing WHAM/Model VII against field measurements of free metal ion concentrations: Model performance and the role of uncertainty in parameters and inputs. *Environ. Chem.*, 8, 501–516, 2011.

Loganathan, B.G. and K. Kannan, Global organochlorine contamination trends: An overview. *Ambio*, 23, 187–191, 1994.

Lomborg, B., *The Skeptical Environmentalist*, Cambridge University Press, Cambridge, United Kingdom, 2001.

Long, E.R., D.D. MacDonald, J.C. Cubbage, and C.G. Ingersoll, Predicting the toxicity of sediment-associated trace metals with simultaneously extracted trace metal: Acid-volatile sulfide concentrations and dry weight-normalized concentrations: A critical comparison. *Environ. Toxicol. Chem.*, 17, 972–974, 1998.

Lorber, M. and L. Phillips, Infant exposure to dioxin-like compounds in breast milk. *Environ. Health Perspect.*, 110, A325–A331, 2002.

Loreau, M., Are communities saturated? On the relationship between α, β and γ diversity. *Eco. Lett.*, 3, 73–76, 2000.

Loreau, M., Linking biodiversity and ecosystems: Towards a unifying ecological theory. *Philos. Trans. R. Soc., B*, 365, 49–60, 2010.

Lotrich, V.A., Summer home range and movements of *Fundulus heteroclitus* (Pisces: Cyprinodontidae) in a tidal creek. *Ecology*, 56, 191–198, 1975.

Lovelock, J.E., *The Ages of Gaia. A Biography of Our Living Earth*, Oxford University Press, Oxford, 1988.

Lovelock, J.E., *Healing Gaia: Practical Medicine for the Planet*, Gaia Books Limited, New York, 1991.

Lovelock, J.E., The living Earth. *Nature*, 426, 769–770, 2003.

Lu, Y., K. Hiza, M. Moto, T. Takeshita, T. Takeuchi, and T. Saito, Genotoxic effects of alpha-endosulfan & beta-endosulfan on human HEPG2 Cells. *Environ. Health Perspt.*, 108, 559–561, 2000.

Lu, B.R. and A.A. Snow, Gene flow from genetically modified rice and its environmental consequences. *Bioscience*, 55, 669–678, 2005.

Lubick, N., Orcas remain burdened by PCBs. *Environ. Sci. Technol.*, 41, 6318, 2007a.

Lubick, N., DDT's resurrection. *Environ. Sci. Technol.*, 41, 6323–6325, 2007b.

Ludvig, W., *Das Rechts-Links Problem im Tierreich und beim Menschen*. Springer, Berlin, Germany, 1932.

Luebke, R.W., P.V. Hodson, M. Fiasal, P.S. Ross, K.A. Grasman, and J. Zelikoff, Aquatic pollution-induced immunotoxicity in wildlife species. *Fund. Appl. Toxicol.*, 37, 1–15, 1997.

Lundberg, P., E. Ranta, and V. Kaitala, Species loss leads to community closure. *Eco. Lett.*, 3, 465–468, 2000.

Lundqvist, M.J., J. Stigler, T. Cedervall, T. Berggård, M.B. Flanagan, I. Lynch, G. Elia, and K. Dawson, The evolution of the protein corona around nanoparticles: A test study. *ACS Nano*, 5, 7503–7509, 2011.

Lundqvist, M., J. Stigler, G. Elia, I. Lynch, T. Cedervall, and K.A. Dawson, Nanoparticle size and surface properties determine the protein corona with possible implications for biological impacts. *Proc. Natl. Acad. Sci. U.S.A.*, 105, 14265–14270, 2008.

Luoma, S.N., Can we determine the biological availability of sediment-bound trace elements? *Hydrobiolia*, 176/177, 379–396, 1989.

Luoma, J.R., New effect of pollutants: Hormone mayhem. *New York Times*, May 24, 1992.

Luoma, S.N. and G.W. Bryan, Factors controlling the availability of sediment-bound lead to the estuarine bivalve, *Scrobicularia plana. J. Mar. Biol. Assoc. U.K.*, 58, 793–802, 1978.

Luoma, S.N. and P.S. Rainbow, *Metal Contamination in Aquatic Environments: Science and Lateral Management*, Cambridge University Press, Cambridge, United Kingdom, 2008.

Lürling, M. and M. Scheffer, Info-disruption: Pollution and the transfer of chemical information between organisms. *Trends Ecol. Evol.*, 22, 374–379, 2007.

Lynch, I. and K.A. Dawson, Protein-nanoparticle interactions. *Nano Today*, 3, 40–47, 2008.

Lyon, R., M. Taylor, and K. Simkiss, Ligand activity in the clearance of metals from the blood of the crayfish (*Austropotamobius pallipes*). *J. Exp. Biol.*, 113, 19–27, 1984.

Lytle, T.F. and J.S. Lytle, Growth inhibition as indicator of stress because of atrazine following multiple exposure of the freshwater macrophyte, *Juncus effusus* L., *Environ. Toxicol. Chem.*, 24, 1198–1203, 2005.

Lyytikäinen, M., P. Hirva, P. Minkkinen, H. Hämäläinen, A.L. Rantalainen, P. Mikkelson, J. Paasivirta, and J.V.K. Kukkonen, Bioavailability of sediment-associated PCDD/Fs and PCDEs: Relative importance of contaminant and sediment characteristics and biological factors. *Environ. Sci. Technol.*, 37, 3926–3934, 2003.

MacDonald, R., D. Mackay, and B. Hickie, Contaminant amplification in the environment. *Environ. Sci. Technol.*, 36, 457A–462A, 2002.

Mackay, D., Correlation of bioconcentration factors. *Environ. Sci. Technol.*, 16, 274–278, 1982.

Mackay, D., *Multimedia Environmental Models: The Fugacity Approach*, Lewis Publishers, Boca Raton, FL, 1991.

Mackay, D., *Multimedia Environmental Models. The Fugacity Approach,* 2nd Edition, Lewis Publishers, Boca Raton, FL, 2001.

Mackay, D., J.A. Arnot, F.A.P.C. Gobas, and D.E. Powell. Mathematical relationships between metrics of chemical bioaccumulation in fish. *Environ. Toxicol. Chem.*, 32, 1459–1466, 2013.

Mackay, D. and S. Paterson, Fugacity revisited. The fugacity approach to environmental transport. *Environ. Sci. Technol.*, 116, 654A–660A, 1982.

Mackenthun, K.M. and J.I. Bregman, *Environmental Regulations Handbook*, Lewis Publishers, Boca Raton, FL, 1992.

Mackey, E.A., R.D. Oflaz, M.S. Epstein, B. Buehler, B.J. Porter, T. Rowles, S.A. Wise, and P.R. Becker, Elemental composition of liver and kidney tissues of rough-toothed dolphin (*Steno bredanensis*). *Arch. Environ. Contam. Toxicol.*, 44, 523–532, 2003.

Mackie, J.A., J.S. Levinton, R. Przeslawski, D. DeLambert, and W. Wallace, Loss of evolutionary resistance by the oligochaete *Limnodrilus hoffmeisteri* to a toxic substance: cost or gene flow? *Evolution*, 64–1, 152–165, 2009.

Macklis, R.M., The great radium scandal. *Sci. Am.*, August 1993, 94–99, 1993.

Macovsky, L.M., The effects of toxicant related mortality upon metapopulation dynamics: A laboratory model. M. S. thesis, Western Washington University, Bellingham, WA, 1999.

Madamanchi, N.R. and R.G. Alscher, Metabolic bases for differences in sensitivity of two pea cultivars to sulfur dioxide. *Plant Physiol.*, 97, 88–93, 1991.

Magurran, A.E., *Ecological Diversity and Its Measurement*, Princeton University Press, Princeton, NJ, 1988.

Majerus, M.E.N., *Melanism: Evolution in Action*, Oxford University Press, Oxford, 1998.

Majerus, M.E.N., Industrial melanism in the peppered moth, *Biston betularia*: An excellent teaching example of Darwinian evolution in action. *Evo. Edu. Outreach.*, 2, 63–74, 2009.

Malhotra, S.S. and D. Hocking, Biochemical and cytological effects of sulphur dioxide on plant metabolism. *New Phytol.*, 76, 227–237, 1976.

Malins, D.C., Identification of hydroxyl radical-induced lesions in DNA base structure: Biomarkers with a putative link to cancer development. *J. Toxicol. Environ. Health*, 40, 247–261, 1993.

Malins, D.C., B.B. McCain, D.W. Brown, S.L. Chan, M.S. Myers, and J.T. Landahl, Chemical pollutants in sediments and diseases of bottom-dwelling fish in Puget Sound, Washington. *Environ. Sci. Technol.*, 18, 705–713, 1984.

Malins, D.C. and G.K. Ostrander, *Aquatic Toxicology: Molecular, Biochemical, and Cellular Perspectives*, CRC Press, Boca Raton, FL, 1994.

Mallatt, J., Fish gill structural changes induced by toxicants and other irritants: A statistical review. *Can. J. Fish. Aquat. Sci.*, 42, 630–648, 1985.

Mallet, J., The evolution of insecticide resistance: Have the insects won? *Trends Ecol. Evol.*, 4, 336–340, 1989.

Malm, O., Gold mining as a source of mercury exposure in the Brazilian Amazon. *Environ. Res.*, 77, 73–78, 1998.

Manahan, S.E., *Fundamentals of Environmental Chemistry*, Lewis Publishers, Chelsea, MI, 1993.

Manahan, S.E., *Fundamentals of Environmental Chemistry*, Taylor & Francis/CRC Press, Boca Raton, FL, 2000.

Mance, G., *Pollution Threat of Heavy Metals in Aquatic Environments*, Elsevier Applied Science, London, United Kingdom, 1987.

Mann, R.M., R.V. Hyne, C.B. Choung, and S.P. Wilson, Amphibians and agricultural chemicals: Review of the risks in a complex environment. *Environ. Pollut.*, 157, 2903–2927, 2009.

Marcial, H.S., A. Hagiwara, and T.W. Snell, Estrogenic compounds affect development of harpacticoid copepod *Tigriopus japonicus*. *Environ. Toxicol. Chem.*, 22, 3025–3030, 2003.

Marciano-Cabral, F., Biology of *Naegleria* spp. *Microbiol. Rev.*, 52, 114–133, 1988.

Marco, G.J., R.M. Hollingworth, and W. Durham, *Silent Spring Revisited*, American Chemical Society, Washington, DC, 1987.

Maret, W. and A. Krezel, Cellular zinc and redox buffering capacity of metallothionein/thionein in health and disease. *Mol. Med.*, 13, 371–375, 2007.

Margoshes, M. and B.L. Vallee, A cadmium protein from equine kidney cortex. *J. Am. Chem. Soc.*, 79, 4813–4814, 1957.

Margulis, L. and J.E. Lovelock, Gaia and geognosy, in *Global Ecology: Towards a Science of the Biosphere*, Rambler, M.B., L. Margulis, and R. Fester, Eds., Academic Press, Boston, MA, 1989.

Markham, A., *A Brief History of Pollution*, Earthscan Publications Ltd, London, United Kingdom, 1995.

Marking, L.L., Toxicity of chemical mixtures, in *Fundamentals of Aquatic Toxicology*, Rand, G.M., and S.R. Petrocelli, Eds., Hemisphere Publishing Corp., Washington, DC, 1985.

Marking, L.L. and V.K. Dawson, Method for assessment of toxicity or efficacy of mixtures of chemicals. *U.S. Fish Wildl. Serv. Invest. Fish Control*, 67, 1–8, 1975.

Maroni, G., J. Wise, J.E. Young, and E. Otto, Metallothionein gene duplications and metal tolerance in natural populations of *Drosophila melanogaster*. *Genetics*, 117, 739–744, 1987.

Marshall, J.S., The effects of continuous gamma radiation on the intrinsic rate of natural increase of *Daphnia pulex*. *Ecology*, 43, 598–607, 1962.

Marshall, E., EPA may allow more lead in gasoline. *Science*, 215, 1375–1378, 1982.

Martin, J.W., K. Kannan, U. Berger, P. deVoogt, J. Field, J.P. Giesy, T. Harner et al, Researchers push for progress in perfluoralkyl analysis. *Environ. Sci. Technol.*, 38, 248A, 2004b.

Martin, J.W., S.A. Mabury, D.M. Whittle, and D.C.G. Muir, Perfluoroalkyl contaminants in a food web from Lake Ontario. *Environ. Sci. Technol.*, 38, 5379–5385, 2004a.

Martineau, D., K. Lemberger, A. Dallaire, P. Labelle, T.P. Lipscomb, P. Michel, and I. Mikaelian, Cancer in wildlife, a case study: Beluga from the St. Lawrence estuary, Québec, Canada. *Environ. Health Perspect.*, 110, 285–292, 2002.

Martinez V.J.L., M. Moreno Frias, A. Garrido Frenich, F. Olea-Serrano, and N. Olea, Determination of endocrine-disrupting pesticides and polychlorinated biphenyls in human serum by GC-ECD and GC-MS-MS and evaluation of contribution to the uncertainty of results. *Anal. Bioanal. Chem.*, 372, 766–775 2002.

Martínez-Jerónimo, F., R. Villaseñor, F. Espinosa, and G. Rios, Use of life-tables and application factors for evaluating chronic toxicity of Kraft mill wastes on *Daphnia magna. Bull. Environ. Contam. Toxicol.*, 50, 377–384, 1993.

Martoja, R. and D. Viale, Accumulation de granules de séléniure mercurique dans le foie d'Odontocètes (Mammiféres, Cétacés): Un mécanisme possible de détoxication du méthylmercure par le sélénium. *C.R. Acad. Sci. D*, 285, 109–112, 1977.

Marubini, E. and M.G. Valsecchi, *Analysing Survival Data from Clinical Trials and Observational Studies*, Wiley, New York, 1995.

Marwood, C.A., K.T.J. Bestari, R.W. Gensemer, K.R. Solomon, and B.M. Greenberg, Cresote toxicity to photosynthesis and plant growth in aquatic microcosms. *Environ. Toxicol. Chem.*, 22, 1075–1085, 2003.

Mason, A.Z. and K.D. Jenkins, Metal detoxification in aquatic organisms, in *Metal Speciation and Bioavailability in Aquatic Systems*, Tessier A. and D.Turner, Eds., J. Wiley & Sons, Chichester, United Kingdom, 1995.

Mason, A.Z. and R.F. Meraz, Cytosolic metal speciation studies by sequential, on-line, SE-IE/HPLC-ICP-MS. *Mar. Environ. Res.*, 46, 573–577, 1998.

Mason, A.Z. and J.A. Nott, The role of intracellular biomineralized granules in the regulation and detoxification of metals in gastropods with special reference to the marine prosobranch *Littorina littorea. Aquat. Toxicol.*, 1, 239–256, 1981.

Mason, R.P., J.R. Reinfelder, and F.M.M. Morel, Uptake, toxicity, and trophic transfer of mercury in a coastal diatom. *Environ. Sci. Technol.*, 30, 1835–1845, 1996.

Mason, R.P. and G.R. Sheu, Role of the ocean in the global mercury cycle. *Global Biogeochem. Cycles*, 16, 1093, 2002.

Mason, A.Z., K. Simkiss, and K.P. Ryan, The ultrastructural localization of metals in specimens of *Littorina littorea* collected from clean and polluted sites. *J. Mar. Biol. Assoc. U.K.*, 64, 699–720, 1984.

Masters, B.A., E.J. Kelly, C.J. Quaife, R.L. Brinster, and R.D. Palmiter, Targeted disruption of metallothionein I and II genes increases sensitivity to cadmium. *Proc. Natl. Acad. Sci. U.S.A.*, 91, 584–588, 1994a.

Maswood, E., Climate panel confirms human role in warming, fights off oil states. *Nature*, 378, 524, 1995.

Mather, K., Genetical control of stability in development. *Heredity*, 7, 297–336, 1953.

Matson, C.W., G. Palatnikov, G. Islamzadeh, T.J. McDonald, R.L. Autenrieth, K.C. Donnelly, and J.W. Bickham, Chromosomal damage in two species of aquatic turtles (*Emys orbicularis* and *Mauremys caspica*) inhabiting contaminated sites in Azerbaijan. *Ecotoxicology*, 14, 1–13, 2005.

Matthews, R.A., W.G. Landis, and G.B. Matthews, The community conditioning hypothesis and its application to environmental toxicology. *Environ. Toxicol. Chem.*, 15, 597–603, 1996.

Matthiessen, P., T. Reynoldson, Z. Billinghurst, D.W. Brassard, P. Cameron, G.T. Chandler, I.M. Davis et al, *Endocrine Disruption in Invertebrates: Endocrinology Testing and Assessment*, SETAC Press, Pensacola, FL, 1999.

Maugh, T.H., It isn't easy being king. *Science*, 203, 637, 1974.

Maurer, B.A. and R.D. Holt, Effects of chronic pesticide stress on wildlife populations in complex landscapes: Processes at multiple scales. *Environ. Toxicol. Chem.*, 15, 420–426, 1996.

Mauri, M., E. Baraldi, and R. Simonini, Effects of zinc exposure on the polychaete *Dinophilus gyrociliatus*: A life-table response experiment. *Aquat. Toxicol.*, 65, 93–100, 2003.

May, R.M., Biological populations with nonoverlapping generations: Stable points, stable cycles, and chaos. *Science*, 186, 645–647, 1974.

May, R.M., *Theoretical Ecology. Principles and Applications*, W.B. Saunders, Philadelphia, PA, 1976a.

May, R.M., Simple mathematical models with very complicated dynamics. *Nature*, 261, 459–467, 1976b.

Mayer, F.L. and P.M. Mehrle, Toxicological aspects of toxaphene in fish: A summary. *Trans. N. Am. Wildl. Nat. Resour. Conf.*, 42, 365–373, 1977.

Mayer, F.L., Residue dynamics of di-2-ethylhexyl phthalate in fathead minnows (*Pimephales promelas*). *J. Fish. Res. Bd. Canada*, 33, 2610–2613, 1976.

Mayne, G.J., C.A. Bishop, P.A. Martin, H.J. Boermans, and B. Hunter, Thyroid function in nestling tree swallows and eastern bluebirds exposed to non-persistent pesticides and p,p'-DDE I apple orchards of southern Ontario, Canada. *Ecotoxicology*, 14, 381–396, 2005.

McAdams, H.H. and A. Arkin, It's a noisy business! Genetic regulation at the nanomolecular scale. *Trends Genet.*, 15, 65–69, 1999.

McBee, K., J.W. Bickham, K.W. Brown, and K.C. Donnelly, Chromosomal aberrations in native small mammals (*Peromyscus leucopus* and *Sigmodon hispidus*) at a petrochemical waste disposal site. I. Standard karyology. *Arch. Environ. Contam. Toxicol.*, 16, 681–688, 1987.

McBee, K. and R.L. Lochmiller, Wildlife toxicology in biomonitoring and bioremediation: Implications for human health, in *Ecotoxicity and Human Health: A Biological Approach to Environmental Remediation*, de Serres, F.J. and Bloom, A.D., Eds., CRC/Lewis Publishers, Boca Raton, 1996.

McBride, W.G., Thalidomide and congenital anomalies. *Lancet*, 2, 1358, 1961.

McCann, K.S., The diversity-stability debate. *Nature*, 405, 228–233, 2000.

McCann, K., A. Hastings, and G.R. Huxel, Weak trophic interactions and the balance of nature, *Nature*, 395, 794–798, 1998.

McCarthy, J.R., R.S. Halbrook, and L.R. Shugart, *Conceptual Strategy for Design, Implementation, and Validation of a Biomarker-Based Biomonitoring Capability*, Oak Ridge National Laboratory, Oak Ridge, TN, 1991.

McCarthy, J.F. and L.R. Shugart, *Biomarkers of Environmental Contamination*, Lewis Publishers, Chelsea, MI, 1990.

McCloskey, D.N., The insignificance of statistical significance. *Am. Sci.*, 272, 32–33, 1995.

McCloskey, J.T., and J.T. Oris, Effect of anthracene and solar ultraviolet radiation exposure on gill ATPase and selected hematologic measurements in the bluegill sunfish. *Aquat. Toxicol.*, 24, 207–218, 1993.

McCloskey, J.T., I.R. Schultz, and M.C. Newman, Estimating the oral bioavailability of methylmercury in channel catfish (*Ictalurus punctatus*). *Environ. Toxicol. Chem.*, 17, 1524–1529, 1998.

McCormick, P.V., J. Cairns, Jr., S.E. Belanger, and E.P. Smith, Response of protistan assemblages to a model toxicant, the surfactant C12-TMAC (dodecyl trimethyl ammonium chloride), in laboratory streams. *Aquat. Toxicol.*, 21, 41–70, 1991.

McDonald, B.G., and P.M. Chapman, PAH phototoxicity: An ecologically irrelevant phenomenon? *Mar. Pollut. Bull.*, 44, 1321–1326, 2002.

McElwee, C.R., *Environmental Law in China. Managing Risk and Ensuring Compliance*, Oxford University Press, Oxford, United Kingdom, 2011.

McFarlane, G.A. and W.G. Franzin, Elevated heavy metals: A stress on a population of white suckers, *Catostomus commersoni*, in Hamell Lake, Saskatchewan. *J. Fish. Res. Board Can.*, 35, 963–970, 1978.

McGraw, J.B. and H. Caswell, Estimation of individual fitness from life-history data. *Am. Nat.*, 147, 47–64, 1996.

McGregor, G.I., *Environmental Law and Enforcement*, Lewis Publishers, Boca Raton, FL, 1994.

McIntosh, A., Trace metals in freshwater sediments: A review of the literature and an assessment of research needs, in *Metal Ecotoxicology. Concepts and Applications*, Newman, M.C. and A.W. McIntosh, Eds., Lewis Publishers, Chelsea, MI, 1991.

McKim, J.M., Early life stage tests, in *Fundamentals of Aquatic Toxicology*, Rand, G.M., and S.R. Petrocelli, Eds., Hemisphere Publishing Corp., Washington, DC, 1985.

McLachlan, J.A., Functional toxicology: A new approach to detect biologically active xenobiotics. *Environ. Health Perspect.*, 101, 386–387, 1993.

McLachlan, M.S., Bioaccumulation of hydrophobic chemicals in agricultural food chains. *Environ. Sci. Technol.*, 30, 252–259, 1996.

Mc Mahon, T.A., N.T. Halstead, S. Johnson, T.R. Raffel, J.M. Romansic, P.W. Crumrine, R.K. Boughton, L.B. Martin, and J.R. Rohr, The fungicide chlorothalonil is nonlinearly associated with corticosterone levels, immunity, and mortality in amphibians. *Environ. Health Perspt.*, 119, 1098–1103, 2011.

McMurry, S.T., Development of an in situ mammalian biomonitor to assess the effect of environmental contaminants on population and community health. Ph.D. Diss., Oklahoma State University, Stillwater, Oklahoma, 1993.

McNeill, J.R., *Something New under the Sun. An Environmental History of the Twentieth-Century World*, W.W. Norton & Co., New York, 2000.

McNeill, K.G. and G.A.D. Trojan, The cesium-potassium discrimination ratio. *Health Phys.*, 4, 109–112, 1960.

Meador J.P., E. Casillas, C.A. Sloan, and U. Varanasi, Comparative bioaccumulation of polycyclic aromatic hydrocarbons from sediment by two infaunal invertebrates. *Mar. Ecol.Prog. Ser.*, 123, 107–124, 1995b.

Meador J.P., J.E. Stein, W.L. Reichert, and U. Varanasi, A review of bioaccumulation of polycyclic aromatic hydrocarbons by marine organisms. *Rev. Environ. Contam. Toxicol.*, 143, 79–165, 1995a.

Medawar, P.B., *The Art of the Soluble*, Methuen, London, United Kingdom, 1967.

Medawar, P.B., *Pluto's Republic*, Oxford University Press, Oxford, 1982.

Medvedev, Z.A., *The Ural and Chernobyl Nuclear Accidents*, seminar presented September 5, 1995 at the University of Georgia's Savannah River Ecology Laboratory.

Mehrle, P.M. and F.L. Mayer, Biochemistry/physiology, in *Fundamentals of Aquatic Toxicology: Methods and Applications*, Rand, G.M. and S.R. Petrocelli, Eds., Hemisphere Publishing Corp., Washington, DC, 1985.

Mekenyan, O.G., G.T. Ankley, G.D. Veith, and D.J. Call, QSARs for photoinduced toxicity: I. Acute lethality of polycyclic aromatic hydrocarbons to *Daphnia magna. Chemosphere*, 28, 567–582, 1994.

Melancon, M.J., Bioindicators used in aquatic and terrestrial monitoring, in *Handbook of Ecotoxicology*, Hoffman, D.J., B.A. Rattner, G.A. Burton, Jr., and J. Cairns, Jr., Eds., CRC Press, Boca Raton, FL, 1995.

Menzel, R., M. Rodel, J. Kulas, and C.E. Steinberg. CYP35: Xenobiotically induced gene expression in the nematode *Caenorhabditis elegans. Arch Biochem Biophys*, 438, 93–102, 2005.

Merilä, J. and M. Bjorklund, Fluctuating asymmetry and measurement error. *Syst. Biol.*, 44, 97–101, 1995.

Mertz, W., The essential trace elements. *Science*, 213, 1332–1338, 1981.

Metcalfe, C.D., V.W. Cairns, and J.D. Fitzsimons, Experimental induction of liver tumors in rainbow trout (*Salmo gairdneri)* by contaminated sediment from Hamilton Harbor, Ontario. *Can. J. Fish. Aquat. Sci.*, 45, 2161–2167, 1988.

Meyer, H.H.D., Biochemistry and physiology of anabolic hormones used for improvement of meat production. *APMIS*, 109, 1–8, 2001.

Meyer, J.S., R.C. Santore, J.P. Bobbitt, L. Debrey, C.J. Boese, P.R. Paquin, H.E. Allen, H.L. Bergman, and D.M. Di Toro, Binding of nickel and copper to fish gills predicts toxicity when water hardness varies, but free-ion activity does not. *Environ. Sci. Technol.*, 33, 913–916, 1999.

Meyers, T.R. and J.D. Hendricks, Histopathology, in *Fundamentals of Aquatic Toxicology. Methods and Applications*, Rand, G.M. and S.R. Petrocelli, Eds., Hemisphere Publishing Corp., Washington, DC, 1985.

Meyers-Schöne, L.R. Shugart, J.J. Beauchamp, and B.T. Walton, Comparison of two freshwater turtle species as monitors of radionuclide and chemical contamination: DNA damage and residue analysis. *Environ. Toxicol. Chem.*, 12, 1487–1496, 1993.

Michaels, D., *Doubt is Their Product*, Oxford University Press, New York, 2008.

Michalska, A.E. and K.H.A. Choo, Targeting and germ-line transmission of a null mutation at the metallothionein I and II loci in mice. *Proc. Natl. Acad. Sci., USA*, 90, 8088–8092, 1993.

Michener, W.K., J.W. Brunt, and S.G. Stafford, *Environmental Information Management and Analysis: Ecosystem to Global Scales*, Taylor & Francis, Ltd., London, United Kingdom, 1994.

Miglioranza,K.S., J.E. Aizpún de Moreno, and V.J. Moreno, Land based sources of marine pollution: Organochlorine pesticide in stream system. *Environ. Sci. Pollut. Res.*, 11, 227–232, 2004a.

Miglioranza, K.S., J.E. De Monero, and V.J. Moreno, Organochlorine pesticide sequestered in the aquatic macrophyte, *Schoenoplectus ediformicus* from a shallow lake in Argentina. *Water Res.*, 38, 1765–1772, 2004b.

Mikol, Y.B., W.R. Richardson, W.H. Van der Schalie, T.R. Shedd, and M.W. Widder, An online real-time biomonitor for contaminant surveillance in water supplies. *J. Am. Water Works Assoc.*, 99, 107–115, 2007.

Millar, I.B. and P.A. Cooney, Urban lead: A study of environmental lead and its significance to school children in the vicinity of a major trunk road. *Atmos. Environ.*, 16, 615–620, 1982.

Miller, R.G., Jr., *Beyond ANOVA, Basics of Applied Statistics*, Wiley, New York, 1986.

Miller, G., C. Hoonhout, E. Sufka, S. Carroll, V. Edirveerasingam, B. Allen, J. Reuter, J. Oris, and M. Lico, Environmental assessment of the impacts of polycyclic aromatic hydrocarbons (PAH) in Lake Tahoe and Donner Lake. A final report to the California State Water Resources Control Board, 79 p, 2003.

Milne, C.J., D.G. Kinniburgh, and E. Tipping, Generic NICA-Donnan model parameters for proton binding by humic substances. *Environ. Sci. Technol.*, 35, 2049–2059, 2001.

Milne, C.J., D.G. Kinniburgh, W.H. Van Riemsdijk, and E. Tipping, Generic NICA-Donnan model parameters for metal-ion binding by humic substances. *Environ. Sci. Technol.*, 37, 958–971, 2003.

Milnes, M.R., A.R. Woodward, A.A. Rooney, and L.J. Guillette, Plasma steroid concentrations in relation to size and age in juvenile alligators from two Florida lakes. *Comp. Biochem. Physiol. A*, 131, 923–930, 2002.

Milodowski, A.E., J.M. West, J.M. Pearce, E.K. Hyslop, I.R. Basham, and P.J. Hooker, Uranium-mineralized microorganisms associated with uraniferous hydrocarbons in southeast Scotland. *Nature*, 347, 465–467, 1990.

Minagawa, M. and E. Wada, Stepwise enrichment of ^{15}N along food chains: Further evidence and the relation between $\delta\,^{15}N$ and animal age. *Geochim. Cosmochim. Acta*, 48, 1135–1140, 1984.

Mineau, P., A. Baril, B.T. Collins, J. Duffe, G. Hoermann, and R. Luttik, Reference values for comparing the acute toxicity of pesticides to birds. *Rev. Environ. Contam. Toxicol.*, 24, 24, 2001.

Mishima, J. and E.P. Odum, Excretion rate of Zn^{65} by *Littorina irrorata* in relation to temperature and body size. *Limnol. Oceanogr.*, 8, 39–44, 1963.

Moe, S.J., Density dependence in ecological risk assessment, in *Population-Level Ecological Risk Assessment*, Barnthouse, L.W., W.R., Jr., Munns, and M.T. Sorensen, Eds., CRC Press, Boca Raton, FL, 2008.

Moe, S.J., K. De Schamphelaera, W.H. Clements, M.T. Sorensen, P.J. Van den Brink, and M. Liess. Combined and interactive effects of global climate change and toxicants on populations and communities. *Environ. Toxicol. Chem.*, 32, 49–61, 2013.

Moe, S.J., N.C. Stenseth, and R.H. Smith, Effects of a toxicant on population growth rates: Sublethal and delayed responses in blowfly populations. *Funct. Ecol.*, 15, 712–721, 2001.

Mohr, S., R. Berghahn, M. Feibicke, S. Meinecke, T. Ottenströer, I. Schmiedling, R. Schmiediche, and R. Schmidt, Effects of the herbicide metazachlor on macrophytes and ecosystem function in freshwater pond and stream mesocosms. *Aquat. Toxicol.*, 82, 73–84, 2007.

Molander, S., H. Blanck, and M. Söderström, Toxicity assessment by pollution-induced community tolerance (PICT), and identification of metabolites in periphyton communities after exposure to 4,5,6-trichloroguaiacol. *Aquat. Toxicol.*, 18, 115–136, 1990.

Møller, A.P., Symmetry, size and stress. *Trends Ecol. Evol.*, 15, 330, 2000.

Møller, V., V.E. Forbes, and M.H. DePledge, Influence of acclimation and exposure temperature on the acute toxicity of cadmium to the freshwater snail *Potamophygus antipodarum* (Hydrobiidae). *Environ. Toxicol. Chem.*, 13, 1519–1524, 1994.

Møller, A.P. and J.P. Swaddle, *Asymmetry, Developmental Stability and Evolution*, Oxford University Press, Oxford, 1997.

Monk, C.D., Effects of short-term gamma irradiation on an old field. *Radiat. Bot.*, 6, 329–335, 1966.

Monkhouse, F.J., *A Dictionary of Geography*, Aldine Publishing Co., Chicago, IL, 1965.

Monserrat, J.M., P.E. MartRnez, L.A. Geractitano, L. Lund Amado, C.M. Gaspar Martins, G.L.L. Pinho, I. Soares Chaves, M. Ferreira-Cravo, J. Ventura-Lima, and A. Bianchini, Pollution biomarkers in estuarine animals: Critical review and new perspectives. *Comp. Biochem. Physiol. C.*, 146, 221–234, 2007.

Monsinjon, T. and T. Knigge, Proteomic applications in ecotoxicology. *Proteomics*, 7, 2997–3009, 2007.

Montiero, L.R. and R.W. Furness, Seabirds as monitors of mercury in the marine environment. *Water Air Soil Pollut.*, 80, 851–870, 1995.

Monteiro, L.R. and R.W. Furness, Accelerated increase in mercury contamination in north Atlantic mesopelagic food chains as indicated by time series of seabird feathers. *Environ. Toxicol. Chem.*, 16, 2489–2493, 1997.

Monteiro, L.R. and R.W. Furness, Kinetics, dose-response, excretion, and toxicity of methylmercury in free-living Cory's shearwater chicks. *Environ. Toxicol. Chem.*, 20, 1816–1823, 2001a.

Monteiro, L.R. and R.W. Furness, Kinetics, dose-response, and excretion of methylmercury in free-living adult Cory's shearwaters. *Environ. Sci. Technol.*, 35, 739–746, 2001b.

Monteiro, L.R., J.P. Granadeiro and R.W. Furness, Relationship between mercury levels and diet in Azores seabirds. *Mar. Ecol. Prog. Ser.*, 166, 259–265, 1998.

Moody, C.A. and J.A. Field, Perfluorinated surfactants and the environmental implications of their use in fire-fighting foams. *Environ. Sci. Technol.*, 34, 3864–3870, 2000.

Moolgavkar, S.H., Carcinogenesis modeling: From molecular biology to epidemiology. *Annu. Rev. Publ. Health*, 151–169, 1986.

Moore, A. and N. Lower, The impact of two pesticides on olfactory-mediated endocrine function in mature male Atlantic salmon (*Salmo salar* L.) parr. *Comp. Biochem. Physiol. B Biochem. Mol. Biol.*, 129, 269–276, 1986.

Moore, M.J. and M.S. Myers, Pathobiology of chemical-associated neoplasia in fish, in *Aquatic Toxicology. Molecular, Biochemical and Cellular Perspectives*, Malins, D.C. and G.K. Ostrander, Eds., CRC Press, Boca Raton, FL, 1994.

Moore M.J., D. Shea, R.E. Hillman, and J.J. Stegeman, Trends in hepatic tumors and hydropic vacuolation, fin erosion, organic chemical and stable isotope ratios in winter flounder, from Massachusetts, USA. *Mar. Pollut. Bull.*, 32, 458–470, 1996.

Moore M.J. and J.J. Stegeman, Hepatic neoplasms in winter flounder, *Pleuronectes americanus* from Boston Harbor, Massachusetts, USA. *Dis. Aquat. Org.*, 20, 33–48, 1994.

Moore, M.N., A. Viarengo, P. Donkin, and A.J.S. Hawkins, Autophagic and lysosomal reactions to stress in the hepatopancreas of blue mussels. *Aquat. Toxicol.*, 84, 80–91, 2007.

Moore, A. and C.P Waring, Mechanistic effects of a triazine pesticide on reproductive endocrine function in mature male Atlantic salmon (*Salmo salar* L.) parr. *Pest. Biochem. Physiol.*, 62, 41–50, 1998.

Mora, M.A., Transboundary pollution: Persistent organochlorine pesticides in migrant birds of the southwestern United States and Mexico. *Environ. Toxicol. Chem.*, 16, 3–11, 1997.

Moreels, D., P. Lodewijks, H. Zegers, E. Rurangwa, N. Vromant, L. Bastiaens, L. Diels, D. Springael, R. Merckx, and F. Ollevier, Effect of short-term exposure to methyl-*tert*-butyl ether and *tert*-butyl alcohol on the hatch rate and development of the African catfish, *Clarias gariepinus. Environ. Toxicol. Chem.*, 25, 514–519, 2006.

Morel, F.M.M., *Principles and Applications of Aquatic Chemistry*, John Wiley & Sons Ltd., NY, 1983.

Moriarty, F., *Ecotoxicology. The Study of Pollutants in Ecosystems*, Academic Press, London, United Kingdom, 1983.

Morrison, H.A., F.A.P.C. Gobas, R. Lazar, and D.G. Haffner, Development and verification of a bioaccumulation model for organic contaminants in benthic invertebrates. *Environ. Sci. Technol.*, 30, 3377–3384, 1996.

Morrison, H.A., F.A.P.C. Gobas, R. Lazar, D.M. Whittle, and D.G. Haffner, Development and verification of a benthic/pelagic food web bioaccumulation model for PCB congeners in western Lake Erie. *Environ. Sci. Technol.*, 31, 3267–3273, 1997.

Morse, J.W. and D. Rickard, Chemical dynamics od sedimentary acid volatile sulfide. *Environ. Sci. Technol.*, 38, 131A–136A, 2004.

Morton, B., The tidal rhythm and rhythm of feeding and digestion in *Cardium edule. J. Mar. Biol. Assoc. U.K.*, 50, 499–512, 1970.

Moser, G.A. and M.S. McLachlan, Modeling digestive tract absorption and desorption of lipophilic organic contaminants in humans. *Environ. Sci. Technol.*, 36, 3318–3325, 2002.

Mottaleb, M.A., W.C. Brumley, S.M. Pyle, and G.W. Sovocool, Determination of a bound musk xylene metabolite in carp hemoglobin as a biomarker of exposure by gas chromatography-mass spectrometry using selected ion monitoring. *J. Anal. Toxicol.*, 28, 581–586, 2004a.

Mottaleb, M.A., W.C. Brumley, and G.W. Sovocool, Nitro musk metabolites bound to carp hemoglobin: Determination by GC with two MS detection modes: EIMS versus electron capture negative ion MS. *Int. J. Environ. Anal. Chem.*, 84, 1069–1078, 2004b.

Mottaleb, M.A., X. Zhoa, L.R. Curtis, and G.W. Sovocool, Formation of nitromusk adducts of rainbow trout hemoglobin for potential use as biomarkers of exposure. *Environ. Toxicol. Chem.*, 67, 315–324, 2004c.

Mougi, A. and M. Kondon, Diversity of interaction types and ecological community stability. *Science*, 337, 349–351, 2012.

Mouneyrac, C., C. Amiard-Triquet, J.C. Amaird, and P.S. Rainbow, Comparison of metallothionein concentrations and tissue distribution of trace metals in crabs (*Pachygrapsus marmoratus*) from a metal-rich estuary, in and out of the reproductive season. *Comp. Biochem. Physiol. C*, 129, 193–209, 2001.

Mount, D.I. and C.E. Stephan, A method for establishing acceptable toxicant limits for fish: Malathion and 2,4-D. *Trans. Am. Fish. Soc.*, 96, 185–193, 1967.

Mouquet, N. and M. Loreau, Coexistence in metacommunities: The regional similarity hypothesis. *Am. Nat.*, 159, 420–426, 2002.

Mouquet, N. and M. Loreau, Community patterns in source-sink metacommunities. *Am. Nat.*, 162, 544–557, 2003.

Mueller, K.K., S. Lofts, C. Fortin, and P.G.C. Campbell, Trace metal speciation predictions in natural aquatic systems: Incorporation of dissolved organic matter (DOM) spectroscopic quality. *Environ. Chem.*, 9, 356–368, 2012.

Muir, D.C.G., N.P. Grift, W.L. Lockhart, P. Wilkinson, B.N. Billeck, and G.J. Brunskill, Spatial trends and historical profiles of organochlorine pesticides in Arctic lake sediments. *Sci. Total Environ.*, 160/161, 447–457, 1995.

Muir, D.C.G., R.J. Norstrom, and M. Simon, Organochlorine contaminants in Arctic marine food chains: Accumulation of specific polychlorinated biphenyls and chlordane-related compounds. *Environ. Sci. Technol.*, 22, 1071–1079, 1988.

Mulholland, P.J., A.V. Palumbo, J.W. Elwood, and A.D. Rosemond, Effect of acidification on leaf decomposition in streams, *J. N. Am. Benthol. Soc.*, 6, 147–158, 1987.

Mulvey, M. and S.A. Diamond, Genetic factors and tolerance acquisition in populations exposed to metals and metalloids, in *Metal Ecotoxicology: Concepts and Applications*, Newman, M.C. and A.W. McIntosh, Eds., Lewis Publishers, Chelsea, MI, 1991.

Mulvey, M., M.C. Newman, A. Chazal, M.M. Keklak, M.G. Heagler, and L.S. Hales, Jr., Genetic and demographic responses of mosquitofish (*Gambusia holbrooki* Girard 1859) populations stressed by mercury. *Environ. Toxicol. Chem.*, 14, 1411–1418, 1995.

Mulvey, M., M.C. Newman, W. Vogelbein, and M.A. Unger, Genetic structure of *Fundulus heteroclitus* from PAH-contaminated and neighboring sites in the Elizabeth and York Rivers. *Aquat. Toxicol.*, 61, 195–209, 2002.

Mulvey, M., M.C. Newman, W. Vogelbein, M.A. Unger, and D.R. Ownby, Genetic structure and mtDNA diversity of *Fundulus heteroclitus* populations from PAH-contaminated and neighboring sites. *Environ. Toxicol. Chem.*, 22, 671–677, 2003.

Mumtaz, M.M., C.T. De Rosa, J. Groten, V.J. Feron, H. Hansen, and P.R. Durkin, Estimation of toxicity of chemical mixtures through modeling of chemical interactions. *Environ. Health Perspect.*, 106, 1353–1360, 1998.

Munkittrick, K.R. and D.G. Dixon, Growth, fecundity, and energy stores of white suckers (*Catostomus commersoni*) from lakes containing elevated levels of copper and zinc. *Can. J. Fish. Aquat. Sci.*, 45, 1355–1365, 1988.

Murdoch, M.H. and P.D.N. Hebert, Mitochondrial DNA diversity of brown bullhead from contaminated and relatively prostine sites in the Great Lakes. *Environ. Toxicol. Chem.*, 8, 1281–1289, 1994.

Murphy, D.L. and J.W. Gooch, Accumulation of *cis* and *trans* chlordane by channel catfish during dietary exposure. *Arch. Environ. Contam. Toxicol.*, 29, 297–301, 1995.

Murphy, P.J. and C.W. Thomas, The synthesis and biological activity of the marine metabolite cylindrospermopsin. *Chem. Soc. Rev.*, 30, 303–312, 2001.

Murray, D.L., *Cultivating Crisis: The Human Cost of Pesticides in Latin America*, University of Texas Press, Austin, TX, 1994.

Murray, D.L., C. Wesseling, M. Keifer, M. Corriols, and S. Henao, Surveillance of pesticide-related illness in the developing world. *Int. J. Occup. Environ. Health.*, 8, 243–248, 2002.

Murry, C.E., R.B. Jennings, and K.A.Reimer, Preconditioning with ischemia: A delay of lethal cell injury in ischemic myocardium. *Circulation*, 74, 1124–1136, 1986.

Musgrave, A., *Common Sense, Science and Scepticism*, Cambridge University Press, Cambridge, United Kingdom, 1993.

Myers, R.A., J.K. Baum, T.D. Shepard, S.P. Powers, and C.H. Peterson, Cascading effects of the loss of apex predatory sharks from a coastal ocean. *Science*, 1846–1850, 2007.

Myers M.S., L.D. Rhodes, and B.B. McCain, Pathologic anatomy and patterns of occurrence of hepatic neoplasms, putative preneoplastic lesions and other idiopathic hepatic conditions in English sole (*Parophrys vetulus*) from Puget Sound, Washington. *J. Natl. Cancer I.*, 78, 333–363, 1987.

Myers, M.S., C.M. Stehr, O.P. Olson, L.L. Johnson, B.B. McCain, S.L. Chan, and U. Varanasi, National Benthic Surveillance Project: Pacific Coast, fish histopathology and relationships between toxicopathic lesions and exposure to chemical contaminants for Cycles I to V (1984–88). U.S. Dept. Commer., NOAA Tech. Memo. NMFS/NWFSC-6, 160 p, 1993.

Myers, M.S., C.M. Stehr, O.P. Olson, L.L. Johnson, B.B. McCain, S.L. Chan, and U. Varanasi, Relationships between toxicopathic hepatic lesions and exposure to chemical contaminants in English sole (*Platichthys stellatus*), and white croaker (*Genyonemus lineatus*) from selected marine sites on the Pacific coast, USA. *Environ. Health Perspect.*, 102, 200–215, 1994.

Nacci, D.E., D. Champlin, and S. Jayaraman, Adaptation of the estuarine fish *Fundulus heteroclitus* (Atlantic killifish) to polychlorinated biphenyls (PCBs). *Estuaries Coast*, 33, 853–864, 2010.

Nacci, D.E., S. Walters, T. Gleason, and W.R. Munns, Using a spatial modeling approach to explore ecological factors relevant to the persistence of an estuarine fish (*Fundulus heteroclitus*) in a PCB-contaminated estuary, in *Demographic Toxicity: Methods in Ecological Risk Assessment*, H.R. Akçakaya, J.D. Stark and T.S. Bridges, Eds., Oxford University Press, Oxford, 2008.

Nadin, P. Ed., *The Use of Plant Protection Products in the European Union, Data 1992–2003*. EUROSTAT Statistical Books, European Communities, Luxembourg, 2007.

Nagel, E., *The Structure of Science. Problems in the Logic of Scientific Explanation*, Harcourt, Brace & World, New York, 1961.

Naimon, J.S., Using expert panels to assess risks of environmental biotechnology applications: A case study of the 1986 Frostban7 risk assessments, in *Risk Assessment in Genetic Engineering. Environmental Release of Organisms*, Levin, M. and H.Strauss, Eds., McGraw-Hill, New York, 1991.

Nakayama, K., Y. Oshima, K. Hiramatsu, Y. Shimasaki, and T. Honjo, Effects of polychlorinated biphenyl on the schooling behavior of Japanese medaka (*Oryzias latipes*). *Environ. Toxicol. Chem.*, 24, 2588–2593, 2005.

Naqvi, S.M., and C. Vaishnavi, Bioaccumulative potential and toxicity of endosulfan insecticide to nontarget animals. *Comp. Biochem. Physiol. C*, 105, 347–361, 1993.

Nash, J.P., D.E. Kime, L.T.M. Van der Ven, P.W. Wester, F. Brion, G. Maack, P. Stahlschmidt-Allner, and C.R. Tyler, Long-term exposure to environmental concentrations of the pharmaceutical ethynylestradiol causes reproductive failure in fish. *Environ. Health Perspect.*, 112, 1725–1733, 2004.

National Academy of Sciences (NAS), *Drinking Water and Health*, Safe Drinking Water Committee, Washington, DC, 1977.

National Research Council, *Risk Assessment in the Federal Government: Managing the Process*, National Academy Press, Washington, DC, 1983.

National Research Council, *Hormonally Active Agents in the Environment*, National Academy Press, Washington, DC, 1999.

National Toxicology Program, *The Ninth Report on Carcinogens*, U.S. Department of Health and Human Services, Public Health Service National Toxicology Program, Washington, DC., 1999.

Natural Resource Damage Assessments, Final Rule. *Federal Register*, 51, 27674–27753, 1986.

Nauen, R. and I. Denholm, Resistance of insect pests to neonicotinoid insecticides: Current status and future prospects. *Arch. Insect Biochem.*, 58, 200–215, 2005.

Nayak, B.N. and M.L. Petras, Environmental monitoring for genotoxicity: In vivo sister chromatid exchange in house mouse (*Mus musculus*). *Can. J. Genet. Cytol.*, 27, 351–356, 1985.

Naz, R.K., Ed., *Endocrine Disruptors: Effects on Male and Female Reproductive Systems*, CRC Press, Boca Raton, FL, 1999.

Neathery, M.W. and W.J. Miller, Metabolism and toxicity of cadmium, mercury, and lead in animals: A review. *J. Diary Sci.*, 58, 1767–1781, 1975.

Nebert, D.W., and T.P. Dalton, The role of cytochrome P450 enzymes in endogenous signalling pathways and environmental carcinogenesis. *Nat. Rev. Cancer*, 6, 947–960, 2006.

Neely, W.B., *Introduction to Chemical Exposure and Risk Assessment*, CRC Press, Boca Raton, FL, 1994.

Neely, W.B., D.R. Bronson, and G.E. Blau, Partition coefficient to measure bioconcentration potential of organic chemicals in fish. *Environ. Sci. Technol.*, 8, 1113–1115, 1974.

Nelson, D.R., Comparison of P450s from human and fugu: 420 million years of vertebrate P450 evolution. *Arch. Biochem. Biophys.*, 409, 18–24, 2003.

Nelson, D.R., Cytochrome P450 nomenclature, 2004. *Methods Mol. Biol. (Clifton, N.J.)*, 320, 1–10, 2006.

Nelson, D.R., J.V. Goldstone, and J.J. Stegeman. The cytochrome P450 genesis locus: The origin and evolution of animal cytochrome P450s. *Philos. Trans. R. Soc. Lond.*, 368, 20120474, 2013.

Nelson, A.L. and E.W. Surber., DDT investigations by the Fish and Wildlife Service in 1946. U.S. Department of the Interior Special Scientific Report 41, Patuxent, MD, 1947.

Nelson, D.R., D.C. Zeldin, S.M. Hoffman, L.J. Maltais, H.M. Wain, and D.W. Nebert, Comparison of cytochrome P450 (CYP) genes from the mouse and human genomes, including nomenclature recommendations for genes, pseudogenes and alternative-splice variants. *Pharmacogenetics*, 14, 1–18, 2004.

NEPC, National Environment Protection (Assessment of Site Contamination) Measure 1999. Schedule B(1) Guideline on the Investigation Levels for Soil and Groundwater. National Environment Protection Council, Adelaide, 16 p, 1999.

NEPC, National Environment Protection (Ambient Air Quality) Measure (amended June 2003). National Environment Protection Council, Adelaide, 20 p, 2003.

NEPC, National Environment Protection (Air Toxics) Measure. National Environment Protection Council, Adelaide, 21 p, 2004.

Neter, J., W. Waserman, and M.H. Kutner, *Applied Linear Statistical Models: Regression, Analysis of Variance, and Experimental Design,* 3rd Edition, Richard D., Ed. Irwin, Homewood, IL, 1990.

Neuhold, J.M., The relationship of life history attributes to toxicant tolerance in fishes. *Environ. Toxicol. Chem.*, 6, 709–716, 1987.

Neutel, A.M., J.A.P. Heesterbeek, and P.C. de Ruiter, Stability in real food webs: Weak links in long loops. *Science*, 296, 1120–1123, 2002.

Neville, C.M. and P.G.C. Campbell, Possible mechanisms of aluminum toxicity in dilute, acidic environment to fingerlings and older life stages of salmonids. *Water Air Soil Pollut.*, 42, 311–327, 1988.

Nevo, E., T. Perl, A. Beiles, and D. Wool, Mercury selection of allozyme genotypes in shrimps. *Experientia*, 37, 1152–1154, 1981.

Nevo, E., T. Shimony, and M. Libni, Thermal selection of allozyme polymorphisms in barnacles. *Nature*, 267, 699–701, 1977.

Nevo, E., T. Shimony, and M. Libni, Pollution selection of allozyme polymorphisms in barnacles. *Experientia*, 34, 1562–1564, 1978.

Newcombe, R.G. and D.G. Altman, Proportions and their differences, in *Statistics with Confidence,* 2nd Edition, Altman, D.G., D. Machin, T.N. Bryant, and M.J. Gardner, Eds., BMJ Books, London, United Kingdom, 2000.

Newman, M.C., *The Comprehensive Cooling Water Report. Volume 2. Water Quality*. SREL UC28-2, Savannah River Ecology Laboratory, Aiken, SC, 600 p, 1986.

Newman, M.C., A statistical bias in the derivation of hardness-dependent metals criteria. *Environ. Toxicol. Chem.*, 10, 1295–1297, 1991.

Newman, M.C., Regression analysis of log-transformed data: Statistical bias and its correction. *Environ. Toxicol. Chem.*, 12, 1129–1133, 1993.

Newman, M.C., *Quantitative Methods in Aquatic Ecotoxicology*, Lewis Publishers, Boca Raton, FL, 1995.

Newman, M.C., Ecotoxicology as a science, in *Ecotoxicology: A Hierarchical Treatment*, Newman, M.C. and C.H. Jagoe, Eds., CRC/Lewis Publishers, Boca Raton, FL, 1996.

Newman, M.C., *Population Ecotoxicology*, Wiley, Chichester, United Kingdom, 2001.

Newman, M.C., "What exactly are you inferring?" A closer look at hypothesis testing. *Environ. Toxicol. Chem.*, 27, 1013–1019, 2008.

Newman, M.C., *Quantitative Ecotoxicology,* 2nd Edition, Taylor & Francis/CRC Press, Boca Raton, FL, 2013.

Newman, M.C., J.J. Alberts, and V.A. Greenhut, Geochemical factors complicating the use of *aufwuchs* to monitor bioaccumulation of arsenic, cadmium, chromium, copper and zinc. *Water Res.*, 19, 111–128, 1985.

Newman, M.C. and M.S. Aplin, Enhancing toxicity data interpretation and prediction of ecological risk with survival time modeling: An illustration using sodium chloride toxicity to mosquitofish (*Gambusia holbrooki*). *Aquat. Toxicol.*, 23, 85–96, 1992.

Newman, M.C. and W.H. Clements, *Ecotoxicology: A Comprehensive Treatment*, Taylor & Francis/CRC Press, Boca Raton, FL, 2008.

Newman, M.C., S.A. Diamond, M. Mulvey, and P. Dixon, Allozyme genotype and time to death of mosquitofish, *Gambusia affinis* (Baird and Girard) during acute toxicant exposure: A comparison of arsenate and inorganic mercury. *Aquat. Toxicol.*, 15, 141–156, 1989.

Newman, M.C. and P.M. Dixon, Ecologically meaningful estimates of lethal effect in individuals, in *Ecotoxicology: A Hierarchical Treatment*, Newman, M.C. and C.H. Jagoe, Eds., CRC/Lewis Publishers, Boca Raton, FL, 1996.

Newman, M.C. and D.A. Evans, Enhancing belief during causality assessments: Cognitive idols or Bayes's theorem?, in *Coastal and Estuarine Risk Assessment*, Newman, M.C., M.H. Roberts, Jr. and R.C. Hale, Eds., CRC/Lewis Publishers, Boca Raton, FL, 2002.

Newman, M.C. and M.G. Heagler, Allometry of metal bioaccumulation and toxicity, in *Metal Ecotoxicology: Concepts and Applications*, Newman, M.C. and A.W. McIntosh, Eds., Lewis Publishers, Chelsea, MI, 1991.

Newman, M.C. and C.H. Jagoe, Ligands and the bioavailability of metals in aquatic environments, in *Bioavailability: Physical, Chemical and Biological Interactions*, Hamelink, J.L., P.F. Landrum, H.L. Bergman, and W.H. Benson, Eds., CRC Press, Boca Raton, FL, 1994.

Newman, M.C. and R.H. Jagoe, Bioaccumulation models with time lags: Dynamics and stability criteria. *Ecol. Modell.*, 1424, 281–286, 1996.

Newman, M.C., M.M.Keklak, and M.S. Doggett, Quantifying animal size effects on toxicity: A general approach. *Aquat. Toxicol.*, 28, 1–12, 1994.

Newman, M.C. and J.T. McCloskey, Predicting relative toxicity and interactions of divalent metal ions: Microtox® bioluminescence assay. *Environ. Toxicol. Chem.*, 15, 275–281, 1966a.

Newman, M.C. and J.T. McCloskey, Time-to-event analyses of ecotoxicology data. *Ecotoxicology*, 5, 187–196, 1996b.

Newman, M.C. and J.T. McCloskey, The individual tolerance concept is not the sole explanation for the probit dose-effect model. *Environ. Toxicol. Chem.*, 19, 520–526, 2000.

Newman, M.C. and A.W. McIntosh, Slow accumulation of lead from contaminated food sources by the freshwater gastropods, *Physa integra* and *Campeloma decisum*. *Arch. Environ. Contam. Toxicol.*, 12, 685–692, 1983.

Newman, M.C. and A.W. McIntosh, Appropriateness of *aufwuchs* as a monitor of bioaccumulation. *Environ. Pollut.*, 60, 83–100, 1989.

Newman, M.C., A.W. McIntosh, and V.A. Greenhut, Geochemical factors complicating the use of *aufwuchs* as a biomonitor for lead levels in two New Jersey reservoirs. *Water Res.*, 17, 625–630, 1983.

Newman, M.C. and S.V. Mitz, Size dependence of zinc elimination and uptake from water by mosquitofish *Gambusia affinis* (Baird and Girard). *Aquat. Toxicol.*, 12, 17–32, 1988.

Newman, M.C., M. Mulvey, A. Beeby, R.W. Hurst, and L. Richmond, Snail (*Helix aspersa*) exposure history and possible adaptation to lead as reflected in shell composition. *Arch. Environ. Contam. Toxicol.*, 27, 346–351, 1994.

Newman, M.C., D.R. Ownby, L.C.A. Mézin, D.C. Powell, T.R.L. Christensen, S.B. Lerberg, B.A. Anderson, and T.V. Padma, Species sensitivity distributions in ecological risk assessment: Distributional assumptions, alternate bootstrap techniques, and estimation of adequate number of species, in *Species Sensitivity Distributions in Ecotoxicology*, Posthuma, L., G.W. Suter II, and T.P. Traas, Eds., CRC/Lewis Publishers, 2002.

Newman, M.C. and M.A. Unger, *Fundamentals of Ecotoxicology,* 2nd Edition, CRC Press, Boca Raton, FL, 2003.

Newman, M.C., X. Xu, A. Condon, and L. Liang, Floodplain methylmercury biomagnification factor higher than that of the contiguous river (South River, Virginia, USA). *Environ. Pollut.*, 159, 2840–2844, 2011.

Newman, M.C., Y. Zhao, and J.F. Carriger, Coastal and estuarine ecological risk assessment: The need for a more formal approach to stressor identification. *Hydrobiologia*, 577, 31–40, 2007.

Newstead, J.L., P.D. Jones, K. Coady, and J.P. Giesy, Avian toxicity reference values for perfluorooctance sulfonate. *Environ. Sci. Technol.*, 39, 9357–9362, 2005.

Newsted, J.L. and J.P. Giesy, Predictive models for photo-induced acute toxicity of polycyclic aromatic hydrocarbons to *Daphnia magna* Strauss (Cladocera: Crustacea). *Environ. Toxicol. Chem.*, 6, 445–461, 1987.

Nichols, J.W., J.M. McKim, M.E. Andersen, M.L. Gargas, H.J. Clewell, III, and R.J. Erickson, A physiologically based toxicokinetic model for the uptake and disposition of waterborne organic chemicals in fish. *Toxicol. Appl. Pharmacol.*, 106, 433–447, 1990.

Nieboer, E. and D.H.S. Richardson, The replacement of the nondescript term "heavy metals" by a biologically and chemically significant classification of metal ions. *Environ. Pollut.*, 1B, 3–26, 1980.

Niederlehner, B.R., J.R. Pratt, A.L. Buikema, Jr., and J. Cairns, Jr., Laboratory tests evaluating the effects of cadmium on freshwater protozoan communities. *Environ. Toxicol. Chem.*, 4, 155–165, 1985.

Niemi, G.J., P. Devore, N. Detenbeck, D. Taylor, A. Lima, J.D. Yount, and R.J. Naiman, Overview of case-studies on recovery of aquatic systems from disturbance. *Environ. Manage.*, 14, 571–587, 1990.

Nierzwicki-Bauer, S.A., C.W. Boylen, L.W. Eichler, J.P. Harrison, J.W. Sutherland, W. Shaw, R.A. Daniels et al, Acidification in the Adirondacks: Defining the biota in trophic levels of 30 diverse acid-impacted lakes. *Environ. Sci. Technol.*, 44, 5721–5727, 2010.

Nihlgård, B., The ammonium hypothesis: An additional explanation to the forest dieback in Europe. *Ambio*, 14, 2–8, 1985.

Nikinmaa, K., How does environmental pollution affect red cell function in fish? *Aquat. Toxicol.*, 22, 227–238, 1992.

Nikli⊠ska, M., M. Chodak, and R. Laskowski, Pollution-induced community tolerance of microorganisms from forest soil organic layers polluted with Zn or Cu. *Appl. Soil. Toxicol.*, 32, 265–272, 2006.

NIOH (National Institute of Occupational Health), Final report of the investigation of unusual illnesses allegedly produced by endosulfan exposure in Padre village of Kasaragod District (N.Kerala), National Institute of Occupational Health, Indian Council for Medical Research, Ahmedabad, India, July 2002.

Nishikawa, J., S. Mamiya, T. Kanayama, T. Nishikawa, F. Shiraishi, and T. Horiguchi, Involvement of the retinoid X receptor in the development of imposex caused by organotins in gastropods. *Environ. Sci. Technol.*, 38, 6271–6276, 2004.

Nissen, P. and A.A. Benson, Arsenic metabolism in freshwater and terrestrial plants. *Physiol. Plant*, 54, 446–450, 1982.

Niyogi, S. and C.M. Wood, Biotic Ligand Model, a flexible tool for developing site-specific water quality guidelines for metals. *Environ. Sci. Technol.*, 38, 6177–6192, 2004.

Nordell, B., Thermal pollution causes global warming. *Global Planet. Change*, 38, 305–312, 2003.

Norton, S.B., D.J. Rodier, J.H. Gentile, W.H. Van Der Schalie, W.P. Wood, and M.W. Slimak, A framework for ecological risk assessment at the EPA. *Environ. Toxicol. Chem.*, 11, 1663–1672, 1992.

Nott, J.A. and A. Nicolaidou, Bioreduction of zinc and manganese along a molluscan food chain. *Comp. Biochem. Physiol. A*, 104, 235–238, 1993.

Nowak, C., D. Jost, C. Vogt, M. Oetken, K. Schwenk, and J. Oehlmann, Consequences of inbreeding and reduced genetic variation on tolerance to cadmium stress in the midge *Chironomus riparius. Aquat. Toxicol.*, 85, 278–284, 2007.

Ntow, W.J., Organochlorine pesticides in water, sediments, crops and human fluids in a farming community in Ghana. *Arch. Environ. Contam. Toxicol.*, 40, 557–563, 2001.

Nyman, A.-M., K. Schirmer, and R. Ashauer, Toxicokinetic-toxicodynamic modelling of survival of *Gammarus pulex* in multiple pulse exposures to propiconazole: Model assumptions, calibration data requirements and predictive power. *Ecotoxicology*, 21, 1828–1840, 2012.

Oaks, J.L., M. Gilbert, M.Z. Virani, R.T. Watson, C.U. Meteyer, B.A. Rideout, H.L. Shivaprasad et al, Diclofenac residues as the cause of vulture population decline in Pakistan. *Nature*, 427, 630–633, 2004.

Oberdörster, G., V. Stone, and K. Donalson, Toxicology of nanoparticles: A historical perspective. *Nanotoxicology*, 1, 2–25, 2007.

Obourn, J.D., S.R. Frame, R.H. Bell, D.S. Longnecker, G.S. Elliott, and J. Cook, Mechanisms for the pancreatic oncogenic effects of the peroxisome proliferator Wyeth-14,643. *Toxicol. Appl. Pharmacol.*, 145, 425–436, 1997.

O'Connor, R.J., Toward the incorporation of spatiotemporal dynamics into ecotoxicology, in *Population Dynamics in Ecological Space and Time*, Rhodes, Jr., O.E., R.K. Chesser, and M.H. Smith, Eds., The University of Chicago Press, Chicago, IL, 1996.

Odin, M., A. Feurtet-Mazel, F. Ribeyre, and A. Boudou, Actions and interactions of temperature, pH and photoperiod on mercury bioaccumulation by nymphs of the burrowing mayfly *Hexagenia rigida*, from the sediment contamination source. *Environ. Toxicol. Chem.*, 13, 1291–1302, 1994.

Odum, E.P., *Fundamentals of Ecology*, W.B. Saunders Co., Philadelphia, PA, 1971.

Odum, E.P., Trends expected in stressed ecosystems. *Bioscience*, 35, 419–422, 1985.

Odum, E.P., Preface, in *Ecotoxicology: A Hierarchical Treatment*, Newman, M.C. and C.H. Jagoe, Eds., CRC/Lewis Publishers, Boca Raton, FL, 1996.

OECD, Current approaches in the statistical analysis of ecotoxicology data: A guidance to application. OECD Environmental Health and Safety Publications Series on Testing and Assessment, No. 54. Environment Directorate, Paris, OECD website: www.oecd.org/document/30/0,2340,en_2649_34377_1916638_1_1_1_1,00.html, 2006.

Ogendi, G.M., W.G. Brumbaugh, R.E. Hannigan, and J.L. Farris, Effects of acid-volatile sulfide on metal bioavailability and toxicity to midge (*Chironomus tentans*) larvae in black shale sediments. *Environ. Toxicol. Chem.*, 26, 325–334, 2007.

Ogilvie, D.M. and D.L. Miller, Duration of a DDT-induced shift in the selected temperature of Atlantic salmon (*Salmo salar*). *Bull. Environ. Contam. Toxicol.*, 16, 86–89, 1976.

O'Halloran, T.V. and V.C. Culotta, Metallochaperones, an intracellular shuttle service for metal ions. *J. Biol. Chem.*, 275, 25057–25060, 2000.

Ohlendorf, H.M., D.J. Hoffman, M.K. Saiki, and T.A. Aldrich, Embryonic mortality and abnormalities of aquatic birds: Apparent impacts of selenium from irrigation drainwater. *Sci. Total Environ.*, 52, 49–63, 1986.

Öhman, L.-O. and S. Sjöberg, Thermodynamic calculations with special reference to the aqueous aluminum system, in *Metal Speciation: Theory, Analysis and Application*, Kramer, J.R. and H.E. Allen, Eds., Lewis Publishers, Chelsea, MI, 1988.

Oke, T.R., City size and the urban heat island. *Atmos. Environ.*, 7, 769–779, 1973.

Okey, A.B., An AH receptor odyssey to the shores of toxicology: The Deichmann lecture. *Toxicol. Sci.*, 98, 5–38, 2007.

Olivieri, G., J. Bodycote, and S. Wolff, Adaptive response of human lymphocytes to low concentrations of radioactive thymidine. *Science*, 223, 594–597, 1984.

Olsen, G.H., E. Sva, J.L. Carroll, L. Camus, W. DeCoen, R. Smolders, H. Øveraas, and K. Hylland, Alterations in the energy budget of Arctic benthic species exposed to oil-related compounds. *Aquat. Toxicol.*, 83, 85–92, 2007.

Olson, C.T. and M.E. Anderson, The acute toxicity of perfluorooctanoic and perfluorodecanoic acids in male rats and effects on tissue fatty acids. *Toxicol. Appl. Pharmacol.*, 70, 362–372, 1983.

Omernik, J.M., Ecoregions in the conterminous United States. *Ann. Assoc. Am. Geogr.*, 77, 118–125, 1987.

Omura, T., Forty years of cytochrome P450. *Biochem. Biophy. Res. Commun.*, 266, 690–698, 1999.

Omura, M., R. Ogata, K. Kubo, Y. Shimasaki, S. Aou, Y. Oshima, A. Tanaka, M. Hirata, Y. Makita, and N. Inoue, Two-generation reproductive toxicity study of tributyltin chloride in male rats. *Toxicol. Sci.*, 64, 224–232, 2001.

Organization for Economic Cooperation and Development, Acute oral toxicity—Up-and-down procedure. OECD Guideline For Testing Of Chemicals, Guideline no. 425, 2001.

Oris, J.T. and J.P. Giesy, The photoenhanced toxicity of anthracene to juvenile sunfish (*Lepomis* spp.). *Aquat. Toxicol.*, 6, 133–146, 1985.

Oris, J.T. and J.P. Giesy, The photo-induced toxicity of polycyclic aromatic hydrocarbons to larvae of the fathead minnow (*Pimephales promelas*), *Chemosphere*, 16, 1396–1404, 1987.

Oris, J.T., A.T. Hall, and J.D. Tylka, Humic acids reduce the photo-induced toxicity of anthracene to fish and daphnia. *Environ. Toxicol. Chem.*, 9(5), 575–583, 1990.

Osano, O., W. Admiraal, and D. Otieno, Developmental disorders in embryos of the frog, *Xenopus laevis* induced by chloroacetanilide herbicides and their degradation products. *Environ. Toxicol. Chem.*, 21, 375–379, 2002.

Osemwengie, L.I. and S. Steinberg, On-site solid-phase extraction and laboratory analysis of ultra-trace synthetic musks in municipal sewage effluent using gas chromatography-mass spectrometry in the full-scan mode. *J. Chromatogr. A*, 932, 107–118, 2001.

O'Shaughnessy, P.T., Parachuting cats and crushed eggs. The controversy over the use of DDT to control malaria, *Am. J. Public Health*, 98, 1940–1948, 2008.

Østby, L., G.W. Gabrielsen, and Å. Krøkje, Cytochrome P4501A induction and DNA adduct formation in glaucous gulls (*Larus hyperboreus*), fed with environmentally contaminated gull eggs. *Ecotoxicol. Environ. Saf.*, 62, 363–37, 2005.

Ostbye, K., S.A. Oxnevad, and L.A. Vollestad, Developmental stability in perch (*Perca fluviatilis*) in acidic aluminium-rich lakes. *Can. J. Zool.*, 75, 919–928, 1997.

Osterberg, C.L., Radiological impacts of releases from nuclear facilities into aquatic environments: USA views, in *Impacts of Nuclear Release into the Aquatic Environment*, International Atomic Energy Agency, Vienna, Austria, 1975.

Otto, E., J.E. Young, and G. Maroni, Structure and expression of a tandem duplication of the *Drosophila* metallothionein gene. *Proc. Natl. Acad. Sci. U.S.A.*, 83, 6025–6029, 1986.

Otvos, J.D., H.R. Engeseth, D.G. Nettesheim, and C.R. Hilt, Interprotein metal exchange reactions of metallothionein. *Experientia Suppl.*, 52, 171–178, 1987.

Otvos, J.D., R.W. Olafson, and I.M. Armitage, Structure of an invertebrate metallothionein from *Scylla serrata. J. Biol. Chem.*, 257, 2427–2431, 1982.

Owen, D.F., Industrial melanism in North American moths. *Am. Nat.*, 95, 227–233, 1961.

Ownby, D., M.C. Newman, M. Mulvey, M. Unger, and W. Vogelbein, Fish (*Fundulus heteroclitus*) populations with different exposure histories differ in tolerance of creosote-contaminated sediments. *Environ. Toxicol. Chem.*, 21, 1897–1902, 2002.

Pacurari, M., V. Castranova, and V. Vallyathan, Single- and multi-wall carbon nanotubes versus asbestos: Are the carbon nanotubes a new health risk to humans? *J. Toxicol. Envir. Health A*. 73, 378–395, 2010.

Pacyna, J.M., The origin of Arctic air pollutants: Lessons learned and future research. *Sci. Total Environ.*, 160/161, 39–53, 1995.

Pacyna, E., J. Pacyna, F. Steenhuisen, and S. Wilson, Global anthropogenic mercury emission inventory for 2000. *Atmos. Environ.*, 40, 4048–4063, 2006.

Pagenkopf, G.K., Gill surface interaction model for trace-metal toxicity to fishes: Role of complexation, pH, and water hardness. *Environ. Sci. Technol.*, 17, 342–347, 1983.

Pain, D.J., Lead in the environment, in *Handbook of Ecotoxicology*, Hoffman, D.J., B.A. Rattner, G.A. Burton, Jr., and J. Cairns, Jr., Eds., CRC Press, Boca Raton, FL, 1995.

Paine, R.T., M.J. Tegner, and E.A. Johnson, Compounded perturbations yield ecological surprises. *Ecosystems*, 1, 535–545, 1998.

Paller, M.H., J.W. Littrell, and E.L. Peters, Ecological half-lives of ^{137}Cs in fishes from the Savannah River Site. *Health Phys.*, 77, 392–402, 1999.

Palm, K., K. Luthman, J. Ros, J. Gråsjo, and P. Artursson, Effect of molecular charge on intestinal transport: pH-dependent transport of cationic drugs. *J. Pharmacol. Exp. Ther.*, 291, 435–443, 1999.

Palmer, A.R., Fluctuating asymmetry analyses: A primer, in *Developmental Instability: Its Origins and Evolutionary Implications*. Markov T.A., Ed., Kluwer, Dordrecht, The Netherlands, 1994.

Palmer, A.R., Waltzing with asymmetry. *Bioscience*, 46, 518–532, 1996.

Palmer, A.R. and C. Strobeck, Fluctuating asymmetry: Measurement, analysis, patterns. *Annu. Rev. Ecol. Syst.*, 17, 391–421, 1986.

Palmer, A.R. and C. Strobeck, Fluctuating asymmetry analyses revisited, in *Developmental Instability: Causes and Consequences*, Polak, M., Ed., Oxford University Press, New York, 2003.

Pan American Health Organization, Epidemiological situation of acute pesticide poisoning in the Central American Isthmus, 1992–2000. *Epidemiol. Bull.*, 23, 1–7, 2002.

Pandolfi, J.M., S.R. Connolly, D.J. Marshall, and A.L. Cohen, Projecting coral reef futures under global warming and ocean acidification. *Science*, 333, 418–422, 2011.

Paquin, P.R., J.W. Gorsuch, S. Apte, G.E. Batley, K.C. Bowles, P.G.C. Campbell, C.G. Delos et al, The biotic ligand model: A historical overview. *Comp. Biochem. Physiol. C*, 133, 3–35, 2002.

Parke, D., Cytochrome P-450 and the detoxification of environmental chemicals. *Aquat. Toxicol.*, 1, 367–376, 1981.

Parkerton, T.F., J.A. Arnot, A.V. Weisbrod, C. Russom, R.A. Hoke, K. Woodburn, T. Traas, M. Bonnell, L.P Burkhard, and M.A. Lampi, Guidance for evaluating in-vivo fish bioaccumulation data. *Integr. Environ. Assess. Manag.*, 4, 139–155, 2008.

Parkinson, A., Biotransformations of xenobiotics, in *Casarett & Doull's Toxicology: The Basic Science of Poisons,* 5th Edition, Klaassen, C.D., Ed., McGraw-Hill, New York, 1996.

Parkinson, A. and B.W. Ogilvie, Biotransformations of xenobiotics, in *Casarett & Doull's Toxicology: The Basic Science of Poisons,* 7th Edition, Klaassen, C.D., Ed., McGraw-Hill, New York, 2008.

Parrott, J.L., C.S. Wood, P. Boutot, and S. Dunn, Changes in growth and secondary sex characteristics of fathead minnows exposed to bleached sulfite mill effluent. *Environ. Contam. Toxicol.*, 22, 2908–2915, 2003.

Parsons, P.A., Fluctuating asymmetry: An epigenetic measure of stress. *Biol. Rev.*, 65, 131–145, 1990.

Parsons, P.A., Inherited stress resistance and longevity: A stress theory of ageing. *Heredity*, 75, 216–221, 1995.

Partridge, G.G., Relative fitness of genotypes in a population of *Rattus norvegicus* polymorphic for warfarin resistance. *Heredity*, 43, 239–246, 1979.

Pastor, N., M. López-Lázaro, J.L. Tella, R. Baos, F. Hiraldo, and F. Cortés, Assessment of genotoxic damage by the comet assay in white storks (*Ciconia ciconia*) after the DoZana ecological disaster. *Mutagenesis*, 16, 219–223, 2001.

Paterson, S. and D. Mackay, A steady-state fugacity-based pharmacokinetic model with simultaneous multiple exposure routes. *Environ. Toxicol. Chem.*, 6, 395–408, 1987.

Patrick, R., Use of algae, especially diatoms, in the assessment of water quality. *ASTM, Special Technical Publication*, 528, 76–95, 1973.

Paustenbach, D.J. and A.J. Madl, The practice of exposure assessment, in *Principles and Methods of Toxicology*, 5th Edition, Hayes, A.W., Ed., CRC/Taylor & Francis Press, Boca Raton, FL, 2008.

Pawel, D.J. and J.S. Puskin, The U.S. Environmental Protection Agency's assessment of risks from indoor radon. *Health Phys.*, 87, 68–74, 2004.

Pawlik-Showrońska, B., J. Pirszel, and M.T. Brown, Concentrations of phytochelatins and glutathione found in natural assemblages of seaweeds depend on species and metal concentrations of the habitat. *Aquat. Toxicol.*, 83, 190–199, 2007.

Pawlowski, S., A. Sauer, J.A. Shears, C.R. Tyler, and T. Braunbeck, Androgenic and estrogenic effects of the synthetic androgen 17a-methyltestosterone on sexual development and reproductive performance in the fathead minnow (*Pimephales promelas*) determined using the gonadal recrudescence assay. *Aquat. Toxicol.*, 68, 277–291, 2004.

Peakall, D., *Animal Biomarkers as Pollution Indicators*, Chapman & Hall, London, United Kingdom, 1992.

Peakall, D.B. and K. McBee, Biomarkers for contaminant exposure and effects in mammals, in *Ecotoxicology of Wild Mammals*, Shore, R.F. and B.A. Rattner, Eds., Wiley, Chichester, United Kingdom, 2001.

Pearson, R.G., Hard and soft acids and bases. *J. Am. Chem. Soc.*, 85, 3533–3539, 1963.

Peltier, W.H. and C.I. Weber, *Methods for Measuring the Acute Toxicity of Effluents to Freshwater and Marine Organisms*, EPA/600/4-85/013, Environmental Monitoring and Support Laboratory, EPA, Cincinnati, OH, 1985.

Pendleton, R.C., R.D. Lloyd, C.W. Mays, and B.W. Church, Trophic level effect on the accumulation of caesium-137 in cougars feeding on mule deer. *Nature*, 204, 708–709, 1964.

Peoples, S.A., The metabolism of arsenic in man and animals, in *Arsenic: Industrial, Biomedical, Environmental Perspectives*, Lederer, W.H. and R.J. Fensterheim, Eds., Van Nostrand Reinhold, New York, 1983.

Pepper, C.B., T.R. Rainwater, S.G. Platt, J.A. Dever, T.A. Anderson, and S.T. McMurry, Organochlorine pesticides in chorioallantoic membranes of Morelet's crocodile eggs from Belize. *J. Wildl. Dis.*, 40, 493–500, 2004.

Peralta-Videa, J.R., L. Zhao, M.L. Lopez-Moreno, G. de la Rosa, J. Hong, and J.L. Gardea-Torresdey, Nanomaterials and the environment: A review for the biennium 2008–2010. *J. Hazard. Mater.*, 186, 1–15, 2011.

Perez, M.H. and W.G. Wallace, Differences in prey capture in grass shrimp, *Palaemonetes pugio*, collected along an environmental impact gradient. *Arch. Environ. Contam. Toxicol.*, 46, 81–89, 2004.

Pérez-Coll, C.S. and J. Herkovits, Lethal and teratogenic effects of naringenin evaluated by means of an amphibian embryo toxicity test (AMPHITOX). *Food Chem. Toxicol.*, 42, 299–306, 2004.

Peterson, B.J. and B. Fry, Stable isotopes in ecosystem studies. *Annu. Rev. Ecol. Syst.*, 18, 293–320, 1987.

Petrelli, G., Figá-Talamanca, R. Tropeano, M. Tangucci, C. Cini, S. Aquilani, L. Gasperini, and P. Mela, Reproductive male-mediated risk: Spontaneous abortion among wives of pesticide applicators. *Eur. J. Epidemiol.*, 16, 391–393, 2000.

Pettis, J.S. and K.S. Delaplane, Coordinated responses to honey bee decline in the USA. *Apidologie*, 2010, 1–8, 2010.

Phelps, K.L. and K. McBee, Ecological characteristics of small mammal communities at a superfund site, *Am. Midl., Nat.*, 161, 57–68, 2009.

Phelps, K.L., and K. McBee, Population parameters of *Peromyscus leucopus* (white-footed deermice) inhabiting a heavy metal contaminated superfund site, *Southwest. Nat.*, 55, 363–373, 2010.

Phillips, D.J.H., The use of biological indicator organisms to monitor trace metal pollution in marine and estuarine environments: A review. *Environ. Pollut.*, 13, 281–317, 1977.

Piatt, J.F., C.J. Lensink, W. Butler, M. Kendziorek, and D.R. Nysewander, Immediate impact of the 'Exxon Valdez' oil spill on marine birds. *Auk*, 10, 387–397, 1990.

Pielou, E.C., *Population and Community Ecology: Principles and Methods*, Gordon and Breach Science Publishers, New York, 1974.

Pimentel, D. and C.A. Edwards, Pesticides and ecosystems. *Bioscience*, 32, 595–600, 1982.

Pimentel, D. and L. Levitan, Pesticides: Amounts applied and amounts reaching pests, *BioScience*, 36, 86–91, 1986.

Pimentel, D., L. McLaughlin, A. Zepp, B. Lakitan, T. Kraus, P. Kleinman, F. Vancini, W.J. Roach, E. Graap, W.S. Keeton, and G. Selig, Environmental and economic effects of reducing pesticide use. *BioScience*, 41, 402–409, 1991.

Pimentel, D., L. Westra, and R.F. Noss, Eds., *Ecological Integrity: Integrating Environment, Conservation, and Health*, Island Press, Washington, DC, 2000.

Pinchot, G., *Breaking New Ground*, Harcourt, Brace & World, New York, 1947.

Pirrone, N., I. Allegrini, G.J. Keeler, J.O. Nriagu, R. Rossmann, and J.A. Robbins, Historical atmospheric mercury emissions and depositions in North America compared to mercury accumulations in sedimentary records. *Atmos. Environ.*, 32, 929–940, 1998.

Piszkiewicz, D., *Kinetics of Chemical and Enzyme-Catalyzed Reactions*, Oxford University Press, New York, 1977.

Plaa, G.L. and M. Charbonneau, Detection and evaluation of chemically induced liver injury, in *Principles and Methods of Toxicology,* 5th Edition, Hayes, A.W., Ed., CRC/Taylor & Francis Press, Boca Raton, FL, 2008.

Plackett, R.L. and P.S. Hewlett, Quantal responses to mixtures of poisons. *J. R. Stat. Soc.*, 14, 141–163, 1952.

Platt, J.R., Strong inference. *Science*, 146, 347–353, 1964.

Playle, R.C. and C.M. Wood, Water chemistry changes in the gill microenviroment of rainbow trout: Experimental observations and theory. *J. Comp. Physiol.*, 159B, 527–537, 1989.

Playle, R.C. and C.M. Wood, Mechanisms of aluminum extraction and accumulation at the gills of rainbow trout, *Oncorhynchus mykiss* (Walbaum) fingerlings. *Aquat. Toxicol.*, 21, 267–278, 1991.

Pojmańska, T. and E. Dzika, Parasites of bream (*Abramis brama* L.) from the Lake Gosławskie (Poland) affected by long-term thermal pollution. *Acta Parasitol. Pol.* 32, 139–161, 1987.

Poland, C., R. Duffini, I. Kinloch, A. Maynard, W.A.H. Wallace, A. Seaton, V. Stones, S. Brown, W. MacNee, and K. Donaldson, Carbon nanotubes introduced into the abdominal cavity of mice show asbestos-like pathogenicity in a pilot study. *Nat. Nanotechnol.*, 3, 423–428, 2008.

Polidoro BA, M.J. Morra, C. Ruepert, and L.E. Castillo, Pesticide sequestration in passive samplers (SPMDs): Considerations for deployment time, biofouling, and stream flow in a tropical watershed. *J Environ. Monit.*, 11, 1866–74, 2009.

Polikarpov, G.G., *Radioecology of Aquatic Organisms*, North-Holland Publishing Co., Amsterdam, The Netherlands, 1966.

Pontasch, K.W., B.R. Niederlehner, and J. Cairns, Jr., Comparisons of single-species, microcosm and field responses to a complex effluent. *Environ. Toxicol. Chem.*, 8, 521–532, 1989.

Posthuma, L., R.F. Hogervorst, E.N.G. Joose, and N.M. Van Straalen, Genetic variation and covariation for characteristics associated with cadmium tolerance in natural populations of the springtail *Orchesella cincta* (L.). *Evolution*, 47, 619–631, 1993.

Posthuma L., G.W. Suter II, and T.P. Traas, Eds., *Species Sensitivity Distributions in Ecotoxicology*, CRC/Lewis Publishers, Boca Raton, FL. 2002.

Postma, J.F., S. Mol, H. Larsen, and W. Admiraal, Life-cycle changes and zinc shortage in cadmium-tolerant midges, *Chironomus riparius* (Diptera), reared in the absence of cadmium. *Environ. Toxicol. Chem.*, 14, 117–122, 1995.

Potts, S.G., J.C. Biesmeijer, C. Kremen, P. Neumann, O. Schweigers, and W.E. Kunin, Global pollinator declines: Trends, impacts and drivers. *TREES*, 25, 345–353, 2010.

Pounds, J.A., M.R. Bustamante, L.A. Coloma, J.A. Consuegra, M.P. Fogden, P.N. Foster, E. La Marca et al, Widespread amphibian extinctions from epidemic disease driven by global warming. *Nature*, 439, 161–167, 2006.

Powell, A., D. Mackay, E. Webster, and J.A. Arnot. Modeling bioaccumulation using characteristic times. *Environ. Toxicol. Chem.*, 28, 272–278, 2009.

Prabhavathi, P.A., P. Padmavathi, and P.P. Reddy, Chromosomal aberrations in the leucocytes of men occupationally exposed to uranyl compounds. *Bull. Environ. Contam. Toxicol.*, 70, 322–327, 2003.

Prasad, M.S., Histochemical observation on crude oil poisoning in the respiratory epithelium of *Puntius sophore. Acta Hydroch. Hydrob.*, 15, 535–539, 2006.

Pratt, J.R. and J. Cairns, Ecotoxicology and the redundancy problem: Understanding effects on community structure and function, in *Ecotoxicology: A Hierarchical Treatment*, Newman, M.C. and C.H. Jagoe, Eds., CRC/Lewis Publishers, Boca Raton, FL, 1996.

Preston, F.W., The commonness, and rarity, of species. *Ecology*, 29, 254–283, 1948.

Preston, B.L., G. Cecchine, and T.W. Snell, Effects of pentachlorophenol on predator avoidance behavior of the rotifer *Brachionus calyciflorus. Aquat. Toxicol.*, 44, 201–212, 1999.

Prevodnik, A.P., J. Gardeström, K. Lilja, T. Elfwing, B. McDonagh, N. Petroviç, M. Tedengren, D. Sheehan, and T. Bollner, Oxidative stress in response to xenobiotics in the blue mussel *Mytilus edulis* L.: Evidence for variation along a natural salinity gradient of the Baltic Sea. *Aquat. Toxicol.*, 82, 63–71, 2007.

Preziosi, P., Natural and anthropogenic environmental estrogens. The scientific basis for risk assessment. Endocrine-disruptors as environmental signalers: An introduction. *Pure Appl. Chem.*, 70, 1617–1631, 1998.

Price, W.J., *Analytical Atomic Absorption Spectrometry*, Heyden & Son, London, United Kingdom, 1972.

Programa Estado de la Nación, Décimo Informe Estado de la Nación. San José, 2004, Costa Rica, 448 p, 2004.

Pronczuk, J., J. Akre, G. Moy, and C. Vallenas, Global perspectives in breast milk contamination: Infectious and toxic hazards. *Environ. Health Perspect.*, 110, A349–A351, 2002.

Prugh, T., *Natural Capital and Human Economic Survival,* 2nd Edition, Lewis Publishers, Boca Raton, FL, 1999.

Prysyazhnyuk, A., V. Gristchenko, Z. Fedorenko, L. Gulak, M. Fuzik, K. Slipenyuk, and M. Tirmarche, Twenty years after the Chernobyl accident: Solid cancer incidence in various groups of Ukrainian population. *Radiat. Environ. Bioph.*, 46, 43–51, 2007.

Pulliam, H.R., G.W. Barrett and E.P. Odum, Bioelimination of tracer ^{65}Zn in relation to metabolic rates in mice, in *Symposium on Radioecology*, Nelson, D.J. and F.C. Evans, Eds., University of Michigan, Ann Arbor, MI, 1967.

Pulliam, H.R. and B.J. Danielson, Source, sinks, and habitat selection: A landscape perspective on population dynamics. *Am. Nat.*, 137, Supplement, S50–S66, 1991.

Putka, G., Research on lead poisoning is questioned. *Wall Street Journal*, March 6, B1, 1992.

Pynnönen, K., D.A. Holwerda, and D.I. Zandee, Occurrence of calcium concretions in various tissues of freshwater mussels, and their capacity for cadmium sequestration. *Aquat. Toxicol.*, 10, 101–114, 1987.

Qiu, J., Pollutants capture the high ground in the Himalayas. *Science*, 339, 1030–1031, 2013.

Quijano, R.F., Risk assessment in a third world reality: An endosulfan case history, *Int. J. Occup. Environ. Health*, 6, 312–317, 2000.

Quijano, R.F., Endosulfan poisoning in Kasargod, Kerala, India: A report on a fact-finding mission, University of the Philippines, College of Medicine, Manila, Philippines, 2002.

Quine, W.V., *Word and Object*, MIT Press, Cambridge, MA, 1960.

Quine, W.V., *Methods of Logic*, Harvard University Press, Cambridge, MA, 1982.

Rago, P.J. and R.M. Dorazio, Statistical inference in life-table experiments: The finite rate of increase. *Can. J. Fish. Aquat. Sci.*, 41, 1361–1374, 1984.

Ragunathan, A.V., Raw cashew nut market goes into a shell. *The Hindu Daily*, April 11, 2012.

Railsback, S.F. and V. Grimm, *Agent-Based and Individual-Based Modeling: A Practical Introduction.* Princeton University Press, Princeto, NJ, 2012.

Raimondo, S., Density dependent functional forms drive compensation in populations exposed to stressors. *Ecol. Modell.*, 265, 149–157, 2013.

Raimondo, S. and C.L. McKenney, Jr., Projected population-level effects of thiobencarb exposure on the mysid, *Americamysis bahia*, and extinction probability in a concentration-decay exposure system. *Environ. Toxicol. Chem.*, 24, 564–572, 2005a.

Raimondo, S. and C.L. McKenney, Jr., Projecting population-level responses of mysids exposed to an endocrine disrupting chemical. *Integr. Comp. Biol.*, 45, 151–157, 2005b.

Raimondo, S. and C.L. McKenney, Jr., From organisms to populations: Modeling aquatic toxicity data across two levels of biological organization. *Environ. Toxicol. Chem.*, 25, 589–596, 2006.

Rainbow, P.S., S.N. Luoma, and W.X. Wang, Trophically available metal: A variable feast. *Environ. Pollut.*, 159, 2347–2349, 2011.

Rainwater, T.R., T.H. Wu, A.G. Finger, J.E. Canas, L. Yu, K.D. Reynolds, G. Coimbatore, B. Barr, S.G. Platt, G.P. Cobb, T.A. Anderson, and S.T. McMurry, Metals and organochlorine pesticides in caudal scutes of crocodiles from Belize and Costa Rica. *Sci. Total Environ.*, 373, 146–156, 2007.

Rambler, M.B., L. Margulis, and R. Fester, Eds., *Global Ecology: Towards a Science of the Biosphere*, Academic Press, Boston, MA, 1989.

Rand, G.M., Behavior, in *Fundamentals of Aquatic Toxicology*, Rand, G.M., and S.R. Petrocelli, Eds., Hemisphere Publishing, Washington, DC, 1985.

Rand, G.M., *Fundamentals of Aquatic Toxicology: Effects, Environmental Fate, and Risk Assessment.* 2nd Edition, Taylor and Francis, Washington DC, 1995.

Rand, G.M. and S.R. Petrocelli, Eds., *Fundamentals of Aquatic Toxicology*, Hemisphere Publishing, Washington, DC, 1985.

Rand, G.M., P.G. Wells, and L.S. McCarty, Introduction to aquatic toxicology, in *Fundamentals of Aquatic Toxicology,* 2nd Edition, Rand, G.M., Ed., Taylor and Francis, Washington, DC, 1995.

Rao, A.S. and R.R. Pillale, The concentrations of pesticides in sediments from Kolleru Lake in India. *Pest Manag. Sci.*, 57, 620–624, 2001.

Rapport, D.J., H.A. Regier, and T.C. Hutchinson, Ecosystem behavior under stress. *Am. Nat.*, 125, 617–640, 1985.

Rasmussen, J.B., D.J. Rowan, D.R.S. Lean, and J.H. Carey, Food chain structure in Ontario lakes determines PCB levels in lake trout (*Salvelinus namaycush*) and other pelagic fish. *Can. J. Fish. Aquat. Sci.*, 47, 2030–2038, 1990.

Ratcliffe, D.A., Decrease in eggshell weight in certain birds of prey. *Nature*, 215, 208–210, 1967.

Ratcliffe, D.A., Changes attributable to pesticides in egg breakage frequency and eggshell thickness in some British birds. *J. Appl. Ecol.*, 7, 67–107, 1970.

Rattner, B.A., P.C. McGowan, N.H. Golden, J.S. Hatfield, P.C. Toschik, R.F. Lukei, Jr., R.C. Hale, I. Schmitz-Afonso, and C.P. Rice, Contaminant exposure and reproductive success of ospreys (*Pandion haliaetus*) nesting in Chesapeake Bay regions of concern. *Arch. Environ. Contam. Toxicol.*, 47, 126–140, 2004.

Rau, G.H., Low $^{15}N/^{14}N$ in hydrothermal vent animals: Ecological implications. *Nature*, 289, 484–485, 1981.

Raukas, A., Past pollution and its remediation in Estonia. *Baltica*, 17, 71–78, 2004.

Rauser, W.E. and E.B. Dumbroff, Effects of excess cobalt, nickel and zinc on the water relations of *Phaseolus vulgaris*. *Environ. Exp. Bot.*, 21, 249–255, 1981.

Ravera, O., The ecological effects of acid deposition. Part 1. An introduction. *Experientia*, 42, 329–330, 1986.

Ray, S., D.W. McLeese, and M.R. Peterson, Accumulation of copper, zinc, cadmium and lead from two contaminated sediments by three marine invertebrates: A laboratory study. *Bull. Environ. Contam. Toxicol.*, 26, 315–322, 1981.

Reddy A.P., J.M. Spitsbergen, C. Mathews, J.D. Hendricks, and G.S. Bailey, Experimental hepatic tumorigenicity by environmental hydrocarbon dibenzo[a,1]pyrene. *J. Environ. Pathol. Toxicol.*,18, 261–269, 1999.

Rees, M., Denial of catastrophic risks. *Science*, 339, 1123, 2013.

Regoli, F. and G. Principato, Glutathione, glutathione-dependent and antioxidant enzymes in mussel, *Mytilus galloprovincialis*, exposed to metals under field and laboratory conditions: Implications for the use of biochemical biomarkers. *Aquat. Toxicol.*, 31, 143–164, 1995.

Reichhardt, T., Environmental GIS: The world in a computer. *Environ. Sci. Technol.*, 30, 340A–343A, 1996.

Reichle, D.E., Relation of body size to food intake, oxygen consumption, and trace element metabolism in forest floor arthropods. *Ecology*, 49, 538–541, 1968.

Reichle, D.E., P.B. Dunaway, and D.J. Nelson, Turnover and concentration of radionuclides in food chains. *Nucl. Safety*, 11, 43–55, 1970.

Reichle, D.E. and R.I. Van Hook, Jr., Radionuclide dynamics in insect food chains. *Manit. Entomol.*, 4, 22–32, 1970.

Reichman, J.R., L.S. Watrud, E.H. Lee, C.A. Burdick, M.A. Bollman, M.J. Storm, G.A. King, and C. Mallory-Smith, Establishment of transgenic herbicide-resistant creeping bentgrass (*Agrostis stolonifera* L.) in nonagronomic habitats. *Mol. Ecol.*, 15, 4243–4255, 2006.

Reimchen, T.E., Parasitism of asymmetrical pelvic phenotypes in sticklebacks. *Can. J. Zool.*, 75, 2084–2094, 1997.

Reinecke, S.A. and A.J. Reinecke, The comet assay as biomarker of heavy metal genotoxicity in earthworms. *Arch. Environ. Contam. Toxicol.*, 46, 208–215, 2004.

Reinert, R.E., L.J. Stone, and W.A. Willford, Effect of temperature on accumulation of methylmercuric chloride and p,p'-DDT by rainbow trout (*Salmo gairdneri*). *J. Fish. Res. Board Can.*, 31, 1649–1652, 1974.

Reinfelder, J.R. and N.S. Fisher, The assimilation of elements ingested by marine copepods. *Science*, 251, 794–796, 1991.

Reinfelder, J.R., W.X. Wang, S.N. Luoma, and N.S. Fisher, Assimilation efficiencies and turnover rates of trace-elements in marine bivalves: A comparison of oysters, clams and mussels. *Mar. Biol.*, 129, 443–452, 1997.

Relyea, R.A., Predator cues and pesticides: A double dose of danger for amphibians. *Ecol. Appl.*, 13, 1515–1521, 2003.

Relyea, R. and J. Hoverman, Assessing the ecology in ecotoxicology: A review and synthesis in freshwater systems. *Eco. Lett.*, 9, 1157–1171, 2006.

Rench, J.D., Environmental Epidemiology, in *Basic Environmental Toxicology*, Cockerham, L.G. and B.S. Shane, Eds., CRC Press, Boca Raton, FL, 1994.

Renwick, A.G., Toxicokinetics, in *Principles and Methods of Toxicology,* 5th Edition, Hayes, A.W., Ed., CRC/ Taylor & Francis Press, Boca Raton, FL, 2008.

Reuber, M.D., The role of toxicity in the carcinogenicity of endosulfan. *Sci. Total Environ.*, 20, 23–47, 1981.

Reuther, R., Mercury accumulation in sediment and fish from rivers affected by alluvial gold mining in the Brazilian Madeira River basin, Amazon. *Environ. Monit. Assess.*, 32, 239–258, 1994.

Reynolds, C.S., Eutrophication and the management of planktonic algae: What Vollenweider couldn't tell us., in *Eutrophication: Research and Application to Water Supply*, Sutcliffe, D.W. and J.G. Jones, Eds., Freshwater Biological Association, London, United Kingdom, 1992.

Rice, P.J., C.D. Drewes, T.M. Klubertanz, S.P. Bradbury, and J.R. Coats, Acute toxicity and behavioral effects of chlorpyrifos, permethrin, phenol, strychnine, and 2,4-dinitrophenol to 30-day-old Japanese medaka (*Oryzias latipes*). *Environ. Toxicol. Chem.*, 16, 696–704, 1997.

Richards, C. and G. Host, Examining land use influences on stream habitats and macroinvertebrates: A GIS approach. *Water Resour. Bull.*, 30, 729–738, 1994.

Richards, C., G.E. Host, and J.W. Arthur, Identification of predominant environmental factors structuring stream macroinvertebrate communities within a large agricultural catchment. *Freshw. Biol.*, 29, 285–294, 1993.

Richardson, P., T. Hideshima, and K. Anderson, Thalidomide: Emerging role in cancer medicine. *Annu. Rev. Med.*, 53, 629–657, 2002.

Rico, A., A.V. Waichman, R. Geber-Correa, and P.J. Van den Brink, Effects of malathion and carbendazim on Amazonian freshwater organisms: Comparison of tropical and temperate species sensitivity distributions. *Ecotoxicology*, 20, 625–634, 2011.

Riddell, D.J., J.M. Culp, and D.J. Baird, Behavioral responses to sublethal cadmium exposure within an experimental aquatic food web. *Environ. Toxicol. Chem.*, 24, 431–441, 2005.

Rieger, M.A., M. Lamond, C. Preston, S.B. Powles, and R.T. Roush, Pollen-mediated movement of herbicide resistance between commercial canola fields. *Science*, 296, 2386–2388, 2002.

Rizak, S.N. and S.E. Hrudey, Misinterpretation of drinking water quality monitoring data with implications for risk management. *Environ. Sci. Technol.*, 40, 5244–5250, 2006.

Roberts, L., Hard choices ahead on biodiversity. *Science*, 241, 1759–1761, 1988.

Roberts, C.M., Connectivity and management of Caribbean coral reefs. *Science*, 278, 1454–1457, 1997.

Roberts, D.K., T.C. Hutchinson, J. Paciga, A. Chattopadhyay, R.E. Jervis, and J. Van Loon, Lead contamination around secondary smelters: Estimation of dispersal and accumulation by humans. *Science*, 186, 1120–1124, 1974.

Robison, S.H., O. Cantoni, and M. Costa, Analysis of metal-induced DNA lesions and DNA-repair replication in mammalian cells. *Mutat. Res.*, 131, 173–181, 1984.

Roch, M., J.A. McCarter, A.T. Matheson, M.J.R. Clark, and R.W. Olafson, Hepatic metallothionein in rainbow trout (*Salmo gairdneri*) as an indicator of metal pollution in the Campbell River system. *Can. J. Fish. Aquat. Sci.*, 39, 1596–1601, 1982.

Rockström, J., W. Steffen, K. Noone, Å. Persson, F.S. Chapin, III, E.F. Lambin, T.M. Lenton, A safe operating space for humanity. *Nature*, 461, 472–475, 2009a.

Rockström, J., W. Steffen, K. Noone, Å. Persson, F.S. Chapin, III, E.F. Lambin, T.M. Lenton, Planetary boundaries: Exploring the safe operating space for humanity. *Ecol. Soc.*, 14, 2009b. Available at http://www ecologyandsociety.org/vol14/iss2/art32/.

Rodgers, B.E. and R.J. Baker, Frequencies of micronuclei in bank voles from zones of high radiation at Chrnobyl, Ukraine. *Environ. Toxicol. Chem.*, 19, 1644–1648, 2000.

Rodgers, B. E., J.K. Wickliffe, C.J. Phillips, R.K. Chesser, and R.J. Baker, Experimental exposure of naïve bank voles (*Clethrionomys glarelus)* to the Chornobyl, Ukraine, environment: A test of radioresistance. *Environ. Toxicol. Chem.*, 20, 1936–1941, 2001.

Rodriguez, R.R., N.T. Basta, S.W. Castell, and L.W. Pace, An *in vitro* gastrointestinal method to estimate bioavailable arsenic in contaminated soils and solid media. *Environ. Sci. Technol.*, 33, 642–649, 1999.

Roesijadi, G., Metallothioneins in metal regulation and toxicity in aquatic animals. *Aquat. Toxicol.*, 22, 81–114, 1992.

Roesijadi, G., Metallothionein and its role in toxic metal regulation. *Comp. Biochem. Physiol. C*, 113, 117–123, 1996.

Roesijadi, G., The basis for increased metallothionein in a natural population *of Crassostrea virginica. Biomarkers*, 4, 467–472, 1999.

Roesijadi, G., R. Bogumil, M. Vasák, and J.H.R. Kägi, Modulation of DNA-binding of a Tramtrack zinc-finger peptide by the metallothionein-thionein conjugate pair. *J. Biol. Chem.*, 273, 17425–17432, 1998.

Roesijadi, G. and P. Klerks, A kinetic analysis of Cd-binding to metallothionein and other intracellular ligands in oyster gills. *J. Exp. Zool.*, 251, 1–12, 1989.

Roesijadi, G. and W.E. Robinson, Metal regulation in aquatic animals: Mechanisms of uptake, accumulation, and release, in *Aquatic Toxicology. Molecular, Biochemical and Cellular Perspectives*, Malins, D.C. and G.K. Ostrander, Eds., CRC Press, Boca Raton, FL, 1994.

Rogers, J.M. and R.J. Kavlock,. Developmental toxicology, in *Casarett & Doull's Toxicology: The Basic Science of Poisons,* 7th Edition, Klaassen, C.D., Ed., McGraw Hill, New York, 2008.

Rohr, J.R. and K.A. McCoy, A qualitative meta-analysis reveals consistent effects of atrazine on freshwater fish and amphibians. *Environ. Health Perspect.*, 118, 20–32, 2010.

Rohr, J.R., A. Swan, T.R. Raffel, and P.J. Hudson, Parasites, info-disruption, and the ecology of fear. *Oecologia*, 159, 447–454, 2009.

Rojas, E., M.C. Lopez, and M. Valverde, Single cell gel electrophoresis assay: Methodology and applications. *J. Chromatogr. B*, 722, 225–254, 1999.

Rolán-Alvarez, E., K. Johnnesson, and J. Erlandsson, The maintenance of a cline in the marine snail *Littorina saxatilus*: The role of home site advantage and hybrid fitness. *Evolution*, 51, 1838–1847, 1997.

Rolff, C., D. Broman, C. Näf, and Y. Zebühr, Potential biomagnification of PCDD/Fs: New possibilities for quantitative assessment using stable isotope trophic position. *Chemosphere*, 27, 461–468, 1993.

Rose, K.C., C.E. Williamson, S.G. Schladow, M. Winder, and J.T. Oris, Patterns of spatial and temporal variability of UV transparency in Lake Tahoe, California-Nevada. *J. Geophys. Res. Biogeosci.*, 114, G00D03, 2009.

Rosner, B., *Fundamentals of Biostatistics,* 6th Edition, Duxbury Press, Belmont, CA, 2006.

Rostkowski, P., N. Yamashita, I.M.K. So, S. Taniyasu, P.K.S. Lam, J. Falandysz, K.-T Lee, Perfluorinated compounds in streams of the Shihwa Industrial Zone and Lake Shihwa, South Korea. *Environ. Toxicol. Chem.*, 25, 2374–2380, 2006.

Roth, D.S., I. Perfecto, and B. Rathcke, The effects of management systems on ground-foraging ant diversity in Costa Rica. *Ecol. Appl.*, 4, 423–436, 1994.

Rousch, W., Putting a price tag on Nature's bounty. *Science*, 276, 1029, 1997.

Roussel, H., S. Joachim, S. Lamothe, O. Palluel, L. Gauthier, and J.M. Bonzom, A long-term copper exposure on freshwater ecosystem using lotic mesocosms: Individual and population responses of three-spined sticklebacks (*Gasterosteus aculeatus*). *Aquat. Toxicol.*, 82, 272–280, 2007a.

Roussel, H., L. Ten-Hage, S. Joachim, R. Le Cohu, L. Gauthier, and J.M. Bonzom, A long-term copper exposure on freshwater ecosystem using lotic mesocosms: Primary producer community responses. *Aquat. Toxicol.*, 81, 168–182, 2007b.

Rowan, D.J. and J.B. Rasmussen, Bioaccumulation of radiocesium by fish: The influence of physicochemical factors and trophic structure. *Can. J. Fish. Aquat. Sci.*, 51, 2388–2410, 1994.

Roy, R. and P.G.C. Campbell, Survival time modeling of exposure of juvenile Atlantic salmon (*Salmo salar*) to mixtures of aluminum and zinc in soft water at low pH. *Aquat. Toxicol.*, 33, 155–176, 1995.

Ruban, G.I., Plasticity of development in natural and experimental populations of Siberian sturgeon *Acipenser baeri* Brandt. *Acta Zool. Fenn.*, 191, 43–46, 1992.

Rudge, D.W., Tut-tut, not so fast. Did Kettlewell really test Tutt's explanation of industrial melanism? *Hist. Philos. Life Sci.*, 32, 493–520, 2010.

Rueter, J.G., Jr. and F.M.M. Morel, The interaction between zinc deficiency and copper toxicity as it affects the silicic acid uptake mechanisms in *Thalassiosira pseudonana. Limnol. Oceanogr.*, 26, 67–73, 1981.

Rule, J.H. and R.W. Alden, III, Cadmium bioavailablity to three estuarine animals in relation to geochemical fractions to sediments. *Arch. Environ. Contam. Toxicol.*, 19, 878–885, 1990.

Ruohtula, M. and J.K. Miettinen, Retention and excretion of ^{203}Hg-labelled methylmercury in rainbow trout. *Oikos*, 26, 385–390, 1975.

Russell, R.W., R. Lazar, and G.D. Haffner, Biomagnification of organochlorines in Lake Erie white bass. *Environ. Toxicol. Chem.*, 14, 719–724, 1995.

Ryan, J.A. and L.E. Hightower, Stress proteins as molecular biomarkers for environmental toxicology. *EXS*, 77, 411–424, 1996.

Saavedra Alvarez, M.M. and D.V. Ellis, Widespread neogastropod imposex in the Northeast Pacific: Implications for TBT contamination surveys. *Mar. Pollut. Bull.*, 21, 244–247, 1990.

Sabins, F.F., Jr., *Remote Sensing: Principles and Interpretation*, W.H. Freeman and Co., New York, 1987.

Saccheri, I.J., F. Rousett, P.C. Watts, P.M. Brakefield, and L.M. Cook, Selection and gene flow on a diminishing cline of melanic peppered moths. *Proc. Natl. Acad. Sci.,* 105, 16212–16217, 2008.

Sagan, L.A., What is hormesis and why haven't we heard about it before? *Health Phys.*, 52, 521–525, 1987.

Saglio, P. and S. Trijasse, Behavioral responses to atrazine and diuron in goldfish. *Arch. Environ. Contam. Toxicol.*, 35, 484–491, 1998.

Saiki, M.K. and R.S. Ogle, Evidence of impaired reproduction by Western mosquitofish inhabiting seleniferous agricultural drainwater. *Trans. Am. Fish. Soc.*, 124, 578–587, 1995.

Saiyed, H., A. Dewan, V. Bhatnagar, U. Shenoy, R. Shenoy, H. Rajmohan, K. Patel, Effect of endosulfan on male reproductive development. *Environ. Health Perspect.*, 111, 1958–1962, 2003.

Saleh, M., A. Kamel, A. Rajab, A. El-Baroty, and A.K. El-Sabae, Regional distribution of organochlorine insecticide residues in human milk from Egypt. *J. Environ. Sci. Health.* B, 31, 241–255, 1996.

Salsburg, D.S., *Statistics for Toxicologists*, Marcel Dekker, New York, 1986.

Samallow, P.B. and M.E. Soule, A case of stress related heterozygote superiority in nature. *Evolution*, 37, 646–649, 1983.

Samson, L., and J. Cairns, A new pathway for DNA in *Escherichia coli. Nature*, 267, 281–283, 1977.

Sanborn, J.R., R.L. Metcalf, W.N. Bruce, and P.Y. Lu, The fate of chlordane and toxaphene in a terrestrial-aquatic model ecosystem. *Environ. Entomol.*, 5, 533–538, 1976.

Sanders, B.M., Stress proteins: Potential as multitiered biomarkers, in *Biomarkers of Environmental Contamination*, McCarthy, J.F. and L.R. Shugart, Eds., Lewis Publishers, Boca Raton, FL, 1990.

Sanders, B.M. and S.D. Dyer, Cellular stress response. *Environ. Toxicol. Chem.*, 13, 1209–1210, 1994.

Sanders, B.M., K.D. Jenkins, W.G. Sunda and J.D. Costlow, Free cupric ion activity in seawater: Effects on metallothionein and growth in crab larvae. *Science*, 222, 53–55, 1983.

Sanders, B.M. and L.S. Martin, Stress proteins as biomarkers of contaminant exposure in archived environmental samples. *Sci. Total Environ.*, 139/140, 459–470, 1993.

Sandheinrich, M.B. and G.J. Atchison, Sublethal toxicant effects on fish foraging behavior: Empirical vs. mechanistic approaches. *Environ. Toxicol. Chem.*, 9, 107–119, 1990.

Sandheinrich, M.B. and K.M. Miller, Effects of dietary methylmercury on reproductive behavior of fathead minnows (*Pimephales promelas*). *Environ. Toxicol. Chem.*, 25, 3053–3057, 2006.

Sandstead, H.H., Interactions that influence bioavailability of essential metals to humans, in *Metal Speciation: Theory, Analysis and Application*, Kramer, J.R. and H.E. Allen, Eds., Lewis Publishers, Chelsea, MI, 1988.

Sang, S. and S. Petrovic, Endosulfan: A review of its toxicity and its effects on the endocrine system, *WWF (World Wild Life Fund: Canada)*, 1999.

Sanghi, R, M.K. Pillai, T.R. Jayalakshmi, and A. Nair, Organochlorine and organo-phosphorous pesticide residues in breast milk from Bhopal and Madhya Pradesh, India. *Hum. Exp. Toxicol.*, 22, 73–76, 2003.

Sargent, J. and R. Seffl, Properties of perfluorinated liquids. *Fed. Proc.*, 29, 1699–1703, 1970.

Sarkar, A., D. Ray, A.N. Shrivastava, and S. Sarker, Molecular biomarkers: Their significance and application in marine pollution monitoring. *Ecotoxicology*, 15, 333–340, 2006.

Sarokin, D. and J. Schulkin, The role of pollution in large-scale population disturbances. Part 1: Aquatic populations. *Environ. Sci. Technol.*, 26, 1476–1484, 1992.

SAS Institute Inc., *SAS Technical Report P-200, SAS/STAT Software: CALIS and LOGISTIC Procedures, Release 6.04*, SAS Institute., Cary, NC, 1990.

Satoh, M., N. Nishimura, Y. Kanayama, A. Naganuma, T. Suzuki, and C. Tohyama, Enhanced renal toxicity by inorganic mercury in metallothionein-null mice. *J. Pharmacol. Exp. Ther.*, 283, 1529–1533, 1997.

Saunders, R.L. and J.B. Sprague, Effects of copper-zinc mining pollution on a spawning migration of Atlantic salmon. *Water Res.*, 1, 419–432, 1967.

Sawyer, S.J., K.A. Gerstner, and G.V. Callard, Real-time PCR analysis of cytochrome P450 aromatase expression in zebrafish: Gene specific tissue distribution, sex differences, developmental programming, and estrogen regulation. *Gen. Comp. Endocr.*, 147, 108–117, 2006.

Schecher, W.D. and D. McAvoy, *MINEQL+: A Chemical Equilibrium Modeling System. (4.5)*, Environmental Research Software, Hallowell, ME, 2001.

Scheffer, M., S.R. Carpenter, T.M. Lenton, J. Bascompte, W. Brock, V, Dakos, J. van de Koppel, Anticipating critical transitions. *Science* 338, 344–348, 2012.

Scheninger, M.J., M.J. CeNiro, and H. Tauber, Stable nitrogen isotope ratios of bone collagen reflect marine and terrestrial components of prehistoric human diet. *Science*, 220, 1381–1383, 1983.

Schieve, M.S., D.D. Weber, M.S. Myers, F.J. Jaques, W.L. Riechert, C.A. Krone, D.C. Malins, Induction of foci of cellular alteration and other hepatic lesions in English sole (*Parophrys vetulus)* exposed to an extract of an urban marine sediment. *Can. J. Fish. Aquat. Sci.*, 48, 1750–1760, 1991.

Schiffer, B., A. Daxenberger, K. Meyer, and H.H. Meyer, The fate of trenbolone acetate and melengestrol acetate after application as growth promoters in cattle: Environmental studies. *Environ. Health Perspect.*, 109, 1145–1151, 2001.

Schindler, D.W., Effects of acid rain on freshwater ecosystems. *Science*, 239, 149–157, 1988.

Schindler, D.W., Ecosystems and ecotoxicology: A personal perspective, in *Ecotoxicology: A Hierarchical Treatment*, Newman, M.C. and C.H. Jagoe, Eds., CRC/Lewis Publishers, Boca Raton, FL, 1996.

Schlezinger, J.J., R.D. White, and J.J. Stegeman, Oxidative inactivation of cytochrome P450 1A (CYP1A) stimulated by 3,3',4,4'-tetrachlorobiphenyl: Production of reactive oxygen by vertebrate CYP1As. *Mol. Pharmacol.*, 56, 588–597, 1999.

Schlueter, M.A., S.I. Guttman, J.T. Oris, and A.J. Bailer, Survival of copper-exposed juvenile fathead minnows (*Pimephales promelas*) differs among allozyme genotypes. *Environ. Toxicol. Chem.*, 10, 1727–1734, 1995.

Schmitt, C.J., M.L. Wildhaber, J.B. Hunn, T. Nash, M.N. Tieger, and B.L. Steadman, Biomonitoring of lead-contaminated Missouri streams with an assay for erythrocyte δ-aminolevulinic acid dehydratase activity in fish blood. *Arch. Environ. Contam. Toxicol.*, 25, 464–475, 1993.

Schnell, J.H., Some effects of neutron-gamma radiation on late summer bird populations. *Auk*, 81, 528–533, 1964.

Schnute, J.T. and L.J. Richards, A unified approach to the analysis of fish growth, maturity, and survivorship data. *Can. J. Fish. Aquat. Sci.*, 47, 24–40, 1990.

Schober, U. and W. Lampert, Effects of sublethal concentrations of the herbicide Atrazin[7] on growth and reproduction of *Daphnia pulex. Bull. Environ. Contam. Toxicol.*, 17, 269–277, 1977.

Schreck, C.B., Physiological, behavioral, and performance indicators of stress. *Am. Fish. Soc. Symp.*, 8, 29–37, 1990.

Schulz, H., Zur Lehre von der Arzneiwirdung. *Virchows Archiv fur Pathologische Anatomie und Physiologie fur Klinishce Medizin*, 108, 423–445, 1887.

Schulz, H., Uber Hefegifte, *Pflugers Archiv fur die gesamte Physiologie des Menschen und der Tiere*, 42, 517–541, 1888.

Schultz, I.R. and W.L. Hayton, Body size and the toxicokinetics of trifluralin in rainbow trout. *Toxicol. Appl. Pharmacol.*, 129, 138–145, 1994.

Schultz, I.R., G. Orner, J.L. Merdink, and A. Skillman, Dose-response relationships and pharmacokinetics of vitellogenin in rainbow trout after intravascular administration of 17 a-ethynylestradiol. *Aquat. Toxicol.*, 51, 305–318, 2001.

Schuster, P.F., D.P. Krabbenhoft, D.L. Naftz, L.D. Cecil, M.L. Olson, J.F. Dewild, D.D. Susong, J.R. Green, and M.L. Abbott, Atmospheric mercury deposition during the last 270 years: A glacial ice core record of natural and anthropogenic sources. *Environ. Sci. Technol.*, 36, 2303–2303, 2002.

Schütt, P. and E.B. Cowling, Waldsterben, a general decline of forests in central Europe: Symptoms, development, and possible causes. *Plant Dis.*, 69, 548–558, 1985.

Schwaiger, J., H. Ferling, U. Mallow, H. Wintermayr, and R.D. Negele, Toxic effects of the non-steroidal anti-inflammatory drug diclofenac. Part 1: Histopathological alterations and bioaccumulation in rainbow trout. *Aquat. Toxicol.*, 68, 141–150, 2004.

Scott, G.R., and K.A. Sloman, The effects of environmental pollutants on complex fish behaviour: Integrating behavioural and physiological indicators of toxicity. *Aquat. Toxicol.*, 68, 369–392, 2004.

Seebaugh, D.R. and W.G. Wallace, Importance of metal-binding proteins in the partitioning of Cd and Zn as trophically available metal (TAM) in the brine shrimp *Artemia franciscana. Mar. Ecol. Prog. Ser.*, 272, 215–230, 2004.

Segner, H. and T. Braunbeck, Cellular response profile to chemical stress, in *Ecotoxicology*, Schüürmann, G. and T. Braunbeck, Eds., Wiley, New York, 1998.

Sekercioglu, C.H., Increasing awareness of avian ecological function. *Trends Ecol. Evol.*, 21, 464–471, 2006.

Sellin Jeffries, M.K., C. Claytor, W. Stubblefield, W. Pearson, and J.T. Oris, A quantitative risk model for polycyclic aromatic hydrocarbon photo-induced toxicity in Pacific herring following the *Exxon Valdez* oil spill. *Environ. Sci. Technol., 47*, 5450–5458, 2013.

Selye, H., Stress and the general adaptation syndrome. *Br. Med. J.*, 4667, 1383–1392, 1950.

Selye, H., *The Stress of Life*, McGraw-Hill, New York, 1956.

Selye, H., The evolution of the stress concept. *Am. Sci.*, 61, 692–699, 1973.

Semenza, G.L., Life with oxygen. *Science*, 318, 62–64, 2007.

Serreze, M.C., M.M. Holland, and J. Stroeve, Perspectives on the Arctic's shrinking sea-ice cover. *Science*, 315, 1533–1536, 2007.

Service, R.F., Rising acidity brings an ocean of trouble. *Science*, 337, 146–148, 2012.

Seth, C.S., P.K. Chaturvedi, and V. Misra, The role of phytochelatins and antioxidants in tolerance to Cd accumulation in *Brassica juncea* L. *Ecotoxicol. Environ. Saf.*, 71, 76–85, 2008.

Settle, D.M. and C.C. Patterson, Lead in albacore: Guide to lead pollution in Americans. *Science*, 207, 1167–1176, 1980.

Shane, B.S., Introduction to ecotoxicology, in *Basic Environmental Toxicology*, Cockerham, L.G. and B.S. Shane, Eds., CRC Press, Boca Raton, FL, 1994.

Shapiro, H. R., *Practical Flow Cytometry*, A. R. Liss Publishing, New York, 2003.

Sharples, F.E., Ecological aspects of hazard identification for environmental uses of genetically engineered organisms, in *Risk Assessment in Genetic Engineering. Environmental Release of Organisms*, Levin, M. and H. Strauss, Eds., McGraw-Hill, New York, 1991.

Shaw, W.H.R., Cation toxicity and the stability of transition-metal complexes. *Nature*, 192, 754–755, 1961.

Shaw-Allen, P.L. and K. McBee, Chromosome damage in wild rodents inhabiting a site contaminated with Aroclor 1254, *Environ.Toxicol.* Chem., 12, 677–684, 1993.

Shea, D., Developing national sediment quality criteria. *Environ. Sci. Technol.*, 22, 1256–1261, 1988.

Shedd, T.R., W.H. van der Schalie, M.W. Widder, D.T. Burton, and E.P. Burrows, Long-term operation of an automated fish biomonitoring system for continuous effluent acute toxicity surveillance. *Bull. Environ. Contam. Toxicol.*, 66, 392–399, 2001.

Shen, D.D., Toxicokinetics, in *Casarett & Doull's Toxicology: The Basic Science of Poisons,* 7th Edition, Klaassen, C.D., Ed., McGraw-Hill, New York, 2008.

Shen, L., F. Wania, Y.D. Lei, C. Teixeira, D.C.G. Muir, and T.F. Bidleman, Atmospheric distribution and long-range transport behavior of organochlorine pesticides in North America. *Environ. Sci. Technol.*, 39, 409–420, 2005.

Shepherd, A., E.R. Ivins, G.A. Valentina, V.R. Barletta, M.J. Bentley, S. Bettadpur, K.H. Briggs, A reconciled estimate of ice-sheet mass balance. *Science* 338, 1183–1189, 2012.

Sherson, J.F., H. Krauter, R.K. Olsson, B. Julsgaard, K. Hammerer, I. Cirac, and E.S. Polzik, Quantum teleportation between light and matter. *Nature*, 443, 557–560, 2006.

Shi, H.H., C.J. Huang, S.X. Zhu, X.J.Yu, and W.Y. Xie, Generalized system of imposex and reproductive failure in female gastropods of coastal waters of mainland China. *Mar. Ecol.Prog. Ser.*, 304, 179–189, 2005.

Shore, P.A., B.B. Brodie, and C.A.M. Hogben, The gastric secretion of drugs: A pH partition hypothesis. *J. Pharmacol. Exp. Ther.*, 119, 361–369, 1957.

Shouqui Cai, The characteristics of development and trend of Chinese contemporary environmental laws. Proceedings of International Conference of Sustainable Environmental and Resources Law. Wuhan, China, 1999.

Shugart, L.R., Quantitation of chemically induced damage to DNA of aquatic organisms by alkaline unwinding assay. *Aquat. Toxicol.*, 13, 43–52, 1988.

Shugart, L.R., Environmental genotoxicology, in *Fundamentals of Aquatic Toxicology: Effects, Environmental Fate, and Risk Assessment,* 2nd Edition, Rand, G.M., Ed., Taylor & Francis, Washington, DC, 1995.

Shugart, L.R., Molecular markers to toxic agents, in *Ecotoxicology: A Hierarchical Treatment*, Newman, M.C. and C.H. Jagoe, Eds., CRC Press, Boca Raton, FL, 1996.

Shugart, L. R., Structural damage to DNA in response to toxicant exposure, in *Genetics and Ecotoxicology*, Forbes, V. E., Ed., Taylor & Francis, Philadelphia, PA, 1999.

Shukla, K.K., C.S. Dombroski, and S.H. Cohn, Fallout ^{137}Cs levels in man over a 12 yr period. *Health Phys.*, 24, 555–557, 1973.

Shunthirasingham, C., T. Gouin, Y.D. Lei, C. Ruepert, L.E. Castillo, and F. Wania, Current use pesticide transport to Costa Rica's high altitude tropical cloud forest. *Environ.Toxicol. Chem.*, 30, 2709–2717, 2011.

Shurin,J.B., E.T. Borer, E.W. Seabloom, K. Anderson, C.A. Blanchette, B. Broitman, S.D. Cooper, and B.S. Halpern, A cross-ecosystem comparison of the strength of trophic cascades. *Ecol. Lett.*, 5, 785–791, 2002.

Sibly, R.M., Effects of pollutants on individual life histories and population growth rates, in Newman, M.C. and C.H. Jagoe, *Ecotoxicology: A Hierarchical Treatment*, CRC Press, Boca Raton, FL, 1996.

Sibly, R.M. and P. Calow, A life-cycle theory of responses to stress. *Biol. J. Linn. Soc.*, 37, 101–116, 1989.

Sigg, L., F. Black, J. Buffle, J. Cao, R. Cleven, W. Davison, J. Galceran, Comparison of analytical techniques for dynamic trace metal speciation in natural freshwaters. *Environ. Sci. Technol.*, 40, 1934–1941, 2006.

Sih, A., A.M. Bell, and J.L. Kerby, Two stressors are far deadlier than one. *Trends Ecol. Evol.*, 19, 274–276, 2004.

Simkiss, K., Calcium, pyrophosphate and cellular pollution. *Trends Biochem. Sci.*, April 1981, 1–3, 1981a.

Simkiss, K., Cellular discrimination processes in metal accumulating cells. *J. Exp. Biol.*, 94, 317–327, 1981b.

Simkiss, K., Lipid solubility of heavy metals in saline solutions. *J. Mar. Biol. Assoc. U.K.*, 63, 1–7, 1983.

Simkiss, K., Ecotoxicants at the cell-membrane barrier, in *Ecotoxicology: A Hierarchical Treatment*, Newman, M.C. and C.H. Jagoe, Eds., CRC/Lewis Publishers, Boca Raton, FL, 1996.

Simkiss, K., S. Daniels, and R.H. Smith, Effects of population density and cadmium on growth and survival of blowflies. *Environ. Pollut.*, 81, 41–45, 1993.

Simkiss, K. and M. Taylor, Cellular mechanisms of metal ion detoxification and some new indices of pollution. *Aquat. Toxicol.*, 1, 279–290, 1981.

Simon, H., Nevada radioiodine levels soar, *The Washington Post*, A1, August 17, 1963.

Simonich, S.I. and R.A. Hites, Global distribution of persistent organochlorine compounds. *Science*, 269, 1851–1854, 1995.

Simpson, G.P., The environmentalist paradox. The World Trade Organization's challenges. *Harvard Int. Rev.*, Winter 2002, 56–61, 2002.

Sinclair, C. and A.A. Hoffmann, Developmental stability as a potential tool in the early detection of salinity stress in wheat. *Int. J. Plant Sci.*, 164, 325–331, 2003.

Sinha,N., R. Narayan, R. Shanker, and D.K. Saxena, Endosulfan induced biochemical changes in testis of rats. *Vet. Hum. Toxicol.*, 37, 547–549, 1995.

Skelly, D.K., Experimental venue and estimation of interaction strength. *Ecology*, 83, 2097–2101, 2002.

Skidmore, J.F. and P.W.A. Tovell, Toxic effects of zinc sulphate on the gills of rainbow trout. *Water Res.*, 6, 217–230, 1972.

Slater, T.F., Free-radical mechanisms in tissue injury. *Biochem. J.*, 222, 1–15, 1984.

Slemr, F., E.G. Brunke, R. Ebinghaus, C. Temme, J. Munthe, I. Wangberg, W. Schroeder, A. Steffen, and T. Berg, Worldwide trend of atmospheric mercury since 1977. *Geophys. Res. Lett.*, 30, 1516, 2003.

Slobodkin, L.B. and D.E. Dykhuizen, Applied ecology, its practice and philosophy, in *Integrated Environmental Management*, Cairns, J, Jr. and T.V. Crawford, Eds., Lewis Publishers, Chelsea, MI, 1991.

Sloman, K.A., D.W. Baker, C.M. Wood, and D.G. McDonald, Social interactions affect physiological consequences of sublethal copper exposure in rainbow trout, *Oncorhynchus mykiss. Environ. Toxicol. Chem.*, 21, 1255–1263, 2003a.

Sloman, K.A., G.R. Scott, Z. Diao, C. Rouleau, C.M. Wood, and D.G. McDonald, Cadmium affects the social behavior of rainbow trout, *Oncorhynchus mykiss. Aquat. Toxicol.*, 65, 171–185, 2003b.

Sluyts, H., F. Van Hoof, A. Cornet and J. Paulussen, A dynamic new alarm system for use in biological early warning system. *Environ. Toxicol. Chem.*, 15, 1317–1323, 1996.

Smith, W.H., *Air Pollution and Forests*, Springer-Verlag New York, New York, 1981.

Smith, R.J., The risks of living near Love Canal. *Science*, 217, 808–811, 1982.

Smith, R.L., *Elements of Ecology,* 2nd Edition, Harper & Row Publishers, New York, 1986.

Smith, D.R., B.J. Crespi, and F.I. Bookstein, Fluctuating asymmetry in the honey bee, *Apis mellifera*: Effects of ploidy and hybridization. *J. Evolution. Biol.*, 10, 551–574, 1997.

Smith, A.L. and R.H. Green, Uptake of mercury by freshwater clams (Family Unionidae). *J. Fish. Res. Board Can.*, 32, 1297–1303, 1975.

Smith, C.J., B.J. Shaw, and R.D. Handy, Toxicity of single walled carbon nanotubes to rainbow trout (*Oncorhynchus mykiss*): Respiratory toxicity, organ pathologies, and other physiological effects. *Aquat. Toxicol.* 82, 94–109, 2007.

Smith, W.E. and A.M. Smith, *Minamata*, Holt, Rinehart and Winston, New York, 1975.

Smith, V.H., G.D. Tilman, and J.C. Nekola. Eutrophication: Impacts of excess nutrient inputs on freshwater, marine, and terrestrial ecosystems. *Environ. Pollut.*, 100, 179–196, 1999.

Smith, E.R., L.T. Tran, and R.V. O'Neill, Regional vulnerability assessment for the mid-Atlantic region: evaluation of integration methods and assessments results. EPA/600/R-03/082, October 2003. EPA Office of Research and Development, Research Triangle Park, NC, 2003.

Smith, S.M., M.E. Wastney, K.O. O'Brien, B.V. Morukov, I.M. Larina, S.A. Abrams, J.E. Davis-Street, V. Oganov, and L.C. Shackelford, Bone markers, calcium metabolism, and calcium kinetics during extended-duration space flight on the Mir Space Station. *J. Bone Miner. Res.* 20, 208–218, 2005.

Snape, J.R., S.J. Maund, D.B. Pickford, and T.H. Hutchinson, Ecotoxicogenomics: The challenge of integrating genomics into aquatic and terrestrial ecotoxicology. *Aquat. Toxicol.*, 67, 143–154, 2004.

Snell, T.W., S.E. Brogdon, and M.B. Morgan, Gene expression profiling in ecotoxicology. *Ecotoxicology*, 12, 475–483, 2003.

Snoeijs, T., T. Dauwe, R. Pinsten, F. Vandesande, and M. Eens, Heavy metal exposure affects the humoral immune response in a free-living small songbird, the Great Tit (*Parus major*). *Arch. Environ. Contam. Toxicol.*, 46, 399–404, 2004.

So, M.K., S. Taniyasu, N. Yamashita, J.P. Giesy, J. Zheng, Z. Fang, S.H. Im, and P.K.S. Lam, Perfluorinated compounds in coastal waters of Hong Kong, South China, and Korea. *Environ. Sci. Technol.*, 38, 4056–4063, 2004.

So, M. K., S. Taniyasu, N. Yamashita, J.P. Giesy, Q. Jiang, K. Chen, and P.K.S. Lam, Health risks in infants associated with exposure to perfluorinated compounds in breast milk from Zhoushan, China. Perfluorinated fatty acids in coastal waters of Hong Kong, South China, and Korea. *Environ. Sci. Technol.*, 40, 2924–2929, 2006.

Sohlenius A.K., K. Andersson, A. Bergstrand, O. Spydevold, and J.W. De Pierre, Hepatic peroxisome proliferation in vitamin A-deficient mice without a simultaneous increase in peroxisomal acyl-CoA oxidase activity. *Biochem. Biophys.*, 1213, 63–74, 1994.

Soimasuo, R., I. Jokinen, J. Kukkoen, T. Petänen, T. Ristola, and A. Oikari, Biomarker responses along a pollution gradient: Effects of pulp and paper mill effluents on caged whitefish. *Aquat. Toxicol.*, 31, 329–345, 1995.

Solomon, G.M. and P.M. Weiss, Chemical contaminants in breast milk: Time trends and regional variability. *Environ. Health Perspect.*, 110, A339–A347, 2002.

Sommer, C., Ecotoxicology and developmental stability as in situ monitor of adaptation. *Ambio*, 25, 374–376, 1996.

Sorkhoh, N., R. Al-Hasan, S. Radwan, and T. Höpner, Self-cleaning of the Gulf. *Nature*, 359, 109, 1992.

Soto, A.M., K.L. Chung, and C. Sonnenschein, The pesticides endosulfan, toxaphene and dieldrin have estrogenic effect on human estrogen-sensitive cells. *Environ. Health Perspect.*, 102, 380–383, 1994.

Soto-Jiménez, M.F., C. Arellano-Fiore, R. Rocha-Velarde, M.E. Jara-Marini, J. Ruelas-Inzunza, and F. Páez-Osuna, Trophic transfer of lead through a model marine four-level food chain: *Tetraselmis suecica, Artemia franciscana, Litopenaeus vannamei*, and *Haemulon scudderi. Arch. Environ. Contam. Toxicol.*, 61, 280–291, 2011.

Souder, W., *On a Farther Shore. The Life and Legacy of Rachel Carson*, Crown Publishers, New York, 2012.

Southam, C.M., and J. Erhlich, Effects of extracts of western red-cedar heartwood on certain wood-decaying fungi in culture. *Phytopathology*, 33, 517–524, 1943.

Spacie, A. and J.L. Hamelink, Bioaccumulation, in *Fundamentals of Aquatic Toxicology*, Rand, G.M. and S.R. Petrocelli, Eds., Hemisphere Publishing Corp., Washington, DC, 1985.

Spacie, A., L.S. McCarty, and G.M. Rand, Bioaccumulation and bioavailability in multiphase systems, in *Fundamentals of Aquatic Toxicology,* 2nd Edition, Rand, G.M., Ed., Taylor & Francis, Washington, DC, 1995.

Sparks, A.K., *Invertebrate Pathology: Noncommunicable Diseases*, Academic Press, New York, 1972.

Sparling, D.W., G.M. Fellers, and L.L. McConnell, Pesticides and amphibian population declines in California, U.S.A. *Environ. Toxicol. Chem.*, 20, 1591–1595, 2001.

Spies, R.B., D.W. Rice, Jr., P.J. Thomas, J.J. Stegeman, J.N. Cross, and J.E. Hose, A field test of correlates of poor reproductive success and genetic damage in contaminated populations of starry flounder, *Platichthys stellatus. Mar. Environ. Res.*, 28, 542–543, 1989.

Spitzer, P.R., R.W. Risebrough, W. Walker II, R. Hernandez, A. Poole, D. Puleston, and I.C.T. Nisbet, Productivity of ospreys in Connecticut–Long Island increases as DDE residues decline. *Science*, 202, 333–335, 1978.

Sprague, J. B., Avoidance of copper-zinc solutions by young salmon in the laboratory. *J. Water Pollut. Con F.*, 36, 990–1004, 1964.

Sprague, J.B., Measurement of pollutant toxicity to fish. I. Bioassay methods for acute toxicity. *Water Res.*, 3, 793–821, 1969.

Sprague, J.B., Measurement of pollutant toxicity to fish. II. Utilizing and applying bioassay results. *Water Res.*, 4, 3–32, 1970.

Sprague, J.B., Measurement of pollutant toxicity to fish. III. Sublethal effects and "safe" concentrations. *Water Res.*, 5, 245–266, 1971.

Sprague, J.B., Current status of sublethal tests of pollutants on aquatic organisms. *J. Fish. Res. Board Can.*, 33, 1988–1992, 1976.

Sprague, J.B., P.F. Elson, and R.L. Saunders, Sublethal copper-zinc pollution in a salmon river: A field and laboratory study. *Int. J. Air Water Pollut.*, 9, 531–543, 1965.

Spromberg, J.A. and W.J. Birge, Modeling the effects of chronic toxicity on fish populations: The influence of life-history strategies. *Environ. Toxicol. Chem.*, 24, 1532–1540, 2005.

Spromberg, J.A., B.M. John, and W.G. Landis, Metapopulation dynamics: Indirect effects and multiple distinct outcomes in ecological risk assessment. *Environ. Toxicol. Chem.*, 17, 1640–1649, 1998.

Spromberg J. A. and N. Scholz, Estimating the future decline of wild Coho salmon populations resulting from early spawner die-offs in urbanizing watersheds of the Pacific Northwest, USA. *Intgr. Environ. Assess. Manag.*, 7, 648–656, 2011.

Squibb, S. and B.A. Fowler, Relationship between metal toxicity to subcellular systems and the carcinogenic response. *Environ. Health Perspect.*, 40, 181–188, 1981.

Srama,R.J., O. Beskid, B. Binkova, I. Chvatalova, Z. Lnenickova, A. Milcova, I. Solansky, Chromosomal aberrations in environmentally exposed population in relation to metabolic and DNA repair gene polymorphisms., 620, 22–33, 2007.

Stacell, M. and D.G. Huffman, Oxytetracycline-induced photosensitivity of channel catfish. *Prog. Fish Cult.*, 56, 211–213, 1994.

Standley. L.J. and B.W. Sweeney, Organochlorine pesticides in stream mayflies and terrestrial vegetation of undisturbed tropical catchments exposed to long-range atmospheric transport. *J. N. Am. Benthol. Soc.*, 14, 38–49, 1995.

Stanley-Horn, D.E., H.R. Matilla, M.K. Sears, G. Dively, R. Rose, R.L. Hellmich, and L. Lewis, Assessing impact of Cry 1Ab-expressing corn pollen on monarch butterfly larvae in field studies. *Proc. Natl. Acad. Sci. U.S.A* 98, 11931–11936, 2001.

Stark, J.D. and J..E. Banks, Population-level effects of pesticides and other toxicants on arthropods. *Annu. Rev. Entomol.*, 48, 505–519, 2003.

Stark, J.D., J.E. Banks, and R.Vargas, How risky is risk assessment: The role that life history strategies play in susceptibility of species to stress. *Proc. Natl. Acad. Sci. U.S.A*, 101, 732–736, 2004.

State of Queensland, Reef Water Quality Protection Plan 2013: Securing the health and resilience of the Great Barrier Reef World Heritage Area and adjacent catchments. Reef Water Quality Protection Secretariat, Brisbane, Queensland, 2013. Available at http://www.reefplan.qld.gov.au/resources/assets/reef-plan-2013.pdf, accessed on 7/10/2013.

Stearns, S.C., *The Evolution of Life Histories*, Oxford University Press, Oxford, 1992.

Stebbing, A.R.D., Hormesis: The stimulation of growth by low levels of inhibitors. *Sci. Total Environ.*, 22, 213–234, 1982.

Steedman, R.J., Modification and assessment of an index of biotic integrity to quantify stream quality in southern Ontario. *Can. J. Fish. Aquat. Sci.*, 45, 492–501, 1988.

Steffen, W., J. Grinewald, P. Crutzen, and J. McNeill, The Anthropocene: Conceptual and historical perspectives, *Philos. Trans. R. Soc. A*, 369, 842–867, 2011.

Stegeman, J.J. and M.E. Hahn, Biochemistry and molecular biology of monooxygenases: Current perspectives on forms, functions, and regulation of cytochrome P450 in aquatic species, in *Aquatic Toxicology: Molecular, Biochemical and Cellular Perspectives*, Malins, D.C. and G.K. Ostrander, Eds., CRC Press, Boca Raton, FL, 1994.

Stein, J.E., W.L. Reichert, M. Nishimoto, and U. Varanasi, Overview of studies on liver carcinogenesis in English sole from Puget Sound; Evidence for a xenobiotic chemical etiology II: Biochemical studies. *Sci. Total Environ.*, 94, 51–69, 1990.

Stenehjem, M., Indecent exposure. *Nat. Hist.*, 9/90, 6–21, 1990.

Stentiford, G.D., M. Longshaw, B.P. Lyons, G. Jones, M. Green, and S.W. Feist, Histopathological biomarkers in estuarine fish species for the assessment of biological effects of contaminants. *Mar. Environ. Res.*, 55, 137–159, 2003.

Stephan, C.E., Methods for calculating an LC50, in *Aquatic Toxicology and Hazard Evaluation.* ASTM STP 634, Mayer, F.L. and J.L. Hamelink, Eds., American Society for Testing and Materials, Philadelphia, PA, 1977.

Stephan, C.E. and J.W. Rogers, Advantages of using regression analysis to calculate results of chronic toxicity tests, in *Aquatic Toxicology and Hazard Assessment: Eighth Symposium.* ASTM STP 891, Bahner, R.C. and D.J. Hansen, Eds., American Society for Testing and Materials, Philadelphia, PA, 1985.

Sterne, J.A.C. and G. Davey Smith, Sifting the evidence: What's wrong with significance tests? *Br. Med. J.*, 322, 226–230, 2001.

Stewart, A.R., S.N. Luoma, C.E. Schekat, M.A. Doblin, and K.A. Hieb, Food web pathway determines how selenium affects aquatic ecosystems: A San Francisco Bay case study. *Environ. Sci. Technol.*, 38, 4519–4526, 2004.

Steward, R.C., Industrial and non-industrial melanism in the peppered moth, *Biston betularia* (L.). *Ecol. Ent.*, 2, 231–243, 1977.

Stirling, I. and C.L. Parkinson, Possible effects of climate warming on selected populations of polar bears (*Ursus maritimus*) in the Canadian Arctic. *Arctic*, 59, 261–275, 2006.

Stocker, T.F., The closing door of climate targets. *Science*, 339, 280–282, 2013.

Stokstad, E., Pollinator diversity declining in Europe. *Science*, 313, 286, 2006.

Stone, D., P. Jepson, P. Kramarz, and R. Laskowski, Time to death in carabid beetles exposed to multiple stressors along a gradient of heavy metals pollution. *Environ. Pollut.*, 113, 239–244, 2001.

Stone, R., Can a father's exposure lead to illness in his children? *Science*, 258, 31, 1992.

Stone, D., P. Jepson, and R. Laskowski, Trends in detoxification enzymes and heavy metals accumulation in ground beetles (Coleoptera: Carabidae) inhabiting a gradient of pollution. *Comp. Biochem. Physiol. C*, 132, 105–112, 2002.

Strode, C., C.S. Wondji, J.P. David, N.J. Hawkes, N. Lumjuan, D.R. Nelson, D. R. Drane, Genomic analysis of detoxification genes in the mosquito *Aedes aegypti. Insect Biochem. Molec.*, 38, 113–123, 2008.

Stroeve, J., M.M. Holland, W. Meier, T. Scambos, and M. Serreze, Arctic sea ice decline: Faster than forecast. *Geophys. Res. Lett.*, 34, L09501, 2007.

Strong, C.R. and S.N. Luoma, Variations in the correlation of body size with concentrations of Cu and Ag in the bivalve *Macoma balthica. Can. J. Fish. Aquat. Sci.*, 38, 1059–1064, 1981.

Stumm, W. and J.J. Morgan, *Aquatic Chemistry. An Introduction Emphasizing Chemical Equilibrium in Natural Waters*, Wiley, New York, 1981.

Stumm, W., L. Sigg, and J.L. Schnoor, Aquatic chemistry of acid deposition. *Environ. Sci. Technol.*, 21, 8–13, 1987.

Sturzenbaum, S.R., O. Georgiev, A.J. Morgan, and P. Kille, Cadmium detoxification in earthworms: From genes to cells. *Environ. Sci. Technol.*, 38, 6283–6289, 2004.

Suedel B.C., J.A. Boraczek, R.K. Peddicord, P.A. Clifford, and T.M. Dillon, Trophic transfer and biomagnification potential of contaminants in aquatic ecosystems. *Rev. Environ. Contam. Toxicol.*, 136, 22–89, 1994.

Sukhotin, A.A., D.L. Lajus, and P.A. Lesin, Influence of age and size on activity and stress resistance in marine bivalve *Mytilus edulis. J. Exp. Mar. Biol. Ecol.*, 284, 129–144, 2003.

Sullivan, J.F., G.J. Atchison, D.J. Kolar, and A.W. McIntosh, Changes in predator-prey behavior of fathead minnows (*Pimephales promelas*) and largemouth bass (*Micropterus salmoides*) caused by cadmium. *J. Fish. Res. Board Can.*, 35, 446–451, 1978.

Surendranath, C., Toxic tales From god's own country, *Down to Earth*, 10, 6, 2001

Sutcliffe, F.E. (Translator), *Discourse on Method and the Meditations by René Descartes*, Penguin Books, London, UK, 1968.

Suter, II, G.W., *Ecological Risk Assessment*, Lewis Publishers, Boca Raton, FL, 1993.

Suter, II., G.W., R.A. Efroymson, B.E. Sample, and D.S. Jones, *Ecological Risk Assessment for Contaminated Sites*, Lewis Publishers, Boca Raton, FL, 2000.

Sward, S., and J. Doyle, Setback for Tahoe jet skiers / Judge rules for protection of lake. *San Francisco Chronicle* 1998-10-07 04:00:00 PDT REGION, 1998. Available at at http://www.sfgate.com/news/article/Setback-For-Tahoe-Jet-Skiers-Judge-rules-for-2986754.php, accessed October 16, 2013.

Sweet, D.H., Organic anion transporter (Slc22a) family members as mediators of toxicity. *Toxicol. Appl. Pharmacol.*, 204, 198–215, 2005.

Sweet, L.I., D.R. Passino-Reader, P.G. Meier, and G.M. Omann, Xenobiotic-induced apoptosis: Significance and potential application as a general biomarker of response. *Biomarkers*, 4, 237–253, 1999.

Szabaldi, E., A model of two functionally antagonistic receptor populations activated by the same agonist. *J. Theor. Biol.*, 69, 101–112, 1977.

Szczypka, M.S. and D.J. Thiele, A cysteine-rich nuclear protein activates yeast metallothionein gene transcription, *Mol. Cell. Biol.*, 9, 421–429, 1989.

Tackett, S.L., Lead in the environment: Effects of human exposure. *Am. Lab. (Fairfield Conn)*, July, 32–41, 1987.

Tagatz, M.E., Effect of mirex on predator-prey interaction in an experimental estuarine ecosystem. *Trans. Am. Fish. Soc.*, 4, 546–549, 1976.

Tanabe, S., POPs: Need for target research on high risk stage. *Mar. Pollut. Bull.*, 48, 609–610, 2004.

Tanaka, Y and J. Nakanishi, Life history elasticity and population-level effect of *p*-nonylphenol on *Daphnia galeata. Ecol. Res.*, 16, 41–48, 2001.

Tanka, Y. and H. Tatsuta, Retrospective estimation of population-level effect of pollutants based on local adaptation and fitness cost of tolerance. *Ecotoxicology*, 22, 795–802, 2013.

Tang, L., Q.P. Liu, H.H. Shi, X.H. Wang, D.M. Zhao, W. Luo, W. Xie, C. Fang, and H.S. Hong, Imposex of *Thais clavigera* and *Cantharus cecillei* as an indicator of marine organotin compounds pollution in coastal waters of Xiamen. *Acta Ecologica Sinica*, 29, 4640–464, 2009.

Tanguy, A. and D. Moraga, Cloning and characterization of a gene coding for a novel metallothionein in the Pacific oyster *Crassostrea gigas* (CgMT2): A case of adaptive response to metal-induced stress?, *Gene*, 273, 123–130, 2001.

Tan-Kristanto, A., A. Hoffmann, R. Woods, P. Batterham, C. Cobbett, and C. Sinclair, Translational asymmetry as a sensitive indicator of cadmium stress in plants: A laboratory test with wild-type and mutant *Arabidopsis thaliana. New Phytol.*, 159, 471–477, 2003.

Tapp, J.S., Eutrophication analysis with simple and complex models. *JWPCF* 50, 484–492, 1978.

Tariq, M.I., S. Afzal, and L. Hussian, Pesticide in shallow groundwater of Bahawalnagar, Mussafargarh, D.G Khan and Rajan Pur districts of Punjab, Pakistan. *Environ. Int.*, 30, 471–479, 2004.

Taub, F.B., Standardized aquatic microcosms. *Environ. Sci. Technol.*, 23, 1064–1066, 1989.

Taussig, H.B., The thalidomide syndrome. *Sci. Am.*, 207, 29–35, 1962.

Tavares P.C., A. Kelly, R. Maia, R.J. Lopes, R. Serrao Santos, M.E. Pereira, A.C. Duarte, and R.W. Furness, Variation in the mobilization of mercury into black-winged stilt chicks in coastal saltpans, as revealed by stable isotopes. *Estuar. Coast. Shelf S.*, 77, 65–76, 2008.

Taylor, A.D., Metapopulations, dispersal, and predator-prey dynamics: An overview. *Ecology*, 71, 429–433, 1990.

Taylor, E.J., J.E. Morrison, S.J. Blockwell, A. Tarr. and D. Pasoe, Effects of lindane on the predator-prey interaction between *Hydra oligactis* Pallas and *Daphnia magna* Strauss. *Arch. Environ. Contam. Toxicol.*, 29, 291–26, 1995.

Teasdale, S., There will come soft rains (War time), in *The Collected Poems of Sara Teasdale*, Macmillan, New York, 1937.

Teather, K., C. Jardine, and K. Gormley, Behavioral and sex ratio modification of Japanese medaka (*Oryzias latipes*) in response to environmentally relevant mixtures of three pesticides. *Environ. Toxicol.*, 20, 110–117, 2005.

Temple, P.J., Dose-response of urban trees to sulfur dioxide. *JAPCA J. Air Waste MA*, 22, 271–274, 1972.

Templeton, D.M., F. Ariese, R. Cornelis, L.G. Danielsson, H. Muntau, H.P. van Leeuwen, and R. Lobinski, Guidelines for terms related to chemical speciation and fractionation of elements. Definitions, structural aspects, and methodological approaches. *Pure Appl. Chem.*, 72, 1453–1470, 2000.

Ten Hoeve, J.E. and M.Z. Jacobson, Worldwide health effects of the Fukushima Daiichi nuclear accident, *Energy Environ. Sci*, 2012.

Terhaar, C.J., W.S. Ewell, S.P. Dziuba, W.W. White, and P.J. Murphy, A laboratory model for evaluating the behavior of heavy metals in an aquatic environment. *Water Res.*, 11, 101–110, 1977.

Terman, A. and U.T. Brunk, Ceroid/lipofuscin formation in cultured human fibroblasts: The role of oxidative stress and lysosomal proteolysis. *Mech. Ageing Dev.*, 104, 277–291, 1998.

Tessier, A., P.G.C. Campbell, and M. Bisson, Sequential extraction procedure for the speciation of particulate trace metals. *Anal. Chem.*, 51, 844–851, 1979.

Tessier, A., P.G.C. Campbell, J.C. Auclair, and M. Bisson, Relationships between partitioning of trace metals in sediments and their accumulation in the tissues of the freshwater mollusc *Elliptio complanata* in a mining area. *Can. J. Fish. Aquat. Sci.*, 41, 1463–1472, 1984.

Tessier, L., G. Vaillancourt, and L. Pazdernik, Temperature effects on cadmium and mercury kinetics in freshwater molluscs under laboratory conditions. *Arch. Environ. Contam. Toxicol.*, 26, 179–184, 1994.

Thanal, Preliminary findings of the survey on the impact of aerial spraying on the people and the ecosystem in Kasargod, Kerala, India, Long term monitoring: The impact of pesticides on the people and ecosystem (LMIPPE). Part 1, Thanal Conservation Action and Information Network, Thiruvananthapuram, Kerala, India, 2001.

Thanal, Preliminary findings of the survey on the impact of aerial spraying of endosulfan on the people and ecosystem in Kasargod, Kerala, India, Long term monitoring: The impact of pesticides on the people and ecosystem (LMIPPE). Part II, Thanal Conservation Action and Information Network, Thiruvananthapuram, Keralam, India, 2002.

Theodorakis, C.W., J.W. Bickham, T. Lamb, P.A. Medica, and T.B. Lyne, Integration of genotoxicity and population genetic analyses in kangaroo *rats (Dipodomys merriami)* exposed to radionuclide contamination at the Nevada Test Site, USA. *Environ. Toxicol. Chem.*, 20, 317–326, 2001.

Thiele, D.J., M.J. Walling, and D.H. Hamer, Mammalian metallothionein is functional in yeast, *Science*, 231, 854–856, 1986.

Thies, M.L., K. Thies, and K. McBee, Organochlorine pesticide accumulation and genotoxicity in Mexican free-tailed bats from Oklahoma and New Mexico. *Arch. Environ. Contam. Toxicol.*, 30, 178–187, 1996.

Thikawa, M. and R. Sawamura, The effects of salinity on pentachlorophenol accumulation and elimination by killifish (*Oryzias latipes*). *Arch. Environ. Contam. Toxicol.*, 26, 304–308, 1994.

Thomann, R.V., Bioaccumulation model of organic chemical distribution in aquatic food chains. *Environ. Sci. Technol.*, 23, 699–707, 1989.

Thomas, P., Molecular and biochemical responses of fish to stressors and their potential use in environmental monitoring. *Am. Fish. Soc. Symp.*, 8, 9–28, 1990.

Thompson, R.D., The changing atmosphere and its impact on planet Earth, in *Environmental Issues in the 1990s*, Mannion, A.M. and S.R. Bowlby, Eds., Wiley, West Sussex, United Kingdom, 1992.

Thompson, H.M., Interactions between pesticides; a review of reported effects and their implications for wildlife risk assessment. *Ecotoxicology*, 5, 59–81, 1996.

Thompson, H.M., Assessing the exposure and toxicity of pesticides to bumblebees (*Bombus* sp.). *Apidologie*, 32, 305–321, 2001.

Thompson, D.R., S. Bearhop, J.R. Speakman, and R.W. Furness, Feathers as a means of monitoring mercury in seabirds: Insights from stable isotope analysis. *Environ. Pollut.*, 101, 193–200, 1998.

Thompson, D.R., R.W. Furness, and S.A. Lewis, Temporal and spatial variation in mercury concentrations in some albatrosses and petrels from the sub-Antarctic. *Polar Biol.*, 13, 239–244, 1993.

Thompson, D.R., R.W. Furness, and P.M. Walsh, Historical the marine ecosystem of the North and northeast Atlantic Ocean as indicated by seabird feathers. *J. Appl. Ecol.*, 29, 79–84, 1992.

Thompson, D.R., K.C. Hamer, and R.W. Furness, Mercury accumulation in great skuas *Catharacta skua* of known age and sex, and its effects upon breeding and survival. *J. Appl. Ecol.*, 28, 672–684, 1991.

Thompson, K.W., A.C. Hendricks, G.L. Nunn, and J. Cairns, Jr., Ventilatory responses of bluegill sunfish to sublethal fluctuating exposures to heavy metals (Zn^{++} and Cu^{++}). *Water Resour. Bull.*, 19, 719–727, 1983.

Thompson, R.A., G.D. Schroder, and T.H. Conner, Chromosomal aberrations in the cotton rat (*Sigmodon hispidus*) exposed to hazardous waste. *Environ. Mol. Mutagen.*, 11, 359–367, 1988.

Thomulka, K.W. and J.H. Lange, A mixture toxicity study employing binary combinations of dinitrobenzene and trinitrobenzene using the bioluminescent marine bacterium *Vibrio harveyi* as the test organism. *Int. J. Environ. Stud.*, 60, 169–178, 2003.

Thoreau, H.D., *Essay on the Duty of Disobedience and Walden*, Reprinted 1968 by Lancer Books, New York, 1854.

Thorrold, S.R., C. Latkoczy, P.K. Swart, and C.M. Jones, Natal homing in a marine fish population. *Science*, 291, 297–299, 2001.

Thurston, R.V., R.C. Russo, and G.A. Vonogradov, Ammonia toxicity to fishes. Effect of pH on the toxicity of the un-ionized ammonia species. *Environ. Sci. Technol.*, 15, 837–840, 1981.

Tice, R.R., B.G. Ormiston, R. Boucher, C.A. Luke, and D.E. Paquette, Environmental biomonitoring with feral rodent species, in *Short-Term Assays in the Analysis of Complex Mixtures II*, Sandhu, S. S., D.M. DeMarini, M.J. Mass, M.M. Moore, and J.L. Mumford, Eds., Plenum Press, New York, 1987.

Tierney, K.B., C.R., Singh, P.S. Ross, and C.J. Kennedy, Relating olfactory neurotoxicity to altered olfactory-mediated behaviors in rainbow trout exposed to three currently-used pesticides. *Aquat. Toxicol.*, 81, 55–64, 2007.

Tietge, J.E., G.W. Holcombe, K.M. Flynn, P.A. Kosian, J.J. Korte, L.E. Anderson, D.C. Wolfe, and S.J. Degitz, Metamorphic inhibition of *Xenopus laevis* by sodium perchlorate: Effects on development and thyroid histology. *Environ. Toxicol. Chem.*, 24, 926–933, 2005.

Tilt, B., Perceptions of risk from industrial pollution in China: A comparison of occupational groups. *Hum. Organ.*, 65, 115–127, 2006.

Timbrell, J., *Principles of Biochemical Toxicology,* 3rd Edition, Taylor & Francis, Philadelphia, PA, 2000.

Tipping, E., S. Lofts, and A.J. Lawlor, Modeling the chemical speciation of trace metals in the surface waters. *Sci. Total Environ.*, 210, 63–77, 1998.

Toft, G., E. Baatrup, and L.J. Guillette, Jr., Altered social behavior and sexual characteristics in mosquitofish (*Gambusia holbrooki*) living downstream of a paper mill. *Aquat. Toxicol.*, 70, 213–222, 2004.

Tolmazin, D., Soviet environmental practices. *Science*, 221, 1136, 1983.

Tom, K.R., M.C. Newman, and J. Schmerfeld, Modeling mercury biomagnification (South River, Virginia USA) to inform river management decision making. *Environ. Toxicol. Chem.* 29, 1013–1020, 2010.

Tomizawa, M. and J.E. Casida, Neonicotinoid insecticide toxicology: Mechanisms of selective action. *Ann. Rev. Pharmacol.*, 45, 247–268, 2005.

Tomlin, C., Ed., *The Pesticide Manual: A World Compendium,* 11th Edition, British Crop Protection Council, Farnham, Surrey, Cambridge, United Kingdom, 2003.

Topashka-Ancheva, M., R. Metcheva, and S. Teodorova, A comparative analysis of the heavy metal loading of small mammals in different regions of Bulgaria II: Chromosomal aberrations and blood pathology. *Ecotoxicol. Environ. Saf.*, 54, 188–193, 2003.

Touart, L.W., The federal insecticide, fungicide, and rodenticide act, in *Fundamentals of Aquatic Toxicology: Effects, Environmental Fate, and Risk Assessment,* 2nd Edition, Rand, G.M., Ed., Taylor & Francis, Washington, DC, 1995.

Toyooka, T. and Y. Ibuki, DNA damage induced by coexposure to PAHs and light. *Environ. Toxicol. Phar.*, 23, 256–263, 2007.

Trabalka, J.R., L.D. Eyman, and S.I. Aurbach, Analysis of the 1957–1958 Soviet nuclear accident. *Science*, 209, 345–353, 1980.

Triebskorn, R., H. Casper, A. Heyd, R. Eikemper, H.R. Köhler, and J. Schwaiger, Toxic effects of the non-steroidal anti-inflammatory drug diclofenac. Part II. Cytological effects in liver, kidney, gills and intestine of rainbow trout (*Oncorhynchus mykiss*). *Aquat. Toxicol.*, 68, 151–166, 2004.

Truhaut, R., Survey of the hazards of the chemical age. *Pure Appl. Chem.*, 21, 419–436, 1970.

Truhaut, R., Ecotoxicology: Objectives, principles and perspectives. *Ecotoxicol. Environ. Saf.*, 1, 151–173, 1977.

Tsigos, C. and G.P. Chrousos, Hypothalamic-pituitary-adrenal axis, neuroendocrine factors and stress. *J. Psychosom. Res.*, 53, 865–871, 2002.

Tsyusko, O.V., M.H. Smith, T.K. Oleksyk, J. Goryanaya, and T.C. Glenn, Genetics of cattails in radioactively contaminated areas around Chornobyl. *Mol. Ecol.*, 15, 2611–2625, 2006.

Tucker, J.D., A. Auletta, M.C. Cimino, K.L. Dearfield, D. Jacobson-Kram, R.R. Tice, and A.V. Carrano, Sister-chromatid exchange: Second report of the Gene-Tox program. *Mutat. Res.*, 297, 101–180, 1993.

Tucker, A.J., C.E. Williamson, and J.T. Oris, Development and application of a UV attainment threshold for the prevention of warm water aquatic invasive species. *Biol. Invasions*, 14, 2331–2342, 2012.

Tucker, A.J., C.E. Williamson, K.C. Rose, J.T. Oris, S. Connelly, M. Olson, and D. Mitchell. Ultraviolet radiation affects invasibility of lake ecosystems by warm water fish. *Ecology*, 91, 882–890, 2010.

Tull-Singleton, S., S. Kimball, and K. McBee, Correlative analysis of heavy metal
bioconcentration and genetic damage in white-footed mice (*Peromyscus leucopus*) from a hazardous waste site. *Bull. Environ. Contam. Toxicol.*, 52, 667–672, 1994.

Turner, D.R., Problems in trace metal speciation modeling, in *Metal Speciation and Bioavailability in Aquatic Systems*, A. Tessier and D.Turner, Eds., Wiley, Chichester, United Kingdom,1995.

Turner, D.R., M. Whitfield, and A.G. Dickson, The equilibrium speciation of dissolved components in freshwater and saltwater at 25° C and 1 atm pressure. *Geochim. Cosmochim. Acta*, 45, 855–881, 1981.

Tutt, J.W., *British Moths*, Routledge, London, 1896.

Tuurala, H. and A. Soivio, Structural and circulatory changes in the secondary lamallae of *Salmo gairdneri* gills after sublethal exposures to dehydroabietic acid and zinc. *Aquat.Toxicol.*, 2, 21–29, 1982.

Tylianakis, J.M., The global plight of pollinators. *Science*, 339, 1532–1533, 2013.

Tyrrell, T., Anthropogenic modification of the oceans. *Philos. Trans. R. Soc. A*, 369, 887–908, 2011.

Ugalde, R., Impact of herbicides used in pineapple plantations over phytoplankton of streams influenced by agricultural lands in the Caribbean zone of Costa Rica. MSc. thesis, University of Bremen, Bremen, Germany, 2007.

Ugedal, O., B. Jonsson, O. Njåstad, and R. Næumanň, Effects of temperature and body size on radiocaesium retention in brown trout, *Salmo trutta. Freshw Biol.*, 28, 165–171, 1992.

Ugolini, F.C. and H. Spaltenstein, Pedosphere, in *Global Biogeochemical Cycles*, Butcher, S.S., R.J. Charlson, G.H. Orians, and G.V. Wolfe, Eds., Academic Press, London, 1992.

Ulmer, D.D., Effects of metals on protein structure, in *Effects of Metals on Cells, Subcellular Elements, and Macromolecules*, Maniloff, J., J.R. Coleman, and M.W. Miller, Eds., Charles C. Thomas Publisher, Springfield, IL, 1970.

UNEP (United Nations Environmental Programme), Regional based assessment of persistent toxic substances—Indian Ocean regional report—chemicals. United Nations Environmental Programme: Global Environment Facility (UNEP–GEF), Châtelaine, Switzerland, 2002a.

UNEP (United Nations Environmental Programme), Regional based assessment of persistent toxic substances—Central America and the Caribbean regional report—chemicals. United Nations Environmental Programme: Global Environment Facility (UNEP–GEF), Châtelaine, Switzerland, 2002b.

UNEP (United Nations Environmental Programme), Regional based assessment of persistent toxic substances—South East Asia and South Pacific regional report—chemicals. United Nations Environmental Programme: Global Environment Facility (UNEP–GEF), Châtelaine, Switzerland, 2002c.

UNEP (United Nations Environmental Programme), Regional based assessment of persistent toxic substances—Europe regional report—chemicals. United Nations Environmental Programme: Global Environment Facility (UNEP–GEF), Châtelaine, Switzerland, 2002d.

UNEP, Regionally Based Assessment of Persistent Toxic Substances. Global Report. UNEP Chemicals, Geneva, 207 p, 2003.

UNEP, The 9 new POPs. An introduction to the nine chemicals added to the Stockholm Convention by the Conference of the Parties at its fourth meeting, UNEP Chemical, 15 p, 2010.

Unger, M.A., M.C. Newman, and G.G. Vadas, Predicting survival of grass shrimp (*P. pugio*) exposed to aromatic compounds derived from spilt oil. *Environ. Toxicol. Chem.* 27, 1802–1808, 2008.

Uno, S., T.P. Dalton, S. Derkenne, C.P. Curran, M.L. Miller, H.G. Shertzer, and D.W. Nebert, Oral exposure to benzo[a]pyrene in the mouse: Detoxication by inducible cytochrome P450 is more important than metabolic activation. *Mol. Pharmacol.*, 65, 1225–1237, 2004.

Unsworth, E.R., K.W. Warnken, H. Zhang, W. Davison, F. Black, J. Buffle, J. Cao, Model predictions of metal speciation in freshwaters compared to measurements by in situ techniques. *Environ. Sci. Technol.*, 40, 1942–1949, 2006.

Urbansky, E.T., Perchlorate as an environmental contaminant. *Environ. Sci. Pollut. Res.Int.*, 9, 187–192, 2002.

USGS (United States Geological Survey), Pesticides in surface waters: Current understanding of distribution and major influences: *U.S. Geological Survey Fact Sheet* FS-039-97, USGS, Washington, DC, 1997.

USGS (United States Geological Survey), Rare earth elements: Critical resources for high technology. USGS Fact Sheet 087-02, USGS, Washington, DC, 2002.

USNIEHS (United States National Institute of Environmental Health Sciences), Final report on continuing disaster of endosulfan in Kasargod and Kannur districts, National Institute of Occupational Health. in: *Environmental Perspective*, United States National Institute of Environmental Health Sciences, Research Triangle Park, NC, 2003.

Vadas, D., The Anthropocene and the international law of the sea. *Philos. Trans. R. Soc. A*, 369, 909–925, 2011.

Vaituzis, Z., Review of effects on monarch butterfly larvae after continuous exposure to Cryl Ab-expressing MON810 and BT11 corn pollen during anthesis. MRID No. 461620-01. Monsanto Co., St. Louis, MO; Syngenta Seeds, Research Triangle Park, NC, and Biose, ID. Memorandum to M. Mendelsohn. Biopesticides and Pollution Division, EPA, 2004.

Valcke, M., F. Chaverri, P. Monge, V. Bravo, D. Mergler, T. Partanen, and C. Wesseling, Pesticide prioritization for a case-control study on childhood leukemia in Costa Rica: A simple stepwise approach. *Environ. Res.*, 97, 335–347, 2005.

Valentine, D.W. and M. Soulé, Effect of p,p'-DDT on developmental stability of pectoral fin rays in the grunion, *Leuresthes tenius. Fish. Bull.*, 71, 921–926, 1973.

Valentine, R. and G.L. Kennedy, Inhalation Toxicology Toxicokinetics, in *Principles and Methods of Toxicology,* 5th Edition, Hayes, A.W., Ed., CRC/Taylor & Francis Press, Boca Raton, FL, 2008.

Valentine D.W., M.E. Soulé, and P. Samolow, Asymmetry analysis in fishes: A possible indicator of environmental stress. *Fish. Bull.NOAA*, 71, 357–370, 1973.

Van Beneden, R.J. and G.K. Ostrander, Expression of oncogenes and tumor suppressor genes in teleost fishes, in: *Aquatic Toxicology: Molecular, Biochemical and Cellular Perspectives*, Malins, D.C. and G.K. Ostrander, Eds., CRC Press, Boca Raton, FL, 1994.

Van Cleef-Toedt, K.A., L.A.E. Kaplan and J.F. Crivello, Killifish metallothionein messenger RNA expression following temperature perturbation and cadmium exposure. *Cell Stress Chaperon.*, 6, 351–359, 2001.

Van Dam, J. W., A.P. Negri, S. Uthicke, and J.F. Mueller, Chemical Pollution on Coral Reefs: Exposure and Ecological Effect, in *Ecological Impacts of Toxic Chemicals*, Sánchez-Bayo, F., P.J. van den Brink, and R.M. Mann, Eds., Bentham Science Publishers, Bussum, The Netherlands, 2011.

Van Dam, R.A., A.J. Harford, and M. S.Warne, Time to get off the fence: The need for definitive international guidance on statistical analysis of ecotoxicity data. *Integr. Environ. Assess. Manag.*, 8, 242–245, 2012.

Van den Berg, M., L. Birnbaum, A.T.C. Bosveld, B. Brunsdtröm, P. Cook, M. Feeley, G.P. Giesy, Toxic equivalency factors (TEFs) for PCBs, PCDDs, PCDFs for human and wildlife. *Environ. Health Perspect.*, 106, 775–792, 1998.

van den Heuvel, M.R., L.S. McCarty, R.P. Lanno, B.E. Hickie, and D.G. Dixon, Effect of total body lipid on the toxicity and toxicokinetics of pentachlorophenol in rainbow trout (*Oncorhynchus mykiss*). *Aquat. Toxicol.*, 20, 235–252, 1991.

Van der Oost, R., J. Beyer, and N.P.E. Vermeulen, Fish bioaccumulation and biomarkers in environmental risk assessment: A review. *Environ. Toxicol. Phar.*, 13, 57–149, 2003.

Van der Vliet, L., L.N. Taylor, and R. Scroggins, NOEC: Notable oversight of enlightened Canadians: A response to van Dam et al. (2012). *Integr. Environ. Assess. Manag.*, 8, 397–398, 2012.

Van Dongen, S. and L. Lens, Symmetry, size and stress. *Trends Ecol. Evol.*, 15, 330, 2000.

Vankar, P.S., R. Mishra, and S. Johnson, Analysis of samples from Padre village from Kasargod District of Kerala for endosulfan residues (Pollution Monitoring Laboratory- Pesticide Residue Monitoring Study CSE/PRM-1/2001). *Centre for Science and Environment*, New Delhi, 2001. http://www.indiaenvironmentportal.org.in/content/324991/analysis-of-samples-from-padre-village-in-kasaragod-district-of-kerala-for-endosulfan-residues/.

Van Leeuwen, C.J., F. Moberts, and G. Niebeek, Aquatic toxicological aspects of dithiocarbamates and related compounds. II. Effects on survival, reproduction and growth of *Daphnia magna. Aquat. Toxicol.*, 7, 165–175, 1985.

Van Nes, E. H. and M. Scheffer, Slow recovery from perturbations as a generic indicator of a nearby catastrophic shift. *Am. Nat.*, 169, 738–747, 2007.

Van Straalen, N.M. and C.A. Denneman, Ecotoxicological evaluation of soil quality criteria. *Ecotoxicol. Environ. Saf.*, 18, 241–251, 1989.

Van Veld, P.A., and D.E. Nacci, Toxicity resistance, in: *The Toxicology of Fishes*, Di Giulio, R.T. and D. E. Hinton, Eds., Taylor & Francis, Boca Raton, FL, 2007.

Van Veld P.A., W.K. Vogelbein, M.K. Cochran, A. Goksoyr, and J.J. Stegeman, Route-specific cellular expression of cytochrome P4501A (CYP1A) in fish (*Fundulus heteroclitus*) following exposure to aqueous and dietary benzo[a]pyrene. *Toxicol. Appl. Pharmacol.*, 142, 348–359, 1997.

Van Veld P.A., W.K. Vogelbein, R. Smolowitz, B.R. Woodin, and J.J. Stegeman, Cytochrome P450IA1 in hepatic lesions of a teleost fish (*Fundulus heteroclitus*) collected from a polycyclic aromatic hydrocarbon-contaminated site. *Carcinogenesis*, 13, 505–507, 1992.

Van Veld, P.A., D.J. Westbrook, B.R. Woodin, R.C. Hale, C.L. Smith, R.J. Huggett, and J.J. Stegeman, Induced cytochrome P-450 in intestine and liver of spot (*Leiostomus xanthurus*) from a polycyclic aromatic hydrocarbon contaminated environment. *Aquat. Toxicol.*, 17, 119–132, 1990.

Van't Hof, A.E., N. Edmonds, M. Dalikova, F. Marec, and I.J. Saccheri, Industrial melanism in British peppered moths has a singular and recent mutational origin. *Science*, 332, 958–960, 2011.

Varanasi, U., M. Nishimoto, W.L. Reichert, and B.T. Le Eberhart, Comparative metabolism of benzo[a]pyrene and covalent binding to hepatic DNA in English sole, starry flounder, and rat. *Cancer Res.*, 46, 3817–3824, 1986.

Vergani, L., M. Grattarola, C. Borghi, F. Dondero, and A. Viarengo, Fish and molluscan metallothioneins: A structural and functional comparison. *FEBS J.*, 272, 6014–6023, 2005.

Verreault, J., U. Berger, and G.W. Gabrielsen, Trends in perfluorinated alkyl substances in Herring gull eggs from two coastal colonies in northern Norway: 1983–2003. *Environ. Sci. Technol.*, 41, 6671–6677, 2007.

Vidas, D., The Anthropocene and the international law of the sea. *Philos. Trans. R. Soc.*, 369, 909–925, 2011.

Víg, É. and J. Nemcsók, The effects of hypoxia and paraquat on the superoxide dismutase activity in different organs of carp, *Cyprinus carpio* L. *J. Fish Biol.*, 35, 23–25, 1989.

Vijver, M.G., C.A.M. Van Gestel, R.P. Lanno, N.M. Van Straalen, and W.J.G.M. Peijnenburg, Internal metal sequestration and its ecotoxicological relevance: A review. *Environ. Sci. Technol.*, 38, 4705–4712, 2004.

Villar, A.J., E.M. Eddy, and R.A. Pedersen, Developmental regulation of genomic imprinting during gametogenesis. *Dev. Biol.*, 172, 264–271, 1995.

Vineis, P., G. Hoek, M. Krzyzanowski, F. Vigna-Taglianti, F. Veglia, L. Airoldi, H. Autrup, Air pollution and risk of lung cancer in a prospective study in Europe. *Int. J. Cancer*, 119, 169–174, 2006.

Vir, A.K., Toxic trade with Africa. *Environ. Sci. Technol.*, 23, 23–25, 1989.

Viswanathan, P.N. and C.R. Krishna Murti, Effects of temperature and humidity on ecotoxicology of chemicals, in: *Ecotoxicology and Climate. SCOPE 38*, Bourdeau, P., J.A. Haines, W. Klein, and C.R. Krishna Murti, Eds., Wiley, Suffolk, United Kingdom, 1989.

Vogel, G. and C. Holden, Field leaps forward with new stem cell advances. *Science*, 318, 1224–1225, 2007.

Vogelbein, W.K. and J.W. Fournie, The ultrastructure of normal and neoplastic exocrine pancreas in the mummichog, *Fundulus heteroclitus. Toxicol. Pathol.*, 22, 248–260, 1994.

Vogelbein W.K., J.W. Fournie, P.S. Cooper, and P.A. Van Veld, Hepatoblastomas in the mummichog, *Fundulus heteroclitus* (Linnaeus), from a creosote-contaminated environment: A histologic, ultrastructural, and immunohistochemical study. *J. Fish Dis.*, 22, 419–431, 1999.

Vogelbein, W.K., J.W. Fournie, P.A. Van Veld, and R.J. Huggett, Hepatic neoplasms in the mummichog *Fundulus heteroclitus* from a creosote-contaminated site. *Cancer Res.*, 50, 5978–5986, 1990.

Vogelbein, W.K. and M.A. Unger, Liver carcinogenesis in a non-migratory fish: The association with polycyclic aromatic hydrocarbon exposure. *Bull. Eur. Assoc. Fish Pathol.*, 26, 11–20, 2006.

Vogelbein W.K., D.E. Zwerner, M.A. Unger, C.L. Smith, and J.W. Fournie, Hepatic and extra-hepatic neoplasms in a teleost fish from a polycyclic aromatic hydrocarbon contaminated habitat in Chesapeake Bay, USA, in: *Spontaneous Animal Tumors: A Survey*, L. Rossi, R. Richardson, and J. Harshbarger, Eds., Press Point di Abbiategrasso, Italy, 1997.

Vollenweider, R.A., Input-output models. With special reference to the phosphorus loading concept in limnology. *Schweiz. Z. Hydrol.*, 37, 53–84, 1975.

Volz, D.C., S.W. Kullman, D.L. Howarth, R.C. Hardman, and D.E. Hinton, Protective response of the ah receptor to ANIT-induced biliary epithelial cell toxicity in see-through medaka. *Toxicol. Sci.*, 102, 262–277, 2008.

Vorkamp, K., F. Riget, M. Glasius, M. Pécseli, M. Lebeuf, and D. Muir, Chlorobenzenes, chlorinated pesticides, coplanar chlorobiphenyls and other organochlorine compounds in Greenland biota. *Sci. Total Environ.*, 331, 157–175, 2004.

Vrijenhoek, R.C., Genetic diversity and fitness in small populations, in: *Conservation Genetics*, Loeschcke, V., J. Tomiuk, and S.K. Jain, Eds., Birkhöuser, Basel, Switzerland, 1994.

Wacholder, S., S. Chanock, M. Garcia-Closas, E. El ghormli, and N. Rothman, Assessing the probability that a positive report is false: An approach for molecular epidemiology studies. *J. Natl. Cancer Inst.*, 96, 434–442, 2004.

Waddington, C.H., *The Strategy of the Genes*, Allen & Unwin, London, United Kingdom, 1957.

Wade, L., Gold's dark side. *Science*, 341, 1448–1449, 2013.

Wagner, C. and H. Løkke, Estimation of ecotoxicological protection levels from NOEC toxicity data. *Water Res.*, 25, 1237–1242, 1991.

Wagner, J.G., *Fundamentals of Clinical Pharmacokinetics*, Drug Intelligence Publications, Hamilton, IL, 1975.

Walker, C.H., Species differences in microsomal monooxygenase activity and their relationship to biological half-lives. *Drug Metab. Rev.*, 7, 295–323, 1978.

Walker, J.S., Nuclear power and the environment: The Atomic Energy Commission and thermal pollution, 1965–1971, *Technol. Cult.*, 30, 964–992, 1989.

Walker, B., Biodiversity and ecological redundancy. *Conserv. Biol.*, 6, 12–23, 1991.

Walker, C.H., S.P. Hopkin, R.M. Sibly, and D.B. Peakall, *Principles of Ecotoxicology,* 2nd Edition, Taylor & Francis, London, 2001

Walker, C.H., S.P. Hopkin, R.M. Sibly, and D.B. Peakall, *Principles of Ecotoxicology,* 4th Edition, CRC Press/Taylor & Francis, Boca Raton, FL, 2012.

Walker, C.H., I. Newton, S.D. Hallam, and M.J.J. Ronis, Activities and toxicological significance of hepatic microsomal enzymes of the kestrel (*Falco tinnunculus*) and sparrowhawk (*Accipiter nisus*). *Comp. Biochem. Physiol. C*, 86, 379–382, 1987.

Wallace, W.G., T.M.H. Brouwer, M. Brouwer, and G.R. Lopez, Alterations in prey capture and induction of metallothioneins in grass shrimp fed cadmium-contaminated prey. *Environ. Toxicol. Chem.*, 19, 962–971, 2000.

Wallace, W.G., B.G. Lee, and S.N. Luoma, Subcellular compartmentalization of Cd and Zn in two bivalves. I. Significance of metal-sensitive fractions (MSF) and biologically detoxified metal (BDM). *Mar. Ecol. Prog. Ser.*, 249, 183–197, 2003.

Wallace, W.G. and G.R. Lopez, Bioavailability of biologically sequestered cadmium and the implications of metal detoxification. *Mar. Ecol.Prog. Ser.*, 147, 149–157, 1997.

Wallace, W.G., G.R. Lopez, and J.S. Levinton, Cadmium resistance in an oligochaete and its effect on cadmium trophic transfer to an omnivorous shrimp. *Mar. Ecol.Prog. Ser.*, 172, 225–237, 1998.

Wallace, W.G. and S.N. Luoma, Subcellular compartmentalization of Cd and Zn in two bivalves. II. Significance of trophically available metal (TAM). *Mar. Ecol.Prog. Ser.*, 257, 125–137, 2003.

Walsh, J.E., Climate of the Arctic marine environment. *Ecol. Appl.*, 18, S3–S22, 2008.

Walters, S., A. Kuhn, M.C. Nicholson, J. Copeland, S.A. Rego, and D.E. Macci, Stressor impacts on common loons in New Hampshire, USA., in: *Demographic Toxicity: Methods in Ecological Risk Assessment*, Akçakaya, H.R., J. D. Stark and T. S. Bridges, Eds., Oxford University Press, Oxford, 2008.

Wang, M. and V. Grimm, Population models in pesticide risk assessment: Lessons for assessing population-level effects, recovery, and alternative exposure scenarios from modeling a small mammal. *Environ. Toxicol. Chem.*, 29, 1292–1300, 2010.

Wang, Y.J., E.A. Mackay, O. Zerbe, D. Hess, P.E. Hunziker, M. Vasák, and J.H.R. Kägi, Characterization and sequential localization of the metal clusters in sea urchin metallothionein. *Biochemistry*, 34, 7460–7467, 1995.

Wang, J., M.C. Newman, X. Xiaoyu, A. Condon, and L. Liang, Floodplain methylmercury biomagnification factor higher and more variable than that of the contiguous South River (Virginia USA). *Ecotoxicol. Environ. Saf.*, 92, 191–198, 2013.

Wang, X.H., L. Tang, M.H.Wong, H.H. Shi, D.M. Zhao, J.S. Zheng, Q.P. Liu, and H.S. Hong, Comparative study on the contamination of organotin compounds and imposex of *Thais clavigera* in the coastal waters of Xiamen and Hong Kong sea area. *Comp. Biochem. Physiol. C*, 148, 463–464, 2008.

Wang, C.R., X.R. Wang, Y. Tian, H.X. Yu, X.Y. Gu, W.C. Du, and H. Zhou, Oxidative stress, defense response, and early biomarkers for lead-contaminated soil in *Vicia faba* seedlings. *Environ. Toxicol. Chem.*, 27, 970–977, 2008.

Wang, C.H., D.L. Willis, and W.D. Loveland, *Radiotracer Methodology in the Biological, Environmental, and Physical Sciences*, Prentice-Hall, Englewood Cliffs, NJ, 1975.

Wang, W.C., Z. Zeng, and T.R. Karl, Urban heat islands in China. *Geophys. Res. Lett.*, 17, 2377–2380, 1990.

Wangen, L.E., Elemental composition of size-fractionated aerosols associated with a coal-fired power plant plume and background. *Environ. Sci. Technol.*, 15, 1080–1088, 1981.

Wania, F. and D. Mackay, Global fractionation and cold condensation of low volatility organochlorine compounds in polar regions. *Ambio*, 22, 10–18, 1993.

Wania, F. and D. Mackay, A global distribution model for persistent organic chemicals. *Sci. Total Environ.*, 160/161, 211–232, 1995.

Wania, F. and D. Mackay, Tracking the distribution of persistent organic pollutants. *Environ. Sci. Technol.*, 30, 390A–396A, 1996.

Warnau, M., G. Ledent, A. Temara, V. Alva, M. Jangoux, and P. Dubois, Allometry of heavy metal bioconcentration in the Echinoid *Paracentrotus lividus*. *Arch. Environ. Contam. Toxicol.*, 29, 393–399, 1995.

Washington Post, Radioactivity here rises sharply, *The Washington Post*, Washington, DC, B7 p., December 12, 1961. B7 p.

Watanabe, M.E., Colony collapse disorder: Many suspects, no smoking gun. *Bioscience*, 58, 384–388, 2008.

Watkins, B. and K. Simkiss, The effect of oscillating temperatures on the metal ion metabolism of *Mytilus edulis*. *J. Mar. Biol. Assoc. U.K.*, 68, 93–100, 1988.

Watrud, L.S., E.H. Lee, A. Fairbrother, C. Burdick, J.R. Reichman, M. Bollman, M. Storm, G. King, and P.K. van de Water, Evidence for landscape-level, pollen-mediated gene flow from genetically modified creeping bentgrass with *CP4 EPSPS* as a marker. *Proc. Natl. Acad. Sci.U.S.A.*, 101, 14533–14538, 2004.

Webb, R.E. and F. Horsfall, Jr., Endrin resistance in the pine mouse. *Science*, 156, 1762, 1967.

Webber, H.M. and T.A. Haines, Mercury effects on predator avoidance behavior of a forage fish, golden shiner (*Notemigonus crysoleucas*). *Environ. Toxicol. Chem.*, 22, 1556–1561, 2003.

Weber, K. and H. Goerke, Persistent organic pollutants (POPs) in Antarctic fish: Levels, patterns, changes. *Chemosphere*, 53, 667–678, 2003.

Weber, C.I., W.H. Peltier, T.J. Norberg-King, W.B. Horning, II., F.A. Kessler, J.R. Menkedick, T.W. Neiheisel, P.A. Lewis, D.J. Klemm, Q.H. Pickering, E.L. Robinson, J.M. Lazorchak, L.J. Wymer, and R.W. Freyberg, *Short-Term Methods for Estimating the Chronic Toxicity of Effluents and Receiving Waters to Freshwater Organisms*, EPA/ 600/4-89/001. Environmental Monitoring Systems Laboratory, EPA, Cincinnati, OH, 315 p, 1989.

Weeks, J.M. and P.S. Rainbow, A dual-labelling technique to measure the relative assimilation efficiencies of invertebrates taking up trace metals from food. *Funct. Ecol.*, 4, 711–717, 1990.

Weijs, L., R.S.H. Yang, A. Covaci, K. Das, and R. Blust, Physiologically based pharmacokinetic (PBPK) models for lifetime exposure to PCB 153 in male and female harbor porpoises (*Phocoena phocoena*): Model development and evaluation. *Environ. Sci. Technol.* 44, 7023–7030, 2010.

Weinstein, J.E., and J.T. Oris, Humic acids reduce the bioaccumulation and photoinduced toxicity of fluoranthene fish. *Environ. Toxicol. Chem.*, 18, 2087–2094, 1999.

Weinstein, J.E., J.T. Oris, and D.H. Taylor, An ultrastructural examination of the mode of UV-induced toxic action of fluoranthene in the fathead minnow, *Pimephales promelas*. *Aquat. Toxicol.*, 39, 1–22, 1997.

Weis, J.S. and J. Perlmutter, Effects of tributyltin on activity and burrowing behavior of the fiddler crab, *Uca pugilator*. *Estuaries*, 10, 342–346, 1987.

Weis, J.S., G. Smith, and C. Santiago-Bass, Predator/prey interactions: A link between the individual level and both higher and lower level effects of toxicants in aquatic ecosystems. *J. Aquat. Ecosyst. Stress Recovery*, 7, 145–153, 2000.

Weis, J.S., G. Smith, T. Zhou, C. Santiago-Bass, and P. Weis, Effects of contaminants on behavior: Biochemical mechanisms and ecological consequences. *BioScience*, 51, 209–217, 2001.

Weis, J.S. and P. Weis, Pollutants as developmental toxicants in aquatic organisms. *Environ. Health Perspect.*, 71, 77–85, 1987.

Weis, J.S. and P. Weis, Effects of environmental pollutants on early fish development. *CRC Crit. Rev. Aquat. Sci.*, 1, 45–73, 1989a.

Weis, J.S. and P. Weis, Effects of embryonic exposure to methylmercury on larval prey-capture ability in the mummichog, *Fundulus heteroclitus*. *Environ. Toxicol. Chem.*, 14, 153–156, 1995.

Welch, W.J., Mammalian stress response: Cell physiology and biochemistry of stress proteins, in: *Stress Proteins in Biology and Medicine*, Morimoto, R.I., A. Tissieres, and C.C. Georgopolis, Eds., Cold Spring Harbor Press, Cold Spring Harbor, New York, 1990.

Wen, B., Y. Liu, X.Y. Hu, and X.Q. Shan, Effect of earthworms (*Eisenia fetida*) on the fractionation and bioavailability of rare earth elements in nine Chinese soils. *Chemosphere*, 63, 1179–1186, 2006.

Weng, G. and Q. Weng, *Remote Sensing of Natural Resources*, Taylor & Francis/CRC Press, Boca Raton, FL, 2013.

Wenning, R.J., R.T. Di Giulio, and E.P. Gallagher, Oxidant-mediated biochemical effects of paraquat in the ribbed mussel, *Geukensia demissa*. *Aquat. Toxicol.*, 12, 157–170, 1988.

Were, F.H., G.N. Kamau, P.M. Shiundu, G.A. Wafula, and C.M. Moturi, Air and blood lead levels in lead acid battery recycling and manufacturing plants in Kenya, *J. Occup Environ. Hyg.*, 9, 340–344, 2012.

Wesseling, C., M. Corriols, and V. Bravo, Acute pesticide poisoning and pesticide registration in Central America. *Toxicol. Appl. Pharmacol.*, 207, S697–S705, 2005.

Westlake, G.F., Behavioral Effects of Industrial Chemicals in Aquatic Animals, in *Hazard Assessment of Chemicals: Current Developments Volume 3*, Saxena, J., Ed., Academic Press, Orlando, FL, 1984.

Weston, D.P. and K.A. Maruya, Predicting bioavailability and bioaccumulation with *in vitro* digestive fluid extraction. *Environ. Toxicol. Chem.*, 21, 962–971, 2002.

Wetzel, R.G., *Limnology*, Saunders College Publishing, Philadelphia, PA, 1982.

Whalen, K. E., V.R. Starczak, D.R. Nelson, J.V. Goldstone, and M.E. Hahn, Cytochrome P450 diversity and induction by gorgonian allelochemicals in the marine gastropod *Cyphoma gibbosum. BMC Ecol.*, 10, 24, 2010.

Wheeler, D.L., An eclectic biologist argues that humans are not evolution's most important result; bacteria are. *Chron. Higher Education*, A23–A24, Sept. 6, 1996.

White, R.M., The great climate debate. *Sci. Am.*, 263, 36–43, 1990.

White, R.M., Preface, in: *The Greening of Industrial Ecosystems*, Allenby, B.R. and D.J. Richards, Eds., National Academy Press, Washington, DC, 1994.

Whitfield, S.M., K.E. Bell, T. Philippi, M. Sasa, F. Bolaños, G. Chaves, J.M. Savage, and M.A. Donnelly, Amphibian and reptile declines over 35 years at La Selva, Costa Rica. *Proc. Natl. Acad. Sci. U.S.A.*, 104, 8352–8356, 2007.

Widmer, R., H. Oswald-Krapf, D. Sinha-Khetriwal, M. Schnellman, and H. Bőni. Global perspectives on e-waste. *Environ. Impact Asses.*, 25, 436–458, 2005.

Wiemeyer, S.M., J.M. Scott, M.P. Anderson, P.H. Bloom, and C.J. Stafford, Environmental contaminants in California condors. *J. Wildl. Manage.*, 52: 238–247, 1988.

Wiener J.G. and P.J. Rago, A test of fluctuating asymmetry in bluegills (*Lepomis macrochirus* Rafinesque) as a measure of pH-related stress. *Environ. Pollut.*, 44, 27–36, 1987.

Wiens, J.A., Spatial scaling in ecology. *Funct. Ecol.*, 3, 385–397, 1989.

Wiig, Ø., Are polar bears threatened? *Science*, 309, 1814–1815, 2005.

Wilkinson, K.J. and P.G.C. Campbell, Aluminum bioconcentration at the gill surface of juvenile Atlantic salmon in acidic media. *Environ. Toxicol. Chem.*, 12, 2083–2095, 1993.

Williams, T. D., A. M. Diab, S. G. George, R. E. Godfrey, V. Sabine, A. Conesa, S. D. Minchin, P. C. Watts, and J. K. Chipman, Development of the GENIPOL European flounder (*Platichthys flesus*) microarray and determination of temporal transcriptional responses to cadmium at low dose. *Environ. Sci. Technol.*, 40, 6479–6488, 2006.

Williams, L.G. and D.I. Mount, Influence of zinc on periphytic communities. *Am. J. Bot.*, 52, 26–34, 1965.

Williams, M.W. and J.E. Turner, Comments on softness parameters and metal ion toxicity. *J. Inorg. Nucl. Chem.*, 43, 1689–1691, 1981.

Williamson, P., Use of ^{65}Zn to determine the field metabolism of the snail *Cepaea nemoralis* L. *Ecology*, 56, 1185–1192, 1975.

Williamson, C.E., B.R. Hargreaves, P.S. Orr, and P.A. Lovera, Does UV play a role in changes in predation and zooplankton community structure in acidified lakes? *Limnol. Oceanogr.*, 44, 774–783, 1999.

Willingham, E., Endocrine-disrupting compounds and mixtures: Unexpected dose-response. *Arch. Environ. Contam. Toxicol.*, 46, 265–269, 2004.

Wilmut, I. and J. Taylor, Primates join the club. *Nature*, 450, 485–486, 2007.

Wilson, J.B., The cost of heavy-metal tolerance: An example. *Evolution*, 42, 408–413, 1988.

Wilson, D.S., Complex interactions in metacommunities, with implications for biodiversity and higher levels of selection. *Ecology*, 73, 1984–2000, 1992.

Wilson, J. D. and W.A. Hopkins, Beyond the wetland: Evaluating the effects of anthropogenic stressors on source-sink dynamics in pond-breeding amphibians. *Conserv. Biol.*, 27, 595–604, 2013.

Wilson, V.S., C. Lambright, J. Ostby, and L.E. Gray, Jr., *In vitro* and *in vivo* effects of 17 b-trenbolone: A feedlot effluent contaminant. *Toxicol. Sci.*, 70, 202–211, 2002.

Windham, G.C., K. Waller, M. Anderson, L. Fenster, P. Mendola, and S. Swan, Chlorination by-products in drinking water and menstrual cycle function. *Environ. Health Perspect.*, 111, 935–941, 2003.

Winge, D.R. and M. Brouwer, Discussion summary. Techniques and problems in metal-binding protein chemistry and implications for proteins in nonmammalian organisms. *Environ. Health Perspect.*, 65, 211–214, 1986.

Winner, R.W., M.W. Boesel, and M.P. Farrell, Insect community structure as an index of heavy-metal pollution in lotic ecosystems. *Can. J. Fish. Aquat. Sci.*, 37, 647–655, 1980.

Winner, R.W., T. Keeling, R. Yeager, and M.P. Farrell, Effect of food type on the acute and chronic toxicity of copper to *Daphnia magna. Freshw. Biol.*, 7, 343–349, 1977.

Winston, G.W. and R.T. Di Giulio, Prooxidant and antioxidant mechanisms in aquatic organisms. *Aquat. Toxicol.*, 19, 137–161, 1991.

Wirgin, I.I., C. Grunald, S. Courtenay, G.L. Kreamer, W.L. Reichert, and J.E. Stein, A biomarker approach to assessing xenobiotic exposure in Atlantic tomcod from the North American Atlantic Coast. *Environ. Health Perspect.*, 102, 764–770, 1994.

Wirgin, I., N.K. Roy, M. Loftus, R.C. Chambers, D.G. Franks, and M.E. Hahn, Mechanistic basis of resistance to PCBs in Atlantic tomcod from the Hudson River. *Science*, 331, 1322–1325, 2011.

Wise, D., J.D. Yarbrough, and R.T. Roush, Chromosomal analysis of insecticide resistant and susceptible mosquitofish. *J. Hered.*, 77, 345–348, 1986.

Witschi, H.R., K.E. Pinkerton, L.S. van Winkle, and J.A. Last, Toxic responses of the respiratory system, in *Casarett & Doull's Toxicology: The Basic Science of Poisons,* 7th Edition, Klaassen, C.D., Ed., McGraw-Hill, New York, 2008.

Wofford, H.W. and P. Thomas, Effect of xenobiotics on peroxidation of hepatic microsomal lipids from striped mullet (*Mugil cephalus*) and Atlantic croaker (*Micropogonus undulatus*). *Mar. Environ. Res.*, 2, 285–289, 1988.

Wojtaszek, B.F., T.M. Buscarini, D.T. Chartland, G.R. Stephenson, and D.G. Thompson, Effect of Release® herbicide on mortality, avoidance response, and growth of amphibian larvae in two forest wetlands. *Environ. Toxicol. Chem.*, 24, 2533–2544, 2005.

Wollenberger, L., L. Dinan, and M. Breitholtz, Brominated flame retardants: Activities in a crustacean development test and in an ecdysteroid screening assay. *Environ. Toxicol. Chem.*, 24, 400–407, 2005.

Woltering, D.M., J.L. Hedtke, and L.J. Weber, Predator-prey interactions of fishes under the influence of ammonia. *Trans. Am. Fish. Soc.*, 107, 500–504, 1978.

Wong, C.K. and P.K. Wong, Life table evaluation of the effects of cadmium exposure on the freshwater cladoceran *Moina macrocopa. Bull. Environ. Contam. Toxicol.*, 44, 135–141, 1990.

Wood, J.M. and H.K. Wang, Microbial resistance to heavy metals. *Environ. Sci. Technol.*, 17, 582A–590A, 1983.

Woodford, J.E., W.H. Karaso, M.W. Meyer, and L. Chambers, Impact of 2,3,7,8-TCDD on survival, growth, and behavior of ospreys breeding in Wisconsin, USA. *Environ. Toxicol. Chem.*, 17, 1323–1331, 1998.

Woods, J.S., M.D. Martin, C.A. Naleway, and D. Echeverria, Urinary porphyrin profiles as a biomarker of mercury exposure: Studies on dentists with occupational exposure to mercury vapor. *J. Toxicol. Environ. Health*, 40, 235–246, 1993.

Woodward, L.A., M. Mulvey, and M.C. Newman, Mercury contamination and population-level responses in chironomids: Can allozyme polymorphism indicate exposure? *Environ. Toxicol. Chem.*, 15, 1309–1316, 1996.

Woodwell, G.M., Effects of ionizing radiation on terrestrial ecosystems. *Science*, 138, 572–577, 1962.

Woodwell, G.M., The ecological effects of radiation. *Sci. Am.*, 208, 2–11, 1963.

Woodwell, G.M., Toxic substances and ecological cycles. *Sci. Am.*, 216, 24–31, 1967.

Woodwell, G.M., The carbon dioxide question. *Sci. Am.*, 238, 34–43, 1978.

Wren, C.D. and H.R. MacCrimmon, Comparative bioaccumulation of mercury in two adjacent freshwater ecosystems. *Water Res.*, 20, 763–769, 1986.

Wren, C.D., H.R. MacCrimmon, and B.R. Loescher, Examination of bioaccumulation and biomagnification of metals in a Precambrian Shield lake. *Water Air Soil Pollut.*, 19, 277–291, 1983.

Wright, P.A. and C.D. Zamuda, Copper accumulation by two bivalue molluscs. Salinity effect is independent of cupric ion activity. *Mar. Environ. Res.*, 23, 1–14, 1987.

Wu, J., J.L Vankat, and Y. Barlas, Effects of patch connectivity and arrangement on animal metapopulation dynamics: A simulation study. *Ecol. Modell.*, 65, 221–254, 1993.

Wu, T.H., T.R. Rainwater, S.G. Platt, S.T. McMurry, and T.A. Anderson, DDE in eggs of two crocodile species from Belize. *J. Agric. Food Chem.*, 48, 6416–6420, 2000.

Wu, T.H., J.E. Cañas, T.R. Rainwater, S.G. Platt, S.T. McMurry, and T.A. Anderson, Organochlorine contaminants in complete clutches of Morelet's crocodile (*Crocodylus moreletii*) eggs from Belize. *Environ. Pollut.*, 144, 151–157, 2006.

Wu, L., J. Chen, K.K. Tanji, and G.S. Banuelos, Distribution and biomagnification of selenium in a restored upland grassland contaminated by selenium from agricultural drain water. *Environ. Toxicol. Chem.*, 14, 733–742, 1995.

Wynne, J. and C. Stringfield, Treatment of lead toxicity and crop stasis in a California condor (*Gymnogyps californianus*). *J. Zoo Anim. Med.*, 38, 588–590, 2007.

Xenopoulos. M.A., P.R. Leavitt, and D.W. Schindler, Ecosystem-level regulation of boreal lake phytoplankton by ultraviolet radiation. *Can. J. Fish. Aquat. Sci.*, 66, 2002–2010, 2009.

Yamada, H., M. Tateishi, and K. Takayanagi, Bioaccumulation of organotin compounds in the red sea bream (*Pagrus major*) by two uptake pathways: Dietary uptake and direct uptake from water. *Environ. Toxicol. Chem.*, 13, 1415–1422, 1994.

Yamaoka, K., T. Nakagawa, and T. Uno, Statistical moments in pharmacokinetics. *J. Pharmacokinet. Biop.*, 6, 547–558, 1978.

Yan, Q.L. and W.X. Wang, Metal exposure and bioavailability to a marine deposit-feeding sipuncula, *Sipunculus nudus. Environ. Sci. Technol.*, 36, 40–47, 2002.

Yang, X., A.P. Gondikas, S.M. Marinokos, M. Auffan, J. Liu, H. Hsu-Kim, and J.N. Meyer, Mechanism of silver nanoparticle toxicity is dependent on dissolved silver and surface coating in *Caenrhabditis elegans. Environ. Sci. Technol.* 46, 1119–1127, 2012.

Yang, R.Q., Q.F. Zhou, J.Y. Liu, and G.B. Jiang, Butyltins compounds in mollusks from Chinese Bohai coastal waters. *Food Chem.*, 97, 637–643, 2006.

Yap, H.H., D. Desaiah, L.K. Cutkomp, and R.B. Koch, Sensitivity of fish ATPases to polychlorinated biphenyls. *Nature*, 233, 61–62, 1971.

Yarbrough, J.D., R.T. Roush, J.C. Bonner, and D.A. Wise, Monogenic inheritance of cyclodiene insecticide resistance in mosquitofish, *Gambusia affinis. Experientia*, 42, 851–853, 1986.

Yeung, L.W.Y, M.K. So, G. Jiang, S. Taniyasu, N. Yamashita, M. Song, Y. Wu, J. Li, J.P. Giesy, and P.K.S. Lam, Perfluorooctanesulfonate and related fluorochemicals in human blood samples from China. *Environ. Sci. Technol.*, 40, 715–720, 2006.

Yongxing, W., W. Xiaorong, and H. Zichun, Genotoxicity of lanthanum (III) and gadolinium (III) in human peripheral blood lymphocytes. *Bull. Environ. Contam. Toxicol.*, 64, 611–616, 2000.

You, C.H., E.A. Mackay, P.M. Gehrig, P.E. Hunziker, and J.H.R. Kagi, Purification and characterization of recombinant *Caenorhabditis elegans* metallothionein. *Arch. Biochem. Biophys.*, 372, 44–52, 1999.

Young, L.B. and H.H. Harvey, Metal concentrations in chironomids in relation to the geochemical characteristics of surficial sediments. *Arch. Environ. Contam. Toxicol.*, 21, 202–211, 1991.

Younglai, E.V., W.G. Foster, E.G. Hughes, K. Trim, and J.F. Jarrell, Levels of environmental contaminants in human follicular fluid, serum and seminal plasma of couples undergoing in vitro fertilization. *Arch. Environ. Contam. Toxicol.* 43, 121–126, 2002.

Yurtseva, A., D. Lajus, F. Artamonova, and A. Makhrov, Effect of hatchery environment on cranial morphology and developmental stability of Atlantic salmon (*Salmo salar* L.) from north-west Russia. *J. Appl. Ichthyology*, 26, 307–314, 2010.

Zagury, G.J., C.M. Neculita, C. Bastien, and L. Deschênes, Mercury fractionation, bioavailability, and ecotxicology in highly contaminated soil from chlor-alkali plants. *Environ. Toxicol. Chem.*, 25, 1138–1147, 2006.

Zakharov, V.M., Future prospects for population phenogenetics. *Sov. Sci. Rev. Sect. F, Physiol. Gen. Biol. Rev.*, 4, 1–79, 1989.

Zakharov, V.M., Analysis of fluctuating asymmetry as a method of biomonitoring at the population level, in: *Bioindications of Chemical and Radioactive Pollution*, Krivolutsky, D.A., Ed., CRC Press, Boca Raton, FL, 1990.

Zakharov, V.M., and E. Krysanov, Eds., *Consequences of the Chernobyl Catastrophe: Environmental Health*, Center for Russian Environmental Policy, Moscow Affiliate of the International "Biotest" Foundation. Moscow, 1996.

Zakharov, V.M. and A.V. Yablokov, Skull asymmetry in the Baltic grey seal: Effects of environmental pollution. *Ambio*, 19, 266–269, 1990.

Zala, S. M. and D. J. Penn, Abnormal behaviours induced by chemical pollution: A review of the evidence and new challenges. *Anim. Behav.*, 68, 649–664, 2004.

Zalasiewicz, J., M. Williams, W. Steffen, and P. Crutzen, The new world of the Anthropocene. *Environ. Sci. Technol.*, 44, 2228–2231, 2010.

Zanette, J., M.J. Jenny, J.V. Goldstone, T. Parente, B.R. Woodin, A.C. Bainy, and J.J. Stegeman. Identification and expression of multiple CYP1-like and CYP3-like genes in the bivalve mollusk *Mytilus edulis. Aquat. Toxicol.*, 128–129, 101–112, 2013.

Zartman, C.E. and A.J. Shaw, Metapopulation extinction thresholds in rain forest remnants. *Am. Nat.*, 167, 177–189, 2006.

Zeng, N., Y. Ding, J. Pan, H. Wang, and J. Gregg, Climate change: The Chinese challenge. *Science*, 319, 730–731, 2008.

Zeng, X., B. Mai, G. Sheng, X. Luo, W. Shao, T. An, and J. Fu, Distribution of polycyclic musks in surface sediments from the Pearl River delta and Macao coastal region, South China. *Environ. Toxicol. Chem.*, 27, 18–23, 2008.

Zhang, Z.X., China is moving away the pattern of "develop first and then treat the pollution." *Energ. Policy*, 35, 3547–3549, 2007.

Zhang, G., X. Fang, X. Guo, L. Li, R. Luo, F. Xu, P. Yang, The oyster genome reveals stress adaptation and complexity of shell formation. *Nature*, 490, 49–54, 2012.

Zhang, H., J. Feng, W. Zhu, C. Liu, S. Xu, P. Shao, D. Wu, W. Yang, and J. Gu, Chronic toxicity of rare-earth elements on human beings. *Biol. Trace Elem. Res.*, 73, 1–17, 2000.

Zhang, Z., J. Huang, G. Yu, and H. Hong, Occurrence of PAHs, PCBs, and organochlorine pesticides in the Tonghui River of Beijing, China. *Environ. Pollut.*, 130, 249–261, 2004.

Zhang, L. and W.X. Wang, Significance of subcellular metal distribution in prey in influencing the trophic transfer of metals in a marine fish. *Limnol. Oceanogr.*, 51, 2008–2017, 2006.

Zhang, G., J.Yan, J.M. Fu, A. Parker, X.D. Li, and Z.S. Wang, Butyltins in sediments and biota from the Pearl River Delta, South China. *Chem. Spec. Bioavailab.*, 14 (Special Issue), 35–42, 2002.

Zhao, Y. and M.C. Newman, Shortcomings of the laboratory-derived median lethal concentrations for predicting mortality in field populations: Exposure duration and latent mortality. *Environ. Toxicol. Chem.*, 23, 2147–2153, 2004.

Zhao, Y. and M.C. Newman, Effects of exposure duration and recovery time during pulsed exposures. *Environ. Toxicol. Chem.*, 25, 1298–1304, 2006.

Zhao, Y. and M.C. Newman, The theory underlying dose-response models influences predictions for intermittent exposures. *Environ. Toxicol. Chem.*, 26, 543–547, 2007.

Ziliak, S.T. and D.N. McCloskey, *The Cult of Statistical Significance: How the Standard Error Costs Us Jobs, Justice, and Lives*, University of Michigan Press, Ann Arbor, MI, 2008.

Zimmermann, S., C.M. Menzel, D. Stüben, H. Taraschewski, and B. Sures, Lipid solubility of the platinum group metals Pt, Pd, and Rh in dependence on the presence of complexing agents. *Environ. Pollut.*, 124, 1–5, 2003.

Zurer, P.S., Chemists solve key puzzle of Anarctic ozone hole. *Chem. Eng. News*, Nov. 30, 25–27, 1987.

Zurer, P.S., Studies on ozone destruction expand beyond Antarctic. *Chem. Eng. News*, 66, 16–25, 1988.

Zurer, P.S., Arctic ozone loss: Fact-finding mission concludes outlook is bleak. *Chem. Eng. News*, 67, 29–31, 1989.

Zvereva, E.L. and M.V. Kozlov, Top-down effects on population dynamics of *Eriocrania* miners (Lepidoptera) under pollution impact: Does an enemy-free space exist? *Oikos*, 115, 413–426, 2006.

Index

B

C

D

F

G

H

M

N

Q

R

S